PRINCIPLES OF Virology

THIRD EDITION

VOLUME I *Molecular Biology*

PRINCIPLES OF Virology

THIRD EDITION

S. J. FLINT
Department of Molecular Biology
Princeton University
Princeton, New Jersey

L. W. ENQUIST
Department of Molecular Biology
Princeton University
Princeton, New Jersey

V. R. RACANIELLO
Department of Microbiology
College of Physicians and Surgeons
Columbia University
New York, New York

A. M. SKALKA
Fox Chase Cancer Center
Philadelphia, Pennsylvania

WASHINGTON, DC

Front cover illustration: A model of the atomic structure of the poliovirus type 1 Mahoney strain. The model has been highlighted by radial depth cuing so that the portions of the model that are farthest from the center are bright. Prominent surface features include a star-shaped mesa at each of the fivefold axes and a propeller-shaped feature at each of the threefold axes. A deep cleft or canyon surrounds the star-shaped feature. This canyon is the receptor-binding site. Courtesy of Robert Grant, Stéphane Crainic, and James Hogle (Harvard Medical School).

Back cover illustration: Progress in the global eradication of poliomyelitis has been striking, as illustrated by maps showing areas of known or probable circulation of wild-type poliovirus in 1988, 1998, and 2008. Dark red indicates the presence of virus. In 1988, the virus was present on all continents except Australia. By 1998, the Americas were free of wild-type poliovirus, and transmission was interrupted in the western Pacific region (including the People's Republic of China) and in the European region (with the exception of southeastern Turkey). By 2008, the number of countries reporting endemic circulation of poliovirus had been reduced to four: Afghanistan, Pakistan, India, and Nigeria.

Address editorial correspondence to ASM Press, 1752 N St. NW, Washington, DC 20036-2904, USA

Send orders to ASM Press, P.O. Box 605, Herndon, VA 20172, USA
Phone: (800) 546-2416 or (703) 661-1593
Fax: (703) 661-1501
E-mail: books@asmusa.org
Online: estore.asm.org

Library of Congress Cataloging-in-Publication Data

Principles of virology / S.J. Flint ... [et al.]. — 3rd ed.
p. ; cm.
Includes bibliographical references and index.
ISBN 978-1-55581-443-4 (pbk. : set) — ISBN 978-1-55581-479-3 (pbk. : v. 1) — ISBN 978-1-55581-480-9 (pbk. : v. 2)
1. Virology. I. Flint, S. Jane. II. American Society for Microbiology.
[DNLM: 1. Viruses. 2. Genetics, Microbial. 3. Molecular Biology. 4. Virology—methods. QW 160 P957 2009]

QR360.P697 2009
579.2—dc22

2008030964

10 9 8 7 6 5 4 3 2 1

ISBN 978-1-55581-479-3

Printed in the United States of America

Illustrations and illustration concepting: Patrick Lane, ScEYEnce Studios
Cover and interior design: Susan Brown Schmidler

We dedicate this book to the students, current and future scientists and physicians, for whom it was written. We kept them ever in mind.

We also dedicate it to our families:

Jonn, Gethyn, and Amy Leedham

Kathy and Brian

Doris, Aidan, Devin, and Nadia

Rudy, Jeanne, and Chris

Oh, be wiser thou!
Instructed that true knowledge leads to love.

William Wordsworth
Lines left upon a Seat in a Yew-tree
1888

Contents

Preface

The enduring goal of scientific endeavor, as of all human enterprise, I imagine, is to achieve an intelligible view of the universe. One of the great discoveries of modern science is that its goal cannot be achieved piecemeal, certainly not by the accumulation of facts. To understand a phenomenon is to understand a category of phenomena or it is nothing. Understanding is reached through creative acts.

A. D. HERSHEY
Carnegie Institution Yearbook 65

The major goal of all three editions of this book has been to define and illustrate the basic principles of animal virus biology. In this information-rich age, the quantity of data describing any given virus can be overwhelming, if not indigestible, for student and expert alike. Furthermore, the urge to write more and more about less and less is the curse of reductionist science and the bane of those who write textbooks meant to be used by students. Consequently, in the third edition, we have continued to distill information with the intent of extracting essential principles, while retaining some descriptions of how the work is done. Our goal is to illuminate process and strategy as opposed to listing facts and figures. We continue to be selective in our choice of topics, viruses, and examples in an effort to make the book readable, rather than comprehensive. Detailed encyclopedic works like *Fields Virology* (2007) have made the best attempt to be all-inclusive, and *Fields* is recommended as a resource for detailed reviews of specific virus families.

What's New

The major change in the third edition is the separation of material into two volumes, each with its unique appendix(es) and general glossary. Volume I covers molecular aspects of the biology of viruses, and Volume II focuses on viral pathogenesis, control of virus infections, and virus evolution. The organization into two volumes follows a natural break in pedagogy and provides considerable flexibility and utility for students and teachers alike. The smaller size and soft covers of the two volumes make them easier for students to carry and work with than the single hardcover volume of earlier editions. The volumes can be used for two courses, or as parts I and II of a one-semester course. While differing in content, the volumes are integrated in style and presentation. In addition to updating the material for both volumes, we have used the new format to organize the material more efficiently and to keep chapter size manageable.

As in our previous edition, we have tested ideas for inclusion in the text in our own classes. We have also received constructive comments and suggestions from other virology instructors and their students. Feedback from students was particularly useful in finding typographical errors, clarifying confusing or complicated illustrations, and pointing out inconsistencies in content.

For purposes of readability, references again are generally omitted from the text, but each chapter ends with an updated and expanded list of relevant books, review articles, and selected research papers for readers who wish to pursue specific topics. In general, if an experiment is featured in a chapter, one or more references are listed to provide more detailed information.

Principles Taught in Two Distinct, but Integrated Volumes

These two volumes outline and illustrate the strategies by which all viruses are propagated in cells, how these infections spread within a host, and how such infections are maintained in populations. The principles established in Volume I enable understanding of the topics of Volume II: viral disease, its control, and the evolution of viruses.

Volume I: the Science of Virology and the Molecular Biology of Viruses

This volume features the molecular processes that take place in an infected host cell. Chapters 1 and 2 discuss the foundations of virology. A general introduction with historical perspectives as well as definitions of the unique properties of viruses is provided first. The unifying principles that are the foundations of virology, including the concept of a common strategy for viral propagation, are then described. Chapter 2 establishes the principle of the infectious cycle with an introduction to cell biology. The basic techniques for cultivating and assaying viruses are outlined, and the concept of the single-step growth cycle is presented.

Chapter 3 introduces the fundamentals of viral genomes and genetics, and it provides an overview of the perhaps surprisingly limited repertoire of viral strategies for genome replication and mRNA synthesis. Chapter 4 describes the architecture of extracellular virus particles in the context of providing both protection and delivery of the viral genome in a single vehicle. In Chapters 5 through 13, we describe the broad spectrum of molecular processes that characterize the common steps of the reproductive cycle of viruses in a single cell, from decoding genetic information to genome replication and production of progeny virions. We describe how these common steps are accomplished in cells infected by diverse but representative viruses, while emphasizing principles applicable to all.

The appendix in Volume I provides concise illustrations of viral life cycles for the main virus families discussed in the text. It is intended to be a reference resource when one is reading individual chapters and a convenient visual means by which specific topics may be related to the overall infectious cycles of the selected viruses.

Volume II: Pathogenesis, Control, and Evolution

This volume addresses the interplay between viruses and their host organisms. Chapters 1 to 7 focus on principles of virus replication and pathogenesis. Chapter 1 provides a brief history of viral pathogenesis and addresses the basic concepts of how an infection is established in a host as opposed to infection of single cells in the laboratory. In Chapter 2, we focus on how viral infections spread in populations. Chapter 3 presents our growing understanding of crucial autonomous reactions of cells to infection and describes how these actions influence the eventual outcome for the host. Chapter 4 provides a virologist's view of immune defenses and their integration with events that occur when single cells are infected. Chapter 5 describes how a particular virus replication strategy and the ensuing host response influence the outcome of infection such that some are short and others are of long duration. Chapter 6 is devoted entirely to the AIDS virus, not only because it is the causative agent of the most serious current worldwide epidemic, but also because of its unique and informative interactions with the human immune defenses. In Chapter 7, we discuss virus infections that transform cells in culture and promote oncogenesis (the formation of tumors) in animals.

Chapters 8 and 9 outline the principles involved in treatment and control of infection. Chapter 8 focuses on vaccines, and chapter 9 discusses the approaches and challenges of antiviral drug discovery. In Chapter 10, the final chapter, we present a foray into the past and future, providing an introduction to viral evolution. We illustrate important principles taught by zoonotic infections, emerging infections, and humankind's experiences with epidemic and pandemic viral infections.

Appendix A summarizes the pathogenesis of common viruses that infect humans in three "slides" (viruses and diseases, epidemiology, and disease mechanisms) for each virus or virus group. This information is intended to provide a simple snapshot of pathogenesis and epidemiology. Appendix B provides a concise discussion of unusual infectious agents, such as viroids, satellites, and prions, that are not viruses but that (like viruses) are molecular parasites of the cells in which they replicate.

Reference

Knipe, D. M., and P. M. Howley (ed. in chief). 2007. *Fields Virology,* 5th ed. Lippincott Williams & Wilkins, Philadelphia, PA.

Acknowledgments

These two volumes of *Principles* could not have been composed and revised without help and contributions from many individuals. We are most grateful for the continuing encouragement from our colleagues in virology and the students who use the text. Our sincere thanks also go to colleagues who have taken considerable time and effort to review the text in its evolving manifestations. Their expert knowledge and advice on issues ranging from teaching virology to organization of individual chapters and style were invaluable, even when orthogonal to our approach, and are inextricably woven into the final form of the book.

We thank the following individuals for their reviews and comments on multiple chapters in both volumes: Nicholas Acheson (McGill University), Karen Beemon and her virology students (Johns Hopkins University), Clifford W. Bond (Montana State University), Martha Brown (University of Toronto Medical School), Teresa Compton (University of Wisconsin), Stephen Dewhurst (University of Rochester Medical Center), Mary K. Estes (Baylor College of Medicine), Ronald Javier (Baylor College of Medicine), Richard Kuhn (Purdue University), Muriel Lederman (Virginia Polytechnic Institute and State University), Richard Moyer (University of Florida College of Medicine), Leonard Norkin (University of Massachusetts), Martin Petric (University of Toronto Medical School), Marie Pizzorno (Bucknell University), Nancy Roseman (Williams College), David Sanders (Purdue University), Dorothea Sawicki (Medical College of Ohio), Bert Semler (University of California, Irvine), and Bill Sugden (University of Wisconsin).

We also are grateful to those who gave so generously of their time to serve as expert reviewers of these or earlier individual chapters or specific topics: James Alwine (University of Pennsylvania), Edward Arnold (Center for Advanced Biotechnology and Medicine, Rutgers University), Carl Baker (National Institutes of Health), Amiya Banerjee (Cleveland Clinic Foundation), Silvia Barabino (University of Basel), Albert Bendelac (University of Chicago), Susan Berget (Baylor College of Medicine), Kenneth I. Berns (University of Florida),

John Blaho (MDL Corporation), Sheida Bonyadi (Concordia University), Jim Broach (Princeton University), Michael J. Buchmeier (The Scripps Research Institute), Hans-Gerhard Burgert (University of Warwick), Allan Campbell (Stanford University), Jim Champoux (University of Washington), Bruce Chesebro (Rocky Mountain Laboratories, National Institute of Allergy and Infectious Diseases), Marie Chow (University of Arkansas Medical Center), Barclay Clements (University of Glasgow), Don Coen (Harvard Medical School), Richard Condit (University of Florida), David Coombs (University of New Brunswick), Michael Cordingley (Bio-Mega/Boehringer Ingelheim), Ted Cox (Princeton University), Andrew Davison (Institute of Virology, MRC Virology Unit), Ron Desrosiers (Harvard Medical School), Robert Doms (University of Pennsylvania), Emilio Emini (Merck Sharp & Dohme Research Laboratories), Alan Engelman (Dana-Farber Cancer Center), Ellen Fanning (Vanderbilt University), Bert Flanagan (University of Florida), Nigel Fraser (University of Pennsylvania Medical School), Huub Gelderblom (University of Amsterdam), Charles Grose (Iowa University Hospital), Samuel Gunderson (European Molecular Biology Laboratory), Pryce Haddix (Auburn University at Montgomery), Peter Howley (Harvard Medical School), James Hoxie (University of Pennsylvania), Frederick Hughson (Princeton University), Clinton Jones (University of Nebraska), Christopher Kearney (Baylor University), Walter Keller (University of Basel), Tom Kelly (Memorial Sloan-Kettering Cancer Center), Elliott Kieff (Harvard Medical School), Elizabeth Kutter (Evergreen State College), Robert Lamb (Northwestern University), Ihor Lemischka (Mount Sinai School of Medicine), Arnold Levine (Institute for Advanced Study), Michael Linden (Mount Sinai School of Medicine), Daniel Loeb (University of Wisconsin), Adel Mahmoud (Princeton University), Michael Malim (King's College London), James Manley (Columbia University), Philip Marcus (University of Connecticut), Malcolm Martin (National Institutes of Health), William Mason (Fox Chase Cancer Center), Loyda Melendez (University of Puerto Rico Medical Sciences Campus), Baozhong Meng (University of Guelph), Edward Mocarski (Emory University), Bernard Moss (Laboratory of Viral Diseases, National Institutes of Health), Peter O'Hare (Marie Curie Research Institute), Radhakris Padmanabhan (University of Kansas Medical Center), Peter Palese (Mount Sinai School of Medicine), Philip Pellett (Cleveland Clinic and Case Western Reserve University), Stuart Peltz (University of Medicine and Dentistry of New Jersey, Robert Wood Johnson Medical School), Roger Pomerantz (Thomas Jefferson University), Glenn Rall (Fox Chase Cancer Center), Charles Rice (Rockefeller University), Jack Rose (Yale University), Barry Rouse (University of Tennessee College of Veterinary Medicine), Rozanne Sandri-Goldin (University of California, Irvine), Nancy Sawtell (Childrens Hospital Medical Center), Priscilla Schaffer (University of Arizona), Robert Schneider (New York University School of Medicine), Christoph Seeger (Fox Chase Cancer Center), Aaron Shatkin (Center for Advanced Biotechnology and Medicine, Rutgers University), Thomas Shenk (Princeton University), Geoff Smith (Wright-Fleming Institute), Greg Smith (Northwestern University), Kathryn Spink (Illinois Institute of Technology), Joan Steitz (Yale University), Victor Stollar (University of Medicine and Dentistry of New Jersey), Wesley Sundquist (University of Utah), John M. Taylor (Fox Chase Cancer Center), Alice Telesnitsky (University of Michigan Medical School), Heinz-Jürgen Thiel (Institut für Virologie, Giessen, Germany), Adri Thomas (University of Utrecht), Gerald Thrush (Western University of Health Sciences), Paula Traktman (Medical College of Wisconsin), James van

Etten (University of Nebraska, Lincoln), Chris Upton (University of Victoria), Luis Villarreal (University of California, Irvine), Herbert Virgin (Washington University School of Medicine), Peter Vogt (The Scripps Research Institute), Simon Wain-Hobson (Institut Pasteur), Gerry Waters (TB Alliance), Robin Weiss (University College London), Sandra Weller (University of Connecticut Health Center), Michael Whitt (University of Tennessee), Lindsay Whitton (The Scripps Research Institute), and Eckard Wimmer (State University of New York at Stony Brook). Their rapid responses to our requests for details and checks on accuracy, as well as their assistance in simplifying complex concepts, were invaluable. All remaining errors or inconsistencies are entirely ours.

Since the inception of this work, our belief has been that the illustrations must complement and enrich the text. Execution of this plan would not have been possible without the support of Jeff Holtmeier (Director, ASM Press) and the technical expertise and craft of our illustrator. The illustrations are an integral part of the exposition of the information and ideas discussed, and credit for their execution goes to the knowledge, insight, and artistic talent of Patrick Lane of ScEYEnce Studios. As noted in the figure legends, many of the figures could not have been completed without the help and generosity of our many colleagues who provided original images. Special thanks go to those who crafted figures tailored specifically to our needs or provided multiple pieces: Mark Andrake (Fox Chase Cancer Center), Edward Arnold (Rutgers University), Bruce Banfield (The University of Colorado), Christopher Basler and Peter Palese (Mount Sinai School of Medicine), Amy Brideau (Peregrine Pharmaceuticals), Roger Burnett (Wistar Institute), Rajiv Chopra and Stephen Harrison (Harvard University), Marie Chow (University of Arkansas Medical Center), Bob Craigie (NIDDK, National Institutes of Health), Richard Compans (Emory University), Friedrich Frischknecht (European Molecular Biology Laboratory), Wade Gibson (Johns Hopkins University School of Medicine), Ramón González (Universidad Autónoma del Estado de Morelos), David Knipe (Harvard Medical School), Thomas Leitner (Los Alamos National Laboratory), Maxine Linial (Fred Hutchinson Cancer Center), Pedro Lowenstein (University of California, Los Angeles), Paul Masters (New York State Department of Health), Rolf Menzel (National Institutes of Health), Thomas Mettenleiter (Federal Institute for Animal Diseases, Insel Reims, Germany), Heather Ongley and Michael Chapman (Oregon Health and Science University), B. V. Venkataram Prasad (Baylor College of Medicine), Botond Roska (Friedrich Miescher Institute, Basel, Switzerland), Michael Rossmann (Purdue University), Alasdair Steven (National Institutes of Health), Phoebe Stewart (Vanderbilt University), Wesley Sundquist (University of Utah), Jose Varghese (Commonwealth Scientific and Industrial Research Organization), Robert Webster (St. Jude's Children's Research Hospital), Thomas Wilk (European Molecular Biology Laboratory), Alexander Wlodawer (National Cancer Institute), and Li Wu (Medical College of Wisconsin).

The collaborative work undertaken to prepare the third edition was facilitated greatly by an authors' retreat at The Institute for Advanced Study, Princeton, NJ, in August 2007. We thank Arnold Levine for making the Biology Library available to us. ASM Press generously provided financial support for this retreat as well as for our many other meetings

We thank all those who guided and assisted in the preparation and production of the book: Jeff Holtmeier (Director, ASM Press) for steering us through the complexities inherent in a team effort, Ken April (Production Manager, ASM Press) for keeping us on track during production, and Susan Schmidler

(Susan Schmidler Graphic Design) for her elegant and creative designs for the layout and cover. We are also grateful for the expert secretarial and administrative support from Trisha Barney and Ellen Brindle-Clark (Princeton University) and Mary Estes and Rose Walsh (Fox Chase Cancer Center) that facilitated preparation of this text. Special thanks go to Ellen Brindle-Clark for obtaining the permissions required for many of the figures.

This often-consuming enterprise was made possible by the emotional, intellectual, and logistical support of our families, to whom the two volumes are dedicated.

I The Science of Virology

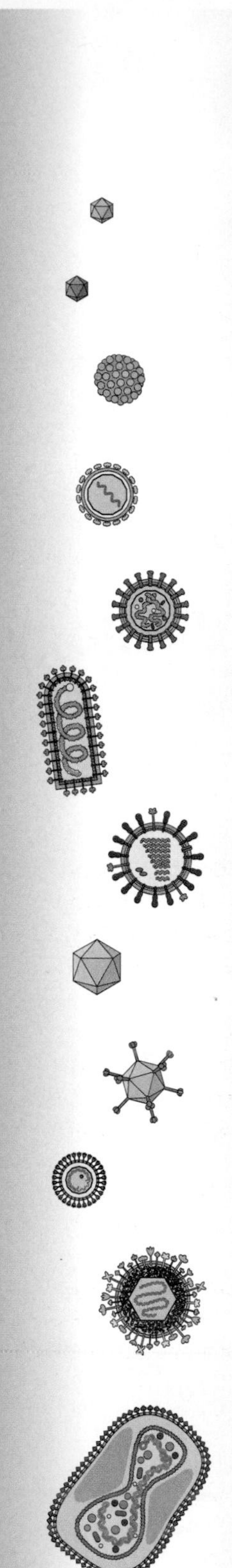

1

Table 1.1 Bacteriophages: landmarks in molecular biology[a]

Year	Discovery (discoverer[s])
1939	One-step growth of viruses (Ellis and Delbrück)
1946	Mixed phage infection leads to genetic recombination (Delbrück)
1947	Mutation and DNA repair (multiplicity reactivation) (Luria)
1952	Transduction of genetic information (Zinder and Lederberg)
1952	DNA, not protein, found to be the genetic material (Hershey and Chase)
1952	Restriction and modification of DNA (Luria)
1955	Definition of a gene (*cis-trans* test) (Benzer)
1958	Mechanisms of control of gene expression by repressors and activators are established (Pardee, Jacob, and Monod)
1958	Definition of the episome (Jacob and Wollman)
1961	Discovery of mRNA (Brenner, Jacob, and Meselson)
1961	Elucidation of the triplet code by genetic analysis (Crick, Barnett, Brenner, and Watts-Tobin)
1961	Genetic definition of nonsense codons as stop signals for translation (Campbell, Epstein, and Bernstein)
1964	Colinearity of the gene with the polypeptide chain (Sarabhai, Stretton, and Brenner)
1966	Pathways of macromolecular assembly (Edgar and Wood)
1974	Vectors for recombinant DNA technology (Murray and Murray, Thomas, Cameron, and Davis)

[a]Sources: T. D. Brock, *The Emergence of Bacterial Genetics* (Cold Spring Harbor Laboratory Press, Cold Spring Harbor, NY, 1990); K. Denniston and L. Enquist, *Recombinant DNA* (*Benchmark Papers in Microbiology*, vol. 15) (Dowden, Hutchinson and Ross, Inc., Stroudsburg, PA, 1981); and C. K. Mathews, E. Kutter, G. Mosig, and P. Berget, *Bacteriophage T4* (American Society for Microbiology, Washington, DC, 1983).

insights into cellular biology and functioning of host defenses. Intensive studies of viruses that infect bacteria, the bacteriophages, laid the foundation of modern molecular biology (Table 1.1), and crystallization of the plant virus tobacco mosaic virus established a landmark in structural biology. Studies of animal viruses established many fundamental principles of cellular function, including the processes of gene replication and transcription and of messenger RNA (mRNA) processing and translation. The study of cancer (transforming) viruses revealed the genetic basis of this disease. It seems clear that studies of viruses will continue to open up such paths of discovery in the future.

Viruses Can Also Be Used To Manipulate Biology

With the development of recombinant DNA technology and our increased understanding of some viral systems, it has become possible to use viral genomes as vehicles for the delivery of selected genes to cells and organisms for both scientific and medical purposes. The use of viral vectors to introduce tagged genes into various cells and organisms to study their function has become a standard method in biology. Viral vectors are also being tested to treat human disease via "gene therapy." Such technological advances also have a dark side, as they make it possible to reprogram viral genomes as agents of bioterrorism. A comprehensive understanding of virus biology and pathogenesis is our only rational defense against such eventualities.

Virus Prehistory

Viruses have been known as distinct biological entities for little more than a century. Consequently, efforts to understand and control these important agents of disease are phenomena of the 20th century. Nevertheless, evidence of viral infection can be found among the earliest recordings of human activity, and methods for combating viral disease were practiced long before the first virus was recognized.

Viral Infections in Antiquity

Reconstruction of the prehistoric past to provide a plausible account of when or how viruses established themselves in human populations is a challenging task. However, extrapolating from current knowledge, we can deduce that some modern viruses undoubtedly were associated with the earliest precursors of mammals and coevolved with humans. Other viruses entered human populations only recently. The last 10,000 years of human history was a time of radical change for humans and our viruses: animals were domesticated, the human population increased dramatically, large population centers appeared, and commerce drove interactions among unprecedented numbers of humans. We can infer from scattered glimpses of ancient history that viruses have long been a part of human experience.

Some viruses that eventually established themselves in human populations were undoubtedly transmitted to early humans from animals, much as still happens today. Early human groups that domesticated and lived with their animals were almost certainly exposed to different viruses than were nomadic hunter societies. Similarly, as many different viruses are **endemic** in the tropics, human societies in that environment must have been exposed to a greater variety of viruses than societies established in temperate climates. When nomadic groups met others with domesticated animals, or individuals from tropical cultures mingled with those from cooler climates, human-to-human contact could have provided new avenues for virus spread. Even so, it seems unlikely that viruses such as those that cause measles or smallpox could have entered a permanent relationship with small groups of early humans. Such highly virulent viruses, as we now know them to be, either kill their hosts or induce lifelong immunity. Consequently, they can survive only where large, interacting host populations, which offer a sufficient number of naive and sensitive hosts, are available for their continued propagation. Such viruses could not have been established in human populations until large, settled communities appeared. Less virulent viruses that enter into a more benign, long-term relationship with their hosts were therefore more likely to be the first to become adapted to replication in the earliest human populations. These viruses include the modern retroviruses, herpesviruses, and papillomaviruses.

Evidence of several viral diseases can be found in ancient records (Fig. 1.1). The Greek poet Homer characterizes Hector as rabid in *The Iliad*. Mesopotamian laws that outline the responsibilities of the owners of rabid dogs date from before 1,000 B.C. Their existence indicates that the communicable nature of this viral disease was already well known by that time. Egyptian hieroglyphs that illustrate what appear to be the consequences of poliovirus infection (a withered leg typical of poliomyelitis [Fig. 1.1B]) or pustular lesions characteristic of smallpox also date from that period. The smallpox virus, which was probably endemic in the Ganges river basin by the fifth century B.C. and subsequently spread to other parts of Asia and Europe, has played an important part in human history. Its introduction into the previously unexposed native populations of Central and South America by colonists in the 15th century led to lethal epidemics, which are considered an important factor in the conquests achieved by a small number of European soldiers. Other viral diseases known in ancient

Figure 1.1 References to viral diseases abound in the ancient literature. (A) An image of Hector from an ancient Greek vase. Courtesy of the University of Pennsylvania Museum (object 30-44-4). **(B)** An Egyptian stele, or stone tablet, from the 18th dynasty (1580–1350 B.C.) depicting a man with a withered leg and the "drop foot" syndrome characteristic of polio. Panel B is reprinted from W. Biddle, *A Field Guide to Germs* (Henry Holt and Co., LLC, New York, NY, 1995; © 1995 by Wayne Biddle), with permission from the publisher.

A

Here this firebrand, rabid Hector, leads the charge.

HOMER, *The Iliad*,
translated by Robert Fagels
(Viking Penguin)

B

times include mumps and, perhaps, influenza. Yellow fever has been described since the discovery of Africa by Europeans, and it has been suggested that this scourge of the tropical trade was the basis for legends about ghost ships, such as the *Flying Dutchman,* in which an entire ship's crew perished mysteriously.

Humans have not only been subject to viral disease throughout much of their history but have also manipulated these agents, albeit unknowingly, for much longer than might be imagined. One classic example is the cultivation of marvelously patterned tulips, which were of enormous value in 17th-century Holland. Such efforts included deliberate spread of a virus (tulip breaking virus or tulip mosaic virus) that we now know causes the striping of tulip petals so highly prized at that time (Fig. 1.2). Attempts to control viral disease have an even more venerable history.

Figure 1.2 ***Three Broken Tulips.*** A painting by Nicolas Robert (1624–1685), now in the collection of the Fitzwilliam Museum, Cambridge, United Kingdom. Striping patterns (color breaking) in tulips were described in 1576 in western Europe and were caused by a viral infection. This beautiful image depicts the remarkable consequences of infection with the tulip mosaic virus. Courtesy of the Fitzwilliam Museum, University of Cambridge.

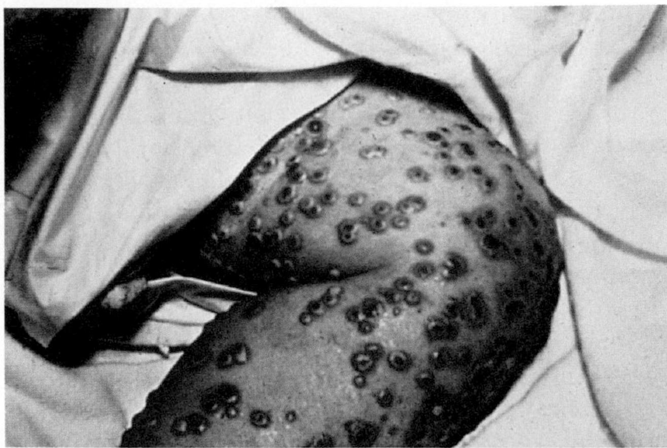

Figure 1.3 Characteristic smallpox lesions in a young victim. Illustrations like these were used to track down individuals infected with the smallpox virus (variola virus) during the World Health Organization campaign to eradicate the disease. Photo courtesy of the Immunization Action Coalition (original source: Centers for Disease Control and Prevention).

The First Vaccines

Measures to control one viral disease have been used with some success for the last millennium. The disease is smallpox (Fig. 1.3), and the practice is called **variolation**, inoculation of healthy individuals with material from a smallpox pustule into a scratch made on the arm. Variolation, which was widespread in China and India by the 11th century, was based on the recognition that smallpox survivors were protected against subsequent bouts of the disease. Variolation later spread to Asia Minor, where its value was recognized by Lady Mary Wortley Montagu, wife of the British ambassador to the Ottoman Empire. She introduced this practice into England in 1721, where it became quite widespread following successful inoculation of children of the royal family. George Washington is said to have introduced variolation among Continental Army soldiers in 1776. However, the consequences of variolation were unpredictable and never pleasant. Serious skin lesions invariably developed at the site of inoculation and might be accompanied by more generalized rash and disease, with a fatality rate of 1 to 2%. From the comfortable viewpoint of a developed country in the 21st century, such a death rate seems unacceptably high. However, in the 18th century, variolation was perceived as a much better alternative than contracting natural smallpox, a disease associated with a fatality rate of 25% in the whole population and 40% in babies and young children.

In the 1790s, Edward Jenner, an English country physician, recognized the principle on which modern methods of viral immunization are based, even though viruses themselves were not to be identified for another 100 years. Jenner himself was variolated as a boy and also practiced this procedure. He was undoubtedly familiar with its effects and risks. Perhaps this experience spurred his great insight upon observing that milkmaids who had been exposed to cowpox (a mild disease in humans) were protected against smallpox. Jenner followed up this astute observation with direct experiments. In 1794 to 1796, he demonstrated that inoculation with extracts from cowpox lesions induced only mild symptoms but protected against the far more dangerous smallpox disease. It is from these experiments with cowpox that we derive the term **vaccination** (*vacca* = cow in Latin). This term was first given its general meaning by Louis Pasteur in 1881 to honor Jenner's accomplishments.

Initially, the only way to propagate and maintain cowpox vaccine was by serial infection of human subjects. This method was eventually banned, as it was associated with transmission of other diseases such as syphilis and hepatitis at significant frequencies. By 1860, the vaccine had been passaged in cows; later, sheep and water buffaloes were also used. While Jenner's original vaccine was based on the virus that causes cowpox, sometime during the human-to-human or cow-to-cow transfers, the poxvirus now called vaccinia virus replaced the cowpox virus. Vaccinia virus is the basis for the modern smallpox vaccine, but its origins are not known. It exhibits little genetic similarity to the viruses that cause cowpox or smallpox or to many of the known members of the poxvirus family. Scientists have recovered the smallpox vaccine used in New York in 1876 and have verified that it contains vaccinia virus and not cowpox virus. Speculation about when and how the switch occurred has produced some fascinating scenarios (Box 1.2).

The first deliberately attenuated viral vaccine was made by Louis Pasteur. In 1885 he inoculated rabbits with material from the brain of a cow infected with rabies virus and then used aqueous suspensions of dried spinal cords from these animals to infect other rabbits. After several such passages, the resulting preparations caused mild disease (i.e., were **attenuated**) yet produced effective immunity against rabies. Safer and more efficient methods for the production of larger quantities of these first vaccines awaited the recognition of viruses as distinctive biological entities and knowledge about their parasitism of cells in their hosts. Indeed, it took almost 50 years to discover the next antiviral vaccines: a vaccine for yellow fever virus was developed in 1935, and an influenza vaccine was available in 1936. These advances became possible only with radical changes in our knowledge of living organisms and of the causes of disease.

BOX 1.2 DISCUSSION

Origin of vaccinia virus

Over the years, at least three hypotheses have been advanced to explain the curious substitution of cowpox virus by vaccinia virus:

1. Recombination of cowpox virus with smallpox virus after variolation of humans
2. Recombination between cowpox virus and animal poxviruses during passage in various animals
3. Genetic drift of cowpox virus after repeated passage in humans and animals

None of these hypotheses has been proven conclusively.

Vaccination against smallpox. The image is from *Public Health Bulletin* vol. 52, no. 3, spring 2003, California County of Orange Health Care Agency (http://www.ochealthinfo.com/newsletters/phbulletin/2003/2003-spring.htm), with permission.

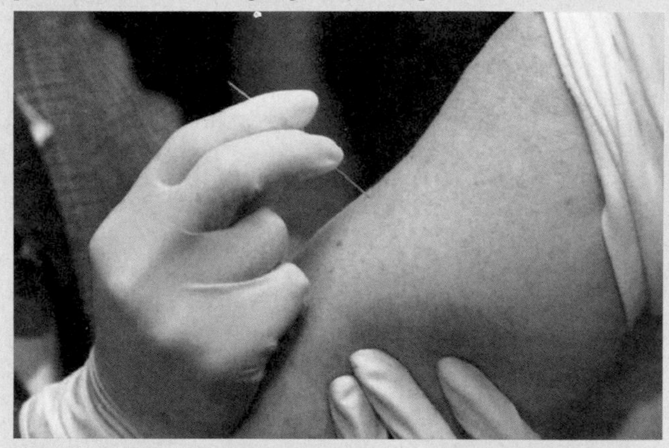

Microorganisms as Pathogenic Agents

The 19th century was a period of revolution in scientific thought, particularly in ideas about the origins of living things. The publication of Charles Darwin's *The Origin of Species* in 1859 crystallized startling (and, to many people, shocking) new ideas about the origin of diversity in plants and animals, until then generally attributed directly to the hand of God. These insights permanently undermined the belief that humans were somehow set apart from all other members of the animal kingdom. From the point of view of the science of virology, the most important changes were in ideas about the causes of disease.

The diversity of macroscopic organisms has been appreciated and cataloged since the dawn of recorded human history. A vast new world of organisms too small to be visible to the naked eye was revealed through the microscopes of Antony van Leeuwenhoek (1632–1723). Among van Leeuwenhoek's vivid and enthusiastic descriptions of living microorganisms as "wee animalcules," seen in such ordinary materials as rain or seawater, are examples of what we now know to be protozoa, algae, and bacteria. By the early 19th century, the scientific community had accepted the existence of microorganisms and turned to the question of their origin—a topic of fierce debate. Some believed that microorganisms arose spontaneously, for example, in decomposing matter, where they were especially abundant. Others held the view that all were generated by the reproduction of like microorganisms, as were macroscopic organisms. The death knell of the spontaneous-generation hypothesis was sounded with the famous experiments of Pasteur. He demonstrated that boiled (i.e., sterilized) medium remained free of microorganisms as long as it was maintained in special flasks with curved, narrow necks designed to prevent entry of airborne microbes (Fig. 1.4). Pasteur also established that particular microorganisms were associated with specific processes, for example, the production of alcohol, lactic acid, or acetic acid (vinegar) by fermentation. This idea was crucial in the development of modern explanations for the causes of disease.

Figure 1.4 Four experiments to challenge the spontaneous-generation hypothesis. The first step in each experiment was to boil the broth medium very thoroughly to destroy all living organisms. Air was then admitted to the flasks to satisfy the believers in spontaneous generation, who insisted that oxygen must be present for life to originate. However, the air admitted to the broth medium was first freed of living organisms in several ingenious ways. Under these conditions the broths remained perfectly sterile; no microorganisms appeared in them, showing that living things could not be generated spontaneously from the lifeless liquid. **(A)** Sterilizing air by chemical treatment. The set of bulbs next to the person's face contained alkali, and the other set contained concentrated acid. Air was drawn in through the acid to inactivate microbes before it reached the broth. **(B)** Sterilizing air by using heat. Air could enter the flask of broth only by passing through the coiled glass tube kept hot by the flame. **(C)** Sterilizing air by physically trapping particles. The aspiration bottle drew air into the flask through the tube containing cotton at the right. The cotton filtered out the microbes in the air, just as the cotton or foam plugs now used in bacteriological culture tubes protect the culture from air contamination. **(D)** Pasteur's famous swan-neck flasks provided passive exclusion of microbes from the sterilized broth. Although the flask was freely open to the air, the broth remained sterile so long as the microbe-bearing dust that collected in the neck did not reach the liquid.

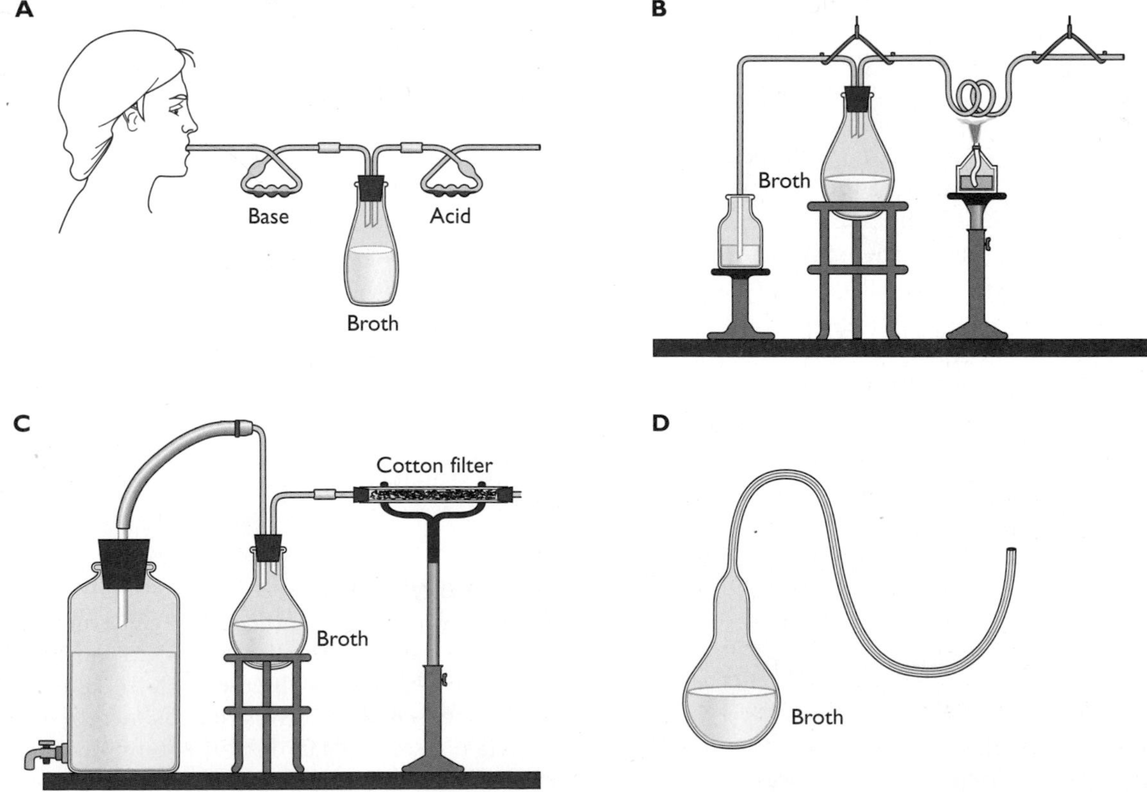

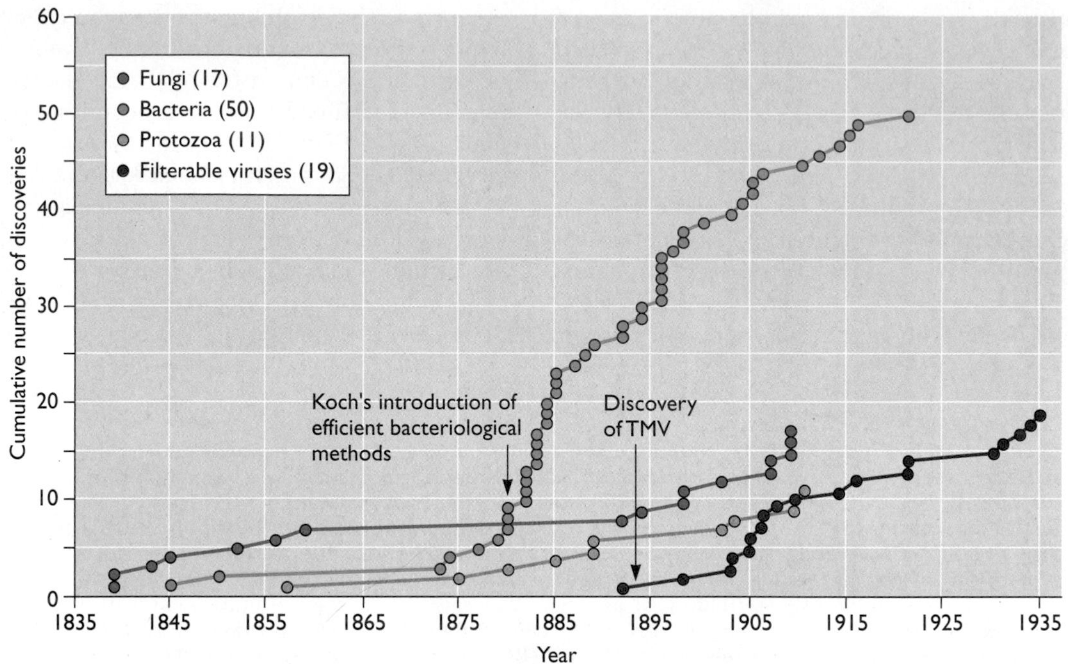

Figure 1.5 The pace of early discovery of new infectious agents. Koch's introduction of efficient bacteriological techniques spawned an explosion of new discoveries of bacterial agents in the early 1880s. Similarly, the discovery of filterable agents launched the field of virology in the early 1900s. Despite an early surge of virus discovery, only 19 distinct human viruses had been reported by 1935. TMV, tobacco mosaic virus. Adapted from K. L. Burdon, *Medical Microbiology* (Macmillan Co., New York, NY, 1939), with permission.

From the earliest times, poisonous air (miasma) was generally invoked to account for **epidemics** of contagious diseases, and there was little recognition of the differences among their causative agents. The association of particular microorganisms, initially bacteria, with specific diseases can be attributed to the ideas of the German physician Robert Koch. He developed and applied a set of criteria for identification of the agent responsible for a specific disease (a **pathogen**). These criteria, **Koch's postulates**, are still applied in the identification of pathogens that can be propagated in the laboratory and tested in an appropriate animal model. The postulates are as follows.

- The organism must be regularly associated with the disease and its characteristic lesions.
- The organism must be isolated from the diseased host and grown in culture.
- The disease must be reproduced when a pure culture of the organism is introduced into a healthy, susceptible host.
- The same organism must be reisolated from the experimentally infected host.

By applying these criteria, Koch demonstrated that anthrax, a common disease of cattle, was caused by a specific bacterium (designated *Bacillus anthracis*) and that a second, distinct bacterial species caused tuberculosis in humans. Guided by these postulates and the methods for the sterile culture and isolation of pure preparations of bacteria developed by Pasteur, Joseph Lister and Koch identified and classified many pathogenic bacteria (as well as yeasts and fungi) during the last part of the 19th century (Fig. 1.5). From these beginnings, investigation into the causes of infectious disease was placed on a secure scientific foundation, the first step toward rational treatment and ultimately control. During the last decade of the 19th century, failures of the paradigm that bacterial or fungal agents are responsible for all diseases led to the identification of a new class of infectious agents—submicroscopic pathogens that came to be called **viruses**.

Discovery of Viruses

The first report of a pathogenic agent smaller than any known bacterium appeared in 1892. The Russian scientist Dimitrii Ivanovsky observed that the causative agent of tobacco mosaic disease was not retained by the unglazed porcelain filters used at that time to remove bacteria from extracts and culture media (Fig. 1.6). Six years later in

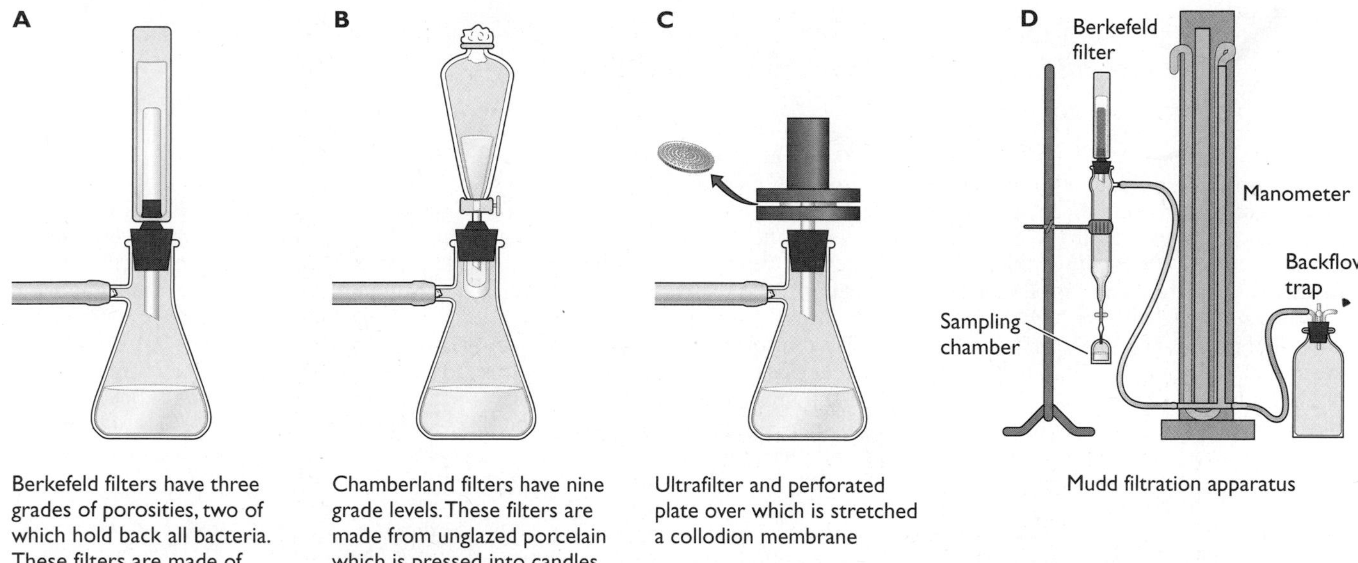

Berkefeld filters have three grades of porosities, two of which hold back all bacteria. These filters are made of diatomaceous earth.

Chamberland filters have nine grade levels. These filters are made from unglazed porcelain which is pressed into candles open at one end.

Ultrafilter and perforated plate over which is stretched a collodion membrane

Mudd filtration apparatus

Figure 1.6 Filter systems used to characterize viruses. Filters for making liquids free of cultivatable organisms were instrumental in the first identification of viruses. Several types of filters were used in the early days of virus research, and four are illustrated here. Berkefeld filters **(A)** and Chamberland filters **(B)** are typical of the "candle" style of filter comprising diatomaceous earth pressed into the shape of a hollow candle open at one end. Only the smallest pore sizes retained bacteria and allowed viruses to pass through. For the Chamberland filter **(B)**, the fluid to be filtered was introduced into the open end by means of a funnel. The filter is cut away to illustrate the hollow center. Filters similar to these were probably used by Ivanovsky, Loeffler, and Frosch to isolate the first plant and animal viruses. **(C)** Example of a collodion membrane "ultrafilter." The thin membrane is the filtering surface and is stretched over a filter plate. **(D)** A Mudd filtration apparatus illustrating the typical setup for filtering fluids. All the filters were designed to operate under negative pressure by connection with a suction pump or aspirator. Positive pressure was also used. The trap (on the right) was essential to catch backflow of liquid when the suction was released. The manometer (in the middle) provided an accurate means of measuring and regulating pressure. The filtrate was collected in a graduated buret, from which measured samples were collected and tested.

Holland, Martinus Beijerinck independently made the same observation. More importantly, Beijerinck made the conceptual leap that because the pathogen was so small that it could pass through filters that trapped all known bacteria, it must be a distinctive agent.

The same year (1898), the German scientists Friedrich Loeffler and Paul Frosch, both former students and assistants of Koch, observed that the causative agent of foot-and-mouth disease was also filterable (Box 1.3). Not only were the tobacco mosaic and foot-and-mouth disease pathogens much smaller than any previously recognized microorganism, but also they replicated **only** in their host organisms. For example, extracts of an infected tobacco plant diluted into sterile solution produced no additional infectious agents until introduced into leaves of healthy plants, which subsequently developed tobacco mosaic disease. The serial transmission of infection by diluted extracts established that these diseases were not caused by a bacterial toxin present in the original preparations derived from infected tobacco plants or cattle. The failure of both pathogens to multiply in solutions that readily supported the growth of bacteria, as well as their dependence on host organisms for reproduction, further distinguished these new agents from pathogenic bacteria. Beijerinck termed the submicroscopic agent responsible for tobacco mosaic disease *contagium vivum fluidum* to emphasize its infectious nature and distinctive reproductive and physical properties. Agents passing through filters that retain bacteria came to be called ultrafilterable viruses, appropriating the term *virus* from the Latin for "poison." This term eventually was simplified to viruses.

The discovery of the first virus, tobacco mosaic virus, is often attributed to the work of Ivanovsky in 1892. However, he did not identify the tobacco mosaic disease pathogen as a distinctive agent, nor was he convinced that its passage through bacterial filters was not the result of some technical failure. It may be more appropriate to attribute the founding of the field of virology to the astute

BOX 1.3

DISCUSSION

The first animal virus discovered remains a scourge today

The first animal virus to be discovered, foot-and-mouth disease virus, infects domestic cattle, pigs, and sheep, as well as many species of wild animals. Although mortality is low, morbidity is high and infected domestic animals lose their commercial value. The virus is highly contagious, and the most common and effective method of control is by the slaughter of entire herds in affected areas.

Outbreaks of foot-and-mouth disease were widely reported in Europe, Asia, Africa, and South and North America in the 1800s. The largest epidemic ever recorded in the United States occurred in 1914. After gaining entry into the Chicago stockyards, the virus spread to more than 3,500 herds in 22 states. This calamity accelerated epidemiologic and disease control programs, eventually leading to the field- and laboratory-based systems maintained by the U.S. Department of Agriculture to protect domestic livestock from foreign animal and plant diseases. Similar control systems have been established in other Western countries, but this virus still presents a formidable challenge throughout the world. A 1997 outbreak of foot-and-mouth disease among pigs in Taiwan resulted in economic losses of greater than $10 billion.

In 2001, an epidemic outbreak in the United Kingdom spread to other countries in Europe and led to the slaughter of more than 3 million infected and uninfected farm animals. The associated economic, societal, and political costs threatened to bring down the British government. Images of mass graves and horrific pyres consuming the corpses of dead animals sensitized the public as never before. It is clear that viruses do not have to infect humans to exact a human cost. It is sobering to realize that we have known about this animal virus longer than any other but have still not learned enough to keep it from exacting its toll.

Murphy, F. A., E. P. J. Gibbs, M. C. Horzinek, and M. J. Studdert. 1999. *Veterinary Virology,* 3rd ed. Academic Press, Inc., San Diego, CA.

insights of Beijerinck, Loeffler, and Frosch, who recognized the distinctive nature of the plant and animal pathogens they were studying more than 100 years ago.

The pioneering work on tobacco mosaic virus and foot-and-mouth disease virus was followed by the identification of viruses associated with specific diseases in many other organisms. Important landmarks from this early period include the identification of viruses that cause leukemias or solid tumors in chickens by Vilhelm Ellerman and Olaf Bang in 1908 and by Peyton Rous in 1911, respectively. The study of viruses associated with cancers in chickens, particularly the Rous sarcoma virus, eventually led to an understanding of the molecular basis of cancer (Volume II, Chapter 7).

The fact that bacteria could also be hosts to viruses was first recognized by Frederick Twort in 1915 and Félix d'Hérelle in 1917. d'Hérelle named them **bacteriophages** because of their ability to lyse bacteria on the surface of agar plates (phage is derived from the Greek for "eating"). In an interesting twist of serendipity, Twort made his discovery of bacterial viruses while testing the smallpox vaccine virus to see if it would grow on simple media. He found bacterial contaminants, some of them exhibiting an unusual "glassy transformation," which proved to be the result of lysis by a bacteriophage. Investigation of bacteriophages provided the foundations for the field of molecular biology, as well as fundamental insights into how viruses interact with their host cells.

The Definitive Properties of Viruses

Throughout the early period of virology, in which many viruses of plants, animals, and bacteria were cataloged, the origin and nature of these distinctive infectious agents were quite controversial. Arguments centered on whether viruses represented parts of a cell that had somehow acquired a new kind of existence or whether they were built from virus-specific components. Little progress was made toward resolving these issues and establishing the definitive properties of viruses until new techniques were developed, which allowed their visualization or propagation in cultured cells.

The Structural Simplicity of Viruses

Dramatic confirmation of the structural simplicity of viruses came in 1935, when crystals of tobacco mosaic virus were obtained by Wendell Stanley. At that time, nothing was known of the structural organization of any biologically important macromolecules, such as proteins and DNA. Indeed, the crucial role of DNA as genetic material had not even been recognized. The ability to obtain an infectious agent in crystalline form, a state that is more generally associated with inorganic material, therefore created much wonder and speculation about whether a virus is truly a life form. In retrospect, it is obvious that the relative ease with which tobacco mosaic virus could be crystallized was a direct result of its structural simplicity and the ability of many particles to associate in regular arrays.

The 1930s saw the introduction of the instrument that rapidly revolutionized virology, the electron microscope. The great magnifying power of this instrument (eventually over 100,000-fold) allowed direct visualization of virus particles for the first time. It has always been an exciting experience for investigators to obtain images of viruses,

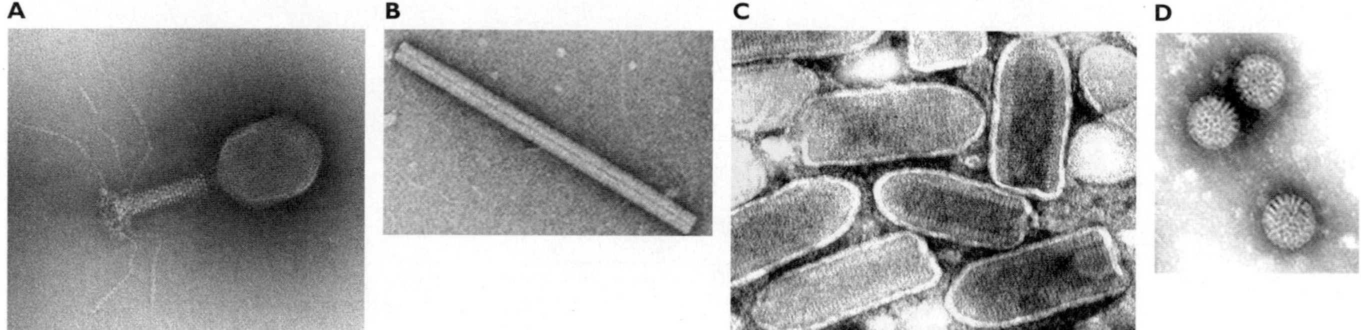

Figure 1.7 Electron micrographs of viruses following negative staining. (A) The complex, nonenveloped virus bacteriophage T4. Note the intricate tail and tail fibers. Courtesy of R. L. Duda, University of Pittsburgh. **(B)** The helical, nonenveloped particle of tobacco mosaic virus. Reprinted from the Universal Virus Database of the International Committee on Taxonomy of Viruses (http://www.ncbi.nlm.nih.gov/ICTVdb/WIntkey/Images/em_tmv.gif), with permission. **(C)** Enveloped particles of the rhabdovirus vesicular stomatitis virus. Courtesy of F. P. Williams, University of California, Davis. **(D)** Nonenveloped, icosahedral human rotavirus particles. Courtesy of F. P. Williams, U.S. Environmental Protection Agency (http://www.epa.gov/nerlcwww/rota.htm).

especially as they often prove to be remarkably elegant in appearance (Fig. 1.7). Images of many different viruses confirmed that these agents are very small (Fig. 1.8) and far simpler in structure than any cellular organism. Many appeared as regular helical or spherical particles. The description of the morphology of virus particles made possible by electron microscopy also opened the way for the first rational classification of viruses.

The Intracellular Parasitism of Viruses

Organisms as Hosts

The fundamental characteristic of viruses is their absolute dependence on a living host for reproduction: they are **obligate parasites**. Transmission of plant viruses such as tobacco mosaic virus can be achieved readily, for example, by applying extracts of an infected plant to a scratch made on the leaf of a healthy plant. Furthermore, as a single infectious particle of many plant viruses is sufficient to induce the characteristic lesion (Fig. 1.9), the concentration of the infectious agent could be measured. Plant viruses were therefore the first to be studied in detail. Viruses of animals could also be propagated under experimental conditions, and methods were developed to quantify these agents by determining the lethal dose. These methods could also be applied to human viruses that were able to infect laboratory animals. The transmission of yellow fever virus to mice by Max Theiler in 1930 was an achievement that led to the isolation of an attenuated strain, still considered one of the safest and most effective ever produced for the vaccination of humans.

After specific viruses and host organisms were identified, it became possible to produce sufficient quantities of virus particles for investigation of their physical and chemical properties. Scientists were also able to determine the consequences of infection for the host. Features such as the incubation period, gross symptoms of infection, and effects on specific tissues and organs were investigated. Laboratory animals remain an essential tool in investigations of the pathogenesis of viruses that cause disease. However, real progress toward understanding the mechanisms of virus replication was made only with the development of tissue and cell culture systems. Among the simplest, but crucial to both virology and molecular biology, were cultures of bacterial cells.

Lessons from Bacteriophages

In the late 1930s and early 1940s, bacteriophages, or "phages," received increased attention as a result of controversy centering on how they were formed. d'Hérelle, one of their discoverers, favored the hypothesis that there was only one phage that attacked all bacteria. This hypothesis was disproved by F. Macfarlane Burnet, who showed that there were a great variety of phages with different physical and biological properties. Nevertheless, the precise nature of these agents remained elusive. John Northrup, a biochemist at the Rockefeller Institute in Princeton, NJ, championed the theory that a phage was a metabolic product of a bacterium. Phage formation was said to be analogous to the autocatalytic formation of enzymes from inactive precursors, a direct rejection of the proposal by d'Hérelle that a phage was a living organism. On the other hand,

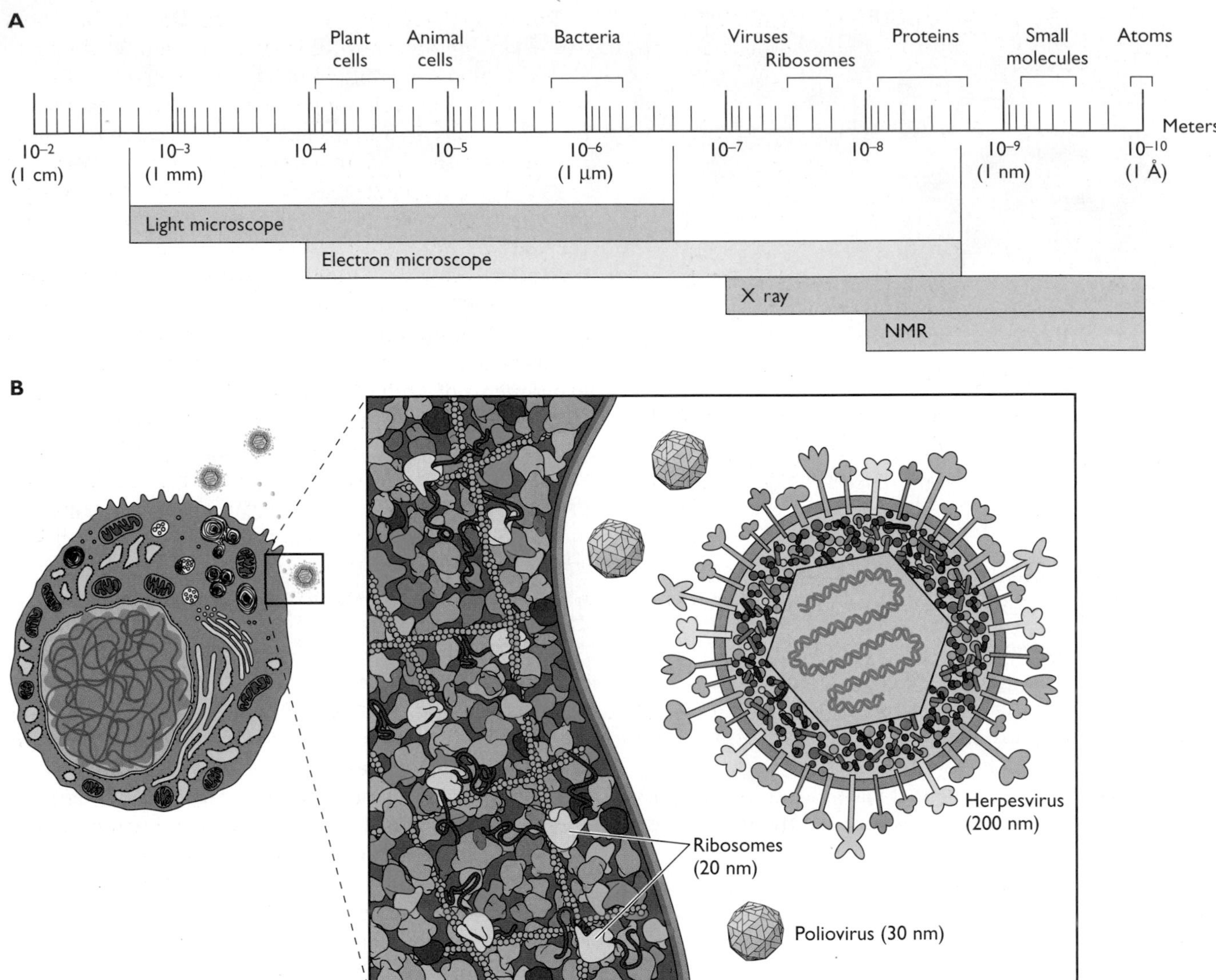

Figure 1.8 Size matters. (A) Sizes of animal and plant cells, bacteria, viruses, proteins, molecules, and atoms are indicated. The resolving powers of various techniques used in virology, including light microscopy, electron microscopy, X-ray crystallography, and nuclear magnetic resonance (NMR) spectroscopy, are indicated. Viruses, which are within the resolving power of the electron microscope, are about 2 orders of magnitude smaller than the smallest bacterium. The units commonly used in descriptions of virus particles or their components are the nanometer (nm [10^{-9} m]) and the angstrom (Å [10^{-10} m]). Adapted from A. J. Levine, *Viruses* (Scientific American Library, New York, NY, 1991); used with permission of Henry Holt and Company, LLC. **(B)** Illustration of the size differences among two viruses and a typical host cell.

Max Delbrück, in his work with Emory Ellis and later with Luria, regarded phages as autonomous, stable, self-replicating entities characterized by heritable traits. According to this paradigm, phages were seen as ideal tools with which to investigate the nature of genes and heredity. Probably the most critical early contribution of Delbrück and Ellis was the perfection of the one-step growth technique for synchronization of the replication of phage, an achievement that allowed analysis of a single cycle of phage growth in a population of bacteria. This approach introduced highly quantitative methods to virology, as well as an unprecedented rigor of analysis. The first experiments showed that phages indeed multiplied in the bacterial host and were liberated in a "burst" by lysis of the cell.

Figure 1.9 Lesions induced by tobacco mosaic virus on an infected tobacco leaf. In 1886, Adolph Mayer first described the characteristic patterns of light and dark green areas on the leaves of tobacco plants infected with tobacco mosaic virus. He demonstrated that the mosaic lesions could be transmitted from an infected plant to a healthy plant by aqueous extracts derived from infected plants. The number of local necrotic lesions that result is directly proportional to the number of infectious particles in the preparation. Photo by J. P. Krausz, from *Lessons in Plant Pathology*, The American Phytopathological Society Education Center (http://www.apsnet.org/education/lessonsPlantPath/TMV/text/symptom.htm), reprinted with permission.

Delbrück was a zealot for phage research and recruited talented scientists to pursue the fundamental issues of what is now known as the field of molecular biology. This group of scientists, working together in what came to be called the "phage school," focused their attention on specific phages of the bacterium *Escherichia coli*. The list of Nobel laureates who were trained as phage workers is a testament to Delbrück's leadership. Progress was rapid, primarily because of the simplicity of the phage infectious cycle. Phages replicate in bacterial hosts, large numbers of which can easily be obtained by overnight culture. By the mid-1950s, it was evident that viruses from bacteria, animals, and plants share many fundamental properties. However, the phages provided a far more tractable experimental system. Consequently, their study had a profound impact on the development of virology.

One critical lesson came from the definitive experiments which established that viral nucleic acid carries genetic information. It was known from studies of the "transforming principle" of *Pneumococcus* by Oswald Avery, Colin MacLeod, and Maclyn McCarty (1944) that nucleic acid was both necessary and sufficient for the transfer of genetic traits of bacteria. However, in the early 1950s, viral protein was still suspected to be an important component of viral heredity. In a brilliantly simple experiment that included the use of a common kitchen item, a food blender, Alfred Hershey and Martha Chase showed that this hypothesis was incorrect (Box 1.4).

Bacteriophages were originally thought to be lethal agents, killing their host cells after infection. In the early 1920s, a previously unknown interaction was discovered, in which the host cell not only survived the infection but also stably inherited the genetic information of the virus. It was also observed that certain bacterial strains not known to be infected could lyse spontaneously and produce bacteriophages after a period of growth in culture. Such strains were called **lysogenic**, and the phenomenon was called **lysogeny**. Studies of lysogeny uncovered many previously unrecognized features of virus-host cell interactions. Recognition of this phenomenon resulted from the work of many scientists, but it began with the elegant experiments of André Lwoff and colleagues at the Institut Pasteur in Paris. Lwoff showed that a viral genome exists in lysogenic cells in the form of a specific genetic element called the **prophage**. This element determined the ability of lysogenic bacteria to produce infectious bacteriophage. Subsequent studies of the *E. coli* phage lambda established a paradigm for one of the many mechanisms of lysogeny, the integration of a phage genome into a specific site on the bacterial chromosome (Box 1.5).

Bacteriophages became inextricably associated with the new field of molecular biology (Table 1.1). Their study also established many fundamental principles. For example, control of the decision to enter a lysogenic or a lytic pathway is encoded in the genome of the virus. The first mechanisms discovered for the control of gene expression, exemplified by the elegant operon theory of Nobel laureates François Jacob and Jacques Monod, were deduced in part from studies of lysogeny by phage lambda. The biology of phage lambda provided a fertile ground for work on gene regulation, but study of virulent T phages (T1 to T7, where T stands for type) of *E. coli* paved the way for many other important advances (Table 1.1). As we shall see, these systems also provided an extensive preview of mechanisms of animal virus replication (Box 1.6).

Animal Cells as Hosts

The culture of animal cells in the laboratory was initially more of an art than a science, restricted to cells that grew out of organs or tissues maintained in nutrient solutions under sterile conditions. The finite life span of such **primary cells**, their dependence for growth on natural media such as lymph, plasma, or chicken embryo extracts, and the technical demands of sterile culture prior to the discovery of antibiotics made reproducible experimentation very difficult. By 1955 the work of many investigators had led to a series of important methodological advances. These included the development of defined media optimal

BOX 1.4

EXPERIMENTS

The Hershey-Chase experiment

By differentially labeling the nucleic acid and protein components of virus particles with radioactive phosphorus (^{32}P) and radioactive sulfur (^{35}S), respectively, Alfred Hershey and Margaret Chase showed that the protein coat of the infecting virus could be removed soon after infection by agitating the bacteria for a few minutes in a blender. In contrast, ^{32}P-labeled phage DNA entered and remained associated with the bacterial cells. As such blended cells produced a normal burst of new virus particles, it was clear that this DNA contained all of the information necessary to produce progeny phages.

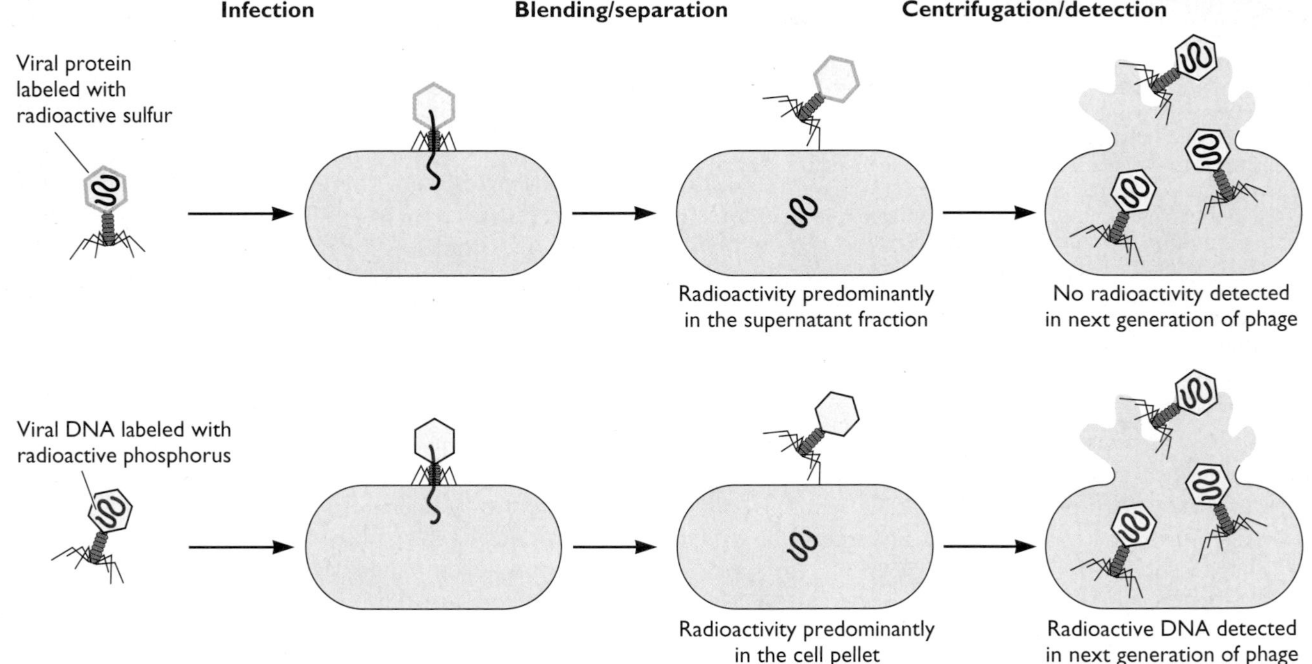

for growth of mammalian cells, incorporation of antibiotics into tissue culture media, and development of immortal cell lines such as the mouse L and human HeLa cells that are still in widespread use. These advances allowed growth of animal cells in culture to become a routine, reproducible exercise.

The availability of well-characterized cell cultures had several important consequences for virology. It allowed the discovery of several new human viruses, such as adenovirus, measles virus, and rubella virus, for which animal hosts were not available. In 1949, John Enders and colleagues used cell cultures to propagate poliovirus, a feat that led to the development of polio vaccines a few years later. Cell culture technology revolutionized the ability to investigate the replication of viruses. Viral infectious cycles could be studied under precisely controlled conditions by employing the analog of the one-step growth cycle of bacteriophages and simple methods for quantification of infectious particles described in Chapter 2. Our current understanding of the molecular basis of parasitism by viruses, the focus of this Volume (I), is based almost entirely on analyses of one-step growth cycles in cultured cells. Such studies established that viruses are **molecular** parasites: their reproduction depends absolutely on their host cell's biosynthetic machinery for synthesis of the components from which they are built. In contrast to cells, viruses do not reproduce by growth and division. Rather, the infecting genome contains information necessary to redirect cellular systems to the production of many copies of all the components needed for the

BOX 1.5

BACKGROUND

Studies of lysogeny established several general principles in virology

Lytic versus Lysogenic Response to Infection

Some bacterial viruses can enter into either destructive (lytic) or relatively benign (lysogenic) relationships with their host cells. Such bacteriophages were called "temperate." In a lysogenic bacterial cell, viral genetic information persists but viral gene expression is repressed. Such cells are called lysogens, and the quiescent viral genome is called a prophage.

Propagation as a Prophage

For some bacteriophages like lambda and Mu (Mu stands for mutator), prophage DNA is integrated into the host genome of lysogens and passively replicated by the host. Virally encoded enzymes, known as integrase (lambda) and transposase (Mu), mediate the covalent insertion of viral DNA into the chromosome of the host bacterium, establishing it as a prophage.

Transposition by Mu establishes an integrated prophage when viral gene expression is repressed, but it also leads to replication of the viral genome during the lytic cycle.

The prophage DNA of other bacteriophages, such as P1, exists as a plasmid, a self-replicating, autonomous chromosome in a lysogen. Both forms of propagation were subsequently identified in certain animal viruses.

Insertional Mutagenesis

Bacteriophage Mu inserts its genome into many random locations on the host chromosome, causing numerous mutations. This process is called insertional mutagenesis and is a phenomenon subsequently observed with retroviruses.

Gene Repression and Induction

Viral gene expression in lysogens is turned off by the action of viral proteins called repressors. Viral gene expression can be turned on when repressors are inactivated (a process called induction). Elucidation of the mechanisms involved set the stage for later investigation of the control of gene expression in experiments with other viruses and their host cells.

Transduction of Host Genes

Viral genomes can pick up cellular genes and deliver them to new cells (a process known as transduction). The process can be generalized, with the acquisition by the virus of any segment from the host chromosome, or specialized, as is the case for viruses that integrate into specific sites in the host chromosome. For example, occasional mistakes in excision of the lambda prophage after induction result in production of unusual progeny phage that have lost some of their own DNA but have acquired the bacterial DNA adjacent to the prophage. As described in Volume II, Chapter 7, the acute transforming retroviruses also arise via capture of proto-oncogenes in the vicinity of their integration as proviruses. These cancer-inducing cellular genes are then transduced along with viral genes during subsequent infection.

intrinsically programmed, *de novo* assembly of new virus particles.

Viruses Defined

Advances in knowledge of the structure of virus particles and the mechanisms by which viruses are reproduced in their host cells have been accompanied by increasingly accurate definitions of these unique agents. The earliest, pathogenic agents, distinguished by their small size and dependence on a host organism for reproduction, emphasized the importance of viruses as agents of disease. We can now provide a much more precise

BOX 1.6

TERMINOLOGY

The episome

In 1958, François Jacob and Elie Wollman realized that lambda prophage, the *E. coli* F sex factor, and the colicinogenic factor had many common genetic properties. This remarkable insight led to the definition of the episome.

An episome is an exogenous genetic element that is not necessary for cell survival. Its defining characteristic is the ability to reproduce in two alternative states, integrating into the host chromosome or by autonomous replication.

Nowadays this term is often applied to genomes that can be maintained in cells by autonomous replication and never integrate, for example, certain viral DNA genomes.

Jacob and Wollman immediately understood that the episome had value in understanding larger problems, including cancer, as revealed by this quotation: "... in the no man's land between heredity and infection, between physiology and pathology at the cellular level, episomes provide a new link and a new way of thinking about cellular genetics in bacteria, and perhaps in mice, men and elephants" (F. Jacob and E. Wollman, *Viruses and Genes: Readings from* Scientific American [W. H. Freeman & Co., New York, NY, 1961]).

BOX 1.7

WARNING

Viruses don't actually "do" anything!

Many have succumbed to the temptation of ascribing various actions and motives to viruses. While remarkably effective in enlivening a lecture or an article, anthropomorphic characterizations are inaccurate and often misleading.

Here are some examples from the anthropomorphic lexicon that should be avoided.

- Viruses cannot think, employ, ensure, synthesize, exhibit, display, destroy, deploy, depend, reprogram, avoid, retain, evade, exploit, generate, etc.
- Infected cells and hosts do many things in the presence of viruses, but viruses themselves are passive agents, totally at the mercy of their environments.

It is exceedingly difficult to purge such anthropomorphic terms from virology communications. Indeed, hours were spent doing so in preparation of this textbook. A good exercise to appreciate the difficulty of the problem may be to find all the examples that were missed.

definition of viruses, elaborating their relationship with the host cell and the important features of virus particles. The definitive properties of viruses are summarized as follows:

- A virus is an infectious, obligate intracellular parasite.
- The viral genome comprises either DNA or RNA.
- Within an appropriate host cell, the viral genome directs the synthesis by cellular systems of many copies of all the viral components.
- Progeny infectious virus particles, called **virions**, are formed by *de novo* self-assembly from the newly synthesized components within the host cell.
- A progeny virion assembled during the infectious cycle is the vehicle for transmission of the viral genome to the next host cell or organism, where disassembly of the virion leads to the beginning of the next infectious cycle.

With these properties in mind, we can accurately place viruses within the evolutionary continuum of biological agents. They are far simpler than even the smallest microorganisms and lack the complex energy-generating and biosynthetic systems necessary for independent existence (Box 1.7). On the other hand, viruses are **not** the simplest biologically active agents: even the smallest virus, built from a very limited genome and a single type of protein, is significantly more complex than other pathogens. Some of these minimalist molecular pathogens such as **viroids**, which are infectious agents of a variety of economically important plants, comprise a single small molecule of RNA. Others, termed **prions**, are thought to be single protein molecules.

Cataloging Animal Viruses

Around 1960, virus classification was a subject of colorful and quite heated controversy (Box 1.8). New viruses were being discovered and studied by electron microscopy. The virus world consisted of a veritable zoo of particles with different sizes, shapes, and compositions (see, for example, Fig. 1.10). Very strong opinions were advanced concerning classification and nomenclature, and opposing

BOX 1.8

TERMINOLOGY

Complexities of viral nomenclature

No consistent system for naming viruses has been established by their discoverers. For example, among the vertebrate viruses, some are named for the associated diseases (e.g., poliovirus, rabies virus), for the specific type of disease they cause (e.g., murine leukemia virus), or for the sites in the body that are affected or from which they were first isolated (e.g., rhinovirus and adenovirus). Others are named for the geographic locations in which they were first isolated (e.g., Sendai virus [Sendai, Japan] and Coxsackievirus [Coxsackie, New York]) or for the scientists who first discovered them (e.g., Epstein-Barr virus). In these cases the virus names are capitalized. Some viruses are even named for the way in which people imagined they were contracted (e.g., dengue, for "evil spirit," and influenza, for the "influence" of bad air). Finally, combinations of the above designations are also used (e.g., Rous sarcoma virus).

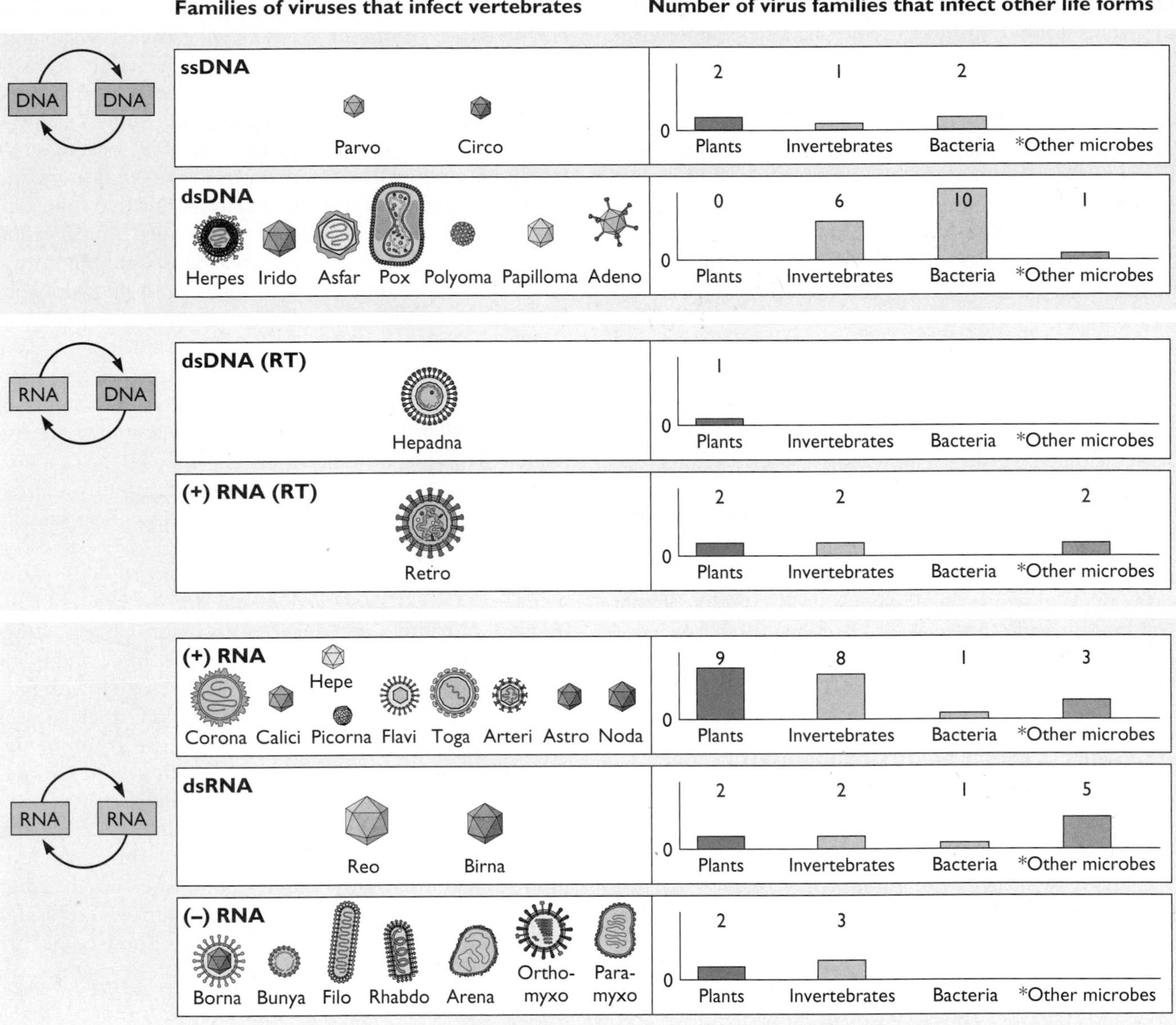

Figure 1.10 Viral families sorted according to the nature of the viral genomes. A wide variety of sizes and shapes are illustrated for the families of viruses that infect vertebrates. Similar diversity exists for the families of viruses that infect other life-forms, but the chart illustrates only the number found to date in each category. As noted, in some categories there are as yet no examples in some life-forms. Adapted from C. M. Fauquet et al. (ed.), *Virus Taxonomy: Classification and Nomenclature of Viruses,* Eighth Report of the International Committee on Taxonomy of Viruses (Academic Press, Inc., San Diego, CA, 2007).

camps developed, as in any controversy involving individuals who tend to focus on differences and those who look for similarities (conventionally known as "splitters" and "lumpers"). Splitters pointed to the inability to infer, from the known properties of viruses, anything about their evolutionary origin or their relationships to one another—the major goal of classical taxonomy. Lumpers maintained that despite such limitations, there were significant practical advantages in grouping isolates with similar properties. Furthermore, it seemed likely that a good classification might actually stimulate fruitful investigation. A major sticking point, however, was finding agreement on **the** properties that should be considered most important in constructing a scheme for virus classification.

The Classical System

In 1962, Lwoff, Robert Horne, and Paul Tournier advanced a comprehensive scheme for the classification of all viruses (bacterial, plant, and animal) under the classical Linnaean hierarchical system consisting of phylum, class, order, family, genus, and species. Although a subsequently formed international committee on the nomenclature of viruses did not adopt this system *in toto*, its designation of families, genera, and species was used for the classification of animal viruses.

One of the most important principles embodied in the system advanced by Lwoff and his colleagues was that viruses should be grouped according to **their** shared properties rather than the properties of the cells or organisms they infect. A second principle was a focus on the nucleic acid genome as the primary criterion for classification. The importance of the genome had become clear when it was inferred from the Hershey-Chase experiment that viral nucleic acid alone can be infectious (Box 1.4). Four characteristics were to be used in the classification of all viruses:

1. Nature of the nucleic acid in the virion (DNA or RNA)
2. Symmetry of the protein shell (**capsid**)
3. Presence or absence of a lipid membrane (**envelope**)
4. Dimensions of the virion and capsid

Genomics, the elucidation of evolutionary relationships by analyses of nucleic acid and protein sequence similarities, is being used increasingly to assign viruses to a particular family and to order members within a family. For example, human herpesvirus 8 was placed in the subfamily *Gammaherpesvirinae* on the basis of sequence analysis of just a small segment of its DNA. Similarly, hepatitis C virus was classified as a member of the family *Flaviviridae* from the sequence of a cloned DNA copy of its genome. However, as our knowledge of molecular properties of viruses and their replication has increased, it has become apparent that **any** comparison based on one or two criteria can be somewhat misleading. For example, *Hepadnaviridae, Retroviridae,* and some plant viruses are classified as different families on the basis of the nature of their genomes, but they are all related by the fact that reverse transcription is an essential step in their reproductive cycles. Moreover, the viral polymerases that perform this task exhibit important similarities in amino acid sequence.

As of the latest report (2005) of the International Committee on Taxonomy of Viruses (ICTV), approximately 40,000 virus isolates from bacteria, plants, and animals had been assigned to one of 3 orders, 73 families, 9 subfamilies, 287 genera, and 1,950 species. Many viruses remain unassigned because they have not yet been characterized adequately, and others are assigned only provisionally. It seems likely that a significant fraction of all existing virus families are now known. However, as we learn more and more about genes, proteins, and reproduction strategies, other relationships will certainly be revealed. Classification refinements can therefore be expected to continue in the future. The ICTV report also includes descriptions of subviral agents (**satellites**, viroids, and prions) and a list of viruses for which information is still insufficient to make assignments. Satellites are composed of nucleic acid molecules that depend for their multiplication on coinfection of a host cell with a helper virus. However, they are not related to this helper. When a satellite encodes the coat protein in which its nucleic acid is encapsidated, it is referred to as a **satellite virus** (e.g., hepatitis delta virus is a satellite virus).

Several years ago, the bacterial virologists who were members of the ICTV agreed to coin similar Latinized family names for the different types of bacteriophages. However, this nomenclature never really took hold, and it has not been widely used by those who do research with bacteriophages. Plant virologists do not classify their viruses into families and genera. Instead, they use group names derived from the prototype virus of each group. For animal viruses, however, the ICTV nomenclature has been applied widely in both the scientific and medical literature, and therefore we adopt it in this text. In this nomenclature, the Latinized virus family names are recognized as starting with capital letters and ending with *-viridae,* as, for example, in the family name *Parvoviridae*. These names are used interchangeably with their common derivatives, as, for example, parvoviruses.

Classification by Genome Type

Because the viral genome carries the entire blueprint for virus propagation, molecular virologists have long considered it the most important characteristic for classification purposes. Therefore, although individual virus families are known by their classical designations, they are more commonly placed in groups according to their genome types, as illustrated in Fig. 1.10. There are seven genome types for all families of viruses, and all seven are represented in viruses that infect vertebrates, although single-stranded RNA genomes are the most numerous. One or more of these types appear to be missing among viruses that infect other life-forms, and the general distributions can vary. For example, double-stranded DNA genomes are most numerous among the bacterial viruses, but there are as yet no known plant viruses with such genomes.

The Baltimore Classification System

The past three decades have seen an enormous increase in knowledge of the molecular biology of animal viruses and cells. Among the most significant advances has been the elucidation of pathways by which viral genomes are expressed. We know that cellular genes are encoded in double-stranded nuclear DNA and that this genetic information is expressed via single-stranded mRNAs. These mRNAs, which are made in the nucleus, are transported to the cytoplasm, where they are translated by ribosomes and associated machinery. This is the so-called central dogma conceptualized by Francis Crick:

$$\text{DNA} \rightarrow \text{RNA} \rightarrow \text{protein}$$

Because viral protein synthesis is completely dependent on the cell's translational machinery, all viruses must direct the synthesis of mRNA to produce proteins. Appreciation of the central role of the translational machinery and of the importance of mRNA molecules in the programming of viral protein synthesis inspired an alternative classification scheme devised by David Baltimore (Fig. 1.11). This classification highlights the obligatory relationship between the viral genome and its mRNA and describes the pathways for formation of mRNA that must be followed by viruses with either RNA or DNA genomes.

By the molecular biologist's convention, mRNA is defined as a **positive [(+)] strand** because it contains immediately translatable information. In the Baltimore classification, a strand of DNA that is of equivalent sequence is also designated a (+) strand. The RNA and DNA complements of (+) strands are designated **negative [(−)] strands**. The principles embodied in this classification have proved to be extremely valuable, especially for viruses with single-stranded RNA genomes. Knowledge of strand polarity provides virologists with immediate insight into the steps that must take place to initiate replication and expression of the viral genome.

Figure 1.11 The Baltimore classification. All viruses must produce mRNA that can be translated by cellular ribosomes. In this classification system, the unique pathways from various viral genomes to mRNA define specific virus classes on the basis of the nature and polarity of their genomes.

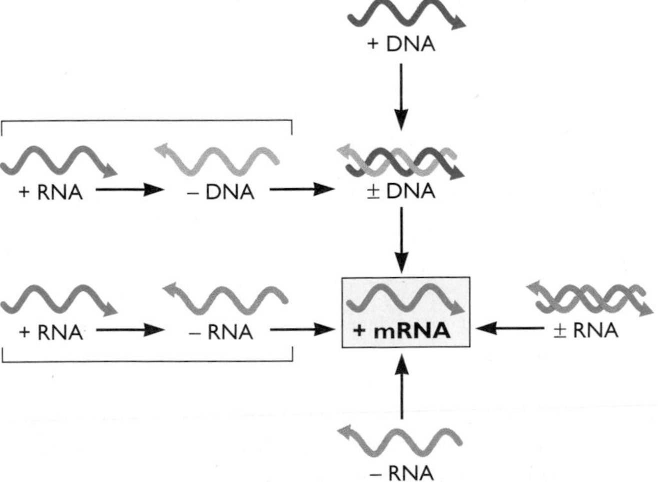

A Common Strategy for Viral Propagation

The basic thesis of this textbook is that **all** viral propagation can be described in the context of three fundamental properties.

- All viral genomes are packaged inside particles that mediate their transmission from host to host.
- The viral genome contains the information for initiating and completing an infectious cycle within a susceptible, permissive cell. An infectious cycle includes attachment and entry of the particle, decoding of genome information, translation of viral mRNA by host ribosomes, genome replication, and assembly and release of particles containing the genome.
- All successful viruses are able to establish themselves in a host population so that virus survival is ensured.

Modern virology is both fascinating and challenging, because of the many and varied ways this strategy is executed. Furthermore, as viruses are obligate molecular parasites, every tactical solution must of necessity tell us something about the host as well as the virus. The intellectual satisfaction of discovering and understanding new principles is as rewarding as the practical consequences of providing solutions to problems of disease.

Perspectives

The viral lifestyle is unique, and understanding virus biology has generated a new appreciation of the molecular biology of host cells and organisms. Some of the important landmarks in animal virology are summarized in Figure 1.12. We have learned that viruses are not primitive entities but, rather, are highly evolved molecular parasitic systems that are subject to checks and

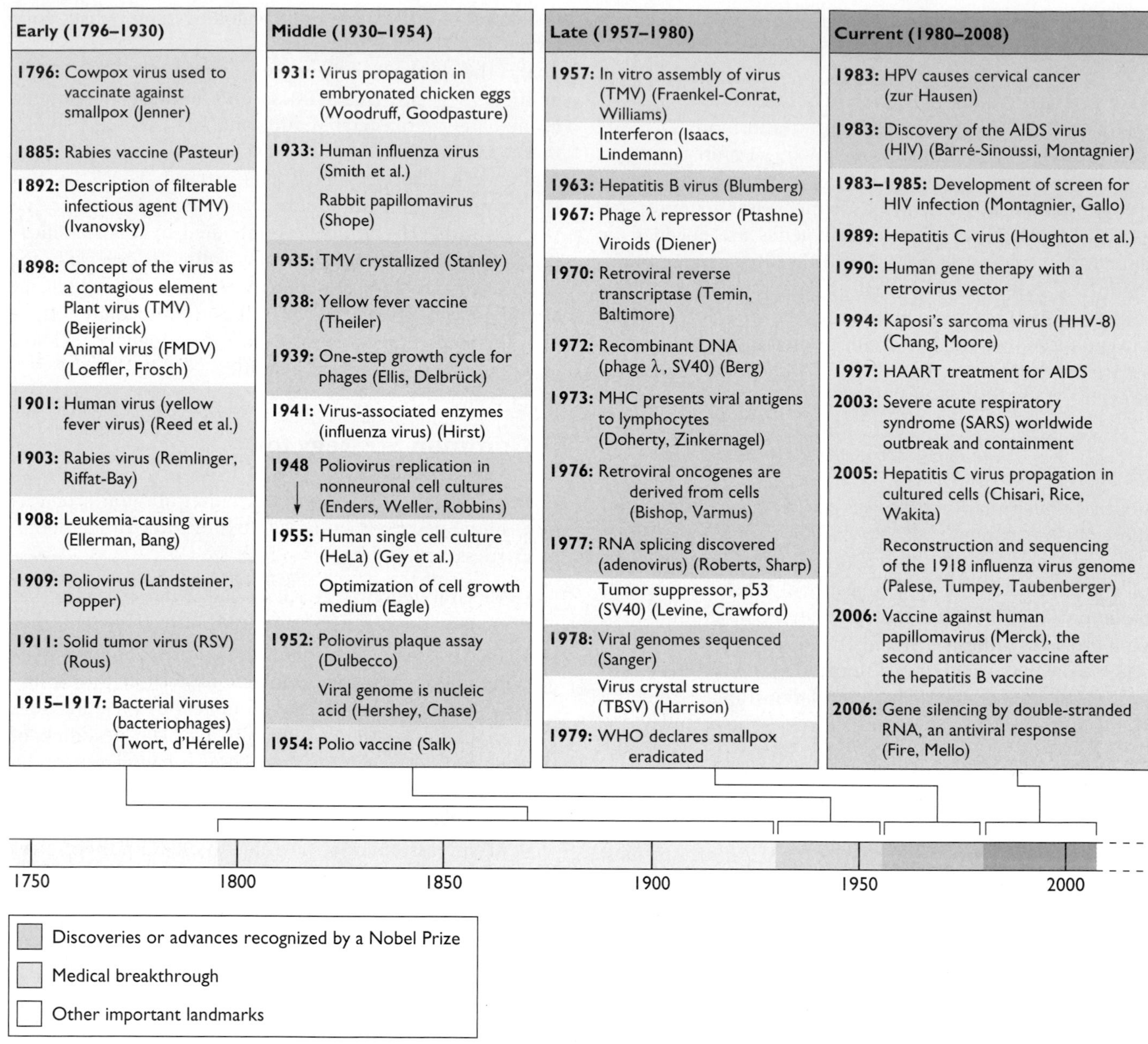

Figure 1.12 Landmarks in the study of animal viruses. Key discoveries and technical advances are listed for each time interval. The pace of discovery has increased exponentially over time. Abbreviations: HAART, highly active antiretroviral therapy; HIV, human immunodeficiency virus; TBSV, tomato bushy stunt virus; TMV, tobacco mosaic virus; SV40, simian virus 40; FMDV, foot-and-mouth disease virus; WHO, World Health Organization; MHC, major histocompatibility complex; HHV-8, human herpesvirus 8; RSV, Rous sarcoma virus.

balances and selected for survival. For example, if viruses are too successful (virulent) and kill all of their hosts, they face annihilation, but if they are too passive they can also be eliminated. Much has been discovered about the biology of viruses and about host defenses against them. Yet the more we learn, the more we appreciate that much is still unknown. It is our conviction that the fundamental principles elaborated in this volume, and its companion Volume II, will serve as a guide to virologists of the future.

References

Books

Barry, J. M. 2005. *The Great Influenza*. Penguin Books, New York, NY.

Brock, T. D. 1990. *The Emergence of Bacterial Genetics.* Cold Spring Harbor Laboratory Press, Cold Spring Harbor, NY.

Brothwell, D., and A. T. Sandison (ed.). 1967. *Diseases in Antiquity.* Charles C Thomas, Publisher, Springfield, IL.

Cairns, J., G. S. Stent, and J. D. Watson (ed.). 1966. *Phage and the Origins of Molecular Biology.* Cold Spring Harbor Laboratory for Quantitative Biology, Cold Spring Harbor, NY.

Creager, A. N. H. 2002. *The Life of a Virus: Tobacco Mosaic Virus as an Experimental Model, 1930–1965.* The University of Chicago Press, Chicago, IL.

Denniston, K., and L. Enquist. 1981. *Recombinant DNA. Benchmark Papers in Microbiology,* vol. 15. Dowden, Hutchinson and Ross, Inc., Stroudsburg, PA.

Fauquet, C. M., M. A. Mayo, J. Maniloff, U. Desselberger, and L. A. Ball (ed.). 2005. *Virus Taxonomy: Classification and Nomenclature of Viruses. Eighth Report of the International Committee on Taxonomy of Viruses.* Academic Press, Inc., San Diego, CA.

Fiennes, R. 1978. *Zoonoses and the Origins and Ecology of Human Disease.* Academic Press, Inc., New York, NY.

Hughes, S. S. 1977. *The Virus: a History of the Concept.* Heinemann Educational Books, London, United Kingdom.

Karlen, A. 1996. *Plague's Progress, a Social History of Man and Disease.* Indigo, Guernsey Press Ltd., Guernsey, Channel Islands.

Knipe, D. M., P. M. Howley, D. E. Griffin, R. A. Lamb, M. A. Martin, B. Roizman, and S. E. Straus (ed.). 2007. *Fields Virology,* 5th ed. Lippincott Williams & Wilkins, Philadelphia, PA.

Luria, S. E. 1953. *General Virology.* John Wiley & Sons, Inc., New York, NY.

Murphy, F. A., C. M. Fauquet, D. H. L. Bishop, S. A. Ghabrial, A. W. Jarvis, and N. Rasmussen. 1997. *Picture Control: the Electron Microscope and the Transformation of Biology in America 1940–1960.* Stanford University Press, Stanford, CA.

Oldstone, M. B. A. 1998. *Viruses, Plagues and History.* Oxford University Press, New York, NY.

Stent, G. S. 1960. *Papers on Bacterial Viruses.* Little, Brown & Co., Boston, MA.

Waterson, A. P., and L. Wilkinson. 1978. *An Introduction to the History of Virology.* Cambridge University Press, London, United Kingdom.

Papers of Special Interest

Baltimore, D. 1971. Expression of animal virus genomes. *Bacteriol. Rev.* **35:**235–241.

Brown, F., J. Atherton, and D. Knudsen. 1989. The classification and nomenclature of viruses: summary of results of meetings of the International Committee on Taxonomy of Viruses. *Intervirology* **30:** 181–186.

Burnet, F. M. 1953. Virology as an independent science. *Med. J. Aust.* **40:**842.

Crick, F. H. C., and J. D. Watson. 1956. Structure of small viruses. *Nature* **177:**473–475.

Lustig, A., and A. J. Levine. 1992. One hundred years of virology. *J. Virol.* **66:**4629–4631.

Murray, N. E. and A. Gann. 2007.What has phage lambda ever done for us? *Curr. Biol.* **17:**R305–R312.

Van Helvoort, T. 1993. *Research Styles in Virus Studies in the Twentieth Century: Controversies and the Formation of Consensus.* Doctoral dissertation. University of Limburg, Maastricht, The Netherlands.

Websites

http://www.ictvonline.org/virustaxonomy.asp?version=2008 *Latest update of virus classification from the ICTV.*

http://www.ncbi.nlm.nih.gov/ICTVdb/ *ICTV-approved virus names and other information as well as links to virus databases.*

http://www.tulane.edu/~dmsander/garryfavweb.html *Has links to almost all the virology sites on the Web.*

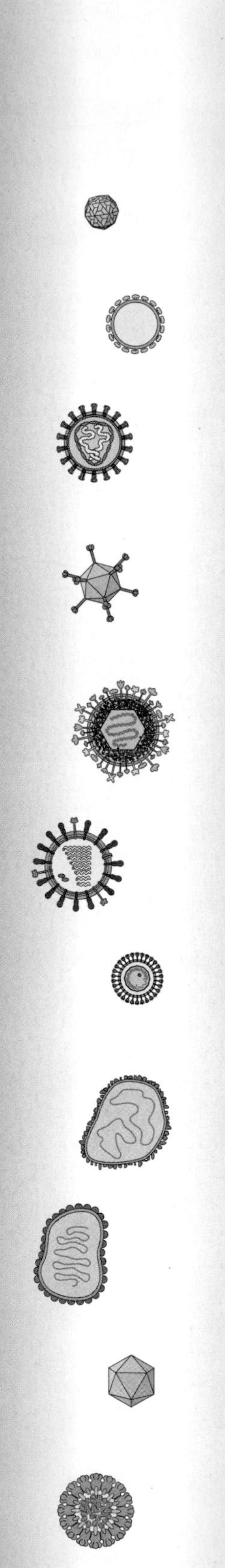

2

The Infectious Cycle

You know my methods, Watson.
SIR ARTHUR CONAN DOYLE

Introduction

Viruses are unique: they are exceedingly small, often made up of nothing more than a nucleic acid molecule within a protein shell, yet when they enter cells, they parasitize the cellular machinery to produce thousands of progeny. This simplicity is misleading: viruses can infect all known life forms, and they comprise a variety of structures and genomes. Despite this complexity, viruses are amenable to study because all viral propagation can be described in the context of three fundamental properties, as described in Chapter 1: all viral genomes are packaged inside particles that mediate their transmission from cell to cell; the viral genome contains the information for initiating and completing an infectious cycle; and all viruses can establish themselves in a host population to ensure virus survival.

The objective of research in virology is to understand how viruses enter individual cells, replicate, and assemble new infectious particles. These studies are usually carried out with cell cultures rather than with animals, because cell cultures provide a much simpler and more homogeneous experimental system. Cell cultures can be infected in such a way as to ensure that a single replication cycle occurs synchronously in every infected cell, the **one-step growth curve**. Because all viral infections take place within the cell, a full understanding of viral life cycles also requires knowledge of cell biology and cellular architecture. In this chapter we review the cell surface (the site at which viruses enter and exit cells), methods for quantifying viral growth, and one-step growth analysis.

The Infectious Cycle

Viruses cannot reproduce extracellularly; the production of new infectious viruses takes place within a cell (Fig. 2.1). Virologists divide the viral infectious cycle into discrete steps to facilitate their study, although in virus-infected cells no such artificial boundaries occur. The infectious cycle comprises attachment and entry of the particle, translation of viral mRNA by host ribosomes, genome

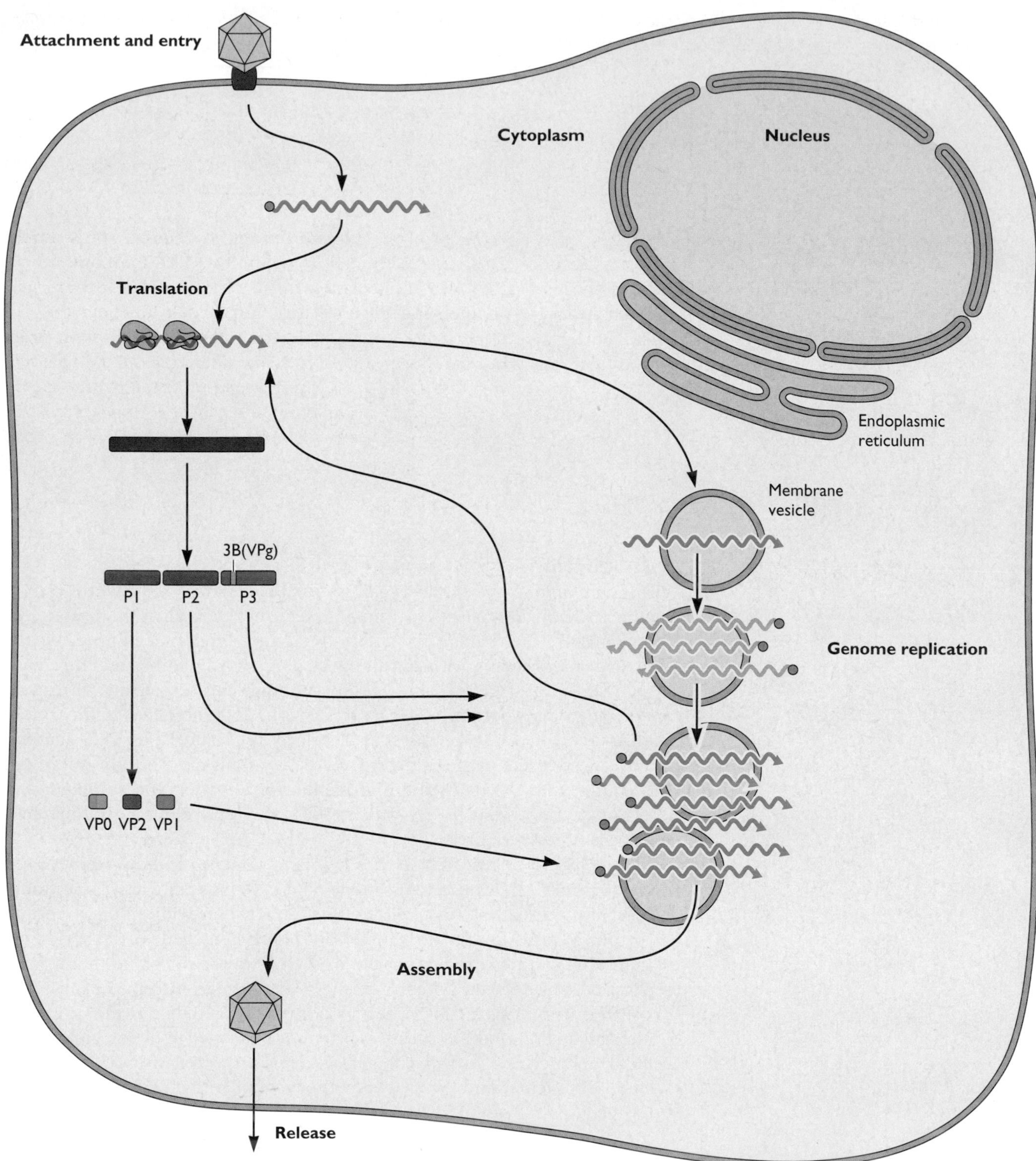

Figure 2.1 The viral infectious cycle. The infectious cycle of poliovirus is shown as an example, illustrating the steps common to all viral life cycles: attachment and entry, translation, genome replication, assembly, and release.

replication, and assembly and release of particles containing the genome. New virus particles produced during the infectious cycle may then infect new cells. The term **virus replication** is another name for the sum total of all events that occur during the infectious cycle.

Because all viral infections begin in a single cell, the virologist can use cultured cells to study stages of the infectious cycle. There are events common to virus replication in animals and in cultured cells, but there are also many important differences. While viruses readily attach to cells in culture, in nature a virus particle must encounter a host, no mean feat for nanoparticles without any means of locomotion. After encountering a host, the virus particle must pass through physical host defenses, such as dead skin, mucous layers, and the extracellular matrix. Host defenses such as antibodies and immune cells, which exist to combat virus infections, are not found in cultured cells. Virus infection of cultured cells has been a valuable tool for understanding viral life cycles, but the differences compared with infection of a living animal must always be considered.

The Cell

Viruses require many different functions of the host cell (Fig. 2.2) for propagation. Cells provide the machinery for translation of viral mRNAs, sources of energy, enzymes for genome replication, and sites of nucleic acid replication and viral assembly. The cellular transport apparatus brings viral genomes to the correct cellular compartment and ensures that viral subunits reach locations where they may be assembled into virus particles. Subsequent chapters include a discussion of cellular functions that are important for individual steps in the viral replication cycle. In the following section we consider in detail the architecture of cell surfaces. The cell membrane merits this special focus because it is not only the portal of entry for all animal viruses, but also the site from which many viruses leave the cell.

Figure 2.2 The mammalian cell. Illustrated schematically are the nucleus and major membrane-bound compartments of the cytoplasm, and components of the cytoskeleton that play important roles in virus replication. The figure is not drawn to scale.

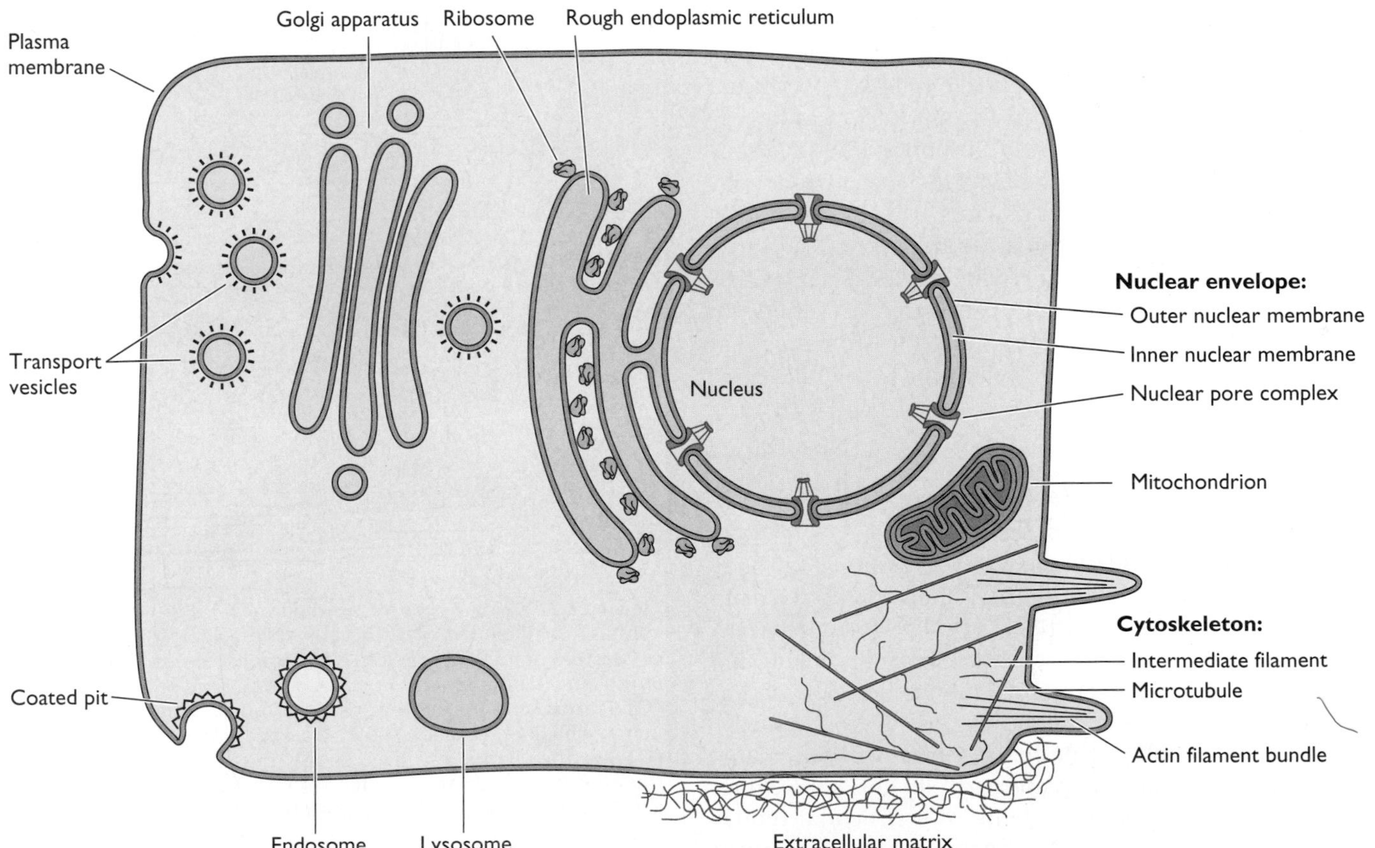

The Architecture of Cell Surfaces

In animals, viral infections usually begin at the epithelial surfaces of the body that are exposed to the environment (Fig. 2.3). Cells cover these surfaces, and the part of these cells exposed to the environment is called the **apical surface**. Conversely, the **basolateral surfaces** of such cells are in contact with adjacent or underlying cells or tissues. These cells exhibit a differential (polar) distribution of proteins and lipids in the plasma membranes that creates the two distinct surface domains. As illustrated in Fig. 2.3, these cell layers differ in thickness and organization. Movement of macromolecules between the cells in the epithelium is prevented by **tight junctions**, which circumscribe the cells at the apical edges of their lateral membranes. Many viral infections are initiated upon entry into epithelial or endothelial cells at their exposed apical surfaces, often by attaching to cell surface molecules specific for this domain. Viruses that both enter and are released at apical membranes can be transmitted laterally from cell to cell without ever transversing the epithelial or endothelial layers; they generally cause localized infections. In other cases, progeny virions are transported to the basolateral surface and released into the underlying cells and tissues, a process that facilitates viral spread to other sites of replication.

There are also more specialized pathways by which viruses reach susceptible cells. For example, some epithelial tissues contain M cells, specialized cells that overlie the collections of lymphoid cells in the gut known as Peyer's patches. M cells function in the transport of intestinal contents to Peyer's patches. Certain viruses, such as poliovirus and human immunodeficiency virus type 1, can also be transported through them by a mechanism called **transcytosis** to gain access to underlying tissues. Such specialized pathways of invasion are considered in Volume II, Chapter 1. Below we describe briefly the structures that surround cells and tissues, as well as the membrane components that mediate the interaction of cells with their environments.

The Extracellular Matrix: Components and Biological Importance

Extracellular matrices hold the cells and tissues of the body together. The extracellular matrix is made up of two main classes of macromolecules (Fig. 2.4). The first class comprises glycosaminoglycans (such as heparan sulfate and chondroitin sulfate), which are unbranched polysaccharides made of repeating disaccharides. Glycosaminoglycans are usually linked to proteins to form **proteoglycans**. The second class of macromolecules in the extracellular matrix consists of fibrous proteins with structural **(collagen** and **elastin)** or adhesive **(fibronectin** and **laminin)**

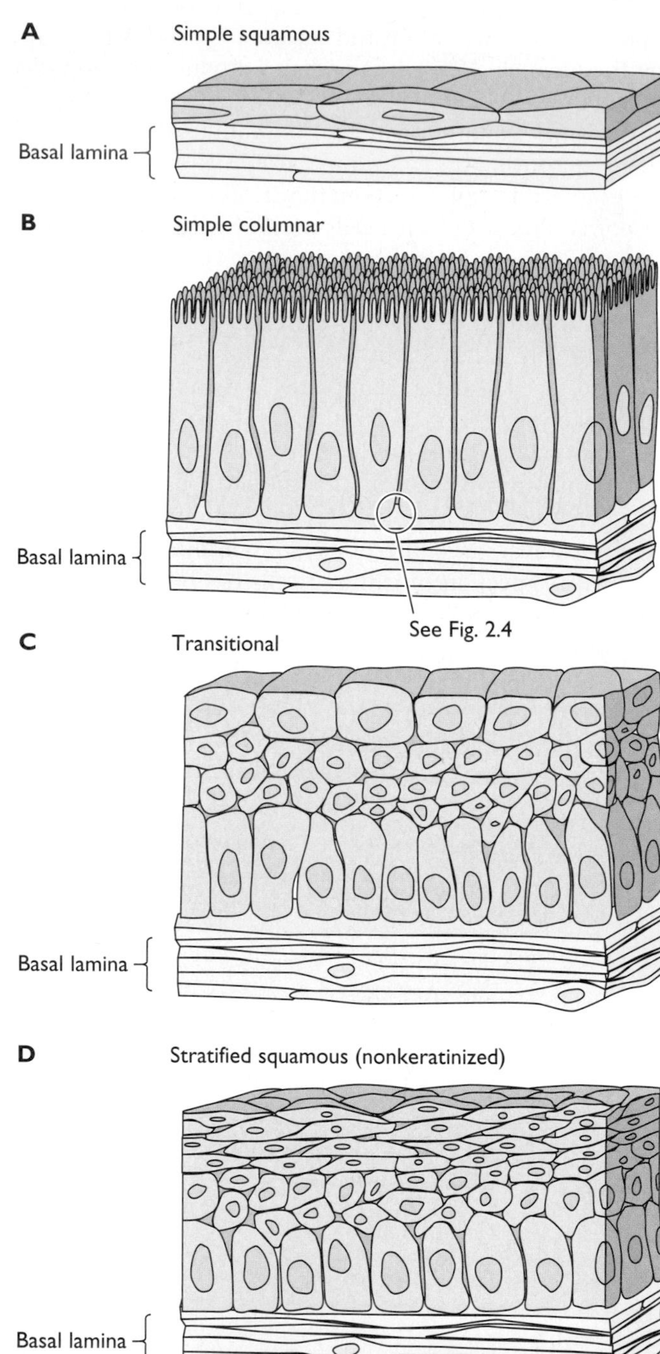

Figure 2.3 Major types of epithelia. (A) Simple squamous epithelium made up of thin cells such as those lining blood vessels and many body cavities. **(B)** Simple columnar epithelium found in the stomach, cervical tract, and small intestine. **(C)** Transitional epithelium, which lines cavities, such as the urinary bladder, that are subject to expansion and contraction. **(D)** Stratified, nonkeratinized epithelium lining surfaces such as the mouth and vagina. Adapted from H. Lodish et al., *Molecular Cell Biology*, 3rd ed. (W. H. Freeman & Co., New York, NY, 1995), with permission.

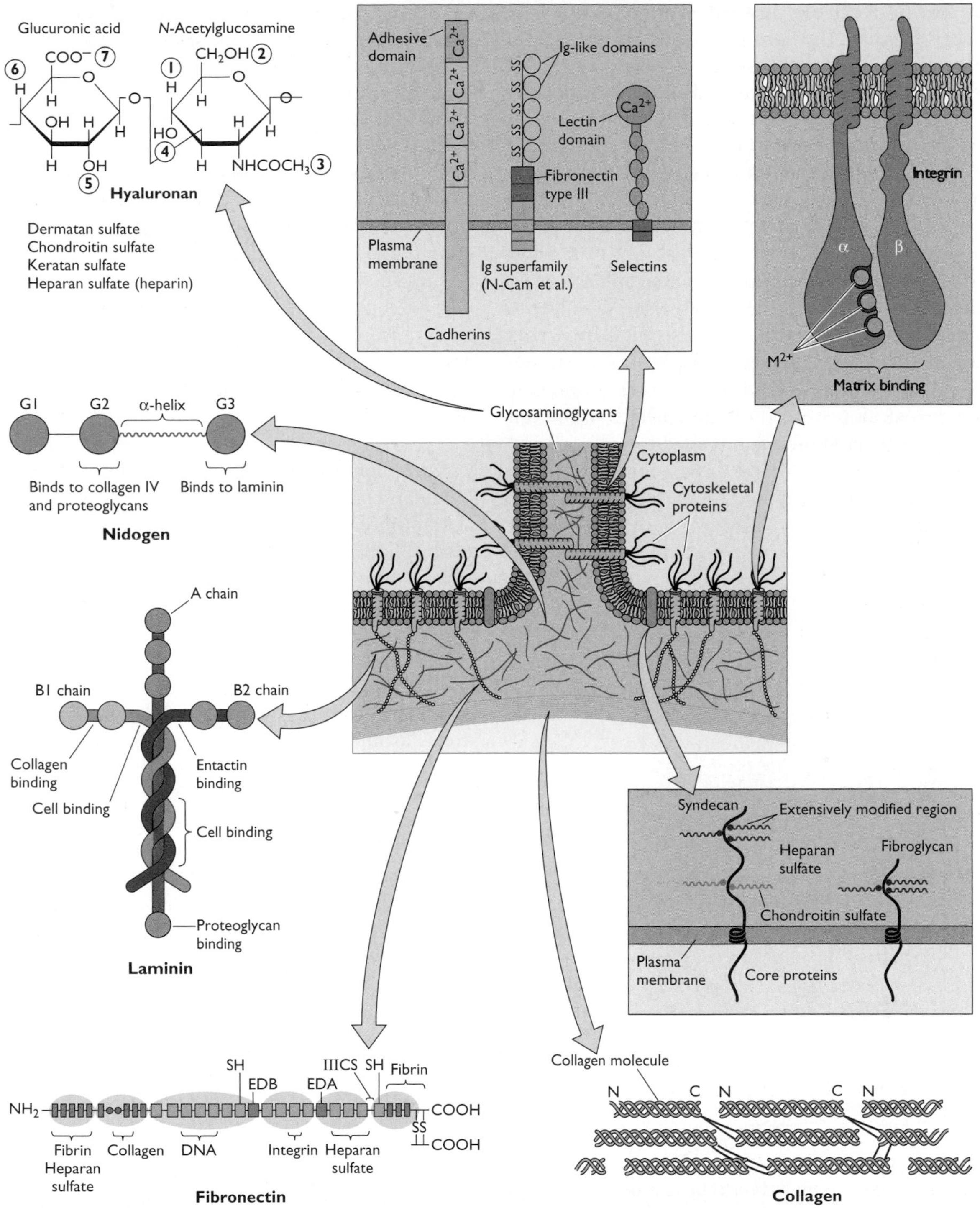

Figure 2.4 Cell adhesion molecules and components of the extracellular matrix. The diagram in the center (expanded from Fig. 2.3B) illustrates the variety of cell surface components that contribute to cell-cell adhesion and attachment to the extracellular matrix. Different glycosaminoglycans such as dermatan sulfate and chondroitin sulfate are produced by covalent linkage of SO_3 at the numbered positions on the hyaluronan molecule. Ig, immunoglobulin; M^{2+}, divalent cation. Adapted from H. Lodish et al., *Molecular Cell Biology*, 3rd ed. (W. H. Freeman & Co., New York, NY, 1995), and G. M. Cooper, *The Cell: a Molecular Approach* (ASM Press, Washington, DC, and Sinauer Associates, Sunderland, MA, 1997), with permission.

functions. The proteoglycan molecules in the matrix form hydrated gels in which the fibrous proteins are embedded, providing strength and resilience to the matrix. The gel provides resistance to compression and allows the diffusion of nutrients between blood and tissue cells. The extracellular matrix of each cell type is specialized for the particular function required, varying in thickness, strength, elasticity, and degree of adhesion.

Most organized groups of cells, like epithelial cells of the skin (Fig. 2.3 and 2.5), are bound tightly on their basal surface to a thin layer of extracellular matrix called the **basal lamina**. This matrix is linked to the basolateral membrane by specific receptor proteins called **integrins** (which are discussed in "Cell Membrane Proteins" below). Integrins are anchored to the intracellular structural network (the **cytoskeleton**) at the inner surface of the cell membrane. The basal lamina is attached to collagen and other material in the underlying loose connective tissue found in many organs of the body (Fig. 2.5). Capillaries, glands, and specialized cells are embedded in this connective tissue. Several viruses gain access to susceptible cells by attaching specifically to components of the extracellular matrix, including some cell adhesion proteins or proteoglycans.

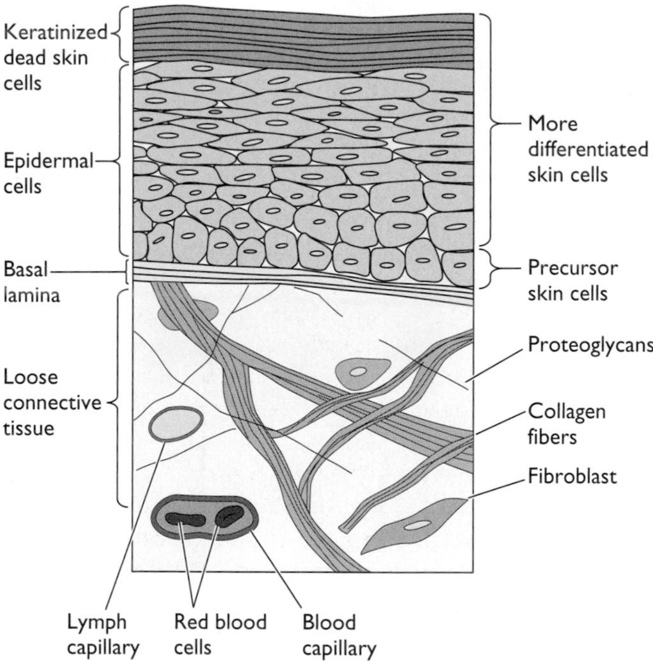

Figure 2.5 Cross section through skin. In this diagram of skin from a pig, the precursor epidermal cells rest on a thin layer of extracellular matrix called the basal lamina. Underneath is loose connective tissue consisting mostly of extracellular matrix. Fibroblasts in the connective tissue synthesize the connective tissue proteins, hyaluronan, and proteoglycans. Blood and lymph capillaries are also located in the loose connective tissue layer. Adapted from H. Lodish et al., *Molecular Cell Biology*, 3rd ed. (W. H. Freeman & Co., New York, NY, 1995), with permission.

Properties of the Plasma Membrane

The plasma membrane of every cell type is composed of a similar phospholipid/glycolipid bilayer, but different sets of membrane proteins and lipids allow the cells of different tissues to carry out their specialized functions. The lipid bilayer consists of molecules that possess both hydrophilic and hydrophobic portions; they are known as **amphipathic** molecules, from the Greek word *amphi* (meaning "on both sides") (Fig. 2.6). They form a sheetlike structure in which polar head groups face the aqueous environment of the cell's cytoplasm (inner surface) or the surrounding environment (outer surface). The polar head groups of the inner and outer leaflets bear side chains with different lipid compositions. The fatty acyl side chains form a continuous hydrophobic interior about 3 nm thick. Hydrophobic interactions are the driving force for formation of the bilayer. However, hydrogen bonding and electrostatic interactions among the polar groups and water molecules or membrane proteins also stabilize the structure.

Thermal energy permits the phospholipid and glycolipid molecules comprising natural cell membranes to rotate freely around their long axes and diffuse laterally. If unencumbered, a lipid molecule can diffuse the length of an animal cell in only 20 s at 37°C. In most cases, phospholipids and glycolipids do not flip-flop from one side of a bilayer to the other, and the outer and inner leaflets of the bilayer remain separate. Similarly, membrane proteins not anchored to the extracellular matrix and/or the underlying structural network of the cell can diffuse rapidly, moving laterally like icebergs in this fluid bilayer. In this way, certain membrane proteins can form functional aggregates. Intracellular organelles such as the nucleus, endoplasmic reticulum, and lysosomes are also enclosed in lipid bilayers, although their composition and physical properties differ.

For many years the plasma membrane was viewed as a uniform and fluid sea, in which lipid and protein components diffused randomly in the plane of the membrane. This simplistic model has been dispelled by experimental findings during the past 15 years. It is now recognized that plasma membranes comprise **microdomains**, regions with distinct lipid and protein composition (Box 2.1). The **lipid raft** is one type of microdomain that is important for virus replication. Lipid rafts are enriched in cholesterol and saturated fatty acids and consequently

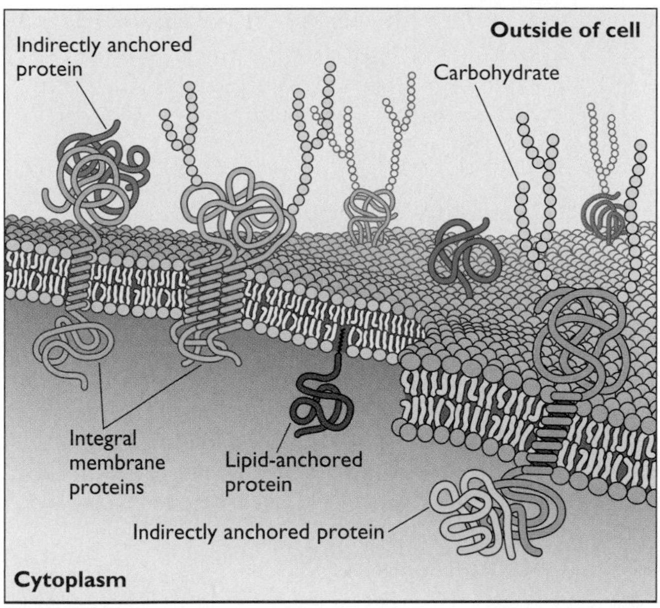

Phospholipid bilayer

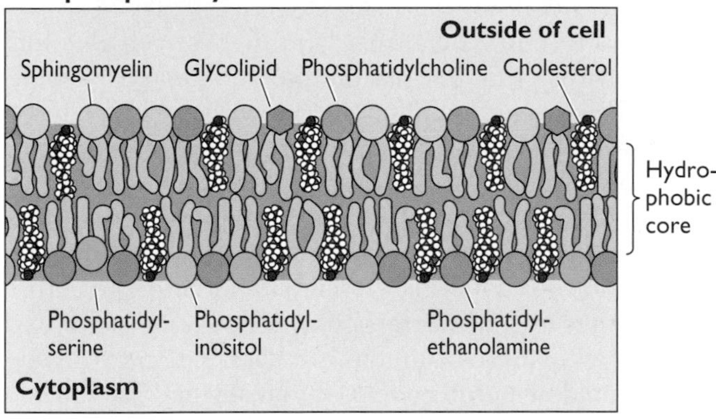

Figure 2.6 The plasma membrane. (Top) Different types of membrane proteins are illustrated. Some integral membrane proteins are transmembrane proteins and are exposed on both sides of the bilayer. **(Bottom)** Lipid components of the plasma membrane. The membrane consists of two layers (leaflets) of phospholipid and glycolipid molecules. Their fatty acid tails converge to form the hydrophobic interior of the bilayer; the polar hydrophilic head groups (shown as balls) line both surfaces. Adapted from G. M. Cooper, *The Cell: a Molecular Approach* (ASM Press, Washington, DC, and Sinauer Associates, Sunderland, MA, 1997), with permission.

are more densely packed and less fluid than other regions of the membrane. The assembly of a variety of viruses takes place at lipid rafts (see Chapter 13). Furthermore, the entry of some viruses requires lipid rafts. For example, virions of human immunodeficiency virus type 1 and Ebola virus enter cells at lipid rafts. Treatment of cells with compounds that disrupt these microdomains blocks entry. One explanation for this requirement might be that cell membrane proteins required for entry are present only in lipid rafts: receptors and coreceptors for human immunodeficiency virus are preferentially located in these domains.

Cell Membrane Proteins

Membrane proteins are classified into two broad categories, **integral membrane proteins** and **indirectly anchored proteins**, names that describe the nature of their interactions with the plasma membrane (Fig. 2.6).

Integral membrane proteins are embedded in the lipid bilayer, because they contain one or more **membrane-spanning domains**, as well as portions that protrude out into the exterior and interior of the cell (Fig. 2.6). Many membrane-spanning domains consist of an α-helix typically 3.7 nm long. It includes 20 to 25 generally hydrophobic or uncharged residues embedded in the membrane, with the hydrophobic side chains protruding outward to interact with the fatty acyl chains of the lipid bilayer. The first and last residues are often positively charged amino acids (lysine or arginine) that can interact with the negatively charged polar head groups of the phospho- or glycolipids to stabilize the membrane-spanning domain. Proteins with membrane spanning domains enable the cell to respond to signals from outside the cell. Such membrane proteins are designed to bind external ligands (e.g., hormones, cytokines, or membrane proteins on the same cell or on other cells) and to signal the occurrence of such interactions to molecules in the interior of the cell. Some proteins with multiple membrane-spanning domains (Fig. 2.6) form critical components of molecular pores or pumps, which mediate the internalization of required nutrients or the expulsion of undesirable material from the cell, or maintain homeostasis with respect to cell volume, pH, and ion concentration.

In many cases, the external portions of membrane proteins are decorated by complex or branched **carbohydrate chains** linked to the peptide backbone. Linkage can be either through nitrogen (**N linked**) in the side chain of asparagine residues or through oxygen (**O linked**) in the side chains of serine or threonine residues. Such membrane **glycoproteins**, as they are called, quite frequently serve as viral receptors. Many viruses attach specifically to one or more of the external components of glycoproteins.

Some membrane proteins do not span the lipid bilayer but are anchored in the inner or outer leaflet by covalently attached hydrocarbon chains (see Chapter 12). Indirectly anchored proteins are bound to the plasma membrane lipid bilayer by interacting either with integral membrane

BOX 2.1 BACKGROUND

Plasma membrane microdomains

According to the Singer-Nicholson fluid mosaic model of membrane structure proposed in 1972, membranes are two-dimensional fluids with proteins inserted into the lipid bilayers (Fig. 2.6). Although the model accurately predicts the general organization of membranes, one of its conclusions has proven incorrect: that proteins and lipids are randomly distributed because they can freely rotate and laterally diffuse within the plane of the membrane. Beginning in the 1990s, the results of a series of experiments indicated that the movement of most proteins in the plasma membrane is partially restricted. In particular, these studies provided evidence for the existence of plasma membrane microdomains that are enriched in glycosphingolipids, cholesterol, glycosylphosphatidylinositol-anchored proteins, and certain intracellular signaling proteins. These microdomains, called lipid rafts, are experimentally defined as being resistant to extraction in cold 1% Triton X-100 and floating in the top half of a 5 to 30% sucrose density gradient. A major component of lipid rafts was found to be caveolin-1, a major coat protein of caveolae. These flask-shaped invaginations of the plasma membrane are involved in the uptake of sphingolipids and integrins, as well as viruses, bacteria, and toxins. Detergent-insoluble microdomains are also present in cells that lack caveolin-1, and it is now known that there are many noncaveolar lipid raft domains in the plasma membrane.

proteins or with the charged sugars of the glycolipids within the membrane lipid. Fibronectin, a protein in the extracellular matrix that binds to integrins (Fig. 2.4), is an example.

Entering Cells

Viral infection is initiated by a collision between the virus particle and the cell, a process that is governed by chance. Therefore, a higher concentration of virus particles increases the probability of infection. However, a virion may not infect every cell it encounters; it must come in contact with the cells and tissues in which it can replicate. Such cells are normally recognized by means of a specific interaction of a virion with a cell surface receptor. This process can be either promiscuous or highly selective, depending on the virus and the distribution of the cell receptor. The presence of such receptors determines whether the cell will be **susceptible** to the virus. However, whether a cell is **permissive** for the replication of a particular virus depends on other, intracellular components found only in certain cell types. Cells must be both susceptible **and** permissive if an infection is to be successful.

In general, viruses have no means of locomotion, but their small size facilitates diffusion driven by Brownian movement. Propagation of viruses is dependent on essentially random encounters with potential hosts and host cells. In this sense, they can be thought of as tiny, opportunistic "Darwinian machines." For such machines, features that increase the probability of favorable encounters are very important. In particular, viral propagation is critically dependent on the production of large numbers of progeny virions with surfaces composed of many copies of structures that enable the virions to attach to susceptible cells.

Successful entry of a virus into a host cell requires that the virus cross the plasma membrane and in some cases the nuclear membrane. The virus particle must also disassemble to make the viral genome accessible in the cytoplasm, and the nucleic acid must be targeted to the correct cellular compartment. These are not simple processes. Cell membranes are not permeable to virus particles. Virions or critical subassemblies are brought across such barriers by specific transport pathways. To survive in the extracellular environment, the viral genome must be encapsidated in a protective coat that shields viral nucleic acid from the variety of potentially harsh conditions that a virion may meet in the environment. For example, mechanical shearing, ultraviolet (UV) irradiation (from sunlight), extremes of pH (in the gastrointestinal tract), dehydration (in the air), and enzymatic attack (in body fluids) are all capable of damaging viral nucleic acids. However, once in the host cell, the protective structures must become sufficiently unstable to release the genome. Virus particles cannot be viewed only as passive vehicles: they must be able to undergo structural transformations that are important for attachment and entry into a new host cell and for the subsequent disassembly required for viral replication.

Making Viral RNA

Although the genomes of viruses come in a number of configurations, they share a common requirement: they must be efficiently copied into progeny genomes for assembly and mRNAs for the synthesis of viral proteins. The synthesis of RNA molecules by RNA viruses is a unique process that has no counterpart in the cell. With the exception of retroviruses, all RNA viruses encode an RNA-dependent RNA polymerase to catalyze the synthesis of mRNAs and genomes. For the majority of DNA viruses and retroviruses, synthesis of mRNA is accomplished by the cellular enzyme that produces mRNA, RNA polymerase II. Much of our current understanding of the mechanisms of cellular transcription comes from study of the transcription of cellular templates. Analysis of the various means of producing viral mRNA has taught us much about the cell itself.

Making Viral Proteins

Because viruses are parasites of translation, all viral RNAs must be translated by the host's cytoplasmic protein-synthesizing machinery (see Chapter 11). Viral infection often results in modification of the host's translational apparatus so that viral mRNAs are selectively translated. The study of such modifications has revealed a great deal about mechanisms of mRNA translation. Analysis of viral translation has also revealed new strategies, such as internal ribosome binding and leaky scanning, that have been subsequently found to occur in uninfected cells.

Making Viral Genomes

Many viral genomes are copied by the cell's synthetic machinery in cooperation with viral proteins (see Chapters 6 through 9). The cell provides nucleotide substrates, energy, enzymes, and other proteins. In all cases, the internal compartmentalization of the cell must be reckoned with, because essential components are found only in the nucleus, are restricted to the cytoplasm, or are present in cellular membranes. Study of the mechanisms of viral genome replication has established fundamental principles of cell biology and nucleic acid synthesis.

Forming Progeny Virions

The various components of a virion—the nucleic acid genome, capsid protein(s), and in some cases envelope proteins—are often synthesized in different cellular compartments. Their trafficking through and among the cell's compartments and organelles requires that they be equipped with the proper homing signals (see Chapter 12). Virion components must be assembled at some central location, and the information for assembly must be preprogrammed in the component molecules (see Chapter 13). The primary sequences of virion proteins contain sufficient information to specify assembly; this property is exemplified by the remarkable *in vitro* assembly of tobacco mosaic virus from coat protein and RNA (Box 2.2).

Viral Pathogenesis

Viruses command our attention because of their association with animal and plant diseases. The process by which viruses cause disease is called **viral pathogenesis**. To study this process, we must investigate not only the relationships of viruses with the specific cells that they infect but also the consequences of infection to the host organism. The nature of viral disease depends on the effects of viral replication on host cells, the responses of the host's defense systems, and the ability of the virus to spread in and among hosts (Volume II, Chapters 1, 4, and 5).

Overcoming Host Defenses

Organisms have evolved many physical barriers to protect themselves from dangers in their environment such as invading parasites. In addition, vertebrates possess an effective immune system to defend against anything recognized as nonself or dangerous. Studies of the interactions between viruses and the immune system are particularly instructive, because of the many viral countermeasures

BOX 2.2 EXPERIMENTS
In vitro assembly of tobacco mosaic virus

The ability of the primary sequence of virion proteins to specify assembly is exemplified by the coat protein of tobacco mosaic virus. Heinz Fraenkel-Conrat and Robley Williams showed in 1955 that purified tobacco mosaic virus RNA and capsid protein, when mixed and incubated for 24 h, assemble into infectious viruses. When examined by electron microscopy, the particles produced in vitro were identical to the rod-shaped virions produced from infected tobacco plants (Figure 1.7B). Neither the purified viral RNA nor the capsid protein was infectious; no virions were observed in these preparations. These results indicate that the viral coat protein contains all the information needed for assembly of an infectious virion. The spontaneous formation of tobacco mosaic virions in vitro from protein and RNA components is the paradigm for self-assembly in biology.

Fraenkel-Conrat, H., and R. C. Williams. 1955. Reconstitution of active tobacco mosaic virus from its inactive protein and nucleic acid components. *Proc. Natl. Acad. Sci. USA* **40:**690–698.

that can defeat this system. Elucidation of these measures teaches us much about the basis of immunity (Volume II, Chapters 3 and 4).

Cultivation of Viruses

Cell Culture

Types of Cell Culture

Although human and other animal cells were first cultured in the early 1900s, contamination with bacteria, mycoplasmas, and fungi initially made routine work with such cultures extremely difficult. For this reason, most viruses were grown in laboratory animals. In 1949, John Enders, Thomas Weller, and Frederick Robbins made the discovery that poliovirus could multiply in cultured cells not of neuronal origin. As noted in Chapter 1, this revolutionary finding, for which these three investigators were awarded the Nobel Prize in physiology or medicine in 1954, led the way to the propagation of many other viruses in cultured cells, the discovery of new viruses, and the development of viral vaccines such as those against poliomyelitis, measles, and rubella. The ability to infect cultured cells synchronously permitted studies of the biochemistry and molecular biology of viral replication. Large-scale growth and purification allowed studies of the composition of virus particles, leading to the solution of high-resolution, three-dimensional structures, as discussed in Chapter 4.

Cell culture is still the most common method for the propagation of animal viruses. To prepare a cell culture, tissues are dissociated into a single-cell suspension by mechanical disruption followed by treatment with proteolytic enzymes. The cells are then suspended in culture medium and placed in plastic flasks or covered plates. As the cells divide, they cover the plastic surface. Epithelial and fibroblastic cells attach to the plastic and form a **monolayer**, whereas blood cells such as lymphocytes settle, but do not adhere. The cells are grown in a chemically defined and buffered medium optimal for their growth. Commonly used cell lines double in number in 24 to 48 h in such media. Most cells retain viability after being frozen at low temperatures (−70 to −196°C).

There are three main kinds of cell cultures (Fig. 2.7). **Primary cell cultures** are prepared from animal tissues as described above. They include several cell types and have a limited life span, usually no more than 5 to 20 cell divisions. The most commonly used primary cell cultures are derived from monkey kidneys, human embryonic amnion, human embryonic kidneys, human foreskins, and chicken or mouse embryos. Such cells are used for experimental virology when the state of cell differentiation is important or when appropriate cell lines are not available. They are also used in vaccine production: for example, live attenuated poliovirus vaccine strains may be propagated in primary monkey kidney cells. Primary cell cultures were mandated for the growth of viruses to be used as human vaccines to avoid contamination of the product with potentially oncogenic DNA from continuous cell lines (see below). Some viral vaccines are now prepared in **diploid cell strains**, which consist of a homogeneous population of a single type and can divide up to 100 times before dying. Despite the numerous divisions, these cell strains retain the diploid chromosome number. The most widely used diploid cells are those established from human embryos, such as the WI-38 strain derived from human embryonic lung.

Continuous cell lines consist of a single cell type that can be propagated indefinitely in culture. These immortal lines are usually derived from tumor tissue or by treating a primary cell culture or a diploid strain with a mutagenic chemical or a tumor virus. Such cell lines often do not resemble the cell of origin; they are less differentiated

Figure 2.7 Different types of cell culture used in virology. Confluent cell monolayers photographed by low-power light microscopy. **(A)** Primary human foreskin fibroblasts; **(B)** established line of mouse fibroblasts (3T3); **(C)** continuous line of human epithelial cells (HeLa [Box 2.3]). The ability of transformed HeLa cells to overgrow one another is the result of a loss of contact inhibition. Courtesy of R. Gonzalez, Princeton University.

A 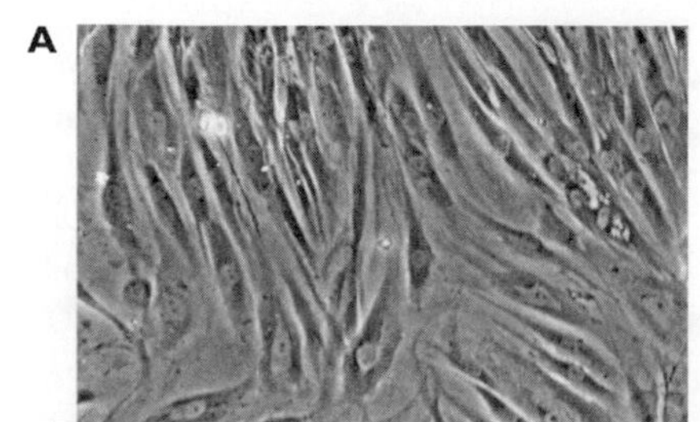

B 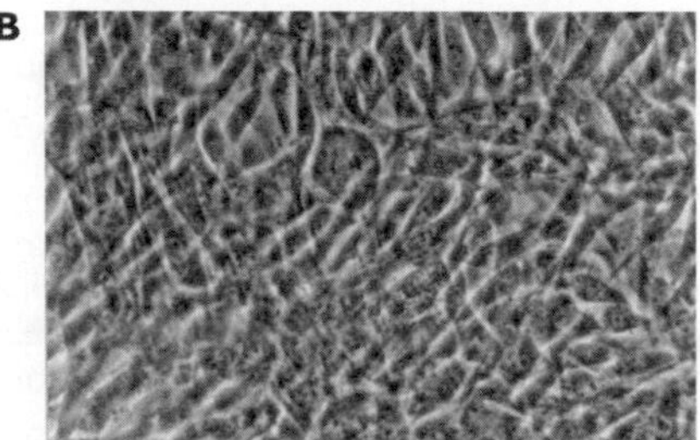

C 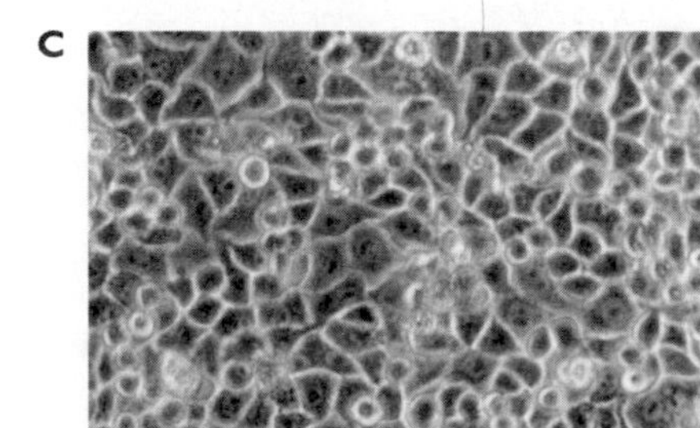

BOX 2.3

BACKGROUND

The cells of Henrietta Lacks

The most widely used continuous cell line in virology, the HeLa cell line, was derived from Henrietta Lacks. In 1951, the 31-year-old mother of five visited a physician at Johns Hopkins Hospital in Baltimore and found that she had a malignant tumor of the cervix. A sample of the tumor was taken and given to George Gey, head of tissue culture research at Hopkins. Dr. Gey had been attempting for years, without success, to produce a line of human cells that would live indefinitely. When placed in culture, Henrietta Lacks's cells propagated as no other cells had before.

On the day in October that Henrietta Lacks died, George Gey appeared on national television with a vial of her cells, which he called HeLa cells. He said, "It is possible that, from a fundamental study such as this, we will be able to learn a way by which cancer can be completely wiped out." Soon after, HeLa cells were used to propagate poliovirus, which was causing poliomyelitis throughout the world, and they played an important role in the development of poliovirus vaccines. Henrietta Lacks's HeLa cells started a medical revolution: not only was it possible to propagate many different viruses in these cells, but also continuous cell lines could be produced from many human tissues. Sadly, the family of Henrietta Lacks did not learn about HeLa cells, or the revolution they started, until 24 years after her death. Her family members were shocked that cells from Henrietta lived in so many laboratories, and hurt that they had not been told that any cells had been taken from her.

The story of HeLa cells is a sad commentary on the lack of informed consent that pervaded medical research in the 1950s. Since then, biomedical ethics have changed greatly, and now there are strict regulations about clinical research: physicians may not take samples from patients without permission.

For additional information, see http://www.jhu.edu/~jhumag/0400web/01.html.

(having lost the morphology and biochemical features that they possessed in the organ), are often abnormal in chromosome morphology and number **(aneuploid)**, and can be tumorigenic (i.e., they produce tumors when inoculated into nude mice). Examples of commonly used continuous cell lines include HeLa (Henrietta Lacks) cells (Box 2.3), derived from human carcinomas, and L and 3T3 cells, derived from mice. Continuous cell lines provide a uniform population of cells that can be infected synchronously for growth curve analyses (see "The One-Step Growth Cycle" below) or biochemical studies of virus replication.

In contrast to cells that grow in monolayers on plastic dishes, others can be maintained in **suspension cultures**, in which a spinning magnet continuously stirs the cells. The advantage of suspension culture is that a large number of cells can be grown in a relatively small volume. This culture method is well suited for applications that require large quantities of virus particles, such as X-ray crystallography or production of vectors.

Initially, it was thought that viruses, as obligatory intracellular parasites, could not replicate outside a living cell. This dictum was nullified in 1991 by the demonstration that infectious poliovirus could be produced in a cell extract of human cells incubated with viral RNA. Despite this finding, most work on viruses is done *in vivo*, using cultured cells, embryonated eggs, or laboratory animals (Box 2.4).

Evidence of Viral Growth in Cultured Cells

Some viruses kill the cells in which they replicate, and the infected cells may eventually detach from the cell culture plate. As more cells are infected, the changes become visible and are called **cytopathic effects** (Table 2.1). Many types of cytopathic effect can be seen with a simple light or phase-contrast microscope at low power,

BOX 2.4

TERMINOLOGY

In vitro and in vivo

The terms "*in vitro*" and "*in vivo*" are common in the virology literature. *In vitro* means "in glass" and refers to experiments carried out in an artificial environment, such as a glass test tube. Unfortunately, the phrase "experiments performed *in vitro*" is used to designate not only work done in the cell-free environment of a test tube but also work done within cultured cells. The use of the phrase *in vitro* to describe living cultured cells leads to confusion and is inappropriate.

In this textbook, descriptions of experiments being carried out *in vitro* signify the absence of cells, e.g., *in vitro* translation.

Table 2.1 Some examples of cytopathic effects of viral infection of animal cells

Cytopathic effect(s)	Virus(es)
Morphological alterations	
Nuclear shrinking (pyknosis), proliferation of membrane	Picornaviruses
Proliferation of nuclear membrane	Alphaviruses, herpesviruses
Vacuoles in cytoplasm	Polyomaviruses, papillomaviruses
Syncytium formation (cell fusion)	Paramyxoviruses, coronaviruses
Margination and breaking of chromosomes	Herpesviruses
Rounding up and detachment of cultured cells	Herpesviruses, rhabdoviruses, adenoviruses, picornaviruses
Inclusion bodies	
Virions in nucleus	Adenoviruses
Virions in the cytoplasm (Negri bodies)	Rabies virus
"Factories" in the cytoplasm (Guarnieri bodies)	Poxviruses
Clumps of ribosomes in virions	Arenaviruses
Clumps of chromatin in nucleus	Herpesviruses

without fixing or staining the cells. These changes include the rounding up and detachment of cells from the culture dish, cell lysis, swelling of nuclei, and sometimes the formation of a group of fused cells called a syncytium (Fig. 2.8). Observation of other cytopathic effects requires high-power microscopy. These cytopathic effects include the development of intracellular masses of virions or unassembled viral components in the nucleus and/or cytoplasm (inclusion bodies), formation of crystalline arrays of viral proteins, membrane blebbing, duplication of membranes, and fragmentation of organelles. The time required for the development of cytopathology varies greatly among animal viruses. For example, depending on the size of the inoculum, enteroviruses and herpes simplex virus can cause cytopathic effects in 1 to 2 days and destroy the cell monolayer in 3 days. In contrast, cytomegalovirus, rubella virus, and some adenoviruses may not produce such effects for several weeks.

The development of characteristic cytopathic effects in infected cell cultures is frequently monitored in diagnostic virology during isolation of viruses from specimens obtained from infected patients or animals. However, cytopathic effect is also of value in the research laboratory: it can be used to monitor the progress of an infection, and it is often one of the phenotypic traits by which mutant viruses are characterized.

Some viruses multiply in cells without causing obvious cytopathic effects. For example, many members of the families *Arenaviridae*, *Paramyxoviridae*, and *Retroviridae* do not cause obvious damage to cultured cells. The growth of such viruses in cells must therefore be assayed using alternative methods, as described in "Assay of Viruses" below.

Embryonated Eggs

Before the advent of cell culture, many viruses were propagated in embryonated chicken eggs (Fig. 2.9). At 5 to 14 days after fertilization, a hole is drilled in the shell and virus is injected into the site appropriate for its replication. This method of virus propagation is now routine only for influenza virus. The robust yield of this virus from chicken eggs has led to their widespread use in research laboratories and for vaccine production.

Laboratory Animals

In the early 1900s, when viruses were first isolated, freezers and cell cultures were not available and it was necessary to maintain virus stocks by continuous passage from animal to animal. This practice not only was inconvenient but also, as we shall see in Volume II, Chapter 8, led to the selection of viral mutants. For example, monkey-to-monkey intracerebral passage of poliovirus selected a mutant that could no longer infect chimpanzees by the oral route, the natural means of infection. Cell culture has largely supplanted the use of animals for propagating viruses, but some viruses, such as Norwalk virus, cannot yet be grown in this way.

Experimental infection of laboratory animals has always been, and will continue to be, obligatory for studying the processes by which viruses cause disease. The use of monkeys in the study of poliomyelitis, the paralytic disease caused by poliovirus, led to an understanding of the basis of this disease and was instrumental in the development

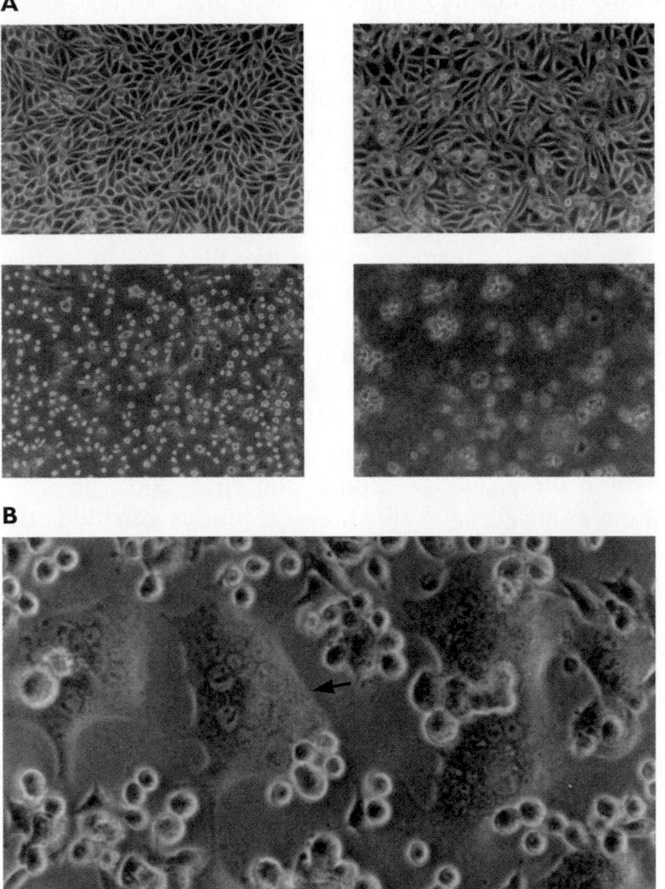

Figure 2.8 Development of cytopathic effect. (A) Cell rounding and lysis during poliovirus infection. (Upper left) Uninfected cells; (upper right) 5 1/2 h after infection; (lower left) 8 h after infection; (lower right) 24 h after infection. **(B)** Syncytium formation induced by murine leukemia virus. The field shows a mixture of individual small cells and syncytia, indicated by the arrow, which are large, multinucleate cells. Courtesy of R. Compans, Emory University School of Medicine.

of a successful vaccine. Similarly, the development of vaccines against hepatitis B virus would not have been possible without experimental studies with chimpanzees. Understanding how the immune system or any complex organ reacts to a virus cannot be achieved without research on living animals. The development of viral vaccines, antiviral drugs, and diagnostic tests for veterinary medicine has also benefited from research on diseases in laboratory animals.

Assay of Viruses

There are two main types of assays for detecting viruses: biological and physical. Because viruses were first recognized by their infectivity, the earliest assays focused on this most sensitive and informative property. However, biological assays such as the plaque assay and end-point titration methods do not measure noninfectious particles. Such particles can be measured by physical assays such as electron microscopy or by immunological methods. Knowledge of the number of noninfectious particles is useful for assessing the quality of a virus preparation.

Measurement of Infectious Units

One of the most important procedures in virology involves measuring the concentration of a virus in a sample, the **virus titer**. This parameter is determined by inoculating serial dilutions of virus into host cell cultures, chicken embryos, or laboratory animals and monitoring for evidence of virus multiplication. The response may be quantitative (as in assays for plaques, fluorescent foci, infectious centers, or transformation) or all-or-none, in which the presence or absence of infection is measured (as in an end-point dilution assay).

Plaque Assay

In 1952, Renato Dulbecco modified the plaque assay developed to determine the titers of bacteriophage stocks for use in animal virology. The plaque assay was adopted

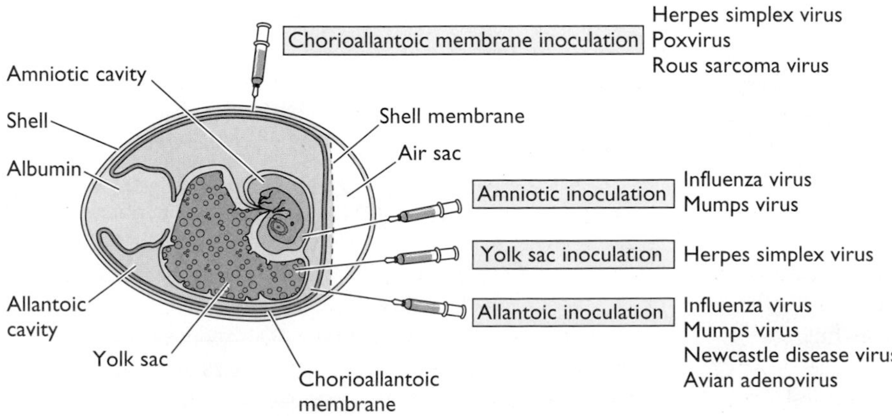

Figure 2.9 Growth of viruses in embryonated eggs. The cutaway view of an embryonated chicken egg shows the different routes by which viruses are inoculated into eggs and the different compartments in which viruses may grow. Adapted from F. Fenner et al., *The Biology of Animal Viruses* (Academic Press, New York, NY, 1974), with permission.

rapidly for reliable determination of the titers of a wide variety of viruses. In this assay, monolayers of cultured cells are incubated with a preparation of virus to allow adsorption to cells. After removal of the inoculum, the cells are covered with nutrient medium containing a supplement, most commonly agar, that results in the formation of a gel. When the original infected cells release new progeny viruses, the spread of viruses to neighboring uninfected cells is restricted by the gel. As a result, each infectious particle produces a circular zone of infected cells, a **plaque**. If the infected cells are damaged, the plaque can be distinguished from the surrounding monolayer. In time, the plaque becomes large enough to be seen with the naked eye (Fig. 2.10). Only viruses that cause visible damage of cultured cells can be assayed in this way.

Figure 2.10 Plaques formed by different animal viruses. Plaque sizes reflect the life cycle of a virus in a particular cell type. **(A)** Photomicrograph of a single plaque formed by pseudorabies virus in Georgia bovine kidney cells. (Left) Unstained cells. (Right) Cells stained with the chromogenic substrate X-Gal (5-bromo-4-chloro-3-indolyl-β-D-galactopyranoside), which is converted to a blue compound by the product of the *lacZ* gene carried by the virus. Courtesy of B. Banfield, Princeton University. **(B)** Different plaque morphology of influenza C virus strains. Monolayers were stained with the vital dye crystal violet. **(C)** Plaques formed by poliovirus on human HeLa cells stained with crystal violet. **(D)** Illustration of the spread of virus from an initial infected cell to neighboring cells, resulting in a plaque.

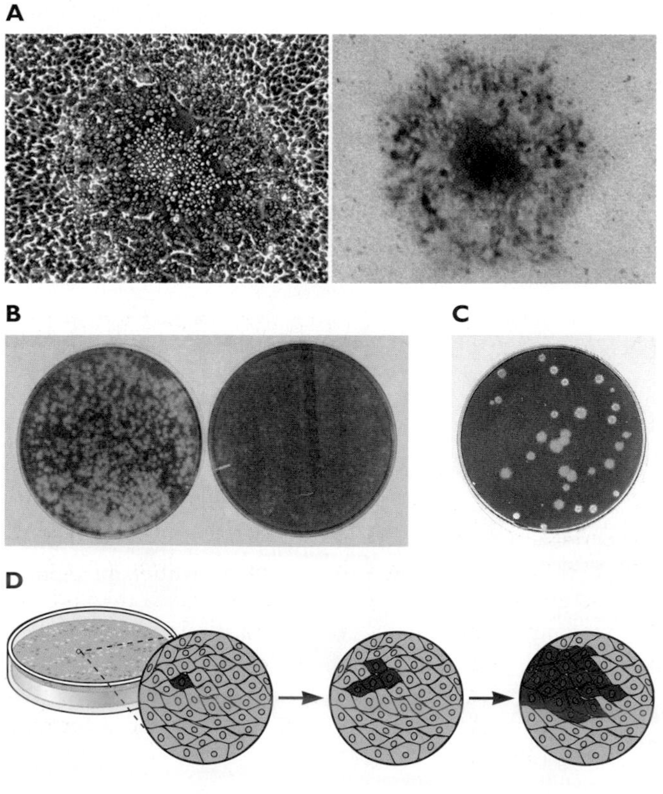

For the majority of animal viruses, there is a linear relationship between the number of infectious virus particles and the plaque count (Fig. 2.11). One infectious particle is therefore sufficient to initiate infection, and the virus is said to infect cells with **one-hit kinetics**. Some examples of **two-hit kinetics**, in which two different types of virus particle must infect a cell to ensure replication, have been recognized. For example, the genomes of some (+) strand RNA viruses of plants consist of two RNA molecules that are encapsidated separately. Both RNAs are required for infectivity. The dose-response curve in plaque assays for these viruses is parabolic rather than linear (Fig. 2.11).

The titer of a virus stock can be calculated in **plaque-forming units (PFU) per milliliter** (Box 2.5) When one infectious virus particle initiates a plaque, the viral progeny within the plaque are clones, and virus stocks prepared from a single plaque are known as **plaque purified**. The tip of a small pipette is plunged into the overlay above the

Figure 2.11 The dose-response curve of the plaque assay. The number of plaques produced by a virus with one-hit kinetics (red) or two-hit kinetics (blue) is plotted against the relative concentration of the virus. In two-hit kinetics, there are two classes of uninfected cells, those receiving one particle and those receiving none. The Poisson distribution can be used to determine the proportion of cells in each class: they are e^{-m} and me^{-m} (Box 2.8). Because one particle is not sufficient for infection, $P(0) = e^{-m}(1 + m)$. At a very low multiplicity of infection, this equation becomes $P(\mathrm{i}) = (1/2)m^2$ (where i = infection) which gives a parabolic curve. Adapted from B. D. Davis et al., *Microbiology* (J. B. Lippincott Co., Philadelphia, PA, 1980), with permission.

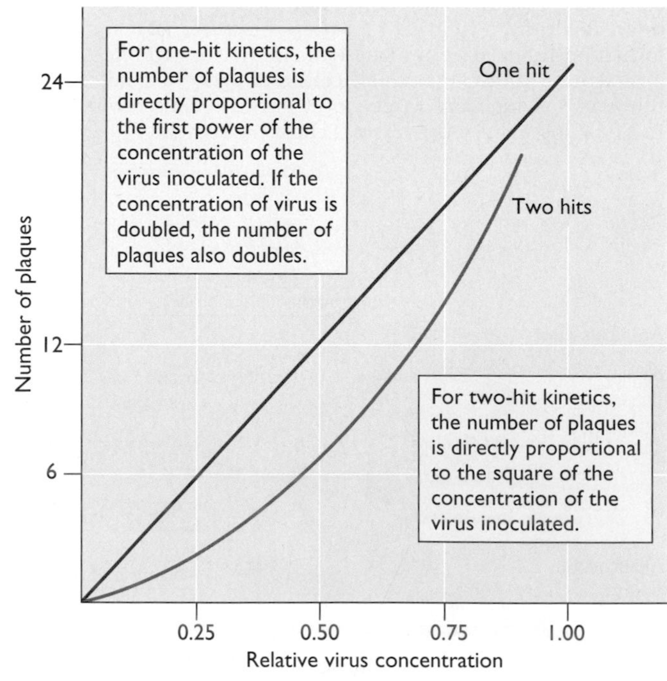

BOX 2.5

METHODS

Calculating virus titer from the plaque assay

To calculate the titer of a virus in plaque-forming units per milliliter, 10-fold serial dilutions of a virus stock are prepared, and 0.1-ml aliquots are inoculated onto susceptible cell monolayers (see figure). After a suitable incubation period, the monolayers are stained and the plaques are counted. To minimize error in calculating the virus titer, only plates containing between 10 and 100 plaques are counted, depending on the area of the cell culture vessel. According to statistical principles, when 100 plaques are counted, the sample titer varies by ±10%. For accuracy, each dilution is plated in duplicate or triplicate (not shown in the figure). Plates with more than 100 plaques are generally not counted because the plaques may overlap, causing inaccuracies. In the example shown in the figure, 17 plaques are observed on the plate produced from the 10^{-6} dilution. Therefore, the 10^{-6} dilution tube contains 17 PFU per 0.1 ml, or 170 PFU per ml, and the titer of the virus stock is 170×10^{6} or 1.7×10^{8} PFU/ml.

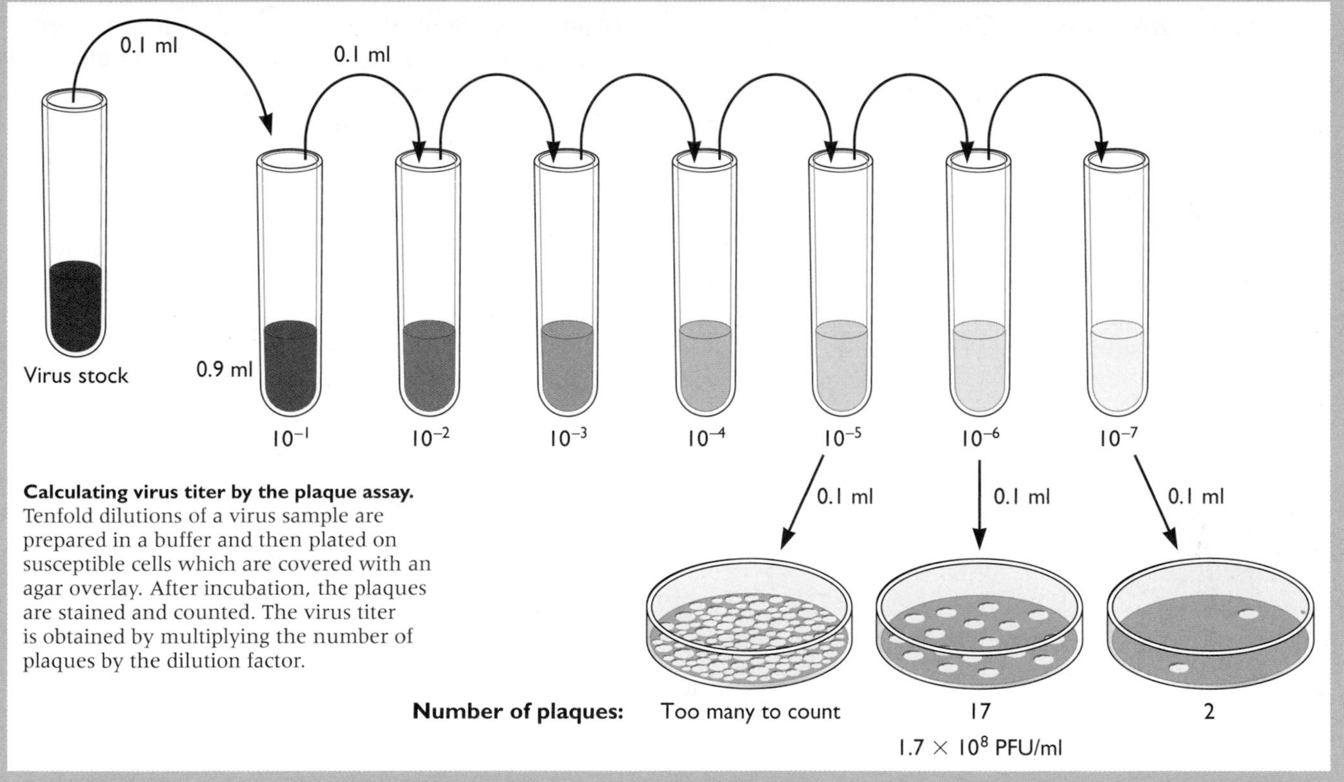

Calculating virus titer by the plaque assay. Tenfold dilutions of a virus sample are prepared in a buffer and then plated on susceptible cells which are covered with an agar overlay. After incubation, the plaques are stained and counted. The virus titer is obtained by multiplying the number of plaques by the dilution factor.

plaque, and the plug of agar containing the virus is recovered. The virus within the agar plug is eluted into buffer and used to prepare virus stocks. To ensure purity, this process is usually repeated at least one more time. Plaque purification is employed widely in virology to establish clonal virus stocks.

Fluorescent-Focus Assay

The fluorescent-focus assay, a modification of the plaque assay, is useful in determining the titers of viruses that do not kill cells. The initial procedure is the same as in the plaque assay. However, after a period sufficient for adsorption and gene expression, cells are permeabilized and incubated with an antibody raised against a viral protein. A second antibody, which recognizes the first, is then added. This second antibody is usually conjugated to a fluorescent indicator, such as fluorescein. The cells are then examined under a microscope at an appropriate wavelength. The titer of the virus stock is expressed in fluorescent-focus-forming units per milliliter.

Infectious-Centers Assay

Another modification of the plaque assay, the infectious-centers assay, is used to determine the fraction of cells in a culture that are infected with a virus. Monolayers of infected cells are suspended before progeny viruses are produced. Dilutions of a known number of infected cells are then plated on monolayers of susceptible cells, which are covered with an agar overlay. The number of plaques that form on the indicator cells is a measure of the number of cells infected in the original population. The fraction of infected cells can therefore be determined. A typical use of the infectious-centers assay is to measure the proportion of infected cells in persistently infected cultures.

Transformation Assay

The transformation assay is useful for determining the titers of some retroviruses that do not form plaques. For example, Rous sarcoma virus transforms chicken embryo cells. As a result, the cells lose their contact inhibition (the property that governs whether cultured cells grow as a single monolayer [see Volume II, Chapter 7]) and become heaped up on one another. The transformed cells form small piles, or **foci**, that can be distinguished easily from the rest of the monolayer (Fig. 2.12). Infectivity is expressed in focus-forming units per milliliter.

End-Point Dilution Assay

The end-point dilution assay provided a measure of virus titer before the development of the plaque assay. It is still used for measuring the titers of certain viruses that do not form plaques or for determining the virulence of a virus in animals. Serial dilutions of a virus stock are inoculated into replicate test units (typically 8 to 10), which can be cell cultures, eggs, or animals. The number of test units that have become infected is then determined for each virus dilution. When cell culture is used, infection is determined by the development of cytopathic effect; in eggs or animals, infection is gauged by death or disease. An example of an end-point dilution assay using cell cultures is shown in Box 2.6. At high dilutions, none of the cell cultures are infected because no infectious particles are delivered to the cells; at low dilutions, every culture is infected. The **end point** is the dilution of virus that affects 50% of the test units. This number can be calculated from the data and expressed as 50% infectious dose (ID_{50}) per milliliter. The first preparation illustrated in Box 2.6 contains 10^5 ID_{50} per ml.

When the end-point dilution assay is used to assess the virulence of a virus or its capacity to cause disease, as defined in Volume II, Chapter 1, the result of the assay can be expressed in terms of 50% lethal dose (LD_{50}) per milliliter or 50% paralytic dose (PD_{50}) per milliliter, end points of death and paralysis, respectively. If the virus titer can be determined separately by plaque assay, the 50% end point determined in an animal host can be related to this parameter. In this way, the effects of the route of inoculation or specific mutations on viral virulence can be quantified.

Efficiency of Plating

The term **relative efficiency of plating** was coined to assign a value to the plaque count determined with the

Figure 2.12 Transformation assay. Chicken cells transformed by two different strains of Rous sarcoma virus are shown. Loss of contact inhibition causes cells to pile up rather than grow as a monolayer. One focus is seen in panel **A**, and three foci are seen in panel **B** at the same magnification. Courtesy of H. Hanafusa, Osaka Bioscience Institute.

A

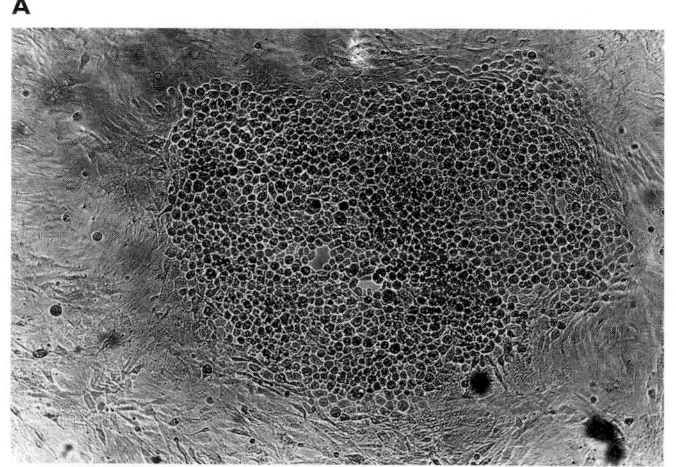

B

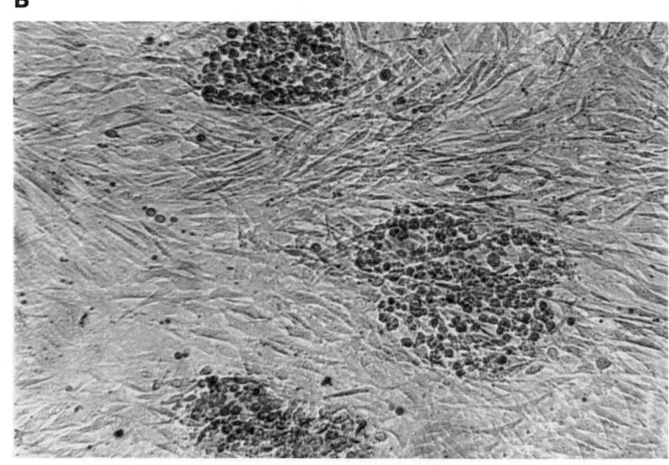

BOX 2.6

METHODS

End-point dilution assays

In the first example, 10 monolayer cell cultures were infected with each virus dilution. After the incubation period, plates that displayed cytopathic effect were scored +. Fifty percent of the cell cultures displayed cytopathic effect at the 10^{-5} dilution, and therefore the virus stock contains 10^5 ID_{50} units.

Virus dilution	Cytopathic effect									
10^{-2}	+	+	+	+	+	+	+	+	+	+
10^{-3}	+	+	+	+	+	+	+	+	+	+
10^{-4}	+	+	–	+	+	+	+	+	+	+
10^{-5}	–	+	+	–	+	–	–	+	–	+
10^{-6}	–	–	–	–	–	–	+	–	–	–
10^{-7}	–	–	–	–	–	–	–	–	–	–

In most cases the 50% end point does not fall on a dilution tested as shown in the example; for this reason, various statistical procedures have been developed to calculate the end point of the titration. In one popular method, the dilution containing the ID_{50} is identified by interpolation between the dilutions on either side of this value. The assumption is made that the location of the 50% end point varies linearly with the log of the dilution. Because the number of test units used at each dilution is usually small, the accuracy of this method is relatively low. For example, if six test units are used at each 10-fold dilution, differences in virus titer of only 50-fold or more can be detected reliably. The method is illustrated in the following example, in which the lethality of poliovirus in mice is the end point. Eight mice were inoculated per dilution.

In the method of Reed and Muench, the results are pooled, as shown in the table. The interpolated value of the 50% end point, which in this case falls between the fifth and sixth dilutions, is calculated to be $10^{-6.5}$. The virus sample therefore contains $10^{6.5}$ LD_{50}s. The LD_{50} may also be calculated as the concentration of the stock virus in PFU per milliliter (1×10^9) times the 50% end-point titer. In the example shown, the LD_{50} is 3×10^2 PFU.

Dilution	Alive	Dead	Total alive	Total dead	Mortality ratio	Mortality (%)
10^{-2}	0	8	0	40	0/40	100
10^{-3}	0	8	0	32	0/32	100
10^{-4}	1	7	1	24	1/25	96
10^{-5}	0	8	1	17	1/18	94
10^{-6}	2	6	3	9	3/12	75
10^{-7}	5	3	8	3	8/11	27

same bacteriophage but with different strains of bacteria. The value is a ratio of viral titers obtained on two different host cells. This number may be more or less than 1, depending on how well the virus grows in different host cells. A very different value is the **absolute efficiency of plating**, which is defined as the plaque titer divided by the number of virus particles in the sample. The **particle-to-PFU ratio**, a term more commonly used today, is the inverse value (Table 2.2). For many bacteriophages, the particle-to-PFU ratio approaches 1, the lowest value that can be obtained. However, for animal viruses this value can be much higher, ranging from 1 to 10,000. These high values have complicated the study of animal viruses. For example, when the particle-to-PFU ratio is high, it is never certain whether properties measured biochemically are in fact those of the infectious particle or those of the noninfectious component.

Although the linear nature of the dose-response curve indicates that a single particle is capable of initiating an

Table 2.2 Particle-to-PFU ratios of some animal viruses

Virus	Particle/PFU ratio
Adenoviridae	20–100
Alphaviridae	
Semliki Forest virus	1–2
Herpesviridae	
Herpes simplex virus	50–200
Orthomyxoviridae	
Influenza virus	20–50
Papillomaviridae	
Papillomavirus	10,000
Picornaviridae	
Poliovirus	30–1,000
Polyomaviridae	
Polyomavirus	38–50
Simian virus 40	100–200
Poxviridae	1–100
Reoviridae	
Reovirus	10

infection (one-hit kinetics) (Fig. 2.11), the high particle-to-PFU ratio for many viruses demonstrates that not all virions are successful. The high value of this ratio is sometimes caused by the presence of noninfectious particles with genomes that harbor lethal mutations or that have been damaged during growth or purification. An alternative explanation is that although all viruses in a preparation are in fact capable of initiating infection, not all of them succeed because of the complexity of the infectious cycle. Failure at any one step in the cycle prevents completion. A high particle-to-PFU ratio does not indicate that most particles are defective but, rather, indicates that they failed to complete the infection.

Measurement of Virus Particles and Their Components

Although the numbers of virus particles and infectious units are often not equal, assays for particle number are frequently used to approximate the number of infectious particles present in a sample. For example, the concentration of viral DNA or protein can be used to estimate the particle number, assuming that the ratio is constant. Biochemical or physical assays are usually more rapid and easier to carry out than assays for infectivity, which may be slow, cumbersome, or not possible. Assays for subviral components also provide information on particle number if the stoichiometry of these components in the virus particle is known.

Imaging Particles

Electron microscopy. With few exceptions, virus particles are too small to be observed directly by light microscopy. However, they can be seen readily in the electron microscope. If a sample contains only one type of virus, the particle count can be determined. First, a virus preparation is mixed with a known concentration of latex beads. The numbers of virus particles and beads are then counted, allowing the concentration of the virus particles in the sample to be determined by comparison.

Live-cell imaging of single fluorescent virions. The discovery of green fluorescent protein revolutionized the study of the cell biology of virus infection. This protein, isolated from the jellyfish *Aequorea victoria*, is a convenient reporter for monitoring transcription or translation, because it is directly visible in living cells without the need for fixation, substrates, or coenzymes. Similar proteins isolated from different organisms, which emit light of different wavelengths, are also widely used in virology. The use of fluorescent proteins has allowed visualization of single virus particles in living cells. The coding sequence for the fluorescent protein is inserted into the viral genome, often fused to the coding region of a virion protein. The fusion protein is incorporated into the viral particle, which is visible in cells by fluorescence microscopy (Fig. 2.13). Using this approach, entry, uncoating, replication, assembly, and egress of single particles can all theoretically be observed in living cells.

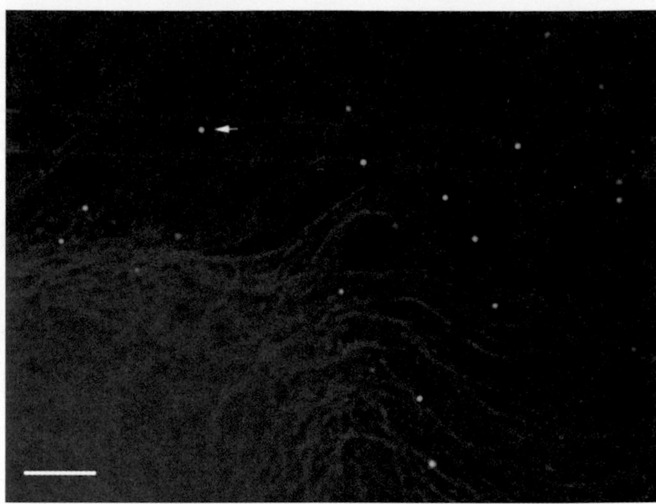

Figure 2.13 Live-cell imaging of single virus particles by fluorescence. Single-virus-particle imaging with green fluorescent protein illustrates microtubule-dependent movement of human immunodeficiency virus type one particles in cells. Rhodamine-tubulin was injected into cells to label microtubules (red). The cells were infected with virus particles that contain a fusion of green fluorescent protein with Vpr. Virus particles can be seen as green dots. Bar, 5 μm. Courtesy of David McDonald, University of Illinois.

Hemagglutination

Members of the *Adenoviridae*, *Orthomyxoviridae*, and *Paramyxoviridae*, among others, contain proteins that can bind to erythrocytes (red blood cells); these viruses can link multiple cells, resulting in formation of a lattice. This property is called **hemagglutination**. For example, influenza viruses contain an envelope glycoprotein called hemagglutinin, which binds to *N*-acetylneuraminic acid-containing glycoproteins on erythrocytes. In practice, twofold serial dilutions of the virus stock are prepared, mixed with a defined quantity of red blood cells, and added to small wells in a plastic tray (Fig. 2.14). Unadsorbed red blood cells tumble to the bottom of the well and form a sharp dot or button. In contrast, agglutinated red blood cells form a diffuse lattice that coats the well. Because the assay is rapid (30 min), it is often used as a quick indicator of the relative quantities of virus particles. However, it is not sufficiently sensitive to detect small numbers of particles.

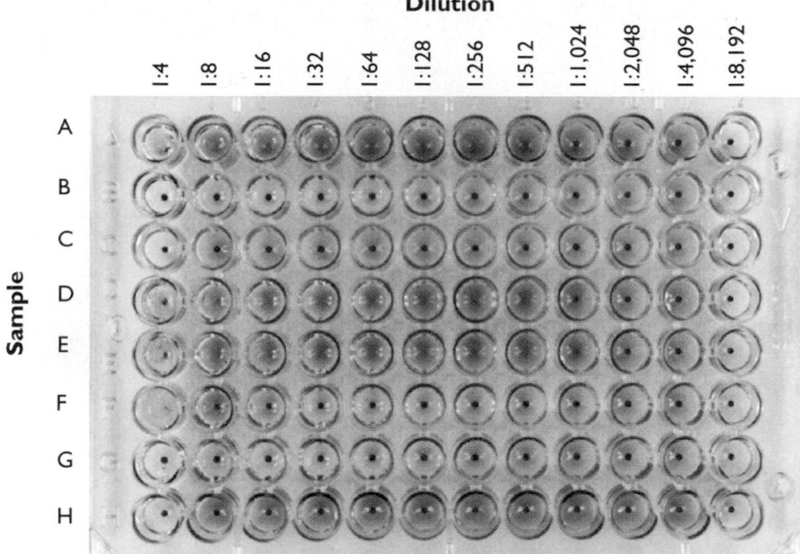

Figure 2.14 Hemagglutination assay. Samples of different influenza viruses were diluted, and a portion of each dilution was mixed with a suspension of chicken red blood cells and added to the wells. After 30 min at 4°C, the wells were photographed. Sample A causes hemagglutination until a dilution of 1:256 and therefore has a hemagglutination titer of 256. Elution of the virus from red blood cells as seen in column 1, rows D and E, is caused by neuraminidase in the virus particle. This enzyme cleaves *N*-acetylneuraminic acid from glycoprotein receptors and elutes bound viruses from red blood cells. Courtesy of C. Basler and P. Palese, Mount Sinai School of Medicine of the City University of New York.

Measurement of Viral Enzyme Activity

Some animal virus particles contain nucleic acid polymerases, which can be assayed by mixing permeabilized particles with radioactively labeled precursors and measuring the incorporation of radioactivity into nucleic acid. This type of assay is used most frequently for retroviruses, many of which do not transform cells or form plaques. The reverse transcriptase incorporated into the virus particle is assayed by mixing cell culture supernatants with a mild detergent (to permeabilize the viral envelope), an appropriate template and primer, and a radioactive nucleoside triphosphate. If reverse transcriptase is present, a radioactive product will be produced by priming on the poly(rC) template. This product can be detected by precipitation or bound to a filter and quantified. Because enzymatic activity is proportional to particle number, this assay allows rapid tracking of virus production in the course of an infection.

Serological Methods

Many virological techniques are based on the specificity of the antibody-antigen reaction. Some of the techniques, such as immunostaining, immunoprecipitation, immunoblotting, and the enzyme-linked immunosorbent assay, are by no means limited to the detection of viruses and viral proteins. All these approaches have been used extensively to study the structures and functions of cellular proteins.

Virus neutralization. When a virus preparation is inoculated into an animal, an array of antibodies is produced. These antibodies can bind to virus particles, but not all of them can block infectivity **(neutralize)**, as discussed in Volume II, Chapter 4. Virus neutralization assays are usually conducted by mixing dilutions of antibodies with virus, incubating them, and assaying for remaining infectivity in cultured cells, eggs, or animals. The end point is defined as the highest dilution of antibody that inhibits the development of cytopathic effect in cells or virus replication in eggs or animals.

Some neutralizing antibodies define **type-specific antigens** on the virus particle. For example, the three **serotypes** of poliovirus are distinguished on the basis of neutralization tests; type 1 poliovirus is neutralized by antibodies to type 1 virus but not by antibodies to type 2 or type 3 poliovirus, and so forth. Neutralization tests have therefore been valuable for virus classification. These antibodies may also be used to map the three-dimensional structure of neutralization antigenic sites on the virion (Box 2.7).

Hemagglutination inhibition. Antibodies against viral proteins with hemagglutination activity can block the ability of virus to bind red blood cells. In this assay, dilutions of antibodies are incubated with virus, and erythrocytes are added as outlined above. After incubation, the hemagglutination inhibition titer is read as the highest dilution of antibody that inhibits hemagglutination. This test is sensitive, simple, inexpensive, and rapid; it is the method of choice for assaying antibodies to any virus that causes hemagglutination. It can be used to detect antibodies to viral hemagglutinin in animal and human sera or to identify the origin of the hemagglutinin of influenza viruses produced in cells coinfected with two parent viruses.

BOX 2.7

DISCUSSION

Neutralization antigenic sites

Knowledge of the antigenic structure of a virus is useful in understanding the immune response to these agents and in designing new vaccination strategies. The use of **monoclonal antibodies** (antibodies of a single specificity made by a clone of antibody-producing cells) in neutralization assays permits mapping of antigenic sites on a virus particle, or of the amino acid sequences that are recognized by neutralizing antibodies. Each monoclonal antibody binds specifically to a short amino acid sequence (8 to 12 residues) that fits into the antibody-combining site. This amino acid sequence, which may be linear or nonlinear, is known as an **epitope**. In contrast, **polyclonal antibodies** comprise the repertoire produced in an animal against the many epitopes of an antigen. Antigenic sites may be identified by cross-linking the monoclonal antibody to the virus and determining which protein is the target of the antibody. Epitope mapping may also be performed by assessing the abilities of monoclonal antibodies to bind synthetic peptides representing viral protein sequences. When the monoclonal antibody recognizes a linear epitope, it may react with the protein in Western blot analysis, facilitating direct identification of the viral protein harboring the antigenic site. The most elegant understanding of antigenic structures has come from the isolation and study of variant viruses that are resistant to neutralization with specific monoclonal antibodies (called **monoclonal antibody-resistant variants**). By identifying the amino acid change responsible for this phenotype, the antibody-binding site can be located and, together with three-dimensional structural information, can provide detailed information on the nature of antigenic sites that are recognized by neutralizing antibodies (see figure).

Locations of neutralization antigenic sites on the capsid of human rhinovirus type 14. Amino acids that change in viral mutants selected for resistance to neutralization by monoclonal antibodies are shown in cyan on a model of the viral capsid. These amino acids are in VP1, VP2, and VP3 on the surface of the virion.

Complement fixation. The complement fixation assay can be used to determine if antibodies against a virus are present in serum. The interaction of viral antigen and antibody can cause complement fixation, which leads to membrane lysis, as discussed in Volume II, Chapter 4. Red blood cells are used as targets, because lysis of their membranes is readily observed (Fig. 2.15).

Because crude preparations of viral antigens are often used in the complement fixation test, this test is not highly specific. The antigen preparations may include proteins or specific epitopes shared by related groups of viruses. As a result, the test detects most serotypes within a given family of viruses and is therefore said to recognize group-specific antigens. For example, the three serotypes of poliovirus share a complement-fixing, group-specific antigen. Complement fixation is not as sensitive as neutralization or hemagglutination inhibition. Nevertheless, it is often the first assay performed on sera from infected patients to identify the family to which the infecting virus belongs.

Viral Growth: the Burst Concept

A fundamental and important concept is that viruses replicate by the assembly of preformed components into particles. The parts are first made in cells and then assembled into the final product. The growth of viruses is very different from the growth of cells, which multiply by binary fission. This simple build-and-assemble strategy is unique to viruses, but the details for members of different virus families are astoundingly different. There are many ways to build a virus particle, and each one tells us something new about virus structure and assembly.

Modern studies of viruses have their origins in the work of Delbrück and colleagues, who studied the T-even bacteriophages starting in 1937. Delbrück believed that these bacteriophages were perfect models for understanding virus replication. He also thought that phages were excellent models for studying the gene: they were self-replicating (a hallmark of a gene), their mutations were inherited, and they were small, easily manipulated entities that grew rapidly.

In the late 1930s, Delbrück focused his attention on the fact that one bacterial cell usually makes hundreds of progeny viruses. The yield from one cell is one viral generation; it was called the **burst** because viruses literally burst from the infected cell. According to the burst concept, an infected cell either produces virus or it does not. Under carefully controlled laboratory conditions, most cells make on average about the same number of bacteriophages per cell. For example, in one of Delbrück's experiments, the number of bacteriophage T4 particles produced from individual

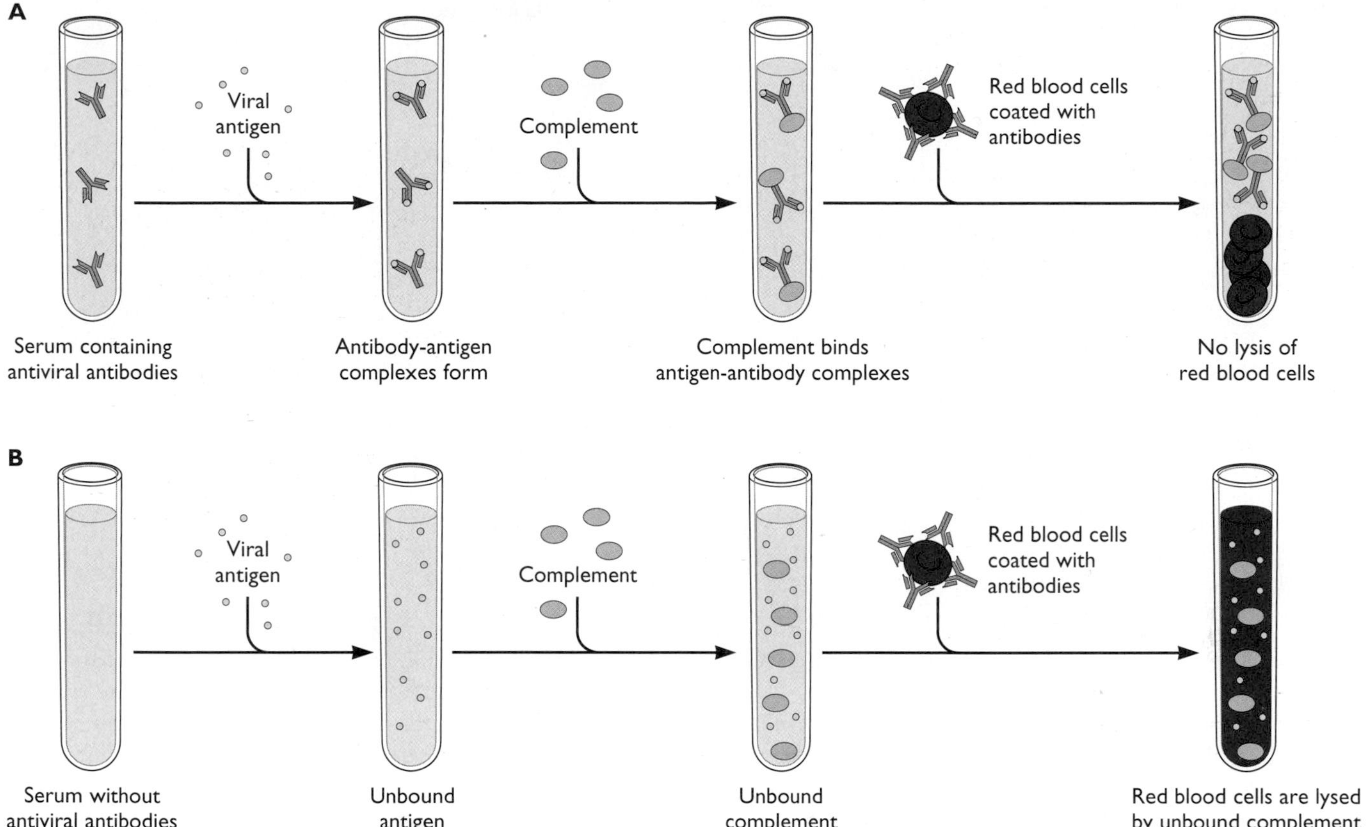

Figure 2.15 Detection of viral antigens by complement fixation. Samples of patients' sera are heated to inactivate endogenous complement and then incubated with preparations of viral antigen and a standardized quantity of guinea pig complement. If an antibody-antigen reaction takes place, complement fixation will occur. The latter is detected by adding sheep red blood cells that have been coated with rabbit anti-red blood cell antibodies. If complement has been fixed by binding of viral antigen to antibody, the red blood cells remain intact **(A)**; if complement is not fixed, the red blood cells are lysed by the action of free complement **(B)**.

single-cell bursts from four *Escherichia coli* cells was 101, 127, 57, and 316. The average burst is 150 particles per cell; if this experiment were done today, using similar experimental conditions, the average burst would be similar. The variation of the burst size from cell to cell is of interest, as it shows that the replication machinery is not precise.

Another important component of the burst concept is that a cell has a finite capacity to produce virus. A number of factors limit the number of particles produced per cell, such as metabolic resources, the number of sites for replication in the cell, the regulation of virion release, and host defenses. The burst principle holds for most viruses. In general, larger cells (e.g., eukaryotic cells) produce more virus particles per cell; yields of 1,000 to 10,000 virions per eukaryotic cell are not uncommon.

The burst concept applies only to viruses that kill the cell after infection, namely, the cytopathic viruses. However, there are viruses that do not kill their host cells; virus particles are produced as long as the cell is alive. There is no burst; instead, there is a continuous release of virus particles. Examples include filamentous bacteriophages, some retroviruses, and hepatitis viruses.

The One-Step Growth Cycle

Initial Concept

The idea that one-step growth analysis can be used to study the single-cell life cycle of viruses originated from the work on bacteriophages by Ellis and Delbrück. In their classic experiment, they added virus particles to a culture of rapidly growing *E. coli* cells; these particles adsorbed quickly to the cells. The infected culture was then diluted, preventing further adsorption of unbound particles. This simple dilution step is the key to the experiment: it reduces further

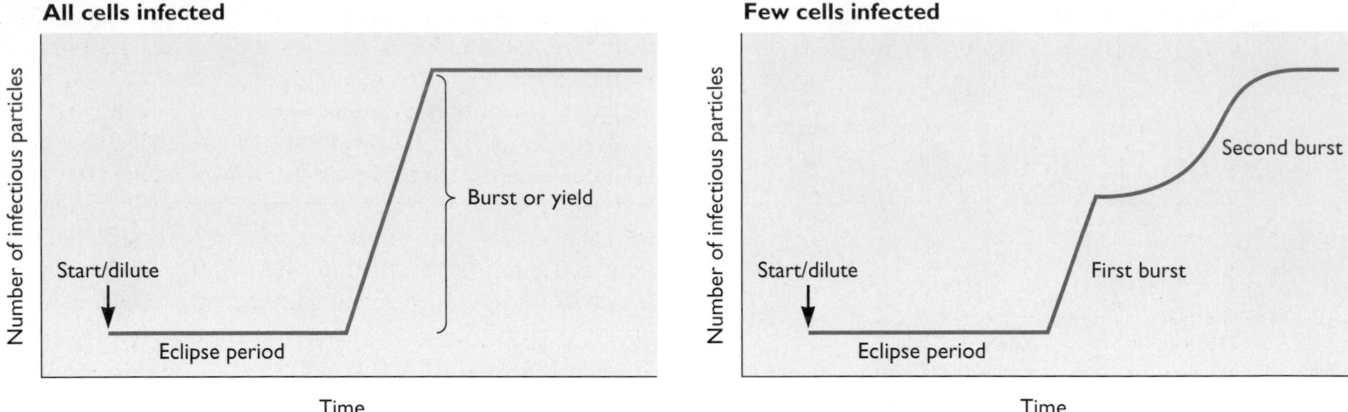

Figure 2.16 One-step growth curves of bacteriophages. The figure shows the growth of a bacteriophage in *E. coli* under conditions when all cells are infected **(A)** and when only a few cells are infected **(B)**.

binding of virus to cells and effectively synchronizes the infection. Samples of the diluted culture were then taken every few minutes and analyzed for the number of infectious bacteriophages. When the results were plotted, several key observations emerged (Fig. 2.16). Numbers of new viruses did not increase in a linear fashion from the start of the infection. There was an initial lag, followed by a rapid increase in virus production, which then plateaued. This single cycle of virus replication produces the "burst" of virus progeny. If the experiment is repeated, so that only a few cells are initially infected, the graph looks different (Fig. 2.16). Instead of a single cycle, there is a stepwise increase in numbers of new viruses with time. Each step represents one cycle of virus infection.

Once the nature of the viral growth cycle was explored using the one-step growth curve, questions emerged about what was happening in the cell before the burst. What was the fate of the incoming virus? Did it disappear? How was more virus produced? These questions were answered by A. Doermann, who in 1951 found an elegant way of looking inside the infected cell. Instead of sampling the diluted culture for virus after various periods of infection, Doermann prematurely lysed the infected cells as the infection proceeded and then assayed the lysate for infectious virus. The results were extremely informative. Immediately after dilution, there was a complete loss, or eclipse, of infectious virus for 10 to 15 min (Fig. 2.16). In other words, input infectious virions disappeared and no new phage particles were detected during this period. The loss of infectivity is a consequence of the release of the genome from the virion, to allow for subsequent transcription of viral genes. Particle infectivity is lost during this phase because the released genome is not infectious under the conditions of the plaque assay. Next, new infectious viruses were detected inside the cell, before they were released into the medium (Fig. 2.16). These were newly assembled virions that had not yet been released by cell lysis. The results of these experiments defined two new terms in virology: the **eclipse period**, the phase in which infectivity is lost when virions are disassembled after penetrating cells, and the **latent period**, the time it takes to replicate, assemble, and release new virus, approximately 20 to 25 min for *E. coli* bacteriophages.

Synchronous infection, the key to the one-step growth cycle, is usually accomplished by infecting cells with a sufficient number of virus particles to ensure that most of the cells are infected rapidly.

One-Step Growth Analysis: a Valuable Tool for Studying Animal Viruses

One-step growth analysis soon became adapted for studying the replication of animal viruses. The experiment begins with removal of the medium from the cell monolayer and addition of virus in a small volume to promote rapid adsorption. One cell monolayer is infected for each time point. After approximately 1 h, the unadsorbed inoculum is removed, the cells are washed, and fresh medium is added. At different times after infection, samples of the cell culture supernatant are collected and the virus titer is determined. The kinetics of intracellular virus production can be monitored by removing the medium containing extracellular particles, scraping the cells into fresh medium, and lysing them by repeated cycles of freeze-thawing. A cell extract is prepared after removal of cellular debris by centrifugation, and the virus titer in the extract is measured.

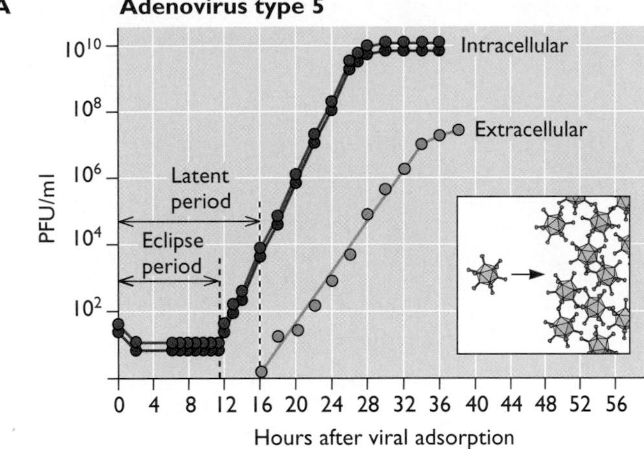

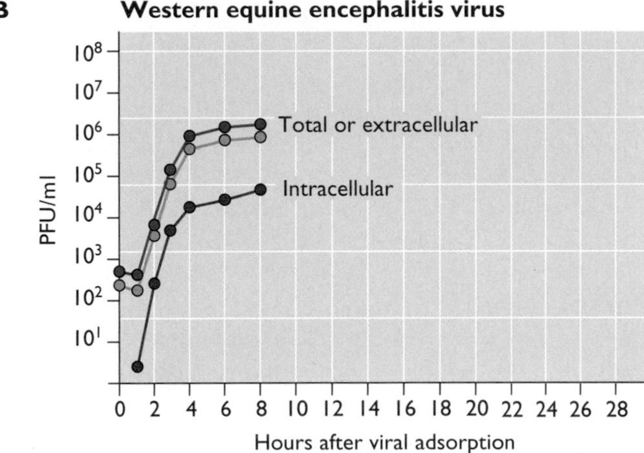

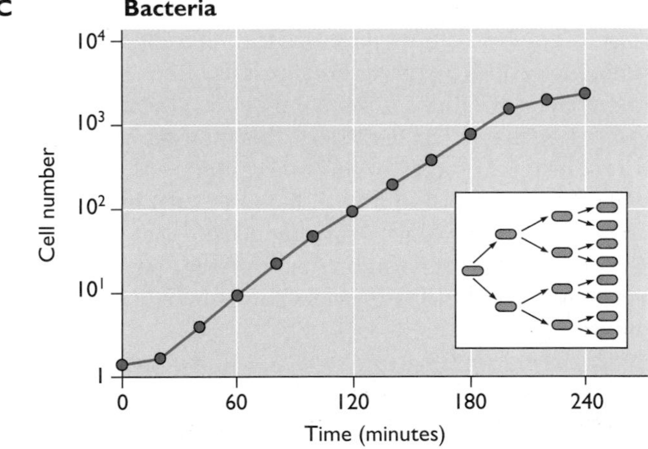

Figure 2.17 One-step growth curves of animal viruses. **(A)** Growth of a nonenveloped virus, adenovirus type 5. The inset illustrates the concept that viruses multiply by assembly of preformed components into particles. **(B)** Growth of an enveloped virus, Western equine encephalitis virus, a member of the *Togaviridae*. This virus acquires infectivity after maturation at the plasma membrane, and therefore little intracellular virus can be detected. The small amounts observed at each time point probably represent released virus contaminating the cell extract. **(C)** Growth curve for a bacterium. The number of bacteria is plotted as a function of time. One bacterium is added to the culture at time zero; after a brief lag, the bacterium begins to divide. The number of bacteria doubles every 20 min until nutrients in the medium are depleted and the growth rate decreases. The inset illustrates the growth of bacteria by binary fission. (A and B) Adapted from B. D. Davis et al., *Microbiology* (J. B. Lippincott Co., Philadelphia, PA, 1980), with permission. (C) Adapted from B. Voyles, *The Biology of Viruses* (McGraw-Hill, New York, NY, 1993), with permission.

The results of a one-step growth experiment establish a number of important features about viral replication. In the example shown in Fig. 2.17A, the first 11 h after infection constitutes the eclipse period, during which the viral nucleic acid is uncoated from its protective shell and no infectious virus can be detected inside cells. The low level of infectivity detected during this period probably results from adsorbed virus that was not uncoated. Beginning at 12 h after adsorption, the quantity of intracellular infectious virus begins to increase, marking the onset of the synthetic phase, during which new virus particles are assembled. During the latent period no extracellular virus can be detected. At 18 h after adsorption, virions are released from cells and found in the extracellular medium. Ultimately, virus production plateaus as the cells become metabolically and structurally incapable of supporting additional replication.

The yield of infectious virus per cell can be calculated from the data collected during a one-step growth experiment (Fig. 2.17). This value varies widely among different viruses and with different virus-host cell combinations. For many viruses, increasing the **multiplicity of infection** (Box 2.8) above a certain point does not increase the yield: cells have a finite capacity to produce new virus particles. In fact, infecting at a very high multiplicity of infection can cause premature cell lysis and decrease virus yields.

The nature of the one-step growth curve can vary dramatically among different viruses. For example, enveloped viruses that mature by budding from the plasma membrane, as discussed in Chapter 13, generally become infectious only as they leave the cell, and therefore little intracellular infectious virus can be detected (Fig. 2.17B). The first one-step growth curves of viruses were prepared for bacteriophages, and the results surprised scientists who had expected that they would resemble the growth curves of bacteria or cultured cells. After a short lag, bacterial cell growth becomes exponential (i.e., each progeny cell is capable of dividing) and follows a straight line (Fig. 2.17C). Exponential growth continues until the nutrients in the medium are exhausted. The one-step growth curves of animal viruses are very different from one another. The curve

BOX 2.8

DISCUSSION

Multiplicity of infection (MOI)

Infection depends on the random collision of cells and virus particles. When susceptible cells are mixed with a suspension of virus, some cells are uninfected and other cells receive one, two, three, etc., particles. The distribution of virus particles per cell is best described by the Poisson distribution:

$$P(k) = e^{-m}m^k/k!$$

In this equation, $P(k)$ is the fraction of cells infected by k virus particles. The multiplicity of infection, m, is calculated from the proportion of uninfected cells, $P(0)$, which can be determined experimentally. If k is made 0 in the above equation, then

$$P(0) = e^{-m} \text{ and } m = -\ln P(0)$$

The fraction of cells receiving 0, 1, and more than one virus particle in a culture of 10^6 cells infected with an MOI of 10 can be determined as follows.

The fraction of cells that receive 0 particles is

$$P(0) = e^{-10} = 4.5 \times 10^{-5}$$

and in a culture of 10^6 cells this equals 45 uninfected cells.

The fraction of cells that receive 1 particle is

$$P(1) = 10 \times 4.5 \times 10^{-5} = 4.5 \times 10^{-4}$$

and in a culture of 10^6 cells, 450 cells receive 1 particle.

The fraction of cells that receive >1 particle is

$$P(>1) = 1 - e^{-m}(m + 1) = 0.9995$$

and in a culture of 10^6 cells, 999,500 cells receive more than 1 particle. [The value in this equation is obtained by subtracting from 1 (the sum of all probabilities for any value of k) the probabilities $P(0)$ and $P(1)$.]

The fraction of cells receiving 0, 1, and more than one virus particle in a culture of 10^6 cells infected with an MOI of 0.001 is

$$P(0) = 99.99\%$$
$$P(1) = 0.0999\% \text{ (for } 10^6 \text{ cells, } 10^4 \text{ are infected)}$$
$$P(>1) = 10^{-6}$$

The MOI required to infect 99% of the cells in a cell culture dish is

$$P(0) = 1\% = 0.01$$
$$m = -\ln (0.01) = 4.6 \text{ PFU per cell}$$

shown in Fig. 2.17A illustrates the pattern observed for a DNA virus with the long latent and synthetic phases typical of many DNA viruses, some retroviruses, and reovirus. For small RNA viruses, the entire growth curve is complete within 6 to 8 h, and the latent and synthetic phases are correspondingly shorter. Counterintuitively, polyomavirus, with one of the smallest genomes of the DNA viruses, has a very long latent period. The basis for these differences is related to the various strategies of gene expression and genome replication (discussed in Chapter 3).

One-step growth curve analysis can provide quantitative information about different virus-host systems. It is frequently employed to study mutant viruses to determine what parts of the replication cycle are affected by a particular genetic lesion. It is also valuable for studying the multiplication of a new virus or viral replication in a new virus-host cell combination.

When cells are infected at a low multiplicity of infection, several cycles of viral replication may occur (Figure 2.16). Growth curves established under these conditions can also provide useful information. For example, if a mutation fails to have an obvious effect on viral replication when infection is done at high multiplicity of infection, the defect may become obvious following a low-multiplicity infection. Because the effect of a mutation in each cycle is multiplied over several cycles, a small effect can be amplified. Defects in the ability of viruses to spread from cell to cell may also be revealed when multiple cycles of replication occur.

Perspectives

The one-step growth analysis is nearly universally used to study virus replication. When millions of cells are infected at a high multiplicity of infection, enough viral nucleic acid or protein can be isolated to allow a study of events during the replication cycle. Synchronous infection is the key to this approach, because under this condition, the same steps of the replication cycle occur in all cells at the same time. Many of the experimental results discussed in subsequent chapters of this book were obtained using one-step growth analysis. The power of this analysis is such that it reports on all stages of the replication cycle in a simple and quantitative fashion. With modest expenditure of time and reagents, virologists can deduce a great deal about viral translation, replication, or assembly. This universal utility is a cardinal feature of a fundamentally important method.

References

Books

Ausubel, F. M., R. Brent, R. E. Kingston, D. D. Moore, J. G. Seidman, J. A. Smith, and K. Struhl. 1998. *Current Protocols in Molecular Biology.* John Wiley & Sons, Inc., New York, NY. Updated frequently: http://www.currentprotocols.com.

Cann, A. J. 2000. *Virus Culture: a Practical Approach.* Oxford University Press, Oxford, United Kingdom.

Cooper, G. M., and R. E. Hausman. 2006. *The Cell: a Molecular Approach.* Sinauer Associates, Sunderland, MA.

Freshney, I. A. 2005. *Culture of Animal Cells: a Manual of Basic Techniques.* Wiley-Liss, Hoboken, NJ.

Harlow, E., and D. Lane. 1998. *Using Antibodies: a Laboratory Manual.* Cold Spring Harbor Laboratory Press, Cold Spring Harbor, NY.

Sambrook, J., and D. Russell. 2001. *Molecular Cloning: a Laboratory Manual,* 3rd ed. Cold Spring Harbor Laboratory Press, Cold Spring Harbor, NY.

Review Articles

Dulbecco, R., and M. Vogt. 1953. Some problems of animal virology as studied by the plaque technique. *Cold Spring Harbor Symp. Quant. Biol.* **18:**273–279.

Laude, A. J., and A. Prior. 2004. Plasma membrane microdomains: organization, function and trafficking. *Mol. Membr. Biol.* **21:**193–205.

Maramorosch, K., and H. Koprowski. *Methods in Virology.* Academic Press, New York, NY. A series of volumes begun in the 1960s that contain review articles on classic virological methods.

Papers of Special Interest

Elliott, G., and P. O'Hare. 1999. Live-cell analysis of a green fluorescent protein-tagged herpes simplex virus infection. *J. Virol.* **73:**4110–4119.

Ellis, E. L., and M. Delbrück. 1939. The growth of bacteriophage. *J. Gen. Physiol.* **22:**365–384.

Reed, L. J., and H. Muench. 1932. A simple method for estimating 50% endpoints. *Am. J. Hyg.* **27:**493–497.

Journals

BioTechniques (http://www.biotechniques.com)

Trends in Biotechnology (http://www.trends.com/tibtech/)

II

Molecular Biology

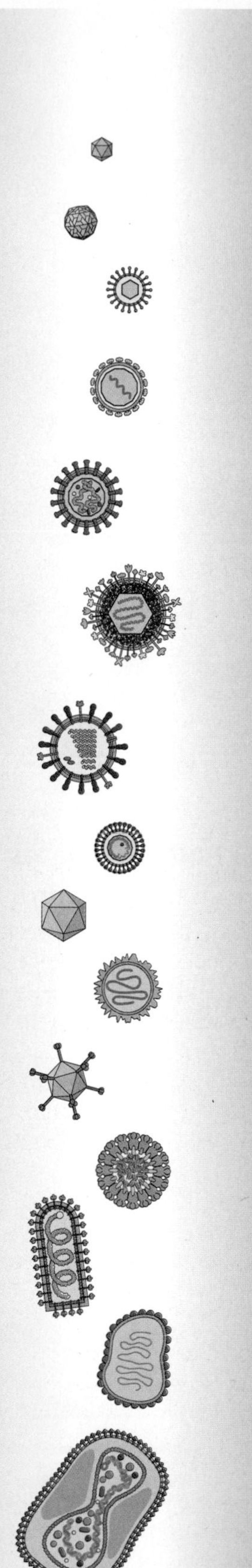

3

Genomes and Genetics

... everywhere an interplay between nucleic acids and proteins; a spinning wheel in which the thread makes the spindle and the spindle the thread.

ERWIN CHARGAFF
as quoted in A. N. H. Creager, The Life of a Virus, University of Chicago Press, 2002

Introduction

The world abounds with uncountable numbers of viruses of apparently overwhelming diversity. Fortunately, as discussed in Chapter 1, taxonomists have devised methods of classifying viruses. As a result, the number of identifiable groups becomes rather manageable (Table 3.1). One of the contributions of molecular biology has been a detailed analysis of the genetic material of representatives of these main virus families. What emerged from these studies was the principle that the **viral genome** is the nucleic acid-based repository of the information needed to build, replicate, and transmit a virus (Box 3.1). These analyses also revealed that the structures of viral genomes are far less complex than is implied by the existence of thousands of distinct entities defined by classical taxonomic methods. In fact, it is possible to organize the known viruses into seven groups, based on the structures of their genomes.

Genome Principles and the Baltimore System

A universal function of viral genomes is to specify proteins. However, viral genomes do not encode the machinery needed to carry out protein synthesis. Consequently, one important principle is that all viral genomes must be copied to produce messenger RNAs (mRNAs) that can be read by host ribosomes. Literally, all viruses are parasites of their host cells' mRNA translation system.

A second principle is that there is unity in diversity; even with immeasurable time, evolution has led to the formation of only seven major types of viral genome. The Baltimore classification system integrates these two principles to construct an elegant molecular algorithm for virologists (see Fig. 1.11). When the bewildering array of viruses is classified by this system, we find fewer than 10 pathways to mRNA. The elegance of the Baltimore system is that by knowing only the nature of the viral genome, one can deduce the basic steps that must take place to produce mRNA. Perhaps more pragmatically, the system simplifies comprehension of the extraordinary life cycles of viruses.

Table 3.1 DNA and RNA viruses according to the report of the International Committee on Taxonomy of Viruses (2005)[a]

			Morphology			
			Isometric[c]		Other[d]	
Type	Genome	No. of families[b]	Enveloped	Naked	Enveloped	Naked
DNA	ds	24	2	9	11	2
	ss	5	0	4	0	1
RNA	ds	8	0	5	1	2
	(+)ss	27	0	14	6	7
	(−)ss	7	0	0	7	0
Subtotal			2	32	25	12
Total		71	34		37	

[a]The known viruses have been grouped into >1,950 species (probably 30,000 to 40,000 total virus isolates), 3 orders, 73 families, 9 subfamilies, 287 genera, subviral agents (satellites, viroids, and prions), and unclassified viruses. These numbers are certainly underestimates as new viruses are discovered regularly.

[b]Not included are the many genera of viruses that have not been assigned to families.

[c]Particles likely to be built with icosahedral symmetry; the latter can be identified only by structural studies, which have not been done for all viruses.

[d]This category comprises nonisometric forms including tailed bacteriophages, pleomorphic, rod shaped, lemon shaped, droplet shaped, spherical, bacilliform, reniform, and filamentous.

The Baltimore system omits the second universal function of viral genomes, i.e., to serve as a template for synthesis of progeny genomes. There is a finite number of nucleic acid copying strategies, each with unique primer, template, and termination requirements. We shall combine this principle with that embodied in the Baltimore system to define seven strategies based on mRNA synthesis **and** genome replication.

For most viruses with DNA genomes, replication and mRNA synthesis present no obvious challenges, as all cells use DNA-based mechanisms. In contrast, animal cells possess no known mechanisms to copy viral RNA templates and to produce mRNA from them. For RNA viruses to survive, their RNA genomes must, by definition, encode a nucleic acid polymerase for genome replication and mRNA synthesis.

BOX 3.1 BACKGROUND

What information is encoded in a viral genome?

Gene products and regulatory signals required for

- replication of the genome
- assembly and packaging of the genome
- regulation and kinetics of the replication cycle
- modulation of host defenses
- spread to other cells and hosts

Information **not** contained in viral genomes:

- no genes or evolutionary relics of genes encoding protein synthesis machinery (e.g., no ribosomal RNA and no ribosomal or translation proteins); note: the genomes of some large DNA viruses contain genes for transfer RNAs (tRNAs), aminoacyl-tRNA synthetases, and genes involved in sugar and lipid metabolism
- no genes encoding proteins of energy metabolism or membrane biosynthesis
- no telomeres (to maintain genomes) or centromeres (to ensure segregation of genomes)

Structure and Complexity of Viral Genomes

By definition, the nucleic acid in the virion is the viral genome. Despite the simplicity of expression strategies, the composition and structures of viral genomes are more varied than those seen in the entire archeal, bacterial, or eukaryotic kingdoms. Nearly every possible method for encoding information in nucleic acid can be found in viruses. Viral genomes can be

- DNA or RNA
- DNA with short segments of RNA
- DNA or RNA with covalently attached protein
- single stranded (+) strand, (−) strand, or ambisense (Box 3.2)
- double stranded
- linear
- circular

BOX 3.2 TERMINOLOGY

Important conventions: plus (+) and minus (−) strands

mRNA is defined as the positive (+) strand, because it contains immediately translatable information. A strand of DNA of the equivalent polarity is also designated as a (+) strand; i.e., if it were mRNA, it would be translated into protein.

The RNA or DNA complement of the (+) strand is the (−) strand. The (−) strand cannot be translated; it must first be copied to make the (+) strand.

- segmented
- gapped

It is clear that the structure of the genome determines the method of replication and packaging. What is the significance of such diversity in genome composition, structure, and replication? Obviously, viral genomes are survivors of constant selective pressure, but can we deduce something from their existence? Is one configuration more advantageous? These are difficult questions to answer. Virus evolution is discussed further in Volume II, Chapter 10.

Figures 3.1 through 3.7 illustrate the seven strategies for expression and replication of viral genomes. In some cases, genomes can enter the replication cycle directly. In others, genomes must first be modified and other viral nucleic acids must participate in the replication cycle. For each strategy, a panel contains the following essential information:

- genome structure
- degree of dependence on host for replication
- gene expression strategies
- noteworthy features of the interaction with the host

Examples of specific viruses in each class are provided.

DNA Genomes

The strategy of having DNA as a viral genome appears at first glance to be simplicity itself: the host genetic system is based on DNA, so viral genome replication and expression could simply emulate the host system. Many surprises await those who believe that this is all such a strategy entails.

Double-Stranded DNA (dsDNA) (Fig. 3.1)

Viral genomes may consist of double-stranded or partially double-stranded nucleic acids. There are 24 families of viruses with such DNA genomes; those that include mammalian viruses are the *Adenoviridae, Herpesviridae, Papillomaviridae, Polyomaviridae*, and *Poxviridae*. These genomes may be linear or circular. mRNA is produced by copying of the genome by host or viral DNA-dependent RNA polymerase. The strategies are shown in Fig. 3.1, with the exception of the *Papillomaviridae*, which have mRNA and replication mechanisms similar to the *Polyomaviridae*.

Gapped DNA (Fig. 3.2)

As the gapped DNA genome is partially double-stranded, the gaps must be filled to produce perfect duplexes. This repair process must precede mRNA synthesis because the host RNA polymerase can transcribe only fully dsDNA. The unusual gapped DNA genome is produced from an RNA template by a virus-encoded enzyme that synthesizes DNA from an RNA template (reverse transcriptase).

Single-Stranded DNA (ssDNA) (Fig. 3.3)

Five families of viruses containing single-stranded DNA genomes have been recognized; the families *Circoviridae*

BOX 3.3 BACKGROUND

RNA synthesis in cells

There are no known enzymes in mammalian cells that can copy the genomes of RNA viruses. However, at least one enzyme, RNA polymerase II, can copy an RNA template. The 1.7-kb circular ssRNA genome of hepatitis delta satellite virus is copied by RNA polymerase II to form multimeric RNAs (see figure). How RNA polymerase II, an enzyme that produces mRNAs from DNA templates, is reprogrammed to copy a circular RNA template is not known.

Hepatitis delta satellite (−) strand genome RNA is copied by RNA polymerase II at the indicated position. The polymerase passes the poly(A) signal (purple box) and the self-cleavage domain (red circle). For more information, see Fig. 6.14. Redrawn from J. M. Taylor, *Curr. Top. Microbiol. Immunol.* **239**:107–122, 1999, with permission.

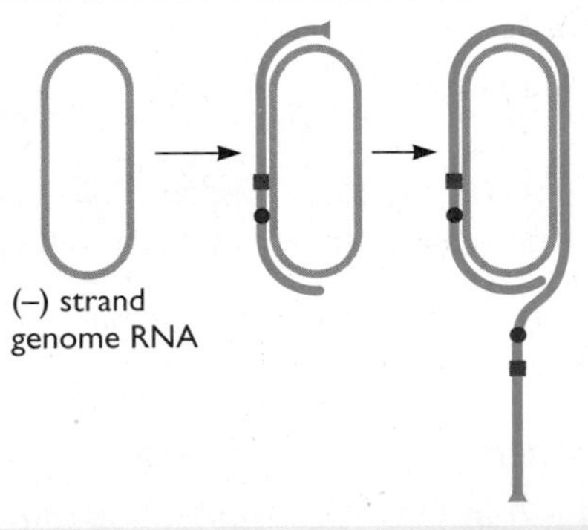

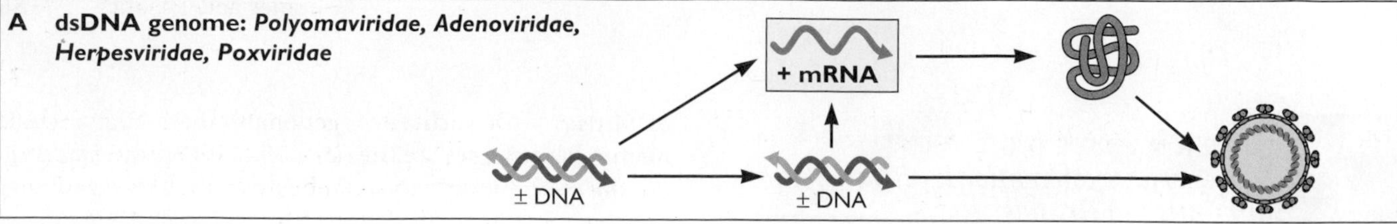

B *Polyomaviridae* (5 kbp)

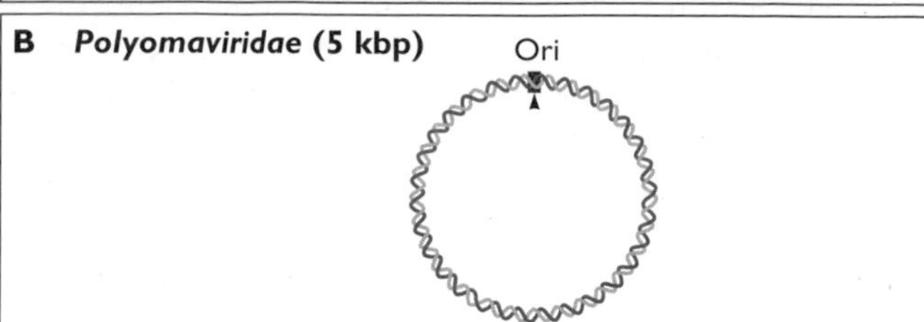

Genome configuration

- Circular DNA genomes, <10 open reading frames

Degree of dependence on host cell

- Dependent on cell for both replication and gene expression
- No viral DNA polymerase
- RNA primers and replication fork to replicate DNA, analog of the cellular mechanism

Gene expression strategies

- Temporal control of early and late gene expression
- Splicing is a major regulatory process
- Other decoding tactics: overlapping genes, leaky scanning

Noteworthy features of the interaction with the host

- Polyomavirus and simian virus 40 encode tumor (T) antigens that modulate cell cycle control

C *Adenoviridae* (36–48 kbp)

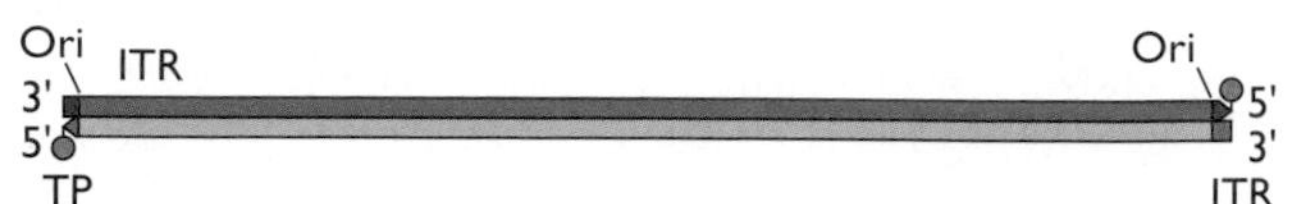

Genome configuration

- Linear DNA, short inverted terminal repeat sequence, protein bound to 5' termini

Degree of dependence on host cell

- Encodes DNA replication system including DNA polymerase
- Protein priming with unusual strand displacement mechanism

Gene expression strategies

- Temporal control (immediate-early, early, intermediate, and late) of gene expression
- 8 pre-mRNAs that are differentially polyadenylated and/or spliced give rise to many mRNAs translated into viral proteins
- Other decoding tactics: overlapping coding and/or control sequences
- Small viral RNAs produced by host RNA polymerase III

Noteworthy features of the interaction with the host

- E1A proteins are powerful transcriptional activators and regulators of the cell cycle

D *Herpesviridae* (120–220 kbp)

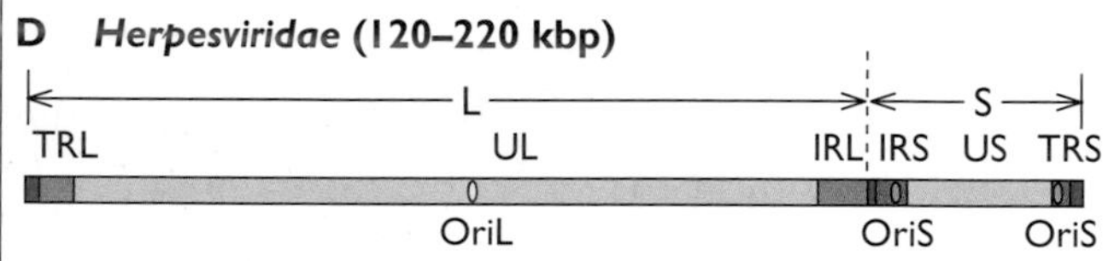

Genome configuration

- Linear DNA
- Genome isomers

Degree of dependence on host cell

- Genomes encode a complete replication system, including a viral DNA polymerase, and enzymes that synthesize DNA precursors
- Genome replication with RNA primers and replication fork

Gene expression strategies

- Temporal regulation; extent of splicing variable among different subfamilies
- Encode a variety of transcriptional and posttranscriptional regulators
- Package activators of transcription and other regulatory proteins in the virion
- Production of many mRNAs from multiple promoters

Noteworthy features of the interaction with the host

- Latent infection

E *Poxviridae* (130–375 kbp)

ITR ITR

Terminal loop

Genome configuration

- Large DNA genome, cross-linked ends
- Long terminal repeats

Degree of dependence on host cell

- Replicates in cytoplasm
- Genome encodes DNA replication system
- Strand displacement DNA synthesis

Gene expression strategies

- Genome encodes a complete mRNA synthesis system including RNA polymerase, capping, and polyadenylation enzymes
- Temporal regulation
- Many viral enzymes packaged in virions

Noteworthy features of the interaction with the host

- Encode many immune modulators

Figure 3.1

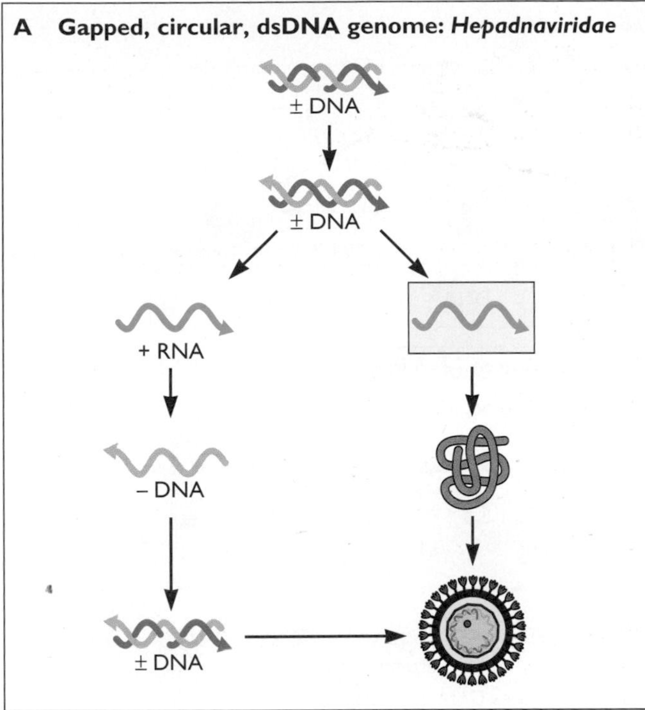

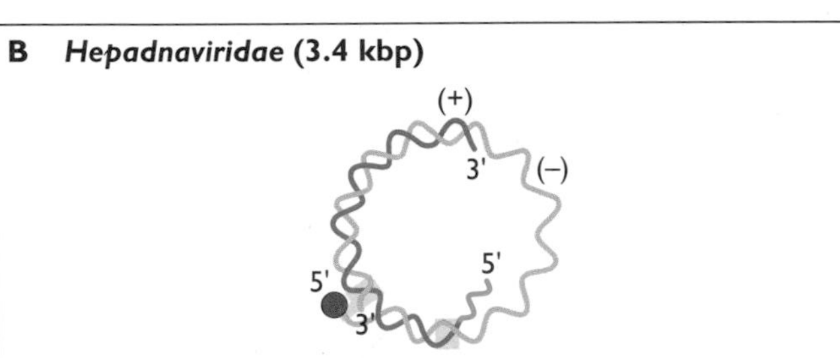

Genome configuration

- Circular DNA genome; full-length but nicked (–) and partial (+) strand
- Reverse transcriptase bound to 5' end of (–) strand; RNA bound to 5' end of (+) strand

Degree of dependence on host cell

- Gapped DNA genome is repaired in nucleus by host enzymes to form covalently closed circular DNA
- Longer-than-genome-length (+) strand mRNA is produced by host RNA polymerase II, and the viral reverse transcriptase copies it to make the gapped dsDNA genome

Gene expression strategies

- Alternative splicing for synthesis of multiple mRNAs
- Overlapping reading frames

Noteworthy features of the interaction with the host

- The X gene product of mammalian members contributes to tumorigenesis

Figure 3.2

and *Parvoviridae* include viruses that infect mammals. ssDNA must be copied into mRNA before proteins can be produced. However, RNA can be made only from a dsDNA template, whatever the sense of the ssDNA. DNA synthesis **must** precede mRNA production in the replication cycles of these viruses. The single-stranded genome is produced by cellular DNA polymerases.

RNA Genomes

Cells have no RNA-dependent RNA polymerases that can replicate the genomes of RNA viruses or make mRNA from RNA (Box 3.3). Two strategies have evolved to solve this problem. One solution is that RNA virus genomes encode RNA-dependent RNA polymerases that produce RNA (both genomes and mRNA) from RNA templates. The other, exemplified by retrovirus genomes, is reverse transcription of the genome to dsDNA, which can be transcribed by host RNA polymerase.

dsRNA (Fig. 3.4)

There are eight families of viruses with dsRNA genomes. The number of dsRNA segments ranges from 1 (*Totiviridae* and *Hypoviridae*, viruses of fungi, protozoa, and plants) to 10 to 12 (*Reoviridae*, viruses of mammals, fish, and plants). While dsRNA contains a (+) strand, it cannot be translated as part of a duplex. The (–) strand of the genomic dsRNA is first copied into mRNAs by a viral RNA-dependent RNA polymerase to produce viral proteins. Newly synthesized mRNAs are encapsidated and then copied to produce dsRNAs.

(+) Strand RNA (Fig. 3.5)

The (+) strand RNA viruses are the most plentiful on this planet; 27 families have been recognized. The families *Arteriviridae, Astroviridae, Caliciviridae, Coronaviridae, Flaviviridae, Picornaviridae,* and *Togaviridae* include viruses that infect mammals. (+) strand RNA genomes usually can be translated directly into protein by host ribosomes. The genome is replicated in two steps. First, the (+) strand genome is copied into a full-length (–) strand. The (–) strand is then copied into full-length (+) strand genomes. In some cases, a subgenomic mRNA is produced.

(+) Strand RNA with DNA Intermediate (Fig. 3.6)

In contrast to other (+) strand RNA viruses, the (+) strand RNA genome of retroviruses is converted to a dsDNA intermediate by viral RNA-dependent DNA polymerase (reverse transcriptase). This DNA then serves as the template for viral mRNA and genome RNA synthesis by cellular enzymes.

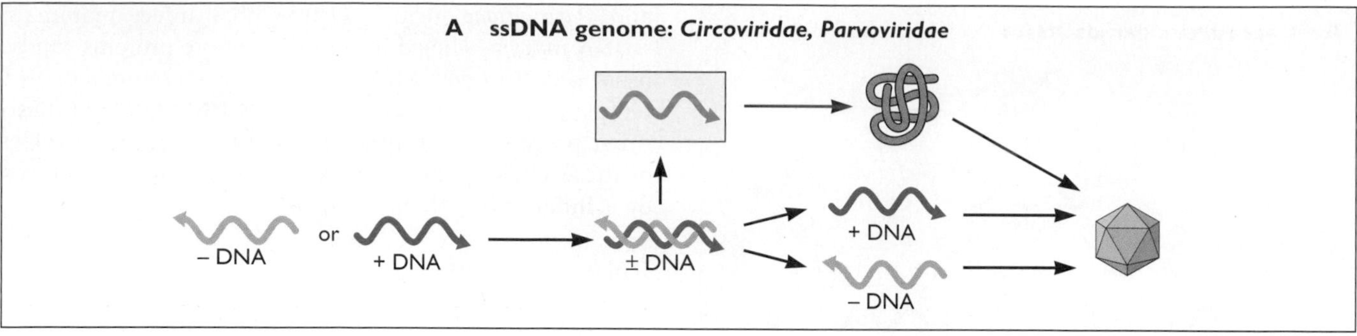

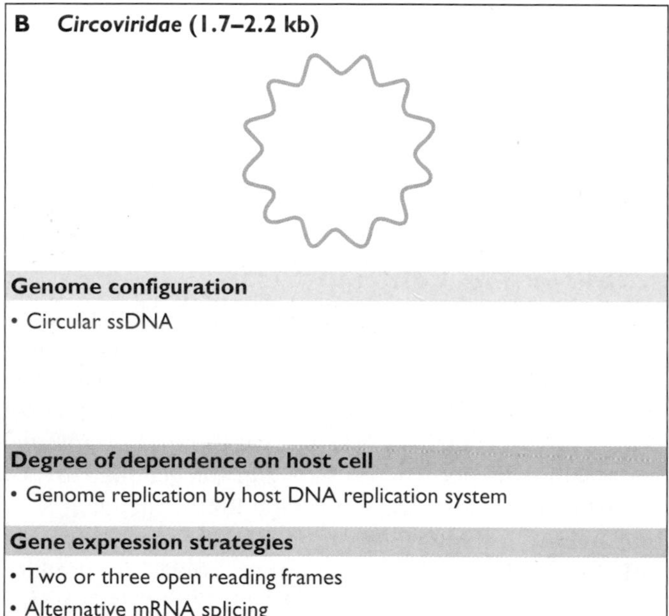

Noteworthy features of the interaction with the host
- Smallest known nondefective genome

C *Parvoviridae* (4–6 kb)

B, B', A, A', A' D, D' A, C, C'

Genome configuration
- Linear with 3' and 5' hairpins; can be both (–) and (+) strands
- 3' hairpin primes DNA synthesis
- NS1 protein covalently attached to 5' end (autonomous parvoviruses)

Degree of dependence on host cell
- Genome replication by host DNA replication system

Gene expression strategies
- Two open reading frames
- Alternative mRNA splicing

Noteworthy features of the interaction with the host
- Replicate only in cells in S phase (autonomous parvoviruses) or cells infected by a helper virus (dependoviruses)

Figure 3.3

(–) Strand RNA (Fig. 3.7)

Viruses with (–) strand RNA genomes are found in seven families. Viruses of this type that can infect mammals are found in the *Bornaviridae, Filoviridae, Orthomyxoviridae, Paramyxoviridae,* and *Rhabdoviridae* families. Unlike (+) strand RNA, (–) strand RNA genomes cannot be translated directly into protein but first must be copied to make (+) strand mRNA by a viral RNA-dependent RNA polymerase. There are no enzymes in the cell that can produce mRNAs from the RNA genomes of (–) strand RNA viruses. These virions therefore contain virus-encoded RNA-dependent RNA polymerases that produce mRNAs from the (–) strand genome. This strand is also the template for the synthesis of full-length (+) strands, which in turn are copied to produce (–) strand genomes. (–) strand RNA viral genomes can be either single molecules (nonsegmented) or segmented.

The genomes of certain (–) strand RNA viruses (e.g., members of the *Arenaviridae* and *Bunyaviridae*) are ambisense: they contain both (+) and (–) strand information on a single strand of RNA (Fig. 3.7C). The (+) sense information in the genome is translated upon entry of the viral RNA into cells. Then, after replication of the RNA genome, new (+) sense information is translated.

What Do Viral Genomes Look Like?

As we have seen, viral genomes exhibit considerable diversity in form and function. Some small RNA and DNA

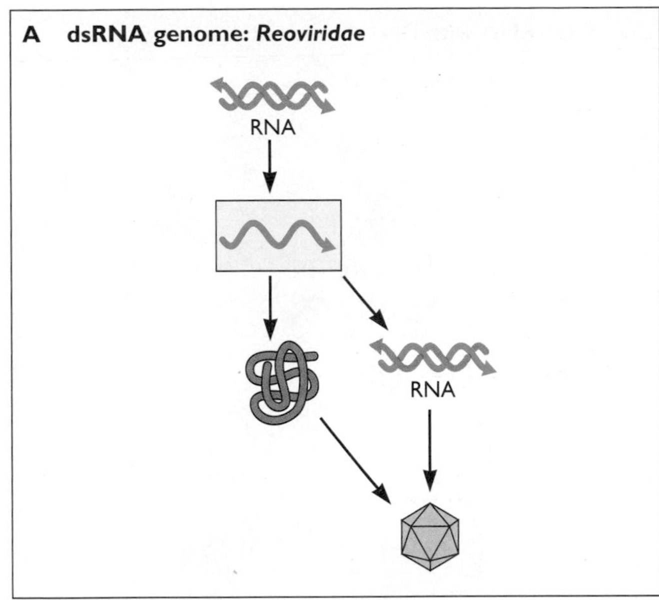

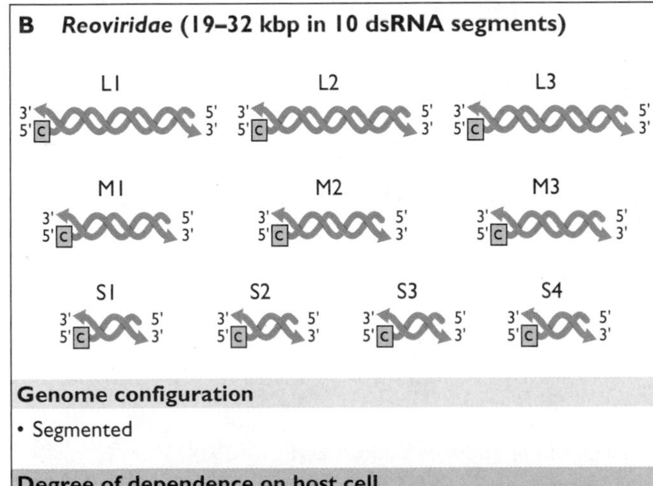

Genome configuration

- Segmented

Degree of dependence on host cell

- Viral RNA polymerase and other enzymes needed for mRNA synthesis are packaged in the particle

Gene expression strategies

- Separate monocistronic mRNA copied from each RNA segment
- mRNAs are capped, lack poly(A)

Noteworthy features of the interaction with the host

- Host lysosomal proteases required for virion uncoating

Figure 3.4

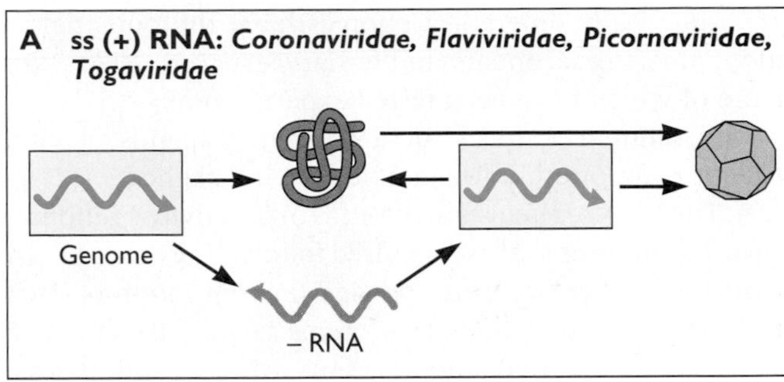

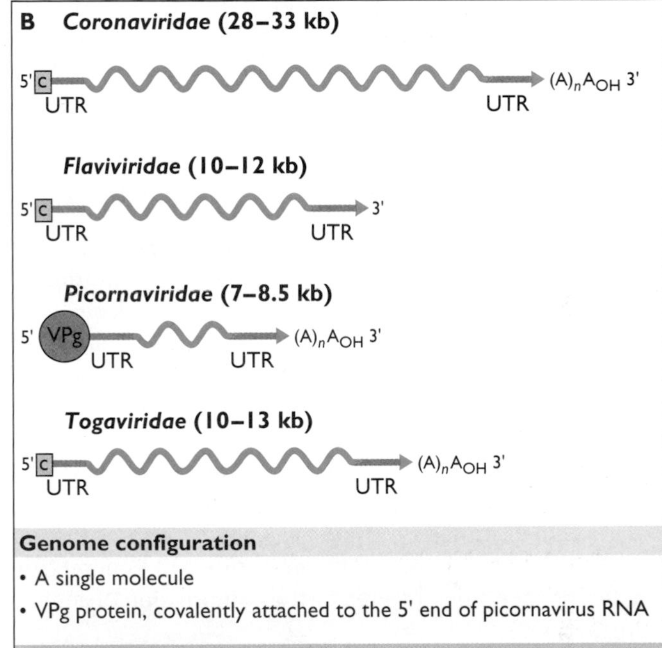

Genome configuration

- A single molecule
- VPg protein, covalently attached to the 5' end of picornavirus RNA

Degree of dependence on host cell

- Encode viral RNA polymerases to replicate genomes and synthesize mRNAs

Gene expression strategies

- Genome is mRNA and is translated immediately after infection
- Genome is copied to make (–) sense RNA, which is copied to make more mRNA and more genomes
- Internal initiation of translation from an IRES (*Picornaviridae* and *Flaviviridae*)
- Production of a polyprotein that is cleaved by proteases into individual proteins

Figure 3.5

genomes enter cells from virus particles as naked molecules of nucleic acid, whereas others are always associated with specialized nucleic acid-binding proteins. A fundamental difference between the genomes of viruses and those of hosts is that while viral genomes often are covered with proteins, they are not bound by histones (polyomaviral genomes are a remarkable exception).

While viral genomes are all nucleic acids, they should **not** be thought of as one-dimensional structures. Virology textbooks (this one included) often draw genomes as straight, one-dimensional lines, but this notation is for illustrative purposes only; physical reality is certain to be dramatically different. Genomes have the potential to adopt amazing secondary and tertiary structures (Fig. 3.8), some of which have been tested experimentally.

The sequences and structures near the ends of viral genomes are often indispensable for viral replication (Fig. 3.9). The DNA sequences at the ends of parvovirus genomes form T structures that are required for priming during DNA synthesis. Other structures needed for replication of RNA and DNA genomes include proteins covalently attached to 5′ ends, inverted and tandem repeats, and tRNAs. Secondary RNA structures may facilitate translation (the internal ribosome entry site (IRES) of picornavirus genomes) and genome packaging (the structured packaging signal of retrovirus genomes).

Coding Strategies

In general, the compact genome of most viruses renders the "one gene, one mRNA" dogma inaccurate. Extraordinary tactics have evolved for information retrieval from viral genomes (Fig. 3.10). Strategies including the production of multiple subgenomic mRNAs, mRNA splicing, RNA editing, and nested transcription units allow the production of multiple proteins from a single viral genome. Further expansion of the coding capacity of the viral genome is achieved by posttranscriptional mechanisms such as polyprotein synthesis, leaky scanning, suppression of termination, and ribosomal frameshifting. In general, the smaller the genome, the greater the compression of genetic information.

What Can Viral Sequences Tell Us?

Knowledge about the physical nature of genomes and coding strategies was first obtained by study of the nucleic acids of viruses. Indeed, DNA sequencing technology was perfected on viral genomes. The first genome of any kind to be sequenced was that of the *Escherichia coli* bacteriophage MS2, a linear ssRNA of 3,569 nucleotides. dsDNA genomes of the largest viruses, such as herpesviruses and poxviruses (vaccinia virus), were sequenced completely by the 1990s. Since then, the sequences of over 1,861 different viral genomes have been determined. Published viral genome sequences can be found at the website http://www.ncbi.nlm.nih.gov/sites/entrez.

Knowing the sequence of viral genomes has many uses, including classification of viruses. Sequence analysis has identified many relationships among diverse viral genomes, providing considerable insight into the origin of viruses. In outbreaks or epidemics of viral infection, even

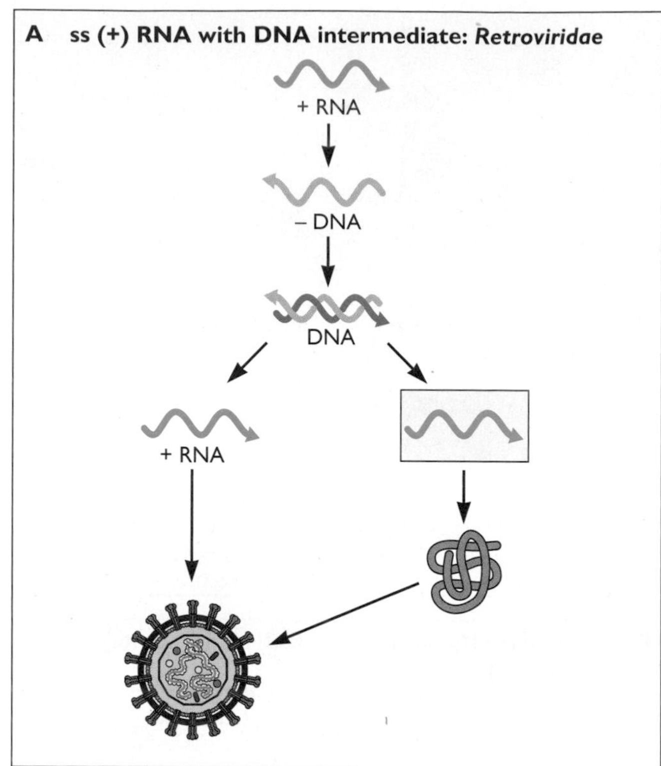

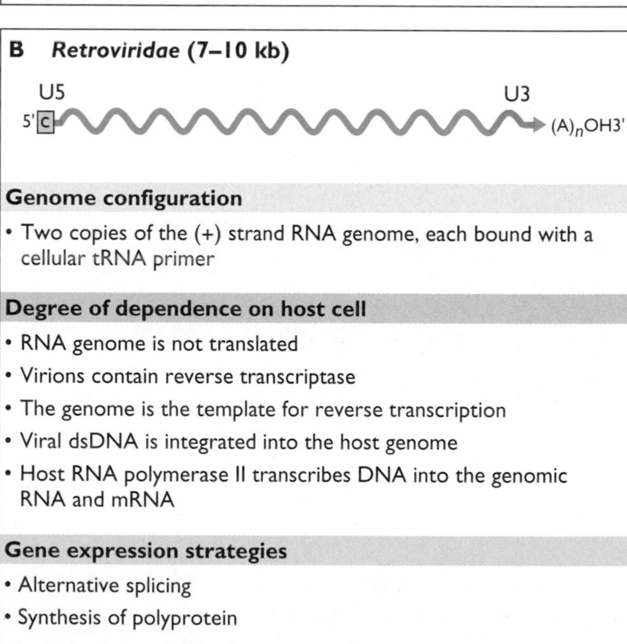

Genome configuration

- Two copies of the (+) strand RNA genome, each bound with a cellular tRNA primer

Degree of dependence on host cell

- RNA genome is not translated
- Virions contain reverse transcriptase
- The genome is the template for reverse transcription
- Viral dsDNA is integrated into the host genome
- Host RNA polymerase II transcribes DNA into the genomic RNA and mRNA

Gene expression strategies

- Alternative splicing
- Synthesis of polyprotein

Noteworthy features of the interaction with the host

- Some cause oncogenic transformation
- Some cause AIDS

Figure 3.6

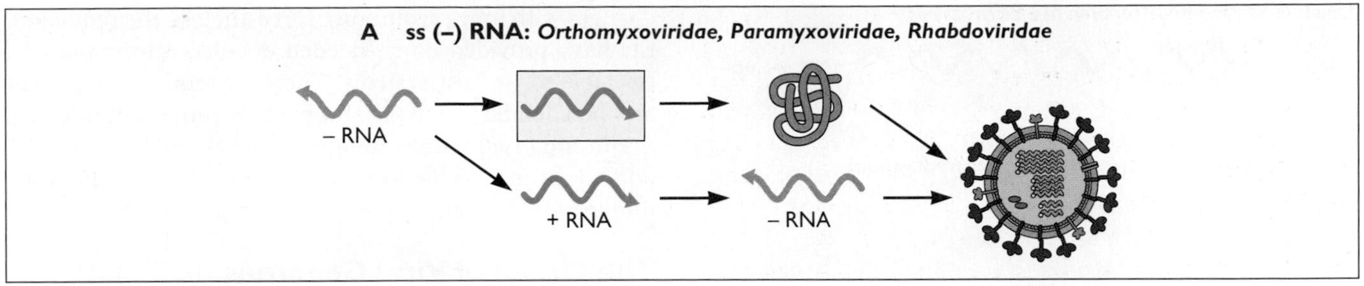

B Segmented genomes: *Orthomyxoviridae* (10–15 kb in 6–8 RNAs)

(–) strand RNA segments

1 3' 5'   2 3' 5'   3 3' 5'   4 3' 5'

5 3' 5'   6 3' 5'   7 3' 5'   8 3' 5'

Nonsegmented genomes: *Paramyxoviridae* (15–16 kb)

3' 5'

***Rhabdoviridae* (13–16 kb)**

3' 5'

Genome configuration

- Nonsegmented or segmented RNA

Degree of dependence on host cell

- Cells cannot copy or translate (–) strand RNA
- RNA-dependent RNA polymerase in virus particle

Gene expression strategies

- Capped, polyadenylated mRNAs produced by viral polymerase
- Cap-stealing mechanism for priming transcription (*Orthomyxoviridae*)
- RNA editing in *Paramyxoviridae* and *Filoviridae*
- Alternative mRNA splicing

Noteworthy features of the interaction with the host

- *Orthomyxoviridae* replicate in the nucleus

C Ambisense (–) strand RNA

***Arenaviridae* (11 kb in 2 RNAs)**
***Bunyaviridae* (12–23 kb in 3 RNAs)**

L RNA 5' 3'

S RNA

Genome configuration

- Segmented

Degree of dependence on host cell

- Cells cannot copy or translate (–) strand RNA
- RNA-dependent RNA polymerase in virus particle

Gene expression strategies

- Cap-stealing mechanism for priming transcription (*Bunyaviridae*)
- (+) and (–) strand information on same strand of RNA

Noteworthy features of the interaction with the host

- Bunyavirus virions form intracellularly by budding in the Golgi apparatus

Figure 3.7

partial genome sequences can provide information on the identity of the virus and its movement in different populations. New viral nucleic acid sequences can be associated with disease and characterized in the absence of standard virological techniques (Volume II, Chapter 10). For example, a new herpesvirus called human herpesvirus 8 was identified by comparing sequences present in diseased and nondiseased tissues.

However, it quickly became clear that a complete understanding of how viruses reproduce cannot be obtained solely from the genome sequence or structure. In retrospect, this fact should have been anticipated. The genome sequence of a virus is at best a biological parts list: it provides some information about the intrinsic properties of a virus (e.g., viral proteins and their composition) but little or nothing about how the virus interacts with cells,

Linear (+) strand RNA genome of a picornavirus

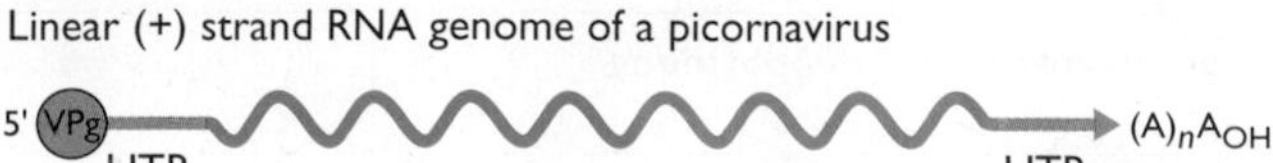

Computer-produced structure of a picornaviral genome

Figure 3.8 Genome structures in cartoons and in real life. (Top) Linear representation of the poliovirus genome. **(Bottom)** Model of the genome predicted by folding algorithms. The arrow points to the 5′ end of the RNA. Adapted from A. C. Palmenberg and J.-Y. Sgro, *Semin. Virol.* **8**:231–241, 1997, with permission. Courtesy of J.-Y. Sgro and A. C. Palmenberg, University of Wisconsin, Madison.

hosts, and populations. Because genomes are so complex, understanding them requires a reductionist analysis, the study of individual components in isolation. Although the reductionist approach is often experimentally the simplest, it is also important to understand how the genome behaves among others (population biology) and how the genome changes with time (evolution). Nevertheless, the reductionists have provided much-needed detailed information for tractable virus-host systems. These systems allow genetic and biochemical analyses and provide models of infection *in vivo* and *in vitro*. Unfortunately, viruses and hosts that are difficult or impossible to manipulate in the laboratory are understudied or ignored.

The Origin of Viral Genomes

Some people think that the origin of the viral genome is an impenetrable mystery, because no fossils are available for study. Others feel that by comparing genomes from diverse viruses, an evolutionary history of viruses may be constructed. This question is discussed further in Volume II, Chapter 10. Another question related to the origin of viruses is why there are both viruses with DNA genomes and viruses with RNA genomes. One possibility is that RNA viruses are relics of the "RNA world," a period during which RNA was **the** information **and** catalytic molecule (no proteins yet existed). During this time, possibly billions of years ago, life could have evolved from RNA, and the earliest organisms might have had RNA genomes. Viruses with RNA genomes might have evolved during this time. Later, DNA replaced RNA as the genetic material, perhaps through the action of reverse transcriptases, enzymes that produce DNA from RNA templates. With the emergence of DNA genomes came the evolution of DNA viruses. However, those with RNA genomes were still evolutionarily competitive, and hence they continue to survive to this day. Although the origins of RNA and DNA viruses will probably never be known, we do know that viruses with either type of nucleic acid are very successful.

The "Big and Small" of Viral Genomes: Does Size Matter?

Currently, the prize for the smallest nondefective animal virus genome goes to the circoviruses, which possess circular, ssDNA genomes of 1.7 to 2.3 kb (Fig. 3.3B). Members of the *Circoviridae* include chicken anemia virus, psittacine beak and feather disease virus, and TT virus, a ubiquitous human virus of no known consequence. The consolation prize goes to the *Hepadnaviridae*, such as hepatitis B virus, which causes hepatitis and liver cancer in millions of people. Its genome comprises 3.2 kb of gapped DNA: one strand is full length, but the complementary strand is

Figure 3.9 Genome structures critical for function. Abbreviations: ITR, inverted terminal repeat; TP, terminal protein; IRES, internal ribosome entry site; pbs and PBS, primer-binding site; DIS, dimerization initiation site; DLS, dimer linkage structure.

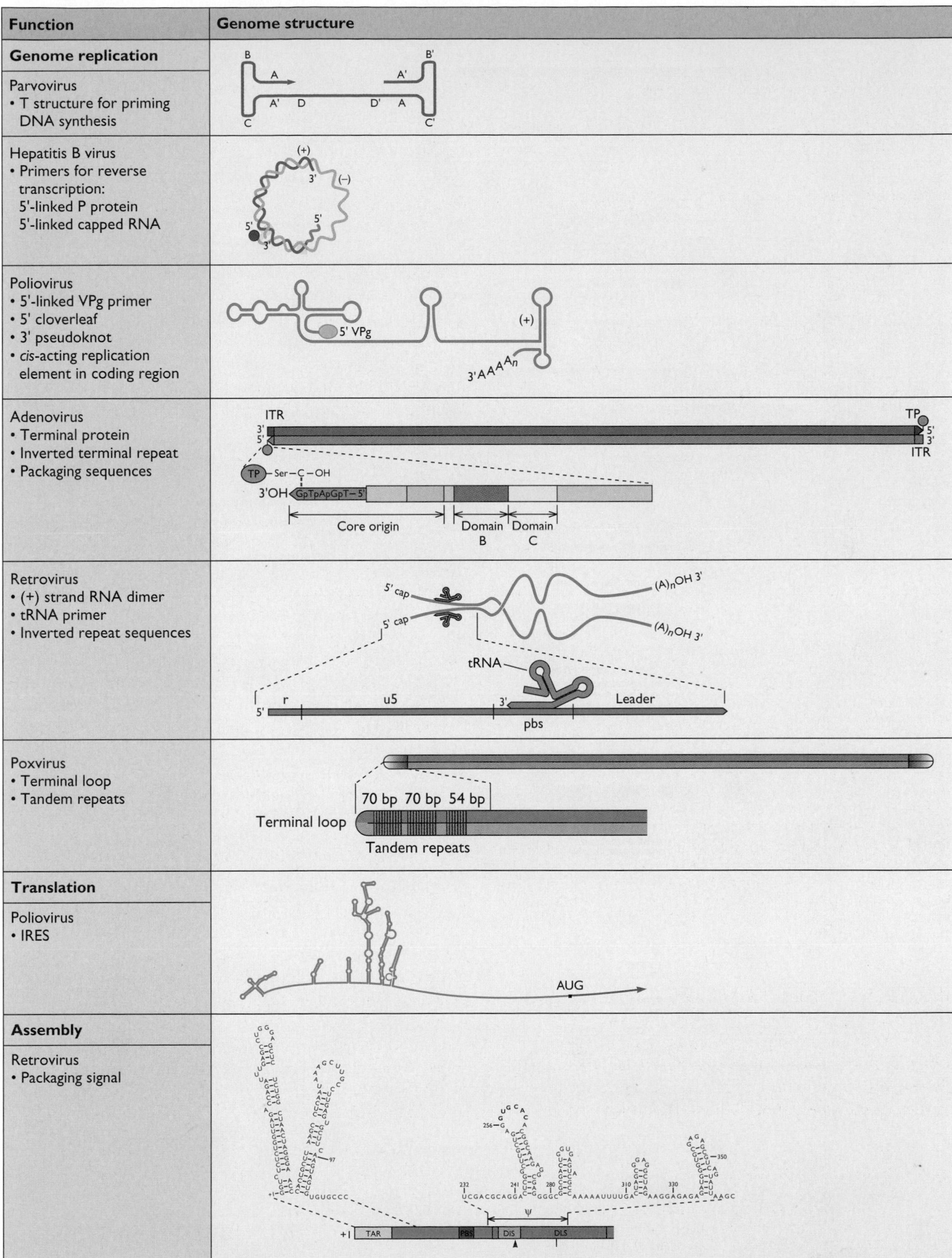

Function
Genome structure
Genome replication
Parvovirus
• T structure for priming DNA synthesis
B
B'
A
A'
A'
D
D'
A
C
C'
Hepatitis B virus
• Primers for reverse transcription:
5'-linked P protein
5'-linked capped RNA
(+)
3'
(–)
5'
5'
3'
Poliovirus
• 5'-linked VPg primer
• 5' cloverleaf
• 3' pseudoknot
• cis-acting replication element in coding region
5' VPg
(+)
3'AAAAn
Adenovirus
• Terminal protein
• Inverted terminal repeat
• Packaging sequences
ITR
3'
5'
TP
5'
3'
ITR
TP
Ser–C–OH
3'OH
GpTpApGpT–5'
Core origin
Domain B
Domain C
Retrovirus
• (+) strand RNA dimer
• tRNA primer
• Inverted repeat sequences
5' cap
5' cap
(A)nOH 3'
(A)nOH 3'
tRNA
r
u5
3'
Leader
5'
pbs
Poxvirus
• Terminal loop
• Tandem repeats
70 bp 70 bp 54 bp
Terminal loop
Tandem repeats
Translation
Poliovirus
• IRES
AUG
Assembly
Retrovirus
• Packaging signal
97
256
232
241
280
310
330
350
UCGACGCAGGA
AAAAAUUUUGA
AAGGAGAGA
AAGC
UGUGCCC
ψ
+1
TAR
PBS
DIS
DLS

Figure 3.9

Mechanism	Diagram	Virus	Chapter(s)	Figures in appendix
Multiple subgenomic mRNAs	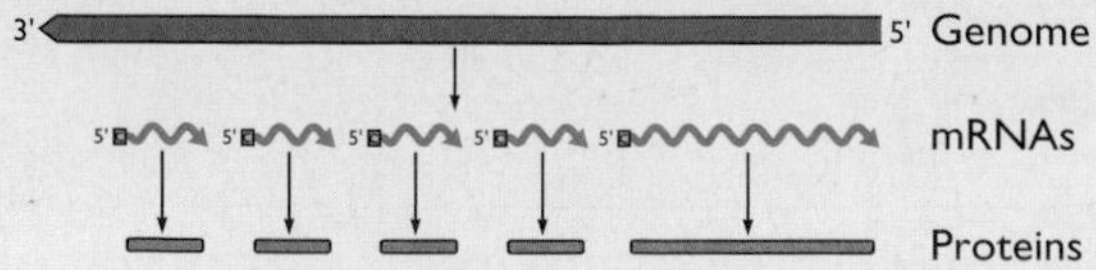 3' 5' Genome; mRNAs; Proteins	*Adenoviridae* *Hepadnaviridae* *Herpesviridae* *Paramyxoviridae* *Poxviridae* *Rhabdoviridae*	8, 10 7, 8 8 6 8 6	1, 2 3, 4 5–7 17, 18 23, 24
mRNA splicing	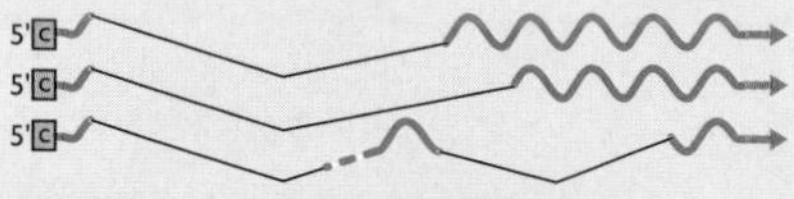 5' C	*Adenoviridae* *Orthomyxoviridae* *Papillomaviridae* *Polyomaviridae* *Retroviridae*	8, 10 10 8, 10 8, 10 7, 10	1, 2 8, 9 15, 16 21, 22
RNA editing	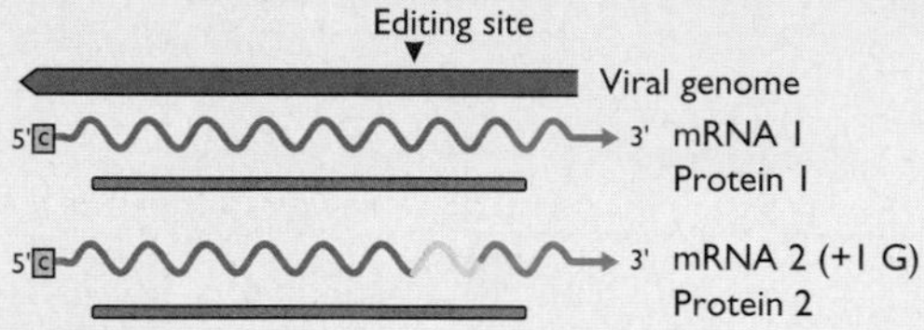 Editing site; Viral genome; 5' C 3' mRNA 1; Protein 1; 5' C 3' mRNA 2 (+1 G); Protein 2	*Paramyxoviridae* *Filoviridae* Hepatitis delta satellite	6, 10 10 10	
Information on both strands	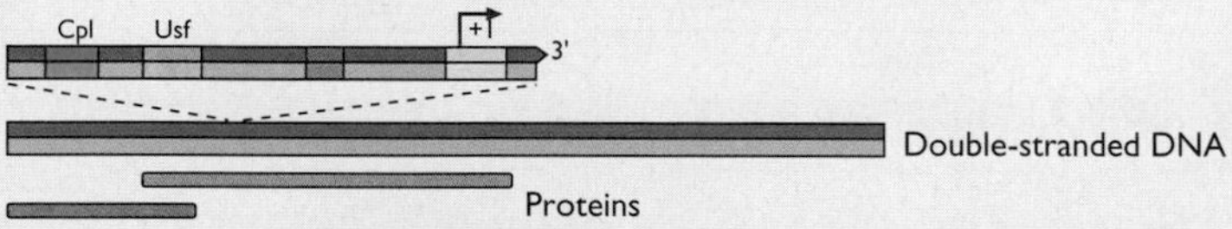 Cpl; Usf; +1; 3'; Double-stranded DNA; Proteins	*Adenoviridae* *Polyomaviridae* *Retroviridae*	8–10 8–10 7	1, 2 15, 16 21, 22
Polyprotein synthesis	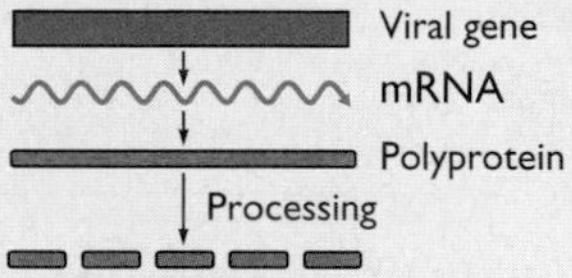 Viral gene; mRNA; Polyprotein; Processing	Alphaviruses *Flaviviridae* *Picornaviridae* *Retroviridae*	6, 11 6, 11 6, 11 6, 11	25, 26 13, 14 21, 22
Leaky scanning	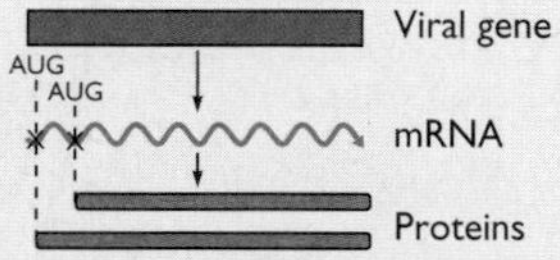 Viral gene; AUG; AUG; mRNA; Proteins	*Orthomyxoviridae* *Paramyxoviridae* *Polyomaviridae* *Retroviridae*	11 11 11 11	8, 9 15, 16 21, 22
Reinitiation	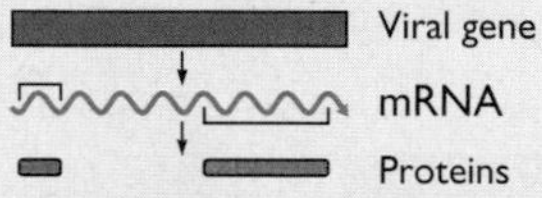 Viral gene; mRNA; Proteins	*Orthomyxoviridae* *Herpesviridae*	11 11	8, 9 5–7
Suppression of termination	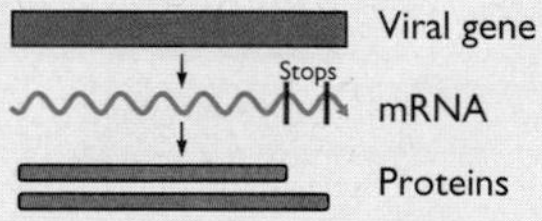 Viral gene; Stops; mRNA; Proteins	Alphaviruses *Retroviridae*	11 11	25, 26 21, 22
Ribosomal frameshifting	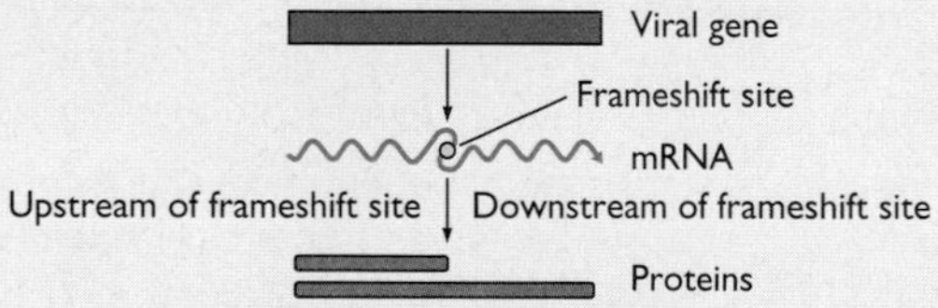 Viral gene; Frameshift site; mRNA; Upstream of frameshift site; Downstream of frameshift site; Proteins	*Astroviridae* *Coronaviridae* *Retroviridae*	11 11 11	 21, 22
IRES	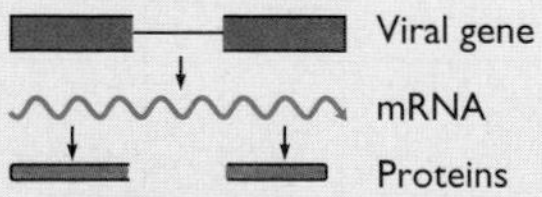 Viral gene; mRNA; Proteins	*Flaviviridae* *Picornaviridae*	11 11	 13, 14
Nested mRNAs	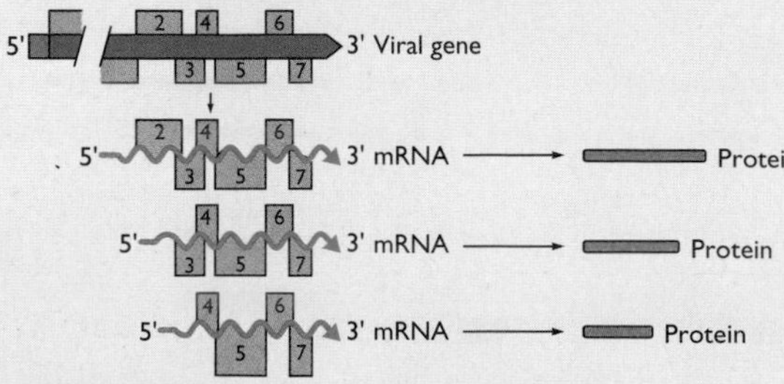 5' 2 4 6 3 5 7 3' Viral gene; 5' 3' mRNA → Protein; 5' 3' mRNA → Protein; 5' 3' mRNA → Protein	*Coronaviridae* *Arteriviridae*	6 6	

Figure 3.10

Table 3.2 Some viral proteins encoded by almost all large DNA viruses

Enzymes
DNA polymerase plus accessory replication proteins
Thymidine kinase
Ribonucleotide reductase
dUTPase
Exonucleases
Proteins that facilitate survival
Many "virulence" gene products to modulate the host's immune response or to foster invasion and spread
Cytokine homologs and receptors
Apoptosis inhibitors
Immune defense system modulators

BOX 3.4 BACKGROUND

Some curious observations

- Most RNA genomes in the world are in capsids (no cells have RNA genomes).
- There are no known animal viruses with "tailed" capsids like those of the bacteriophages.
- Most plant viruses have RNA genomes. DNA viruses of plants exist, of course (e.g., curious geminiviruses and the strange cauliflower mosaic virus).
- Most RNA viruses can cross species barriers; multiple hosts are common. For example, hosts of West Nile virus include birds, horses, and humans.
- DNA viruses tend to be more species specific than RNA viruses.
- Circular RNA viral genomes are rare, but circular DNA genomes abound.

incomplete (Fig. 3.2). Although it is not really a virus, honorable mention goes to hepatitis delta satellite virus, with a 1.7-kb single-stranded but highly base-paired circular RNA genome. This agent depends on hepatitis B virus to provide an envelope for transmission of its genome. Hepatitis B and delta satellite viruses infect hundreds of millions of people around the world.

The largest known virus genome, a DNA molecule of some 1,200 kbp, is that of mimivirus, a virus that infects amoebae. The largest RNA virus genome, 31 kb, is characteristic of some coronaviruses (Fig. 3.5). Despite detailed analyses, there is no evidence that one size is more advantageous than another. All viral genomes have evolved under relentless selection, so extremes of size must provide advantages. One feature distinguishing large genomes from smaller ones is the presence of many genes encoding proteins for viral replication, nucleic acid metabolism, and evasion of the host defense systems (Table 3.2). In other words, these large viruses have sufficient coding capacity to escape some restrictions of host cell biochemistry. The smallest viable cell genome is thought to contain less than 300 genes (predicted from bacterial genome sequences). Remarkably, this number of genes is smaller than the genetic content of large viral DNA genomes. Nevertheless, the big viruses are **not** cells; their replication absolutely requires host translation machinery, as well as host cell systems to make membranes and generate energy.

The factors that limit the size of viral genomes are largely unknown. There are cellular DNA and RNA molecules which are much longer than those found in virus particles. Consequently, the rate of nucleic acid synthesis is not likely to be limiting. For some viruses, the capsid volume (Box 3.4) might limit the size of viral genomes. There is a penalty inherent in having a large genome: a huge particle must be provided, and this is not a simple matter. The 150-kb herpes simplex virus genome resides in a $T = 16$ icosahedral nucleocapsid built from multiple copies of four proteins. In addition, the virion carries multiple copies of 20 or 30 tegument proteins and more than 15 membrane proteins in the viral envelope. Not evident in the inventory of the particle itself are the gene products required to assemble large, complicated capsids. In the case of herpes simplex virus, 50 to 60 gene products are needed to build the final particle to house the genome, yet there are only 84 known open reading frames. In other words, 75% of the viral genetic information is required to build the capsid. The largest known DNA viral genomes, those of mimivirus (1,200 kbp) and phycodnaviruses (330 kbp), are housed in the biggest capsids constructed with icosahedral symmetry. Although the principles of icosahedral symmetry are quite flexible in allowing a wide range of capsid sizes, it is possible that building a very large and stable capsid that can also come apart to release the viral genome is beyond the intrinsic properties of macromolecules.

One solution to the capsid size problem is to exploit helical symmetry. Particles built with helical symmetry can

Figure 3.10 Information retrieval from viral genomes. Different strategies for decoding the information in viral genomes are depicted. IRES, internal ribosome entry site.

in principle accommodate very large genomes, e.g., baculoviruses with DNA genomes up to 180 kbp.

In cells, DNA molecules are much longer than RNA molecules. RNA molecules are less stable than DNA, but in the cell much of the RNA is meant to be used for the synthesis of proteins and therefore need not exceed the size needed to specify the largest polypeptide. However, this constraint does not apply to viral genomes. Yet, the largest viral single-molecule RNA genomes, the 27- to 31-kb (+) strand RNAs of the coronaviruses, are dwarfed by the largest (1,200-kbp) DNA virus genomes. Susceptibility of RNA to nuclease attack might limit the size of viral RNA genomes, but there is little direct support for this hypothesis. The most likely explanation is that, as far as we know, there are no enzymes that can correct errors introduced during RNA synthesis. RNA polymerases, like their DNA counterparts, make mistakes. DNA polymerases can eliminate errors during polymerization, a process known as proofreading, and the errors can also be corrected after synthesis is complete. Such processes are not available during RNA synthesis. The average error frequencies for RNA genomes are about 1 misincorporation in 10^4 or 10^5 nucleotides polymerized. In an RNA viral genome of 10 kb, a mutation frequency of 1 in 10^4 would produce about one mutation in every replicated genome. Hence, very long viral RNA genomes, perhaps longer than 32 kb, would sustain too many lethal mutations. Even the 7.5-kb genome of poliovirus exists at the edge of viability: treatment of the virus with the RNA mutagen ribavirin causes a >99% loss in infectivity after a single round of replication.

Genetic Analysis of Viruses

The application of genetic methods to study the structure and function of animal viral genes and proteins began with development of the plaque assay by Dulbecco in 1952. This assay permitted the preparation of clonal stocks of virus, the measurement of virus titers, and a convenient system for studying viruses with conditional lethal mutations. Although a limited repertoire of classical genetic methods was available, the mutants that were isolated were invaluable in elucidating many aspects of infectious cycles and of cell transformation. Contemporary methods of genetic analysis based on recombinant DNA technology confer an essentially unlimited scope for genetic manipulation; in principle, any viral gene of interest can be mutated, and the precise nature of the mutation can be predetermined by the investigator. Much of the large body of information about viruses and their lifestyles that we now possess can be attributed to the power of these methods.

Classical Genetic Methods

Spontaneous and Induced Mutations

In the early days of experimental virology, mutant viruses could be isolated only by screening stocks for interesting phenotypes, for none of the tools that we now take for granted, such as restriction endonucleases, efficient DNA sequencing methods, or molecular cloning procedures, were developed until the mid- to late 1970s. RNA virus stocks usually contain a high proportion of mutants, and it is only a matter of devising the appropriate selection conditions (e.g., high or low temperature or exposure to drugs that inhibit viral growth) to select mutants with the desired phenotype from the total population. For example, the live attenuated poliovirus vaccine strains developed by Albert Sabin are mutants that were selected from a virulent virus stock (Volume II, Fig. 8.7). RNA virus mutants resistant to neutralization with monoclonal antibodies are often isolated from stocks at a frequency of 1 in 10^5 PFU. Mutation frequencies of RNA virus genomes are on the order of 1 misincorporation in 10^4 to 10^5 nucleotides polymerized, compared with 1 misincorporation in 10^8 to 10^{11} nucleotides incorporated for DNA genomes. This difference has been attributed to a lack of proofreading and error-correcting abilities in enzymes that replicate RNA.

The low spontaneous mutation rate of DNA viruses necessitated random mutagenesis by exposure to a chemical **mutagen**. Mutagens such as nitrous acid, hydroxylamine, and alkylating agents chemically modify the nucleic acid in preparations of virus particles, resulting in changes in base pairing during subsequent replication, and the substitution of an incorrect nucleotide. Mutagens such as base analogs, intercalating agents, or ultraviolet (UV) light are applied to the infected cell and cause changes in the viral genome during replication. Such agents introduce mutations more or less at random. Some mutations are lethal under all conditions, while others have no effect and are said to be silent. To increase the chances of obtaining viruses with a single genetic change, selective screens were applied after exposure to the lowest concentration of mutagen that produced a useful number of mutants. Under such conditions, many of the virus particles in the mutagenized population contain **wild-type** genomes (Box 3.5), making identification of mutants laborious. Some investigators have used chemicals to mutagenize RNA genomes, despite their high mutation rate.

To facilitate identification of mutants, the population must be screened for a phenotype that can be identified easily in a plaque assay. One such phenotype is temperature-sensitive growth of the virus. Virus mutants with this phenotype reproduce well at low temperatures but poorly or not at all at high temperatures. The permissive and nonpermissive temperatures are typically 39 and 33°C,

BOX 3.5

TERMINOLOGY

What is wild type?

Terminology can be confusing. Virologists often use terms such as strains, variants, and mutants to designate a virus that differs in some heritable way from a parental or wild-type virus. In conventional usage, the wild type is defined as the original (often laboratory-adapted) virus from which mutants are selected and which is used as the basis for comparison. A wild-type virus may **not** be identical to a virus isolated from nature. In fact, the genome of a wild-type virus may include numerous mutations accumulated during propagation in the laboratory. For example, the first isolate of poliovirus obtained in 1909 probably is very different from the virus we call wild type today.

We distinguish carefully between laboratory wild types and new virus isolates from the natural host. The latter are called field isolates or clinical isolates.

respectively, for viruses that replicate in mammalian cells. Other commonly sought phenotypes are changes in plaque size or morphology, drug resistance, antibody resistance, and host range (that is, loss of the ability to multiply in certain hosts or host cells). The nomenclature used to identify viral and cellular genes, essential for describing mutations, is explained in Box 3.6.

Mapping Mutations

Before the advent of recombinant DNA technology, it was extremely difficult for investigators to determine the locations of mutations in viral genomes. The **marker rescue** technique (described in "Introducing Mutations into the Viral Genome" below) was a solution to this problem, but before it was developed, other, less satisfactory approaches were exploited.

Recombination mapping can be applied to both DNA and RNA viruses. Recombination results in genetic exchange between genomes within the infected cell. For viruses with unimolecular genomes, the frequency of recombination between two mutations increases with the physical distance separating them. In practice, cells are coinfected with two mutants, and the frequency of recombination is calculated by dividing the titer of phenotypically wild-type virus obtained under restrictive conditions (e.g., high temperature) by the titer obtained under permissive conditions (e.g., low temperature). The recombination frequency between pairs of mutants is determined, allowing the mutations to be placed on contiguous maps. Although a location can be assigned for each mutation relative to others, this approach does not result in a physical map of the actual location of the base change in the genome.

In the case of RNA viruses with segmented genomes, the technique of **reassortment** allowed the assignment of mutations to specific genome segments. When cells are coinfected with both mutant and wild-type viruses, the progeny includes **reassortants** that inherit RNA segments from either parental virus. The origins of the RNA segments can be deduced from their migration patterns during gel electrophoresis (Fig. 3.11) or by nucleic acid

BOX 3.6

TERMINOLOGY

Genetic nomenclature

From the earliest days of genetic analysis, it has been customary to use abbreviations to designate gene names. Unfortunately, no standard nomenclature has been established for the genes of mammalian cells and their viruses. To facilitate our discussion of genes and mutations, we use the following conventions. Abbreviations for cellular genes are given in lowercase, italicized letters (e.g., *pvr* for poliovirus receptor gene). In abbreviations for cellular proteins, the first letter is capitalized and the remainder are lowercase (e.g., Pvr for poliovirus receptor protein). We use the names of some cellular proteins, and all viral genes and proteins, that do not conform to these conventions: they have become firmly entrenched in the scientific literature, and have come to have a life of their own. This nomenclature is not consistent among viruses, and in some cases the gene and protein names do not relate to one another (e.g., the membrane glycoprotein gD of herpes simplex virus is encoded by the US6 gene). The reader should consult the appendix in this volume to avoid confusion about such names.

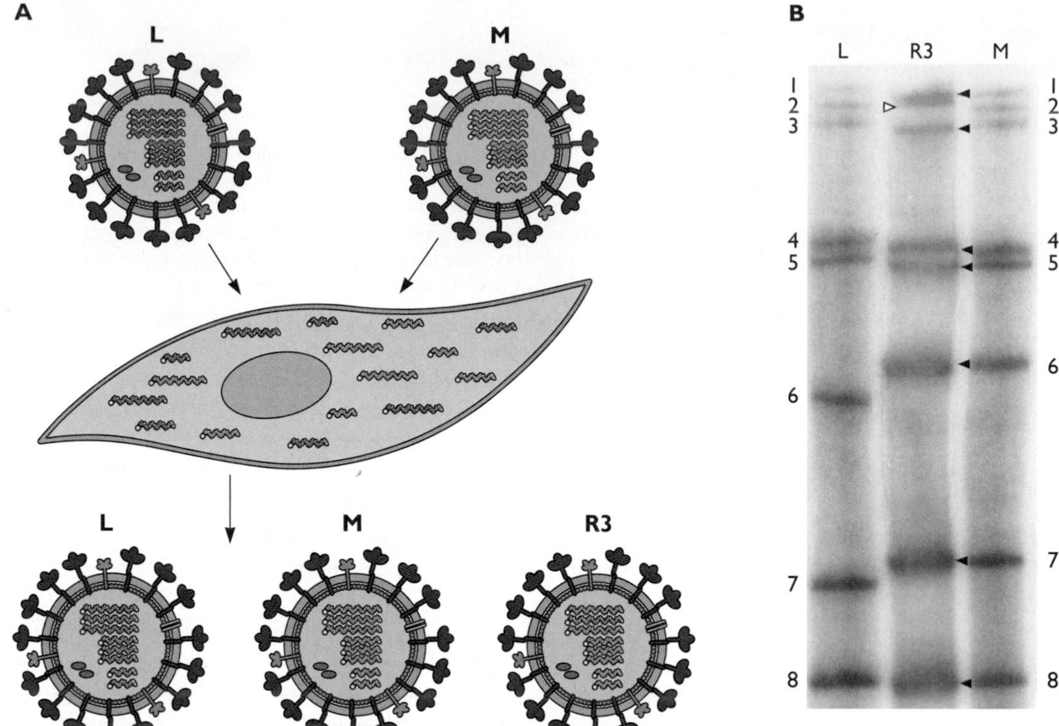

Figure 3.11 Reassortment of influenza virus RNA segments. (A) Progeny viruses of cells that are coinfected with two influenza virus strains, L and M, include both parents and viruses that derive RNA segments from them. Recombinant R3 has inherited segment 2 from the L strain and the remaining seven segments from the M strain. **(B)** ^{32}P-labeled influenza virus RNAs were fractionated in a polyacrylamide gel and detected by autoradiography. Migration differences of parental viral RNAs (M and L) permitted identification of the origin of RNA segments in the progeny virus R3. Panel B is reprinted from V. R. Racaniello and P. Palese, *J. Virol.* **29**:361–373, 1979, with permission.

hybridization. By analyzing a panel of such reassortants, the segment responsible for the phenotype can be identified. When the protein products of each RNA segment are later identified, the mutation can be assigned unambiguously.

Functional Analysis

The term **complementation** describes the ability of gene products from two different nonreplicating mutant viruses to interact functionally in the same cell, permitting viral replication. If the mutations are in separate genes, each virus is able to supply a functional gene product, allowing both viruses to replicate. If the two viruses carry mutations in the same gene, no replication will occur. In this way, the members of collections of mutants obtained by chemical mutagenesis were initially organized into complementation groups defining separate viral functions. In theory, there can be as many complementation groups as genes.

Complementation can be distinguished from recombination or reassortment by examining the progeny produced by coinfected cells. True complementation yields only the two parental mutants, while wild-type genomes also result from recombination or reassortment.

Engineering Mutations into Viral Genomes

Infectious DNA Clones

Recombinant DNA techniques have made it possible to introduce any kind of mutation anywhere in the genome of most animal viruses, whether that genome comprises DNA or RNA. The holy grail of virology today is the **infectious DNA clone,** a dsDNA copy of the viral genome that is carried on a bacterial plasmid. Infectious DNA clones, or *in vitro* transcripts derived from them, can be introduced into cultured cells by **transfection** (Box 3.7) to recover infectious virus. This approach is a modern validation of the Hershey-Chase experiment described in Chapter 1.

BOX 3.7

TERMINOLOGY

DNA-mediated transformation and transfection

The introduction of foreign DNA into cells is called DNA-mediated transformation to distinguish it from the oncogenic transformation of cells caused by tumor viruses and other insults. The term "transfection" (transformation-infection) was coined to describe the production of infectious virus after transformation of cells by viral DNA, first demonstrated with bacteriophage lambda. Unfortunately, the term "transfection" is now routinely used to describe the introduction of any DNA or RNA into cells. In this textbook, we use the correct nomenclature: the term "transfection" is restricted to the introduction of viral DNA or RNA into cells with the goal of obtaining virus replication.

The availability of site-specific bacterial restriction endonucleases, DNA ligases, and an array of methods for mutagenesis has made it possible to manipulate these infectious clones at will. Infectious DNA clones also provide a stable repository of the viral genome, which is particularly important for vaccine strains.

DNA viruses. Current genetic methods for the study of most viruses with DNA genomes are based on the infectivity of viral DNA. When deproteinized viral DNA molecules are introduced into permissive cells by transfection, they generally initiate a complete infectious cycle, although the infectivity (i.e., number of plaques per microgram of DNA) may be low. For example, the infectivity of deproteinized human adenoviral DNA is between 10 and 100 plaque-forming units (PFU) per μg. When the genome is isolated by procedures that do not degrade the covalently attached terminal protein, infectivity is increased by 2 orders of magnitude, probably because this protein participates in the assembly of initiation complexes on the viral origins of replication.

The complete genomes of polyomaviruses, papillomaviruses, and adenoviruses can be cloned in plasmid vectors, and such DNA is infectious under appropriate conditions. The DNA genomes of herpesviruses and poxviruses are too large to insert into conventional bacterial plasmid vectors, but they can be cloned in vectors that accept larger insertions (e.g., cosmids and bacterial artificial chromosomes). The plasmids containing these cloned herpesvirus genomes are infectious. Poxvirus DNA is not infectious, because the viral promoters cannot be recognized by cellular DNA-dependent RNA polymerase. Poxvirus DNA is infectious when early functions (viral DNA-dependent RNA polymerase and transcription proteins) are provided by a helper virus.

RNA viruses. *(+) strand RNA viruses.* The genomic RNA of retroviruses is copied into a dsDNA form by reverse transcriptase early during infection, a process described in Chapter 7. Such DNA is infectious when introduced into cells, as are molecularly cloned forms inserted into bacterial plasmids.

Introduction of a plasmid containing cloned poliovirus DNA into cultured mammalian cells results in the production of progeny virus (Fig. 3.12A). The mechanism by which cloned poliovirus DNA initiates infection is not known, but it has been suggested that the DNA enters the nucleus, where it is transcribed by cellular DNA-dependent RNA polymerase from cryptic, promoter-like sequences on the plasmid. The resulting (+) strand RNA transcripts initiate an infectious cycle. During replication, the extra terminal nucleotide sequences must be removed or ignored, because the viruses that are produced contain RNA with the authentic 5′ and 3′ termini.

The genomic RNA of poliovirus has a higher specific infectivity (10^6 PFU per μg) than does cloned DNA (10^3 PFU per μg). By incorporating promoters for bacteriophage T7 DNA-dependent RNA polymerase in plasmids containing poliovirus DNA, full-length (+) strand RNA transcripts can be synthesized *in vitro*. The specific infectivity of such RNA transcripts resembles that of genomic RNA. Infectious DNA clones have been constructed for many (+) strand RNA viruses, including members of the *Arteriviridae, Caliciviridae, Coronaviridae, Flaviviridae, Picornaviridae,* and *Togaviridae.*

(–) strand RNA viruses. Genomic RNA of (–) strand RNA viruses is not infectious, because it can be neither translated nor copied into (+) strand RNA by host cell RNA polymerases, as discussed in Chapter 6. Two different experimental approaches have been used to develop infectious DNA clones of these viral genomes (Fig. 3.12B and C).

The recovery of influenza virus from cloned DNA is achieved by an expression system in which cloned DNA copies of the eight RNA segments of the viral genome are inserted between two promoters (Fig. 3.12B). When eight plasmids carrying DNA for each viral RNA segment

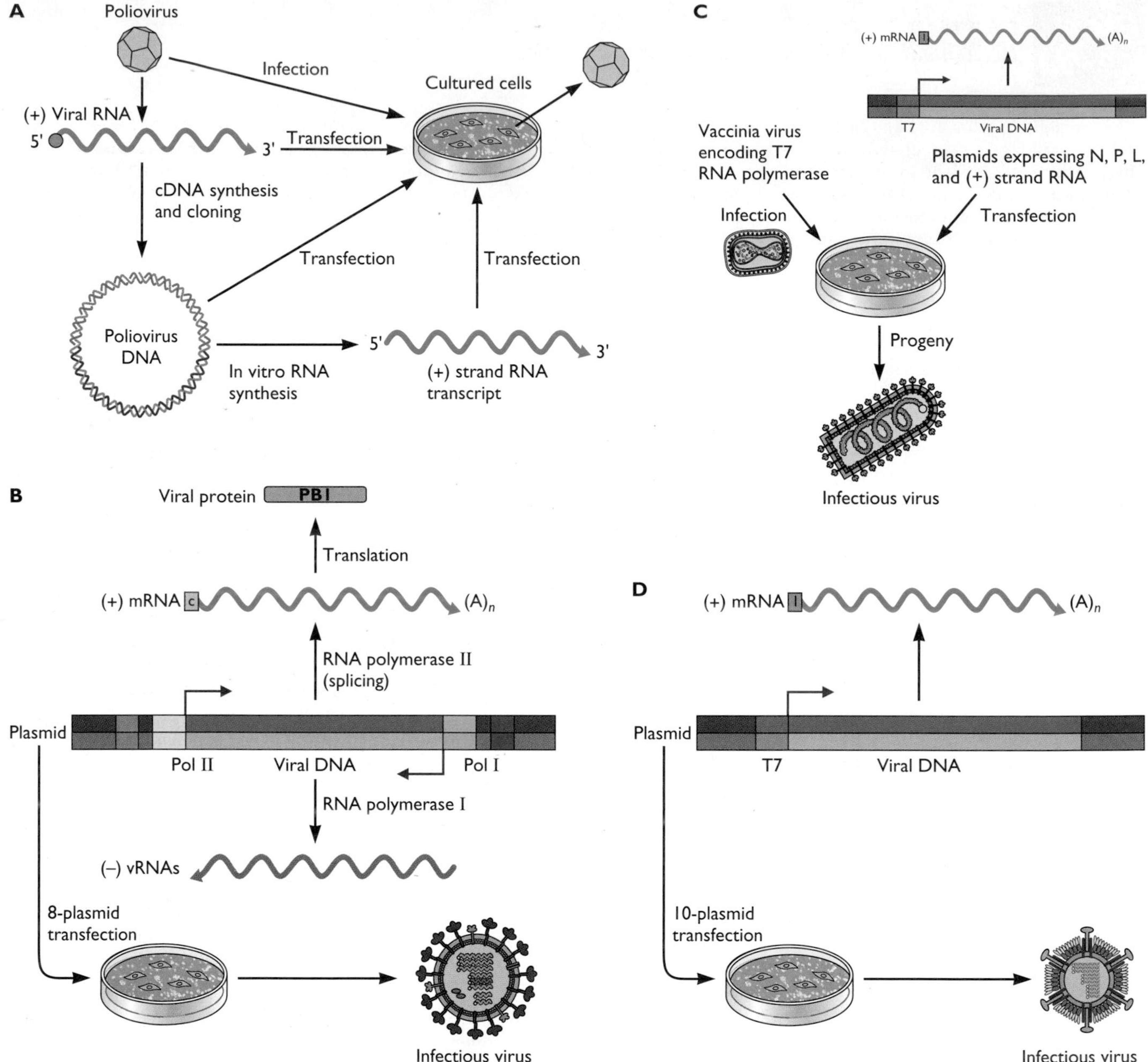

Figure 3.12 Genetic manipulation of RNA viruses. (A) Recovery of infectivity from cloned DNA of (+) strand RNA genomes as exemplified by genomic RNA of poliovirus, which is infectious when introduced into cultured cells by transfection. A complete DNA clone of the viral RNA, carried in a plasmid, is also infectious, as are RNAs derived by *in vitro* transcription of the full-length DNA. **(B)** Recovery of influenza viruses by transfection of cells with eight plasmids. Cloned DNA of each of the eight influenza virus RNA segments is inserted between an RNA polymerase I promoter (Pol I, green) and terminator (brown), and an RNA polymerase II promoter (Pol II, yellow) and a polyadenylation signal (red). When the plasmids are introduced into mammalian cells, (–) strand viral RNA (vRNA) molecules are synthesized from the RNA polymerase I promoter, and mRNAs are produced by transcription from the RNA polymerase II promoter. The mRNAs are translated into viral proteins, and infectious virus is produced from the transfected cells. For clarity, only one cloned viral RNA segment is shown. Adapted from E. G. Hoffmann et al., *Proc. Natl. Acad. Sci. USA* **97:**6108–6113, 2000, with permission. **(C)** Recovery of infectious virus from cloned DNA of viruses with a (–) strand RNA genome. Cells are infected with a vaccinia virus recombinant that synthesizes T7 RNA polymerase and transformed with plasmids that encode a full length (+) strand copy of the viral

(continued on next page)

BOX 3.8

TERMINOLOGY

Operations on DNA and on protein

A mutation is a change in DNA or RNA comprising base changes and nucleotide additions, deletions, and rearrangements. When mutations occur in open reading frames, they can be manifested as changes in the synthesized proteins. For example, one or more base changes in a specific codon may produce a single amino acid substitution, a truncated protein, or no protein. The terms "mutation" and "deletion" are often used incorrectly, or ambiguously to describe alterations in proteins. In this textbook, these terms are used to describe genetic changes, and the terms "amino acid substitution" and "truncation" are used to describe protein alterations.

are introduced into cells, infectious influenza virus is produced.

For (–) strand RNA viruses with a nonsegmented genome, such as vesicular stomatitis virus (a rhabdovirus), the full-length (–) strand genomic RNA is not infectious, because it cannot be translated into protein or copied into mRNA by the host cell. When the full-length (–) strand is introduced into cells that produce viral proteins required for production of mRNA, no infectious virus is recovered. However, when a full-length (+) strand RNA is transfected into cells that synthesize the vesicular stomatitis virus nucleocapsid protein, phosphoprotein, and polymerase, the (+) strand RNA is copied into (–) strand RNAs. These RNAs initiate an infectious cycle, leading to the production of new virus particles.

dsRNA viruses. Genomic RNA of dsRNA viruses is not infectious because the (+) strand cannot be translated. The recovery of reovirus from cloned DNA is achieved by an expression system in which cloned DNA copies of the 10 RNA segments of the viral genome are inserted under the control of an RNA polymerase promoter (Fig. 3.12D). When 10 plasmids carrying DNA for each viral dsRNA segment are introduced into cells, infectious reovirus is produced.

Types of Mutation

Recombinant DNA techniques allow the introduction of many kinds of mutation at any desired site in cloned DNA (Box 3.8). Indeed, provided that the sequence of the segment of the viral genome to be mutated is known, there is little restriction on the type of mutation that can be introduced. **Deletion mutations** can be used to remove an entire gene to assess its role in replication, to produce truncated gene products, or to assess the functions of specific segments of a coding sequence. Noncoding regions can be deleted to identify and characterize regulatory sequences such as promoters. **Insertion mutations** can be made by the addition of unrelated sequences or sequences derived from a closely related virus. **Substitution mutations**, which can correspond to one or more nucleotides, are often made in coding or noncoding regions. Included in the latter class are **nonsense mutations**, in which a termination codon is introduced, and **missense mutations**, in which a single nucleotide or a codon is changed, resulting in the production of a protein with a single amino acid substitution. The introduction of a termination codon is frequently exploited to cause truncation of a membrane protein so that it is secreted or to eliminate the synthesis of a protein without changing the size of the viral genome or mRNA. Substitutions are used to assess the roles of specific

Figure 3.12 *(continued)* genome RNA and proteins required for viral RNA synthesis (N, P, and L proteins). Production of RNA from these plasmids is under the control of the bacteriophage T7 RNA polymerse promoter (brown). Because bacteriophage T7 RNA transcripts are uncapped, an internal ribosome entry site (I) is included so the mRNAs will be translated. After the plasmids are transfected into cells, the (+) strand RNA is copied into (–) strands, which in turn are used as templates for mRNA synthesis and genome replication. The example shown is for viruses with a single (–) strand RNA genome (e.g., rhabdoviruses and paramyxoviruses). A similar approach has been demonstrated for Bunyamwera virus, with a genome comprising three (–) strand RNAs. **(D)** Recovery of infectious virus from cloned DNA of dsRNA viruses. Cloned DNA of each of the 10 reovirus dsRNA segments is inserted under the control of a bacteriophage T7 RNA polymerase promoter (brown). Because bacteriophage T7 RNA transcripts are uncapped, an internal ribosome entry site (I) is included so the mRNAs will be translated. Cells are infected with a vaccinia virus recombinant that synthesizes T7 RNA polymerase and transformed with all 10 plasmids. For clarity, only one cloned viral RNA segment is shown.

nucleotides in regulatory sequences or of amino acids in protein function, such as polymerase activity or binding of a viral protein to a cell receptor. An example of a longer substitution is the exchange of nucleic acid sequences encoding a specific protein sequence among virus strains that differ in host range. The construction of such hybrid viruses can be used to assign a biological function to a specific gene product and to determine whether it functions independently of the viral genetic background.

Introducing Mutations into the Viral Genome

Mutations can be introduced rapidly into a viral genome when it is cloned in its entirety. Mutagenesis is usually carried out on cloned subfragments of the viral genome, which are then substituted into full-length cloned DNA. The final step is introduction of the mutagenized DNA into cultured cells by transfection. This approach has been applied to cloned DNA copies of RNA and DNA viral genomes.

For the large DNA viruses such as adenoviruses, herpesviruses, and poxviruses, it is not always practical to replace cloned and mutagenized subfragments of the viral DNA because of the lack of convenient, unique restriction enzyme sites. A less direct, but effective, method for introducing mutations into these viral DNA genomes is **marker transfer** (where a mutation is marked or traced by its phenotype). The marker transfer method is applicable to any DNA virus provided that the target sequence is available as a DNA fragment in a vector that allows cloning and amplification in *E. coli.* Cells are transfected with the viral genome together with a DNA fragment containing the desired mutation. Recombination between the DNAs produces viral genomes containing the desired mutation. The use of marker transfer to introduce mutations into the genome of a DNA virus is shown in Fig. 3.13.

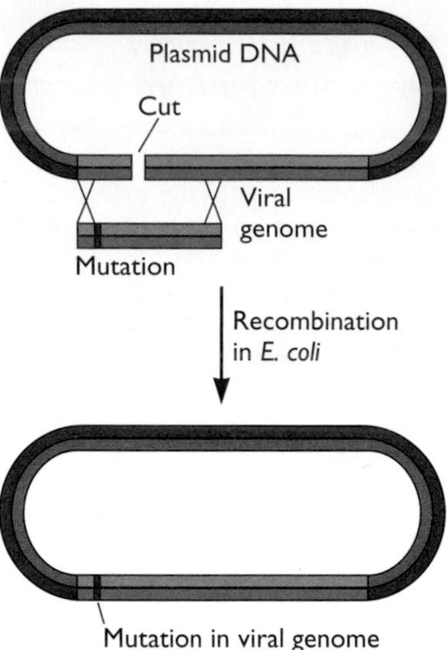

Figure 3.13 Introduction of mutations into a DNA viral genome. A plasmid carrying the entire viral DNA genome is cleaved with a suitable restriction endonuclease to produce a single cut in the area of the desired mutation. This cleaved plasmid is introduced into *E. coli* cells together with a shorter DNA containing the desired mutation. The cleaved plasmid is unable to replicate in *E. coli.* Only plasmids produced by homologous recombination between the adenoviral sequences in the cleaved plasmid and on the DNA fragment will be propagated in *E. coli*. The resulting recombinant plasmids contain a full-length copy of the viral DNA genome with the desired mutation.

Introduction of mutagenized viral nucleic acid into cultured cells by transfection may produce one of several possible results. The mutation may have no effect on viral replication; it may have a subtle effect that is discovered only on subsequent study; it may impart a readily detectable phenotype; or it may prevent the production of infectious virus. In the last-mentioned case, it is necessary to complement the defect in cell lines that have been engineered to produce specific viral proteins.

When the sequences of interest control viral gene expression or replication, complementation is not possible. This limitation is not as prohibitive as it might appear. One approach that can be used to study control regions is to introduce mutations that impair but do not eliminate virus replication. For example, certain mutations introduced into the major late promoter of human adenoviruses reduce virus yield more than 30-fold. An alternative approach is to study the effects of lethal mutations in cloned copies of the viral genome carried in plasmids or bacterial artificial chromosomes.

Mutations introduced into a specific gene might have unanticipated **polar** effects on other genes that are not altered in sequence. Insertions of the *lacZ* gene into the coding sequence for the gG gene of pseudorabies virus impaired cell-to-cell spread of the virus. It was initially concluded that this phenotype is the result of loss of gG gene expression. However, introduction of a nonsense mutation that prevents gG synthesis produced viruses with no defects in cell-to-cell spread. It was then discovered that large DNA insertions in US4 cause dramatic reductions in proteins encoded in an upstream gene, US3. These findings emphasize the risk of unanticipated changes in gene expression when inserting new sequences into the viral genome.

Reversion Analysis

The phenotypes caused by mutation can **revert** in one of two ways: by change of the original mutation to the wild-type sequence or by acquisition of a mutation at a second site, either in the same gene or in a different gene.

BOX 3.9

DISCUSSION

The raison d'être of reversion analysis: is the observed phenotype due to the mutation?

The Problem

The requirement for construction and analysis of revertants is well understood in genetic analysis. The assertion that a phenotype arises from the mutation and the conclusion that the mutated gene mediates the function cannot stand without careful reversion analysis.

In genetic analysis of viruses, mutations are made *in vitro* by a variety of techniques, all of which can introduce unexpected mutations. Errors can be introduced during cloning, during polymerase chain reactions (pcr), and during sequencing. Unexpected mutations often arise during the introduction of viral DNA and plasmid DNA into the eukaryotic cell. The process of recombination when viral DNA meets plasmid DNA in the eukaryotic cell is not error-proof. Tens of thousands of DNA copies are forced into a cell (transfection), and only a tiny portion give rise to the desired genome. Viral DNA isolated from cells or virions contains nicks, gaps, and even RNA. Isolation of large, linear viral DNA by standard micropipetting techniques usually shears the DNA into a population of subgenomic fragments that must be repaired and rejoined in the eukaryotic cells. Less than perfect repair fidelity is a potential source of undesired, second-site mutations. Indeed, linear DNA in eukaryotic cells can be attacked at the ends, and the ssDNA that results from such an attack can invade at nicked sites in linear DNA. In addition, the transfection process itself induces a variety of stress responses that might activate repair pathways.

Some Solutions

- Construct more than one mutant by using totally different sources of DNA. It is unlikely that an unlinked mutation with the same phenotype would occur twice.
- Marker rescue. All local DNA around the mutation is replaced with parental DNA. If the mutation indeed causes the phenotype, the wild-type phenotype should be restored in the rescued virus.
- Ectopic expression of the wild-type protein in the mutant background. If the wild-type phenotype is restored (complemented), then the probability is high that the phenotype arises from the mutation. The merit of this method over marker rescue is that the latter shows only that unlinked mutations are unlikely to be the cause of the phenotype.

All of these approaches have limitations, and it is prudent to use more than one.

Reversion analysis is an integral part of genetics: when confidence is high that the phenotype observed is due to the known mutation, research progresses rapidly.

Phenotypic reversion caused by second-site mutation is known as **suppression**, or **pseudoreversion**, to distinguish it from reversion at the original site of mutation. Reversion has been studied since the beginnings of classical genetic analysis (Box 3.9). When suspected pseudorevertants are crossed with wild-type viruses, the mutant phenotype should be observed in the progeny as a result of segregation of the original mutation from the suppressor mutation. In the modern era of genetics, cloning and sequencing techniques can be used to demonstrate suppression and to identify the nature of the suppressor mutation (see below). The identification of suppressor mutations is a powerful tool for studying protein-protein and protein-nucleic acid interactions. Some suppressor mutations complement changes made at several sites, whereas **allele-specific** suppressors complement only a specific change. The allele specificity of second-site mutations provides evidence for physical interactions among proteins and nucleic acids.

Phenotypic revertants can be isolated either by propagating the mutant virus under restrictive conditions or, with mutants exhibiting nonconditional phenotypes, by searching for wild-type properties. For DNA viruses, chemical mutagenesis may be required to produce revertants, but this is not necessary for RNA viruses, which spawn mutants at a higher frequency. Next, nucleotide sequence analysis is used to determine if the original mutation is still present in the genome of the revertant. The presence of the original mutation indicates that reversion has occurred by second-site mutation. The objective is to identify a fragment that contains the suppressor mutation and is small enough (~1 kb) that its nucleotide sequence can be determined. A similar analysis can be carried out for RNA viruses, except that DNA clones must first be derived from the genomes of phenotypic suppressors. The final step is introduction of the suspected suppressor mutation into the genome of the original mutant virus to confirm its effect. Several specific examples of suppressor analysis are provided below.

Some mutations within the origin of replication (Ori) of simian virus 40 reduce viral DNA replication and induce the formation of small plaques. Pseudorevertants of Ori mutants were isolated by random mutagenesis of mutant viral DNA followed by introduction into cultured cells and selection of viruses that form large plaques. The second-site mutations that suppressed the replication defects were localized to a specific region within the gene for large T antigen. These results indicated that a specific domain of large T antigen interacts with the Ori sequence during viral replication.

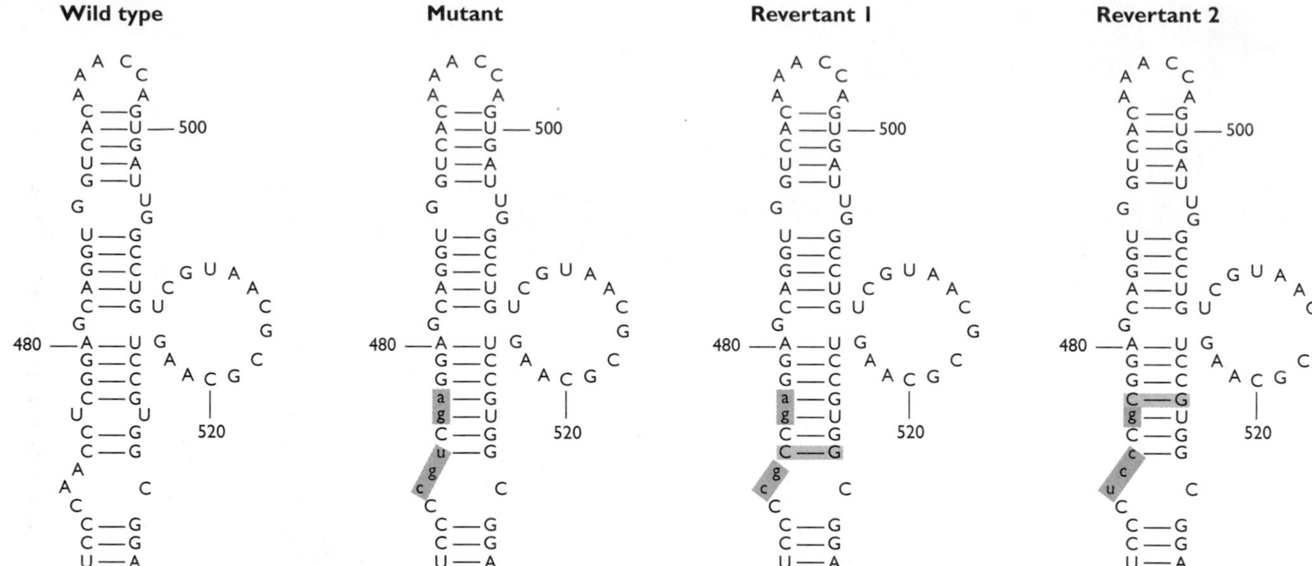

Figure 3.14 Effect of second-site suppressor mutations on predicted secondary structure in the 5' untranslated region of poliovirus (+) strand RNA. Diagrams of the region between nucleotides 468 and 534, which corresponds to stem-loop V (Chapter 11), are shown. These include, from left to right, sequences of wild-type poliovirus type 1, a mutant containing the nucleotide changes highlighted in orange, and two phenotypic revertants. Two C:G base pairs present in the wild-type parent and destroyed by the mutation are restored by second-site reversion (blue shading). Adapted from A. A. Haller et al., *J. Virol.* **70**:1467–1474, 1996, with permission.

The 5′ untranslated region of the poliovirus genome contains extensive RNA secondary-structure features that are important for RNA translation and replication, as discussed in Chapters 6 and 11. Disruption of such secondary structure by substitution of an 8-nucleotide sequence produces a virus that replicates poorly and readily gives rise to pseudorevertants that replicate more efficiently (Fig. 3.14). Nucleotide sequence analysis of two pseudorevertants demonstrated that they contain base changes that restore the disrupted secondary structure. These results confirm that the RNA secondary structure is important for the biological activity of this untranslated region.

Genetic Interference by Double-Stranded RNA

RNA interference (RNAi) has become a powerful and widely used tool for analyzing gene function. In this technique, duplexes of 21-nucleotide RNA molecules, called **small interfering RNAs (siRNAs)**, corresponding to the gene to be silenced are synthesized chemically or by transcription reactions. siRNAs or plasmids which encode them are introduced into cultured cells by transformation, and they efficiently inhibit the production of specific proteins by causing sequence-specific mRNA degradation. The functions of specific viral or cellular proteins during infection can therefore be studied by using this technique (Fig. 3.15). The mechanism of mRNA degradation mediated by siRNA is discussed in Chapter 10.

Engineering Viral Genomes: Viral Vectors

Naked DNA can be introduced into cultured animal cells as complexes with calcium phosphate or lipid-based reagents or by electroporation. Such DNA can support expression of its gene products transiently or stably from an integrated or episomal copy. Introduction of DNA into cells is a routine method in virological research and is also employed for certain clinical applications, such as the production of a therapeutic protein or a vaccine. However, this approach is not suitable for certain applications. For example, one goal of gene therapy is to deliver a gene to patients who either lack the gene or carry defective versions of it (Table 3.3). An approach to this problem is to create a cell line that synthesizes the gene product. After infusion into patients, the cells can become permanently established in tissues. If the primary cells to be used are limiting in a culture (e.g., stem cells), it is not practical to select and amplify the rare cells that receive naked DNA. Recombinant viruses carrying foreign genes can infect a greater percentage of cells and thus facilitate generation of the desired cell lines. These viral vectors have also found widespread use in the research laboratory. A complete understanding of the

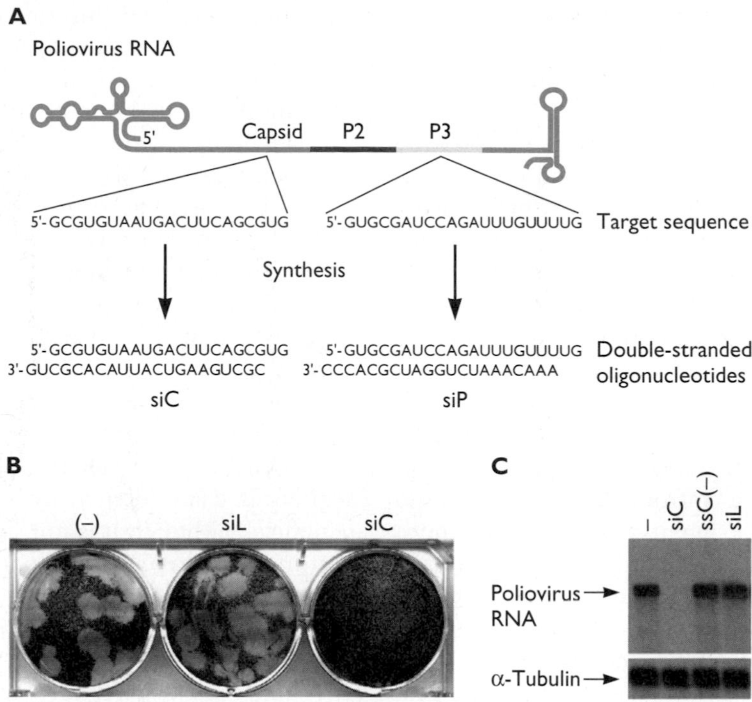

Figure 3.15 Inhibition of poliovirus replication by siRNA. siRNAs were introduced into cells by transformation, and the cells were then subjected to poliovirus infection. **(A)** Location of siRNAs siC and siP on a map of the poliovirus RNA genome. **(B)** Inhibition of plaque formation by siRNA siC. The number of plaques is not reduced in untreated cells (–) or when siRNA from *Renilla* luciferase is used (siL). Plaque formation was also inhibited with siP (not shown). **(C)** Northern blot analysis of RNA from poliovirus-infected cells 6 h after infection. Poliovirus RNA replication is blocked by siC but not by the (–) strand of siC RNA, ssC(–), or siL. The blot was rehybridized with a DNA probe directed against α-tubulin to ensure that all lanes contained equal amounts of RNA. Adapted from L. Gitlin et al., *Nature* **418:**430–434, 2002, with permission.

Table 3.3 Some genetic diseases that might be treated using viral vectors containing human genes

Disease	Defect	Incidence	Target cell
Severe combined immunodeficiency	Adenosine deaminase (25% of patients)	Rare	Bone marrow cells or T lymphocytes
Hemophilia A	Factor VII deficiency	1 in 10,000 males	Liver, muscle, fibroblasts, bone marrow cells
Hemophilia B	Factor IX deficiency	1 in 30,000 males	Liver, muscle, fibroblasts, bone marrow cells
Familial hypercholesterolemia	Deficiency of low-density lipoprotein receptor	1 in 1,000,000	Liver
Cystic fibrosis	Defective salt transport in lung epithelium	1 in 3,000 whites	Lung airways
Hemoglobinopathies and thalassemias	Defects in α- or β globin gene	1 in 600 in specific ethnic groups	Bone marrow precursors of red blood cells
Gaucher's disease	Defect in glucocerebrosidase	1 in 450 Ashkenazi Jews	Bone marrow cells, macrophages
α_1-Antitrypsin deficiency, inherited emphysema	α_1-Antitrypsin not produced	1 in 3,500	Lung or liver cells
Duchenne muscular dystrophy	Dystrophin not produced	1 in 3,000 males	Muscle cells

structure and function of viral vectors requires knowledge of viral genome replication, which is discussed in subsequent chapters for selected viruses and is summarized in the appendix in this volume.

The design requirements for viral vectors include the inclusion of an appropriate promoter, maintenance of genome size within the packaging limit of the particle, and elimination of viral virulence, the capacity of the virus to cause disease. Expression of foreign genes from viral vectors may be controlled by homologous or heterologous promoters and enhancers chosen to support efficient transcription (e.g., the human cytomegalovirus immediate-early transcriptional control region), depending on the goals of the experiment. Such genes can be built directly into the viral genome or introduced by recombination in cells, as described above (see "Introducing Mutations into the Viral Genome"). The recipient viral genome generally carries deletions and sometimes additional mutations. Deletion of some viral sequences is often required to overcome the limitations on the size of viral genomes that can be packaged in virions. For example, adenoviral DNA molecules more than 105% of the normal length are packaged very poorly. As this limitation would allow only 1.8 kbp of exogenous DNA to be inserted into the genome, adenovirus vectors often include deletions of the E3 gene (which is not essential for growth in cells in culture) and of the E1A and E1B transcription units, which encode proteins that can be provided by complementing cell lines.

When viral vectors are designed for therapeutic purposes, it is essential to prevent replication and destruction of target host cells. The deletions necessary to accommodate a foreign gene may contribute to such disabling of the vector. For example, the E1A protein-coding sequences that are invariably deleted from adenovirus vectors are necessary for efficient transcription of viral early genes; in their absence, viral yields from cells in culture are reduced by about 3 to 6 orders of magnitude (depending on the cell type). Removal of E1A coding sequences from adenovirus vectors is therefore doubly beneficial, although it is not sufficient to ensure that the vector cannot replicate or induce damage in a host animal. As discussed in detail in Volume II, Chapter 8, production of viruses that do not cause disease can be more difficult to achieve.

A summary of viral vectors is presented in Table 3.4, and examples are discussed below.

Table 3.4 Some viral vectors

Virus	Insert size	Integration	Duration of expression	Advantages	Potential disadvantages
Adeno-associated virus	~4.5–9 (?) kb	Low efficiency	Long	Nonpathogenic, episomal, infects nondividing cells	Immunogenic, toxicity, small packaging limit
Adenovirus	2–38 kb	No	Short	Efficient gene delivery, infects nondividing cells	Transient, immunogenic
Alphavirus	~5 kb	No	Short	Broad host range, high level expression	Virulence
Epstein-Barr virus	~120 kb	No; episomal	Long	High capacity, episomal, long-term expression	
Gammaretrovirus	1–7.5 kb	Yes	Shorter than formerly	Stable integration	May rearrange genome, insertional mutagenesis, require cell division
Herpes simplex virus	~30 kb	No	Long in central nervous system, short elsewhere	Infects nondividing cells; neurotropic, large capacity	Virulence, persistence in neurons, immunogenic
Lentivirus	7–18 kb	Yes	Long	Stable integration; infects nondividing and terminally differentiated mammalian cells	Insertional mutagenesis
Poliovirus	~300 bp for helper-free virus; ~3 kb for defective virus	No	Short	Excellent mucosal immunity	Limited capacity; reversion to neurovirulence
Rhabdovirus	Unknown	No	Short	High-level expression, rapid cell killing	Virulence, highly cytopathic
Vaccinia virus	At least ~25 kb, probably ~75–100 kb	No	Short	Wide host range, ease of isolation, large capacity, high-level expression	Transient, immunogenic

DNA Virus Vectors

One goal of gene therapy is to introduce genes into terminally differentiated cells. Such cells normally do not divide, they cannot be cultured *in vitro*, and the organs of which they are a part cannot be infused with virus-infected cells. DNA virus vectors have been developed to overcome some of these problems.

Adenovirus vectors were originally developed for the treatment of cystic fibrosis because of the tropism of the virus for the respiratory epithelium. Adenovirus can infect terminally differentiated cells, but only transient expression is provided, as this viral DNA is not integrated into host cell DNA. Adenoviruses carrying the cystic fibrosis transmembrane conductance regulator gene, which is defective in patients with this disease, have been used in clinical trials. Many other gene products with therapeutic potential have been produced from adenovirus vectors in a wide variety of cell types. In the earliest adenovirus vectors that were designed, foreign genes were inserted into the E1 and/or E3 regions. As these vectors had limited capacity, vectors that contain minimal adenovirus sequences were designed (Fig. 3.16). This strategy allows 27 to 38 kb of foreign sequence to be introduced into the vector. In addition, elimination of most viral genes reduces the host immune response to viral proteins, simplifying multiple immunizations. Considerable efforts have been made to modify the adenovirus capsid to target the vectors to different cell types. The fiber protein, which mediates adenovirus binding to cells, has been altered by insertion of ligands that bind particular cell surface receptors. Such alterations could increase the cell specificity of adenovirus attachment and the efficiency of gene transfer and could decrease the amount of virus that is administered.

Figure 3.16 Adenovirus vectors. High-capacity adenovirus vectors are produced by inserting a foreign gene and promoter into the viral E1 region, which has been deleted. The E3 region also has been deleted. Two *loxP* sites for cleavage by the Cre recombinase have been introduced into the adenoviral genome (black arrowheads). Infection of cells that produce Cre leads to excision of sequences flanked by the *loxP* sites. The result is a "gutless" vector that contains only the origin-of-replication-containing inverted terminal repeats (ITR), the packaging signal (yellow), the viral E4 transcription unit (orange), and the transgene with its promoter (green). Additional DNA flanking the foreign gene must be inserted to allow packaging of the viral genome (not shown). Adapted from A. Pfeifer and I. M. Verma, *in* D. M. Knipe et al. (ed.), *Fields Virology*, 4th ed. (Lippincott Williams & Wilkins, Philadelphia, PA, 2001), with permission.

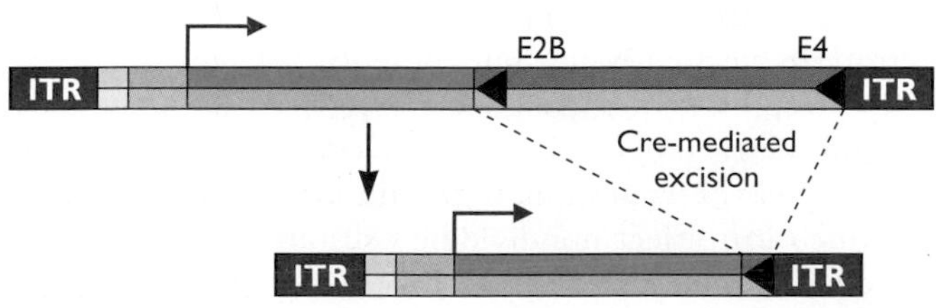

Adeno-associated virus has attracted much attention as a vector for gene therapy, because the virus integrates into the host genome. Genomes packaged into recombinant viruses can integrate at numerous sites in the human genome or replicate as an episome. The vectors persist, in some cases with high levels of expression, in many different tissues. There has been increasing interest in these vectors to target therapeutic genes to smooth muscle, a tissue that is highly susceptible and supports sustained high-level expression of foreign genes. Although the first-generation adeno-associated virus vectors were limited in the size of inserts that could be transferred, other systems have been developed to overcome the limited genetic capacity (Fig. 3.17). The cell specificity of adeno-associated virus vectors has been altered by inserting receptor-specific ligands into the capsid. In addition, many new viral serotypes have been identified that vary in their tropism and ability to trigger immune responses.

Vectors based on the genome of Epstein-Barr virus, a gammaherpesvirus, replicate as episomes in the cell. This property confers the advantage of long-term expression of foreign genes. Episomal replication of the vector is maintained by the origin for plasmid maintenance, OriP, whose activity is dependent on Epstein-Barr virus nuclear antigen 1 (EBNA-1) (see Chapter 10). Vectors containing OriP and encoding EBNA-1 can accept at least 120 kbp of foreign DNA.

Vaccinia virus and other animal poxvirus vectors offer the advantages of a wide host range, a genome that accepts very large fragments, high expression of foreign genes, and relative ease of preparation. Foreign DNA is usually inserted into the viral genome by homologous recombination, using an approach similar to that described for marker transfer. Because of the relatively low pathogenicity of the virus, vaccinia virus recombinants have been considered candidates for human and animal vaccines (Volume II, Fig. 8.11).

RNA Virus Vectors

A number of RNA viruses have also been developed as vectors for foreign gene expression (Table 3.4). Because poliovirus is such an excellent mucosal immunogen, it has been an attractive candidate for the delivery of antigens to these surfaces. Alphavirus vectors have been shown to be capable of expressing foreign genes. These (+) strand RNA viruses produce subgenomic mRNAs from which efficient gene expression can be obtained. In one type of vector,

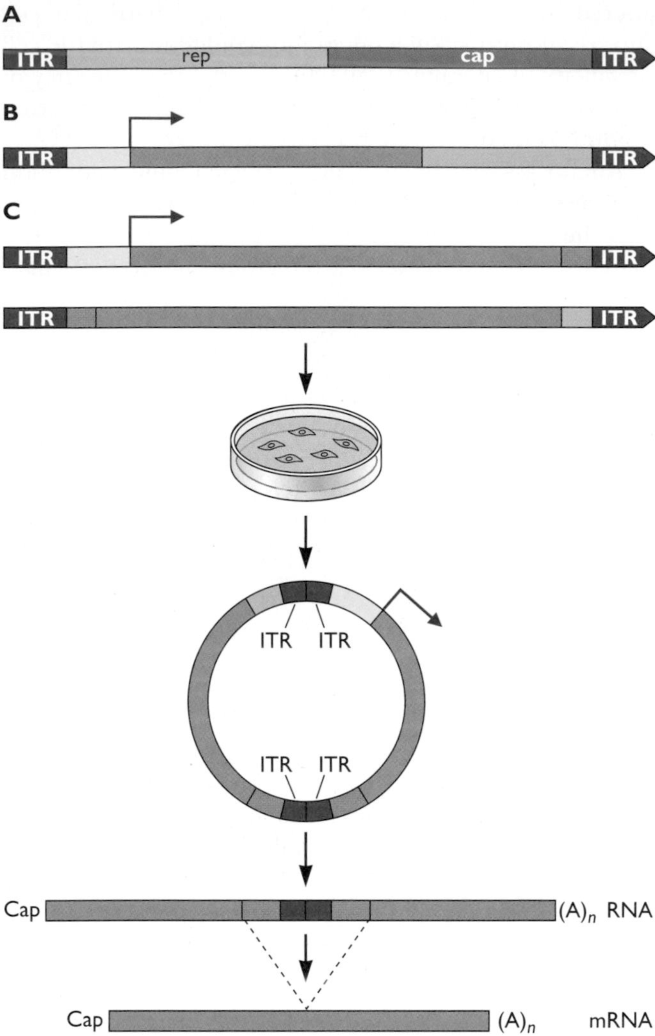

Figure 3.17 Adeno-associated virus vectors. (A) Map of the genome of wild-type adeno-associated virus. The viral DNA is single stranded and flanked by two inverted terminal repeats (ITR); it encodes capsid (blue) and nonstructural (orange) proteins. **(B)** In one type of vector, the viral genes are replaced with the transgene (pink) and its promoter (yellow) and a poly(A) addition signal (green). These DNAs are introduced into cells which have been engineered to produce capsid proteins, and the vector genome is encapsidated into virus particles. A limitation of this vector structure is that only 4.1 to 4.9 kb of foreign DNA can be packaged efficiently. To overcome this limitation, a two-plasmid expression system was developed. **(C)** The transgene of interest (pink) is divided between two adeno-associated virus vectors. One plasmid contains the promoter, the 5′ part of the transgene, and a 5′ splice site (purple), and the second plasmid contains a 3′ splice site, the remainder of the transgene, and the polyadenylation site. In cells, the recombinant viral genomes form circular multimers that bring the divided transgene together in a head-to-tail orientation. A long RNA is then produced that undergoes splicing to form a functional mRNA that can be translated into protein. Adapted from A. Pfeifer and I. M. Verma, *in* D. M. Knipe et al. (ed.), *Fields Virology*, 4th ed. (Lippincott Williams & Wilkins, Philadelphia, PA, 2001), and Z. Yan et al., *Proc. Natl. Acad. Sci. USA* **97:**6716–6721, 2000, with permission.

the foreign gene replaces those encoding the viral structural proteins. As a result, this vector requires the presence of a helper virus to provide the missing proteins, so that the vector genome can be packaged into virus particles. An alternative arrangement is to place the foreign gene under the control of a second, subgenomic RNA promoter, leaving intact the promoter that controls synthesis of the structural proteins. This vector replicates without a helper and produces infectious particles.

The development of an infectious DNA clone of vesicular stomatitis virus, a (−) strand RNA virus, has led to an approach to therapy for human immunodeficiency virus type 1 infection. Recombinant vesicular stomatitis virus particles that lack the viral glycoprotein G, required for attachment to host cells, have been produced. The G gene is replaced by genes encoding two cellular proteins required for the attachment of human immunodeficiency virus type 1 to cells (CD4 and CXCr4) (see Chapter 5). The recombinant vesicular stomatitis viruses cannot infect normal cells. However, they can infect and lyse cells infected with human immunodeficiency virus type 1, because these cells synthesize viral glycoproteins on the surface. This approach has the potential of limiting the number of infected cells in the host.

Retroviruses have enjoyed great popularity as vectors (Fig. 3.18) because their replication cycles include the integration of a dsDNA copy of viral RNA into the cell genome, a topic of Chapter 7. The integrated provirus remains permanently in the cell's genome and is passed on to progeny during cell division. This feature of retroviral vectors results in permanent modification of the genome of the infected cell. The choice of the envelope glycoprotein carried by retroviral vectors has a significant influence on their tropism. The vesicular stomatitis virus G glycoprotein is often used because it confers a wide tissue tropism. Retrovirus vectors can be targeted to specific cell types by using other viral envelope proteins.

One problem with the use of retroviruses in correcting genetic deficiencies is that only a few cell types can be infected by the commonly used murine retroviral vectors, and these can integrate their DNA efficiently only in actively dividing cells. Often the cells that are targets of gene therapy, hepatocytes, and muscle cells, do not divide. This problem can be circumvented if ways can be found to induce such cells to divide before being infected with the retrovirus. Another important limitation of the murine retrovirus vectors is the phenomenon of gene silencing, which represses foreign gene expression in many cells. An alternative approach is to use viral vectors that contain sequences from human immunodeficiency virus type 1, which can infect nondividing cells and is less severely affected by gene silencing.

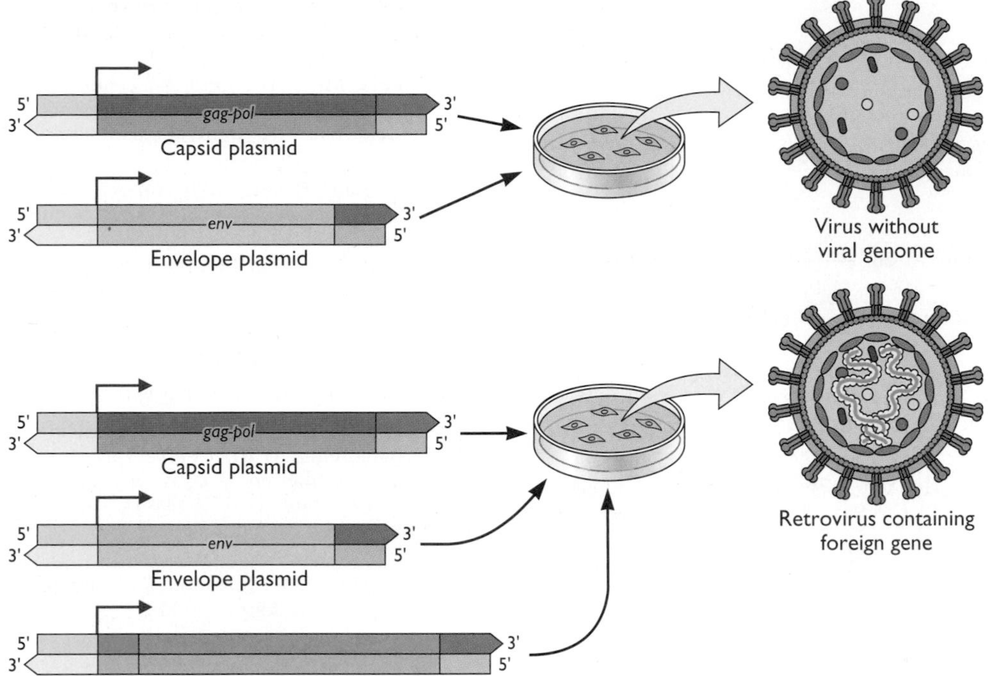

Figure 3.18 Retroviral vectors. The minimal viral sequences required for retroviral vectors are 5′- and 3′-terminal sequences (yellow and blue, respectively) that control gene expression and packaging of the RNA genome. The foreign gene (pink) and promoter (green) are inserted between the viral sequences. To package this DNA into viral particles, it is introduced into cultured cells with plasmids that encode viral proteins required for encapsidation. No wild-type viral RNA is present in these cells. If these plasmids alone are introduced into cells, virus particles that do not contain viral genomes are produced. When all three plasmids are introduced into cells, retrovirus particles are formed that contain only the recombinant vector genome and no wild-type particles. The host range of the recombinant vector can be controlled by the type of envelope protein. Envelope protein from amphotropic retroviruses allows the recombinant virus to infect human and mouse cells. The vesicular stomatitis virus glycoprotein G allows infection of a broad range of cell types in many species and also permits concentration with simple methods. Adapted from A. Pfeifer and I. M. Verma, *in* D. M. Knipe et al. (ed.), *Fields Virology*, 4th ed. (Lippincott Williams & Wilkins, Philadelphia, PA, 2001), with permission.

Perspectives

The information in this chapter can be used as a "road map" for negotiating this book and for planning a virology course. Figures 3.1 through 3.7 serve as the points of departure for detailed analyses of the principles of virology. These figures illustrate seven strategies based on viral mRNA synthesis and genome replication, with references to chapters containing information about the replication and pathogenesis of members of each class. The material in this chapter can be used to structure individual reading or to design a virology course based on specific viruses or groups of viruses while adhering to the overall organization of this textbook by function. Refer to this chapter and the figures to find answers to questions about specific virus families or individual species. For example, Fig. 3.5 provides information about (+) strand RNA viruses and Fig. 3.10 indicates specific chapters in which these viruses are discussed.

Since the earliest days of experimental virology, genetic analysis has proven invaluable for studying the function of the viral genome. Initially, methods were developed to produce viral mutants by chemical or UV mutagenesis followed by screening for readily identifiable phenotypes. Because it was not possible to identify the genetic changes in such mutants, it was difficult to associate proteins with virus-specific processes. This limitation vanished with the development of infectious DNA clones of viral genomes: it suddenly was possible to introduce defined mutations into any region of the viral genome. This complete genetic toolbox provides countless possibilities for studying the function of the viral genome, limited only by the creativity and enthusiasm of the investigator.

References

Books

Casjens, S. 1985. *Virus Structure and Assembly.* Jones and Bartlett Publishers, Inc., Boston, MA.

Griffiths, A. J. F., W. M. Gelbart, R. C. Lewontin, and J. H. Miller. 2002. *Modern Genetic Analysis.* W. H. Freeman & Co., New York, NY.

Review Articles

Alba, R., A. Bosch, and M. Chillon. 2005. Gutless adenovirus: last-generation adenovirus for gene therapy. *Gene Ther.* **12:**S18–S27.

Baltimore, D. 1971. Expression of animal virus genomes. *Bacteriol. Rev.* **35:**235–241.

Conzelmann, K.-K., and G. Meyers. 1996. Genetic engineering of animal RNA viruses. *Trends Microbiol.* **4:**386–393.

Domingo, E. 2000. Viruses at the edge of adaptation. *Virology* **270:**251–253.

Drake, J. W., and J. J. Holland. 1999. Mutation rates among RNA viruses. *Proc. Natl. Acad. Sci. USA* **96:**13910–13913.

Flotte, T. R. 2007. Gene therapy: the first two decades and the current state-of-the-art. *J. Cell. Physiol.* **213:**301–305.

Hino, S. 2002. TTV, a new human virus with single stranded circular DNA genome. *Rev. Med. Virol.* **12:**151–158.

Joyce, G. F. 2002. The antiquity of RNA-based evolution. *Nature* **418:**214–221.

La Scola, B., S. Audic, C. Robert, L. Jungang, X. de Lamballerie, M. Drancourt, R. Birtles, J.-M. Claverie, and D. Raoult. 2003. A giant virus in amoebae. *Science* **299:**2033.

McGeoch, D. J., and A. J. Davison. 1995. Origins of DNA viruses, p. 67–75. *In* A. J. Giffs, C. H. Calisher, and F. Garcia-Arenal (ed.), *Molecular Basis of Virus Evolution.* Cambridge University Press, Cambridge, United Kingdom.

Suzan-Monti, M., B. La Scola, and D. Raoult. 2006. Genetic and evolutionary aspects of *Mimivirus. Virus. Res.* **117:**145–155.

Tidona, C. A., and G. Darai. 2000. Iridovirus homologues of cellular genes—implications for the molecular evolution of large DNA viruses. *Virus Genes* **21:**77–81.

Van Etten, J. L., M. V. Graves, D. G. Müller, W. Boland, and N. Delaroque. 2002. *Phycodnaviridae*—large DNA algal viruses. *Arch. Virol.* **147:**1479–1516.

Papers of Special Interest

Almazan, F., J. M. Gonzalez, Z. Penzes, A. Izeta, E. Calvo, J. Plana-Duran, and L. Enjuanes. 2000. Engineering the largest RNA virus genome as an infectious bacterial artificial chromosome. *Proc. Natl. Acad. Sci. USA* **97:**5516–5521.

Bridgen, A., and R. M. Elliott. 1996. Rescue of a segmented negative-strand RNA virus entirely from cloned complementary DNAs. *Proc. Natl. Acad. Sci. USA* **93:**15400–15404.

Chang, Y., E. Cesarman, M. S. Pessin, F. Lee, J. Culpepper, D. M. Knowles, and P. S. Moore. 1994. Identification of herpesvirus-like DNA sequences in AIDS-associated Kaposi's sarcoma. *Science* **266:**1865–1869.

Crotty, S., C. E. Cameron, and R. Andino. 2001. RNA virus error catastrophe: direct molecular test by using ribavirin. *Proc. Natl. Acad. Sci. USA* **98:**6895–6900.

Demmin, G. L., A. C. Clase, J. A. Randall, L. W. Enquist, and B. W. Banfield. 2001. Insertions in the gG gene of pseudorabies virus reduce expression of the upstream Us3 protein and inhibit cell-to-cell spread of virus infection. *J. Virol.* **75:**10856–10869.

Dmitriev, I., V. Krasnykh, C. R. Miller, M. Wang, E. Kashentseva, G. Mikheeva, N. Belousova, and D. T. Curiel. 1998. An adenovirus vector with genetically modified fibers demonstrates expanded tropism via utilization of a coxsackievirus and adenovirus receptor-independent cell entry mechanism. *J. Virol.* **72:**9706–9713.

Elbashir, S. M., J. Harborth, W. Lendeckel, A. Yalcin, K. Weber, and T. Tuschl. 2001. Duplexes of 21-nucleotide RNAs mediate RNA interference in cultured mammalian cells. *Nature* **411:**494–498.

Goff, S. P., and P. Berg. 1976. Construction of hybrid viruses containing SV40 and lambda phage DNA segments and their propagation in cultured monkey cells. *Cell* **9:**695–705.

Graham, F. L., J. Rudy, and P. Brinkley. 1989. Infectious circular DNA of human adenovirus type 5: regeneration of viral DNA termini from molecules lacking terminal sequences. *EMBO J.* **8:**2077–2085.

Hoffmann, E., G. Neumann, Y. Kawaoka, G. Hobom, and R. G. Webster. 2000. A DNA transfection system for generation of influenza A virus from eight plasmids. *Proc. Natl. Acad. Sci. USA* **97:**6108–6113.

Kobayashi, T., A. A. R. Antar, K. W. Boehme, P. Danthi, E. A. Eby, K. M., Guglielmi, G. H. Holm, E. M. Johnson, M. S. Maginnis, S. Naik, W. B. Skelton, J. D. Wetzel, G. J. Wilson, J. D. Chappell, and T. S. Dermody. 2007. A plasmid-based reverse genetics system for animal double-stranded RNA viruses. *Cell Host Microbe* **1:**147–157.

Luytjes, W., M. Krystal, M. Enami, J. D. Pavin, and P. Palese. 1989. Amplification, expression, and packaging of a foreign gene by influenza virus. *Cell* **59:**1107–1113.

Mackett, M., G. L. Smith, and B. Moss. 1982. Vaccinia virus: a selectable eukaryotic cloning and expression vector. *Proc. Natl. Acad. Sci. USA* **79:**7415–7419.

Palmenberg, A. C., and J. Y. Sgro. 1997. Topological organization of picornaviral genomes: statistical prediction of RNA structural signals. *Semin. Virol.* **8:**231–241.

Palese, P., and J. L. Schulman. 1976. Mapping of the influenza virus genome: identification of the hemagglutinin and the neuraminidase genes. *Proc. Natl. Acad. Sci. USA* **73:**2142–2146.

Post, L. E., and B. Roizman. 1981. A generalized technique for deletion of specific genes in large genomes: alpha gene 22 of herpes simplex virus 1 is not essential for growth. *Cell* **25:**227–232.

Racaniello, V. R., and D. Baltimore. 1981. Cloned poliovirus complementary DNA is infectious in mammalian cells. *Science* **214:**916–919.

Saiki, R. K., D. H. Gelfand, S. Stoffel, S. J. Scharf, R. Higuchi, G. T. Horn, K. B. Mullis, and H. A. Erlich. 1988. Primer-directed enzymatic amplification of DNA with a thermostable DNA polymerase. *Science* **239:**487–491.

Sanger, F., A. R. Coulson, T. Friedmann, G. M. Air, B. G. Barrell, N. L. Brown, J. C. Fiddes, C. A. Hutchison III, P. M. Slocombe, and M. Smith. 1978. The nucleotide sequence of bacteriophage phiX174. *J. Mol. Biol.* **25:**225–246.

Schnell, M. J., T. Mebatsion, and K. K. Conzelmann. 1994. Infectious rabies viruses from cloned cDNA. *EMBO J.* **13:**4195–4203.

Smith, G. A., and L. W. Enquist. 2000. A self-recombining bacterial artificial chromosome and its application for analysis of herpesvirus pathogenesis. *Proc. Natl. Acad. Sci. USA* **97:**4873–4878.

van Dinten, L. C., J. A. den Boon, A. L. M. Wassenaar, W. J. M. Spaan, and E. J. Snijder. 1997. An infectious arterivirus cDNA clone: identification of a replicase point mutation that abolishes discontinuous mRNA transcription. *Proc. Natl. Acad. Sci. USA* **94:**991–996.

Wu, Z., A. Asokan, J. C. Grieger, L. Govindasamy, M. Agbandje-McKenna, and R. J. Samulski. 2006. Single amino acid changes can influence titer, heparin binding, and tissue tropism in different adeno-associated virus serotypes. *J. Virol.* **80:**11393–11397.

Yan, Z., Y. Zhang, D. Duan, and J. F. Engelhardt. 2000. Trans-splicing vectors expand the utility of adeno-associated virus for gene therapy. *Proc. Natl. Acad. Sci. USA* **97:**6716–6721.

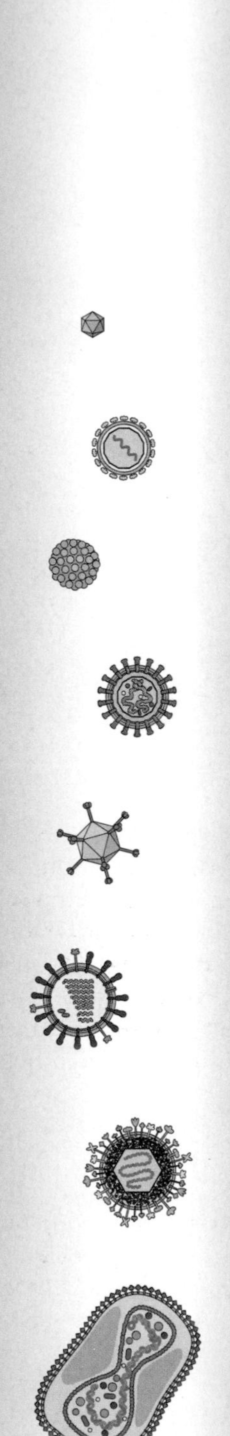

4

Structure

In order to create something that functions properly—a container, a chair, a house—its essence has to be explored, for it should serve its purpose to perfection, i.e. it should fulfill its function practically and should be durable, inexpensive and beautiful.

WALTER GROPIUS
Neue Arbeiten der Bauhaus Werk Stäten, Bauhaus Books, no. 7

Introduction

Virus particles are elegant assemblies of viral, and occasionally cellular, macromolecules. They are marvelous examples of architecture on the molecular scale, with forms beautifully adapted to their functions. Virus particles come in many sizes and shapes (Fig. 4.1; also see Fig. 1.7) and vary enormously in the number and nature of the molecules from which they are built. Nevertheless, they fulfill common functions and are built according to general principles that apply to them all. These properties are described in subsequent sections, in which we also discuss some examples of the architectural detail characteristic of members of different virus families.

Functions of the Virion

Virus particles are designed for effective transmission of the nucleic acid genome from one host cell to another within a single animal or among host organisms. A primary function of the **virion**, an infectious virus particle, is therefore protection of the genome, which can be damaged irreversibly by a break in the nucleic acid or by mutation during passage through hostile environments. During its travels, a virus particle may encounter a variety of potentially lethal chemical and physical agents, including proteolytic and nucleolytic enzymes, extremes of pH or temperature, and various forms of natural radiation. In all virus particles, the nucleic acid is sequestered within a sturdy barrier formed by extensive interactions among the viral proteins that comprise the protein coat. Such protein-protein interactions maintain surprisingly stable capsids: many virus particles composed of only protein and nucleic acid survive exposure to large variations in the temperature, pH, or chemical composition of their environment. Some, such as certain picornaviruses, are even resistant to strong detergents such as sodium dodecyl sulfate. The highly folded nature of coat proteins and their dense packing within virions render them largely inaccessible to proteolytic enzymes. Some viruses also possess an **envelope**, derived from cellular membranes, into which viral glycoproteins have been inserted.

A

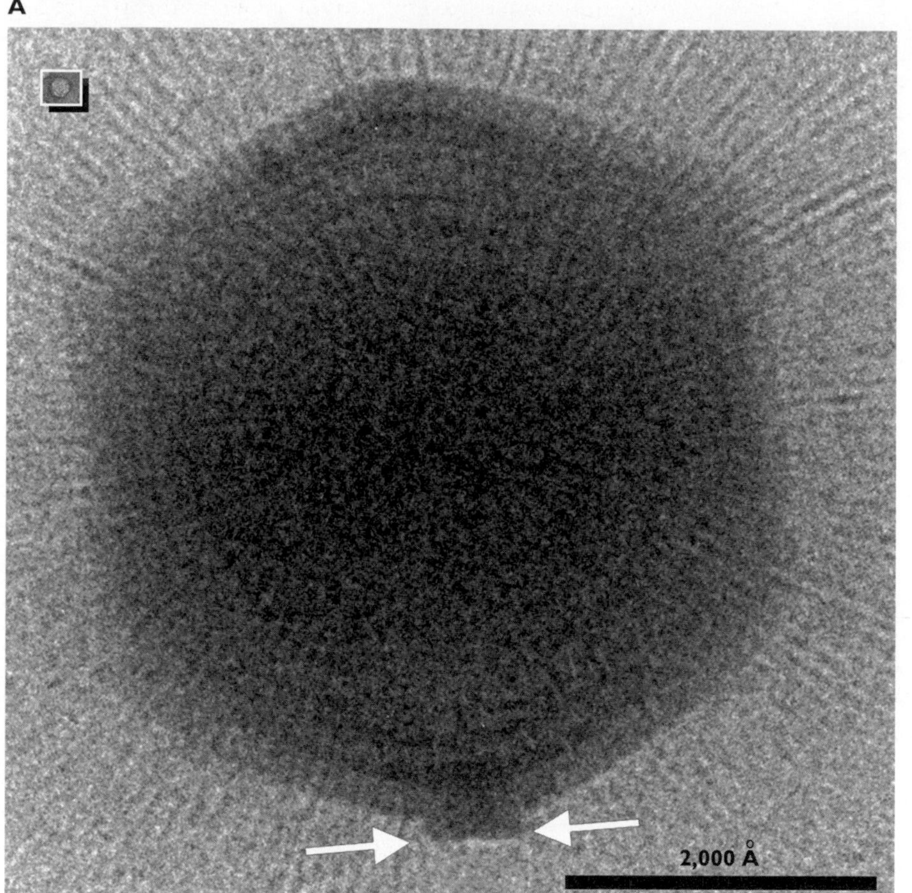

B

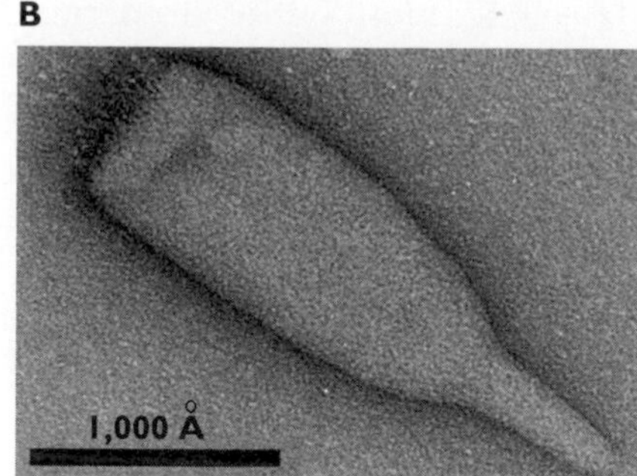

Figure 4.1 Variation in the size and shape of virus particles. (A) Cryo-electron micrographs of mimivirus and, in the inset (upper left), the parvovirus adeno-associated virus type 4 shown to scale relative to one another to illustrate the ~50-fold range in diameter among viruses that appear roughly spherical. Rod-shaped viruses also exhibit considerable variation in size, ranging in length from less than 200 nm to over 2,000 nm. Adapted from C. Xiao et al., *J. Mol. Biol.* **353**:493–496, 2005, and E. Pardon et al., *J. Virol.* **79**:5047–5058, 2005, respectively, with permission. Courtesy of M. G. Rossmann, Purdue University, and M. Agbandje-McKenna, University of Florida, Gainesville. **(B)** Complex shape of acidianus bottle virus isolated from a hot spring in Italy. The mimivirus particle (A) is also structurally complex: a large number of long, closely packed filaments project from its surface and a unique, protruding structure (white arrows) is present at one vertex. Adapted from M. Häring et al., *J. Virol.* **79**:9904–9911, 2005, with permission. Courtesy of D. Prangishvili, Institut Pasteur.

The envelope adds not only a protective lipid membrane but also an external layer of protein and sugars formed by the glycoproteins. Like the cellular membranes from which they are derived, viral envelopes are impermeable and block entry of chemicals or enzymes in aqueous solution into virus particles.

To protect the nucleic acid genome, virus particles must be stable structures. However, virions must also attach to an appropriate host cell and deliver the genome to the interior of that cell, where the particle is at least partially disassembled. The protective function of virus particles depends on very stable intermolecular interactions among their components during assembly, egress from the virus-producing cell, and transmission. On the other hand, these interactions must be reversed readily during entry and uncoating in a new host cell. In only a few cases do we understand the molecular mechanisms by which these apparently paradoxical requirements are met. Nevertheless, it is clear that contact of a virion with the appropriate cell surface receptor, or exposure to a specific intracellular environment, can trigger substantial conformational changes. Virus particles are therefore **metastable structures** that have not yet attained the minimum free energy conformation. The latter state can be attained only when an unfavorable energy barrier is surmounted, following induction of the irreversible conformational transitions associated with attachment and

Table 4.1 Functions of virion proteins

Protection of the genome
Assembly of a stable protective protein shell
Specific recognition and packaging of the nucleic acid genome
In many virions, interaction with host cell membranes to form the envelope
Delivery of the genome
Specific binding to external receptors of the host cell
Transmission of specific signals that induce uncoating of the genome
Induction of fusion with host cell membranes
Interaction with internal components of the infected cell to direct transport of the genome to the appropriate site
Other interactions with the host
With cellular components for transport to intracellular sites of assembly
With cellular components to ensure an efficient infectious cycle
With the host immune system

entry. Virions are **not** simply inert structures. Rather, they are molecular machines that play an active role in delivery of the nucleic acid genome to the appropriate host cell and initiation of the reproductive cycle.

Specific functions of virion proteins associated with protection and delivery of the genome are summarized in Table 4.1.

Nomenclature

Virus architecture is described in terms of **structural units** of increasing complexity, from the smallest biochemical unit (the polypeptide chain) to the infectious particle (or virion). These terms, which are used throughout this text, are defined in Table 4.2. Although virus particles are complex assemblies of macromolecules exquisitely suited for protection and delivery of viral genomes, they are constructed according to the general principles of biochemistry and protein structure.

Methods for Studying Virus Structure

The method most widely used for the examination of virus structure and morphology is electron microscopy. This technique, which has been applied to viruses since the middle decades of the last century, traditionally relied on negative staining of purified virus particles (or of sections of infected cells) with an electron-dense material, such as uranyl acetate or phosphotungstate. It can yield quite detailed and often beautiful images (Fig. 1.7; see the appendix) and provided the first rational basis for the classification of viruses.

The greatest contrast between virus particle and stain occurs where portions of the folded protein chain protrude from the virion surface. Consequently, surface knobs or projections, termed morphological units, are the main features identified by negative-contrast electron microscopy. However, these structures are often formed by multiple proteins and so their organization does not necessarily correspond to that of the individual proteins that form the capsid shell. Even when virus structure is well preserved and a high degree of contrast can be achieved, the minimal size of an object that can be distinguished by classical electron microscopy, its **resolution**, is limited to 50 to 75 Å. Detailed structural interpretation of images of negatively stained virus particles, which possess dimensions on a scale of a few hundred to a few thousand angstroms, is therefore impossible. Cryo-electron microscopy, in which samples are rapidly frozen and examined at very low temperatures in a hydrated, vitrified (noncrystalline, glasslike) state, preserves native structure. Because samples are not stained, this technique allows direct visualization of the contrast inherent in the virus particle. When combined with computerized mathematical methods of image analysis and three-dimensional reconstruction (Fig. 4.2), cryo-electron

Table 4.2 Nomenclature used in description of virus structure

Term	Synonym	Definition
Subunit (protein subunit)		Single, folded polypeptide chain
Structural unit	Asymmetric unit	Unit from which capsids or nucleocapsids are built; may comprise one protein subunit or multiple, different protein subunits
Morphological unit	Capsomere	Surface structures (e.g., knobs, projections, clusters) seen by electron microscopy; as these structures do not necessarily correspond to structural units (see the text), the term is restricted to descriptions of electron micrographs of viruses
Capsid	Coat	The protein shell surrounding the nucleic acid genome
Nucleocapsid	Core	The nucleic acid-protein assembly packaged within the virion; used when this complex is a discrete substructure of a complex particle
Envelope	Viral membrane	The host cell-derived lipid bilayer carrying viral glycoproteins
Virion		The infectious virus particle

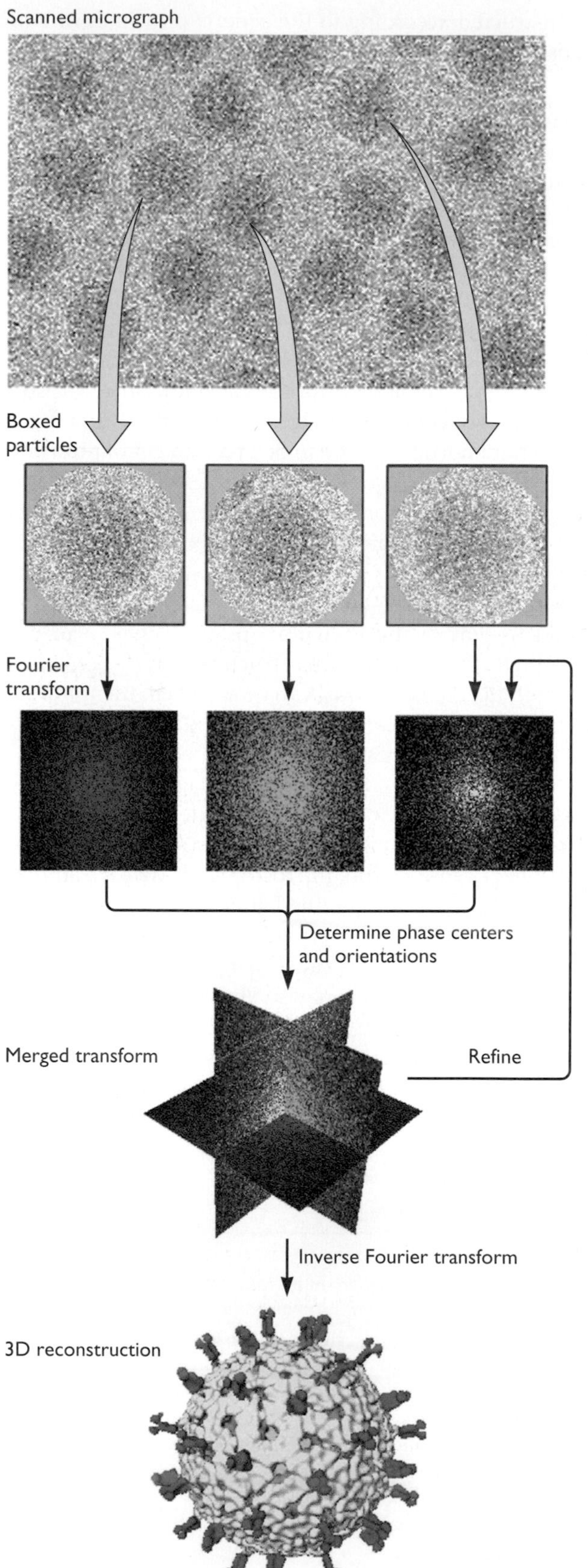

Concentrated preparations of purified virus particles are prepared for cryo-electron microscopy by rapid freezing on an electron microscope grid so that a glasslike, noncrystalline water layer is produced. This procedure avoids sample damage that can be caused by crystallization of the water or by chemical modification or dehydration during conventional negative-contrast electron microscopy. The sample is maintained at or below −160°C during all subsequent operations. Fields containing sufficient numbers of vitrified virus particles are identified by transmission electron microscopy at low magnification (to minimize sample damage from the electron beam) and photographed at high resolution (top).

These electron micrographs can be treated as two-dimensional projections (Fourier transforms) of the particles. Three-dimensional structures can be reconstructed from such two-dimensional projections by mathematically combining the information given by different views of the particles. For the purpose of reconstruction, the images of different particles are treated as different views of the same structure.

For reconstruction, micrographs are digitized for computer processing. Each particle to be analyzed is then centered inside a box, and its orientation is determined by application of programs that orient the particle on the basis of its icosahedral symmetry. In cryo-electron tomography, a series of images are collected with the sample at different angles to the electron beam and combined computationally to reconstruct a three-dimensional structure. The advantage of this approach is that no assumptions about the symmetry of the structure are required. The parameters that define the orientation of the particle must be determined with a high degree of accuracy, for example, to within 1° for even a low-resolution reconstruction (~40 Å). These parameters are improved in accuracy (**refined**) by comparison of different views (particles) to identify common data.

Once the orientations of a number of particles sufficient to represent all parts of the asymmetric unit have been determined, a low-resolution three-dimensional reconstruction is calculated from the initial set of two-dimensional projections by using computerized algorithms.

This reconstruction is refined by including data from additional views (particles). The number of views required depends on the size of the particle and the resolution sought. The reconstruction is initially interpreted in terms of the external features of the virus particle. Various computational and computer graphics procedures have been developed to facilitate interpretation of internal features. Courtesy of B. V. V. Prasad, Baylor College of Medicine.

And is it not true that even the small step of a glimpse through the microscope reveals to us images that we should deem fantastic and over-imaginative if we were to see them somewhere accidentally, and lacked the sense to understand them.

Paul Klee, *On Modern Art,* translated by Paul Findlay (London, United Kingdom, 1948)

Figure 4.2 Cryo-electron microscopy and image reconstruction illustrated with images of rotavirus.

A

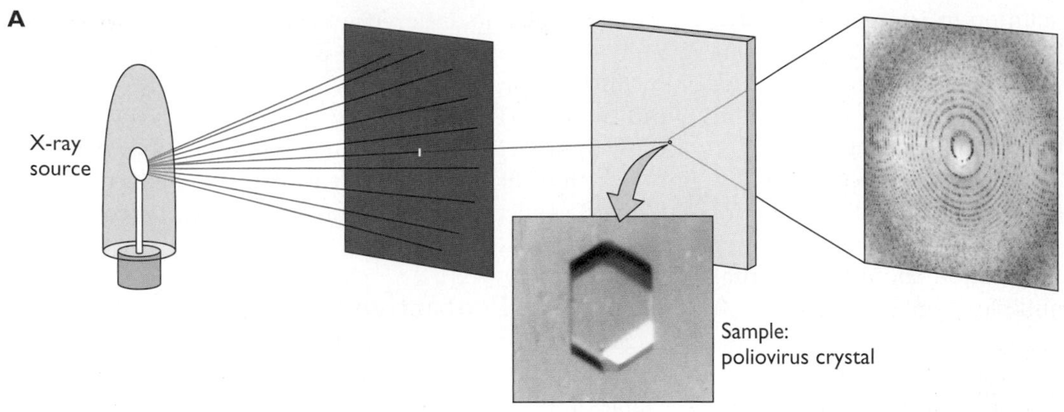

A picture of a section of the diffraction pattern generated by the poliovirus crystal.

B

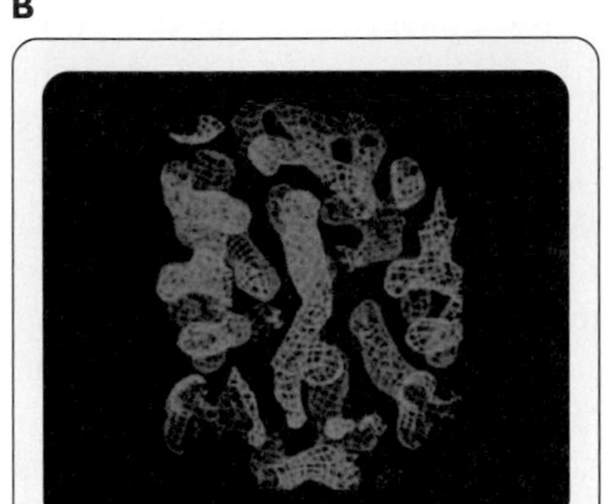

A section of the poliovirus electron density map showing part of the region around the fivefold axis of symmetry.

C

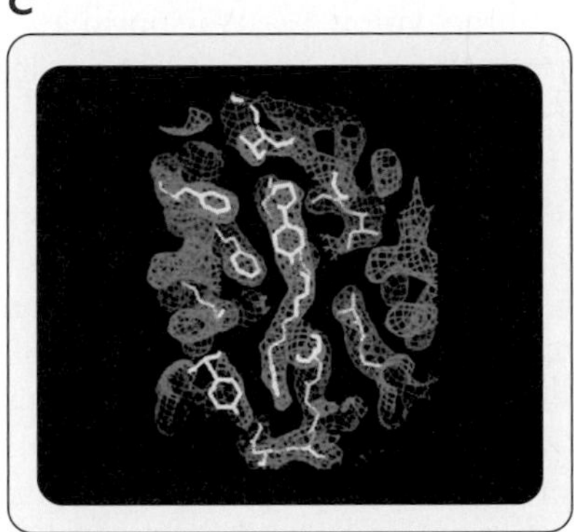

The same section of the map through one plane of the virus particle with segments of the structure built to fit the electron density. The double-ring structure with the long aliphatic chain (center) is an antiviral drug that is bound to the poliovirus particles in the crystal.

D

A portion of the virus structure shown as a ribbon diagram, with the three proteins that form the surface, VP1, VP2, and VP3, colored blue, yellow, and red, respectively.

Figure 4.3 Determination of virus structure by X-ray diffraction. (A) A virus crystal is composed of virus particles arranged in a well-ordered three-dimensional lattice. When the crystal is bombarded with a monochromatic X-ray beam, each atom within the virus particle scatters the radiation. Interactions of the scattered rays with one another form a diffraction pattern that is recorded. Each spot contains information about the position and the identity of all atoms in the crystal. The locations and intensities of the spots are stored electronically. Determination of the three-dimensional structure of the virus from the diffraction pattern requires information that is lost in the X-ray diffraction experiment. This missing information (the phases of the diffracted rays) can be retrieved by collecting the diffraction information from otherwise identical (isomorphous) crystals in which the phases have been systematically perturbed by the introduction of heavy metal atoms at known positions. Comparison of the two diffraction patterns yields the phases. This process is called **multiple isomorphous replacement**. Alternatively, if the structure of a related molecule is known, the diffraction pattern collected from the crystal can be interpreted by using the phases from the known structure as a starting point and subsequently using computer algorithms to calculate iteratively the actual values of the phases. This method is known as **molecular replacement**. Once the phases are known, the intensities and spot positions from the diffraction pattern are used to calculate the locations of the atoms within the crystal, again by using computer programs. **(B)** The product of this mathematical analysis is an electron density map, which is a map of the location of electron-dense atoms (C, N, O, and S), as well as of the covalent bonds within and between amino acids in the virus. **(C)** A model of the peptide backbone linking amino acids within proteins is built by tracing the electron density through each section of the map and specifying the location of each atom. The amino acids are identified by the shapes of the electron density of the side chains and by using the primary sequence of the proteins. **(D)** When the entire structure is built, the completed images can be reassembled and visualized in various representations. Courtesy of J. Hogle, Harvard Medical School, and M. Chow, University of Arkansas for Medical Sciences.

microscopy can improve resolution to 10 to 20 Å. Indeed, sufficient resolution (~6 Å) was attained in recent applications of these techniques to identify α-helices in structural proteins of adenovirus particles, in the core protein of hepatitis B virus, and in the transmembrane domains of viral glycoproteins. Within the past decade, cryo-electron microscopy has become a standard tool of structural biology. Its application to virus particles has provided a wealth of previously inaccessible information about the external and internal structures of multiple members of at least 20 virus families.

The diameter of an α-helix in a protein is on the order of 10 Å, so even the highest resolution attained by these sophisticated techniques cannot reveal the molecular interactions that cement structural units in the virion. Such information can be obtained by X-ray crystallography (Fig. 4.3), provided that the virus yields crystals suitable for X-ray diffraction. It has been known for more than 60 years that simple plant viruses can be crystallized, and the first high-resolution virus structure determined was that of tomato bushy stunt virus. Since this feat was accomplished in 1978, high-resolution structures of a number of increasingly larger animal viruses have been determined, placing our understanding of the principles of capsid structure on a firm molecular foundation. Although they cannot capture dynamic processes, atomic-level descriptions of specific viruses, or of individual viral proteins, have also greatly improved our knowledge of mechanisms of attachment and entry of virions and of virus assembly. They have also provided new opportunities for the design of antiviral drugs.

Not all viruses can be examined directly by X-ray crystallography: some do not form suitable crystals, and the larger viruses lie beyond the power of the current procedures by which X-ray diffraction spots are converted into a structural model (Fig. 4.3). Nevertheless, individual viral proteins can be examined by this method and by multidimensional nuclear magnetic resonance (NMR) techniques (Box 4.1). The latter methods, which allow structural models to be constructed from knowledge of the distances between specific atoms in a polypeptide chain (Fig. 4.4), can be applied to proteins in solution, a significant advantage. At present, NMR methods can be applied to only relatively small proteins (20 to 30 kilodaltons [kDa]), but their power is being expanded rapidly.

High-resolution structures of individual proteins have been particularly important in illuminating mechanisms of attachment and entry of enveloped viruses. However, even more valuable is the more recent development of methods in which high-resolution structures of individual viral proteins are combined with cryo-electron microscopy reconstructions of intact virus particles. For example, in difference imaging, the structures of individual proteins are in essence subtracted from the reconstruction of the particle to yield new structural information (Fig. 4.5). In the past few years, this powerful approach has provided fascinating new views of interactions of viral envelope proteins embedded in lipid bilayers, and even of internal surfaces and components of virus particles.

Building a Protective Coat

Regardless of their structural complexity, all virions contain at least one protein coat, the capsid or nucleocapsid (Table 4.2), that encases and protects the nucleic acid genome. As first pointed out by Francis Crick and James Watson in 1956, most viruses appear to be rod shaped or spherical under the electron microscope. Because the coding capacities of viral genomes are limited, these authors proposed that construction of capsids from a small number of subunits by using helical symmetry (rod-shaped viruses) or the symmetry of Platonic polyhedra (e.g., tetrahedron or icosahedron) (spherical viruses) would minimize the genetic cost of encoding structural proteins. Such genetic economy dictates that capsids and nucleocapsids be built from identical copies of a small number of viral proteins with structural properties that permit regular and repetitive interactions among them. These protein molecules are arranged to provide maximal contact and noncovalent bonding among subunits and structural units. The repetition of such interactions among a limited number of proteins results in a regular structure, with symmetry that is determined by the spatial patterns of the interactions. In fact, the protein coats of all but a few viruses display **helical** or **icosahedral symmetry**.

Helical Structures

The **nucleocapsids** of some enveloped animal viruses, as well as certain plant viruses and bacteriophages, are rodlike or filamentous structures with helical symmetry. Helical symmetry is described by the number of structural units per turn of the helix, μ, the axial rise per unit, ρ, and the pitch of the helix, P, given by the formula

$$P = \mu \times \rho$$

A characteristic feature of a helical structure is that any volume can be enclosed simply by varying the length of the helix. Such a structure is said to be **open.** In contrast, capsids with icosahedral symmetry (described below) are **closed** structures of fixed internal volume.

From a structural point of view, the best-understood helical nucleocapsid is that of tobacco mosaic virus, the first virus to be identified. The virus particle comprises a single molecule of (+) strand RNA, about 6.4 kb in length,

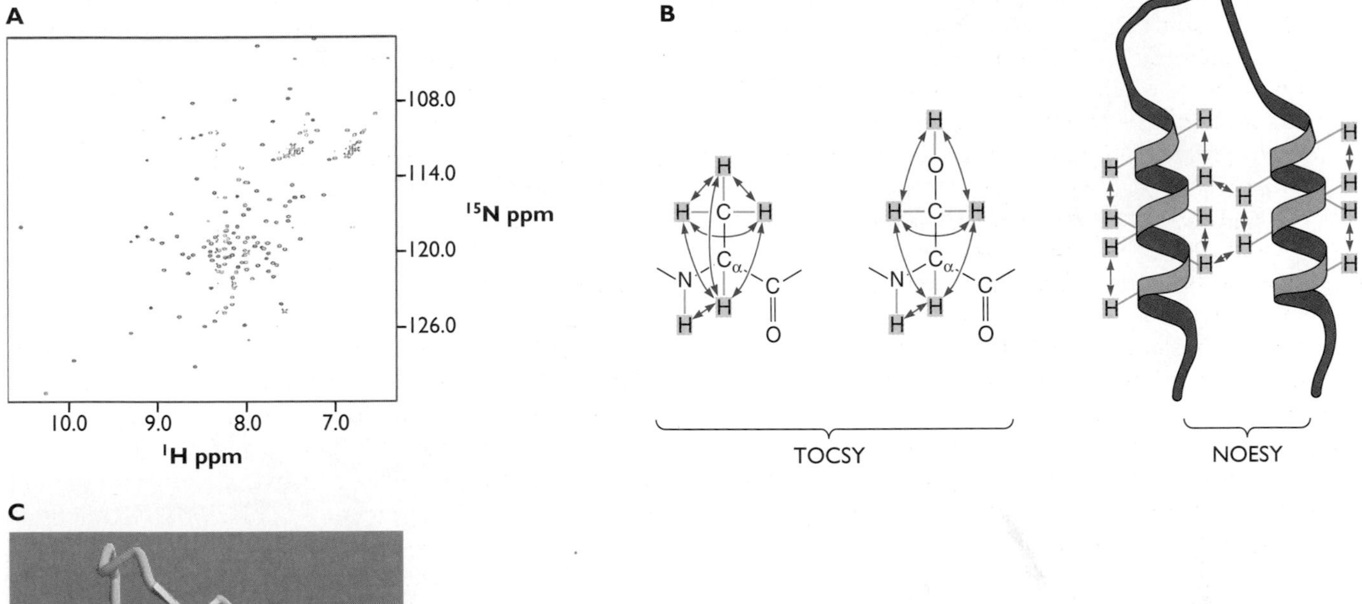

Figure 4.4 Information obtained from multidimensional nuclear magnetic resonance (NMR) spectroscopy. (A) Example showing part of a two-dimensional $^{1}H^{15}N$ heteronuclear, NMR spectrum of the human immunodeficiency virus type 1 matrix (MA) protein. The horizontal and vertical axes show the radiation absorbed by covalently linked protons and nitrogen atoms, respectively, expressed relative to reference signals and termed the **chemical shifts**. These signals are observed when the protein is placed in a strong magnetic field and exposed to radiofrequency pulses. The chemical shift is different for each atom and is determined in part by the molecular environment of its nucleus. The different atoms in a complex molecule like the MA protein therefore exhibit different chemical shifts in this kind of experiment. The initial problem is to resolve and assign the very large number of signals generated by even a small protein (~1,000 for a protein of 15 kDa). The assignment is typically made by using multidimensional experiments, which allow individual protons to be resolved on the basis of several different chemical shift indices, ^{15}N and ^{1}H in the spectrum shown. **(B)** Proton-proton correlations obtained from $^{1}H,^{1}H$ total correlation spectroscopy (TOCSY), and nuclear Overhauser effect spectroscopy (NOESY) NMR spectra. TOCSY reveals correlations between protons that are covalently connected via only one or two additional atoms, and therefore identifies interactions among protons within the same amino acid in a protein; hydrogen atoms in adjacent amino acid are connected by at least three other atoms. A TOCSY experiment therefore generates a characteristic set of linked signals for each amino acid, because each bears a unique side chain. NOESY identifies correlations between protons that are closer than 5 Å in space, regardless of whether they are closely linked in the primary sequence. Secondary-structure elements, such as α-helices, generate characteristic sets of NOE signals. NOE signals also provide information about the tertiary structure of the protein because they place constraints on the distances between specific pairs of hydrogen atoms in the protein. **(C)** When sufficient NOE constraints have been collected and assigned, it is possible to produce structural models of the protein that are consistent with these constraints, as shown for the human immunodeficiency virus type 1 MA protein. (A and C) Courtesy of M. F. Summers, University of Maryland, Baltimore County, and W. I. Sundquist, University of Utah. (B) Adapted from C. Branden and J. Tooze, *An Introduction to Protein Structure* (Garland Publishing, Inc., New York, NY, 1991), with permission.

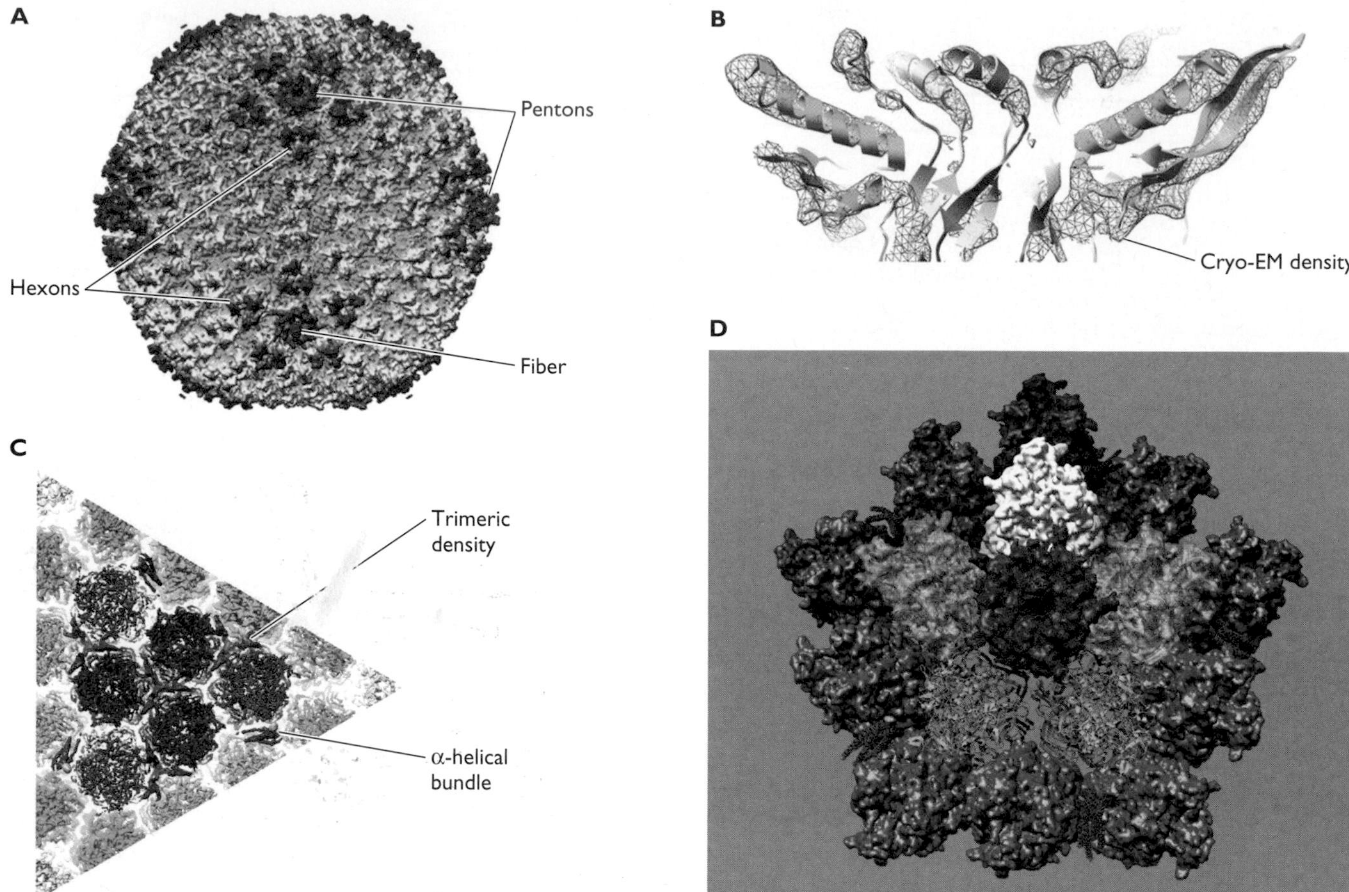

Figure 4.5 Difference mapping illustrated by a 6-Å-resolution reconstruction of adenovirus. (A) Surface view along a twofold symmetry axis of the 6-Å-resolution cryo-electron microscopy reconstruction of a derivative of adenovirus type 5 carrying the Ad35 fiber. As this fiber is flexible, only a short portion protruding from the penton base is visible in the reconstruction. The density is radially color coded (red = 596 Å; blue = 316 Å). **(B)** Comparison of α-helices of the penton base in the cryo-electron microscopic density and crystal structure of this protein bound to a fiber peptide (ribbon). The excellent agreement shown established that α-helices could be detected reliably in the 6-Å cryo-electron microscopy reconstruction. **(C)** Portion of the cryo-electron microscopy difference map corresponding to the surface of one icosahedral face of the capsid. The crystal structures of the penton base (yellow) and the hexon (green, cyan, blue, and magenta at different positions) at appropriate resolution were docked within the cryo-electron microscopic density at 6-Å resolution. The cryo-electron microscopic density that does not correspond to these structural units (the difference map) is shown in red. At this resolution, the difference map revealed four trimeric structures located between neighboring hexons and three bundles of coiled-coiled α-helices. The former were previously assigned to protein IX. Predictions of secondary structure and formation of α-helical coiled coils indicated that the helical bundles at the edges of the icosahedral face are formed by the C-terminal half of protein IX. **(D)** One vertex region of the capsid illustrating the α-helical coiled coil assigned to the C-terminal domain of protein IX (magenta). The penton base and fiber are in red; peripentonal hexons are in white, gray, or cyan (depending on how they are represented); and other hexons are in blue. Adapted from S. D. Saban et al., *J. Virol.* **80**:12049–12059, 2006, with permission. Courtesy of Phoebe Stewart, Vanderbilt University Medical Center.

BOX 4.1

DISCUSSION

Human immunodeficiency virus type 1, a virus understood in great structural detail

This causative agent of acquired immunodeficiency syndrome (AIDS) was identified about 25 years ago. As illustrated in the figure, the structures of most of the proteins present in the complex virions have been solved at high resolution by one or more of the methods described in the text. The need to develop effective drugs for the treatment of AIDS has provided an important impetus for structural studies of human immunodeficiency virus type 1 proteins.

Structures of human immunodeficiency virus type 1 proteins. The virion is depicted at the center, with the structures of the component proteins illustrated on either side in surface representation. RNA is green. Adapted from H. Berman et al., *Am. Sci.* **90**:350–359, 2002, with permission.

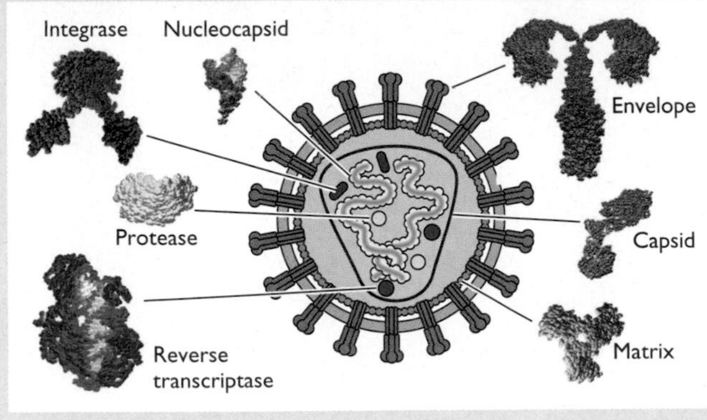

enclosed within a helical protein coat (Fig. 4.6A; see also Fig. 1.7). The coat is built from a single protein that folds into an extended structure shaped like a Dutch clog. Repetitive interactions among coat protein subunits form disks that have been likened to lock washers, which in turn assemble as a long, rodlike, right-handed helix with 16.3 coat protein molecules per turn. In the interior of the helix, each coat protein molecule binds three nucleotides of the RNA genome. The coat protein molecules therefore engage in **identical**, equivalent interactions with one another and with the genome, allowing the construction of a large, stable structure from multiple copies of a single protein subunit.

The virions of several families of animal viruses with (–) strand RNA genomes, including paramyxoviruses, rhabdoviruses, and orthomyxoviruses, contain internal structures with helical symmetry encased within an envelope. In all cases, these structures contain an RNA molecule, many copies of an RNA-packaging protein (designated NP or N), and the viral RNA polymerase and associated enzymes responsible for synthesis of mRNA. Despite common helical symmetry and similar composition, the internal components of these (–) strand RNA viruses exhibit considerable diversity in morphology and organization. Like tobacco mosaic virus particles, the nucleocapsids of paramyxoviruses and rhabdoviruses contain a single molecule of RNA (of about 15 and 11 kb, respectively) tightly associated with the nucleocapsid protein. Nucleocapsids of paramyxoviruses, such as Sendai virus, are long, filamentous structures in which the RNA and NP protein form a left-handed helix with a hollow core (Fig. 4.6B). In contrast, nucleocapsids of rhabdoviruses such as vesicular stomatitis virus are squat, bullet-shaped structures closed at one end (Fig. 4.6C). Furthermore, an additional virion protein is essential to maintain their organization. Vesicular stomatitis virus nucleocapsids released from within the virion envelope retain the dimensions and morphology observed in intact particles, but become highly extended and filamentous once the matrix (M) protein is also removed (Fig. 12.22). The inherent flexibility of the resulting ribonucleoprotein, which contains the RNA-packaging nucleocapsid (N) protein, as well as the viral polymerase (L) and phosphoprotein (P), has precluded high-resolution structural studies. However, it has been possible to determine the X-ray crystal structure of a ring-like N protein-RNA complex containing 10 molecules of the N protein bound to a 90-nucleotide RNA (Fig 4.7). In this complex, each N protein molecule binds to 9 nucleotides of RNA, which is largely sequestered within cavities within each protein. Furthermore, each N protein makes extensive and regular contacts with neighboring N molecules, exactly as predicted from first considerations by Crick and Watson.

The internal components of influenza A virus particles differ more radically. In the first place, they comprise not a single nucleocapsid but, rather, multiple ribonucleoproteins, one for each molecule of the segmented RNA genome present in the virion (Appendix, Fig. 8). The viral NP protein organizes each RNA into a helical structure analogous to the nucleocapsids of viruses with nonsegmented (–) strand RNA genomes. However, with the exception of terminal sequences, the RNA in these ribonucleoproteins is fully accessible to solvent. This property suggests that the RNA is wound **around** a helical core formed by the NP protein rather than being sequestered within a helical structure. Each ribonucleoprotein is further folded into a compact, circular conformation as a result of the binding of the virion enzyme complex (P proteins) to specific sequences conserved at the 5′ and 3′ ends of each (–) strand RNA molecule.

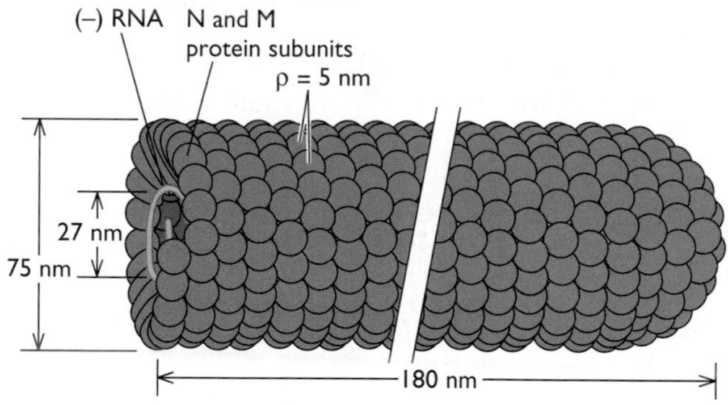

Figure 4.6 Virus structures with helical symmetry. (A) Tobacco mosaic virus. The structure of this virus has been determined at high resolution by X-ray diffraction of fibers or oriented gels of the particles. The single genomic RNA molecule and the single coat protein form an extended right-handed helix with 16.3 protein subunits per turn, μ, and an axial rise per residue, ρ, of 0.14 nm, for a pitch, *P*, of 2.3 nm. The lock-washer-like broken-disk organization of the proteins in a single turn of the helix can be seen in the expanded view. **(B and C)** The structures of Sendai virus (a paramyxovirus) and vesicular stomatitis virus (a rhabdovirus) nucleocapsids are based on electron microscopy of virions or nucleocapsids released from them. The Sendai virus nucleocapsid resembles the tobacco mosaic virus particle in diameter and the presence of a hollow core, but it is in the form of a left-handed helix with distinctive helical parameters ($\mu = 13$, $\rho = 0.41$ nm, $P = 5.3$ nm). The vesicular stomatitis virus nucleocapsid is also hollow and comprises about 30 turns of uniform diameter followed by 5 or 6 turns of decreasing diameter.

Capsids or Nucleocapsids with Icosahedral Symmetry

General Principles

Icosahedral symmetry. An icosahedron is a solid with 20 triangular faces and 12 vertices related by two-, three-, and fivefold axes of rotational symmetry (Fig. 4.8A). In a few cases, virus particles can be readily seen to be icosahedral (for examples of such particles, see Fig. 4.15 and 4.25). However, most closed capsids and nucleocapsids **look** spherical, and they often possess prominent surface structures or viral glycoproteins in the envelope that do not conform to the underlying icosahedral symmetry of the capsid shell. Nevertheless, the symmetry with which the structural units interact is that of an icosahedron.

In solid geometry, each of the 20 faces of an icosahedron is an equilateral triangle (the triangle is one of the three basic units that can form solid structures approximating

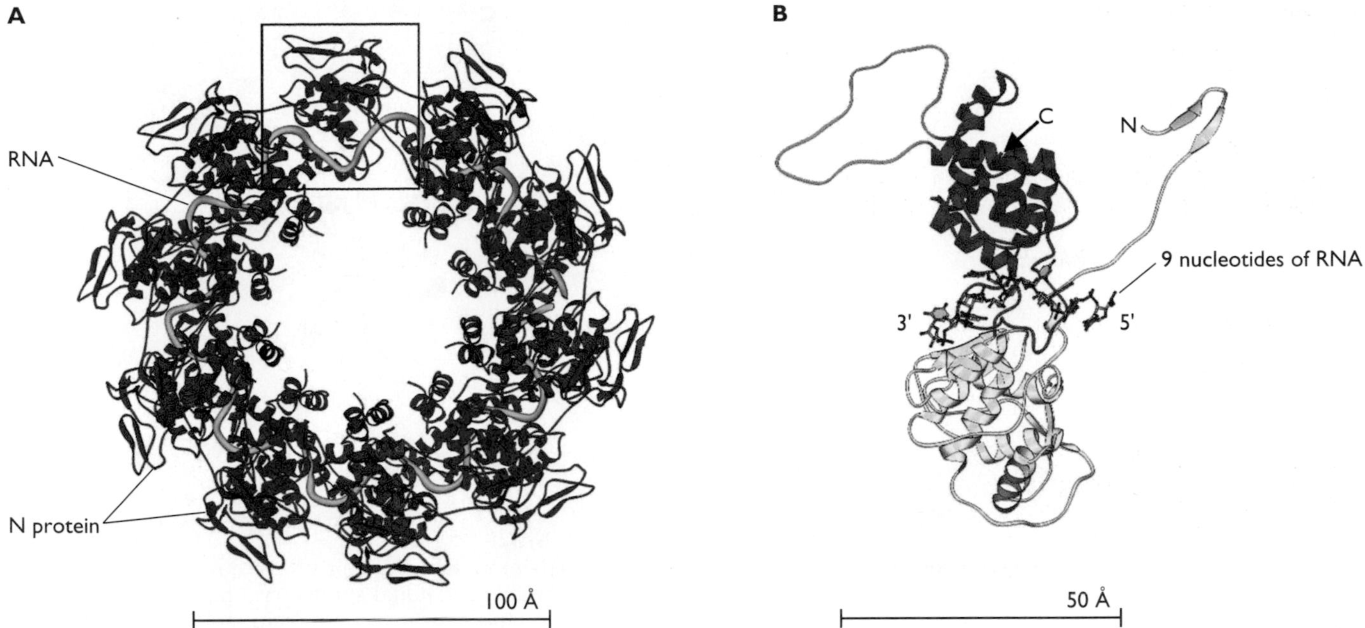

Figure 4.7 Structure of a ribonucleoprotein-like complex of vesicular stomatitis virus. (A) The structure of the decamer of the N protein bound to RNA is shown, with alternating monomers in the ring colored red and blue and the RNA ribose-phosphate backbone depicted as a green tube. To allow visualization of the RNA, the C-terminal domain of the boxed monomer is not shown. The decamer was isolated by dissociation of the viral P protein from RNA-bound oligomers formed when the N and P proteins were synthesized in *Escherichia coli*. **(B)** A single N protein molecule bound to 9 nucleotides of RNA is shown, looking out from the inside of the ring and perpendicular to the view in panel A. The RNA, depicted as a ball-and-stick model, is tightly bound at the interface between the N- and C-terminal lobes of the N protein, which are colored yellow and orange, respectively. The N-terminal extension and the extended loop in the C-terminal lobe contribute to the extensive interactions among neighboring N monomers. Courtesy of M. Luo, University of Alabama at Birmingham. Adapted from T. J. Green et al., *Science* **313**:357–360, 2006, with permission.

a sphere). Five such triangles interact at each of the 12 vertices of the icosahedron (Fig 4.8A). In the simplest protein shells, a trimer of a single viral protein (the **subunit**) corresponds to each triangular face of the icosahedron: as shown in Fig 4.8B, such trimers interact with one another at the five-, three-, and twofold axes of rotational symmetry that define an icosahedron. As an icosahedron has 20 faces, 60 identical subunits (3 per face × 20 faces) is the minimum number needed to build a capsid or nucleocapsid with icosahedral symmetry. Such well-defined symmetry poses a number of interesting questions about the organization and assembly of icosahedral viruses. This property may also prove to have great practical value (Box 4.2).

Large capsids and quasiequivalent bonding. In the simple icosahedral packing arrangement shown in Fig. 4.8B, each of the 60 subunits (**structural** or **asymmetric units**) consists of a single molecule in a structurally identical environment. Consequently, all subunits interact with their neighbors in an identical (or **equivalent**) manner, just like the subunits of a helical nucleocapsid such as that of tobacco mosaic virus. The size of a protein shell constructed according to this simplest icosahedral design is therefore determined by the size of its protein subunits. As the viral proteins that form such closed shells are generally less than ~100 kDa in molecular mass, the size of the viral genome that can be accommodated in this type of particle is restricted severely. In fact, the capsids or nucleocapsids of the majority of animal viruses are built from many more than 60 subunits and can house quite large genomes. In 1962, Donald Caspar and Aaron Klug developed a theoretical framework accounting for the structural properties of larger particles with icosahedral symmetry. This theory has had enormous influence on the way virus structures are described and interpreted.

The triangulation number *T*. A crucial idea introduced by Caspar and Klug was that of **triangulation**, the description of the triangular face of a large icosahedral structure in terms of its subdivision into smaller triangles, termed **facets**. In the example shown in Fig. 4.9, four such

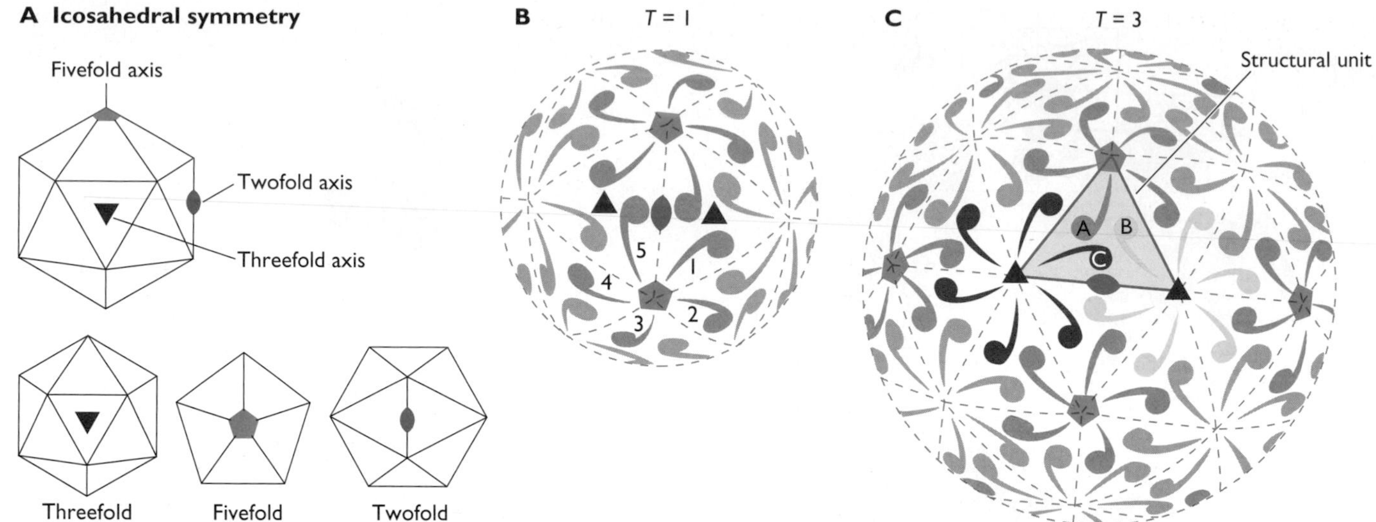

Figure 4.8 Icosahedral packing in simple structures. (A) An icosahedron, which comprises 20 equilateral triangular faces characterized by positions of five-, three-, and twofold rotational symmetry. The three views at the bottom illustrate these positions. **(B and C)** A comma represents a single protein molecule, and axes of rotational symmetry are indicated as in panel A. In the simplest and ideal case, $T = 1$ **(B)**, the protein molecule forms the structural unit, and each of the 60 molecules is related to its neighbors by the two-, three-, and fivefold rotational axes that define a structure with icosahedral symmetry. In such a simple icosahedral structure, the interactions of all molecules with their neighbors are identical. In the $T = 3$ structure **(C)** with 180 identical protein subunits, there are three modes of packing of a subunit (shown in orange, yellow, and purple): the structural unit (outlined in blue) is now the asymmetric unit, which, when replicated according to 60-fold icosahedral symmetry, generates the complete structure. The orange subunits are present in pentamers, formed by tail-to-tail interactions, and interact in rings of three (head to head) with purple and yellow subunits, and in pairs (head to head) with a purple or a yellow subunit. The purple and yellow subunits are arranged in rings of six molecules (by tail-to-tail interactions) that alternate in the particle. Despite these packing differences, the bonding interactions in which each subunit engages are similar, that is, **quasiequivalent:** for example, all engage in tail-to-tail and head-to-head interactions. Adapted from S. C. Harrison et al., *in* B. N. Fields et al. (ed.), *Fundamental Virology* (Lippincott-Raven, New York, NY, 1995), with permission.

facets (each equivalent to one face of the simplest structure shown in Fig. 4.8B) are combined to assemble a larger face of an icosahedrally symmetric structure from the same homotrimer (structural unit). This process is described by the triangulation number, *T*, which gives the number of structural units (small "triangles") per face. Because the minimum number of subunits required is 60, the total number of subunits in the structure is 60*T*.

Quasiequivalence. A second cornerstone of the theory developed by Caspar and Klug was the proposition that when a capsid contains more than 60 subunits, each subunit occupies a **quasiequivalent position**; that is, the noncovalent bonding properties of subunits in different structural environments are **similar** (but not identical, as is the case for the simplest, 60-subunit structure). This property is illustrated in Fig. 4.8C for a particle with 180 identical subunits. In the small, 60-subunit structure, 5 subunits make fivefold symmetric contact at each of the 12 vertices (Fig. 4.8B). In the larger structure with 180 subunits, this arrangement is retained at the 12 vertices, but the additional subunits, arranged with sixfold symmetry, are interposed between the fivefold-symmetric clusters (Fig. 4.8C). In such a structure, each subunit can be present in one of three **different** structural environments (designated A, B, or C in Fig. 4.8C). Nevertheless, all subunits bond to their neighbors in similar (**quasiequivalent**) ways, for example, via head-to-head and tail-to-tail interactions (Fig. 4.8C). Likewise, in a $T = 4$ structure (Fig. 4.9), the packing interactions of the 240 subunits are very similar. It is important to appreciate that there is not a simple and direct relationship between structural (asymmetric) units and icosahedral faces (Box 4.3). However, the *T* number defines the number of subunits in a face.

Icosahedrally symmetric structures can be formed from quasiequivalent subunits for only certain values of *T* (Box 4.4). Virus structures corresponding to various values of *T*, some very large, have been described (Table 4.3). The triangulation number and quasiequivalent bonding among subunits describe the structural properties of many simple

BOX 4.2 METHODS
Nanoconstruction with virus particles

Nanochemistry is the synthesis and study of well-defined structures with dimensions of 1 to 100 nm. Nano-building blocks span the size range between molecules and materials. Molecular biologists study nanochemistry, nanostructures, and molecular machines including the ribosome, the photosynthetic center, and membrane-bound signaling complexes. Icosahedral virus particles are proving to be precision building blocks for nanochemistry. The $T = 1$ icosahedral cowpea mosaic virus particle is 30 nm in diameter, and its atomic structure is known in detail. Grams of virions can be prepared easily from kilograms of infected leaves. Insertional mutagenesis is straightforward, and precise amino acid changes can be introduced. As illustrated in the figure, cysteine residues inserted in the capsid protein provide functional groups for chemical attachment of 60 precisely placed molecules.

High local concentrations of the attached chemical agent, coupled with precise placement, enable rather remarkable nanoconstruction. For example, virus surfaces can be patterned with metal nanoparticles that may function as a conducting "wire" for electronics. In addition, the propensity of virions for self-organization into two- and three-dimensional lattices leads to well-ordered arrays of 10^{13} particles in a 1-mm^3 crystal.

Viruses are not just for infections any more! They will provide a rich source of building blocks for applications spanning the worlds of molecular biology and materials science.

Nam, K. T., D. W. Kim, P. J. Yoo, C. Y. Chiang, N. Meethong, P. T. Hammond, Y. M. Chiang, and A. M. Belcher. 2006. Virus-enabled synthesis and assembly of nanovirus for lithium ion battery electrodes. *Science* **312:**885–888.

Wang, Q., T. Lin, L. Tang, J. E. Johnson, and M. G. Finn. 2002. Icosahedral virus particles as addressable nanoscale building blocks. *Angew. Chem. Int. Ed. Engl.* **41:**459–462.

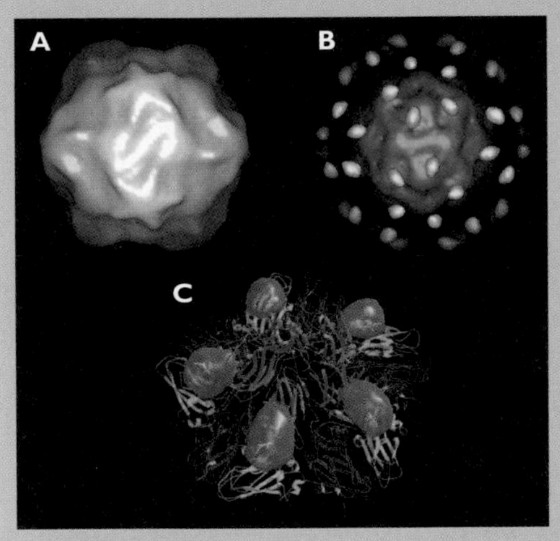

Gold particles attached to cowpea mosaic virus. Cryo-electron microscopy was performed on derivatized cowpea mosaic virus with a cysteine residue inserted on the surface of each of the 60 subunits and to which nanogold particles with a diameter of 1.4 nm were chemically linked. **(A)** Three-dimensional reconstruction of such derivatized virus particles at 29-Å resolution. **(B)** Difference electron density map obtained by subtracting the density of unaltered cowpea mosaic virus at 29 Å from the density map shown in panel A. This procedure reveals both the genome (shown in green) and the gold nanoparticles. **(C)** A section of the difference map imposed on the atomic model of cowpea mosaic virus. The positions of the gold indicate that it is attached at the sites of the introduced cysteine residues. Courtesy of M. G. Finn and J. Johnson, The Scripps Research Institute.

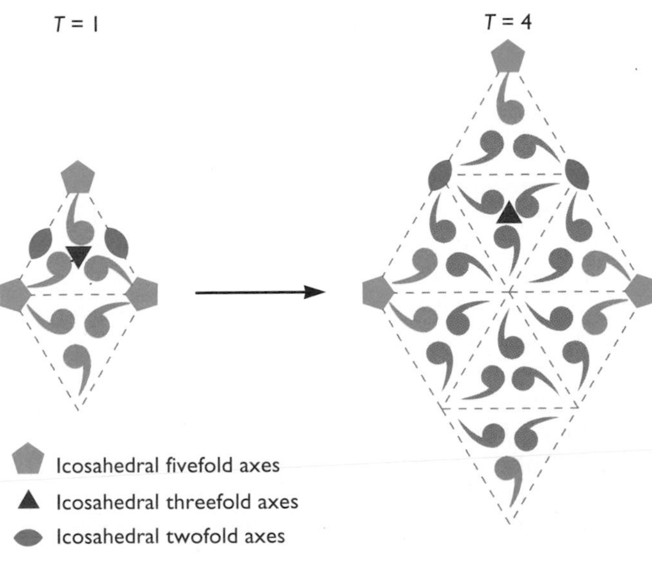

Figure 4.9 $T = 4$ triangulation of an icosahedral face. Two faces of a $T = 1$ icosahedron, each composed of three protein subunits, are shown flattened into a single plane on the left. The two-, three-, and fivefold rotational axes that define icosahedral symmetry are indicated on the top face. The combination of four identical faces (now facets) to form a $T = 4$ face is shown on the right. The icosahedral symmetry axes are in the same relative positions as in the $T = 1$ face. In such a $T = 4$ structure, the size of the face, which is defined by the length of one of its sides, is twice that in the $T = 1$ structure. Adapted from S. Casjens, *in* S. Casjens (ed.), *Virus Structure and Assembly* (Jones and Bartlett Publishers Inc., Boston, MA, 1985), with permission.

BOX 4.3

TERMINOLOGY

Structural (asymmetric) units and faces of virus particles with icosahedral symmetry: not one and the same

The structural (asymmetric) unit of an icosahedral capsid is the **smallest** unit that can give rise to the complete structure when replicated according to the rules of icosahedral symmetry. Therefore

- the capsid must contain 60 copies of the structural unit
- five structural units must interact around each of the axes of fivefold rotational symmetry.

In the $T = 1$ structure shown in Fig. 4.8, the structural unit is a single protein subunit, whereas it is a homotrimer in the $T = 3$ structure.

An icosahedron has 20 faces. Nevertheless, there is **not** a simple relationship between a structural (asymmetric) unit and a face, as shown in the figure.

On the other hand, the ***T*** number defines the number of subunits in each face:

T number	No. of subunits (60 × T)	No. of subunits/face [(60 × T)/20]
1	60	3
3	180	9
4	240	12

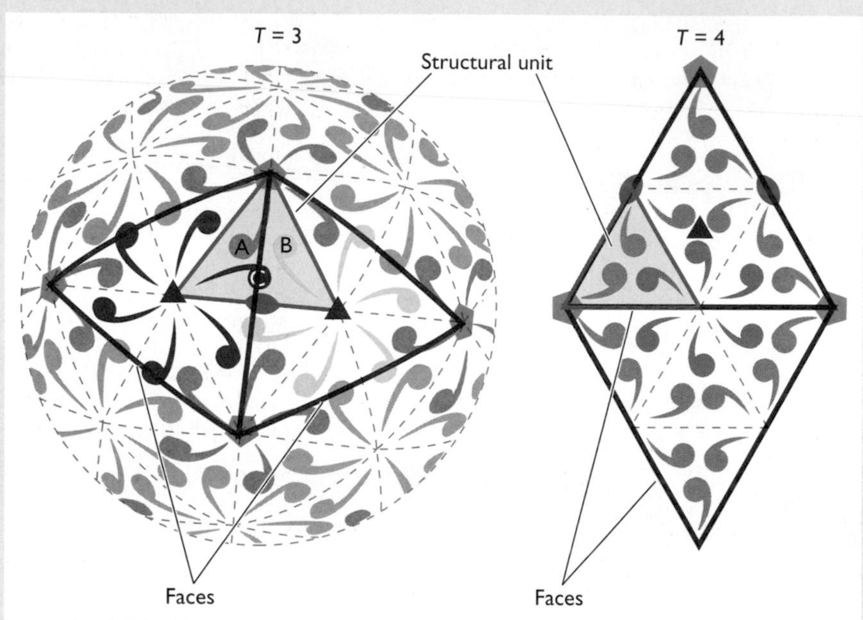

Structural units and faces of icosahedral virus particles. A $T = 3$ particle and part of a $T = 4$ particle are depicted as in Fig. 4.8 and 4.9, respectively. Structural asymmetric units are outlined in blue, and faces are shown in red.

viruses with icosahedral symmetry. However, it is now clear that the structures adopted by specific segments of capsid proteins can govern the packing interactions of identical subunits. This property indicates that interactions among chemically identical protein subunits must be regulated during assembly, with appropriate conformational switching of flexible segments. Such large conformational differences between small regions of chemically identical (and otherwise quasiequivalent) subunits were not anticipated in early considerations of virus structure. This omission is not surprising, for these principles were formulated when little was known about the structural properties of proteins. As we discuss in the next sections, the structural properties of both small and more complex viruses can depart radically from the constraints imposed by quasiequivalent bonding. For example, the capsid of the small polyomavirus simian virus 40 is built from 360 subunits, corresponding to the $T = 6$ triangulation number excluded by the rules formulated by Caspar and Klug (Box 4.4). Moreover, a capsid stabilized by covalent joining of subunits to form viral "chain mail" has been described (Box 4.5). Our current view of icosahedrally symmetric virus structures is therefore one that includes greater diversity in the mechanisms by which stable capsids can be formed than was anticipated by the pioneers in this field.

Structurally Simple Capsids

Several nonenveloped animal viruses are small enough to be amenable to high-resolution structural studies by X-ray crystallography. We have chosen four examples, the parvovirus adenovirus-associated virus 2, the nodavirus Nodamura virus, the picornavirus poliovirus, and the polyomavirus simian virus 40, to illustrate the molecular foundations of icosahedral architecture.

Structure of adeno-associated virus 2: classic $T = 1$ icosahedral design. The parvoviruses are very small animal viruses, with particles of ~25 nm in diameter that encase single-stranded DNA genomes of less than 5 kb. These small naked nucleocapsids are built from 60 copies of a single subunit organized according to $T = 1$ icosahedral symmetry. The structure of adenovirus-associated virus type 2, a member of the dependovirus subgroup of parvoviruses that require a helper virus for replication (Appendix, Fig. 11), has been determined at high resolution by X-ray crystallography. The subunits that form the nucleocapsid contain a three-dimensional motif, the **β-barrel jelly roll** that is conserved in the structural proteins of unrelated viruses. In adenovirus-associated virus particles, the interactions among neighboring subunits are mediated by loops that connect the β-strands in the β-barrel. Some of these loops are long and contain additional β-strands or α-helices (Fig. 4.10A). As illustrated in Fig. 4.10B, the interactions among them result in formation of a particle with classic $T = 1$ icosahedral symmetry. The prominent projections near the threefold axes of rotational symmetry (Fig. 4.10B), which have been implicated in receptor binding of adenovirus-associated virus type 2, are formed by extensive interdigitation among loops from adjacent subunits.

Structure of Nodamura virus: typical $T = 3$ icosahedral design. The nodaviruses are small (290 to 320 Å in diameter), nonenveloped viruses that infect mammals, fish, and insects. The capsid of Nodamura virus (Fig. 4.11A) encases the (+) strand RNA genome, which comprises two molecules of RNA. It is built from 180 copies of a single coat protein, organized according to a $T = 3$ quasiequivalent design: the 60 structural units contain one copy of each of three types of coat protein subunit that are defined by occupancy of structurally distinct environments (A, B, and C in Fig. 4.11B). The A subunits are arranged as pentamers around the 12 axes of fivefold rotational symmetry, whereas the B and C subunits alternate in hexameric rings around the axes of threefold symmetry (Fig. 4.11B; compare with Fig. 4.8B). Despite the differences in subunit packing, interactions among A, B, and C subunits within a structural unit are very similar. However, adjacent subunits in neighboring asymmetric units engage in either flat or bent contacts (Fig. 4.11B). This difference is the result of the presence of a short, ordered protein segment only in the C subunits and of a duplex segment of the RNA genome at the flat contacts. These structural features hold the neighboring subunits farther apart than at the bent contacts, so that the asymmetric units form a flat, diamond-shaped structure (Fig. 4.11B). Whether internal surfaces of adjacent subunits interact with ordered segments of the RNA genome is therefore a crucial determinant of the quasiequivalent, icosahedral architecture of Nodamura virus (and other members of the *Nodaviridae*).

Structure of poliovirus: a pseudo $T = 3$ structure. As their name implies, the picornaviruses are among the smallest of animal viruses. The approximately 300-Å-diameter poliovirus particle is composed of 60 copies of each of four virally encoded structural proteins, VP1, VP2, VP3, and VP4 (VP = virion protein). These proteins form a closed capsid encasing the (+) strand RNA genome of about 7.5 kb and its covalently attached 5′-terminal protein, VPg (Appendix, Fig. 13). Our understanding of the structure of the *Picornaviridae* took a quantum leap in 1985 with the determination of high-resolution structures of human rhinovirus 14 (genus *Rhinovirus*) and poliovirus (genus *Enterovirus*).

The heteromeric structural unit of the poliovirus capsid contains one copy each of VP1, VP2, VP3, and VP4 (Fig. 4.12A). VP4 lies on the inner surface of the protein shell formed by VP1, VP2, and VP3. Although the three latter proteins are not related in amino acid sequence, all contain a central β-sheet structure termed a **β-barrel jelly roll**. The organization of this β-barrel domain is illustrated schematically in Fig. 4.12B, for comparison with the actual structures of VP1, VP2, and VP3. It is a wedge-shaped structure comprising two antiparallel β-sheets. One of the β-sheets forms one wall of the wedge, while the second, sharply twisted β-sheet forms both the second wall and the floor (Fig. 4.12B). The β-barrel domains of VP1, VP2, and VP3 are folded in the same way; that is, they possess the same **topology**. The differences among these proteins are therefore restricted largely to the loops that connect β-strands and to the N- and C-terminal segments that extend from the central β-barrel domains (Fig. 4.12B).

The β-barrel jelly rolls of these picornaviral proteins are similar in structure to the core domains of capsid proteins of a number of plant, insect, and vertebrate (+) strand RNA viruses such as tomato bushy stunt virus and Nodamura virus. This property was entirely unanticipated. Even more remarkably, this relationship is not restricted to small RNA viruses: the major capsid proteins of the DNA-containing parvoviruses, polyomaviruses, and adenoviruses also contain such β-barrel domains. In fact, the β-barrel jelly roll continues to be the structural motif most commonly encountered as the structures of additional capsid proteins are determined. It is well established that the three-dimensional structures of cellular proteins have been highly conserved during evolution, even though there may be very little amino acid sequence identity. For example, all globins possess a common three-dimensional structure based on eight α-helices, even though the only residues conserved are those important for function. One interpretation of the common occurrence of the β-barrel jelly roll domain in viral capsid proteins is therefore that seemingly unrelated modern viruses (e.g., picornaviruses and adenoviruses) share some portion of their evolutionary history. It is also possible that this structural domain represents one of a limited number commensurate with

BOX 4.4

BACKGROUND

The triangulation number, T, and how it is determined

A $T = 1$

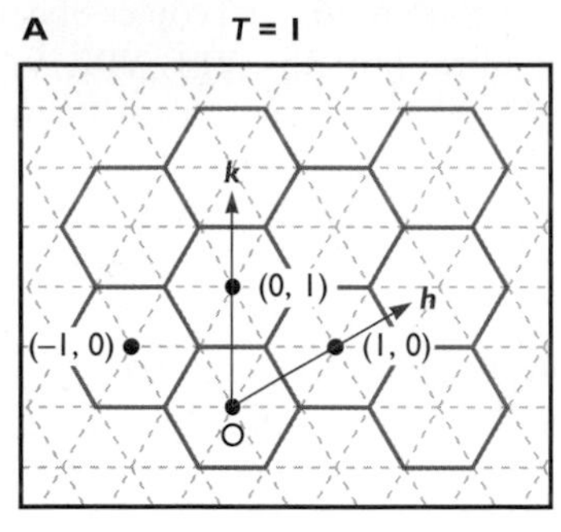

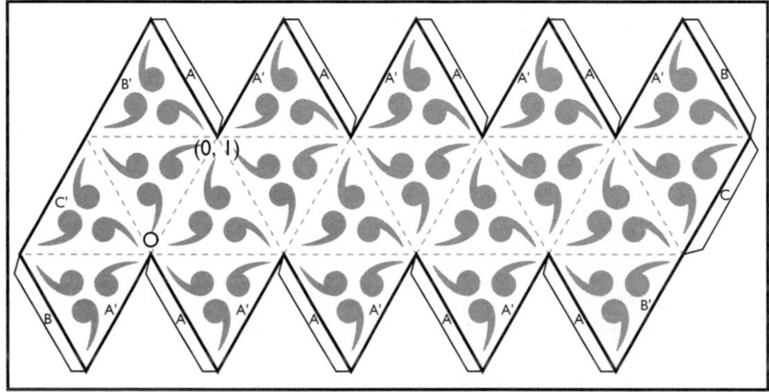

B $T = 3$

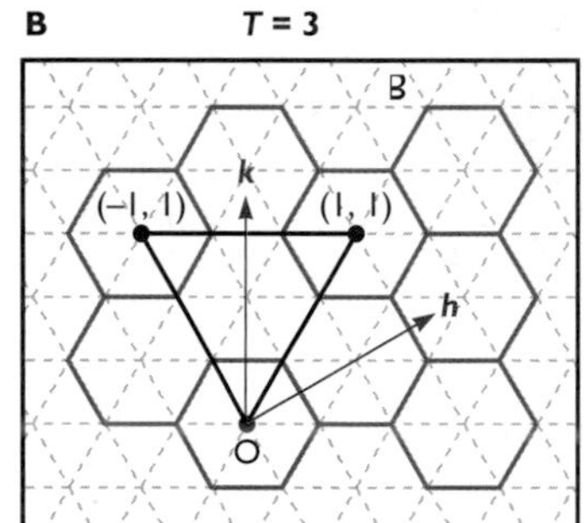

In developing their theories about virus structure, Caspar and Klug used graphic illustrations of capsid subunits, such as the net of flat hexagons shown at the top of panels **A** and **B**. Each hexagon represents a hexamer, with identical subunits shown as equilateral triangles. When all subunits assemble into such hexamers, the result is a flat sheet, or lattice, which can never form a closed structure. However, curvature can be introduced into the hexagonal net by converting specific hexamers to pentamers. As an icosahedron has 12 axes of fivefold symmetry, 12 pentamers must be introduced to form a closed structure with icosahedral symmetry. If 12 adjacent hexamers are converted to pentamers **(A)**, an icosahedron of the minimal size possible for the net is formed. This structure is built from 60 equilateral-triangle asymmetric units and corresponds to a $T = 1$ icosahedron. Larger structures with icosahedral symmetry are built by converting 12 **nonadjacent** hexamers to pentamers at precisely spaced and regular intervals. To illustrate this, we use nets in which an origin (O) is fixed and the positions of all other hexamers are defined by the coordinates along the axes labeled h and k, where h and k are any integers greater than or equal to zero (that is, positive integers). The hexamer (h, k) is therefore defined as that reached from the origin (O) by h steps in the direction of the h axis and k steps in the direction of the k axis. In the $T = 1$ structure, $h = 1$ and $k = 0$ (or $h = 0$ and $k = 1$), and adjacent hexamers are converted to pentamers (A, top left). To construct a model of an icosahedron when $h = 1$ and $k = 1$ **(B)**, we generate one of its faces in the net by conversion of the origin (O) and (1, 1) hexamers to pentamers. The lattice point (1, 1) shown is reached by one step in each direction. The third hexamer replaced (–1, 1) is that identified by threefold symmetry to complete the large, marked equilateral triangle (black) representing the face of an icosahedron formed when $h = 1$ and $k = 1$ (**B**, top left). The resulting quasiequivalent lattice (**B**, bottom panel) and the $T = 1$ lattice (**A**, bottom panel) can be folded to form the icosahedra shown, by excision from the sheet and sequential joining of the edges marked A to D with those marked A′ to D′, respectively.

The triangulation number, *T*, is the number of asymmetric units per face on icosahedron constructed in this way. It can be shown, for example by geometry, that

$$T = h^2 + hk + k^2$$

Therefore, when both *h* and *k* are 1 **(B)**, $T = 3$, and each face of the icosahedron contains three asymmetric units. The total number of units, which must be $60T$ (see the text), is 180.

The integers *h* and *k* describe the spacing and spatial relationships of pentamers in the lattice and of fivefold vertices in the corresponding icosahedron. For example, when $h = 1$ and $k = 0$, or when $h = 0$ and $k = 1$, pentamers are in direct contact with other pentamers (see, e.g., Fig. 4.8B), but when $h = 1$ and $k = 1$, each pentamer is separated from neighboring pentamers by one hexamer (Fig. 4.8C). The values of *h* and *k* are therefore determined by inspection of electron micrographs of virus particles or their constituents **(C)**. For example, in the bacteriophage p22 capsid (top), one pentamer is separated from another by two steps (from one structural unit to the next) along the *h* axis and one step along the *k* axis, as illustrated for the bottom left pentamer shown. Hence, $h = 2$, $k = 1$, and $T = h^2 + hk + k^2 = 7$. In contrast, pentamers of the herpes simplex virus type 1 (HSV-1) nucleocapsid (bottom) are separated by four and zero steps along the directions of the *h* and *k* axes, respectively. Therefore, $h = 4$, $k = 0$, and $T = 16$. Cryo-electron micrographs of bacteriophage P22 and herpes simplex virus type 1 courtesy of B. V. V. Prasad and W. Chiu (Baylor College of Medicine), respectively. (A and B) Adapted from Fig. 2 of J. E. Johnson and A. J. Fisher, *in* R. G. Webster and A. Granoff (ed.), *Encyclopedia of Virology*, 3rd ed. (Academic Press, London, England, 1994), with permission.

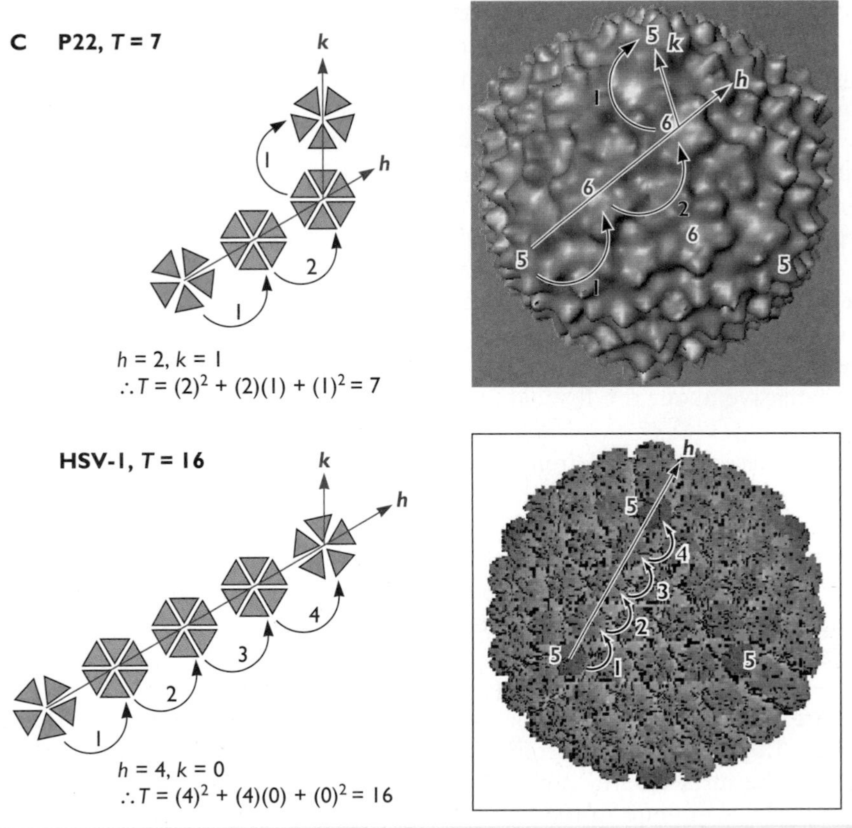

Table 4.3 *T* numbers of representative viruses

Triangulation no. (T)[a]	Family	Member(s)	60T	Protein(s) that forms the capsid or nucleocapsid shell
1	*Parvoviridae*	Canine parvovirus	60	60 copies of VP2
3	*Nodaviridae*	Black beetle virus	180	180 copies of coat protein
3 (pseudo)	*Picornaviridae*	Poliovirus, human rhinovirus	180	60 copies each of VP1, VP2, and VP3
4	*Alphaviridae*	Ross River virus	240	240 copies of C protein
7	*Polyomaviridae*	Simian virus 40, polyomavirus	420	360 copies of VP1
13	*Reoviridae*	Reovirus (outer shell)	780	60 copies of λ2, 600 copies of μ1
16	*Herpesviridae*	Herpes simplex virus type 1	960	960 copies of VP5
25	*Adenoviridae*	Adenovirus type 2	1,500	720 copies of protein II, 60 copies of protein III
219	Unassigned	*Phaeocystis pouchetii* virus	13,050	Major capsid protein (hexamers)

[a]The triangulation number, T, is a description of the face of an icosahedron indicating the number of equilateral triangles into which each face is divided following the laws of solid geometry (Box 4.4). The total number of subunits is equal to $60T$. The simplest assembly is $T = 1$. A $T = 3$ icosahedron requires 180 subunits in sets of three; a $T = 4$ icosahedron requires 240 subunits; a $T = 7$ icosahedron requires 420 subunits; and so forth! The 60 structural units of picornavirus capsids each contain four different polypeptides, and the capsid is described as a pseudo $T = 3$ structure. Note that papovaviruses, reoviruses, and adenoviruses do not contain the number of subunits predicted.

packing of proteins to form a sphere and is an example of convergent evolution. The remarkable similarities in the detailed structures of the capsid proteins of human adenoviruses and the bacteriophage PRD1 (Box 4.6), provide support for the first hypothesis.

The overall similarity in shape of the β-barrel domains of poliovirus VP1, VP2, and VP3 facilitates both their interaction with one another to form the 60 structural units of the capsid and the packing of these structural units in this structure. How well these interactions are tailored to form a protective shell is illustrated by the space-filling model of the capsid shown in Fig. 4.13A and B: the extensive interactions among the β-barrel domains of adjacent proteins form a dense, rigid protein shell around a central cavity in which the genome resides. The packing of the β-barrel domains is reinforced by a network of protein-protein contacts on the inside of the capsid. These interactions are particularly extensive about the fivefold axes (Fig. 4.13C).

One of several important lessons learned from high-resolution structures of picornaviruses is that their design does not conform strictly to the principle of **quasiequivalence**. For example, despite the topological identity and geometric similarity of the central domains of the poliovirus proteins that form the capsid shell, the subunits do not engage in quasiequivalent bonding: interactions among VP1 molecules around the fivefold axes are not chemically or structurally equivalent to those in which VP2 or VP3 engage (Fig. 4.13A and B). These differences account for the characteristic features of the surface of the capsid. The interaction of five VP1 molecules, unique to the fivefold axes, results in a prominent protrusion extending to about 25 Å from the capsid shell (Fig. 4.13A and B). The resulting structure appears as a steep-walled plateau encircled by a valley or cleft. In the capsids of many picornaviruses, these depressions, which contain the receptor-binding sites, are so deep that they have been termed **canyons**.

An alternative icosahedral design: structure of simian virus 40. The capsids of the small DNA polyomaviruses simian virus 40 and polyomavirus, about 500 Å in diameter, are organized according to a rather different design. The structural unit is a pentamer of the major structural protein, VP1 (Fig. 4.14A). The capsid is built from 72 such structural units engaged in one of two kinds of interaction, which are described by the number of neighbors that surround any particular pentamer. Twelve structural units occupy the 12 positions of fivefold rotational symmetry, in which each is surrounded by five neighbors. Each of the remaining 60 pentamers is surrounded by 6 neighbors at positions of sixfold rotational symmetry in the capsid (Fig. 4.14A). Consequently, the 72 pentamers of simian virus 40 do not engage in identical or quasiequivalent interactions. Rather, they occupy a number of different local environments in the capsid, because of differences in packing around the five- and sixfold axes (Fig. 4.14A).

Like the three poliovirus proteins that form the capsid shell, simian virus 40 VP1 contains a large central β-barrel jelly roll domain, in this case with an N-terminal arm and a long C-terminal extension (Fig. 4.14B). However, the arrangement and packing of VP1 molecules bear little resemblance to the organization of poliovirus capsid proteins. In the first place, the VP1 β-barrels in each pentamer project outward from the surface of the capsid to a distance of about 50 Å (Fig. 4.14A), in sharp contrast to those of the poliovirus capsid proteins, which tilt along the surface of the capsid shell. As a result, the surface of simian virus 40 is much more "bristly" than that of poliovirus

BOX 4.5 EXPERIMENTS

Viral chain mail: stabilization of the bacteriophage HK97 capsid by formation of covalently bonded, interlinked subunit rings

The mature capsid of the tailed, double-stranded DNA bacteriophage HK97 is a $T = 7$ structure built from hexamers and pentamers of a single viral protein, Gp5. It is formed in a multistep assembly and maturation pathway, like the virions of many other DNA bacteriophages (e.g., lambda and T4) and some animal viruses (e.g., herpesviruses). The first hints of the remarkable and unprecedented mechanism of stabilization of this structure came from biochemical experiments, which showed that

- a previously unknown covalent protein-protein linkage forms in the final reaction in the assembly of the HK97 capsid: the side chain of a lysine in every Gp5 subunit forms a covalent bond with an asparagine in an adjacent subunit. Consequently, **all** subunits are joined covalently to each other;
- this reaction is **autocatalytic**, depending only on Gp5 subunits organized in a particular conformational state: the capsid is enzyme, substrate, and product;
- HK97 mature particles are extraordinarily stable and cannot be disassembled into individual subunits by boiling in sodium dodecyl sulfate: it was therefore proposed that the cross-linking also interlinks the subunits from adjacent structural units to catenate rings of hexamers and pentamers.

The determination of the structure of the HK97 capsid to 3.6-Å resolution by X-ray crystallography has confirmed the formation of such capsid "chain mail" (see figure). The HK97 capsid is the first example of a protein catenane (an interlocked ring). This unique structure has been shown to increase the stability of the virus particle, and it may be of particular advantage as the capsid shell is very thin. The delivery of the DNA genome to host cells via the tail of the particle obviates the need for capsid disassembly and reversal of its covalent subunit-subunit bonds.

Duda, R. L. 1998. Protein chainmail: catenated protein in viral capsids. *Cell* **94:**55–60.

Wikoff, W. R., L. Liljas, R. L. Duda, H. Tsuruta, R. W. Hendrix, and J. E. Johnson. 2000. Topologically linked protein rings in the bacteriophage HK97 capsid. *Science* **289:**2129–2133.

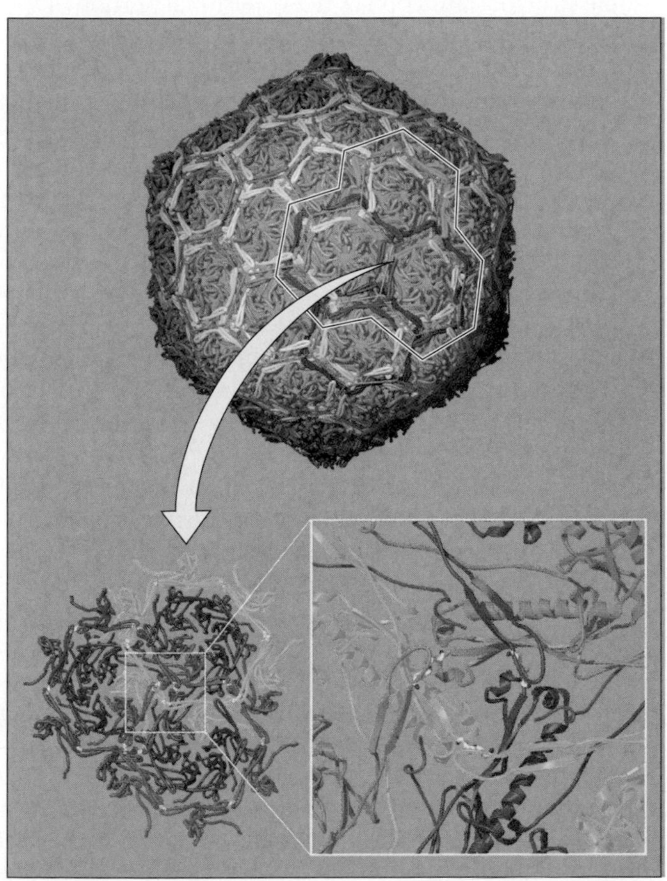

Chain mail in the bacteriophage HK97 capsid. The exterior of the HK97 capsid is shown at the top, with hexamers and pentamers of the Gp5 protein in gray. The segments of subunits that are cross-linked into rings are colored the same, to illustrate the formation of catenated rings of subunits. The cross-linking is shown in the more detailed view below, down a quasithreefold axis with three pairs of cross-linked subunits. The K-N isopeptide bonds are shown in yellow. The cross-linked monomers (shown in blue) loop over a second pair of covalently joined subunits (green), which in turn cross over a third pair (magenta). Adapted from W. R. Wikof et al., *Science* **289:**2129–2133, 2000, with permission. Courtesy of J. Johnson, The Scripps Research Institute.

A

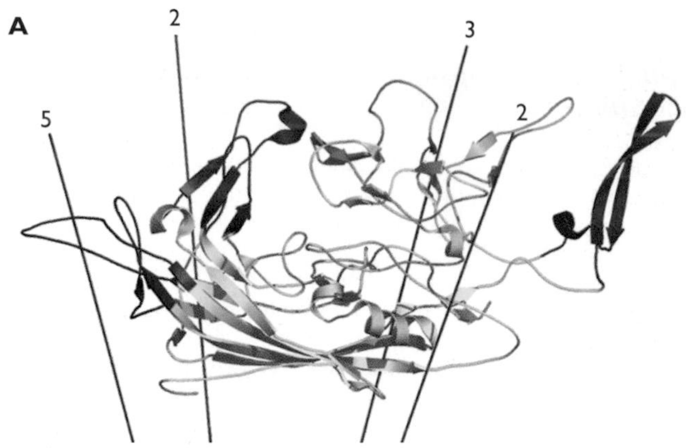

B

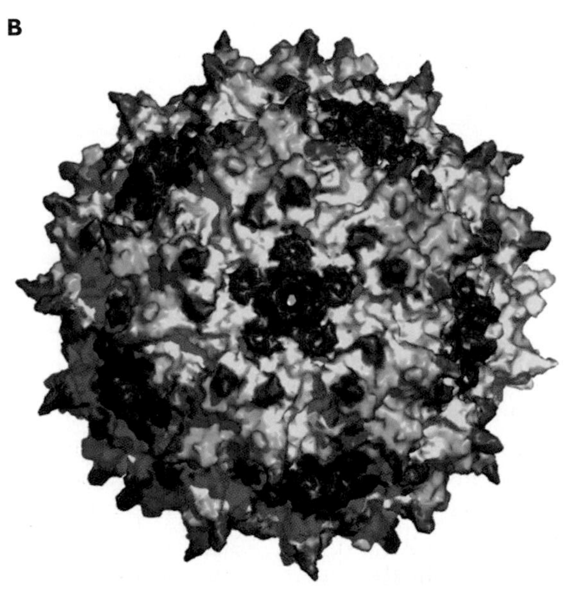

Figure 4.10 Structure of the parvovirus adeno-associated virus 2. (A) Ribbon diagram of the single subunit of the $T = 1$ particle. This comprises the C-terminal domain (about 130 amino acids) common to the VP1, VP2, and VP3 proteins, which are encoded within overlapping reading frames. As the VP1-VP2-VP3 ratio is 1:1:10 in the capsid, most subunits are contributed byVP3. The regions of the subunit that interact around the five-, three-, and twofold axes (indicated) of icosahedral symmetry are shown in blue, green, and yellow, respectively. The red segments form peaks that cluster around the threefold axes. (B) Surface view of the 3-Å-resolution structure determined by X-ray crystallography of purified virions. The regions of the single subunits from which the capsid is built are colored as in panel A. Courtesy of Michael Chapman, Florida State University. Adapted from Q. Xie et al., *Proc. Natl. Acad. Sci USA* **99:**10405–10410, 2002, with permission.

(compare Fig. 4.14A and 4.13B). Furthermore, the VP1 molecules present in adjacent pentamers in the simian virus 40 capsid do not make extensive contacts via the surfaces of their β-barrel domains. Rather, stable interactions among pentamers are mediated by their N- and C-terminal arms (Fig. 4.14). The packing of VP1 pentamers in both pentameric and hexameric arrays in the simian virus 40 capsid requires different contacts among these structural units, depending on their local environment. In fact, there are just three kinds of interpentamer contact, which are the result of alternative conformations and noncovalent interactions of the long C-terminal arms of VP1 molecules (Fig. 4.14A). The same capsid design is also exhibited by human papillomaviruses, with further strengthening of the interactions among adjacent pentamers by the formation of intermolecular disulfide bonds.

Each VP1 pentamer contains a tapering cavity that is narrowest toward the outer surface of the virion. This aperture contains a common sequence of VP2 or VP3, which associates tightly with the VP1 pentamer. Both these proteins are internal and make no contribution to the outer surface of the virion. VP2 plays an important role during entry, but why VP3 is present in virions is not known. This protein is identical in sequence to the C-terminal segment of the larger VP2 protein (Appendix, Fig. 17). It is possible that the virus particle is simply too small to accommodate 72 molecules of VP2 (as well as the DNA genome) in its interior. VP3 could serve the same structural function as the C-terminal portion of VP2, presumably stabilizing VP1 pentamers but occupying less internal space.

Simian virus 40 and poliovirus capsids differ in their surface appearance, in the number of structural units, and in the ways in which the structural units interact. Despite such differences, these virions share important features, including modular organization of the proteins that form the capsid shell and a common β-barrel domain as the capsid building block. Neither poliovirus nor simian virus 40 capsids conform to strict quasiequivalent construction: all contacts made by all protein subunits are not similar, and in the case of simian virus 40, the majority of VP1 **pentamers** are packed in **hexameric** arrays. Nevertheless, close packing with icosahedral symmetry is achieved by limited variations of the contacts, either among topologically similar, but chemically distinct, surfaces (poliovirus) or made by a flexible arm (simian virus 40).

Structurally simple icosahedral capsids or nucleocapsids in more complex virions. Several viruses that are structurally more sophisticated than those described in the previous sections nevertheless possess simple protein coats built from one or a few structural proteins. The complexity comes from the additional protein and lipid layers in which the capsid is enclosed. For example, togaviruses such as Semliki Forest virus and Ross River virus (genus *Alphavirus*), contain a $T = 4$ icosahedral capsid built from a single protein (Table 4.3) within an envelope.

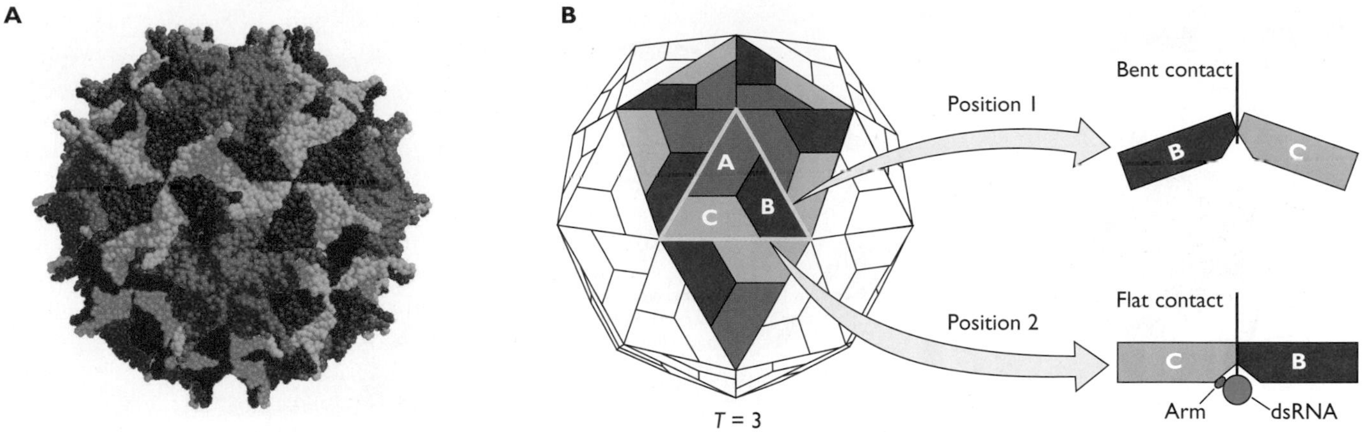

Figure 4.11 Structure of Nodamura virus. The structure of the Nodamura virus particle determined by X-ray crystallography is shown in panel **A** and summarized schematically in panel **B**. The coat protein subunits that occupy A, B, and C structural environments are colored yellow, red, and green, respectively. In panel B, the structural (asymmetric) unit is outlined in blue. The interactions between subunits at the sides of the structural units (position 1) are extensive and similar, and the asymmetric units make contact at an angle of 144°. In contrast, flat contacts are made between B and C subunits at position 2. This difference is the result of interaction of a 10-bp segment of double-stranded RNA with the internal surfaces of the subunits, and the ordering of a specific N-terminal segment of the protein only in the C subunits. (A) Courtesy of P. Natarajan and J. Johnson, The Scripps Research Institute and Virus Particle Explorer (VIPER). See V. S. Reddy et al., *J. Virol.* **75**:11943–11947, 2001.

Structurally Complex Capsids

Some naked viruses are considerably larger and more elaborate than the small RNA and DNA viruses described in the previous section. The characteristic feature of such virus particles is the presence of proteins devoted to specialized structural or functional roles. Despite such complexity, reasonably detailed pictures of the organization of this type of virion can be constructed by using combinations of biochemical, structural, and genetic methods. The well-studied human adenovirus and members of the *Reoviridae* family exemplify this approach.

Adenovirus. The most striking morphological features of the adenovirus particle (maximum diameter, 1,500 Å) are the well-defined icosahedral appearance of the capsid and the presence of long fibers at the 12 vertices (Appendix A, Fig. 1A). Each fiber, which terminates in a distal knob that binds to the adenoviral receptor, is attached to 1 of the 12 penton bases located at positions of fivefold symmetry in the capsid. The remainder of the shell is built from 240 additional subunits, the hexons, each of which is a trimer of viral protein II (Table 4.4). Formation of this capsid therefore depends on nonequivalent interactions among subunits: the hexons that surround pentons occupy a different bonding environment than those surrounded entirely by other hexons.

The intact adenovirus particle has not been studied by X-ray crystallography, but high-resolution structures have been determined for the penton base, the fiber, and the hexon. Each hexon subunit (Table 4.4) contains two β-barrel domains, each with the topology of the β-barrels of the simpler RNA and DNA viruses described in the previous section (Fig. 4.15A). As the two β-barrels are very similar in structure, the hexon trimer exhibits pseudohexagonal symmetry, a property that facilitates its close packing in the capsid. In the trimer, the β-barrel domains of the three monomers are packed together in a hollow base from which rise three towers formed by intertwining loops from each monomer (Fig. 4.15B). The interactions among monomers are very extensive, particularly in the tower. Consequently, once the trimer has formed, the monomers cannot be dissociated easily, and the hexon is extremely stable.

The adenovirus particle contains seven additional structural proteins (Table 4.4), The presence of so many proteins and the large size of the virion have made elucidation of adenovirus structure a challenging problem. One approach that has proved generally useful in the study of complex viruses is the isolation and characterization of discrete subviral particles. For example, adenovirus particles can be dissociated into a core structure that contains the DNA genome, groups of nine hexons, and pentons. Analysis of the composition of such subassemblies, together with identification of virion proteins that contact one another (by cross-linking methods), identified two classes of virion proteins in addition to the major capsid proteins described

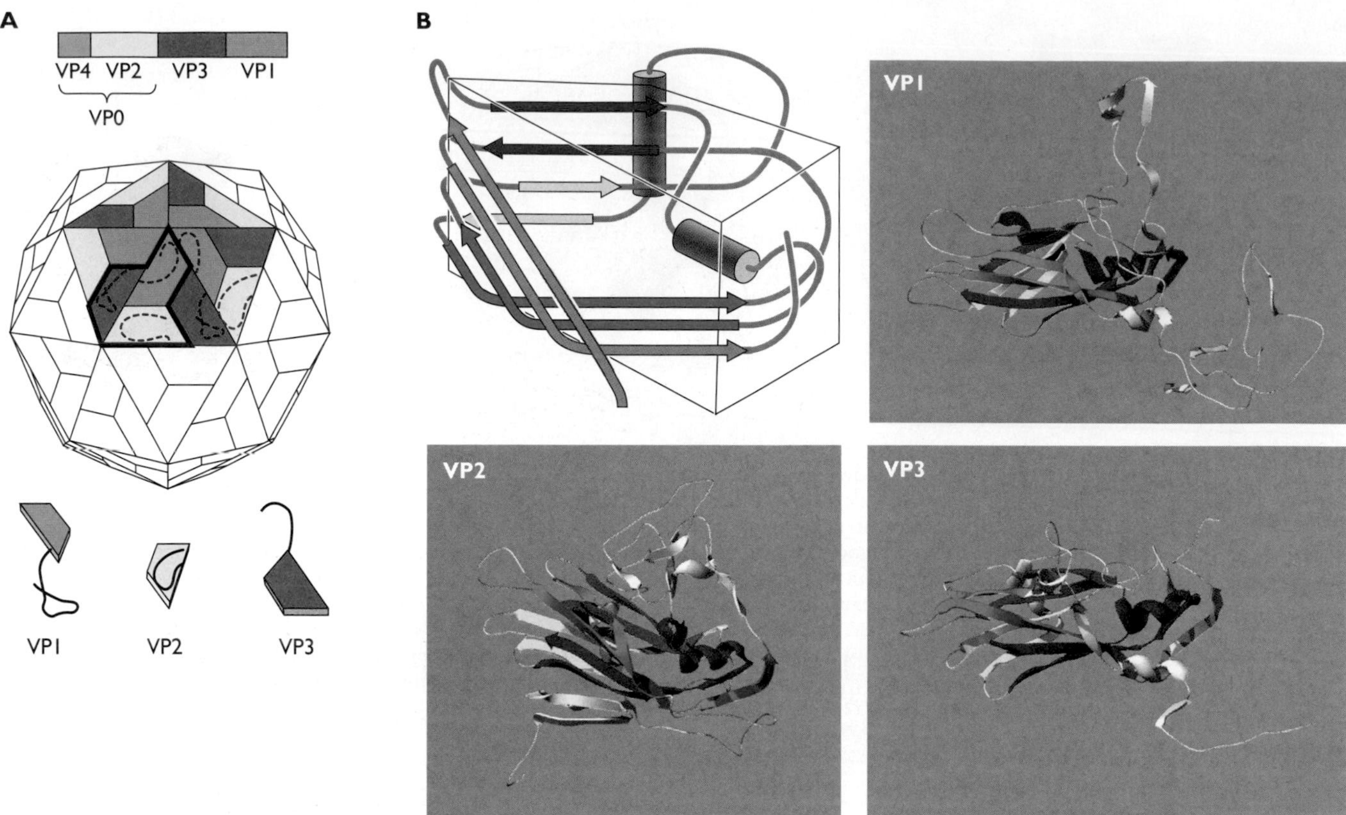

Figure 4.12 Packing and structures of poliovirus proteins. (A) The packing of the 60 VP1-VP2-VP3 structural units, represented by wedge-shaped blocks corresponding to their β-barrel domains. Note that the structural unit (outlined in black) contributes to two adjacent faces of an icosahedron rather than corresponding to a facet. When virions are assembled, VP4 is covalently joined to the N terminus of VP2. It is located on the inner surface of the capsid shell (see Fig. 4.13A). **(B)** The topology of the polypeptide chain in a β-barrel jelly roll is shown at the top left. The β-strands, indicated by arrows, form two antiparallel sheets juxtaposed in a wedgelike structure. The two α-helices (purple cylinders) that surround the open end of the wedge are also conserved in location and orientation in these proteins. As shown, the VP1, VP2, and VP3 proteins each contain a central β-barrel jelly roll domain. However, the loops that connect the β-strands in this domain of the three proteins vary considerably in length and conformation, particularly at the top of the β-barrel, which, as represented here, corresponds to the outer surface of the capsid. The N- and C-terminal segments of the protein also vary in length and structure. The very long N-terminal extension of VP3 has been truncated in this representation. Adapted from J. M. Hogle et al., *Science* **229**:1358–1365, 1985, with permission.

above. One comprises the proteins present in the core, such as protein VII, the major DNA-binding protein (Table 4.4). The remaining proteins are associated with either individual hexons or the groups of hexons that form an icosahedral face of the capsid (Table 4.4), suggesting that they stabilize the structure. Protein IX has been clearly identified as capsid "cement": a mutant virus that lacks the protein IX coding sequence produces the typical yield of virions, but these particles are much less heat stable than wild-type virions.

The locations of protein IX and other minor proteins with hexons and/or pentons have been visualized more recently by difference imaging (Fig. 4.5) and constructions of a quasiatomic model of the virion (Fig. 4.14C). In the latter approach, the crystal structures of the hexon and penton base were built into a 10-Å resolution map of the intact particle obtained by cryo-electron microscopy to reveal density associated with minor capsid proteins on both the exterior and internal surfaces of the capsid. The interactions of minor capsid proteins with the major structural units are extensive (see, e.g., Fig. 4.5), and clearly contribute to stabilizing the virus particle. During assembly, interactions among hexons and other major structural proteins must be relatively weak, so that incorrect associations can

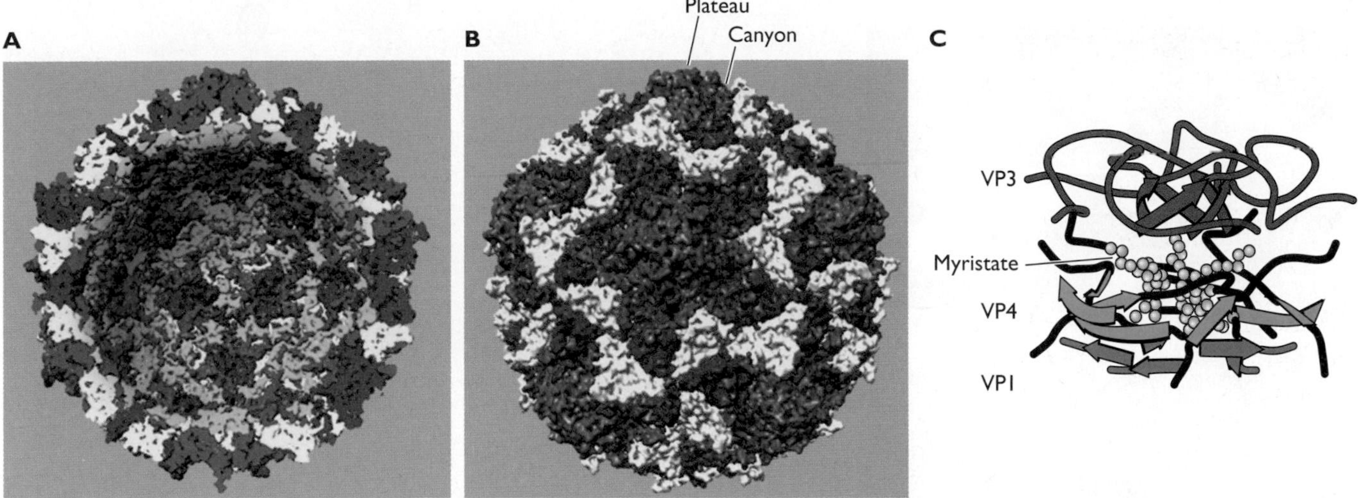

Figure 4.13 Interactions among the proteins of the poliovirus capsid. (A) Space-filling representation of the particle, with four pentamers removed from the capsid shell and VP1 in blue, VP2 in yellow, VP3 in red, and VP4 in green. Note the large central cavity in which the RNA genome resides, the dense protein shell formed by packing of the VP1, VP2, and VP3 β-barrel domains, and the interior location of VP4, which decorates the inner surface of the capsid shell. **(B)** Space-filling representation of the exterior surface showing the packing of the β-barrel domains of VP1, VP2, and VP3. Interactions among the loops connecting the upper surface of the β-barrel domains of these proteins create the surface features of the virion, such as the plateaus at the fivefold axes, which are encircled by a deep cleft or canyon. The virion is also stabilized by numerous interactions among the proteins on the inner side of the capsid. **(C)** These internal contacts are most extensive around the fivefold axes, where the N termini of five VP3 molecules are arranged in a tube-like, parallel β-sheet. The N termini of VP4 molecules carry chains of the fatty acid myristate (gray), which are added to the protein posttranslationally. The lipids mediate interaction of the β-sheet formed by VP3 N termini with a second β-sheet structure, containing strands contributed by both VP4 (green) and VP1 (blue) molecules. This internal structure is not completed until the final stages of, or after, assembly of virus particles, when proteolytic processing liberates VP2 and VP4 from their precursor, VP0. This reaction therefore stabilizes the capsid. Panels A and B were created from a PDB file.

be reversed and corrected. However, the assembled virion must be stable enough to survive passage from one host to another. It has been proposed that the incorporation of stabilizing proteins like protein IX allows these paradoxical requirements to be met.

Reoviruses. Reoviruses are naked $T = 13$ icosahedral particles, 700 to 900 Å in diameter, containing the 10 to 12 segments of the double-stranded genome and the enzymatic machinery to synthesize viral mRNA. Both adenoviruses and reoviruses are built from multiple proteins. However, reovirus particles exhibit an unusual architecture: they contain multiple protein shells. The particles of human reovirus (genus *Orthoreovirus*) contain eight proteins organized in two concentric shells, with spikes projecting from the inner layer through and beyond the outer layer at each of the 12 vertices (Fig. 4.16A). Members of the genus *Rotavirus,* which includes the leading causes of severe infantile gastroenteritis in humans (Volume II, Appendix A, Fig. 22), contain three nested protein layers, with 60 projecting spikes (Fig. 4.16B). Although differing in architectural detail, reovirus particles have common structural features, including an unusual design of the innermost protein shell.

Removal of the outermost protein layer, a process thought to occur during entry into a host cell (see Chapter 5), yields an inner core structure, comprising one shell (orthoreoviruses) or two (rotaviruses and members of the genus *Orbivirus,* such as bluetongue virus). These structures also contain the genome and virion enzymes and synthesize viral mRNAs under appropriate conditions *in vitro*. High-resolution structures have been obtained for bluetongue virus (Fig. 4.16C) and human reovirus cores, the largest viral assemblies yet to be examined by X-ray crystallography. The thin inner layer contains 120 copies of a single protein, termed VP3 in bluetongue virus and λ1 in human reovirus. These proteins are not related in their primary sequences, but they nevertheless have similar topological features and the same plate-like shape. Moreover, in both cases, the dimeric proteins occupy

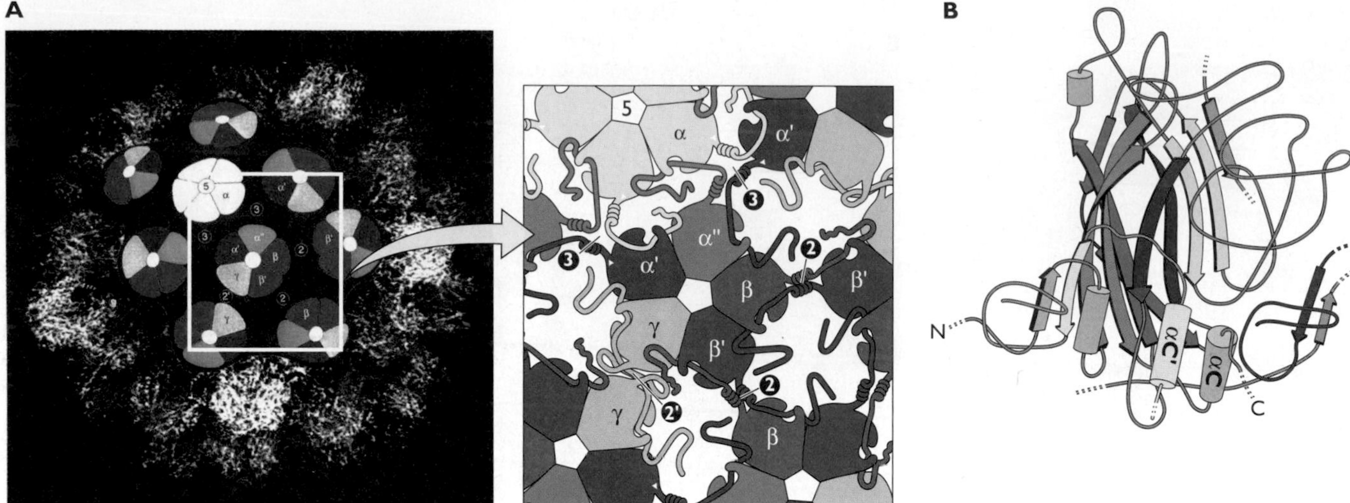

Figure 4.14 Structural features of the simian virus 40 virion. (A) View of the simian virus 40 virion showing the organization of VP1 pentamers. The 12 5-coordinated pentamers are shown in white, and the 60 pentamers present in hexameric arrays are colored. The three types of interpentameric clustering that allow 5-coordinated and 6-coordinated association of pentamers are shown in the schematic overlay and at the right. The VP1 subunits shown in white, purple, and green form a threefold cluster, designated 3. Those shown in red and blue engage in one kind of twofold interaction, designated 2, and the yellow subunits form a second kind of twofold cluster, labeled 2′. C-terminal extensions are shown as lines with coils representing the αC-helices shown in panel B. The three types of interpentamer clustering are accommodated by variation in interpentamer contacts made by the C-terminal extensions. In the threefold cluster, the C-terminal α-helices from the three interacting VP1 molecules form a three-stranded α-helical bundle. A two-chain α-helical bundle is formed at one kind of twofold cluster (2, between red and blue subunits), whereas at the second kind of twofold cluster (2′, yellow subunits) the subunits are packed so closely that there is no space to accommodate an α-helical structure. **(B)** The topology of the VP1 protein shown in a ribbon diagram, with the strands of the β-barrel jelly roll colored as in Fig. 4.12B. This β-barrel domain is radial to the capsid surface. The C-terminal arm and α-helix (orange) of the VP1 subunit invades a neighboring pentamer (not shown). The C-terminal arm and C α-helix shown in gray (αC) is the invading arm from a different neighboring pentamer (not shown), which is clamped in place by extensive interactions of its β-strand with the N-terminal segment of the subunit shown. The subunit shown also interacts with the N-terminal arm from its anticlockwise neighbor in the same pentamer (dark gray) and with the C-terminal arm of the pentamer that invades that neighbor (black). (A [left] and B) From R. C. Liddington et al., *Nature* **354:**278–284, 1991, with permission. Courtesy of S. C. Harrison, Harvard University.

one of two different structural environments, and to do so they adopt one of two distinct conformational states (Fig. 4.16C, right). Because of this arrangement, the A and B dimers are **not** quasiequivalent: virtually all contacts in which A and B monomers engage are very different. However, these differences allow the formation of VP3 assemblies with either five- or threefold rotational symmetry and hence of an icosahedral shell (Fig. 4.16C, right). The VP3 shell of bluetongue virus abuts directly on the inner surface of the middle layer, which comprises trimers of a single protein (VP7) organized into a classical $T = 13$ lattice (Fig. 4.16C, left). A large number of different (nonequivalent) contacts between VP3 and VP7 structural units weld the two layers together. These properties of reoviruses illustrate that a quasiequivalent structure is not the **only** solution to the problem of building large viral particles: viral proteins that interact with each other and with other proteins in multiple ways can provide an effective alternative.

Packaging the Nucleic Acid Genome

A definitive property of a virion is the presence of a nucleic acid genome. Incorporation of the genome requires its discrimination from a large population of cellular nucleic acid and its packaging. These processes are described in Chapter 13. The volumes of closed capsids or nucleocapsids are finite. Consequently, accommodation of viral genomes necessitates a high degree of condensation and compaction. A simple analogy illustrates vividly the scale of this problem; packing of the ~150-kbp DNA genome of herpes simplex virus type 1 into the viral nucleocapsid is equivalent to stuffing some 10 ft of wire into a tennis

BOX 4.6

EXPERIMENTS

Ancient evolutionary relationships deduced from structural comparisons

The specific and unusual properties of the major capsid proteins of human adenovirus and the bacteriophage PRD1 make a compelling case for common ancestry of these viruses. The adenovirus hexons form the faces of the icosahedral capsid (Fig. 4.15). This structural unit is not formed by six monomers arranged with hexagonal symmetry, but instead is a trimer. However, each hexon monomer contains two β-barrel jelly roll domains, labeled P1 and P2 in the figure below, such that the hexon exhibits pseudohexagonal symmetry. As the figure illustrates, the same arrangement is seen in the major capsid protein, P3, of bacteriophage PRD1. Moreover, the hexon and P3 monomers have similar connections within and between the β-barrel domains. These human and bacterial viruses also have in common $T = 25$ capsids, an arrangement not seen in any other virus family, and a structural unit built from distinct proteins at the positions of fivefold symmetry, from which attachment proteins project. They also share features of their genome organization and mechanism of viral DNA synthesis.

Several other large viruses, including the algal virus *Paramecium bursaria* chlorella virus and *Sulfobus* turreted icosahedral virus with an archeal host, are also built from major coat proteins with the same double-barrel fold. It is difficult to escape the conclusion that these modern viruses evolved from an ancient common ancestor.

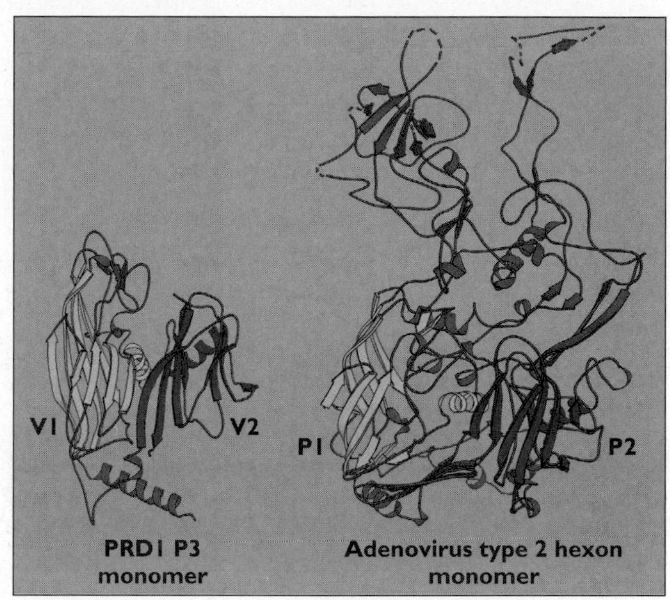

β-Barrel jelly rolls. The human adenovirus type 2 hexon (right) and bacteriophage protein P3 (left) monomers are shown, with the two β-barrel jelly rolls present in each colored green and blue. Adapted from R. W. Hendrix, *Curr. Biol.* **9**:R914–R917, 1999, with permission. Courtesy of R. Burnett, The Wistar Institute.

Benson, S. D., J. K. H. Bamford, D. H. Bamford, and R. M. Burnett. 1999. Viral evolution revealed by bacteriophage PRD1 and human adenovirus coat protein structures. *Cell* **98**:825–833.

Benson, S. D., J. K. H. Bamford, D. H. Bamford, and R. M. Burnett. 2004. Does common architecture reveal a viral lineage spanning all three domains of life? *Mol. Cell* **16**:673–685.

Hendrix, R. W. 1999. Evolution: the long evolutionary reach of viruses. *Curr. Biol.* **9**:R914–R917.

ball! In addition, packaging of nucleic acids is an intrinsically unfavorable process because of loss of entropy as the nucleic acid becomes highly constrained in conformation. In some cases, the energy required to achieve packaging is provided, at least in part, by specialized viral proteins that harness the energy released by hydrolysis of ATP to drive the insertion of DNA. In many others, the binding of viral nucleic acids to capsid or nucleocapsid proteins appears to provide sufficient energy. Such interactions also help to neutralize the negative charge of the sugar-phosphate backbone, a prerequisite for close juxtaposition of genome sequences. Three mechanisms for condensing, and presumably organizing, nucleic acid molecules within capsids or nucleocapsids can be distinguished (Table 4.5) and are described in the following sections.

Direct Contact of the Genome with a Protein Shell

In the simplest arrangement, the viral nucleic acid makes direct contact with the protein(s) that forms a protective shell of the particle (Table 4.5). Proteins on the inner surfaces of the icosahedral capsids of many small RNA viruses interact with the viral genome. As we have seen, the interior surface of the poliovirus capsid can be described in detail. Nevertheless, we possess no structural information about the arrangement of the RNA genome, for the nucleic acid is not visible in the X-ray structure. This property indicates that the RNA genome lacks the symmetry of the virion and does not adopt the identical conformation in every virus particle. In contrast, in other small viruses with icosahedral symmetry, segments of the RNA or DNA genomes are highly ordered. For example, in the $T = 3$

Table 4.4 Specialization of adenovirus type 2 structural proteins[a]

Protein	Molecular mass (kDa)	No. of copies	Location	Function
II	109,677	720	Hexon (trimer)	Formation of capsid shell
III	63,296	60	Penton base (pentamer)	Formation of capsid shell; entry
IV	61,960	36	Fiber (trimer)	Attachment to host cell
IIIa	63,287	74 ± 1	Inner capsid surface below the penton base	Stabilization of capsid
VI	23,449	342 ± 4	Hexon-associated, inner capsid surface	Stabilization of capsid; entry
VIII	14,539	211 ± 2	Hexon-associated, inner capsid surface	Stabilization of capsid
IX	14,339	247 ± 2	Outer surface of groups-of-nine hexons; edges of icosahedral faces	Stabilization of capsid
V	41,631	157 ± 1	Core, outer surface	Packaging of DNA genome
VII	19,412	835 ± 20	Core, bound to DNA	Packaging of DNA genome
μ	2,441	120 ± 1	Core	Packaging of DNA genome

[a]In addition to the structural proteins listed, the adenovirus virion contains two copies of pTP, one covalently linked to each 5′ end of the viral genome, the viral L3 protease, and several small proteins generated upon cleavage of proteins that enter the virion as precursors. Only one of these, the C-terminal extension from pVI, which activates the L3 protease, has been ascribed a function.

nodavirus Nodamura virus (Fig. 4.11), double-stranded segments of the RNA genome bind to subunits of adjacent structural units at the icosahedral twofold axes. And in virions of the small bacteriophage MS2, the entire (+) strand RNA genome appears to be icosahedrally ordered in two connected and concentric internal shells (Fig. 4.17A).

Use of the same protein or proteins both to package the genome and to build a capsid allows efficient utilization of limited genetic capacity. It is therefore an advantageous arrangement for viruses with small genomes. However, this mode of genome packing is also characteristic of some more complex viruses, notably rotaviruses and herpesviruses. The genome of rotaviruses comprises 11 segments of double-stranded RNA located within the innermost of the three protein shells of the virion. About 25% of the RNA (more than 4,000 base pairs [bp]) is highly ordered, forming a dodecahedral structure in which RNA helices are in close contact with the interior surface of the inner nucleocapsid (Fig. 4.17B).

Figure 4.15 Structural features of adenovirus particles. (A and B) Structure of the hexon. The monomer **(A)** is shown as a ribbon diagram, with gaps indicating regions that were not defined in the X-ray crystal structure at 2.9-Å resolution, and the trimer **(B)** is shown as a space-filling model with each monomer in a different color. The monomer contains two β-barrel jelly roll domains colored green and blue in panel A. The trimers are stabilized by extensive interactions within both the base and the towers. From M. M. Roberts et al., *Science* **232:**1148–1151, 1986, and F. K. Athappilly et al., *J. Mol. Biol.* **242:**430–455, 1994, with permission. Courtesy of J. Rux, S. Benson, and R. M. Burnett, The Wistar Institute. **(C)** Quasiatomic model of the capsid of human adenovirus type 5, viewed down a threefold axis. The model was made by fitting the crystal structures of the hexon (light blue) and the penton base (dark blue) into a 10-Å-resolution structure obtained by cryo-electron microscopy. Adapted from C. M. S. Fabry et al., *EMBO J.* **24:**1045–1645, 2005, with permission. Courtesy of G. Schoen, CNRS-Université Joseph Fourier, Grenoble, France.

A

B

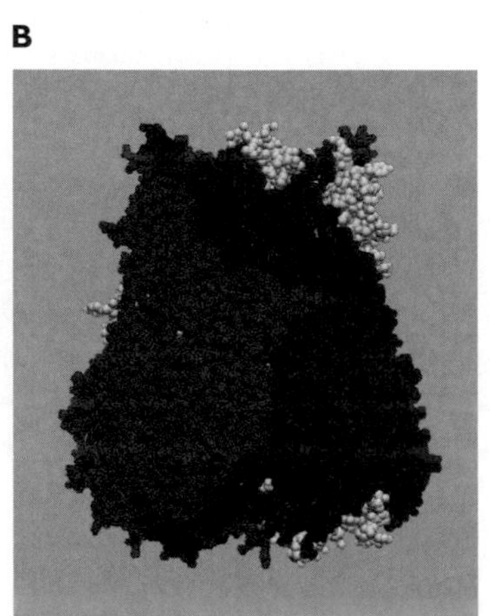

C

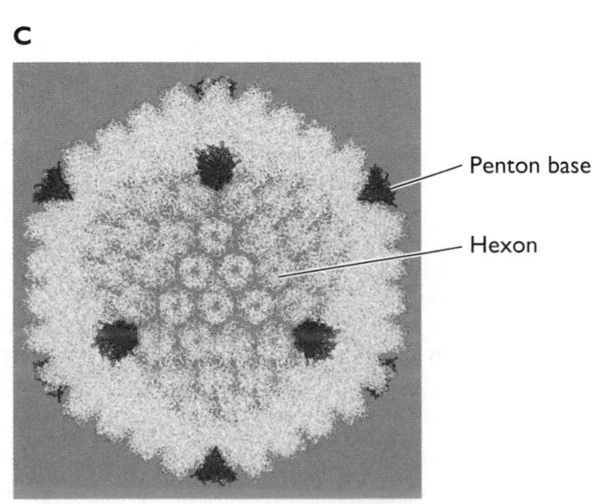

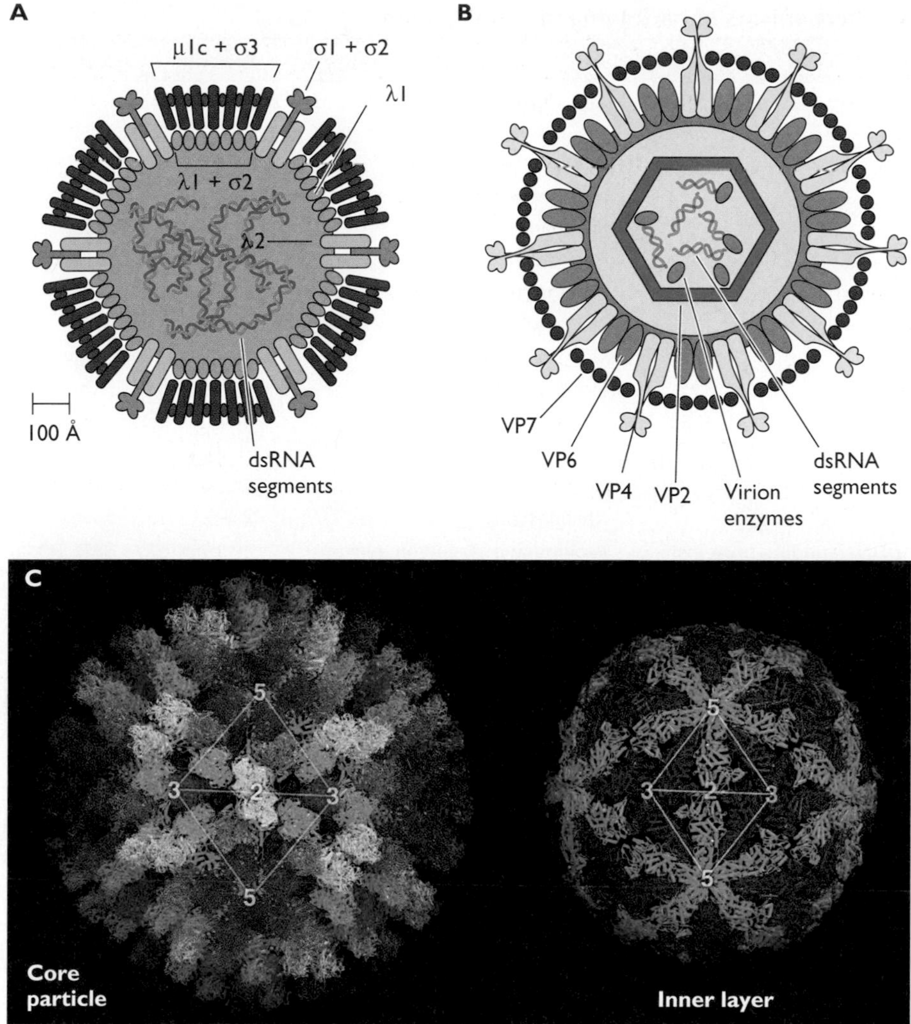

Figure 4.16 Structures of members of the *Reoviridae*. The organization of mammalian reovirus **(A)** and rotavirus **(B)** particles is shown schematically to indicate the locations of proteins, deduced from the protein composition of virions and of subviral particles that can be readily isolated from them. **(C)** X-ray crystal structure of the core of bluetongue virus, a member of the *Orbivirus* genus of the *Reoviridae*, showing the core particle (left) and the inner scaffold (right). Trimers of VP7 project radially from the outer layer of the core particle. Each icosahedral asymmetric unit, two of which are indicated by the white lines, contains 13 copies of VP7 arranged as five trimers colored red, orange, green, yellow, and blue, respectively. The outer layer is organized with classical $T = 13$ icosahedral symmetry. As shown on the right, the inner layer is built from VP3 dimers that occupy one of two completely different structural environments, designated A and B and colored green and red, respectively. A-type monomers span the icosahedral twofold axes and interact in rings of five around the icosahedral fivefold axes. In contrast, B-type monomers are organized as triangular "plugs" around the threefold axes. Differences in the interactions among monomers at different positions allow close packing to form the closed shell, the equivalent of a $T = 2$ lattice. As might be anticipated, VP7 trimers in pentameric or hexameric arrays in the outer layer make different contacts with the two classes of VP3 monomer in the inner layer. Nevertheless, each type of interaction is extensive and, in total, these contacts compensate for the symmetry mismatch between the two layers of the core. The details of these contacts suggest that the inner shell both defines the size of the virus particle and provides a template for assembly of the outer $T = 13$ structure. From J. M. Grimes et al., *Nature* **395:**470–478, 1998, with permission. Courtesy of D. I. Stuart, University of Oxford.

Table 4.5 Mechanisms of packaging the viral genome

Mechanism	Example(s)	Protein(s) contacting nucleic acid
By direct contact with nucleocapsid or capsid proteins	Alphavirus	C
	Herpesvirus	?
	Parvovirus	Ca
	Picornavirus	VP4
	Reovirus	λ1
By specialized viral nucleic acid-binding proteins	Adenovirus	VII, V, μ
	Orthomyxovirus	NP
	Poxvirus	L4R, A10L
		A3L, A4L
		F17R, I1
	Retrovirus	NC
	Rhabdovirus	N
By cellular DNA-binding proteins	Papillomavirus, polyomavirus	Histones H2A, H2B, H3, and H4

One of the most surprising properties of the large herpesviral nucleocapsid (described in "Complex Viruses" below) is the absence of internal proteins associated with viral DNA: despite intense efforts, no such core proteins have been identified. The viral genome has not yet been visualized in the herpesviral nucleocapsid. In contrast, cryo-electron microscopy has allowed visualization of the large, double-stranded DNA genome of bacteriophage T4, which is organized in closely opposed concentric layers (Fig. 4.18). This arrangement illustrates graphically the remarkably dense packing needed to accommodate such large viral DNA genomes in closed structures of fixed dimensions. This type of organization must require neutralization of the negative charges of the sugar-phosphate backbone. In herpesviral nucleocapsids, such neutralization might be accomplished by proteins that form the inner surface of the nucleocapsid, or by incorporation of small, positively charged, cellular molecules like spermine and spermidine.

Packaging by Specialized Virion Proteins

In many other virus particles, the genome is associated with specialized nucleic acid-binding proteins, such as the nucleocapsid proteins of (−) strand RNA viruses and retroviruses, or the core proteins of adenoviruses described above. An important function of such proteins is to condense and protect viral genomes. Consequently, they do not recognize specific nucleic acid sequences but rather bind nonspecifically to RNA or DNA genomes. This mode of binding is exemplified by the organization of the vesicular stomatitis virus N protein-RNA complexes, in which 9 nucleotides of RNA are tightly but nonspecifically bound in a cavity formed between the two domains of each N protein molecule (Fig. 4.7). These protein-RNA interactions both sequester the RNA within the protein ring and organize it into a helical structure. Electron micrographs of nucleocapsids of other RNA viruses, including members of the *Paramyxoviridae* and *Filoviridae*, suggest that organization of RNA genomes into helical ribonucleoproteins by two-domain RNA-binding proteins may be a common genome packaging mechanism.

Electron microscopy of cores released from adenovirus particles suggested that the internal nucleoprotein is also organized in some regular fashion. However, this structure has proved difficult to study in detail because the core is not stable once released from virions, nor have structures of core proteins been determined. The fundamental DNA packaging unit is a multimer of protein VII, which appears as beads on a string of adenoviral DNA when other core proteins are removed. Protein VII and the other core proteins are basic, as would be expected for proteins that bind to a negatively charged DNA molecule without sequence specificity.

Retrovirus virions contain approximately 2,000 molecules of the nucleocapsid (NC) protein that bind to the two copies of the encapsidated (+) strand RNA genome (Appendix, Fig. 21). The NC protein is also responsible for recognition of a specific packaging signal in the RNA during assembly and therefore binds both specifically and nonspecifically to the genome. Like other viral genome-binding proteins, NC is positively charged and, in most retroviruses, contains at least one copy of a well-characterized nucleic acid-binding motif. The structures of human immunodeficiency virus type 1 NC and of NC bound to the RNA packaging signal determined by NMR methods indicate that a long, N-terminal helix rich in basic residues

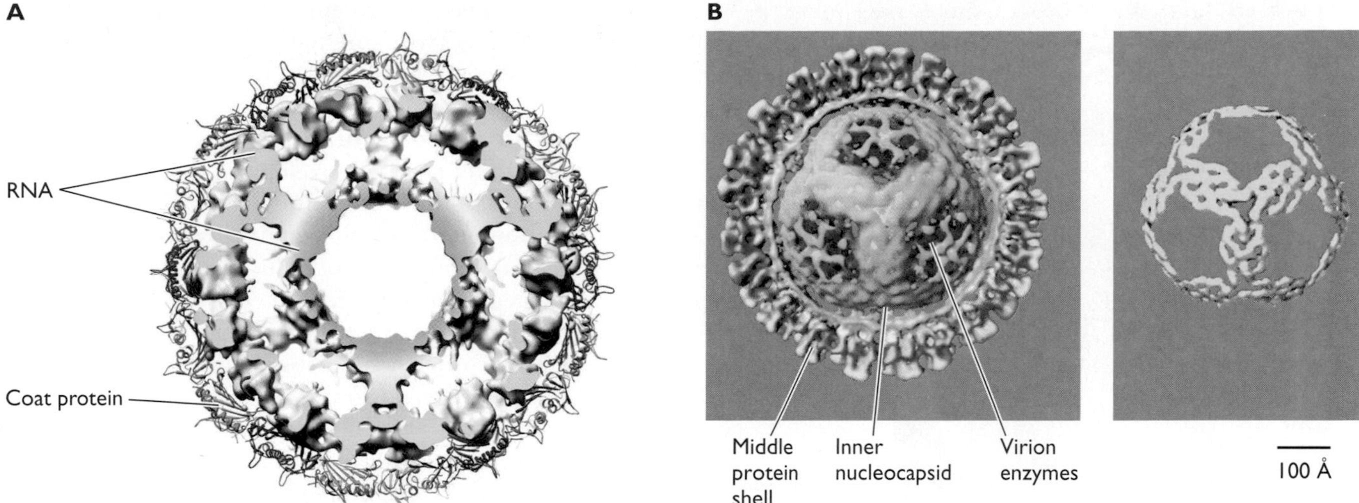

Figure 4.17 Structural organization of RNA genomes. (A) Single-stranded bacteriophage MS2 RNA. This structure (9-Å resolution) was determined by cryo-electron microscopy and difference imaging. Shown is a 40-Å section through the virion, viewed down the threefold axis. The crystal structure of the coat protein dimers is shown in cartoon form, with different conformers of the coat protein dimmer colored blue, green, or red. The regions of the cryo-electron microscopy map corresponding to the RNA genome are shown as radially colored density ranging from pale blue ($r \sim 108$ Å) to pink ($r \sim 42$ Å). One RNA shell lies immediately beneath, and makes close contacts with, the inner surface of the capsid protein shell. It is connected around the fivefold axes to a second, inner shell of RNA. It has been estimated that this ordered RNS represents ~90% of the genome. From K. Tropova et al., *J. Mol. Biol.* **375:**824–836, 2008, with permission. Courtesy of N. Ranson, University of Leeds, Leeds, United Kingdom. **(B)** Double-stranded rotaviral RNA. This structure was determined by cryo-electron microscopy, image reconstruction, and difference imaging of infectious double-layered particles and various virus-like particles. The 19-Å-resolution structure of the double-layered particle is shown on the left. The largely enclosed double-stranded RNA, which is packed around the structures formed by the virion enzymes (red), is shown in yellow. The dodecahedral shell of ordered RNA, in which each strand is about 20 Å in diameter as expected for RNA in double-stranded helices is shown on the right. From B. V. V. Prasad et al., *Nature* **382:**471–473, 1996, with permission. Courtesy of B. V. V. Prasad, Baylor College of Medicine.

interacts nonspecifically with the RNA. Whether this interaction is responsible for coating the entire RNA genome in the virion and how NC molecules condense the genome are not yet known. Retroviral ribonucleoproteins are encased within a protein shell built from the capsid (CA) protein to form an internal core. Although they contain the same components, these cores vary considerably in morphology (Fig. 4.19). Studies of the structures of the human immunodeficiency virus type 1 CA protein and the assemblies it forms *in vitro* (Box 4.7) indicate that this protein determines the conical shape of the core of this retrovirus.

Packaging by Cellular Proteins

The final mechanism of packaging the viral genome, by cellular proteins, is unique to polyomaviruses, such as simian virus 40, and papillomaviruses. The circular, double-stranded DNA genomes of these viruses are organized into nucleosomes that contain the four cellular core histones, H2A, H2B, H3, and H4. These genomes are organized within the virion (and in infected cells) like cellular DNA in chromatin to form a minichromosome. This packaging mechanism is elegant, with two major advantages: none of the limited viral genetic information needs to be devoted to DNA-binding proteins, and the viral genome, which is transcribed by cellular RNA polymerase II, enters the infected cell nucleus as a nucleoprotein closely resembling the cellular templates for this enzyme.

In each simian virus 40 particle, the 20 or so nucleosomes that package the viral genome condense the DNA by a factor of approximately 7. Within the virion, the minichromosome must be further compacted, presumably as a result of its interactions with the internal proteins of the capsid, VP2 and VP3, and perhaps the N-terminal arms of VP1 that lie on the interior surface. Both VP2 and VP3 can bind nonspecifically to DNA as well as to cellular histones, and all three proteins associate with the minichromosome during virion assembly.

Although three different ways of condensing and organizing genomic nucleic acids within virions can be distinguished readily (Table 4.5), few of these packaging

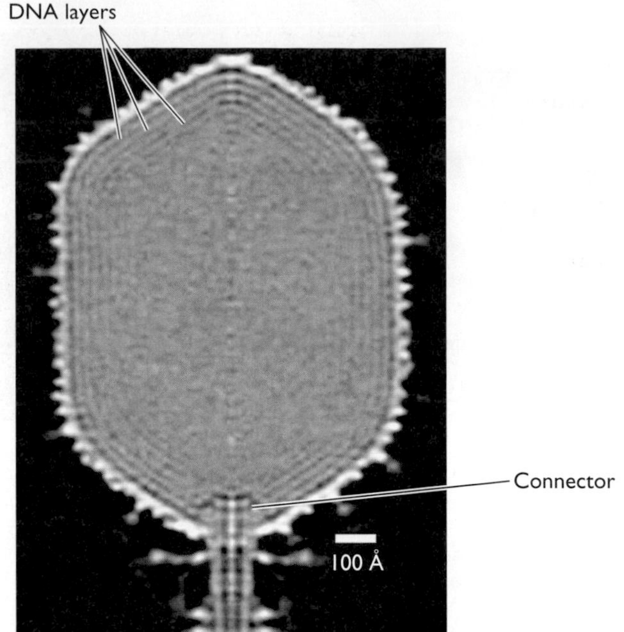

Figure 4.18 Dense packing of the double-stranded DNA genome in the head of bacteriophage T4 DNA. The central section of a 22-Å cryo-electron microscopy reconstruction of the head of bacteriophage T4 viewed perpendicular to the fivefold axis is shown. The concentric layers seen underneath the capsid shell have been attributed to the viral DNA genome. The connector, which is derived from the portal structure by which the DNA genome enters the head during assembly, connects the head to the tail. Adapted from A. Fokine et al., *Proc. Natl. Acad. Sci. USA* **101**:6003–6008, 2004, with permission. Courtesy of M. Rossmann, Purdue University.

arrangements are understood in detail. High-resolution structural descriptions of virion interiors, and of the nucleoproteins that reside within them, would undoubtedly provide important insights into mechanisms of condensation of the nucleic acid. Such structural information might also improve our understanding of the advantages conferred by the various packaging mechanisms described above.

Viruses with Envelopes

Many viruses contain structural elements in addition to the capsids or nucleocapsids described previously. All such virus particles possess an envelope formed by a viral protein-containing membrane derived from the host cell, but they vary considerably in size, morphology, and complexity. Furthermore, viral membranes differ in lipid composition, the number of proteins they contain, and their location. The envelopes form the outermost layer of enveloped animal viruses, but in bacteriophages of the PRD1 family the membrane lies **beneath** an icosahedral capsid. Typical features of viral envelopes and their proteins are described in the next section, to set the stage for consideration of the structures of envelope proteins and the various ways in which they interact with internal components of the virion (Fig. 4.20).

Viral Envelope Components

The foundation of all viral envelopes is a lipid membrane acquired from the host cell during assembly. The precise lipid composition is variable, for viral envelopes can be derived from different kinds of cellular membrane. Embedded in the membrane are viral proteins, the great majority of which are **glycoproteins** that carry covalently linked sugar chains, or **oligosaccharides** (Fig. 4.21). Sugars are added to the proteins posttranslationally, during transport to the cellular membrane at which progeny virions assemble. Intra- or interchain disulfide bonds, another common chemical feature of these proteins, are also acquired during transport to assembly sites. These covalent bonds stabilize the tertiary or quaternary structures of viral glycoproteins (Table 4.6).

Envelope Glycoproteins

Viral glycoproteins are **integral membrane proteins** firmly embedded in the lipid bilayer by a short **membrane-spanning domain** (Fig. 4.21). The membrane-spanning domains of viral proteins are hydrophobic α-helices of sufficient length to span the lipid bilayer. They generally separate large external domains that are decorated with oligosaccharides from smaller internal domains (Fig. 4.21). The former contain binding sites for cell surface virus receptors, major antigenic determinants, and sequences that mediate fusion of viral with cellular membranes during entry. Internal domains, which make contact with other components of the virion, are often essential for virus assembly.

With few if any exceptions, the structures formed by viral membrane glycoproteins are oligomeric. These oligomers vary considerably in composition. Some comprise multiple copies of a single protein, but in many cases each subunit contains two or more protein chains (Table 4.6). The subunits are held together by noncovalent interactions and disulfide bonds. On the exterior of the virion, these oligomers form surface projections, often called spikes. Because of their critical roles in initiating infection, the structures of many viral glycoproteins have been determined.

The hemagglutinin (HA) protein of human influenza A virus is a trimer of disulfide-linked HA1 and HA2 molecules. This protein contains a globular head with a top surface that is projected about 135 Å from the viral membrane by a long stem (Fig. 4.22A). The latter is formed and stabilized by the coiling of α-helices present in each monomer. The membrane-distal globular domain contains the binding site for the virus receptor. This important

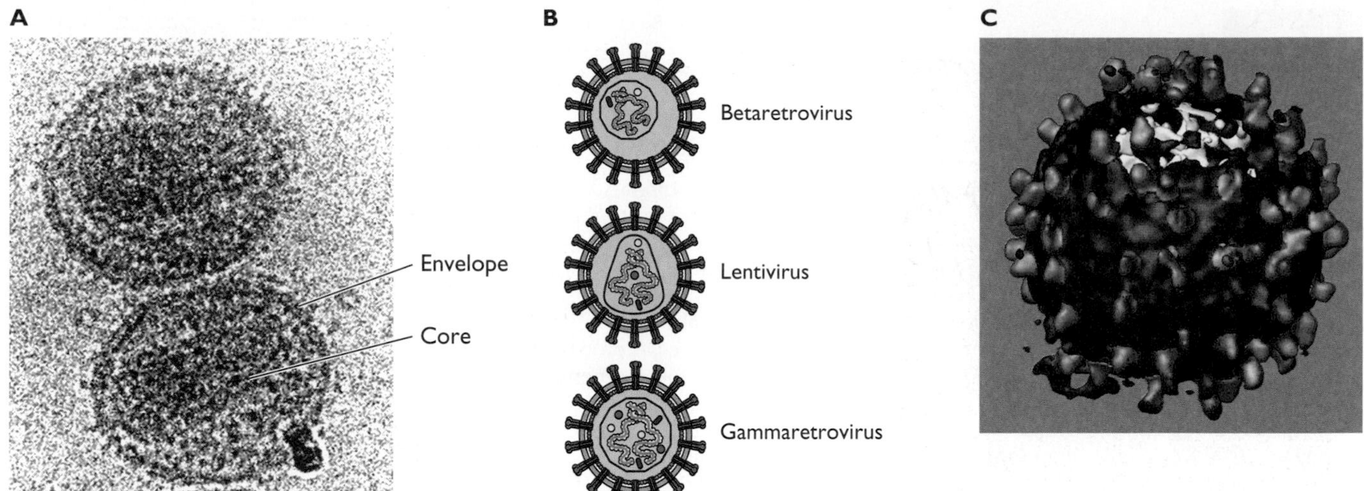

Figure 4.19 Morphology of retroviruses. (A) Cryo-electron micrograph of mature human immunodeficiency virus type 1 showing the elongated internal cores of these particles. Courtesy of T. Wilk, European Molecular Biology Laboratory. **(B)** Variations in the morphology of retroviruses shown schematically. Although retrovirus particles are assembled from the same components, some contain roughly spherical cores, whereas the cores of lentiviruses like human immunodeficiency virus type 1 are elongated and conical. In beta retroviruses the position of the capsid is variable and in some case is not central. The tertiary structures of retroviral CA proteins are highly conserved, and all examined form hexameric arrays in the absence of other viral proteins. It has therefore been proposed that the differences in morphology among retroviral capsids result from alternative positioning of the 12 pentamers required to form a closed lattice (see Box 4.6). **(C)** Cryo-electron tomogram of a Moloney murine leukemia virus particle, showing the dense but irregular packing of Env glycoprotein spikes (magenta) on the lipid bilayer (purple) of the viral envelope. Adapted from F. Förster et al., *Proc. Natl. Acad. Sci. USA* **102**:4729–4734, 2005. Courtesy of F. Förster, Max-Planck-Institut für Biochemie, Martinsried, Germany.

functional region is located more than 100 Å away from the lipid membrane of influenza virus particles. Other viral glycoproteins that mediate cell attachment and entry, such as the E protein of the flavivirus tick-borne encephalitis virus, are quite different in structure; the external domain of E protein is a flat, elongated dimer that would lie on the surface of the viral membrane rather than projecting from it (Fig. 4.22B). Despite their lack of common structural features, both the HA protein and the E protein are primed for dramatic conformational change to allow entry of internal virion components into a host cell.

The particles of the majority of other enveloped viruses contain one or two glycoproteins with typical properties. For example, the surfaces of all retroviruses are covered by a dense array of projections (Fig. 4.19C) formed by the Env proteins TM (transmembrane) and SU (surface unit). However, some enveloped viruses contain a larger collection of glycoproteins. One of the several remarkable features of the virions of herpesviruses is the large number of envelope proteins: so far, more than 12 viral glycoproteins have been identified (Table 4.7). As expected, some of them are important for attachment of the virus to host cells and for entry.

The high-resolution viral glycoprotein structures mentioned above are those of the large external domains of the proteins cleaved from the viral envelope by proteases. This treatment facilitated crystallization, but of course precluded determination of the structure of membrane-spanning or internal segments of the proteins, both of which play important structural or functional roles. Membrane-spanning domains can contribute to the stability of oligomeric glycoproteins, as in influenza virus HA, and transmit signals from the exterior to the interior of the virion. Internal domains can participate in anchoring the envelope to internal virion structures (Fig. 4.20). Recent improvements in resolution achieved by application of cryo-electron microscopy and image reconstruction (Fig. 4.5) have allowed visualization of these segments of glycoproteins of some enveloped viruses. This important advance has provided much previously inaccessible information, as discussed in "Simple Enveloped Viruses: Direct Contact of External Proteins with the Capsid or Nucleocapsid" below.

Other Envelope Proteins

The envelopes of some more complex viruses, including orthomyxoviruses, herpesviruses, and poxviruses,

BOX 4.7 EXPERIMENTS
A fullerene cone model of the human immunodeficiency virus type 1 core

(A) Purified human immunodeficiency virus type 1 CA-NC protein self-assembles into cylinders and cones when incubated with a segment of the viral RNA genome *in vitro*. Although the RNA facilitates formation of these structures, it is not essential. The cones assembled *in vitro* are capped at both ends, and many appear very similar in dimensions and morphology to cores isolated from viral particles.

The very regular appearance of the synthetic CA-NC cones suggested that, despite their asymmetry, they are constructed from a regular, underlying lattice (analogous to the lattices that describe structures with icosahedral symmetry discussed in Box 4.4). In fact, these human immunodeficiency virus type 1 cores can be modeled using the geometric principles that describe cones formed by carbon. Such elemental carbon cones comprise helices of hexamers closed at each end by caps of buckminsterfullerene, which are structures that contain pentamers surrounded by hexamers. As in structures with icosahedral symmetry (Box 4.4), the positions of pentamers determine the geometry of cones. However, in cones, pentamers are present **only** in the terminal caps. **(B)** The human immunodeficiency virus type 1 cones formed *in vitro* and isolated from mature virions can be modeled as a fullerene cone assembling on a curved hexagonal lattice with five pentamers at the narrow end of the cone, as shown in the expanded view. The wide end would be closed by an additional 7 pentamers (because 12 pentamers are required to form a closed structure from a hexagonal lattice). In this type of structure, the cone angle at the narrow end can adopt only one of five allowed angles, determined by the number of pentamers. A narrow cap with five pentamers, as in the model shown in panel B, should exhibit a cone angle of 19.2°. Approximately 90% of all the synthetic CA-NC cores examined met this prediction, consistent with the fullerene cone model. *In vitro*, the purified CA protein can form tubes, which are built from CA hexamers. The N-terminal domain of the CA protein, which is essential for capsid assembly, forms the hexameric rings, and adjacent hexamers interact via the C-terminal dimerization domain. In tubes, the planes of hexameric rings are parallel to the helix axis. Electron microscopy and image reconstruction of authentic cores established that the structures are also constructed from hexameric rings of the CA protein. **(C)** As illustrated by the yellow CA hexamers, to form cones the planes of the hexameric rings must be tilted with respect to the cone axis. Consequently, hexamers are organized in spirals that change gradually in pitch, diameter, and eventually helical handedness. This model requires that CA hexameric rings adopt many different orientations. The flexible hinge between the N- and C-terminal domains of each CA monomer and loose packing of adjacent rings of hexamers appear to provide the necessary conformational flexibility. Pentamers are shown in red. (A and C) From B. K. Ganser et al., *Science* **283:**80–83, 1999, with permission. Courtesy of W. Sundquist.

Briggs, J. A. G., T. Wik, R. Welker, H.-G. Kraüsslich, and S. D. Fuller. 2003. Structural organization of authentic, mature HIV-1 virions and cores. *EMBO J.* **22:**1707–1715.

Ganser, B. K., S. Li, V. Y. Klishko, J. T. Finch, and W. I. Sundquist. 1999. Assembly and analysis of conical models for the HIV-1 core. *Science* **283:**80–83.

Ganser-Pormillos, B. K., and U. K. von Schwedler. 2004. Assembly properties of the human immunodeficiency virus type 1 CA protein. *J. Virol* **78:**2545–2552.

Li, S., C. P. Hill, W. I. Sundquist, and J. T. Finch. 2000. Image constructions of helical assemblies of the HIV-1 CA protein. *Nature* **407:**409–413.

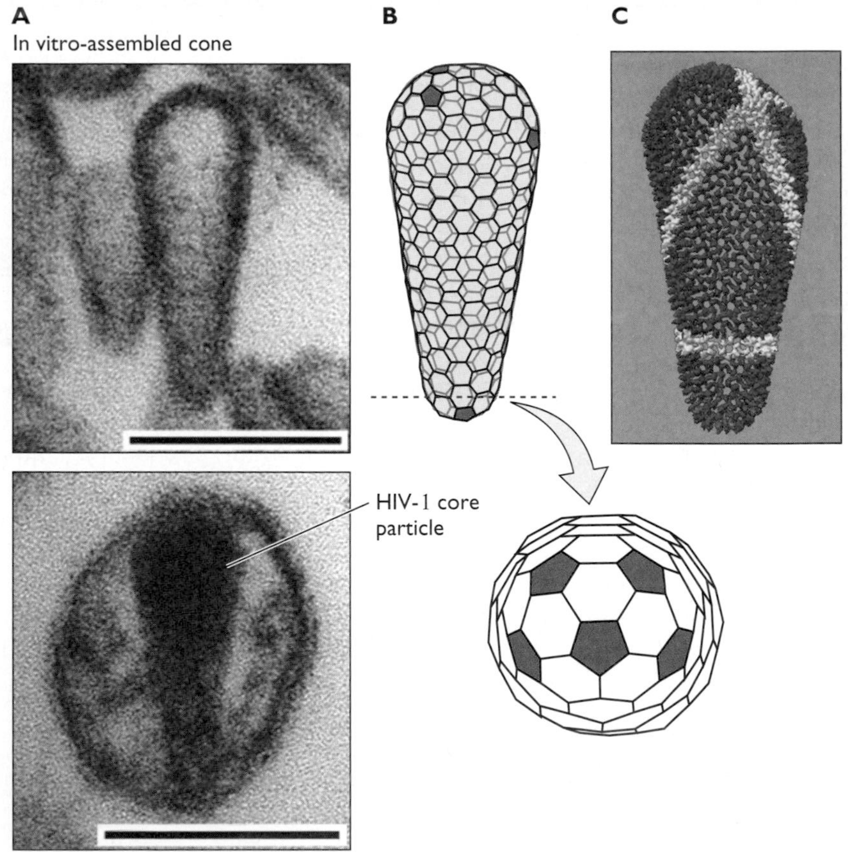

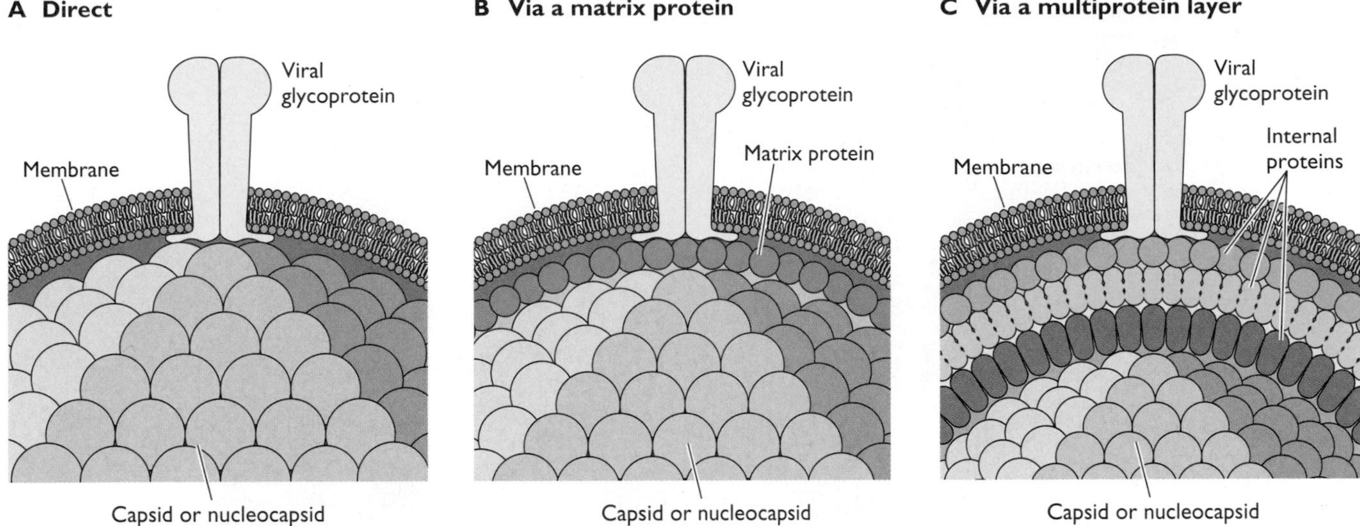

Figure 4.20 Schematic illustration of three modes of interaction of capsids or nucleocapsids with envelopes of virus particles.

contain integral membrane proteins that lack large external domains or possess multiple membrane-spanning segments (Table 4.7). Among the best characterized of these is the influenza A virus M2 protein. This small (97-amino-acid) protein is a minor component of virions, estimated to range from 14 to 68 copies per particle. In the viral membrane, two disulfide-linked M2 dimers associate to form a noncovalent tetramer that functions as an ion channel. The M2 ion channel is the target of the influenza virus inhibitor amantadine (Volume II, Fig. 9.11). The effects of this drug, as well as of mutations in the M2 coding sequence, indicate that M2 can play important roles in entry and assembly by controlling the pH of the local environments in which virus particles and HA molecules are present.

Figure 4.21 Structural and chemical features of a typical viral envelope glycoprotein shown schematically. The protein is inserted into the lipid bilayer via a single membrane-spanning domain. This segment separates a larger external domain, decorated with N-linked oligosaccharides (purple) and containing disulfide bonds (green), from a smaller internal domain.

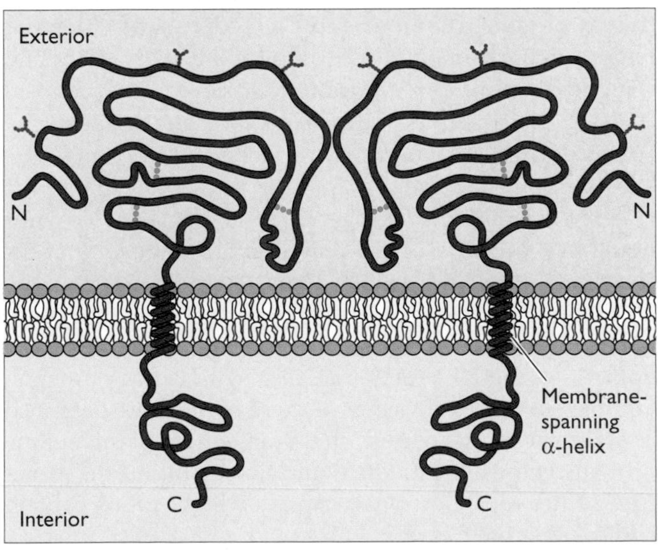

Table 4.6 Oligomeric structures of some viral membrane proteins[a]

Virus	Protein	Quaternary structure
Alphavirus		
Semliki Forest virus	E1, E2, E3	$(E1E2E3)_3$
Ross River virus	E1, E2	$(E1E2)_3$
Herpesvirus		
Herpes simplex virus type 1	gH, gL	$(gHgL)_?$
Orthomyxovirus		
Influenza virus	HA	$(HA1\text{-}HA2)_3$
	NA	$(NA\text{-}NA)_2$
	M2	$(M2\text{-}M2)_2$
Retrovirus		
Avian sarcoma virus	Env	$(SU\text{-}TM)_3$
Rhabdovirus		
Vesicular stomatitis virus	G	$(G)_3$

[a]The best-predicted oligomeric structures of the viral membrane proteins listed are shown in column 3, with hyphens indicating disulfide-bonded protein chains. The proteins that comprise many of the heteromeric subunits listed (e.g., influenza virus HA, retroviral Env) are produced by proteolytic processing of precursors during transport to the cell surface.

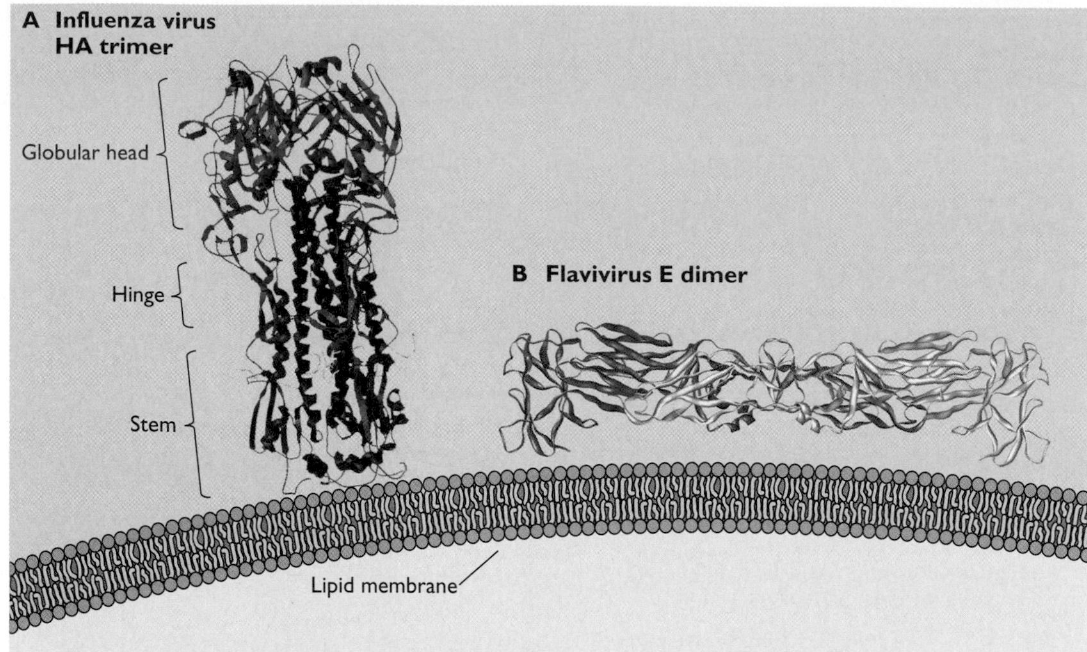

Figure 4.22 Structures of extracellular domains of viral glycoproteins. (A) X-ray crystal structure of the influenza virus HA glycoprotein trimer. Each monomer comprises HA1 (blue) and HA2 (red) subunits covalently linked by a disulfide bond. Adapted from J. Chen et al., Cell **95**:409–417, 1998, with permission. **(B)** X-ray structure of the tick-borne encephalitis virus (a flavivirus) E protein dimer, with the subunits shown in orange and yellow. Adapted from F. A. Rey et al., *Nature* **375**:291–298, 1995, with permission.

Simple Enveloped Viruses: Direct Contact of External Proteins with the Capsid or Nucleocapsid

In the simplest enveloped viruses, exemplified by (+) strand RNA alphaviruses such as Semliki Forest and Ross River viruses, the envelope directly abuts an inner nucleocapsid containing the (+) strand RNA genome. This inner protein layer is a $T = 4$ icosahedral shell built from 240 copies of a single capsid (C) protein arranged as hexamers and pentamers. The outer glycoprotein layer also contains 240 copies of the envelope proteins (Table 4.6). They cover the surface of the particle, such that the lipid membrane is not exposed on the exterior. Strikingly, the glycoproteins are also organized into a $T = 4$ icosahedral shell (Fig. 4.23A), as a result of binding of their internal domains to the C-protein subunits of the underlying nucleocapsid.

The structure of Sindbis virus has been determined by cryo-electron microscopy and image reconstruction to 9-Å resolution (Fig. 4.23A), the highest yet achieved for an enveloped virus. The structures of the E1 and C proteins of the related Semliki Forest virus have been solved at high resolution. The organization of the alphavirus envelope, including the transmembrane anchoring of the outer glycoprotein layer to structural units of the nucleocapsid, can therefore be described with unprecedented precision (Fig. 4.23). The transmembrane segments of the E1 and E2 glycoproteins form a pair of tightly associated α-helices, with the cytoplasmic domain of E2 in close opposition to a cleft in the capsid protein (Fig. 4.23C and D). On the outer surface of the membrane, the external portions of these glycoproteins, together with the E3 protein, form an unexpectedly elaborate structure: a thin $T = 4$ icosahedral protein layer (called the skirt) covers most of the membrane and supports the spikes, which are hollow, three-lobed projections (Fig. 4.23B and C).

The structures formed by external domains of membrane (E) proteins of the important human pathogens West Nile virus and dengue virus *(Flaviviridae)* are quite different: they lie flat on the particle surface, rather than forming protruding spikes. Nevertheless, the external domains of these E proteins are also icosahedrally ordered, and the envelopes of viruses of these families are described as **structured**. In contrast, the arrangement of membrane proteins generally exhibits little relationship to the structure of the capsid or nucleocapsid) when virions contain additional protein layers.

Table 4.7 Some herpes simplex virus type 1 virion proteins

Location	Protein	Gene	Function or properties
Nucleocapsid	VP5	UL19	Major capsid protein; forms both hexamers and pentamers
	VP19C/VP23	UL38/UL18	The heterotrimer ("triplex") connecting VP5 subunits on surface
	VP24	UL26	Protease; synthesized as a precursor
	VP26	UL35	Caps VP5 hexons but not pentons
	Portal	UL6	Present at one vertex; required for entry of DNA
Tegument	VP1-3	UL36	Very large (~273 kDa)
	VP16	UL48	Abundant structural protein; activator of IE gene transcription
	VP18.8	UL13	Protein kinase
	VP22	UL49	Hyperacetylation and stabilization of microtubules
	Vhs	UL41	Host shutoff factor
	US11p	US11	Myristylated protein; envelopment and transport of nascent virions
Envelope	gB	UL27	Fusion; binds heparan sulfate; binds tegument proteins, e.g., VP16
	gC	UL44	Attachment protein; binds heparan sulfate
	gD	US6	Binds cell surface receptors; entry
	gE/gI	US8/US7	Heterodimer; binds Fc domain of immunoglobulin G; cell-cell spread
	gH/gL	UL22/UL1	Heterodimer; fusion
	gG	US4	Not known
	gJ	US5	Not known
	gK	UL53	Virus-induced cell fusion; egress
	gM/gN	UL10/UL49.5	Heterodimer; function not known
	US9p	US9	Function not known for HSV-1; no large external domain
		UL20	Egress; probably contains multiple membrane-spanning domains

Enveloped Viruses with an Additional Protein Layer

Virions of several enveloped viruses contain an additional protein layer that mediates interactions of the genome-containing structure with the viral envelope. In the simplest case, a single viral structural protein, termed the matrix protein, welds an internal ribonucleoprotein to the envelope (Fig 4.20B). This arrangement is found in members of several groups of (−) strand RNA viruses (Appendix, Fig. 8 and 23). Retrovirus particles also contain an analogous, membrane-associated matrix protein, which makes contact with an internal capsid in which the viral ribonucleoprotein is encased.

Because the internal capsids or nucleocapsids of these more complex enveloped viruses are not in direct contact with the envelope, the organization and symmetry of the internal structure are not necessarily evident from the external appearance of the surface glycoprotein layer. Nor does the organization of these proteins reflect the symmetry of the capsid or nucleocapsid. For example, the outer surface of all retroviruses appears as a dense, roughly spherical array of projecting knobs or spikes, regardless of whether the internal core is spherical or cone shaped (Fig. 4.19). Likewise, influenza virus particles, which contain helical nucleocapsids, are generally roughly spherical particles 800 to 1,200 Å in diameter (Appendix, Fig. 8) although long, filamentous forms are common in clinical isolates. In general, the interior architecture of these enveloped viruses cannot be described in detail. However, high-resolution structures have been obtained for several matrix proteins. In conjunction with the results of *in vitro* assays for lipid binding and mutational analyses, such structural information allows molecular modeling of matrix protein-envelope interactions.

Internal proteins that mediate contact with the viral envelope are not embedded within the lipid bilayer, but rather bind to its inner face. Such viral proteins are targeted to, and interact with, membranes by means of specific signals, which are described in more detail in Chapter 12. For example, a posttranslationally added fatty acid chain is important for membrane binding of the MA proteins of most retroviruses. The human immunodeficiency virus type 1 MA protein was the first viral peripheral membrane protein for which a high-resolution structure was determined, initially by NMR methods (Fig. 4.3). Subsequent analysis by X-ray crystallography established that MA is a trimer (Fig. 4.24). Each MA molecule comprises a compact, globular domain of α-helices capped by a

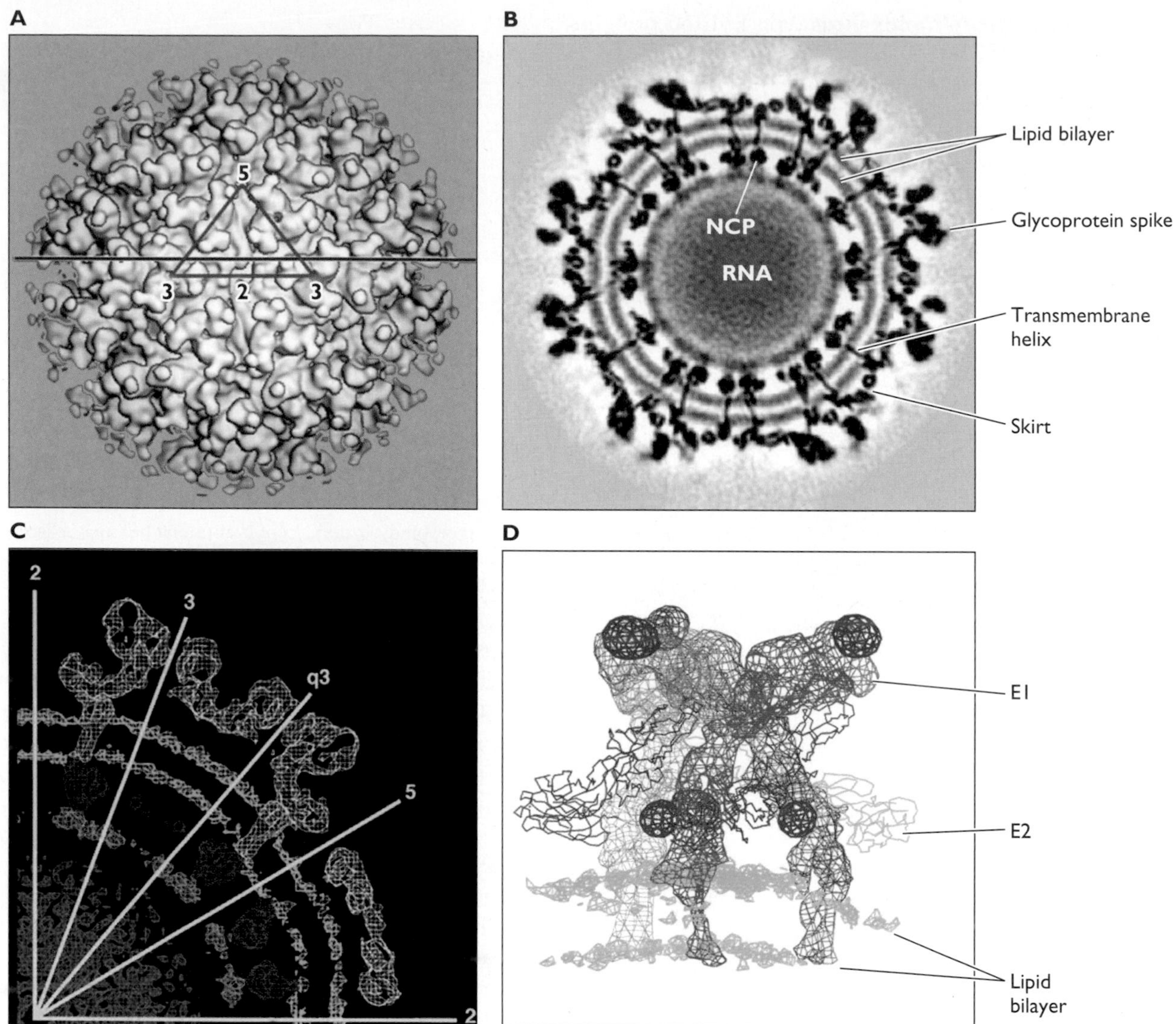

Figure 4.23 Structure of a simple enveloped virus, Sindbis virus. (A) The surface structure of Sindbis virus, a member of the alphavirus genus of the *Togaviridae*, at 20-Å resolution determined by cryo-electron microscopy. The boundaries of the structural (asymmetric) unit are demarcated by the red triangle, on which the icosahedral five-, three-, and twofold axes of rotational symmetry are indicated. This outer surface is organized as a $T = 4$ icosahedral shell studded with 80 spikes, each built from three copies of each of the transmembrane glycoproteins E1 and E2. These spikes are connected by a thin, external protein layer, termed the skirt. **(B)** Cross section through the density map at 11-Å resolution, along the black line shown in panel A. The lipid bilayer of the viral envelope is clearly defined at this resolution, as are the transmembrane domains of the glycoproteins. **(C)** Different layers of the particle, based on the fitting of a high-resolution structure of the E1 glycoprotein into a 9-Å reconstruction of the virus particle. The nucleocapsid (red) surrounds the genomic (+) strand RNA. The RNA is the least well-ordered feature in the reconstruction, although segments (orange) lying just below the capsid (C) protein appear to be ordered by interaction with this protein. The C protein penetrates the inner leaflet of the lipid membrane, where it interacts with the cytoplasmic domain of the E2 glycoprotein (blue). The membrane is spanned by rod-like structures that are connected to the skirt by short stems. **(D)** The structure of the E1 and E2 glycoproteins, obtained by fitting the crystal structure of the closely related Semliki Forest virus E1 glycoprotein into the 11-Å density map and assigning density unaccounted for to the E2 glycoprotein. The view shown is around a quasi-threefold symmetry axis (q3 in panel C), with the three E2 glycoprotein molecules in a trimeric spike colored light blue, brown, and purple and the E1 molecules shown as backbone traces colored red, green and dark blue. The E1 glycoprotein is largely tangential to the surface of the particle. The portions of the proteins that cross the lipid bilayer are helical, twisting around one another in a left-handed coiled coil. Courtesy of Michael Rossmann, Purdue University. Adapted from W. Zhang et al., *J. Virol.* **76:**11645–11658, 2002, with permission.

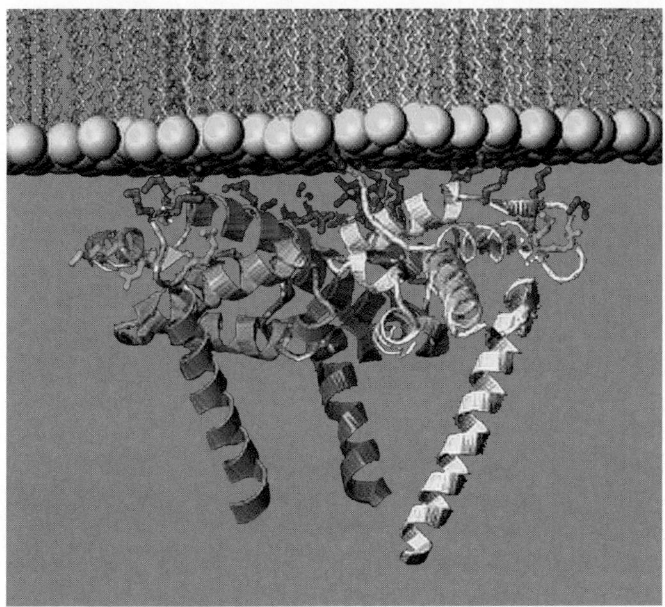

Figure 4.24 Model of the interaction of human immunodeficiency virus type 1 MA protein with the membrane. The membrane is shown with the polar head groups of membrane lipids in yellow. This model is based on the X-ray crystal structure of recombinant MA protein synthesized in *E. coli* and consequently lacking the N-terminal (myristate) 14-carbon fatty acid normally added in human cells. The three monomers in the MA trimer are shown in different colors. Basic residues in the β-sheet that caps the globular α-helical domain are magenta or green. Substitution of those shown in magenta impairs replication of the virus in cells in culture. The position of the myrsitate (red) was modeled schematically. From C. P. Hill et al., *Proc. Natl. Acad. Sci. USA* **93**:3099–3104, 1996, with permission. Courtesy of C. P. Hill and W. I. Sundquist, University of Utah.

β-sheet that contains positively charged amino acids that are also necessary for membrane binding. As illustrated in the model of MA oriented on a membrane shown in Fig. 4.24, the basic residues form a positively charged surface, positioned for interaction with phospholipid head groups on the inner surface of the envelope. The matrix proteins of (−) strand RNA viruses such as vesicular stomatitis virus and influenza virus also contain positively charged domains required for membrane binding, despite having three-dimensional folds that are quite different from those of retroviral MA proteins (and from one another).

Complex Viruses

Virus particles that house large DNA genomes are structurally far more complex than any considered in previous sections. Such virions comprise obviously distinct components with different symmetries and/or multiple layers. We illustrate these properties using as examples bacteriophage T4, herpes simplex virus type 1, and vaccinia virus.

Bacteriophage T4

Bacteriophage T4, which has been studied for over 50 years, is the classic example of a structurally complex virus that lacks an envelope. The T4 virion, which is built from about 50 of the proteins encoded in the ~170-kbp double-stranded DNA genome, is a structurally elegant machine tailored for active delivery of the genome to host cells. The most striking feature is the presence of morphologically distinct and functionally specialized structures, notably the head containing the genome and a long tail that terminates in a baseplate from which six long tail fibers protrude (Fig. 4.25A). A set of short fibers are also attached to the baseplate, while yet another set, termed whiskers, decorate the junction between head and tail.

The head of the mature T4 particle, an elongated prolate icosahedron, is built from hexamers of a single viral protein (gp23*). In contrast to the other capsids or nucleocapsids considered so far, two *T* numbers are needed to describe the organization of gp23* in the two end structures ($T = 13$) and in the elongated midsection ($T = 20$). As in adenoviral capsids, the pentamers that occupy the vertices contain a different viral protein, and additional proteins reside on the outer or inner surfaces of the icosahedral shell (Fig. 4.25B). One of the 12 vertices is occupied by a unique structure termed the connector, which joins the head to the tail. Such structures are derived from the nanomachine that pulls DNA into immature heads termed the **portal**. Portals are a characteristic feature of the nucleocapisds of other families of DNA-containing bacteriophages, as well as of herpesviruses.

In contrast to the head, the ~100-nm-long tail, which comprises two protein layers, exhibits helical symmetry (Fig 4.25A). The outer layer is a contractile sheath that functions in injection of the viral genome into host cells. The tail is connected to the head via a hexameric ring and at its head-distal end to a complex, dome-shaped structure termed the baseplate that contains at least 16 different proteins (Fig 4.25C). Both long and short tail fibers project from the baseplate. The former, which are long and bent, are the primary receptor-binding structures of bacteriophage T4. As discussed in Chapter 5, remarkable conformational changes induced upon receptor binding by the tips of the long fibers are transmitted via the baseplate to initiate injection of the DNA genome.

Herpesviruses

Virions of the *Herpesviridae* contain many more proteins than any animal virus described in previous sections, and exhibit a number of unusual architectural features. Over half of the more than 80 genes of herpes simplex virus type 1 encode proteins found in virus particles (Table 4.7), which are correspondingly large, about 2,000 Å in

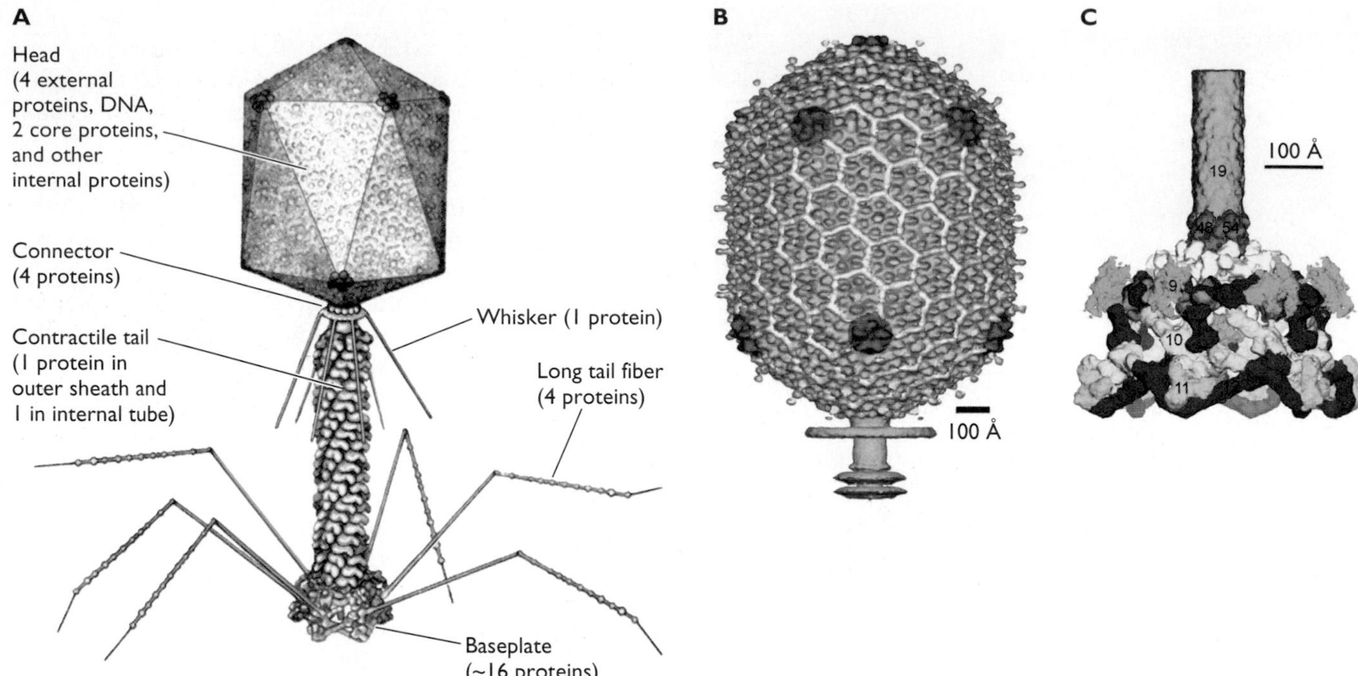

Figure 4.25 Morphological complexity of bacteriophage T4. (A) A Model of the virion. Adapted from P.G. Leiman et al., *Cell Mol. Life Sci.* **60**:2356–2370, 2003, with permission. **(B)** structure of the head (22-Å resolution) determined by cryo-electron microscopy, with the major capsid proteins shown in blue (gp23*) and magenta (gp24*), the protruding noc protein in yellow, the protein that binds between gp23* subunits in white, and the beginning of the tail in green. Adapted from A. Fokine et al., *Proc. Natl. Acad. Sci. USA* **101**:6003–6008, 2004, with permission. **(C)** Side view of the baseplate connected to the inner tube of the tail (protein 19), obtained by fitting the X-ray structures of individual baseplate proteins (numbered and shown in different colors) in a 12-Å-resolution reconstruction of the baseplate-tail tube complex. The baseplate is a complex dome-shaped structure constructed on a hexameric base. Adapted from V. A. Kostyachenko et al., *Nat. Struct. Biol.* **10**:688–693, 2003, with permission. (B and C) Courtesy of M. Rossmann, Purdue University.

diameter. These proteins are components of the envelope (Table 4.7) from which glycoprotein spikes project, or of two distinct internal structures. The latter are the nucleocapsid surrounding the DNA genome and the protein layer encasing this structure, called the **tegument** (Fig. 4.26A).

A single protein (VP5) forms both the hexons and the pentons of the T = 16 icosahedral nucleocapsid of herpes simplex virus type 1 (Fig. 4.26B). Like the structural units of the smaller simian virus 40 capsid, these VP5-containing assemblies make direct contact with one another. However, the large (~1,500-Å-diameter) herpesviral nucleocapsid is stabilized by additional proteins, VP19C and VP23. These two proteins form triplexes that link the major structural units (Fig. 4.26B). Although apparently a typical and quite simple icosahedral shell, this viral nucleocapsid is in fact an asymmetric structure: in infectious virions, 1 of the 12 vertices is occupied not by a VP5 penton but by a unique portal (Fig. 4.26C). This structure comprises 12 copies of the UL6 protein (Table 4.7). The portal assembled from the UL6 protein made in insect cells is a squat hollow cylinder that is wider at one end and surrounded by a two-tiered ring at the wider end (Fig. 4.26C). The asymmetry of the herpesviral nucleocapsid and the incorporation of the portal have important implications for the mechanism of assembly (see Chapter 13).

The tegument contains at least 13 viral proteins, viral RNAs, and cellular components. Descriptions of this structure have changed considerably with the application of increasingly sophisticated methods of electron microscopy (Box 4.8). It is generally agreed that specific tegument proteins are icosahedrally ordered, as a result of direct contacts with the structural units of the nucleocapsid (Fig. 4.26D). However, some tegument proteins are **not** uniformly distributed around the nucleocapsid. Rather, they are concentrated on one side of the capsid, where they form a well-defined cap-like structure (Fig 4.26E). As this totally unanticipated asymmetry of herpesviral particles has been

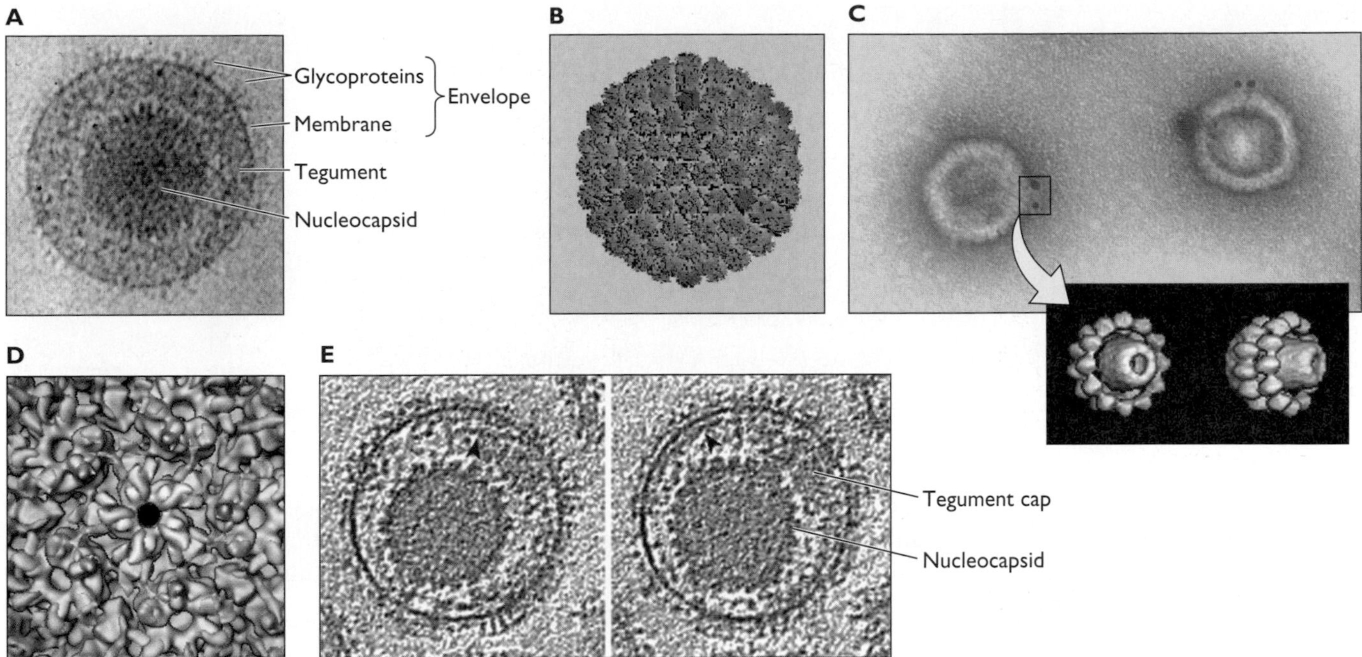

Figure 4.26 Structural features of herpesvirus particles. (A) Electron micrograph of a frozen, hydrated herpes simplex virus type 1 virion. From F. J. Rixon, *Semin. Virol.* **4:**135–144, 1993, with permission. Courtesy of F. Rixon, Institute of Virology, Glasgow, United Kingdom, and W. Chiu, Baylor College of Medicine. **(B)** Reconstruction of the herpes simplex virus type 1 nucleocapsid (8.5-Å resolution), with VP5 hexamers and pentamers colored blue and red, respectively, and the triplexes that reinforce the connections among these structural units in green. VP5 hexamers, but not pentamers, are capped by a hexameric ring of VP26 protein molecules (not shown). Adapted from Z. H. Zhou et al., *Science* **288:**877–880, 2000, with permission. Courtesy of W. Chiu, Baylor College of Medicine. **(C)** The single portal of herpes simplex virus type 1 nucleocapsids visualized by staining with an antibody specific for the viral UL6 protein conjugated to gold beads. The gold beads are electron dense and appear as dark spots in the electron micrograph. They are present at a single vertex in each nucleocapsid, which therefore contains one portal. The inset shows a 16-Å reconstruction of the UL6 protein portal based on cryo-electron microscopy. Adapted from W. W. Newcomb et al., *J. Virol.* **75:**10923–10932, 2001, and B. L. Trus et al., *J. Virol.* **78:**12668–12671, 2004, with permission. Courtesy of A. C. Steven, National Institutes of Health. **(D)** Interactions of two tegument proteins with the simian cytomegalovirus nucleocapsid. Tegument proteins that bind to hexons plus pentons and to triplexes are shown in blue and red, respectively. These proteins were visualized by cryo-electron microscopy, image reconstruction (to 22-Å resolution), and difference mapping of nucleocapsids purified from the nucleus and cytoplasm of virus-infected cells. The latter carry the tegument, but the former do not. Adapted from B. L. Trus et al., *J. Virol.* **73:**2181–2192, 1999, with permission. Courtesy of A. C. Steven, National Institutes of Health. **(E)** Two slices through a cryo-electron tomogram of a single herpes simplex virus type 1 particle, showing the eccentric tegument cap. Adapted from K. Grunewald et al., *Science* **302:**1396–1398, 2003, with permission. Courtesy of A. C. Steven, National Institutes of Health.

viewed only at low resolution, the molecular organization of the cap is not yet understood. Herpesvirus particles completely lacking certain tegument proteins appear morphologically normal and are fully infectious, suggesting that this layer may be structurally plastic.

Poxviruses

Like bacteriophage T4 and herpesvirus particles, those of poxviruses such as vaccinia virus comprise multiple, distinct structural elements. Two forms of infectious particles, termed mature virions and enveloped extracellular virions, are produced in vaccinia virus-infected cells (see Chapter 13). Mature virions are large, enveloped structures (~330 × 360 × 125 nm) comprising at least 75 proteins that appear in the electron microscope as brick- or barrel-shaped (Fig. 4.27A). Whether one or two envelopes are present has been the subject of long-standing debate. However, there is now a growing consensus for the presence of just a single membrane. At least 25 membrane proteins, most of which contain one or two membrane-spanning domains, have been identified. An unusual feature of these viral membranes is that they are not glycosylated.

A number of internal structures have been observed by examination of thin sections through purified particles

BOX 4.8

METHODS

Evolution of descriptions of the organization of the herpesviral tegument with technical advances

~1975 to 1993: An amorphous tegument, an irregular structure with no obvious organization or symmetry. This view was based on negative staining of purified virions, a technique that we now know resulted in distortion of the tegument (and envelope).

1993 to 2003: A regular, uniformly organized tegument revealed by cryo-electron microscopy (see, e.g., Fig. 4.26A), which avoids artifacts associated with negative staining. However, image reconstruction assumed icosahedral symmetry. Consequently, structures with different, or no, symmetry cannot be visualized.

2003 to present: A regular structure that is asymmetrically organized within virions to form a cap on one side of the nucleocapsid (Fig. 4.26E). This view is based on cryo-electron tomography, in which no assumptions about symmetry are made during image reconstruction.

(Fig 4.27B). These features include the core wall, which surrounds the central core that contains the ~200-kbp DNA genome. Remarkably, the core contains at least 30 enzymes with many different activities. The outer surface of the core wall abuts the inner surface of the virion membrane, except where the two central masses termed lateral bodies are located (Fig 4.27B). Although viral proteins that contribute to these various structures have been identified, our understanding of vaccinia virus architecture remains at low resolution.

Other Components of Virions

Some virus particles comprise only the nucleic acid genome and structural proteins necessary for protection and delivery into a host cell. However, many contain additional viral proteins or other components, which are generally present at much lower concentrations but play essential or important roles in establishing an efficient infectious cycle (Table 4.8).

Virion Enzymes

Many types of virus particle contain enzymes necessary for synthesis of viral nucleic acids. Such enzymes generally catalyze reactions unique to virus-infected cells, such as synthesis of viral mRNA from an RNA template or of viral DNA from an RNA template. However, virions of vaccinia virus contain a DNA-dependent RNA polymerase, analogous to cellular RNA polymerases, as well as several enzymes that modify viral RNA transcripts (Table 4.8). This complement of enzymes is necessary because transcription

Figure 4.27 Structural features of the poxvirus vaccinia virus. (A) Electron micrograph of a negatively stained purified virion, showing the protrusions termed surface tubular elements. From S. Willon et al., *Virology* **214:**503–511, 1995. Courtesy of S. Dales, University of Western Ontario, London, Canada. **(B)** Electron micrograph of frozen section of a purified virion. Adapted from M. Hollinshead et al., *J. Virol.* **73:**1503–1517, 1999, with permission. Courtesy of D. J. Vaux, University of Oxford, Oxford, England.

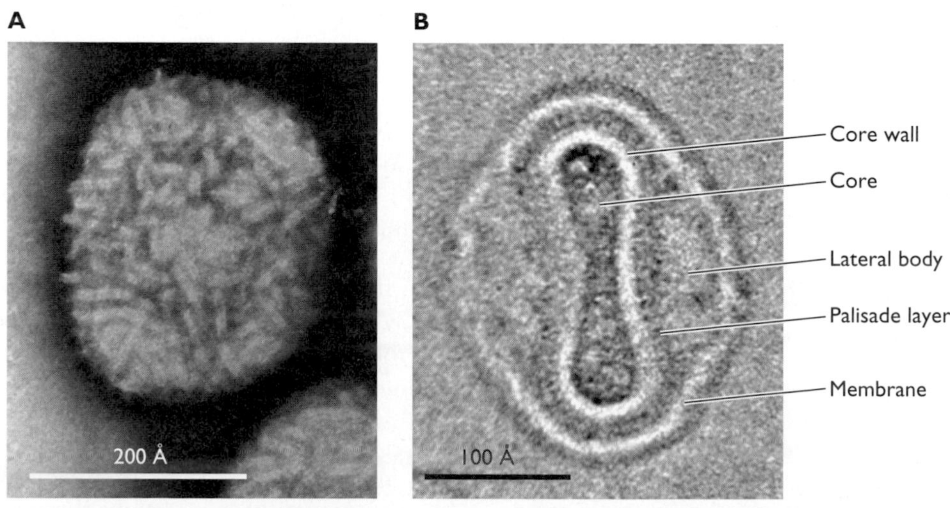

Table 4.8 Some virion enzymes

Virus	Protein	Function(s)
Adenovirus		
Human adenovirus type 2	L3 23k	Protease, production of infectious virions
Herpesvirus		
Herpes simplex virus type 1	VP24	Protease, capsid maturation for genome encapsidation
	UL13	Protein kinase
	Vhs	RNase
Orthomyxovirus		
Influenza A virus	P proteins	RNA-dependent RNA polymerase; synthesis of viral mRNA and vRNA; cap-dependent endonuclease
Poxvirus		
Vaccinia virus[a]	DNA-dependent RNA polymerase (8 subunits)	Synthesis of viral mRNA
	Poly(A) polymerase (2 subunits)	Synthesis of poly(A) on viral mRNA
	Capping enzyme (2 subunits)	Addition of 5′ caps to viral pre-mRNA
	DNA topoisomerase	Sequence-specific nicking of viral DNA
	Proteases 1 and 2	Virion morphogenesis
	Glutaredoxin	Thiol transferase; nonessential
Reovirus		
Reovirus type 1	λ2	Guanyl transferase
	λ3	Double-stranded RNA-dependent RNA polymerase
Retrovirus		
Avian sarcoma virus	Pol	Reverse transcriptase, proviral DNA synthesis
Human immunodeficiency virus type 1	IN	Integrase, integration of proviral DNA into the cellular genome
	PR	Protease, production of infectious virions
Rhabdovirus		
Vesicular stomatitis virus	L	RNA-dependent RNA polymerase, synthesis of viral mRNA and vRNA

[a]Vaccinia virions contain at least 17 enzymes, only a few of are listed.

of the viral double-stranded DNA genome takes place in the cytoplasm of infected cells, whereas cellular DNA-dependent RNA polymerases and the RNA-processing machinery are restricted to the nucleus. Other types of enzyme found in virions include integrase, cap-dependent endonuclease, and proteases (Table 4.8). The proteases, which eliminate covalent connections among specific proteins from which virions assemble, are necessary for the production of infectious particles.

Other Viral Proteins

More complex virions may also contain additional viral proteins that are not enzymes, but nonetheless are important for an efficient infectious cycle (Table 4.9). Among the best characterized examples of this class are several tegument proteins of herpesviruses. The VP16 protein activates transcription of viral immediate-early genes to initiate the viral program of gene expression. In contrast to the majority of viral proteins discussed in this section, VP16 is present at high concentration, because it is also necessary for virion assembly. Other herpesvirus tegument proteins induce the degradation of cellular mRNA, or block the cellular mechanism by which viral proteins are presented to the host's immune system. Complex retroviruses like human immunodeficiency virus type 1 also contain additional proteins required for efficient viral replication in certain cell types, such as Nef and Vpr. These proteins are discussed in Volume II, Chapter 6.

Nongenomic Viral Nucleic Acid

The **viral** nucleic acids present in virions have, by definition, been considered to be genomes. It was therefore quite a surprise to find other, functional viral nucleic acids within virus particles. This unexpected property was first

Table 4.9 Some additional viral components of virions

Virus	Component	Function
Proteins		
Herpesvirus		
Herpes simplex virus type 1	VP16	Structural protein, activation of IE gene transcription
Human cytomegalovirus	pp65	Inhibition of presentation of viral IE proteins as antigens
Poxvirus		
Vaccinia virus	VETF	Binds to early promoters; essential for their transcription
	RAP94	Associated with virion RNA polymerase, specificity factor
	I6L protein	DNA packaging; binds ends of genome
Retrovirus		
Human immunodeficiency virus type 1	Vpr	Required for efficient infection in some cell types
	Nef	
Other viral nucleic acids		
Herpesvirus		
Human cytomegalovirus	5 late mRNAs	?

reported for human cytomegalovirus, a betaherpesvirus with one of the largest DNA genomes known, and an important human pathogen (Volume II, Appendix A, Fig. 9). As discussed in more detail below, it can be difficult to distinguish macromolecules that are assembled specifically into enveloped virus particles from those that are incorporated nonspecifically. However, of the 200 or so viral mRNAs synthesized in human cytomegalovirus-infected cells, only 4 can be readily detected in purified virions. Moreover, these same viral mRNAs were found within cells very soon after infection in the presence of a drug that prevents synthesis of **all** mRNAs. These viral mRNAs must therefore be packaged specifically. The virion-delivered mRNAs, which probably reside within the tegument, are translated in the cytoplasm of newly infected cells. Neither the functions of the viral proteins made from these viral mRNAs nor the mechanism by which the mRNAs enter assembling virus particles is yet understood.

Cellular Macromolecules

Virus particles can also contain cellular macromolecules that play important roles during the infectious cycle (Table 4.9), such as the cellular histones that package polyomaviral and papillomaviral DNAs. Because they are formed by budding, enveloped viruses can readily incorporate cellular proteins and other macromolecules. Cellular glycoproteins may not be excluded from the membrane from which the viral envelope is derived. Moreover, as a bud enlarges and pinches off during virus assembly, internal cellular components may be trapped within it. In addition, enveloped viruses are generally more difficult to purify than naked viruses. As a result, preparations of these viruses may be contaminated with vesicles formed from cellular membranes during virus purification. Consequently, it can be difficult to distinguish cellular components specifically incorporated into enveloped virus particles from those trapped randomly or copurifying with the virus. Nevertheless, in some cases it is clear that cellular molecules are important components of virus particles: these molecules are reproducibly observed in virions at a specific stoichiometry and can be shown to play essential or important roles in the infectious cycle. The cellular components captured in retrovirus particles have been particularly well characterized, but cellular proteins are also present in other virions. For example, the herpesviral tegument contains substantial quantities of actin, as well as other cytoskeletal proteins (Table 4.9).

The primer for initiation of synthesis of the (−) strand DNA during reverse transcription is invariably a cellular transfer RNA (tRNA). This RNA is incorporated into virions by virtue of its binding to a specific sequence in the RNA genome and to reverse transcriptase. A variety of cellular proteins are also present in some retroviral particles. One of the most unusual properties of human immunodeficiency virus type 1 is the presence of cellular cyclophilin A, a **chaperone** that assists or catalyzes protein folding. This protein is a member of a ubiquitous family of peptidyl-prolyl isomerases, which catalyze the intrinsically slow *cis-trans* isomerization of peptide bonds preceding proline residues in newly synthesized proteins. Cyclophilin A is the major cytoplasmic member of this family. It is incorporated within human immunodeficiency virus type 1 particles via specific interactions with the central portion of (CA) the capsid protein, and it catalyzes isomerization of a single Gly-Pro bond in CA. Although incorporation of cyclophilin A is not a prerequisite for assembly, particles that lack this cellular chaperone have reduced infectivity. It is thought that in human cells, cyclophilin provides protection against an intrinsic antiviral defense mechanism, but this mechanism has not been identified.

Cellular membrane proteins, such as Icam-1 and Lfa1 (see Chapter 5), can also be incorporated in the viral envelope and can contribute to attachment and entry of human immunodeficiency virus type 1 particles. They may also influence pathogenesis (see Volume II, Chapter 6). Other cellular proteins assembled into viral particles, such as actin found in herpesviral virions, may facilitate assembly or budding reactions (see Chapter 13).

The majority of cellular components present in virus particles serve to facilitate virus replication, a property exemplified by the cellular tRNA primers for retroviral reverse transcription. However, it has become clear recently that incorporation of cellular components can also provide antiviral defense. As discussed in Volume II, Chapters 3 and 6, packaging of a cellular enzyme that converts cytosine to uracil (Apobec3) into retrovirus particles at the end of one infectious cycle leads to degradation and hypermutation of viral DNA synthesized early in the next cycle of infection.

It is clear from these examples that virus particles contain a surprisingly broad repertoire of molecular functions that are delivered to their host cells (Tables 4.8 and 4.9). This repertoire is undoubtedly larger than we presently appreciate, for the precise molecular functions and/or roles in the infectious cycles of many virion components have yet to be established.

Perspectives

Virus particles are among the most elegant and visually pleasing structures found in nature, properties emphasized by the images presented in this chapter. Now that many structures of particles or their components have been examined, we can appreciate the surprisingly diverse architectures exhibited by virions. Nevertheless, the simple principles of their construction proposed over 50 years ago remain pertinent: with few exceptions, the capsid or nucleocapsid shells that encase and protect nucleic acid genomes are built from a small number of protein subunits arranged with helical or icosahedral symmetry.

The detailed views of nonenveloped virions provided by X-ray crystallography emphasize just how well these protein shells provide protection of the genome during passage from one host cell or organism to another. They have also identified several mechanisms by which identical or nonidentical subunits can interact to form icosahedrally symmetric structures. More complex virus particles, which may contain additional protein layers, a lipid envelope carrying viral proteins, and enzymes or other proteins necessary to initiate the infectious cycle, cannot be described in the exquisite detail provided by a high-resolution structure. For many years we possessed only schematic views of these structures, deduced from negative-contrast electron microscopy and biochemical or genetic methods of analysis. Within the past decade, the development and refinement of cryo-electron microscopy and techniques of image reconstruction have revolutionized structural studies of larger and more complex viruses. To cite but a few examples, these methods have yielded remarkable views of genome organization and interactions among envelope and internal proteins. The recent improvement in the resolution that can be achieved by these techniques to <10 Å, the power of difference imaging methods, and the now routine ability to produce large quantities of viral proteins promise many new insights into the structure, molecular organization, and function of virus particles. Such information will improve our understanding, presently quite limited, of mechanisms of packaging of nucleic acid genomes and of tethering envelopes (when present) to internal structures. Some surprises are undoubtedly in store. But we can predict with some confidence that future studies will reinforce and elaborate the general principles of virus structure described here.

References

Chapters in Books

Baker, T. S., and J. E. Johnson. 1997. Principles of virus structure determination, p. 38–79. *In* W. Chiu, R. M. Burnett, and R. L. Garcea (ed.), *Structural Biology of Viruses*. Oxford University Press, New York, NY.

Chow, M., R. Basavappa, and J. M. Hogle. 1997. The role of conformational transitions in poliovirus pathogenesis, p. 157–187. *In* W. Chiu, R. M. Burnett, and R. L. Garcea (ed.), *Structural Biology of Viruses*. Oxford University Press, New York, NY.

Garcea, R. L., and R. C. Liddington. 1997. Structural biology of polyomaviruses, p. 187–208. *In* W. Chiu, R. M. Burnett, and R. L. Garcea (ed.), *Structural Biology of Viruses*. Oxford University Press, New York, NY.

Reviews

Baker, T. S., N. H. Olson, and S. D. Fuller. 1999. Adding the third dimension to virus life cycles: three-dimensional reconstruction of icosahedral viruses from cryo-electron micrographs. *Microbiol. Mol. Biol. Rev.* **63:**862–922.

Benson, S. D., J. K. H. Bamford, D. H. Bamford, and R. M. Burnett. 2004. Does common viral architecture reveal a viral lineage spanning all three domains of life? *Mol. Cell* **16:**673–685.

Chapman, M. S., V. L. Giranda, and M. G. Rossmann. 1990. The structures of human rhinovirus and mengo virus: relevance to function and drug design. *Semin. Virol.* **1:**413–427.

Chiu, W., and F. J. Rixon. 2002. High resolution structural studies of complex icosahedral viruses: a brief overview. *Virus Res.* **82:**9–17.

Condit, R. C., N. Moussatche, and P. Traktman. 2006. In a nutshell: structure and assembly of the vaccinia virion. *Adv. Virus Res.* **66:**31–124.

Leiman, P. G., S. Kanamara, V. V. Mesyanzkinov, E. Arisaka, and M. G. Rossmann. 2003. Structure and morphogenesis of bacteriophage T4. *Cell Mol. Life Sci.* **60:**2356–2370.

Stabbs, G. 1990. Molecular structures of viruses from the tobacco mosaic virus group. *Semin. Virol.* **1:**405–512.

Strauss, J. H., and E. G. Strauss. 2001. Virus evolution: how does an enveloped virus make a regular structure? *Cell* **105:**5–8.

Vellinga, J., S. Van der Heijdt, and R. C. Hoeben. 2005. The adenovirus capsid: major progress in minor proteins. *J. Gen. Virol.* **86:**1581–1588.

Wilk, T., and S. D. Fuller. 1999. Towards the structure of human immunodeficiency virus: divide and conquer? *Curr. Opin. Struct. Biol.* **9:**231–243.

Papers of Special Interest

Theoretical Foundations

Caspar, D. L. D., and A. Klug. 1962. Physical principles in the construction of regular viruses. *Cold Spring Harbor Symp. Quant. Biol.* **27:**1–22.

Crick, F. H. C., and J. D. Watson. 1956. Structure of small viruses. *Nature* **177:**473–475.

Structures of Nonenveloped Viruses

Brenner, S., and R. W. Horne. 1959. A negative staining method for high resolution electron microscopy of viruses. *Biochim. Biophys. Acta* **34:**103–110.

Harrison, S. C., A. Olson, C. E. Schutt, F. K. Winkler, and G. Bricogne. 1978. Tomato bushy stunt virus at 2.9 Å resolution. *Nature* **276:**368–373.

Hogle, J. M., M. Chow, and D. J. Filman. 1985. Three-dimensional structure of poliovirus at 2.9 Å resolution. *Science* **229:**1358–1365.

Liddington, R. C., Y. Yan, H. C. Zhao, R. Sahli, T. L. Benjamin, and S. C. Harrison. 1991. Structure of simian virus 40 at 3.8 Å resolution. *Nature* **354:**278–284.

Prasad, B. V. V., R. Rothnagel, C. Q.-Y. Zeng, J. Jakana, J. A. Lawton, W. Chiu, and M. K. Estes. 1996. Visualization of ordered genomic RNA and localization of transcriptional complexes in rotavirus. *Nature* **382:**471–473.

Reinisch, K. M., M. L. Nibert, and S. C. Harrison. 2000. Structure of the reovirus core at 3.6 Å resolution. *Nature* **404:**960–967.

Rossman, M. G., E. Arnold, and J. W. Erickson. 1985. Structure of a common cold virus and functional relationship to other picornaviruses. *Nature* **317:**145–153.

Saban, S. D., M. Silvestry, G. R. Nemerow, and P. L. Stewart. 2006. Visualization of α-helices in a 6Å resolution cryelectron microscopy structure of adenovirus allows refinement of capsid protein assignments. *J. Virol.* **80:**12049–12059.

Structures of Enveloped Viruses

Briggs, J. A. G., T. Wilk, R. Walker, H. G. Kräusslich, and S. D. Fuller. 2003. Structural organization of authentic, mature HIV-1 virions and cores. *EMBO J.* **22:**1707–1715.

Cheng, R. H., R. G. Kuhn, N. H. Olson, M. G. Rossmann, H.-K. Choi, T. J. Smith, and T. S. Baker. 1995. Nucleocapsid and glycoprotein organization in an enveloped virus. *Cell* **80:**621–630.

Lescar, J., A. Roussel, M. W. Wien, J. Navaza, S. D. Fuller, G. Wengler, G. Wengler, and F. A. Rey. 2001. The fusion glycoprotein shell of Semliki Forest virus: an icosahedral assembly primed for fusogenic activation at endosomal pH. *Cell* **105:**137–148.

Mancini, E. J., M. Clarke, B. E. Gowen, T. Rulten, and S. D. Fuller. 2000. Cryo-electron microscopy reveals the functional organization of an enveloped virus, Semliki Forest virus. *Mol. Cell* **5:**255–266.

Newcomb, W. W., R. M. Juhas, D. R. Thomsen, F. L. Homa, A. D. Burch, S. K. Weller, and J. C. Brown. 2001. The UL6 gene product forms the portal for entry of DNA into the herpes simplex virus capsid. *J. Virol.* **75:**10923–10932.

Wynne, S. A., R. A. Crowther, and A. G. W. Leslie. 1999. The crystal structure of the human hepatitis B virus capsid. *Mol. Cell* **3:**771–780.

Zhou, Z. H., M. Dougherty, J. Jakana, J. He, F. J. Rixon, and W. Chiu. 2000. Seeing the herpesvirus capsid at 8.5 Å resolution. *Science* **288:**877–880.

Structures of Virion Proteins

Malashkevich, V. N., B. J. Schneider, M. L. McNally, M. A. Milhollen, J. X. Pang, and P. S. Kim. 1999. Core structure of the envelope glycoprotein GP2 from Ebola virus at 1.9 Å resolution. *Proc. Natl. Acad. Sci. USA* **96:**2662–2667.

Massiah, M. A., M. R. Starick, C. Paschall, M. F. Summers, A. M. Christensen, and W. I. Sundquist. 1994. Three-dimensional structure of the human immunodeficiency virus type 1 matrix protein. *J. Mol. Biol.* **244:**198–223.

Rey, F. A., F. X. Heinz, C. Mandl, C. Kunz, and S. C. Harrison. 1995. The envelope glycoprotein from tick-borne encephalitis virus at 2 Å resolution. *Nature* **375:**291–298.

Wilson, J. A., T. S. Skehel, and D. C. Wiley. 1981. Structure of the haemagglutinin membrane glycoprotein of influenza virus at 3 Å resolution. *Nature* **289:**366–373.

Zubieta, C., G. Schoehn, J. Chroboczek, and S. Cusak. 2005. The structure of the human adenovirus 2 penton. *Mol. Cell* **17:**121–135.

Other Components of Virions

Bresnahan, W. A., and T. Shenk. 2000. A subset of viral transcripts packaged within human cytomegalovirus particles. *Science* **288:**2373–2376.

Mariani, R., D. Chen, B. Schrofelbauer, F. Navamo, R. Konig, B. Bollman, C. Munk, H. Numerk-McMahon, and N. R. Landau. 2003. Species-specific exclusion of APOBEC3G from HIV-1 virions by vif. *Cell* **114:**21–31.

Poon, D. T. K., L. V. Coren, and D. E. Ott. 2000. Efficient incorporation of HLA class II onto human immunodeficiency virus type 1 requires envelope glycoprotein packaging. *J. Virol.* **74:**3918–3923.

Rizzuto, C. D., and J. G. Sodroski. 1997. Contribution of virion ICAM-1 to human immunodeficiency virus infectivity and sensitivity to neutralization. *J. Virol.* **71:**4847–4851.

Thali, M., A. Bukovsky, E. Kondo, B. Rosenwirth, C. T. Walsh, J. Sodroski, and H. G. Göttlinger. 1994. Functional association of cyclophilin A with HIV-1 virions. *Nature* **372:**363–365.

Websites

http://viperbd.scripps.edu/ *Virus Particle Explorer*

http://www.virology.net/Big_Virology/BVHomePage.html *The Big Picture Book of Viruses*

http://virology.wisc.edu/virusworld/ *Virus World*

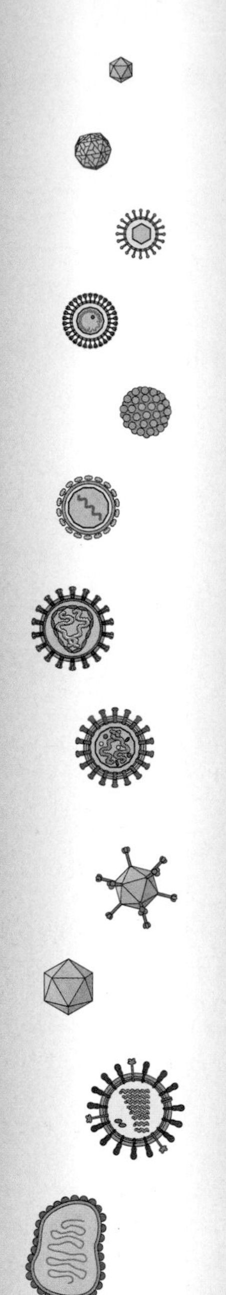

5

Attachment and Entry

Who hath deceived thee so often as thyself?
BENJAMIN FRANKLIN

Introduction

Because viruses are obligate intracellular parasites, the viral genome must enter a cell for the viral replication cycle to occur. The physical properties of the virion are obstacles to this seemingly simple goal. Virions are too large to diffuse passively across the plasma membrane. Furthermore, the viral genome is encapsidated in a stable coat that shields the nucleic acid as it travels through the harsh extracellular environment. These impediments must all be overcome during the process of viral entry into cells. When viruses encounter the surface of a susceptible host cell, a series of events lead to entry of the viral genome into the cytoplasm or nucleus. The first step in entry is adherence of virus particles to the plasma membrane, an interaction mediated by binding to a specific **receptor** molecule on the cell surface.

The cellular receptor plays an important role in **uncoating**, the process by which the viral genome is exposed, so that gene expression and genome replication can begin. Interaction of the virus particle with its cell receptor may initiate conformational changes that prime the capsid for uncoating. Alternatively, the cell receptor may direct the virion into endocytic pathways, where uncoating may be triggered by low pH or by the action of proteases. These steps bring the genome into the cytoplasm, where the genomes of most RNA-containing viruses replicate. The genomes of viruses that replicate in the nucleus are brought to that location by cellular transport pathways. Viruses that replicate in the nucleus include all DNA-containing viruses except poxviruses, RNA-containing retroviruses, influenza viruses, and Borna disease virus.

Early studies of virus entry into host cells, from the 1950s until the late 1970s, led to the view that viruses enter by an entirely passive process: virus particles attach to the cell surface, are taken up into the cell, and release their genomes, which are then replicated. No active role for the receptor in uncoating was envisioned. Beginning in the 1980s, the techniques of cellular, molecular, and structural biology were applied to elucidate the earliest events in viral infection. It is now understood that virus entry into cells is

not a passive process but, rather, relies on viral usurpation of normal cellular processes, including endocytosis, membrane fusion, vesicular trafficking, and transport into the nucleus. Because of the limited functions encoded by viral genomes, virus entry into cells absolutely depends on cellular processes.

Attachment of Viruses to Cells

General Principles

Infection of cells by many, but not all, viruses requires binding to a receptor on the cell surface. Exceptions include viruses of yeasts and fungi, which have no extracellular phases, and plant viruses, which are thought to enter cells through openings produced by mechanical damage, such as those caused by farm machinery or insects. In some cases, the receptor is the only cell surface molecule required for entry into cells. In others, binding to a cellular receptor is not sufficient for infection: an additional cell surface molecule, or **coreceptor**, is required for entry (Box 5.1).

The cell receptor may determine the **host range** of a virus, i.e., its ability to infect a particular animal or cell culture. For example, poliovirus infects primates and primate cell cultures but not mice or mouse cell cultures. Mouse cells synthesize a protein that is homologous to the poliovirus receptor but sufficiently different that poliovirus cannot attach to it. In this example, the poliovirus receptor is **the** determinant of poliovirus host range. However, production of the receptor in a particular cell type does **not** ensure that virus replication will occur. Some primate cell cultures produce the poliovirus receptor but cannot be infected. The restricted host range of the virus in such cells is most probably due to a block in viral replication beyond the attachment step. Cell receptors can also be determinants of tissue **tropism**, the predilection of a virus to invade and replicate in a particular cell type. However, there are many other determinants of tissue tropism. For example, the sialic acid residues on membrane glycoproteins or glycolipids, which are receptors for influenza virus, are found on many tissues, yet viral replication in the host is restricted. The basis for such restriction is discussed in Volume II, Chapter 1.

Our understanding of the earliest interactions of viruses with cells comes almost exclusively from analysis of synchronously infected cells in culture. The initial association of virions with cells is probably via electrostatic forces, as they are sensitive to low pH or high concentrations of salt. Subsequent high-affinity binding relies mainly on hydrophobic and other short-range forces whose strength and specificity are governed primarily by the conformations of the interacting viral and cellular interfaces. Although the affinity of a receptor for a single virus particle is low, the presence of multiple receptor-binding sites on the virion and the fluid nature of the plasma membrane allows engagement of multiple cell receptors. Consequently the avidity of virus binding to cells is usually very high. Virion binding can usually occur at 4°C (even though entry does not) as well as at body temperature (e.g., 37°C). Infection of cultured cells can therefore be synchronized by allowing binding to take place at a low temperature and then shifting the cells to a physiological temperature to allow the initiation of subsequent steps.

The first steps in virus attachment are governed largely by the probability that a virion and a cell will collide, and therefore by the concentrations of free virions and host cells. The rate of attachment can be described by the equation

$$dA/dt = k[V][H]$$

where $[V]$ and $[H]$ are the concentrations of virions and host cells, respectively, and k is a rate constant. Values of k for animal viruses vary greatly from a maximal value that represents the limits of diffusion to one that is as much as 5 orders of magnitude lower. It can be seen from this

BOX 5.1

TERMINOLOGY
Receptors and coreceptors

By convention, the first cell surface molecule that is found to be essential for virus binding is called its **receptor**. Sometimes, such binding is not sufficient for entry into the cell. When binding to another cell surface molecule is needed, that protein is called a **coreceptor**. For example, human immunodeficiency virus binds to cells via a receptor, CD4, and then requires interaction with a second cell surface protein such as CXCR4, the coreceptor.

In practice, the use of receptor and coreceptor can be confusing and inaccurate. A particular cell surface molecule that is a coreceptor for one virus may be a receptor for another. Furthermore, as is the case for the human immunodeficiency viruses, binding only to the coreceptor may be sufficient for entry of some members. Distinguishing receptors and coreceptors by the order in which they are bound is difficult to determine experimentally and is likely to be influenced by cell type and multiplicity of infection. Furthermore, some viruses can infect cells that synthesize only the coreceptor. Usage of the terms "receptor" and "coreceptor" is convenient when describing virus entry, but the appellations may not be entirely accurate.

equation that if a mixture of viruses and cells is diluted after virions have been allowed to attach, subsequent binding is greatly reduced. For example, a 100-fold dilution of the mixture reduces the attachment rate 10,000-fold (i.e., $1/100 \times 1/100$). Dilution can be used to prevent subsequent virus adsorption and hence to synchronize an infection.

Identification of Cell Receptors for Virus Particles

Early investigations of viral receptors exploited a variety of enzymes to characterize the cell surface components that are required for virus attachment. The first cell receptor discovered, sialic acid, which binds influenza virus, was identified because the enzyme neuraminidase removes this carbohydrate from cells and blocks virus attachment. In a similar way, experiments with proteases showed that many receptors are proteins. These types of analyses provided the first clues concerning the chemical nature of cell surface components to which virions become attached. It was also possible to determine whether different viruses share receptors, by determining whether saturating cells with one kind of virion prevented binding of a second.

Despite these approaches, identification of cell receptors for viruses languished because biochemical purification of these molecules proved difficult. As late as 1985, only one cell receptor, the sialic acid receptor of influenza viruses, had been identified unequivocally. The development of three crucial technologies rapidly changed this situation. The first, production of monoclonal antibodies, provided a powerful means of isolating and characterizing individual cell surface proteins. Hybridoma cell lines which secrete monoclonal antibodies that block virus attachment are obtained after immunizing mice with intact cells. Such antibodies can be used to purify the receptor protein by immunoprecipitation (Box 5.2) or affinity chromatography.

A second technology that advanced the cell receptor field was the development of DNA-mediated transformation. This method was crucial for isolating genes encoding receptors following introduction of DNA from susceptible cells into nonsusceptible cells (see Fig. 5.1). Cells that acquire DNA encoding the receptor and carry the corresponding protein on their surface are able to bind virus specifically. Clones of such cells are recognized and selected, for example, by the binding of receptor-specific

BOX 5.2

METHODS

Immunoprecipitation

Immunoprecipitation depends on the interaction of specific antibodies with proteins in solubilized extracts of cells or tissues (see figure). The antibody-protein complexes are isolated, and the proteins are dissociated from the complex and fractionated by electrophoresis in polyacrylamide gels. If the antibody is sufficiently specific, it may be possible to identify the protein that is bound by the antibody. For example, if the antibody used blocks virus attachment and is directed against a cell membrane protein, immunoprecipitation can provide information on the size of the protein. To identify the protein, it can be extracted from the gel and a partial amino acid sequence can be determined.

Isolation of proteins by immunoprecipitation. Cells are lysed with a detergent to solubilize proteins. Antibodies directed against the desired protein are coupled to beads and then added to the cell lysate. The beads are removed by centrifugation and washed free of protein not bound by the antibody. The bound proteins can then be fractionated by gel electrophoresis and visualized by staining. The heavy and light chains of the antibody molecules are not shown. The numbers next to the gel on the right are molecular masses in kilodaltons. SDS-PAGE, sodium dodecyl sulfate-polyacrylamide gel electrophoresis.

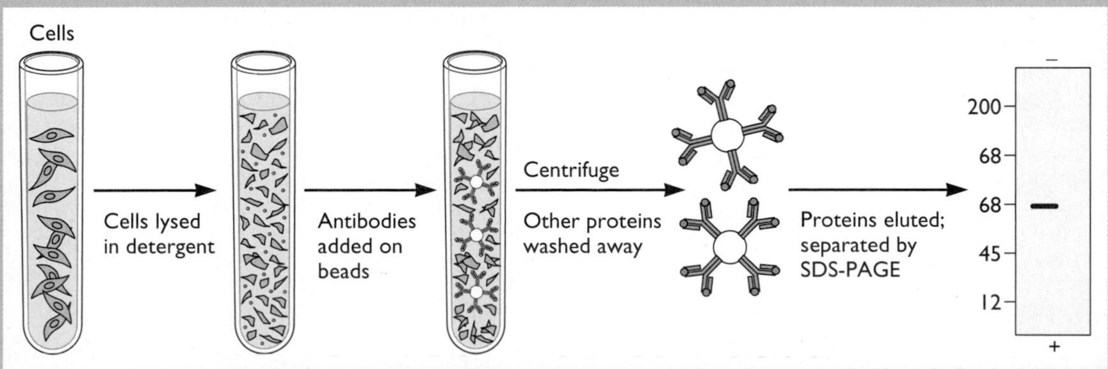

monoclonal antibodies. The receptor genes can then be isolated from these selected cells by using a third technology, molecular cloning. Although these technologies have led to the identification of many cell receptors for viruses, each method has associated uncertainties (Box 5.3).

The availability of cloned receptor genes has made it possible to investigate the details of receptor interaction with viruses by site-directed mutagenesis. Receptor proteins can be synthesized in heterologous systems and purified, and their properties can be studied *in vitro*, while animal cells producing altered receptor proteins can be used to test the effects of alterations on virus attachment. Because of their hydrophobic membrane-spanning domains, many of these cell surface proteins are relatively insoluble and difficult to work with. Soluble forms obtained by the expression of genes encoding truncated proteins lacking the membrane-spanning domain have been essential for structural studies of receptor-virus interactions. Cloned receptor genes have also been used to produce transgenic mice that synthesize receptor proteins. Such transgenic animals can serve as useful models in the study of human viral diseases.

Examples of Cell Receptors

Many different cell surface molecules can serve as receptors for the attachment of viruses. Some viruses attach to more than one, and viruses of different families share some receptors. Diverse molecules that serve as receptors for viruses are discussed below.

The Cell Receptor for Poliovirus, CD155

Members of the enterovirus genus of the *Picornaviridae* include human polioviruses, coxsackieviruses, echoviruses, and enteroviruses. These viruses are stable at acidic pH and multiply in the gastrointestinal tract. However, they also can replicate in other tissues, such as nerve, heart, and muscle. The cell receptor for poliovirus, CD155, was identified by using a DNA transformation and cloning strategy (Fig. 5.1). It was well known that mouse cells cannot be infected with poliovirus because they do not produce the cell receptor. Introduction of infectious poliovirus RNA into mouse tissue cell cultures leads to poliovirus replication, indicating that there is no intracellular block to virus multiplication. Cloning of the human gene from receptor-positive mouse cells established that the CD155 glycoprotein is a member of the immunoglobulin (Ig) superfamily (Fig. 5.2). The first of the three Ig-like domains is essential and sufficient for receptor function, but efficient (wild-type) binding requires all three. A structure of CD155 bound to poliovirus has been determined (Fig. 5.3A). This structure confirmed the results of mutational studies, which indicate that only domain 1 of CD155 contacts the viral capsid.

As mouse cells are permissive for poliovirus replication and as susceptibility appeared to be limited **only** by the absence of CD155, a small-animal model for the disease was developed by producing transgenic mice that synthesize this receptor. Inoculation of CD155 transgenic mice with poliovirus by various routes produces paralysis, tremors, and death, as is observed in human poliomyelitis. These CD155-synthesizing mice were the first new animal model created by transgenic technology for the study of viral disease. Similar approaches have subsequently led to animal models for viral diseases caused by measles virus and echoviruses.

The observation that even some human cells that make CD155 are resistant to poliovirus infection prompted the search for a second cellular protein that could regulate

BOX 5.3 BACKGROUND

Criteria for identifying cell receptors for viruses

The combination of monoclonal antibodies, molecular cloning, and DNA-mediated transformation provides a powerful approach for identifying cellular proteins that are receptors for viruses. Each method has associated uncertainties. For example, a monoclonal antibody that blocks virus attachment might recognize not the receptor but a closely associated membrane protein (see "The Cell Receptor for Poliovirus, CD155"). To prove that the protein recognized by the monoclonal antibody **is** a receptor, the protein must be isolated and its DNA must be cloned and introduced into cells to demonstrate that it can confer virus-binding activity. Any of the approaches outlined in Fig. 5.1 can result in identification of a cellular gene that encodes a putative receptor. However, the encoded protein might not be a receptor but may modify another cellular protein so that it can serve as a receptor. Proof that the DNA codes for a receptor would come from the identification of a monoclonal antibody that blocks virus attachment and is directed against the encoded protein.

For some viruses, synthesis of the receptor on cells leads to binding but not infection. In such cases a coreceptor is required, either for internalization or for membrane fusion. The techniques of molecular cloning also can be used to identify coreceptors. For example, production of CD4 on mouse cells leads to binding of human immunodeficiency virus type 1 but not infection, because fusion of viral and cell membranes does not occur. To identify the coreceptor, a DNA clone was isolated from human cells that allowed membrane fusion catalyzed by viral SU protein in mouse cells synthesizing CD4.

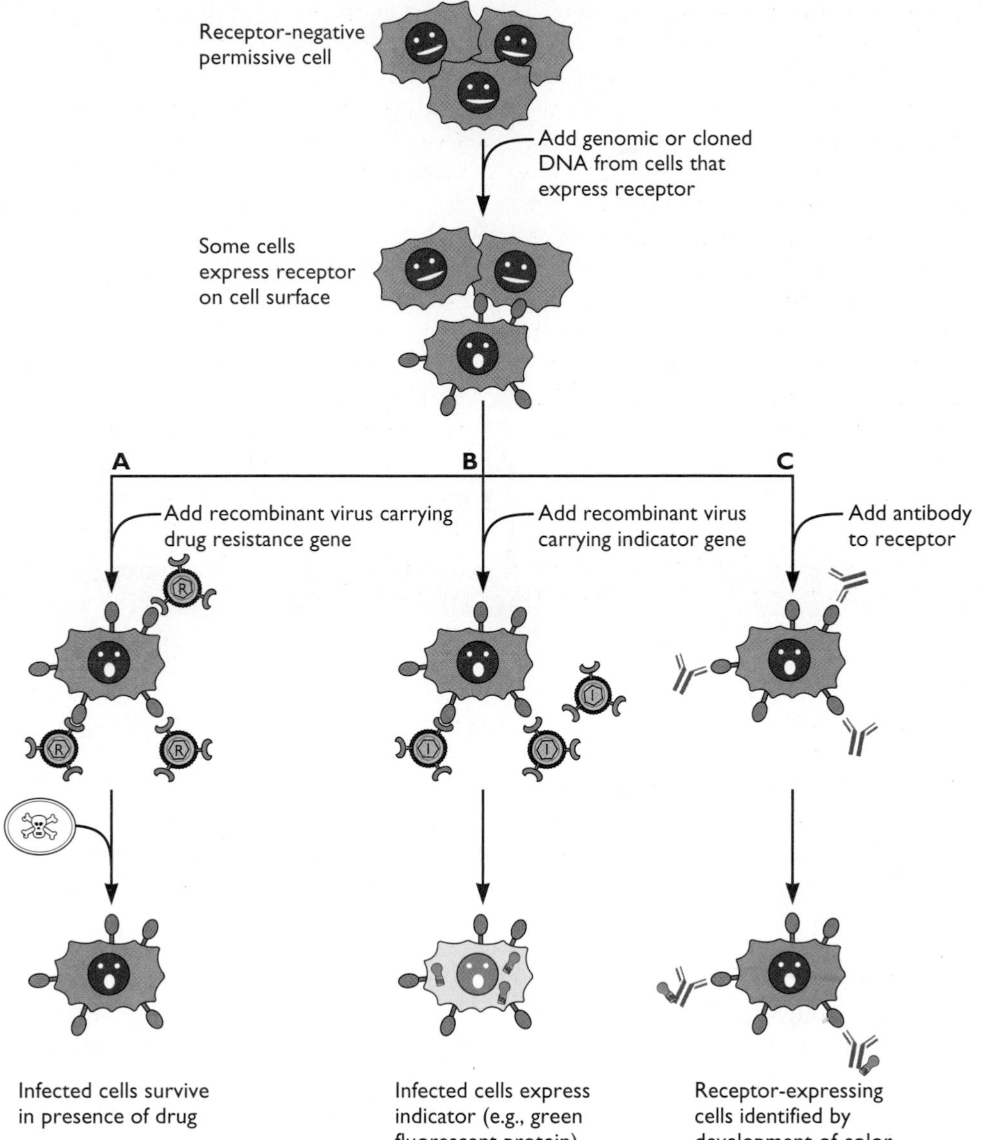

Figure 5.1 Experimental strategies for identification and isolation of genes encoding cell receptors for viruses. Genomic DNA or pools of DNA clones from cells known to synthesize the receptor are introduced into receptor-negative permissive cells. A small number of recipient cells express the receptor. Three different strategies for identifying such rare receptor-expressing cells are outlined. **(A)** The cells are infected with a virus that has been engineered so that it carries a gene encoding drug resistance. Cells that express the receptor will become resistant to the drug. This strategy works only for viruses that persist in cells without killing them. **(B)** For lytic viruses, an alternative is to engineer the virus to express an indicator, such as green fluorescent protein or β-galactosidase. Cells that make the correct receptor and become infected with such viruses can be distinguished by a color change, such as green in the case of green fluorescent protein. **(C)** The third approach depends on the availability of an antibody directed against the receptor, which binds to cells that express the receptor gene. Bound antibodies can be detected by an indicator molecule. When complementary DNA (cDNA) cloned in a plasmid is used as the donor DNA, pools of individual clones (usually 10,000 clones per pool) are prepared and introduced individually into cells. The specific DNA pool that yields receptor-expressing cells is then subdivided, and the screening process is repeated until a single receptor-encoding DNA is identified.

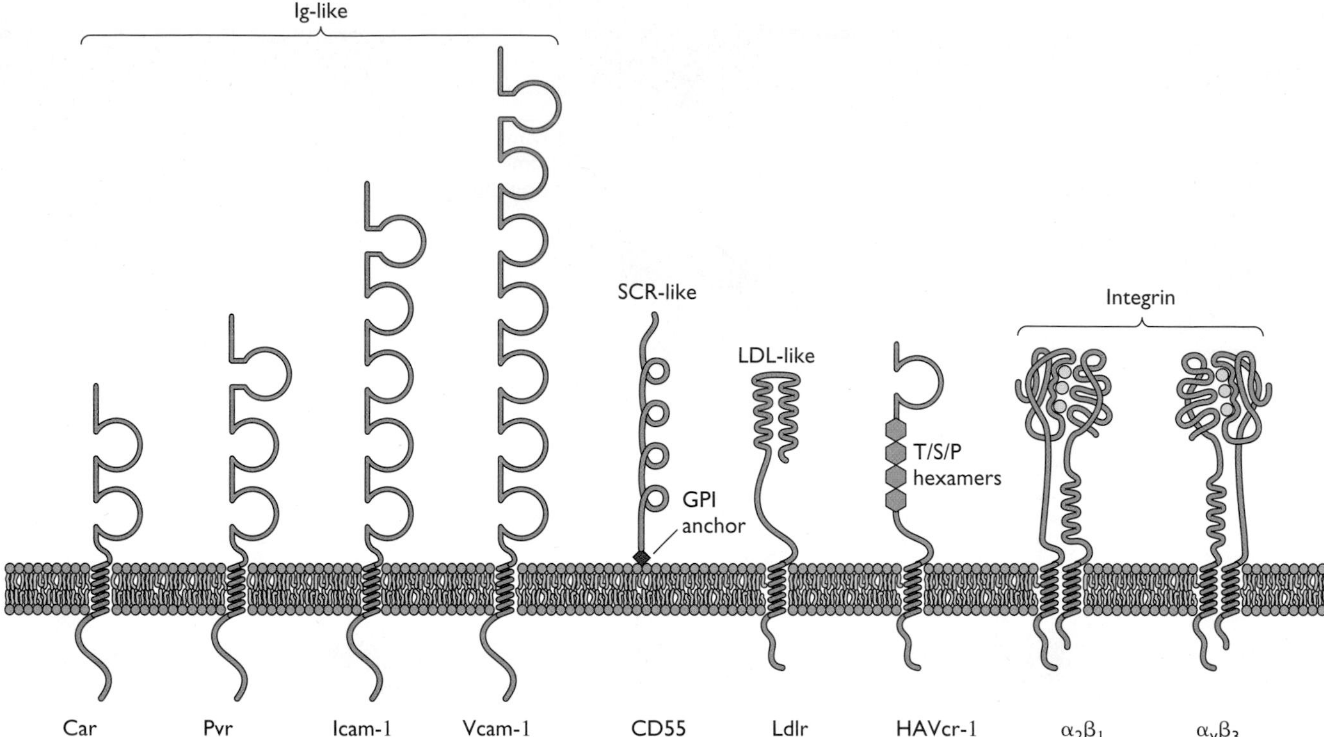

Figure 5.2 Cell receptors for picornaviruses. A schematic diagram of the cell proteins that function as receptors for different picornaviruses is shown. The different domains (Ig-like, short consensus repeat-like [SCR-like], low-density lipoprotein-like [LDL-like], and threonine/serine/proline [T/S/P]) are labeled. Car, coxsackievirus and adenovirus receptor; Vcam-1, vascular cell adhesion molecule 1; Ldlr, low-density lipoprotein receptor. Adapted from D. J. Evans and J. W. Almond, *Trends Microbiol.* **6**:198–202, 1998, with permission.

virus entry. One candidate was the lymphocyte homing receptor, CD44, which was thought to be a coreceptor for poliovirus, because a monoclonal antibody against it blocks poliovirus binding to cells. This multifunctional 100-kDa membrane glycoprotein normally helps direct the migration of lymphocytes to the lymph nodes and regulates lymphocyte adhesion and other functions. However, CD44 is not a receptor for poliovirus, nor is it required for poliovirus infection of cells that produce CD155. It is thought that CD155 and CD44 are associated in the plasma membrane and that anti-CD44 antibodies may block poliovirus attachment by sterically hindering the poliovirus-binding site on CD155. These results emphasize that monoclonal antibodies that block virus binding do not necessarily identify the receptor on cells. The resistance to poliovirus infection of certain cells that carry CD155 is determined by the type I interferon response (see Volume II, Chapter 3).

The Cell Receptor for Rhinovirus, Intercellular Adhesion Molecule 1

Members of the *Rhinovirus* genus of the *Picornaviridae* are unstable below pH 5 to 6 and multiply primarily in the upper respiratory tract. Up to 50% of all human common colds are caused by members of the major group of rhinoviruses. The cell surface receptor for these rhinoviruses (~90 serotypes) was identified by screening monoclonal antibodies for their ability to block rhinovirus infection. When such a monoclonal antibody was identified, it was used to isolate a 95-kDa cell surface glycoprotein by affinity chromatography. Amino acid sequence analysis of the purified protein, which bound to rhinovirus *in vitro*, identified it as the integral membrane protein intercellular adhesion molecule 1 (Icam-1). The cell receptor for the remaining rhinovirus serotypes is the low-density lipoprotein receptor.

Icam-1 is a member of the Ig superfamily. It has five domains that are homologous to one another and to the constant domains of antibody molecules. Each domain is stabilized by intrachain disulfide bonds. Icam-1 is found on the surface of many cell types, including those of the nasal epithelium, the normal entry site for rhinoviruses. The natural ligand of Icam-1 is another integral membrane glycoprotein, the integrin known as lymphocyte function-associated antigen 1 (Lfa-1). The normal function of Icam-1

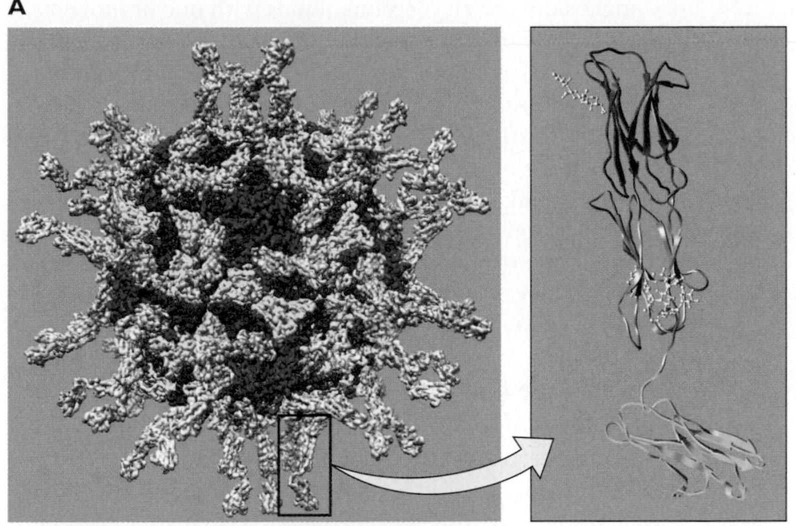

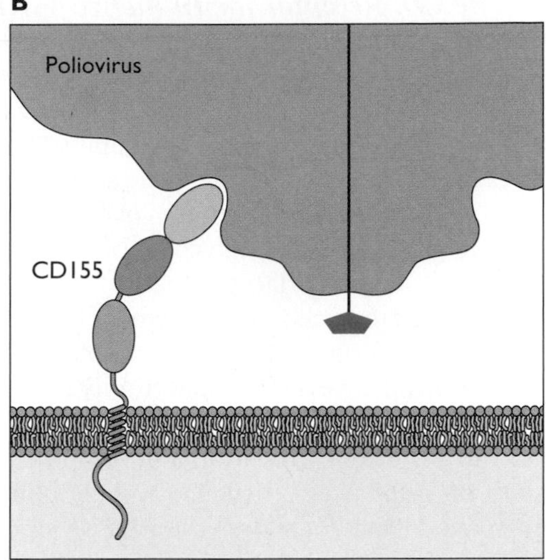

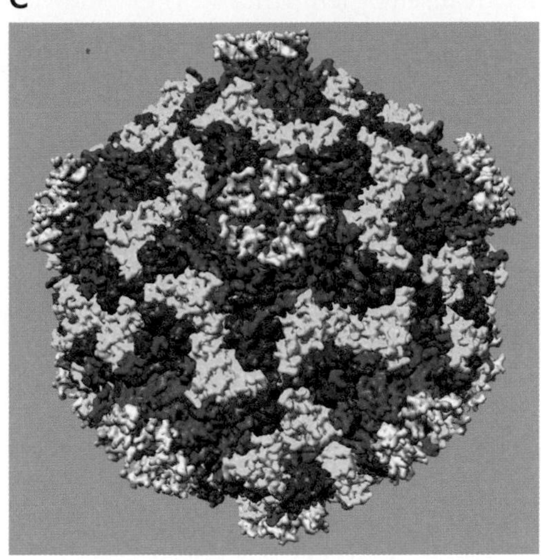

Figure 5.3 Picornavirus-receptor interactions. (A) Structure of poliovirus bound to a soluble form of CD155 (gray), derived by cryo-electron microscopy and image reconstruction. Capsid proteins are color coded (VP1, blue; VP2, yellow; VP3, red). One CD155 molecule is shown as a ribbon model in the panel to the right, with each Ig-like domain in a different color. The first Ig-like domain of CD155 (magenta) binds in the canyon of the viral capsid. **(B)** Depiction of CD155 binding to the canyon of the poliovirion. Adapted from Fig. 3e of D. M. Belnap et al., *Proc. Natl. Acad. Sci. USA* **97**:73–78, 2000, with permission. **(C)** Structure of human rhinovirus type 2 bound to a soluble form of low-density lipoprotein receptor (gray). The receptor binds on the plateau at the fivefold axis of symmetry of the capsid.

is to bind Lfa-1 on the surface of lymphocytes and promote a wide variety of immunological and inflammatory responses. Mediators of inflammation induce increased synthesis of Icam-1. Therefore, the initial reaction of host defenses to rhinoviruses, which leads to the production of such mediators, might actually induce the appearance of Icam-1 in nearby cells and thereby enhance subsequent spread of the virus.

Mutational analyses of cloned Icam-1 DNA established that the binding sites for rhinovirus and Lfa-1 are located in domains 1 and 2, with critical contact points in both cases mapping to domain 1, the most membrane distal and accessible. Although the binding sites for rhinovirus and Lfa-1 partially overlap, they are distinct. Amino acids in first N-terminal domains of CD155 and Icam-1 that are crucial for virus binding are different, demonstrating the diverse interactions that may occur even among structurally related viruses and receptors.

For other picornaviruses, one type of cell receptor is not enough for infection. Decay-accelerating protein (CD55), a member of the complement cascade, is the cell receptor for many enteroviruses (Fig. 5.2), but infection also requires the presence of a coreceptor. For example, coxsackievirus A21 can bind to cell surface decay-accelerating protein, but this interaction does not lead to infection unless Icam-1 is also present. It is thought that virions first bind to decay-accelerating protein and then interact with Icam-1 to allow cell entry. A similar process, with different receptors and coreceptors, is likely to occur during infection with other enteroviruses.

The Cell Receptor for Influenza Virus, Sialic Acid

The family *Orthomyxoviridae* comprises the three genera of influenza viruses, A, B, and C. Influenza A viruses are the best-studied members of this family. This virus binds to negatively charged, terminal sialic acid moieties present in oligosaccharide chains that are covalently attached to cell surface glycoproteins or glycolipids. The presence of sialic acid on most cell surfaces accounts for the ability of influenza virions to attach to many types of cell. The interaction of influenza virus with individual sialic acid moieties is of low affinity. However, the opportunity for multiple interactions among the numerous hemagglutinin (HA) molecules on the surface of the virion and multiple sialic acid residues on cellular glycoproteins and glycolipids results in a high overall avidity of the virus particle for the cell surface. The surfaces of influenza viruses were shown in the early 1940s to contain an enzyme that, paradoxically, removes the receptors for attachment from the surface of cells. Later this enzyme was identified as the virus-encoded envelope protein neuraminidase, which cleaves the glycoside linkages of sialic acids (Fig. 5.4). This enzyme is required for release of virions bound to the surfaces of infected cells, facilitating virus spread through the respiratory tract (Volume II, Chapter 9).

Glycolipids, Unusual Cell Receptors for Polyomaviruses

The family *Polyomaviridae* includes simian virus 40, mouse polyomavirus, and human BK virus. These viruses are unusual because they bind to ganglioside cell receptors. Gangliosides are glycosphingolipids with one or more sialic acids linked to a sugar chain. There are over 40 known gangliosides, which differ in the position and number of sialic acid residues. Simian virus 40, polyomavirus, and BK virus bind to three different types of ganglioside. Structural studies have revealed that sialic acid linked to galactose by an α(2,3) linkage binds to a pocket on the surface of the polyomavirus capsid. Gangliosides are highly concentrated in lipid rafts (Chapter 2, Box 2.1) and participate in signal transduction, two properties that play roles during polyomavirus entry into cells.

CD4, the Cell Receptor for Human Immunodeficiency Virus Type 1

Animal retroviruses have long been of interest because of their ability to cause a variety of serious diseases, especially cancers (caused by oncogenic retroviruses) and neurological disorders (caused by lentiviruses). The worldwide acquired immunodeficiency syndrome (AIDS) epidemic has focused enormous attention on the lentivirus human immunodeficiency virus type 1 and its close relatives. The cell surface receptors of this virus have been among the most intensively studied and currently are the best understood.

The cell receptor for human immunodeficiency virus type 1 is CD4 protein, a 55-kDa rodlike molecule that is a member of the Ig superfamily and has four Ig-like domains. A variety of techniques have been used to identify the site of interaction with human immunodeficiency virus type 1, including site-directed mutagenesis and X-ray crystallographic studies of a complex of CD4 and the viral attachment protein SU (Fig. 5.5). The interaction site

Figure 5.4 Sialic acid receptors for influenza viruses. An integral membrane glycoprotein is shown at left; the arrows point to terminal sialic acid units that are attachment sites for influenza virus. The structure of a terminal sialic acid moiety that is recognized by the viral envelope protein hemagglutinin is shown at right. Sialic acid is attached to galactose by an α(2,3) linkage in the example shown; certain HA subtypes preferentially bind to molecules with an α(2,6) linkage. The site of cleavage by the influenza virus envelope protein neuraminidase is indicated. The sialic acid shown is *N*-acetylneuraminic acid, which is the preferred receptor for influenza A and B viruses. These viruses do not bind to 9-*O*-acetyl-*N*-neuraminic acid, the receptor for influenza C viruses.

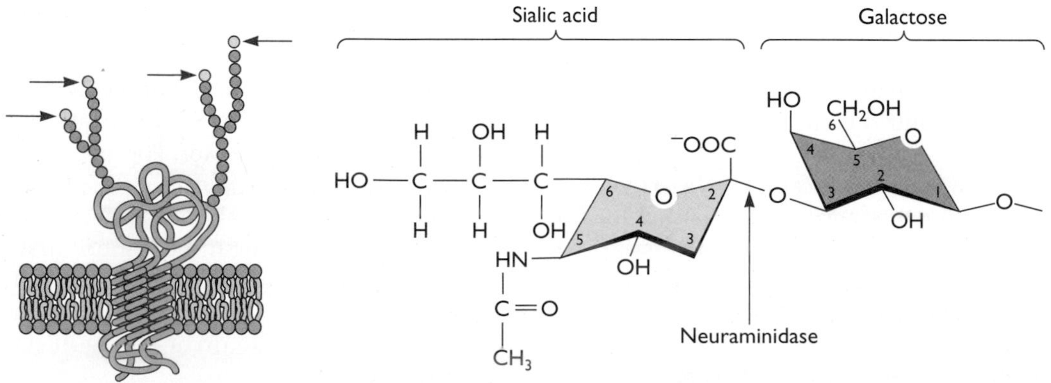

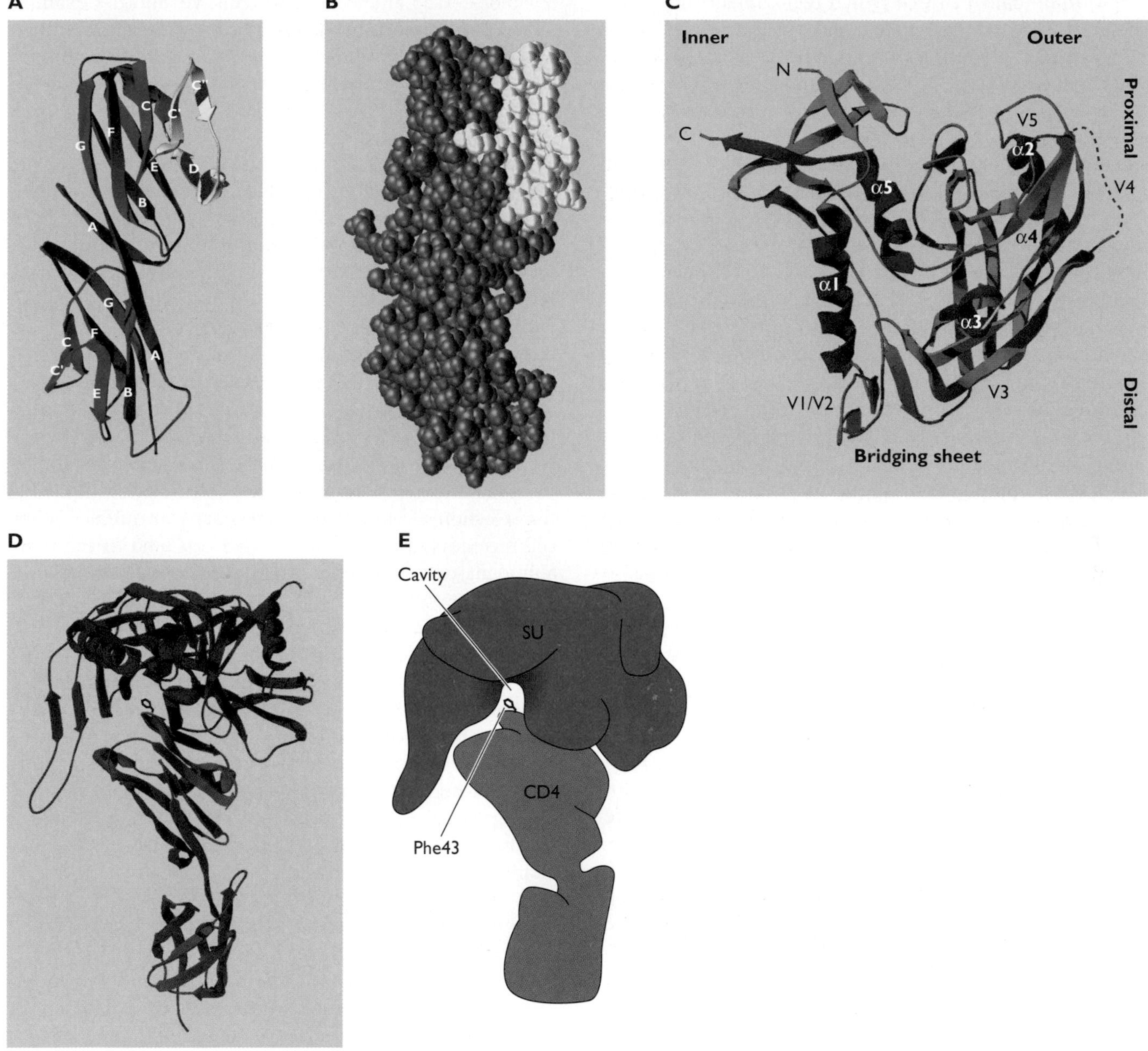

Figure 5.5 Interaction of human immunodeficiency virus type 1 SU with its cell receptor, CD4. **(A)** Ribbon model of the backbone carbons of CD4 domains 1 and 2 (residues 1 to 182), derived from X-ray crystallography. **(B)** Space-filling model of CD4. Shown in yellow in panels A and B are residues that interact with human immunodeficiency virus type 1 during attachment, as revealed by mutagenesis. **(C)** Ribbon diagram of a core of SU, derived from X-ray crystallographic data. This modified SU binds CD4 with an affinity comparable to that of the native protein. α-helices are red, β-strands are magenta, and β-strand 15, which forms an antiparallel β-sheet with strand C'' of CD4, is yellow. **(D)** Ribbon diagram of SU (red) bound to CD4 (brown), derived from X-ray crystallographic data. The side chain of CD4 Phe43 is shown. **(E)** Cartoon of the CD4-SU complex. Mutagenesis has identified CD4 Phe43 as a residue critical for binding to SU. Phe43 is shown penetrating the hydrophobic cavity of SU. This amino acid, which makes 23% of the interatomic contacts between CD4 and SU, is at the center of the interface and appears to stabilize the entire complex. Adapted from J. Wang et al., *Nature* **348:**411–418, 1990 (A and B), and P. D. Kwong et al., *Nature* **393:**648–659, 1998 (D), with permission.

for SU in domain 1 of CD4 is in a region analogous to the site in CD155 that binds to poliovirus. Because two viruses with entirely different virion architectures bind to analogous surfaces of these Ig-like domains, some feature of this region seems likely to be especially advantageous for virus attachment.

One of the first strategies to be considered for the treatment of AIDS was the use of soluble CD4 protein, which lacks the membrane-spanning domain, to inhibit viral infection. The rationale for such treatment was that soluble CD4 should bind virus particles and block their attachment to CD4 on host cell surfaces. Although inhibition of viral infectivity could be demonstrated in cell culture experiments, clinical trials gave disappointing results. This failure can be attributed in part to the fact that each viral envelope includes many copies (~30) of the structure that binds CD4. Consequently, relatively high concentrations in serum would be required to block all of them. This problem was further compounded by the short half-life of soluble CD4 in the blood. Furthermore, human immunodeficiency virus can also be spread from cell to cell by fusion, a process that is not readily blocked by circulating, soluble CD4.

Cell Receptors for Adenoviruses

The results of competition experiments indicated that members of two different virus families, group B coxsackieviruses and adenoviruses, share a cell receptor. This receptor is a 46-kDa member of the Ig superfamily called Car (Coxsackievirus and adenovirus receptor). Binding to this receptor is not sufficient for infection by most adenoviruses. Interaction with a coreceptor, the α_v integrin $\alpha_v\beta_3$ or $\alpha_v\beta_5$, is required for uptake of the capsid into the cell by receptor-mediated endocytosis. An exception is adenovirus type 9, which can infect hematopoietic cells after binding directly to α_v integrins. Adenoviruses of subgroup B bind CD46, which is also a cell receptor for some strains of measles virus, an enveloped member of the *Paramyxoviridae*.

Cell Receptors for Alphaherpesviruses

The alphaherpesvirus subfamily of the *Herpesviridae* includes herpes simplex virus types 1 and 2, pseudorabies virus, and bovine herpesvirus. Initial contact of these viruses with the cell surface is made by low-affinity binding to glycosaminoglycans (preferentially heparan sulfate), abundant components of the extracellular matrix. This interaction concentrates virus particles near the cell surface and facilitates subsequent attachment to an integral membrane protein, which is required for entry into the cell. Members of at least two different protein families serve as entry receptors for alphaherpesviruses. One of these families, the nectins, comprises the poliovirus receptor CD155 and related proteins, yet another example of receptors shared by different viruses. When members of these two protein families are not present, 3-O-sulfated heparan sulfate can serve as an entry receptor for alphaherpesviruses.

Alternative Receptors

Some examples of the use of alternative cell surface molecules as receptors for the same virus have already been discussed. Two additional examples illustrate how receptor usage depends on the nature of the virus isolate or the cell line. Infection with foot-and-mouth disease virus type A12 requires the RGD-binding integrin $\alpha_v\beta_3$. However, the receptor for the O strain of foot-and-mouth disease virus, which has been extensively passaged in cell culture, is not integrin $\alpha_v\beta_3$ but cell surface heparan sulfate. On the other hand, the type A12 strain cannot infect cells that lack integrin $\alpha_v\beta_3$, even if heparan sulfate is present. In a similar way, adaptation of Sindbis virus to cultured cells has led to the selection of variants that bind heparan sulfate. When cell receptors are rare, viruses that can bind to the more abundant glycosaminoglycan are readily selected.

How Virions Attach to Receptors

Animal viruses have multiple receptor-binding sites on their surfaces. Of necessity, one or more of the capsid proteins of nonenveloped viruses specifically interacts with the cell receptor. Receptor-binding sites for enveloped viruses are surface glycoproteins that have been incorporated into their cell-derived membranes. Although the details vary among viruses, most virus-receptor interactions follow one of several mechanisms illustrated by the best-studied examples described below.

Nonenveloped Viruses Bind via the Capsid Surface or Projections

Attachment via surface features: canyons and loops. The RNA genomes of picornaviruses are protected by capsids made up of four virus-encoded proteins, VP1, VP2, VP3, and VP4, arranged with icosahedral symmetry (see Fig. 4.12). Three-dimensional structures have been determined by X-ray crystallography for at least one member of each of the five picornavirus genera. The picornavirus capsid is built from 60 subunits arranged as 12 pentamers (see Fig. 4.12A). Each subunit contains the four capsid proteins in an identical arrangement, with portions of the first three exposed on the surface. Although the arrangement is similar, the surface architecture of the three exposed proteins varies among the family members, a property that accounts for the different serotypes and the different modes of interaction that can take place with cell receptors. For example, the capsids of rhinoviruses

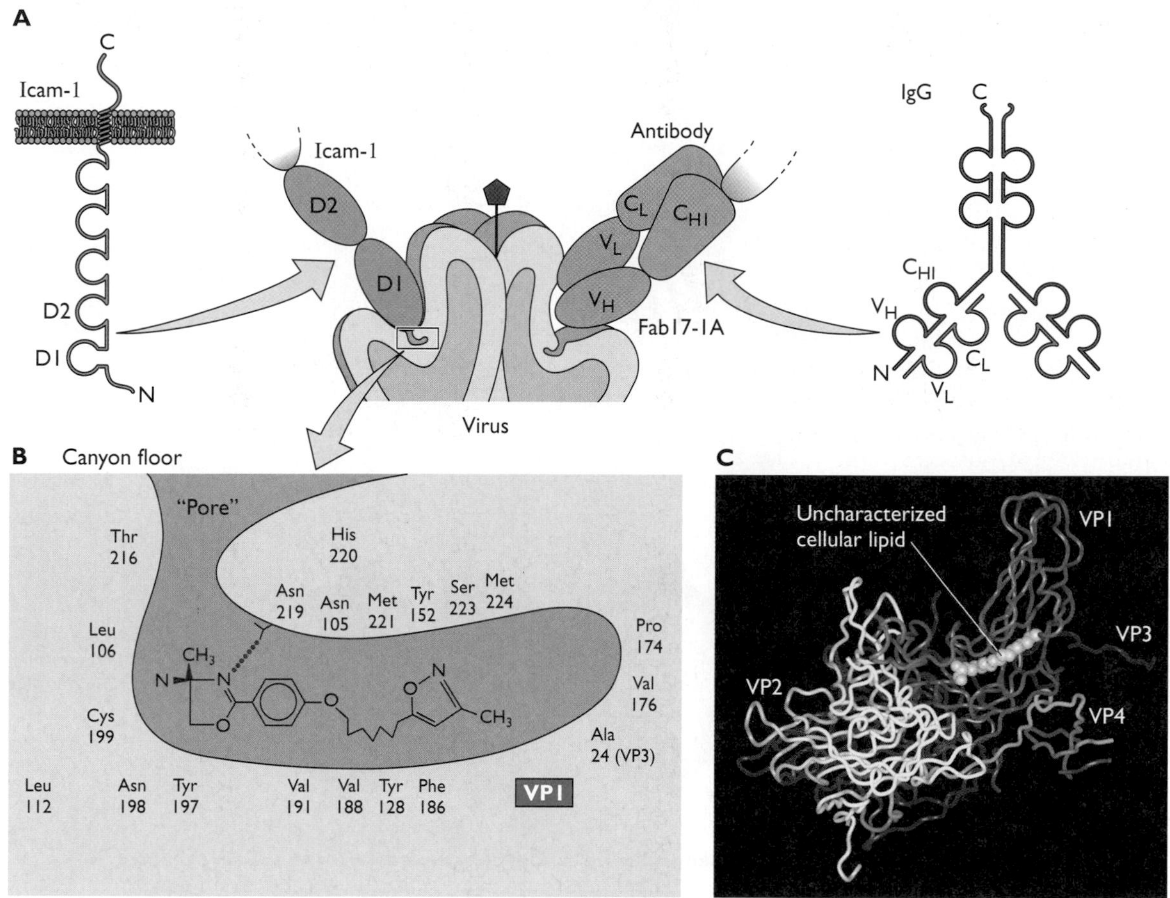

Figure 5.6 Receptor, antibody, and drug binding to the picornavirus capsid. (A) Schematic diagram of the canyon in the human rhinovirus capsid. The domain structure of the cell receptor Icam-1 is illustrated at the left, and the model in the center shows the tip of domain 1 dipping into the canyon. The Fab portion of the antibody contacts a good deal of the canyon but not residues at the deepest regions. Antibodies that bind to the virus in this manner neutralize viral infectivity by blocking entry of receptor into the canyon. **(B)** Location of a WIN compound in a hydrophobic pocket below the canyon floor. **(C)** Location of lipid, possibly sphingosine, in the capsid of poliovirus type 1. Shown is a protomer consisting of one copy of VP1 (blue), VP2 (yellow), VP3 (red), and VP4 (green). The lipid, shown as gray spheres, is bound in the hydrophobic tunnel beneath the canyon floor. Adapted from T. Smith et al., *Nature* **383**:350–354, 1996 (A), and J. Badger et al., *Proc. Natl. Acad. Sci. USA* **85**:3304–3308, 1988 (B), with permission.

and some enteroviruses such as poliovirus have deep canyons surrounding the 12 fivefold axes of symmetry (Fig. 5.3), whereas cardioviruses and aphthoviruses lack this feature.

The canyons in the capsids of some rhinoviruses and enteroviruses are the sites of interaction with cell surface receptors (Fig. 5.3). Amino acids that line the canyons are more highly conserved than any others on the viral surface, and their substitution can alter the affinity of binding to cells. Poliovirus bound to a receptor fragment comprising CD155 domains 1 and 2 has been visualized in reconstructed images from cryo-electron microscopy. The results indicate that the first domain of CD155 binds to the central portion of the canyon in an orientation oblique to the surface of the virion (Fig. 5.3B).

Although canyons are present in the capsid of rhinovirus type 2, they are not the binding sites for the cellular receptor, low-density lipoprotein receptor. Rather, the binding site on the capsid is located on the star-shaped plateau at the fivefold axis of symmetry (Fig. 5.3C). Sequence and structural comparisons have revealed why different rhinovirus serotypes bind distinct receptors. The key amino acid that interacts with a negatively charged cluster of low-density lipoprotein receptor is a lysine of VP1 conserved in all rhinoviruses that bind this receptor. This lysine is not found in VP1 of rhinoviruses that bind Icam-1.

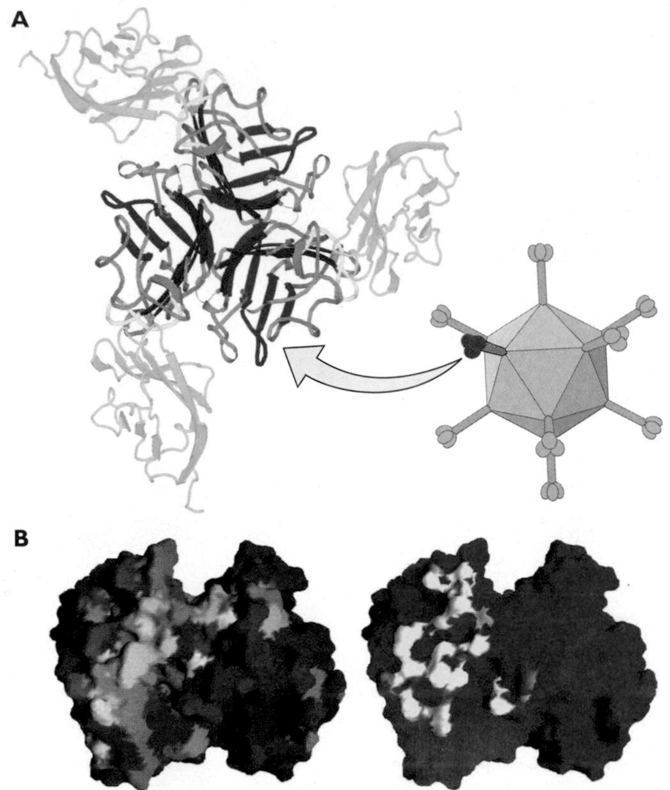

Figure 5.7 Structure of the adenovirus 12 knob complexed with the Car receptor. **(A)** Ribbon diagram of the knob-Car complex as viewed down the axis of the viral fiber. The trimeric knob is in the center. The AB loop of the knob protein, which contacts Car, is in yellow. The first Ig-like domains of three Car molecules bound to the knob are colored blue. **(B)** Surface models of the interface between the knob and Car domain 1. Both models depict two knob monomers and are viewed looking down at the interface with Car. (Left) Conservation of amino acids among all known adenovirus knob protein sequences is represented by a color scale from white (conserved amino acids) to red (nonconserved amino acids). The white strip of conserved amino acids is covered when Car is bound. (Right) Amino acids are colored according to whether they do (yellow) or do not (red) contact Car. From M. C. Bewley et al., *Science* **286:**1579–1583, 1999, with permission.

The canyons of picornaviruses were at first thought to be too deep and narrow to permit entry of antibody molecules with adjacent Ig domains. It was hypothesized that this physical barrier would allow amino acids crucial for receptor interactions to be hidden from the immune system. However, X-ray crystallographic analyses of a specific rhinovirus-antibody complex have shown that the antibody penetrates deep into the canyon, as does Icam-1: the shape of the canyon is not likely to play a role in immune escape (Fig. 5.6).

For picornaviruses with capsids that do not have prominent canyons, including coxsackievirus A and foot-and-mouth disease virus, attachment is to VP1 surface loops that include RGD motifs recognized by their integrin receptors. Alteration of the RGD sequence in VP1 of foot-and-mouth disease virus blocks virus binding.

Attachment via protruding fibers. The nonenveloped DNA-containing adenoviruses are much larger than picornaviruses, and their icosahedral capsids are more complex, comprising at least 10 different proteins. Electron microscopy shows that adenovirus particles have fibers protruding from each pentamer (Fig. 5.7; see the appendix, Fig. 1A). The fibers are composed of homotrimers of the adenovirus fiber protein and are anchored in the pentameric penton base protein; both proteins have roles to play in virus attachment and uptake.

For many adenovirus serotypes, attachment via the fibers is necessary but not sufficient for infection. A region comprising the N-terminal 40 amino acids of each subunit of the fiber protein is bound noncovalently to the penton base. The central shaft region is composed of repeating

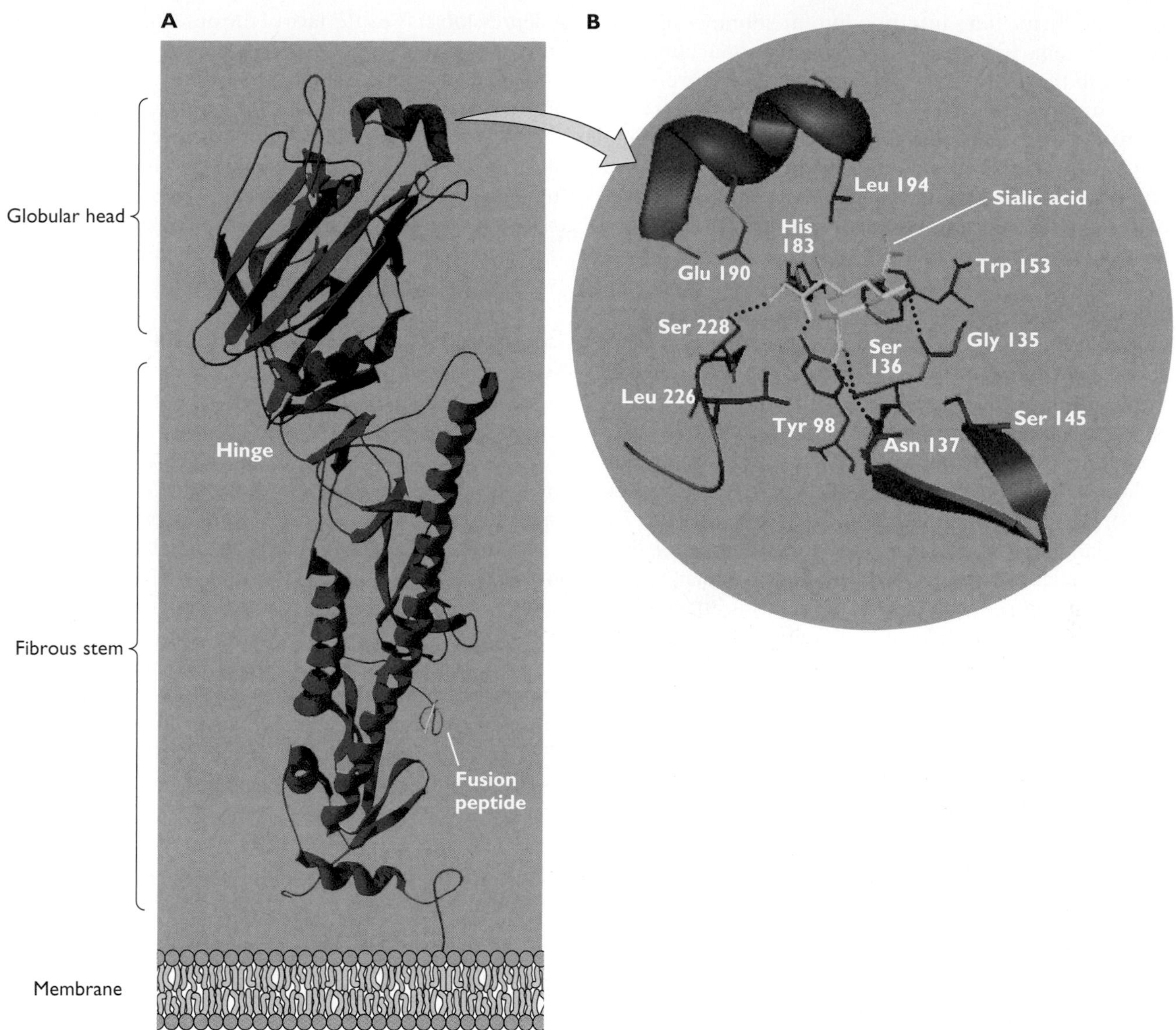

Figure 5.8 Structure of a monomer of influenza virus HA protein and details of the receptor-binding site. (A) HA monomer modeled from the X-ray crystal structure of the natural trimer. HA1 (blue) and HA2 (red) subunits are held together by a disulfide bridge as well as by many noncovalent interactions. The fusion peptide at the N terminus of HA2 is indicated (yellow). **(B)** Close-up of the receptor-binding site with a bound sialic acid molecule. Side chains of the conserved amino acids that form the site and hydrogen-bond with the receptor are included.

motifs of approximately 15 amino acids; the length of the shaft in different serotypes is determined by the number of these repeats. The three constituent shaft regions appear to form a rigid triple-helical structure in the trimeric fiber. The C-terminal 180 amino acids of each subunit interact to form a terminal knob. Genetic analyses and competition experiments indicate that determinants for the initial, specific attachment to host cell receptors reside in this knob. The structure of this receptor-binding domain bound to Car reveals that surface loops of the knob contact one face of Car (Fig. 5.7).

Enveloped Viruses Bind via Transmembrane Glycoproteins

As noted above, the lipid membranes of enveloped viruses originate from the cells they infect. The process of virion assembly includes insertion into membranes of specific viral proteins that carry membrane-spanning domains analogous to those of cellular integral membrane proteins. Attachment sites (i.e., viral ligands) on one or more of these envelope proteins bind to specific cell receptors. The two best-studied examples of enveloped virus attachment and its consequences are provided by the interactions of influenza A virus and the retrovirus human immunodeficiency virus type 1 with their cell receptors.

Influenza virus HA. Influenza virus HA is the viral glycoprotein that binds to the cell receptor sialic acid. The HA monomer is synthesized as a precursor that is glycosylated and subsequently cleaved to form HA1 and HA2 subunits. Each HA monomer consists of a long, helical stalk anchored in the membrane by HA2 and topped by a large HA1 globule, which includes the sialic acid-binding pocket (Fig. 5.8). While attachment of all influenza A virus strains requires sialic acid, strains vary in their affinities for different sialyloligosaccharides. For example, human virus strains are preferentially bound by sialic acids attached to galactose via an $\alpha(2,6)$ linkage, the major sialic acid present on human respiratory epithelium (Fig. 5.4). Avian virus strains bind preferentially to sialic acids attached to galactose via an $\alpha(2,3)$ linkage, the major sialic acid in the duck gut epithelium. Amino acids in the sialic acid-binding pocket of HA (Fig. 5.8) determine which sialic acid is preferred and can therefore determine viral host range. An example is the origin of the 1918 influenza virus strain, which may have evolved from an avian virus. It is thought that an amino acid change in the sialic acid-binding pocket of the avian HA allowed it to recognize the $\alpha(2,6)$-linked sialic acids that predominate in human cells.

The envelope glycoprotein of human immunodeficiency virus type 1. When examined by electron microscopy, the envelopes of human immunodeficiency virus type 1 and other retroviruses appear to be studded with "spikes" (see Fig. 4.19). These structures are composed of trimers of the single viral envelope glycoprotein. The spikes bind the cell receptor CD4 (Fig. 5.5). The monomers of the spike protein are synthesized as heavily glycosylated precursors that are cleaved by a cellular protease to form SU and TM. The latter is anchored in the envelope by a single membrane-spanning domain and remains bound to SU by numerous noncovalent bonds.

The atomic structure of a complex of human immunodeficiency virus type 1 SU, a two-domain fragment of CD4, and a neutralizing antibody against SU has been determined by X-ray crystallography (Fig. 5.5). The polypeptide of SU is folded into an inner and an outer domain linked by an antiparallel four-stranded "bridging sheet." A depression at the interface of the outer and inner domains and the bridging sheet forms the binding site for CD4. The CD4-binding site in SU is a deep cavity, and the opening of this cavity is occupied by CD4 amino acid Phe43, which is critical for SU binding. Comparison with the structure of SU in the absence of CD4 indicates that receptor binding induces conformational changes in SU. These changes expose binding sites on SU for the chemokine receptors, which are required for fusion of viral and cell membranes (see "Uncoating at the Plasma Membrane" below).

Endocytosis of Virions by Cells

Many viruses enter cells by the same pathways by which cells take up macromolecules. The plasma membrane, the limiting membrane of the cell, permits nutrient molecules to enter and waste molecules to leave, thereby ensuring an appropriate internal environment. Water, gases, and small hydrophobic molecules such as ethanol can freely traverse the lipid bilayer, but most metabolites and certain ions (Ca^{2+}, H^+, K^+, and Na^+) cannot diffuse through the membrane. These essential components enter the cell by specific transport processes. Integral membrane proteins are responsible for the transport of ions, sugars, and amino acids, while proteins and large particles are taken into the cell by phagocytosis or endocytosis. The former process (Fig. 5.9) is nonspecific, which means that any particle or molecule can be taken into the cell.

Figure 5.9 Mechanisms for the uptake of macromolecules from extracellular fluid. During phagocytosis, large particles such as bacteria or cell fragments that come in contact with the cell surface are engulfed by extensions of the plasma membrane. Phagosomes ultimately fuse with lysosomes, resulting in degradation of the material within the vesicle. Macrophages use phagocytosis to ingest bacteria and destroy them. Endocytosis comprises the invagination and pinching off of small regions of the plasma membrane, resulting in the nonspecific internalization of molecules (pinocytosis or fluid-phase endocytosis) or the specific uptake of molecules bound to cell surface receptors (receptor-mediated endocytosis). Adapted from J. Darnell et al., *Molecular Cell Biology* (Scientific American Books, New York, NY, 1986), with permission.

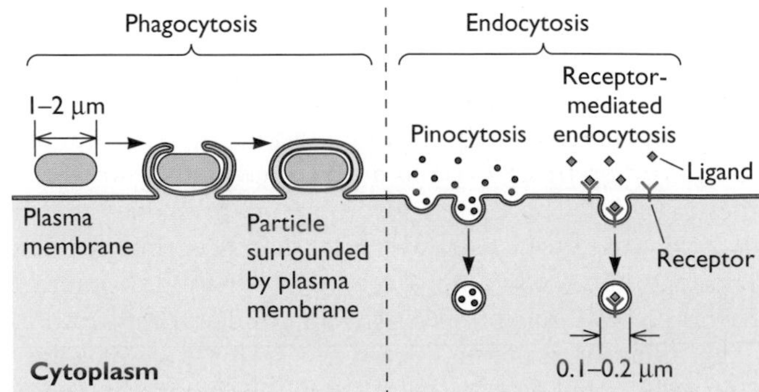

Specific molecules are selectively taken into cells from the extracellular fluid by **receptor-mediated endocytosis** (Fig. 5.9 and 5.10); this is also the mechanism of entry of many viruses. Ligands in the extracellular medium bind to cells via specific plasma membrane receptor proteins. The receptor-ligand complex diffuses along the membrane until it reaches an invagination that is coated on its cytoplasmic surface by a cagelike lattice composed of the fibrous protein clathrin (Fig. 5.10). Such clathrin-coated pits can comprise as much as 2% of the surface area of a cell, and some receptors are clustered over these areas even in the absence of their ligands. Following the accumulation of receptor-ligand complexes, the clathrin-coated pit invaginates and then pinches off to form a clathrin-coated vesicle containing the ligand-receptor complex. Within a few seconds, the clathrin coat is lost and the vesicles fuse with small, smooth-walled vesicles located near the cell surface, called early **endosomes.** The lumen of early endosomes is mildly acidic (pH 6.5 to 6.0), a result of energy-dependent transport of protons into the interior of the vesicles by a membrane proton pump. The contents of the early endosome are then transported via endosomal carrier vesicles to late endosomes located close to the nucleus. The lumen of late endosomes is more acidic (pH 6.0 to 5.0). Some ligands dissociate from their receptors in the acidic environment of the endosome, and the receptors are recycled to the cell surface by transport vesicles that bud from the endosome and fuse with the plasma membrane. Late endosomes in turn fuse with **lysosomes,** which are vesicles containing a variety of enzymes that degrade sugars, proteins, nucleic acids, and lipids. Ligands that reach the lysosomes are degraded by enzymes for further use of their constituents. In some cases, the entire ligand-receptor complex travels to the lysosomal compartment, where it is degraded. Viruses usually enter the cytoplasm from the early or late endosomes, and a few enter from lysosomes.

Clathrin-mediated endocytosis is a continuous but regulated process. For example, the uptake of vesicular stomatitis virus into cells may be influenced by over 90 different cellular protein kinases. Influenza virus and reovirus particles are taken into cells, not into preexisting pits but mainly by clathrin-coated pits that form after virus binds to the cell surface. It is not known how virus binding to the plasma membrane induces the formation of the clathrin-coated pit.

Although uptake of most viruses occurs by the clathrin-mediated endocytic pathway, other pathways are also involved. These include caveolin- (or raft-mediated) and clathrin-independent endocytosis (Fig. 5.10). The caveolar pathway requires cholesterol (a major component of lipid rafts). Three types of caveolar endocytosis have been identified. Endocytosis by caveolin 1-containing **caveolae** is observed in cells infected with simian virus 40 and polyomavirus. Dynamin 2-dependent, noncaveolar, raft-mediated endocytosis occurs during echovirus and rotavirus infection, while dynamin-independent, noncaveolar, raft-mediated endocytosis is also observed during simian virus 40 and polyomavirus infection. Caveolae are distinguished from clathrin-coated vesicles by their flask-like shape, their size (50 to 70 nm in diameter), the absence of a clathrin coat, and the presence of a marker protein called caveolin. In the uninfected cell, caveolae participate in transcytosis, signal transduction, and uptake of membrane components and extracellular ligands. When a virus particle binds the caveolae, a signal transduction pathway involving tyrosine phosphorylation is activated. Such signaling is required for pinching off of the vesicle, which then moves within the cytoplasm. Disassembly of filamentous actin also occurs, presumably to facilitate movement of the vesicle deeper into the cytoplasm. There it fuses with the **caveosome**, a larger membranous organelle that contains caveolin (Fig. 5.10). In contrast to endosomes, the pH of the caveosome lumen is neutral. Some viruses (e.g., echovirus type 1) penetrate the cytoplasm from the caveosome. Others (simian virus 40, polyomavirus, coxsackievirus B3) are sorted to the endoplasmic reticulum by a transport vesicle that lacks caveolin. These viruses enter the cytoplasm by a process mediated by thiol oxidases present in the lumen of the endoplasmic reticulum and by a component of the protein degradation pathway present in the membrane.

The study of virus entry by endocytosis can be confusing because some viruses may enter cells by multiple routes, depending on cell type and multiplicity of infection. For example, herpes simplex virus can enter cells by three different routes and influenza A virus may enter cells by both clathrin-dependent and clathrin-independent pathways.

Membrane Fusion

The formation of vesicles during the process of endocytosis requires the fusion of cell membranes. For example, during endocytosis, fusion produces the intracellular vesicle following invagination of a small region of the plasma membrane (Fig. 5.10). Membrane fusion also takes place during many other cellular processes, such as cell division, myoblast fusion, and exocytosis.

Membrane fusion must be regulated in order to maintain the integrity of the cell and its intracellular compartments. Consequently, membrane fusion does not occur spontaneously but proceeds by specialized mechanisms mediated by proteins. The two membranes must first come into close proximity. This reaction is mediated by interactions of integral membrane proteins that protrude from the lipid bilayers, a targeting protein on one membrane and a docking protein on the other. The next step, fusion,

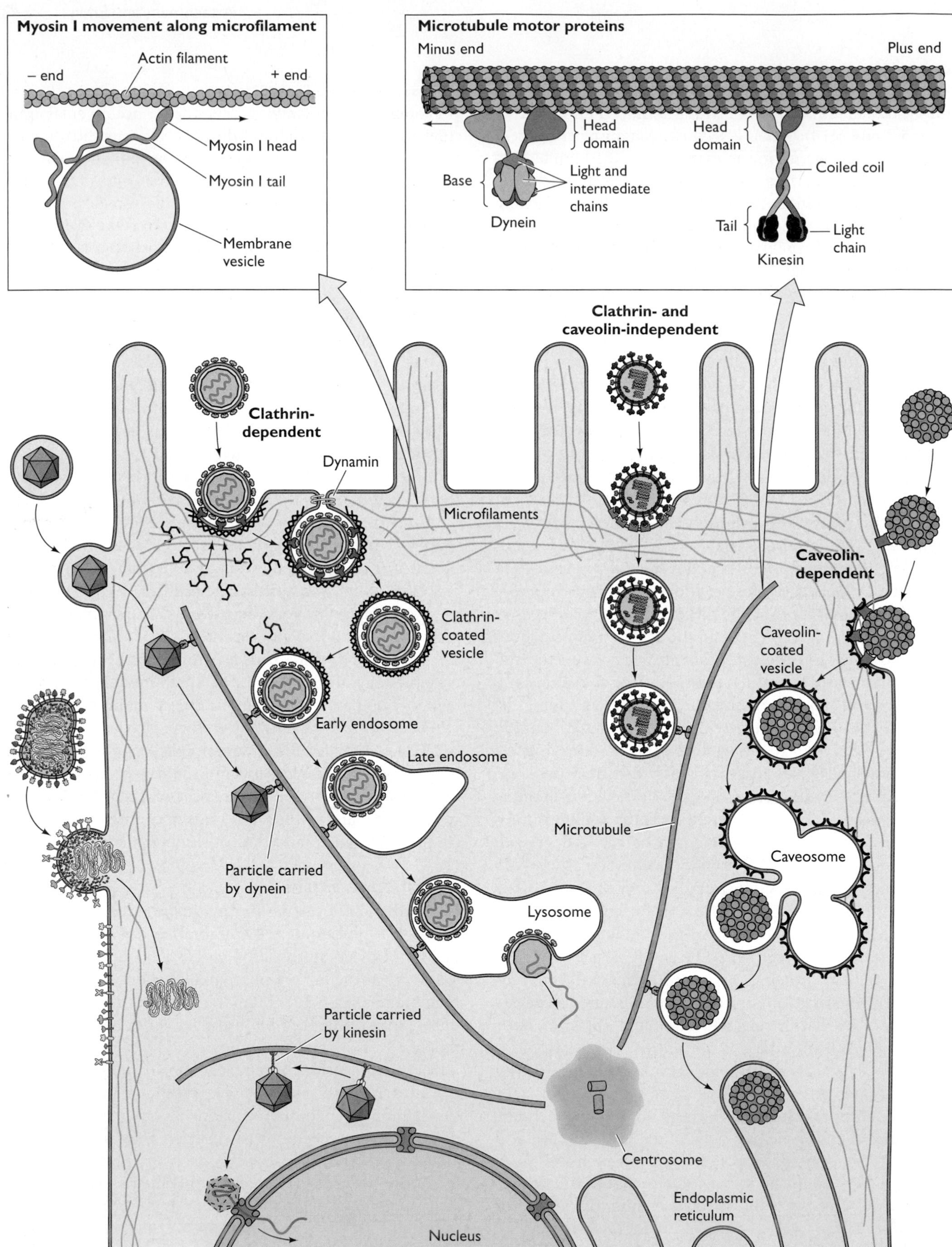

Myosin I movement along microfilament
Actin filament
– end
+ end
Myosin I head
Myosin I tail
Membrane vesicle
Microtubule motor proteins
Minus end
Plus end
Head domain
Base
Light and intermediate chains
Dynein
Head domain
Coiled coil
Tail
Light chain
Kinesin
Clathrin- and caveolin-independent
Clathrin-dependent
Dynamin
Microfilaments
Caveolin-dependent
Clathrin-coated vesicle
Caveolin-coated vesicle
Early endosome
Late endosome
Microtubule
Caveosome
Particle carried by dynein
Lysosome
Particle carried by kinesin
Centrosome
Endoplasmic reticulum
Nucleus

Figure 5.10

requires an even closer approach of the membranes, to within 1.5 nm of each other. This step depends on the removal of water molecules from the membrane surfaces, an energetically unfavorable process. A multisubunit protein complex is thought to provide the energy required for such a close approach in mammalian cells. The complex formed by targeting and docking proteins recruits additional proteins that induce fusion of the two membranes. After fusion occurs, the complex dissociates until needed once again. As individual components of the complex lack fusion activity, fusion can be regulated by assembly and disassembly.

The precise mechanism by which lipid bilayers fuse is not completely understood, but the action of fusion proteins is thought to result in the formation of an opening called a **fusion pore**, allowing exchange of material across the membranes. Much of our understanding about membrane fusion reactions comes from studies using individual viral proteins that promote membrane fusion (Box 5.4). Membrane fusion reactions catalyzed by such proteins appear to be less complex than those mediated by cellular proteins, for in most cases a single viral gene product is sufficient. This simplicity may be a consequence of the fact that abundant quantities of viral fusion proteins are produced during infection. Cell fusion proteins are far less abundant and must therefore be recycled, a requirement that is best accomplished by assembling and disassembling a multisubunit protein complex.

The membranes of enveloped viruses fuse with those of the cell as a first step in delivery of the viral nucleic acid. Viral fusion may occur either at the plasma membrane or from within an endosome or other vesicle. The membranes of the virus and the cell are first brought into close contact by interaction of a viral glycoprotein with a cell receptor. The same viral glycoprotein, or a different viral integral membrane protein, then catalyzes the fusion of the juxtaposed membranes. As described in the following sections, virus-mediated fusion must be regulated to prevent viruses from aggregating or to ensure that fusion does not occur in the incorrect cellular compartment. In some cases, fusogenic potential is masked until the fusion protein interacts with other integral membrane proteins. In others, low pH is required to expose fusion domains. The activity of fusion proteins may also be regulated by cleavage of a precursor. This requirement probably prevents premature activation of fusion potential during virus assembly. Cleavage also generates the metastable states of viral glycoproteins that can subsequently undergo the conformational rearrangements required for fusion activity.

Movement of Virions and Subviral Particles within Cells

Virions and subviral particles move within the host cell during entry and egress (Chapters 12 and 13). However, movement of molecules larger than 500 kDa does not occur by passive diffusion, because the cytoplasm is crowded with organelles, high concentrations of proteins, and the cytoskeleton (Box 5.5). Rather, viruses and their components are transported via the actin and microtubule cytoskeletons. Such movement can be visualized in live cells by using fluorescently labeled virions (Chapter 2).

The cytoskeleton is a dynamic network of protein filaments that extends throughout the cytoplasm. It is composed of three types of filament—microtubules, intermediate filaments, and microfilaments (Fig. 5.10).

Figure 5.10 Virus entry and movement in cells. Examples of genome uncoating at the plasma membrane are shown on the left side of the cell. Fusion at the plasma membrane releases the nucleocapsid into the cytoplasm. In some cases, the subviral particle is transported on microtubules toward the nucleus, where the nucleic acid is released. Uptake of virions by clathrin-dependent endocytosis commences with binding to a specific cell surface receptor. The ligand-receptor complex diffuses into an invagination of the plasma membrane coated with the protein clathrin on the cytosolic side (clathrin-coated pits). The coated pit further invaginates and pinches off, a process that is facilitated by the GTPase dynamin. The resulting coated vesicle then fuses with an early endosome. Endosomes are acidic, as a result of the activity of vacuolar proton ATPases. Virion uncoating ususally occurs from early or late endosomes. Late endosomes then fuse with lysosomes. Virions may enter cells by a dynamin- and caveolin-dependent endocytic pathway (right side of the cell). This pathway brings virions to the endoplasmic reticulum via the caveosome, a pH-neutral compartment. Clathrin- and caveolin-independent endocytic pathways of viral entry have also been described (center of cell). Movement of endocytic vesicles within cells occurs on microfilaments (inset, top left) or microtubules (inset, top right), components of the cytoskeleton. Microfilaments are two-stranded helical polymers of the ATPase actin. They are dispersed throughout the cell but are most highly concentrated beneath the plasma membrane, where they are connected via integrins and other proteins to the extracellular matrix. Transport along microfilaments is accomplished by myosin motors. Microtubules are 25-nm hollow cylinders made of the GTPase tubulin. They radiate from the **centrosome** to the cell periphery. Movement on microtubules is carried out by kinesin and dynein motors. Insets adapted from G. M. Cooper, *The Cell: a Molecular Approach* (ASM Press, Washington, DC, and Sinauer Associates, Sunderland, MA, 1997), with permission.

BOX 5.4

EXPERIMENTS

Membrane fusion proceeds through a hemifusion intermediate

Fusion is thought to proceed through a hemifusion intermediate in which the outer leaflets of two opposing bilayers fuse (see figure), followed by fusion of the inner leaflets and the formation of a fusion pore. Direct evidence that fusion proceeds via a hemifusion intermediate has been obtained with influenza virus HA (see figure). **(Left)** Cultured mammalian cells expressing wild-type HA are fused with erythrocytes containing two different types of fluorescent dye, one in the cytoplasm and one in the lipid membrane. Upon exposure to low pH, HA undergoes conformational change and the fusion peptide is inserted into the erythrocyte membrane. The green dye is transferred from the lipid bilayer of the erythrocyte to the bilayer of the cultured cell. The HA trimers tilt, causing reorientation of the transmembrane domain and generating stress within the hemifusion diaphragm. Fusion pore formation relieves the stress. The red dye within the cytoplasm of the erythrocyte is then transferred to the cytoplasm of the cultured cell. **(Right)** An altered form of HA was produced, lacking the transmembrane and cytoplasmic domains and with membrane anchoring provided by linkage to a glycosylphosphatidylinositol (GPI) moiety. Upon exposure to low pH, the HA fusion peptide is inserted into the erythrocyte membrane, and green dye is transferred to the membranes of the mammalian cell. When the HA trimers tilt, no stress is transmitted to the hemifusion diaphragm because no transmembrane domain is present, and the diaphragm becomes larger. Fusion pores do not form, and there is no mixing of the contents of the cytoplasm, indicating that complete membrane fusion has not occurred. These results prove that hemifusion, or fusion of only the inner leaflet of the bilayer, can occur among whole cells. The findings also demonstrate that the transmembrane domain of the HA polypeptide plays a role in the fusion process.

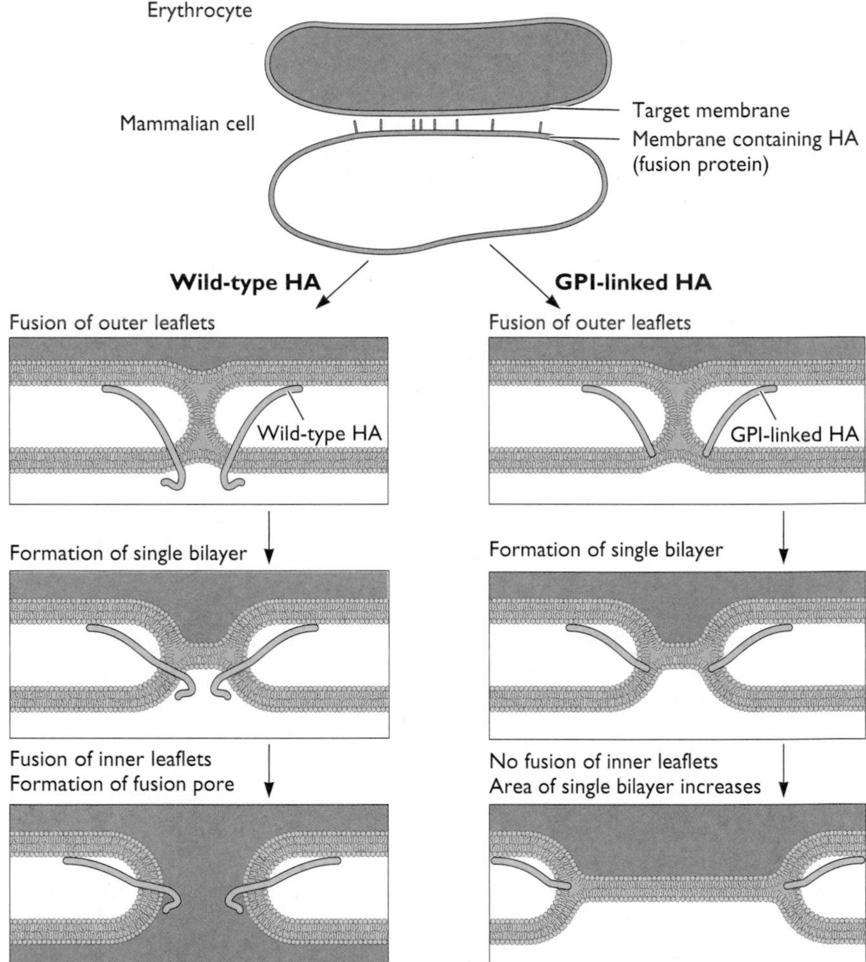

Glycosylphosphatidylinositol-anchored influenza virus HA induces hemifusion. (Left) Model of the steps of fusion mediated by wild-type HA. **(Right)** Effect on fusion by an altered form of HA lacking the transmembrane and cytoplasmic domains. Adapted from G. B. Melikyan et al., *J. Cell Biol.* **131**:679–691, 1995, with permission.

Microtubules are organized in a polarized manner, with minus ends situated at the microtubule-organizing center and plus ends located at the cell periphery. This arrangement permits directed movement of cellular and viral components over long distances. Actin filaments (microfilaments) typically assist in virus movement close to the plasma membrane.

Transport along actin filaments is accomplished by myosin motors, and movement on microtubules is carried out by kinesin and dynein motors (Fig. 5.10). Hydrolysis of adenosine triphosphate (ATP) provides the energy for the motors to move their cargo along cytoskeletal tracks. Dyneins and kinesins participate in movement of viral components during both entry (see "Mechanisms of

BOX 5.5

EXPERIMENTS

Passive diffusion cannot account for intracellular movement of virion components

The crowded cytoplasm of a cell. The image shows the interior of a eukaryotic cell, starting at the cell surface, which is studded with membrane proteins. A small portion of the cytoplasm is shown, illustrating the profusion of cytoskeletal elements, ribosomes, and other small molecules. Adapted from David S. Goodsell (http://www.scripps.edu/pub/goodsell/gallery/patterson.html), with permission.

In aqueous solutions, molecules can move rapidly by diffusion, a process of random motion produced by collision with other molecules in the solution. Under ideal conditions, diffusion coefficients typically range from 10^{-6} to 10^{-8} cm^2/s. However, the intracellular milieu is far from such an ideal: the very high intracellular protein concentrations (up to 300 mg/ml), the presence of numerous organelles, and the cytoskeletal networks (see figure) severely restrict diffusion of molecules with molecular mass greater than 500 kDa. Measurements of diffusion coefficients of beads microinjected into cells, of cytoplasmic vesicles, and of DNA molecules indicate that these values are from 5- to 1,000-fold lower in the cytoplasm than in aqueous solution. As shown in the table, such estimates indicate that viral particles (or the components to be assembled into progeny virions) could not reach the appropriate intracellular destinations by passive diffusion within even a few years, let alone the few hours or days that comprise infectious cycles.

Estimated rates of transport of viral components by diffusion[a]

Viral component	Time to travel 10 μm[b,c]	
	In H_2O (s)	In cytoplasm (h)
Poliovirus capsid	3.85	0.5
Herpes simplex virus nucleocapsid	14.6	2.0
Vaccinia virus intracellular mature virion	35.0	4.9

[a]Adapted from B. Sodeik, *Trends Microbiol.* **8**:465–472, 2000, with permission.

[b]The length of a typical human cell. Note the different timescales for H_2O and cytoplasm.

[c]Diffusion constants were calculated by a formula that considers the radius of the virus particle and the viscosity of water at room temperature. The assumption was made that diffusion constants in the cytoplasm would be 500 times lower than in water.

Uncoating" below) and egress (Chapters 12 and 13). In some cases, the actin cytoskeleton is remodeled during entry and egress, for example, when viruses bud from the plasma membrane.

There are two basic ways for viruses to travel within the cell—within a membrane vesicle such as an endosome, which interacts with the cytoskeletal transport machinery, or directly in the cytoplasm (Fig. 5.10). In the latter case, some form of the virus particle must bind directly to the transport machinery. The cytoplasmic domain of CD155, the cellular receptor for poliovirus, binds the light chain of the motor protein dynein. This interaction might target endocytic vesicles containing CD155 to the microtubule network, allowing transport of the viral capsid in the cytoplasm. After leaving endosomes, the subviral particles derived from adenoviruses and parvoviruses are transported along microtubules to the nucleus. Although adenovirus particles have an overall net movement toward the nucleus, they exhibit bidirectional plus- and minus-end-directed microtubule movement. Adenovirus binding to cells activates two different signal transduction pathways that increase the net velocity of minus-end-directed capsid motility. The signaling pathways are therefore required for efficient delivery of the viral genome to the nucleus. It is not yet known how viral subviral particles are loaded onto and released from the microtubules to move to the nuclear pore complex, where the viral genomes enter the nucleus.

Some viruses move along the surfaces of cells prior to entry, often to locate a clathrin-coated pit. If the cell receptor is rare or inaccessible, virions may first bind to more abundant or accessible receptors, such as carbohydrates, and then migrate to receptors that allow entry into the cell. For example, after binding, polyomavirus particles move laterally ("surf") on the plasma membrane for 5 to 10 s and then are internalized. Virions can be visualized moving along the plasma membrane toward the cell body on **filopodia**, thin extensions of the plasma membrane (Fig. 5.10). Virions move along filopodia by an actin-dependent mechanism. Filopodial bridges mediate cell-to-cell spread of a retrovirus in cultured cells. The filopodia originate from uninfected cells and contact infected cells with their tips. The interaction of the viral envelope glycoprotein on the surface of infected cells with the receptor on uninfected cells stabilizes the interaction.

Virions move along the outside of the filipodial bridge to the uninfected cell. Virion transport is a consequence of actin-based movement of the viral receptor toward the uninfected cell.

The intricate mechanisms by which the genomes of viruses move in eukaryotic cells are in stark contrast to the simple injection of the bacterial genome into the host cell (Box 5.6). During this process, the bacteriophage particle remains on the surface of the bacterium.

Virus-Induced Signaling via Cell Receptors

Binding of virions to cell receptors not only concentrates the particles on the cell surface but also may activate signaling pathways that facilitate virus entry and movement within the cell or produce cellular responses that enhance virus propagation and/or affect pathogenesis.

Signaling triggered by binding of coxsackievirus B3 to its cellular receptor makes receptors accessible for virus entry. The coxsackievirus and adenovirus receptor, Car, is not present on the apical surface of epithelial cells that line the intestinal and respiratory tracts. This membrane protein is a component of tight junctions and is inaccessible to virions. To enter epithelial cells, group B coxsackieviruses bind a receptor, CD55, which is present on the apical surface. Virus binding to CD55 activates Abl kinase, which in turn triggers Rac-dependent actin rearrangements. These changes allow virus movement to the tight junction, where it can bind Car and enter cells.

Signaling is essential for the entry of simian virus 40 into cells. Binding of this virus to its glycolipid cell receptor, GM1 ganglioside, causes activation of tyrosine kinases. The signaling that ensues causes reorganization of actin filaments, internalization of the virus in caveolae, and transport of the caveolar vesicles to the endoplasmic reticulum. The activities of nearly 80 cellular protein kinases regulate the entry of this virus into cells.

Interactions between human immunodeficiency virus type 1 SU and CD4 have been implicated in virus-induced cell killing. Both CD4 and human immunodeficiency virus type 1 coreceptor molecules are coupled via their

BOX 5.6 DISCUSSION
The bacteriophage DNA injection machine

The mechanisms by which the bacteriophage genome enters the bacterial host are unlike those for viruses of eukaryotic cells. One major difference is that the bacteriophage particle remains on the surface of the bacterium as the nucleic acid passes into the cell. The DNA genome of some bacteriophages is packaged under high pressure (up to 870 lb/in^2) in the capsid and is injected into the cell in a process that has no counterpart in the entry process of eukaryotic viruses. The complete structure of bacteriophage T4 illustrates this remarkable process (see figure). To initiate infection, the tail fibers attach to receptors (black) on the surface of *Escherichia coli*. Binding causes a conformational change in the baseplate, which leads to contraction of the sheath. This movement drives the rigid tail tube through the outer membrane, using a needle at the tip. When the needle touches the peptidoglycan layer in the periplasm, the needle dissolves and three lysozyme domains in the baseplate are activated. These disrupt the peptidoglycan layer of the bacterium, allowing DNA to enter.

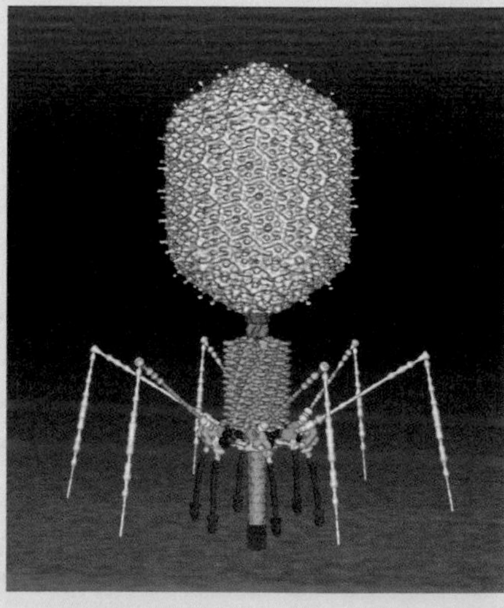

Structure of bacteriophage T4. A model of the 2,000-Å bacteriophage as produced from electron microscopy and X-ray crystallography. Components of the virion are color coded: virion head (beige), tail tube (pink), contractile sheath around the tail tube (green), baseplate (multicolored), and tail fibers (white and magenta). In the illustration, the virion contacts the cell surface, and the tail sheath is contracted prior to DNA release into the cell. From P. G. Leitman et al., *Cell* **118:**419–430, 2004, with permission. Courtesy of Michael Rossmann, Purdue University.

cytoplasmic domains to intracellular signaling pathways. The normal role of CD4 is to bind to the major histocompatibility complex class II-peptide complex on antigen-presenting cells and stabilize its interaction with the T-cell receptor. This interaction leads to activation and differentiation of the T cell by means of a protein kinase ($p56^{lck}$) associated with the cytoplasmic domain of CD4 at the inner leaflet of the plasma membrane (Volume II, Fig. 4.11). The chemokine receptors also signal interaction with their ligand, affecting cellular gene expression. The binding of SU to human $CD4^{+}$ T cells is followed by signaling through chemokine receptors and induction of apoptosis. It has been reported that interactions between SU and chemokine receptors on neuronal cells induce apoptosis. The destruction of cytotoxic T cells by macrophages has also been attributed to such interactions. Such effects may explain the depletion of cytotoxic T cells and the neurological disorders that are symptoms of AIDS.

Mechanisms of Uncoating

Uncoating is the release of viral nucleic acid from its protective protein coat and/or lipid envelope, although in most cases the liberated nucleic acid is still associated with viral proteins. For enveloped viruses, uncoating occurs when viral and cellular membranes fuse, either at the plasma membrane or within intracellular vesicles. Nonenveloped viruses typically enter the cell by endocytosis, and the genome is released from intracellular transport vesicles or while docked at the nuclear pore complex.

Uncoating at the Plasma Membrane

The particles of many enveloped viruses, including members of the family *Paramyxoviridae* such as Sendai virus and measles virus, fuse directly with the plasma membrane at neutral pH. These virions bind to cell surface receptors via a viral integral membrane protein (Fig. 5.11). Once the viral and cell membranes have been closely juxtaposed by this receptor-ligand interaction, fusion is induced by a second viral glycoprotein known as fusion (F) protein, and the viral nucleocapsid is released into the cell cytoplasm.

F protein is a type I integral membrane glycoprotein (the N terminus lies outside the viral membrane) with similarities to influenza virus HA in its synthesis and structure. It is a homotrimer that is synthesized as a precursor called F0 and cleaved during transit to the cell surface by a host cell protease to produce two subunits, F1 and F2, held together by disulfide bonds. The newly formed N-terminal 20 amino acids of the F1 subunit, which are highly hydrophobic, form a region called the **fusion peptide** because it inserts into target membranes to initiate fusion. Viruses with the uncleaved F0 precursor can be produced in cells that lack the protease responsible for its cleavage. Such virus particles are noninfectious; they bind to target cells but the viral genome does not enter. Cleavage of the F0 precursor is necessary for fusion, not only because the fusion peptide is made available for insertion into the plasma membrane, but also to generate the metastable state of the protein that can undergo the conformational rearrangements needed for fusion.

Because cleaved F-protein-mediated fusion can occur at neutral pH, it must be controlled, both to ensure that virus particles fuse with only the appropriate cell and to prevent aggregation of newly assembled virions. The fusion peptide of F1 is buried between two subunits of the trimer in the pre-fusion protein. Conformational changes in F protein lead to refolding of the protein, assembly of an α-helical coiled coil, and movement of the fusion peptide toward the cell membrane (Fig. 5.11). Such movement of the fusion peptide has been described in atomic detail by comparing structures of the F protein before and after fusion.

The trigger that initiates conformational changes in the F protein is not known. The results of experiments in which hemagglutinin-neuraminidase (HN) and F glycoproteins are synthesized in cultured mammalian cells indicate that the fusion activity of F protein is absent or inefficient if HN is not present. It has therefore been hypothesized that an interaction between HN and F proteins is essential for fusion. It is thought that binding of HN protein to its cellular receptor induces conformational changes, which in turn trigger conformational change in the F protein, exposing the fusion peptide and making the protein fusion competent (Fig. 5.11). The requirement for HN protein in F fusion activity has been observed only with certain paramyxoviruses, including human parainfluenza virus type 3 and mumps virus.

As a result of fusion of the viral and plasma membranes, the viral nucleocapsid, which is a ribonucleoprotein (RNP) consisting of the (–) strand viral RNA genome and the viral proteins L, NP, and P, is released into the cytoplasm (Fig. 5.11). Once in the cytoplasm, the L, NP, and P proteins begin the synthesis of viral messenger RNAs (mRNAs), a process discussed in Chapter 6. Because members of the *Paramyxoviridae* replicate in the cytoplasm, fusion of the viral and plasma membranes achieves uncoating and delivery of the viral genome to this cellular compartment in a single step.

Fusion of human immunodeficiency virus type 1 with the plasma membrane requires participation not only of the cell receptor CD4 but also of an additional cellular protein. These proteins are cell surface receptors for small molecules produced by many cells to attract and stimulate cells of the immune defense system at sites of infection; hence these small molecules are called **chemotactic cytokines** or **chemokines**. The chemokine receptors on such cells

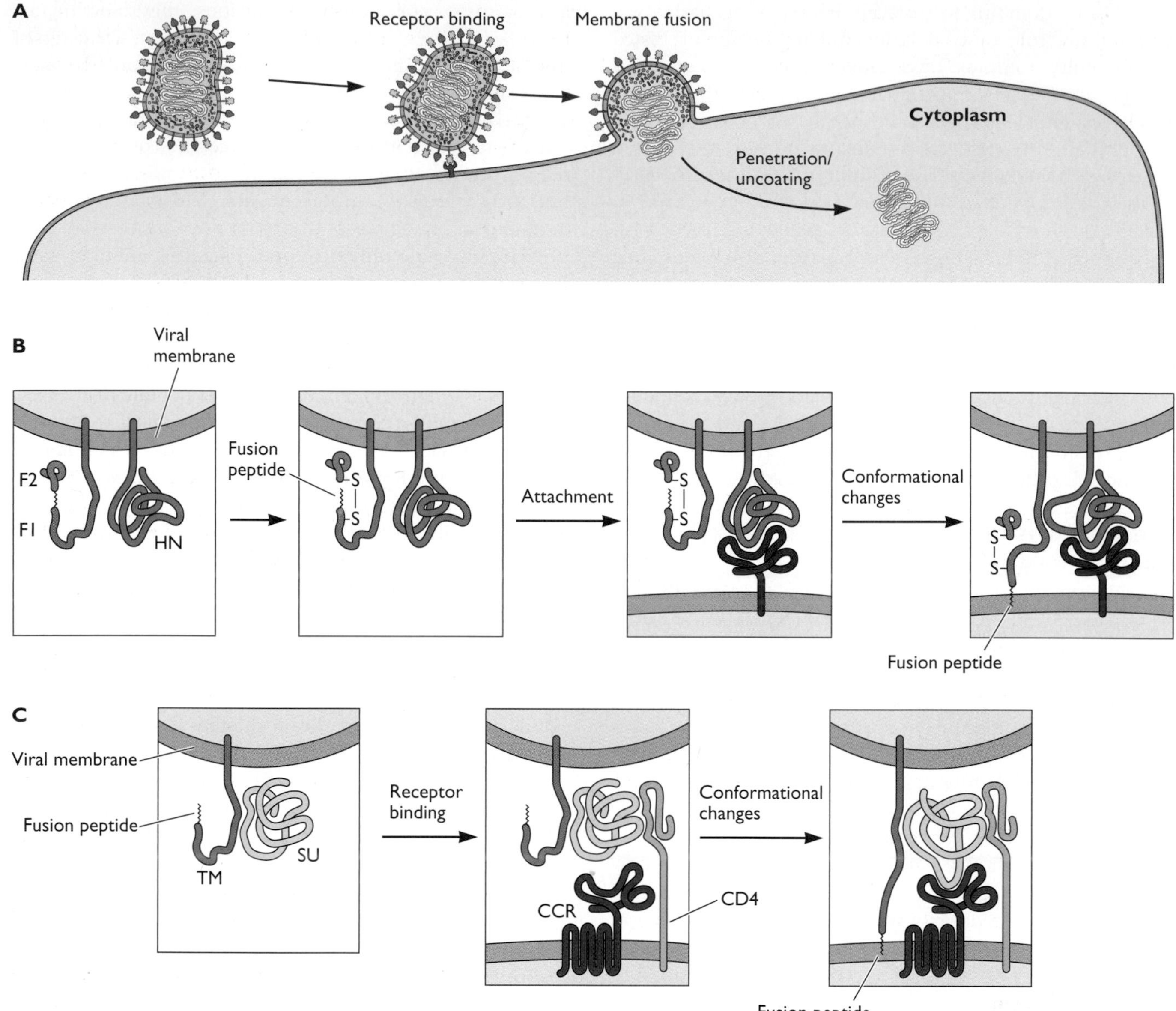

Figure 5.11 Penetration and uncoating at the plasma membrane. (A) Overview. Entry of a member of the *Paramyxoviridae*, which bind to cell surface receptors via the HN, H, or G glycoprotein. The fusion protein (F) then catalyzes membrane fusion at the cell surface at neutral pH. The viral nucleocapsid, as RNP, is released into the cytoplasm, where RNA synthesis begins. The mechanism by which contacts between the viral nucleocapsid and the M protein, which forms a shell beneath the lipid bilayer, are broken to facilitate release of the nucleocapsid is not known. **(B)** Model for F-protein-mediated membrane fusion. Binding of HN to the cell receptor (red) induces conformational changes in HN that in turn induce conformational changes in the F protein, moving the fusion peptide from a buried position nearer to the cell membrane. **(C)** Model of the role of chemokine receptors in human immunodeficiency virus type 1 fusion at the plasma membrane. For simplicity, the envelope glycoprotein is shown as a monomer, although trimer and tetramer forms have been reported. Binding of SU to CD4 exposes a high-affinity chemokine receptor-binding site on SU. The SU-chemokine receptor interaction leads to conformational changes in TM that expose the fusion peptide and permit it to insert into the cell membrane, catalyzing fusion in a manner similar to that proposed for influenza virus (cf. Fig. 5.12 and 5.13).

comprise a large family of proteins with seven membrane-spanning domains and are coupled to intracellular signal transduction pathways. There are two major coreceptors for human immunodeficiency virus type 1 infection. CXCr4 (a member of a family of chemokines characterized by having their first two cysteines separated by a single amino acid) appears to be a specific coreceptor for virus strains that infect T cells preferentially. The second is CCr5, a coreceptor for the macrophage-tropic strains of the virus. The chemokines that bind to this receptor activate both T cells and macrophages, and the receptor is found on both types of cell. Individuals who are homozygous for deletions in the CCr5 gene and produce nonfunctional coreceptors have no discernible immune function abnormality, but they appear to be resistant to infection with human immunodeficiency virus type 1. Even heterozygous individuals seem to be somewhat resistant to the virus. Other members of the CC chemokine receptor family (CCr2b and CCr3) were subsequently found to serve as coreceptors for the virus.

Attachment to CD4 appears to create a high-affinity binding site on SU for CCr5. The atomic structure of SU bound to CD4 revealed that binding of CD4 induces conformational changes that expose binding sites for chemokine receptors (Fig. 5.11). Studies of CCr5 have shown that the first N-terminal extracellular domain is crucial for coreceptor function, suggesting that this sequence might interact with SU.

Human immunodeficiency virus type 1 TM mediates envelope fusion with the cell membrane. The high-affinity SU-CCr5 interaction may induce conformational changes in TM to expose the fusion peptide, placing it near the cell membrane, where it can catalyze fusion (Fig. 5.11). Such changes are similar to those that influenza virus HA undergoes upon exposure to low pH. X-ray crystallographic analysis of fusion-active human immunodeficiency virus type 1 TM revealed that its structure is strikingly similar to that of the low-pH fusogenic form of HA (see "Acid-Catalyzed Membrane Fusion" below).

Certain isolates of human immunodeficiency virus types 1 and 2 and simian immunodeficiency virus enter cells independently of CD4 via chemokine receptors. Given the large number of members of the chemokine receptor family and the ability of human immunodeficiency virus type 1 to interact with these proteins, it is possible that CD4-independent, chemokine receptor-mediated infection may occur with some frequency.

Uncoating during Endocytosis

Acid-Catalyzed Membrane Fusion

Many enveloped viruses undergo fusion within an endosomal compartment. The entry of influenza virus from the endosomal pathway is one of the best-understood viral entry mechanisms. At the cell surface, the virus attaches to sialic acid-containing receptors via the viral HA glycoprotein (Fig. 5.12). The virus-receptor complex is then internalized by the clathrin-dependent receptor-mediated endocytic pathway. When the endosomal pH reaches approximately 5.0, HA undergoes an acid-catalyzed conformational rearrangement, exposing a fusion peptide. The viral and endosomal membranes then fuse, allowing penetration of the viral RNP (vRNP) into the cytoplasm. Because influenza virions have a low pH threshold for fusion, uncoating occurs in late endosomes. Viruses with a high pH threshold (pH 6.5 to 6) undergo fusion with the membranes of early endosomes in the periphery of the cytoplasm.

The fusion reaction mediated by the influenza virus HA protein is a remarkable event when viewed at atomic resolution (Fig. 5.13). In native HA, the fusion peptide is joined to the three-stranded coiled-coil core by which the HA monomers interact via a 28-amino-acid sequence that forms an extended loop structure buried deep inside the molecule, about 100 Å from the globular head. In contrast, in the low-pH HA structure, this loop region is transformed into a three-stranded coiled coil. In addition, the long α-helices of the coiled coil bend upward and away from the viral membrane. The result is that the fusion peptide has moved a great distance toward the endosomal membrane (Fig. 5.13). Despite these dramatic changes, HA remains trimeric and the globular heads can still bind sialic acid. In this conformation, HA holds the viral and endosome membranes 100 Å apart, too distant for the fusion reaction to occur. To bring the viral and cellular membranes closer, it is thought that the top of the acid-induced coiled coil splays apart, spreading into the lipid bilayer (Fig. 5.12). The stems of the HA tilt, further facilitating close contact of the membranes.

In contrast to cleaved HA, the precursor HA0 is stable at low pH and cannot undergo structural changes. How does cleavage of HA produce a protein capable of fusion only at acidic pH? Cleavage of the covalent bond between HA1 and HA2 might simply allow movement of the fusion peptide, which is restricted in the uncleaved molecule. Another possibility is suggested by the observation that cleavage of HA is accompanied by movement of the fusion peptide into the cavity in HA (Fig. 5.13). This movement buries ionizable residues of the fusion peptide, perhaps setting the low-pH "trigger." It should be emphasized that after cleavage, the N terminus of HA2 is tucked into the hydrophobic interior of the trimer (Fig. 5.13). This rearrangement presumably buries the fusion peptide so that newly synthesized virions do not aggregate and lose infectivity.

When the structure of influenza virus HA is compared with those of the TM proteins of two retroviruses, the

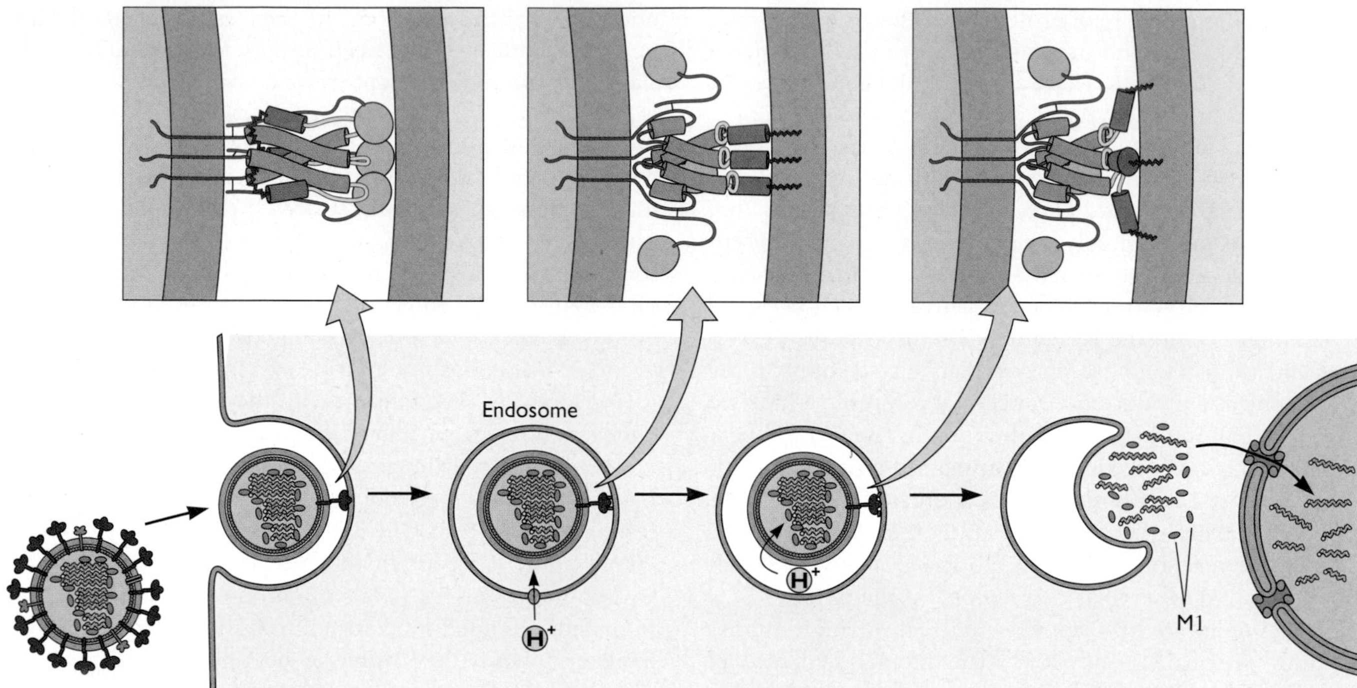

Figure 5.12 Influenza virus entry. The globular heads of native HA mediate binding of the virus to sialic acid-containing cell receptors. The virus-receptor complex is endocytosed, and import of H+ ions into the endosome acidifies the interior. Upon acidification, the viral HA undergoes a conformational rearrangement that produces a fusogenic protein. The loop region of native HA (yellow) becomes a coiled coil, moving the fusion peptides (red) to the top of the molecule near the cell membrane. At the viral membrane, the long α-helix (purple) packs against the trimer core, pulling the globular heads to the side. The long coiled coil splays into the cell membrane, bringing it closer to the viral membrane so that fusion can occur. Not shown is the tilting of HA that occurs. To allow release of vRNP into the cytoplasm, the H^+ ions in the acidic endosome are pumped into the virion interior by the M2 ion channel. As a result, vRNP is primed to dissociate from M1 after fusion of the viral and endosomal membranes. The released vRNPs are imported into the nucleus through the nuclear pore complex via a nuclear localization signal-dependent mechanism (see "Import of Influenza Virus Ribonucleoprotein" below). Adapted from C. M. Carr and P. S. Kim, *Science* **266:**234–236, 1994, with permission.

F protein of simian virus 5 and Gp2 of Ebola virus, remarkable similarities become apparent (Fig. 5.14). In all five cases, the fusion peptides are presented to membranes on top of a three-stranded coiled coil. Such a scaffold is a common feature of viral type I membrane fusion proteins: they have a region of high α-helical content and a 4-3 heptad repeat of hydrophobic amino acids, characteristic of coiled coils, next to the N-terminal fusion peptide.

The envelope proteins of alphaviruses and flaviviruses exemplify a different class of viral fusion protein (type II fusion proteins). These viral proteins contain an internal fusion peptide and are tightly associated with a second viral protein. Proteolytic cleavage of the second protein converts the fusion protein to a metastable state that can undergo structural rearrangements at low pH to promote fusion. In contrast, the fusion peptide of the influenza virus HA is adjacent to the cleavage point and becomes the N terminus of the mature fusion protein. The envelope proteins of alphaviruses and flaviviruses do not form coiled coils, as do type I fusion proteins. Rather, they contain predominantly β-barrels that are thought to tilt toward the membrane at low pH, thereby exposing the fusion peptide (Fig. 5.15).

The membrane fusion mediated by the envelope protein of the alphavirus Semliki Forest virus exhibits several unusual features. This process requires the presence of cholesterol in the cell membrane, which is not needed for fusion mediated by other viral proteins. Why cholesterol is needed for fusion is not understood. In contrast to the situation with other viruses, proteolytic cleavage of E1 is not required to produce a fusogenic protein. However, protein processing may control fusion potential in another way. In the endoplasmic reticulum, E1 protein is associated with the precursor of E2, called p62. In this heterodimeric form, p62-E1, E1 protein cannot be activated for fusion by mildly

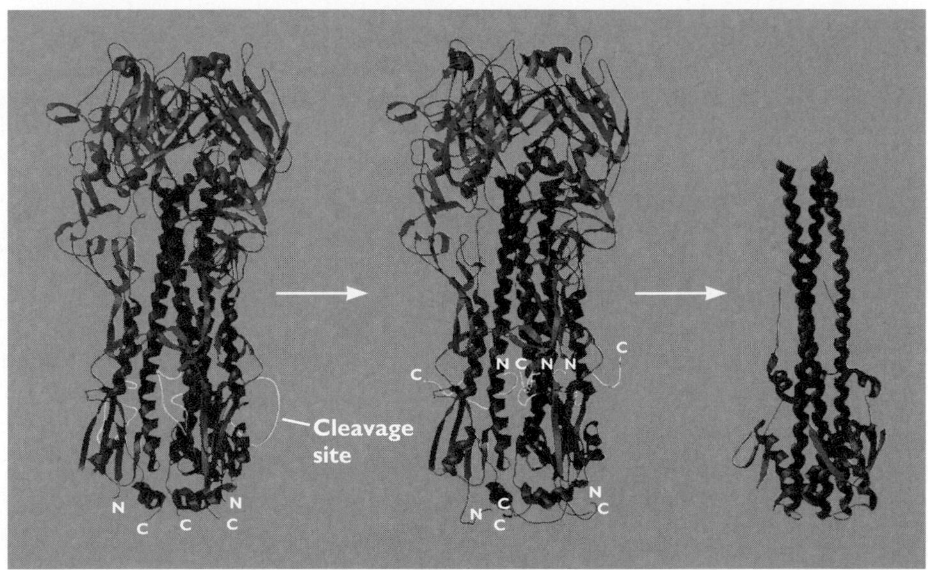

Figure 5.13 Cleavage- and low-pH-induced structural changes in the extracellular domains of influenza virus HA. (Left) Structure of the uncleaved HA0 precursor extracellular domain at neutral pH. HA1 subunits are blue, HA2 subunits are red, residues 323 of HA1 to 12 of HA2 are yellow, and the locations of some of the N and C termini are indicated. The viral membrane is at the bottom, and the globular heads are at the top. The cleavage site between HA1 and HA2 is in a loop adjacent to a deep cavity. **(Middle)** Structure of the cleaved HA trimer at neutral pH. Cleavage of HA0 generates new N and C termini, which are separated by 20 Å. The N and C termini visible in this model are labeled. The cavity is now filled with residues 1 to 10 of HA2, part of the fusion peptide. **(Right)** Structure of the low-pH trimer. The protein used for crystallization was treated with proteases, and therefore the HA1 subunit and the fusion peptide are not present. This treatment is necessary to prevent aggregation of HA at low pH. At neutral pH the fusion peptide is close to the viral membrane, linked to a short α-helix, and at acidic pH this α-helix is reoriented toward the cell membrane, carrying with it the fusion peptide. The structures are aligned on a central α-helix that is unaffected by the conformational change. Adapted from J. Chen et al., *Cell* **95**:409–417, 1998, with permission.

acidic conditions. Only after p62 has been cleaved to E2 can low pH induce disruption of E1-E2 heterodimers and formation of fusion-active E1 homotrimers.

Release of Viral Ribonucleoprotein

The genomes of many enveloped RNA viruses are present as vRNP in the virus particle. One mechanism for release of vRNP during virus entry has been identified by studies of influenza virus. Each influenza virus vRNP is composed of a segment of the RNA genome bound by nucleoprotein (NP) molecules at about 10- to 15-nucleotide intervals and the virion RNA polymerase. This complex interacts with viral M1 protein, an abundant virion protein that underlies the viral envelope and provides rigidity (Fig. 5.12). The M1 protein also contacts the internal tails of HA and neuraminidase proteins in the viral envelope. This arrangement presents two problems. Unless M1-vRNP interactions are disrupted, vRNPs might not be released into the cytoplasm. Furthermore, the vRNPs must enter the nucleus, where mRNA synthesis occurs. However, vRNP cannot enter the nucleus if M1 protein remains bound, because this protein masks a nuclear localization signal (see "Import of Influenza Virus Ribonucleoprotein" below).

The influenza virus M2 protein, the first viral protein discovered to be an ion channel, probably provides the solution to both problems. The virion envelope contains a small number (14 to 68) of molecules of M2 protein, which form a homotetramer. When purified M2 was reconstituted into synthetic lipid bilayers, ion channel activity was observed, indicating that this property requires only M2 protein. The M2 protein channel is structurally much simpler than other ion channels and is the smallest channel discovered to date.

The M2 ion channel is activated by the low pH of the endosome before HA-catalyzed membrane fusion occurs. As a result, protons enter the interior of the virus particle. It has been suggested that the reduced pH of the virion interior leads to conformational changes in the M1 protein, thereby disrupting M1-vRNP interactions. When fusion between the viral envelope and the endosomal membrane subsequently occurs, vRNPs are released into the cytoplasm free of M1 and can then be imported into

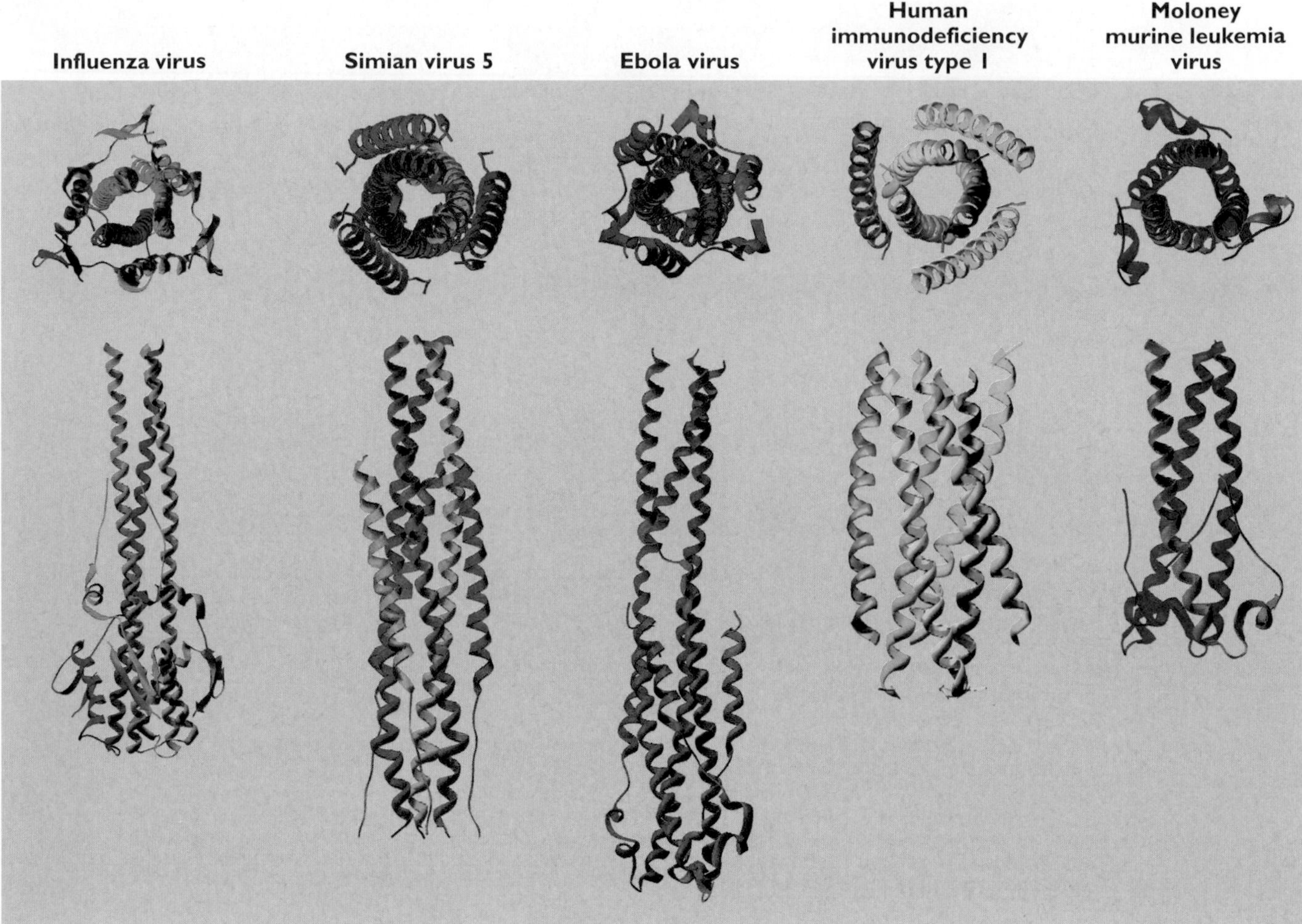

Figure 5.14 Similarities among five viral fusion proteins. (Top) View from the top of the structures. **(Bottom)** Side view. The structure shown for HA is the low-pH, or fusogenic, form. The structure of simian virus 5 F protein is of peptides from the N- and C-terminal heptad repeats. Structures of retroviral TM proteins are derived from interacting human immunodeficiency virus type 1 peptides and a peptide from Moloney murine leukemia virus and are presumed to represent the fusogenic forms because of structural similarity to HA. In all three molecules, fusion peptides would be located at the membrane-distal portion (the tops of the molecules in the bottom view). All present fusion peptides to cells on top of a central three-stranded coiled coil supported by C-terminal structures. Adapted from K. A. Baker et al., *Mol. Cell* **3**:309–319, 1999, with permission.

the nucleus (Fig. 5.12). Support for this model comes from studies with the anti-influenza virus drug **amantadine,** which specifically inhibits M2 ion channel activity (Volume II, Fig. 9.11). In the presence of this drug, influenza virus particles can bind to cells, enter endosomes, and undergo HA-mediated membrane fusion, but vRNPs are not released from the endosomal membrane.

Receptor Priming for Low-pH Fusion: Two Entry Mechanisms Combined

During the entry of avian leukosis virus into cells, virion binding to the cell receptor primes the viral fusion protein for low-pH-activated fusion. Avian leukosis virus, like many other simple retroviruses, was believed to enter cells at the plasma membrane in a pH-independent mechanism resembling that of members of the *Paramyxoviridae* (Fig. 5.11). It is now known that binding of the viral membrane glycoprotein Env-A to the cellular receptor Tva induces conformational rearrangements that convert Env-A from a native metastable state that is insensitive to low pH to a second metastable state. In this state, exposure of Env-A to low pH within the endosomal compartment leads to membrane fusion and release of the viral capsid.

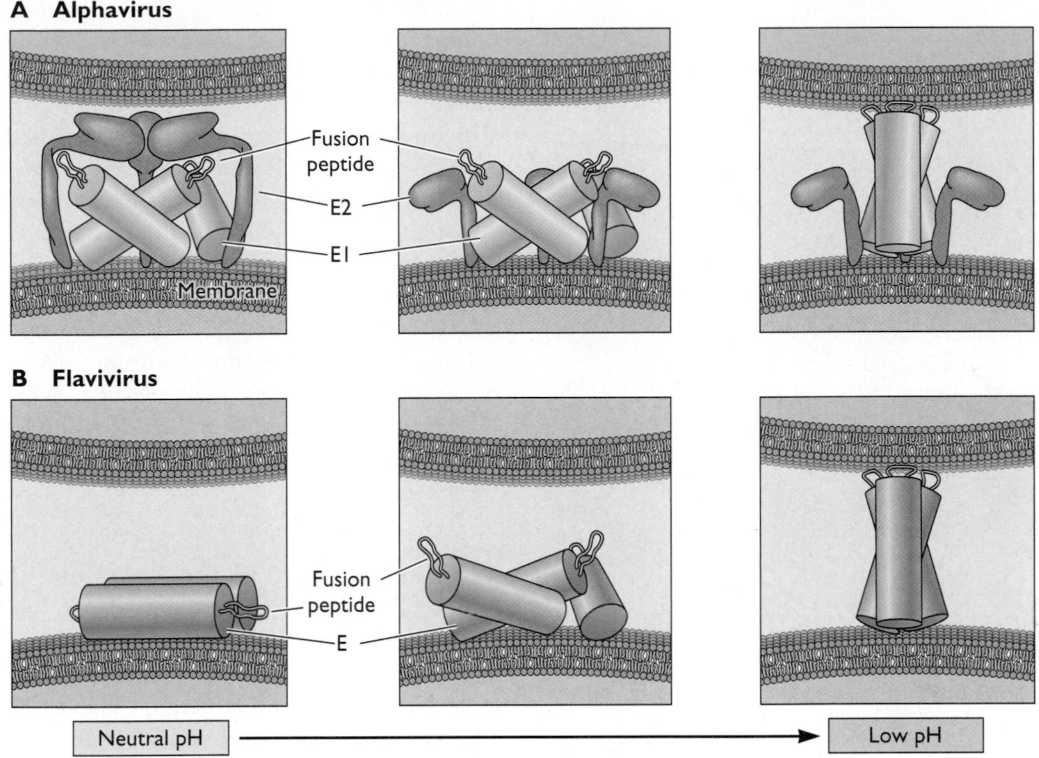

Figure 5.15 Models for low-pH-induced movement of alphavirus and flavivirus glycoproteins. Low pH causes conformational changes in the viral glycoproteins to produce the fusion-active forms. **(A)** In alphavirus virions, the fusion peptide in E1 is masked by E2. Low pH leads to disruption of E1-E2 dimers, exposing the fusion peptide. **(B)** In flavivirus virions, the fusion peptide is buried in dimers of the fusion glycoprotein E. At low pH, the dimers are disrupted, the proteins rotate to form trimers, and the fusion peptide is directed toward the cell membrane. Adapted from R. J. Kuhn et al., *Cell* **108:**717–725, 2002, with permission.

Uncoating in the Cytoplasm by Ribosomes

Some enveloped RNA-containing viruses, such as Semliki Forest virus, contain nucleocapsids that are disassembled in the cytoplasm by pH-independent mechanisms. The nucleocapsid of this virus is an icosahedral shell composed of a single viral protein, C protein, which encloses the (+) strand viral RNA. This structure is surrounded by an envelope containing viral glycoproteins called E1 and E2, which are arranged as heterodimers clustered into groups of three, each cluster forming a spike on the virus surface.

Fusion of the viral and endosomal membrane exposes the nucleocapsid to the cytoplasm (Fig. 5.16). The viral RNA within this structure is sensitive to digestion with RNase, suggesting that the nucleocapsid is permeable. Crystallographic studies of the nucleocapsid of Sindbis virus, a closely related alphavirus, confirm the presence of holes ranging in diameter from 30 to 60 Å. These holes permit the entry of small proteins such as RNase (25 to 40 Å in diameter) into the nucleocapsid. To begin translation of (+) strand viral RNA, the nucleocapsid must be disassembled, a process mediated by an abundant cellular component—the ribosome. Each ribosome binds three to six molecules of C protein, causing them to detach from the nucleocapsid. This process occurs while the nucleocapsid is attached to the cytoplasmic side of the endosomal membrane (Fig. 5.16) and ultimately results in disassembly. The uncoated viral RNA remains associated with cellular membranes, where translation and replication begin.

Disrupting the Endosomal Membrane

Adenoviruses are composed of a double-stranded DNA genome packaged in an icosahedral capsid made up of at least 10 structural proteins, as described in Chapter 4. Internalization of most adenovirus serotypes by receptor-mediated endocytosis requires attachment of fiber to an integrin or Ig-like cell surface receptor and binding of the penton base to a second cell receptor, the cellular vitronectin-binding integrins $\alpha_v\beta_3$ and $\alpha_v\beta_5$. Attachment is mediated by RGD sequences in each of the five subunits of

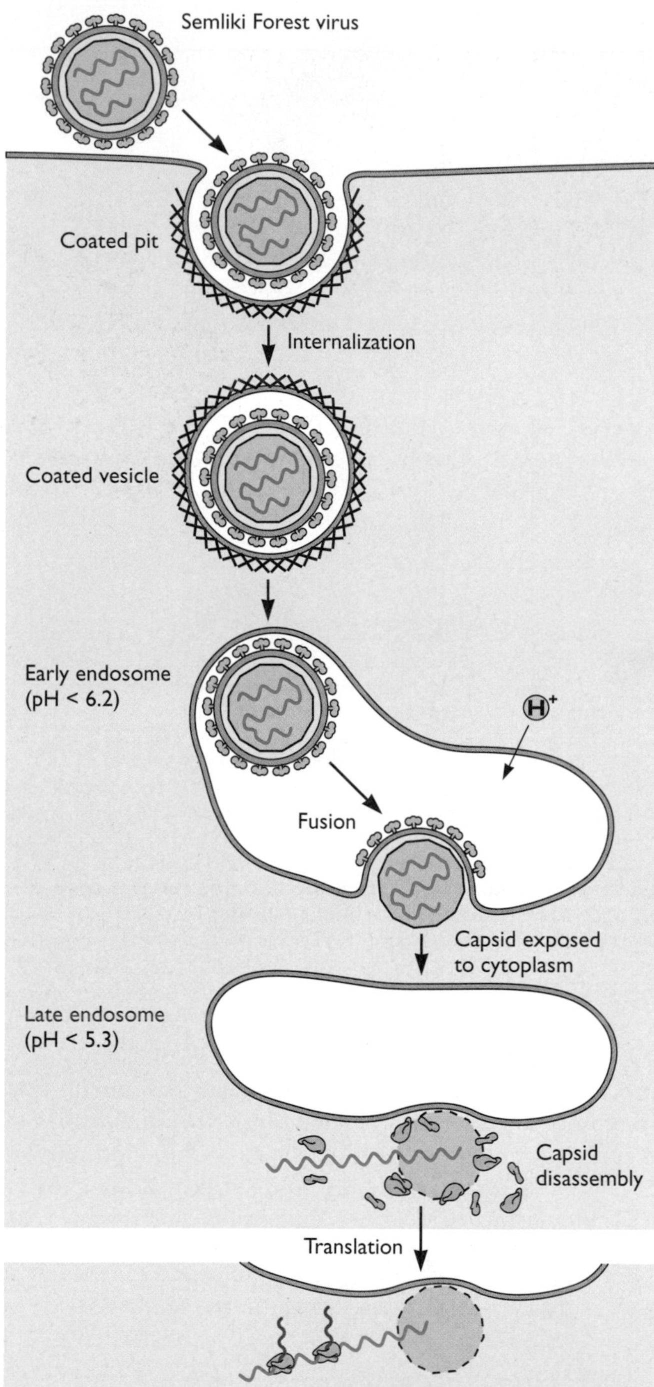

Figure 5.16 Entry of Semliki Forest virus into cells. Semliki Forest virus enters cells by clathrin-dependent receptor-mediated endocytosis, and membrane fusion is catalyzed by acidification of endosomes. Fusion results in exposure of the viral nucleocapsid to the cytoplasm, although the nucleocapsid remains attached to the cytosolic side of the endosome membrane. Cellular ribosomes then bind the capsid, disassembling it and distributing the capsid protein throughout the cytoplasm. The viral RNA is then accessible to ribosomes, which initiate translation. Adapted from M. Marsh and A. Helenius, *Adv. Virus Res.* **36**:107–151, 1989, with permission.

the adenovirus penton base that mimic the normal ligands of cell surface integrins. As the virus particle is transported via the endosomes from the cell surface toward the nuclear membrane, it undergoes multiple uncoating steps by which structural proteins are removed sequentially (Fig. 5.17). As the endosome becomes acidified, the viral capsid is destabilized, leading to release of proteins from the capsid. Among these is protein VI, which causes disruption of the endosomal membrane, thereby delivering the remainder of the particle into the cytoplasm. An N-terminal amphipathic α-helix of protein VI is probably responsible for its pH-dependent membrane disruption activity. This region of the protein appears to be masked in the native capsid by the hexon protein. The liberated subviral particle then docks onto the nuclear pore complex (see "Import of DNA Genomes" below).

Forming a Pore in the Endosomal Membrane

The genome of the nonenveloped picornaviruses is transferred across the cell membrane by a different mechanism, as determined by structural information at the atomic level and complementary genetic and biochemical data obtained from studies of cell entry. The interaction of poliovirus with its Ig-like cell receptor, CD155, leads to major conformational rearrangements in the virus particle (Fig. 5.18A). These altered (A) particles are missing the internal capsid protein VP4, and the N terminus of capsid protein VP1 is on the surface rather than on the interior. Because of the latter change, A particles are hydrophobic and possess an increased affinity for membranes compared to the native virus particle. It is thought that the exposed lipophilic N terminus of VP1 inserts into the cell membrane, forming a pore that allows transport of viral RNA into the cytoplasm (Fig. 5.18B). In support of this model, ion channel activity can be detected when A particles are added to lipid bilayers.

The fate of VP4 is not known, but the study of a virus with an amino acid change in VP4 indicates that this protein is required for an early stage of cell entry. Mutant virus particles can bind to target cells and convert to altered particles, but are blocked at a subsequent, unidentified step. During poliovirus assembly, VP4 and VP2 are part of the precursor VP0, which remains uncleaved until the viral RNA has been encapsidated. The cleavage of VP0 during poliovirus assembly therefore primes the capsid for uncoating by separating VP4 from VP2.

In cultured cells, release of the poliovirus genome occurs from within early endosomes located close (within 100 to 200 nm) to the plasma membrane (Fig. 5.18A). Uncoating is dependent upon actin and tyrosine kinases, possibly for movement of the capsid through the network of actin filaments (Fig. 5.10), but not on dynamin, clathrin,

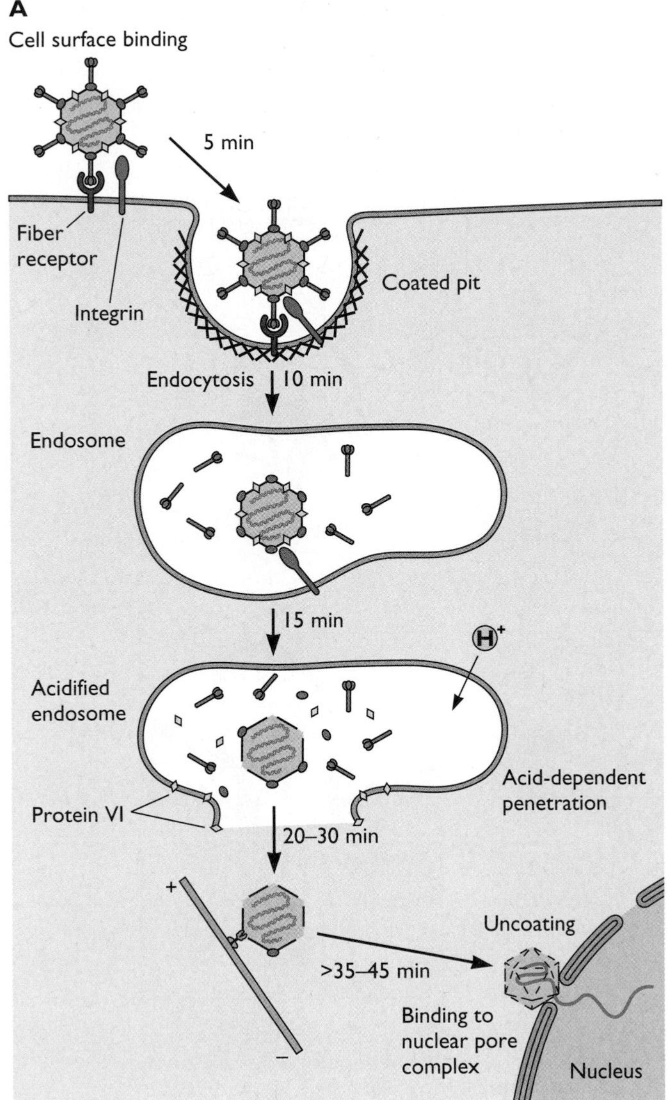

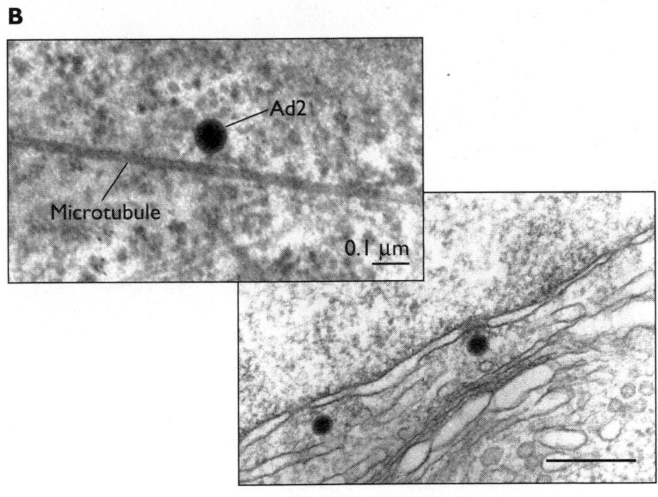

Figure 5.17 Stepwise uncoating of adenovirus. (A) Adenoviruses bind the cell receptor via the fiber protein. Interaction of the penton base with an integrin receptor leads to internalization by endocytosis. Low pH in the endosome causes destabilization of the capsid and release of protein VI. The hydrophobic N terminus of protein VI disrupts the endosome membrane, leading to release of a subviral particle into the cytoplasm. The capsid is transported in the cytoplasm along microtubules and docks onto the nuclear pore complex. **(B)** Electron micrograph of adenovirus type 2 particles bound to a microtubule (top) and bound to the cytoplasmic face of the nuclear pore complex (bottom). Bar in bottom panel = 200 nm. (A) Adapted from U. F. Greber et al., *Cell* **75**:477–486, 1993, and L. C. Trotman et al., *Nat. Cell Biol.* **3**:1092–1100, 2001, with permission. (B) Reprinted from U. F. Greber et al., *Trends Microbiol.* **2**:52–56, 1994, with permission. Courtesy of Ari Helenius, Urs Greber, and Paul Webster, University of Zurich.

caveolin, or flotillin (a marker protein for clathrin- and caveolin-independent endocytosis), endosome acidification, or microtubules. The trigger for RNA release from early endosomes is not known but is clearly dependent on prior interaction with CD155. This conclusion derives from the finding that antibody-poliovirus complexes can bind to cells that produce Fc receptors but cannot infect them. As the Fc receptor is known to be endocytosed, these results suggest that interaction of poliovirus with CD155 is required to induce conformational changes in the particle that are required for uncoating.

A critical regulator of the receptor-induced structural transitions of poliovirus appears to be a hydrophobic tunnel located below the surface of each structural unit (Fig. 5.18). The tunnel opens at the base of the canyon and extends toward the fivefold axis of symmetry. In poliovirus type 1, each tunnel is occupied by a natural ligand thought to be a molecule of sphingosine. Similar lipids have been observed in the capsids of other picornaviruses. Because of the symmetry of the capsid, each virion may contain up to 60 lipid molecules.

The lipids are thought to contribute to the stability of the native virus particle by locking the capsid in a stable conformation. Consequently, removal of the lipid is probably necessary to endow the particle with sufficient flexibility to permit the RNA to leave the shell. These conclusions come from the study of antiviral drugs known as WIN compounds (named after Sterling-Winthrop, the pharmaceutical company at which they were discovered). These compounds displace the lipid and fit tightly in the hydrophobic tunnel

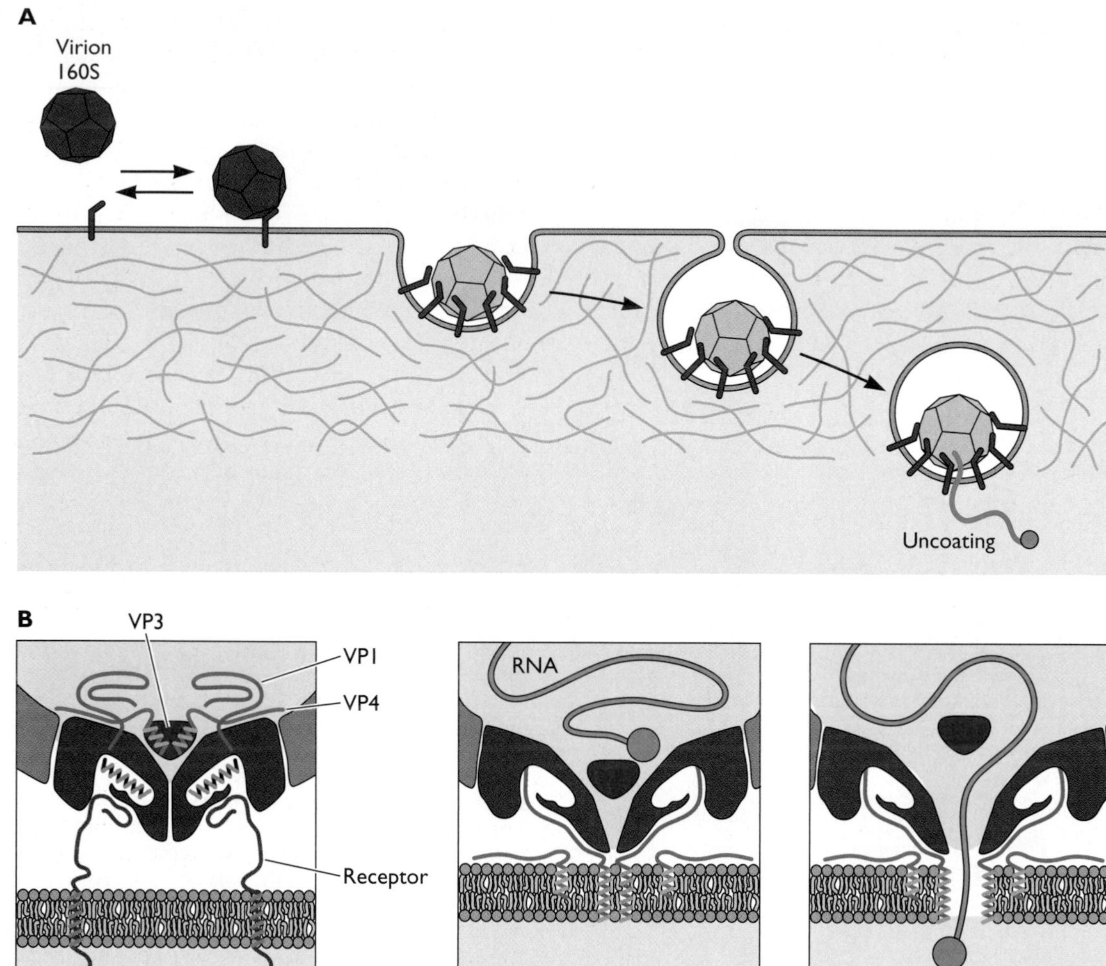

Figure 5.18 Model for poliovirus entry into cells. (A) Overview. The native virion (160S) binds to its cell receptor, CD155, and at temperatures higher than 33°C undergoes a receptor-mediated conformational transition resulting in the formation of altered (A) particles. The viral RNA, shown as a curved green line, leaves the capsid from within early endosomes close to the plasma membrane. **(B)** Model of the formation of a pore in the cell membrane after poliovirus binding. 1, Poliovirus (shown in cross section, with capsid proteins purple) binds to CD155 (brown). 2, A conformational change leads to displacement of the pocket lipid (black). The pocket may be occupied by sphingosine in the capsid of poliovirus type 1. The hydrophobic N termini of VP1 (blue) are extruded and insert into the plasma membrane. 3, A pore is formed in the membrane by the VP1 N termini, through which the RNA is released from the capsid into the cytosol. Adapted from J. M. Hogle and V. R. Racaniello, p. 71–83, *in* B. L. Semler and E. Wimmer (ed.), *Molecular Biology of Picornaviruses* (ASM Press, Washington, DC, 2002), with permission.

(Fig. 5.6). Polioviruses containing bound WIN compounds can bind to the cell receptor, but A particles are not produced. WIN compounds may therefore inhibit poliovirus infectivity by preventing the receptor-mediated conformational alterations required for uncoating. The properties of poliovirus mutants that cannot replicate in the absence of WIN compounds underscore the role of the lipids in uncoating. These drug-dependent mutants spontaneously convert to altered particles at 37°C, in the absence of the cell receptor, probably because they do not contain lipid in the hydrophobic pocket. The lipids are therefore viewed as switches, because their presence or absence determines whether the virus is stable or will be uncoated. The interaction of the virus particle with its receptor probably initiates structural changes in the virion that lead to the release of lipid. Consistent with this hypothesis is the observation that CD155 docks onto the poliovirus capsid just above the hydrophobic pocket.

Some picornaviruses enter cells by a pH-dependent pathway. For example, foot-and-mouth disease virus enters cells by receptor-mediated endocytosis. At a pH of approximately 6.5, the viral capsid dissociates to pentamers, releasing viral RNA. Dissociation of the capsid is probably a consequence of protonation of multiple histidine residues that line the pentamer interface and confer stability to the capsid at neutral pH. Consistent with this entry mechanism, antibody-coated foot-and-mouth disease virus can bind to and infect cells that carry Fc receptors, in contrast to findings with poliovirus. This result suggests that the cell receptor for foot-and-mouth disease virus does not induce uncoating-related changes in the virus particle.

Uncoating in the Lysosome

Most viruses that enter cells by receptor-mediated endocytosis leave the pathway before the vesicles reach the lysosomal compartment. This departure is not surprising, for lysosomes contain proteases and nucleases that would degrade virus particles. However, these enzymes play an important role during the uncoating of members of the *Reoviridae*, an event that takes place in lysosomes.

Orthoreoviruses are naked icosahedral viruses containing a double-stranded RNA genome of 10 segments. The viral capsid is a double-shelled structure composed of eight different structural proteins. These viruses bind to cell receptors via protein σ1 and are internalized into cells by endocytosis (Fig. 5.19A). Infection of cells by reoviruses is sensitive to bafilomycin A1, indicating that acidification of endosomes is required for entry. Low pH activates lysosomal proteases, which then modify several virion proteins, enabling the virus to cross the vesicle membrane. One viral outer capsid protein is cleaved and another is removed from the particle, producing an infectious subviral particle. These subviral particles penetrate the lysosome membrane and escape into the cytosol by a mechanism that is not yet understood. Isolated infectious subviral particles cause cell membranes to become permeable to toxins and produce pores in artificial membranes. These particles can initiate an infection by penetrating the plasma membrane, entering the cytoplasm directly. Their infectivity is not sensitive to bafilomycin A1, further supporting the idea that these particles are primed for membrane entry and do not require further acidification for this process. The core particles generated from infectious subviral particles after penetration into the cytoplasm carry out viral mRNA synthesis.

Import of Viral Genomes into the Nucleus

The replication of most DNA viruses, and some RNA viruses including retroviruses and influenza viruses, begins in the cell nucleus. The genomes of these viruses must therefore be imported from the cytoplasm into the nucleus. One way to accomplish this movement is via the cellular pathway for protein import into the nucleus. An alternative, observed in cells infected by some retroviruses, is to enter the nucleus during cell division. At this time in the cell cycle, cellular chromatin becomes accessible to virus particles. This strategy restricts infection to cells that undergo mitosis.

Many subviral particles are too large to pass through the nuclear pore complex. There are several strategies to overcome this limitation (Fig. 5.20). The influenza virus genome, which consists of eight segments that are each small enough to pass through the nuclear pore complex, is uncoated in the cytoplasm. Adenovirus subviral particles dock onto the nuclear pore complex and are disassembled by the import machinery, allowing the viral DNA to pass into the nucleus. Herpes simplex virus capsids also dock onto the nuclear pore but remain largely intact, and the nucleic acid is injected into the nucleus through a portal in the virion.

The cellular genome is highly compacted in the nucleus, and it is not understood how viral DNAs are imported against this steep gradient. The DNA of some bacteriophages is packaged in the virion at high pressure, which provides sufficient force to insert the viral DNA genome into the bacterial cell. However, no similar mechanism is known for animal viruses. Furthermore, because transport through the nuclear pore complex depends upon hydrophobic interactions with nucleoporins, the charged and hydrophilic viral nucleic acids would have difficulty passing through the pore. How the nuclear import machinery overcomes these obstacles is not known.

Nuclear Localization Signals

Proteins that reside within the nucleus are characterized by the presence of specific nuclear targeting sequences. Such **nuclear localization signals** are both necessary for nuclear localization of the proteins in which they are present and sufficient to direct heterologous, nonnuclear proteins to enter this organelle. Nuclear localization signals identified by these criteria share a number of common properties: they are generally fewer than 20 amino acids in length, they are not removed after entry of the protein into the nucleus, and they are usually rich in basic amino acids. Despite these similarities, no consensus nuclear localization sequence can be defined.

Most nuclear localization signals belong to one of two classes, simple or bipartite sequences (Fig. 5.21). A particularly well characterized example of a simple nuclear localization signal is that of simian virus 40 large T antigen, which comprises five contiguous basic residues flanked by a single hydrophobic amino acid (Fig. 5.21). This sequence is

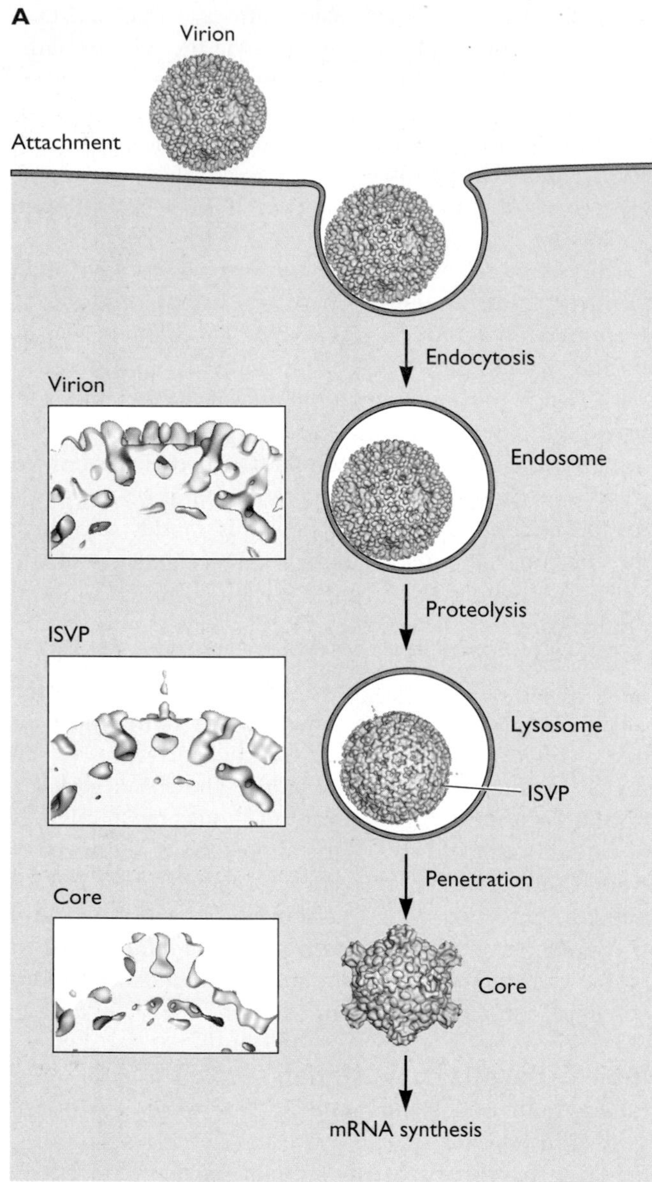

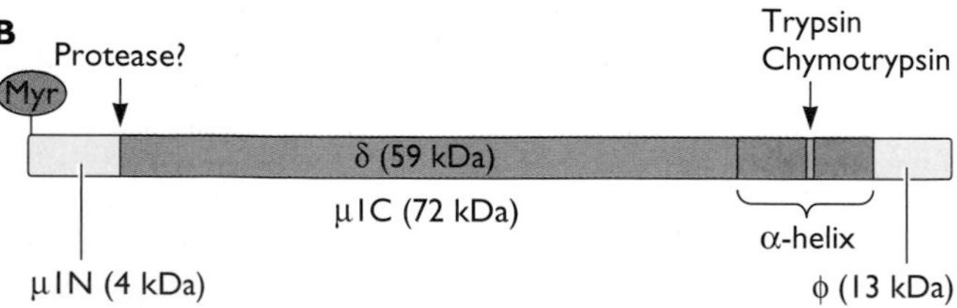

Figure 5.19 Entry of reovirus into cells. (A) The different stages in cell entry of reovirus. After the attachment of σ1 protein to the cell receptor, the virus particle enters the cell by receptor-mediated endocytosis. Proteolysis in the late endosome produces the infectious subviral particle (ISVP), which may then cross the lysosomal membrane and enter the cytoplasm as a core particle. The intact virion is composed of two concentric, icosahedrally organized protein capsids. The outer capsid is made up largely of σ3 and μ1. The dense core shell is formed mainly by λ1 and σ2. In the ISVP, 600 σ3 subunits have been released by proteolysis, and the σ1 protein changes from a compact form to an extended flexible fiber. The μ1 protein, which is thought to mediate interaction of the ISVP with membranes, is present as two cleaved fragments, μ1N and μ1C (see schematic of μ1 in panel B). The N terminus of μ1N is modified with myristate, suggesting that the protein functions in the penetration of membranes. A pair of amphipathic α-helices flank a C-terminal trypsin/chymotrypsin cleavage site at which μ1C is cleaved by lysosomal proteases. Such cleavage may release the helices to facilitate membrane penetration. The membrane-penetrating potential of μ1C in the virion may be masked by σ3; release of the σ3 in ISVPs might then allow μ1C to interact with membranes. The core is produced by the release of 12 σ1 fibers and 600 μ1 subunits. In the transition from ISVP to core, domains of λ2 rotate upward and outward to form a turretlike structure. (Insets) Close-up views of the emerging turretlike structure as the virus progresses through the ISVP and core stages. This structure may facilitate the entry of nucleotides into the core and the exit of newly synthesized viral mRNAs. **(B)** Schematic of the μ1 protein, showing locations of myristate, protease cleavage sites, and amphipathic α-helices. Virus images reprinted from K. A. Dryden et al., *J. Cell Biol.* **122:**1023–1041, 1993, with permission. Courtesy of Norm Olson and Tim Baker, Purdue University.

sufficient to relocate the enzyme pyruvate kinase, normally found in the cytoplasm, to the nucleus. Many other viral and cellular nuclear proteins contain short, basic nuclear localization signals, but these signals are not identical in primary sequence to the T-antigen signal. The presence of a nuclear localization signal is all that is needed to target a macromolecular substrate for import into the nucleus. Even gold particles with diameters as large as 26 nm are readily imported following their microinjection into the cytoplasm, as long as they are coated with proteins or peptides containing a nuclear localization signal.

The Nuclear Pore Complex

The nuclear envelope is composed of two typical lipid bilayers separated by a lumenal space (Fig. 5.22). Like all other cellular membranes, it is impermeable to macromolecules such as proteins. However, the nuclear pore complexes that stud the nuclear envelopes of all eukaryotic cells provide aqueous channels that span both the inner and outer nuclear membranes for exchange of small molecules, macromolecules, and macromolecular assemblies between nuclear and cytoplasmic compartments. Numerous experimental techniques, including direct visualization of gold particles attached to proteins or RNA molecules as they are transported, have established that nuclear proteins enter and RNA molecules exit the nucleus by transport through the nuclear pore complex. The functions of the nuclear pore complex in both protein import and RNA export are far from completely understood, not least because this important cellular machine is large (molecular mass,

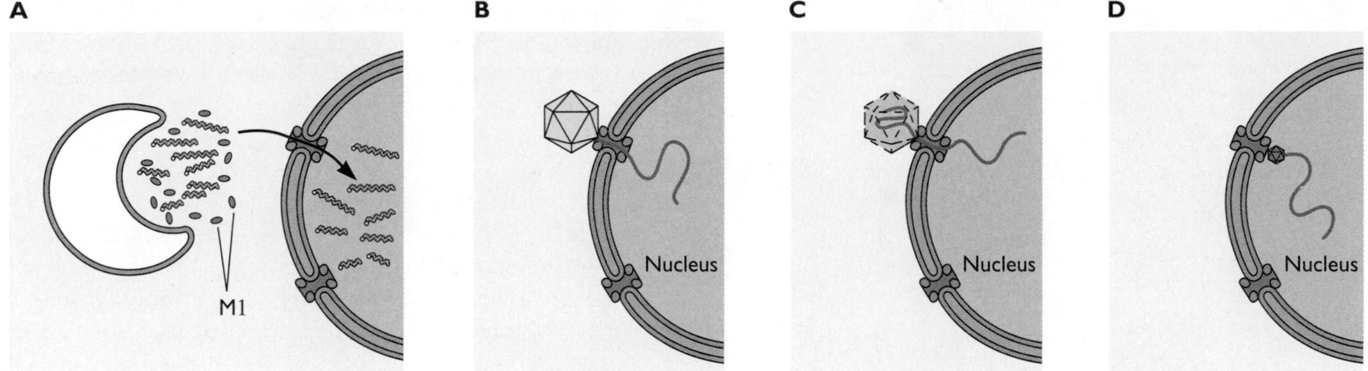

Figure 5.20 Different strategies for entering the nucleus. (A) Each segment of the influenza virus genome is small enough to be transported through the pore complex. **(B)** The herpes simplex virus type 1 capsid docks onto the nuclear pore complex and is minimally disassembled to allow transit of the viral DNA into the nucleus. **(C)** The adenovirus subviral particle is substantially dismantled by the nuclear import machinery, allowing transport of the viral DNA into the nucleus. **(D)** The capsids of some viruses (parvovirus and hepadnavirus) are small enough to enter the nuclear pore complex without disassembly.

approximately 124×10^3 kDa in vertebrates), built from many different proteins, and architecturally complex (Fig. 5.22). In comparison, ribosomes, which consist of ~82 proteins and 4 RNA molecules, have a molecular mass of 4.2×10^3 kDa.

The nuclear pore complex allows passage of cargo in and out of the nucleus by either passive diffusion or facilitated translocation. Passive diffusion does not require interaction between the cargo and components of the nuclear pore complex, and becomes inefficient as molecules approach 20 to 40 kDa in mass. Objects as large as several megadaltons can pass through nuclear pore complexes by facilitated translocation. This process requires specific interactions between the cargo and components of the nuclear pore complex and is therefore selective.

The Nuclear Import Pathway

Import of a protein into the nucleus via nuclear localization signals occurs in two distinct, and experimentally separable, steps (Fig. 5.22C). A protein containing such a signal first binds to a soluble cytoplasmic receptor protein. This complex then engages with the cytoplasmic surface of the nuclear pore complex, in a reaction often called docking, and is translocated through the nuclear pore complex into the nucleus. In the nucleus, the complex is disassembled, releasing the protein cargo.

Different groups of proteins are imported into the nucleus by specific receptor systems. In what is known as the "classical system" of import, cargo proteins containing basic nuclear localization signals bind to the cytoplasmic nuclear localization signal receptor protein importin-α

Figure 5.21 Nuclear localization signals. The general form and a specific example of simple and bipartite nuclear localization signals are shown in the one-letter amino acid code, where X is any residue. Bipartite nuclear targeting signals are defined by the presence of two clusters of positively charged amino acids separated by a spacer region of variable sequence. Both clusters of basic residues, which often resemble the simple targeting sequences of proteins like simian virus 40 T antigen, are required for efficient import of the proteins in which they are found. The subscript indicates either length (3–7) or composition (e.g., 3/5 means at least 3 residues out of 5 are basic).

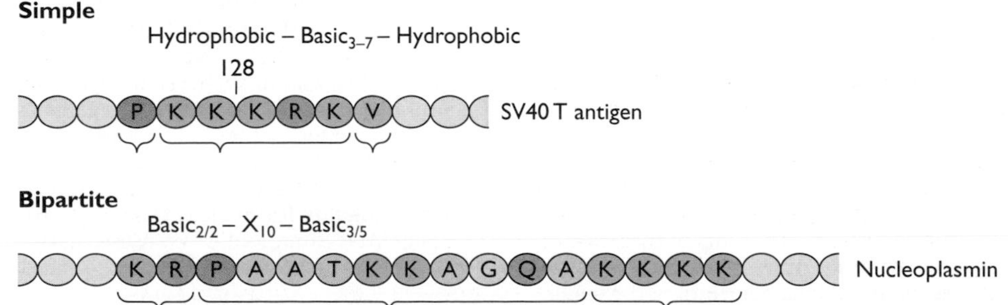

A

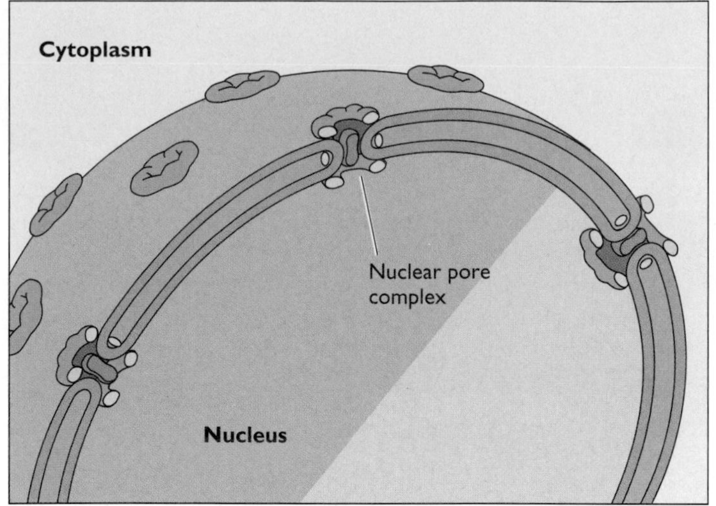

B

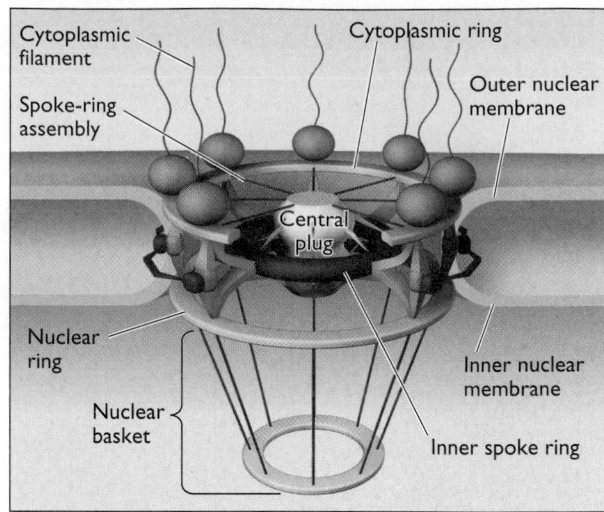

C

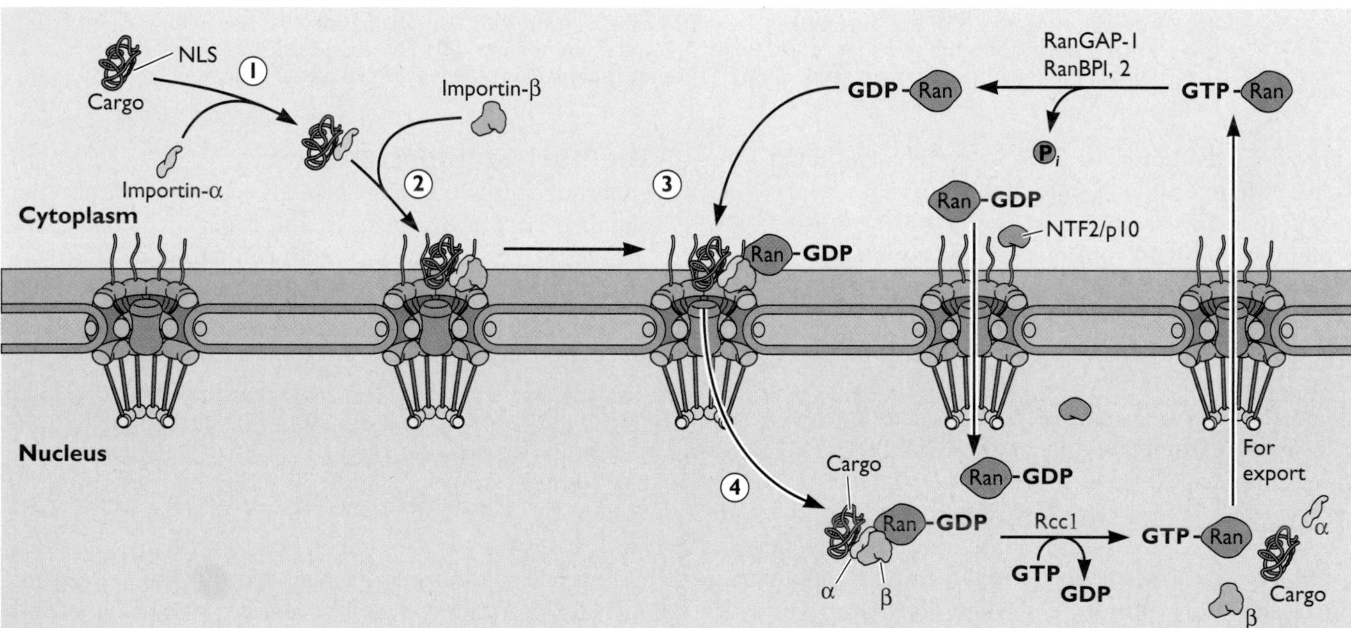

Figure 5.22 Structure and function of the nuclear pore complex. (A) Overview of the nuclear membrane, showing the topology of the nuclear pore complexes. **(B)** Schematic drawing of the nuclear pore complex, showing the spoke-ring assembly at its waist and its attachment to cytoplasmic filaments and the nuclear basket. The latter comprises eight filaments, extending 50 to 100 nm from the central structure and terminating in a distal annulus. The nuclear pore channel is shown containing the transporter. **(C)** An example of the classical protein import pathway for proteins with a simple nuclear localization signal (NLS). This pathway is illustrated schematically from left to right. Cytoplasmic and nuclear compartments are shown separated by the nuclear envelope studded with nuclear pore complexes. In step 1, a nuclear localization signal on the cargo (red) is recognized by importin-α. In step 2, importin-β binds the cargo–importin-α complex and docks onto the nucleus, probably by associating initially with nucleoporins present in the cytoplasmic filaments of the nuclear pore complex. Translocation of the substrate into the nucleus (step 4) requires additional soluble proteins, including the small guanine nucleotide-binding protein Ran (step 3). A Ran-specific guanine nucleotide exchange protein (Rcc1) and a Ran-GTPase-activating protein (RanGap-1) are localized in the nucleus and cytoplasm, respectively. The action of RanGAP-1, with the accessory proteins RanBp1 and RanBp2, maintains cytoplasmic Ran in the GDP-bound form. When Ran is in the GTP-bound form, nuclear import cannot occur. Following import, the complexes are dissociated when Ran-GDP is converted to Ran-GTP by Rcc1. Ran-GTP participates in export from the nucleus. The nuclear pool of Ran-GDP is replenished by the action of the transporter Ntf2/p10, which efficiently transports Ran-GDP from the cytoplasm to the nucleus. Hydrolysis of Ran-GTP in the cytoplasm and GTP-GDP exchange in the nucleus therefore maintain a gradient of Ran-GTP/Ran-GDP. The asymmetric distribution of RanGap-1 and Rcc1 allows for the formation of such a gradient. This gradient provides the driving force and directionality for nuclear transport. (B) Adapted from Q. Yang, M. P. Rout, and C. W. Akey, *Mol. Cell* **1:**223–234, 1998, with permission.

(Fig. 5.22C). This complex then binds importin-β, which mediates docking with the nuclear pore complex by binding to members of a family of nucleoporins. Some of these nucleoporins are found in the cytoplasmic filaments of the nuclear pore complex (Fig. 5.22), which associate with import substrates as seen by electron microscopy. The complex is translocated to the opposite side of the nuclear envelope, where the cargo is released. Other importins can bind cargo proteins directly without the need for an adapter protein. A monomeric receptor called transportin mediates the import of heterogeneous nuclear RNA-binding proteins that contain glycine- and arginine-rich nuclear localization signals. Transportin is related to importin-β, as are other monomeric receptors that mediate nuclear import of ribosomal proteins.

Release of cargo occurs when the importins associate with a small guanosine triphosphate (GTP)-binding protein termed Ran in the GTP form (Fig. 5.22). How these components work together to move the import substrate through the channel of the nuclear pore complex, a distance of more than 100 nm, is not yet well understood. It is clear that a single translocation through the nuclear pore complex does not require energy consumption. However, maintenance of a gradient of the guanosine nucleotide-bound forms of Ran, with Ran-GDP and Ran-GTP concentrated in the cytoplasm and nucleus, respectively, is absolutely essential for continued transport. For example, conversion of Ran-GDP to Ran-GTP in the nucleus, catalyzed by the guanine nucleotide exchange protein Rcc-1, promotes dissociation of imported proteins from importins (Fig. 5.22).

Import of Influenza Virus Ribonucleoprotein

Influenza virus is among the few RNA-containing viruses that replicate in the cell nucleus. After vRNPs separate from M1 and are released into the cytosol, they are rapidly imported into the nucleus (Fig. 5.12). Such import depends on the presence of a nuclear localization signal in the NP protein, a component of vRNPs. Naked viral RNA does not dock onto the nuclear pore complex, nor is it taken up into the nucleus, but in the presence of NP the viral RNA can enter this organelle.

Import of DNA Genomes

The capsids of many DNA-containing viruses are larger than 26 nm and cannot be imported into the nucleus from the cytoplasm. One mechanism for crossing the nuclear membrane involves docking onto the nuclear pore complex, followed by delivery of the viral DNA into the nucleus. Adenoviral and herpesviral DNAs are transported into the nucleus via this mechanism. Partially disassembled adenovirus capsids dock onto the nuclear pore complex through interactions with the filament protein Can/Nup214 (Fig. 5.23). Small quantities of histone H1 from the nucleus bind to hexon proteins on the nuclear side of the viral capsid. The H1 import proteins importin-β and importin-7 recognize H1 bound to hexon, and promote further disassembly. These interactions also promote conformational changes that allow viral protein VII and the viral DNA to associate with transportin. The protein VII-viral DNA complex is imported into the nucleus, where viral transcription begins.

Herpesvirus capsids also dock onto the nuclear pore complex, but undergo only limited disassembly. The viral

Figure 5.23 Uncoating of adenovirus at the nuclear pore complex. After release from the endosome, the partially disassembled capsid docks onto the nuclear pore complex-filament protein Can/Nup214. Histone H1 from the nucleus (green ovals) binds to hexon. Importin-β and importin-7 bind histone H1, leading to further disassembly of the capsid. Once there is sufficient dismantling, the viral DNA, bound to protein VII, is delivered into the nucleus by the import protein transportin.

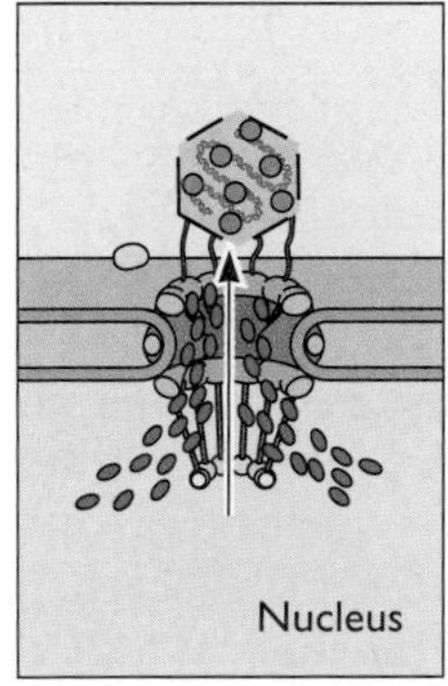

Histone H1 contacts capsid-hexon

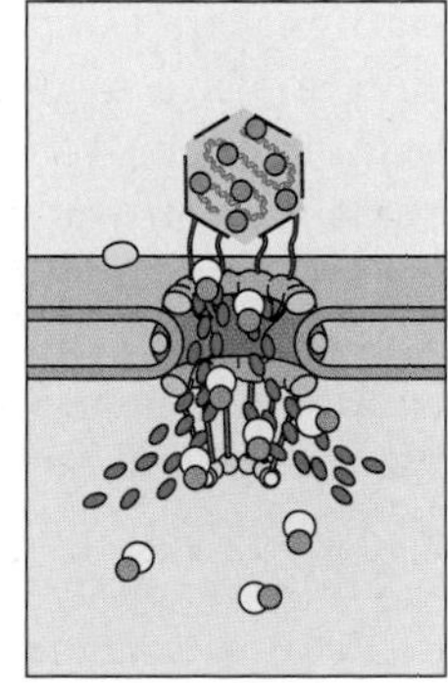

Importin-7 and importin-β bind histone H1

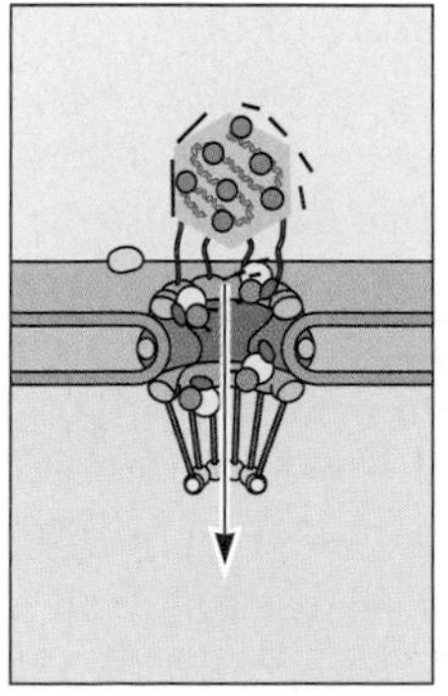

Capsid disassembly

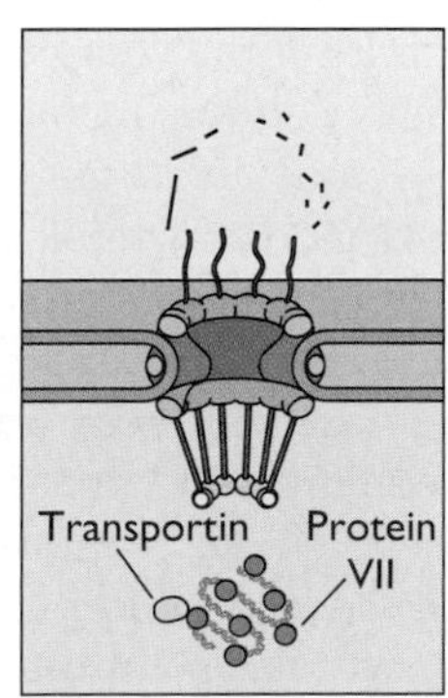

Import of DNA

DNA probably exits through one of the pentameric faces of the capsid and passes through the nuclear pore complex.

Only the smallest capsids can enter the nuclear pore complex without disassembly. The capsids of parvoviruses and hepatitis B virus can be observed intact within the central channel of the complex. Uncoating takes place within the nuclear basket (Fig. 5.22B).

Import of Retroviral Genomes

Fusion of retroviral and plasma membranes releases the viral core into the cytoplasm. The retroviral core consists of the viral RNA genome, coated with NC protein, and the enzymes reverse transcriptase (RT) and integrase (IN), enclosed by CA protein. Retroviral DNA synthesis commences in the cytoplasm, within the nucleocapsid core, and after 4 to 8 h of DNA synthesis the preintegration complex, comprising viral DNA, IN, and other proteins, localizes to the nucleus. There the viral DNA is integrated into a cellular chromosome, and viral transcription begins. The mechanism of nuclear import of the preintegration complex is poorly understood, but it is quite clear that this structure is too large (~60S) to pass through the nuclear pore complex. The betaretrovirus Moloney murine leukemia virus can efficiently infect only dividing cells. These and other observations suggest that exposure of chromatin that occurs during mitosis is essential to allow efficient entry of the preintegration complex of this retrovirus into the nucleus.

In contrast to Moloney murine leukemia virus, human immunodeficiency virus type 1 can replicate in nondividing cells. The preintegration complex of this virus, and probably other lentiviruses, must therefore be transported into an intact nucleus. The exact mechanism by which the DNA of these retroviruses enter the nucleus is still unclear. There is evidence for participation of various viral proteins that contain nuclear localization signals (e.g., Vpr, MA, and IN). Others discount the role of these proteins in import and suggest that breakdown of CA is critical. Such controversy may stem from the complex nature of the import mechanism, possibly comprising more than one pathway.

Avian sarcoma and leukosis viruses, like Moloney murine leukemia virus, do not replicate in nondividing cells. However, it was recently shown that the DNA of these avian retroviruses can be integrated in cell cycle-arrested cells and during interphase in cycling cells, implying a mitosis-independent mechanism of nuclear import. IN protein of these viruses contains a nuclear localization signal in the C-terminal domain which, when fused to heterologous cytoplasmic proteins, can direct them to the nucleus. This protein may have a role in the nuclear import of the preintegration complex of these avian retroviruses.

Perspectives

Since the last edition of this textbook, it has become clear that there are many pathways for virus entry into cells. Clathrin- and dynamin-dependent endocytosis is no longer the sole entry pathway known; other routes are caveolin-dependent endocytosis and clathrin- and caveolin-independent endocytosis. The road used seems to depend on the virus, the cell type, and the conditions of infection. Do additional entry pathways exist that bring viruses into cells? What is the significance of multiple pathways used by the same virus? What pathways of viral entry operate in living animals?

The notion that endocytosis is an unregulated process has been shattered. Of particular interest has been the application of high-throughput small interfering RNA (siRNA) screens to identify host protein kinases that regulate clathrin- and caveolin-mediated endocytosis. The results indicate that vesicular stomatitis virus entry is regulated by 92 kinases while simian virus 40 entry is regulated by 80; 36 kinases are common to both virus entry pathways. Curiously, these 36 kinases have the opposite effects on entry of the two viruses studied. It will be important to determine whether such patterns are common to other virus infections and to identify other cellular genes that regulate these uptake pathways.

The development of single-particle tracking methods has advanced considerably in the past 5 years. As a consequence, our understanding of the routes that viruses travel once they are inside the cell has improved markedly. The role of cellular transport pathways in bringing viruses to the point of replication within the cell is beginning to be clarified. Yet many questions remain. How are viruses transported on the cytoskeletal network? What are the precise virus-host interactions needed? Do viral proteins regulate such transport? What are the signals for a virus to attach to and detach from microtubules and filaments?

It has become clear that virus binding to the cell surface leads to major alterations in cell activities, effects mediated by signal transduction. Virus binding induces the formation of pits, pinching off of vesicles, and rearrangement of actin filaments to facilitate vesicle movement. The precise signaling pathways required need to be elucidated. Such efforts may identify specific targets for inhibiting virus movement in cells.

The genomes of many viruses replicate in the nucleus. Incoming viral genomes enter this cellular compartment by transport through the nuclear pore complex. Studies of adenovirus import into the nucleus have revealed an active role for components of the nuclear pore complex in subviral particle disassembly. What is the molecular basis for this process? What other proteins are involved, and how general is the process? Can it be interrupted therapeutically?

How does the hydrophilic viral DNA pass through the hydrophobic pore, against a steep gradient of nucleic acid in the nucleus? Nuclear import of the lentivirus genome is barely understood. What signal allows transport of the preintegration complex through the nuclear pore?

Nearly all the conclusions discussed in this chapter were derived from studies of viral infection in cultured cells. How viruses attach to and enter cells of a living animal remains an uncharted territory. Methods are being developed to study virus entry in whole animals, and the results will be important for understanding how viruses spread and breach host defenses to reach target cells.

References

Reviews

Brandenburg, B. and X. Zhuang. 2007. Virus trafficking—learning from single-virus tracking. *Nat. Rev. Microbiol.* **5**:197–208.

Damm, E.-M., and L. Pelkmans. 2006. Systems biology of virus entry in mammalian cells. *Cell. Microbiol.* **8**:1219–1227.

Fassati, A. 2006. HIV infection of non-dividing cells: a divisive problem. *Retrovirology* **3**:73.

Greber, U., and M. Way. 2006. A superhighway to virus infection. *Cell* **124**:741–754.

Greber, U., and M. Fornerod. 2005. Nuclear import in viral infections. *Curr. Top. Microbiol. Immunol.* **285**:109–138.

Hogle, J. M. 2002. Poliovirus cell entry: common structural themes in viral cell entry pathways. *Annu. Rev. Microbiol.* **56**:677–702.

Lamb, R. A., R. G. Paterson, and T. S. Jardetzky. 2005. Paramyxovirus membrane fusion: lessons from the F and HN atomic structures. *Virology* **344**:30–37.

Marsh, M., and A. Helenius. 2006. Virus entry: open sesame. *Cell* **124**:729–740.

Pietiäinen, V. M., V. Marjomäki, J. Heino, and T. Hyypiä. 2005. Viral entry, lipid rafts and caveosomes. *Ann. Med.* **37**:394–403.

Radtke, K., K. Dohner, and B. Sodeik. 2006. Viral interactions with the cytoskeleton: a hitchhiker's guide to the cell. *Cell. Microbiol.* **8**:387–400.

Weissenhorn, W., A. Hinz, and Y. Gaudin. 2007. Virus membrane fusion. *FEBS Lett.* **581**:2150–2155.

Papers of Special Interest

Baker, K. A., R. E. Dutch, R. A. Lamb, and T. S. Jardetzky. 1999. Structural basis for paramyxovirus-mediated membrane fusion. *Mol. Cell* **3**:309–319.

Belnap, D. M., D. J. Filman, B. L. Trus, N. Cheng, F. P. Booy, J. F. Conway, S. Curry, C. N. Hiremath, S. K. Tsang, A. C. Steven, and J. M. Hogle. 2000. Molecular tectonic model of virus structural transitions: the putative cell entry states of poliovirus. *J. Virol.* **74**:1342–1354.

Belnap, D. M., B. M. McDermott, Jr., D. J. Filman, N. Cheng, B. L. Trus, H. J. Zuccola, V. R. Racaniello, J. M. Hogle, and A. C. Steven. 2000. Three-dimensional structure of poliovirus receptor bound to poliovirus. *Proc. Natl. Acad. Sci. USA* **97**:73–78.

Brandenburg, B., L. Y. Lee, M. Lakadamyali, M. J. Rust, X. Zhuang, and J. M. Hogle. Imaging poliovirus entry in live cells. *PLoS Biol.* **5**:1543–1555.

Carr, C. M., C. Chaudhry, and P. S. Kim. 1997. Influenza hemagglutinin is spring-loaded by a metastable native conformation. *Proc. Natl. Acad. Sci. USA* **94**:14306–14313.

Chandran, K., D. L. Farsetta, and M. L. Nibert. 2002. Strategy for nonenveloped virus entry: a hydrophobic conformer of the reovirus membrane penetration protein micro 1 mediates membrane disruption. *J. Virol.* **76**:9920–9933.

Chen, J., K. H. Lee, D. A. Steinhauer, D. J. Stevens, J. J. Skehel, and D. C. Wiley. 1998. Structure of the hemagglutinin precursor cleavage site, a determinant of influenza pathogenicity and the origin of the labile conformation. *Cell* **95**:409–417.

Chipman, P. R., M. Agbandje-McKenna, S. Kajigaya, K. E. Brown, N. S. Young, T. S. Baker, and M. G. Rossmann. 1996. Cryo-electron microscopy studies of empty capsids of human parvovirus B19 complexed with its cellular receptor. *Proc. Natl. Acad. Sci. USA* **93**:7502–7506.

Dohner, K., A. Wolfstein, U. Prank, C. Echeverri, D. Dujardin, R. Vallee, and B. Sodeik. 2002. Function of dynein and dynactin in herpes simplex virus capsid transport. *Mol. Biol. Cell* **13**:2795–2809.

Dragic, T., V. Litwin, G. P. Allaway, S. R. Martin, Y. Huang, K. A. Nagashima, C. Cayanan, P. J. Maddon, R. A. Koup, J. P. Moore, and W. A. Paxton. 1996. HIV-1 entry into $CD4^+$ cells mediated by the chemokine receptor CC-CKR-5. *Nature* **381**:667–673.

Ebert, D. H., J. Deussing, C. Peters, and T. S. Dermody. 2002. Cathepsin L and cathepsin B mediate reovirus disassembly in murine fibroblast cells. *J. Biol. Chem.* **277**:24609–24617.

Ewers, H., A. E. Smith, I. F. Sbalzarini, H. Lilie, P. Koumoutsakos, and A. Helenius. 2005. Single-particle tracking of murine polyoma virus-like particles on live cells and artificial membranes. *Proc. Natl. Acad. Sci. USA* **102**:15110–15115.

Follis, K. E., S. J. Larson, M. Lu, and J. H. Nunberg. 2002. Genetic evidence that interhelical packing interactions in the gp41 core are critical for transition of the human immunodeficiency virus type 1 envelope glycoprotein to the fusion-active state. *J. Virol.* **76**:7356–7362.

Frick, M., N. A. Bright, K. Riento, A. Bray, C. Merrified, and B. J. Nichols. 2007. Coassembly of flotillins induces formation of membrane microdomains, membrane curvature, and vesicle budding. *Curr. Biol.* **17**:1151–1156.

Han, X., J. H. Bushweller, D. S. Cafiso, and L. K. Tamm. 2001. Membrane structure and fusion-triggering conformational change of the fusion domain from influenza hemagglutinin. *Nat. Struct. Biol.* **8**:715–720.

He, Y., V. D. Bowman, S. Mueller, C. M. Bator, J. Bella, X. Peng, T. S. Baker, E. Wimmer, R. J. Kuhn, and M. G. Rossmann. 2000. Interaction of the poliovirus receptor with poliovirus. *Proc. Natl. Acad. Sci. USA* **97**:79–84.

Herbein, G., U. Mahlknecht, F. Batliwalla, P. Gregersen, T. Pappas, J. Butler, W. A. O'Brien, and E. Verdin. 1998. Apoptosis of $CD8^+$ T cells is mediated by macrophages through the interaction of human immunodeficiency virus gp120 with chemokine receptor CXCR4. *Nature* **395**:189–194.

Hesselgesser, J., D. Taub, P. Basker, M. Greenberg, J. Hoxie, D. L. Kolson, and R. Horuk. 1998. Neuronal apoptosis induced by human immunodeficiency virus-1 gp120 and the chemokine Sdf-1a is mediated by the chemokine receptor CXCR4. *Curr. Biol.* **8**:595–598.

Hewat, E. A., E. Neumann, J. F. Conway, R. Moser, B. Ronacher, T. C. Marlovits, and D. Blaas. 2000. The cellular receptor to human rhinovirus 2 binds around the 5-fold axis and not in the canyon: a structural view. *EMBO J.* **19**:6317–6325.

Kuhn, R. J., W. Zhang, M. G. Rossmann, S. V. Pletnev, J. Corver, E. Lenches, C. T. Jones, S. Mukhopadhyay, P. R. Chipman, E. G. Strauss, T. S. Baker, and J. H. Strauss. 2002. Structure of dengue virus: implications for flavivirus organization, maturation, and fusion. *Cell* **108**:717–725.

Kwong, P. D., R. Wyatt, J. Robinson, R. W. Sweet, J. Sodroski, and W. A. Hendrickson. 1998. Structure of an HIV gp120 envelope glycoprotein in complex with the CD4 receptor and a neutralizing human antibody. *Nature* **393:**648–659.

Lescar, J., A. Roussel, M. W. Wien, J. Navaza, S. D. Fuller, G. Wengler, G. Wengler, and F. A. Rey. 2001. The fusion glycoprotein shell of Semliki Forest virus: an icosahedral assembly primed for fusogenic activation at endosomal pH. *Cell* **105:**137–148.

Mabit, H., M. Y. Nakano, U. Prank, B. Saam, K. Dohner, B. Sodeik, and U. F. Greber. 2002. Intact microtubules support adenovirus and herpes simplex virus infections. *J. Virol.* **76:**9962–9971.

McDonald, D., M. A. Vodicka, G. Lucero, T. M. Svitkina, G. G. Borisy, M. Emerman, and T. J. Hope. 2002. Visualization of the intracellular behavior of HIV in living cells. *J. Cell Biol.* **159:**441–452.

Mothes, W., A. L. Boerger, S. Narayan, J. M. Cunningham, and J. A. Young. 2000. Retroviral entry mediated by receptor priming and low pH triggering of an envelope glycoprotein. *Cell* **103:**679–689.

Mueller, S., X. Cao, R. Welker, and E. Wimmer. 2002. Interaction of the poliovirus receptor CD155 with the dynein light chain Tctex-1 and its implication for poliovirus pathogenesis. *J. Biol. Chem.* **277:**7897–7904.

Ojala, P. M., B. Sodeik, M. W. Ebersold, U. Kutay, and A. Helenius. 2000. Herpes simplex virus type 1 entry into host cells: reconstitution of capsid binding and uncoating at the nuclear pore complex in vitro. *Mol. Cell. Biol.* **20:**4922–4931.

Pelkmans, L., J. Kartenbeck, and A. Helenius. 2001. Caveolar endocytosis of simian virus 40 reveals a new two-step vesicular-transport pathway to the ER. *Nat. Cell Biol.* **3:**473–483.

Pelkmans, L., D. Puntener, and A. Helenius. 2002. Local actin polymerization and dynamin recruitment in SV40-induced internalization of caveolae. *Science* **296:**535–539.

Pelkmans, L., E. Fava, H. Grabner, M. Hannus, B. Habermann, E. Krausz, and M. Zerial. 2005. Genome-wide analysis of human kinases in clathrin- and caveolae/raft-mediated endocytosis. *Nature* **436:**78–86.

Pietiäinen, V., V. Marjomäki, P. Upla, L. Pelkmans, A. Helenius, and T. Hyypiä. 2004. Echovirus 1 endocytosis into caveosomes requires lipid rafts, dynamin II, and signaling events. *Mol. Biol. Cell* **15:**4911–4925.

Rietdorf, J., A. Ploubidou, I. Reckmann, A. Holmstrom, F. Frischknecht, M. Zettl, T. Zimmermann, and M. Way. 2001. Kinesin-dependent movement on microtubules precedes actin-based motility of vaccinia virus. *Nat. Cell Biol.* **3:**992–1000.

Russell, C. J., T. S. Jardetzky, and R. A. Lamb. 2001. Membrane fusion machines of paramyxoviruses: capture of intermediates of fusion. *EMBO J.* **20:**4024–4034.

Ryu, S.-E., P. D. Kwong, A. Truneh, T. G. Porter, J. Arthos, M. Rosenberg, X. Dai, N.-H. Xuong, R. Axel, R. W. Sweet, and W. A. Hendrickson. 1990. Crystal structure of an HIV-binding recombinant fragment of human CD4. *Nature* **348:**419–426.

Schelhaas, M., J. Malmstrom, L. Pelkmans, J. Haugstetter, L. Ellgaard, K. Grunewald, and A. Helenius. 2007. Simian virus 40 depends on ER protein folding and quality control factors for entry into host cells. *Cell* **131:**516–529.

Suomalainen, M., M. Y. Nakano, K. Boucke, S. Keller, and U. F. Greber. 2001. Adenovirus-activated PKA and p38/MAPK pathways boost microtubule-mediated nuclear targeting of virus. *EMBO J.* **20:**1310–1319.

Trotman, L. C., N. Mosberger, M. Fornerod, R. P. Stidwill, and U. F. Greber. 2001. Import of adenovirus DNA involves the nuclear pore complex receptor CAN/Nup214 and histone H1. *Nat. Cell Biol.* **3:**1092–1100.

Weis, W., J. H. Brown, S. Cusack, J. C. Paulson, J. J. Skehel, and D. C. Wiley. 1988. Structure of the influenza virus hemagglutinin complexed with its receptor, sialic acid. *Nature* **333:**426–431.

Wickham, T. J., P. Mathias, D. A. Cheresh, and G. R. Nemerow. 1993. Integrins alpha v beta 3 and alpha v beta 5 promote adenovirus internalization but not virus attachment. *Cell* **73:**309–319.

Wiethoff, C. M., H. Wodrich, L. Gerace, and G. R. Nemerow. 2005. Adenovirus protein VI mediates membrane disruption following capsid disassembly. *J. Virol.* **79:**1992–2000.

Xing, L., K. Tjarnlund, B. Lindqvist, G. G. Kaplan, D. Feigelstock, R. H. Cheng, and J. M. Casasnovas. 2000. Distinct cellular receptor interactions in poliovirus and rhinoviruses. *EMBO J.* **19:**1207–1216.

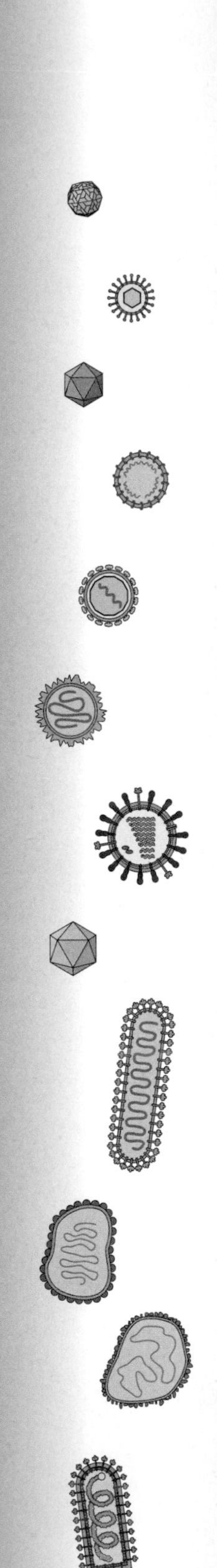

6

Synthesis of RNA from RNA Templates

Truth is ever to be found in the simplicity, and not in the multiplicity and confusion of things.
SIR ISAAC NEWTON

Introduction

The genomes of RNA viruses come in a number of conformations, including unimolecular or segmented, single stranded of (+) or (−) polarity, double stranded, and circular. These structurally diverse viral RNA genomes share a common requirement: they must be efficiently copied within the infected cell to provide both genomes for assembly into progeny virions and messenger RNAs (mRNAs) for the synthesis of viral proteins. The synthesis of these RNA molecules by RNA viruses is a unique process that has no parallel in the cell. The genomes of all RNA viruses except retroviruses (see below) encode an **RNA-dependent RNA polymerase** (Box 6.1) to catalyze the synthesis of new genomes and mRNAs.

The virions of RNA viruses with (−) strand and double-stranded RNA genomes must contain the RNA polymerase, because the incoming viral RNA can be neither translated nor copied by the cellular machinery. The deproteinized genomes of (−) strand and double-stranded RNA viruses are therefore noninfectious. In contrast, viral particles containing a (+) strand RNA genome lack a virion polymerase; the deproteinized RNAs of these viruses **are** infectious because they are translated in cells to produce, among other viral proteins, the viral RNA polymerase.

The mechanisms by which viral mRNA is made and the RNA genome is replicated in cells infected by RNA viruses appear even more diverse than the structure and organization of viral RNA genomes (Fig. 6.1). For example, the genomes of both picornaviruses and alphaviruses are single molecules of (+) strand genomic RNA, but the strategies for the production of viral RNA are quite different. Nevertheless, each mechanism of viral RNA synthesis meets two essential requirements common to the infectious cycles of all these viruses: (i) during replication the RNA genome must be copied from one end to the other with no loss of nucleotide sequence; and (ii) viral mRNAs must be produced that can be translated efficiently by the cellular protein synthetic machinery.

BOX 6.1

TERMINOLOGY

What should we call RNA polymerases and the processes they catalyze?

Historically, viral RNA-dependent RNA polymerases were given two different names depending on their activities during infection. The term **replicase** was used to describe the enzyme that copies the viral RNA to produce additional genomes, while the enzyme that produces mRNA was called **transcriptase.** In some cases this terminology indicates true differences in the enzymes that carry out synthesis of functionally different RNAs. For example, for some RNA viruses, genomic replication and mRNA synthesis are the **same** reaction. For double-stranded RNA viruses, mRNA synthesis produces templates that can also be used for genomic replication. However, these terms can also be inaccurate and misleading, and so are not used here.

The production of mRNAs from viral RNA templates is often designated **transcription.** However, this term refers to a specific process, the copying of genetic information carried in DNA into RNA. Consequently, it is not used here to describe synthesis of the mRNAs of viruses with RNA genomes. Similarly, use of the term **promoter** is reserved to designate sequences controlling transcription of DNA templates.

In this chapter we consider the mechanisms of viral RNA synthesis, the mechanism for switching from mRNA production to genome replication, and how the process of RNA-directed RNA synthesis leads to genetic diversity. Much of our understanding of viral RNA synthesis comes from experiments done with purified components. Because it is possible that events proceed differently in infected cells, the results of such *in vitro* studies are used to build hypothetical models for the different steps in RNA synthesis. While many models exist for each reaction, those presented in this chapter were selected because they are consistent with experimental results obtained in different laboratories.

The general principles of RNA synthesis are illustrated with a few viruses as examples. Members of another family of RNA viruses, the *Retroviridae,* are discussed in Chapter 7. Retroviruses encode an RNA-dependent DNA polymerase, and therefore mRNA synthesis and genome replication are very different from those of other RNA viruses.

The Nature of the RNA Template

Secondary Structures in Viral RNA

RNA molecules are not simple linear chains but can form secondary structures that have important functions. Viral RNA genomes contain secondary-structure elements such as base-paired **stem regions, hairpin loops, bulge loops, interior loops,** and **multibranched loops** (Fig. 6.2). An **RNA pseudoknot** is formed when a single-stranded loop region base pairs with a complementary sequence outside the loop. Such structures are important for RNA synthesis, translation, and assembly.

The first step in identifying a structural feature in RNA is to scan the nucleotide sequence with computer programs designed to fold the RNA into energetically stable structures. Comparative sequence analysis can provide evidence for RNA secondary structures. For example, comparison of the RNA sequences of several related viruses might establish that the structure, but not the sequence, of a stem-loop is conserved. Evidence for RNA structure comes from experiments in which RNAs are treated with enzymes or chemicals that attack single- or double-stranded regions specifically. The results of such analyses can confirm that predicted stem regions are base paired while loop regions are unpaired. Structures of RNA hairpins and pseudoknots have been determined by X-ray crystallography or nuclear magnetic resonance (Fig. 6.2C).

Naked or Nucleocapsid RNA

The genomes of (−) strand viruses are organized into nucleocapsids in which protein molecules, including the RNA-dependent RNA polymerase and accessory proteins, are bound to the genomic RNAs at regular intervals (Fig. 6.3). These tightly wound ribonucleoprotein complexes are very stable and resistant to RNase. The RNA polymerases of (−) strand viruses copy viral RNAs **only** when they are present in the nucleocapsid. For example, vesicular stomatitis virus genomic RNA is a template for RNA polymerase only when it is bound to the nucleocapsid protein N. In contrast, the genomes of (+) strand RNA viruses are not coated with viral proteins in the virion. This structural difference is consistent with the fact that mRNAs are produced from the genomes of (−) strand RNA viruses upon cell entry whereas the genomes of (+) strand RNA viruses are translated.

The viral nucleoproteins are cooperative, single-stranded RNA-binding proteins, as are the single-stranded nucleic acid-binding proteins required during DNA-directed DNA and RNA synthesis. Their function during replication is to keep the RNA single stranded and prevent base pairing between the template and product, so that additional rounds of RNA synthesis can occur. The genomes of many (+) strand RNA viruses encode helicases that serve a similar function (see "Unwinding the RNA Template" below). In addition to its enzymatic activity, polioviral $3D^{pol}$ is a cooperative single-stranded RNA-binding protein and can unwind RNA duplexes without the hydrolysis of ATP

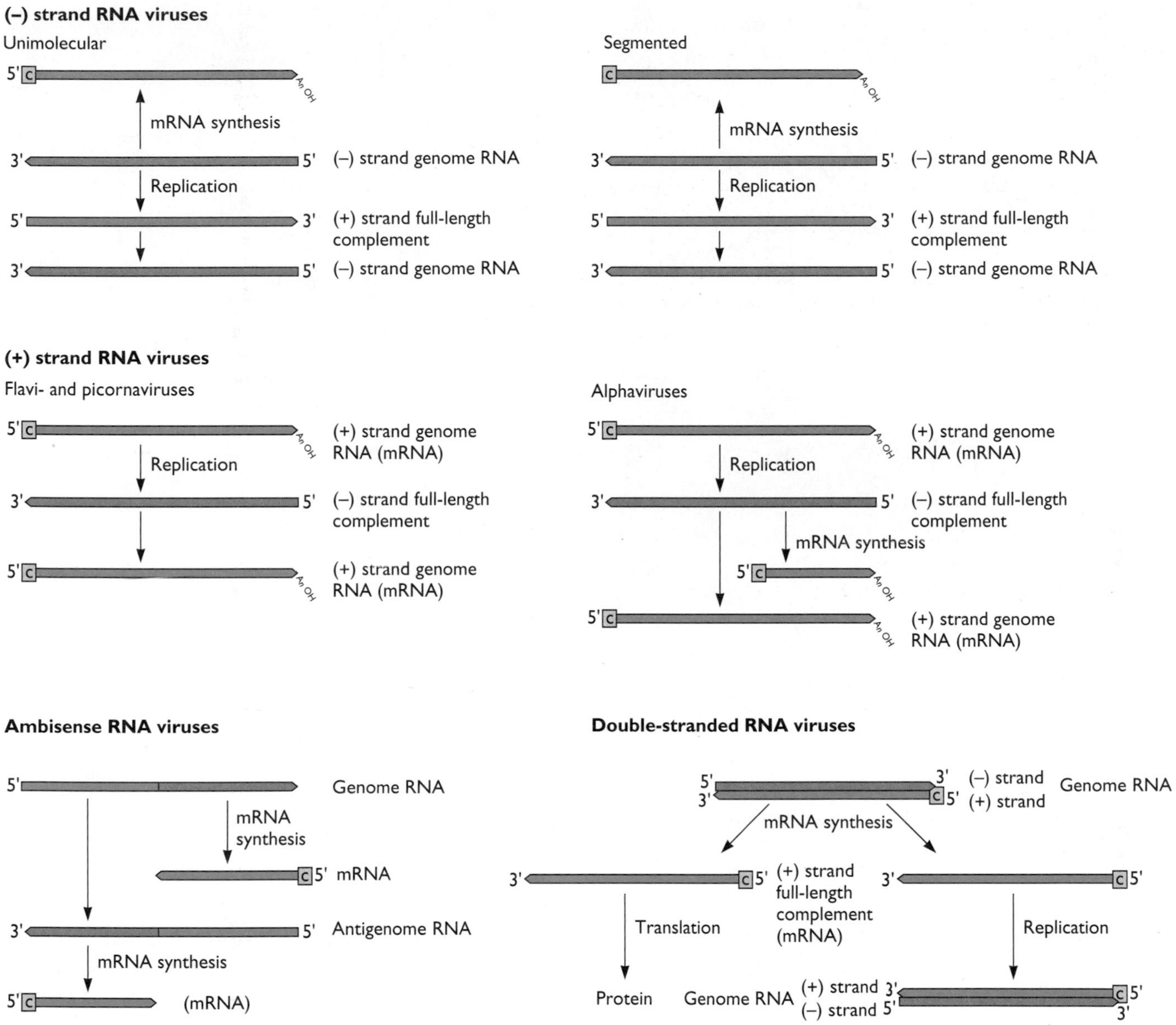

Figure 6.1 Strategies for replication and mRNA synthesis of RNA virus genomes are shown for representative virus families. Picornaviral genomic RNA is linked to VPg at the 5′ end. The (+) genomic RNA of some flaviviruses does not contain poly(A). Only one RNA segment is shown for segmented (–) strand RNA viruses.

characteristic of helicase-mediated unwinding. Polioviral RNA polymerase is therefore functionally similar to the RNA-binding nucleoproteins of (–) strand viruses.

The RNA Synthesis Machinery

Identification of RNA-Dependent RNA Polymerases

The first evidence for a viral RNA-dependent RNA polymerase emerged in the early 1960s from studies with mengovirus and poliovirus, both (+) strand RNA viruses. In these experiments, extracts were prepared from virus-infected cells and incubated with the four ribonucleoside triphosphates (adenosine triphosphate [ATP], uridine triphosphate [UTP], cytosine triphosphate [CTP], and guanosine triphosphate [GTP]), one of which was radioactively labeled. The incorporation of nucleoside monophosphate into RNA was then measured. Infection with mengovirus or poliovirus led to the appearance of a cytoplasmic enzyme that could synthesize viral RNA in the presence of actinomycin D, a drug that was known to inhibit cellular DNA-directed RNA synthesis. Lack of sensitivity to actinomycin

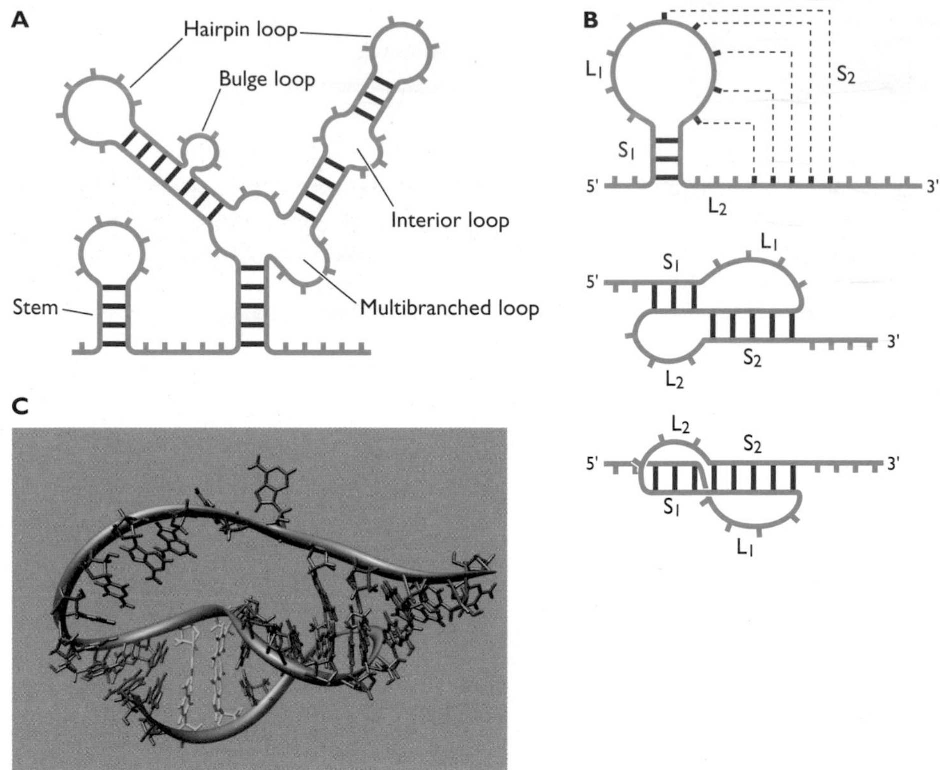

Figure 6.2 RNA secondary structure. (A) Schematic of different structural motifs in RNA. Red bars indicate base pairs; green bars indicate unpaired nucleotides. **(B)** Schematic of a pseudoknot. (Top) Stem 1 (S_1) is formed by base pairing in the stem-loop structure, and stem 2 (S_2) is formed by base pairing of nucleotides in the loop with nucleotides outside the loop. (Middle) A different view of the formation of stems S_1 and S_2. (Bottom) Coaxial stacking of S_1 and S_2 resulting in a quasicontinuous double helix. **(C)** Structure of a pseudoknot as determined by X-ray crystallography. The sugar backbone is highlighted with a green tube. Stacking of the bases in the areas of S_1 and S_2 can be seen. From Protein Data Bank file 1l2x. Adapted from C. W. Pleij, *Trends Biochem. Sci.* **15:**143–147, 1990, with permission.

D suggested that the enzyme was virus specific and could copy RNA from an RNA template and not from a DNA template. This enzyme was presumed to be an RNA-dependent RNA polymerase. Several years later, similar assays were used to demonstrate that the virions of (–) strand viruses and of double-stranded RNA viruses contain an RNA-dependent RNA polymerase that synthesizes mRNAs from the (–) strand RNA present in the virus particles.

The initial discovery of a putative RNA polymerase in poliovirus-infected cells was followed by attempts to purify the enzyme and show that it can copy viral RNA. Because polioviral genomic RNA contains a 3′ poly(A) sequence, polymerase activity was measured with a poly(A) template and an oligo(U) **primer**. After several fractionation steps, a poly(U) polymerase that could copy polioviral genomic RNA in the presence of an oligo(U) primer was purified from infected cells. Poly(U) polymerase activity coincided with a single polypeptide, now known to be the polioviral RNA polymerase $3D^{pol}$ (see Appendix, Fig. 13, for a description of this nomenclature). Purified $3D^{pol}$ RNA polymerase cannot copy polioviral genomic RNA in the absence of a primer; it is therefore a template- and primer-dependent enzyme.

Assays for RNA polymerase activity have been used to demonstrate virus-specific enzymes in virions or in extracts of cells infected with a wide variety of RNA viruses. Amino acid sequence alignments (see "Sequence Relationships among RNA Polymerases" below) can be used to identify viral proteins with motifs characteristic of RNA-dependent RNA polymerases. These approaches have been used to identify the L proteins of paramyxoviruses and bunyaviruses, the PB1 protein of influenza viruses, and the nsP4 protein of alphaviruses as candidate RNA polymerases. When the genes encoding these polymerases are expressed in cells, the proteins that are produced can copy viral RNA templates.

RNA-directed RNA synthesis follows a set of universal rules that differ slightly from those followed by DNA-dependent DNA polymerases. RNA synthesis initiates and

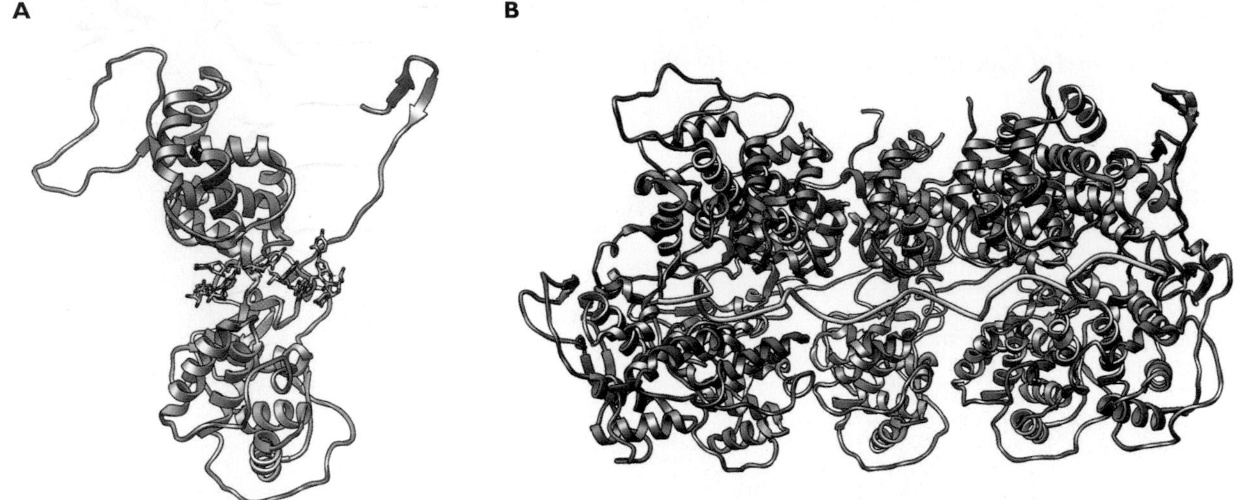

Figure 6.3 Structure of vesicular stomatitis N protein bound to RNA. (A) Ribbon diagram of the N protein monomer bound to a 9-nucleotide RNA. The RNA (stick representation) is bound in a groove located between N- and C-terminal lobes of the protein. **(B)** Structure of five N molecules bound to RNA. The ribose-phosphate backbone of the RNA is shown as a yellow tube. From Protein Data Bank file 2qvj.

terminates at specific sites in the template and is catalyzed by virus-encoded polymerases, but viral accessory proteins and even host cell proteins may also be required. Like cellular DNA-dependent RNA polymerases, some RNA-dependent RNA polymerases can initiate RNA synthesis *de novo*. Others require a primer with a free 3′-OH end to which nucleotides complementary to the template strand are added. Some RNA primers are protein linked, while others bear a 5′ cap structure (the cap structure is described in Chapter 10). A comparison of the structures and sequences of polynucleotide polymerases (see below) has led to the hypothesis that all polymerases catalyze synthesis by a mechanism that requires two metals (Box 6.2). RNA is usually synthesized by template-directed, stepwise incorporation of ribodeoxynucleoside monophosphates (NMPs) into the 3′-OH end of the growing RNA chain, which undergoes **elongation** in the 5′ → 3′ direction. Examples of nontemplated synthesis of viral RNA in cells infected with viruses of the *Paramyxoviridae* and *Filoviridae* have been described (Chapter 10).

Sequence Relationships among RNA Polymerases

The amino acid sequences of viral RNA polymerases have been compared to identify conserved regions and to provide information about the evolution of these enzymes. Although polymerases have very different amino acid sequences, four common motifs (A to D) have been identified in all polymerases (Fig. 6.3). Motif C includes a Gly-Asp-Asp sequence conserved in the RNA polymerases of most (+) strand RNA viruses. It was suggested that this sequence is part of the active site of the enzyme. In support of this hypothesis, alterations in this sequence in polioviral $3D^{pol}$ and many other viral polymerases produce an inactive enzyme. Evidence that a viral protein is an RNA polymerase is considerably strengthened when this 3-amino-acid sequence is found (Box 6.3).

Motifs A to D are also present in the sequences of polymerases that copy DNA templates (Fig. 6.4). These sequence comparisons indicate that all four classes of nucleic acid polymerases have a similar core catalytic domain (the palm domain [see below]) and most probably evolved from a common ancestor.

Three-Dimensional Structure of RNA-Dependent RNA Polymerases

Determination of the crystal structures of more than 10 RNA-dependent RNA polymerases has confirmed the hypothesis that all polynucleotide polymerases are similar structurally. The shapes of all four types of polymerases resemble a right hand consisting of a palm, fingers, and a thumb, with the active site of the enzyme located in the palm (see Fig. 7.12). This shape provides the correct arrangement of substrates and metal ions at the catalytic site. The structures of RNA-dependent RNA polymerases differ in detail from those of other polymerases, presumably to accommodate different templates and priming mechanisms.

All the available structures of RNA-dependent RNA polymerases show that the enzymes adopt closed structures

BOX 6.2 BACKGROUND
Two-metal mechanism of catalysis by polymerases

All polynucleotide polymerases are thought to catalyze synthesis by a two-metal mechanism that requires two conserved aspartic acid residues, one in motif A and one in motif C (see figure). The carboxylate groups of these amino acids coordinate two divalent metal ions, shown as Mg^{2+} in the figure. One metal ion promotes deprotonation of the 3′-OH group of the nascent strand, and the other ion stabilizes the transition state at the α-phosphate of NTP and facilitates the release of pyrophosphate (PP_i).

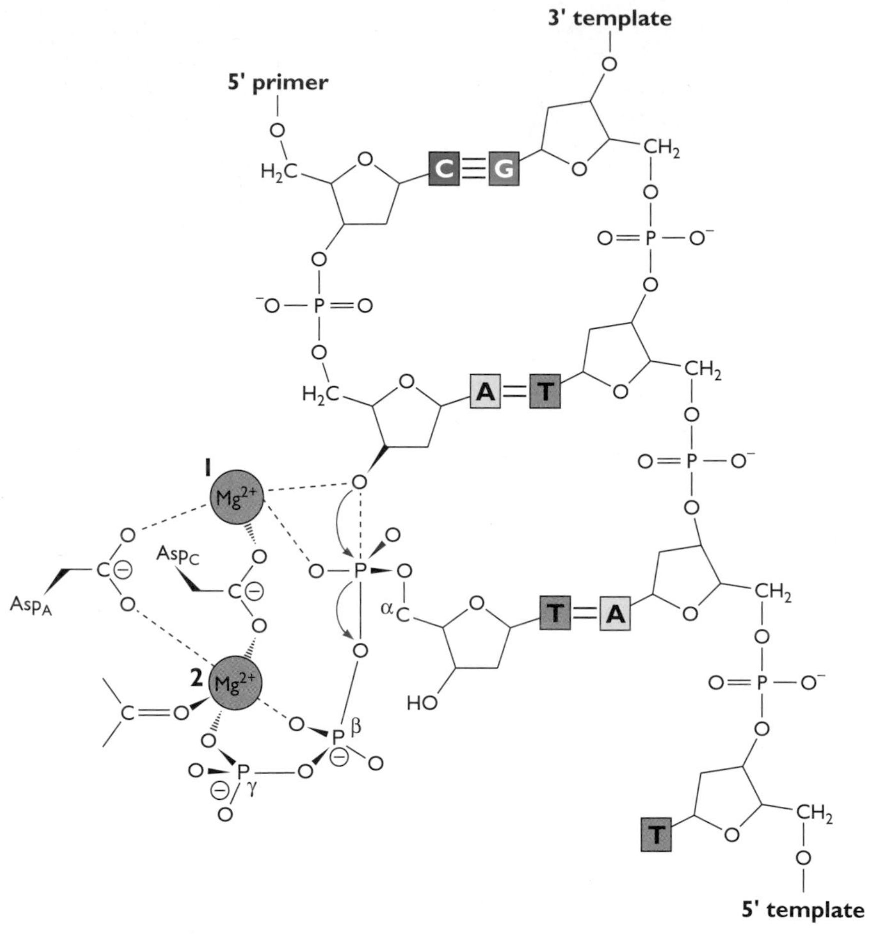

Two-metal mechanism of polymerase catalysis. Red arrows indicate the net movement of electrons.

in which the active site is completely encircled (Fig. 6.5). In contrast, structures of other polynucleotide polymerases resemble an open hand (Fig. 6.6). The closed structure, which is formed by interactions between the fingers and thumb domains, creates a nucleoside triphosphate (NTP) entry tunnel at the back of the enzyme and a template-binding site at the front. Residues within motif F, a conserved region unique to RNA-dependent RNA polymerases (Fig. 6.4), form the NTP entry tunnel.

The palm domain of RNA-dependent RNA polymerases is structurally similar to that of other polymerases and contains the four motifs (A to D) that are conserved in all polymerases (Fig. 6.6). The motifs confer specific functions, such as nucleotide recognition and binding (A and B), phosphoryl transfer (A and C), and mediation of the structure of the palm domain (D). The fifth motif, E, which is present in RNA-dependent, but not in DNA-dependent, polymerases, lies between the palm and thumb domains. It binds the nucleotide primer.

RNA-dependent RNA polymerases prefer to incorporate NTPs over deoxyribonucleoside triphosphates (dNTPs). NTP recognition by poliovirus $3D^{pol}$ is regulated by Asp238, which forms a hydrogen bond with the ribose 2′-OH (Fig. 6.7). dNTPs are not bound because Asp238 cannot form a hydrogen bond with 2′-deoxyribose. An Asp is present at this position in all RNA-dependent RNA polymerases.

BOX 6.3

BACKGROUND

The Gly-Asp-Asp sequence of RNA polymerase motif C

The Asp-Asp sequence of motif C is also conserved in RNA-dependent DNA polymerases of retroviruses and in RNA polymerases of double-stranded RNA and segmented (–) strand viruses. The RNA polymerases of nonsegmented (–) strand viruses contain Gly-Asp-Asn instead of Gly-Asp-Asp. Mutational studies have shown that this sequence in the L protein of vesicular stomatitis virus is essential for RNA synthesis. The RNA polymerase of birnavirus, an insect virus with a double-stranded RNA genome, has Ala-Asp-Asn instead of Gly-Asp-Asp. Substitution of Ala-Asp-Asn with Gly-Asp-Asp produces an RNA polymerase with increased synthetic activity. This observation has led to the suggestion that Ala-Asp-Asn may have been selected during the evolution of these birnaviruses to reduce pathogenicity and facilitate virus spread.

A Tyr at this position in RNA-dependent DNA polymerase is responsible for discriminating against NTPs and selecting dNTPs. Motif C of $3D^{pol}$ contains the Asp-Asp sequence conserved in RNA-dependent polymerases; the first Asp is also conserved in DNA-dependent polymerases. The two Asp residues of motif C and the conserved Asp238 of motif A form a cluster that coordinates the triphosphate moiety of the NTP and the metal ions required for catalysis (Fig. 6.7).

The binding site on the RNA polymerase for the RNA template has been revealed by structural studies of foot-and-mouth disease virus $3D^{pol}$ complexed with a molecule of RNA that serves as both template and primer (Fig. 6.8). The phosphodiester backbone of the template RNA contacts basic amino acid residues in a channel on the polymerase. These contacts propel the template toward the active site. The template is near the fingers domain, while the phosphodiester backbone of the primer interacts with an α-helix of the thumb domain. Amino acids in motifs C and E of the palm domain stabilize the 3′ end of the primer to facilitate elongation. The double-stranded region of the nucleic acid, which mimics the double-stranded product, runs from the active site to the protein C terminus. This structure indicates that the product would exit the polymerase from the large central cavity.

Mechanisms of RNA Synthesis

Initiation

The requirement for a primer in the initiation step of nucleic acid synthesis varies among the different classes of polymerases (Fig. 6.9). All DNA polymerases are primer-dependent enzymes, while DNA-dependent RNA polymerases initiate RNA synthesis *de novo*. Some RNA-dependent RNA polymerases can initiate RNA synthesis without a primer, while others require a primer. Nucleic acid synthesis by these RNA polymerases is initiated by a protein primer or an oligonucleotide cleaved from the 5′ end of cellular mRNA.

De Novo Initiation

The requirements for *de novo* initiation are an RNA-dependent RNA polymerase, an RNA template, the initiating NTP, and a second NTP. The first phosphodiester bond is made between the 3′-OH of the initiating NTP and the second NTP. Elongation usually follows immediately.

Figure 6.4 Protein domain alignments for the four categories of nucleic acid polymerases. Numbers at the top are from the $3D^{pol}$ amino acid sequence. Sequence and structure motifs in each polymerase category are shaded green, and the alignments between the different polymerases are shown in green. Motif F is found only in RNA-dependent RNA polymerases. Adapted from J. L. Hansen et al., *Structure* **5**:1109–1127, 1997, with permission. Courtesy of S. Schultz, University of Colorado, Boulder.

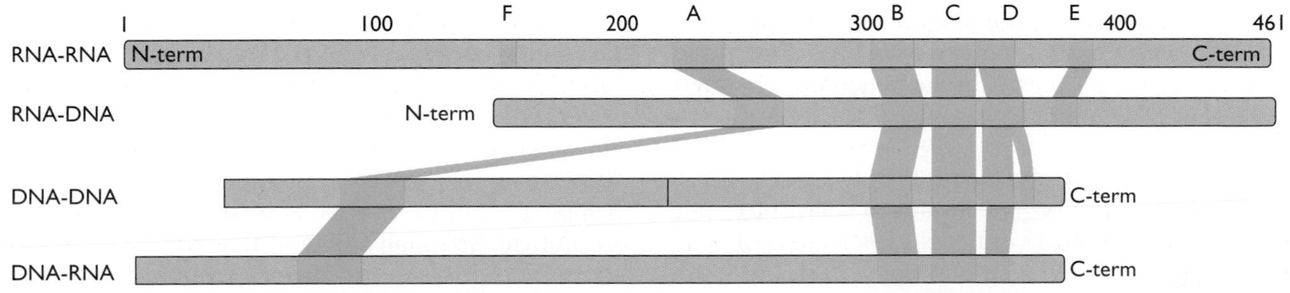

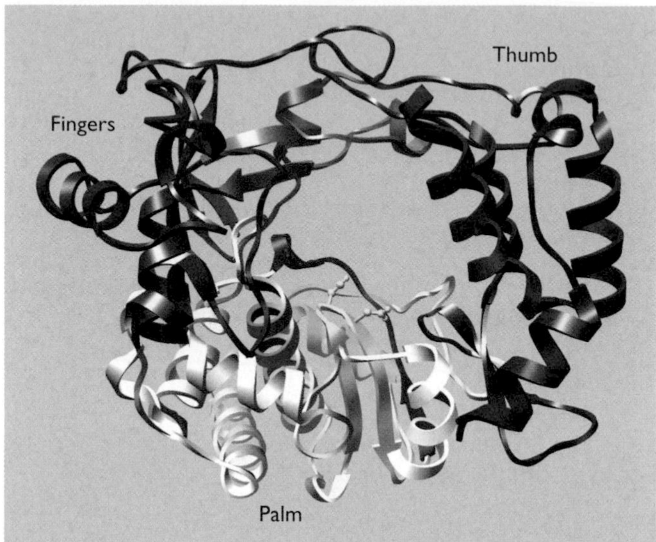

Figure 6.5 Structure of polioviral 3D^{pol}. The thumb, palm, and fingers domains are labeled. Fingers and thumb domains are blue. Conserved structure/sequence motifs are colored as follows: motif A, red; B, green; C, yellow; D, cyan; E, purple; F, orange. From Protein Data Bank file 1ra6.

In most cases initiation takes place at the exact 3′ end of the template, but it may instead occur internally, for example during replication of the genomes of some (−) strand RNA viruses, such as bunyaviruses, arenaviruses, and nairoviruses (Fig. 6.9). Initiation begins at an internal C, and after extension of a few nucleotides, the daughter strand is shifted in the 3′ direction so that the 5′-terminal G residue is not base paired with the template strand. Because the daughter strand slips, this mechanism is called "prime and realign."

The structures of RNA polymerases that initiate *de novo* have a less accessible active site compared with polymerases that initiate with a primer. The thumb domains of some RNA polymerases, such as the hepatitis C virus and bacteriophage ϕ6 RNA polymerases, are large and partially block the active site (Fig. 6.10). This conformation produces two positively charged tunnels that allow access of RNA template and NTP. The NTP tunnel is formed by the fingers and palm domains. The template tunnel, which is formed by the thumb and fingers, is wide enough to accommodate single-stranded but not double-stranded RNA. A β-hairpin that protrudes from the thumb domain toward the active site of the hepatitis C virus RNA polymerase appears to be important for correct positioning of the 3′ terminus of the viral RNA.

Primer-Dependent Initiation

Protein priming. A protein-linked oligonucleotide serves as a primer for RNA synthesis by some RNA polymerases. Protein priming also occurs during DNA replication of adenoviruses and certain DNA-containing bacteriophages (Chapter 9). Picornaviral 3D^{pol} is a primer-dependent enzyme that does not copy viral RNA *in vitro* without an oligo(U) primer. Polioviral genomic RNA, as well as newly synthesized (+) and (−) strand RNAs, are covalently linked at their 5′ ends to the 22-amino acid protein VPg (Fig. 6.11A), suggesting that VPg might function as a primer for RNA synthesis. This hypothesis was supported by the discovery of a uridylylated form of the protein, VPg-pUpU, in infected cells. VPg can be uridylylated *in vitro* by 3D^{pol} and then can prime the synthesis of VPg-linked poly(U) from a poly(A) template RNA. The template for uridylylation of VPg is an RNA hairpin, the *cis*-acting replication element (cre), located in the coding region of picornaviruses (Fig. 6.11B and C). VPg-pUpU is likely to serve as a primer for viral RNA synthesis in infected cells (Fig. 6.12).

Structures of the RNA polymerases of different picornaviruses indicate that the active site is more accessible than in polymerases with a *de novo* mechanism of initiation. The thumb domains of picornaviral polymerases are small, which leaves a wide central cavity that can accommodate the template primer complex (Fig. 6.8) and the protein primer (Fig. 6.13). Binding of VPg in the central cavity is mediated by the interaction of the protein with motif F and portions of the fingers and thumb domains. This interaction places the third amino acid of VPg, tyrosine, which is linked to the first U, in the active site of 3D^{pol}.

Protein priming by the birnavirus RNA polymerase VP1 is unusual because the primer is the polymerase, not a separate protein. VP1 has self-guanylylation activity, which occurs in the absence of a template. The guanylylation site is a serine located approximately 23 Å from the catalytic site of the polymerase. The long distance between them suggests that guanylylation may be carried out at a second active site. After two G residues are added to VP1, it binds to the conserved CC sequence at the terminus of the viral RNA template to initiate nucleotide synthesis. The 5′ ends of mRNAs and genomic double-stranded RNAs produced by this reaction are therefore linked to a VP1 molecule.

Priming by capped RNA fragments. Influenza virus mRNA synthesis is blocked by treatment of cells with the fungal toxin α-amanitin at concentrations that inhibit cellular DNA-dependent RNA polymerase II. In contrast, the toxin does not affect viral mRNA synthesis *in vitro*. This surprising finding demonstrated that the viral RNA polymerase is dependent on a host nuclear function. Inhibition by α-amanitin is explained by the requirement for a continuous supply of newly synthesized cellular RNA polymerase II transcripts to provide primers for viral mRNA synthesis. Nuclear RNA polymerase II transcripts are cleaved by a virus-encoded, cap-dependent endonuclease that is part

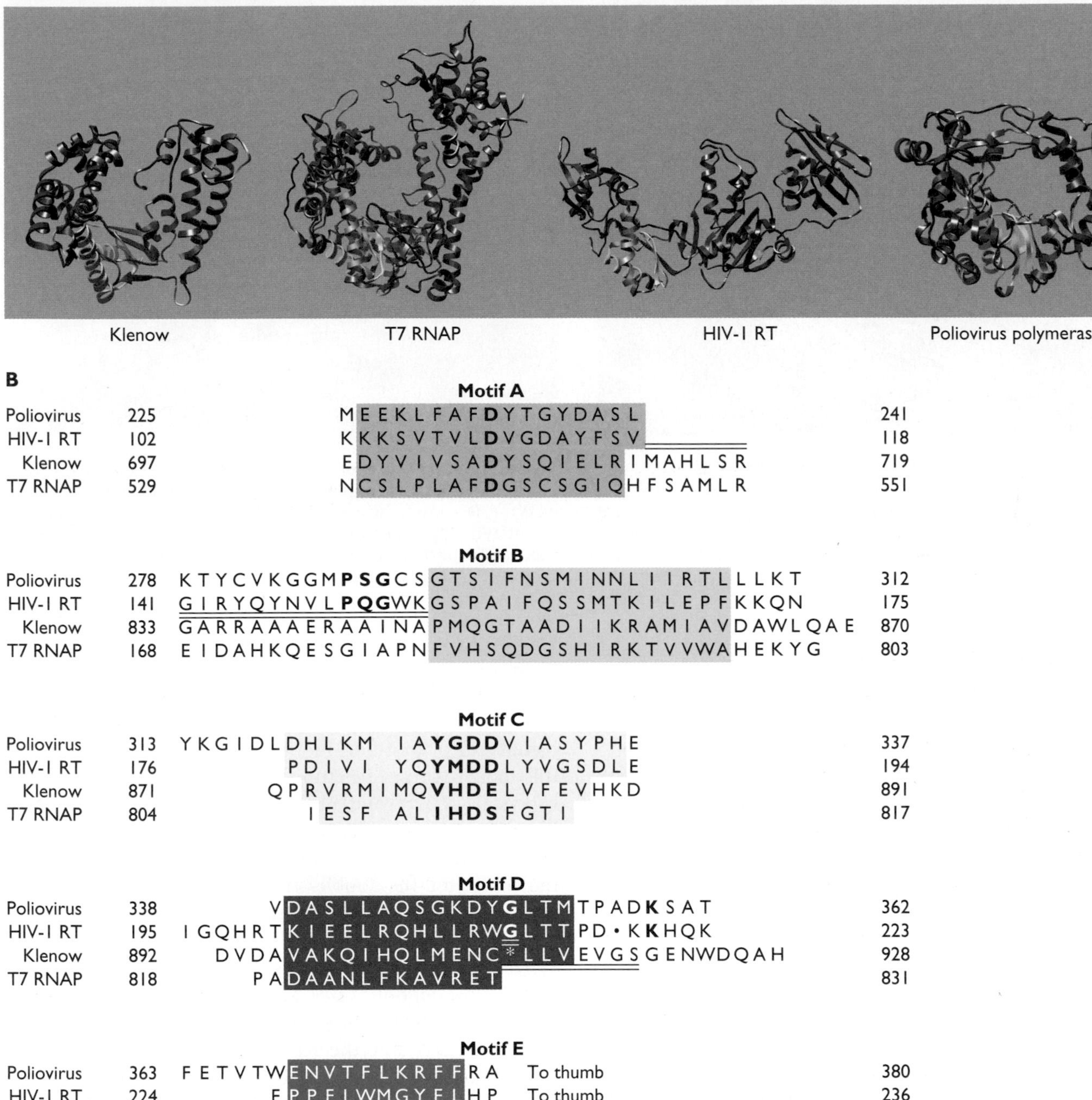

Figure 6.6 Polymerase structure and sequence motifs in representative structures of each of the four types of nucleic acid polymerases. (A) Ribbon diagrams of the polymerase domain of the large (Klenow) fragment of *Escherichia coli* DNA polymerase I, a DNA-dependent DNA polymerase; T7 RNA polymerase (T7 RNAP), a DNA-dependent RNA polymerase; human immunodeficiency virus type 1 reverse transcriptase (HIV-1 RT), an RNA-dependent DNA polymerase; and polioviral 3D^{pol}, an RNA-dependent RNA polymerase. The thumb domain is at the right, and the fingers domain is at the left. The conserved structure/sequence motifs A, B, C, D, and E are red, green, yellow, cyan, and purple, respectively. From Protein Data Bank files 1qsl, 1s77, 3hvt, and 1ra6. **(B)** Sequence alignments based on crystal structures. Boxes are color coded to match the motifs in panel A. The most highly conserved amino acids are in bold type. Double lines are regions in which structures differ. Adapted from J. L. Hansen et al., *Structure* **5:**1109–1127, 1997, with permission. Courtesy of S. Schultz, University of Colorado, Boulder.

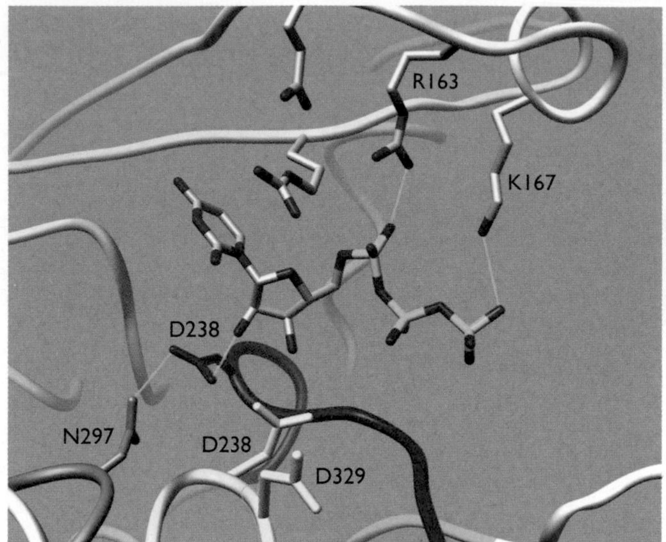

Figure 6.7 Structure of UTP bound to poliovirus 3D^pol^. The NTP bridges the fingers (top) and palm (bottom) domains. The base is stacked with Arg174 from the fingers. Hydrogen bonds are shown as cyan lines. The Asp238 of motif A, which is conserved in all RNA-dependent RNA polymerases, hydrogen bonds with the 2′-OH of the ribose moiety; this interaction discriminates NTPs from dNTPs. Asp328 and Asp329, which coordinate Mg^{2+}, are also labeled. Produced from Protein Data Bank file 2im2.

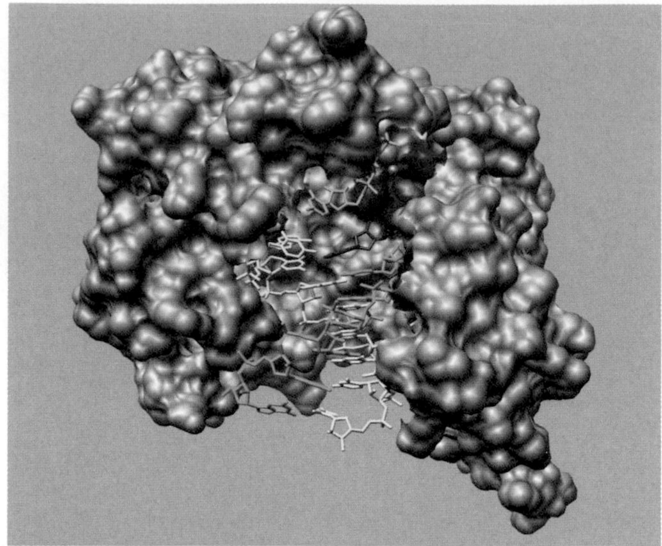

Figure 6.8 Structure of foot-and-mouth disease virus 3D^{pol} complexed with a template-primer RNA, ATP, and UTP. The enzyme is shown in a surface representation (gray). Parts of the enzyme have been removed to reveal the substrate cavities. Catalytic aspartates are red. The Mg^{2+} ion is an orange ball. The bound RNA is shown as a stick representation; the template is yellow, and the primer is green. The ATP, which is incorporated into the primer, is purple, and the UTP is cyan. From Protein Data Bank file 2e9z.

of the RNA polymerase (Fig. 6.14B). The resulting 10- to 13-nucleotide capped fragments serve as primers for the initiation of viral mRNA synthesis.

Bunyaviral mRNA synthesis is also primed with capped fragments of cellular RNAs. In contrast to that of influenza virus, bunyaviral mRNA synthesis is not inhibited by α-amanitin because it occurs in the cytoplasm, where there are many capped cellular RNAs.

Elongation

All nucleic acid synthesis begins with the formation of a complex of polymerase, template-primer, and initiating NTP, followed by a conformational change that reorients the triphosphate moiety. This change leads to phosphoryl transfer, and incorporation of the nucleoside monophosphate into the 3′ terminus of the primer or the growing chain. During synthesis by T7 RNA polymerase, the NTP first binds in a preinsertion site. The enzyme then undergoes conformational changes that push the NTP into the active site for catalysis. NTP has also been observed bound in a preinsertion site in the fingers domain of picornaviral 3D^{pol} (Fig. 6.7); no major movements of domains are observed when NTPs are bound. However, 3D^{pol} is known to undergo conformational changes after binding template RNA and the initiating NTP. Therefore, it is hypothesized that movements of the fingers domain occur that push NTPs into the catalytic site. The structures of elongation complexes of foot-and-mouth disease virus 3D^{pol} support this model. These structures were obtained by incubating cocrystals of 3D^{pol} and a template-primer RNA with different NTPs. For example, when ATP was added to the cocrystals, it was hydrolyzed and incorporated into the primer molecule, and the double-stranded RNA was translocated. The 3′ end of the template strand, the newly incorporated adenine, was positioned at the active site of the enzyme. When UTP was added to these cocrystals, it became positioned close to, but not at, the active site. Presumably this structure represents the state just before the fingers domain pushes the UTP into the active site. These structures show that Asp245 (Fig. 6.6, motif A) and Asn307 (motif B) play an important role in recognition of the NTP and positioning the ribose in its binding pocket.

Template Specificity

Viral RNA-dependent RNA polymerases must select viral templates in a vast excess of cellular mRNAs and then initiate correctly to ensure accurate RNA synthesis. Different mechanisms that contribute to template specificity have been identified. Initiation specificity may be regulated by the affinity of the RNA polymerase for the initiating nucleotide. For example, the RNA polymerase of bacteriophage ϕ6 prefers

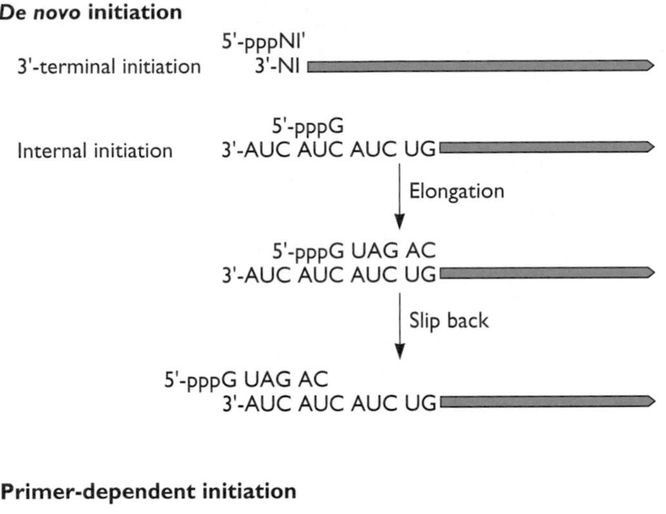

Figure 6.9 Mechanisms of initiation of RNA synthesis. *De novo* initiation may occur at the 3′ end of the viral RNA or from an internal base. When a primer is required, it may be a capped or protein-linked oligonucleotide.

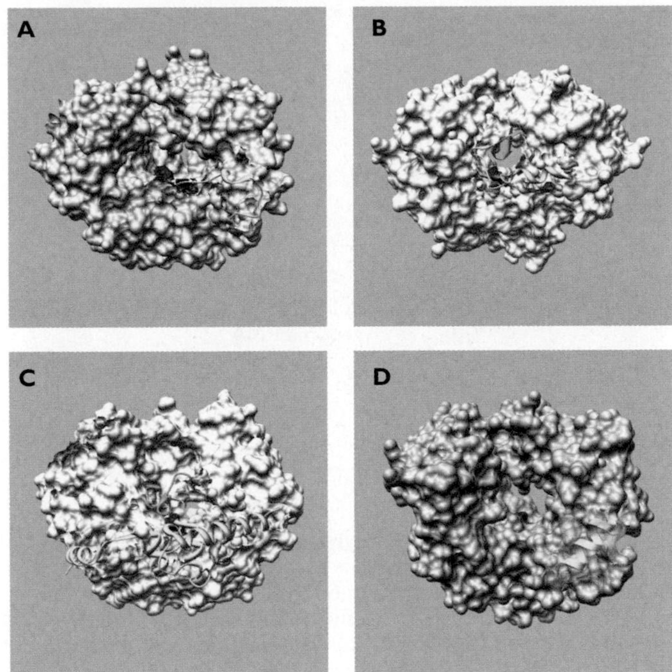

Figure 6.10 Partially blocked active site of flavivirus and bacteriophage ϕ6 RNA polymerases. RNA polymerase of Norwalk virus **(A)**, hepatitis C virus **(B)**, bacteriophage ϕ6 **(C)**, and poliovirus **(D)** are shown in surface representation. The carboxy terminus is shown in ribbon representation and colored yellow. Loop insertions of hepatitis C virus and bacteriophage ϕ6 enzymes are green. From Protein Data Bank files 1sh2, 1c2p, 1hhs, and 1ra6.

3′-terminal C, while the N3 and C4 amino groups of the first C of the template are important for bovine viral diarrhea virus RNA synthesis. Reovirus RNA polymerase prefers a G at the second position of the template RNA. This preference is controlled by hydrogen bonding of carbonyl and amino groups of the G with two amino acids of the enzyme.

Template specificity may also reside in the recognition of RNA sequences or structures at the 5′ and 3′ ends of viral RNAs by viral proteins. RNA synthesis initiates specifically within a polypyrimidine tract in the 3′ untranslated region of hepatitis C virus RNA. The 3′ noncoding region of polioviral genomic RNA contains an RNA pseudoknot structure that is conserved among picornaviruses (Fig. 6.11B). Protein 3AB-3CD binds this structure and may direct the polymerase to that site for the initiation of (−) strand RNA synthesis (Box 6.4). Polioviral 3CD^{pro} protein plays an important role in viral RNA synthesis by participating in the formation of a ribonucleoprotein at the 5′ end of the (+) strand RNA. The 3CD protein is a precursor of the 3C^{pro} protease and 3D^{pol}. Protein 3CD, together with a cellular protein known as poly(rC)-binding protein 2, binds to a cloverleaf structure in the viral RNA (Fig. 6.12). The RNA-binding domain of 3CD is contained within the 3C portion of the protein, and alterations within this domain inhibit complex formation and RNA synthesis without affecting viral protein processing.

Internal RNA sequences may confer initiation specificity to RNA polymerases. The *cis*-acting replication elements (cre) in the coding sequence of poliovirus protein 2C and rhinovirus capsid protein VP1 contain short RNA sequences that are required for RNA synthesis. These sequences are binding sites for 3CD^{pro} and serve as a template for the priming of viral RNA synthesis by VPg protein (Fig. 6.12).

During mRNA synthesis by influenza virus polymerase, sequences at the RNA termini play an important role in ensuring that the 5′ ends of newly synthesized influenza virus mRNAs are not cleaved and used as primers (Fig. 6.15). If such cleavage were to occur, there would be no net synthesis of viral mRNAs. The three P proteins form a multisubunit assembly that can neither bind to capped primers nor synthesize mRNAs. Addition of a sequence corresponding to the 5′-terminal 11 nucleotides of the viral RNA, which is highly conserved in all eight genome segments, activates the cap-binding activity of the P proteins. The PB1 protein binds this RNA sequence and activates the cap-binding PB2 subunit, probably by conformational change. Concomitantly with activation of cap binding, the P proteins acquire the ability to bind to a conserved sequence at the 3′ ends of genomic RNA segments. This second interaction activates the endonuclease that cleaves

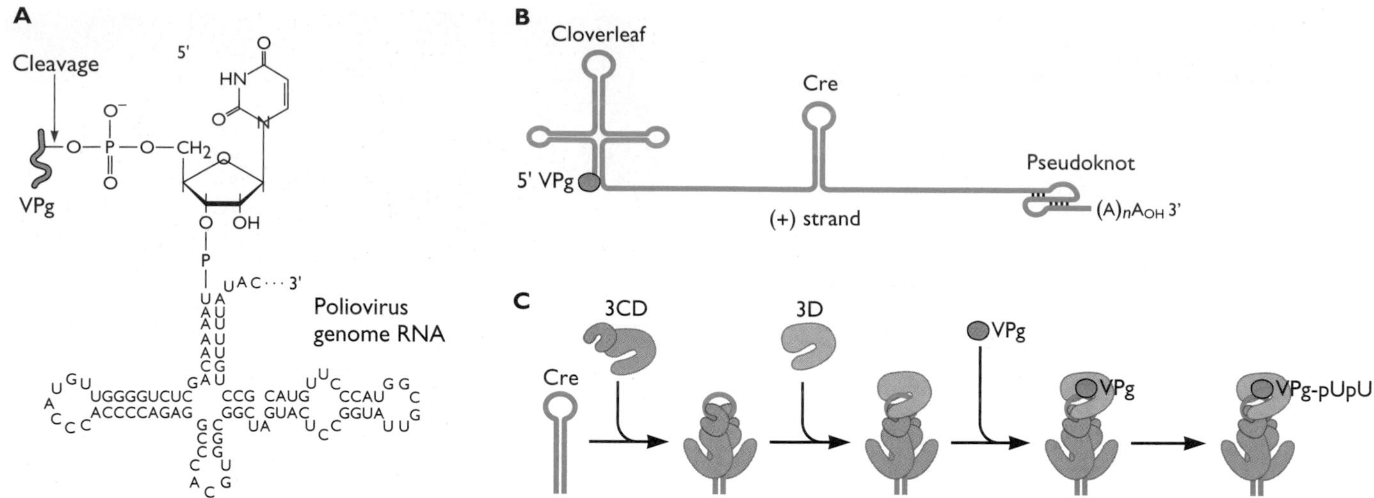

Figure 6.11 Uridylylation of VPg. (A) Linkage of VPg to polioviral genomic RNA. Polioviral RNA is linked to the 22-amino-acid VPg (orange) via an *O*4-(5′-uridylyl)-tyrosine linkage. This phosphodiester bond is cleaved at the indicated site by a cellular enzyme to produce the viral mRNA containing a 5′-terminal pU. **(B)** Structure of the poliovirus (+) strand RNA template, showing the 5′ cloverleaf structure, the internal cre (*cis*-acting replication element) sequence, and the 3′ pseudoknot. **(C)** Model for assembly of the VPg uridylylation complex. Two molecules of 3CD bind to cre. The 3C dimer melts part of the stem. $3D^{pol}$ binds to the complex by interactions between the back of the thumb domain and the surface of 3C. VPg then binds the complex and is linked to two U moieties in a reaction templated by the cre sequence.

capped host cell RNAs, producing the primers for viral mRNA synthesis. Such binding to two sites in the nascent viral mRNA blocks access of a second P protein and protects newly synthesized viral mRNA from endonucleolytic cleavage by P proteins.

Protein-protein interactions play roles in directing the RNA polymerase to the RNA template. The vesicular stomatitis virus RNA polymerase for mRNA synthesis consists of the P protein and the L protein, the catalytic subunit. The P protein binds both the L protein and the complex of N and the (−) strand RNA. In this way the P protein brings the L polymerase to the RNA template. Cellular general initiation proteins have a similar function in bringing RNA polymerase II to the correct site to initiate transcription of DNA templates.

Unwinding the RNA Template

Base-paired regions in viral RNA must be disrupted to permit copying by RNA-dependent RNA polymerase. RNA helicases, which are encoded in the genomes of many RNA viruses, are thought to unwind the genomes of double-stranded RNA viruses as well as the secondary structures in template RNAs. They also prevent extensive base pairing between template RNA and the nascent complementary strand. The RNA helicases of several viruses that are important human pathogens, including the flaviviruses hepatitis C virus and dengue virus, have been studied extensively because these proteins are potential targets for chemotherapeutic intervention. To facilitate the development of new agents that inhibit these helicases, their three-dimensional structures have been determined by X-ray crystallography. These molecules comprise three domains that mediate hydrolysis of NTPs and RNA binding (Fig. 6.16). Between the domains is a cleft that is large enough to accommodate single-stranded but not double-stranded RNA. Unwinding of double-stranded RNA probably occurs as one strand of RNA passes through the cleft and the other passes outside of the molecule.

The bacteriophage ϕ6 RNA polymerase can separate the strands of double-stranded RNA without the activity of a helicase. Examination of the structure of the enzyme suggests how such melting might be accomplished. This RNA polymerase has a plow-like protuberance around the entrance to the template channel that is thought to separate the two strands of the double-stranded RNA, allowing only one to enter the channel.

Role of Cellular Proteins

Host cell components required for viral RNA synthesis were initially called "host factors" because nothing was known about their chemical composition. Evidence that cellular proteins are essential components of a viral RNA polymerase first came from studies of the bacteriophage Qβ. The RNA-dependent RNA polymerase of this virus

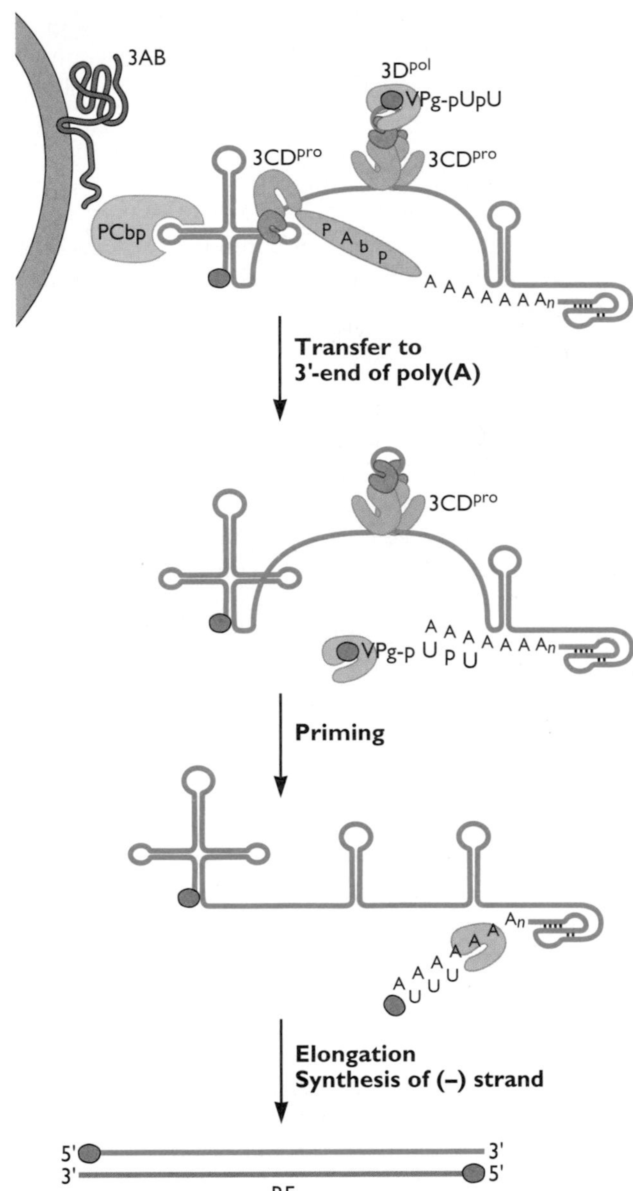

Figure 6.12 Poliovirus (−) strand RNA synthesis. The precursor of VPg, 3AB, contains a hydrophobic domain and is a membrane-bound donor of VPg. A ribonucleoprotein complex is formed when poly(rC)-binding protein 2 (PCbp2) and $3CD^{pro}$ bind the cloverleaf structure located within the first 108 nucleotides of (+) strand RNA. The ribonucleoprotein complex interacts with poly(A)-binding protein 1 (PAbp1), which is bound to the 3′ poly(A) sequence, bringing the ends of the genome into close proximity. Protease $3CD^{pro}$ cleaves membrane-bound 3AB, releasing VPg and 3A. VPg-pUpU is synthesized by $3D^{pol}$ using the sequence AAACA of cre as a template, transferred to the 3′ end of the genome, and used by $3D^{pol}$ as a primer for RNA synthesis. Modified from A. V. Paul, p. 227–246, *in* B. L. Semler and E. Wimmer (ed.), *Molecular Biology of Picornaviruses* (ASM Press, Washington, DC, 2002), with permission.

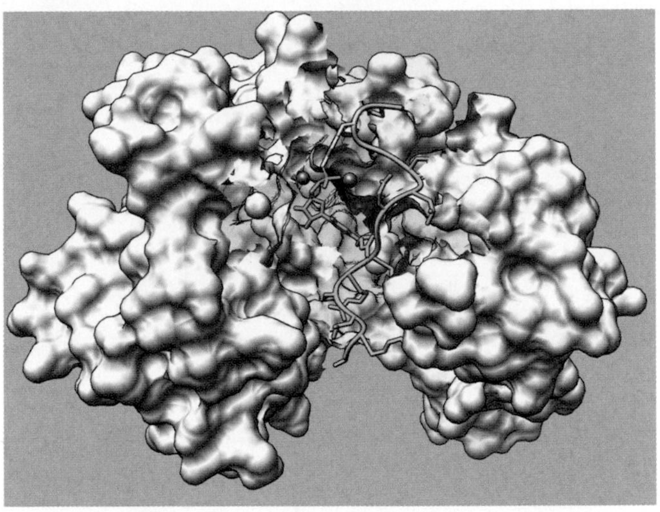

Figure 6.13 Structure of VPg bound to $3D^{pol}$ of foot-and-mouth disease virus. The polymerase is shown by surface representation. VPg protein is shown in cyan, tyrosine-3 of VPg is yellow, and the UMP linked to VPg is green. Catalytic aspartates are red, and Mg^{2+} ions are orange balls. From Protein Data Bank file 2f8e.

is a multisubunit enzyme, consisting of a 65-kDa virus-encoded protein and three host proteins, ribosomal protein S1 and the translation elongation proteins EF-Tu and EF-Ts. Proteins S1 and EF-Tu contain RNA-binding sites that enable the RNA polymerase to recognize the viral RNA template. The 65-kDa viral protein exhibits no RNA polymerase activity in the absence of the host proteins, but has sequence and structural similarity to known RNA-dependent RNA polymerases.

Polioviral RNA synthesis also requires host cell proteins. When purified polioviral RNA is incubated with a cytoplasmic extract prepared from uninfected permissive cells, the genomic RNA is translated and the viral RNA polymerase is made. If guanidine hydrochloride is included in the reaction mixture, the polymerase assembles on the viral genome but RNA synthesis is not initiated. The RNA polymerase-template assembly can be isolated free of guanidine, but RNA synthesis does not occur unless a new cytoplasmic extract is added, indicating that soluble cellular proteins are required for initiation. A similar conclusion comes from studies in which polioviral RNA was injected into oocytes derived from the African clawed toad *Xenopus laevis*. Polioviral RNA cannot replicate in *Xenopus* oocytes unless it is coinjected with a cytoplasmic extract from human cells. These observations can be explained by the requirement of the viral RNA polymerase for a mammalian protein that is absent in toad oocytes.

Host cell poly(rC)-binding protein is required for polioviral RNA synthesis. This protein binds to a cloverleaf

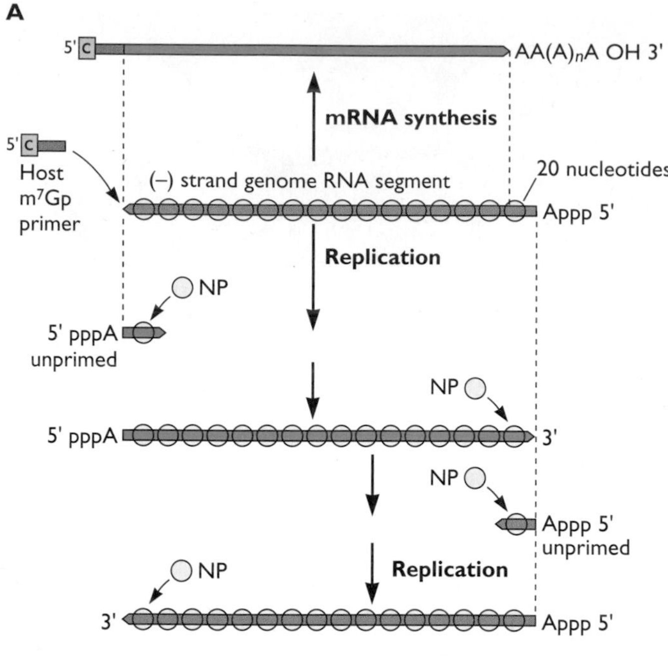

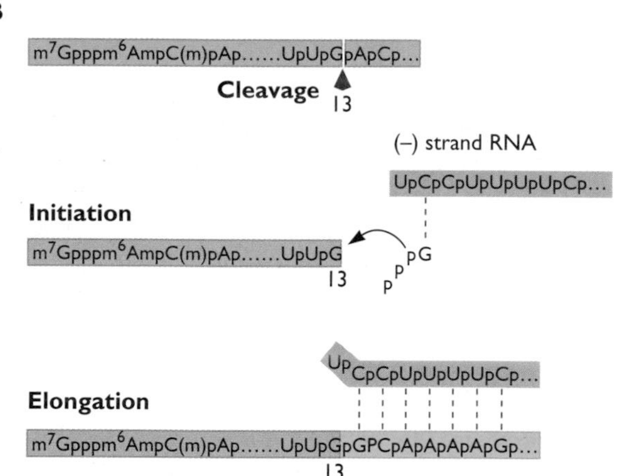

Figure 6.14 Influenza virus RNA synthesis. (A) Viral (–) strand genomes are templates for the production of either subgenomic mRNAs or full-length (+) strand RNAs. The switch from viral mRNA synthesis to genomic RNA replication is regulated by both the number of nucleocapsid (NP) protein molecules and the acquisition by the viral RNA polymerase of the ability to catalyze initiation without a primer. Binding of the NP protein to elongating (+) strands enables the polymerase to read to the 5′ end of genomic RNA. **(B)** Capped RNA-primed initiation of influenza virus mRNA synthesis. Capped RNA fragments cleaved from the 5′ ends of cellular nuclear RNAs serve as primers for viral mRNA synthesis. The 10 to 13 nucleotides in these primers do not need to hydrogen bond to the common sequence found at the 3′ ends of the influenza virus genomic RNA segments. The first nucleotide added to the primer is a G residue templated by the penultimate C residue of the genomic RNA segment; this is followed by elongation of the mRNA chains. The terminal U residue of the genomic RNA segment does not direct the incorporation of an A residue. The 5′ ends of the viral mRNAs therefore comprise 10 to 13 nucleotides plus a cap structure snatched from host nuclear pre-mRNAs. Adapted from S. J. Plotch et al., *Cell* **23:**847–858, 1981, with permission.

structure that forms in the first 108 nucleotides of (+) strand RNA (Fig. 6.12). Formation of a ribonucleoprotein composed of the 5′ cloverleaf, 3CD, and poly(rC)-binding protein is essential for the initiation of viral RNA synthesis. Interaction of poly(rC)-binding protein with the cloverleaf facilitates the binding of viral protein 3CD to the opposite side of the same cloverleaf.

Another candidate for a host protein that is essential for polioviral RNA synthesis is poly(A)-binding protein 1. This protein interacts with poly(rC)-binding protein 2, $3CD^{pro}$, and the 3′-poly(A) tail of poliovirus RNA, bringing together the ends of the viral genome (Fig. 6.12). Formation of this circular ribonucleoprotein complex is required for (–) strand RNA synthesis.

Host cell cytoskeletal proteins participate in paramyxoviral RNA synthesis. Measles virus and Sendai virus mRNA synthesis *in vitro* is stimulated by tubulin, the major structural component of microtubules of the cell's cytoskeleton, and is inhibited by anti-β-tubulin antibodies. In contrast, mRNA synthesis of human parainfluenza virus type 3 and respiratory syncytial virus requires cellular actin, which forms the microfilaments of the cytoskeletal network. Actin and tubulin might serve as anchoring sites on the cytoskeleton for the viral RNA polymerase. In support of this hypothesis, it has been shown that the incoming viral ribonucleoproteins associate with the cytoskeletal framework, where they are active in RNA synthesis. Assembling the RNA polymerase on the cytoskeleton may ensure high local concentrations of replication components and hence increase the rates or efficiencies of replication reactions.

Why Are There Unequal Amounts of (–) and (+) Strands?

Different concentrations of (+) and (–) strands are produced in infected cells. For example, in cells infected with poliovirus, genomic RNA is produced at 100-fold higher concentrations than its complement. There are different explanations for these observations. RNA genomes and their complementary strands might have different stabilities, or the two strands might be synthesized by different mechanisms that vary in efficiency.

Viral (–) strand RNA is approximately 20 to 50 times more abundant than (+) strand RNA in cells infected with vesicular stomatitis virus. It was suggested that the asymmetry is a consequence of more efficient initiation of RNA synthesis at the 3′ end of (+) strand RNA than at the 3′ end of (–) strand RNA. An elegant proof of this hypothesis came from the construction and study of a rabies virus genome with identical initiation sites at the 3′ ends of both (–) and (+) strand RNAs. In cells infected with this virus, the ratio of (–) to (+) strands is 1:1.

BOX 6.4

DISCUSSION

Does the 3′ noncoding region of poliovirus RNA confer template specificity?

The isolation of infectious polioviruses without the entire 3′ noncoding region of the viral RNA has led to the suggestion that the template specificity imparted by terminal structures of RNA might be of greater importance early in infection. Early in infection, the 3′ pseudoknot structure might facilitate template selection when few viral polymerase molecules are available and membrane association has not yet provided highly concentrated areas of replication components. Later in infection, determinants of template selection by the polymerase might include the membrane association of the RNA polymerase, and the fact that translation of $3D^{pol}$, the most 3′-terminal gene, would position the polymerase at the 3′ end of the genome, ready for initiation.

Figure 6.15 Activation of the influenza virus RNA polymerase by specific virion RNA sequences. Binding of the PB1 protein to this RNA sequence induces the PB2 protein to bind to the cap of a cellular RNA. When the 3′-terminal sequence of genomic RNA binds to a second amino acid sequence in the PB1 protein, the polymerase acquires the activity to cleave the capped cellular RNA 10 to 13 nucleotides from the cap. The RNA polymerase can then carry out initiation and elongation of mRNAs. p, polymerase active site. 5′, and 3′ indicate the binding sites for the 5′ and 3′ ends, respectively, of (−) strand genomic RNA. Blue indicates an inactive site, and red indicates an active site. The polymerase is bound to both the 5′ and 3′ ends of the genomic RNA, with the capped RNA primer associated with the PB2 protein. Adapted from D. M. Knipe et al. (ed.), *Fields Virology*, 4th ed. (Lippincott Williams & Wilkins, Philadelphia, PA, 2001); P. Rao et al., *EMBO J.* **22**:1188–1198, 2003; and S. R. Shih and R. M. Krug, *Virology* **226**:430–436, 1996, with permission.

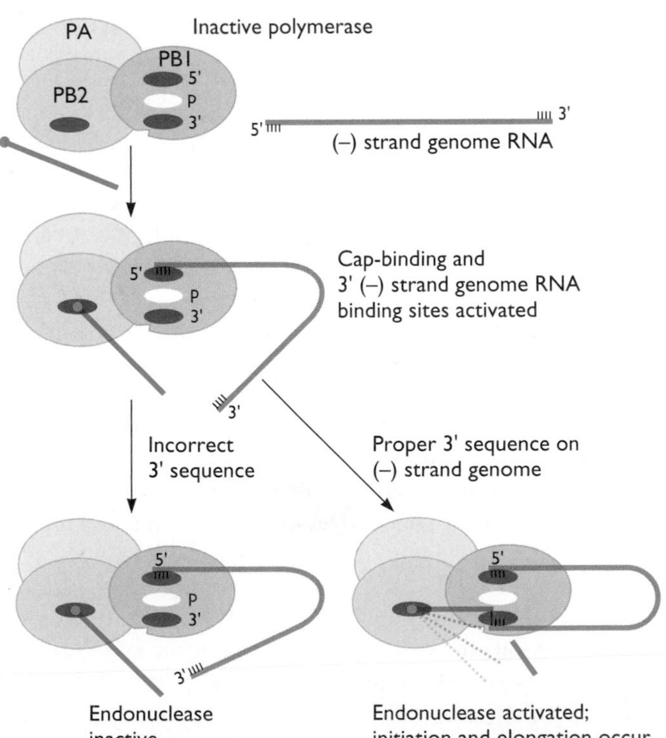

In alphavirus-infected cells, the abundance of genomic RNA is explained by the fact that (−) strand RNAs are synthesized only for a short time early in infection. The RNA polymerase that catalyzes (−) strand RNA synthesis is produced only during this period. The synthesis of (+) strands continues for much longer and leads to accumulation of mRNA and (+) strand genomic RNA.

Do Ribosomes and RNA Polymerases Collide?

The genomic RNA of (+) strand viruses can be translated in the cell, and the translation products include the viral RNA polymerase. At a certain point in infection, the RNA polymerase copies the RNA in a 3′ → 5′ direction while ribosomes traverse it in a 5′ → 3′ direction (Fig. 6.17), raising the question of whether the viral polymerase avoids

Figure 6.16 Structure of a viral RNA helicase. The RNA helicase of yellow fever virus is shown in surface representation, colored red, white, or blue depending on the distance of the amino acid from the center of the molecule. A model for melting of double-stranded RNA is shown. From Protein Data Bank file 1yks.

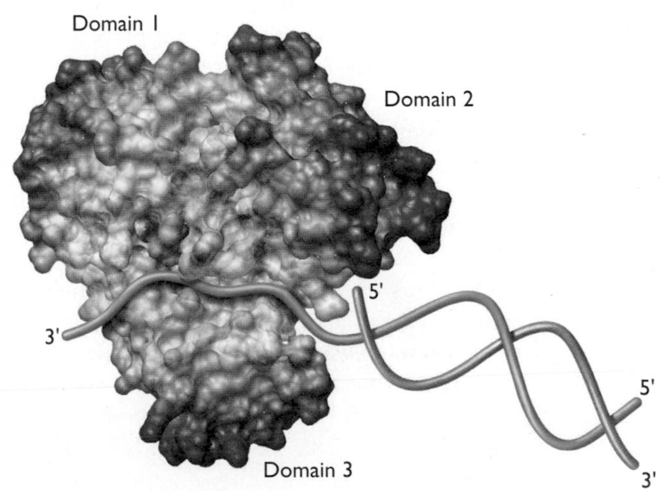

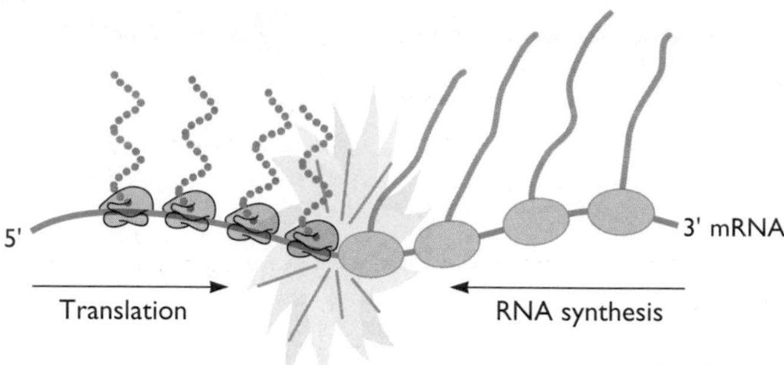

Figure 6.17 Ribosome-RNA polymerase collisions. A strand of viral RNA is shown, with ribosomes translating in the 5′ → 3′ direction and RNA polymerase copying the RNA chains in the 3′ → 5′ direction. Ribosome-polymerase collisions would occur in cells infected with (+) strand RNA viruses unless mechanisms exist to avoid simultaneous translation and replication.

collisions with ribosomes. When ribosomes are frozen on polioviral RNA by using inhibitors of protein synthesis, replication is inhibited. In contrast, when ribosomes are released, replication of the RNA increases. These results suggest that ribosomes must be cleared from viral RNA before it can serve as a template for (−) strand RNA synthesis; in other words, replication and translation cannot occur simultaneously.

The interactions of viral and cellular proteins with the cloverleaf structure in the polioviral 5′ untranslated region might determine whether the RNA genome is translated or replicated. In this model, interaction of cellular poly(rC)-binding protein 2 with the 5′ untranslated region initially stimulates translation in infected cells. Once protein $3CD^{pro}$ has been synthesized, it cleaves poly(rC)-binding protein, which prevents it from binding and leads to reduced translation. The cleaved poly(rC)-binding protein can still bind to the cloverleaf (Fig. 6.12) and promote viral RNA synthesis.

Restricting translation and RNA synthesis to distinct compartments may prevent collisions of ribosomes and polymerases. Viral mRNA synthesis takes place in the reovirus capsid, where the enzymes responsible for this process are located. The viral mRNAs are exported to the cytoplasm for translation. Synthesis of retroviral RNAs occurs in the cell nucleus, where translation does not take place. The architecture of membranous replication complexes of (+) strand viruses may favor RNA synthesis and exclude translation.

Even if mechanisms exist for controlling whether the polioviral genome is translated or replicated, some ribosome-RNA polymerase collisions are likely to occur. This conclusion is drawn from the isolation of a polioviral mutant with a genome that contains an insertion of a 15-nucleotide sequence from 28S ribosomal RNA (rRNA).

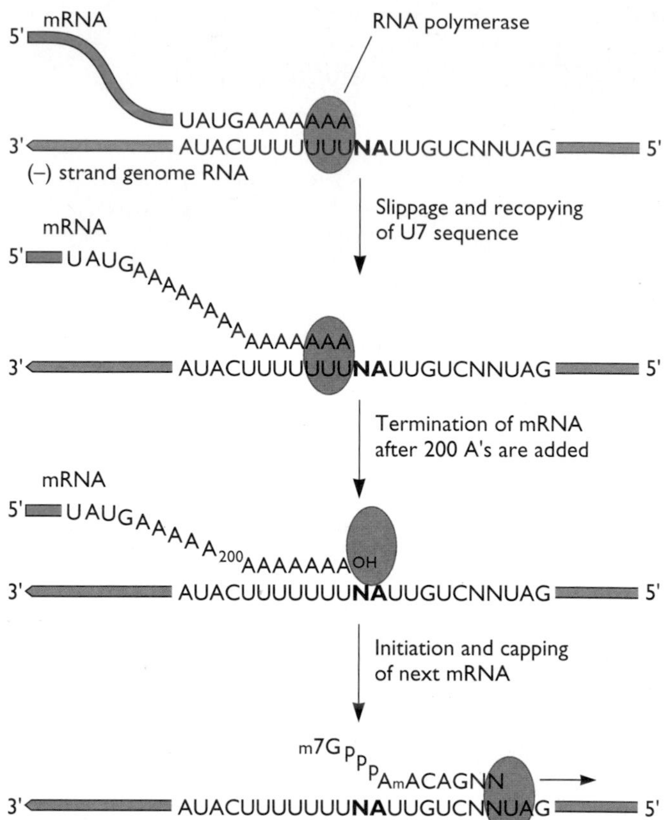

Figure 6.18 Poly(A) addition and termination at an intergenic region during vesicular stomatitis virus mRNA synthesis. Copying of the last seven U residues of an mRNA-encoding sequence is followed by slipping of the resulting seven A residues in the mRNA off the genomic sequence, which is then recopied. This process continues until approximately 200 A residues are added to the 3′ end of the mRNA. Termination then occurs, followed by initiation and capping of the next mRNA. The dinucleotide NA in the genomic RNA is not copied.

After colliding with a ribosome, the RNA polymerase apparently copied 15 nucleotides of rRNA before returning to the viral RNA template.

Synthesis of Poly(A)

The mRNAs synthesized during infection by most RNA viruses contain a 3′-poly(A) sequence, as do the vast majority of cellular mRNAs (exceptions are arenaviruses and reoviruses). The poly(A) sequence is encoded in the genome of (+) strand viruses. For example, polioviral (+) strand RNAs contain a 3′ stretch of poly(A), approximately 62 nucleotides in length, which is required for infectivity. The (−) strand RNA contains a 5′ stretch of poly(U), which is copied to form this poly(A).

Another mechanism for poly(A) addition is reiterative copying of, or "stuttering" at, a short U sequence in the (−) strand template. After initiation, vesicular stomatitis virus

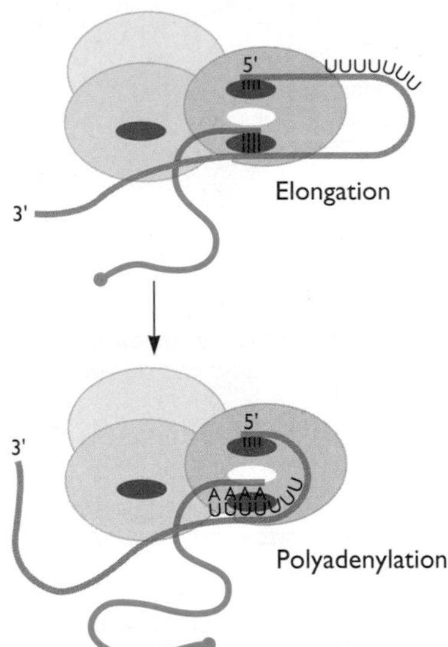

Figure 6.19 Moving-template model for influenza virus mRNA synthesis. During RNA synthesis, the polymerase remains bound to the 5′ end of the genomic RNA, and the 3′ end of the genomic RNA is threaded through (or along the surface of) the polymerase as the PB1 protein catalyzes each nucleotide addition to the growing mRNA chain. This threading process continues until the mRNA reaches a position on the genomic RNA that is close to the binding site of the polymerase. At this point the polymerase itself blocks further mRNA synthesis, and reiterative copying of the adjacent U_7 tract occurs. After about 150 A residues are added to the 3′ end of the mRNA, mRNA synthesis terminates. Adapted from D. M. Knipe et al. (ed.), *Fields Virology*, 4th ed. (Lippincott Williams & Wilkins, Philadelphia, PA, 2001); P. Rao et al., *EMBO J.* **22**:1188–1198, 2003; and S. R. Shih and R. M. Krug, *Virology* **226**:430–436, 1996, with permission.

mRNAs are elongated until the RNA polymerase reaches a conserved stop-polyadenylation signal [3′-AUACU$_7$-5′] located in each intergenic region (Fig. 6.18). Poly(A) (approximately 150 nucleotides) is added by reiterative copying of the U stretch, followed by termination. Polyadenylation is achieved by a similar mechanism during influenza virus mRNA synthesis.

Some models for RNA synthesis invoke a stationary template and moving polymerase. If instead we imagine a stationary enzyme and moving template, a mechanism becomes apparent for synthesis of poly(A) by the influenza virus polymerase (Fig. 6.19). The RNA polymerase specifically binds the 5′ end of (−) strand RNA and remains bound at this site of each genomic RNA segment throughout mRNA synthesis. The genomic RNAs would be threaded through the polymerase in a 3′ → 5′ direction as mRNA synthesis proceeds. Eventually the template would be unable to move, causing reiterative copying of the U residues.

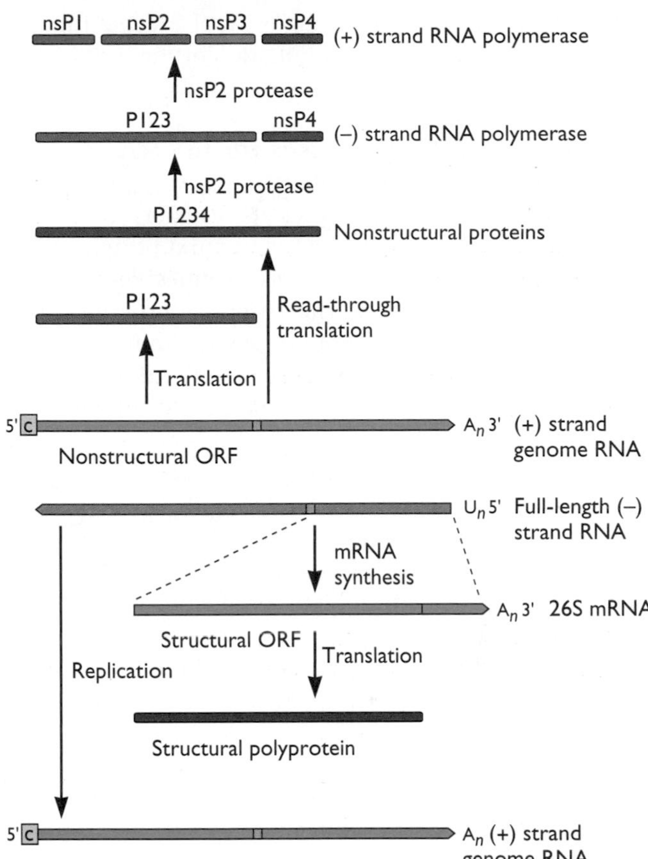

Figure 6.20 Genome structure and expression of an alphavirus, Sindbis virus. The 11,703-nucleotide Sindbis virus genome contains a 5′-terminal cap structure and a 3′-poly(A) tail. A conserved RNA secondary structure at the 3′ end of (+) strand genomic RNA is thought to control the initiation of (−) strand RNA synthesis. At early times after infection, the 5′ region of the genomic RNA (nonstructural open reading frame [ORF]) is translated to produce two nonstructural polyproteins: P123, whose synthesis is terminated at the first translational stop codon (indicated by the box), and P1234, produced by an occasional (15%) readthrough of this stop codon. The P1234 polyprotein is proteolytically cleaved to produce the enzymes that catalyze the various steps in genomic RNA replication: the synthesis of a full-length (−) strand RNA, which serves as the template for (+) strand synthesis, and either full-length genomic RNA or subgenomic 26S mRNA. The 26S mRNA, shown in expanded form, is translated into a structural polyprotein (p130) that undergoes proteolytic cleavage to produce the virion structural proteins. The 26S RNA is not copied into a (−) strand because a functional initiation site fails to form at the 3′ end.

The Switch from mRNA Production to Genome RNA Synthesis

Exact replicas of the RNA genome must be made for assembly of infectious viral particles. However, the mRNAs of most RNA viruses are **not** complete copies of the viral RNA. The replication cycle of these viruses must therefore include a switch from mRNA synthesis to the production of full-length genomes. The majority of mechanisms for this switch regulate either the initiation or the termination of RNA synthesis.

Different RNA Polymerases for mRNA Synthesis and Genome Replication

The mechanism of mRNA synthesis of other (+) strand RNA viruses allows structural and nonstructural proteins (generally needed in greater and lesser quantities, respectively) to be produced separately. The latter are produced from full-length (+) strand (genomic) RNA, while structural proteins are translated from subgenomic mRNA(s). This strategy is a feature of the replication cycles of coronaviruses, caliciviruses, and alphaviruses. Translation of the Sindbis virus (+) strand RNA genome yields the nonstructural proteins that synthesize a full-length (−) strand (Fig. 6.20). Such RNA molecules contain not only a 3′-terminal sequence for initiation of (+) strand RNA synthesis but also an internal initiation site, used for production of a 26S subgenomic mRNA.

Alphaviral genome and mRNA synthesis is regulated by the sequential production of three RNA polymerases with different template specificities. All three enzymes are derived from the nonstructural polyprotein P1234 and contain the complete amino acid sequence of this precursor (Fig. 6.20). The covalent connections among the segments (P1, P2, P3, and P4) of the polyprotein are successively broken, with ensuing alterations in RNA polymerase specificity (Fig. 6.21). It seems likely that each proteolytic cleavage induces a conformational change in the polymerase that alters its template specificity.

The genes of RNA viruses with a nonsegmented (−) strand RNA genome are expressed by the production of subgenomic mRNAs in infected cells. The switch from subgenomic mRNA to full-length (+) strand synthesis may be accomplished by producing RNA polymerases with different specificities. For example, polymerase composed of one molecule of L protein associated with four molecules of P protein is thought to carry out vesicular stomatitis virus mRNA synthesis, while genome replication would be accomplished by an LN-(P) 4 assembly (Fig. 6.22). The ratio of the two polymerases at different times after infection is regulated by viral protein synthesis, and by the phosphorylation state of the P protein, which controls its ability to form oligomers.

Suppression of Intergenic Stop-Start Reactions by Nucleocapsid Protein

The transition from mRNA to genome RNA synthesis in cells infected with vesicular stomatitis virus requires not only different RNA polymerases but also the viral nucleocapsid (N) protein. Individual mRNAs are produced by a

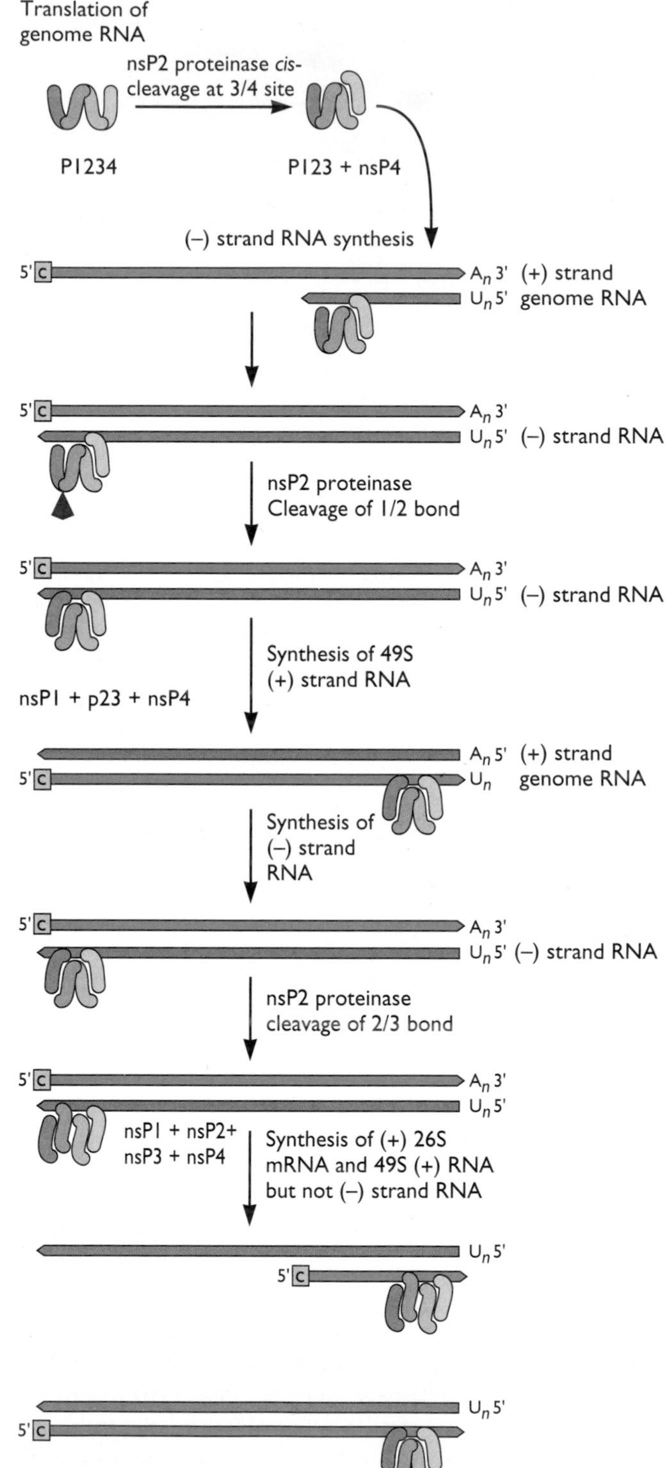

Figure 6.21 Three RNA polymerases with distinct specificities in alphavirus-infected cells. These RNA polymerases contain the entire sequence of the P1234 polyprotein and differ only in the number of proteolytic cleavages in this sequence. Adapted from J. H. Strauss and E. G. Strauss, *Microbiol. Rev.* **58**:491–562, 1994, with permission.

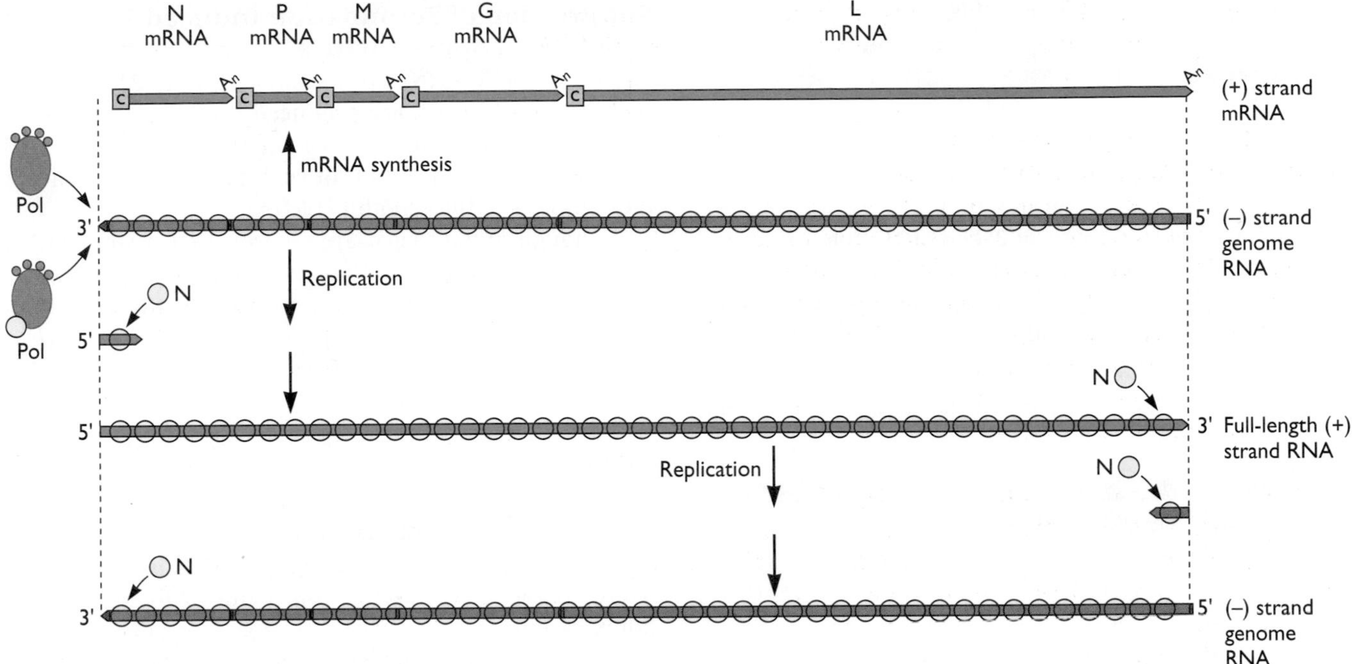

Figure 6.22 Vesicular stomatitis viral RNA synthesis. Viral (–) strand genomes are templates for the production of either subgenomic mRNAs or full-length (+) strand RNAs. The switch from mRNA synthesis to genomic RNA replication is mediated by two RNA polymerases and by the N protein. mRNA synthesis initiates at the beginning of the N gene, near the 3′ end of the viral genome. Poly(A) addition is a result of reiterative copying of a sequence of seven U residues present in each intergenic region. Chain termination and release occur after approximately 150 A residues have been added to the mRNA. The RNA polymerase then initiates synthesis of the next mRNA at the conserved start site 3′UUGUC . . . 5′. This process is repeated for all five viral genes. Synthesis of the full-length (+) strand begins at the exact 3′ end of the viral genome and is carried out by the RNA polymerase L-N-(P)4. The (+) strand RNA is bound by the viral nucleocapsid (N) protein, which is complexed with the P protein in a 1:1 molar ratio. The N-P complexes bind to the nascent (+) strand RNA, allowing the RNA polymerase to read through the intergenic junctions at which polyadenylation and termination take place during mRNA synthesis.

series of initiation and termination reactions as the RNA polymerase moves down the viral genome (Fig. 6.23). This start-stop mechanism accounts for the observation that 3′-proximal genes must be copied before downstream genes (Box 6.5); the viral RNA polymerase is unable to initiate synthesis of each mRNA independently. To produce a full-length (+) strand RNA, the stop-start reactions at intergenic regions must be suppressed. Suppression depends on the synthesis of the N and P proteins. The P protein maintains the N protein in a soluble form so that it can encapsidate the newly synthesized RNA. The N-P complexes bind to leader RNA and cause antitermination, signaling the polymerase to begin processive RNA synthesis. Additional N protein molecules then associate with the (+) strand RNA as it is elongated, and eventually bind to the seven A bases in the intergenic region. This interaction prevents the seven A bases from slipping backward along the genomic RNA template and therefore blocks reiterative copying of the seven U bases in the genome. Consequently, RNA synthesis continues through the intergenic region. The number of N-P protein complexes in infected cells therefore regulates the relative efficiencies of mRNA synthesis and genome RNA replication. The copying of full-length (+) strand RNAs to (–) strand genomic RNAs also requires the binding of N-P protein complexes to elongating RNA molecules. Newly synthesized (–) strand RNAs are produced as nucleocapsids that can be readily packaged into progeny viral particles.

The (–) strand RNA genome of paramyxoviruses is copied efficiently only when its length in nucleotides is a multiple of 6. This requirement, called the **rule of six**, is probably a consequence of the association of each N monomer with exactly six nucleotides. Assembly of the nucleocapsid begins with the first nucleotide at the 5′ end of the RNA and continues until the 3′ end is reached. If the genome length is not a multiple of 6, then the 3′ end of the genome will not be precisely aligned with the last N monomer. Such misalignment reduces the efficiency of initiation of RNA synthesis at the 3′ end. Curiously, although the

RNA polymerase binds at 3' end of N gene

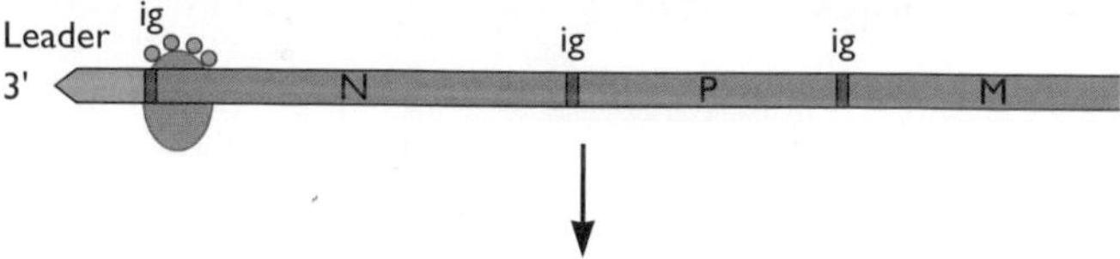

Initiation of mRNA synthesis at 3' end of N gene

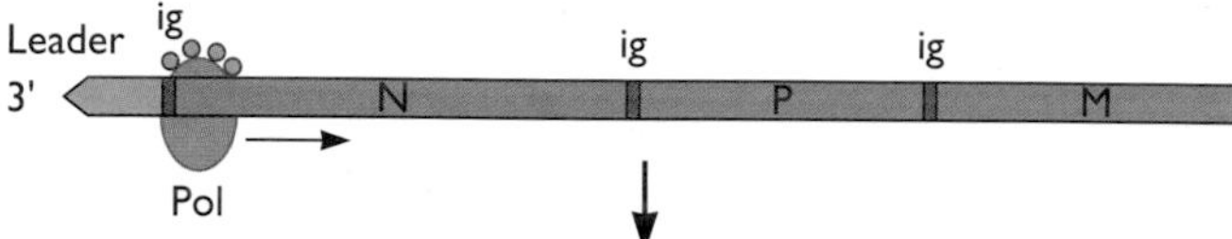

Synthesize N mRNA and terminate at intergenic region (ig)

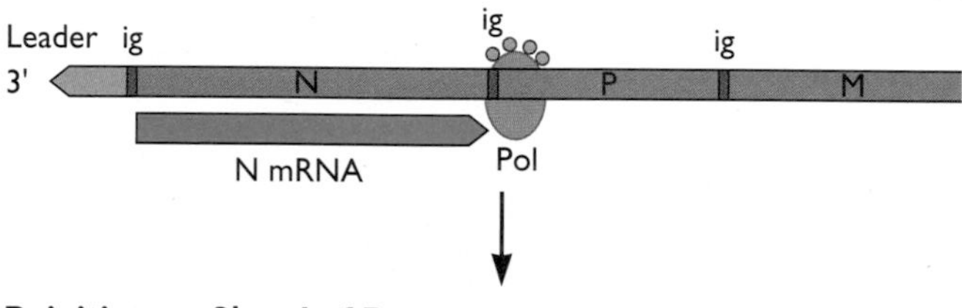

Reinitiate at 3' end of P gene

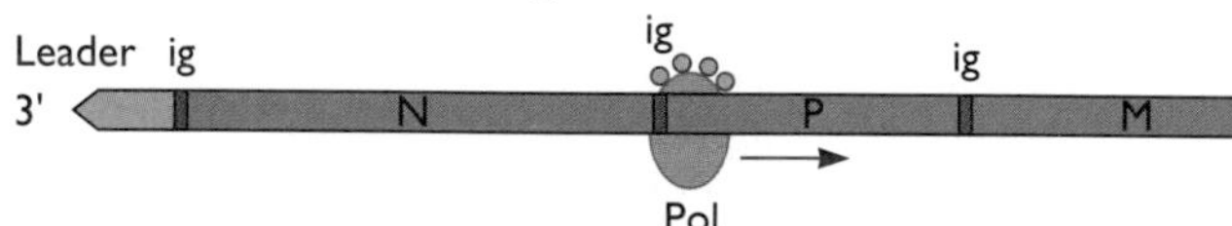

Figure 6.23 Stop-start model of vesicular stomatitis mRNA synthesis. The RNA polymerase (Pol) initiates RNA synthesis at the 3′ end of the N gene. After synthesis of the N mRNA, RNA synthesis terminates at the intergenic region, followed by reinitiation at the 3′ end of the P gene. This process continues until all five mRNAs are synthesized. Reinitiation does not occur after the last mRNA (the L mRNA) is synthesized, and as a consequence the 59 5′-terminal nucleotides of the vesicular stomatitis virus genomic RNA are not copied. Only a fraction of the polymerase molecules successfully make the transition from termination to reinitiation of mRNA synthesis at each intergenic region.

rhabdovirus N protein binds nine nucleotides of RNA, the genome length need not be a multiple of this number for efficient copying.

The influenza virus NP protein also regulates the switch from viral mRNA to full-length (+) strand synthesis (Fig. 6.14). The RNA polymerase for genome replication reads through the polyadenylation and termination signals used for mRNA production only if NP is present. This protein is thought to bind nascent (+) strand transcripts and block poly(A) addition by a mechanism analogous to that described for vesicular stomatitis virus N protein. Copying of (+) strand RNAs into (−) strand RNAs also requires NP protein. Intracellular concentrations of NP protein are therefore an important determinant of whether mRNAs or full-length (+) strands are synthesized.

Suppression of Termination Induced by a Stem-Loop Structure

The subgenomic mRNAs of certain (−) strand RNA viruses, such as arenaviruses, are produced when the RNA polymerase terminates at a stem-loop structure in the viral RNA template. Suppression of such termination results in the synthesis of full-length (+) RNAs.

Although arenaviruses are considered (−) strand RNA viruses, their genomic RNA is in fact **ambisense**: mRNAs are produced both from (−) strand genomic RNA and from complementary full-length (+) strands. The arenavirus genome comprises two RNA segments, S (small) and L (large) (Fig. 6.24). Shortly after infection, a virion-associated RNA polymerase synthesizes mRNAs from the 3′ region of the S and L genomic RNA segments. Synthesis of each mRNA terminates at a stem-loop structure. These mRNAs, which are translated to produce the nucleocapsid (NP) protein and RNA polymerase (L) protein, respectively, are the only viral RNAs made during the first several hours of infection. Later in infection, the block imposed by the stem-loop structure is overcome, permitting the synthesis of full-length S and L (+) RNAs. It was initially thought that melting of the stem-loop structure by the NP protein allowed the transcription termination signal to be bypassed. It now seems more likely that two different RNA polymerases are made in infected cells, one that produces mRNAs and a second that synthesizes full-length copies of the genome. The finding that viral mRNAs are capped while genomes are not is consistent with this hypothesis.

Different Templates Used for mRNA Synthesis and Genome Replication

A distinctive feature of the infectious cycle of double-stranded RNA viruses such as reovirus is the production of mRNAs and genomic RNAs from distinct templates in different viral particles. Because the viral genomes are double stranded, they cannot be translated by the cell. Therefore, the first step in infection is the production of mRNAs from each viral RNA segment by the virion-associated RNA polymerase λ3 (Fig. 6.25). Reoviral mRNAs contain 5′ cap structures but lack 3′ poly(A) sequences.

Although the reovirus RNA polymerase resembles a right hand, with thumb, fingers, and palm domains, it has distinctive features not observed in other enzymes. The reovirus RNA polymerase is a cube-like structure, with a catalytic site in the center that is accessible by four tunnels. One tunnel allows template entry, one serves for the exit of newly synthesized double-stranded RNA, a third permits exit of mRNA, and a fourth is for substrate entry. A priming loop is present in the palm domain that is not observed in this region of other RNA polymerases. This loop supports stacking of the initiating NTP, then retracts into the palm

BOX 6.5

EXPERIMENTS

Mapping gene order by UV irradiation

The effects of ultraviolet (UV) irradiation provided insight into the mechanism of vesicular stomatitis virus mRNA synthesis. In these experiments, vesicular stomatitis virus particles were irradiated with UV light and the effect on the synthesis of individual mRNAs was determined. UV light causes the formation of pyrimidine dimers in RNA that block passage of the RNA polymerase. In principle, larger genes require less UV irradiation to inactivate mRNA synthesis and have a larger **target size.** The dosage of UV irradiation needed to inactivate synthesis of the N mRNA corresponded to the predicted size of the N gene, but this was not the case for the other viral mRNAs. The target size of each other mRNA was the sum of its size plus the size of other genes located 3′ to it. For example, the UV target size of the L mRNA is the size of the entire viral genome. These results indicate that these mRNAs are synthesized sequentially, in the 3′ → 5′ order in which their genes are arranged in the viral genome: N-P-M-G-L.

Ball, L. A., and C. N. White. 1976. Order of transcription of genes of vesicular stomatitis virus. *Proc. Natl. Acad. Sci. USA* **73:**442–446.

Vesicular stomatitis virus mRNA map and UV map. The genome is shown as a dark green line at the top, and the N, P, M, G, and L genes and their relative sizes are indicated. The 47-nucleotide leader RNA is encoded at the 3′ end of the genomic RNA. The leader and intergenic regions are shown in orange. The RNAs encoded at the 3′ end of the genome are made in larger quantities than the RNAs encoded at the 5′ end of the genome. UV irradiation experiments determined the size of the vesicular stomatitis virus genome (UV target size) required for synthesis of each of the viral mRNAs. The UV target size of each viral mRNA corresponded to the size of the genomic RNA sequence encoding the mRNA plus all of the genomic sequence 3′ to this coding sequence. The transition from reiterative copying and termination to initiation is not perfect, and only about 70 to 80% of the polymerase molecules accomplish this transition at each intergenic region. Such inefficiency accounts for the observation that 3′-proximal mRNAs are more abundant than 5′-proximal mRNAs.

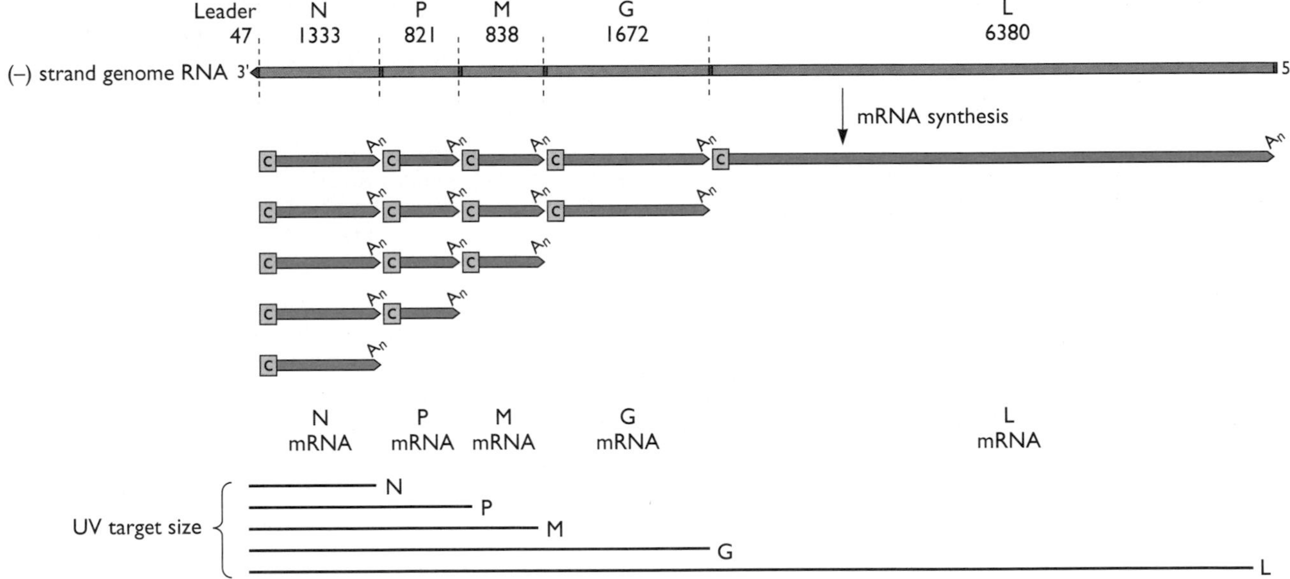

and fits into the minor groove of the double-stranded RNA product. This movement assists in the transition between initiation and elongation, and also allows the newly synthesized RNA to exit the polymerase.

In the reovirus core, the λ3 polymerase molecules are attached to the inner shell at each fivefold axis, below an RNA exit pore. Viral mRNAs are produced by the polymerase inside the viral particle and then extruded into the cytoplasm through this pore. Attachment of the polymerase molecules to the pores ensures that the mRNA can leave the particle, without depending upon diffusion, which would be very inefficient. Examination of the structure of actively transcribing rotavirus, a member of the *Reoviridae*, has allowed a three-dimensional visualization of how mRNAs are released from the particle (Box 6.6). Viral (+) strand RNAs that will serve as templates for (–) strand RNA synthesis are first packaged into newly assembled subviral particles (Fig. 6.25). Each (+) strand RNA is then copied just once within the subviral particle to produce double-stranded RNA.

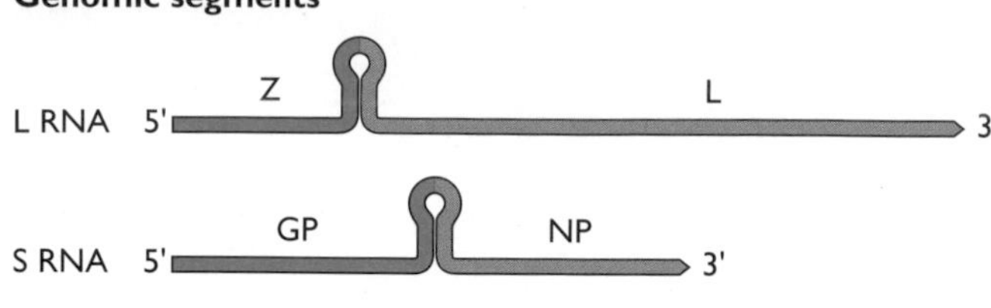

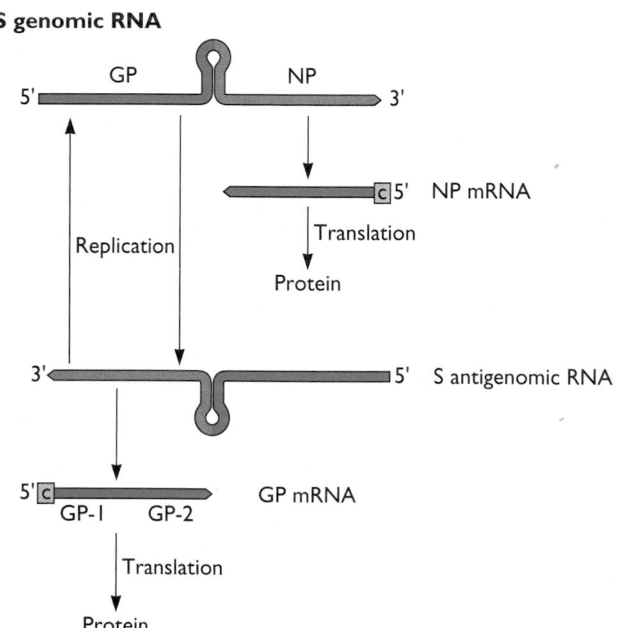

Figure 6.24 Arenavirus RNA synthesis. Arenaviruses contain two genomic RNA segments, L (large) and S (small) (top). At early times after infection, only the 3′ region of each of these segments is copied to form mRNA: the N mRNA from the S genomic RNA and the L mRNA from the L genomic RNA. Copying of the remainder of the S and L genomic RNAs may be blocked by a stem-loop structure in the genomic RNAs. After the S and L genomic RNAs are copied into full-length strands, their 3′ regions are copied to produce mRNAs: the glycoprotein precursor (GP) mRNA from S RNA and the Z mRNA (encoding an inhibitor of viral RNA synthesis) from the L RNA. Only RNA synthesis from the S RNA is shown in detail. Adapted from D. M. Knipe et al. (ed.), *Fields Virology*, 4th ed. (Lippincott Williams & Wilkins, Philadelphia, PA, 2001), with permission.

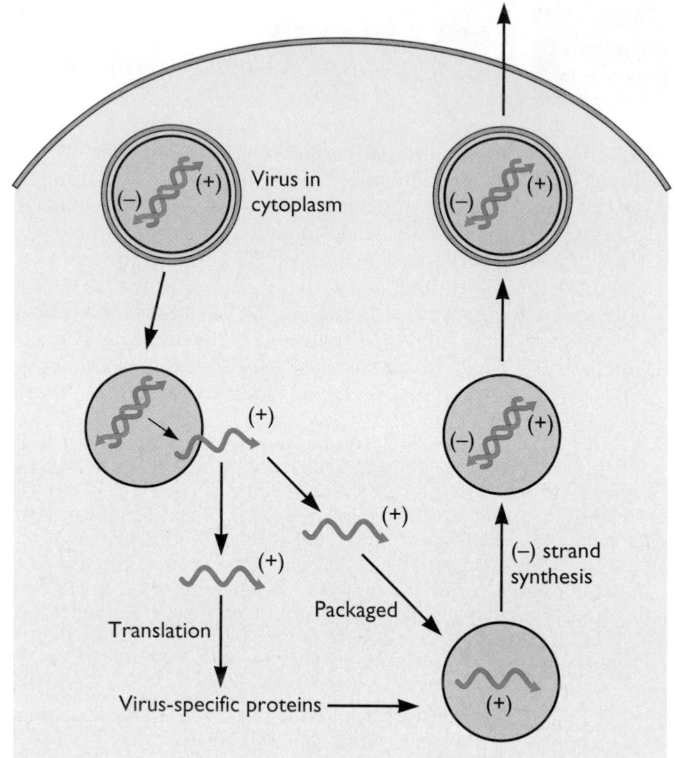

Figure 6.25 mRNA synthesis and replication of double-stranded RNA genomes. These processes occur in subviral particles containing the RNA templates and necessary enzymes. During cell entry, the virion passes through the lysosomal compartment, and proteolysis of viral capsid proteins activates the RNA synthetic machinery. Single-stranded (+) viral mRNAs, which are synthesized in parental subviral particles, are extruded into the cytoplasm, where they serve either as mRNAs or as templates for the synthesis of (–) RNA strands. In the latter case, viral mRNAs are first packaged into newly assembled subviral particles in which the synthesis of (–) RNAs to produce double-stranded RNAs occurs. These subviral particles become infectious particles. Only 1 of the 10 to 12 double-stranded RNA segments of the reoviral genome is shown.

Members of different families of double-stranded RNA viruses carry out RNA synthesis in diverse ways. Replication of the genome of bacteriophage ϕ6 (3 double-stranded RNA segments) and birnaviruses (2 double-stranded RNA segments) is semiconservative, whereas that of reoviruses (10 to 12 double-stranded RNA segments) is conservative (Fig. 6.26). During conservative replication, the double-stranded RNA that exits the polymerase must be melted, so that the newly synthesized (+) strand is released and the template (–) strand reanneals with the original (+) strand. In reovirions, each double-stranded RNA segment is attached to a polymerase complex, by interaction of the 5′ cap structure with a cap-binding site on the RNA polymerase. Attachment of the 5′ cap to the polymerase facilitates insertion of the 3′ end of the (–) strand into the template channel. This arrangement allows very efficient reinitiation of RNA synthesis in the crowded core of the virion. The RNA polymerase of bacteriophage ϕ6 and birnaviruses do not have such a cap-binding site, as would be expected for enzymes that copy both strands of the double-stranded RNA segments. This strategy appears less efficient, but may be acceptable when the genome consists of only two or three double-stranded RNA segments.

BOX 6.6

EXPERIMENTS

Release of mRNA from rotavirus particles

Rotaviruses, the most important cause of gastroenteritis in children, are large icosahedral viruses made of a three-shelled capsid containing 11 double-stranded RNA segments. The structure of this virus reveals that a large portion of the viral genome (~25%) is ordered within the particle and forms a dodecahedral structure (see Fig. 4.17). In this structure, the RNAs interact with the inner capsid layer and pack around the RNA polymerase located at the fivefold axis of symmetry. Further analysis of rotavirus particles in the process of synthesizing mRNA has shown that newly synthesized molecules are extruded through the capsid through several channels located at the fivefold axes (see figure). Multiple mRNAs are released at the same time from such particles. On the basis of these observations, it has been suggested that each double-stranded genomic RNA segment is copied by an RNA polymerase located at a fivefold axis of symmetry. This model may explain why no double-stranded RNA virus with more than 12 genomic segments—the maximum number of fivefold axes—has ever been identified.

Lawton, J. A., M. K. Estes, and B. V. Prasad. 1997. Three-dimensional visualization of mRNA release from actively transcribing rotavirus particles. *Nat. Struct. Biol.* **4:**118–121.

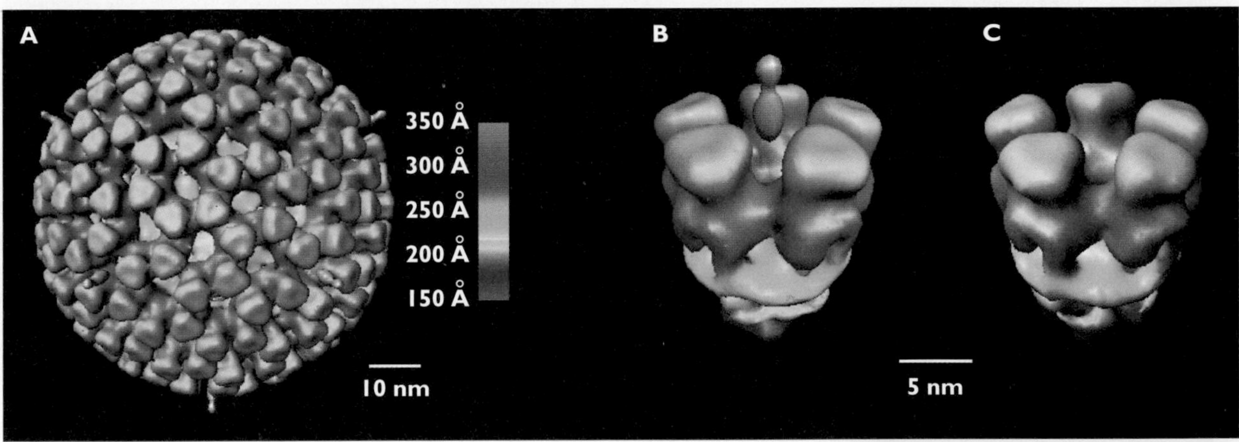

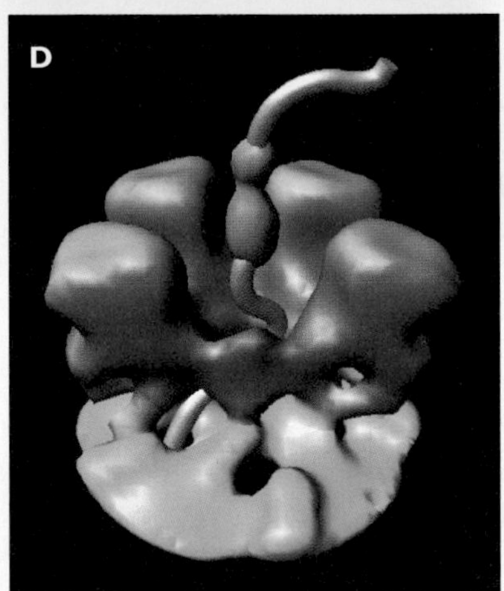

Three-dimensional visualization of mRNA release from rotavirus particles synthesizing mRNA. (A) Structure of a rotavirus particle in the process of synthesizing mRNA. The capsid is depth-cued according to the color chart. Parts of newly synthesized mRNA that are ordered, and therefore structurally visible, are shown in pink at the fivefold axes of symmetry. **(B)** Close-up view of the channel at the fivefold axis and the visible mRNA. The mRNA is surrounded by five trimers of capsid protein VP6. **(C)** Close-up view of the channel at the fivefold axis of a particle not in the process of synthesizing mRNA. **(D)** Model of the pathway of mRNA transit through the capsid. One VP6 trimer has been omitted for clarity. The green protein is VP2, and the mRNA visible in the structure is shown in pink. The gray tube represents the possible path of an mRNA molecule passing through the VP2 and VP6 layers through the channel. Courtesy of B. V. V. Prasad, Baylor College of Medicine. Reprinted from J. A. Lawton et al., *Nat. Struct. Biol.* **4:**118–121, 1997, with permission.

Suppression of Polyadenylation

Another mechanism for switching from mRNA synthesis to genomic replication is to suppress a poly(A) addition signal; an example occurs in hepatitis delta satellite virus-infected cells. Host RNA polymerase II initiates viral mRNA synthesis at a position on the genome near the beginning of the delta antigen-coding region (Boxes 6.7 and 6.8). Once the polymerase has moved past a polyadenylation signal and the self-cleavage domain (Box 6.9), the 3′-poly(A) end of the mRNA is produced by host cell enzymes. The

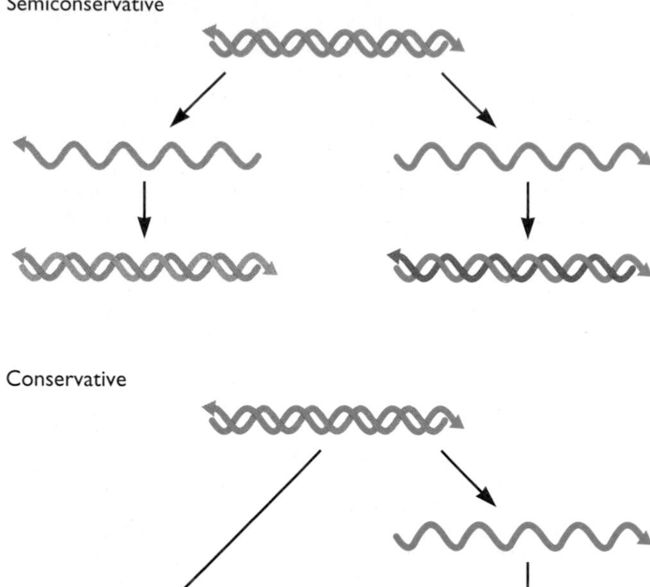

Figure 6.26 Two mechanisms for copying nucleic acids. During semiconservative replication, both strands of nucleic acid serve as templates for the synthesis of new strands (shown in red). In contrast, only one strand is copied during conservative replication.

RNA downstream of the poly(A) site is not degraded, in contrast to that of other mRNA precursors made by RNA polymerase II, but is elongated until a complete full-length (+) strand is made. The poly(A) addition site in this full-length (+) RNA is not used. The delta antigen bound to the rodlike RNA may block access of cellular enzymes to the poly(A) signal, thereby inhibiting polyadenylation.

The Same Template Used for mRNA Synthesis and Genome Replication

No switch from mRNA to genomic RNA synthesis is needed when the only viral mRNA made in infected cells is identical to the (+) strand genome, as occurs in cells infected with picornaviruses (Fig. 6.1). Newly synthesized polioviral (+) strand RNA molecules can serve as RNA templates for further genomic RNA replication, as mRNAs for the synthesis of viral proteins, or as genomic RNAs to be packaged into progeny virions. Because picornaviral mRNA is identical in sequence to the viral RNA genome, all RNAs needed for the reproduction of these viruses can be made by a simple set of RNA synthesis reactions (Fig. 6.1). Such simplicity comes at a price, because there can be no regulation of individual gene products at the level of mRNA synthesis. However, polioviral gene expression can be controlled by the rate and extent of polyprotein processing. For example, $3CD^{pro}$, the precursor of the viral polymerase $3D^{pol}$, cannot polymerize RNA. Rather, $3CD^{pro}$ is a protease that cleaves at certain Gln-Gly amino acid pairs in the polyprotein. Regulating the processing of $3CD^{pro}$ controls the concentration of RNA polymerase.

Cellular Sites of Viral RNA Synthesis

Genome and mRNA synthesis of most RNA viruses occurs in the cytoplasm of the cell, invariably in specific structures such as the nucleocapsids of (–) strand RNA viruses, subviral particles of double-stranded RNA viruses, and membrane-bound replication complexes for (+) strand RNA viruses. Membrane-bound replication complexes of different viruses are morphologically diverse, and the membranes originate from various cellular compartments. Alphaviral RNA synthesis occurs on the cytoplasmic surface of endosomes and lysosomes, and polioviral RNA polymerase is located on the surfaces of small, membranous vesicles that assemble into higher-order structures.

The membrane vesicles observed early in poliovirus-infected cells are thought to originate from two different cellular pathways. One source appears to be the endoplasmic reticulum, specifically vesicles whose production is regulated by proteins of coat protein complex II (CopII) (Chapter 12). Unlike vesicles produced from the endoplasmic reticulum in uninfected cells, those in poliovirus-infected cells do not fuse with the Golgi and therefore accumulate in the cytoplasm. The vesicles produced during poliovirus infection bear several hallmarks of autophagosomes, including a double-membrane morphology and colocalization with protein markers of these vesicles. Synthesis of poliovirus 2BC and 3A proteins in uninfected cells leads to production of such autophagosomes. Similar double-membrane vesicles are observed during infection with a variety of (+) strand RNA viruses, indicating that they may serve as a general replication platform. However, neither the production of CopII-coated vesicles nor the formation of autophagosomes is blocked by the fungal metabolite brefeldin A, which inhibits poliovirus RNA synthesis. The cellular target of this drug is the Arf GTPases, proteins that regulate the formation of secretory vesicles. The structures upon which poliovirus RNA is produced may therefore also require membrane vesicles that depend upon Arf for their formation.

It is thought that membrane association of viral replication complexes ensures high local concentrations of replication components and hence increases the rates or efficiencies of replication reactions. As we have seen, this property may contribute to the specificity of polioviral $3D^{pol}$ for viral RNA templates. Membrane association may also have other functions, such as allowing efficient packaging of progeny RNA into virions, or providing lipid components

BOX 6.7

DISCUSSION

Unique mechanisms of mRNA and genome synthesis of the hepatitis delta satellite virus

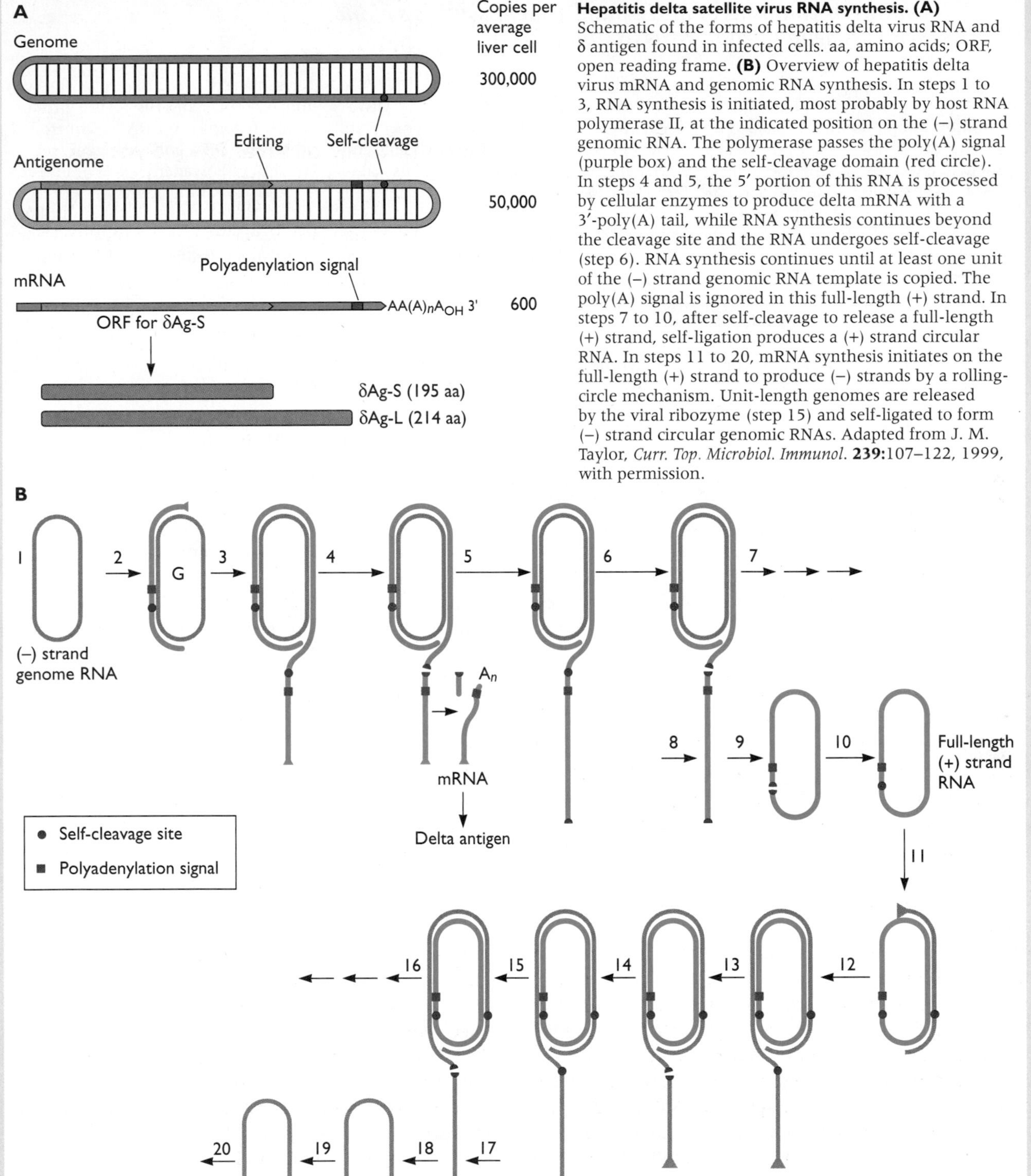

Hepatitis delta satellite virus RNA synthesis. (A) Schematic of the forms of hepatitis delta virus RNA and δ antigen found in infected cells. aa, amino acids; ORF, open reading frame. **(B)** Overview of hepatitis delta virus mRNA and genomic RNA synthesis. In steps 1 to 3, RNA synthesis is initiated, most probably by host RNA polymerase II, at the indicated position on the (–) strand genomic RNA. The polymerase passes the poly(A) signal (purple box) and the self-cleavage domain (red circle). In steps 4 and 5, the 5′ portion of this RNA is processed by cellular enzymes to produce delta mRNA with a 3′-poly(A) tail, while RNA synthesis continues beyond the cleavage site and the RNA undergoes self-cleavage (step 6). RNA synthesis continues until at least one unit of the (–) strand genomic RNA template is copied. The poly(A) signal is ignored in this full-length (+) strand. In steps 7 to 10, after self-cleavage to release a full-length (+) strand, self-ligation produces a (+) strand circular RNA. In steps 11 to 20, mRNA synthesis initiates on the full-length (+) strand to produce (–) strands by a rolling-circle mechanism. Unit-length genomes are released by the viral ribozyme (step 15) and self-ligated to form (–) strand circular genomic RNAs. Adapted from J. M. Taylor, *Curr. Top. Microbiol. Immunol.* **239:**107–122, 1999, with permission.

(continued)

BOX 6.7

DISCUSSION

Unique mechanisms of mRNA and genome synthesis of the hepatitis delta satellite virus (continued)

The strategy for synthesis of the hepatitis delta satellite virus genome is apparently unique among animal viruses. The viral RNAs are synthesized by host cell DNA-dependent RNA polymerase II (Box 6.8), and the hepatitis delta virus RNAs are RNA catalysts, or **ribozymes** (Box 6.9). The genome of hepatitis delta virus is a 1,700-nucleotide (−) strand circular RNA, the only RNA with this structure that has been found in animal cells. As approximately 70% of the nucleotides are base paired, the viral RNA is folded into a rodlike structure (see figure). Surprisingly, all hepatitis delta virus-specific RNA synthesis is catalyzed by cellular DNA-dependent RNA polymerase II, which can also utilize RNA as a template. Genomic RNA replication requires reactions catalyzed by ribozymes (Box 6.9) residing in both the hepatitis delta virus genomic (−) strand RNA and its full-length (+) strand RNA copy. Both these hepatitis delta virus RNAs catalyze self-cleavage in the absence of any protein and may also catalyze self-ligation.

All hepatitis delta satellite virus RNAs are synthesized in the nucleus by RNA polymerase II (see figure). The switch from mRNA synthesis to the production of full-length (+) RNA is controlled by suppression of a poly(A) signal (see "Suppression of Polyadenylation"). Full-length (−) and (+) strand RNAs are copied by a rolling-circle mechanism, and ribozyme self-cleavage releases linear monomers. Subsequent ligation of the two termini by the same ribozyme produces a monomeric circular RNA. The hepatitis delta virus ribozymes are therefore needed to process the intermediates of rolling-circle RNA replication.

or physical support to the replication complex. The surfaces of membranous replication complexes isolated from poliovirus-infected cells appear to be coated with two-dimensional arrays of polymerase. These arrays are formed by interaction of $3D^{pol}$ molecules in a head-to-tail fashion. Surface catalysis is known to have several advantages, including a higher probability of collision among reactants, an increase in substrate affinity from clustering of multiple binding sites, and retention of reaction products. The last property would facilitate multiple rounds of copying (+) and (−) strand RNA templates.

Viral RNA polymerases are recruited to membranous replication complexes in different ways. Polioviral $3D^{pol}$ is targeted to membrane vesicles by binding to the viral membrane protein 3AB (Fig. 6.12). In contrast, a C-terminal, transmembrane segment of the hepatitis C virus RNA polymerase, NS5b, is responsible for attachment of the enzyme to cellular membrane replication complexes.

BOX 6.8

DISCUSSION

RNA-dependent RNA polymerase II

The mRNAs produced during hepatitis delta satellite virus infection of cells have typical properties of DNA-dependent RNA polymerase II products, including a 5′ cap and 3′-poly(A) tail. Production of these satellite mRNAs is also sensitive to α-amanitin, an inhibitor of DNA-dependent RNA polymerase II. Furthermore, the RNA genome of plant viroids can be copied by plant DNA-dependent RNA polymerase II. Based on these observations, it was suggested that the RNA genome of hepatitis delta satellite virus is copied by RNA polymerase II. Recently, experimental support for this hypothesis has been obtained. When purified mammalian RNA polymerase II was incubated with NTPs and an RNA template-primer, an RNA product was produced. Similar results were obtained when the antigenome of hepatitis delta satellite virus was used in the reaction. Structural studies revealed that the RNA template-product duplex occupies the same site on the enzyme as the DNA-RNA hybrid during transcription. When transcription protein IIS was added to the reaction mixture, the satellite genome was cleaved, and the new 3′ end was used as a primer. Compared with DNA-dependent RNA synthesis, RNA-dependent RNA synthesis by RNA polymerase II was slower and less processive. These properties may explain why the enzyme can copy only short RNA templates.

The ability of DNA-dependent RNA polymerase II to copy an RNA template provides a missing link in molecular evolution. This activity supports the hypothesis that an ancestor of RNA polymerase II copied RNA genomes that are thought to have existed during the ancient RNA world. During the transition from RNA to DNA genomes, this enzyme evolved to copy DNA templates. Today these enzymes can still copy small RNAs such as the genome of hepatitis delta satellite virus.

Chang, J., X. Nie, H. E. Chang, Z. Han, and J. Taylor. 2008. Transcription of hepatitis delta virus RNA by RNA polymerase II. *J. Virol.* **82:**1118–1127.

Lehmann, E., F. Brueckner, and P. Cramer. 2007. Molecular basis of RNA-dependent RNA polymerase II activity. *Nature* **450:**445–449.

Rackwitz, H. R., W. Rohde, and H. L. Sanger. 1981. DNA-dependent RNA polymerase II of plant origin transcribes viroid RNA into full-length copies. *Nature* **291:**297–301.

BOX 6.9

BACKGROUND

Ribozymes

A **ribozyme** is an enzyme in which RNA, not protein, carries out catalysis. The first ribozyme discovered was the group I intron of the ciliate *Tetrahymena thermophila*. Other ribozymes have since been discovered, including RNase P of bacteria, group II self-splicing introns, hammerhead RNAs of viroids and satellite RNAs, and the ribozyme of hepatitis delta virus. Ribozymes are very diverse in size, sequence, and the mechanism of catalysis. For example, the hepatitis delta satellite virus ribozyme (see figure) catalyzes a transesterification reaction that produces products with 2′,3′-cyclic phosphate and 5′-OH termini. Only an 85-nucleotide sequence is required for activity of this ribozyme, which can cleave optimally with as little as a single nucleotide 5′ to the site of cleavage.

By joining the 85-nucleotide fragment to other upstream sequences, accurate 3′ ends of heterologous RNA transcripts synthesized *in vitro* can be obtained. This property proved crucial for producing infectious RNAs from cloned DNA copies of the genomes of (–) strand RNA viruses. Ribozymes can also be used to target the cleavage of other RNAs and are therefore being tested as antiviral and antitumor agents.

Kruger, K., P. J. Grabowski, A. J. Zaug, J. Sands, D. E. Gottschling, and T. R. Cech. 1982. Self-splicing RNA: autoexcision and autocyclization of the ribosomal RNA intervening sequence of Tetrahymena. *Cell* **31:**147–157.

Westhof, E., and F. Michel. 1998. Ribozyme architectural diversity made visible. *Science* **282:**251–252.

Whelan, S. P., L. A. Ball, J. N. Barr, and G. T. Wertz. 1995. Efficient recovery of infectious vesicular stomatitis virus entirely from cDNA clones. *Proc. Natl. Acad. Sci. USA* **92:**8388–8392.

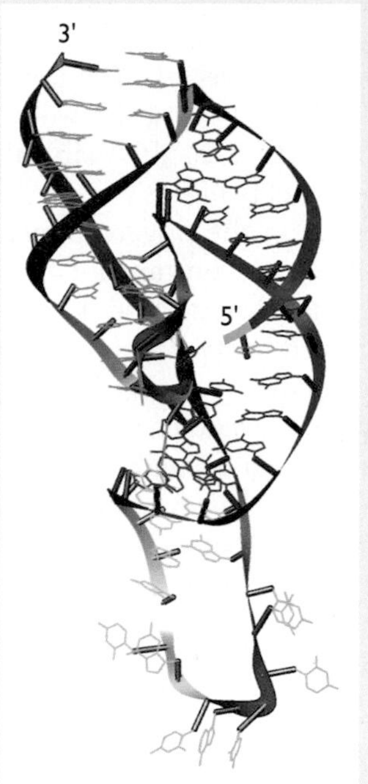

Crystal structure of the hepatitis delta satellite virus ribozyme. The RNA backbone is shown as a ribbon. The two helical stacks are shown in red and blue, and unpaired nucleotides are grey. The 5′ nucleotide, which marks the active site, is green. Produced from Protein Data Bank file 1cx0.

Polioviral $3D^{pol}$ cannot by itself associate with membranes but is brought to the replication complex by binding to protein 3AB. When the membrane association of this protein is disrupted by amino acid changes, viral RNA synthesis is inhibited. The hydrophobic domain of 3AB can be substituted for the C-terminal transmembrane segment of NS5b with little effect on RNA polymerase activity, indicating that membrane association is the sole function of this sequence.

Origins of Diversity in RNA Virus Genomes

Misincorporation of Nucleotides

All nucleic acid polymerases insert incorrect nucleotides during chain elongation. DNA-directed DNA polymerases have **proofreading** capabilities in the form of exonuclease activities that can correct such mistakes. RNA-dependent RNA polymerases do not possess proofreading ability. The result is that error frequencies in RNA replication can be as high as one misincorporation per 10^3 to 10^5 nucleotides polymerized, whereas the frequency of errors in DNA replication is about 10,000-fold lower. Many of these polymerization errors cause lethal amino acid changes, while other mutations may appear in the genomes of infectious virions. This phenomenon has led to the realization that RNA virus populations are **quasispecies**, or mixtures of many different genome sequences. The errors introduced during RNA replication have important consequences for viral pathogenesis and evolution, as discussed in Volume II, Chapter 10. Because RNA virus stocks are mixtures of genotypically different viruses, specific viral mutants may be isolated directly from such stocks. For example, live attenuated poliovirus vaccine strains are viral mutants that were isolated from an unmutagenized stock of wild-type virus.

Fidelity of copying by RNA-dependent RNA polymerases is determined by how the template, primer, and NTP interact at the active site. The results of structural and genetic studies of picornaviral $3D^{pol}$ have led to a hypothesis about how fidelity is ensured during RNA synthesis. Nucleotide binding occurs in two steps: first, the NTP is bound in such a way that the ribose cannot interact properly with the Asp of motif A and the Asn of motif B (Fig. 6.6). Next, if the NTP is correctly base paired with the template strand, there is a conformational change in the enzyme which reorients the triphosphate and allows phosphoryl transfer to occur. This conformational change requires reorientation of the Asp and Asn residues, which would stabilize the position of the ribose in the binding pocket. This conformational change is thought to be a key fidelity checkpoint for the picornaviral RNA polymerase. This model derives from the structures of 3D^{pol} bound to a template primer and NTP (Fig. 6.7) and the study of an altered poliovirus 3D^{pol} with higher fidelity than the wild-type enzyme. The increased fidelity of this enzyme, which has a single amino acid change in the fingers domain, is due to a change in the equilibrium constant for the conformational change. Although this amino acid is remote from the active site, it is involved in hydrogen bonding to motif A, which, as discussed above, is important in holding the NTP in a catalytically appropriate conformation. Of great interest is the observation that a similar interaction between fingers and motif A can be observed in RNA polymerases from a wide variety of viruses. This mechanism of regulating fidelity may therefore be conserved in all RNA-dependent RNA polymerases.

These studies also provide mechanistic information on how ribavirin, an antiviral compound, causes lethal mutagenesis of picornaviruses. The structure of foot-and-mouth disease virus 3D^{pol} bound to ribavirin shows the compound positioned in the active site of the enzyme, adjacent to the 3′ end of the primer. The ribose of ribavirin is bound in the pocket, indicating that it has bypassed the fidelity checkpoint and has caused the conformational change that holds the analog in a position ready for catalysis. Therefore, ribavirin is a mutagen because it can bypass the fidelity checkpoint and be inserted into incorrect positions in newly synthesized RNA molecules.

The RNA polymerase of members of the *Nidovirales* (Box 6.10) may have evolved to allow faithful replication of the large (up to 32 kb) RNA genomes. The RNA synthesis machinery includes proteins not found in other RNA viruses, such as a U-specific endonuclease. It has been suggested that these proteins are part of a repair mechanism similar to the proofreading activity associated with DNA replication.

Segment Reassortment and RNA Recombination

Reassortment is the exchange of entire RNA molecules between genetically related viruses with segmented genomes. In cells coinfected with two different influenza viruses, the eight genome segments of each virus replicate. When new progeny viruses are assembled, they can package RNA segments from **either** parental virus. Because reassortment involves the simple exchange of RNA segments, it can occur at high frequencies.

In contrast to reassortment, recombination is the exchange of nucleotide sequences among different genomic RNA molecules (Fig. 6.27A). Recombination, a feature of many RNA viruses, is an important mechanism for producing new genomes with selective growth advantages. This process has shaped the RNA virus world by rearranging genomes or moving functional parts of RNA molecules among different viruses. RNA recombination was first discovered in cells infected with poliovirus and was subsequently found to occur with other (+) and (–) strand RNA viruses. The frequency of recombination can be relatively high: it has been estimated that 10 to 20% of polioviral genomic RNA molecules recombine in one growth cycle. RNA recombination is not limited to the research laboratory. For example, recombinants among the three serotypes of poliovirus are readily isolated from the feces of individuals immunized with the live (Sabin) vaccine. Such recombinants may possess an improved ability to replicate in the human alimentary tract and have a selective advantage over the parental viruses. Recombination can also lead to the production of pathogenic viruses (Box 6.11).

Polioviral recombination is predominantly **base pairing dependent**: it occurs between nucleotide sequences that have a high percentage of nucleotide identity (Fig. 6.27B). Other viral genomes undergo **base-pairing-independent** recombination between very different nucleotide sequences. RNA recombination is coupled with the process of genomic RNA replication: it occurs by template exchange during (–) strand synthesis, as first demonstrated in poliovirus-infected cells. The RNA polymerase first copies the 3′ end of one parental (+) strand and then exchanges one template for another at the corresponding position on a second parental (+) strand (Fig. 6.27B). Template exchange in poliovirus-infected cells occurs predominantly during (–) strand synthesis, presumably because the concentration of (+) strands is 100-fold higher than that of (–) strands. This template exchange mechanism of recombination is also known as **copy-choice**. The exact mechanism of template exchange is not known, but it might be triggered by pausing of the polymerase during chain elongation (Fig. 6.27).

If the RNA polymerase skips sequences during template switching, deletions may occur. These RNAs can replicate if they contain the appropriate signals for the initiation of RNA synthesis. Subgenomic RNAs replicate more rapidly than full-length RNA and therefore compete for the components of the RNA synthesis machinery. Because of these properties, they are called **defective interfering RNAs**. Such RNAs can be packaged into viral particles only in the presence of a **helper virus** that provides viral proteins.

Defective interfering particles accumulate during the replication of both (+) and (–) strand RNA viruses.

BOX 6.10

BACKGROUND

Synthesis of nested subgenomic mRNAs

An unusual pattern of mRNA synthesis occurs in cells infected with members of the families *Coronaviridae* and *Arteriviridae*, in which subgenomic mRNAs that form a 3′-coterminal nested set with the viral genome are synthesized (see figure). These viral families were combined into the order *Nidovirales* to reflect this property (*nidus* is Latin for nest).

The subgenomic mRNAs of these viruses are composed of a leader and a body that are synthesized from noncontiguous sequences at the 5′ and 3′ ends, respectively, of the viral (+) strand genome. The leader and body are separated by a conserved junction sequence encoded both at the 3′ end of the leader and at the 5′ end of the mRNA body. Subgenome-length (–) strands are produced when the template loops out as the polymerase completes synthesis of the leader RNA (see figure). These (–) strand subgenome-length RNAs then serve as templates for mRNA synthesis.

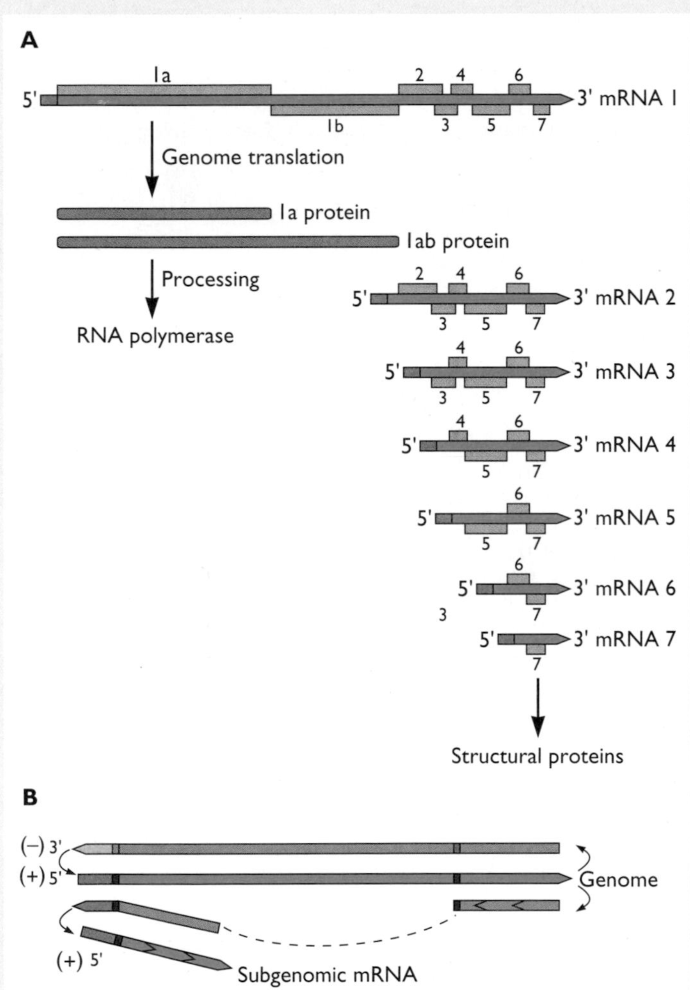

Nidoviral genome organization and expression. (A) Organization of open reading frames. The (+) strand viral RNA is shown at the top, with open reading frames as boxes. The genomic RNA is translated to form polyproteins 1a and 1ab, which are processed to form the RNA polymerase. Structural proteins are encoded by six nested mRNAs that share a common 5′ leader sequence (orange box). **(B)** Model of the synthesis of nested mRNAs. Discontinuous transcription occurs during (–) strand RNA synthesis. Most of the (+) strand template is not copied, probably because it loops out as the polymerase completes synthesis of the leader RNA. The resulting (–) strand RNAs, with leader sequences at the 3′ ends, are then copied to form mRNAs. Adapted from E. J. Snijder et al., *J. Gen. Virol.* **79:**961–979, 1998, with permission.

Production of these viruses requires either a high multiplicity of infection or serial passaging, conditions that are achieved readily in the laboratory but rarely in nature. It is not known whether defective interfering viruses generally play a role in viral pathogenesis. However, some recombination reactions that lead to the appearance of cytopathic bovine viral diarrhea viruses delete viral RNA sequences rather than inserting cellular RNA sequences (Box 6.11). Such deletions create a new protease cleavage site at the N terminus of the NS3 protein, and the defective

RNA recombination

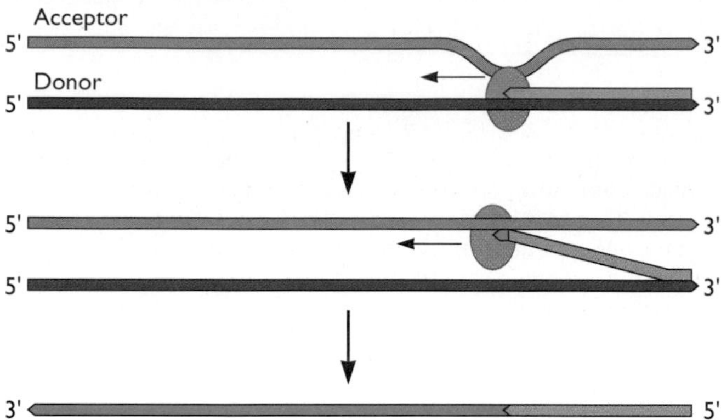

Class 1: Base pairing dependent

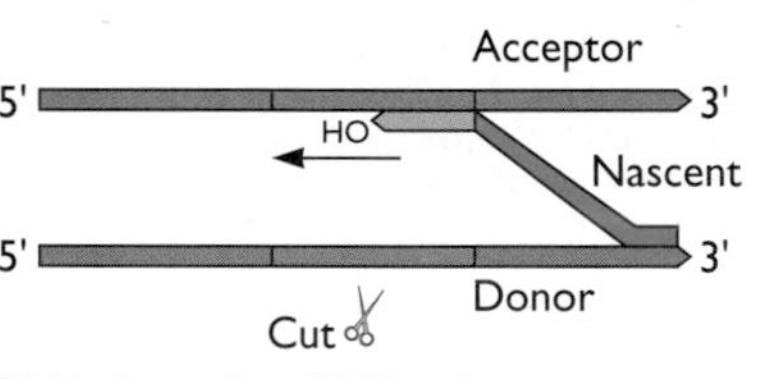

Class 2: Base pairing independent

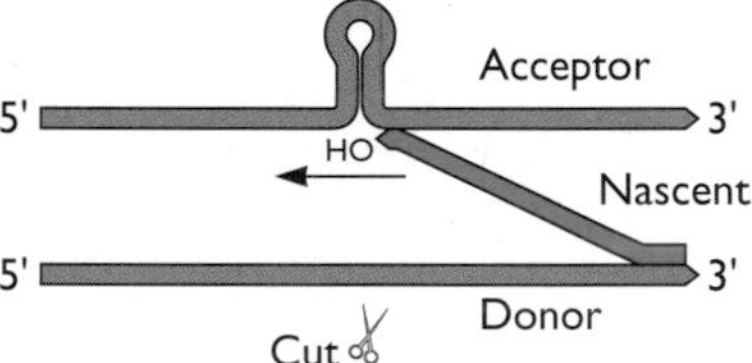

Class 3: Base pairing assisted

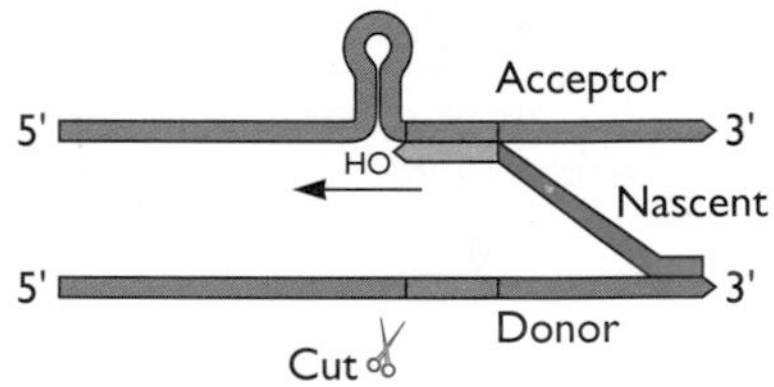

Figure 6.27 RNA recombination. (Top) Schematic representation of RNA recombination occurring during template switching by RNA polymerase, or copy choice. Two parental genomes are shown. The RNA polymerase (purple oval) has copied the 3′ end of the donor genome and is switching to the acceptor genome. The resulting recombinant molecule is shown. **(Bottom)** Three classes of RNA recombination. All three classes require RNA polymerase-mediated template exchange. Events that occur after the exchange are shown by an arrow. The hairpin symbolizes various RNA features required for class 2 and 3 recombination. In base-pairing-dependent recombination, substantial sequence similarity between parental RNAs is required and is the major determinant of recombination. In base-pairing-independent recombination, sequence similarity is not required but may be present. Recombination may be determined by other RNA features, such as RNA polymerase-binding sites, secondary structures, and heteroduplex formation between parental RNAs. Base-pairing-assisted recombination combines features of class 1 and class 2 recombination. Sequence similarity influences the frequency or site of recombination, but additional RNA features are required. Adapted from P. D. Nagy and A. E. Simon, *Virology* **235**:1–9, 1997, with permission.

interfering viruses also cause severe gastrointestinal disease in livestock.

RNA Editing

Diversity in RNA viral genomes is also achieved by RNA editing. Viral mRNAs can be edited either by insertion of a nontemplated nucleotide during synthesis or by alteration of the base after synthesis. Examples of RNA editing have been documented in members of the *Paramyxoviridae* and *Filoviridae* and in hepatitis delta satellite virus. This process is described in Chapter 10.

Perspectives

In a previous edition of this textbook, we wrote, "The crystal structure of polioviral 3D^{pol} hints at the beautiful pictures of other viral RNA polymerases and their associated proteins that will be forthcoming." The structural biologists have made us appear clairvoyant: since then, the three-dimensional structures of many RNA-dependent RNA polymerases have been resolved. These structures underscore the relationship of these enzymes to other nucleic acid polymerases, but also underscore structural differences that have evolved to accommodate the wide diversity of

BOX 6.11

DISCUSSION

RNA recombination leading to the production of pathogenic viruses

A remarkable property of pestiviruses, members of the *Flaviviridae*, is that RNA recombination produces viruses that cause disease. Bovine viral diarrhea virus causes a usually fatal gastrointestinal disease. Infection of a fetus with this virus during the first trimester is noncytopathic, but RNA recombination produces a cytopathic virus that causes severe gastrointestinal disease after the animal is born.

Pathogenicity of bovine viral diarrhea virus is associated with the synthesis of a nonstructural protein, NS3, encoded by the recombinant cytopathic virus (see figure). The NS3 protein cannot be produced in cells infected by the noncytopathic parental virus because its precursor, the NS2-3 protein, is not proteolytically processed. In contrast, NS3 **is** synthesized in cells infected by cytopathic bovine viral diarrhea virus because base-pairing-independent RNA recombination reactions add an extra protease cleavage site in the viral polyprotein, precisely at the N terminus of the NS3 protein (see figure). This cleavage site can be created in several ways. One of the most frequent is insertion of a cellular RNA sequence coding for ubiquitin, which targets cellular proteins to a degradative pathway. Insertion of ubiquitin at the N terminus of NS3 permits cleavage of NS2-3 by any member of a widespread family of cellular proteases. This recombination event provides a selective advantage because pathogenic viruses outgrow nonpathogenic viruses. Why cytopathogenicity is associated with release of the NS3 protein, which is thought to be part of the enzymatic system for genomic RNA replication, is not known.

Retroviruses acquire cellular genes by recombination, and the resulting viruses can have lethal disease potential, as described in Volume II, Chapters 7 and 10.

Pathogenicity of bovine viral diarrhea virus is associated with production of the NS3 protein. Two cytopathic viruses, Osloss and CP1, in which the ubiquitin sequence (UCH) has been inserted at different sites, are shown. In Osloss, UCH has been inserted into the NS2-3 precursor, and NS3 is produced. In CP1, a duplication has also occurred such that an additional copy of NS3 is present after the UCH sequence. Adapted from D. M. Knipe et al. (ed.), *Fields Virology*, 4th ed. (Lippincott Williams & Wilkins, Philadelphia, PA, 2001), with permission.

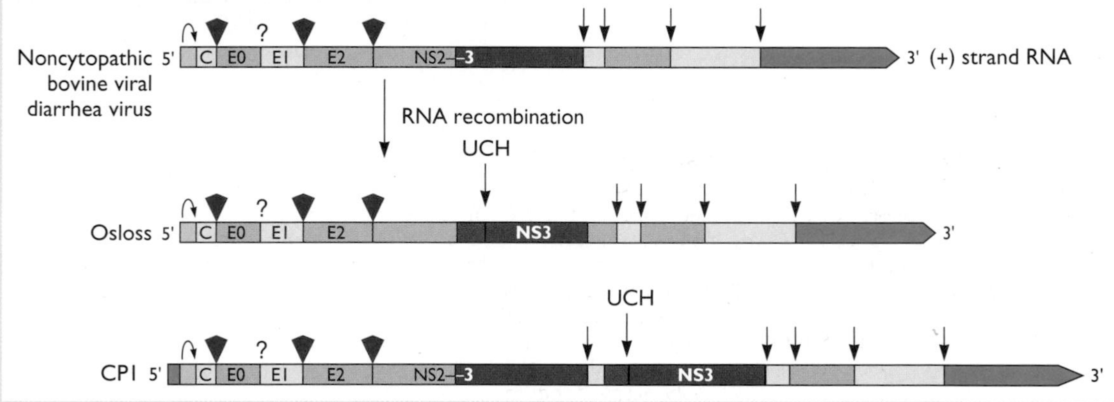

RNA genome configurations. We expect that forthcoming structures will detail the conformational movements that occur during the switch between initiation and elongation, highlighting the changes that occur as the polymerase moves from an open to a closed conformation. Our understanding of how RNA polymerases alter between synthesis of mRNA and genome replication is largely hypothetical; additional structures should provide insight into the mechanism of this crucial aspect of the infectious cycle.

The nature and function of host proteins that are required for viral RNA synthesis remain obscure, as does the reason why RNA synthesis occurs in certain cellular compartments. Some kinds of membranes are the sites of RNA replication for different viruses, but the reasons for the specificity are unknown. Is the membrane simply a platform for RNA synthesis? Viral RNA-directed RNA synthesis is a unique process and therefore an excellent target for antiviral intervention. As our understanding of RNA catalysis improves, new ways of limiting viral infections will be revealed.

References

Reviews

Barr, J. N., S. P. Whelan, and G. W. Wertz. 2002. Transcriptional control of the RNA-dependent RNA polymerase of vesicular stomatitis virus. *Biochim. Biophys. Acta* **1577:**337–353.

Belov, G. A., and E. Ehrenfeld. 2007. Involvement of cellular membrane traffic proteins in poliovirus replication. *Cell Cycle* **6:**36–38.

Brierley, I., S. Pennell, and R. J. Gilbert. 2007. Viral RNA pseudoknots: versatile motifs in gene expression and replication. *Nat. Rev. Microbiol.* **5:**598–610.

Cowton, V. M., D. R. McGivern, and R. Fearns. 2006. Unravelling the complexities of respiratory syncytial virus RNA synthesis. *J. Gen. Virol.* **87:**1805–1821.

Ferrer-Orta, C., A. Arias, C. Escarmis, and M. Verdaguer. 2006. A comparison of viral RNA-dependent RNA polymerases. *Curr. Opin. Struct. Biol.* **16:**27–34.

Horikami, S. M., and S. A. Moyer. 1995. Structure, transcription, and replication of measles virus. *Curr. Top. Microbiol. Immunol.* **191:**35–50.

Kääriäinen, L., and T. Ahola. 2002. Functions of alphavirus nonstructural proteins in RNA replication. *Prog. Nucleic Acid Res. Mol. Biol.* **71:**187–222.

Nagy, P. D., and A. E. Simon. 1997. New insights into the mechanisms of RNA recombination. *Virology* **235:**1–9.

Neumann, G., G. G. Brownlee, E. Fodor, and Y. Kawaoka. 2004. Orthomyxovirus replication, transcription, and polyadenylation. *Curr. Top. Microbiol. Immunol.* **283:**121–43.

Nowakowski, J., and I. Tinoco. 1997. RNA structure and stability. *Semin. Virol.* **8:**153–165.

Paul, A. V. 2002. Possible unifying mechanism of picornavirus genome replication, p. 227–246. *In* B. L. Semler and E. Wimmer (ed.), *Molecular Biology of Picornaviruses.* ASM Press, Washington, DC.

Salonen, A., T. Ahola, and L. Kääriäinen. 2005. Viral RNA replication in association with cellular membranes. *Curr. Top. Microbiol. Immunol.* **285:**139–173.

Sawicki, S. G., D. L. Sawicki, and S. G. Siddell. 2007. A contemporary view of coronavirus transcription. *J. Virol.* **81:**20–29.

Van Dijk, A. A., E. V. Makeyev, and D. H. Bamford. 2004. Initiation of viral RNA-dependent RNA polymerization. *J. Gen. Virol.* **85:**1077–1093.

Vulliémoz, D., and L. Roux. 2001. "Rule of six": how does the Sendai virus RNA polymerase keep count? *J. Virol.* **75:**4506–4518.

Papers of Special Interest

Ago, H., T. Adachi, A. Yoshida, M. Yamamoto, N. Habuka, K. Yatsunami, and M. Miyano. 1999. Crystal structure of the RNA-dependent RNA polymerase of hepatitis C virus. *Structure* **7:**1417–1426.

Andino, R., G. E. Rieckhof, P. L. Achacoso, and D. Baltimore. 1993. Poliovirus RNA synthesis utilizes an RNP complex formed around the 5′-end of viral RNA. *EMBO J.* **12:**3587–3598.

Arnold, J. J., M. Vignuzzi, J. K. Stone, R. Andino, and C. E. Cameron. 2005. Remote site control of an active site fidelity checkpoint in a viral RNA-dependent RNA polymerase. *J. Biol. Chem.* **280:**25706–25716.

Baltimore, D., and R. M. Franklin. 1963. Properties of the mengovirus and poliovirus RNA polymerases. *Cold Spring Harbor Symp. Quant. Biol.* **28:**105–108.

Barik, S. 1992. Transcription of human respiratory syncytial virus genome RNA in vitro: requirement of cellular factor(s). *J. Virol.* **66:**6813–6818.

Barton, D. J., and J. B. Flanegan. 1997. Synchronous replication of poliovirus RNA: initiation of negative-strand RNA synthesis requires the guanidine-inhibited activity of protein 2C. *J. Virol.* **71:**8482–8489.

Barton, D. J., B. J. Morasco, and J. B. Flanegan. 1999. Translating ribosomes inhibit poliovirus negative-strand RNA synthesis. *J. Virol.* **73:**10104–10112.

Beaton, A. R., and R. M. Krug. 1986. Transcription antitermination during influenza viral template RNA synthesis requires the nucleocapsid protein and the absence of a 5′ capped end. *Proc. Natl. Acad. Sci. USA* **83:**6282–6286.

Brown, D., and L. Gold. 1996. RNA replication by Q beta replicase: a working model. *Proc. Natl. Acad. Sci. USA* **93:**11558–11562.

Butcher, S. J., J. M. Grimes, E. V. Makeyev, D. H. Bamford, and D. I. Stuart. 2001. A mechanism for initiating RNA-dependent RNA polymerization. *Nature* **410:**235–240.

Charini, W. A., S. Todd, G. A. Gutman, and B. L. Semler. 1994. Transduction of a human RNA sequence by poliovirus. *J. Virol.* **68:**6547–6552.

Cianci, C., L. Tiley, and M. Krystal. 1995. Differential activation of the influenza virus polymerase via template RNA binding. *J. Virol.* **69:**3995–3999.

Cuconati, A., A. Molla, and E. Wimmer. 1998. Brefeldin A inhibits cell-free, de novo synthesis of poliovirus. *J. Virol.* **72:**6456–6464.

D'Abramo, C. M., J. Deval, C. E. Cameron, L. Cellai, and M. Götte. 2006. Control of template positioning during de novo initiation of RNA synthesis by the bovine viral diarrhea virus NS5B polymerase. *J. Biol. Chem.* **281:**24991–24998.

Das, T., M. Mathur, A. K. Gupta, G. M. Janssen, and A. K. Banerjee. 1998. RNA polymerase of vesicular stomatitis virus specifically associates with translation elongation factor-1 alphabetagamma for its activity. *Proc. Natl. Acad. Sci. USA* **95:**1449–1454.

Dasgupta, A., M. H. Baron, and D. Baltimore. 1979. Poliovirus replicase: a soluble enzyme able to initiate copying of poliovirus RNA. *Proc. Natl. Acad. Sci. USA* **76:**2679–2683.

Deval, J., C. M. D'Abramo, Z. Zhao, S. McCormick, D. Coutsinos, S. Hess, M. Kvaratskhelia, and M. Götte. 2007. High resolution footprinting of the hepatitis C virus polymerase NS5B in complex with RNA. *J. Biol. Chem.* **282:**16907–16916.

Ferrer-Orta, C., A. Arias, R. Perez-Luque, C. Escarmis, E. Domingo, and N. Verdaguer. 2004. Structure of foot-and-mouth disease virus RNA-dependent RNA polymerase and its complex with a template-primer RNA. *J. Biol. Chem.* **279:**47212–47221.

Ferrer-Orta, C., A. Arias, R. Agudo, R. Pérez-Luque, C. Escarmis, E. Domingo, and N. Verdaguer. 2006. The structure of a protein primer-polymerase complex in the initiation of genome replication. *EMBO J.* **25:**880–888.

Ferrer-Orta, C., A. Arias, R. Perez-Luque, C. Escarmis, E. Domingo, and N. Verdaguer. 2007. Sequential structures provide insights into the fidelity of RNA replication. *Proc. Natl. Acad. Sci. USA* **104:**9463–9468.

Finke, S., and K. K. Conzelmann. 1997. Ambisense gene expression from recombinant rabies virus: random packaging of positive- and negative-strand ribonucleoprotein complexes into rabies virions. *J. Virol.* **71:**7281–7288.

Gamarnik, A. V., and R. Andino. 1998. Switch from translation to RNA replication in a positive-stranded RNA virus. *Genes Dev.* **12:**2293–2304.

Gamarnik, A. V., and R. Andino. 1996. Replication of poliovirus in Xenopus oocytes requires two human factors. *EMBO J.* **15:**5988–5998.

Giachetti, C., and B. L. Semler. 1991. Role of a viral membrane polypeptide in strand-specific initiation of poliovirus RNA synthesis. *J. Virol.* **65:**2647–2654.

Gupta, S., B. P. De, J. A. Drazba, and A. K. Banerjee. 1998. Involvement of actin microfilaments in the replication of human parainfluenza virus type 3. *J. Virol.* **72:**2655–2662.

Gupta, A. K., D. Shaji, and A. K. Banerjee. 2003. Identification of a novel tripartite complex involved in replication of vesicular stomatitis virus genome RNA. *J. Virol.* **77:**732–738.

Hagen, M., T. D. Chung, J. A. Butcher, and M. Krystal. 1994. Recombinant influenza virus polymerase: requirement of both 5′ and 3′ viral ends for endonuclease activity. *J. Virol.* **68:**1509–1515.

Hansen, J. L., A. M. Long, and S. C. Schultz. 1997. Structure of the RNA-dependent RNA polymerase of poliovirus. *Structure* **5:**1109–1122.

Herold, J., and R. Andino. 2001. Poliovirus RNA replication requires genome circularization through a protein-protein bridge. *Mol. Cell* **7:**581–591.

Holmes, D. E., and S. A. Moyer. 2002. The phosphoprotein (P) binding site resides in the N terminus of the L polymerase subunit of Sendai virus. *J. Virol.* **76:**3078–3083.

Horikami, S. M., S. Smallwood, and S. A. Moyer. 1996. The Sendai virus V protein interacts with the NP protein to regulate viral genome RNA replication. *Virology* **222:**383–390.

Houben, K., D. Marion, N. Tarbouriech, R. W. Ruigrok, and L. Blanchard. 2007. Interaction of the C-terminal domains of Sendai virus N and P proteins: comparison of polymerase-nucleocapsid interactions within the paramyxovirus family. *J. Virol.* **81:**6807–6816.

Jackson, W. T., T. H. Giddings Jr., M. P. Taylor, S. Mulinyawe, M. Rabinovitch, R. R. Kopito, and K. Kirkegaard. 2005. Subversion of cellular autophagosomal machinery by RNA viruses. *PLoS Biol.* **3:** e156.

Kirkegaard, K., and D. Baltimore. 1986. The mechanism of RNA recombination in poliovirus. *Cell* **47:**433–443.

Khromykh, A. A., H. Meka, K. J. Guyatt, and E. G. Westaway. 2001. Essential role of cyclization sequences in flavivirus RNA replication. *J. Virol.* **75:**6719–6728.

Klumpp, K., R. W. Ruigrok, and F. Baudin. 1997. Roles of the influenza virus polymerase and nucleoprotein in forming a functional RNP structure. *EMBO J.* **16:**1248–1257.

Kopek, B. G., G. Perkins, D. J. Miller, M. H. Ellisman, and P. Ahlquist. 2007. Three-dimensional analysis of a viral RNA replication complex reveals a virus-induced mini-organelle. *PLoS Biol.* **5:**e220.

Lawton, J. A., M. K. Estes, and B. V. Prasad. 1997. Three-dimensional visualization of mRNA release from actively transcribing rotavirus particles. *Nat. Struct. Biol.* **4:**118–121.

Lawton, J. A., C. Q. Zeng, S. K. Mukherjee, J. Cohen, M. K. Estes, and B. V. Prasad. 1997. Three-dimensional structural analysis of recombinant rotavirus-like particles with intact and amino-terminal-deleted VP2: implications for the architecture of the VP2 capsid layer. *J. Virol.* **71:**7353–7360.

Lemm, J. A., A. Bergqvist, C. M. Read, and C. M. Rice. 1998. Template-dependent initiation of Sindbis virus RNA replication in vitro. *J. Virol.* **72:**6546–6553.

Lemm, J. A., T. Rumenapf, E. G. Strauss, J. H. Strauss, and C. M. Rice. 1994. Polypeptide requirements for assembly of functional Sindbis virus replication complexes: a model for the temporal regulation of minus- and plus-strand RNA synthesis. *EMBO J.* **13:**2925–2934.

Li, M. L., B. C. Ramirez, and R. M. Krug. 1998. RNA-dependent activation of primer RNA production by influenza virus polymerase: different regions of the same protein subunit constitute the two required RNA-binding sites. *EMBO J.* **17:**5844–5852.

Li, M. L., P. Rao, and R. M. Krug. 2001. The active sites of the influenza cap-dependent endonuclease are on different polymerase subunits. *EMBO J.* **20:**2078–2086.

Li, T., and A. K. Pattnaik. 1999. Overlapping signals for transcription and replication at the 3′ terminus of the vesicular stomatitis virus genome. *J. Virol.* **73:**444–452.

Lohmann, V., F. Korner, J. Koch, U. Herian, L. Theilmann, and R. Bartenschlager. 1999. Replication of subgenomic hepatitis C virus RNAs in a hepatoma cell line. *Science* **285:**110–113.

Lyle, J. M., E. Bullitt, K. Bienz, and K. Kirkegaard. 2002. Visualization and functional analysis of RNA-dependent RNA polymerase lattices. *Science* **296:**2218–2222.

Marcotte, L. L., A. B. Wass, D. W. Gohara, H. B. Pathak, J. J. Arnold, D. J. Filman, C. E. Cameron, and J. M. Hogle. 2007. Crystal structure of poliovirus 3CD protein: virally encoded protease and precursor to the RNA-dependent RNA polymerase. *J. Virol.* **81:**3583–3596.

McKnight, K. L., and S. M. Lemon. 1998. The rhinovirus type 14 genome contains an internally located RNA structure that is required for viral replication. *RNA* **12:**1569–1584.

Molla, A., A. V. Paul, and E. Wimmer. 1991. Cell-free, de novo synthesis of poliovirus. *Science* **254:**1647–1651.

Momose, F., T. Naito, K. Yano, S. Sugimoto, Y. Morikawa, and K. Nagata. 2002. Identification of Hsp90 as a stimulatory host factor involved in influenza virus RNA synthesis. *J. Biol. Chem.* **277:**45306–45314.

Mullin, A. E., R. M. Dalton, M. J. Amorim, D. Elton, and P. Digard. 2004. Increased amounts of the influenza virus nucleoprotein do not promote higher levels of viral genome replication. *J. Gen. Virol.* **85:**3689–3698.

Palese, P., M. B. Ritchey, and J. L. Schulman. 1977. P1 and P3 proteins of influenza virus are required for complementary RNA synthesis. *J. Virol.* **21:**1187–1195.

Pan, J., V. N. Vakharia, and Y. J. Tao. 2007. The structure of a birnavirus polymerase reveals a distinct active site topology. *Proc. Natl. Acad. Sci. USA* **104:**7385–7390.

Parsley, T. B., J. S. Towner, L. B. Blyn, E. Ehrenfeld, and B. L. Semler. 1997. Poly(rC) binding protein 2 forms a ternary complex with the 5′-terminal sequences of poliovirus RNA and the viral 3CD proteinase. *RNA* **3:**1124–1134.

Pattnaik, A. K., L. Hwang, T. Li, N. Englund, M. Mathur, T. Das, and A. K. Banerjee. 1997. Phosphorylation within the amino-terminal acidic domain I of the phosphoprotein of vesicular stomatitis virus is required for transcription but not for replication. *J. Virol.* **71:**8167–8175.

Paul, A. V., J. H. van Boom, D. Filippov, and E. Wimmer. 1998. Protein-primed RNA synthesis by purified poliovirus RNA polymerase. *Nature* **393:**280–284.

Perera, R., S. Daijogo, B. L. Walter, J. H. Nguyen, and B. L. Semler. 2007. Cellular protein modification by poliovirus: the two faces of poly(rC)-binding protein. *J. Virol.* **81:**8919–3892.

Plotch, S. J., M. Bouloy, I. Ulmanen, and R. M. Krug. 1981. A unique cap(m7GpppXm)-dependent influenza virion endonuclease cleaves capped RNAs to generate the primers that initiate viral RNA transcription. *Cell* **23:**847–858.

Prasad, B. V., R. Rothnagel, C. Q. Zeng, J. Jakana, J. A. Lawton, W. Chiu, and M. K. Estes. 1996. Visualization of ordered genomic RNA and localization of transcriptional complexes in rotavirus. *Nature* **382:**471–473.

Pritlove, D. C., L. L. Poon, E. Fodor, J. Sharps, and G. G. Brownlee. 1998. Polyadenylation of influenza virus mRNA transcribed in vitro from model virion RNA templates: requirement for 5′ conserved sequences. *J. Virol.* **72:**1280–1286.

Qanungo, K. R., D. Shaji, M. Mathur, and A. K. Banerjee. 2004. Two RNA polymerase complexes from vesicular stomatitis virus-infected cells that carry out transcription and replication of genome RNA. *Proc. Natl. Acad. Sci. USA* **101:**5952–5957.

Rao, P., W. Yuan, and R. M. Krug. 2003. Crucial role of CA cleavage sites in the cap-snatching mechanism for initiating viral mRNA synthesis. *EMBO J.* **22:**1188–1198.

Rust, R. C., L. Landmann, R. Gosert, B. L. Tang, W. Hong, H. P. Hauri, D. Egger, and K. Bienz. 2001. Cellular COPII proteins are involved in production of the vesicles that form the poliovirus replication complex. *J. Virol.* **75:**9808–9818.

Sawicki, D. L., and S. G. Sawicki. 1994. Alphavirus positive and negative strand RNA synthesis and the role of polyproteins in formation of viral replication complexes. *Arch. Virol. Suppl.* **9:**393–405.

Silvestri, L. S., Z. F. Taraporewala, and J. T. Patton. 2004. Rotavirus replication: plus-sense templates for double-stranded RNA synthesis are made in viroplasms. *J Virol.* **78:**7763–7774.

Stillman, E. A., and M. A. Whitt. 1998. The length and sequence composition of vesicular stomatitis virus intergenic regions affect mRNA levels and the site of transcript initiation. *J. Virol.* **72:**5565–5572.

Tao, Y., D. L. Farsetta, M. L. Nibert, and S. C. Harrison. 2002. RNA synthesis in a cage—structural studies of reovirus polymerase lambda3. *Cell* **111:**733–745.

Testa, D., P. K. Chanda, and A. K. Banerjee. 1980. Unique mode of transcription in vitro by vesicular stomatitis virus. *Cell* **21:**267–275.

Thompson, A. A., R. A. Albertini, and O. B. Peersen. 2006. Stabilization of poliovirus polymerase by NTP binding and fingers-thumb interactions. *J. Mol. Biol.* **366:**1459–1474.

Todd, S., J. S. Towner, D. M. Brown, and B. L. Semler. 1997. Replication-competent picornaviruses with complete genomic RNA 3′ noncoding region deletions. *J. Virol.* **71:**8868–8874.

Tolskaya, E. A., L. I. Romanova, M. S. Kolesnikova, A. P. Gmyl, A. E. Gorbalenya, and V. I. Agol. 1994. Genetic studies on the poliovirus 2C protein, an NTPase: a plausible mechanism of guanidine effect on the 2C function and evidence for the importance of 2C oligomerization. *J. Mol. Biol.* **236:**1310–1323.

Tomar, S., R. W. Hardy, J. L. Smith, and R. J. Kuhn. 2006. Catalytic core of alphavirus nonstructural protein nsP4 possesses terminal adenylyltransferase activity. *J. Virol.* **80:**9962–9969.

Towner, J. S., T. V. Ho, and B. L. Semler. 1996. Determinants of membrane association for poliovirus protein 3AB. *J. Biol. Chem.* **271:**26810–26818.

Walter, B. L., T. B. Parsley, E. Ehrenfeld, and B. L. Semler. 2002. Distinct poly(rC) binding protein KH domain determinants for poliovirus translation initiation and viral RNA replication. *J. Virol.* **76:**12008–12022.

Wang, Y. F., S. G. Sawicki, and D. L. Sawicki. 1994. Alphavirus nsP3 functions to form replication complexes transcribing negative-strand RNA. *J. Virol.* **68:**6466–6475.

Whelan, S. P., and G. W. Wertz. 2002. Transcription and replication initiate at separate sites on the vesicular stomatitis virus genome. *Proc. Natl. Acad. Sci. USA* **99:**9178–9183.

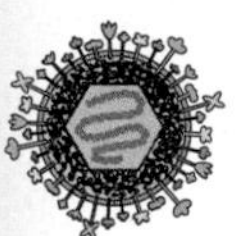

7

Reverse Transcription and Integration

> *"One can't believe impossible things,"*
> *said Alice.*
> *"I dare say you haven't had much practice," said the Queen. "Why, sometimes I've believed as many as six impossible things before breakfast."*
>
> Lewis Carroll
> *Alice in Wonderland*

Retroviral Reverse Transcription

Discovery

In 1970, back-to-back reports in the scientific journal *Nature* from the laboratories of Howard Temin and David Baltimore provided the first concrete evidence for the existence of an RNA-directed DNA polymerase activity in retrovirus particles. The pathways to this discovery were quite different in the two laboratories. In Temin's case, the discovery came about through attempts to understand how this group of RNA-containing viruses could permanently alter the heredity of cells, as they do in the process of oncogenic transformation. Temin proposed that retroviral RNA genomes become integrated into the host cell's chromatin in a DNA form, an idea supported by the observation that purified cellular DNA polymerases can use RNA as a template under certain reaction conditions in vitro. Furthermore, studies of bacterial viruses such as bacteriophage lambda had established a precedent for viral DNA integration into host DNA (Box 7.1). However, with the technology available at the time, it was a difficult hypothesis to test, and attempts by Temin and others to demonstrate the existence of such a phenomenon in infected cells were generally met with skepticism. Baltimore's entrée into the problem of reverse transcription came from his interest in virion-associated polymerases, in particular one that he had just discovered to be present in vesicular stomatitis virions, a virus with a (−) strand RNA genome. It occurred to Baltimore and Temin independently that retrovirus particles might also contain the sought-after RNA-dependent DNA polymerase. As subsequent experiments showed, this was indeed the case, and the retrovirus particles themselves yielded up the enzyme activity that had earlier eluded Temin. In 1975, Temin and Baltimore were awarded the Nobel Prize in physiology or medicine for their independent discoveries of retroviral reverse transcriptase (RT).

BOX 7.1 DISCUSSION

Bacteriophage lambda, a paradigm for the joining of viral and host DNAs

In 1962, Allan Campbell proposed an elegant, but at the time revolutionary, model for site-specific integration of DNA of the bacteriophage lambda into chromosome of its host, *Escherichia coli*. The model was deduced from the fact that different linkage maps could be constructed for viral genomes at different stages in its life cycle. One linkage map, that of the prophage, was obtained from the study of lysogenic bacteria. A different linkage map was obtained by measuring recombination frequencies in the phage yield from infected cells (see part A of figure).

Campbell proposed that these unique features could be explained by a model for integration in which the incoming, linear double-stranded DNA phage genome must first circularize. Subsequent recombination between a specific, internal sequence in the phage genome (called *attP*) and a specific sequence in the bacterial chromosome (called *attB*) would produce an integrated viral genome, with a linkage map that was a circular permutation of that of the linear phage genome, as had been observed (see part B of figure).

Induction of the integrated provirus and subsequent viral DNA replication would proceed by a reversal of this site-specific integration reaction, a process called excision. The model explained how abnormal excision (arrows a and b) could give rise to the observed rare, specialized transducing phages that lacked certain viral genes and carried either the *gal* or *bio* genes of the host. These were the first cellular "cloned" DNAs to be identified.

Although this model seems obvious today, it was not obvious in the 1960s. An alternative model in which the linear viral DNA was attached by a partial binding or "synapse" with the bacterial chromosome was favored by a number of investigators. However, shortly after Campbell's elaboration of his model, circular molecules of lambda phage DNA were detected in infected cells, and the linear DNA extracted from purified virus particles was found to possess short complementary single-strand extensions, "cohesive ends," that could promote circle formation. Other predictions of the model were also validated in several laboratories, and viral and cellular proteins that catalyzed integration and excision were identified.

Lambda DNA integration remains an important paradigm for understanding the molecular mechanisms of DNA recombination and the parameters that influence the joining of viral and host DNAs.

Campbell, A. M. 1962. Episomes. *Adv. Genet.* **11**:101–145.

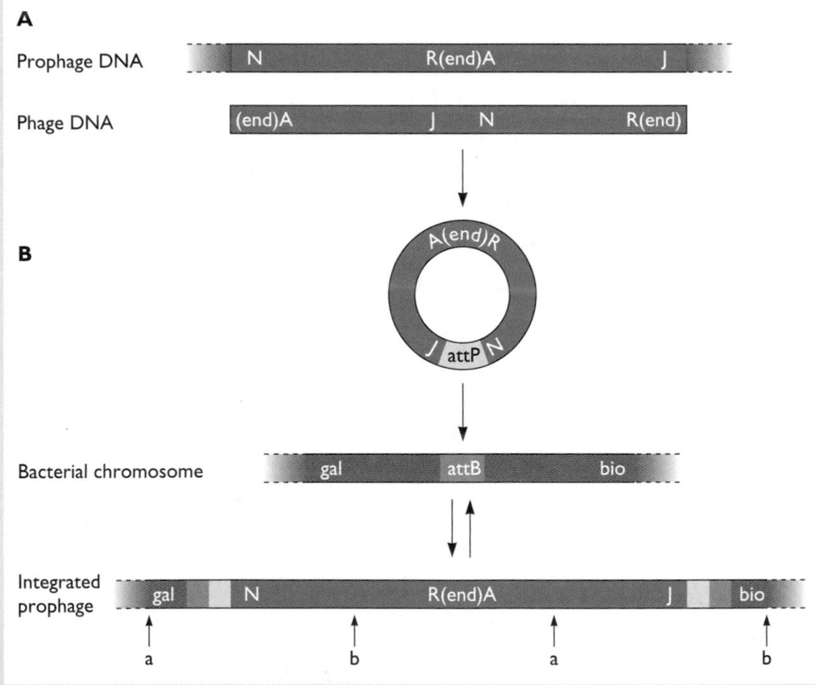

Impact

The immediate impact of the discovery of RT was to amend the then-accepted central dogma of molecular biology, that the transfer of genetic information is unidirectional: DNA → RNA → protein. It was now apparent that there could also be a "retrograde" flow of information from RNA to DNA, and the name **retroviruses** eventually came to replace the earlier designation of RNA tumor viruses. In the years following this revision of dogma, many additional reverse transcription reactions have been discovered.

Furthermore, as Temin hypothesized, study of RT has contributed to our understanding of cancer. As described in Volume II, Chapter 7, the study of oncogenic retroviruses has provided a framework for current concepts of the genetic basis of this disease. Study of reverse transcription (and integration) has also allowed us to understand the persistence of retroviral infections and aspects of the pathogenesis of acquired immunodeficiency syndrome (AIDS) that is caused by the human immunodeficiency virus. Finally, RT itself, first purified from virions and now produced in bacteria, has become an indispensable tool in molecular biology, for example allowing experimentalists to capture cellular messenger RNAs (mRNAs) as complementary DNAs (cDNAs), which can then be amplified, cloned, and expressed by well-established methodologies. For such reasons, we devote an entire chapter to these very important reactions.

The Pathways of Reverse Transcription

Significant insight into the mechanism of reverse transcription can be obtained by comparing the amino acid sequences of RTs with those of other enzymes that catalyze similar reactions. For example, RTs share (with the RNA and DNA polymerases of both prokaryotes and eukaryotes) certain sequence motifs in regions known to include critical active-site residues (see Fig. 6.4). It is hypothesized, therefore, that these enzymes employ similar catalytic mechanisms for nucleic acid polymerization reactions. Like DNA polymerases, viral RTs cannot initiate DNA synthesis *de novo*, but require a specific primer. In this chapter, we provide a detailed description of priming and reverse transcription for retroviral and hepadnaviral enzymes. But it should be noted that even as arcane and distinct from each other as these two systems may appear, they do not exhaust the repertoire for reverse transcription reactions that have evolved in nature. A wide variety of primers, as well as sites and modes of initiation, are used by the RTs of other retroelements.

Much of what has been learned about reverse transcription in retroviruses comes from the analysis of intermediates in the reaction pathway that have been identified in extracts from infected cells. Reverse transcription intermediates have also been detected in **endogenous reactions**, which take place within purified virus particles, using the encapsidated viral RNA template. It was amazing to discover that intermediates and products virtually identical to those made in infected cells can actually be synthesized in purified virions; all that is required is treatment with a mild detergent to permeabilize the envelope and addition of the metal cofactor and deoxyribonucleoside triphosphate (dNTP) substrates. The fidelity and robustness of the endogenous reaction suggests that the reverse transcription system is poised for action as soon as the virus enters the cell. Indeed, small amounts of viral DNA can be detected in purified human immunodeficiency virus type 1 particles, presumably having been synthesized using substrates picked up from the infected cell during budding. Retroviral reverse transcription intermediates have also been analyzed in totally reconstituted reactions with purified enzymes and model RNA templates.

Retroviral RT is the only protein required to accomplish all the diverse steps in the pathway described below. However, as the reactions that take place inside cells are significantly more efficient than those observed in either endogenous or reconstituted systems, it is unlikely that all the essential features have been reproduced.

Essential Components

Genomic RNA. Retrovirus particles contain two copies of the RNA genome held together by multiple regions of base pairing. (See Box 7.2 for labeling conventions.) This RNA sediments in a 70S complex, as expected for a dimer of 35S genomes. Partial denaturation and electron microscopic analyses of the 70S complex indicate that the most stable pairing is via sequences located near the 5′ ends of the two genomes (Fig. 7.1). Retroviral genomes can be thought of as being annealed head to head, an arrangement that may discourage the encapsidation of multiples larger than two (i.e., concatemers). The 70S RNA complex also includes two molecules of a specific cellular transfer RNA (tRNA) that serves as a primer for the initiation of reverse transcription (discussed below).

Despite the fact that two genomes are encapsidated, only one integrated copy of the viral DNA typically is detected after infection with single virion. Therefore, retroviral virions are said to be **pseudodiploid**. Why should such a feature have been selected during evolution? One popular notion is that the availability of two RNA templates can help retroviruses survive extensive damage to their genomes. At least parts of both genomes can, and typically are, used as templates during the reverse transcription process, accounting for the high rates of genetic recombination in these viruses. Presumably, being able to patch together

BOX 7.2 TERMINOLOGY

Conventions for designating sequences in nucleic acids

For clarity, lowercase designations are used throughout this chapter to refer to RNA sequences; uppercase designations identify the same or complementary sequences in DNA (e.g. pbr in RNA; PBR in DNA).

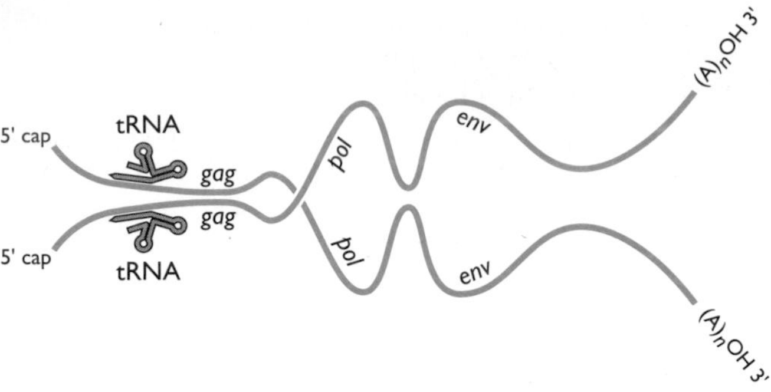

Figure 7.1 The diploid retroviral genome. The diploid genome includes the following, from 5′ to 3′: the m^{7}Gppp capping group; the coding regions for viral structural proteins and enzymes; *gag*, *pol*, and *env*; and the 3′-poly(A) sequence. The cell-derived primer tRNA is also shown. Points of contact represent multiple short regions of complementary base pairing. From J. M. Coffin, p. 1767–1848, *in* B. N. Fields et al. (ed.), *Fields Virology*, 3rd ed. (Lippincott-Raven, Philadelphia, PA, 1996), with permission.

one complete DNA copy from two randomly damaged RNA genomes would provide survival value; the genetic recombination produced during the process may also contribute to survival. Nevertheless, genetic experiments have shown that the use of two RNA templates is not an essential feature of the reverse transcription process. Therefore, all of the known steps in the reverse transcription pathway can take place on a single genome.

Like the genomes of (–) strand RNA viruses, the retroviral genome is coated along its length by a viral nucleocapsid protein (NC), with approximately one molecule for every 10 nucleotides. This small basic protein can bind nonspecifically to both RNA and DNA and promote the annealing of nucleic acids. Biochemical experiments suggest that NC may facilitate template exchanges and perform a role in reverse transcription similar to that of prokaryotic single-stranded-DNA-binding (SSB) proteins. In the synthesis of DNA catalyzed by bacterial DNA polymerases, the single-stranded-DNA-binding proteins enhance **processivity** (the efficiency of elongation). The ability of NC first to organize RNA genomes within the virion, and then to facilitate reverse transcription within the infected cell, may account for some of the differences in efficiency observed when comparing reactions reconstituted *in vitro* with those that take place in a natural infection.

Primer tRNA. In addition to the viral genome, retroviral virions contain a collection of cellular RNAs. These include approximately 100 copies of a nonrandom sampling of tRNAs, some 5S rRNA, 7S RNA, and traces of cellular mRNAs. We do not know how most of these cellular RNAs become incorporated into virus particles, and most have no obvious function. However, one particular tRNA molecule does have a critical role, that of serving as primer for the initiation of reverse transcription. The tRNA primer is positioned on the template genome during virus assembly via interactions with both RT and viral RNA, probably facilitated by NC. The primer tRNA is partially unwound and hydrogen-bonded to complementary sequences near the 5′ end of each RNA genome in a region called the **primer-binding site (pbs)** (Fig. 7.2). The reverse transcriptases of all retroviruses studied to date are primed by one of only a few classes of cellular tRNAs. Most mammalian retroviral RTs rely on either tRNA^{Pro}, tRNA^{Lys3}, or tRNA^{Lys1,2} for this function.

In addition to the 3′-terminal 18 nucleotides that anneal to the pbs, other regions in the tRNA primer contact the RNA template and modulate reverse transcription. The template-primer interaction has been studied extensively in reconstituted reactions with RNA and RT of avian sarcoma/leukosis virus. In these *in vitro* analyses, the ability of the viral RNA to form stem-loop structures, and specific interactions between the primer tRNA^{Trp} and one of these loops, appears to be critical for reverse transcription (Fig. 7.2). Similar interactions have been reported for human immunodeficiency virus RNA and its primer. Although the interactions are likely to be significant biologically, we do not yet know how RTs recognize structural features in these template-primer complexes.

Reverse transcriptase. Each retrovirus particle contains 50 to 100 molecules of RT. Retroviral RTs probably function as dimers, but the number of dimers in each virion that are actually engaged in reverse transcription is not known. As noted previously, results from studies with purified virions suggest that viral DNA synthesis can begin as soon as the viral envelope is removed. With the three orthoretroviruses studied most extensively (avian sarcoma/leukosis virus, murine leukemia virus, and human immunodeficiency virus type 1), DNA synthesis takes place mainly in the cytoplasm, in a subviral structure called the RT complex. Enzymes of these three retroviruses are used as examples throughout this chapter.

Retroviral RTs are complex molecular machines with moving parts and multiple activities. The distinct catalytic activities brought into play at various stages in the pathway of reverse transcription include RNA-directed and DNA-directed DNA polymerization, DNA unwinding, and the hydrolysis of RNA in RNA-DNA hybrid by RNase H. The first three activities reside within the polymerase domain, with RNase H activity comprising a separate domain. In digesting RNA-DNA hybrids, the RNase H functions primarily as an endonuclease, producing oligoribonucleotides of 2 to 15 nucleotides. The activities of the RNase H domain

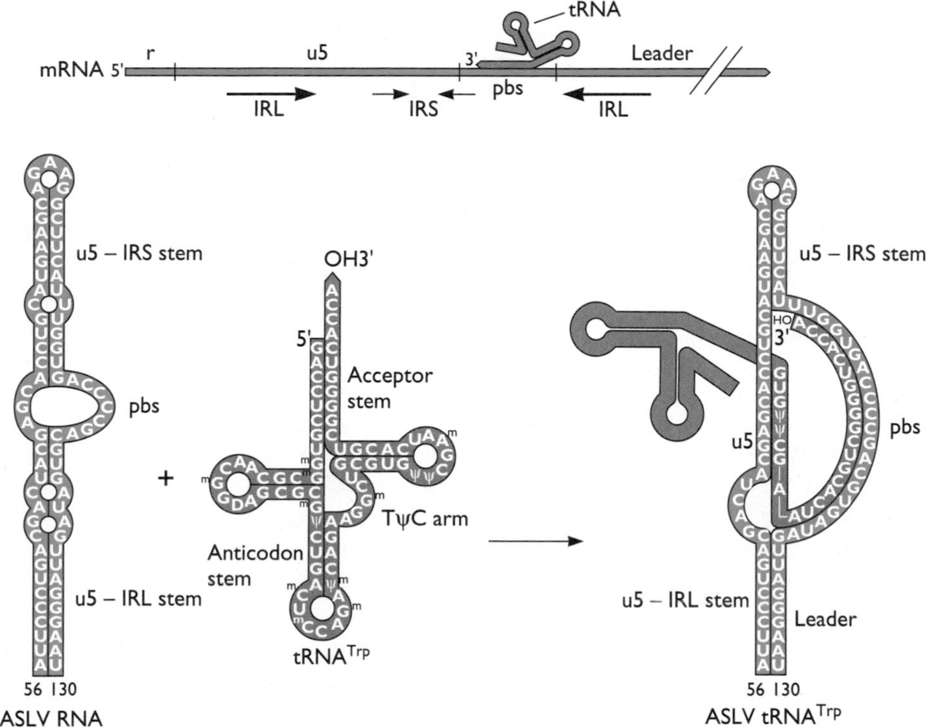

Figure 7.2 Primer tRNA binding to the retroviral RNA genome. (Top) Linear representation of the 5′ terminus of retroviral RNA, indicating locations of the r, u5, and leader regions. A tRNA primer is shown schematically annealed to the pbs. Two inverted-repeat (IR) sequences that flank the pbs are represented by arrows. **(Bottom)** Avian sarcoma/leukosis virus RNA can form an extended hairpin structure around the pbs in the absence of primer tRNA (left). Primer $tRNA^{Trp}$ is shown in the cloverleaf structure (middle). Modified bases are indicated. Viral RNA annealed to $tRNA^{Trp}$, with flanking u5-leader and u5-IR stem structures (right). The TψC arm of the primer and u5 RNA also form hydrogen bonds. Bottom diagram is from J. Leis et al., p. 33–47, *in* A. M. Skalka and S. P. Goff (ed.), *Reverse Transcriptase* (Cold Spring Harbor Laboratory Press, Cold Spring Harbor, NY, 1993), with permission.

of RT degrade the genomic RNA after it has been copied into cDNA, form the primer for (+) strand DNA synthesis from the genomic RNA, and, finally, remove this primer and the tRNA primer from the 5′ ends of the nascent viral DNA strands.

Critical Reactions in Reverse Transcription

Initiation of (–) strand DNA synthesis. Based on our understanding of DNA synthesis, the simplest way of copying an RNA template to produce DNA would be to start at its 3′ end and finish at its 5′ end. It was therefore somewhat of a shock for early researchers to discover that retroviral reverse transcription in fact starts near the 5′ end of the viral genome—only to run out of template after little more than about 100 nucleotides (Fig. 7.3). As we will see later, this counterintuitive strategy for initiation of DNA synthesis allows the duplication and translocation of critical transcription and integration signals encoded in both the 5′ and 3′ ends of the genomic RNA, called u5 and u3, respectively.

The 5′ end of the genome RNA is digested by the RNase H domain of RT, after (or as) it is copied to form (–) strand DNA. The short (ca. 100-nucleotide) DNA product of this first reaction, which includes the tRNA primer, accumulates in large quantities in the endogenous and reconstituted systems and is called **(–) strong-stop DNA.** For simplicity, and because genetic analyses indicate that this can occur, the reactions illustrated in Fig. 7.3 to 7.6 are shown as taking place on a single RNA genome. Although a template exchange involving the 5′ end of one RNA genome and 3′ end of the second RNA genome in the virion might also occur, the principles illustrated and the final end products would be the same.

The first template exchange. In the next step (Fig. 7.4), the 3′ end of the RNA genome is engaged as a template via hydrogen bonding between the R sequence in the (–) strong-stop DNA and the complementary r sequence upstream of the poly(A) tail. This reaction, which we call

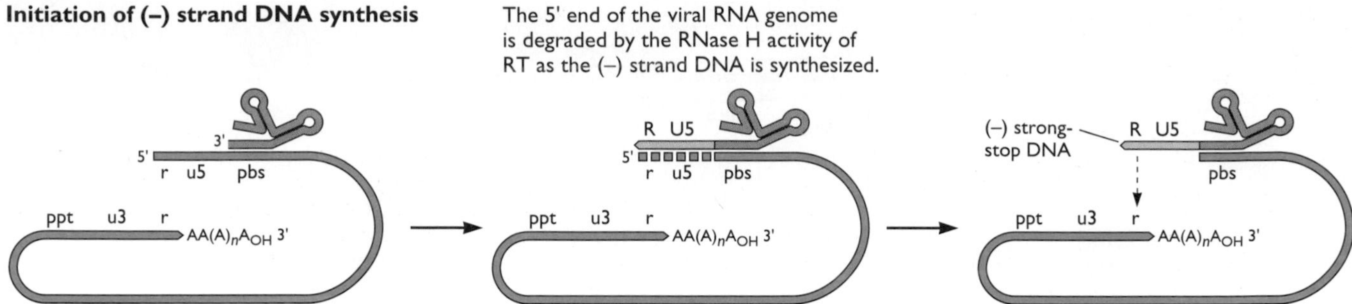

Figure 7.3 Retroviral reverse transcription: initiation of (−) strand DNA synthesis from the tRNA primer.

the **first template exchange,** corresponds to the substitution of one end of the RNA for another to be copied by the RT "machine" (Box 7.3). It has been suggested that base pairing internal to the R and r sequences, followed by branch migration, may allow this exchange to take place without complete digestion of the copied 5′ end of the viral RNA. As (–) strong-stop DNA is barely detectable in infected cells, this first template exchange must be efficient *in vivo*. Once the 3′ end of the genome RNA is engaged, the RNA-dependent DNA polymerase activity of RT can continue copying all the way to the 5′ end of the template, with the RNase H activity digesting the RNA template in its wake.

Initiation of (+) strand DNA synthesis. Among the early products of digestion of genomic RNA by RNase H is a fragment comprising a **polypurine tract (ppt)** of approximately 13 to 15 nucleotides. This RNA fragment is especially important as it serves as the primer for (+) strand DNA synthesis, which begins even before (–) strand DNA synthesis is completed (Fig. 7.4). As illustrated in Fig. 7.5, synthesis of (+) strand DNA proceeds to the nearby end of the (–) strand DNA template and terminates after copying the first 18 nucleotides of the primer tRNA, when it encounters a modified base that cannot be copied by RT. This product is called **(+) strong-stop DNA**. The (–) strand DNA synthesis continues to the end of the viral DNA template, which includes the pbs sequence that had been annealed to the tRNA primer. The production of (+) strong-stop DNA and the converging (–) strand DNA synthesis disengage the template ends. The product is a (–) strand of viral DNA comprising the equivalent of an entire genome (but in permuted order) annealed to the (+) strong stop DNA (Fig 7.5). After removal of the tRNA primer by the RNase H, the single-stranded 3′ end of the (+) strong stop DNA becomes available for annealing to complementary sequences in the single-stranded pbs at the 3′ end of the (–) strand DNA.

The second template exchange. The next steps in the pathway of reverse transcription are facilitated by a **second template exchange** in which annealing of the complementary pbs sequences provides a circular DNA template for continued polymerization by RT (Fig. 7.6, top). Synthesis of the (+) strand DNA can now continue, using the (–) strand DNA as a template. The (–) strand DNA synthesis also continues using the strand-displacement activity of RT, a reaction that opens the DNA circle. Synthesis stops when RT reaches the terminus of each template strand (Fig. 7.6,

Figure 7.4 Retroviral reverse transcription: first template exchange, mediated by annealing of short terminal repeat sequences.

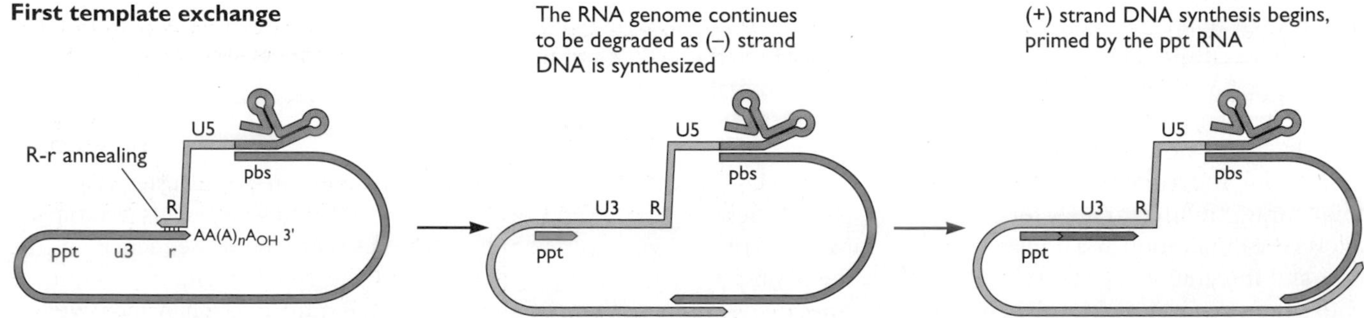

BOX 7.3 TERMINOLOGY
Enzymes can't jump!

The exchange of one template for another to be copied by either DNA or RNA polymerases is sometimes referred to as enzyme "jumping." This inappropriate term comes from a too literal reading of cartoon illustrations of the process, in which the templates to be exchanged may be opposite ends of the nucleic acid or different nucleic acid molecules. In actuality, such enzyme movement is quite improbable, and use of this terminology can cloud thinking about these processes. In almost all cases, these polymerases are components of large complexes with architecture designed to bring different parts of the template, or different templates, close to each other. Although some dynamic changes must occur for the proteins to accommodate template exchanges, it seems likely that most of the "movement" is done by the flexible nucleic acid molecules (see, e.g., Fig. 7.3).

left). The final product is a linear, DNA duplex copy of the viral genome with a **long terminal repeat (LTR)**, of critical *cis*-acting signals at either end. This linear form of viral DNA is the major product of reverse transcription found in the nucleus of infected cells.

Small quantities of two circular DNA products are also invariably present in the nucleus. These are nonfunctional, dead-end products; their presumed origin is illustrated in Fig. 7.6 (right). The smaller of the two, a circle with only one LTR, can arise either from a failure of strand displacement synthesis of RT, or by recombination between the terminal LTR sequences in the linear molecule. The circle with two LTRs is presumed to arise by ligation of the ends of the linear viral DNA. Because formation of this product requires a nuclear enzyme, DNA ligase, and it is easy to detect by polymerase chain reaction (PCR) techniques, the two-LTR circles have been used as convenient markers for the transport of viral DNA into the nucleus.

Retroviral reverse transcription has been called "destructive replication," as there is no net gain of genomes, but rather a substitution of one double-stranded DNA for two molecules of single, (+) strand RNA. However, by this rather complex but elegant pathway, RT not only makes a linear DNA copy of the retroviral genome to be integrated, but also produces the LTRs that contain signals necessary for transcription of the integrated DNA, which is called the **provirus**. The promoter in the upstream LTR is now in the appropriate location for synthesis of progeny RNA genomes and viral mRNAs by host cell RNA polymerase II, and the downstream LTR contains signals for polyadenylation of the mRNA. Integration also ensures subsequent replication of the provirus via the host's DNA synthesis machinery as the cell divides.

Reverse transcription promotes recombination. The above description of reverse transcription has been idealized, for clarity. Analyses of reaction intermediates show that RT pauses periodically during synthesis, presumably at some sequences or breaks that impede copying. If a break is encountered in one RNA template, synthesis can be completed by utilization of the second RNA genome. Such internal template exchanges, known to occur even in the absence of breaks, probably proceed via the same steps outlined for the first template exchange. Indeed, internal exchanges that take place during RNA-directed DNA synthesis are estimated to be the source of at least half of the genetic recombination that occurs in retroviruses, a mechanism known as **copy choice**.

Figure 7.5 Retroviral reverse transcription: (+) strand DNA synthesis primed from ppt RNA.

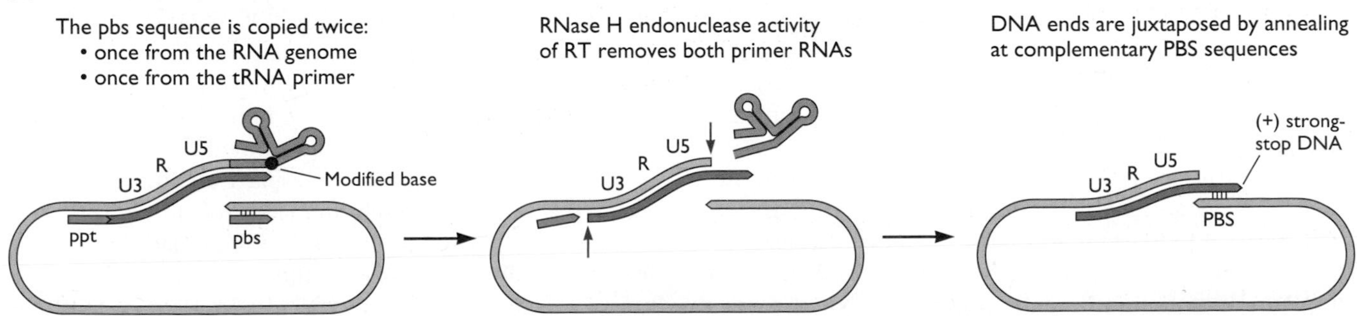

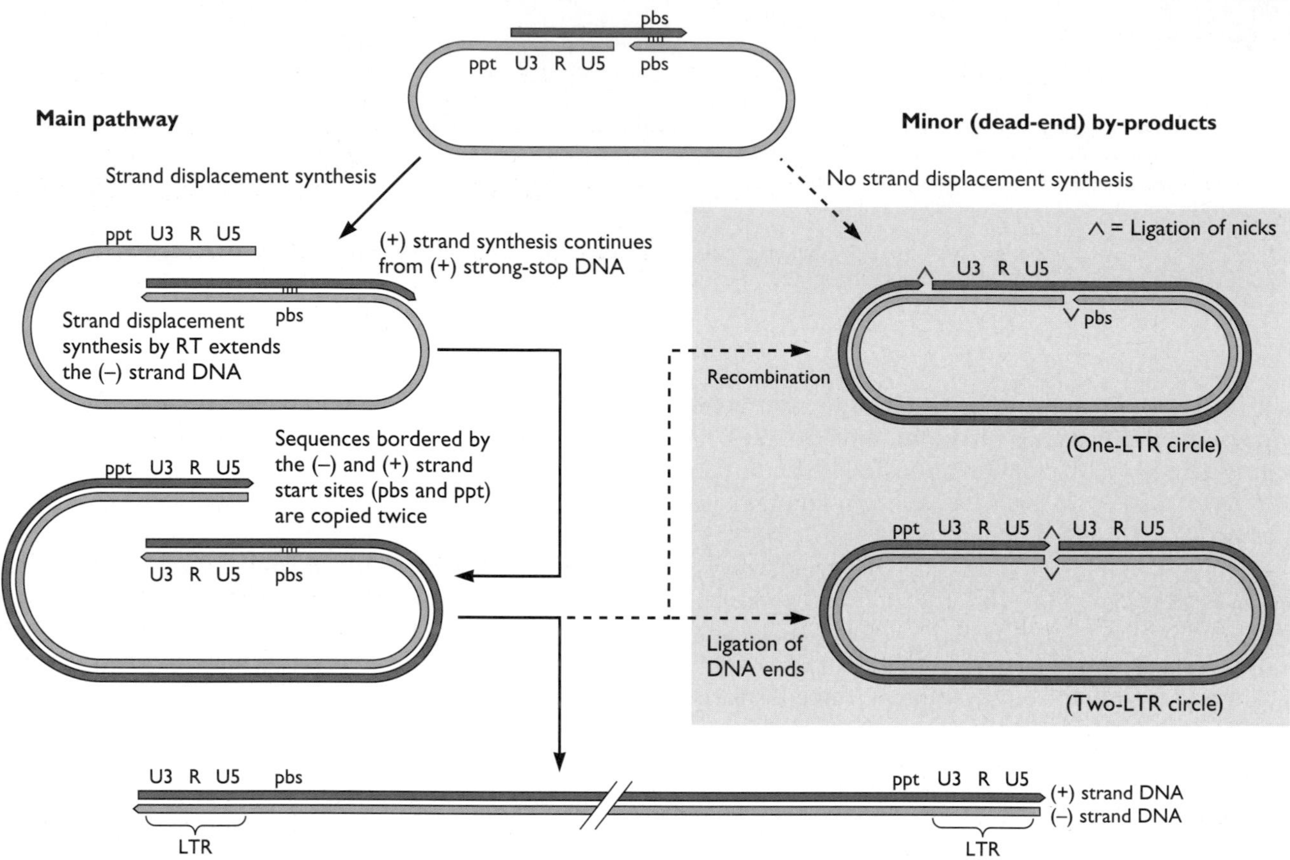

Figure 7.6 Retroviral reverse transcription: the second template exchange and formation of the final linear DNA product. The second exchange is facilitated by annealing of pbs sequences in (+) and (–) strands of retroviral DNA.

In addition to the ppt, other oligoribonucleotides products of RNase H digestion of the RNA template can also serve as primers for DNA-directed DNA polymerization by RT. Human immunodeficiency virus contains a second, specific internal polypurine tract for priming (+) strand synthesis, which is located near the center of the genome. With both the human virus and avian sarcoma/leukosis viruses, (+) strand DNA synthesis can also be initiated at several additional locations. On the other hand, there is little or no evidence of (+) strand initiation at sites other than the 3′ ppt during murine leukemia virus reverse transcription. As illustrated in Fig. 7.7, RT intermediates of murine retroviruses (mouse mammary tumor virus and murine leukemia virus) contain large gaps in the DNA (+) strand. In contrast, avian sarcoma/leukosis virus (+) strands contain single-stranded tails formed by strand displacement synthesis through internal sites of initiation on the DNA template. Human immunodeficiency virus DNA also has a uniquely positioned tail generated by strand displacement synthesis through the internal polypurine tract. This apparent abundance of single-stranded DNA, opposite gaps and in tails, could provide additional opportunities for recombination. It is estimated that approximately half of the genetic recombination that occurs during retroviral reverse transcription takes place during (+) strand synthesis. One proposed mechanism, the **strand displacement-assimilation model**, is illustrated in Fig. 7.8, together with the (–) strand synthesis-dependent mechanism of copy choice.

It is not known how or exactly when linear DNA products with (+) strand discontinuities or branches are processed. From *in vitro* studies it appears unlikely that strand displacement synthesis by RT can displace all downstream (+) strand segments completely. Indeed, a role in transport into the nucleus has been proposed for the displaced (+) strand segment ("flap") that includes the central ppt sequence in human immunodeficiency type 1 DNA (Fig. 7.7). It may

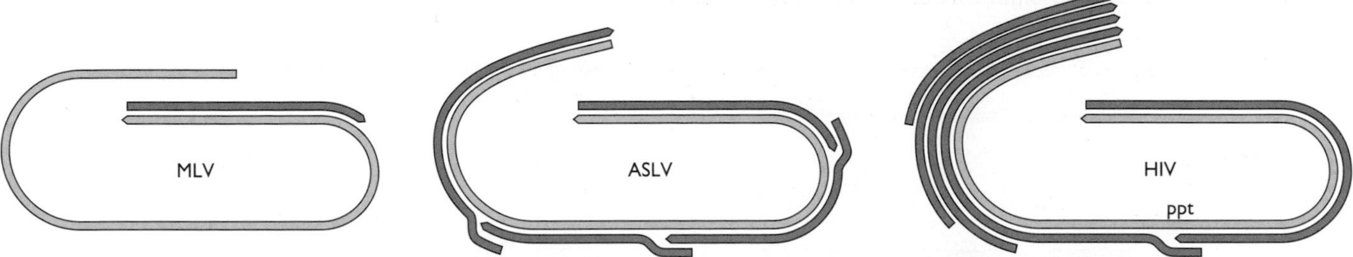

Figure 7.7 Schematic representation of the differences in the structure of retroviral DNAs extracted from infected cells. Most unintegrated murine leukemia virus (MLV) DNA molecules extracted from infected cells have a full-length, continuous DNA (−) strand but only short fragments of the (+) strand. The (+) strands of avian sarcoma/leukosis virus (ASLV) DNA can be initiated at several (apparently random) locations. These are elongated past adjacent initiation sites to generate long, single-stranded tails. Human immunodeficiency virus type 1 (HIV) DNA has two classes of (+) strands. The (+) strands are continuous to a point just beyond the internal polypurine tract (ppt), where they appear to encounter a block to further elongation. Downstream of the internal ppt, there appear to be many possible initiation sites that give rise to a collection of (+) strands of different lengths.

be that cellular repair enzymes in the nucleus participate in the final process. Theoretically, such repair could occur even after the viral DNA is joined to the host DNA, as such joining can proceed as soon as the duplex LTR ends are formed. Indeed, studies with murine leukemia virus show that the first step in the integration process, processing of the blunt ends of the LTRs by integrase (IN), takes place in the cytoplasm, before the viral DNA enters the nucleus. Furthermore, nucleoprotein complexes isolated from the cytoplasmic fraction of human immunodeficiency virus-infected cells are competent to join viral DNA to exogenously supplied target DNA, despite the fact that their (+) DNA strands are not yet continuous.

General Properties and Structure of Retroviral Reverse Transcriptases

Domain Structure and Variable Subunit Organization

The RTs of retroviruses are encoded in the *pol* genes. Despite the sequence homologies and similar organization of coding sequences in their *pol* regions, each of the three Pol polyproteins studied most extensively, those of avian sarcoma/leukosis, murine leukemia, and human immunodeficiency viruses, is processed differently (Fig. 7.9). The mature, functional RT of avian sarcoma virus is a heterodimer, the larger subunit of which includes IN. Murine leukemia virus RT functions as a monomer or homodimer from which IN is absent. As with the avian sarcoma/leukosis virus, the human immunodeficiency virus type 1 RT is a heterodimer. However, in this case only one subunit includes the RNase H domain, and neither includes IN sequences. Little is known about the subunit compositions of RTs from most other retroviruses. In the absence of more structural data, it is difficult to gauge the significance of this apparent structural diversity. However, it seems likely that in the confines of the subviral nucleoprotein complex in which reverse transcription takes place in vivo, the Pol-derived proteins of all retroviruses work together and in similar ways.

Distinctive Catalytic Properties

DNA polymerization is slow. The biochemical properties of RT have been studied with enzymes purified from virions or synthesized in bacteria, and using model templates and primers. Kinetic analyses have revealed an ordered reaction pathway for DNA polymerization similar to that of other polymerases. Like cellular polymerases and nucleases, RTs require divalent cations as cofactors (physiologically, most likely Mg^{2+}). The rate of elongation by RT on natural RNA templates in vitro is 1 to 1.5 nucleotides per second, approximately 1/10 the rate of other eukaryotic DNA polymerases. Assuming that RT begins to act immediately upon viral entry, the long time required to produce a genome's equivalent of retroviral DNA after infection, approximately 4 h for ca. 9,000 nucleotides, supports the view that reverse transcription is a relatively slow process *in vivo*.

In reactions *in vitro*, the rate of dissociation of the enzyme from the template-primer decreases considerably after addition of the first nucleotide, suggesting that initiation and elongation are distinct steps in reverse transcription, as is the case during DNA synthesis by DNA-dependent DNA polymerases. Contrary to most other DNA polymerases, retroviral RTs do not remain attached to their template-primers during a large number of successive deoxyribonucleotide additions; this property is described as "poor processivity."

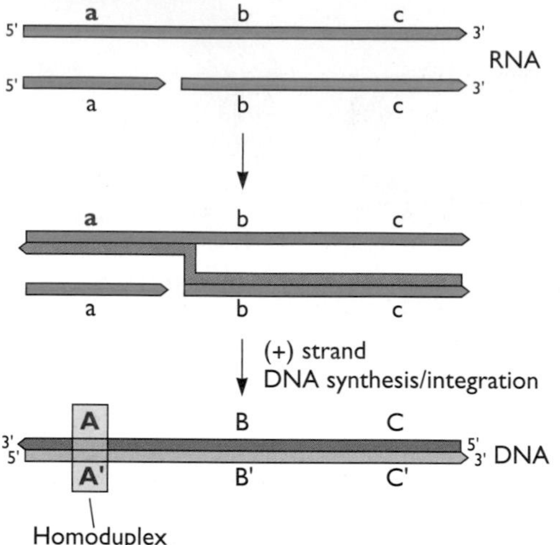

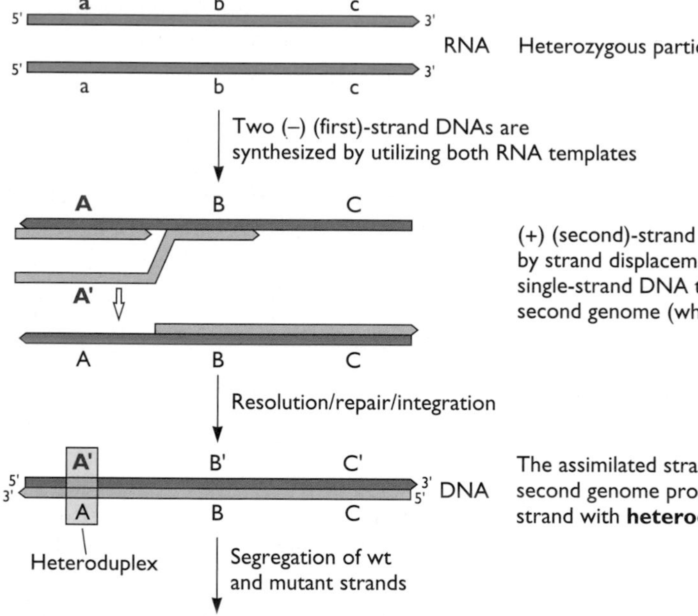

Figure 7.8 Two models for recombination during reverse transcription. Virtually all retroviral recombination occurs between coencapsidated genomes at the time of reverse transcription. The copy choice model **(A)** postulates a mechanism for genetic recombination during (–) strand, DNA synthesis; the strand displacement-assimilation model **(B)** proposes a mechanism for recombination during (+) strand DNA synthesis. These two models are not mutually exclusive, and there is experimental support for both. Viral genetic markers, arbitrarily labeled a, b, and c, are indicated to illustrate recombination. Single recombinations are shown for simplicity, focusing on the a allele, with the mutant form in red. However, multiple crossover events are frequently observed. The boxes highlight the region of recombination. **(A)** Recombination during (–) strand synthesis: copy choice. A break in the RNA is represented as a gap. Exchanges between RNA templates may occur at RNA breaks, as shown, but such breaks may not be required. Introducing intentional breaks, e.g., by gamma irradiation, does not seem to increase the frequency of such recombination. The copy choice mechanism predicts only one DNA homoduplex product from two RNA molecules. **(B)** Recombination during (+) strand synthesis: strand assimilation. (+) strand DNA synthesis initiates at internal sites with RNA primers produced by partial RNase H digestion. The strand assimilation model requires that (+) strand DNA synthesis take place on two templates and predicts formation of a heteroduplex as a consequence of recombination of (+) strands. One DNA molecule is shown for simplicity. Adapted from R. Katz and A. M. Skalka, *Annu. Rev. Genet.* **24:**409–445, 1990, with permission.

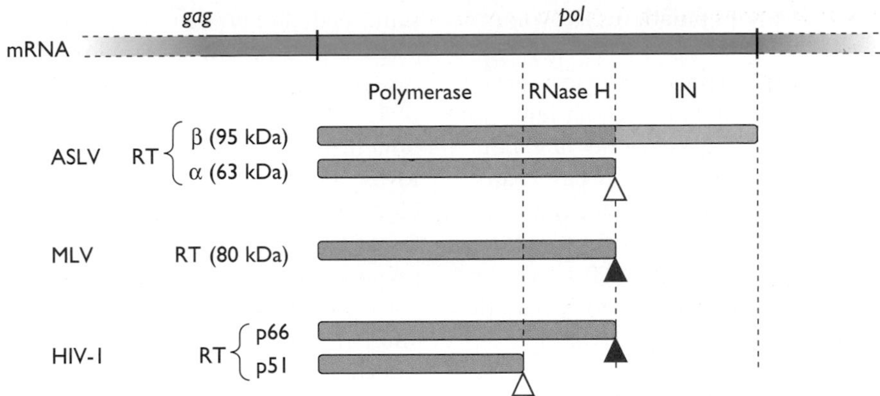

Figure 7.9 Domain and subunit relationships of RTs of different retroviruses. The organization of *pol* domains in retroviral mRNA is shown at the top. The protein products of the three indicated retroviruses are shown below, with arrows pointing to the sites of proteolytic processing that produce the observed diversity of RT subunit composition. Open red arrows indicate partial (asymmetric) processing, and solid red arrows indicate complete processing. ASLV, avian sarcoma/leukosis virus; MLV, murine leukemia virus; HIV-1, human immunodeficiency virus type 1.

Fidelity is low. Retroviral genomes, like those of other RNA viruses, accumulate mutations at much higher rates than do cellular genes. RTs lack an editing activity (resembling the 3′ → 5′ exonuclease of *Escherichia coli* DNA polymerase I, which is capable of excising mispaired nucleotides), and have been shown to be error prone *in vitro*. RTs are presumed therefore to contribute to the high *in vivo* mutation rate of retroviruses.

In vitro, RT errors include not only misincorporations but also rearrangements such as deletions and additions (Fig. 7.10). Misincorporations by human immunodeficiency virus type 1 RT can occur as frequently as 1 per 70 copies at some positions, and as infrequently as 1 per 10^6 copies at others. Many types of genetic experiments have been conducted in attempts to determine the error rates of RTs in a single replication cycle within a cell. The general conclusion is that such rates are also quite high, with reported misincorporations in the range of 1 per 10^4 to 1 per 10^6 nucleotides polymerized, in contrast to 1 per 10^7 to 1 per 10^{11} for cellular DNA replication. As retroviral genomes are approximately 10^4 nucleotides in length, this rate can translate to approximately one lesion per retroviral genome per replication cycle. This high mutation rate explains, in part, the difficulties inherent in treating AIDS patients with inhibitors of RT or other viral proteins; a large population of mutant viruses preexist in every chronically infected individual, some encoding drug-resistant proteins. These mutants can replicate in the presence of a drug and

Figure 7.10 Mutational intermediates for base substitution and frameshift errors. Misincorporated nucleotides are indicated in red. Slippage and dislocations are presumed to be mediated by looping out of nucleotides in the template. Only single-nucleotide dislocations are shown here, but large dislocations leading to deletions are also possible. From K. Bebenek and T. A. Kunkel, p. 85–102, *in* A. M. Skalka and S. P. Goff (ed.), *Reverse Transcriptase* (Cold Spring Harbor Laboratory Press, Cold Spring Harbor, NY, 1993), with permission.

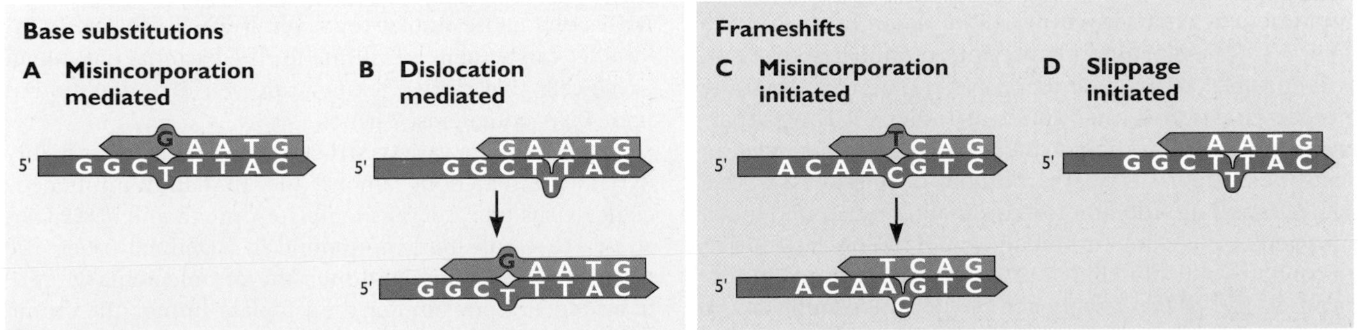

quickly comprise the bulk of the population (see Volume II, Chapters 6 and 9).

Several unique activities of RTs are also likely to contribute to their high error rates. The avian sarcoma/leukosis and human immunodeficiency virus enzymes are both proficient at extending mismatched terminal base pairs, such as those that result from nontemplated addition (Fig. 7.10A). This process facilitates incorporation of the mismatched nucleotide into the RT product. RTs also seem to allow a certain type of slippage within homopolymeric runs, in which one or more bases are extruded on the template strand (Fig. 7.10B and C). Mispairing occurs after the next deoxyribonucleotide is added and the product strands attempt to realign with the template. Deletions can also be produced by this mechanism (Fig. 7.10D). Both deletions and insertions are also known to occur during reverse transcription within an infected cell, apparently because template exchanges can take place within short sequence repeats (e.g., 4 or 5 nucleotides) that are not in homologous locations on the two RNA templates. This mechanism is of major importance in the capture of oncogenes (see Volume II, Chapter 7).

RNase H. The RNase H of RT also requires a divalent cation, (most likely Mg^{2+}, which is abundant in cells). Like other RNase H enzymes, present in all prokaryotic and eukaryotic cells, the RNase H of RT digests only RNA that is annealed to cDNA. RNase H cleaves phosphodiester bonds to produce 5′-PO_4 and 3′-OH ends; the latter can be extended directly by the RT. Biochemical studies have uncovered three distinguishable activities for RNase H of human immunodeficiency virus type 1 and murine leukemia virus; recessed 3′ DNA end-directed and recessed 5′ RNA end-directed endonucleases, both of which cleave the RNA strand 12 to 20 nucleotides from the recessed end, and an internal endonuclease that is not end directed. A loose consensus site for all three activities has also been observed.

Structure of RT

Although RTs from avian and murine retroviruses have been studied extensively *in vitro*, the importance of human immunodeficiency virus type 1 RT as a target for drugs to treat AIDS has focused intense interest and resources on this enzyme. As a consequence, we know more about the human viral protein than about any other RT. Three aspartic acid residues in the polymerase region are included in conserved motifs in a large number of polymerases (see Fig. 6.4) and are thought to coordinate the required metal ions and contribute to binding deoxyribonucleoside triphosphates and subsequent catalysis. As illustrated in Figure 7.9, the primary sequence of the p51 subunit is the same as that of p66, minus the RNase H domain. However, as discussed below, the analogous portions of these subunits are arranged quite differently in RT.

The first high-resolution structure of human immunodeficiency virus type 1 RT to be solved was that of the RNase H domain. As illustrated by the ribbon models in Fig. 7.11, the general structure of this domain is almost superimposable on that of *E. coli* RNase H. Included in the regions of structural homology are seven residues that are components of conserved motifs present in all retroviral and bacterial RNases H, clustered at what is thought to be the site at which the metal ion cofactors are bound. The results of site-directed mutagenesis have confirmed that many of the conserved residues are important for the activities of both *E. coli* and human retroviral RNase H.

The high-resolution structure of the intact human immunodeficiency virus type 1 RT heterodimer was obtained for complexes with the nonnucleoside inhibitor nevirapine. Subsequently, several additional structures were determined, including RT with no bound inhibitors, and RT complexed with a short RNA-DNA duplex or a DNA duplex and a deoxynucleotide substrate (Fig. 7.12). The most surprising finding to come from the analysis of the first RT crystals was the **structural asymmetry** in the subunits. Results of biochemical studies showed that the two subunits also perform different functions in the heterodimeric enzyme. The catalytic functions are contributed by the larger subunit, p66, whereas the role of p51 appears to be mainly structural. In the crystal structures, the two subunits are nestled on top of each other, with an extensive subunit interface. The p66 polymerase domain is divided into three subdomains denoted "finger," "palm," and "thumb" by analogy to the convention used for describing the topology of the *E. coli* DNA polymerase I Klenow fragment, described in Chapter 6 (Fig. 6.3 and 6.4). A fourth subdomain lies between the remainder of the polymerase and the RNase H domain, and is therefore called the "connection" domain. It contains the major contacts between the two subunits of RT. The extended thumb of p51 contacts the RNase H domain of p66, an interaction that appears to be required for RNase H activity. Not only are human immunodeficiency virus type 1 RT and *E. coli* DNA polymerase similar topologically, but also this retroviral RT can actually substitute for the bacterial enzyme in *E. coli* cells that carry a temperature-sensitive mutation in their DNA polymerase I gene.

Analysis of the various structural models predicts highly dynamic interactions among the human immunodeficiency virus type 1 RT, its template-primer, and dNTP substrates. The substrates are bound in a defined order: the template-primer first, and then the complementary dNTP to be added. Upon binding the template-primer, the thumb

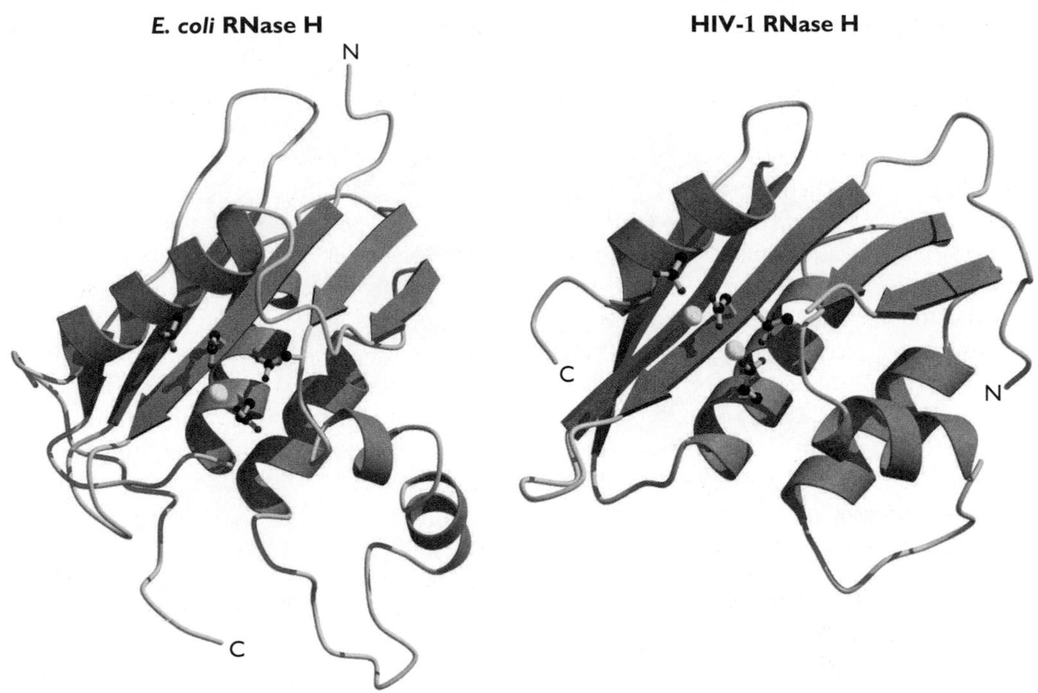

Figure 7.11 Ribbon diagrams based on the crystal structures of *E. coli* and human immunodeficiency virus type 1 (HIV-1) RNase H. The highly conserved acidic residues are shown as ball-and-stick models, with metal ion complexes indicated by the large yellow spheres. Courtesy of Mark Andrake, Fox Chase Cancer Center.

is moved away from the fingers, allowing contacts between the fingers and the 5′ extension of the template. Another conformational change takes place when the dNTP is bound. This substrate interacts directly with two fingertip residues; the interaction may induce closure of the binding pocket. This conformational change facilitates attack of the 3′-OH of the primer on the α-phosphate of the incoming dNTP. After this addition, the fingertips may resume their open position, allowing the diphosphate product to be released and the template-primer to be translocated, so that the next dNTP can be accepted. Such translocation may be driven, in part, by the energy released upon hydrolysis of the previous dNTP substrate.

Figure 7.13 is a schematic drawing of an RNA-DNA heteroduplex bound in the cleft region of the human immunodeficiency virus type 1 RT. It illustrates how the RNA strand is fed into the polymerizing site from the left. As a new DNA chain is synthesized by addition of deoxyribonucleotides to the primer, the template RNA is translocated in stepwise fashion to the RNase H site at the right, where it can be degraded. This schematic is consistent with biochemical evidence for polymerase-coupled RNase H activity that results in cleavage of the RNA template about once for every 50 to 100 dNTPs polymerized. The distance between the polymerizing and RNase H sites can account for the length (ca. 18 nucleotides) of the terminal RNase H oligoribonucleotide product.

Production of the p51 subunit of the human immunodeficiency virus type 1 RT, which possesses identical amino acid sequences to but has both structure and function that are distinct from those of the p66 subunit (Fig. 7.9), appears to be an excellent example of viral genetic economy. Results of cross-linking experiments suggest that the p51 subunit may perform a unique function in the RT heterodimer, that of binding the tRNA primer. The C terminus of the p51 subunit, at the end of the connection domain, is buried within the N-terminal β-sheet of the RNase H domain of p66. This arrangement suggests a model for proteolytic processing in which a p66 homodimer intermediate in the reaction is also arranged asymmetrically and the RNase H domain of the subunit destined to become p51 is unfolded. This mechanism would account for asymmetric cleavage by the viral protease (Fig. 7.9). Without comparable structural data for other retroviral RTs, we can only wonder if an analogous processing strategy applies to these proteins. The asymmetric processing of the avian sarcoma/leukosis virus Pol protein suggests that this may be the case for the alpharetroviruses as well (Fig. 7.9).

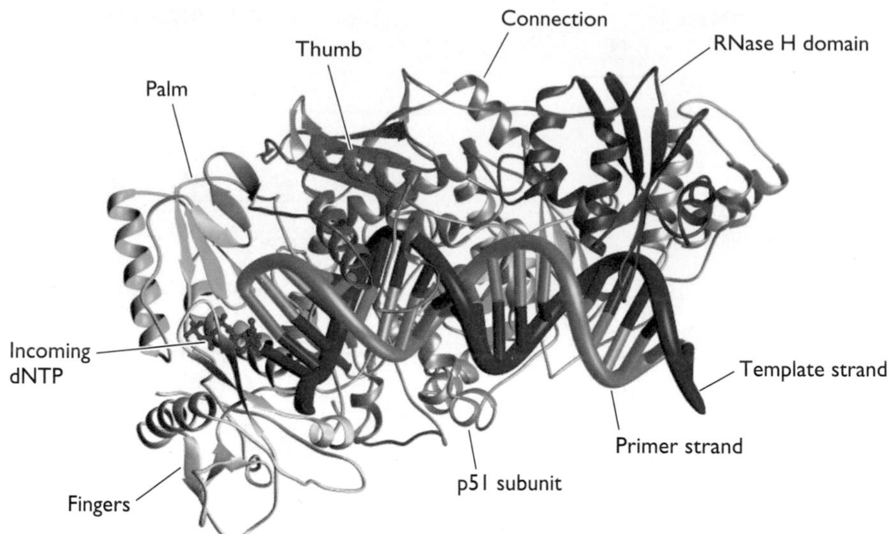

Figure 7.12 A ribbon representation of human immunodeficiency virus type 1 RT in complex with DNA template/primer. dNTP, deoxyribonucleoside triphosphate. Courtesy of Rajiv Chopra and Stephen Harrison, Harvard University.

There Are Many Other Examples of Reverse Transcription

When it was first discovered, RT was thought to be a peculiarity of retroviruses. We now know that other animal viruses, the hepadnaviruses, and some plant viruses, such as the caulimoviruses, also replicate by producing duplex genomic DNA via an mRNA intermediate. All are classified therefore as **retroid viruses**. In fact, the discovery of RT activity in some strains of myxobacteria and *E. coli* places the evolutionary origin of this enzyme before the separation of prokaryotes and eukaryotes. As it is now widely held that the evolving biological world was initially based on RNA molecules, as both catalysts and genomes, the development of the modern (DNA) stage of evolution would have required an RT activity. If so, the retroid viruses may also be viewed as living fossils, shining the first dim light into an ancient evolutionary passageway from the primordial world (Volume II, Chapter 10).

Figure 7.13 Model for a DNA-RNA hybrid bound to human immunodeficiency virus type 1 RT. The RNA template-DNA product duplex is shown lying in a cleft. The polymerase active site and the putative RNase H active site are indicated. Me^{2+} signifies a divalent metal ion. As illustrated, the RNA template enters from the left, and is degraded at the RNase H active site after being copied into DNA at the polymerase active site. The newly synthesized DNA strand exits to the right. Adapted from L. A. Kohlstaedt et al., p. 223–250, *in* A. M. Skalka and S. P. Goff (ed.), *Reverse Transcriptase* (Cold Spring Harbor Laboratory Press, Cold Spring Harbor, NY, 1996), with permission.

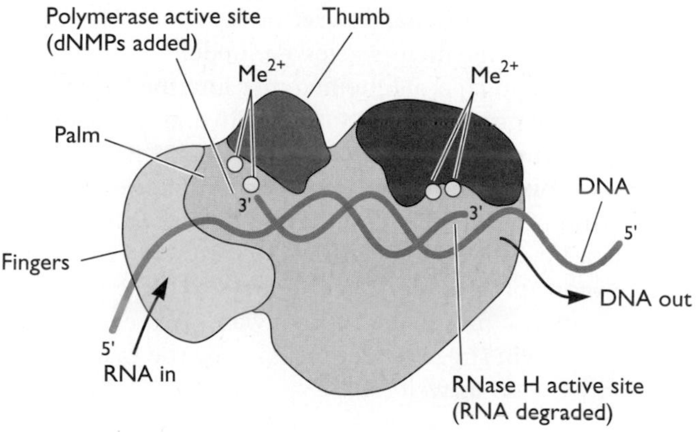

Since the discovery of RT in retroviruses, additional RT-related sequences have been found in cellular genomes. Some of these sequences are contained in cellular constituents known generally as **retroelements**, which are derived from retroviruses. During the retroviral life cycle (Appendix, Fig. 20), the double-stranded DNA molecule synthesized by reverse transcription is integrated into the genomes of infected animal cells by a second retroviral enzyme, integrase (discussed in the following section). In some cases, retroviral DNA may be integrated into the DNA of germ line cells in a host organism. These integrated DNAs are then passed on to future generations in Mendelian fashion as **endogenous proviruses**. Such proviruses are often replication defective, a property that may facilitate coexistence with their hosts. Almost 50% of the human genome is now known to comprise mobile genetic elements, including endogenous proviruses and other retroelements such as retrotransposons, retroposons, and processed pseudogenes, all of which have been accumulated during evolution (Box 7.4). Chromosomal

BOX 7.4 DISCUSSION *Retroelements*

Retrotransposons are dispersed widely in nature. Their gene content and arrangements are similar to those of retroviruses. Like retroviral proviruses, integrated retrotransposon DNAs have LTRs that include signals for transcription by cellular RNA polymerase II and short direct repeats of host DNA at their borders. RT sequence comparisons (Fig. 7.14) have allowed classification of these elements into two families. Most retrotransposons are distinguished from retroviruses by lack of an extracellular phase; they have no *env* gene, and hence the virus-like particles formed intracellularly are noninfectious. However, in the family *Metaviridae*, members of one genus do include open reading frames corresponding to *env*, and at least one of these elements, the *Drosophila* gypsy, produces infectious extracellular particles. Phylogenetic comparisons of LTR-retrotransposons from invertebrates provide evidence that several of these elements have acquired *env* sequences via genetic recombination with both RNA and DNA viruses. These results support the view that retrotransposons are retroviral progenitors. An alternative possibility, but with less phylogenetic support, is that retrotransposons are degenerate forms of retroviruses.

Retroposons, also called non-LTR retrotransposons, are dispersed widely in the DNA of human cells. Like retroviral proviruses and retrotransposons, they are flanked by direct repeats of host DNA at their boundaries. Although they lack LTRs, some contain internal promoters for transcription by cellular RNA polymerase III. All retroposons have A-rich stretches at one terminus, presumed to be derived by reverse transcription of the 3′-poly(A) tails in their RNA intermediates. Retroposons comprise fairly long stretches of related sequences (up to 6 kb). These long interspersed repeat elements are commonly called LINEs. Most LINEs encode RT-related sequences, but they often contain large deletions and translational stop codons and are therefore defective. The sole LINE element of humans, called L-1, comprises approximately 20% of the human genome and, although most L-1 elements are dead, approximately 80 to 100 are transposition competent. Genetic analyses have documented a number of disease-causing genetic lesions resulting from L-1-mediated retrotranspositional events.

Retrosequences, another type of reverse-transcribed sequence, are found as short interspersed repeat elements on the order of 300 bp long. These sequences, called SINEs, comprise approximately 11% of the human genome. SINEs have no known open reading frames and are retrotransposed in *trans* by L-1.

Processed pseudogenes have no introns (hence "processed"), and their sequences are related to exons in functional genes that map elsewhere in the genome. Like retroposons, they include long A-rich stretches at one end and lack LTRs. However, they contain no promoter for transcription, and no RT. Generally, they are thought to arise from rare reverse transcription of cellular mRNAs catalyzed by the RTs of retroviruses or nondefective LINEs. The table summarizes the defining characteristics of the major classes of retroelements.

Malik, H. S., S. Henikoff, and T. H. Eickbush. 2000. Poised for contagions: evolutionary origins of the infectious abilities of invertebrate retroviruses. *Genome Res.* **10:**1037–1318, 2000.

Characteristics of retroelements resident in eukaryotic genomes

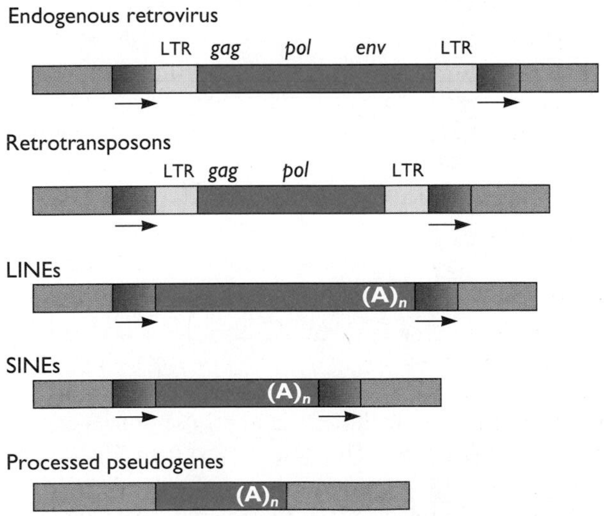

Endogenous retrovirus: LTR *gag* *pol* *env* LTR

Retrotransposons: LTR *gag* *pol* LTR

LINEs: (A)$_n$

SINEs: (A)$_n$

Processed pseudogenes: (A)$_n$

Designation	Characteristic	Example	Copy no.
Endogenous retroviruses	RT, LTR (internal Pol II promoter), and *env*	HERVs (human)	1–10^2
Retrotransposons	RT, LTR (internal Pol II promoter)	Ty3 (yeast)	10^2–10^4
Retroposons (LINEs)	RT, internal Pol III promoter, A-rich sequence at end	LINE 1 (human)	10^4–10^5
Retrosequences (SINEs)	A-rich sequence at end, internal Pol III promoter, but no RT	*Alu* (human)	10^5–10^6
Processed pseudogenes	A-rich sequence at end, no internal promoter, no RT	β-Tubulin (human)	1–10^2

telomeres are also formed via reverse transcription by an enzyme known as telomerase. Comparisons of the predicted RT-related amino acid sequences of representative retroelements provide clues about their relatedness and hints to their possible origin (Fig. 7.14).

Retroviral DNA Integration Is a Unique Process

The **integrase (IN)** of retroviruses and the related retrotransposons catalyzes specific and efficient insertion of the DNA product of RT into the host cell DNA. This activity is unique in the eukaryotic world. Some retroposons also contain IN-related sequences (Fig. 7.14). Establishment of an integrated copy of the genome is a critical step in the life cycle of retroviruses, as this reaction ensures stable association of viral DNA with the host cell's genome. The integrated **proviral DNA** is transcribed by cellular RNA polymerase II to produce the viral RNA genome and the mRNAs required to complete the replication cycle.

IN is encoded in the 3′ region of the retroviral *pol* gene (Fig. 7.9), and the mature protein is produced, in most cases, by viral protease (PR)-mediated processing of the Gag-Pol polyprotein precursor. During progeny virus assembly, all three viral enzymes (PR, RT, and IN) are incorporated into the virion core. Virus particles contain equimolar quantities of RT and IN (ca. 50 to 100 molecules per virion). The viral

Figure 7.14 Schematic representation of the eukaryotic retroid family phylogeny. The unrooted family tree derived from comparison of deduced RT amino acid sequences is displayed at the left. It is separated into two major branches (I and II), based primarily on the presence of long terminal repeats (LTRs) that contain important regulatory sequences in the DNA versions of some of its members. The table compares DNA sequences and deduced open reading frames encoded in all family members with those sequences that are characteristic of retroviruses. The region that defines membership in the retroid family, the RT, is highlighted in red. The *cis*-acting sequences include the following: LTRs, long terminal repeats flanking the retroviral genome; PBS, tRNA primer-binding site; and poly(A), 3′-poly(A) tract in RNA or DNA form. The open reading frames (yellow shaded area) are as follows: CA, capsid protein; NC, nucleocapsid; PR, protease; RT, reverse transcriptase domain; T, "tether" region connecting the RT and RNase H domains; RH, RNase H domain; H/C, histidine and cysteine motif in the N-terminal portion of the integrase; IN, integrase; Env, envelope protein. From M. A. McClure, *in* A. M. Skalka and S. P. Goff (ed.), *Reverse Transcriptase* (Cold Spring Harbor Laboratory Press, Cold Spring Harbor, NY, 1996), with permission.

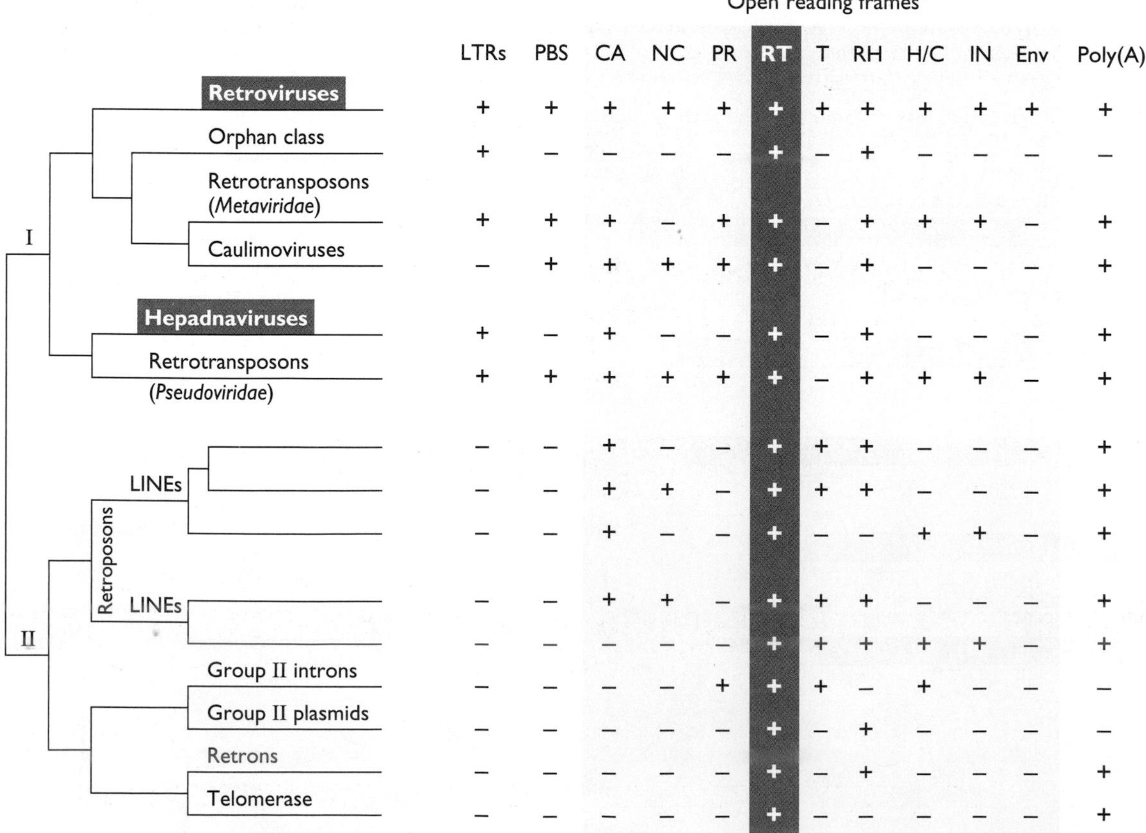

	LTRs	PBS	CA	NC	PR	RT	T	RH	H/C	IN	Env	Poly(A)
Retroviruses	+	+	+	+	+	+	+	+	+	+	+	+
Orphan class	+	–	–	–	–	+	–	+	–	–	–	–
Retrotransposons (*Metaviridae*)	+	+	+	–	+	+	–	+	+	+	–	+
Caulimoviruses	–	+	+	+	+	+	–	+	–	–	–	+
Hepadnaviruses	+	–	+	–	–	+	–	+	–	–	–	+
Retrotransposons (*Pseudoviridae*)	+	+	+	+	+	+	–	+	+	+	–	+
LINEs	–	–	+	–	–	+	+	+	–	–	–	+
	–	–	+	+	–	+	+	+	–	–	–	+
	–	–	+	–	–	+	–	–	+	+	–	+
LINEs	–	–	+	+	–	+	+	+	–	–	–	+
	–	–	–	–	–	+	+	+	+	+	–	+
Group II introns	–	–	–	–	+	+	+	–	+	–	–	–
Group II plasmids	–	–	–	–	–	+	–	+	–	–	–	–
Retrons	–	–	–	–	–	+	–	+	–	–	–	+
Telomerase	–	–	–	–	–	+	–	–	–	–	–	+

DNA product of RT is the direct substrate for IN, and genetic and biochemical studies indicate that these enzymes function in concert within infecting particles. As already noted, IN sequences actually are included in one of the subunits of avian sarcoma/leukosis virus RT, and gentle extraction of murine leukemia virus particles yields RT-IN complexes. However, as with RT, virtually nothing is known about the molecular organization of IN within virions.

The first insights into the mechanism of the integration process came in the early 1980s, when it was established that proviral DNA is flanked by LTRs and is colinear with the unintegrated viral DNA and the RNA genome (Fig. 7.15). Nucleotide sequencing of cloned retroviral DNAs and host-virus DNA junctions revealed several unique features of the process. Both viral and cellular DNAs undergo characteristic changes; viral DNA is cropped, usually by 2 bp from each end, and a short duplication of host DNA flanks the provirus on either end. Finally, the proviral ends of all retroviruses comprise the same dinucleotide: 5′-TG . . . CA-3′. This dinucleotide is often embedded in an extended, imperfect inverted repeat that can be as long as 20 bp for some viral genomes. The fact that the length of the host cell DNA duplication is characteristic of the virus, provided the first clue that a viral protein must play a critical role in the integration process.

The inverted terminal repeat, conserved terminal dinucleotide sequence, and flanking direct repeats of host DNA were strikingly reminiscent of features observed earlier in a number of bacterial transposons and the *E. coli* bacteriophage Mu (for "mutator"). Homologies to the predicted amino acid sequences in a portion of the retroviral IN were also found in the transposases of certain bacterial insertion sequences and transposons such as Tn*5*. This observation suggested that, like RT, IN probably evolved before the divergence of prokaryotes and eukaryotes. These similarities predicted what is now known to be a common mechanism for retroviral DNA integration and DNA transposition.

Integrase-Catalyzed Steps in the Integration Process

A generally accepted model for the IN-catalyzed reactions has been developed on the basis of results from many different types of experiment, including studies of infected cells and reconstituted systems (Box 7.5). The two ends of viral DNA

Figure 7.15 Characteristic features of retroviral integration. Unintegrated linear DNA of the avian retrovirus avian sarcoma/leukosis virus (top) after reverse transcription has produced blunt-ended LTRs (Fig. 7.3). The break in the bottom (+) strand indicates that this strand may include discontinuities, whereas the top (−) strand must be continuous (Fig. 7.7). Two base pairs (AA·TT) are lost from both termini upon completion of the integration process, and a 6-bp "target site" in host DNA (pink, indicated by an arrow) is duplicated on either side of the proviral DNA. The integrated proviral DNA (middle) includes short, imperfect inverted repeats at its termini, which end with the conserved 5′-TG...CA-3′ sequence; these repeats are embedded in the LTR, which is itself a direct repeat. The gene order is identical in unintegrated and proviral DNA, and is colinear with that in the viral RNA genome (bottom), for which a provirus serves as a template (described more fully in Chapter 8).

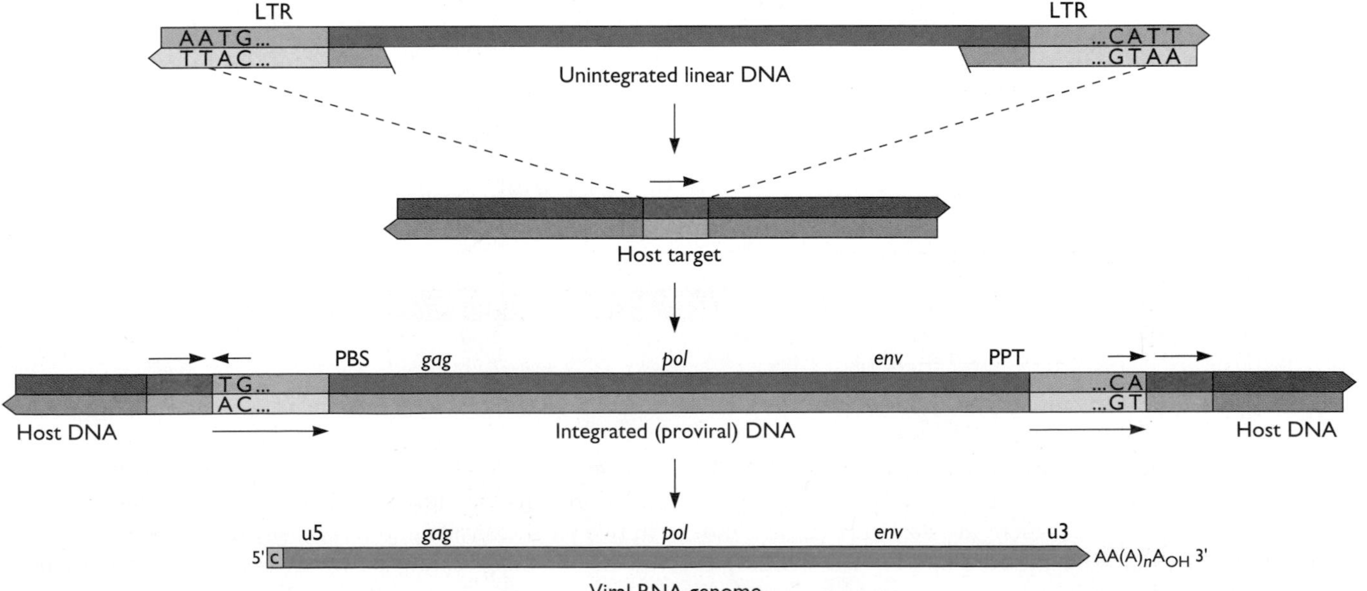

BOX 7.5

BACKGROUND

Model in vitro reactions uncover catalytic mechanisms

The development of simple *in vitro* assays for the processing and joining steps catalyzed by IN marked an important turning point for investigation of the biochemistry of these reactions. With such assays, it was discovered that retroviral IN protein is both necessary and sufficient for catalysis, that no exogenous source of energy (ATP or an ATP-generating system) is needed, and that the only required cofactor is a divalent cation, Mn^{2+} or Mg^{2+}. Use of simple substrates and other biochemical analyses with purified IN protein helped to delineate the sequence and structural requirements for DNA recognition.

The substrates in the simplest assays comprise short duplex DNAs (ca. 25 bp) whose sequences correspond to one retroviral DNA terminus, and are labeled with ^{32}P (red asterisk). The use of such model viral DNA substrates showed that IN can also catalyze an apparent reversal of the joining reaction, which has been called disintegration. Such assays also showed that while the processing, joining, and disintegration reactions produce different products, their underlying chemistry is the same. All comprise a nucleophilic attack on a phosphorus atom by the oxygen in an OH group, and result in cleavage of a phosphodiester bond in the DNA backbone. In **processing (A)**, the OH comes from a water molecule. In **joining (B)**, the OH is derived from the processed 3′ end of the viral DNA, and the result is a direct transesterification. In **disintegration (C)**, also a direct transesterification, a 3′-OH end in the interrupted duplex attacks an adjacent phosphorus atom, forming a new phosphodiester bond and releasing the overlapping DNA. The products of all these reactions can be distinguished by gel electrophoresis.

Although assays with short, single-viral-end model substrates have been invaluable in elucidation of the catalytic mechanisms of IN, they are limited in that the major products represent "half-reactions" in which only one viral end is processed and joined to a target. More recently, conditions for efficient, **concerted processing and joining (D)** of two viral DNA ends to a target DNA have been described, with a variety of specially designed "miniviral" model DNA substrates. The concerted reaction has been shown to be more efficient with longer (150- to 900-bp) DNA "donor" fragments that include viral DNA sequences at a terminus. After preincubation, excess plasmid DNA is added as target, and the concerted joining of two donor fragments produces a linear DNA product that can be detected after electrophoresis in an agarose gel.

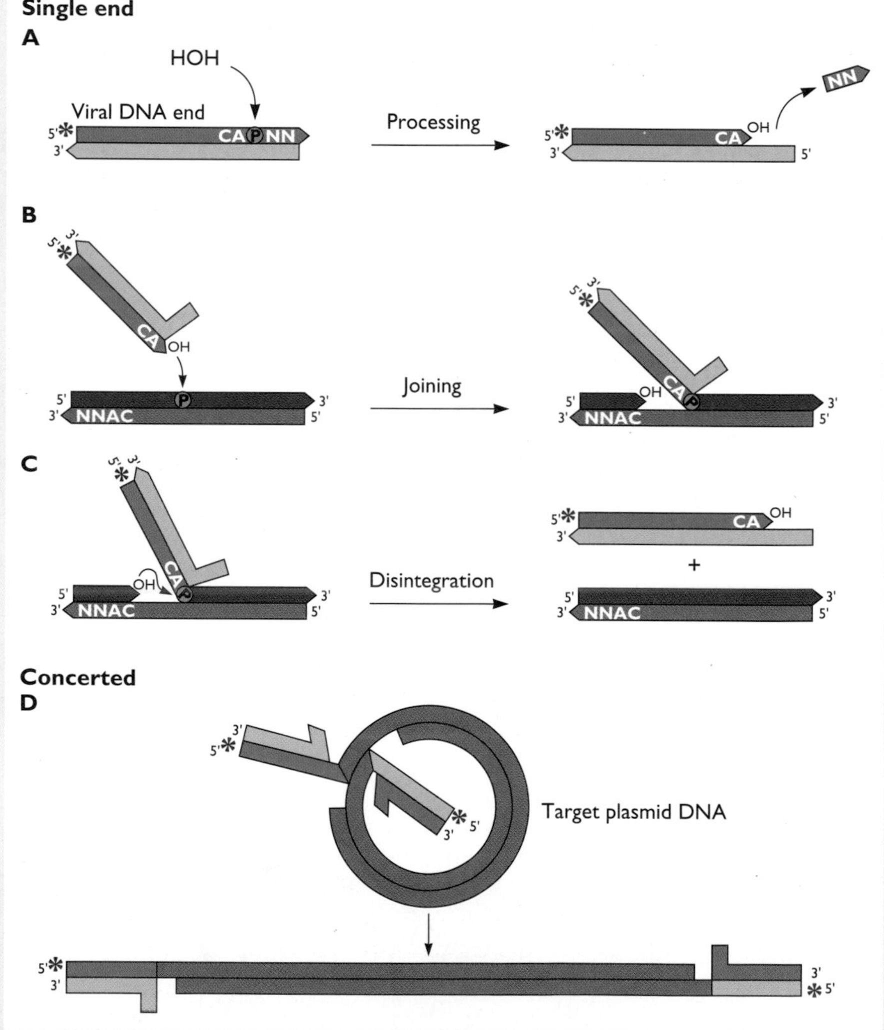

are recognized specifically and nicked, and the new 3′ ends are then joined covalently to the host DNA in a relatively sequence-independent manner, at staggered nicks also introduced by IN. The enzyme accomplishes these multiple activities as a multimer in a large nucleoprotein complex, probably assisted by other viral components as well as cellular proteins. As illustrated in Fig. 7.16, IN catalysis occurs in two biochemically and temporally distinct steps.

The first step catalyzed by IN is a processing reaction, which requires duplex ends and therefore can take place

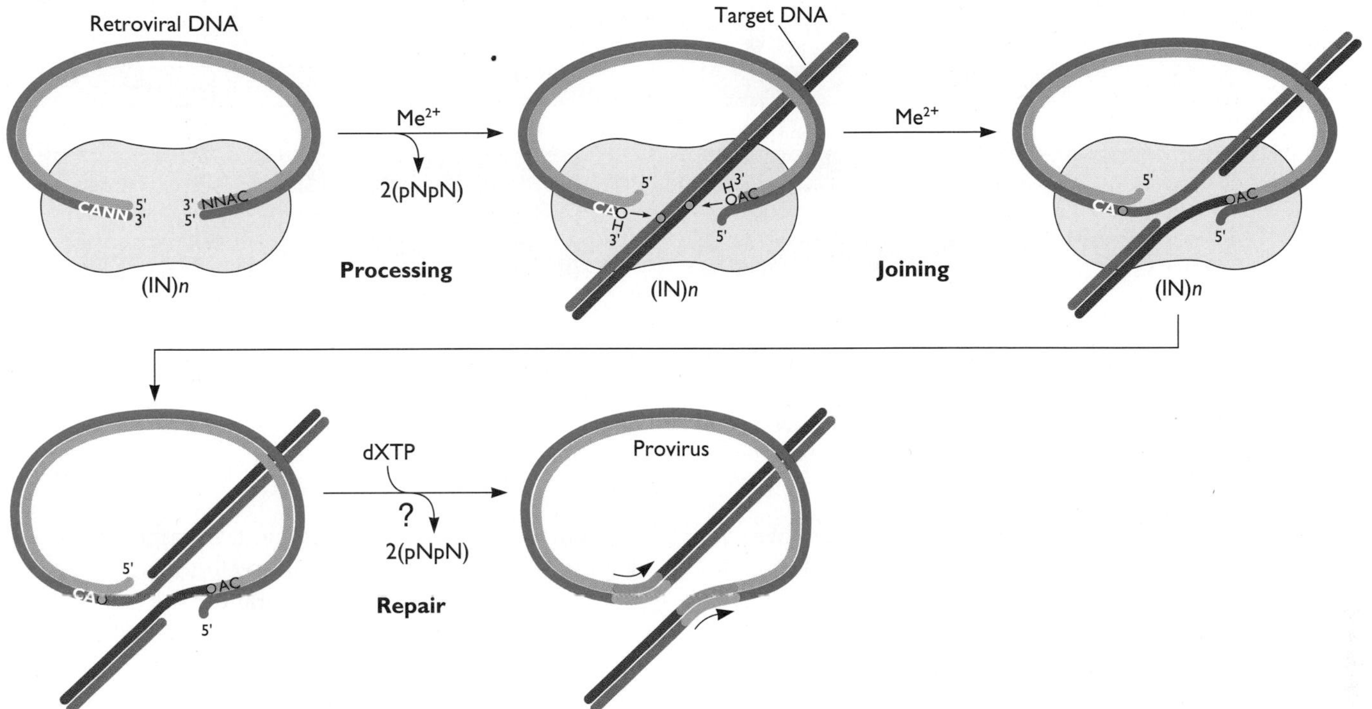

Figure 7.16 Three steps in the retroviral DNA integration process. Endonucleolytic nicking adjacent to the conserved dinucleotide near each DNA end results in the removal of a terminal dinucleotide, with formation of a new, recessed 5′...CA_{OH}-3′ end that will be joined to target DNA in the second step of the IN-catalyzed reaction. Both processing and joining reactions require a divalent metal, Mg^{2+} or Mn^{2+}. The region shaded in beige represents an IN multimer [(IN)*n*]. Results of site-directed mutagenesis of viral DNA ends have shown that the conserved 5′...CA_{OH}-3′ dinucleotide is essential for correct and efficient integration. The small gold circles represent the phosphodiester bonds cleaved and reformed in the joining reaction. The final step in the integration process is a repair reaction that utilizes host enzymes. Adapted from N. D. Grindley and A. E. Leschziner, *Cell* **83:**1063–1066, 1995, with permission.

only when synthesis of the ends of the unintegrated linear viral DNA is complete. This requirement prevents the integration of defective molecules, with imperfect ends. A specific sequence in the duplex viral DNA and close proximity of this sequence to a terminus are critical, but sequences upstream also affect the efficiency of the reaction (Fig. 7.17). It has been shown that the processing reaction can take place in the cytoplasm of an infected cell before viral DNA enters the nucleus, within a subviral structure commonly referred to as the **preintegration complex** (described below). Although there is strong evidence for sequence specificity for processing *in vivo,* only limited sequence-specific binding of purified IN protein to retroviral DNA has been detected in reconstituted systems. It seems likely, therefore, that some structural features or interactions among components within the preintegration complex help to place IN at its site of action near the viral DNA termini. The second step catalyzed by IN is a concerted cleavage and ligation reaction in which the two newly processed 3′ viral DNA ends are joined to staggered (4- to 6-bp) phosphates at the target site in host DNA (Fig. 7.16). The product of the joining step is a **gapped intermediate** in which the 5′-PO_4 ends of the viral DNA are not linked to the 3′-OH ends of host DNA.

Host Proteins Are Recruited for Repair of the Integration Intermediate

Retroviral DNA integration creates a discontinuity in the host cell chromatin, and repair of this damage is required to complete the integration process (Fig. 7.15). As with double-strand breaks produced by ionizing radiation or genotoxic drugs, retroviral DNA integration promotes rapid phosphorylation of the histone variant H2AX in the vicinity of the integration site and recruitment of proteins of the DNA damage-sensing pathways. Various lines of evidence indicate that components of the nonhomologous end-joining DNA repair pathway (DNA-dependent protein kinase, ligase IV, and Xrcc4) are required for postintegration repair. In cells that are defective in any of these components, retroviral DNA integration is essentially a lethal

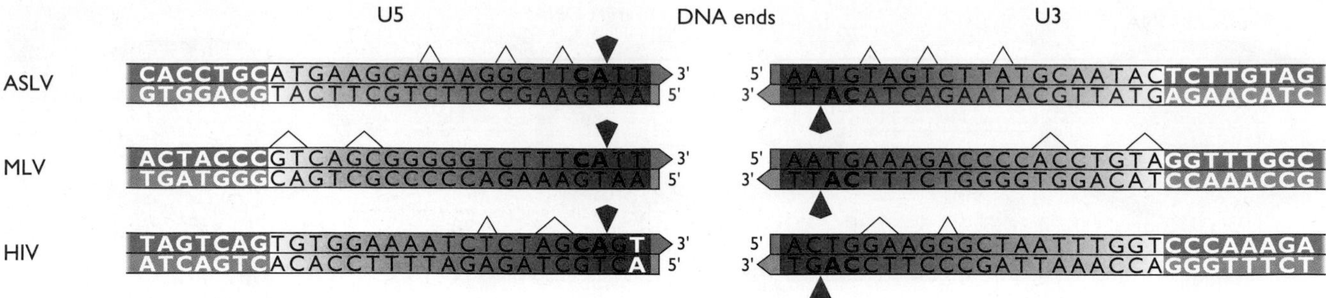

Figure 7.17 Nucleotide sequences at the termini of retroviral DNAs. Sequences are arranged with juxtaposed LTR termini. Red-shaded sections are important for integration. The most intensely shaded regions appear to be most critical, and the conserved 5′...CA...3′ is essential for efficient and accurate integration. Triangles above the sequences mark positions where base pairs are not repeated. Red arrowheads indicate the location of the nick produced during processing by IN.

event, triggering either cell cycle arrest or programmed cell death. It is likely that other host proteins play a role both in postintegration repair and in reconstitution of chromatin structure at the site of integration.

Properties of the Preintegration Complex

Rapidly sedimenting nucleoprotein complexes have been isolated from the cytoplasm of cells infected with avian sarcoma/leukosis virus, murine leukemia virus, and human immunodeficiency virus type 1. These complexes appear to be compact, but the viral DNA associated with them is accessible to nucleases. The nucleoprotein complexes contain IN and viral DNA in a form that can be joined to exogenously provided plasmid or bacteriophage DNA. Such *ex vivo* reactions exhibit all the features expected for products of authentic integration. Other viral proteins, among them RT, capsid, and nucleocapsid proteins, have also been reported to be present in the preintegration complexes of murine leukemia and human immunodeficiency viruses, but no clearly defined role in the integration process has yet been assigned to any of them. Despite intense efforts, the mechanism by which retroviral preintegration complexes enter the nucleus is still unclear, and variable (often contrasting) results have generated a good deal of controversy that has yet to be resolved (see Chapter 5).

Multiple Parameters Affect Selection of Host DNA Target Sites

Many sites in host DNA can be targets for retroviral DNA integration, and *in vitro* studies have revealed little sequence specificity for such sites, although the patterns exhibited by different retroviral INs with the same DNA target are not identical. However, such studies have established a preference for integration into DNA sequences that are intrinsically bent or underwound as a consequence of being wrapped around a nucleosome.

Recent advances in genomics and transcriptional profiling have provided a wealth of both technical and informational resources for studying integration site selection by retroviruses following infection of cultured cells. Results from a number of investigators who, collectively, have mapped thousands of integration sites in human and other cell lines have identified weak consensus sequences for host target sites. As illustrated for the two cases shown in Figure 7.18, the sequence patterns preferred for integration of proviruses of different genera are distinct. Although all of the consensus sequences studied to date form weak palindromes, the significance of this observation is still unclear. The symmetry in the patterns is consistent with the idea that IN complexes function as symmetrical multimers in the preintegration complex (discussed in "Higher-Order Structure in the Preintegration Complex" below). However, this statistically derived feature is also consistent with a mechanism whereby a preferred, nonpalindromic sequence can be used for integration of viral DNA in either direction. The pattern observed after insertion of viral DNA into purified human DNA by isolated preintegration complexes of human immunodeficiency virus type 1 is similar to that seen after infection of cultured cells by this virus (Fig. 7.18). It seems likely, therefore, that recognition by IN proteins depends on structural features that are shaped by primary DNA sequence.

These same large-scale, global analyses have also shown that all human chromosomes are targets for integration, but different retroviruses display distinguishable preferences for defined chromosomal features (Table 7.1). For example, human immunodeficiency virus type 1 shows the highest preference for integration anywhere inside of genes, especially in highly transcribed genes, whereas murine leukemia virus integrates preferentially in and near transcription start sites. The avian sarcoma/leukosis virus also shows a slight preference for integrating within

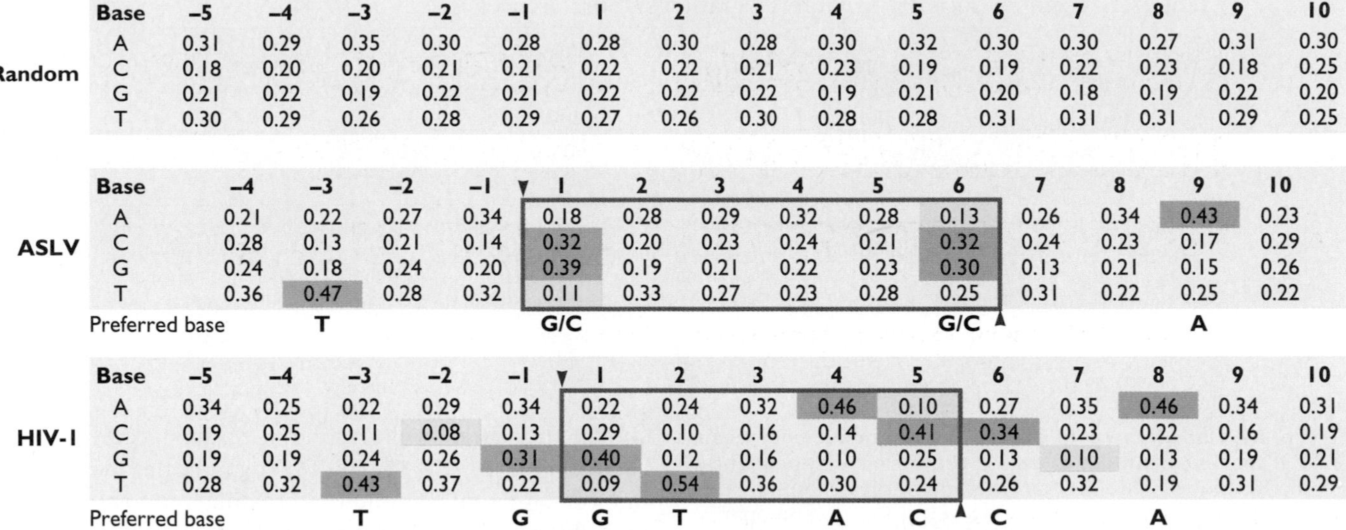

Random

Base	−5	−4	−3	−2	−1	1	2	3	4	5	6	7	8	9	10
A	0.31	0.29	0.35	0.30	0.28	0.28	0.30	0.28	0.30	0.32	0.30	0.30	0.27	0.31	0.30
C	0.18	0.20	0.20	0.21	0.21	0.22	0.22	0.21	0.23	0.19	0.19	0.22	0.23	0.18	0.25
G	0.21	0.22	0.19	0.22	0.21	0.22	0.22	0.22	0.19	0.21	0.20	0.18	0.19	0.22	0.20
T	0.30	0.29	0.26	0.28	0.29	0.27	0.26	0.30	0.28	0.28	0.31	0.31	0.31	0.29	0.25

ASLV

Base	−4	−3	−2	−1	1	2	3	4	5	6	7	8	9	10
A	0.21	0.22	0.27	0.34	0.18	0.28	0.29	0.32	0.28	0.13	0.26	0.34	0.43	0.23
C	0.28	0.13	0.21	0.14	0.32	0.20	0.23	0.24	0.21	0.32	0.24	0.23	0.17	0.29
G	0.24	0.18	0.24	0.20	0.39	0.19	0.21	0.22	0.23	0.30	0.13	0.21	0.15	0.26
T	0.36	0.47	0.28	0.32	0.11	0.33	0.27	0.23	0.28	0.25	0.31	0.22	0.25	0.22
Preferred base		T			G/C					G/C			A	

HIV-1

Base	−5	−4	−3	−2	−1	1	2	3	4	5	6	7	8	9	10
A	0.34	0.25	0.22	0.29	0.34	0.22	0.24	0.32	0.46	0.10	0.27	0.35	0.46	0.34	0.31
C	0.19	0.25	0.11	0.08	0.13	0.29	0.10	0.16	0.14	0.41	0.34	0.23	0.22	0.16	0.19
G	0.19	0.19	0.24	0.26	0.31	0.40	0.12	0.16	0.10	0.25	0.13	0.10	0.13	0.19	0.21
T	0.28	0.32	0.43	0.37	0.22	0.09	0.54	0.36	0.30	0.24	0.26	0.32	0.19	0.31	0.29
Preferred base			T		G	G	T		A	C	C		A		

Figure 7.18 Palindromic consensus sequences at retroviral integration sites. The frequency of each base at each position around the integration sites was calculated, where 1 equals 100%. Integration occurs between positions −1 and 1 on the top strand. Colored positions have statistically different frequencies of bases from that of randomly generated sequences shown at the top. Bases with a greater than 10% increase of frequency at a position are blue, and bases with a greater than 10% decrease of frequency at a position are yellow. The preferred bases are listed below. Inferred duplicated target sites are in the blue box, and joining to the 3′ ends of viral DNA occurs at positions labeled by arrows. The symmetry of the palindromic patterns is centered on the duplicated target sites. Adapted from X. Wu et al., *J. Virol.* **79:**5211–5214, 2005.

genes, but not necessarily those that are highly transcribed, and it displays no preference for transcription start sites. These observations suggest that the interaction of pre-integration complexes with different chromatin-bound proteins promotes integration into specific chromosomal locations. Indeed, recent studies have shown that interaction of lentiviral IN proteins with a 75-kDa transcriptional coactivator, (mis)named lens epithelium-derived growth factor **(Ledgf)**, is a critical component of integration site selection for these viruses. The efficiency of integration of human immunodeficiency virus type 1 is greatly reduced in cells in which Ledgf has been depleted by treatment with small interfering RNA (siRNA), or in which the gene has been deleted. Furthermore, the pattern of preference for various chromosomal features is altered in the small percentage of Ledgf-deficient cells in which integration does occur (Table 7.1). Exactly how this protein promotes such preferential integration is still unclear. Interactions that may be responsible for the preferences observed with other retroviral genera have also yet to be defined. Because integration of the yeast transposon Ty3 into sites upstream of polymerase III promoters has been linked to interaction of the Ty3 IN protein with RNA polymerase III transcription initiation factors, it is possible that a promoter-specific protein could play a role in the preferential integration of murine leukemia virus DNA near transcription start sites.

Integrase Structure and Mechanism

IN Proteins Are Composed of Three Structural Domains

Retroviral IN proteins are approximately 300 amino acids in length, comprising three domains (Fig. 7.19). Attempts to determine the three-dimensional structure of a full-length IN by X-ray crystallography have so far proved unsuccessful. However, considerable insight has been obtained from analyses of single- and two-domain polypeptides.

The **N-terminal domain**, comprising approximately the first 50 amino acids, is characterized by two pairs of invariant, Zn^{2+}-chelating histidine and cysteine residues (HHCC motif). The bound Zn^{2+} ion stabilizes a helix-turn-helix structural motif that is almost identical in topology to the DNA-binding domain of the bacterial *trp* repressor protein. The **catalytic core domain**, included in a central region of approximately 150 amino acids, is characterized by a constellation of three invariant acidic amino acids, the last two separated by 35 residues, the D,D(35)E motif. Results of site-directed mutagenesis experiments have

Table 7.1 Comparison of retroviral integration site preferences

Site or region	% Integration[a]					
	Human cells[b]				Mouse cells[c]	
	Random	ASLV	MLV	HIV	$HIV^{LEDGF^{+/-}}$	$HIV^{LEDGF^{-/-}}$
Within genes	26	42	40	60–70	62	44
Transcription start sites	5	8	20	10	6	17

[a]HIV, human immunodeficiency virus; ASLV, avian sarcoma/leukosis virus; MLV, murine leukemia virus.

[b]Percentages are approximates for integration into human cells and are from A. Narezkina et al., *J. Virol* **78:**11656–11663, 2004.

[c]Percentages for mouse embryo fibroblasts are from M. C. Shun et al., *Genes Dev.*, **21:**1767–1768, 2007. Calculations performed in this study indicated HIV gene usage in mouse $LEDGF^{-/-}$ cells at ~8% above random, which was ~3% less than ASV/MLV gene usage in human cells.

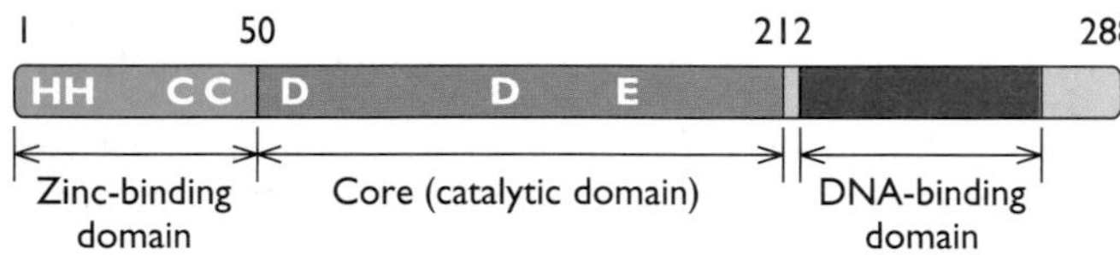

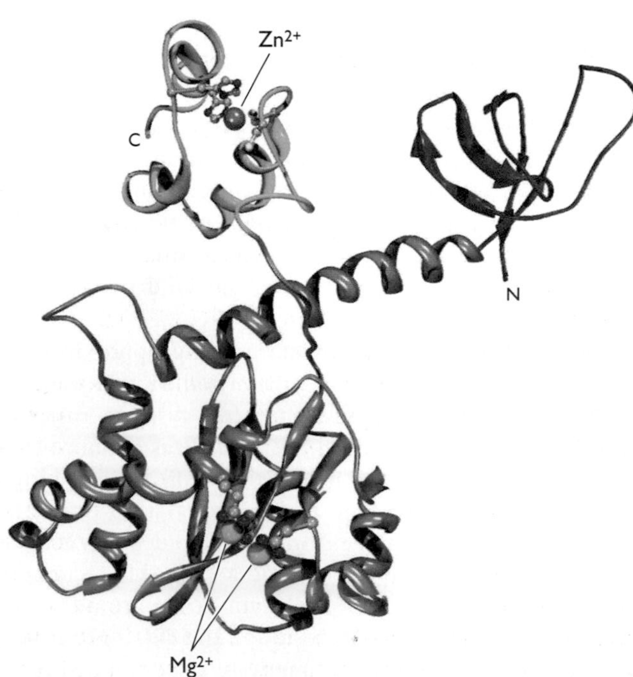

Figure 7.19 Linear map of human immunodeficiency virus type 1 IN and a model showing the three independently folding domains. Numbers at the top indicate amino acid residues starting with 1 at the N terminus. Evolutionarily conserved amino acids are indicated in the single-letter code. In the ribbon model below, conserved amino acids are in ball-and-stick representation with metal ions represented by orange (Zn^{2+}) or green (Mn^{2+} or Mg^{2+}) balls. The conserved Glu residue in the core domain is on an unstructured region in the X-ray structure used for the model. It is presumed to chelate the second metal ion together with one of the conserved Asp residues, as illustrated in Fig. 7.20. Coordinates are from Protein Data Bank codes 1K6Y and 1EX4; figure prepared by Mark Andrake, Fox Chase Cancer Center.

shown that these acidic amino acids are required for all catalytic functions of IN. This finding indicates that there is one catalytic center for both processing and joining. Solution of the crystal structures of the catalytic core domains of the human and avian viral IN proteins was an important milestone in the study of integration. These core domain structures established that IN proteins are members of a large superfamily of nucleases and recombinases that includes the RNase H domain of the human virus RT (Fig. 7.20). The sequence of the 80- to 100-amino-acid **C-terminal domain** is the least conserved among IN proteins from different retroviral genera. However, despite their sequence differences, the three-dimensional structures of this domain are quite similar in the two examples analyzed to date: the human immunodeficiency virus type 1 and the avian leucosis/sarcoma virus IN proteins. This domain contains critical DNA-binding activity and multimerization determinants. A molecular model of a full-length IN protein, which illustrates the structural features of each domain, is shown in Figure 7.19.

Structures of two-domain proteins comprising the catalytic core and either the N- or C-terminal domains of several retroviral IN proteins have been solved by X-ray crystallography. A model of a full-length IN dimer, based on a consolidation of the two-domain structures, is shown in Fig. 7.21. Although no C-terminal domain interactions are depicted in this model, previous biochemical studies indicate that this domain contributes to formation of multimers. C-terminal interactions are included in several computer-derived molecular models of IN tetramers, but to date none of these models has been fully validated.

Higher-Order Structure in the Preintegration Complex

The joining of viral to host DNA comprises a dynamic, multistep reaction in which two separate DNA duplexes are bound to an IN complex, cut, and recombined with a DNA target. We have yet to understand completely how the substrates and protein are organized and acted upon

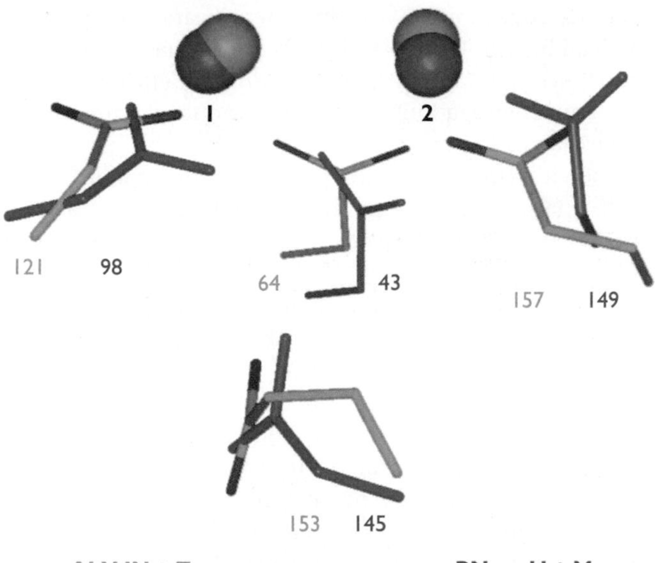

Figure 7.20 Metal complexes in the human immunodeficiency virus type 1 RNase H and the avian sarcoma/leukosis virus (ALV) IN catalytic core domain. The orientation of the side chains and the positions of the metals in the catalytic core domain of human immunodeficiency virus type 1 and avian sarcoma/leukosis virus IN proteins are superimposable on those of metal ion complexes obtained with the RNase H domain of the human immunodeficiency virus type 1 RT (Fig. 7.11). In the comparison shown here, the side chains of critical acidic amino acids form a tripod that interacts with two divalent cations in almost superimposable positions. The IN catalytic core also binds a single Mn^{2+} or Mg^{2+} ion in position 1 with no change in the position of the side chains of D64 and D121. The binding of two metal ions at the active sites of the viral RNase H and IN proteins is consistent with a two-metal mechanism of catalysis by this superfamily, as proposed for the human retroviral RT polymerase activity.

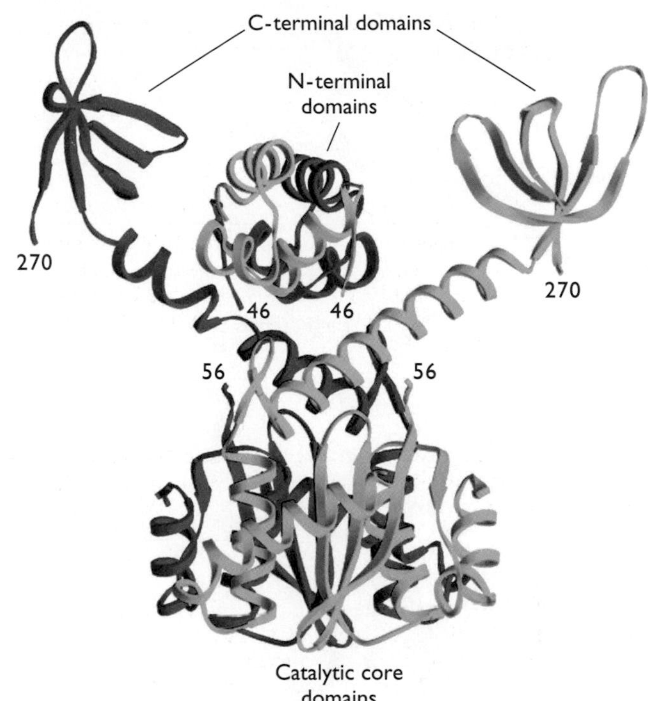

Figure 7.21 Model of a dimer of human immunodeficiency virus type 1 integrase. This ribbon model is derived from the superposition of the nearly identical core domain dimers in the crystal structures of the core with the N-terminal domain (Protein Data Bank code 1K6Y), and the core with the C-terminal domain (Protein Data Bank code 1EX4). The domains comprising the individual monomers are shown in green and blue. Courtesy of Robert Craigie, National Institutes of Health.

in any recombination reaction. However, because details of the structure of IN and substrate DNAs in the preintegration complex are important for the development of antiviral drugs, many investigators have focused on this question.

Organization of DNA ends. In most models of the IN-catalyzed reactions, viral DNA ends are placed near one another, as presumed to be necessary for their concerted processing and joining to a target site. The presence of inverted repeats at these ends (Fig. 7.17) suggests that they are recognized by a multimeric protein with a twofold axis of symmetry. *In vitro* experiments indicate that the invariant dinucleotides at each viral DNA end are bound to IN at a fixed distance from each other during the concerted processing reaction. The optimal distance between them is related to the distance between the staggered cuts in target DNA made during joining. These and other experiments also indicate that the viral DNA ends do not remain double stranded when bound at the active site of the enzyme, but that the two strands are partially unwound and distorted.

A multimeric form of IN is necessary for activity. The results of early biochemical analyses and *in vitro* complementation studies indicated that IN functions as a multimer. A reversible equilibrium among monomeric, dimeric, and tetrameric forms of IN is observed in the absence of DNA substrate. A conservative estimate of the intracapsid concentration of IN gives a value (~150 μM) high enough for most of the IN protein in virions to be in the form of dimers and tetramers. Recent *in vitro* studies indicate that an IN tetramer is stabilized by interaction with a pair of viral DNA ends, and various lines of evidence suggest that each end is held mainly through contacts with C-terminal domain residues in one IN monomer, and then positioned and acted upon by the catalytic core domain of another. In this model, the catalytic core domains of only two of the four subunits in the IN tetramer would provide catalytic function (as illustrated in Fig. 7.22). DNA footprinting experiments with preintegration complexes isolated from

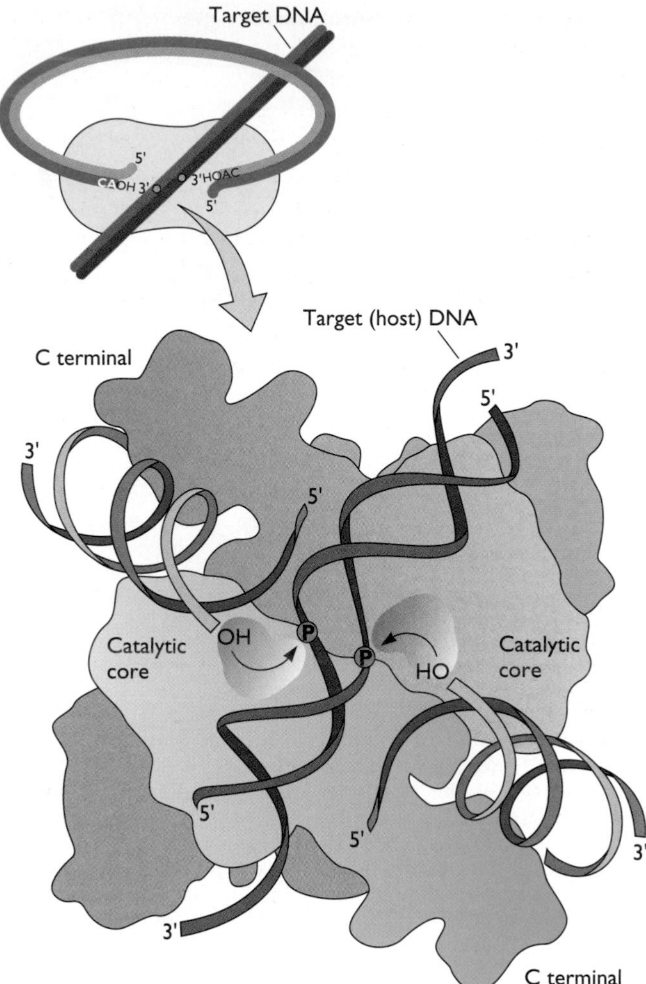

Figure 7.22 Tetramer model for functional integrase. The active sites of only two IN monomers (represented in green and yellow) participate in catalysis during the concerted joining reaction illustrated. As suggested by various lines of evidence, the distal end of viral DNA is held by the C-terminal domain of one monomer and the terminus is acted upon by the catalytic core domain of a second monomer. The other two subunits likely contribute to the stability of the complex. Binding of the target DNA to basic residues at the dimer-dimer interface may help to stabilize the tetramer.

Moloney murine leukemia virus-infected cells suggest that several hundred base pairs of the viral DNA ends are protected by association with IN and therefore that many multimers may be bound. The significance of this association is not yet known.

Host Proteins That May Regulate the Integration Reaction

Two general approaches have been used to identify host proteins that are important for retroviral DNA integration: analysis of the cellular components in purified preintegration complexes, and identification of proteins that can bind to IN. Most of these studies have focused on the clinically important human immunodeficiency virus, but some investigations have been conducted with murine leukemia virus.

The viral DNA within the preintegration complex is not itself a target for joining; such autointegration reactions would be suicidal. Analysis of preintegration complexes of murine leukemia virus showed that this restriction depends on the presence of an 89-amino-acid cellular protein called **barrier-to-autointegration factor (Baf).** This small protein forms dimers in solution, binds to DNA, and can produce intermolecular bridges that compact the DNA. It has been proposed that such compaction prevents autointegration. As purified virions do not contain this cellular protein, it must be acquired from the cytoplasm of a newly infected host cell (Fig. 7.23). The homologous human protein has been shown to block autointegration in human immunodeficiency virus preintegration complexes. Baf was recently shown to bind to a specific domain (the Lem domain) in **lamina-associated polypeptide 2α (Lap2α)**, another cellular protein that accumulates in the preintegration complex of murine leukemia virus. Baf also binds to the Lem domain of **emerin**, a component of the inner nuclear membrane. Transcribed chromatin tends to be associated with the nuclear lamina, and the interaction of Baf with emerin has been proposed to promote access of the preintegration complex to chromatin after the complex has entered the nucleus. However, as deletion of either Lap2α or emerin from mouse cells is reported to have little or no effect on the replication of murine leukemia virus or human immunodeficiency virus type 1, neither interaction can be essential for integration. Another host protein, the high-mobility-group, nonhistone DNA-bending protein **Hmga1**, is present in human immunodeficiency preintegration complexes and was originally thought to play a role in targeting integration to chromatin. However, because cells that lack Hmga1 have no obvious defect in human immunodeficiency virus type 1 replication, this protein is also either redundant or not essential for integration.

The first host protein reported to bind specifically to human immunodeficiency virus type 1 IN was called IN-interacting protein 1 **(Ini-1)**. It was later found to be a core component of the Swi/Snf chromatin remodeling complex. After infection, Ini-1 translocates from the nucleus to the preintegration complex in the cytoplasm. Although the idea that Ini-1 might play a role in targeting the preintegration complex to chromatin initially seemed attractive, no apparent integration defect has been observed in cells that lack this protein. The only human immunodeficiency virus IN-binding protein that has a verified role in mediating access to host chromatin is Ledgf (Fig. 7.23).

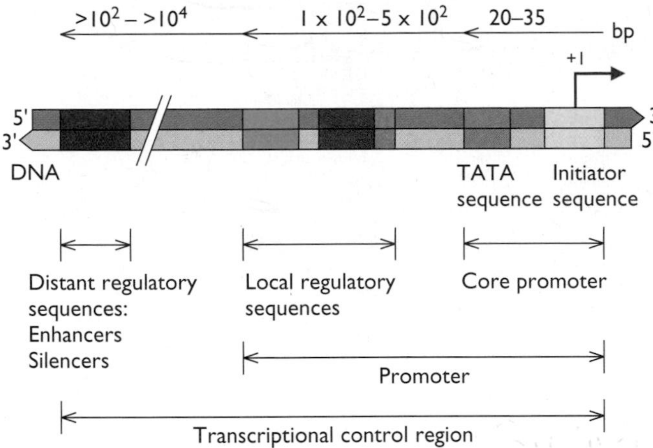

Figure 8.1 RNA polymerase II transcriptional control elements. The site of initiation is represented by the red arrow drawn in the direction of transcription on the nontranscribed DNA strand, a convention used throughout this text. The core promoter comprises the minimal sequence necessary to specify accurate initiation of transcription. The TATA sequence is the binding site for TfIId (Box 8.4), and the initiator is a sequence sufficient to specify initiation at a unique site. The activity of the core promoter is modulated by local regulatory sequences typically found within a few hundred base pairs of the initiation site. The location of these sequences upstream of the TATA sequence as shown is common, but such sequences can also lie downstream of the initiation site (see, e.g., Fig. 8.3). Distant regulatory sequences that stimulate (enhancers) or repress (silencers) transcription are present in a large number of transcriptional control regions.

initiate transcription as soon as they are recruited to a promoter.

Regulation of RNA Polymerase II Transcription

Numerous patterns of gene expression are necessary for eukaryotic life: some RNA polymerase II transcription units must be expressed in all cells, whereas others are transcribed only during specific developmental stages, or in specialized differentiated cells. Many others must be maintained in a silent ground state, from which they can be activated rapidly in response to specific stimuli, and to which they can be returned readily. Transcription of viral genes is also regulated during the single-cell life cycles of most of the viruses considered in this chapter. Large quantities of viral proteins for assembly of progeny virions must be made within a finite (and often short) infectious cycle. Consequently, some viral genes must be transcribed at high rates. As noted previously, in many cases viral genes are transcribed in a specific temporal sequence. Such regulated transcription is achieved in part by means of cellular control mechanisms, for example, signal transduction cascades that transmit specific environmental stimuli to the transcriptional machinery, or cellular proteins that repress transcription. However, viral proteins are generally critical components of the circuits that establish orderly transcription of viral genes.

BOX 8.2

EXPERIMENTS

Mapping of a human adenovirus type 2 initiation site and accurate transcription in vitro

When cellular RNA polymerase II was identified in 1969, investigators had access to only preparations of total cellular DNA. Furthermore, nothing was known about the organization of eukaryotic transcription units. The genomes of smaller DNA viruses, such as simian virus 40 and human adenovirus type 2, were therefore perceived as a valuable resource for investigation of mechanisms of transcription, a view that inspired many investigators to enter virology. Indeed, it was detailed information about a particular adenoviral transcription unit that finally allowed biochemical studies of the mechanism of initiation. In 1978, the site at which major late transcription begins was mapped precisely, by determining the sequence of the 5′ end of the RNA transcript. This knowledge was exploited to develop a simple assay for accurate initiation of transcription, the "runoff" assay (Box 8.3). Purified RNA polymerase II produced no specific transcripts in the runoff assay, but unfractionated nuclear extracts of human cells were shown to contain all the components necessary for accurate initiation of transcription.

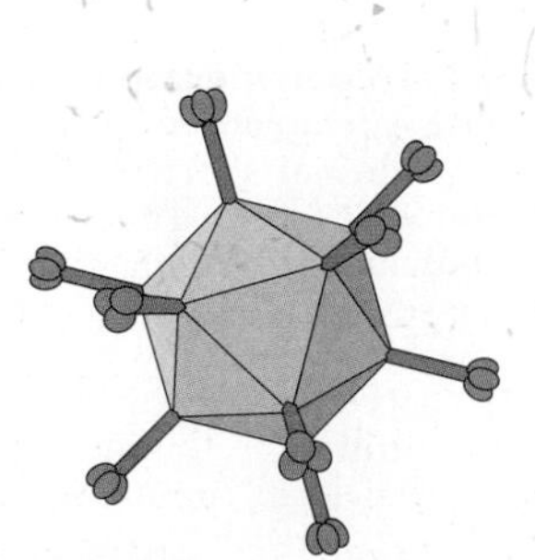

Weil, P. A., D. S. Luse, J. Segall, and R. G. Roeder. 1979. Selective and accurate initiation of transcription at the Ad2 major late promotor in a soluble system dependent on purified RNA polymerase II and DNA. *Cell* **18:**469–484.

Ziff, E. B., and R. M. Evans. 1978. Coincidence of the promoter and capped 5′ terminus of RNA from the adenovirus 2 major late transcription unit. *Cell* **15:**1463–1475.

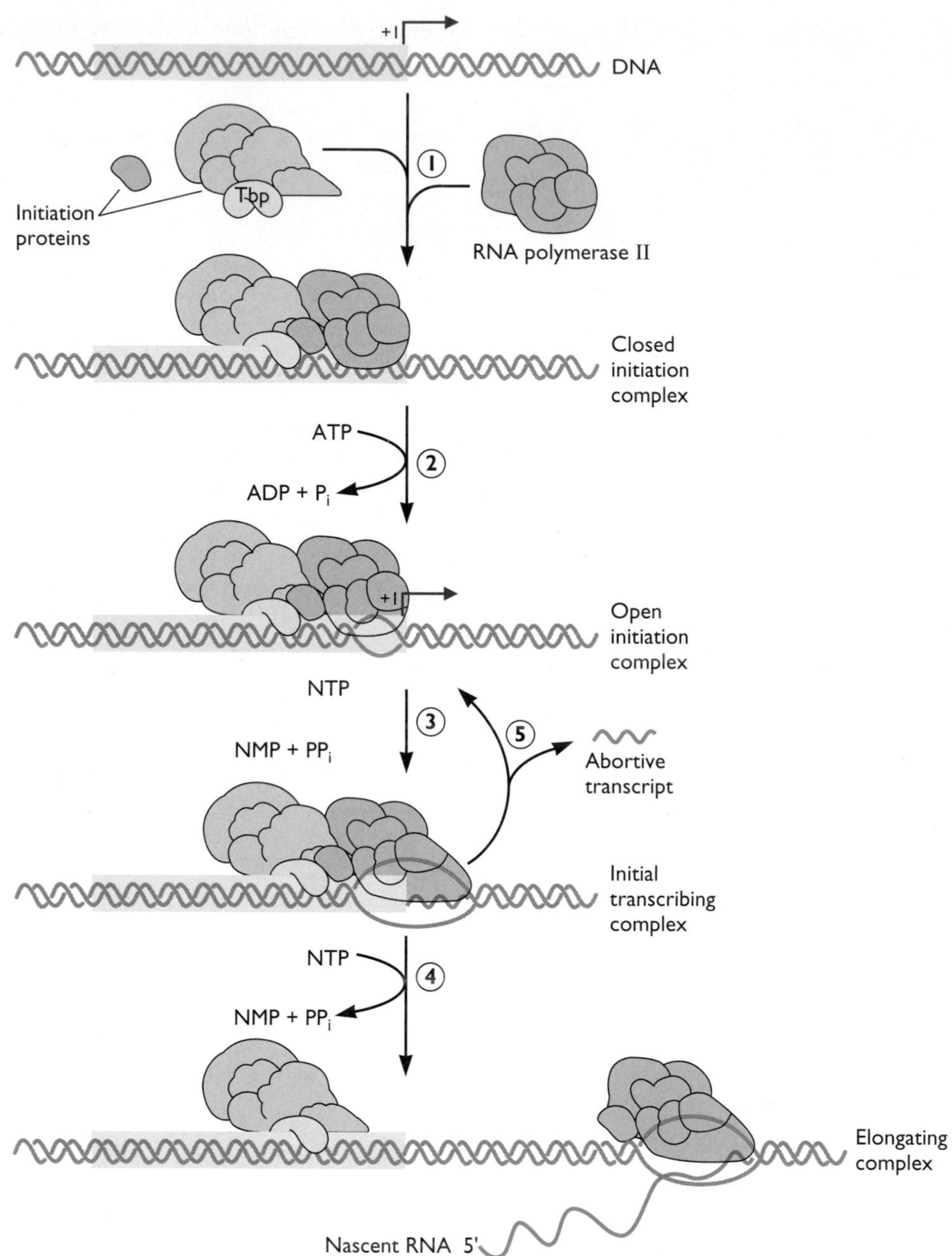

Figure 8.2 Initiation of transcription by RNA polymerase II. Assembly of the closed initiation complex (step 1) is followed by unwinding of the DNA template in the region spanning the site of initiation (step 2). RNA polymerase II then synthesizes an initial short transcript (less than 10 to 15 nucleotides) by template-directed incorporation of nucleotides (step 3). The initial transcribing complex is thought to be conformationally strained, because RNA polymerase II remains in contact with promoter-bound initiation proteins as it synthesizes short RNAs. The severing of these contacts allows the transcribing complex to escape from the promoter and proceed with elongation (step 4). This promoter clearance step is often inefficient, with abortive initiation (step 5) predominating. In the latter process, initial transcripts are cleaved and released, reforming the open initiation complex. The elongating transcriptional complex contains some but not all of the proteins that form the preinitiation complex, as well as proteins that stimulate elongation.

BOX 8.3

METHODS

Assays for the activity of RNA polymerase II promoters

(A) *In vitro* transcription assay. In this simple assay, linear DNA templates are prepared by restriction endonuclease cleavage (black arrow), a known distance, *x* bp, downstream of the initiation site (+1). When the template is incubated with the transcriptional machinery and nucleoside triphosphate (NTP) substrates, transcription initiated at position +1 continues until the transcribing complex "runs off" the linear template. Specific transcription is therefore assayed as the production of ^{32}P-labeled RNA *x* nucleotides in length. This runoff transcription assay is convenient and can be used to assess both specificity and efficiency of transcription. However, it can be applied only to linear DNA templates and often suffers from high background, for example, nonspecific initation of transcription at the ends of the template. These problems can be circumvented by the use of circular templates and an indirect assay for specific transcripts, for example, copying of an unlabeled transcript by reverse transcriptase to form a labeled, complementary DNA. Such *in vitro* transcription assays differ from transcription under normal intracellular conditions in several important ways. Transcription templates are typically provided as naked circular or linear DNA molecules, physical states that may not resemble those of cellular or viral genes within cells. Another important parameter that may be altered is the relative, as well as the absolute, concentrations of the proteins necessary for transcription. **(B)** Transient-expression assay. A segment of DNA containing the transcriptional control region of interest (yellow) is ligated to the coding sequence (orange) of an enzyme not synthesized in the recipient cells to be used (luciferase in this example), and RNA-processing signals, such as those specifying polyadenylation, shown by the green box at the end of the coding sequence. Plasmids containing such chimeric reporter genes are introduced into cells in culture by any one of several methods, including electroporation and incubation with synthetic vesicles containing the plasmid DNA. The proportion of cells that will take up foreign DNA varies with a number of parameters, including cell type. Within a cell that takes up the reporter gene, the DNA enters the nucleus, where the transcriptional control region directs transcription of chimeric RNA. The RNA is exported from the nucleus following processing, and translated on cytoplasmic polyribosomes. The activity of the luciferase enzyme is then assayed, generally 48 h after introduction of the reporter gene. Note that this indirect measure of transcription assumes that it is **only** the activity of the transcriptional control region that determines the concentration level of the enzyme. Alternatively, the concentration of the chimeric reporter RNA can be measured.

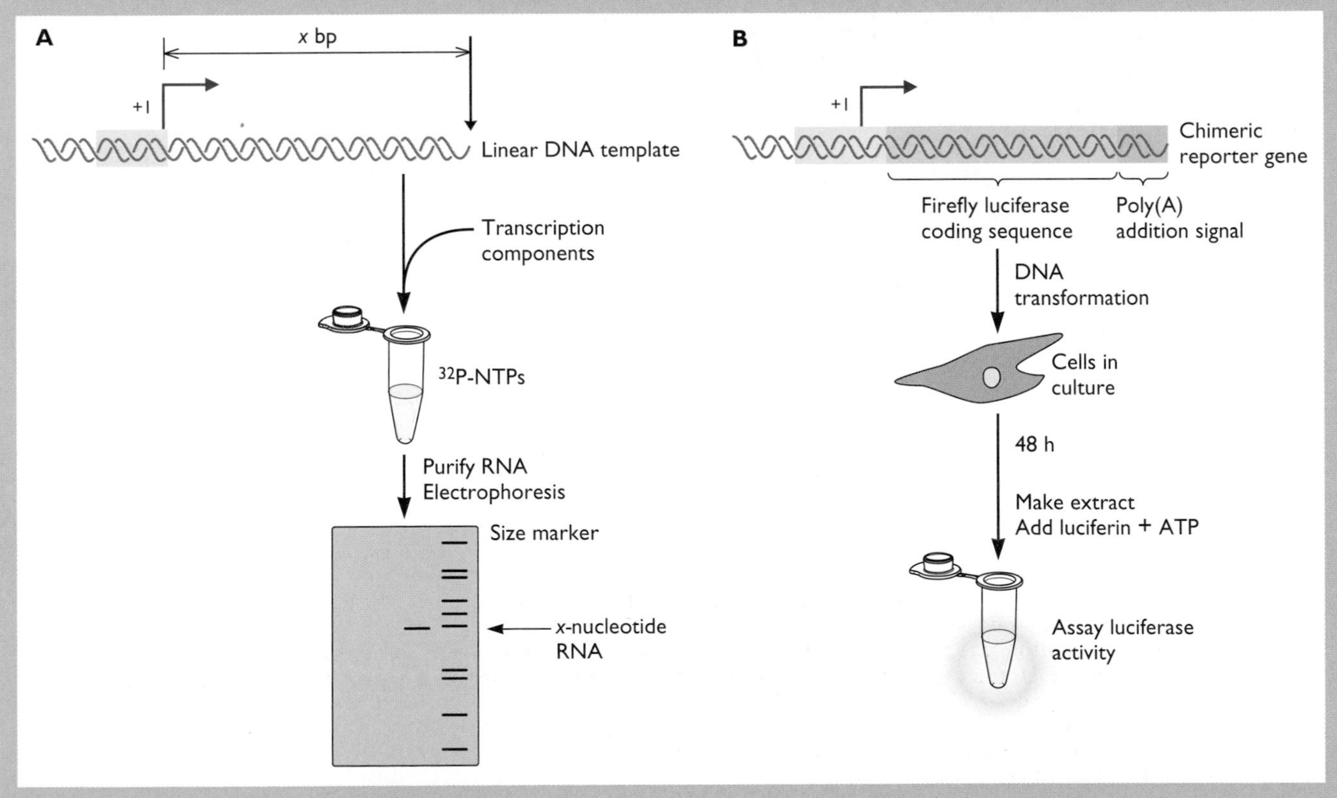

BOX 8.4

BACKGROUND

The RNA polymerase II closed initiation complex

The closed initiation complex is shown on a promoter that contains both a TATA and an initiator sequence (e.g., the adenovirus major late promoter). The TfIId protein contains a subunit that recognizes the TATA sequence (TATA-binding protein [Tbp]) and 8 to 10 additional subunits, termed Tbp-associated proteins (Tafs). X-ray crystal structures of DNA-bound Tbp, such as that of *Arabidopsis thaliana*, bound to the adenoviral major late TATA sequence shown in the inset, revealed that this protein induces sharp bending of the DNA. One popular hypothesis is that such bending facilitates interaction among proteins bound to local regulatory sequences located upstream of the TATA sequence and the basal transcriptional machinery. TfIId is required for transcription from all RNA polymerase II promoters. It can recognize those that lack TATA sequences (see, e.g., Fig. 8.3) by binding of a Taf to an initiator or an internal sequence, or it can be recruited to the promoter by interactions with proteins bound to specific sequences near the initiation site. The largest subunit of TfIIh supplies DNA-dependent ATPase and helicase activities essential for transcription. Two others are components of a kinase that can phosphorylate the C-terminal segment of the largest subunit of RNA polymerase II. The preinitiation complex assembles via an ordered pathway *in vitro*. In the cell, however, many of these proteins are associated with RNA polymerase II prior to encounter with a promoter.

Buratowski, S., S. Hahn, L. Guarente, and P. A. Sharp. 1989. Five intermediate complexes in transcription initiation by RNA polymerase II. *Cell* **56:**549–561.

Kim, J. L., D. B. Nikolov, and S. K. Burley. 1993. Co-crystal structure of TBP recognizing the minor groove of a TATA element. *Nature* **365:** 520–527.

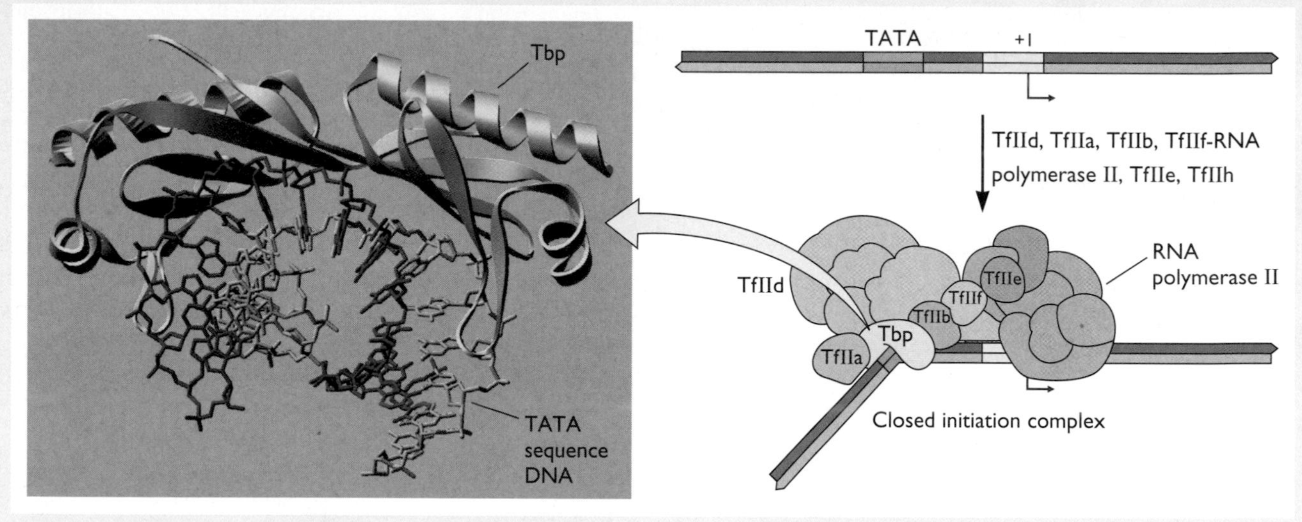

Figure 8.3 Variations in core RNA polymerase II promoter architecture. Variations in promoter architecture are illustrated using four viral promoters represented as in Fig. 8.1. The TATA or initiator sequences of the different promoters are not identical in DNA sequence. In the case of the simian virus 40 (SV40) late transcription unit, each of the sites of initiation is included within a DNA sequence resembling an initiator of another promoter. It has not been shown experimentally that all actually function as autonomous initiator sequences. The relative frequencies with which different initiation sites in a single promoter are used are indicated by the thickness of the red arrows. Ad2, adenovirus type 2.

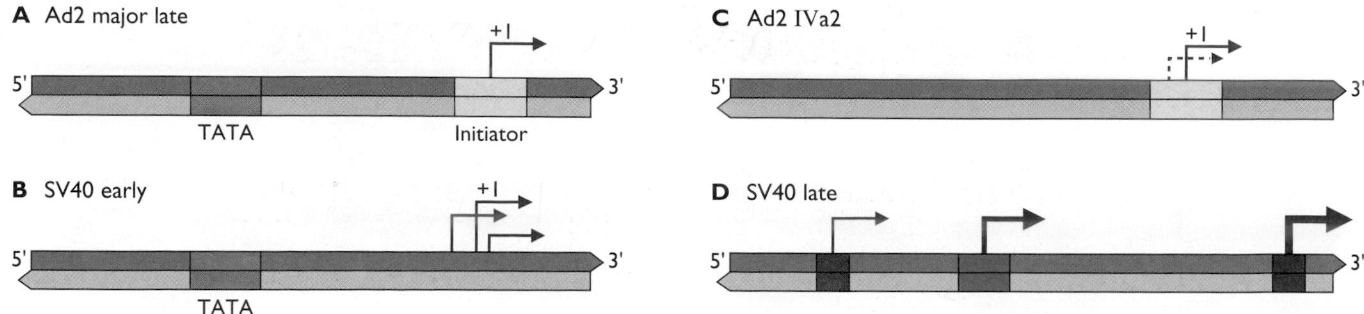

Recognition of Local and Distant Regulatory Sequences

Both local and distant sequences (Fig. 8.1) can control transcription from core promoters. However, in many cases, local sequences are sufficient for proper transcriptional regulation. These local regulatory sequences are recognized by sequence-specific DNA-binding proteins (Fig. 8.4), a property first demonstrated with the simian virus 40 early promoter. An enormous number of sequence-specific proteins that regulate transcription are now known, many first identified through analyses of viral promoters (Fig. 8.4). Unfortunately, the nomenclature applied to these regulatory proteins (Table 8.4) presents serious difficulties for both writer and reader, for it is unsystematic and idiosyncratic (Box 8.5).

Proper regulation of transcription of many viral and cellular genes also requires more distant regulatory sequences in the DNA template, which possess properties that were entirely unanticipated. The first example was discovered in the genome of simian virus 40 and was termed an **enhancer**, because it stimulated transcription to a large degree. Enhancers are defined by their position- and orientation-independent stimulation of transcription of homologous and heterologous genes over distances as great as 10,000 bp in the genome. Despite these unusual properties, enhancers are built with binding sites for the proteins that recognize local promoter sequences.

The Simian Virus 40 Enhancer: a Model for Viral and Cellular Enhancers

The majority of viral DNA templates described in this chapter contain enhancers of transcription that are recognized by cellular DNA-binding proteins. The simian virus 40 enhancer has been studied intensively, in part because it was the first such regulatory sequence to be identified. Its properties and mechanism of action are considered characteristic of many enhancers, whether of viral or cellular origin.

The simian virus 40 enhancer displays the hierarchical organization typical of these regulatory regions (Box 8.6). It is built from three units, termed enhancer elements, which are subdivided into smaller sequences recognized by DNA-binding proteins (Fig. 8.5). The DNA-binding proteins that interact with this viral enhancer are differentially produced in different cell types. For example, Nf-κb and certain members of the octamer-binding protein (Obp) family are specific to cells of lymphoid origin, and their binding sites are necessary for enhancer activity in these cells. Other elements of the simian virus 40 enhancer, such as the activator protein 1 (Ap-1)-binding sites, confer responsiveness to cellular signaling pathways. The combination of different DNA sequences ensures the activity of the enhancer, and therefore initiation of the viral infectious cycle, in many different cellular environments. This property is exhibited by several other viral enhancers, including those of avian retroviruses (Table 8.5). In contrast, some viral templates for RNA polymerase II transcription contain

Figure 8.4 Local regulatory sequences of three viral transcriptional control regions. The TATA sequences, initiator sequences, and sites of transcription initiation are depicted as in Fig. 8.3. The local regulatory sequences of each promoter, which are recognized by the cellular DNA-binding proteins listed (see Table 8.4), are drawn to the scale shown at the bottom, where +1 is the major initiation site. The black arrows below the Ad2 E2 early promoter indicate the orientation of the E2f-binding sites.

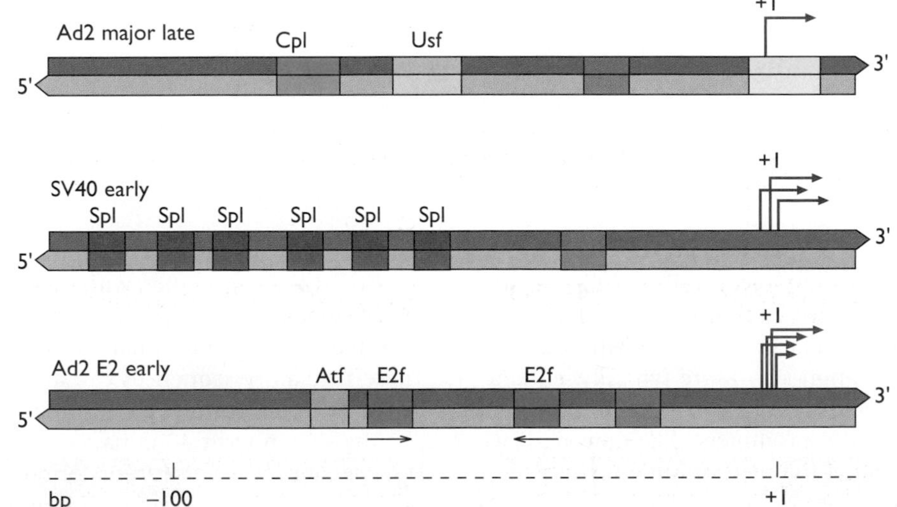

Table 8.4 Some cellular transcriptional regulators that recognize specific DNA sequences

Abbreviation	Full name	Characteristics	Viral or cellular promoters/enhancers recognized
Ap-1	Activator protein 1	Dimers of various basic-leucine zipper proteins, including c-Fos and c-Jun; regulated by dimerization and phosphorylation; mediate transcriptional responses to a variety of extracellular stimuli	Simian virus 40 enhancer
Atf-2	Activating transcription factor 2	Member of large Atf/Creb family; basic-leucine zipper proteins, related to Ap-1 family; mediate transcriptional responses to second-messenger cyclic AMP	Adenovirus type 2 E2E, E3, and E4 promoters; virus-responsive enhancer of human beta interferon gene.
E2f	E2 factor	Member of family of dimeric activators; activity regulated by association with Rb and Rb-related proteins; important in cell cycle regulation	Adenovirus type 2 E2E promoter
Ebps, e.g., A1/Ebp and C/Ebp-α	CCAAT/enhancer-binding proteins	Members of basic leucine-zipper family; C/Ebp α enriched in hepatocytes	Rous sarcoma virus enhancer; hepatitis B virus enhancer
Gata-3	GATA (binding protein) 3	Member of family defined by conserved Zn finger DNA-binding domain; expressed in T cells and specific neurons of the central nervous system	Human immunodeficiency virus type 1 enhancers
Ibp	Initiator-binding protein	Member of nuclear hormone receptor superfamily; binds via Zn fingers as dimers	Simian virus 40 major late promoter
Nf-κb	Nuclear factor κB	Dimer of p50-p65 Rel family proteins; activity regulated by sequestration in cytoplasm (Fig. 8.11)	Human immunodeficiency virus type 1 core enhancer; simian virus 40 enhancer
Nf-Il6	Nuclear factor for Il6 expression	Member of basic-leucine protein family	Rous sarcoma virus and human immunodeficiency virus type 1 enhancers
Oct-1	Octamer-binding protein 1	Ubiquitously expressed member of Obp family	Herpes simplex virus type 1 IE promoters, in complex with VP16 and Hcf (see the text)
Sp1	Stimulatory protein 1	Binds as monomer via Zn finger DNA-binding domain; first sequence-specific transcription factor identified	Simian virus 40 E promoter
Srf	Serum response factor	Member of Mads box DNA-binding domain family; mediates transcriptional response to serum growth factors	Rous sarcoma virus enhancer/ promoter; interacts with HTLV-1 Tax
Tef-1	Transcriptional enhancer factor 1	Complex DNA binding domains; no independent activation domain identified; requires limiting coactivator	Simian virus 40 enhancer; human papillomavirus type 16 E6/E7 promoter
Usf	Upstream stimulatory factor	Contains basic, helix-loop-helix and leucine zipper domains; binds as dimer	Adenovirus type 2 major late promoter

enhancers that are active only in a specific tissue (for example, hepatitis B virus DNA enhancer in liver cells) or only in the presence of inducers or viral proteins (Table 8.5).

The simian virus 40 enhancer is located within 200 bp of the transcription initiation site. More typically, enhancers are found thousands, or tens of thousands of base pairs up- or downstream of the promoters that they regulate. The most popular model of the mechanism by which these sequences exert remote control of transcription, the DNA-looping model, invokes interactions among enhancer-bound proteins and the transcriptional components assembled at the promoter, with the intervening DNA looped out. Compelling evidence in favor of this model has been collected by using the simian virus 40 enhancer (Box 8.7). These regulatory sequences can also facilitate access of the transcriptional machinery to chromatin templates. For example, the simian virus 40 enhancer lies within, and contains DNA sequences necessary for formation of, a nucleosome-free region of the viral genome in infected cells. Enhancers can, therefore, stimulate RNA polymerase II transcription by

BOX 8.5

TERMINOLOGY

The idiosyncratic nomenclature for sequence-specific DNA-binding proteins that regulate transcription

When proteins that bind to specific promoter sequences to regulate transcription by RNA polymerase II were first identified, no rules for naming mammalian proteins (or the genes encoding them) were in place. Consequently, the names given by individual investigators were based on different properties of the protein.

- Some names indicate the function of the regulator, e.g., the glucocortoid receptor, Gr.
- Some names indicate the promoter sequence to which the protein binds, e.g., cyclic AMP response element (CRE)-binding protein, Creb.
- Some names are based on the promoter in which binding sites for the regulator were first identified, e.g., adenovirus E2 transcription factor, E2f.
- Some names report some very general property of the regulator, e.g., stimulatory protein 1, Sp1 (the first sequence-specific activator to be identified), and upstream stimulatory factor, Usf.

Such inconsistency, coupled with the universal use of acronyms, can mystify rather than inform: the historical origins of the names of transcriptional regulators are not known to most readers. The subsequent recognition that many "factors" are members of families of closely related proteins compounds such difficulties.

multiple molecular mechanisms. The primary effect of these mechanisms is to increase the probability that the gene to which an enhancer is linked will be transcribed.

Proteins That Regulate Transcription Share Common Properties

All viral templates transcribed by cellular RNA polymerase II contain transcriptional control regions directly recognized by cellular regulatory proteins (see, e.g., Fig. 8.4). Indeed, the transcriptional programs of retroviruses with simple genomes are executed by the cellular transcriptional machinery alone. Consequently, cellular, sequence-specific, transcriptional regulators play pivotal roles in expression of viral genes and its regulation. However, the genomes of many viruses also encode transcriptional regulatory proteins, some of which resemble cellular proteins that recognize local promoter or more distant enhancer sequences. The cellular and viral DNA-binding proteins necessary for transcription from viral DNA templates share a number of common properties. Their most characteristic feature is modular organization: they are built from discrete structural and functional domains (Fig. 8.6). The basic modules are a DNA-binding domain and an activation domain, which function as independent units. Other common properties are binding to DNA as dimers, and membership in families of related proteins that share the same types of DNA-binding and dimerization domains (Fig. 8.6).

Regulation of transcription by sequence-specific DNA-binding proteins usually requires additional proteins termed **coactivators**. In general, these proteins cannot bind specifically to DNA, nor can they modulate transcription on their own. However, once recruited to a promoter by interaction with a DNA-bound sequence-specific regulator, they dramatically augment (or damp) transcriptional responses. Coactivators can cooperate with multiple, sequence-specific transcriptional activators and stimulate transcription from many promoters, but are not required at all. A common feature of many coregulators is their ability to alter the structure of nucleosomal templates for transcription, either directly or by interaction with appropriate enzymes. For example, several coactivators, including p300/Cbp, are histone acetyltransferases that catalyze the addition of acetyl groups

BOX 8.6

DISCUSSION

Typical properties of the simian virus 40 enhancer

- An enhancer is composed of multiple units, termed **enhancer elements** (e.g., the B1, B2, and A elements of the simian virus 40 enhancer) (Fig. 8.5).
- Enhancer elements operate synergistically: a single element has little activity, but the complete set stimulates transcription more than 100-fold.
- Multimerization of a single inactive enhancer element can create an active enhancer.
- Enhancer activity is relatively insensitive to the orientation or position of individual elements: stimulation of transcription generally depends on the presence of multiple elements, rather than on the exact way in which these are arranged.
- Enhancer elements comprise multiple sequences recognized by sequence-specific DNA-binding proteins that can also bind to promoter sequences (Fig. 8.5, 8.7, and 8.10).

Table 8.5 Some viral enhancers and their recognition

Virus	Enhancer location	Enhancer-binding proteins		Enhancer properties and functions
		Cellular	Viral	
Adenovirus				
Human adenovirus type 2	Enhancer 1, sequences repeated at −300 and −200 of the E1A gene	E2f, Ets family members	None	Broad cell type specificity
Hepadnavirus				
Hepatitis B virus	Enhancer I, adjacent to X gene promoter	Nf-κb, Ap-1, Nf-1, C/Ebp, hepatocyte nuclear factor 3 (Hnf3), Hnf4	None	Strong specificity for hepatocytes, because C/Ebp, Hnf3 and Hnf4 are specific for, or enriched in, these cells; activity may be increased by the viral X protein
Herpesvirus				
Human cytomegalovirus	Immediate-early proximal enhancer, 613 to −70 of major immediate-early transcription unit	Nf-κb, Ap-1, Creb, Srf, Ets family member Elk1, Rxr	None	Can function as a strong basal enhancer and an inducible enhancer activated by signal transduction pathways
Papillomavirus				
Human papillomavirus type 16	"Constitutive" enhancer in the long control region (LCR)	Nf-κb, Ap-1, Tef-1, Tef-2, Oct-1, Nf-1	None	Epithelial cell specific; Ap-1-binding sites confer responsiveness to epidermal growth factor
	E2 protein-dependent enhancers formed by E2-binding sites	None	E2 protein dimers	Active only in cells synthesizing the viral E2 protein; activates transcription from all early genes
Polyomavirus				
Simian virus 40	Between early and late promoters; tandem copies of a 72-bp repeat sequence	Nf-κb, Ap-1, Tef-1, octamer family members	None	Active in many mammalian cell types; activated by signal transduction pathways that converge on Ap-1
Retrovirus				
Human immunodeficiency virus type 1	Core enhancer, −95 to −50 of viral transcription unit	Nf-κb, Ets-1	None	Active only in cells in which Nf-κb is activated (e.g., T cells exposed to various growth factors)
Rous sarcoma virus	LTR enhancer, −250 to −130 of the viral transcription unit	bZip proteins, Srf, Y-box-binding proteins	None	Active in many cell types

to specific lysine residues in histones. Such enzymes, and the histone deacetylases associated with corepressors, help establish the histone codes that distinguish transcriptionally active from inactive chromatin (Box 8.8). A second class of coactivators, exemplified by members of the Swi/Snf family, contain ATP-dependent chromatin remodeling enzymes that alter the way in which DNA is bound to the histone octamer in a nucleosome. It is thought that the coordinated action of these two types of enzyme helps remove the nucleosomal barriers to transcription described previously.

As noted above, the RNA polymerase II system can mediate many patterns of transcription. Such plasticity stems in part from the variety in the nature of core RNA polymerase II promoters, and in the constellations of sequence-specific proteins and coactivators that regulate their activity. Equally important is the power of the transcriptional machinery to integrate signals from multiple, promoter-bound regulators. The transcriptional machinery must also be able to sense and respond to developmental and environmental cues. The proteins that control transcription are therefore frequently regulated by mechanisms that govern their activity, availability, or intracellular concentration. These mechanisms include regulation of the phosphorylation (or other modification) of specific amino acids, which can determine how

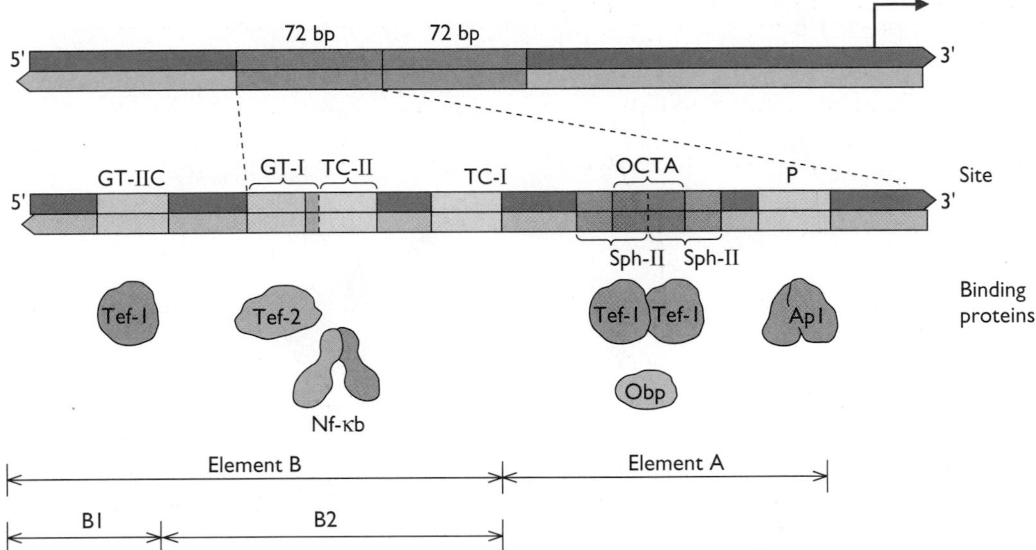

Figure 8.5 Organization of the simian virus 40 enhancer. The positions of the 72-bp repeat region containing the enhancer elements are shown relative to the early promoter at the top. Shown to scale below are functional DNA sequence units of the early promoter-distal 72-bp repeat and its 5′ flanking sequence, which forms part of enhancer element B, and the proteins that bind to them. All the protein-binding sites shown between the expansion lines are repeated in the promoter-proximal 72-bp repeat. The complete enhancer contains one copy of the enhancer element B1 and two directly repeated copies of the enhancer elements B2 and A. Some enhancer elements are built from repeated binding sites for a single sequence-specific protein. For example, cooperative binding of Tef-1 to the two Sph-II sequences forms a functional enhancer element. Such cooperative binding renders enhancer activity sensitive to small changes in the concentration of a single protein. A second class of enhancer elements comprise sequences bound by two different proteins, as illustrated by the enhancer element B GT-IIC and GT-I sequences. Binding of Tef-1 and Tef-2 is not cooperative, but these proteins interact once bound to DNA to form an active enhancer element. A third kind of enhancer element forms an active enhancer element upon oligomerization of a single sequence. The TC-II sequence recognized by Nf-κb functions in this manner.

well a protein binds to DNA, its **oligomerization** state, or the properties of its regulatory domain(s). In some cases, the intracellular location of a sequence-specific DNA-binding protein, or its association with inhibitory proteins within the nucleus, is controlled. Autoregulation of expression of the genes encoding transcriptional regulators is also common. This brief summary illustrates the varied repertoire of mechanisms available for regulation of transcription of viral templates by RNA polymerase II. Not surprisingly, virus-infected cells provide examples of all items on this menu, with the added zest of virus-specific mechanisms.

Transcription of Viral DNA Templates by the Cellular Machinery Alone

Retroviral transcription is characterized by production of a single viral transcript, which serves as both the genome for assembly of progeny virions and the source of viral mRNA species. In cells infected by many retroviruses, the components of the cellular transcriptional machinery described in the previous section complete the viral transcriptional program without the assistance of **any** viral proteins. The proviral DNA created by reverse transcription and integration (Table 8.3) comprises a single RNA polymerase II transcription unit organized into chromatin, exactly like the cellular templates for transcription. Such integrated proviral DNA is a permanent resident in the cellular genome. Because the genomes of these retroviruses do not encode transcriptional regulators, the rate at which proviral DNA is transcribed is determined by the constellation of cellular transcription proteins present in an infected cell. This rate may be influenced by the nature and growth state of the infected cell, as well as by the organization of cellular chromatin containing the proviral DNA. Transcription of viral genetic information can occur throughout the lifetime of the host cell, indeed even in descendants of the cell initially infected. This strategy for transcription of viral DNA is exemplified by avian sarcoma and leukosis viruses, such as Rous-associated viruses.

The long terminal repeat (LTR) of these proviral DNAs contains a compact enhancer located immediately upstream of the viral promoter (Fig. 8.7). As noted previously, the close proximity of the avian proviral LTR enhancer to promoter

BOX 8.7 EXPERIMENTS
Mechanisms of enhancer action

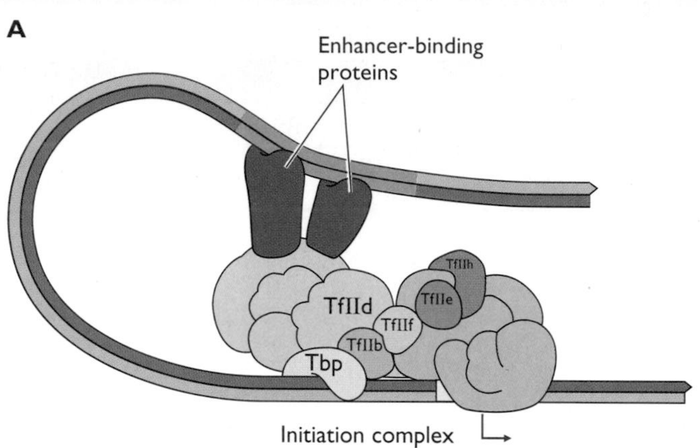

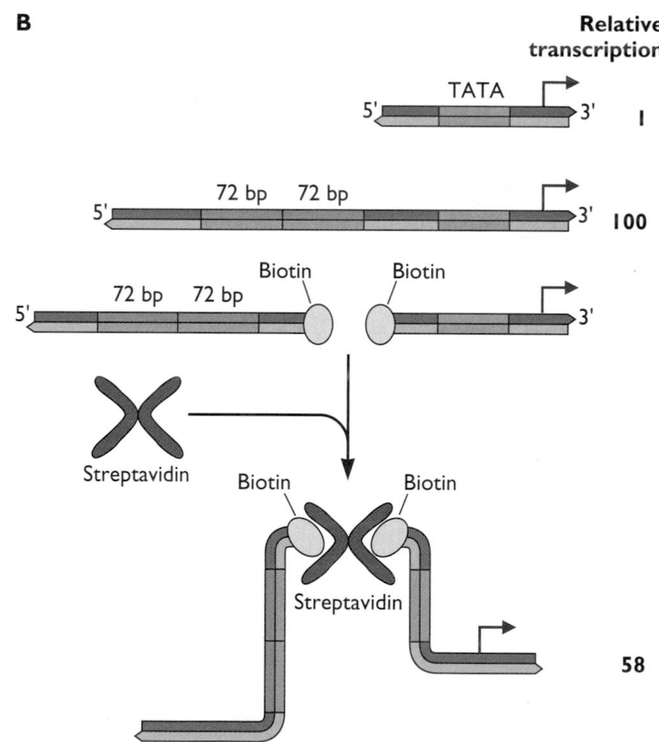

(A) The DNA-looping model postulates that proteins bound to a distant enhancer, here shown upstream of a gene (orange), interact directly with components of the transcription initiation complex, with the intervening DNA looped out. These interactions of proteins bound to a distal enhancer are analogous to those that can take place when the enhancer is located close to the promoter. Such interactions might stabilize the initiation complex and therefore stimulate transcription. **(B)** An enhancer noncovalently linked to a promoter via a protein bridge is functional. When placed immediately upstream of the rabbit β-globin gene promoter, the simian virus 40 enhancer stimulates specific transcription from circular plasmids *in vitro* by a factor of 100. In the experiment summarized here, the enhancer and promoter were separated by restriction endonuclease cleavage. Under this condition, the enhancer cannot stimulate transcription. Biotin was added to the ends of each DNA fragment by incorporation of biotinylated UTP. Biotin binds the protein streptavidin noncovalently, but with extremely high affinity (K_d, 10^{-15} M). Because streptavidin can bind four molecules of biotin, its addition to the biotinylated DNA fragments allows formation of a noncovalent protein "bridge" linking the enhancer and the promoter. Under these conditions, the viral enhancer stimulates *in vitro* transcription from the rabbit β-globin promoter almost as efficiently as when present in the same DNA molecule, as summarized in the column shown on the right. Because this result indicates that an enhancer can stimulate transcription when present in a separate DNA molecule (i.e., in *trans*), it rules out models in which enhancers are proposed to serve as entry sites for RNA polymerase II (or other components of the transcription machinery): such a mechanism requires that RNA polymerase II slide along the DNA from the enhancer to the promoter, a passage that would be blocked by the protein bridge. The results of this experiment are therefore consistent with the looping model shown in panel A.

Muller, H. P., J. M. Sogo, and W. Schappner. 1989. An enhancer stimulates transcription in trans when attached to the promoter via a protein bridge. *Cell* **58:**767–777.

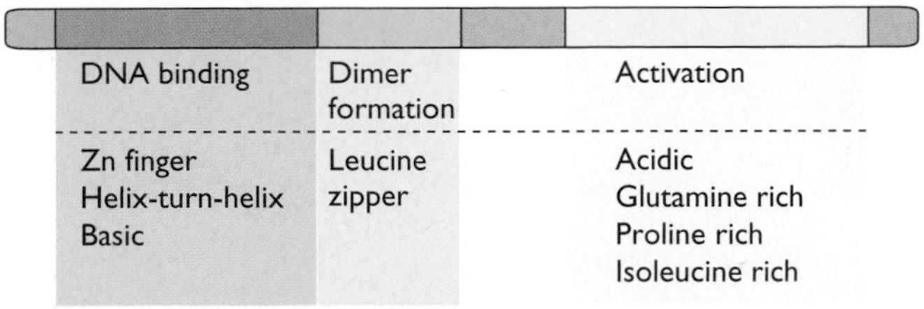

DNA binding	Dimer formation	Activation
Zn finger Helix-turn-helix Basic	Leucine zipper	Acidic Glutamine rich Proline rich Isoleucine rich

Figure 8.6 Modular organization of sequence-specific transcriptional activators. Common functional domains of eukaryotic transcriptional regulators are shown at the top, with some of the types of each domain listed below. DNA-binding and activation domains are defined by their structure (e.g., Zn finger or helix-turn-helix) and chemical makeup (e.g., acidic, glutamine rich), respectively. Transcriptional activators are often more complex than illustrated here. They can contain two activation domains, as well as regulatory domains, such as ligand-binding domains.

sequences is typical of viral genomes (for examples, see Fig. 8.5 and 8.10). The avian and mammalian serum response proteins that bind to the enhancer also bind to a specific sequence in the promoter (Fig. 8.7). This arrangement emphasizes the fact that enhancer and promoter sequences cannot be distinguished by the kinds of protein that recognize them. The other proteins that bind to this enhancer are all members of the basic-leucine zipper family of proteins (Fig. 8.6). Such proteins share a "leucine zipper" dimerization motif located immediately adjacent to a DNA-binding domain rich in basic residues (Fig. 8.8). As with many other transcriptional regulators, dimerization is essential for DNA binding, in this case to align the adjacent basic regions of the monomers to form the DNA-binding surface (Fig. 8.8).

The most remarkable property of the avian retroviral transcriptional control region is that it is active in many different cell types of both the natural avian hosts and mammals. This unusual feature can be explained by the widespread distribution of the cellular proteins that bind to it. The enhancer- and promoter-binding proteins, presumably assembled on the LTR transcriptional control region in different combinations in different cell types, allow efficient transcription of the provirus in both avian and mammalian cells. This property of the LTR enhancers/promoters of avian retroviruses has been exploited in the development of viral vectors.

Despite these properties, transcription of proviral DNA is not an inevitable consequence of integration, but can be blocked or impaired by specific cellular proteins. The avian retroviral integrase binds to the cellular protein Daxx, which becomes associated with proviral DNA. This association leads to repression of viral transcription. It is thought that once it has bound to chromatin containing the proviral template, Daxx recruits histone deacetylases (corepressors). As discussed in Volume II, Chapter 3, such epigenetic silencing of proviral transcription is but one example of general antiviral defense mechanisms.

Because the LTRs are direct repeats of one another (Fig. 8.7), transcription directed by the 3′ LTR extends into cellular DNA, and cannot contribute to the expression of retroviral genetic information. In fact, the transcriptional control region of the 3′ LTR is normally inactivated by a process called **promoter occlusion**: the passage of transcribing complexes initiating at the 5′ LTR through the 3′ LTR prevents recognition of the latter by enhancer- and promoter-binding proteins. Occasionally, transcription from the 3′ LTR does occur, with important consequences for the host cell (see Volume II, Chapter 7).

Absolute dependence on cellular components for the production of viral transcripts avoids the need to devote limited viral genetic information to transcriptional regulatory proteins. Nevertheless, such a strategy is not the rule.

Viral Proteins That Regulate RNA Polymerase II Transcription

Patterns of Regulation

Transcription of many viral DNA templates by the RNA polymerase II machinery results in the synthesis of large quantities of viral transcripts (in some cases, more than 10^5 copies of individual mRNA species per cell) in relatively short periods. Such bursts of transcription are elicited by transcriptional regulatory proteins. Viral proteins that stimulate RNA polymerase II transcription establish one of two kinds of regulatory circuit. The first is a **positive autoregulatory loop**, epitomized by transcription of human immunodeficiency virus type 1 proviral DNA (Fig. 8.9A). A viral activating protein stimulates the rate of transcription, but does not alter the nature of viral proteins made in infected cells. The second is a **transcriptional cascade**, in which different viral transcription units are activated in a fixed sequence (Fig. 8.9B). This mechanism, which ensures that different classes of viral proteins are made during different periods of the infectious cycle, is characteristic of viruses with DNA genomes. The participation of viral regulatory proteins presumably confers a measure of control lacking when the transcriptional program is executed solely by cellular components. The following sections describe some well-studied examples of regulatory circuits established by viral proteins.

The Human Immunodeficiency Virus Type 1 Tat Protein Autoregulates Transcription

Like those of their simpler cousins, the proteins of retroviruses with complex genomes are encoded in a single proviral transcription unit controlled by an LTR enhancer and promoter. However, in addition to the common structural proteins and enzymes, these genomes encode auxiliary proteins, including transcriptional regulators. Some of these regulatory proteins, such as the Tax protein of

BOX 8.8

DISCUSSION

The histone code hypothesis

In eukaryotic cells, genomic DNA is organized and highly compacted by histones and many other chromosomal proteins in chromatin. Transcriptionally active DNA is present as less condensed euchromatin. Although it has been known for decades that the nucleosomal histones present in euchromatin are enriched in acetylated residues, the complexity and regulatory importance of posttranslational modification of nucleosomal histone has become apparent only within the past decade.

The N-terminal tails of the four core histones of the nucleosome (H2A, H2B, H3, and H4) are subject to acetylation, methylation, and phosphorylation of specific residues. Panel A of the figure summarizes modifications of this segment of histone H3. The N-terminal segment of histone H4 can also be methylated or acetylated at several positions, while H2A and H2B can be modified by acetylation, phosphorylation and ubiquitinylation. The large number of modifications results in a much greater number of possible combinations. For example, over 150 combinations present in different molecules of histone H3 in human cells have been identified by mass spectrometry, just 2 of which are shown in panel B.

A Sites of modification

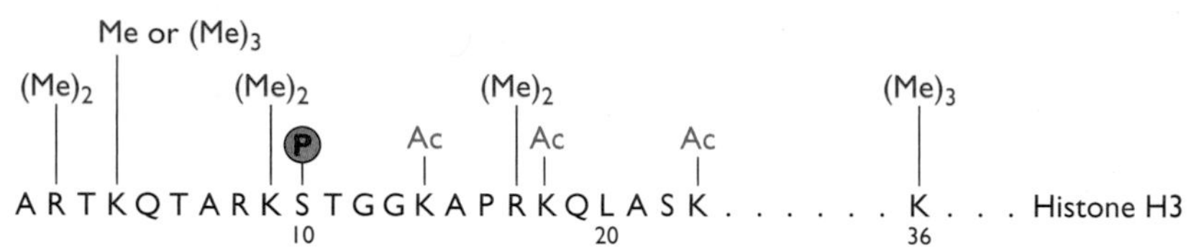

B Some observed combinations

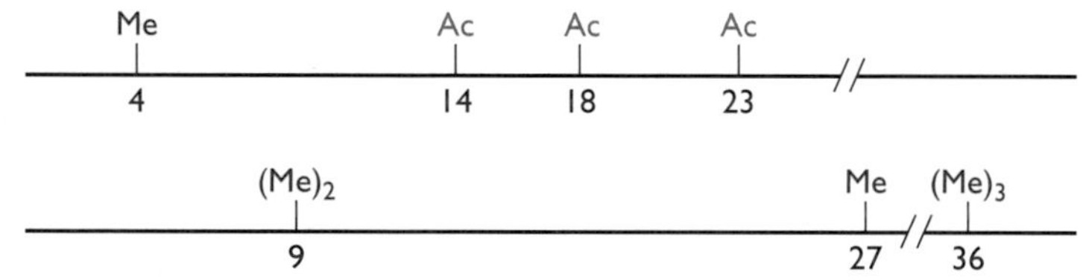

It was initially proposed that particular combinations of posttranscriptionally modified histones identify transcriptionally active or inactive DNA. Consistent with this "histone code" hypothesis, some combinations, such as S10 phosphorylation plus K9 and/or K14 acetylation of histone H3, are characteristic of transcriptionally active genes. Conversely, methylation of K9 in this histone has been implicated in formation of heterochromatin. These modified amino acids serve as recognition sites for proteins that modify histones, remodel nucleosomes, or facilitate transcription by other mechanisms.

Although the idea of a simple code of histone posttranslational modifications has great appeal, it is now clear that it may be more appropriate to consider this a complex "language": for example, the same modification can recruit either activators or repressors of transcription, probably depending on the context of other posttranscriptional modifications. Furthermore, histone modifications are dynamic, changing, for example, during transcriptional elongation, or from one transcriptional cycle to another.

Berger, S. L. 2007. The complex language of chromatin regulation during transcription. *Nature* **447:**407–412.

Garcia, B. A., J. J. Pesavento, Z. A. Mizzen, and N. L. Kelleher. 2007. Pervasive combinational modification of histone H3 in human cells. *Nat. Methods* **4:**487–489.

Jenuwein, T., and C. D. Allis. 2001. Translation the histone code. *Science* **293:**1074–1079.

Mellor, J. 2006. Dynamic nucleosomes and gene transcription. *Trends Genet.* **22:**320–329.

human T-lymphotropic virus type 1, resemble activators of other virus families, and stimulate transcription from a wide variety of viral and cellular promoters (Table 8.6). Others, exemplified by the transactivator of transcription (Tat) of human immunodeficiency virus type 1, are unique sequence-specific activators of transcription: they recognize an RNA element in nascent transcripts.

The positive feedback loop that is established once a sufficient concentration of Tat has accumulated in an infected cell is simplicity itself (Fig. 8.9A). Cellular proteins initially direct transcription of the proviral DNA in infected cells at some basal rate. Among the processed products of viral primary transcripts are spliced mRNA species from which the Tat protein is synthesized. This protein is imported into the nucleus, where it stimulates transcription of the proviral template upon binding to its RNA recognition site in nascent viral transcripts. However, the molecular mechanisms that establish this autostimulatory loop are sophisticated and unusual. Their elucidation has been an important area of research, because the Tat protein is essential for virus propagation and represents a good target for antiviral therapy.

Cellular Proteins Recognize the Human Immunodeficiency Virus Type 1 LTR Transcriptional Control Region

The Tat protein is necessary for efficient human immunodeficiency virus type 1 transcription, but it is not sufficient: cellular proteins that bind to the LTR enhancer

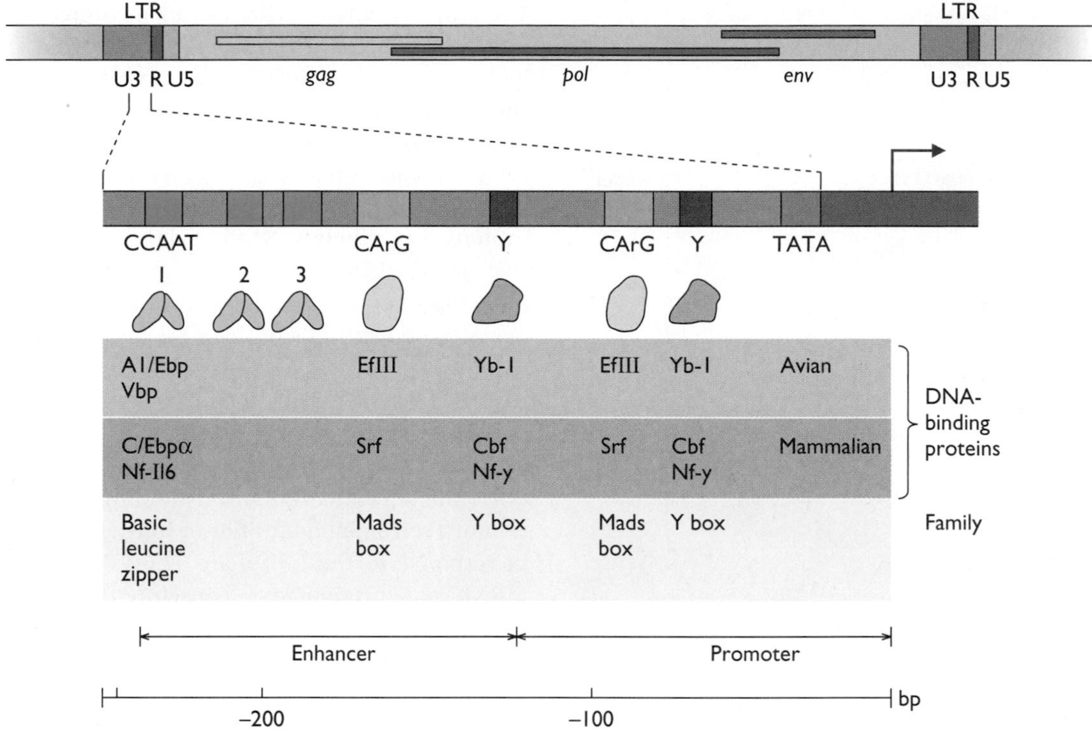

Figure 8.7 The transcriptional control region of an avian retrovirus. The proviral DNA of an avian leukosis virus is shown at the top. The enhancer and promoter present in the U3 regions of the LTRs are drawn to scale below. Each of the multiple CCAAT, CArG, and Y-box sequences, which are required for maximally efficient transcription, is recognized by the avian and mammalian proteins listed below. The chicken protein EfIII is the avian homolog of mammalian serum response factor (Srf), which plays an important role in the activation of transcription in response to serum growth factors. These proteins are members of a large, widespread family defined by a conserved sequence motif within the DNA-binding domain (the Mads box) and named for four of the originally identified members.

and promoter are also required. These proteins support a low level of proviral transcription before Tat is made in infected cells.

In contrast to avian retroviruses, human immunodeficiency virus type 1 propagates efficiently in only a few cell types, notably $CD4^+$ T lymphocytes and cells of the monocytic lineage. Viral reproduction (i.e., transcription) in infected T cells in culture is stimulated by T-cell growth factors, indicating that viral transcription requires cellular components available only in such stimulated T cells. Indeed, the failure of the virus to propagate efficiently in unstimulated T cells correlates with the absence of active forms of particular enhancer-binding proteins. The distribution of cellular enhancer-binding proteins is therefore an important determinant of the host range of retroviruses with both simple and complex genomes. The difference is that the transcription of proviral DNA of avian retroviruses depends on proteins that are widely distributed, whereas human immunodeficiency virus type 1 transcription requires proteins that are found in only a few cell types, or active only under certain conditions.

Within the LTR, the promoter is immediately preceded by two important regulatory regions (Fig. 8.10). The promoter-proximal enhancer core is essential for enhancer function in transient-expression assays (Box 8.3) and virus reproduction in T cells. The upstream enhancer region is also necessary for efficient viral transcription in both peripheral blood lymphocytes and certain T-cell lines. Both the core and the upstream enhancers are densely packed with binding sites for cellular proteins, many of which are enriched in the types of cell in which the virus can reproduce (Fig. 8.10). For example, the human Gata-3 and Ets-1 proteins, which stimulate viral transcription in transient-expression assays, are restricted to cells of the T-lymphocyte and hematopoietic lineages, respectively. However, it is not known whether these proteins are important for transcription of the provirus in infected cells. Transient-expression assays do not reproduce physiological conditions (Box 8.9). Consequently, a positive result in this kind of assay establishes only that a certain protein **can** stimulate transcription, not that it normally does so.

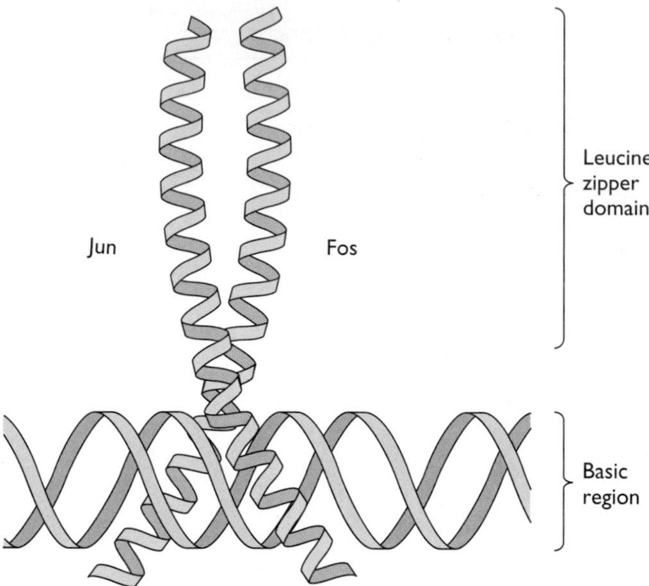

Figure 8.8 The structure of a basic-leucine zipper domain bound specifically to DNA. The model of the leucine zipper and adjacent basic region of the Jun-Fos heterodimer (Ap1) bound to a specific recognition sequence in DNA is based on the structure determined by X-ray crystallography. The leucine zipper forms an α-helical coiled coil, both in crystals and in solution. In the protein-DNA complex, the basic (DNA-binding) region is also α-helical, but in the absence of DNA it is disordered: DNA binding induces a major conformational change in proteins of this class. The α-helical regions containing the basic amino acids that make specific contact with DNA fit snugly into successive major grooves. Consequently, the basic-leucine zipper dimerization and DNA-binding domains have been likened to DNA "forceps." Adapted from J. N. M. Glover and S. C. Harrison, *Nature* **373**:257–261, 1995, with permission.

The mechanisms by which viral transcription is regulated by cellular pathways are illustrated by the cellular transcriptional activator Nf-κb, which plays a critical role in relication of human immunodeficiency virus type 1 in T cells (as well as in cells infected by several other viruses) (Fig. 8.11). Unstimulated T cells display no Nf-κb activity, because the protein is retained in inactive form in the cytoplasm by binding of inhibitory proteins of the Iκb family. Upon T-cell activation by treatment with any one of several growth factors, Iκb proteins rapidly disappear from the cytoplasm, with the concomitant appearance of active Nf-κb in the nucleus. Phosphorylation of Iκb at specific sites, the result of signaling pathways activated by growth factors at the cell surface, targets the inhibitor for degradation by the cytoplasmic multiprotease complex (the **proteasome**) (Fig. 8.11). Consequently, Nf-κb is free for transit to the nucleus, where it can bind to its recognition sites within the viral LTR core enhancer (Fig. 8.10 and 8.11). This pathway can account for the induction of human immunodeficiency virus type 1 transcription observed when T cells are stimulated. The severe, or complete, inhibition of virus reproduction (transcription) in normal human $CD4^+$ lymphocytes, caused by mutations in the Nf-κb-binding site, emphasizes the importance of activation of Nf-κb in the infectious cycle of this virus. Nf-κb is also indispensable for viral transcription in macrophages, which, when differentiated, contain nuclear pools of constitutively active protein. Nevertheless, Nf-κb and the other cellular proteins that act via LTR enhancer- or promoter-binding sites do not support efficient expression of viral genes. This process depends on synthesis of the viral Tat protein (see the next section).

The efficient production of progeny human immunodeficiency virus type 1 genomes by transcription of the provirus depends ultimately on the constellation or activation state of cellular enhancer-binding proteins. Transcription of even small quantities of full-length viral transcripts in response to these proteins allows the synthesis of Tat mRNA and protein, and consequently activation of the positive, autoregulatory loop (Fig. 8.9A). The provirus can be considered dormant in cells that do not contain the necessary enhancer-binding proteins, or in which these proteins are inactive. However, it is important to keep in mind that such transcriptional inactivity is **not** the cause of the clinical latency characteristic of human immunodeficiency virus type 1 infections. In clinical latency, few, if any, symptoms are manifested. Nevertheless, virus is produced continuously (Volume II, Chapter 6), because the positive, autoregulatory circuit is triggered whenever infected cells can support LTR enhancer-dependent transcription of proviral DNA.

The Unique Mechanisms by Which the Tat Protein Regulates Transcription

Tat recognizes an RNA structure. Stimulation of human immunodeficiency virus type 1 transcription by Tat requires an LTR sequence, termed the **transactivation response (TAR)** element, which lies within the transcription unit (Fig. 8.12A). This sequence is active only in the sense orientation and only when located close to the initiation site of the promoter, properties that distinguish it from enhancer elements (see Box 8.4). The observation that mutations that disrupted the predicted secondary structure of TAR RNA inhibited Tat-dependent transcription suggested that the TAR element is recognized as RNA. Indeed, the Tat protein binds specifically to a trinucleotide bulge and adjacent base pairs in the stem of the TAR RNA stem-loop structure (Fig. 8.12B). Such specific binding requires an arginine-rich basic region and adjacent sequences of Tat. Binding of Tat to this region of TAR induces a local conformational rearrangement in the RNA, resulting in formation of a more stable and compact structure (Fig. 8.12C), and is therefore energetically

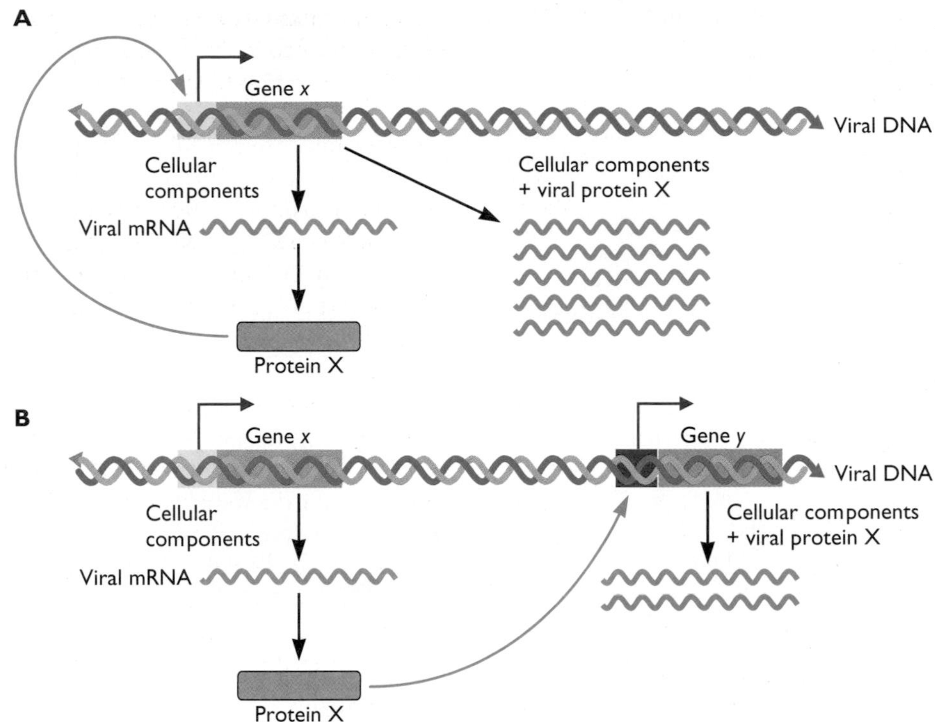

Figure 8.9 Mechanisms of stimulation of transcription by viral proteins. Cellular transcriptional components acting alone transcribe the viral gene encoding protein X. Once synthesized and returned to the nucleus, viral protein X can stimulate transcription either of the same transcription unit **(A)** or of a different one **(B)**. In either case, viral protein X acts in concert with components of the cellular transcriptional machinery.

favorable. Tat may also make contact with residues within the TAR loop (Fig. 8.12B). Recognition of a transcriptional control sequence as RNA by a regulatory protein remains unique to Tat proteins.

Tat stimulates transcriptional elongation. Binding of Tat to TAR RNA stimulates production of viral RNA as much as 100-fold. In contrast to many cellular and viral proteins that stimulate transcription by RNA polymerase II, the Tat protein has little effect on initiation. Rather, it greatly improves elongation. Complexes that initiate transcription in the absence of Tat elongate poorly, and many terminate transcription within 60 bp of the initiation site (Fig. 8.13A). Consequently, in the absence of Tat, full-length transcripts of proviral DNA account for no more than 10% of the total. This property resolves the paradox of why the human immunodeficiency virus type 1 LTR enhancer and promoter do not support efficient viral RNA synthesis: they direct initiation effectively, but the transcription complexes formed cannot carry out sustained transcription of the proviral genome over long distances. The Tat protein overcomes such poor **processivity** of elongating complexes to allow efficient production of full-length viral transcripts.

How Tat stimulates transcriptional elongation. A search for cellular proteins that stimulate viral transcription when bound to the N-terminal region of Tat (Fig. 8.12C) identified the human Ser/Thr kinase p-Tefb (positive-acting transcription factor b), which was known to stimulate elongation of cellular transcripts. This cellular protein is essential for Tat-dependent stimulation of processive viral transcription both *in vitro* and in infected cells. One subunit of the p-Tefb heterodimer is a **cyclin**, cyclin T. Cyclins are so named because members of the family accumulate during specific periods of the cell cycle. Cyclin T regulates the activity of the second subunit of p-Tefb, cyclin-dependent kinase 9 (Cdk9). The active kinase phosphorylates an unusual domain at the C terminus of the largest subunit of RNA polymerase II, which is essential for Tat-dependent stimulation of viral transcription. This domain comprises multiple, tandem copies of a heptapeptide sequence rich in Ser and Thr residues. The C-terminal domain of RNA polymerase II present in preinitiation complexes is hypophosphorylated, and its hyperphosphorylation has been implicated in successful promoter clearance and transcriptional elongation.

Tat and p-Tefb bind **cooperatively** to the TAR RNA stem-loop, that is, with higher affinity than either protein alone, and with greater specificity. This property is the result of the

Table 8.6 Some viral transcriptional regulators that bind to specific DNA sequences

Virus	Protein	Properties	Functions
Adenovirus			
Human adenovirus type 2	IVa_2	Binds to intragenic sequence in the major late promoter; late-phase specific	In conjunction with a viral L4 protein, stimulates transcription from major late promoter
Herpesviruses			
Herpes simplex virus type 1	ICP_4	Typical domain organization; one domain binds to a degenerate consensus sequence	Stimulates transcription from early and late promoters; represses transcription from all IE promoters by binding to promoter sequences
Epstein-Barr virus	Zta	bZIP protein; synthesis and activity regulated by multiple mechanisms (see the text)	Essential activator of early gene transcription; commits to lytic infection (see the text)
Papillomavirus			
Bovine papillomavirus type 1	E2	Typical domain organization; binds as dimers; can bind TfIIb and TfIId	Stimulates transcription from several promoters as enhancer-binding protein; necessary for replication of viral DNA *in vivo*
Poxvirus			
Vaccinia virus	VETF	Binds to TA-rich early promoter sequences as heterodimer; DNA-dependent ATPase	Essential for recognition of early promoters by the viral RNA polymerase

interaction of the cyclin T1 subunit of p-Tefb with nucleotides within the TAR RNA loop (Fig. 8.12B) that are not contacted by Tat, but are nevertheless crucial for stimulation of transcription. Assembly of the ternary complex containing TAR RNA, Tat, and p-Tefb induces conformational changes that activate the Cdk9 kinase of the latter protein. Once associated with transcription complexes, this enzyme phosphorylates the C-terminal domain of RNA polymerase II to stimulate elongation of transcription (Fig. 8.13A). This model was developed using *in vitro* assays and other simplified experimental systems. However, the results of experiments in which p-Tefb was inhibited in infected cells, as well as of genetic experiments, have established that p-Tefb is essential for stimulation of viral transcription by Tat *in vivo*.

Figure 8.10 The transcriptional control region of human immunodeficiency virus type 1. The organization of the U3 region of the proviral long terminal repeat is shown to scale. Proteins that bind to the enhancer or promoter sequences indicated are shown above or below the DNA. The upstream enhancer, which stimulates viral transcription in infected cells, contains binding sites for several proteins that are enriched in cells in which the virus reproduces. For example, the Lef protein is produced primarily in lymphocytes. Upon binding, this protein induces a large (130°) bend in DNA. The cellular enhancer-binding proteins listed have been shown to modulate the efficiency of transcription from the viral promoter in experimental situations. However, not all their binding sites are well conserved in different viral isolates, and only a few have been shown to contribute to transcription in infected cells.

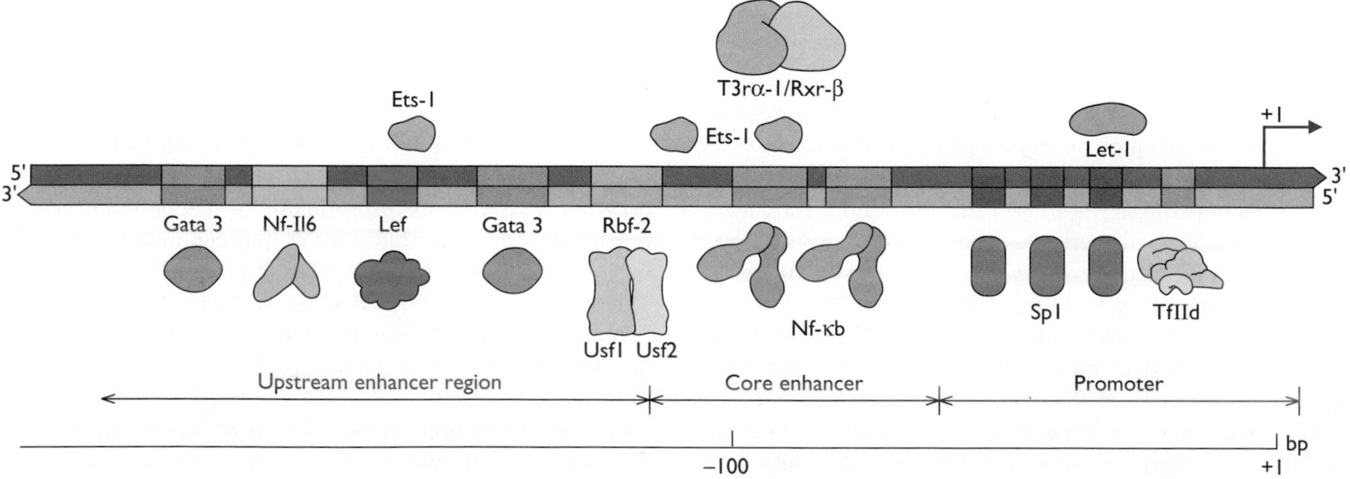

BOX 8.9

WARNING

Caution: transient-expression assays do not reproduce conditions within virus-infected cells

Transient-expression assays (Box 8.3, panel B of the figure) provide a powerful, efficient way to investigate regulation of transcription. Advantages include the following:

- simplicity and sensitivity of assays for reporter gene activity
- ready analysis of mutated promoters to identify DNA sequences needed for the action of the regulatory protein
- application with chimeric fusion proteins and synthetic promoters to avoid transcriptional responses due to endogenous cellular proteins
- simplification of complex transcriptional regulatory circuits to focus on the activity of a single protein

Despite these advantages, transient-expression assays do not necessarily tell us how transcription is regulated in virus-infected cells, for they do not reproduce normal intracellular conditions. Important differences include the following:

- association of transcriptional templates with different proteins: the exogenous DNA may associate with cellular histones, but viral DNA may be packaged by virus-specific proteins
- abnormally high concentrations of exogenous template DNA: concentrations of reporter genes as high as 10^6 copies per cell are not unusual; this value is significantly greater than even the maximal concentrations of viral DNA molecules attained toward the end of an infectious cycle
- abnormally high concentrations of the regulatory protein as a result of its deliberate overproduction
- because of these high concentrations of template and protein, potential interactions of the viral regulator with template, or components of the cellular transcriptional machinery, that would not take place in an infected cell
- absence of viral components that might negatively or positively modulate the activity of the protein under study

The last three caveats apply to any experiment in which a viral protein is overproduced, for example, for investigation of its interactions with other proteins.

Human cells contain several proteins that stimulate elongation by RNA polymerase II, and others that inhibit this process. There have been several reports of Tat-dependent phosphorylation of such proteins, suggesting that additional mechanisms contribute to efficient elongation of viral transcription.

Tat also facilitates nucleosome remodeling. As discussed previously, integrated proviral DNA templates for transcription are organized in chromatin, by association with nucleosomes (and many other host cell proteins). Although proviral DNA is integrated preferentially into transcriptionally active genes of the host cell (Chapter 7), efficient transcription requires reorganization of nucleosomes. Nucleosomes are located at specific positions on the LTR promoter of integrated DNA. The promoter and enhancers are nucleosome free, and hence accessible to the transcriptional activators described above. In contrast, the nucleosome that is located immediately downstream of the site of initiation of transcription must be repositioned, or otherwise reorganized, to allow elongation of proviral DNA transcription. The results of recent experiments implicate Tat in inducing remodeling of this nucleosome.

In addition to binding to the cyclin T1 subunit of p-Tefb, Tat can bind to specific subunits of ATP-dependent chromatin-remodeling enzymes of the Swi/Snf family, as well as to several histone acetyltransferases, including p300/Cbp and pCaf. The data currently available are consistent with a model in which binding of Tat to TAR RNA recruits not only p-Tefb, but also Swi/Snf enzymes, which then alter the position or structure of the downstream nucleosome to promote elongation of viral transcription (Fig. 8.13B). Although some details are not yet clear, the inhibition of Tat-dependent transcription induced by small interfering RNA-mediated knockdown of specific subunits of Swi/Snf enzymes provides compelling evidence for the important contribution of this function of Tat to transcription of integrated proviral DNA.

The ability of Tat to bind to cellular proteins is governed by posttranslational modification. For example, acetylation at Lys28 (by Pcaf) promotes binding to p-Tefb, whereas acetylation at Lys50 prevents this interaction. The latter modification also prevents association of Tat with some Swi/Snf subunits, but allows binding to others (Fig. 8.13B).

Why such unusual transcriptional regulation? At this juncture, it is difficult to appreciate the value of the intricate transcriptional program of human immunodeficiency virus type 1. Those of many other viruses are executed successfully by transcriptional regulators that operate, directly or indirectly, via specific DNA sequences, and Tat can stimulate transcription effectively when made to bind to DNA experimentally. Why, then, is efficient transcription of the proviruses mediated by this unique, RNA-dependent mechanism? Binding of Tat to nascent viral RNA close to the site at which many transcriptional

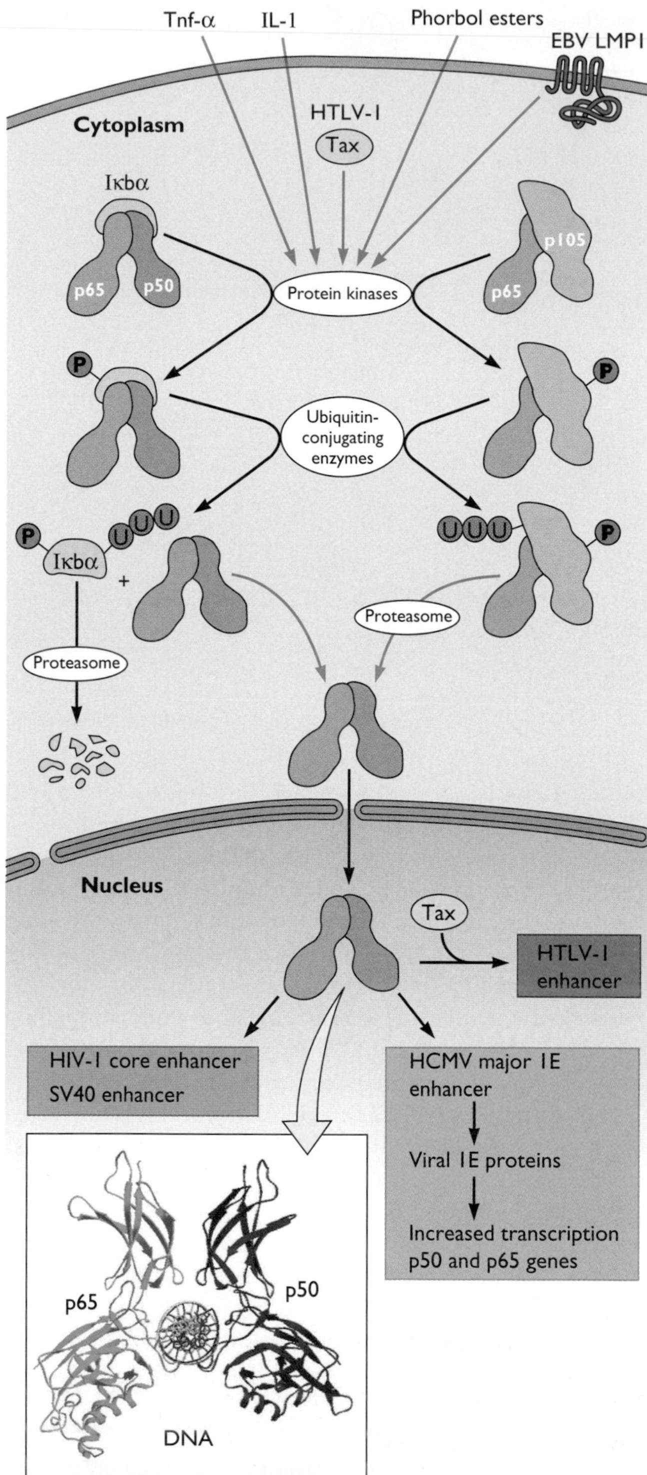

Figure 8.11 The cellular transcriptional regulator Nf-κB and its participation in viral transcription. The members of the Nf-κb Rel protein family are defined by the presence of the Rel homology region, which contains DNA-binding and dimerization motifs, and a nuclear localization signal. The p65 protein of the p50-p65 heterodimer **(left)** also contains an acidic activation domain at its C terminus. p50 is synthesized as an inactive precursor, p105 **(right)**. The p105-p65 heterodimer is one of two forms of inactive Nf-κb found in the cytoplasm (e.g., of unstimulated T cells). The second consists of mature Rel heterodimers, p50-p65, associated with an inhibitory protein such as Iκb-α (left), which blocks the nuclear localization signals of the p50 and p65 proteins. The C-terminal segment of p105 functions like Iκb, with which it shares sequences, to block nuclear localization signals and retain Nf-κb in the cytoplasm. Exposure of the cells to any of several growth factors, indicated by tumor necrosis factor alpha (Tnf-α), interleukin-1 (IL-1), and phorbol esters, results in activation (green arrows) of protein kinases that phosphorylate specific residues of Iκb or p105. Upon phosphorylation, Iκb dissociates and is recognized by the system of enzymes that adds branched chains of ubiquitin (Ub) to proteins targeted for degradation. It is then degraded by the proteasome, a multicatalytic protease that degrades polyubiquinylated proteins. Specific p105 cleavage by the proteasome also produces the p50-p65 dimer. Free Nf-κb produced by either mechanism can translocate to the nucleus because its nuclear localization signals are now accessible. In the nucleus of uninfected cells, Nf-κb binds to specific promoter sequences to stimulate transcription via the p65 activation domain. Viral transcriptional control regions to which Nf-κb binds and some viral proteins that induce activation (green arrows) of Nf-κb are indicated. The X-ray crystal structure of a p50-p65 heterodimer bound specifically to DNA is shown in the inset. The structure is viewed down the helical axis of DNA with the two strands in purple and yellow and with the p50 and p65 subunits in green and red, respectively. The dimer makes extensive contact with DNA via protein loops. HCMV, human cytomegalovirus; HIV-1, human immunodeficiency virus type 1; HTLV-1, human T-lymphotropic virus type 1; SV40, simian virus 40; EBV, Epstein-Barr virus. From F. E. Chen et al., *Nature* **391**:410–413, 1998, with permission. Courtesy of G. Ghosh, University of California, San Diego.

complexes pause or stall (Fig. 8.13A) could provide a particularly effective way to recruit the cellular proteins that stimulate processive transcription and induce nucleosome remodeling. Or it may be that regulation of transcription via an RNA sequence is a legacy from some ancestral viral-host cell interaction in an RNA world.

The Transcriptional Cascades of DNA Viruses

Common Features

The transcriptional strategies characteristic of the productive infectious cycles of viruses with DNA genomes exhibit a number of common features. The most striking is the transcription of viral genes in a reproducible and precise temporal sequence. Prior to initiation of genome replication, during **immediate-early** and **early** phases, infected cells synthesize viral proteins necessary for viral DNA synthesis, efficient gene expression, or other regulatory functions. Transcription of the **late** genes, most of which encode structural proteins, requires genome replication (Fig. 8.14). This property ensures coordinated production of the DNA genomes and the structural proteins from which progeny virus particles are assembled. Another common feature is the control of the transitions from one transcriptional stage to the next by both viral proteins

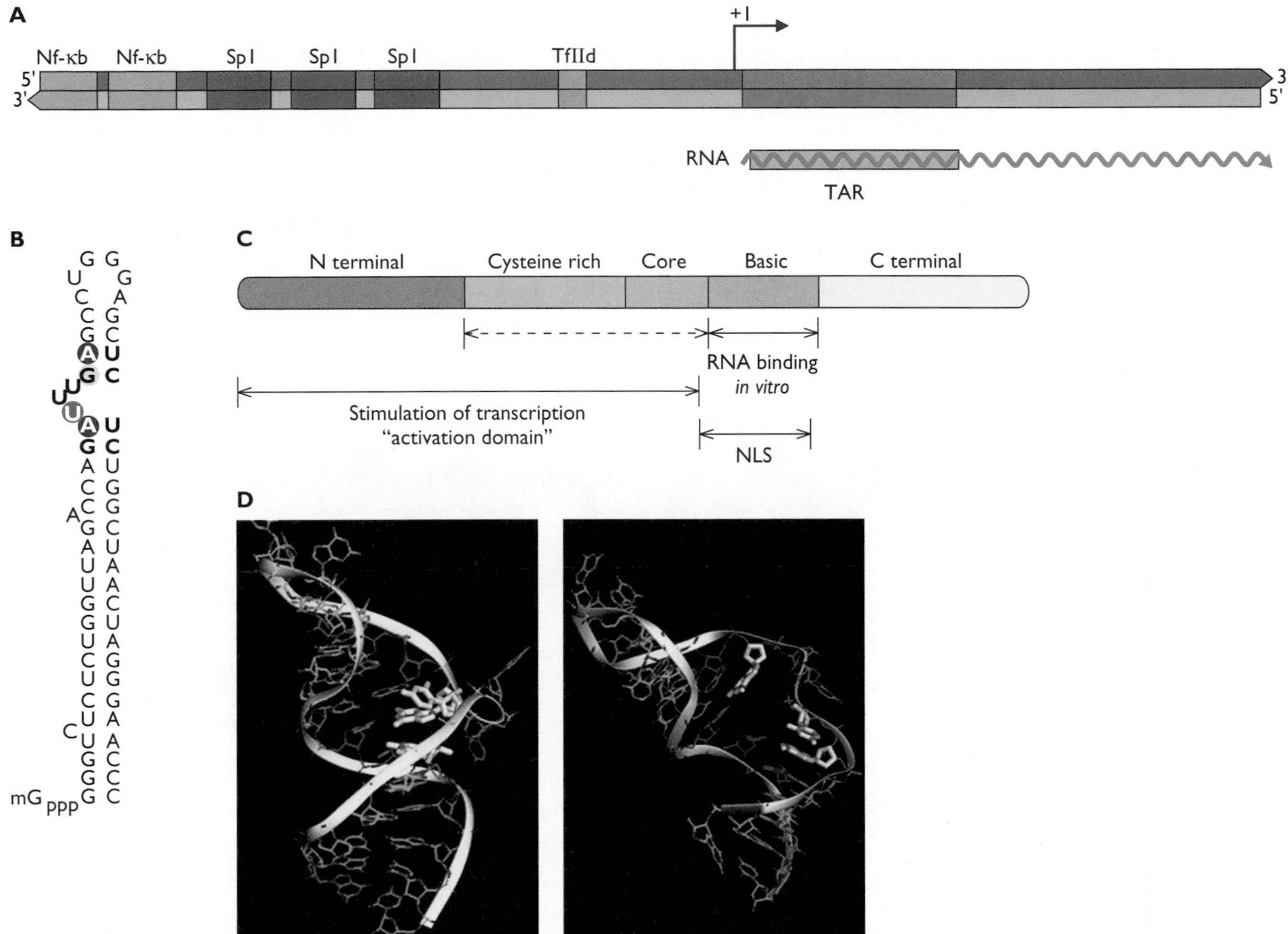

Figure 8.12 Human immunodeficiency virus type 1 TAR and the Tat protein. (A) The region of the viral genome spanning the site of transcription initiation is drawn to scale, with the core enhancer and promoter depicted as in Fig. 8.10. The DNA sequence lying just downstream of the initiation site (pink) negatively regulates transcription. Transcription of the proviral DNA produces nascent transcripts that contain the TAR sequence (tan box). **(B)** The TAR RNA hairpin extends from position +1 to position +59 in nascent viral RNA. Sequences important for recognition of TAR RNA by the Tat protein are colored. Optimal stimulation of transcription by Tat requires not only this binding site in TAR but also the terminal loop. **(C)** The Tat protein is made from several different, multiply spliced mRNAs (Appendix, Fig. 21B) and therefore varies in length at its C terminus. The regions of the protein are named for the nature of their sequences (basic, cysteine rich) or greatest conservation among lentiviral Tat proteins (core). Experiments with fusion proteins containing various segments of Tat and a heterologous RNA-binding domain identified the N-terminal segment indicated as sufficient to stimulate transcription. The basic region contains the nuclear localization signal (NLS). The basic region alone can bind specifically to RNA containing the bulge characteristic of TAR RNA. However, high-affinity binding, effective discrimination of wild-type TAR from mutated sequences *in vitro*, and RNA-dependent stimulation of transcription within cells require additional, N-terminal regions of the protein, shown by the dashed arrow. The sequences required for stimulation of transcription and specific recognition of TAR RNA in cells do not form discrete domains of the protein. **(D)** Major groove views of structures of a free TAR RNA corresponding to the apical stem and loop regions (but with a truncated stem) (left) and of the same RNA when bound to the Tat peptide (right) were determined by nuclear magnetic resonance methods. The residues shown in yellow are A22, U23, and G26, which are colored red, blue, and yellow, respectively, in panel B. Note the substantial conformational change in the trinucleotide bulge region on binding of the Tat peptide. From F. Aboul-ela et al., *J. Mol. Biol.* **253**:313–332, 1995, and F. Aboul-ela et al., *Nucleic Acids Res.* **24**:3974–3981, 1996, with permission. Courtesy of M. Afshar, RiboTargets, and J. Karn, MRC Laboratory of Molecular Biology.

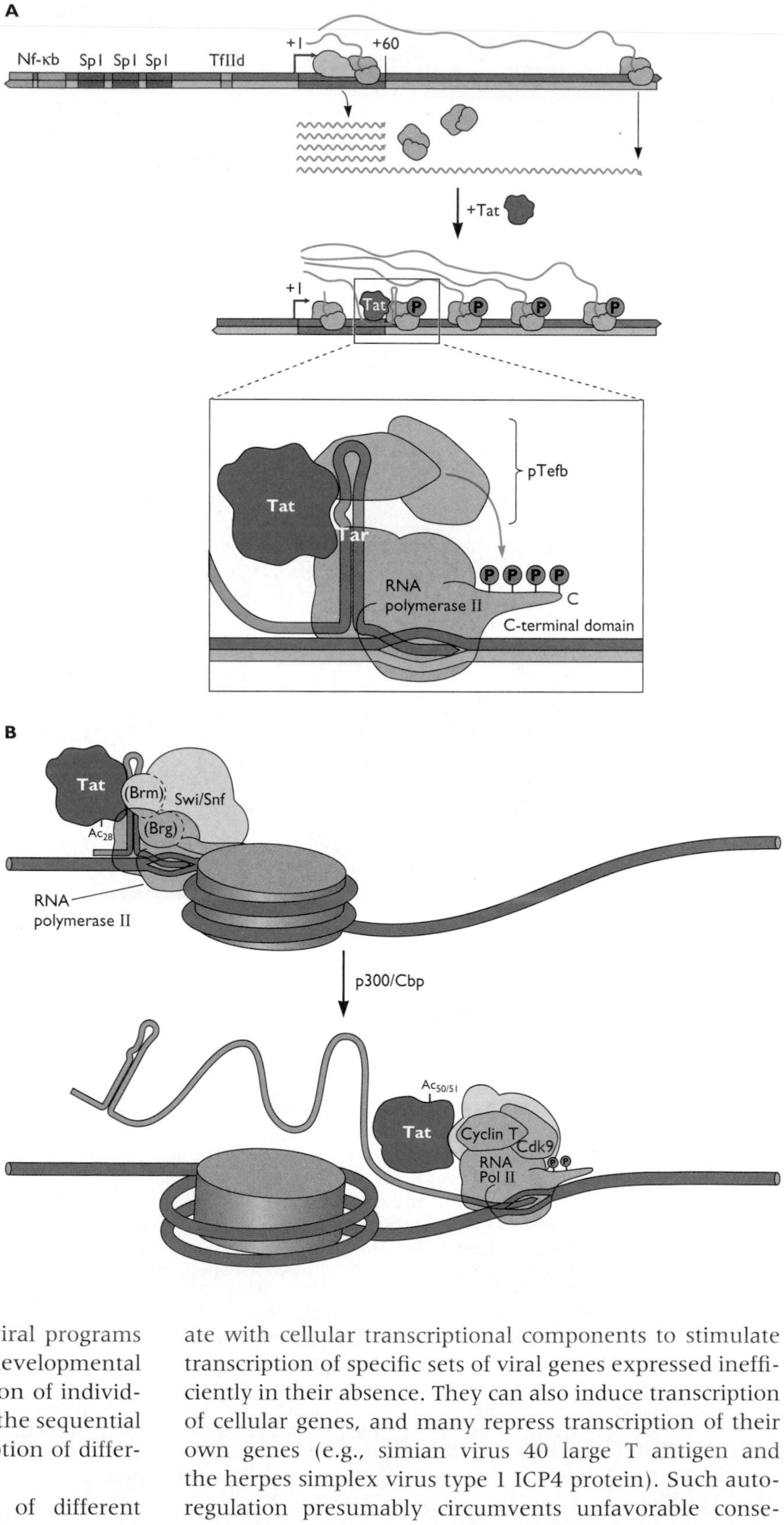

Figure 8.13 Mechanisms of stimulation of transcription by the human immunodeficiency virus type 1 Tat protein. (A) Model for the stimulation of elongation. The regulatory sequences flanking the site of initiation of transcription are depicted as in Fig. 8.10. In the absence of Tat, transcription complexes are poorly processive, and the great majority (9 of 10) terminate within 60 bp of the initiation site, releasing transcription components and short transcripts. Production of the Tat protein upon translation of mRNAs spliced from rare, full-length transcripts allows all transcriptional complexes to pass through the elongation blocks to synthesize full-length viral RNA. Binding of Tat to TAR in conjunction with the cyclin T subunit of p-Tefb leads to phosphorylation (P) of the C-terminal domain of the largest subunit of RNA polymerase II by the Cdk9 kinase subunit. As a result of this modification, transcriptional complexes become competent to carry out highly processive transcription. **(B)** Model for nucleosome remodeling. The initial transcript of proviral DNA (blue line) is depicted as in panel A. The nucleosome located a short distance downstream of the initation site blocks transcriptional elongation. When Tat bound to TAR is acetylated at Lys28, it binds to a particular subunit (Brm) of the Swi/Snf chromatin remodeling protein. Acetylation of Tat at Lys50 and Lys51 by the histone acetylases p300/Cbp induces dissociation of Tat from TAR, presumably the result of conformational change, and binding of Tat to a different subunit (Brg-1) of Swi/Snf. Remodeling of the nucleosome by Swi/Snf relieves the block to transcriptional elongation.

and genome replication (Fig. 8.14). Such viral programs closely resemble those that regulate many developmental processes in animals, in both the transcription of individual genes in a predetermined sequence, and the sequential action of proteins that regulate the transcription of different sets of genes.

The transcriptional activating proteins of different viruses exhibit distinctive properties. However, all cooperate with cellular transcriptional components to stimulate transcription of specific sets of viral genes expressed inefficiently in their absence. They can also induce transcription of cellular genes, and many repress transcription of their own genes (e.g., simian virus 40 large T antigen and the herpes simplex virus type 1 ICP4 protein). Such autoregulation presumably circumvents unfavorable consequences of excess concentrations of these viral proteins, for

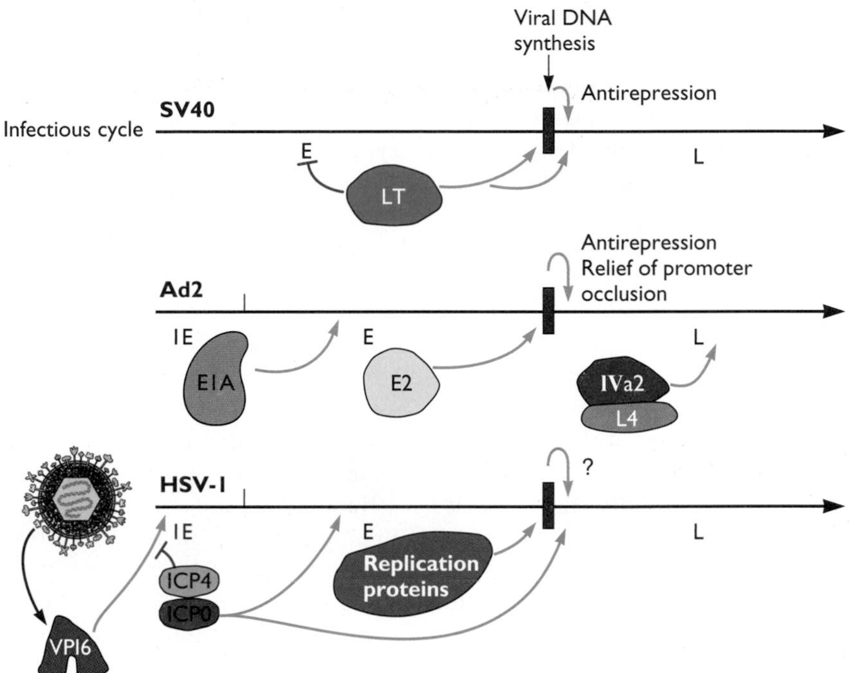

Figure 8.14 Important features of the simian virus 40 (SV40), human adenovirus type 2 (Ad2), and herpes simplex virus type 1 (HSV-1) transcriptional programs. The transcriptional programs of these three viruses are depicted by the horizontal time lines, on which the onset of viral DNA synthesis is indicated by the purple boxes. For comparative purposes only, the three reproductive cycles are represented by lines of equal length. The immediate-early (IE), early (E), and late (L) transcriptional phases are indicated, as are viral proteins that participate in regulation of transcription. Stimulation of transcription by these proteins and effects contingent on viral DNA synthesis in infected cells are indicated by green and orange arrows, respectively. Red bars indicate negative regulation of transcription.

example, unnecessary use of the resources of the infected cell, or inappropriate transcription of cellular genes.

Simian Virus 40

The transcriptional programs established in cells infected by viruses with small DNA genomes are quite simple: they comprise only two phases regulated by a single viral protein. For example, the genome of simian virus 40 contains only one early and one late transcription unit (Appendix, Fig. 13B), each of which encodes more than one protein. This type of organization reduces the genetic information that must be devoted to transcription punctuation marks and regulatory sequences, a significant advantage when genome size is limited by packaging constraints. The price for such a transcriptional strategy is heavy dependence on the host cell's RNA-processing systems to generate multiple mRNAs by differential polyadenylation and/or splicing of a single primary transcript (Chapter 10). Expression of the simian virus 40 early transcription unit by cellular enhancer- and promoter-binding proteins (Fig. 8.5) leads to the synthesis of large T antigen in infected cells. This multifunctional viral protein both induces initiation of viral DNA synthesis and activates late transcription (Fig. 8.14). Consequently, T-antigen synthesis in a permissive host cell leads to progression to the late phase of infection, when the segment of the viral genome encoding structural proteins is transcribed.

Human Adenovirus

Adenoviral genomes contain eight RNA polymerase II transcription units, which encode more than 40 proteins and are transcribed in four discrete phases (Appendix, Fig. 1B). At least three viral regulators of transcription, as well as viral DNA synthesis, govern the transitions from one phase to the next (Fig. 8.14). Upon entry of the genome into the nucleus, a single viral gene (E1A) is transcribed under the control of typical enhancers. If functional E1A proteins cannot be made, progression of the infectious cycle beyond this immediate-early phase is severely impaired. The E1A proteins, which regulate transcription by multiple mechanisms, are necessary for efficient transcription of all early transcription units. Among this set is the E2 gene, which encodes the proteins required for viral DNA synthesis and entry into the late phase of infection (Fig. 8.14). The adenoviral late phase is marked not only by transcription of late phase-specific genes, but also by an increased rate of initiation from the major late promoter that is active during the early phase of infection. Late genes of the more complex DNA viruses therefore are defined as those that attain their maximal rates of transcription following viral DNA synthesis. No known activator of late transcription is included among the adenoviral early gene products. Instead, the synthesis of progeny adenoviral DNA molecules is indirectly coordinated with production of the protein components that will encapsidate them in virions (Fig. 8.14).

Herpes Simplex Virus Type 1

The organization of the herpes simplex virus type 1 genome demands an entirely different transcriptional strategy. The more than 80 known genes of this virus are, with few exceptions, expressed as individual transcription units. Furthermore, splicing of primary transcripts is the exception, rather than the rule. In contrast to those of simian virus 40 or human adenoviruses, herpes simplex virus type 1 genes encoding proteins that participate in the same process are not clustered but are scattered throughout the genome (Appendix, Fig. 5B). The basic distinction of early and late phases is maintained in the herpesviral transcriptional program (Fig. 8.14), but temporal control of the activity of more than 80 viral promoters is obviously more complicated. In fact, the potential for finely tuned regulation is much greater when the viral genome comprises a large number of independent transcription units.

Initial expression of both polyomaviral and adenoviral genes is directed by enhancers that operate efficiently with cellular proteins. In contrast, a viral activating protein is imported into cells infected by herpes simplex virus type 1 (Fig. 8.14). This virion structural protein, VP16, is necessary for efficient transcription of viral immediate-early genes. This simple device might seem to guarantee transcription of these genes in **all** infected cells. However, this is not the case, because VP16 functions only in conjunction with specific cellular proteins. Stimulation of immediate-early gene transcription by VP16 is of considerable importance for the success of the viral reproductive cycle.

Herpesviral immediate-early gene products resemble adenoviral E1A proteins in performing a number of regulatory functions. The regulatory scheme of herpes simplex virus type 1 is, however, considerably more complex, for this viral genome carries five immediate-early genes (Appendix, Fig. 5B). The products of two, ICP4 and ICP0, are transcriptional regulators. The ICP4 protein, which is essential for progression beyond the immediate-early phase of infection, is regarded as the major transcriptional activator. It stimulates transcription of both early and late genes, and also acts as a repressor of immediate-early transcription. As for the simpler DNA viruses, transcription of herpesviral late genes is governed by viral DNA replication (Fig. 8.14). As noted previously, two classes of late genes can be distinguished: some are transcribed only following synthesis of progress genomes and others are transcribed at maximal rates during the late phase. More subtle distinctions among the large number of late genes may be made as their transcriptional regulation becomes better understood (Box 8.10).

This summary of three DNA virus transcriptional programs illustrates the increasing sophistication of the circuits that control transcription with increasing genome size, a feature of little surprise. Underlying such diversity are the common themes of the central role of virus-encoded transcriptional regulators and the coordination of transcriptional control with viral DNA synthesis. Regulation of initiation is the mainstay of the transcriptional control programs of these DNA viruses. Nevertheless, expression of some genes is controlled at other steps in the transcription cycle, such as termination.

Viral Proteins That Stimulate Transcription Are Linchpins of Transcriptional Cascades

In this section, we focus on a few well-studied viral regulators to illustrate general principles of their operation, or fundamental insights into cellular processes that have been gained through their study.

Some viral transcriptional regulators are close relatives of well-studied cellular proteins that bind to specific DNA sequences in promoters or enhancers (Table 8.6). These viral proteins possess discrete DNA-binding domains, some with sequences that conform to motifs characteristic of cellular DNA-binding proteins, and activation domains that interact with cellular initiation proteins (Table 8.6). These properties are described in more detail for one such protein, the Epstein-Barr virus Zta protein, in the next section.

Sequence-specific DNA-binding proteins play ubiquitous roles in the transcription of cellular genes by RNA polymerase II, so it is not surprising that viral DNA genomes transcribed by this enzyme encode analogous proteins. However, viral transcriptional regulators that possess no intrinsic ability to bind specifically to DNA are equally common (Table 8.7). The preponderance of such viral proteins, quite unexpected when they were first characterized, was a strong indication that host cells also contain proteins that control transcription without themselves binding to DNA. Many such proteins (e.g., coactivators) have now been recognized. Two examples, the herpes simplex virus type 1 VP16 and adenovirus E1A proteins, illustrate the diversity of mechanisms by which this class of viral proteins regulates transcription.

BOX 8.10

EXPERIMENTS

Global analysis of herpesviral gene expression using DNA microarrays

Individual viral RNAs can be detected and quantified by several methods that exploit hybridization to DNA representing specific viral sequences, such as RNase protection or primer extension. Application of these methods to **every** RNA encoded in a large DNA genome, like that of herpes simplex virus type 1, would be extremely laborious, time-consuming, and expensive. Furthermore, it is difficult to examine more than a few RNAs simultaneously. Hybridization to DNA microarrays provides a powerful tool with which to determine and quantify expression of all viral genes in a single experiment, and to compare expression under different conditions.

In this method, illustrated in panel A of the figure, RNA isolated from infected cells is converted to complementary DNA (cDNA) by reverse transcription in the presence of a fluorescently labeled deoxyribonucleoside triphophate (dNTP). Dyes that fluoresce at different wavelengths are generally incorporated into different samples. For example, mock-infected-cell cDNA could be labeled with a cyanine 5-labeled dNTP (red) and infected-cell cDNA could be labeled with cyanine-3 dNTP (green), or cDNAs synthesized from wild-type and mutant virus-infected cell RNA could be labeled with the different dyes. The labeled cDNAs are then mixed and hybridized to a DNA micorarry, which contain DNAs representing numerous genes. Microarrays for cellular genomes are made using cDNAs or synthetic oligonucleotides, and contain DNA representing tens of thousands of genes. Microarrays for viral genomes, such as those used in the experiments described below, typically are made with oligonucleotides, and each viral gene is represented by multiple sequences. After hybridization, the microarrays are washed and exposed to

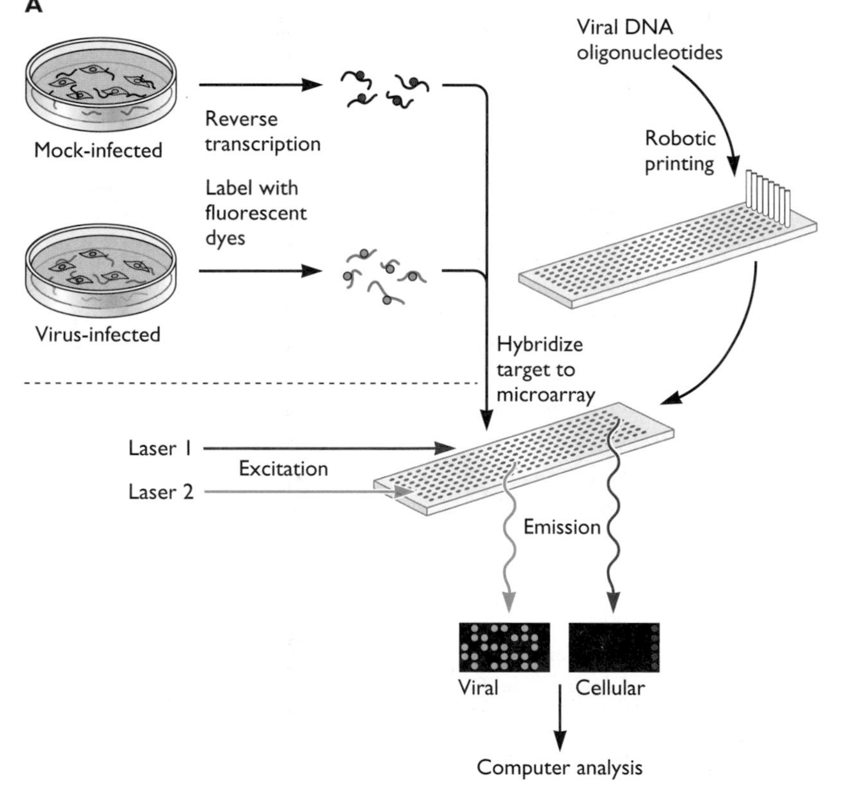

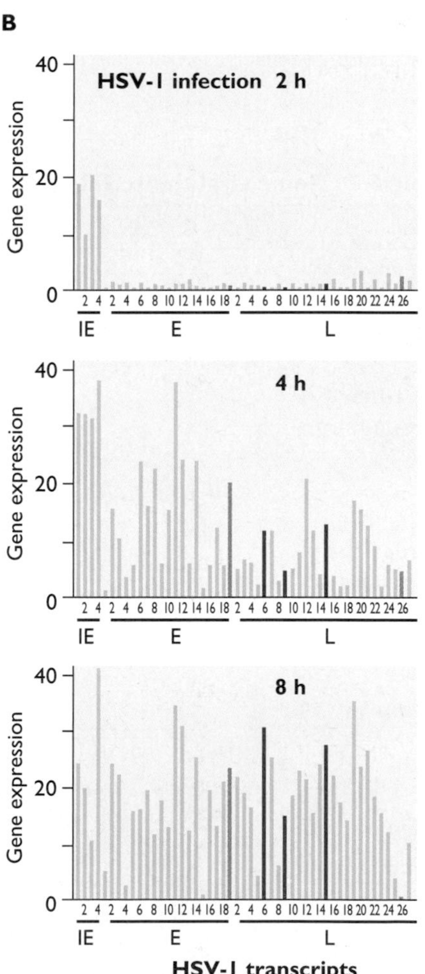

(continued)

BOX 8.10

EXPERIMENTS

Global analysis of herpesviral gene expression using DNA microarrays (continued)

laser light of appropriate wavelengths. The fluorescent signals of each DNA spot are detected, quantified and recorded by an automatic scanner. After subtraction of background and application of various other quality control procedures, they can then be analyzed in a number of different ways.

The results of an experiment in which microarrays were used to compare expression of herpes simplex virus type 1 genes at various times after infection are illustrated in panel B. In this experiment, each viral gene was represented by multiple spots on the array and a number of independently prepared samples was analyzed for each condition of interest, so that levels of expression of individual genes could be quantified reliably. The median signals for each gene examined at immediate-early (IE), early (E), and late (L) times of infection were then compared.

As expected, only IE genes were expressed efficiently 2 h after infection, whereas late RNAs predominated by 8 h. However, this global comparison revealed considerable variation in the kinetics and efficiency of expression of late genes. The concentrations of some late RNAs increased significantly between 4 and 8 h after infection (red), whereas other late RNAs reached close to maximal levels by 4 h (orange) or actually decreased in concentration between 4 and 8 h (pink).

In this experiment (as is currently typical), total RNA was isolated for microarray analysis. Consequently, the results do not establish whether changes in viral RNA concentration are the result of alterations in rates of transcription, mRNA processing, or degradation of the RNA. Nevertheless, the results shown indicate that there are several distinct patterns of expression of herpesviral late genes, and hence multiple mechanisms that govern production of late mRNAs.

The concentrations of RNA encoded by herpes simplex virus type 1 IE, E and L genes are shown in arbitrary units. Adapted from W.C. Yang et al., *J. Virol.* **76:**12758–12774, 2002, with permission.

Stringley, S. W., J. J. Garcia Ramirez, S. A. Aguilar, K. Simmen, R. M. Sandri-Goldin, P. Ghazal, and E. K. Wagner. 2000. Global analysis of herpes simplex virus type 1 transcription using an oligonucleotide-based DNA microarray. *J. Virol.* **74:**9916–9927.

Table 8.7 Some viral transcriptional regulators that lack sequence-specific DNA-binding activity

Virus	Protein	Functions and properties
Adenovirus		
Human adenovirus type 2	289R and 243R E1A proteins	Larger one stimulates early gene transcription in infected cells; both bind to the Rb protein releasing cellular E2f, and to the cellular coactivator p300
Hepadnavirus		
Hepatitis B virus	X	Required for efficient virus reproduction *in vivo*; stimulation of RNA polymerase II transcription depends on cellular, sequence-specific activators; may also stimulate transcription indirectly via activation of cellular signaling pathways
Herpesviruses		
Herpes simplex virus type 1	VP16	Stimulates transcription from IE promoters via potent acidic activation domain; achieves promoter specificity by interaction with the cellular Oct-1 protein
	ICP0	Cooperates with ICP4 to stimulate transcription of E and L genes in infected cells; important for reactivation from latency
Epstein-Barr virus	EBNA-2	Essential for B-cell transformation; induces transcription of viral LMP-1 and LMP-2, and of some cellular genes via specific DNA sequence; sequence recognition achieved by interaction with the cellular recombination binding protein Jκ
Polyomavirus		
Simian virus 40	Large T antigen[a]	Stimulates late-gene transcription; can bind to Tbp, TfIIb, and Tef-1 via different segments; contains no activation domain; may stabilize initiation complexes
Retrovirus		
Human T-lymphotropic virus type 1	Tax	Stimulates transcription from the viral LTR transcriptional control region; can bind directly to the basic region of bZIP proteins to stabilize dimers and increase their affinity for DNA; can regulate transcription indirectly by inducing degradation of IκB

[a]The sequence-specific DNA-binding activity of large T antigen (see Fig. 9.3) is not required for stimulation of transcription by this protein.

The Epstein-Barr virus Zta protein: a sequence-specific DNA-binding protein that induces entry into the productive cycle. When the gammaherpesvirus Epstein-Barr virus infects B lymphocytes, only a few viral genes are transcribed (the latent state). The products of these genes maintain the viral genome via replication from a latent phase-specific origin of replication (Fig 8.15A), modulate the immune system, and alter the growth properties of the cells. Treatment of such latently infected lymphocytic cells in culture with a variety of agents induces entry into the productive cycle. Epstein-Barr virus also enters this cycle upon infection of epithelial cells. Virus replication begins with synthesis of three viral proteins that regulate gene expression. However, just one of these, the transcriptional regulator Zta (also known as ZEBRA, Z, or EB-1), is sufficient to interrupt latency and induce entry into the productive cycle.

The Zta protein exhibits many properties characteristic of the cellular proteins that recognize local promoter sequences: it is a modular, sequence-specific DNA-binding protein that belongs to the basic-leucine zipper family (Table 8.6). This dimerization and DNA-binding domain mediates binding of Zta to viral promoters that contain its recognition sequence. The discrete activation domain, which can bind directly to cellular initiation proteins, such as subunits of TfIId, is thought to facilitate the assembly of preinitiation complexes.

The availability or activity of Zta is regulated by numerous mechanisms. Expression of the viral gene encoding Zta is controlled by cellular DNA-binding proteins that stimulate or repress transcription (Fig. 8.15B). Many of these cellular proteins are targets of signal transduction cascades that induce entry of Epstein-Barr virus into the productive cycle in latently infected B cells. In addition, the Zta protein itself is a positive autoregulator of transcription. The availability of Zta mRNA for translation is also regulated, in part, by annealing of Zta pre-mRNA to the complementary transcripts of the viral EBNA-1 gene (Fig. 8.15A). Finally, phosphorylation of Zta modulates its DNA-binding activity, whereas binding of cellular proteins, including the p53 tumor suppressor protein and the p65 subunit of Nf-κb, prevents Zta from stimulating transcription.

The net effect of these regulatory mechanisms, which depends on the type and on the proliferation and differentiation states of the Epstein-Barr virus-infected cell, determines whether active Zta protein is available, and consequently whether infection leads to latent or productive replication. Entry into the latter cycle appears to be an inevitable consequence of production of active Zta: this protein stimulates transcription from the promoters of its own gene and other early genes and plays an important role in replication from the lytic origins.

The herpes simplex virus type 1 VP16 protein: sequence-specific activation of transcription via a cellular DNA-binding protein. The herpesviral VP16 protein enters infected cells in the virion (Fig. 8.14). It has taught us much about mechanisms by which transcription by RNA polymerase II can be stimulated. Furthermore, its unusual mode of promoter recognition illustrates the importance of conformational change in proteins during formation of DNA-bound complexes.

The VP16 protein lacks a DNA-binding domain. Its acidic activation domain is one of the most potent known and has been exploited to investigate mechanisms of stimulation of transcription. Chimeric proteins in which this activation domain is fused to heterologous DNA-binding domains strongly stimulate transcription from promoters that contain the appropriate binding sites. When part of such fusion proteins, the VP16 acidic activation domain can stimulate several reactions required for initiation of transcription (Fig. 8.16A). It can also increase the rate of transcriptional elongation and stimulate transcription from chromatin templates (Fig. 8.16B). These properties established that a single protein can regulate RNA polymerase II transcription by multiple molecular mechanisms.

The VP16 protein is the founding member of a class of viral regulators that activate transcription from promoters that contain a specific consensus sequence, yet possess no sequence-specific DNA-binding activity. The 5′-flanking regions of viral immediate-early genes contain at least one copy of a consensus sequence, 5′TAATGARAT3′ (where R is a purine), which is necessary for VP16-dependent activation of their transcription. This sequence is bound by VP16 only in association with at least two cellular proteins, octamer-binding protein 1 (Oct-1) and host cell factor (Hcf) (Fig. 8.17). The Oct-1 protein is a ubiquitous transcriptional activator named for its recognition of a consensus DNA sequence termed the octamer motif. This protein and VP16 can associate to form a ternary (three-component) complex on the 5′TAATGARAT3′ sequence, but the second cellular protein, Hcf, is necessary for stable, high-affinity binding. Hcf is thought to regulate transcription, as it can bind to several sequence-specific transcriptional regulators and proteins that modulate chromatin structure. The VP16 protein and Hcf form a heteromeric complex in the absence of Oct-1 or DNA. An important function of Hcf therefore appears to be induction of a conformational change in VP16, to allow its stable binding to Oct-1 on the immediate-early promoters (Fig. 8.17).

One of the most remarkable features of the mechanism by which the VP16 protein is recruited to immediate-early promoters is its specificity for Oct-1. This protein is a member of a family of related transcriptional regulators defined by a common DNA-binding motif called the POU-homeodomain.

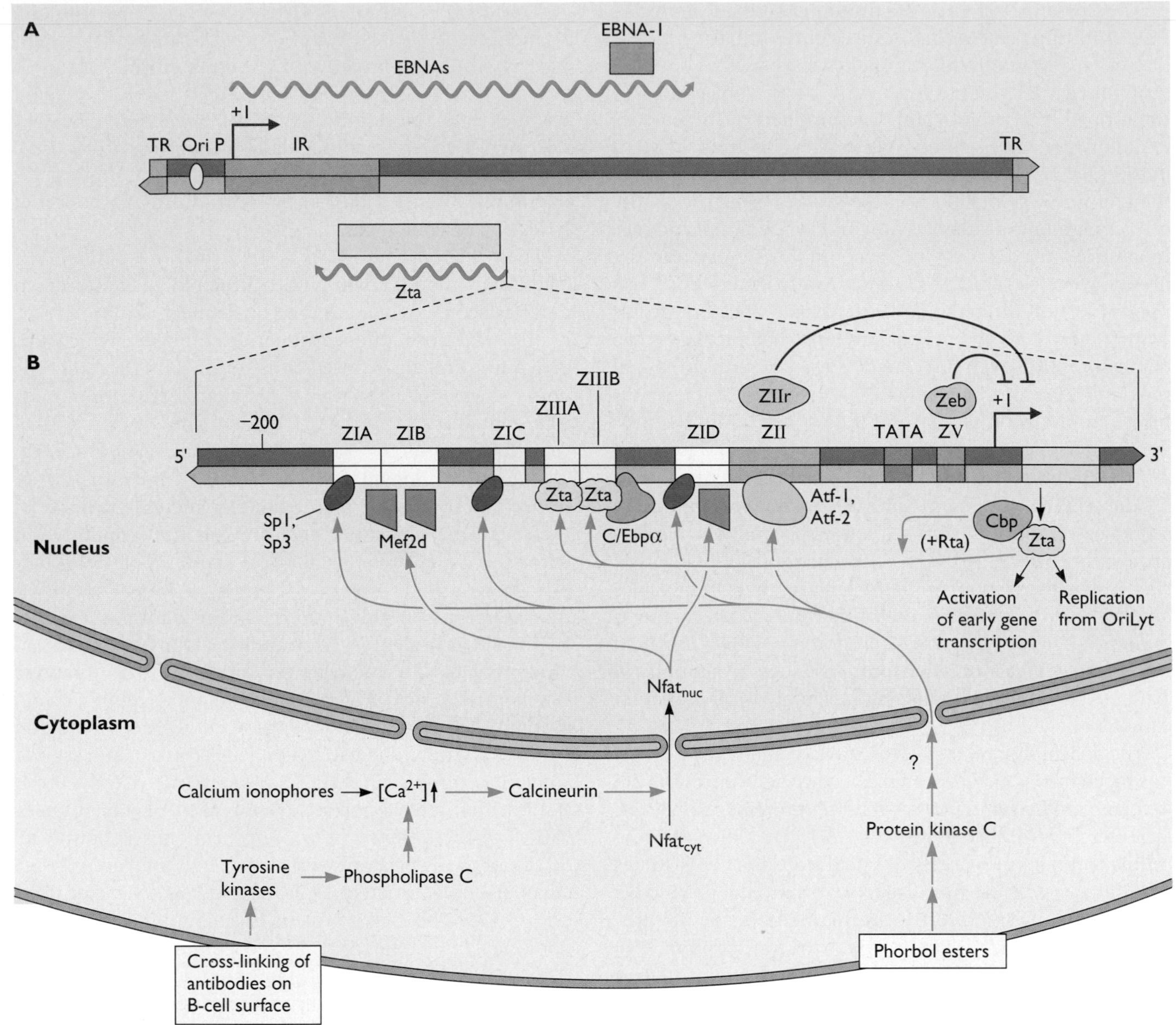

Figure 8.15 Organization and regulation of the Epstein-Barr virus Zta gene promoter. (A) Organization of the transcription units that contain the coding sequence for the Epstein-Barr virus nuclear antigen (EBNA) proteins (an ~100-kb transcription unit), Zta, and Rta. The locations of the genomic terminal (TR) and internal (IR) repeat sequences, the origin of replication for plasmid maintenance (OriP), and the coding sequences for the EBNA-1 and Zta proteins are indicated. The Zta protein is synthesized from spliced mRNAs processed from the primary transcripts shown. **(B)** Sequences that regulate transcription from the Zp promoter of the Zta gene are shown to scale and in the conventional 5′→3′ direction. The cellular or viral proteins that operate via these sequences are shown below. The Atf and Mef2d proteins are members of the basic-leucine zipper and Mads families, respectively. During latent infection, binding of cellular transcriptional repressors (for example, to the ZII and ZV sequences) prevents transcription from the Zp promoter (red bars). Signals that induce entry into the productive cycle activate transcription of the Zta gene via effects on cellular transcriptional activators (green arrows). Such pathways are illustrated with signaling initiated by cross-linking of antibodies of B cell surfaces, and treatment of cells with phorbol esters. Once Zta protein is produced, it stimulates transcription further by binding to the ZIII sequences in association with the cellular coactivator Cpb. It also stimulates transcription from the Zp promoter indirectly, by increasing the intracellular concentration of another cellular activator C/Ebpα.

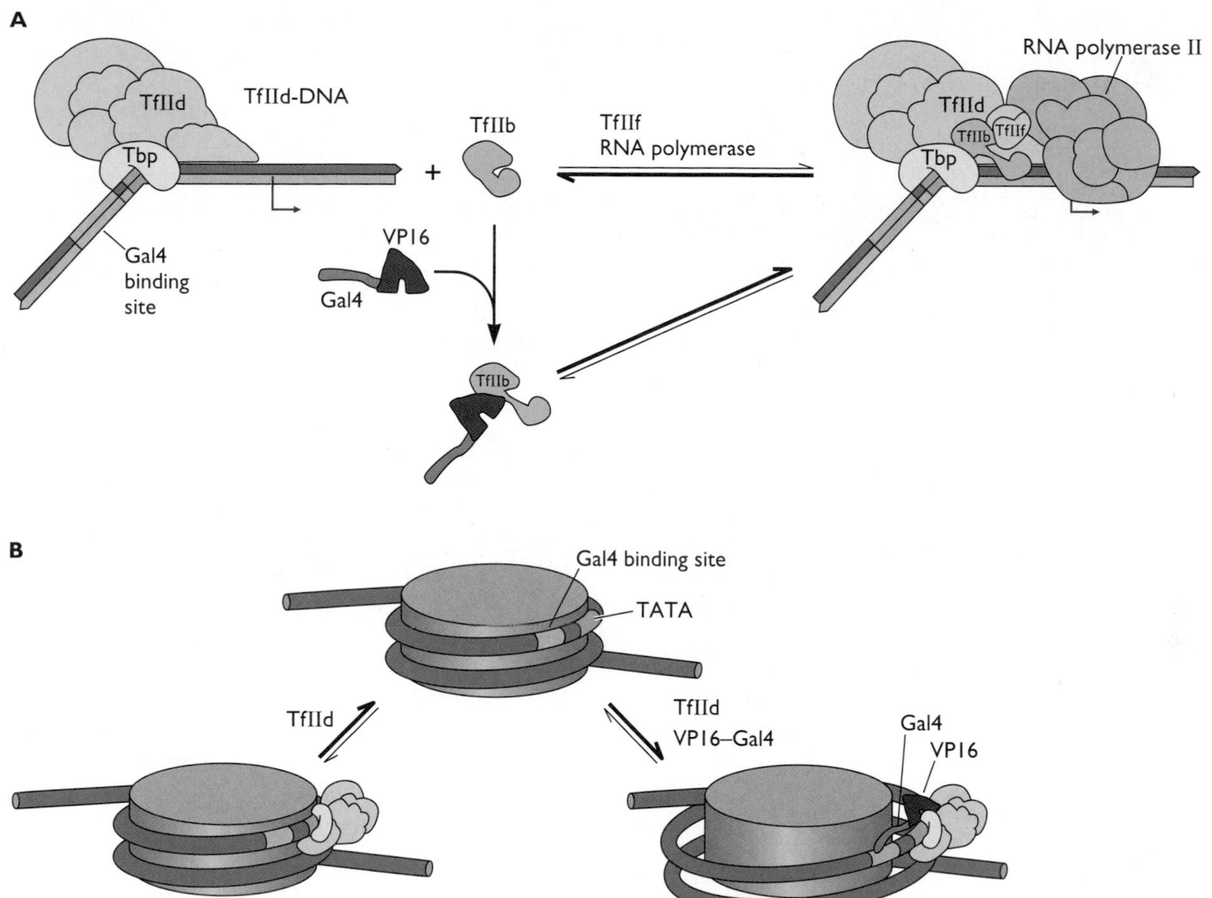

Figure 8.16 Models for transcriptional activation by the herpes simplex virus type 1 VP16 protein. (A) Induction of conformational change in TfIIb. In native TfIIb, the N- and C-terminal domains associate with one another such that internal segments of the protein that interact with TfIIf-RNA polymerase II are inaccessible. Binding of the acidic activation domain of VP16, for example as a chimera with the DNA-binding domain of the yeast protein Gal4, disrupts this intermolecular association of TfIIb domains, exposing its binding sites for TfIIf and RNA polymerase II. Consequently, formation of the preinitiation complex containing TfIIb, TfIIf, and RNA polymerase II is now a more favorable reaction. The VP16 activation domain can also bind directly to subunits of TfIId, TfIIf, and TfIIh. These interactions may also favor formation of the closed initiation complex, or they may be related to the ability of the activation domain to stimulate other reactions, such as promoter clearance. **(B)** Alleviation of transcriptional repression by nucleosomes. Many activators, including the acidic activation domain of VP16, stimulate transcription from nucleosomal DNA templates to a much greater degree than they do transcription from naked DNA. This property is the result of their ability to alleviate repression of transcription by nucleosomes, an activity therefore termed antirepression. Organization of DNA into a nucleosome can block access of proteins to their DNA-binding sites, as illustrated for binding of TfIId to a TATA sequence (left). Association of the acidic activation domain of VP16 with the template alters the interaction of the DNA with the nucleosome to allow TfIId access to the TATA sequence (right), presumably as a result of recruitment of ATP-dependent chromatin remodeling enzymes and/or histone acetyltransferases.

The VP16 protein distinguishes Oct-1 from all other members of this family, including Oct-2, which binds to exactly the same DNA sequence as Oct-1. In fact, VP16 detects a **single** amino acid difference in the exposed surfaces of DNA-bound Oct-1 and Oct-2 homeodomains. Such discrimination is not only remarkable but also of biological importance, because only the Oct-1 protein is synthesized ubiquitously.

The GARAT segment of TAATGARAT-containing elements of immediate-early genes is indispensable for the assembly of VP16 into the DNA-bound complex and therefore defines promoters that are recognized by the viral protein. Nevertheless, the DNA-binding domain of Oct-1 recognizes the GARAT motif (Fig. 8.17). This DNA sequence is therefore the crucial effector of assembly of

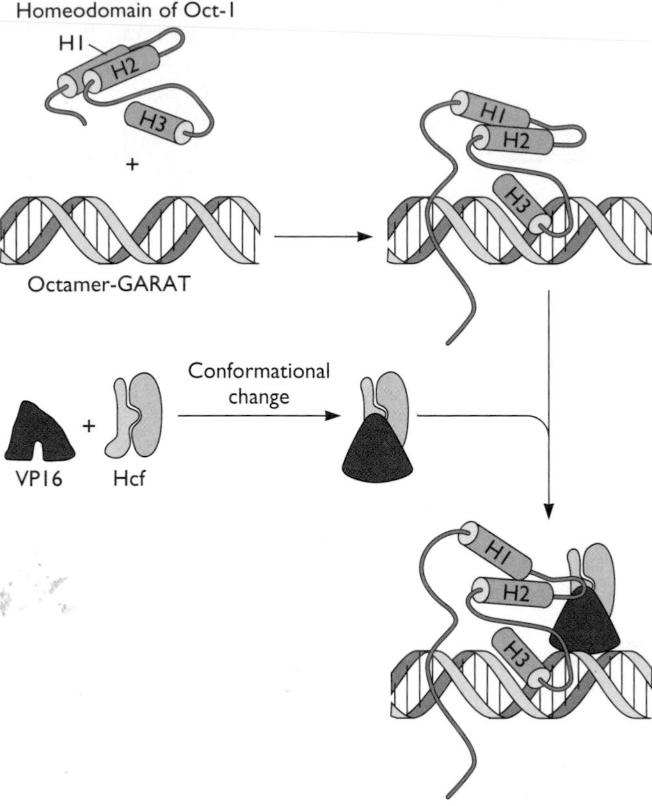

Figure 8.17 Conformational changes and recruitment of VP16 to herpes simplex virus type 1 promoters. Binding of the Oct-1 homeodomain to DNA containing the GARAT sequence and of VP16 to Hcf induces conformational changes that allow specific recognition of GARAT-bound Oct-1 by VP16. This mechanism ensures that the VP16 protein is recruited only to promoters that contain the GARAT sequence, that is, viral immediate-early promoters.

the quaternary complex. The mechanism by which VP16, in conjunction with Hcf and Oct-1, recognizes herpesviral immediate-early promoters illustrates the important contributions of induced conformational changes to transcriptional specificity.

The VP16 protein interacts with cellular Hcf and Oct-1 proteins via its N-terminal region. Its C-terminal region contains the acidic activation domain discussed previously. The results of chromatin immunoprecipitation experiments suggest that stimulation of IE gene transcription in infected cells is mediated by several of the biochemical activities exhibited by the acidic activation domain in simplified experimental systems (Box 8.11).

The incorporation of the VP16 protein into virions at the end of one infectious cycle is an effective way to ensure transcription of viral genes and initiation of a new cycle in a new host cell. Nevertheless, some features of this mechanism are not fully appreciated, in particular the benefits conferred by the indirect mechanism by which VP16 recognizes immediate-early promoters. One advantage over direct DNA binding may be the opportunity to monitor the growth state of the host cell that is provided by the requirement for binding to Hcf, a protein that is important for proliferation of uninfected cells. The dependence on Hcf may also contribute to the establishment of latent infections in neurons (see "Entry into one of two alternative transcriptional programs" below).

The VP16 activation domain (as well as Hcf) is necessary for efficient virus reproduction when infected cells contain one or a few viral genomes. However, the need for VP16 is circumvented, at least partially, when a large number of genomes enter infected cell nuclei (Box 8.12). Adenovirus mutants lacking the coding sequence for E1A CR3 (see the next section) exhibit the same phenotype.

Adenoviral E1A proteins: regulation of transcription of multiple mechanisms. Two E1A proteins, synthesized from differentially spliced mRNAs (Fig. 8.18), are synthesized during the immediate-early phase of adenovirus infection. These two proteins share all sequences except for an internal segment (conserved region 3 [CR3]) that is unique to the larger protein. Nevertheless, they differ considerably in their regulatory potential, because the CR3 segment is primarily responsible for stimulation of transcription of viral early genes. As the larger E1A protein neither binds specifically to DNA nor depends on a specific promoter sequence, it is often considered the prototypical example of viral proteins that stimulate transcription by indirect mechanisms, some of which are listed in Table 8.7.

The CR3 segment of the larger E1A protein comprises an N-terminal zinc finger motif followed by 10 amino acids that are highly conserved (Fig 8.19A). The latter mediate binding of the E1A protein to such cellular, sequence-specific activators as Atf-2, and Sp1, and hence association with the viral promoters. The zinc finger motif is essential for stimulation of transcription by the E1A protein in infected cells. It binds with exceptionally high affinity to a single component (Med23) of the human mediator complex, which contains at least 20 different subunits and is essential for regulation of transcription by RNA polymerase II. This interaction stimulates assembly of preinitiation complexes (see Fig. 8.2) in *in vitro* reactions, and is required for stimulation of transcription in infected cells: CR3-dependent stimulation of viral E2 transcription is observed in mouse embryonic stem cells, but not in mutant cells, homozygous for deletion of the *Med23* gene.

Adenoviral E1A proteins activate transcription by a second mechanism, which is mediated by the conserved regions of the N-terminal exon, CR1 and CR2 (Fig. 8.18). This function was elucidated through studies of

BOX 8.11

EXPERIMENTS

In vivo *functions of the VP16 acidic activation domain*

The chromatin immunoprecipitation assay (Box 8.1) has been used to compare the proteins associated with herpes simplex virus type 1 immediate-early (IE) promoters in cells infected by the wild-type virus, or a mutant encoding VP16 that lacks the acidic activation domain. Cross-linked DNA was immunoprecipated with antibodies to VP16, RNA polymerase, or several other cellular proteins. The concentrations of viral promoter DNA present in such immunoprecipates was then assessed by using PCR. The results summarized in the table suggest that the activation domain stimulates IE gene transcription in infected cells by mechanisms observed in simplified experimental systems.

Association of RNA polymerase II and Tbp with the viral promoters depended on synthesis of VP16 containing an activation domain, consistent with stimulation of assembly of initiation complexes. This domain is also necessary for efficient recruitment of histone acetyltransferases (Cbp) and ATP-dependent remodeling proteins (Brg-1), as well as loss of histone H3, suggesting that it also allows remodeling of chromatin templates in infected cells.

Herrera, F. J., and S. J. Triezenberg. 2004. VP16-dependent association of chromatin-modifying coactivators and underpresentation of histones at immediate-early gene promoters during herpes simplex virus infection. *J. Virol.* **78:**9689–9696.

Promoter-bound proteins	VP16 acidic activation domain	
	Not present	Present
VP16	++	++
Oct-1	++	++
RNA polymerase II	–	++
Tbp	–	++
Cbp	+	++
Brg-1	–	++
Histone H3	++	–

transformation and immortalization by E1A gene products. The CR1 or CR2 segments interact with several cellular proteins. Two of these, Rb and p300 (Fig. 8.18), are of special relevance to transcriptional regulation. The Rb protein is the product of the cellular retinoblastoma susceptibility gene, a tumor suppressor that plays a crucial role in cell cycle progression (Volume II, Chapter 7). The discovery that Rb binds to adenoviral E1A proteins, as well as to transforming proteins of papillomaviruses and polyomaviruses, led to elucidation of the mechanism of action of this important cellular protein.

In uninfected cells, Rb binds to cellular E2f proteins, sequence-specific transcriptional activators originally discovered because they bind to the human adenovirus type 2 E2 early promoter (Fig. 8.4). Such E2f-Rb complexes possess the specific DNA-binding activity characteristic of E2f, but Rb represses transcription (Fig. 8.19B). The CR1 and CR2 regions of the E1A proteins bind to the same

BOX 8.12

DISCUSSION

How does high multiplicity of infection overcome the need for viral activators of early transcription?

Mutants of human adenovirus type 5 (Ad5) that do not encode the larger (289R) E1A protein and of herpes simplex virus type 1 deleted for the coding sequence for the VP16 activation domain grow very poorly in cells infected at a low multiplicity of infection, e.g., 1 PFU/cell. However, when the multiplicity of infection is increased, these mutants do replicate, albeit more slowly and to lower yields than the wild-type viruses. How does increasing the concentration of infecting virus overcome the need for the viral activators of transcription?

The answer is thought to lie in the number of genomes that reach infected cell nuclei. Because of the high particle-to-PFU ratio typical of these viruses (see Chapter 2), we cannot infer that a single genome enters the nucleus when the multiplicity of infection is 1 PFU/cell. Nevertheless, increasing the multiplicity of infection, for example to 100 PFU/cell, obviously results in entry of a much larger number of viral DNA molecules. Such high DNA template concentrations presumably compensate for the low affinity with which cellular transcriptional regulators bind to viral proteins in the absence of the E1A or VP16 proteins. Consequently, early or immediate-early genes **are** transcribed, although less efficiently than in cells containing the viral activators.

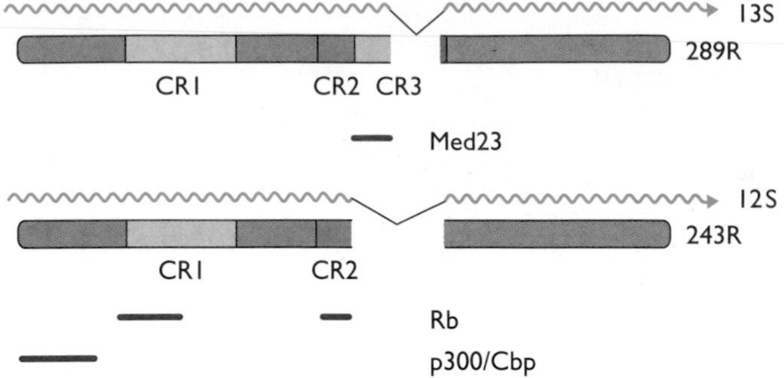

Figure 8.18 Features of the adenovirus type 2 E1A proteins. Primary transcripts of the immediate-early E1A gene are alternatively spliced to produce the abundant 13S and 12S mRNAs. As such splicing does not change the translational reading frame, the E1A proteins are identical, except for an internal segment of 46 amino acids unique to the larger protein. The three most highly conserved regions are designated CR1, CR2, and CR3. The regions of the E1A proteins necessary for interaction with the Rb protein, p300 Cbp, and Med23 are indicated (red lines).

segments of Rb as does E2f. Binding of the E1A proteins to Rb therefore releases E2f proteins from association with this repressor to allow transcription from E2f-dependent promoters. During the early phase of infection, E2f proteins are essential for efficient transcription of the gene that encodes the proteins required for viral DNA synthesis. Sequestration of Rb by the E1A proteins therefore ensures production of replication proteins and progression into the late phase of the infectious cycle (Fig. 8.14).

The N-terminal sequences common to the two E1A proteins also bind directly to the cellular coactivators p300 and Cbp (Fig. 8.18). These proteins are coactivators that interact with, and mediate stimulation of transcription by, the basic-leucine zipper protein Creb and steroid/thyroid hormone receptors. p300/Cbp are histone acetylases and bind to other histone acetylases. Modification of histones by these enzymes alters the structure of transcriptionally active chromatin and stimulates multiple reactions in transcription. The E1A proteins disrupt the interaction of p300/Cbp with Creb, and compete for their binding to other histone acetylases. This activity of E1A proteins has been implicated in repression of enhancer-dependent transcription, and is required for induction of cell proliferation and transformation.

The multiplicity of mechanisms by which the E1A proteins engage with components of the cellular transcriptional machinery is one of their most interesting features. Regulation by multiple mechanisms may prove to be a definitive property of viral proteins that cannot bind directly to DNA (Table 8.7). For example, the human T-lymphotropic virus type 1 Tax protein stimulates transcription by binding to specific cellular members of the basic-leucine zipper family, and by activating Nf-κb. Tax can activate the signaling cascade that induces degradation of Iκb and translocation of Nf-κb from the cytoplasm to the nucleus (Fig. 8.11). This mechanism illustrates the potential for intervention of viral regulators in cellular pathways well upstream of any transcriptional activator.

Coordination of Transcription of Late Genes with Viral DNA Synthesis

Large quantities of viral structural proteins must be made to ensure efficient assembly of virions. Such a high concentration of foreign proteins is undoubtedly injurious to the host cell, and could therefore interfere with cellular processes needed for virus replication. In the case of the viruses under consideration here, this potential problem is avoided, because structural proteins are synthesized only during the late phase of infection. As replication of viral DNA genomes is in full swing by this time, virion assembly is the only step then necessary to complete the reproductive cycle. This pattern of viral structural protein synthesis results from the dependence of late-gene transcription on viral DNA replication: drugs or mutations that inhibit viral DNA synthesis in infected cells also block efficient expression of late genes. In fact, late genes are defined experimentally as those that are not transcribed when DNA synthesis is blocked. Despite their importance, the mechanisms by which activation of transcription can be integrated with viral DNA synthesis remain incompletely understood.

Titration of cellular repressors. The most obvious consequence of genome replication in cells infected by DNA viruses is the large increase in concentration of viral DNA molecules. Even in experimental situations, infected cells contain a relatively small number of copies of the viral genome during the early phase of infection, typically 1 to 100 copies per cell depending on the multiplicity of infection. As soon as viral DNA synthesis begins, this number rises geometrically to values as high as hundreds of thousands of viral DNA molecules per infected cell nucleus. At such high concentrations, viral promoters can compete effectively for components of the cellular transcription machinery.

The increase in DNA template concentration also titrates cellular transcriptional repressors that bind to specific sequences of certain viral late promoters. For example, the simian virus 40 major late promoter contains multiple binding sites for a cellular repressor that belongs to the steroid/thyroid hormone receptor superfamily (Fig. 8.20). Viral DNA replication increases the concentration of the late promoter above that at which every copy can be bound by the repressor, and therefore allows this promoter to become active (Fig. 8.20). This "antirepression" mechanism directly coordinates activation of transcription of late genes with

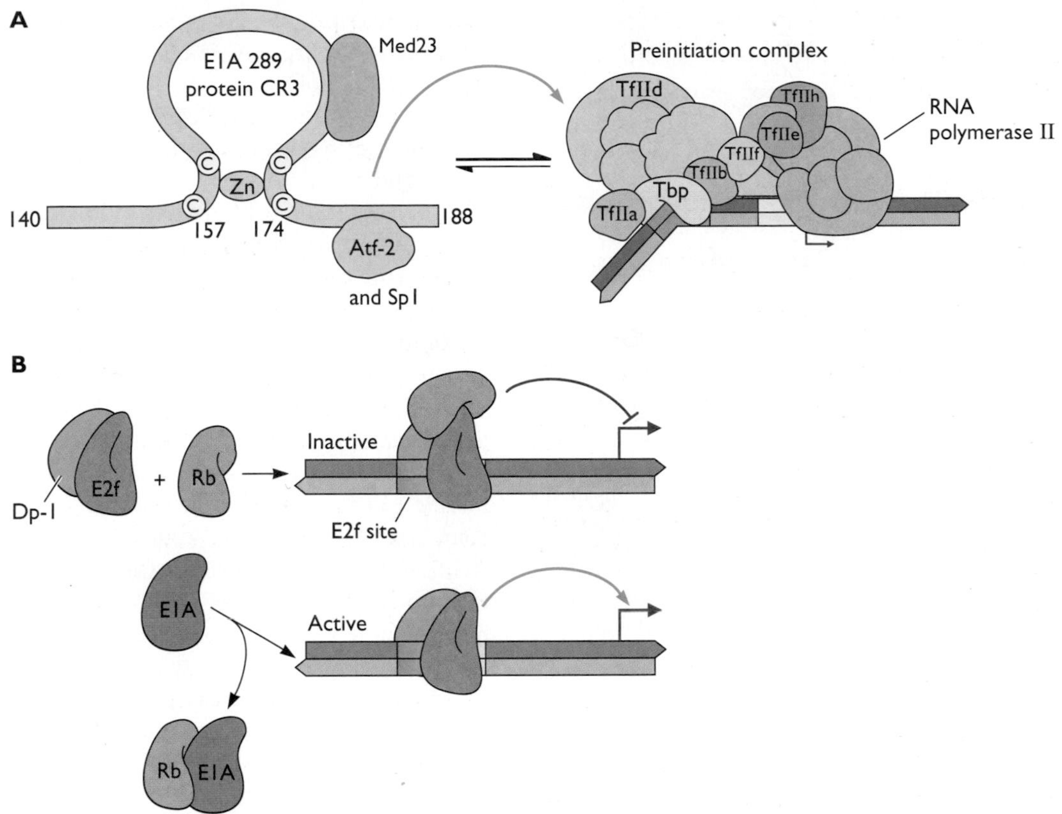

Figure 8.19 Indirect stimulation of transcription by adenoviral E1A proteins. (A) Interactions of the E1A CR3 sequences with components of the RNA polymerase II transcriptional machinery. The C-terminal segment of CR3 interacts with several cellular activators that bind to specific DNA sequences, as indicated, as well as with particular Taf subunits of TfIId. The Zn finger motif is required for tight binding to the Med23 subunit of the Mediator complex. This interaction stimulates assembly of closed preinitiation complexes. The exceptionally high affinity binding of CR3 to Med23 may also facilitate reinitiation. **(B)** Model of competition between E1A proteins and E2f for binding to Rb protein. The E2f transcriptional activators are heterodimers of a member of the E2f protein family (described in Chapter 9) and a related differentiation-regulated transcription factor protein, such as Dp-1. The binding of E2f to its recognition sites in specific promoters is not inhibited by association with the Rb protein, but Rb represses transcription from E2f-dependent promoters (top). Adenoviral E1A proteins made in infected (or transformed) cells bind to Rb and disrupt the E2f-Rb interaction. Consequently, Rb is removed from association with E2f, which can then stimulate transcription.

viral DNA synthesis, and is highly efficient. Consequently, it is not surprising that the same mechanism regulates transcription of the adenoviral IVa$_2$ gene, and hence the switch to the late phase transcriptional program (Fig. 8.14).

Viral activators of late-gene transcription. Although viral DNA synthesis is sufficient for activation of transcription of some viral late genes (e.g., the adenoviral IVa$_2$ gene), in most cases efficient transcription from late promoters also requires one or more viral proteins. For example, maximally efficient transcription from the simian virus 40 major late promoter depends on the viral early gene product large T antigen. This protein controls simian virus 40 late transcription both directly, as an activating protein (Fig. 8.14; Table 8.7), and indirectly, as a result of its essential functions in viral DNA synthesis (Chapter 9). Among the more complex DNA viruses, the coupling of replication to transcription of late genes can be more indirect. As noted above, late phase-specific transcription of the adenoviral IVa$_2$ gene is controlled by viral DNA synthesis-dependent titration of a cellular repressor. However, the IVa$_2$ protein is itself a sequence-specific activator of transcription. Once synthesized in infected cells, it cooperates with a viral L4 protein that also binds to specific promoter sequences, and stimulates the rate of initiation of transcription from the major late promoter at least 20-fold. Synthesis of adenoviral DNA therefore initiates a transcriptional cascade in which late promoters are activated sequentially (Fig. 8.14).

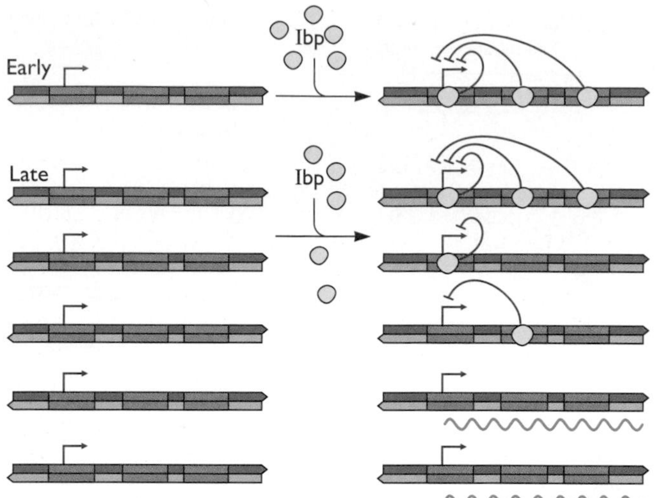

Figure 8.20 Cellular repressors regulate of the activity of the simian virus 40 late promoter. The sequence surrounding the simian virus 40 major late initiation site (the thickest arrow in Fig. 8.3D) contains three binding sites for the cellular repressor termed initiator-binding protein (Ibp), which contains members of the steroid/thyroid receptor superfamily. During the early phase of infection, the concentration of Ibp relative to that of the viral major late promoter is sufficiently high to allow all Ibp-binding sites in the viral genomes to be occupied. The concentration of Ibp does not change during the course of infection. However, as viral DNA synthesis takes place in the infected cell, the concentration of the major late promoter becomes sufficiently high that not all Ibp-binding sites can be occupied. Consequently, the major late promoter becomes accessible to cellular transcription components. Although we generally speak of "activation" of late gene transcription, this DNA replication-dependent mechanism is, in fact, one of escape from repression.

Transcription from simple promoters. Transcription of herpesviral late genes also requires viral DNA replication and synthesis of viral activators of transcription, for example, ICP4 and ICP0 in the case of herpes simplex virus type 1. However, the promoters of true late genes (those that are not expressed during the earlier phases) also share an unusually simple structure: they comprise typical TATA and initiator sequences and one additional element located between positions +10 and +40. How this simple promoter structure facilitates transcription of late genes has not been established. Nevertheless, this property might relate to the presence of a unique form of RNA polymerase II during the late phase of infection. The hypophosphorylated and hyperphosphorylated forms of the enzyme present in uninfected cells are replaced by a species in which the C-terminal domain of the large subunit is phosphorylated only on the second serine of the repeated YSPTSPS sequence. This modification has been implicated in efficient promoter clearance, whereas phosphorylation of the first serine facilitates transcriptional elongation by pTEFb. The herpesviral late genes that are transcribed by this infected cell-specific form of RNA polymerase II lack introns and are considerably shorter than typical cellular genes. Moreover, they are probably not associated with nucleosomes. Consequently, pTefb and regulators that modify nucleosome structure may be dispensable for transcription of these genes, and the simple structure of the promoters sufficient for recruitment of the components that mediate their transcription.

Regulation of termination. Viral DNA synthesis can also affect transcriptional termination, a regulatory mechanism illustrated by the human adenovirus type 2 major late transcription unit. During the early phase of infection, major late transcription terminates within a region in the middle of the transcription unit, such that no elongating complexes reach the L4 poly(A) addition site (Fig. 8.21). As discussed in Chapter 10, such restricted transcription is coupled with preferential utilization of specific RNA-processing signals to produce a single major late mRNA and protein during the early phase. Viral DNA synthesis is necessary to induce transcription of distal segments of the transcription unit to a termination site close to the right-hand end of the viral genome (Fig. 8.21). The inefficient termination of early transcription at many sites spread over about 12 kbp of DNA, and the fact that only replicated viral DNA molecules can support complete transcription, are features that suggest an unusual regulatory mechanism. One hypothesis is that alterations in template structure upon viral DNA synthesis may contribute to the regulation of termination of major late transcription (Box 8.13).

Figure 8.21 DNA replication-dependent alterations in termination of human adenovirus type 2 major late transcription. The segment of the viral genome from 15 to 100 map units containing the major late transcription unit is shown to scale **(top)**. The five sites at which primary major late transcripts are polyadenylated, designated L1 to L5, are indicated by the vertical arrows. During the early phase, major late transcripts terminate at multiple sites beyond the L1 site, as indicated by the dashed line, and none extend beyond the position indicated by the green arrowhead. The mechanisms responsible for this termination, which takes place within a DNA segment of 11 to 12 kb, are not understood. Following synthesis of both viral DNA and late proteins, all regions of the major late transcription unit are transcribed with equal efficiency to a termination site near the right-hand end of the genome.

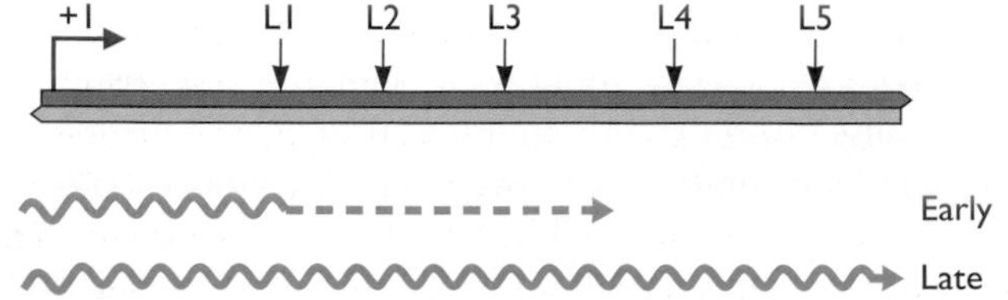

BOX 8.13

DISCUSSION

Unusual viral templates for RNA polymerase II transcription?

In contrast to the genomes of papillomaviruses and polyomaviruses, adenoviral and herpesviral DNAs are not packaged by cellular nucleosomes to form regular, repeating structures in either virions or infected cells. The significance of the unusual nature of these transcriptional templates is not fully appreciated, but there are hints that this property might contribute to regulation of viral gene expression.

In virions, adenoviral DNA is packaged within the core by specialized viral DNA-binding proteins (Appendix, Fig. 1A). The major core protein, protein VII, is retained when the genome enters the infected cell nuclei and remains associated with viral DNA throughout the early phase. The results of recent experiments indicate that transcription of viral DNA is necessary and sufficient to induce dissociation of protein VII from DNA by the end of the early phase. Furthermore, because protein VII is the product of a late gene, viral DNA synthesis must necessarily be accompanied by reorganization of the viral nucleoproteins. Indeed, the templates for major late transcription during the late phase of infection appear protein free in the electron microscope, and by application of other assays. It is therefore possible that production of protein-free viral DNA molecules contributes to alterations in termination of major late transcription that accompany entry into the late phase (see "Coordination of Transcription of Late Genes with Viral DNA Synthesis").

Herpes simplex virus DNA is not associated with dedicated viral packaging proteins in virions. It has been reported that viral DNA in infected cells is associated with histones, but not organized in the form of a regular array. This conclusion is based largely on results of chromatin immunoprecipitation assays (Box 8.1). At this juncture, it remains unclear whether such histone-associated viral DNA molecules serve as templates for transcription. For example, the chromatin immunoprecipitation assay does not establish the fraction of viral molecules that are associated with a particular histone. Furthermore, experiments to investigate whether RNA polymerase II and histones are present on the same or different herpesviral DNA molecules have not been reported. Consequently, whether changes in the structure of DNA templates for transcription accompany viral DNA synthesis and entry into the late phase of infection is not known. On the other hand, there is accumulating evidence that changes in the nucleoprotein organization of herpesviral DNA are important in establishment of latency, and reactivation from this state (see the text).

Chatterjee, P. D., M. E. Vayda, and S. J. Flint. 1986. Adenoviral protein VII packages intracellular viral DNA throughout the early phase of infection. *EMBO J.* **5:**1633–1644.

Chen, J., N. Morral, and D. A. Engel. 2007. Transcription releases protein VII from adenovirus chromatin. *Virology* **369:**411–422.

Xue, Y., J. S. Johnson, D. A. Ornelles, J. Libberman, and D. A. Engel. 2005. Adenovirus protein-VII functions throughout early phase and interacts with cellular proteins SET and PP32. *J. Virol.* **79:**2474–2483.

Availability and structure of templates. Newly replicated DNA molecules must be partitioned for such fates as entry into additional replication cycles, service as templates for transcription, or assembly into virions. This process must be regulated so that different fates predominate at different times in the infectious cycle. Although transcription of all viral DNA molecules made in infected cells seems to be a simple mechanism to ensure efficient transcription of late genes, no more than 5 to 10% of the large numbers that accumulate are transcriptionally active. In the case of simian virus 40, synthesis of viral DNA molecules is coordinated with assembly into nucleosomes, and transcriptional activity can be ascribed to establishment of the open chromatin region in minichromosomes described above. Analogous mechanisms do not appear to be operating in cells infected by adenoviruses or herpesviruses (Box 8.13). It is not known whether a subset of these viral DNA molecules is also marked for transcriptional activity by some unknown molecular device. However, one important parameter governing the concentration of transcriptional templates must be the relative concentrations of viral DNA molecules and the proteins that package them for assembly into virions, because packaging and transcription are mutually exclusive.

Entry into One of Two Alternative Transcriptional Programs

As discussed in Chapter 1, studies of bacteriophage lambda led to the discovery that some viral infections result in maintenance of the viral genome for long periods in infected cells, rather than in viral replication. Whether lambda enters this lysogenic state or the lytic cycle is determined by the outcome of the opposing actions of two viral proteins that repress transcription (Box 8.14). This regulatory mechanism, which was among the first to be elucidated in detail, established the importance of repression of transcription of specific genes, and general paradigms for transcriptional switches. Several animal viruses can establish a similar pattern of infection. For example, **latent infection** is a characteristic feature of herpesvirus infection of specific types of host cell. As in bacteriophage lambda lysogeny, latent infections are characterized by both lack of efficient expression of many viral genes that are transcribed during productive infection, and activation of a unique, latent-phase

BOX 8.14

DISCUSSION

Two bacteriophage lambda repressors govern the outcome of infection

As summarized in panel A of the figure, infection of *Escherichcia coli* by bacteriophage lambda leads to either synthesis of progeny virions and lysis of the host cell (lytic infection) or stable integration of the viral genome into that of the host cell (lysogenic infection). During lysogeny, lytic genes are not expressed. The actions of two repressors of transcription encoded within the viral genome, the cI (lambda) repressor and Cro, make a major contribution to the lytic/lysogeny "decision." The regulatory circuits by which these proteins govern expression of lambda lytic and lysogeny genes were among the first to be understood in detail.

The region of the lambda genome containing the *cI* repressor and *cro* genes is illustrated at the top of panel B. These coding sequences are flanked by genes encoding proteins that regulate transcription during lytic infection (e.g., N), or that are required during establishment of lysogeny (e.g., *int*, which encodes an integrase). Although both repressors bind to the operator sequences O_R and O_L adjacent to the right (P_R) and left (P_L) promoters, respectively, events at O_R are critical in determining the outcome of infection. The expanded view of the region of the genome containing O_R and P_R indicates the three binding sites for the repressors, and the two promoters from which the *cI* gene is expressed, the promoters for repressor establishment and for repressor maintenance, P_{RE} and P_{RM}, respectively.

When the lambda genome enters a host cell, transcription from the P_R and P_{RE} promoters by the bacterial RNA polymerase leads to synthesis of the cI repressor and Cro. The highest-affinity binding site for the cI repressor in O_R is O_{R1}, but this dimeric protein binds cooperatively to O_{R1} and O_{R2}. As these binding sites overlap sequences of the P_R promoter essential for binding of *E. coli* RNA polymerase, transcription of *cro* (and other rightward lytic genes) is repressed (red bar). Transcription from P_L is blocked in the same way by binding of the cI repressor to O_{L1} and O_{L2}. The N-terminal domain of cI repressor bound to O_{R2} contacts the subunit of RNA polymerase that binds to the nearby P_{RM} promoter. This interaction stimulates the formation of an open initiation

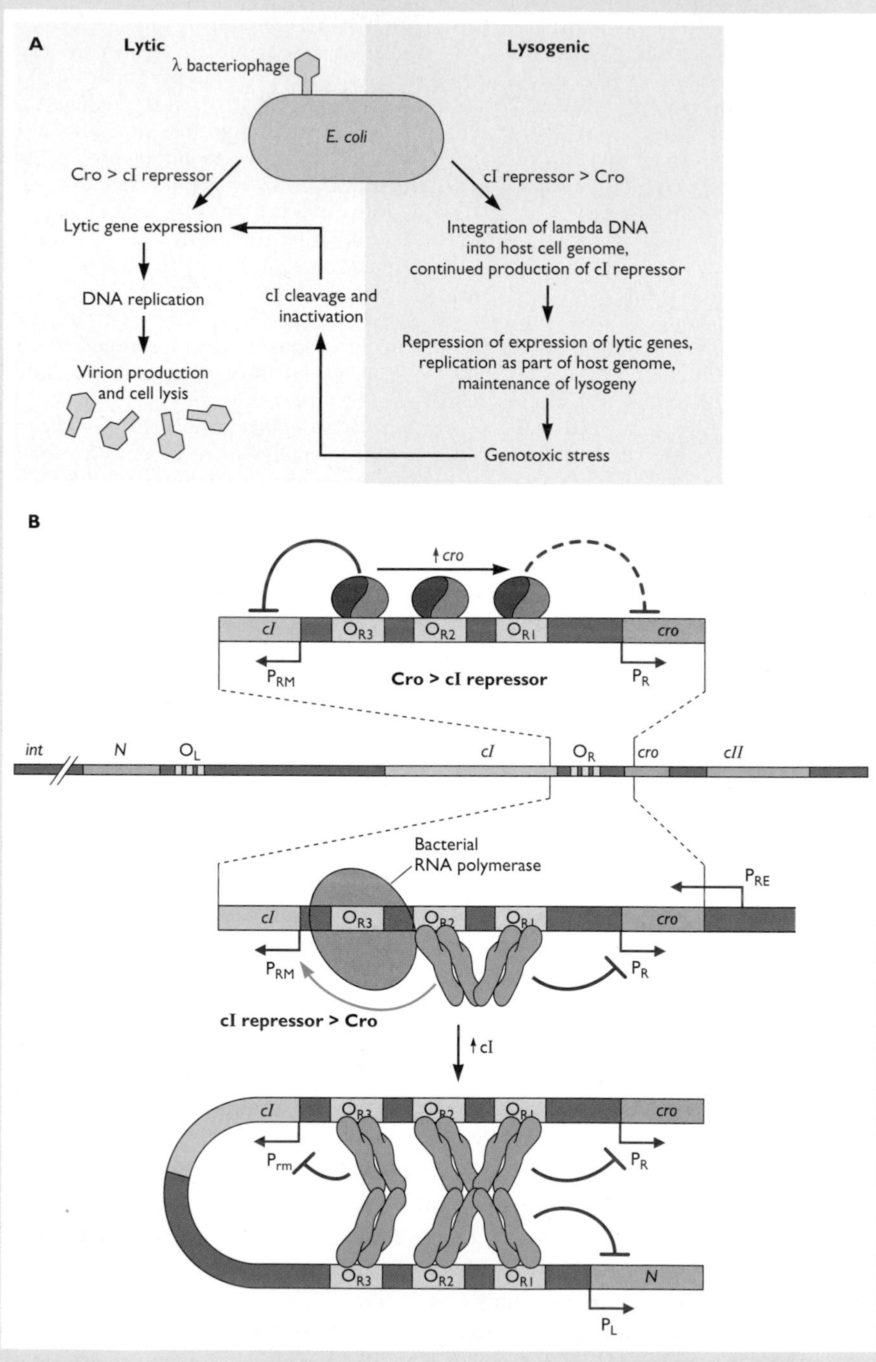

(continued)

complex at the P_{RM} promoter, and hence transcription of the *cI* gene (green arrow). Consequently, as expression of lytic genes is repressed, the concentration of cI repressor is increased to a value some 10-fold higher than that compatible with expression of lytic genes. The cI repressor has only low affinity for the O_{R3}-binding site. However, cooperative interactions occur between dimers bound to the O_L and O_R sites, to facilitate binding to O_{R3} and repression of transcription from P_{RM}. Because of such cooperative binding, whether cI repressor stimulates or blocks its own synthesis is very sensitive to cI concentration, and repressor concentration is maintained within a narrow range.

Although Cro binds to the same O_R sites as the cI repressors, it has the highest affinity for O_{R3}. It therefore occupies this site preferentially, and then binds to O_{R2}, to block binding of RNA polymerase to the P_{RM} promoter. Consequently, the cI repressor does not attain the concentrations necessary for establishment (and maintenance) of lysogeny. Binding of Cro to O_{R2} and O_{R1} leads to weak repression of transcription from P_R (and from P_L by an analogous mechanism). This function of Cro favors lytic infection, for example, by reducing production of the cII protein, a transcriptional regulator that promotes lysogeny by activating transcription of the *cI* gene from P_{RE}, and of the integrase gene.

It has been known for many years that environmental conditions and the activities of particular host cell gene products influence the outcome of lambda infection. How these cues are integrated with one another and into the regulatory network comprising the cI repressor/Cro switch and other lambda regulators remains incompletely understood.

Dodd, B., K. E. Shearwin, and J. B. Egan. 2005. Revisited gene regulation in bacteriophage λ. *Curr. Opin. Genet. Dev.* **15:**145–152.

Ptashne, M., A. Jeffrey, A. D. Johnson, R. Maurer, B. J. Meyer, C. O. Pabo, T. M. Roberts, and R. T. Sauer. 1980. How λ repressor and Cro work. *Cell* **19:**1–11.

transcriptional program. Furthermore, whether a herpesvirus infection is latent or lytic, as well as reentry into the productive cycle from latency (**reactivation**), is governed by mechanisms that regulate transcription.

In the case of the gammaherpesvirus Epstein-Barr virus, the availability and activity of a single viral protein determine whether an infection is latent or lytic. As described in a previous section, transcription of the gene encoding the transcriptional activator Zta is repressed by cellular proteins when B cells are initially infected. This protein is necessary for transcription of viral early genes, as well as viral DNA replication during the lytic cycle. Consequently, a latent infection ensues until the infected cell is exposed to conditions that activate transcription of the Zta gene. As Zta also represses transcription of the genes expressed in latently infected cells, it can be viewed as a simple regulatory switch. In contrast, more complex mechanisms appear to determine the outcome of infection by the alphaherpesviruses, which establish latent infections in neurons.

During latent infection of neurons by herpes simplex virus type 1, a set of specific RNAs, termed the **latency-associated transcripts (LATs)**, are the only viral products synthesized in large quantities (Fig. 8.22). The 2.0- and 1.5-kb LATs accumulate to 40,000 to 100,000 copies within the nucleus, lack poly(A) tails, and are not linear molecules. Indeed, all properties established to date indicate that they are stable introns produced by splicing of precursor RNA, as discussed in Chapter 10. These noncoding RNAs have been reported to inhibit apoptosis, interfere with expression of interferon genes, and stimulate expression of the protein chaperone Hsp70, by mechanisms that are not yet well understood. Such functions presumably promote the survival of latently infected neurons.

The LATs are also important for establishing a latent infection. Their stable production in neuronal cells suppresses replication of the virus, and synthesis of the immediate-early gene products needed for progression through the productive infectious cycle. These observations raise the intriguing question of how viral RNAs with the properties of introns excised from pre-mRNAs repress expression of these viral genes. The 2.0-kb LAT RNA has been reported to be associated with cellular ribosomes and splicing proteins, suggesting that it might alter splicing of immediate-early pre-mRNAs or production of cellular proteins that modulate viral gene expression. However, the ability of LATs to modulate chromatin structure is likely to make a major contribution to inhibition of transcription of lytic genes. As a latent infection is established, lytic genes become organized by nucleosomes that carry modifications associated with repression of transcription. In contrast, the LAT gene is associated with nucleosomes containing histones with modifications characteristic of actively transcribed genes, at least in part because it is bounded by specialized DNA sequences (insulators) that serve as boundaries between different types of chromatin. When the LAT region is deleted, repressive chromatin does not form efficiently on lytic genes. The viral LATs may function like the growing list of cellular RNAs

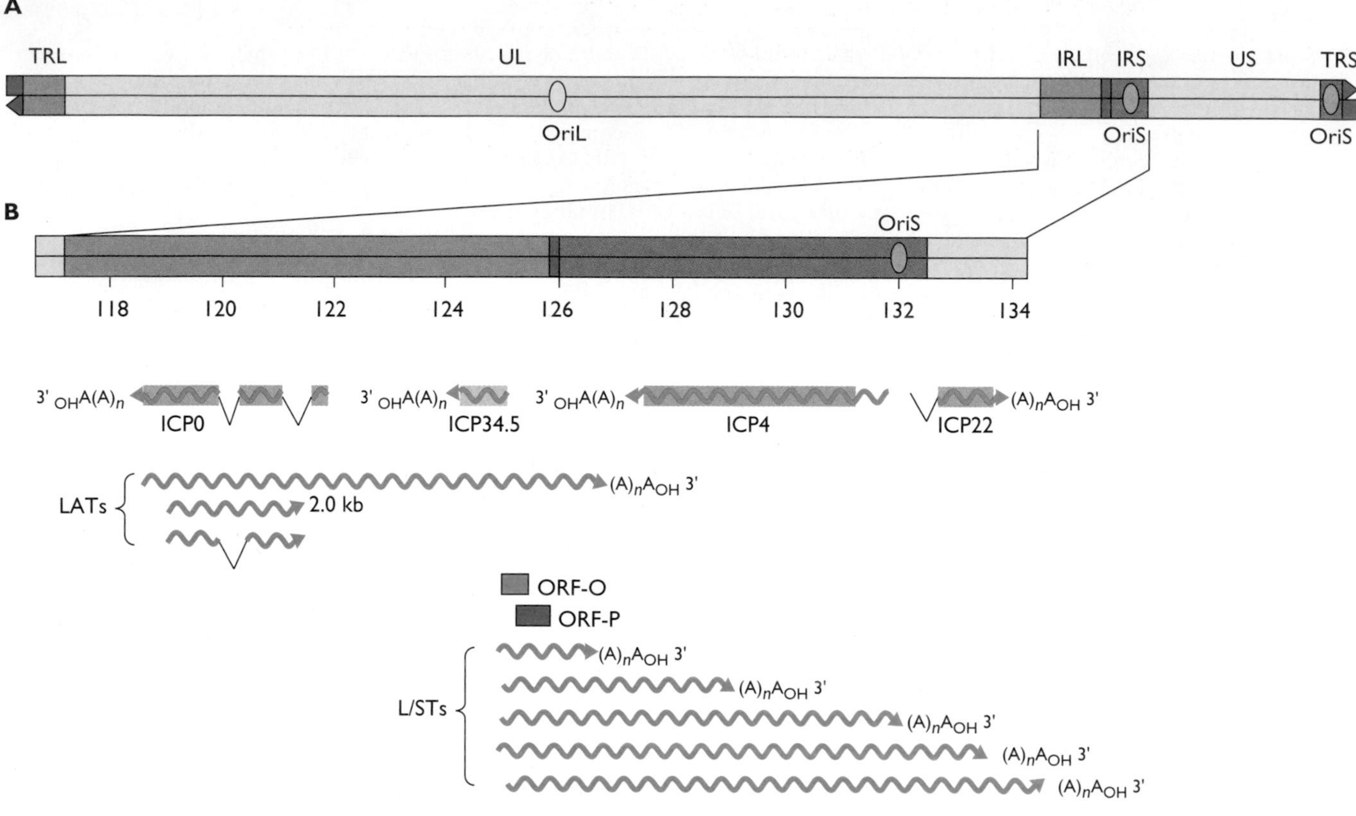

Figure 8.22 The latency-associated transcripts of herpes simplex virus type 1. (A) Diagram of the herpes simplex virus type 1 genome, showing the unique long and short segments, UL and US, respectively, the terminal repeat (TRL and TRS) and internal repeat (IRL and IRS) sequences, and the origins of replication, OriL and OriS. **(B)** Expanded map of the region shown, with the scale in kilobase pairs. This region encodes immediate-early proteins ICP0, ICP4, and ICP22, which play important roles in establishing a productive infection. Below are shown the locations of sequences encoding the latency-associated transcripts (LATs) and the transcripts named L/STs. The arrows indicate the direction of transcription, and $(A)_n A_{OH}$ indicates 3′ poly(A) sequences where known. Open reading frames, including ORF-O and ORF-P, are shown as boxes. The evidence currently available suggests that ORF-P (and perhaps ORF-O) may control LAT RNA synthesis.

that are known to regulate chromatin structure. The products of other RNAs made in smaller quantities in latently infected neurons, such as the L/ST RNAs (Fig. 8.22), can block the binding of the immediate-early ICP4 protein to promoter sequences. These observations suggest that once latency-associated genes are expressed in an infected neuron, the transcriptional cascade characteristic of productive infection is repressed.

As we have seen, the virion protein VP16 activates transcription of viral immediate-early genes and entry into the productive cycle transcriptional cascade. This viral protein enters infected neurons, but its ability to activate transcription is blocked by at least two mechanisms. In these cells, the essential VP16 cofactor Hcf is localized largely in the cytoplasm, sequestered from viral genomes and VP16 that enter infected cell nuclei. In addition, Hcf binds to Zhangfei, a cellular protein that is a strong repressor of transcription. Zhangfei is synthesized in sensory neurons (a natural site of latency), but not in most other cell types. Inhibition of production of the immediate-early ICP0 protein is also likely to be important, particularly to prevent reactivation from latency (when VP16 is not present). This protein is a powerful activator of transcription in simplified experimental systems, and is required for efficient viral replication in cells infected at low multiplicity and for efficient reactivation from latency. Among the several functions of ICP0 (see Volume II, Chapter 3) is activation of immediate-early transcription (in cooperation with ICP4), probably mediated at least in part by changes in histone modification. Whether the effects of LATs described previously are

Table 8.8 Viral RNA polymerase III transcription units

Virus	RNA polymerase III transcript	Function
Adenovirus		
Human adenovirus type 2	VA-RNA I	Prevention of activation of RNA-dependent protein kinase
	VA-RNA II	Not known
Herpesviruses		
Epstein-Barr virus	EBER 1 EBER 2	Always made in latently infected cells; implicated in transformation and oncogenesis
Herpesvirus saimiri	HSVR 1–5	Degradation of certain cellular mRNAs
Retrovirus		
Moloney murine leukemia virus	Let	Stimulation of transcription of specific cellular genes (e.g., CD4 and class I major histocompatibility complex genes)

sufficient to ensure that ICP0 and ICP4 cannot activate transcription of early genes remains to be established.

Transcription of Viral Genes by RNA Polymerase III

Several of the viruses considered in this chapter contain RNA polymerase III transcription units in their genomes (Table 8.8). The first, and still best understood, example is the gene encoding human adenovirus type 2 virus-associated RNA I (VA-RNA I). The VA-RNA I gene specifies an RNA product that ameliorates the effects of a host cell defense mechanism, and has also been implicated in RNA interference (Volume II, Chapter 3). It contains a typical, intragenic promoter that has been widely used in studies of initiation of transcription by RNA polymerase III.

RNA Polymerase III Transcribes the Adenoviral VA-RNA Genes

The VA-RNA I Promoter

The human adenovirus type 2 genome contains two VA-RNA genes located very close to one another (Appendix, Fig. 1B). The VA-RNA I promoter is described here, for it is the more thoroughly characterized. Transcription of the VA-RNA I gene depends on two intragenic sequences, the A and B boxes (Fig. 8.23A). Sequences upstream of sites of initiation are not necessary for promoter activity, but do modulate the specificity or efficiency of initiation. As in the RNA polymerase II system, the essential promoter sequences are binding sites for accessory proteins necessary for promoter recognition. The internal sequences are recognized by the RNA polymerase III-specific initiation protein TfIIIc, which binds to the promoter to begin assembly of an initiation complex that also contains TfIIIb and the enzyme (Fig. 8.23B). This pathway of initiation was elucidated by using *in vitro* assays. We can be confident that this same pathway operates in adenovirus-infected cells, for there is excellent agreement between the effects of A and B box mutations on VA-RNA I synthesis *in vitro* and in mutant virus-infected cells.

Regulation of VA-RNA Gene Transcription

The two VA-RNA genes are initially transcribed at similar rates, but during the late phase of infection, production of VA-RNA I is accelerated greatly. Such preferential transcription is the result of competition of the VA-RNA I promoter with the intrinsically much weaker transcriptional control region of the VA-RNA II gene for a limiting component of the RNA polymerase III transcriptional machinery. Repression of VA-RNA I transcription may account for the similar rates at which the two genes are transcribed during the early phase. The control of transcription of VA-RNA genes emphasizes the fact that transcription by RNA polymerase III can, and must, be regulated, although the mechanisms are probably less elaborate than those that govern transcription by RNA polymerase II. The properties of other viral RNA polymerase III transcription units illustrate the kinship of the RNA polymerase II and III systems (Fig. 8.24).

Inhibition of the Cellular Transcriptional Machinery in Virus-Infected Cells

Inhibition of cellular transcription in virus-infected cells offers several advantages. Cellular resources, such as substrates for RNA synthesis, can be devoted to the exclusive

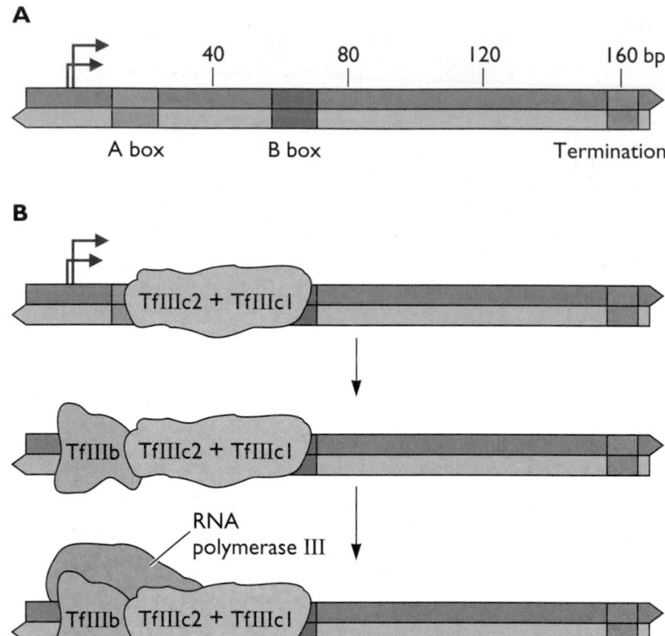

Figure 8.23 Organization and recognition of the human adenovirus type 2 VA-RNA I RNA polymerase III promoter. (A) The VA-RNA I gene is depicted to scale, in base pairs. The intragenic A and B box sequences are essential for efficient VA-RNA I transcription, and are closely related to the consensus A and B sequences of cellular tRNA genes. The VA-RNA termination site sequences are also typical of those of cellular genes transcribed by RNA polymerase III. **(B)** Assembly of an initiation complex on the VA-RNA I promoter is initiated by binding of TfIIIc2 to the intragenic promoter sequences. This reaction is stimulated by TfIIIc1, which makes contact with the A box. The TfIIIc-DNA complex is recognized by TfIIIb, which in turn allows recruitment of the polymerase. All mammalian RNA polymerase III initiation proteins contain multiple subunits. The TATA-binding protein is an essential component of TfIIIb.

production of viral mRNAs (and, in many cases, RNA genomes), and competition between viral and cellular mRNAs for components of the translational machinery is minimized. The essential participation of cellular transcriptional systems in the infectious cycles of most viruses considered in this chapter precludes inactivation of this machinery. However, posttranscriptional mechanisms allow selective expression of adenoviral and herpesviral genes (Chapter 10). Furthermore, herpes simplex virus type 1 infection induces inhibition of transcription of many cellular genes. Selective transcription of viral genes is accompanied by loss of the forms RNA polymerase II present in uninfected cells. Degradation of the hypophosphorylated form by the proteasome correlates with inhibition of transcription of cellular genes. As discussed previously, the unusual organization of viral late promoters probably facilitates selective transcription of late genes by the modified transcriptional machinery of the host cell. The poxviruses induce rapid inhibition of synthesis of all classes of cellular RNA in infected cells. Such inhibition requires viral proteins, but these have not been identified.

Replication of the majority of viruses with RNA genomes requires neither the cellular transcriptional machinery nor its RNA products, and is accompanied by inhibition of cellular mRNA synthesis. Among the best-characterized examples is the inhibition of transcription by RNA polymerase II characteristic of poliovirus-infected cells. The defect in RNA polymerase II transcription in extracts of poliovirus-infected cells can be explained by the fact that $3C^{pro}$ cleaves the Tbp subunit of TfIId at several sites. This modification eliminates the DNA-binding activity of Tbp and hence transcription by RNA polymerase II. The TATA-binding protein is also a subunit of initiation proteins that function with RNA polymerase III (TfIIIb) and RNA polymerase I. Its cleavage by $3C^{pro}$ in poliovirus-infected cells therefore appears to be a very efficient way to inhibit all cellular transcriptional activity. As poliovirus yields are reduced in cells that synthesize an altered form of Tbp resistant to cleavage by $3C^{pro}$, it is clear that inhibition of cellular transcription is necessary for optimal virus replication.

Two gene products of the rhabdovirus vesicular stomatitis virus have been implicated in inhibition of cellular transcription. The first is the viral leader RNA described in Chapter 6, which inhibits transcription by both RNA polymerase II and RNA polymerase III *in vitro* and is primarily responsible for the rapid inhibition of cellular RNA synthesis in infected cells. Following synthesis in the cytoplasm, the leader RNA enters the nucleus. The question of how short RNA molecules inhibit DNA-dependent RNA transcription cannot yet be answered, although *in vitro* experiments suggest that binding of a cellular protein to specific sequences within the RNA may be important. The viral M protein is also a potent inhibitor of transcription by RNA polymerase II, even in the absence of other viral gene products. This activity may become important later in infection, when replication of genome RNA predominates over transcription, such that less leader RNA is produced.

Unusual Functions of Cellular Transcription Components

In the preceding sections, we concentrated on the similarities among the mechanisms by which viral and cellular DNA are transcribed. Even though all mechanisms of regulation of expression of viral genes by the host cell's RNA polymerase II or RNA polymerase III cannot be described in molecular detail, it seems clear that the majority are not unique to viral systems. It is therefore an axiom of

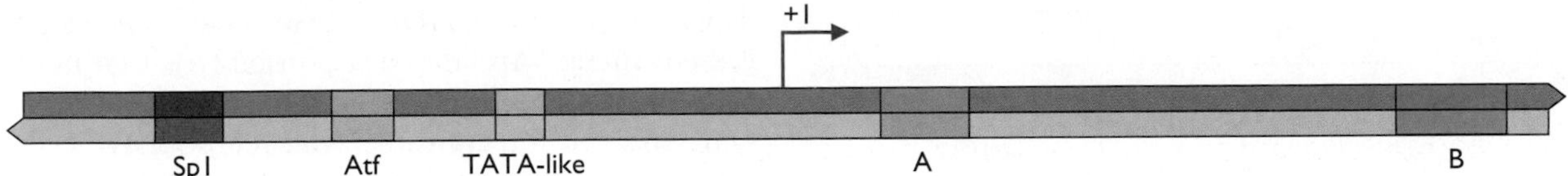

Figure 8.24 Some viral RNA polymerase III genes contain both external and internal promoter elements. The 5′ end of the Epstein-Barr virus EBER-2 transcription unit is shown to scale. This gene contains typical intragenic A and B box sequences. However, efficient transcription by RNA polymerase III also depends on the 5′ flanking sequence, which includes binding sites for the RNA polymerase II stimulatory proteins Spl and Atf. The TATA-like sequence is important for efficient transcription, and essential for specifying transcription by RNA polymerase III.

molecular virology that **every** mechanism by which viral transcription units are expressed by cellular components, or by which their activity is regulated, will prove to have a normal cellular counterpart. However, virus-infected cells also provide examples of novel functions or activities of cellular proteins that mediate transcription.

One example of the latter kind of mechanism is the production of hepatitis delta satellite virus RNA from an **RNA** template by RNA polymerase II, described in Chapter 6. The RNA of viroids, infectious agents of plants, is synthesized in the same manner. Such RNA-dependent RNA synthesis by RNA polymerase II from a specific RNA promoter is one of the most remarkable interactions of a viral genome with the cellular transcriptional machinery. No cellular analog of this reaction is yet known. Even more divergent functions of cellular transcriptional components in virus-infected cells are illustrated by the participation of the RNA polymerase III initiation proteins TfIIIb and TfIIIc in integration of the yeast retroid element Ty3 (see Chapter 7). Given the large repertoire of molecular and biochemical activities displayed by components of the eukaryotic transcriptional machinery, it seems likely that other unusual activities of these cellular proteins in virus-infected cells will be discovered.

A Viral DNA-Dependent RNA Polymerase

The DNA genomes of viruses considered in preceding sections replicate in the nucleus of infected cells, where the cellular transcriptional machinery resides. In contrast, poxviruses such as vaccinia virus are reproduced exclusively in the cytoplasm of their host cells. This feat is possible because the genomes of these viruses encode the components of a transcription and RNA-processing system that produces viral mRNAs with the hallmarks of cellular mRNA 5′ caps and 3′ poly(A) tails. This system, which is carried into infected cells within virions, includes a DNA-dependent RNA polymerase responsible for transcription of all vaccinia virus genes. A striking feature of the viral enzyme is its structural and functional resemblance to cellular RNA polymerases.

The vaccinia virus RNA polymerase, like cellular RNA polymerases, is a complex, multisubunit enzyme built from the products of at least eight genes. The amino acid sequences of several of these subunits, including the two largest and the smallest, are clearly related to subunits of RNA polymerase II. The viral enzyme transcribes all classes of vaccinia virus genes expressed at different times in the infectious cycle (early, intermediate, and late). Like the cellular enzymes, vaccinia virus RNA polymerase recognizes promoters by cooperation with additional proteins. For example, formation of initiation complexes on vaccinia virus early promoters is mediated by two viral proteins, VETF and RAP94, which are responsible for the recognition of promoter sequences and recruitment of the RNA polymerase, respectively (Fig. 8.25). These viral proteins are functional analogs of the cellular RNA polymerase II initiation proteins TfIId and TfIIf (Box 8.3). However, the vaccinia virus transcriptional machine is not analogous to its cellular counterpart in every respect. Cellular RNA polymerase II generally transcribes far beyond the sites at which the 3′ ends of mature cellular or viral mRNAs are produced by processing of the primary transcript, and does not terminate transcription at simple, discrete sites. In contrast, transcription of the majority of vaccinia virus early genes **does** terminate at discrete sites, 20 to 50 bp downstream of specific T-rich sequences in the template. Termination requires the viral termination protein, which is also the viral mRNA-capping enzyme (see Chapter 10). The 3′ ends of the viral mRNAs correspond to sites of transcription termination. This viral mechanism is considerably simpler than the cellular counterpart.

The viral RNA polymerase and the several other proteins necessary for transcription of early genes enter host cells within vaccinia virus particles. Subsequently, viral gene expression is controlled by the ordered synthesis of viral proteins that permit sequential recognition of intermediate and late promoters. For example, transcription of intermediate genes requires synthesis of the viral RPO30 gene product (a subunit of the viral polymerase) and a second viral protein termed VITF3, while late transcription depends on synthesis of several intermediate gene

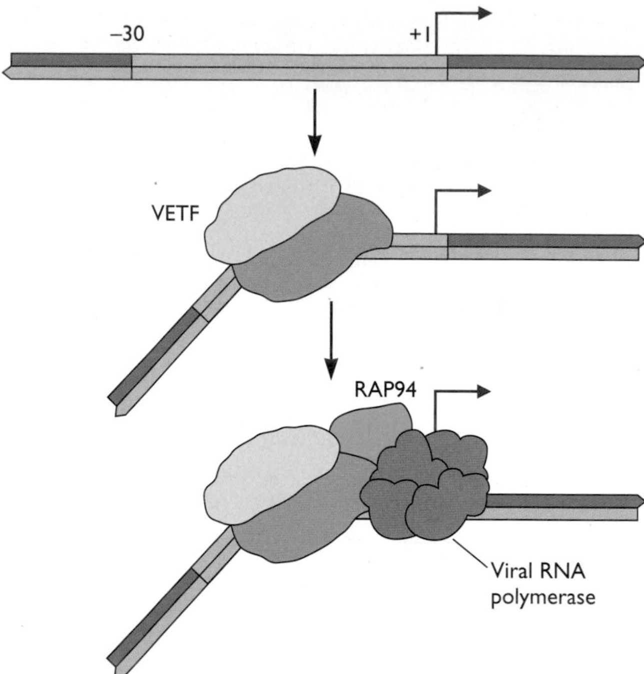

Figure 8.25 Assembly of an initiation complex on a vaccinia virus early promoter. Vaccinia virus early promoters contain of an AT-rich sequence (tan) immediately upstream of the site of initiation. Vaccinia virus RNA polymerase cannot recognize these (or any other) viral promoters in the absence of other viral proteins. The vaccinia virus early transcription protein (VETF) is necessary for early promoter recognition and must bind before the viral RNA polymerase. This heteromeric protein binds specifically to early promoters and induces DNA bending. It also possesses DNA-dependent ATPase activity. VETF and the second protein necessary for early promoter specificity, RAP94, enter infected cells in virions. The RAP94-RNA polymerase complex associates with early promoter-bound VETF to form a functional initiation complex. Assembly of these vaccinia virus initiation complexes is therefore analogous to, although simpler than, formation of RNA polymerase II initiation complexes (Box 8.4).

products. The viral genome also encodes several proteins that regulate elongation during transcription of late genes. Activation of intermediate and late gene transcription also requires viral DNA replication. Transcription of vaccinia virus genetic information is therefore regulated by mechanisms similar to those operating in cells infected by other DNA viruses, even though the transcriptional machinery is viral in origin.

Surprisingly, the vaccinia virus transcription system is not entirely self-contained: a cellular protein is essential for transcription of viral intermediate genes. This protein, termed Vitf2, is located in the nucleus of uninfected cells, but is present in both the cytoplasm and the nucleus of infected cells. As a significant number of vaccinia virus genes encode proteins necessary for transcription, such dependence on a cellular protein must confer some special advantage. An attractive possibility is that interaction of the viral transcriptional machinery with a cellular protein serves to integrate the viral reproductive cycle with the growth state of its host cell. The identification of Vitf2 as a heterodimer of proteins that are produced in greatest quantities in proliferating cells is consistent with this hypothesis.

Perspectives

It is difficult to exaggerate the contributions of viral systems to the elucidation of mechanisms of transcription and its regulation in eukaryotic cells. The organization of RNA polymerase II promoters considered typical was first described for viral transcriptional control regions, enhancers were first discovered in viral genomes, and many important cellular regulators of transcription were identified by virtue of their specific binding to viral promoters. Perhaps even more importantly, efforts to elucidate the molecular basis of regulatory circuits crucial to viral infectious cycles have established general principles of eukaryotic transcriptional regulation. These include the importance of proteins that do not recognize DNA sequences directly, and the ability of a single transcriptional regulator to modulate multiple components of the machinery. The insights into regulation of elongation by RNA polymerase II gained from studies of the human immunodeficiency virus type 1 Tat protein emphasize the intimate relationship of viral proteins with cellular components that make viral systems such rich resources for the investigation of eukaryotic transcription.

The identification of cellular and viral proteins necessary for transcription of specific viral genes has allowed many regulatory circuits to be traced. For example, the tissue distribution or the availability of particular cellular activators that bind to specific viral DNA sequences can account for the tropism of individual viruses, or conditions under which different transcriptional programs (latent or lytic) can be established. Furthermore, the mechanisms that allow sequential expression of viral genes are quite well established. Regardless of whether regulatory circuits are constructed of largely cellular or mostly viral proteins, these transcriptional cascades share such mechanistic features as sequential production of viral activators, and integration of transcription of late genes with synthesis of viral DNA. Our ability to describe these regulatory mechanisms in some detail represents enormous progress.

The models for the individual regulatory processes described in this chapter were developed by using convenient and powerful experimental systems. It is clear that such simplified systems (e.g., *in vitro* transcription reactions and transient-expression assays) do not reproduce all features characteristic of infected cells, even in the simplest cases, and cannot address such issues as how transcription of specific genes can be coupled with replication of the viral genome. It is therefore crucial that the models be tested in virus-infected cells, even though it is more difficult to elucidate the molecular functions and mechanisms of action of transcriptional components *in vivo*. Many viral regulatory proteins perform multiple functions, a property that can confound genetic analysis. Moreover, the study of individual reactions, such as binding of a protein to a specific promoter sequence, or formation of preinitiation complexes, is technically more demanding. Nevertheless, viral *cis*-acting sequences and regulatory proteins remain more amenable to genetic analyses of their function in the natural context than do their cellular counterparts. In conjunction with increasingly powerful and sensitive methods for examining intracellular processes, such as the chromatin immunoprecipitation assay, continued efforts to exploit such genetic malleability will eventually establish how transcription of viral DNA templates is mediated and regulated within infected cells. Such information not only will address outstanding virological questions, but also should allow current models of the fundamental cellular process of transcription to be refined.

References

Books

McKnight, S. L., and K. Yamamoto (ed.). 1992. *Transcriptional Regulation*. Cold Spring Harbor Laboratory Press, Cold Spring Harbor, NY.

Book Chapters

Berk, A. J. 2007. Adenoviridae: the viruses and their replication, p. 2355–2394. *In* D. M. Knipe and P. M. Howley (ed.), *Fields Virology*, 5th ed. Lippincott Williams & Wilkins, Philadelphia, PA.

Imperiale, M. J., and E. O. Major. 2007. Polyomaviridae, p. 2263–2298. *In* D. M. Knipe and P. M. Howley (ed.), *Fields Virology*, 5th ed. Lippincott Williams & Wilkins, Philadelphia, PA.

Kieff, E., and A. B. Rickinson. 2007. Epstein-Barr Virus and its replication, p. 2655–2700. *In* D. M. Knipe and P. M. Howley (ed.), *Fields Virology*, 5th ed. Lippincott Williams & Wilkins, Philadelphia, PA.

Roizman, B., D. M. Knipe, and R. J. Whitley. 2007. Herpes simplex viruses, p. 2501–2601. *In* D. M. Knipe and P. M. Howley (ed.), *Fields Virology*, 5th ed. Lippincott Williams & Wilkins, Philadelphia, PA.

Reviews

Broyles, S. S. 2003. Vaccinia virus transcription. *J. Gen. Virol.* **84:**2293–2303.

Everett, R. D. 1999. A surprising role for the proteasome in the regulation of herpesvirus infection. *Trends Biochem. Sci.* **24:**293–295.

Flint, J., and T. Shenk. 1997. Viral transactivating proteins. *Annu. Rev. Genet.* **31:**177–212.

Greenblatt, J. 1997. RNA polymerase II holoenzyme and transcriptional activation. *Curr. Opin. Cell Biol.* **9:**310–319.

Hassan, A. H., K. E. Neely, M. Vignali, J. C. Reese, and J. L. Workman. 2001. Promoter targeting of chromatin-modifying complexes. *Front. Biosci.* **6:**D1054–1064.

Jenuwein, T., and C. D. Allis. 2002 Translating the histone code. *Science* **293:**1074–1080.

Karn, J. 1999. Tackling Tat. *J. Mol. Biol.* **293:**235–254.

Khoury, G., and P. Grüss. 1983. Enhancer elements. *Cell* **33:** 313–314.

Naar, A. M., B. D. Lemon, and R. Tjian. 2001. Transcriptional coactivator complexes. *Annu. Rev. Biochem.* **70:**475–501.

Preston, C. M. 2000. Repression of viral transcription during herpes simplex virus latency. *J. Gen. Virol.* **81:**1–19.

Roeder, R. G. 1996. The role of general initiation factors in transcription by RNA polymerase II. *Trends Biochem. Sci.* **21:**327–335.

Ruddell, A. 1995. Transcription regulatory elements of the avian retroviral long terminal repeat. *Virology* **206:**1–7.

Sinclair, A. S. 2003. bZIP proteins of gamma herpesviruses. *J. Gen. Virol.* **84:**1941–1949.

Workman, J. L. 2006. Nucleosome displacement in transcription. *Genes Dev.* **20:**2009–2017.

Wysocka, J., and W. Herr. 2003. The herpes simplex virus VP16-induced complex: the makings of a regulatory switch. *Trends Biochem. Sci.* **28:**294–304.

Papers of Special Interest

Viral RNA Polymerase II Promoters and the Cellular Transcriptional Machinery

Cramer, P., D. A. Bushnell, and R. D. Kornberg. 2001. Structural basis of transcription: RNA polymerase II at 2.8 Ångstrom resolution. *Science* **292:**1863–1876.

Dynan, W. S., and R. Tjian. 1983. The promoter-specific transcription factor Sp1 binds to upstream sequences in the SV40 early promoter. *Cell* **35:**79–87.

Grüss, P., R. Dhar, and G. Khoury. 1981. Simian virus 40 tandem repeated sequences as an element of the early promoter. *Proc. Natl. Acad. Sci. USA* **78:**943–947.

Hu, S. L., and J. L. Manley. 1981. DNA sequences required for initiation of transcription in vitro from the major late promoter of adenovirus 2. *Proc. Natl. Acad. Sci. USA* **78:**820–824.

Matsui, T., J. Segall, P. A. Weil, and R. G. Roeder. 1980. Multiple factors required for accurate initiation of transcription by purified RNA polymerase II. *J. Biol. Chem.* **255:**11992–11996.

Yamamoto, T., B. de Crombrugghe, and I. Pastan. 1980. Identification of a functional promoter in the long terminal repeat of Rous sarcoma virus. *Cell* **22:**787–797.

Viral Enhancers

Banerji, J., S. Rusconi, and W. Schaffner. 1981. Expression of the α-globin gene is enhanced by remote SV40 DNA sequences. *Cell* **27:**299–308.

Fromental, C., M. Kanno, H. Nomiyama, and P. Chambon. 1988. Cooperativity and hierarchical levels of functional organization in the SV40 enhancer. *Cell* **54:**943–953.

Moreau, P., R. Hen, B. Wasylyk, R. Everett, M. P. Gaub, and P. Chambon. 1981. The SV40 72 base repeat has a striking effect on gene expression both in SV40 and other chimeric recombinants. *Nucleic Acids Res.* **9:**6047–6068.

Nabel, G. J., and D. Baltimore. 1987. An inducible transcription factor that activates expression of human immunodeficiency virus in T cells. *Nature* **326:**711–713.

Spalholz, P. A., Y.-C. Yang, and P. M. Howley. 1985. Transactivation of a bovine papillomavirus transcriptional regulatory element by the E2 gene product. *Cell* **42:**183–191.

Yee, J. 1989. A liver-specific enhancer in the core promoter of human hepatitis B virus. *Science* **246:**658–670.

RNA-Dependent Stimulation of Human Immunodeficiency Virus Type 1 Transcription by the Tat Protein

Brès, V., H. Tagami, J. M. Pèloponèse, E. Loret, K. T. Jeang, Y. Nakatani, S. Emiliani, M. Benkirane, and R. E. Kiernan. 2002. Differential aceyltation of Tat coordinates its interactions with the coactivators cyclin T1 and VCAF. *EMBO J.* **21:**6811–6819.

Dingwall, C., J. Ernberg, M. J. Gait, S. M. Green, S. Heaphy, J. Karn, M. Singh, and J. J. Shinner. 1990. HIV-1 Tat protein stimulates transcription by binding to a U-rich bulge in the stem of the TAR RNA structure. *EMBO J.* **9:**4145–4153.

Feinberg, M. B., D. Baltimore, and A. D. Frankel. 1991. The role of Tat in the human immunodeficiency virus life cycle indicates a primary effect on transcriptional elongation. *Proc. Natl. Acad. Sci. USA* **88:**4045–4049.

Mancebo, H. S. Y., G. Lee, J. Flygare, J. Tomassini, P. Luu, Y. Zhu, J. Peng, C. Blau, D. Hazuda, D. Price, and O. Flores. 1997. P-TEFb kinase is required for HIV Tat transcriptional activation in vivo and in vitro. *Genes Dev.* **11:**2633–2644.

Parada, C. A., and R. G. Roeder. 1996. Enhanced processivity of RNA polymerase II triggered by Tat-induced phosphorylation of its carboxy-terminal domain. *Nature* **384:**375–378.

Tréand, C., I. du Chéné, V. Brès, R. Kiernan, R. Benarous, M. Benkirane, and S. Emiliani. 2006. Requirement of SWI/SNF chromatin-remodeling complex in Tat-mediated activation of the HIV-1 promoter. *EMBO J.* **25:**1690–1699.

Wei, P., M. F. Garber, S.-M. Fang, W. H. Fischer, and K. A. Jones. 1998. A novel CDK9-associated C-type cyclin interacts directly with HIV-1 Tat and mediates its high-affinity, loop-specific binding to TAR RNA. *Cell* **92:**451–462.

Regulation of Transcription of DNA Genomes by Viral Proteins

Ali, H., G. LeRoy, G. Bridge, and S. J. Flint. 2007. The adenoviral L4 33 kDa protein binds to intragenic sequences of the major late promoter required for late phase-specific stimulation of transcription. *J. Virol.* **81:**1327–1338.

Bandara, L. R., and N. B. La Thangue. 1991. Adenovirus E1A prevents the retinoblastoma gene product from complexing with a cellular transcription factor. *Nature* **351:**494–497.

Batterson, W., and B. Roizman. 1983. Characterization of the herpes simplex virion-associated factor responsible for the induction of alpha genes. *J. Virol.* **46:**371–377.

Berk, A. J., F. Lee, T. Harrison, J. F. Williams, and P. A. Sharp. 1979. Pre-early adenovirus 5 gene product regulates synthesis of early viral messenger RNAs. *Cell* **17:**935–944.

Chellappan, S. P., S. Hiebert, M. Mudryj, J. M. Horowitz, and J. R. Nevins. 1991. The E2F transcription factor is a cellular target for the Rb protein. *Cell* **65:**1053–1061.

Chi, T., and M. Carey. 1993. The ZEBRA activation domain: modular organization and mechanism of action. *Mol. Cell. Biol.* **13:**7045–7055.

Jones, N., and T. Shenk. 1979. An adenovirus type 5 early gene function regulates expression of other early viral genes. *Proc. Natl. Acad. Sci. USA* **76:**3665–3669.

Keller, J. M., and J. C. Alwine. 1984. Activation of the SV40 late promoter: direct effects of T antigen in the absence of viral DNA replication. *Cell* **36:**381–389.

Lai, J.-S., M. A. Cleary, and W. Herr. 1992. A single amino acid exchange transfers VP16-induced positive control from the Oct-1 to the Oct-2 homeo domain. *Genes Dev.* **6:**2058–2065.

Stevens, J. L., G. T. Cantin, G. Wang, A. Shevchenko and A. J. Berk. 2002. Transcription control by E1A and MAP kinase pathway via Sur2 Mediator subunit. *Science* **296:**755–758.

Tribouley, C., P. Lutz, A. Staub, and C. Kédinger. 1994. The product of the adenovirus intermediate gene IVa_2 is a transcriptional activator of the major late promoter. *J. Virol.* **68:**4450–4457.

Wiley, S. R., R. J. Kraus, F. Zuo, E. E. Murray, K. Loritz, and J. E. Mertz. 1993. SV40 early-to-late switch involves titration of cellular transcriptional repressors. *Genes Dev.* **7:**2206–2219.

Wilson, A. C., R. W. Freeman, H. Goto, T. Nishimoto, and W. Herr. 1997. VP16 targets an amino-terminal domain of HCF involved in cell cycle progression. *Mol. Cell. Biol.* **17:**6139–6146.

Establishing Latent or Lytic Infections

Akhova, O., M. Bainbridge, and V. Misra. 2005. The neuronal host cell-factor binding protein Zhangfei inhibits herpes simplex virus replication. *J. Virol.* **79:**14708–14718.

Atanasia, D., J. R. Kent, J. J. Gartner, and N. W. Fraser. 2006. The stable 2-kb LAT intron of herpes simplex stimulates the expression of heat shock proteins and protects cells from stress. *Virology* **350:** 26–33.

Deshmane, S. L., and N. W. Fraser. 1989. During latency, herpes simplex virus type I DNA is associated with nucleosomes in a chromatin structure. *J. Virol.* **63:**943–947.

Jin, L., W. Peng, G.-C. Perng, D. J. Brick, A. B. Nesburn, C. Jones, and S. L. Wechsler. 2003. Identification of herpes simplex virus type 1 latency-associated transcript sequences that both inhibit apoptosis and enhance the spontaneous reaction phenotype. *J. Virol.* **77:**6556–6561.

Kraus, R. J., J. G. Perrigove, and J. E. Mertz. 2003. ZEB negatively regulates the lytic switch BZLF1 gene promoter of Epstein-Barr virus. *J. Virol.* **77:**199–207.

Lomonte, P., J. Thomas, P. Texiev, C. Caron, S. Khochbin, and A. L. Epstein. 2004. Functional interaction between class II histone deacetylases and ICP0 of herpes simplex virus type 1. *J. Virol.* **78:**6744–6757.

Wang, Q.-Y., C. Zhou, K. E. Johnson, R. C. Colgrove, D. M. Coen, and D. M. Krupe. 2005. Herpesviral latency-associated gene promotes assembly of heterochromatin on viral lytic gene promoters in latent infection. *Proc. Natl. Acad. Sci. USA* **102:**16055–16059.

The Poxviral Transcriptional System

Hagler, J., and S. Shuman. 1992. A freeze-frame view of eukaryotic transcription during elongation and capping of nascent RNA. *Science* **255:**983–986.

Kates, J. R., and B. R. McAuslan. 1967. Poxvirus DNA-dependent RNA polymerase. *Proc. Natl. Acad. Sci. USA* **58:**134–141.

Latner, D. R., J. M. Thompson, P. D. Gershon, C. Storrs, and R. C. Condit. 2002. The positive transcription elongation factor activity of the vaccinia virus J3 protein is independent from its (nucleoside-2'-O-)methyltransferase and poly(A) polymerase stimulatory functions. *Virology* **301:**64–80.

Passurelli, A. L., G. R. Kovacs, and B. Moss. 1996. Transcription promoter: requirement for the product of the A2L intermediate stage gene. *J. Virol.* **70:**4444–4450.

Rosales, R., G. Sutter, and B. Moss. 1994. A cellular factor is required for transcription of vaccinia viral intermediate-stage genes. *Proc. Natl. Acad. Sci. USA* **91:**3794–3798.

Transcription of Viral Genes by RNA Polymerase III

Fowlkes, D. M., and T. Shenk. 1980. Transcriptional control regions of the adenovirus VA1 RNA gene. *Cell* **22:**405–413.

Howe, E., and M.-D. Shu. 1993. Upstream basal promoter element important for exclusive RNA polymerase III transcription of EBER2 gene. *Mol. Cell. Biol.* **13:**2656–2665.

Huang, W., R. Pruzan, and S. J. Flint. 1994. In vivo transcription from the adenovirus E2 early promoter by RNA polymerase III. *Proc. Natl. Acad. Sci. USA* **91:**1265–1269.

Komano, J., S. Maruo, K. Kurozumi, T. Oda, and K. Takada. 1999. Oncogenic role of Epstein-Barr virus-encoded RNAs in Burkett's lymphoma cell line Akata. *J. Virol.* **73:**9827–9831.

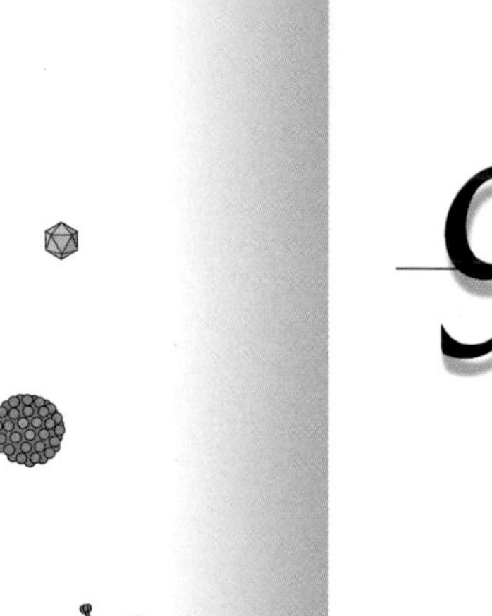

9

Genome Replication Strategies: DNA Viruses

The more the merrier.
ANONYMOUS

Introduction

The genomes of DNA viruses come in a considerable variety of sizes and shapes, from small single-stranded to large double-stranded molecules that may be linear or circular (Table 9.1). Whatever their physical nature, viral DNA molecules must be replicated efficiently within an infected cell to provide genomes for assembly into progeny virions. Such replication invariably requires the synthesis of at least one viral protein and often expression of several viral genes. Consequently, viral DNA synthesis cannot begin immediately upon arrival of the genome at the appropriate intracellular site, but rather is delayed until viral replication proteins have attained a sufficient concentration. Initiation of viral DNA synthesis typically leads to many cycles of replication, and the accumulation of very large numbers of newly synthesized DNA molecules. However, more long-lasting virus-host cell interactions, such as latent infections, are also common, both in nature and in the laboratory. In these circumstances, the number of viral DNA molecules made is strictly controlled.

Replication of all DNA, from the genome of the simplest virus to that of the most complex vertebrate cell, follows a set of universal rules: (i) DNA is always synthesized by template-directed, stepwise incorporation of deoxynucleoside monophosphates (dNMPs) from deoxynucleoside triphosphate (dNTP) substrates into the 3′-OH end of the growing DNA chain; (ii) each parental strand of a duplex DNA template is copied by base pairing to produce two daughter molecules identical to one another and to their parent (**semiconservative replication**); (iii) replication of DNA begins and ends at specific sites in the template, termed **origins** and **termini**, respectively; and (iv) DNA synthesis is catalyzed by DNA-dependent DNA polymerases, but many accessory proteins are required for initiation or elongation. In contrast to all DNA-dependent, and many RNA-dependent, RNA polymerases, **no** DNA polymerase can initiate template-directed DNA synthesis *de novo*. All require a **primer** with a free 3′-OH end to which dNMPs complementary to those of the template strand are added.

Table 9.1 Cellular and viral proteins used in synthesis of viral DNA

Virus	Genome	Viral origin-binding protein(s)	DNA polymerase(s)	Effects of infection on cellular DNA synthesis
Parvovirus				
Adeno-associated virus type 2	4,680 bp, linear, single (−) or (+) strands	Rep 78/68	Cellular	None
Papovavirus				
Simian virus 40	5,243 bp, closed, double-stranded circle	LT	Cellular, DNA polymerases α and δ	Induction of entry into S phase and cellular DNA synthesis
Adenovirus				
Human adenovirus type 2	35,937 bp, linear, double stranded	DNA polymerase preterminal protein complex	Viral	Induction of entry into S phase in quiescent cells; inhibition of cellular DNA synthesis in actively growing cells
Herpesviruses				
Herpes simplex virus type 1	~150 kbp, linear, double stranded, 4 isomers	UL9	Viral	Inhibition of cellular DNA synthesis
Epstein-Barr virus	172 kbp, linear, double stranded			Immortalization of latently infected B cells
OriP		EBNA-1	Cellular	
OriLyt		Zta	Viral	Induction of cell cycle arrest and inhibition of cellular DNA synthesis by Zta
Poxvirus				
Vaccinia virus	~200 kbp, linear, double stranded with closed terminal loops	None known	Viral	Inhibition of cellular DNA synthesis

The genomes of RNA viruses must encode enzymes that catalyze RNA-dependent RNA or DNA synthesis. In contrast, those of DNA viruses can be replicated by the cellular machinery. Indeed, replication of the smaller DNA viruses, such as parvoviruses and polyomaviruses, requires but a single viral replication protein, and the majority of reactions are carried out by cellular proteins (Table 9.1). This strategy avoids the need to devote limited viral genetic coding capacity to enzymes and other proteins required for DNA synthesis. In contrast, the genomes of all larger DNA viruses encode DNA polymerases and additional replication proteins (Table 9.1). In the extreme case, exemplified by poxviruses, the viral genome encodes a complete DNA synthesis system, and is replicated in the cytoplasm of its host cells.

There is also considerable variety in the mechanism of priming of viral DNA synthesis. In some cases, DNA synthesis is initiated with RNA primers, the mechanism by which cellular genomes are replicated. In others, unusual structural features of the genome or viral proteins provide primers. Despite such distinctions, the replication strategies of different viral DNAs are based on the only two mechanisms of double-stranded DNA synthesis known (Box 9.1) and on common molecular principles. For example, the genomes of polyomaviruses and herpesviruses, which are quite different in size and structure, are replicated by the cellular replication machinery and viral replication proteins, respectively (Table 9.1). Nevertheless, synthesis of these two DNAs is initiated by the same priming mechanism, and the herpesviral replication machinery carries out the same biochemical reactions as the host proteins that mediate synthesis of simian virus 40 DNA.

DNA Synthesis by the Cellular Replication Machinery: Lessons from Simian Virus 40

Our current understanding of the intricate reactions by which both strands of a typical double-stranded DNA template are copied in eukaryotic cells is based on *in vitro* studies of simian virus 40 DNA synthesis. In this section, we focus on the cellular replication machinery and the molecular functions of its components that have been established by using this viral system. However, we first briefly discuss general features of eukaryotic DNA replication, and why simian virus 40 proved to be a priceless resource for those seeking to understand this process.

Eukaryotic Replicons

General Features

The complete replication of large eukaryotic genomes within the lifetime of an actively growing cell depends on their organization into smaller units of replication termed **replicons** (Fig. 9.1): at the maximal rate of replication

BOX 9.1

BACKGROUND

The two mechanisms of synthesis of double-stranded viral DNA molecules

Replication of double-stranded nucleic acids proceeds by **either** synthesis of daughter strands from a replication fork, **or** strand displacement. No other replication mechanisms are known.

Among viral genomes, only those of certain double-stranded DNA viruses are synthesized via a replication fork. Replication of viral double-stranded RNAs **never** proceeds via this mechanism.

DNA synthesis via a replication fork is **always** initiated from an RNA primer. In contrast, strand displacement synthesis of viral DNA **never** requires an RNA primer.

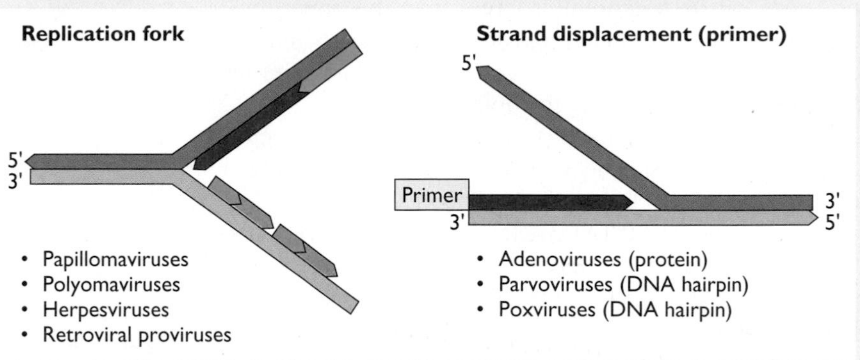

Parental DNA, RNA primers, and newly synthesized DNA are shown in blue, green, and red/pink, respectively. The primer indicated by the tan box can be a DNA structure or a protein.

Figure 9.1 Properties of replicons. (A) Electron micrographs of replicating simian virus 40 DNA, showing the "bubbles" of replicating DNA, in which the two strands of the template are unwound. These linear DNA molecules were obtained by restriction cleavage of viral enzyme DNA that had replicated to different degrees in infected cells. They are arranged in order of increasing degree of replication to illustrate the progressive movement of the two replication forks from a single origin of replication. From G. C. Fareed et al., *J. Virol.* **10**:484–491, 1972, with permission. **(B)** Bidirectional replication from an origin. Newly synthesized DNA is shown in red and pink, a convention used throughout the text.

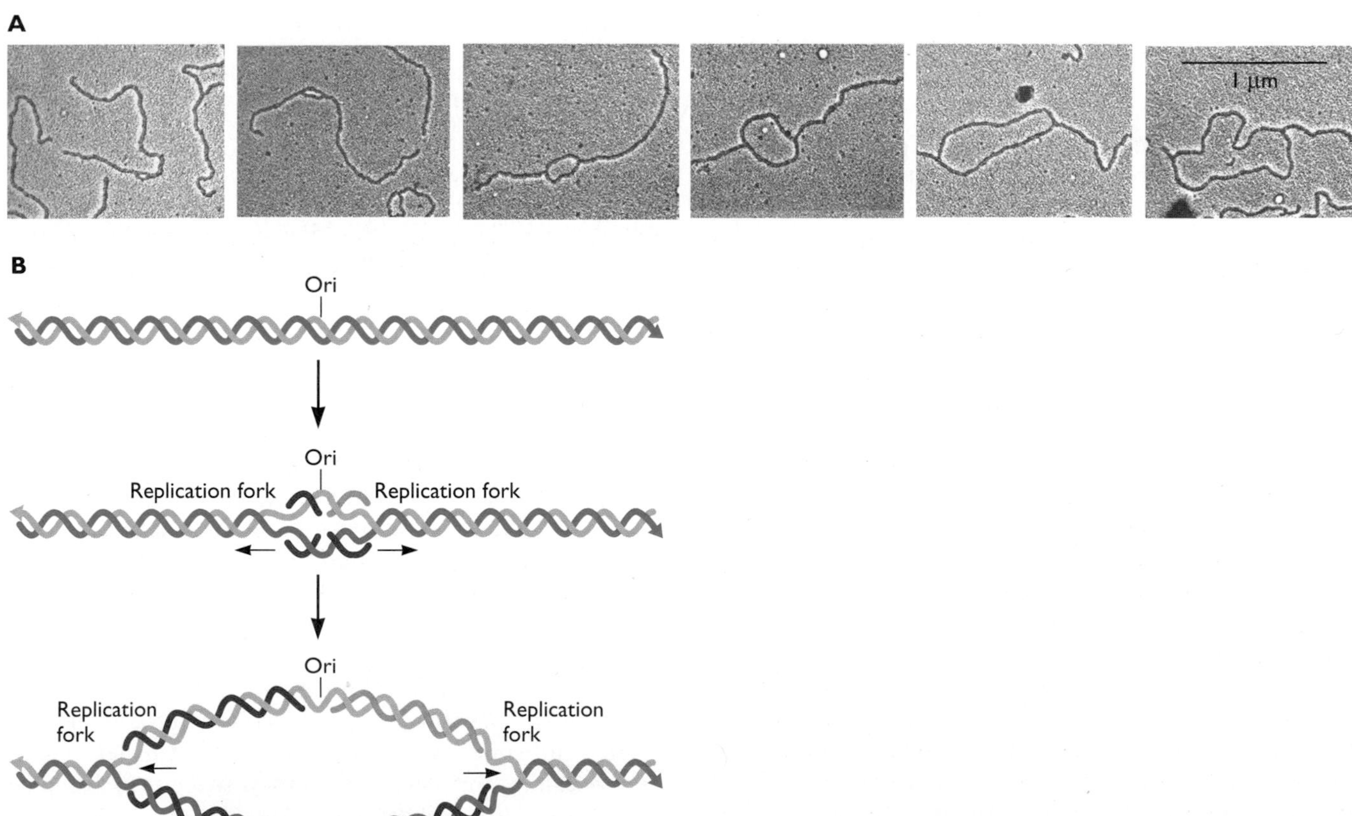

observed *in vivo*, a typical human chromosome could not be replicated as a single unit in less than 10 days! Each chromosome therefore contains many replicons, ranging in length from about 20 to 300 kbp. With some exceptions, no viral DNA genome is larger than the upper limit reported for such cellular replicons. Nevertheless, all but the smallest contain two or three origins (see "Properties of Viral Replication Origins" below).

Each replicon contains a single origin at which replication begins. The sites at which nascent DNA chains are being synthesized, the ends of "bubbles" seen in the electron microscope (Fig. 9.1A), are termed **replication forks**. In bidirectional replication, two replication forks are established at a single origin and move away from it as the new DNA strands are synthesized (Fig. 9.1B). However, as DNA must be synthesized in the 5′ → 3′ direction, only one of the two parental strands can be copied continuously from a primer deposited at the origin. The long-standing conundrum of how the second strand is synthesized was solved with the elucidation of the discontinuous mechanism of synthesis (Fig. 9.2A). Primers for DNA synthesis are synthesized at multiple sites, such that this new DNA strand is made initially as short, discontinuous segments.

The discontinuous mechanism of DNA synthesis creates a special problem at the ends of linear DNAs, where excision of the terminal primer creates a gap at the 5′ end of the daughter DNA molecules (Fig. 9.2B). In the absence of a mechanism for completing synthesis of termini, discontinuous DNA synthesis would lead to an intolerable loss of genetic information. In chromosomal DNA, specialized elements, called **telomeres**, at the ends of each chromosome prevent loss of end sequences. These structures comprise simple, repeated sequences maintained by reverse transcription of an RNA template, which is an essential component of the ribonucleoprotein enzyme **telomerase**. Complete replication of all sequences of linear viral DNA genomes (Table 9.1) is achieved by a variety of elegant mechanisms.

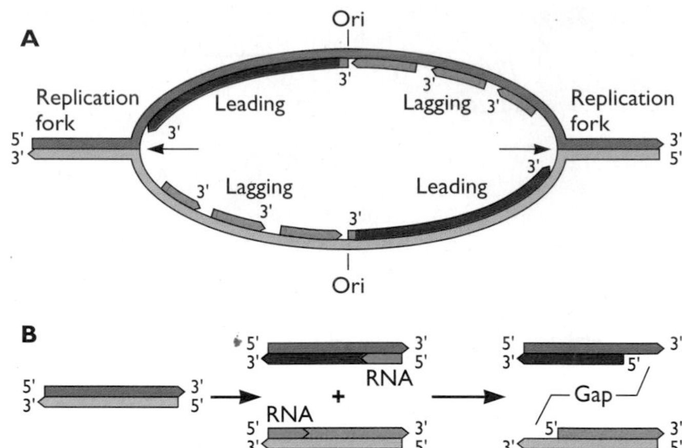

Figure 9.2 Semidiscontinuous DNA synthesis from a bidirectional origin. (A) Semidiscontinuous synthesis of the daughter strands. Synthesis of the RNA primers (green) at the origin allows initiation of continuous copying of one of the two strands on either side of the origin in the replication bubble. The second strand cannot be made in the same way (see the text). The nascent DNA population contains many small molecules termed **Okazaki fragments** in honor of the investigator who first described them. The presence of short segments of RNA at the 5′ ends of Okazaki fragments indicated that the primers necessary for DNA synthesis are molecules of RNA. With increasing time of replication, these small fragments are incorporated into long DNA molecules, indicating that they are precursors. It was therefore deduced that the second nascent DNA strand is synthesized discontinuously, also in the 5′ → 3′ direction. Because synthesis of this strand cannot begin until the replication fork has moved some distance from the origin, it is called the **lagging strand**, while the strand synthesized continuously is termed the **leading strand**. Complete replication of the lagging strand requires enzymes that can remove RNA primers, repair the gaps thus created, and ligate the individual DNA fragments to produce a continuous copy of the template strand. **(B)** Incomplete synthesis of the lagging strand. When a DNA molecule is linear, removal of the terminal RNA primer from the 5′ end of the lagging strand creates a gap that cannot be repaired by any DNA-dependent DNA polymerase.

Origins of Cellular Replication

Identification of replication origins of eukaryotes (other than budding yeasts) has proved to be one of the most frustrating endeavors of modern molecular biology. A large body of evidence indicates that replication initiates at specific sites *in vivo*. With the availability of cloned genomic DNA segments, these origins can be identified and mapped with some precision. The difficulty lies in their functional identification. *Saccharomyces cerevisiae* origins can be assayed readily, because they support replication of small plasmids and hence their maintenance as episomes. All yeast origins behave as such **autonomously replicating sequences**. In contrast, this simple functional assay generally has failed to identify analogous mammalian sequences, even when applied to DNA segments containing origins mapped *in vivo*. Conversely, many long DNA fragments (including prokaryotic DNAs) support replication of plasmids in mammalian cells, but apparently direct random initiation of DNA replication. The paradox of why DNA sequences that support plasmid replication are so difficult to find has yet to be fully resolved. However, there is accumulating evidence that metazoan DNA replication initiates within long DNA segments (several kilobase pairs or more), and that their function can depend on far more distant sequences, or epigenetic changes. Despite the considerable differences in size and complexity of fungal and mammalian origins of replication, the origin recognition machinery is highly conserved among eukaryotes.

The Origin of Simian Virus 40 DNA Replication

Because the identification of functional origins of most eukaryotic genomes was so difficult, progress in elucidation of the mechanisms of origin-dependent DNA synthesis came through study of viral origins, notably that of simian virus 40. This origin, which supports bidirectional replication, was the first viral control sequence to be located on a physical map in which the reference points are restriction endonuclease cleavage sites (Fig. 9.1A; Box 9.2). We now possess a detailed picture of this origin (Fig. 9.3), and of the binding sites for the simian virus 40 origin recognition protein, large T antigen (LT).

A 64-bp sequence, the core origin, which lies between the initiation sites of early and late transcription, is sufficient for initiation of simian virus 40 DNA synthesis in infected cells. This sequence contains four copies of a pentanucleotide-binding site for LT, flanked by an AT-rich element and a 10-bp imperfect palindrome (Fig. 9.3). Additional sequences within this busy control region of the viral genome increase the efficiency of initiation of DNA synthesis from the core origin.

Cellular Replication Proteins and Their Functions during Simian Virus 40 DNA Synthesis

Eukaryotic DNA Polymerases

It has been known for over 50 years that eukaryotic cells contain DNA-dependent DNA polymerases. Mammalian cells contain several such nuclear enzymes, which are distinguished by such properties as sensitivities to various inhibitors and their degree of **processivity**, the number of nucleotides incorporated into a nascent DNA chain per initiation from a primer carrying a 3′-OH group. These characteristics can be readily assayed in *in vitro* reactions with artificial template-primers, such as gapped or

BOX 9.2 EXPERIMENTS

Mapping of the simian virus 40 origin of replication

As illustrated in panel **A** (left), exposure of simian virus 40-infected monkey cells to [^{3}H]thymidine ([^{3}H]dT) for a period less than the time required to complete one round of replication (e.g., 5 min) results in labeling of the growing points of replicating DNA. If replication proceeds from a specific origin (Ori) to a specific termination site (T), the DNA replicated last will be preferentially labeled in the population of completely replicated molecules (panel **A**, right). The distribution of [^{3}H]thymidine among the fragments of completely replicated viral DNA generated by cleavage with HindII and HindIII is shown in panel **B**. The simian virus 40 genome is represented as cleaved within the G fragment, and relative distances are given with respect to the junction of the A and C fragments. The observation of two decreasing gradients of labeling that can be extrapolated (dashed lines) to the same region of the genome confirmed that simian virus 40 replication is bidirectional (Fig. 9.1A) and allowed location of the origin on the physical map of the viral genome. Modified from K. J. Danna and D. Nathans, *Proc. Natl. Acad. Sci. USA* **69**:3097–3100, 1972, with permission.

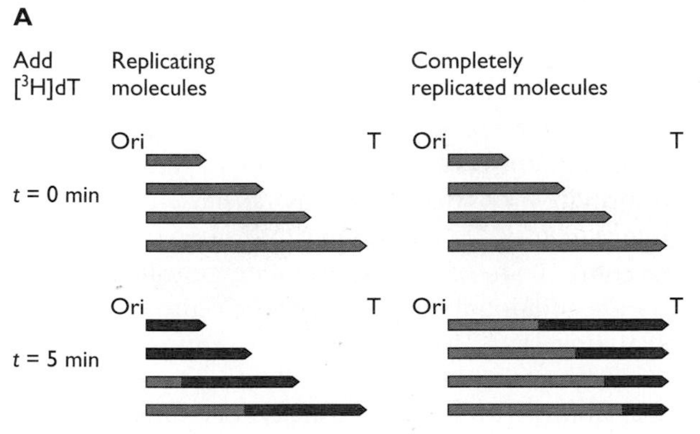

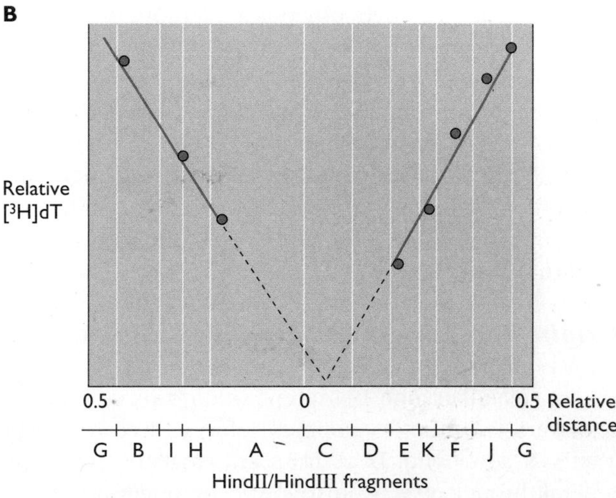

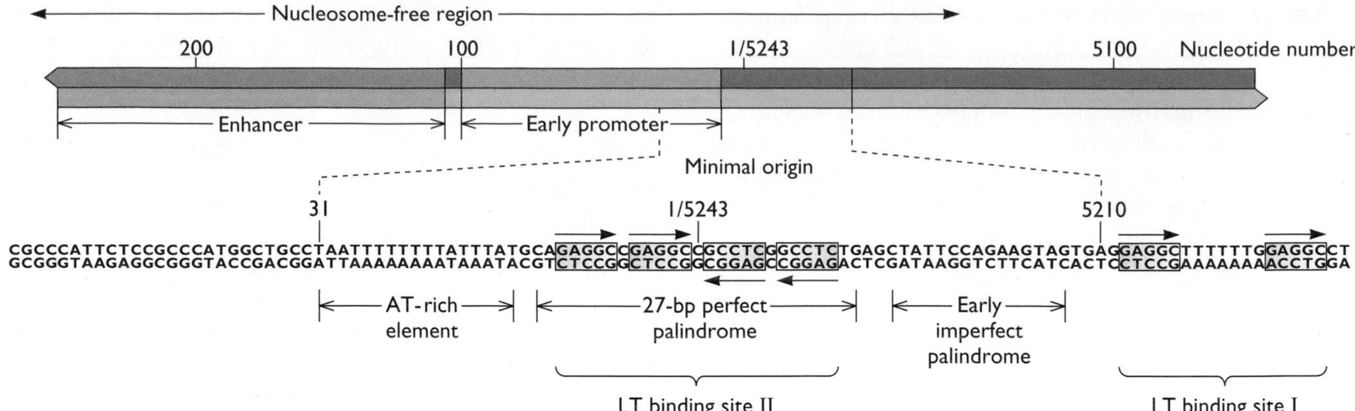

Figure 9.3 The origin of simian virus 40 DNA replication. The positions of the minimal origin necessary for simian virus 40 DNA replication *in vivo* and *in vitro* and of the enhancer and early promoter (see Chapter 8) in the viral genome are indicated. The sequence of the minimal origin is shown below, with the pentameric LT-binding sites ($G^A_C GGC$) in yellow. The AT-rich element and early imperfect palindrome, as well as LT-binding site II, are essential for replication. A second LT-binding site (site I) modestly stimulates replication *in vivo*. Other sequences, including the enhancer and Sp1-binding sites of the early promoter, increase the efficiency of viral DNA replication at least 10-fold. The activation domains (see Chapter 8) of these transcriptional regulators might help recruit essential replication proteins to the origin. Alternatively, the binding of transcriptional activators might induce remodeling of chromatin in the vicinity of the origin. This possibility is consistent with the fact that, as indicated at the top, the region of the genome containing the origin and transcriptional control regions is nucleosome free in a significant fraction (about 25%) of minichromosomes in infected cells.

nicked DNA molecules. However, distinguishing their roles in DNA replication was impossible until simian virus 40 origin-dependent DNA synthesis was reproduced *in vitro* (see the next section). The requirements for viral DNA synthesis *in vitro* and genetic analyses (performed largely with yeasts) identified DNA polymerases α, δ, and ε as replication enzymes. Only DNA polymerase α is associated with priming activity. A heteromeric **primase** is tightly bound to DNA polymerase α, and copurifies with it.

One of the most striking properties of these replicative DNA polymerases is their obvious evolutionary relationships to prokaryotic and viral enzymes. Six regions are highly conserved in both primary amino acid sequence and relative position in a wide variety of these DNA polymerases. As discussed in Chapters 6 and 7, all template-directed nucleic acid polymerases share several sequence motifs and probably a similar core structure, indicating that important features of the catalytic mechanisms are also common to all these enzymes.

Origin-Dependent Viral DNA Replication In Vitro

Studies of eukaryotic DNA replication took a quantum leap forward with the development of a cell-free system for synthesis of adenoviral DNA from an exogenous template. This breakthrough was soon followed by origin-dependent replication of simian virus 40 DNA *in vitro*. Because cellular components are largely responsible for simian virus 40 DNA synthesis, this system proved to be the watershed in the investigation of eukaryotic DNA replication: it allowed the identification of previously unknown cellular replication proteins, and elucidation of the mechanism of DNA synthesis.

Mechanism of Simian Virus 40 DNA Synthesis

Origin recognition and unwinding. The first step in simian virus 40 DNA synthesis is the recognition of the origin by LT, the major early gene product of the virus. This viral protein can bind to the core origin to form a hexamer on any of the pentanucleotide repeats (Fig. 9.3). However, initiation of viral DNA synthesis requires the flanking sequences of the minimal origin and ATP-dependent formation of two LT hexamers that encase the DNA sequence (Fig. 9.4). The double hexamers bind over the palindromic pairs of pentanucleotide repeats in the origin, with additional non-sequence-specific contacts with the flanking DNA. In reactions that require ATP but not its hydrolysis, the LT hexamers undergo conformational change and, in turn, induce structural transitions in the AT-rich and the early imperfect palindrome sequences that flank the LT-binding sites (Fig. 9.4). The next reaction, LT-induced unwinding of the origin, requires cellular replication protein A (Rp-A), which possesses single-stranded-DNA-binding activity and binds specifically to LT

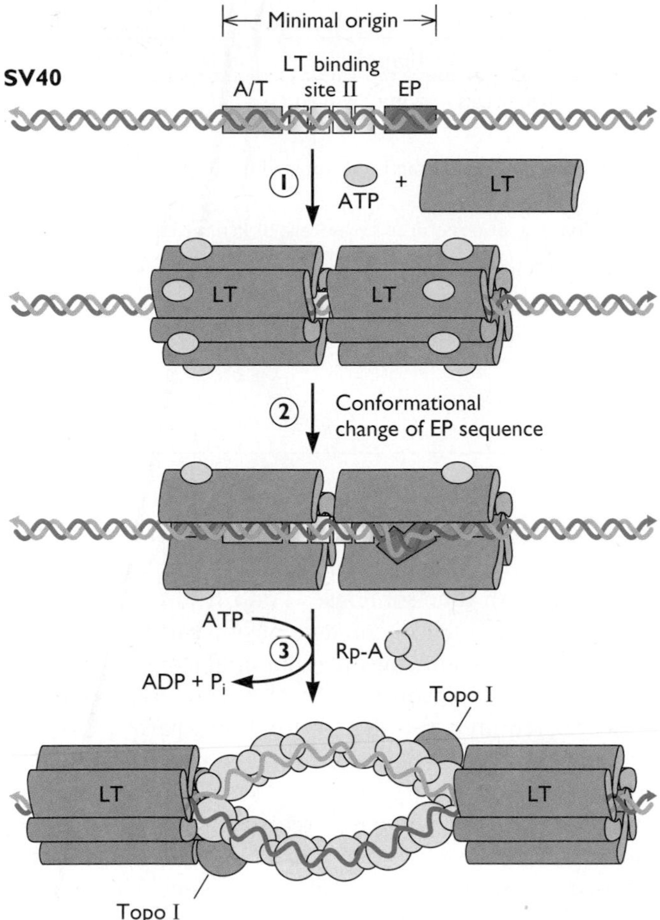

Figure 9.4 Model of the recognition and unwinding of the simian virus 40 (SV40) origin. In the presence of ATP, two hexamers bind to the origin via the pentanucleotide LT-binding sites (step 1). Binding of LT hexamers protects the flanking AT-rich (A/T) and early palindrome (EP) sequences of the minimal origin from DNase I digestion and induces conformational changes, for example distortion of the early palindrome (step 2). Stable unwinding of the origin requires the cellular, single-stranded-DNA-binding protein replication protein A (Rp-A), which binds to LT. LT helicase activity, in concert with Rp-A and topoisomerase I, progressively unwinds the origin (step 3). For simplicity, the two LT hexamers are shown moving apart, but they probably remain in contact (Box 9.3).

(Table 9.2). In concert with Rp-A and cellular topoisomerase I, the intrinsic 3′ → 5′ helicase activity of LT harnesses the energy of ATP hydrolysis to unwind DNA bidirectionally from the core origin (Fig. 9.4). Formation of the LT complex at the simian virus 40 origin resembles assembly reactions at well-characterized bacterial origins, such as *Escherichia coli* OriC, or the origin of phage λ, in which multimeric protein structures form on AT-rich sequences. Furthermore, formation of hexamers around DNA is a property common to several viral and cellular replication proteins.

Leading-strand synthesis. Binding of DNA polymerase α-primase to both LT and Rp-A at the simian virus 40 origin completes assembly of the **presynthesis complex**, and sets the stage for the initiation of leading-strand synthesis (Fig. 9.5). The primase synthesizes the RNA primers of the leading strand at each replication fork, while DNA polymerase α extends them to produce short fragments. The 3′-OH ends of these DNA fragments are then bound by cellular replication factor C (Rf-C), proliferating-cell nuclear antigen (Pcna), and DNA polymerase δ (Fig. 9.5). Pcna, which is highly conserved, is the processivity factor for DNA polymerase δ: it is required for synthesis of long DNA chains from a single primer and essential for simian virus 40 DNA replication *in vitro*. This mammalian protein is the functional analog of the β subunit of *E. coli* DNA polymerase III and phage T4 gene 45 product. These remarkable proteins are **sliding clamps**, which track along the DNA template and serve as movable platforms for DNA polymerases (Fig. 9.6). As sliding clamp proteins are closed rings, Pcna cannot load onto DNA molecules that lack ends, such as the circular simian virus 40 genome or chromosomal domains of genomic DNA. This essential, ATP-dependent step is carried out by the clamp-loading protein Rf-C, which induces transient opening of the Pcna ring. Because Rf-C binds to the 3′-OH ends of DNA fragments, it places the processivity protein at the replication forks (Fig. 9.5). Binding of these two proteins inhibits DNA polymerase α, and induces dissociation of the DNA polymerase α-primase complex. Subsequent binding of DNA polymerase δ completes assembly of a multiprotein structure capable of leading-strand synthesis by continuous copying of the parental template strand.

DNA replication can neither initiate nor proceed without the action of enzymes that unwind the strands of the DNA template. In the simian virus 40 *in vitro* replication system, LT is the helicase responsible for unwinding of the origin, and remains associated with each of the two replication forks, unwinding the template during elongation (Fig. 9.5).

Lagging-strand synthesis. The first Okazaki fragment of the lagging strand is synthesized by the DNA polymerase α-primase complex (Fig. 9.5, step 4). The lagging strand is also synthesized by DNA polymerase δ, and transfer of the 3′ end of the first Okazaki fragment to this enzyme is thought to proceed as on the leading strand. The lagging-strand template is then copied **toward** the origin of replication. Consequently, synthesis of the lagging strand requires initiation by DNA polymerase α-primase at multiple sites progressively further from the origin (Fig. 9.5). The mechanisms by which leading- and lagging-strand synthesis are coordinated are not fully understood. If the replication

Table 9.2 Cellular proteins required for simian virus 40 DNA replication

Protein	Synonym(s)	Contacts	Functions
Rp-A	Rf-A ssB	Primase LT	Binds to single-stranded DNA; origin unwinding in cooperation with LT
DNA polymerase α-primase	Polα/primase	LT	Synthesis of RNA primers and Okazaki fragments on leading and lagging strands
Rf-C	Activator 1	Pcna	ATP-dependent clamp loading; also required for release of Pcna from DNA
Pcna		Rf-C	Sliding clamp
DNA polymerase δ		Pcna	Processive synthesis of leading and lagging strands when bound to Pcna
Rnase H1			Endonucleotytic degradation of RNA base-paired with DNA; removal of RNA primers
Fen1			5′→3′ exonuclease; removal of RNA primers
DNA ligase I			Sealing of daughter DNA fragments

machinery tracked along an immobile DNA template, the complexes responsible for leading- and lagging-strand synthesis would have to move in opposite directions. Furthermore, lagging-strand synthesis would require repeated assembly of priming complexes at sites that do not contain LT-binding sequences. A more attractive alternative is that the DNA template is spooled through an immobile replication complex that contains all the proteins necessary for synthesis of both daughter strands. This mechanism would allow simultaneous copying of the template strands in opposite directions by a single replication machine present at each fork (Fig. 9.7). Consistent with this idea, replication of chromosomal DNA occurs at fixed sites in the nucleus. Moreover, structures indicative of DNA spooling have been observed in the electron microscope during the initial, LT-dependent unwinding of the simian virus 40 origin (Box 9.3). While a model invoking an immobile replication machine possesses the virtues of elegance and parsimony, it remains to be firmly established.

Termination and resolution. Because the circular simian virus 40 DNA genome possesses no termini, its replication does not lead to gaps in the strands made discontinuously. Nevertheless, additional cellular proteins are needed for the production of two daughter molecules from the circular template. These essential components of the simian virus 40 replication system are cellular enzymes that alter the topology of DNA, topoisomerases I and II. These enzymes, which differ in their catalytic mechanisms, and functions in the cell, reverse the winding of one duplex DNA strand around another (**supercoiling**). Because they remove supercoils, topoisomerases are said to **relax** DNA. In a closed circular DNA molecule the unwinding of duplex DNA at the origin and subsequently at the replication forks is necessarily accompanied by supercoiling of the remainder of the DNA (Fig. 9.8A). If not released, the torsional stress so introduced would act as a brake on movement of the replication forks, eventually bringing them to a complete halt. Both topoisomerases I and II can remove such torsional stress to allow movement of SV40 DNA replication forks. These enzymes play an analogous role during replication of chromosomal DNA *in vivo*. Topoisomerase II is also required specifically for the separation of the viral daughter molecules from late replication intermediates. A cycle of simian virus 40 DNA synthesis produces two interlocked (catenated) circular DNA molecules that can be separated only when one DNA molecule is passed through a double-strand break in the other. The break is then resealed (Fig. 9.8B). Topoisomerase II catalyzes this series of reactions.

Replication of chromatin templates. The simian virus 40 genome is associated with cellular nucleosomes both in the virion and in infected cell nuclei. It is therefore replicated as a minichromosome, in which the DNA is wrapped around nucleosomes. This arrangement raises the question of how the replication machinery rapidly copies a DNA template that is tightly bound to nucleosomal histones. A similar problem is encountered during the replication of many viral RNA genomes, when the template RNA is packaged by viral RNA-binding proteins in a ribonucleoprotein. The mechanisms by which replication complexes circumvent the nucleosomal barriers to movement are not understood in detail. Nevertheless, as discussed in Chapter 8, numerous proteins that couple ATP hydrolysis to remodeling of nucleosomal DNA have been identified. The organization of the simian virus 40 genome into a minichromosome also implies that viral DNA replication must be coordinated with binding of newly synthesized DNA to cellular nucleosomes. In fact, new nucleosomes are deposited at viral replication forks, a reaction that is catalyzed by the essential human protein chromatin assembly factor 1.

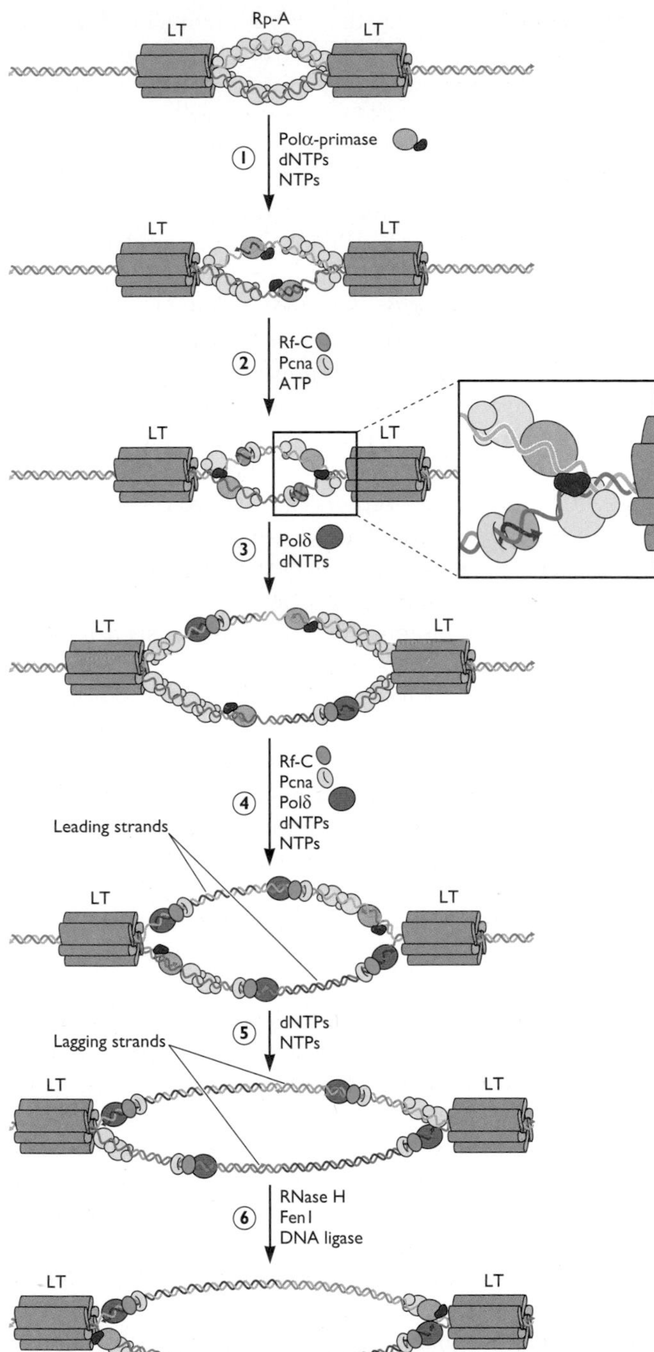

Figure 9.5 Synthesis of leading and lagging strands. The DNA polymerase α-primase responsible for the synthesis of Okazaki fragments binds specifically to both the Rp-A and LT proteins assembled at the origin in the presynthesis complex. Once bound, this enzyme complex synthesizes leading-strand RNA primers that are then extended as DNA (step 1). The 3′-OH group of the nascent RNA-DNA fragment (about 30 nucleotides in total length) is then bound by replication factor C (Rf-C) in a reaction that requires ATP but not its hydrolysis. Rf-C allows ATP-dependent opening of the proliferating-cell nuclear antigen (Pcna) (Fig. 9.6) ring and its loading onto the template (step 2). DNA polymerase δ then binds to the Pcna/Rf-C complex (step 3). This replication complex is competent for continuous and highly processive synthesis of the leading strands (steps 4 and 5). Lagging-strand synthesis begins with synthesis of the first Okazaki fragment by DNA polymerase α-primase (step 3). DNA polymerase δ is recruited as during leading strand synthesis, and produces a lagging-strand segment (step 5). The multiple DNA fragments produced by discontinuous lagging-strand synthesis are sealed by removal of the primers by RNase H (an enzyme that specifically degrades RNA hybridized to DNA) and the 5′ → 3′ exonuclease Fen1, repair of the resulting gaps by DNA polymerase δ, and joining of the DNA fragments by DNA ligase I (step 6).

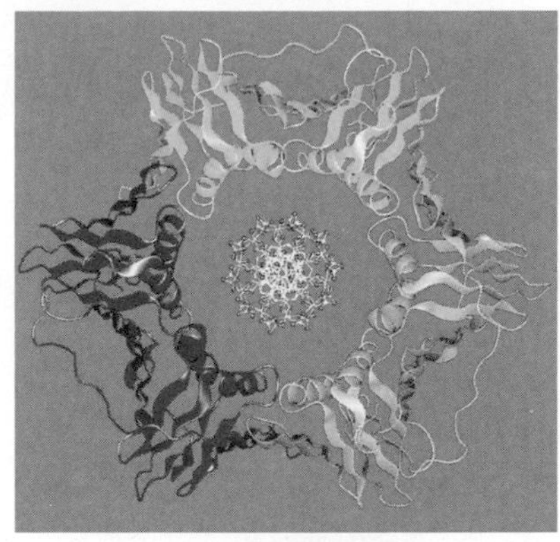

Figure 9.6 Structure of proliferating cell nuclear antigen, Pcna. The protein trimer is represented schematically in ribbon form, with the three subunits colored red, yellow, and orange. Each monomer contains two topologically identical domains. A model of double-stranded B form DNA is shown in the geometric center of the structure to illustrate how the closed ring formed by tight association of the three subunits might encircle DNA. The α-helices from the three subunits line the central channel, forming a surface of positively charged residues favorable for interaction with the negatively charged phosphodiester backbone of DNA. The nonspecificity of such interactions and the orientation of the internal α-helices appear ideal for strong but nonspecific contact with the phosphate groups of the DNA backbone, as modeled in the figure. From T. S. R. Krishna et al., *Cell* **79:**1233–1243, 1994, with permission. Courtesy of J. Kuriyan, Rockefeller University.

Summary. Analysis of simian virus 40 replication *in vitro* has identified essential cellular replication proteins, led to molecular descriptions of crucial reactions in the complex process of DNA synthesis, and provided new insights into chromatin assembly. The detailed understanding of the reactions completed by the cellular DNA replication machinery laid the foundation for elucidation of the mechanisms by which other animal viral DNA genomes are replicated, and of some of the intricate circuits that regulate DNA synthesis and its initiation.

Mechanisms of Viral DNA Synthesis

The replication of all viral DNA genomes within infected cells comprises reactions analogous to those necessary for simian virus 40 DNA synthesis, namely, origin recognition

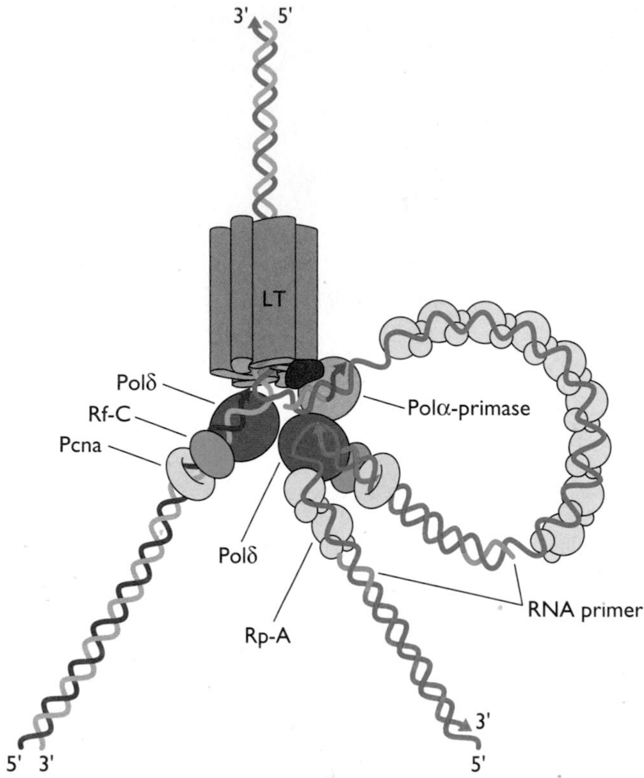

Figure 9.7 A hypothetical simian virus 40 replication machine. A replication machine containing all proteins necessary for both continuous synthesis of the leading strand and discontinuous synthesis of the lagging strand would assemble at each replication fork. Spooling of a loop of the template DNA strand for discontinuous synthesis would allow the single complex to copy the two strands in opposite directions. Pol, polymerase.

and assembly of a presynthesis complex, priming of DNA synthesis, elongation, termination, and often resolution of the replication products. However, the mechanistic problems associated with each of the steps in DNA synthesis are solved by a variety of virus-specific mechanisms.

Priming and Elongation

Synthesis of viral DNA molecules is initiated by a number of unusual mechanisms in which not only RNA, but also DNA and even protein molecules function as primers. Because each of the viral priming strategies has profound consequences for the mechanism of elongation, priming and elongation are considered together.

Synthesis of Viral RNA Primers by Cellular or Viral Enzymes

The standard method of priming is synthesis of a short RNA molecule by a specialized primase. As we have seen, cellular DNA polymerase α-primase synthesizes all RNA primers needed for replication of both template strands of polyomaviral genomes. A similar mechanism operates at certain origins of some herpesviruses, such as that directing replication of the episomal Epstein-Barr viral genome in latently infected cells (see "Regulation of Replication via Different Viral Origins: Epstein-Barr Virus" below). The integrated proviral genomes of retroviruses are also replicated via RNA primers synthesized by the cellular primase at the origin of the cellular replicon into which the provirus is inserted. In actively dividing cells, proviral DNA is therefore replicated once per cell cycle by the cellular replication machinery. Less common among DNA viruses is synthesis of RNA primers by viral proteins. However, this mechanism is characteristic of herpesviral replication in productively infected cells. Such productive replication has been best characterized in cells infected by herpes simplex virus type 1 when a viral primase synthesizes RNA primers. This primase is a heterotrimer of the products of the viral UL5, UL8, and UL25 genes.

An inevitable consequence of DNA synthesis from RNA primers by either cellular or viral DNA polymerases is that one of the two parental strands must be copied discontinuously. In the case of linear templates, a specialized mechanism is also necessary to complete synthesis of the lagging strand. In contrast, other viral DNA genomes are replicated by means of alternative priming mechanisms that eliminate the need for discontinuous synthesis.

Priming via DNA: Specialized Structures in Viral Genomes

Self-priming of viral DNA synthesis via specialized structures in the viral genome is a hallmark of all *Parvoviridae*, among the smallest DNA viruses that replicate in animal cells. This virus family includes the dependoviruses, such as adeno-associated viruses, and the autonomous parvoviruses, such as minute virus of mice. Synthesis of all parvoviral DNAs exhibits a number of unusual features, the most striking being self-priming. This mechanism is illustrated here with adeno-associated virus.

The adeno-associated virus genome is a small molecule (<5 kb) of single-stranded, linear DNA that carries **inverted terminal repetitions** (ITRs). Genomic DNA is of both (+) and (–) polarity, for both strands are encapsidated, but in separate virus particles. As illustrated in Fig. 9.9A, palindromic sequences within the central 125 nucleotides of the ITR base pair to form T-shaped structures. Formation of this structure at the 3′ end of either single strand of viral DNA provides an ideal template-primer for initiation of the first cycle of viral DNA synthesis (Fig. 9.9B). Experimental evidence for such **self-priming** includes the dependence of adeno-associated virus DNA synthesis on self-complementary sequences within the ITR. Following

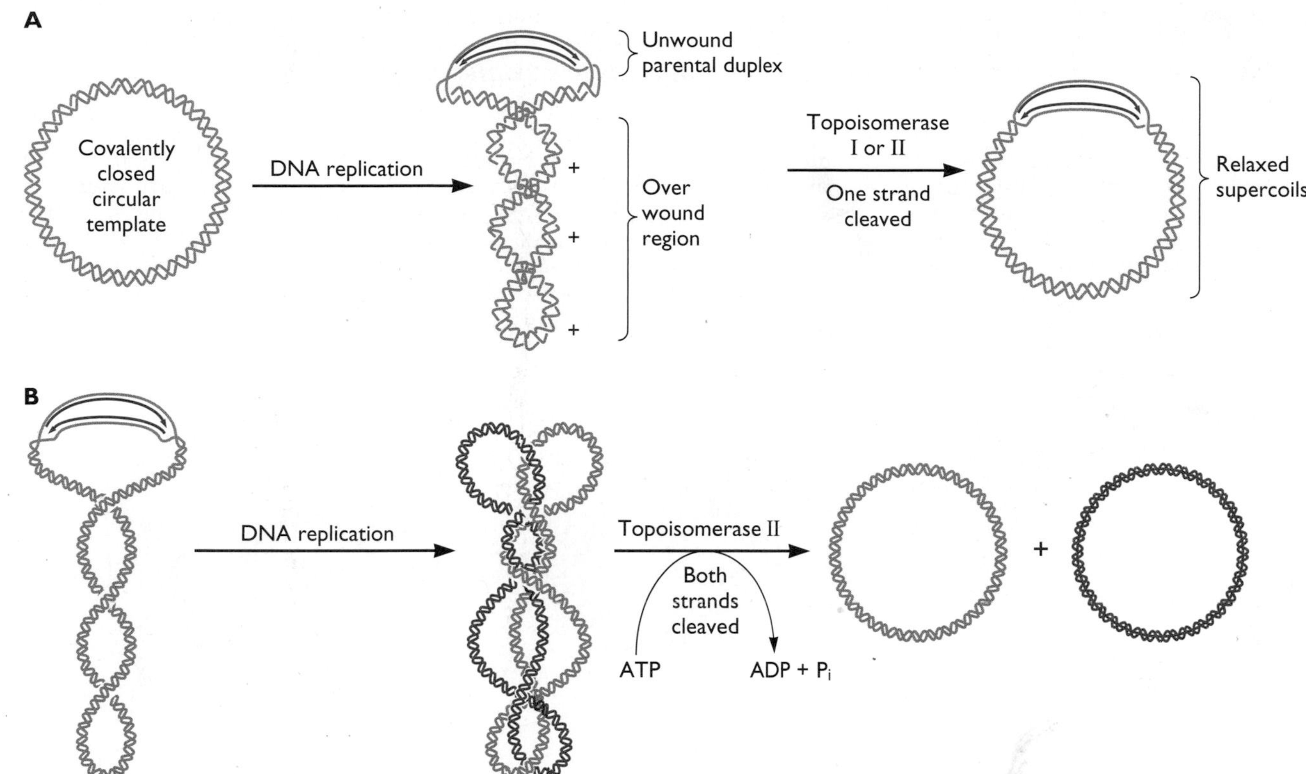

Figure 9.8 Function of topoisomerases during simian virus 40 DNA replication. (A) Relief from overwinding. Unwinding of the template DNA at the origin and two replication forks leads to overwinding (positive supercoiling) of the DNA ahead of the replication forks (middle). Either topoisomerase I or topoisomerase II can remove the supercoils to allow continued movement of the replication fork. **(B)** Decatenation of replication products. Separation of interlocked daughter molecules (middle) requires topoisomerase II, which makes a double-strand break in DNA, passes one double strand over the other to unwind one turn, and reseals the DNA in reactions that require hydrolysis of ATP.

recognition of the viral DNA primer, the single template strand of an infecting genome can be copied by a continuous mechanism, analogous to leading-strand synthesis during replication of double-stranded DNA templates. In subsequent cycles of replication, the same 3′-terminal priming structures form in the duplex replication intermediate produced in the initial round of synthesis (Fig. 9.9B). Adeno-associated virus DNA synthesis is therefore always continuous, and requires cellular DNA polymerase δ, Rf-c and Pcna, but not DNA polymerase α-primase.

On the other hand, a specialized mechanism **is** necessary to complete replication, by copying of the sequences that form the priming structure: the initial product retains the priming hairpin and is largely duplex DNA in which parental and daughter strands are covalently connected (Fig. 9.9.B, step 2). This step is achieved by nicking of the intermediate within the parental DNA strand at a specific site. The new 3′-OH end liberated in this way primes continuous synthesis to the end of the DNA molecule (Fig. 9.9B, step 4). The nick is introduced by the related viral proteins Rep 78 and Rep 68 (Rep 78/68) (Box 9.4). These proteins are site- and strand-specific endonucleases, which bind to, and cut at, specific sequences within the ITR. During this **terminal resolution** process, Rep 78/68 becomes covalently linked to the cleaved DNA at the site that will become the 5′ terminus of the fully replicated molecule (Fig. 9.9B). Following the synthesis of a duplex of the genomic DNA molecule (the **replication intermediate**), formation of the 3′-terminal priming hairpin allows continuous synthesis of single-stranded genomes by a strand displacement mechanism, with re-formation of the replication intermediate (Fig. 9.9B, steps 6 and 7).

Rep 78 and Rep 68 are similar to simian virus 40 LT in several respects, and can be considered origin recognition proteins (Table 9.3). They are the only viral gene products necessary for parvoviral DNA synthesis. In addition to recognizing and cleaving the terminal resolution site, these proteins provide the ATP-dependent, 3′ → 5′ helicase

BOX 9.3

EXPERIMENTS

Unwinding of the simian virus 40 origin leads to spooling of DNA

Visualization by electron microscopy of structures formed during LT-dependent unwinding from the simian virus 40 origin *in vitro* suggested that the two hexamers remain in contact as DNA is unwound. **(A)** LT bound to the origin, as a characteristic bilobed structure (the double hexamer shown in Fig. 9.4); **(B)** unwinding intermediates; **(C)** the intermediate at the bottom right in panel B at higher magnification. This intermediate contains a bilobed LT complex connecting the two replication forks, so the single-stranded DNA (ssDNA) is looped out as "rabbit ears." The formation of such structures containing a dimer of the LT hexamer, in which each monomer is bound to a replication fork, stimulates the helicase activity of LT. This property supports the view that the DNA template is spooled through an immobile replication machine (see the text). dsDNA, double-stranded DNA. From R. Wessel et al., *J. Virol.* **66:**804–815, 1992, with permission. Courtesy of H. Stahl, Universität des Saarlandes.

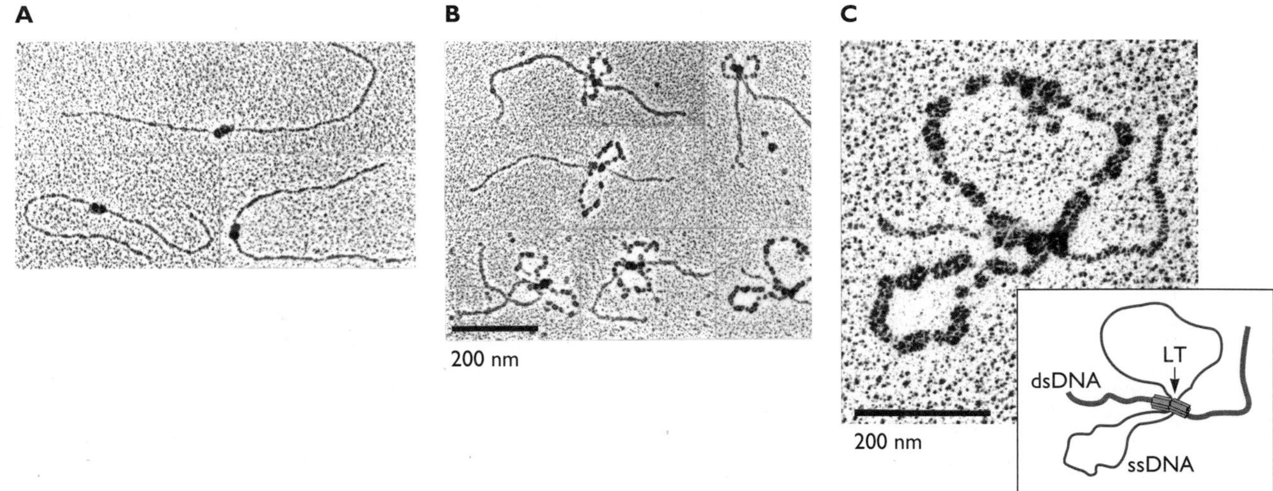

activity needed for unwinding of the replicated ITR and re-formation of the priming hairpin (Fig. 9.9, step 5). Whether cellular helicases are also required, for example, when double-stranded replication intermediates serve as templates, is not yet clear.

Priming via self-complementary terminal sequences of the viral genome eliminates the need for proteins that make RNA primers, or are themselves primers for DNA synthesis (see the next section). Such a mechanism would therefore seem especially advantageous for viruses with small genomes, such as the parvoviruses. Nevertheless, a similar self-priming mechanism is thought to initiate replication of the large double-stranded DNA genomes of poxviruses such as vaccinia virus.

Protein Priming

Judged by the criterion of simplicity, the most effective mechanism of initiation of DNA synthesis is via a protein primer. Nevertheless, this mechanism is rare, restricted to some bacteriophages (e.g., ϕ29 and PRD1) and to hepadnaviruses and adenoviruses among DNA viruses that infect animal cells. The replication of some viral RNA genomes is also initiated from a protein primer, notably the VPg protein of poliovirus discussed in Chapter 6. Here, we use adenoviral DNA replication to illustrate the mechanism of protein priming.

The primer for human adenoviral DNA synthesis is the precursor to the terminal protein, which is covalently attached to the 5′ ends of the linear, double-stranded DNA genomes. In a template-dependent reaction, the adenoviral DNA polymerase covalently links the α-phosphoryl group of dCMP to the hydroxyl group of a specific serine residue in this preterminal protein (Fig. 9.10). The 3′-OH group of the protein-linked dCMP then primes synthesis of daughter viral DNA strands by the viral DNA polymerase. The nucleotide is added to the preterminal protein **only** when this protein primer is assembled with the DNA polymerase into preinitiation complexes at the origins of replication. As the origins lie at the ends of the linear genome, each template strand is copied continuously from one end to the other (Fig. 9.10). The parental template strand initially displaced is copied by the same mechanism, following formation of a duplex stem upon annealing of an ITR sequence. This unusual strand displacement mechanism therefore results in semiconservative replication, even though the two parental strands of viral DNA are not copied at the same replication fork.

Properties of Viral Replication Origins

As we have seen, origins of replication contain the sites at which viral DNA synthesis begins and can be defined experimentally as the minimal DNA segment necessary for initiation of replication in cells or *in vitro* reactions. Viral origins of replication support initiation of DNA synthesis by a variety of mechanisms, including some with no counterpart in cellular DNA synthesis. Nevertheless, they are discrete DNA segments that contain sequences recognized by viral origin recognition proteins to seed assembly of multiprotein complexes. When initiation is by self-priming, DNA sequences essential for replication include those needed to form and maintain a specific secondary structure in the template, as well as the sequence at which replication intermediates are cleaved for complete copying

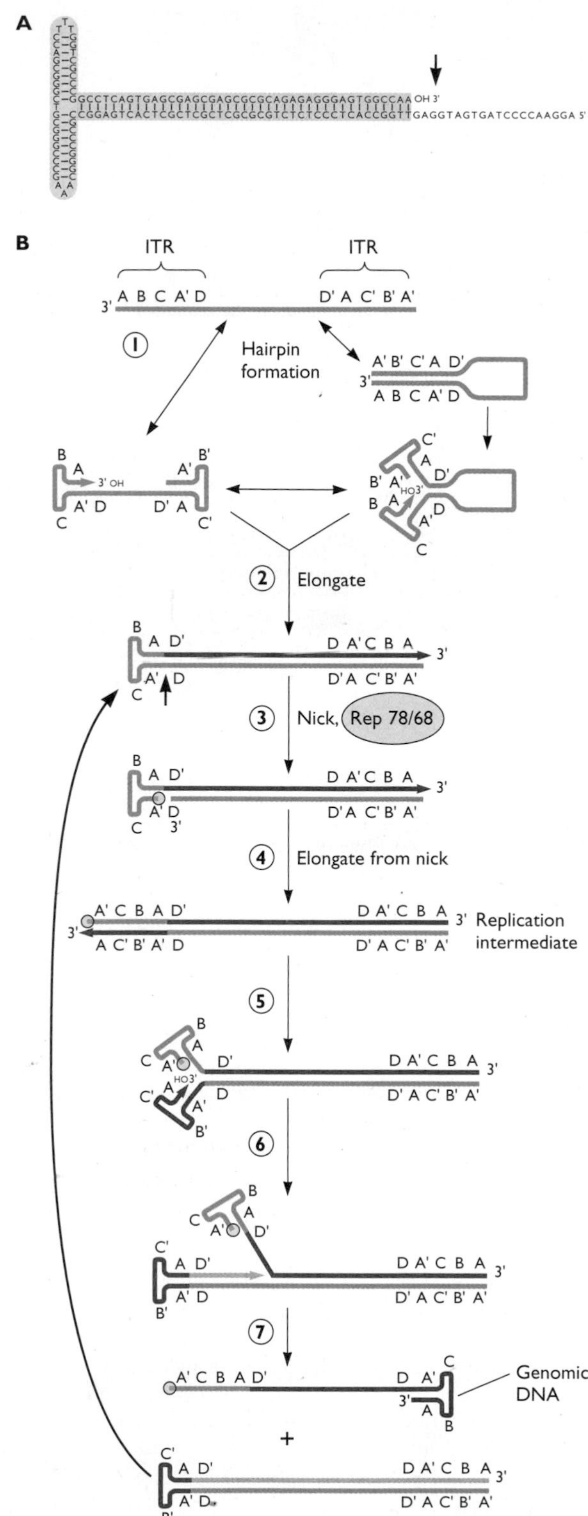

Figure 9.9 Replication of parvoviral DNA. (A) Sequence and secondary structure of the adeno-associated virus type 2 inverted terminal repetition (ITR). A central palindrome (tan background) is flanked by a longer palindrome (light blue background) within the ITR. When these bases pair at the 3′ end of the genome, a T-shaped structure in which the internal duplex stem terminates in a free 3′-OH group (arrow) is formed. **(B)** Model of adeno-associated virus DNA replication. The ITRs are represented by 3′ABCA′D5′ and 5′A′B′C′AD′3′. Formation of the 3′-terminal hairpin provides a primer-template (step 1, left). However, it is possible that the self-complementary terminal repeat sequences first base pair to form a "panhandle," a single-stranded circle stabilized by the duplex terminal sequence (step 1, right). Such a structure could explain the repair of deletions or mutations within one ITR observed when the other is intact. In either case, elongation from the 3′-OH group of the hairpin allows continuous synthesis (red) to the 5′ end of the parental strand (step 2). To complete copying of the parental strand, a nick to generate a new 3′ OH is introduced at the specific **terminal resolution site** (marked by the arrow) by the viral Rep 78/68 proteins. Elongation from the nick results in copying of sequences that initially formed the self-priming hairpin to form the double-stranded replication intermediate (step 4). However, the parental strand then contains newly replicated DNA (red) at its 3′ end. As a result, the ITR of the parental strand is no longer the initial sequence but rather its complement. This palindromic sequence is therefore present in populations of adeno-associated virus DNA molecules in one of two orientations. Indeed, such sequence heterogeneity provided an important clue for elucidation of the mechanism of viral DNA synthesis. The newly replicated 3′ end of the replication intermediate can form the same terminal hairpin structure (step 5) to prime a new cycle of DNA synthesis (step 6) with displacement of a molecule of single-stranded genomic DNA, and the formation of the incompletely replicated molecule initially produced (step 7). The latter molecule can undergo additional cycles of replication as in steps 3 and 4.

BOX 9.4

BACKGROUND

Organization of coding sequences in the adeno-associated virus genome

The viral genome is depicted at the top, with the locations of the ITR, the promoters for transcription by RNA polymerase II (p5, p19, and p40), and the single site of polyadenylation (vertical red arrow) indicated. The two coding regions within the viral genome are termed Rep (replication) and Cap (capsid). As shown below, each is expressed as multiple related proteins, which are synthesized from differentially spliced mRNAs transcribed from the different promoters. The two smaller products of the Rep region, Rep 52 and Rep 40, are not required for viral DNA synthesis, but facilitate encapsidation of single-stranded genomic DNA molecules during assembly of virus particles.

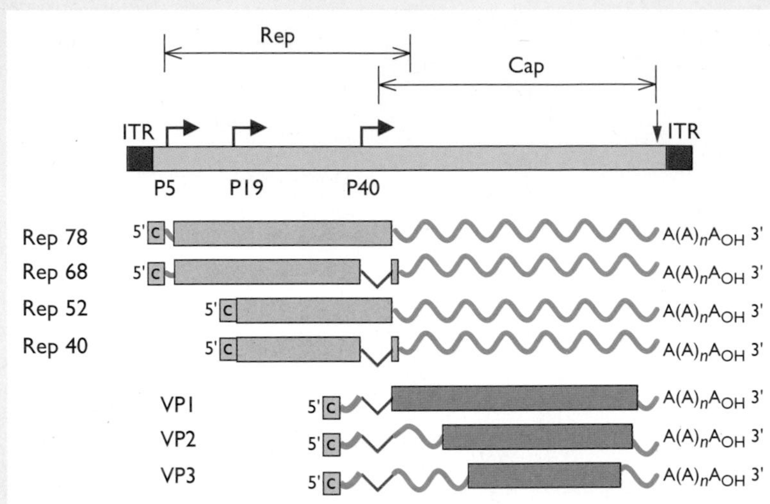

Table 9.3 Viral origin recognition proteins

Virus	Protein(s)	DNA-binding properties	Other activities and functions
Parvovirus			
Adeno-associated virus	Rep 78/68	Binds to specific sequences in ITR; binds as hexamer	Site- and strand-specific endonuclease; ATPase and ATP-dependent helicase; transcriptional regulator
Papovaviruses			
Simian virus 40	LT	Binds cooperatively to origin site II to form double hexamer; distorts origin; DNA binding regulated by phosphorylation at specific sites	DNA-dependent ATPase and 3′→5′ helicase; binds to cellular Rp-A and polymerase α-primase; represses early and activates late transcription; binds cellular Rb protein to induce progression through the cell cycle
Bovine papillomavirus type 5	E1	Binds origin with low affinity; binds strongly and cooperatively in presence of E2 protein	Binding to E2 essential for viral DNA replication; DNA-dependent ATPase and helicase
	E2	Binds as dimer to specific sequences in origin	Regulates transcription by binding to viral enhancers
Adenovirus			
Human adenovirus type 2	Pre-TP-DNA polymerase	Binds to origins; association stimulated by cellular transcriptional activators Nf-1 and Oct-1	Priming of DNA synthesis via addition of dCMP to pre-TP by the DNA polymerase; continuous synthesis of both strands of viral genome
Herpesviruses			
Herpes simplex virus type 1	UL9	Binds cooperatively to specific sites in viral origins; distorts DNA to which it binds	ATPase and 3′→5′ helicase; binds UL29 protein, UL8 subunit of viral primase, and UL42 processivity protein
Epstein-Barr virus	EBNA-1	Binds as dimer to multiple sites in two clusters (FR and DS) in viral OriP	Binding to FR sequences required for maintenance of episomal viral genomes; stimulates transcription from viral promoters

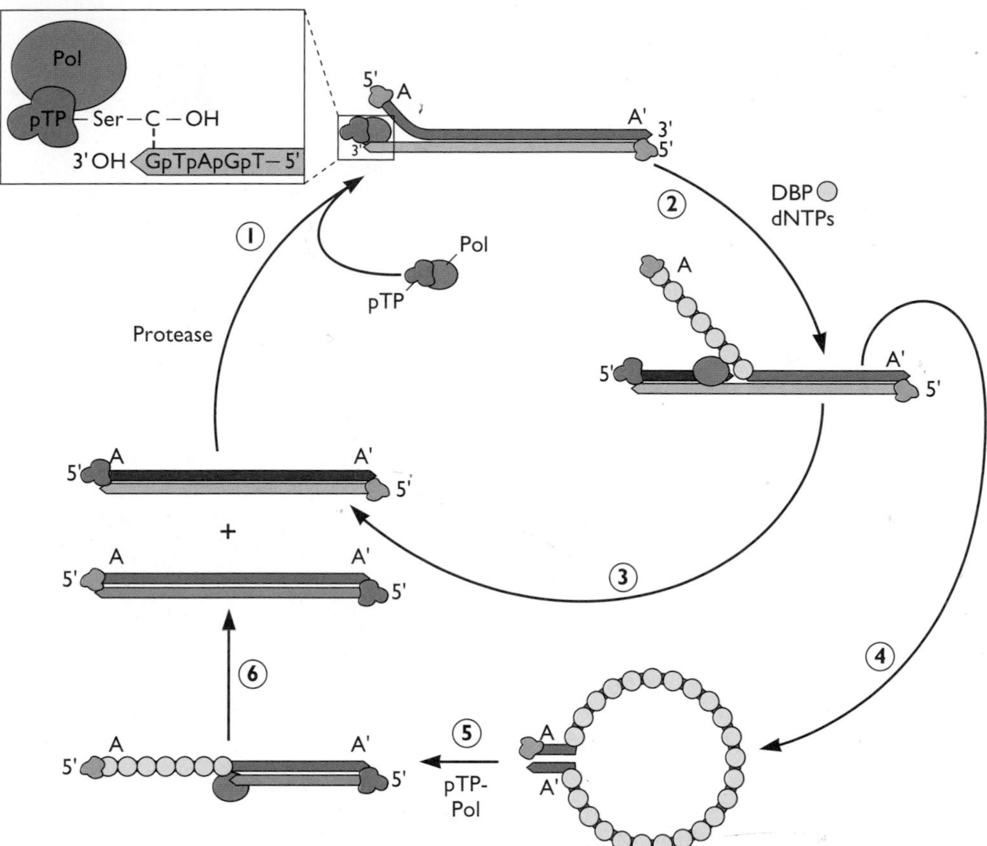

Figure 9.10 Replication of adenoviral DNA. Assembly of the viral preterminal protein (pTP) and DNA polymerase (Pol), into a preinitiation complex at each terminal origin of replication activates covalent linkage of dCMP to a specific serine residue in pTP by the DNA polymerase (step 1). The free 3′-OH group of preterminal protein-dCMP primes continuous synthesis in the 5′ → 3′ direction by Pol (step 2). This reaction also requires the viral E2 single-stranded-DNA-binding protein (DBP), which coats the displaced second strand of the template DNA molecule, and a cellular topoisomerase. As the terminal segments of the viral genome comprise an inverted repeat sequence (A and A′), there is an origin at each end, and both parental strands can be replicated by this displacement mechanism. Reannealing of the complementary terminal sequences of the parental strand initially displaced forms a short duplex stem identical to the terminus of the double-stranded genome (step 3). The origin re-formed in this way directs a new cycle of protein priming and continuous DNA synthesis (steps 4 and 5).

of the parental strand (Fig. 9.9A). Even though viral origins may be unconventional, they exhibit a number of common features, as do the proteins that recognize them.

Number of Origins

In contrast to papillomaviral and polyomaviral DNAs, the genomes of the larger DNA viruses contain not one, but two or three origins. As noted above, the two identical adenoviral origins at the ends of the linear genome are the sites of assembly of complexes containing the viral DNA polymerase and protein primer (Fig. 9.10). The genomes of herpesviruses, such as Epstein-Barr virus and herpes simplex virus type 1, contain three origins of replication. Different functions can be ascribed to the different Epstein-Barr virus origins: a single origin (OriP) allows maintenance of episomal genomes in latently infected cells, while two others (OriLyt) support replication of the viral genome during productive infection. The advantage of three origins, two copies of OriS and one copy of OriL (Fig. 9.11), to herpes simplex viruses is less clear: a full complement of these origins is not necessary for efficient viral DNA synthesis, at least in cells in culture. The two types of origin possess considerable nucleotide sequence similarity, but differ in their organization, and can be distinguished functionally. For example, OriL is activated when differentiated neuronal cells are exposed to a glucocorticoid hormone, but OriS is repressed. As glucocorticoids are produced in response to stress, a condition that reactivates

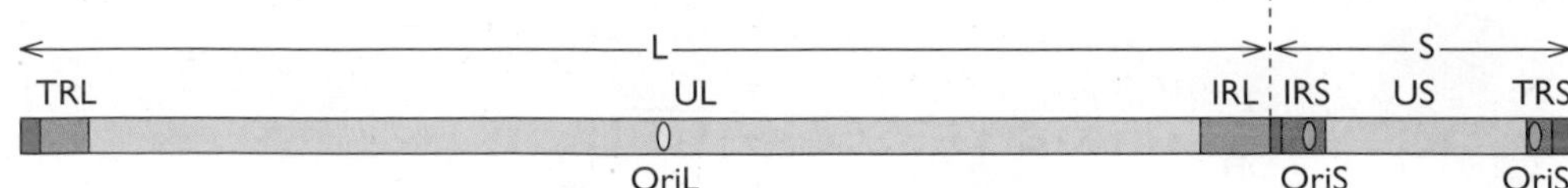

Figure 9.11 Features of the herpes simplex virus type 1 genome. The long (L) and short (S) regions of the viral genome that are inverted with respect to one another in the four genome isomers (Fig. 9.20) are indicated at the top. Each segment comprises a unique sequence (UL or US) flanked by internal and terminal repeated sequences (IR and TR). The locations of the two identical copies of OriS, in repeated sequences, and of the single copy of OriL are indicated.

a latent herpes simplex virus type 1 infection, it has been suggested that replication from OriL may be particularly important during the transition to a productive infection. An unusual feature of these viral origins is that they are required only during the initial stage of replication, which begins early in infection. Subsequent viral DNA synthesis is origin independent.

A recently discovered, and unanticipated, feature of the small adeno-associated virus genome is the presence of a second origin that can support amplification of viral DNA in the absence of the ITR origin described previously. This origin, which includes the p5 promoter (Box 9.4), also contains a binding site for Rep 78/68 and a sequence resembling the terminal resolution site that can be cleaved by the viral protein (Fig. 9.12). The function of this origin during adeno-associated virus infection in the presence of a helper virus is not yet clear. However, it may contribute to viral genome integration in cells infected in the absence of a helper (see "Integrated Parvoviral DNA Can Replicate as Part of the Cellular Genome" below).

Figure 9.12 Sequence features of the adeno-associated virus type 2 ITR and p5 origins. The Rep 78/68-binding sites in the two origins are shown in pink, and the terminal resolution sites (TRS) are shown in purple. The p5 origin TRS, like that of the ITR origin, has been shown to be cleaved by the viral protein (red arrowheads). The TATA box (green) that overlaps the Rep binding site in the p5 origin is required for replication from this origin *in vivo*. Adapted from D. L. Glauser et al., *J. Virol.* 79:12218–12230, 2005, with permission.

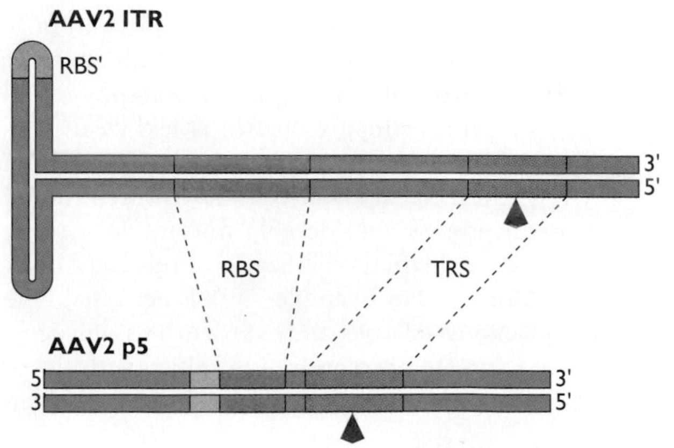

Viral Replication Origins Share Common Features

Even though the origins of replication of double-stranded DNA viruses are recognized by different proteins and support different mechanisms of initiation, they exhibit a number of common features (Fig. 9.13). The most prominent of these is the presence of AT-rich sequences. In general, AT base pairs contain only two hydrogen bonds, whereas GC pairs interact via three such noncovalent bonds. The less stable AT-rich sequences are therefore thought to facilitate the unwinding of origins that is necessary for initiation of viral DNA synthesis from either RNA or protein primers. The best-characterized viral origins of DNA replication comprise a minimal essential core origin flanked by sequences that are dispensable, but nonetheless significantly increase replication efficiency. These stimulatory sequences contain binding sites for cellular transcriptional activators (Fig. 9.13). Yet other viral origins, those of papillomaviruses and parvoviruses, and OriLyt of Epstein-Barr virus, include binding sites for viral proteins that are both transcriptional regulators and essential replication proteins. And all three herpes simplex virus type 1 origins lie between sites at which transcription of viral genes is initiated (Fig. 9.13). A close relationship between origin sequences and those that regulate transcription is therefore a second general feature.

Recognition of Viral Replication Origins

The paradigm for viral origin recognition are the polyomaviral LT proteins. We therefore describe the properties of simian virus 40 LT in more detail as a prelude to discussion of other viral proteins with similar functions.

Properties of Simian Virus 40 LT

Functions and organization. The LT proteins of polyomaviruses are remarkable proteins that provide essential replication functions and play several other important roles in the infectious cycle. As we have seen, simian virus 40 LT is both necessary and sufficient for recognition of the viral origin, and also supplies the helicase activity necessary for origin unwinding and perhaps movement of the replication fork. The LT proteins make a major contribution to the species specificity of polyomaviruses. Although the

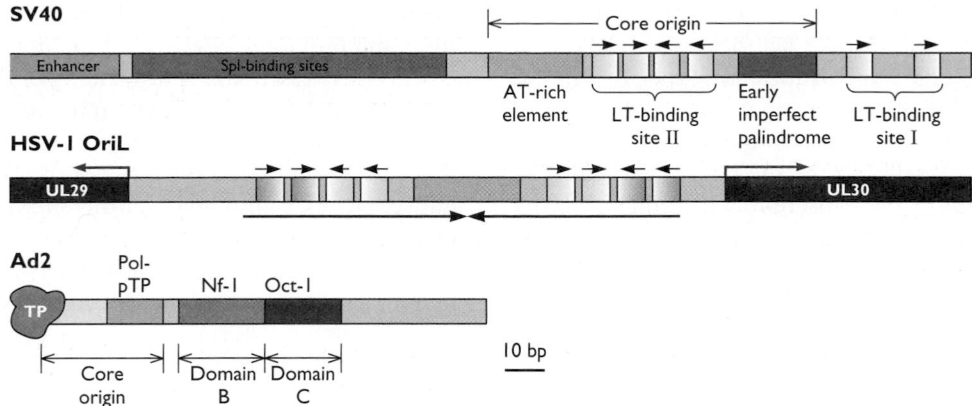

Figure 9.13 Common features of viral origins of DNA replication. The single simian virus 40 (SV40) origin, herpes simplex virus type 1 (HSV-1) OriL (Fig. 9.11), and adenovirus type 2 (Ad2) origin (Fig. 9.10), are illustrated to scale, emphasizing common features. Binding sites for origin recognition proteins and AT-rich sequences are indicated in yellow and green, respectively, and sites of initiation of transcription are indicated by jointed red arrows. Herpes simplex virus OriL, which is a perfectly symmetrical palindrome as indicated by the arrows, is located within the transcriptional control regions of the divergently transcribed UL29 and UL30 genes. The two copies of OriS (Fig. 9.11) are very similar in sequence to OriL. The terminal sequence of the adenoviral origin designated the core origin functions inefficiently in the absence of the adjacent binding site for the transcriptional activator nuclear factor 1 (Nf-1).

genomes of simian virus 40 and mouse polyomavirus are closely related in organization and sequence, they replicate only in simian and murine cells, respectively. Such host specificity is largely the result of species-specific binding of LT to the largest subunit of DNA polymerase α of the host cell in which the virus will replicate (Table 9.3). Although the precise mechanism remains to be determined, assembly of preinitiation complexes competent for unwinding of the origin does not take place when the LT of one polyomavirus binds to the origin of another.

Figure 9.14 Functional organization of simian virus 40 LT. LT is represented to scale by the blue bar. Indicated are the sequences required for binding to the DNA polymerase α-primase complex (Polα), to the cellular chaperone Hsc70, to the cellular retinoblastoma (Rb) and p53 proteins, the origin of replication (origin DNA binding), and single-stranded (ss) DNA. Also shown are segments necessary for the helicase and ATPase activities, hexamer assembly at the origin, the nuclear localization signal (NLS), and a C-terminal sequence necessary for virion production but not viral DNA synthesis (Host range). The region that binds to Hsc70 lies within an N-terminal segment termed the J domain, because it shares sequences and functional properties with the *E. coli* protein DnaJ, a chaperone that assists the folding and assembly of proteins and is required during replication of bacteriophage λ. The chaperone functions of the J domain and Hsc70 are essential for replication, and seem likely to assist assembly or rearrangement of the preinitiation complex. Below are shown the two regions of the protein in which sites of phosphorylation are clustered, indicating modifications that have been shown to inhibit (red) or activate (green) the replication activity of LT.

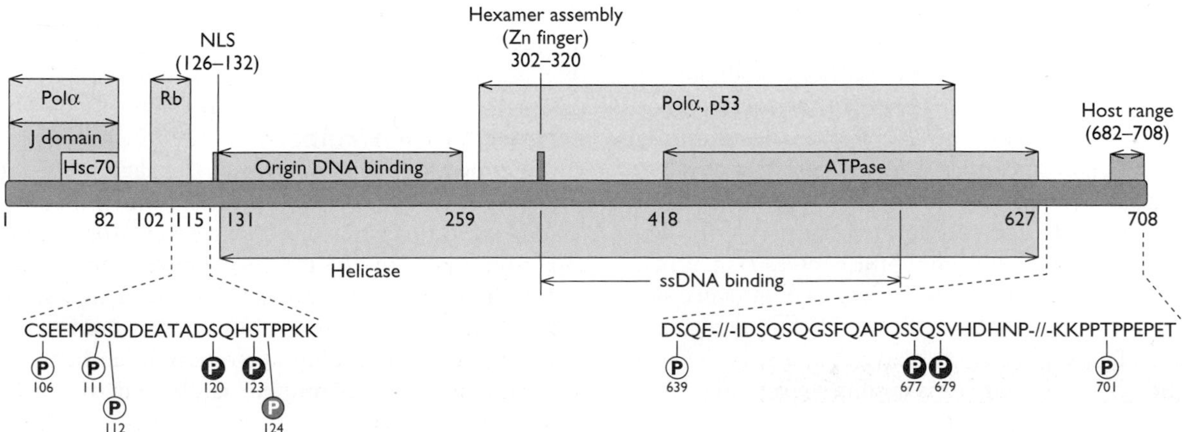

LT proteins also ensure that the cellular components needed for simian virus 40 DNA synthesis are available in the host cell. By binding and sequestering specific cellular proteins, LT perturbs mechanisms that control cell proliferation and can induce infected cells to enter S phase when they would not normally do so (see below). In addition, LT carries out a number of transcriptional regulatory functions during the infectious cycle, including regulation of its own synthesis and activation of late gene expression.

Sequences of simian virus 40 LT necessary for its numerous activities have been mapped by analysis of the effects of specific alterations in the protein on virus replication in infected cells, DNA synthesis *in vitro*, or the individual biochemical activities of the protein. The properties of such altered proteins indicate that LT contains discrete structural and functional domains, such as the minimal domain for specific binding to sites I and II at the viral origin (Fig. 9.14). However, the activities of such functional regions defined by genetic and biochemical methods may be influenced by distant sites, as discussed in the next section.

High-resolution structures of the LT origin-binding and helicase domains have been determined by X-ray crystallography. The former forms a hexameric spiral that contains a gap (Fig. 9.15A). The central channel of the spiral is more than wide enough, some 30A, to accommodate double-stranded DNA and is lined by residues required for origin binding. In protein crystals, a C-terminal segment of LT that possesses helicase activity also forms a hexameric structure, comprising a two-tiered ring encasing a central channel of varying diameter (Fig. 9.15B). The mechanism by which the helicase domain couples ATP hydrolysis to unwinding of the origin is not fully understood. However, this reaction requires nonspecific binding of LT to double- and single-stranded DNA, and a β-hairpin structure in the helicase domain that is conserved among several viral and cellular proteins with helicase activity (Box 9.5).

Regulation of LT synthesis and activity. As the simian virus 40 early enhancer and promoter are active in many cell types (Chapter 8), the viral early gene encoding LT is transcribed efficiently as soon as the viral chromosome enters the nucleus. The spliced early mRNA encoding LT is the predominant product of processing of these early transcripts. Although production of LT is not regulated during the early phase of infection, its activity is tightly controlled.

Simian virus 40 LT is phosphorylated at multiple Ser and Thr residues, most of which are located within one of two clusters near the N and C termini of the protein (Fig. 9.14). Although the significance of phosphorylation at every site is not known, it is clear that specific modifications regulate the ability of LT to support viral DNA synthesis. For example, phosphorylation of Thr124 is absolutely required. This modification specifically stimulates binding of LT to origin site II and promotes assembly of the double hexamer (Fig. 9.4). It is also essential for unwinding of DNA from the origin. As Thr124 does not lie within the minimal origin-binding domain (Fig. 9.14), such regulation of DNA-binding activity is thought to be the result of conformational change induced by phosphorylation at this site. The best candidate for the protein kinase that phosphorylates Thr124 is cyclin-dependent kinase 2 (Cdk2) associated with cyclin A. The accumulation and activity of this complex are strictly regulated as growing cells traverse the cell cycle, but the mechanisms by which LT ensures that cellular replication proteins are made in simian virus 40-infected cells also result in production of active Cdk2 (see "Viral Proteins Can Induce Synthesis of Cellular Replication Proteins" below).

Viral Origin Recognition Proteins Share Several Properties

Other viral origin recognition proteins share with simian virus 40 LT the ability to bind specifically to DNA sequences within the appropriate origin of replication. They also bind to other replication proteins (although these may be viral or cellular), and several possess the biochemical activities exhibited by LT (Table 9.3). For example, the herpes simplex virus type 1 UL9 protein, which recruits viral rather than cellular replication proteins, binds cooperatively to specific sites that flank the AT-rich sequences of the viral origins (Fig. 9.12), which are then distorted, and it possesses an ATP-dependent helicase activity that unwinds DNA in the 3′ → 5′ direction (Table 9.4). The adeno-associated virus Rep 78/68 protein possesses these same activities (Table 9.4), but is also the site-specific endonuclease that is essential for terminal resolution (Fig. 9.9). The endonuclease active site and the domain that mediates sequence-specific binding adjacent to the terminal resolution site includes a large region very similar in structure to the origin-binding domains of simian virus 40 LT and the papillomavirus E1 protein (Fig. 9.16). Such structural homology is remarkable, as there is no amino acid identity among the three viral proteins. As would be expected, a unique endonuclease active site is present in Rep 78/68 (Fig. 9.16). In addition, this protein contains a second, discrete DNA-binding domain that mediates its binding to a stem-loop structure at the tip of one of the hairpin arms. This interaction greatly stimulates cleavage at the terminal resolution site, by a mechanism that is not yet clear.

In many respects, the herpesviral UL9 protein is a typical origin-binding protein (Table 9.4). However, it is required only during the initial stage of viral DNA synthesis: later in the infectious cycle, replication requires neither the viral origins nor the origin-binding protein. It has been reported recently that the UL9 protein is cleaved by the cellular

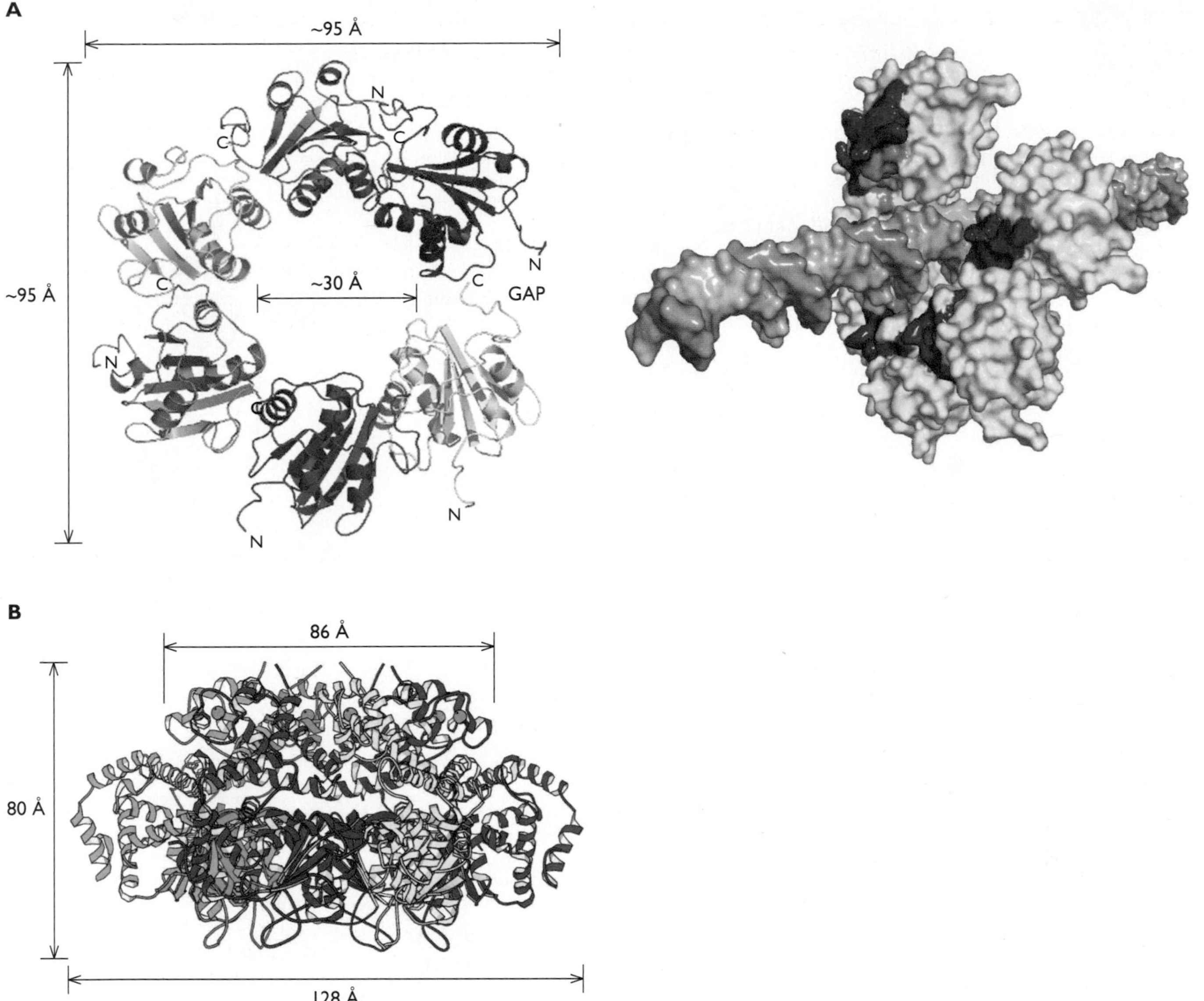

Figure 9.15 Structure of origin-binding and helicase domains of simian virus 40 large T antigen, determined by X-ray crystallography. (A) The structure of the origin-binding domain (amino acids 131 to 260) hexamer is shown in ribbon form on the left, with each subunit a different color. The hexamer forms a left-hand spiral with a gap. In the model of the hexamer shown in surface representation bound to DNA (right), the DNA is gray, with the palindromic LT-binding sequences (Fig. 9.3) in cyan and magenta. The DNA-binding regions of LT are colored red and purple. This model, and the results of mutational analysis, indicate that in the double hexamer of the full-length protein the origin-binding domains in the two hexamers interact with one another. Adapted from G. Meinke et al., *J. Virol.* **80:**4304–4312, 2006, with permission. Courtesy of Andrew Bohm, Tufts University School of Medicine. **(B)** The LT helicase domain (residues 251 to 627) hexamer is shown in ribbon form with each monomer in a different color. The hexameric ring comprises two tiers of different diameters around a central channel. Consistent with the DNA-binding activity of this LT fragment, the central channel is lined with positively charged residues. A large chamber in the middle of the central channel is wide enough (~67 Å in diameter) to accommodate DNA strands separated during DNA unwinding. The diameters of the channel on either side of this chamber are sufficient to accommodate double-stranded DNA in the smaller tier (~25 Å), but only single-stranded DNA (~15 Å) in the larger tier. The sizes of these channels were observed to differ significantly in different crystal forms of the LT hexamer and under different buffer conditions. This property, and structural features of LT monomers and the interfaces that form the two-tier hexamer, have led to the proposal that during DNA unwinding, the central channel might expand and constrict like the diaphragm (iris) of a camera. Adapted from D. Li et al., *Nature* **423:**512–518, 2003, with permission. Courtesy of Xiaojiang Chen, University of Colorado Health Science Center.

BOX 9.5

EXPERIMENTS

A conserved β-hairpin in the simian virus 40 LT helicase domain and unwinding of the origin

The β-hairpin of the LT helicase **(bottom)** is conserved in the polyomavirus and papillomavirus origin-binding proteins, and is present in cellular replicative helicases (Mcm proteins [see Fig 9.23]). Alanine substitution of the lysine and histidine residues shown resulted in complete inhibition of simian virus 40 DNA synthesis *in vitro*. Analysis of effects of the substitutions on interactions of LT established that the β-hairpin residues are required for unwinding of the origin. Furthermore, such unwinding is necessary for oligomerization of the helicase domain. These observations support a model of the action of LT **(top)**, in which initial unwinding of DNA in the EP region of the origin (Fig. 9.3) allows assembly of the hexamer, by coordinated binding of the helicase β-hairpins around a single strand of origin DNA. Adapted from A. Kumar et al., *J. Virol.* **81:**4808–4818, 2007, with permission.

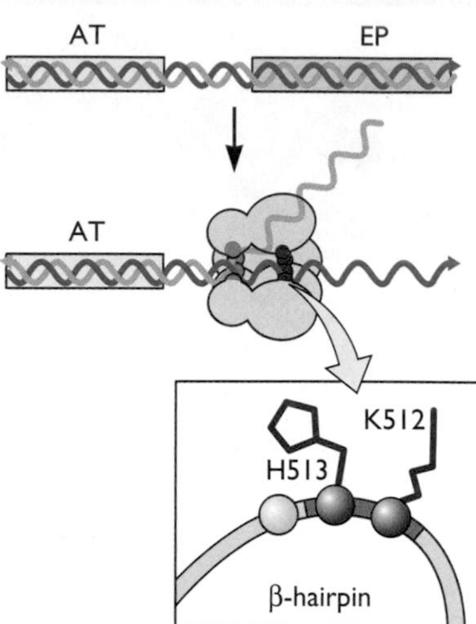

Table 9.4 Viral proteins that participate in genome replication

Virus	Protein	Properties and functions
Adenovirus		
Human adenovirus type 2	DNA polymerase	Initiates replication by covalent linkage of dCMP to the protein primer; completes synthesis of all daughter strands
	Preterminal protein	Protein primer for DNA synthesis
	Terminal protein	Facilitates origin unwinding
	Single-stranded-DNA-binding protein	Binds displaced single-stranded DNA; confers processivity to viral DNA polymerase
Herpesvirus		
Herpes simplex virus type 1	UL9 protein	Origin recognition
	UL30 protein (DNA polymerase)	Synthesis of viral DNA; associated 3′→5′ exonuclease for proofreading; target of several antiherpesviral drugs
	UL5, UL8, and UL52 proteins	Heterotrimer with primase and helicase activities; UL8 protein binds to the viral DNA polymerase
	UL42 protein	Processivity protein for viral DNA polymerase
	UL29 protein (ICP8)	Single-stranded-DNA-binding protein essential for viral DNA synthesis, binds DNA cooperatively; binds to the UL42 protein
Poxvirus		
Vaccinia virus	DNA polymerase	Synthesis of vaccinia virus DNA; associated 3′→5′ exonuclease activity for proofreading
	DNA type I topoisomerase	? essential; present in virions; structurally and functionally similar to cellular counterpart
	A20R protein	Component of DNA polymerase complex

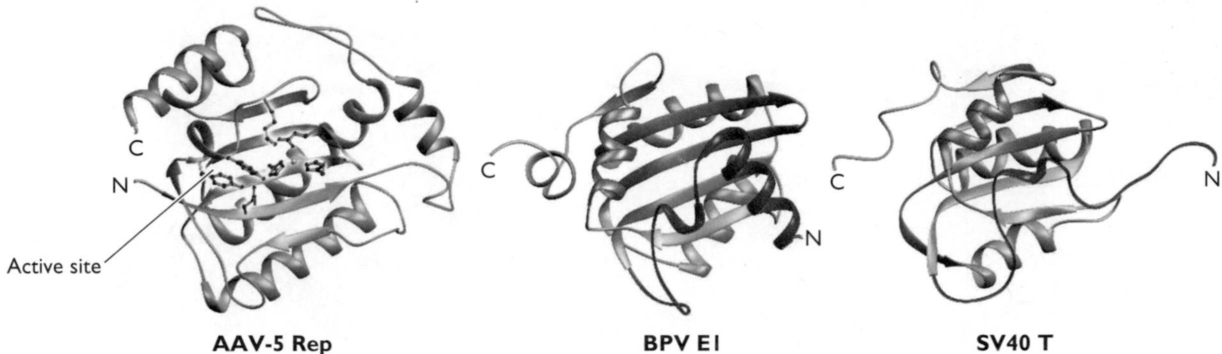

Figure 9.16 Structural homology among DNA-binding domains of viral origin recognition proteins. The X-ray crystal structures of the adenovirus-associated virus type 5 Rep 68 endonuclease domain and the bovine papillomavirus E1 and simian virus 40 LT origin-binding domains are shown in ribbon form. Each protein contains a central antiparallel β-sheet flanked by three α-helices. However, the Rep protein includes a cleft on one surface of the β-sheet that contains the endonuclease active site (residues shown in ball-and-stick). In the other two viral proteins, no cleft is present, as this region is occupied by N-terminal extensions (red). Adapted from A. Hickman et al., *Mol. Cell* **10**:327–337, 2002. Courtesy of Alison Hickman, National Institutes of Health.

protease cathepsin B following the onset of viral DNA synthesis in infected cells. Such cleavage might inactivate the protein, or the stable cleavage product, which retains the origin-binding domain, might act as a dominant negative inhibitor of origin-dependent replication.

Although recognition of viral origins of replication by a single viral protein is common, it is not universal. The human papillomavirus E1 protein possesses the same activities as simian virus 40 LT (Table 9.4), to which it is related in sequence, organization, and structure. The topologies of the origin-binding (Fig. 9.16) and the helicase-ATPase domains of the two viral proteins are strikingly similar. Nevertheless, E1 cannot support papillomaviral DNA replication in infected cells: a second viral protein, the E2 transcriptional regulator, is also necessary. Essential sequences of the minimal origin of replication of papillomaviral genomes include adjacent binding sites for both the E1 and E2 proteins (Fig. 9.17A). The E1 protein binds to DNA, but with only low specificity for origin sequences. In contrast, the E1 and E2 proteins bind cooperatively, and the specificity and the affinity of the E1-DNA interaction are increased significantly. However, the E2 protein is required only transiently: once a specific E1-E2 protein complex has assembled on the origin, the E2 protein dissociates. The results of structural studies of the E1 helicase domain bound to the N-terminal segment of E2 suggest that the latter protein binds to the E1 surface that mediates interaction among subunits in hexameric assemblies (Fig. 9.17B). Binding of ATP to E1 appears to induce conformational change that leads to dissociation of E2. Additional molecules of E1 then associate (Fig. 9.17A). The final product is an E1 hexamer assembled on, and probably closed around, single-stranded DNA at the origin. The E2 protein therefore serves as an initiation protein and loads the E1 helicase.

The adenoviral origins of replication are also recognized by two viral proteins, the preterminal protein and viral DNA polymerase. In this case, the proteins associate as they are synthesized in the cytoplasm of infected cells and, once within the nucleus, bind specifically to a conserved sequence within the minimal origin of replication (Fig. 9.13). Assembly of the viral proteins into a preinitiation complex at the origin is stimulated by their direct interaction with two cellular transcriptional regulators that bind to adjacent sequences. These viral and cellular proteins are sufficient to reconstitute initiation of adenoviral DNA synthesis from templates that carry the mature terminal protein covalently linked to their 5′ ends, like the viral genomes that enter infected cell nuclei (Fig. 9.10). The terminal protein present on such natural adenoviral DNA templates facilitates unwinding of the duplex termini early in the initiation reaction, and may be important for directing viral DNA molecules to the specialized nuclear sites at which replication takes place (see below). Once the first few nucleotides have been incorporated, the DNA polymerase must disassociate from the preterminal protein to allow elongation of the daughter DNA strand (Fig. 9.10). The structure of the ϕ29 DNA polymerase bound to its priming terminal protein suggests that such dissociation is the result of conformational change induced by displacement of the priming domain from the catalytic site in the polymerase (Box 9.6).

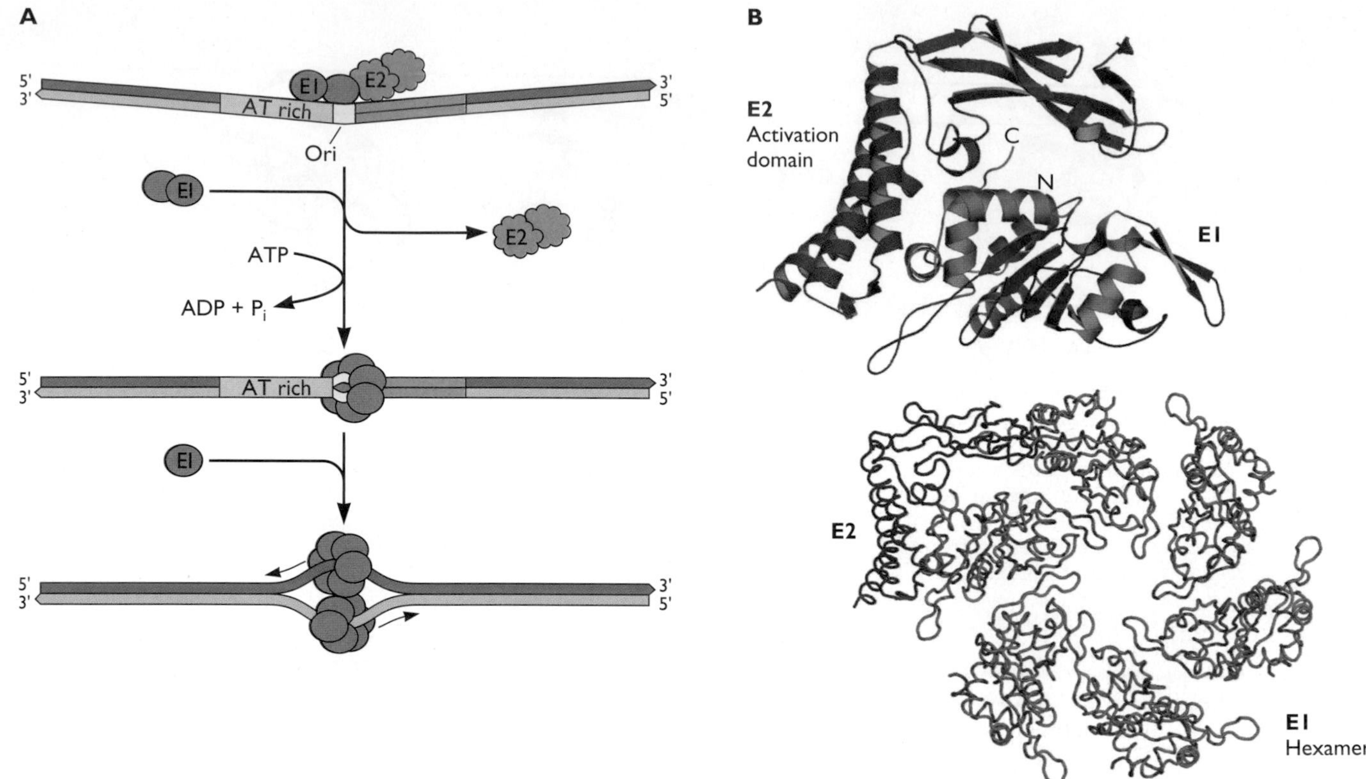

Figure 9.17 Origin loading of the papillomaviral E1 initiation protein by the viral E2 protein. **(A)** Schematic model. The sequence features of the minimal origin of replication of bovine papillomavirus type 1 are depicted as in Fig. 9.13. This origin contains an essential binding site for the viral E2 protein, a sequence-specific transcriptional regulator. The model of the origin loading of the viral E1 by the E2 protein is based on *in vitro* studies of the interactions of these proteins with the origin. The E1 and E2 proteins, which are both homodimers, bind cooperatively to the viral origin, with specificity and affinity far greater than that exhibited by the E1 protein alone. When ATP is hydrolyzed (presumably by the ATPase of the E1 protein), the $(E1)_2(E2)_2$-Ori complex is destabilized, the E2 dimers are displaced, and additional E1 molecules bind. In the resulting E1-Ori complex, the DNA is distorted and becomes partially single stranded. Finally, a larger E1 structure (probably a hexameric ring) is assembled on single-stranded DNA. **(B)** The X-ray crystal structure of the E2 activation domain bound to the E1 ATPase/helicase domain is shown in ribbon form, with E1 and E2 in blue and red, respectively. The overlay of E2 and the E1 hexamer (below) illustrates how association with E2 blocks the E1 surface that mediates hexamer assembly. Hence, E2 must dissociate prior to E1 assembly. Consistent with this model, the E1 and E2 proteins form a 1:1 complex in the absence of ATP, but in the presence of ATP E1 forms of a high-molecular-mass assembly that contains no E2. Adapted from E. Abbate et al., *Genes Dev.* **18:**1981–1986, 2004, with permission. Courtesy of Eric Abbate and Michael Botchan, University of California, Berkeley.

Viral DNA Synthesis Machines

In addition to origin recognition proteins, larger viral DNA genomes encode DNA polymerases and other essential replication proteins (Table 9.4). The simplest viral replication apparatus is that of adenoviruses, which comprises the preterminal protein primer and DNA polymerase and only one other protein, a single-stranded-DNA-binding protein. The latter protein stimulates initiation, and the switch from initiation to elongation. It is essential during elongation, when it coats the displaced strands of the template DNA molecule (Fig. 9.10). Cooperative binding of this protein to single-stranded DNA stimulates the activity of the viral DNA polymerase as much as 100-fold. It also induces highly processive DNA synthesis by this viral enzyme, and progression of replication forks over quite large distances. Remarkably, no ATP hydrolysis is required. The crystal structure of a large segment of the DNA-binding protein indicates that this protein multimerizes via a C-terminal hook (Fig. 9.18A). The formation of long protein chains results in cooperative, high-affinity binding to single-stranded DNA, and is the driving force for ATP-independent unwinding of the duplex template (Fig. 9.18B). Other single-stranded-DNA-binding proteins, such as the herpes simplex virus type 1 UL8 protein and cellular replication

BOX 9.6

DISCUSSION

Model for the transition between initiation and elongation during protein-primed DNA synthesis

Association of the adenoviral DNA polymerase with the preTP primer is necessary for catalysis of covalent linkage of the priming dCMP to preTP (see the text). However, this interaction must be reversed following initiation to allow processive elongation by the enzyme. Clues about how this transition occurs have come from structural studies of the bacteriophage ϕ29 replication proteins.

Replication of the linear, double-stranded ϕ29 genome is initiated by protein priming from origins at the ends of the genome. The phage DNA polymerase (Pol) and priming terminal protein (TP) form a heterodimer and the enzyme catalyzes linkage of the priming nucleotide to TP, just as in adenoviral DNA synthesis (see Fig. 9.10). The structure of the ϕ29 Pol-TP dimer has been determined by X-ray crystallography. In this complex, the TP priming domain lies in the site occupied by the DNA template-primer in a model of the elongating enzyme. The loop that contains the Ser to which the priming nucleotide is attached lies closest to the Pol active site. The priming domain is connected to a domain that makes extensive contacts with the DNA polymerase via a hinge.

The results of modeling studies indicate that up to 6 or 7 nucleotides can be added to the nascent DNA while TP maintains close contacts with DNA polymerase: motion about the hinge allows displacement of the priming domain while the intermediate domain maintains contact with Pol (see figure). However, this mechanism cannot accommodate further translocation of the priming domain. Rather, the intermediate domain of TP and Pol must dissociate, presumably as a result of addition structural changes. Consequently, the DNA polymerase is released for elongation, as illustrated for incorporation of eight dNMPs in the figure.

Kamtekar, S., A. J. Berman, J. Wang, J. M. Lazaro, M. de Vega, L. Blanco, M. Salas, and T. A. Steitz. 2006. The ϕ29 DNA polymerase: protein primer structure suggests a model for the initiation to elongation transition. *EMBO J.* **25:**1335–1343.

The 3′-OH group of the priming nucleotide attached to TP and nascent DNA are shown in green and red, respectively, with newly incorporated dNMPs indicated by yellow circles. Adapted from S. Kamtekar et al., *EMBO J.* **25:**1335–1343, 2006, with permission.

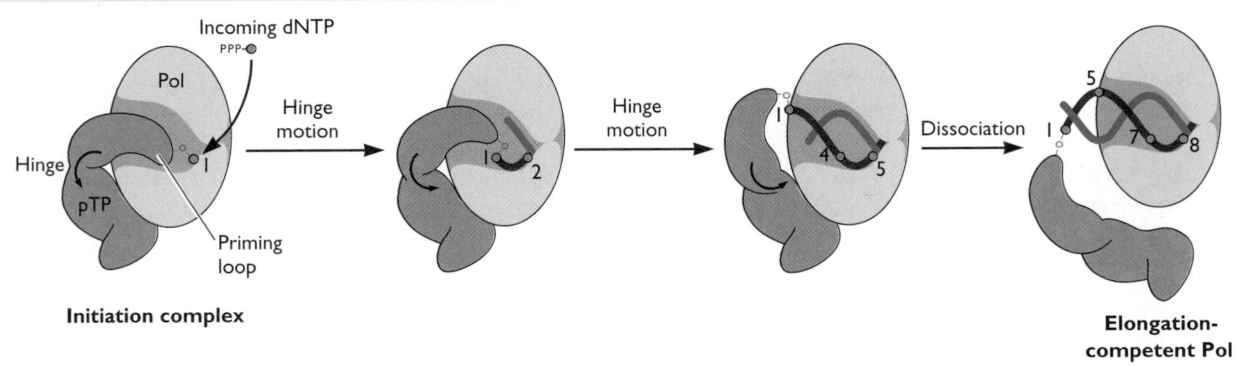

protein A, may destabilize double-stranded DNA helices by a similar mechanism.

Complete copying of adenoviral DNA templates also requires a cellular topoisomerase, such as topoisomerase I, which relieves overwinding of the template upon extensive replication, as during simian virus 40 DNA synthesis. The simplicity of the adenoviral DNA synthesis system stands in stark contrast to the complexity of the cellular replication machine that is assembled at simian virus 40 replication forks. Such simplicity can be attributed to the protein-priming mechanism, which allows continuous synthesis of full-length daughter DNA strands (Fig. 9.10).

Other viral replication systems include a larger number of accessory replication proteins. Herpes simplex virus type 1 genes encoding essential replication proteins have been identified by both genetic methods and a DNA-mediated transformation assay for the gene products necessary for plasmid replication directed by a viral origin (Fig. 9.19). Replication from a herpes simplex virus type 1 origin requires not only the viral DNA polymerase and origin recognition protein, but also five other viral proteins (Table 9.4). These proteins are functional analogs of essential components of the cellular replication machinery. The products of the UL5, UL8, and UL52 genes form a viral primase, which also functions as a helicase, and the UL42 protein is a processivity factor. Although the herpesviral proteins listed in Table 9.5 provide a large repertoire of replication functions, they do not support viral DNA synthesis *in vitro*. Not all the additional viral and/or cellular proteins needed to reconstitute herpesviral DNA synthesis have

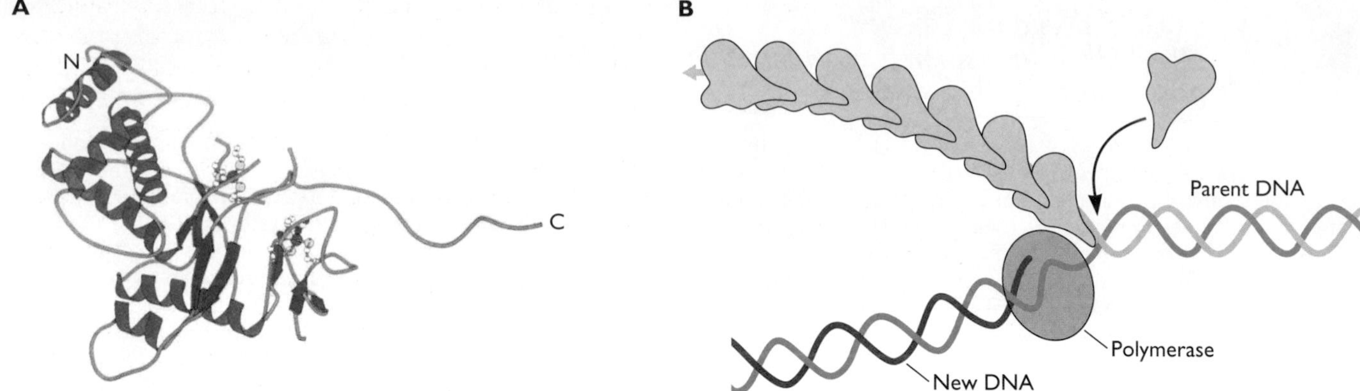

Figure 9.18 Crystal structure of the adenoviral single-stranded-DNA-binding protein. (A) Ribbon diagram of the C-terminal nucleic acid-binding domain (amino acids 176 to 529) of the human adenovirus type 5 protein, showing the two sites of Zn^{2+} (red atom) coordination. The most prominent feature is the long (~40-Å) C-terminal extension. This C-terminal extension of one protein molecule invades a cleft between two α-helices in its neighbor in the protein array formed in the crystal. Deletion of the C-terminal 17 amino acids of the DNA-binding protein fragment eliminates cooperative binding of the protein to DNA, indicating that the interaction of one molecule with another via the C-terminal hook is responsible for cooperativity in DNA binding. From P. A. Tucker et al., *EMBO J.* **13**:2994–3002, 1994, with permission. Courtesy of P. C. van der Vliet, Utrecht University. **(B)** Model of unwinding of double-stranded adenoviral DNA by cooperative interactions among the viral single-stranded-DNA-binding protein.

yet been identified. However, cellular topoisomerase II is essential for replication in infected cells.

Resolution and Processing of Viral Replication Products

Several of the viral DNA replication mechanisms described in preceding sections yield products that do not correspond to the parental viral genome. As we have seen, replication of simian virus 40 DNA yields two interlocked, double-stranded, circular DNA molecules that must be separated by cellular topoisomerase II. Such resolution is required whenever circular templates (e.g., episomal papillomavirus or Epstein-Barr virus DNA) are replicated as monomers. In other cases, replication yields multimeric DNA molecules, from which linear genomes of fixed length and sequence must be processed for packaging into virions. This situation is exemplified by the herpes simplex virus type 1 genome.

The products of herpesviral DNA synthesis are head-to-tail concatemers containing multiple copies of the viral genome. It is well established that the linear viral DNA genomes that enter infected cell nuclei at the start of a productive infection are converted rapidly to "endless" molecules; that is, that the DNA termini are joined together. This reaction requires cellular DNA ligase IV, which normally mediates joining of nonhomologous DNA ends during a cellular repair process. On the other hand, there is no consensus as to whether joining of the herpes simplex viral DNA produces unit-length circles (as found in latently infected cells) or linear concatemers: the large size of the viral genome and the presence of repeated sequences at the termini (Fig. 9.20) has made it difficult to provide an unambiguous experimental demonstration of which type of "endless" structure is formed (Box 9.7). This distinction is of more than esoteric interest, as the configuration of the parental DNA has profound implications about the mechanism of viral DNA synthesis. When a genome is circular, concatemers can be synthesized by the rolling-circle mechanism (Box 9.8), as during replication of the double-stranded DNA genome of bacteriophage lambda. In contrast, recombination is required to produce longer-than-unit-length genomes during replication of a linear template.

However they are made, the concatemeric products of herpesviral DNA synthesis are not simple linear molecules. Rather, they comprise a mixture of structures, including branched (Y- and X-shaped) molecules. A second characteristic feature of herpes simplex virus DNA replication is that it is accompanied by a high degree of recombination. Indeed, conversion of the genome from one of its four isomers to another (Fig. 9.20) occurs by the time that newly replicated DNA can first be detected in infected cells. Such isomerization is the result of recombination between repeated viral DNA sequences (see "Recombination of Viral Genomes" below). These properties suggest that recombination may be an essential reaction during herpes simplex virus type 1 DNA synthesis.

Linear herpes simplex virus type 1 DNA molecules with termini identical to those of the infecting genome are

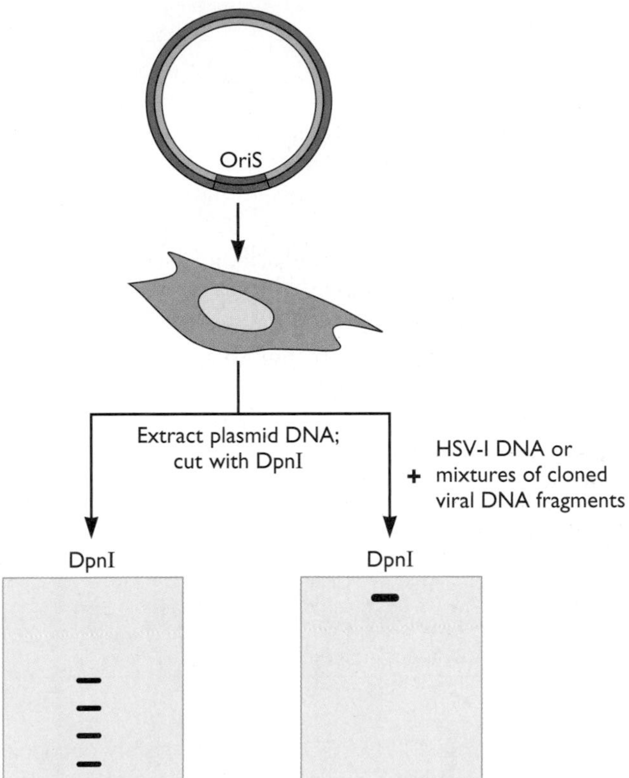

Figure 9.19 DNA-mediated transformation assay for essential herpes simplex virus type 1 replication proteins. A plasmid carrying a viral DNA fragment spanning OriS is introduced into monkey cells permissive for herpesvirus replication. In the absence of viral proteins **(left)**, the plasmid DNA is not replicated and retains the methyl groups added to A residues in a specific sequence by the *E. coli dam* methylation system. As these sequences include the recognition site for the restriction endonuclease DpnI, which cleaves only such methylated DNA, the unreplicated plasmid DNA remains sensitive to DpnI cleavage. When all viral genes encoding proteins required for OriS-dependent replication are also introduced into the cells **(right)**, the plasmid is replicated. Because the newly replicated DNA is not methylated at DpnI sites, it cannot be cleaved by this enzyme. Resistance of the plasmid to DpnI cleavage therefore provides a simple assay for plasmid replication, and hence for the identification of viral proteins required for replication from OriS.

liberated from concatemeric replication products by cleavage at specific sites within the *a* repeats (Fig. 9.11). As described in Chapter 13, such cleavage is coupled with encapsidation of viral DNA molecules during virion assembly.

Mechanisms of Exponential Viral DNA Replication

The details of the mechanisms by which replication of viral DNA genomes is achieved vary considerably from one virus to another. Nevertheless, each of these strategies generally results in efficient replication of viral DNA. Production of 10^4 to 10^5 viral genomes, or more, per infected cell is not uncommon, as the products of one cycle of replication are recruited as templates for the next. Such **exponential** viral DNA synthesis sets the stage for assembly of a large burst of progeny virions. In this section, we discuss regulatory mechanisms that ensure efficient viral DNA synthesis. These regulatory circuits impinge primarily on the expression of cellular or viral genes that encode the proteins that carry out viral DNA synthesis.

Viral Proteins Can Induce Synthesis of Cellular Replication Proteins

With few exceptions, virus reproduction is studied by using established cell lines that are permissive for the virus of interest. Such immortal or transformed cell lines proliferate indefinitely, and differ markedly from the cells in which viruses reproduce in nature. For example, highly differentiated cells, such as neurons or the outer cells of an epithelium, do not divide and are permanently in a specialized resting state, termed the **G_0 state**. Many other cells in an organism divide only rarely, or only in response to specific stimuli, and therefore spend much of their lives in G_0. Such cells do not contain many of the components of the cellular replication machinery, and are characterized by generally low rates of synthesis of RNAs and proteins. Virus reproduction entails the synthesis of large quantities of viral nucleic acids (and proteins), often at a high rate. Consequently, the resting state would not seem to provide a hospitable environment. Nevertheless, viruses often replicate successfully within cells infected when they are in G_0. In some cases, such as replication of several herpesviruses in neurons, the replication machinery is encoded within the viral genome. Infection by other viruses stimulates resting or slowly growing cells to abnormal activity, by disruption of cellular circuits that restrain cell proliferation. This strategy is characteristic of polyomaviruses and adenoviruses. The discovery that infection by these viruses disrupts the same cellular cell growth control circuits was quite unanticipated, and of the greatest importance in elucidating the roles of critical regulators of cell proliferation, such as the cellular retinoblastoma (Rb) protein.

Functional Inactivation of the Rb Protein

Loss or mutation of both copies of the cellular retinoblastoma (*rb*) gene is associated with the development of tumors of the retina in children and young adults. Because it is the **loss** of normal function that leads to tumor formation, *rb* is defined as a **tumor suppressor gene**. As discussed in Volume II, Chapter 7, the Rb protein is an important component of the regulatory program that ensures that cells grow, duplicate their DNA, and divide in an orderly manner. In particular, the Rb protein controls entry into the period of

Table 9.5 Viral enzymes of nucleic acid metabolism

Virus	Protein	Functions
Herpesvirus		
Herpes simplex virus type 1	Thymidine kinase (UL23 protein, ICP36)	Phosphorylates thymidine and other nucleosides; essential for efficient replication in animal hosts
	Ribonucleotide reductase ($\alpha_2\beta_2$ dimer of UL39 and U40 proteins)	Reduces ribose to deoxyribose in ribonucleotides; essential in nondividing cells
	dUTPase (UL50 protein)	Hydrolyzes dUTP to dUMP, preventing incorporation of dUTP into DNA and providing dUMP for conversion to dTMP
	Uracil DNA glycosylase	Corrects insertion of dUTP or deamination of C in viral DNA
	Alkaline nuclease (UL12 protein)	Required for production of infectious DNA
Poxvirus		
Vaccinia virus	Thymidine kinase	Phosphorylates thymidine; required for efficient virus reproduction in animal hosts
	Thymidylate kinase	Phosphorylates TMP
	Ribonucleotide reductase, dimer	Reduces ribose to deoxyribose in ribonucleotides; essential in nondividing cells
	dUTPase	Hydrolyzes dUTP to dUMP (see above)
	DNase	Has nicking-joining activity; present in virion cores
	D4R protein	Uracil DNA glycosylase

the cell cycle in which DNA is synthesized, the **S phase**, from the preceding (G_1) phase. Our current appreciation of the critical participation of this protein in the control of cell cycle progression, and of the mechanism by which it operates, stems from the discovery that Rb binds directly to the two adenoviral E1A proteins (see Chapter 8).

In uninfected cells in the G_1 phase, the Rb protein is bound to cellular transcriptional regulators of the E2f family. These complexes, which bind to specific promoters via the DNA-binding activity of E2f, function as repressors of transcription (Fig. 9.21A). Binding of adenoviral E1A proteins (or of simian virus 40 LT or E7 proteins of highly oncogenic human papillomaviruses) to Rb releases E2f from this association and sequesters Rb. The E2f proteins therefore become available to stimulate transcription of cellular genes that encode proteins that participate directly or indirectly in DNA synthesis, or in control of cell cycle progression (Fig. 9.21A; Volume II, Chapter 7).

The benefits conferred by activation of E2f when a polyomavirus or an adenovirus infects a cell that is proliferating slowly, or in G_0, are virus specific. While the advantages for viral DNA replication can be deduced, it is important

Figure 9.20 Isomers of the herpes simplex virus type 1 genome. The organization of the unique and repeated sequences of the viral genome are depicted at the top, as in Fig. 9.11. This orientation is defined as the prototype (P) genome isomer. The other three isomers differ, with respect to the P form, in the orientation of S (IS), in the orientation of L (IL), or in both S and L (ISIL). These differences are illustrated using HindIII fragments. The unusual isomerization of this viral genome was deduced from the presence of fragments that span the terminal or internal inverted repeat sequences at 0.5 and 0.25 molar concentrations, respectively, in such HindIII digests, and examination of partially denatured DNA in the electron microscope.

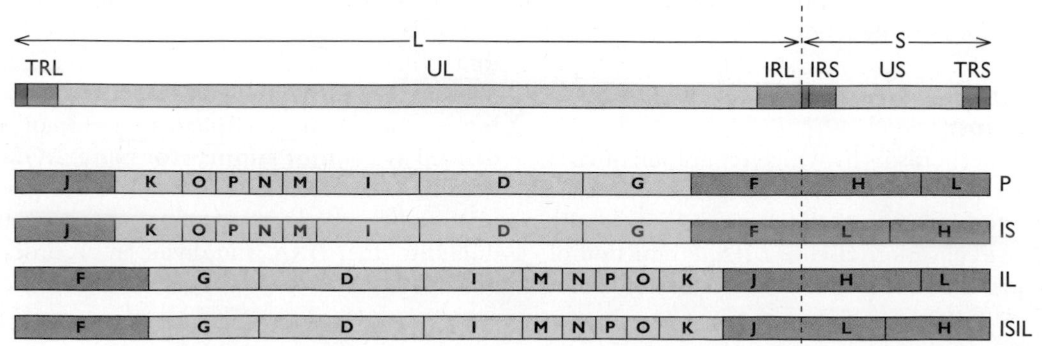

BOX 9.7

DISCUSSION

Circularization or concatemerization of the herpes simplex virus type 1 genome in productively infected cells?

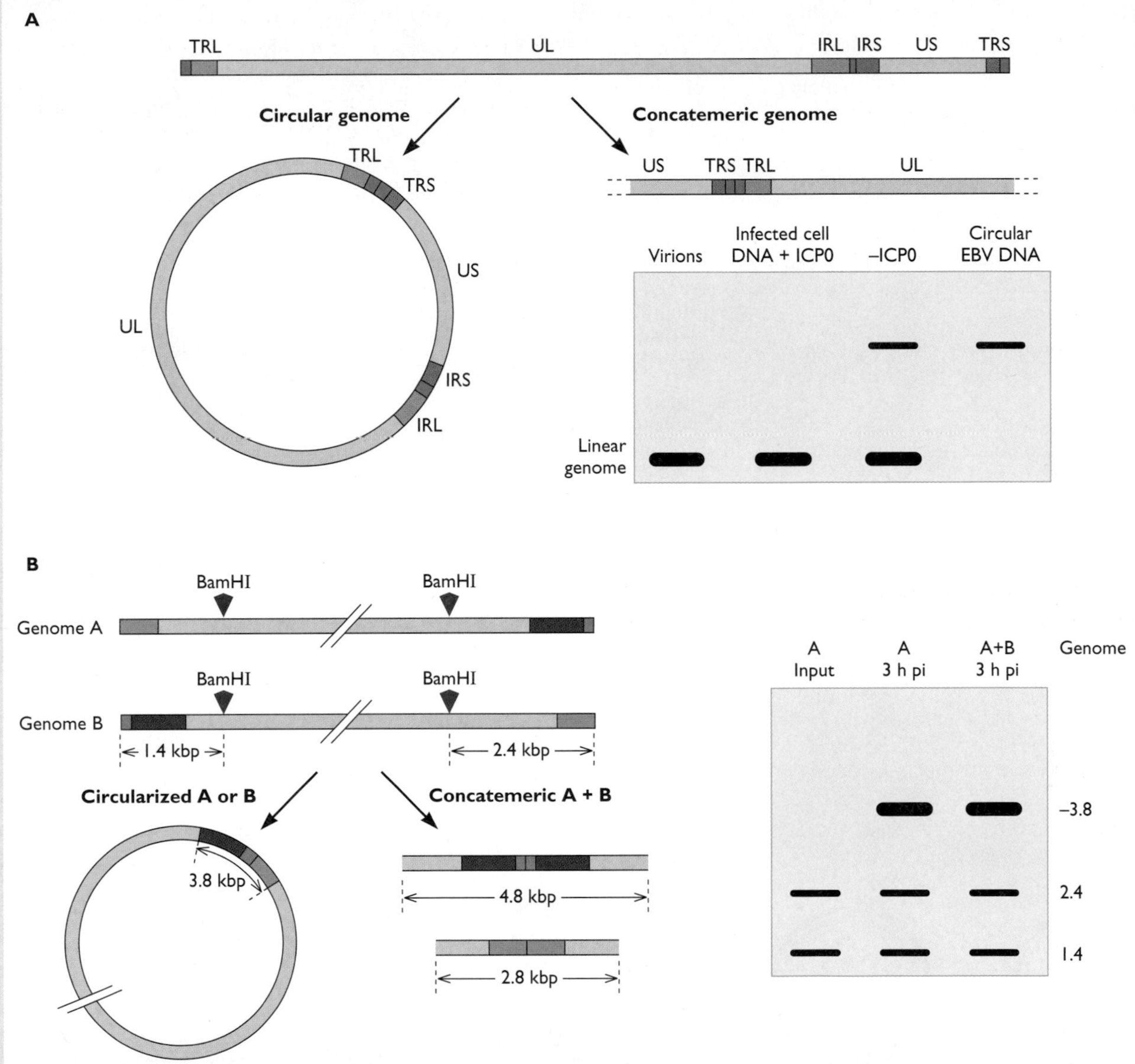

(A) Evidence for genome concatemerization. These experiments exploited a method of electrophoresis in which infected cells (or nuclei) are lysed in the initial portion of the gel (to minimize damage of the DNA) prior to electrophoresis for a very long period. The viral DNA was then detected by hybridization to labeled viral DNA, following transfer to a membrane. As summarized in the figure, a fraction of infecting viral DNA molecules was detected as a species of low mobility that migrated in similar fashion to circular Epstein-Barr virus DNA. Such molecules were detected only in the absence of the immediate-early gene product ICP0. No such molecules were observed in cells infected by wild-type virus, in the absence or presence of an inhibitor of viral DNA synthesis. These observations led to the conclusion that the herpesviral genome circularized in the absence of ICP0 (as in latently infected cells), but not during productive infection. Consequently, linear viral DNA molecules would serve as the templates for initial viral DNA synthesis. **(B) Evidence for a circular genome.** These experiments exploited mutant viruses that lacked all a repeats (the packaging signal) and contained a minimal packaging signal at an ectopic site. As illustrated, cleavage of the linear mutant genomes designated A and B by BamHI yields terminal fragments of 1.4 and 2.4 kb. Circularization of either genome generates a new, unique junction fragment of 3.8 kb, whereas head-to-tail concatemerization of A with B would produce distinguishable BamHI fragments. The fragment characteristic of circularization was detected within 1 h of infection by either virus in cells that synthesized ICP0. In contrast, neither of the junction fragments diagnostic of concatemers were observed in cells coinfected by the two viruses. These observations led to the conclusion that the genome circularizes very soon after entry into infected cell nuclei, and in this form serves as a template for initial viral DNA synthesis.

(continued)

BOX 9.7

DISCUSSION

Circularization or concatemerization of the herpes simplex virus type 1 genome in productively infected cells? (continued)

It has been known for some time that the linear herpes simplex virus type 1 DNA that enters nuclei of productively infected cells rapidly adopts a new conformation in which the termini are fused: restriction endonuclease cleavage of viral DNA recovered shortly after infection established that cleavage products generated from free ends decrease in concentration as those characteristic of joined ends increase. However, the origin of the latter fragments has been difficult to determine, for several reasons.

- Formation of either unit length circles or concatemers results in loss of free termini.
- The presence of an internal inverted copy of the joined terminal repeats precludes the use of an assay based on detection of joined termini.
- Conventional methods for separation and identification of linear and circular DNA molecules by electrophoresis cannot be applied to the large herpesviral genome.
- Because of its large size, the herpesviral genome may be easily damaged during extraction from infected cells.
- Under conditions that facilitate detection of viral DNA, high multiplicity of infection, the majority of infecting DNA molecules may be neither transcribed nor replicated.

These different conclusions have yet to be reconciled. Complications of approach A in the figure include the difficulty of interpreting electrophoretic mobility in terms of specific DNA structures and the possible trapping of molecules with complicated structure in the wells of the gel. Approach B (see figure) appears straightforward. However, the mutant genomes lack sequences present in wild-type DNA, and the authors could not detect circular genomes by the electrophoretic method.

Jackson, S. A., and N. A. DeLuca. 2003 Relationship of herpes simplex virus genome configuration to productive and persistent infections. *Proc. Natl. Acad. Sci. USA* **100:**7871–7876.

Strang, B. L., and N. D. Stow. 2005 Circularization of the herpes simplex virus type 1 genome upon lytic infection. *J. Virol.* **79:**12487–12494.

BOX 9.8

BACKGROUND

Rolling-circle replication

The rolling-circle replication mechanism of DNA synthesis was discovered during studies of the replication of the single-stranded DNA genome of bacteriophage ϕX174. However, it also operates during replication of double-stranded genomes, such as that of bacteriophage lambda.

Rolling-circle replication is initiated by introduction of a nick that creates a 3′-OH end into one strand of a double-stranded circular DNA. One strand of the template is copied continuously, and multiple times, while the displaced strand is copied discontinuously. As shown, this mechanism produces genome concatemers.

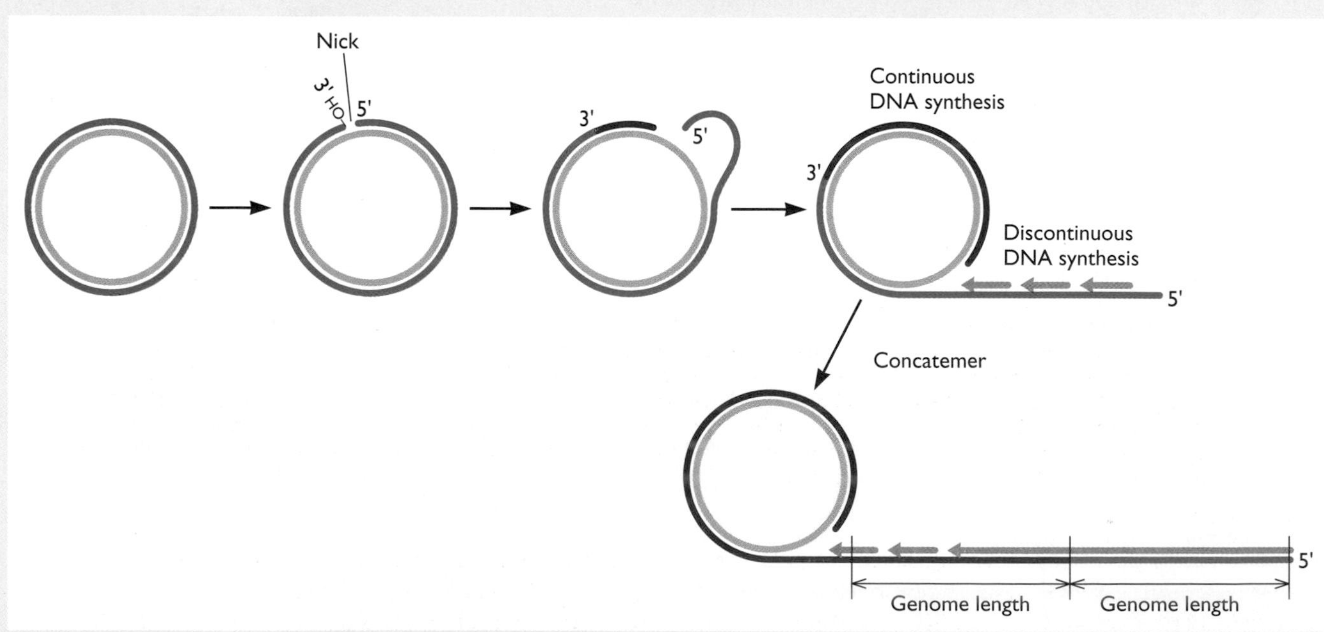

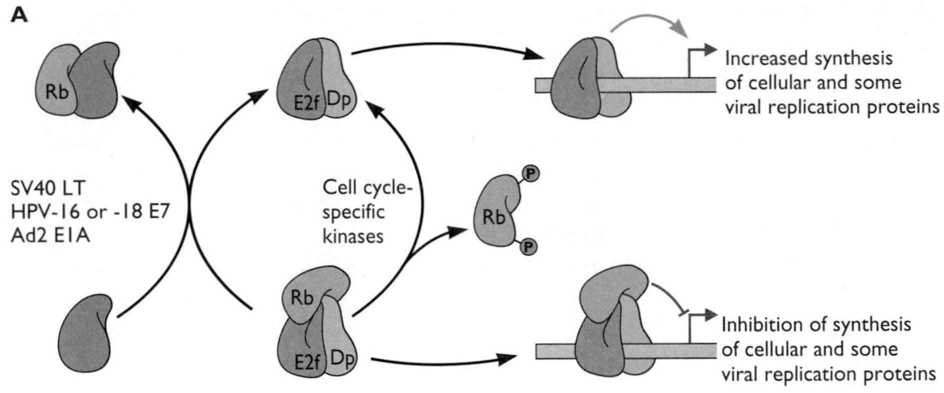

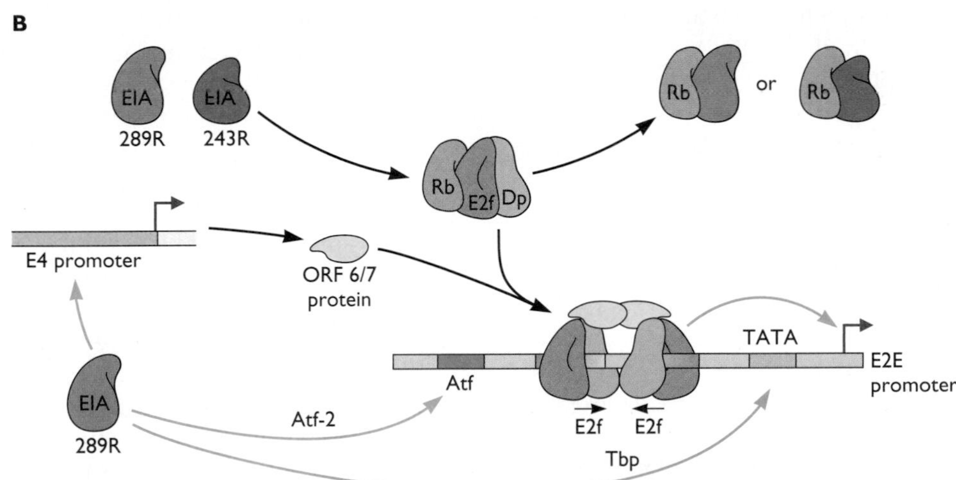

Figure 9.21 Regulation of production of cellular and viral replication proteins. (A) Model for the abrogation of the function of the Rb protein by viral proteins. E2f transcriptional regulators are heterodimeric proteins, each containing one E2f and one Dp subunit. E2f dimers stimulate transcription of cellular genes encoding replication proteins, histones, and proteins that allow passage through the cell cycle (green arrow). Binding of Rb protein does not prevent promoter recognition by E2f. However, Rb protein represses transcription (red bar). Phosphorylation of Rb protein at specific sites induces its dissociation from E2f, and activates transcription of cellular genes expressed in S phase. The adenoviral E1A proteins, simian virus 40 LT, and the E7 proteins of certain human papillomaviruses (types 16 and 18) bind to the region of Rb protein that contacts E2f to disrupt Rb-E2f complexes and activate E2f-dependent transcription. **(B)** Stimulation of transcription from the adenoviral E2 early promoter by E1A proteins. The E2E promoter-binding sites for the cellular Atf, E2f, and TfIId proteins are necessary for E2E transcription in infected cells. The inversion of the two E2f sites (arrows) and their precise spacing are essential for assembly of an E2f-DNA complex unique to adenovirus-infected cells, in which the E4 Orf6/7 protein is bound to each E2f heterodimer. Binding of the E4 protein promotes cooperative binding of E2f, and increases the lifetime of E2f-DNA complexes. The availability of the cellular E2f and viral E4 Orf6/7 proteins is a result of the action of immediate-early E1A proteins: either the 243R or 289R protein can sequester unphosphorylated Rb to release active E2f from Rb-E2f complexes, and the 289R protein stimulates transcription from the E4 promoter. This larger E1A protein can also stimulate transcription from the E2E promoter directly.

to stress that essentially all studies of the cell growth control functions of LT, E7, or E1A proteins have been directed toward elucidation of mechanisms of transformation. For example, it has been shown that the smaller adenoviral E1A protein is necessary for efficient viral DNA synthesis in quiescent human cells. However, it has not been established that the E1A protein plays a critical role in activation of cellular E2f, as assumed in the model described below.

E2f and simian virus 40 DNA synthesis. Expression of the cellular genes encoding DNA polymerases, accessory replication proteins, histones, and enzymes that synthesize substrates for DNA synthesis is tightly controlled during the cell cycle: these proteins are made only just before they are needed in S phase. The transcriptional control regions of many of these genes contain E2f-binding sites, which are required for their efficient transcription. Synthesis of simian virus 40 LT is therefore thought to induce production of the cellular proteins necessary for viral DNA synthesis, and hence to ensure efficient replication of the viral genome regardless of the growth state of the host cell. The ability of simian virus 40 LT both to induce synthesis of components of the cellular replication machinery and to initiate viral DNA synthesis seems likely to coordinate viral replication with the entry of the host cell into S phase. Indeed, as discussed previously, LT replication functions are regulated by phosphorylation, probably by the cyclin A–cyclin-dependent kinase 2 that accumulates as cells enter S phase.

E2f and adenoviral DNA synthesis. Activation of E2f in adenovirus-infected cells is believed to promote viral DNA synthesis in two ways: stimulation of transcription of the cellular genes encoding enzymes that make substrates for DNA synthesis, such as thymidine kinase and dihydrofolate reductase, and activation of production of all three viral replication proteins. The viral DNA polymerase, preterminal protein primer, and DNA-binding protein are encoded within the E2 gene, which is transcribed from an early promoter that contains two binding sites for E2f (Fig. 9.21B). In fact, these critical cellular regulators derive their name from these E2-binding sites, which are necessary for efficient E2 transcription during the early phase of infection. As discussed previously, the viral E1A proteins disrupt Rb-E2f complexes and sequester Rb. They also regulate transcription from the E2 promoter by two other mechanisms (Fig. 9.21B). These E1A-dependent regulatory mechanisms presumably operate synergistically, to allow synthesis of the viral mRNAs encoding replication proteins in quantities sufficient to support numerous cycles of viral DNA synthesis. A posttranscriptional regulatory mechanism may allow production of the appropriate concentrations of the three E2 replication proteins (Box 9.9).

Synthesis of Viral Replication Machines and Accessory Enzymes

The DNA genomes of several viruses, exemplified by that of herpes simplex virus type 1, encode large cohorts of proteins that participate directly or indirectly in viral DNA synthesis. Two classes of such replication proteins are synthesized during productive infection. The first comprises the viral DNA polymerase and other proteins responsible for viral DNA synthesis discussed previously (Table 9.4). The second class contains enzymes, analogous to host cell enzymes, that catalyze reactions by which dNTPs are synthesized (Table 9.5). For example, viral thymidine kinase and ribonucleotide reductase are synthesized in herpesvirus-infected cells. In general, members of this second class of proteins are dispensable for replication in proliferating cells in culture and in animal hosts. However, herpes simplex viruses that lack thymidine kinase or ribonucleotide reductase genes cannot replicate in neurons: such terminally differentiated cells are permanently withdrawn from the cell cycle, and do not make enzymes that produce substrates for DNA synthesis.

Efficient synthesis of all herpes simplex virus type 1 replication proteins is primarily the result of the viral transcriptional cascade described in Chapter 8. Expression of the genes encoding these viral proteins, which are early (β) genes, is regulated by products of immediate-early genes. These immediate-early proteins operate transcriptionally (e.g., ICP0 and ICP4) or posttranscriptionally (e.g., ICP27) to induce synthesis of viral replication proteins at concentrations sufficient to support exponential replication of viral DNA. Whether the multiple origins of herpes simplex virus type 1 and other herpesviruses increase the rate of replication will not be clear until the viral replication mechanism has been better characterized.

Viral DNA Replication Independent of Cellular Proteins

One method guaranteed to ensure replicative success of a DNA virus, regardless of the growth state of the host cell, is encoding of **all** necessary proteins in the viral genome. On the other hand, this mechanism is genetically expensive, which may be the reason why it is restricted to the viruses with the largest DNA genomes, such as the poxvirus vaccinia virus. The genome of this virus, which is replicated in the cytoplasm at specialized foci called **viral factories**, encodes all the proteins needed for viral DNA synthesis. The current catalog of such proteins includes a DNA polymerase and several accessory replication proteins (Table 9.4). The vaccinia virus genome also encodes several other enzymes that would be expected to participate in DNA synthesis or in resolution of replication products, including a type I topoisomerase and a DNase with endonucleolytic activity, as well as several enzymes for synthesis of dNTPs (Table 9.5). None of the latter appear to be essential for virus reproduction in actively growing cells. However, several enzymes, such as the thymidine kinase, are necessary for efficient virus propagation in quiescent cells or animal hosts, where they presumably increase less than optimal substrate pools. Much remains to be learned about the mechanism by which the vaccinia virus genome is replicated and the functions of viral replication proteins.

BOX 9.9

DISCUSSION

Production of appropriate quantities of the adenoviral replication proteins by differential polyadenylation?

The three adenoviral replication proteins, Pol, pTP, and DBP, are all encoded within the E2 transcription unit of the viral genome. However, they are needed in very different quantities during viral DNA synthesis. The replication of one molecule of viral DNA requires four molecules each of Pol and the pTP primer (Fig. 9.10). Once the daughter strands have been synthesized, Pol dissociates and can catalyze additional replication cycles. This protein is therefore required at **catalytic** concentrations. In contrast, the pTP remains covalently attached to the 5′ end of each newly synthesized DNA strand and is incorporated into progeny virus particles. This protein is therefore required at a greater concentration than Pol. The DBP coats the single-stranded DNA displaced during viral DNA synthesis (Fig. 9.10). This protein, which interacts with about 12 nucleotides, binds cooperatively to such DNA to form a long chain (Fig. 9.18). Thousands of molecules of DBP are needed during replication of a single molecule of the viral genome.

As shown in the figure, E2 primary transcripts can be polyadenylated at either a promoter-proximal (DBP mRNA) or a promoter-distal (Pol and pTP mRNAs) site. The former appears to be utilized much more frequently than the latter: DBP mRNA accumulates to 10-to 20-fold-higher concentrations than the Pol and pTP mRNAs. In this way, the replication proteins can be made in the appropriate relative concentrations, even though they are encoded in the same transcription unit.

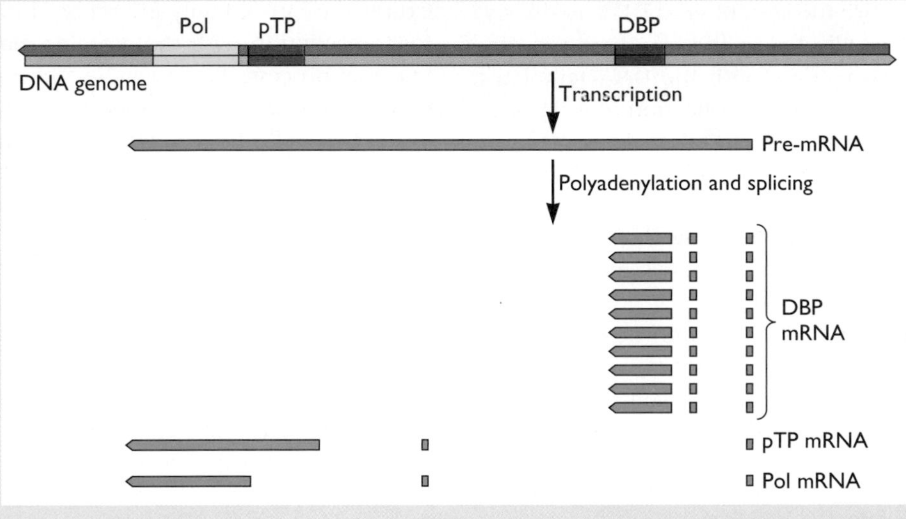

Delayed Synthesis of Virion Structural Proteins Prevents Premature Packaging of DNA Templates

During productive infection by DNA viruses, each cycle of replication increases the number of DNA molecules that can be copied in the subsequent cycle. This increase in the pool of replication templates, a doubling in each cycle of the simpler viruses like simian virus 40 and adenovirus, undoubtedly makes an important contribution to rapid amplification of genomes. Progeny viral DNA molecules are eventually encapsidated during the assembly of new virus particles, and consequently become unavailable to serve as templates for DNA synthesis. However, assembly and sequestration of the genome are delayed with respect to initiation of viral DNA synthesis, because transcription of late genes encoding virion proteins cannot begin until viral DNA has been replicated (Chapter 8).

Inhibition of Cellular DNA Synthesis

When viral DNA replication is carried out largely by viral proteins, cellular DNA synthesis is often inhibited, presumably to increase the availability of critical substrates. Infection by the larger DNA viruses (herpesviruses and poxviruses) induces severe inhibition of synthesis of cellular DNA. Cellular DNA synthesis is also blocked when adenoviruses infect proliferating cells in culture, despite the elaborate mechanisms by which these viruses induce quiescent host cells to reenter the cell cycle. Inhibition of cellular DNA synthesis in cells infected by these DNA viruses was described in some of the earliest studies of their infectious cycles. However, very little is known about the mechanisms that shut down this cellular process.

There is accumulating evidence that inhibition of cellular DNA synthesis is an active process, rather than an indirect result of passive competition between viral and cellular DNA polymerases for the finite pools of dNTP substrates. For example, infection of proliferating cells by adenovirus or the herpesvirus human cytomegalovirus has been reported to induce cell cycle arrest. Synthesis of the Epstein-Barr virus Zta protein, a sequence-specific transcriptional regulator and origin-binding protein, also arrests cells at a point in the cell cycle prior to S phase, such that cellular DNA synthesis is precluded. In this case, arrest is the result of increased concentrations of cellular proteins that negatively regulate progression through the cell cycle, such as the Rb protein.

Viral DNAs Are Synthesized in Specialized Intracellular Compartments

A common, probably universal, feature of cells infected by viruses with DNA genomes is the presence of virus-specific structures that are the sites of viral DNA synthesis. Vaccinia virus DNA is replicated in the cytoplasm, in discrete viral factories that contain both the DNA templates and viral replication proteins. The replication of viral DNA genomes within infected cell nuclei also takes place in specialized compartments, which can be visualized as distinctive, infected cell-specific structures containing viral proteins (Fig. 9.22). Such structures, known as **replication centers** or **replication compartments**, have been best characterized in human cells infected by adenovirus or herpes simplex virus type 1. They contain newly synthesized viral DNA and all the viral proteins necessary for viral DNA synthesis, as well as other viral and cellular proteins.

Figure 9.22 Nuclear replication compartments of herpes simplex virus type 1-infected monkey cells. The locations of the viral ICP8 single-stranded-DNA-binding protein and newly replicated DNA were visualized 6 h after infection by indirect immunofluorescence with antibodies against ICP8 and bromodeoxyuridine incorporated into DNA, respectively. Newly synthesized viral DNA and ICP8 colocalize to a limited number of globular patches, which contain all seven known viral replication proteins. Courtesy of Lindsey Silva and David Knipe, Harvard Medical School.

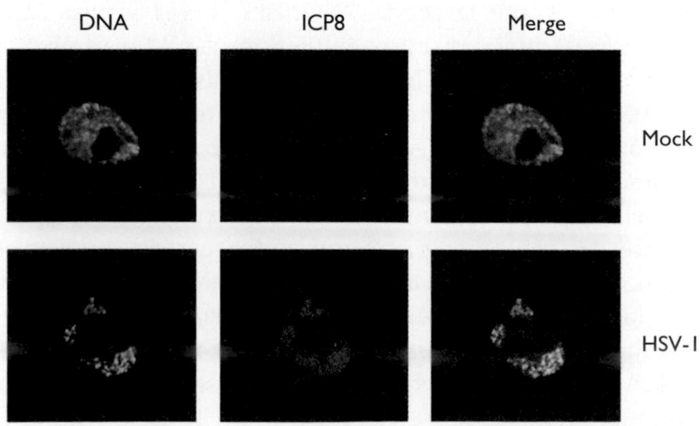

Among the cellular proteins associated with herpes simplex virus type 1 replication centers are several enzymes, such as DNA polymerases α and γ, and topoisomerase II, and numerous DNA repair and recombination proteins.

The localization of both the templates for viral DNA synthesis and the replication proteins at a limited number of sites undoubtedly facilitates exponential viral DNA replication: this arrangement increases the local concentrations of proteins that must interact with one another, or with viral origin sequences or replication forks, favoring such intermolecular interactions by the law of mass action. In addition, the high local concentrations of replication templates and proteins are likely to allow for efficient recruitment of the products of one replication cycle as templates for the next.

Viral replication centers also serve as foci for viral gene expression, presumably in part by concentrating templates for transcription with the proteins that carry out or regulate this process. For example, the herpes simplex virus type 1 immediate-early ICP4 and ICP27 proteins, as well as the host cell's RNA polymerase II, are recruited to these nuclear sites. Similarly, an adenoviral early protein complex necessary for selective export of viral late mRNAs from the nucleus (Chapter 10) is associated with the periphery of viral replication centers, as is nascent viral RNA.

Viral replication centers do not assemble at random sites, but rather are formed by viral colonization of specialized niches within mammalian cell nuclei. When they enter the nucleus, infecting adenoviral or herpes simplex virus type 1 genomes (as well as those of papillomaviruses and polyomaviruses) localize to preexisting nuclear bodies that contain specific cellular proteins, the promyelocytic leukemia proteins (Pmls). These structures are therefore called **Pml bodies**, or nuclear domain 10 structures, a name derived from the average number present in most cells. Viral proteins then induce reorganization of Pml bodies as viral replication centers are established (Fig. 9.23). For example, the herpes simplex virus type 1 ICP0 protein causes disruption of these cellular structures. This viral protein is an E3 ubiquitin ligase. Such enzymes catalyze addition of polyubiquitin chains to proteins, a modification that targets them for degradation by the proteasome. ICP0 induces the degradation of several components of Pml bodies, including some Pml proteins. Others, including specific isoforms of the Pml protein, Daxx, and many proteins that participate in repair, recombination, and chromatin remodeling, become associated with viral replication centers. Recruitment of specific proteins present in Pml bodies to viral replication centers also occurs in adenovirus-infected cells, and requires the E4 Orf 3 protein.

The association of replication centers of different DNA viruses with constituents of the same intranuclear

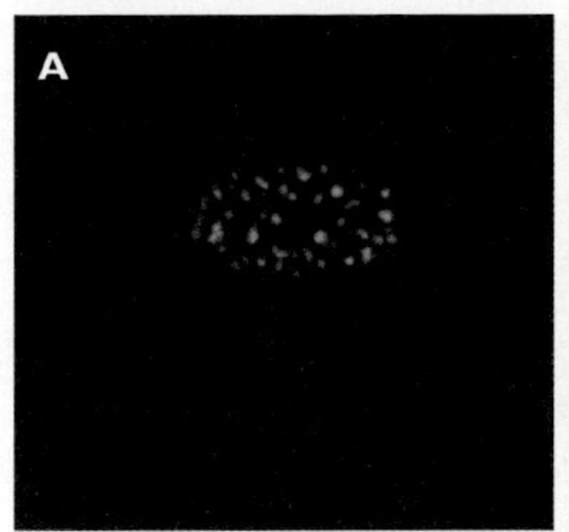

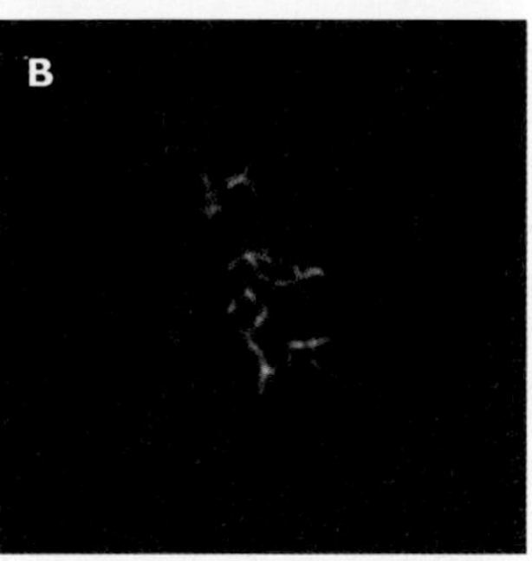

Figure 9.23 Reorganization of Pml bodies by the adenoviral E4 Orf3 protein. Plasmids for expression of human Pml isoforms I to VI linked to a FLAG tag were introduced in *Pml*-null mouse embryo fibroblasts in the absence **(A)** or presence **(B)** of an expression vector for the viral E4 Orf3 protein. The Pml (green) and E4Orf3 (red) proteins were visualized by using indirect immunofluorescence. All Pml isoforms formed Pml-like bodies alone, but only structures containing PmlI were disrupted and reorganized by the E4 Orf 3 protein. Adapted from A. Hoppe et al., *J. Virol.* **80**:3042–3049, 2006, with permission. Courtesy of K. Leppard, University of Warwick, Coventry, United Kingdom.

structures suggests that reorganization of host cell nuclei facilitates viral DNA synthesis. The discovery that viral DNA genomes home to Pml bodies has stimulated characterization of their constituents, but much remains to be learned about the molecular functions of these nuclear structures. There is accumulating evidence that Pml bodies represent a form of intrinsic antiviral defense (Volume II, Chapter 3). However, other advantages conferred by the degradation or dispersal of Pml body proteins are likely to be virus specific. For example, the human papillomavirus type 18 E6 protein induces proteasomal degradation of a Pml isoform (Pml-IV) that causes primary human cells to become senescent (a state in which cellular proteins required for replication of the viral genome are not made). In cells infected by adenovirus, alterations in specific Pml components block disadvantageous concatemerization of the viral genome (see below). In contrast, herpesviral DNA synthesis may require cellular repair and recombination proteins that become relocalized to viral replication centers. Furthermore, it has been reported recently that formation of herpesviral replication compartments represents but an early step in the massive alterations in nuclear architecture necessary for egress of newly formed nucleocapsids (see Chapter 13).

Limited Replication of Viral DNA

Exponential replication of viral DNA is the typical pattern when the majority of DNA viruses infect cells in culture. Nevertheless, several can establish long-term relationships with their hosts and host cells, in which the number of genomes produced is limited. Various mechanisms that effect such copy number control are described in this section.

Integrated Parvoviral DNA Can Replicate as Part of the Cellular Genome

The adeno-associated viruses reproduce only in cells coinfected with a helper adenovirus or herpesvirus. Although the latter viruses are widespread in hosts infected by adeno-associated viruses, the chances that a particular host cell will be infected simultaneously by two viruses are very low. The strategy of exploiting other viruses to provide functions for efficient expression of the genetic information of adeno-associated viruses would therefore appear to be potentially lethal for individual virus particles. In fact, this is not the case, for adeno-associated virus can survive in the absence of a helper virus by an alternative mechanism: its genome becomes integrated into that of the host cell, and is replicated as part of a cellular replicon.

This program for long-term survival of the adeno-associated virus genome depends on expression of its regulatory region (Rep) (Box 9.4). The two larger proteins encoded by this region, Rep 78/68, are multifunctional and control all phases of the viral life cycle (Table 9.3). When helper virus proteins, such as adenoviral E1A, E1B, and E4 proteins, allow synthesis of large quantities of Rep 78/68, adeno-associated virus DNA is replicated by the mechanism described previously. In the absence of helper functions, only very small quantities of Rep 78/68 are made, even in healthy, dividing cells. Consequently, there is little viral DNA synthesis, and the genome is integrated into that of the host cell. The latter reaction is mediated by Rep 78/68.

One of the most unusual features of this integration reaction is that it occurs preferentially near one end of human chromosome 19. It was believed for many years that integration required the recognition of the viral ITR origin (Fig. 9.12) by Rep 78/68. However, the observation that integration of DNA molecules containing only the ITR was exceedingly inefficient led to the identification of a viral sequence that increased the frequency of site-specific integration by up to 100-fold. This sequence corresponds to the viral origin that overlaps the p5 promoter (Fig. 9.12) and is, in fact, sufficient for integration. The Rep 78/68 protein can bind simultaneously to both viral and human chromosomal 19 DNA sequences required for integration, at least *in vitro*. The current model of integration therefore proposes that its specificity is the result of such simultaneous binding to the two DNA molecules by Rep 78/68. Rep mediates nonhomologous recombination reactions that result in integration of the viral genome, with concomitant, large deletions or duplications of cellular DNA.

Following high-multiplicity infection of cells in culture, as many as 40% of the cells contain the integrated adeno-associated virus genome. However, the results of more recent studies indicate that site-specific integration is very rare when the virus infects humans or laboratory animals. Rather, the viral genome can persist in various tissues, for example tonsil and lung in humans, as a double-stranded, circular episome. The mechanisms by which such episomes are produced from the linear, single-stranded DNA genome are not yet clear.

Regulation of Replication via Different Viral Origins: Epstein-Barr Virus

During herpesviral latent infections, the viral genome is stably maintained at low concentrations, often for long periods (Volume II, Chapter 5). Furthermore, replication of viral and cellular genomes can be coordinated. This pattern is characteristic of human B cells latently infected by Epstein-Barr virus. Many such cell lines have been established from patients with Burkitt's lymphoma, and this state is the usual outcome of infection of B cells in culture. Characteristic features of latent Epstein-Barr virus infection include expression of only a small number of viral genes, the presence of a finite number of viral genomes, and replication from a specialized origin. Because replication from this origin, which is not active in lytically infected cells, is responsible for maintenance of episomal viral genomes, it is termed the **origin for plasmid maintenance** (OriP).

The Epstein-Barr virus genome is maintained in nuclei of latently infected cells as a stable circular **episome**, present at 10 to 50 copies per cell. For example, one Burkitt's lymphoma cell line (Raji) has carried about 50 copies per cell of episomal viral DNA for over 40 years. When Epstein-Barr virus infects a B cell, the linear viral genome circularizes by a mechanism that is not well understood. The circular viral DNA is then amplified during S phase of the host cell to the final concentration listed above. Such replication is by the cellular DNA polymerases and accessory proteins that synthesize simian virus 40 DNA. However, it also requires OriP (Fig. 9.24A), and the viral protein that binds specifically to it, EBNA-1 (Table 9.3), which is invariably synthesized in latently infected cells. In contrast to exponential replication, such amplification of the episomal viral genome is limited to a few cycles. Following such limited amplification, the viral DNA is duplicated once per cell cycle, in S phase, such that its concentration is maintained as the host lymphocyte divides. The EBNA-1 protein and OriP are sufficient for both such once-per-cell cycle replication and orderly segregation of viral genomes to daughter cells.

The availability of cellular replication proteins only in late G_1 and S can account for the timing of Epstein-Barr virus replication in latently infected cells. However, this property **cannot** explain why each genome is replicated only once in each cell cycle. Such controlled initiation of replication is analogous to the tight control of initiation of replication from cellular origins, each of which also fires **once and only once** in each S phase. The mechanisms that control such once-per-cycle firing of eukaryotic origins, a process termed **replication licensing**, were initially elucidated in budding yeasts, which contain compact origins of replication. Mammalian homologs of the yeast origin recognition complex (Orc) and proteins that regulate initiation of DNA synthesis, such as Mcm and Cdc6, have been identified in all other eukaryotes examined. The human Orc proteins are associated with OriP and can bind to EBNA-1. Experimental manipulations that reduce the concentrations of specific Orc proteins severely inhibit OriP-dependent replication. Human Mcm proteins are also associated with OriP during G_1 phase. However, they are not present at OriP during G_2, as would be expected if replication from OriP were licensed by the mechanism shown in Fig. 9.24B. Overproduction of a protein that prevents recruitment of Mcm inhibits the OriP-dependent replication complex (Fig. 9.24B). This observation provides strong support for the conclusion that synthesis of viral DNA genomes in latently infected cells is governed by the mechanisms that ensure once-per-cell-cycle firing of cellular origins.

In addition to EBNA-1-binding sites, OriP contains three copies of a nonameric sequence that resemble repeated sequences present in telomeres. In the presence of EBNA-1, several cellular proteins bind to the repeats. Such cellular proteins include telomerase-associated poly(ADP-ribose) polymerases and telomere repeat-binding protein 2. These protein regulate OriP-dependent replication negatively and positively, respectively, but how they do so is not yet clear.

Orderly segregation of episomal viral DNA molecules during mitosis requires binding of EBNA-1 to its high affinity sites in the family of repeat (FR) sequences of OriP (Fig. 9.24A). Direct observation of episomal viral genomes by *in situ* hybridization has established that these DNA molecules become tethered to the cellular sister chromatids that are separated during mitosis. Tethering of viral DNA chromosomes, and their partitioning during mitosis, is mediated by an N-terminal sequence of EBNA-1 that contains two domains that can bind directly to AT-rich DNA. In metaphase chromosomes, regions of uncondensed (that is, accessible) AT-rich DNA are found between segments of the genome that are highly condensed. Any derivative of EBNA-1 that contains two such AT-hook domains (even if they are derived from cellular proteins) binds to chromosomes and supports maintenance of OriP-containing episomes in a host cell population.

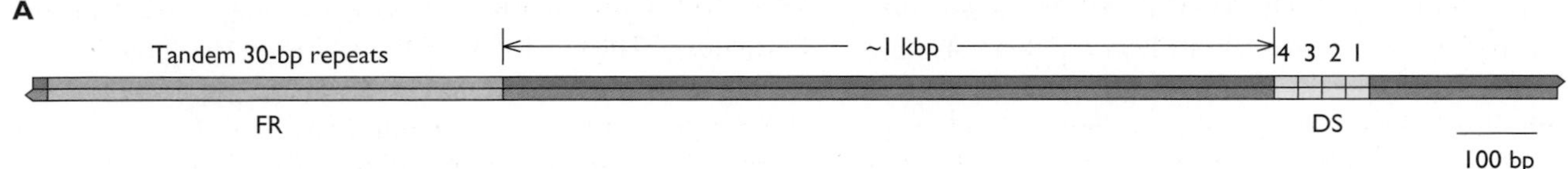

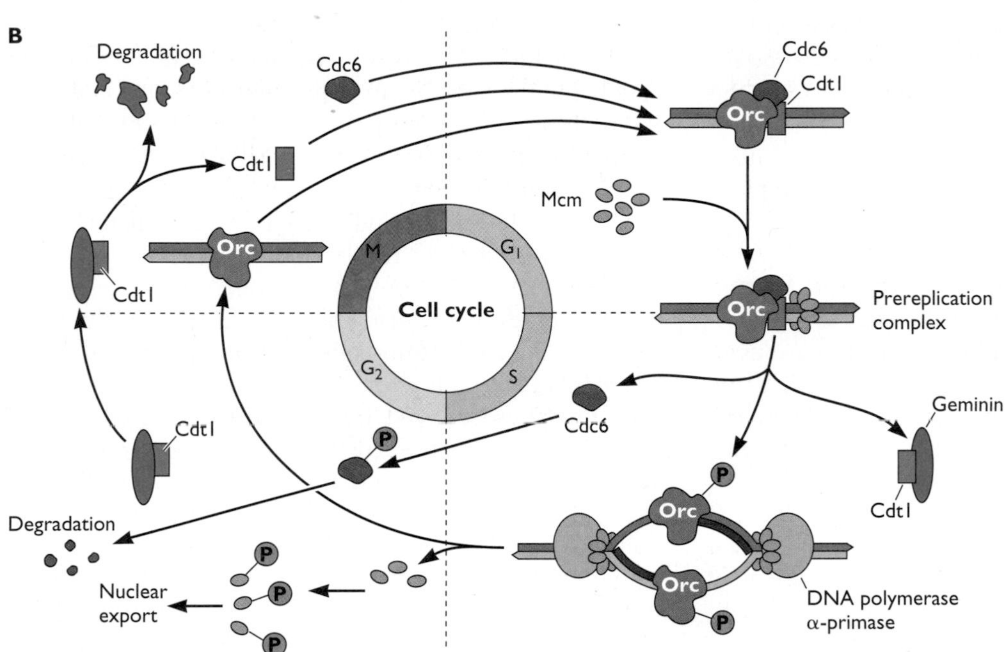

Figure 9.24 Licensing of replication from Epstein-Barr virus OriP. (A) Organization of EBNA-binding sites, shown to scale. The dyad symmetry (DS) sequence, which comprises four binding sites (1 to 4) for EBNA-1 dimers, is the site of initiation of DNA synthesis. The activity of the DS origin is regulated by adjacent sequences recognized by cellular telomere-binding proteins (see the text) and stimulated by the family of repeat (FR) sequence. The mechanism of such stimulation is not clear. However, binding of EBNA-1 to both DS and FR sequences, with the intervening 1 kbp of DNA looped out, appears to be important. Binding of EBNA-1 to FR sequences is also necessary for maintenance of episomal viral DNA in latently infected B cells. **(B)** The multiprotein origin recognition complex (Orc) is present throughout the cell cycle and is associated with replication origins. However, initiation of DNA synthesis requires loading of the hexameric minichromosome maintenance complex (Mcm), which provides helicase activity. It is the recruitment of Mcm that is controlled during the cell cycle to set the timing of the initiation of DNA synthesis in S phase. This reaction requires two proteins, Cdc6 and Cdt1. The concentrations and activities of both are tightly controlled during the cell cycle. As cells complete mitosis and enter G_1, Cdc6 and Cdt1 accumulate in the nucleus, where they associate with DNA-bound Orc. These interactions permit loading of Mcm at the G_1-to-S-phase transition, and subsequently of components of the DNA synthesis machinery, such as Rpa and DNA polymerase α-primase. The latter step requires phosphorylation of specific components of the prereplication complex by cyclin-dependent kinases that accumulate during the G_1-to-S-phase transition (Volume II, Chapter 7). Reinitiation of DNA synthesis is prevented by several mechanisms. A cyclin-dependent kinase that accumulates during the G_2 and M phases phosphorylates both Mcm proteins and Cdc6. This modification induces nuclear export of the former and degradation of the latter. In addition, the protein called geminin is present in the nucleus from S until M phase (when it is degraded). This protein binds to Cdt1, sequestering it from interaction with Cdc6 and Orc. As a consequence of such regulatory mechanisms, the prereplication complex can form **only** in the G_1 phase, ensuring firing of the origin once per cell cycle.

As a latent infection is established, the Epstein-Barr virus genome becomes increasingly methylated at C residues present in CG dinucleotides, although sequences that must function in latently infected cells, such as OriP, generally escape this modification. Such DNA methylation is associated with repression of transcription, and contributes to inhibition of expression of viral genes. The viral genome also becomes packaged by cellular nucleosomes and is therefore replicated as a circular minichromosome, much like that of simian virus 40. Replication of the Epstein-Barr virus genome once per cell cycle persists unless conditions that induce entry into the viral productive cycle are encountered. The critical step for this transition is activation of transcription of the viral genes encoding the transcriptional activators Zta and Rta (Chapter 8). These proteins induce expression of the viral early genes that encode the viral DNA polymerase and other proteins necessary for replication from OriLyt. In addition, Zta appears to be the viral OriLyt recognition protein. Consequently, once this protein is made in an Epstein-Barr virus-infected cell, its indirect and direct effects on viral DNA synthesis ensure a switch from OriP-dependent to OriLyt-dependent replication, and progression through the infectious cycle.

Controlled and Exponential Replication from a Single Origin: the Papillomaviruses

Three different modes of viral DNA replication are associated with papillomavirus infection (Fig. 9.25). Entry of a papillomaviral genome into the nucleus of a host cell initiates a period of amplification of the circular genome, just as during the early stages of latent infection by Epstein-Barr virus. Replication continues until a moderate number of viral genomes (~50 to 100) accumulates in the cell. A maintenance replication pattern, in which the complement of viral episomes is duplicated on average once per cell cycle, is then established. The mechanism that governs the switch from amplification to maintenance replication is not known. In natural human papillomavirus infections, these two types of replication take place in the proliferating basal cells of an epithelium. They can also be reproduced in cells in culture transformed by these viruses.

The single viral origin and the viral E1 and E2 proteins that bind to specific origin sequences (Fig. 9.17; Table 9.3) are necessary for both the initial amplification of the papillomavirus genome and its maintenance for long periods at a more-or-less constant concentration. Initial studies of bovine papillomavirus indicated that such maintenance replication is not the result of strict, once-per-cell-cycle replication of viral DNA. Rather, replication of individual viral episomes occurs at random, taking place on average once per cell cycle. Subsequent studies of human papillomavirus DNA replication in different epithelial cell lines, including those derived from naturally infected cervical epithelia, have established that the viral genome can be replicated by both random and strict, once-per-cell-cycle mechanisms (Box 9.10). Which mode of replication prevails is determined by both the host cell and the concentration of the viral E1 protein: at high concentrations, this

Figure 9.25 Regulation of papillomaviral DNA replication in epithelial cells. The outer layers of the skin are shown as depicted in Fig. 2.5. The virus infects proliferating basal epithelial cells, to which it probably gains access after wounding. The double-stranded, circular viral DNA genome is imported into the infected cell nucleus and initially amplified to a concentration of 50 to 100 copies per cell. This concentration of viral DNA episomes is maintained by further limited replication as the basal and parabasal cells of the epithelium divide **(maintenance replication)**. As cells move to the outer layers of the epidermis and differentiate, productive replication of the viral genome to thousands of copies per cell takes place.

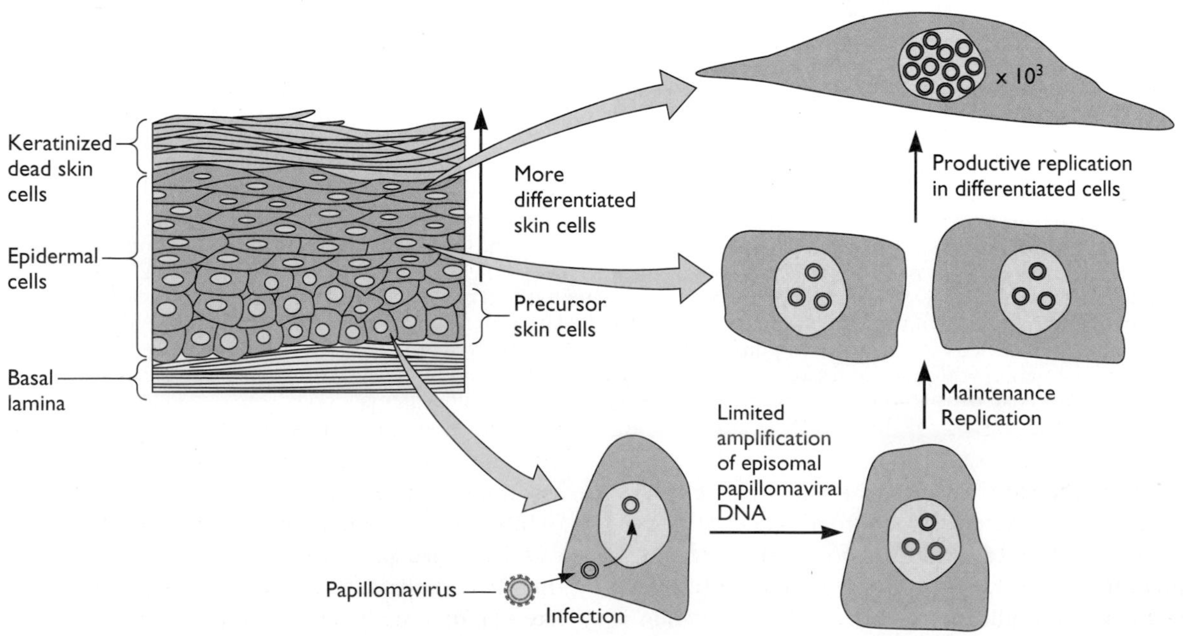

BOX 9.10

EXPERIMENTS

Distinguishing one-per-cell cycle from random replication of human papillomavirus DNA

In once-per-cell cycle replication, each molecule of episomal viral DNA is replicated just once per cell cycle during S phase. In random replication, some DNA molecules are replicated several times in a single cell cycle, some are replicated once, and some do not replicate. As illustrated in the figure, these mechanisms can be distinguished by the densities of the DNA molecules synthesized when cells are incubated with the dense analog of thymidine, bromodeoxyuridine (BUdR), for a period less than the time required to complete one cell cycle.

Results obtained when this method was applied to W12 cervical keratinocytes that contain human papillomavirus type 16 DNA are shown schematically in the figure. In these cells, viral DNA replication is by the once-per-cell cycle mechanism: no HH DNA could be detected **(left)**. When a vector for expression of the viral E1 protein was introduced, random replication of the viral DNA ensued **(right)**.

Hoffman, R., B. Hirt, V. Bechtold, P. Beard, and K. Raj. 2006. Different modes of human papillomavirus DNA replication during maintenance. *J. Virol.* **80:**4431–4439.

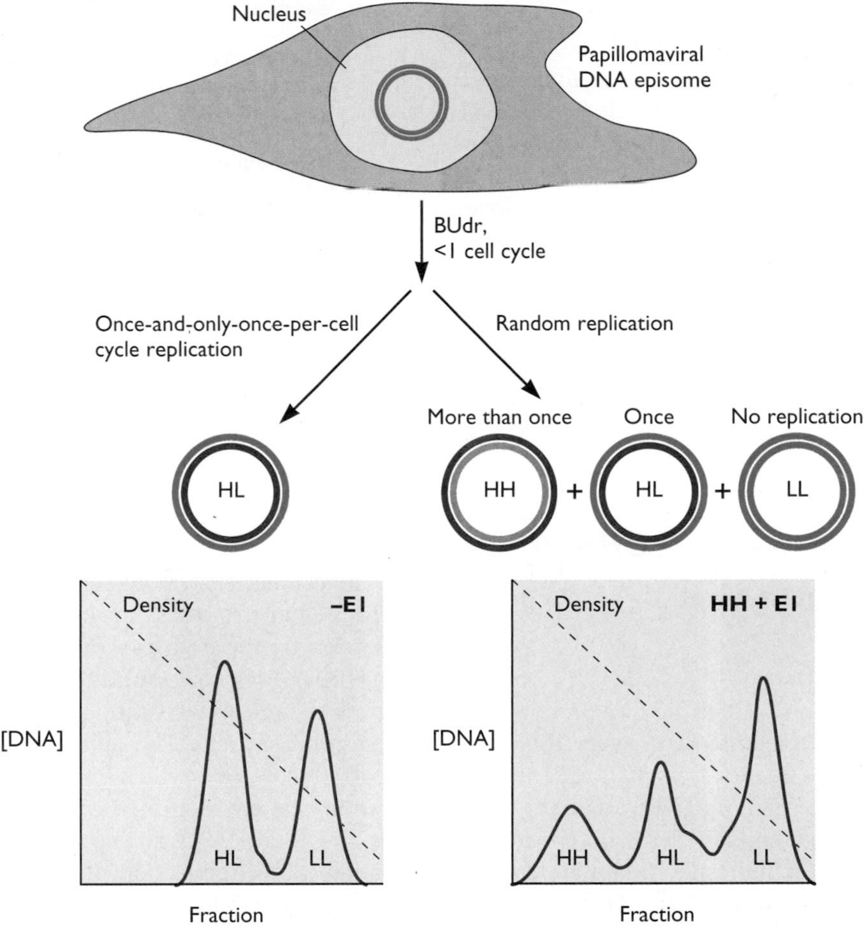

protein converts once-per-cell cycle viral DNA synthesis to random-choice replication.

Stable maintenance of the viral genome requires an additional sequence, called the **minichromosome maintenance element**, which is composed of multiple binding sites for the E2 protein. When bound by the viral protein, the minichromosome maintenance element is attached to mitotic chromosomes and remains associated with them during all stages of mitosis. This association is mediated by binding of E2 to the C-terminal domain of the cellular bromodomain-containing protein 4 (Brd 4), an acetylated histone H4-binding protein that interacts with mitotic chromosomes. The C-terminal domain of Brd4 acts as a dominant negative inhibitor of E2 binding: when overproduced, it induces release of viral DNA from mitotic chromosomes and loss of the viral genome from cells transformed by bovine papillomavirus type 1. This protein also appears to mediate binding to mitotic chromosomes of episomal DNA of human herpesvirus type 8, a gammaherpesvirus that is associated with various human tumors (Volume II, Chapter 7).

Remarkably, the final stage of papillomaviral DNA replication, production of high concentrations of the viral genome for assembly into progeny virions, is restricted to nondividing, differentiated epithelial cells, such as terminally differentiated keratinocytes formed as basal cells move toward the surface of an epithelium (Fig. 9.25). Cell culture systems that support such so-called productive replication have now been developed, and should allow elucidation of the mechanism of this poorly understood process. The viral E7 protein is necessary for productive replication and induction of synthesis of the cellular replication proteins needed for viral DNA synthesis, such as DNA polymerase α and Pcna, by the mechanism shown in Fig. 9.21A. This protein perturbs the program of keratinocyte differentiation. The viral E5 and E1^E4 proteins, which are produced at high concentrations following differentiation, also contribute to maintenance of proliferative competence in infected keratinocytes. The function of these viral proteins therefore appears to be to establish a cellular environment favorable for productive replication of the papillomaviral genome.

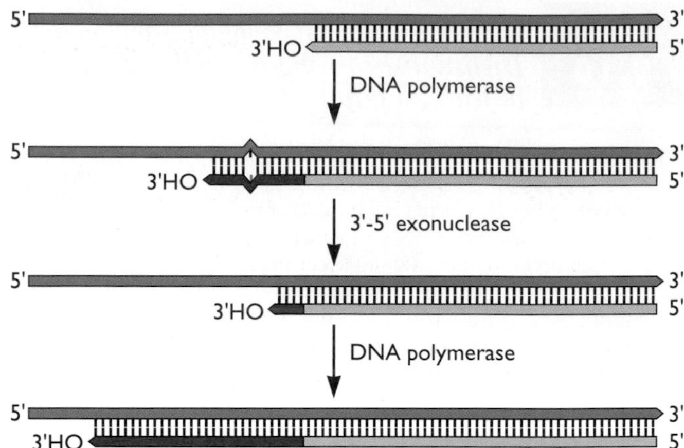

Figure 9.26 Proofreading during DNA synthesis. If permanently fixed into the genome, mispaired bases would result in mutation. However, the majority are removed by the proofreading activity of replicative DNA polymerases. A mismatch at the 3′-OH terminus of the primer-template during DNA synthesis activates the 3′ →5′ exonuclease of all replicative DNA polymerases, which excises the mismatched region to create a perfect duplex for further extension. In the best-characterized case, DNA polymerase I of *E. coli*, the rate of extension from a mismatched nucleotide is much lower than when a correct base pair is formed at the 3′ terminus of the nascent strand. This low rate of extension allows time for spontaneous unwinding (breathing) of the new duplex region of the DNA and transfer of the 3′ end to the 3′ → 5′ exonuclease site for removal of the mismatched nucleotide. Because preferential excision of mismatched nucleotides is the result of differences in the **rate** at which the polymerase can add the next nucleotide, this mechanism is called **kinetic proofreading**.

Origins of Genetic Diversity in DNA Viruses

Fidelity of Replication by Viral DNA Polymerases

Proofreading Mechanisms

Cellular DNA replication is a high-fidelity process with an error rate of only about one mistake in every 10^9 nucleotides incorporated. Such fidelity, which is essential to maintain the integrity of the genome, is based on the accurate pairing of substrate and template deoxyribonucleotide bases prior to synthesis of each phosphodiester bond. Nonstandard base pairs can form quite readily. However, DNA synthesis requires perfect base pairing between the nascent and template strands, and DNA synthesis does not begin if the terminal nucleotide, or the preceding region of the primer-template, is mismatched. In such circumstances, the mismatched base in the primer strand is excised by a 3′ → 5′ exonuclease present in all replicative DNA polymerases until a perfectly base-paired primer-template is created (Fig. 9.26). Replicative DNA polymerases are therefore self-correcting enzymes, removing errors made in newly synthesized DNA as replication continues.

The cellular DNA polymerases that replicate small viral DNA genomes possess such proofreading exonucleases. Infection by these viruses (e.g., papillomavirus and polyomavirus) does not result in inhibition of cellular protein synthesis, and indeed may induce expression of cellular replication proteins. The cellular mechanisms of mismatch repair (Fig. 9.27), which correct errors in the daughter strand of newly replicated DNA missed during proofreading, are therefore available to operate on progeny viral genomes. The replication of the genomes of these small DNA viruses is therefore likely to be as accurate as that of the genome of their host cells.

Proofreading by Viral DNA Polymerases

The question of how accurately viral DNA is replicated by viral DNA polymerases, such as those of adenoviruses, herpesviruses, and poxviruses, has received relatively little attention. However, each of these viral enzymes possesses an intrinsic 3′ → 5′ exonuclease that preferentially excises mismatched nucleotides from duplex DNAs *in vitro*. These viral enzymes contain short, conserved sequences related

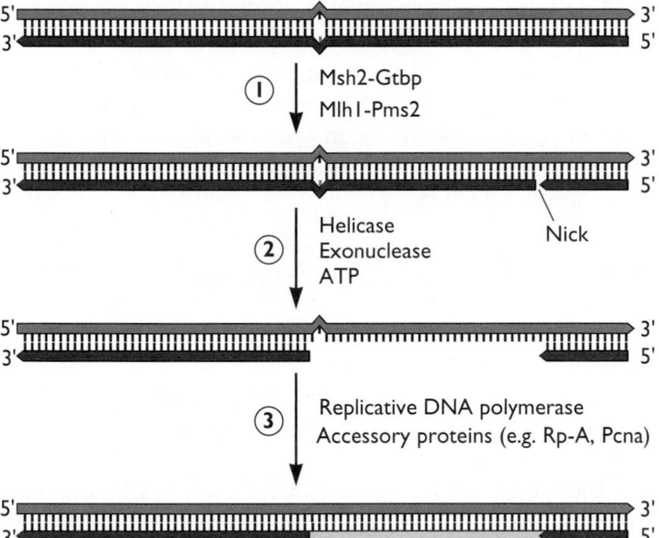

Figure 9.27 Mismatch repair in newly synthesized DNA. This activity requires recognition of the newly synthesized strand containing a misincorporated nucleotide and introduction of a nick (step 1), exonucleolytic degradation of the new strand from the nick as it is unwound (step 2), and resynthesis of DNA to repair the gap (step 3). The long-patch repair system is responsible for removal of mismatches introduced during DNA synthesis in eukaryotic cells. The mammalian Msh2-Gtbp and Mlh1-Pms2 heterodimers are composed of homologs of the well-characterized *E. coli* mismatch repair proteins MutS and MutL, respectively. This relationship suggests that they participate in steps 1 and 2. The importance of this repair system is illustrated by the predisposition to tumor development induced by mutations in the human genes encoding MutS or MutL homologs.

to those that flank the exonuclease active site residues of well-characterized cellular DNA polymerases. Indeed, mutations that impair the exonuclease activity of the herpes simplex virus type 1 DNA polymerase greatly increase the mutation rate.

At this juncture, relatively little is known about the effects of infection by the larger DNA viruses on the production or function of cellular mismatch repair proteins that normally back up proofreading by replicative DNA polymerases. As expression of cellular genes and cellular DNA synthesis are generally inhibited in cells infected by these viruses, it is possible that such mismatch repair proteins are not present in the concentrations necessary for effective surveillance and repair of newly synthesized viral DNA. Indeed, infection of primary human fibroblasts by human cytomegalovirus (a betaherpesvirus) reduces the activity of an enzyme important for excision of alkylated bases. On the other hand, it has been reported that the herpes simplex virus type 1 UL8 and UL42 replication proteins interact with two proteins that participate in mismatch repair, Pcna and Msh6. More detailed information about the rates at which viral DNA polymerases introduce errors during DNA synthesis *in vitro*, and the rates of mutation of viral DNA genomes during productive infection, would help establish the role of cellular repair systems in maintaining the integrity of these viral genomes. Similarly, the contributions of viral enzymes that could prevent or repair DNA damage, such as the dUTPase and uracil DNA glycosylases of herpesviruses and poxviruses (Table 9.5), remain to be established.

Inhibition of Repair of Double-Stranded Breaks in DNA

Exposure of mammalian cells to ultraviolet (UV) or infrared light, as well as stalling or collapse of replication forks, can produce double-stranded breaks in the DNA genome. Such lesions are potentially lethal, so it is not surprising that they elicit powerful and sensitive damage-sensing and response systems. Proteins that recognize double-stranded DNA ends initiate signaling to effector proteins that halt progression through the cell cycle (to allow time for repair) and that repair the broken ends (Fig. 9.28). The DNA ends are sealed either by nonhomologous end joining or homologous recombination repair (Fig. 9.28). In the former process, which is prone to errors, double-stranded ends of DNA are simply joined together. This important repair pathway is blocked in cells infected by several DNA viruses.

The products of adenoviral DNA synthesis are unit-length copies of the linear viral genome that require no processing (Fig. 9.10). However, accumulation of these viral DNA molecules requires inactivation of nonhomologous end joining. In the absence of the viral E4 Orf 3 and Orf 6 proteins, newly synthesized viral DNA forms concatemers far too large to be packaged into progeny virions. Accumulation of such multimeric DNA molecules depends on the cellular MreII-Rad50-Nsb1 repair complex, which normally accumulates at sites of DNA damage (Fig. 9.28), and is necessary for repair of double-stranded breaks. In adenovirus-infected cells, the protein components of this complex become redistributed within nuclei, and are then degraded by the proteasome. These alterations are induced by the Orf 3 and Orf 6 proteins, respectively. When the E4 proteins cannot be made, the cellular repair proteins accumulate in viral replication centers. These E4 gene products therefore appear to prevent triggering of a double-strand-break repair mechanism by the accumulation of linear viral DNA genomes in infected cell nuclei.

Infection by alpha-, beta-, or gammaherpesviruses induces phosphorylation of the Atm and Chk2 proteins (Fig. 9.28). These modifications are signatures of an active

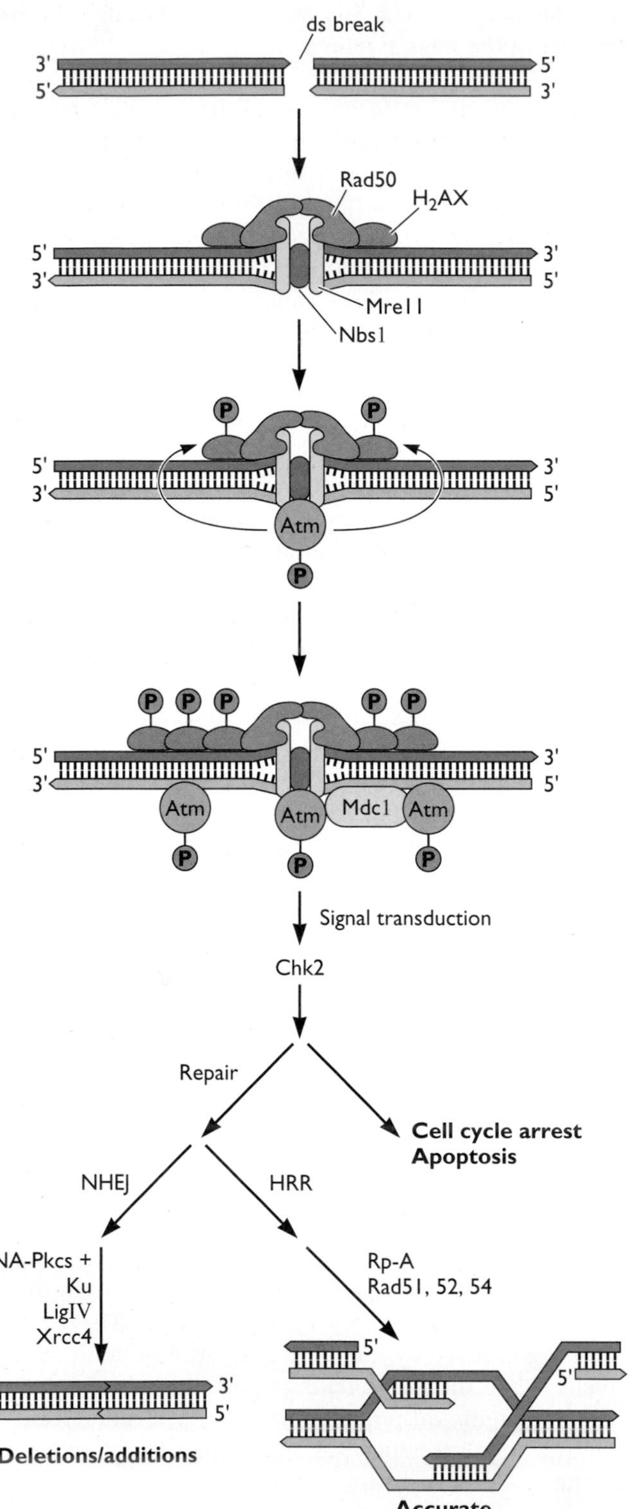

Figure 9.28 Detection of double-stranded breaks in DNA. Induction of a double-stranded break in the DNA genome triggers rapid accumulation of the MRN complex at the break. This complex contains two copies each of the Mre11 and Rad50 proteins, which move from the cytoplasm into the nucleus, and one of the Nbs1 protein. Mre11 possesses 3′ → 5′ exonuclease, single-stranded DNA endonuclease, and helicase activities. It is thought that these activities unwind the DNA ends at the site of the break, allowing recruitment of the large protein kinase ataxia telangiectasia mutated, Atm. This kinase then becomes activated, perhaps by conformational change and autophosphorylation, and phosphorylates substrates such as the variant histone H2AX. This modification allows amplification of the signal via binding of additional MRN complexes and of mediator of DNA damage checkpoint protein 1, Mdc1. Both this protein and Nsb1 bind phosphorylated H2AX. The Mdc1 protein transduces the signal via additional protein kinases (e.g., Chk2) and other proteins to induce such responses as cell cycle arrest and DNA repair. The two major repair pathways are nonhomologous end joining (NHEJ) and homologous recombination repair (HRR).

DNA damage response which is thought to be triggered by free DNA ends present in the viral genome and replication intermediates. The mechanisms and consequences of activation of this cellular response for viral replication are not understood. However, in the case of cells infected by human cytomegalovirus, the signaling pathway is blocked upstream of effectors.

Recombination of Viral Genomes

General Mechanisms of Recombination

Genetic recombination is an important source of genetic variation in populations. It also makes a major contribution to repair of breaks in a DNA genome (Fig. 9.28) and can rescue replication when this process has stalled at unfavorable sequences (or chromosomal sites) in the template. Much of our understanding of the mechanisms of recombination is based on studies of bacterial viruses, such as bacteriophage lambda. Similar principles apply to recombination of DNA genomes of animal viruses.

Two types of recombination are generally recognized: site specific and homologous. In **site-specific recombination**, exchange of DNA takes place at short DNA sequences that are specifically recognized by proteins that catalyze recombination. These sequences may be present in only one or both of the DNA sequences that are recombined in this way. Much more common during reproduction of DNA viruses is **homologous recombination**, the exchange of genetic information between **any** pair of related DNA sequences.

Viral Genome Recombination

The integration of adeno-associated virus DNA into the cellular genome, and its excision when conditions are appropriate, are the result of **site-specific** recombination reactions mediated by the Rep 78/68 viral proteins, which bind to specific sequences in both DNA molecules.

In contrast, integration of retroviral DNA (Chapter 7) is site specific only for the viral DNA.

All viral DNA genomes undergo homologous recombination. Indeed, viral recombination is favored by the large numbers of viral DNA molecules present in productively infected cells, and their concentration within specialized replication compartments: the initial step in recombination, pairing of homologous sequences with one another, depends on random collision and is therefore concentration dependent. Furthermore, the structures of replication intermediates, or the nicking of viral DNA during replication or packaging that yields DNA ends, can facilitate recombination among viral DNA molecules. The formation of replication compartments can also result in the concentrations of cellular proteins that participate in recombination (and repair) with viral genomes, as, for example, observed in cells infected by herpes simplex virus type 1.

The ease with which viral DNA sequences can recombine is an important factor in the evolution of these viruses. It is also of great benefit to the experimenter, facilitating introduction of specific mutations into the viral genome or construction of viral vectors (see Chapter 3). Conversely, recombination may be necessary for productive replication of some viral genomes (those of herpesviruses [see below]) or stimulation of viral DNA synthesis, for example, from the nicks that initiate recombination.

As viral genomes do not generally encode homologous recombination proteins, it is thought that this process is catalyzed by host cell enzymes. One exception is the herpes simplex virus type 1 alkaline nuclease (Table 9.5). This enzyme is a $5' \rightarrow 3'$ exonuclease with homology to the Red α component of the bacteriophage lambda recombinase (Box 9.11). In conjunction with the viral single-stranded DNA-binding protein (ICP8), the alkaline nuclease can mediate the exchange of strands between two DNA molecules *in vitro*. The precise role of recombination mediated by these viral proteins during infection is not known. However, this process is important for production of normal, infectious genomes: the viral DNA synthesized and packaged in cells infected by mutants that do not direct synthesis of the active nuclease contains structural abnormalities and is poorly infectious.

Although recombination among animal viral DNA sequences has been widely exploited in the laboratory, the mechanisms have not received much attention. One exception is adenovirus recombination. Recombination between markers in two adenovirus genomes exhibits many properties typical of this process, such as the dependence of recombination frequency on the distance between the markers. However, several features suggest that recombination is coupled with viral DNA synthesis, because the initial invasion is by the single strands of viral DNA displaced during replication of the adenoviral genome (Fig. 9.10). In particular, this mechanism accounts for the **polarity** of adenovirus recombination, the decreasing gradient of recombination frequency with distance of the recombining sequences from the ends of the viral genome. Another important exception is the homologous recombination of DNA sequences of some herpesviruses, including herpes simplex virus type 1, which is responsible for isomerization of the genome.

Populations of viral DNA molecules purified from herpes simplex virus type 1 virions contain four isomers of the genome, defined by the relative orientations of the two unique sequence segments (L and S) with respect to one another (Fig. 9.20). These unique sequences are flanked by several inverted repeats, including the conserved *a* sequence. The viral DNA population isolated from a single plaque contains all four isomers at equimolar concentrations, suggesting that a single virus particle containing just one genome isomer gives rise to all four by recombination between repeated DNA sequences. It was initially thought that the *a* sequences mediate these recombination reactions by a site-specific mechanism, for example, because insertion of an *a* sequence at an ectopic site induces additional inversions. However, it is now clear that an *a* sequence can act as a hot spot for recombination simply because it contains the viral packaging signal: double-stranded DNA breaks that promote recombination are made within this sequence during cleavage of replication products for packaging of the viral genome. In fact, the *a* sequences are dispensable for production of all four genome isomers at the normal frequency, and recombination between **any** of the inverted repeat sequences in the viral genome promotes inversion of the L and S segments. Such homologous recombination takes place during viral DNA synthesis and requires the viral replication machinery.

Despite some 30 years of study, the function of the unusual isomerization of the genome of herpes simplex virus type 1 and certain other herpesviruses remains enigmatic. Isomerization is not absolutely essential for virus replication in cells in culture, because viruses "frozen" as a single isomer by deletion of internal inverted repeats are viable. On the other hand, the reduced yield of such viruses, and the presence of the inverted repeat sequences in **all** strains of herpes simplex virus type 1 examined, emphasizes the importance of the repeated sequences. It may be that these sequences themselves fulfill some beneficial function (as yet unknown). Recombinational isomerization would then be a secondary result of the presence of multiple, inverted copies of these sequences in the viral genome. Alternatively, isomerization might be a consequence of an important role

BOX 9.11

DISCUSSION

Replication and recombination/repair are two sides of the same coin: earliest insights from bacteriophage λ

In the early 1970s, studies of the replication of bacteriophage λ showed that mutants defective in viral recombination genes (*redα⁻* or *redβ⁻*, *gam⁻*) synthesize DNA at only half to one-third the wild-type rate, the concatemers typical of late DNA synthesis were on average shorter than usual, and viral bursts were only 30 to 40% of wild-type levels. The role of Gam was explained by its inhibition of the cellular RecBCD nuclease, which would be expected to destroy free concatemer ends. However, the role of Red proteins was not so readily apparent. Furthermore, the fact that viral *red⁻* mutants failed to plate at all on certain cells, for example those that were deficient in host DNA polymerase A or ligase, suggested a critical role for recombination and repair functions in λ DNA replication.

An elegant series of genetic and biochemical experiments led to a model (shown here) for the transition from circle to rolling-circle replication, which proposed a mechanism by which viral recombination or host DNA repair functions might produce new replication forks when encountering damage induced by a single-strand break.

It was suggested at the time that the principles illustrated in this model might very well be applicable to cellular DNA metabolism. The idea that recombination could generate a replication origin was novel at the time, but current schemes for the repair of stalled replication forks in both prokaryotic and eukaryotic cells incorporate the very same ideas elaborated from studies of λ over 30 years ago.

Enquist, L.W., and A. Skalka. 1973. Replication of bacteriophage lambda DNA dependent on the function of host and viral genes. I. Interaction of *red, gam,* and *rec. J. Mol. Biol.* **75:**185–212.

Skalka, A. 1974. A replicator's view of recombination (and repair), p. 421–432. *In* R. F. Grell (ed.), *Mechanisms in Recombination*. Plenum Press, New York, NY.

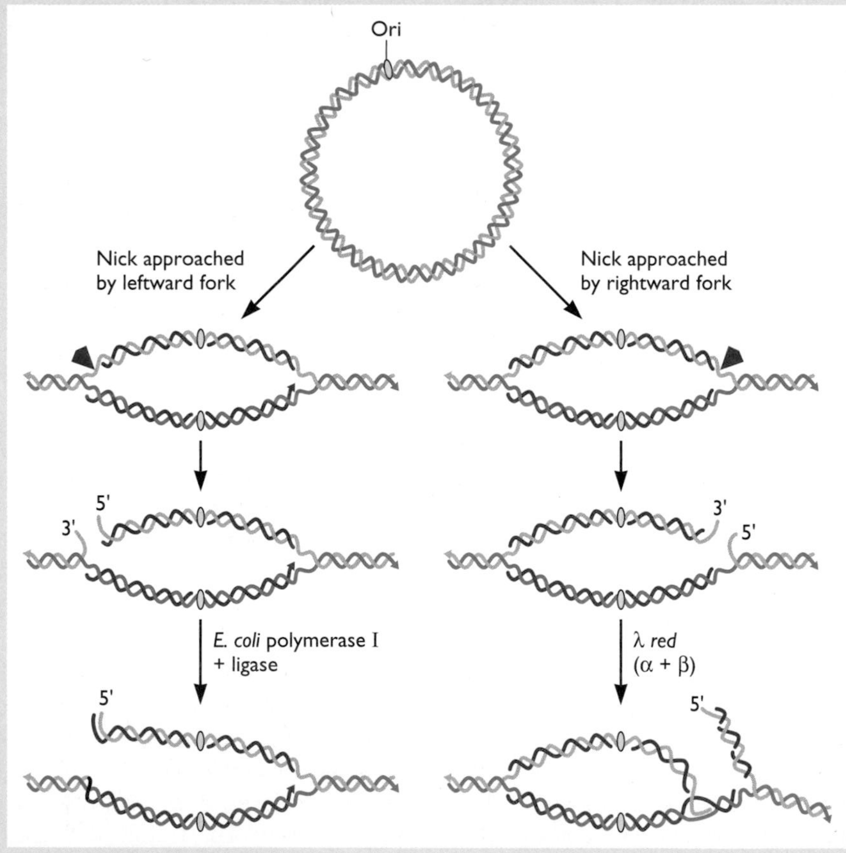

for recombination in replication of the viral genome, as during replication of bacteriophage lambda (Box 9.11). The switch to origin-independent replication later in the infectious cycle is consistent with this idea, but a definitive answer is not yet available. The value of the unusual isomerization of the herpes simplex virus type 1 genome may become clearer as understanding of the mechanism of viral DNA synthesis improves.

Perspectives

Our understanding of mammalian replication proteins and the intricate reactions they carry out during DNA synthesis would still be rudimentary were it not for the simian virus 40 origin of replication. This relatively simple viral DNA sequence, which was initially well characterized by genetic methods, supports origin-dependent replication *in vitro* when cellular proteins are supplemented with a single viral protein, LT.

The mechanism of synthesis of this small viral DNA genome has also provided the conceptual framework within which to appraise the considerable diversity in replication of viral DNA genomes. One parameter that varies considerably is the degree of dependence on the host cell's replication machinery. In contrast to those of papillomaviruses, parvoviruses, and polyomaviruses, the genomes of the larger DNA viruses (herpesviruses and poxviruses) encode the components of a complete DNA synthesis system as well as accessory enzymes responsible for the production of dNTP substrates. Nevertheless, replication of **all** viral DNA genomes requires proteins that carry out the reactions first described for simian virus 40 DNA synthesis, notably an origin recognition protein(s), one or more DNA polymerases, proteins that confer processive DNA synthesis, origin-unwinding and helicase proteins, and, usually, proteins that synthesize, or serve as, primers.

Replication of viral DNA genomes ranges from simple, continuous synthesis of both strands of a linear, double-stranded DNA template (adenovirus) to baroque (and not well understood) mechanisms that produce DNA concatemers (herpesviruses). These various replication strategies represent alternative mechanisms for circumventing the inability of all known DNA-dependent DNA polymerases to initiate DNA synthesis *de novo*. In some cases, initiation of viral DNA synthesis requires RNA primers and the lagging strand is synthesized discontinuously, but in others, the priming mechanism leads to continuous synthesis of all daughter DNA strands from protein or DNA sequence primers.

Efficient reproduction of DNA viruses requires the rapid production of very large numbers of progeny viral DNA molecules for assembly of viral particles. One factor contributing to such exponential replication is the efficient production of the proteins that mediate or support DNA synthesis, be they viral or cellular in origin. However, it is also likely that viral DNA replication at specialized intracellular sites, a common feature of cells infected by these viruses, contributes to efficient viral DNA synthesis. Further exploration of this incompletely understood phenomenon should shed new light on host cell biology, in particular the structural and functional compartmentalization of the nucleus. Also far from well understood are the cues that set the stage for the alternative mode of limited replication characteristic of some DNA viruses, when the number of replication cycles and their timing with respect to the host cell cycle are governed by the nature (e.g., alphaherpesviruses) or differentiation state (papillomaviruses) of the host cell. Elucidation of the mechanisms that result in such close integration of viral DNA synthesis with the physiological state of the host cell seems certain to provide important insights into both host cell control mechanisms and the long-term relationships these viruses can establish with their hosts.

References

Books

Kornberg, A., and T. Baker. 1992. *DNA Replication*, 2nd ed. W. H. Freeman and Company, New York, NY.

Chapters in Books

Berns, K., and C. R. Parrish. 2007. Parvoviridae, p. 2437–2477. *In* D. M. Knipe and P. M. Howley (ed.), *Fields Virology*, 5th ed. Lippincott Williams & Wilkins, Philadelphia, PA.

Reviews

Beaud, G. 1995. Vaccinia virus DNA replication: a short review. *Biochimie* **77:**774–779.

Bell, S. P. 2002. The origin recognition complex: from simple origins to complex functions. *Genes Dev.* **16:**659–672.

Boehmer, P. E., and I. R. Lehman. 1997. Herpes simplex virus DNA replication. *Annu. Rev. Biochem.* **66:**347–384.

Everett, R. D. 2001. DNA viruses and viral proteins that interact with PML nuclear bodies. *Oncogene* **20:**7266–7273.

Fanning, E. 1994. Control of SV40 DNA replication by protein phosphorylation: a model for cellular DNA replication? *Trends Cell. Biol.* **4:**250–255.

Kolodner, R. D., and G. T. Marsischky. 1999. Eukaryotic DNA mismatch repair. *Curr. Opin. Genet. Dev.* **9:**89–96.

Lindahl, T., and R. D. Wood. 1999. Quality control by DNA repair. *Science* **286:**1897–1905.

Machida, Y. J., J. L. Hamlin, and A. Dulta. 2005. Right place, right time, and only once: replication initiation in metazoans. *Cell* **123:** 13–24.

Marintcheva, B., and S. K. Weller. 2001. A tale of two HSV-1 helicases: roles of phage and animal virus helicases in DNA replication and recombination. *Prog. Nucleic Acid Res. Mol. Biol.* **70:**77–118.

Sternland, A. 2003. Initiation of DNA replication: lesson from viral initiation proteins. *Nat. Rev. Mol. Cell Biol.* **4:**777–785.

Sugden, B. 2002. In the beginning: a viral origin exploits the cell. *Trends Biochem. Sci.* **27:**1–3.

Van der Vliet, P. C. 1995. Adenovirus DNA replication. *Curr. Top. Microbiol. Immunol.* **199**(Part 2):1–30.

Waga, S., and B. Stillman. 1998. The DNA replication fork in eukaryotic cells. *Annu. Rev. Biochem.* **67:**721–751.

Weinberg, R. A. 1995. The retinoblastoma protein and cell cycle control. *Cell* **81:**323–330.

Zhang, Y., J. Zhou, and C. V. K. Lim. 2006. The role of NBS1 in DNA double strand break repair, telomere stability and cell cycle control checkpoint. *Cell Res.* **16:**45–54.

Papers of Special Interest

Viral Replication Mechanisms and Replication Proteins

Challberg, M. D. 1986. A method for identifying the viral genes required for herpesvirus DNA replication. *Proc. Natl. Acad. Sci. USA* **83:** 9094–9098.

Challberg, M. D., S. V. Desidero, and T. J. Kelly. 1980. Adenovirus DNA replication in vitro: characterization of a protein covalently linked to nascent DNA strands. *Proc. Natl. Acad. Sci. USA* **77:**5105–5109.

Earl, P. L., E. V. Jones, and B. Moss. 1986. Homology between DNA polymerases of poxviruses, herpesviruses and adenoviruses: nucleotide sequence of the vaccinia virus DNA polymerase gene. *Proc. Natl. Acad. Sci. USA* **83:**3659–3663.

Fixman, E. D., G. S. Hayward, and S. D. Hayward. 1995. Replication of Epstein-Barr virus OriLyt: lack of a dedicated virally encoded origin-binding protein and dependence on Zta in cotransfection assays. *J. Virol.* **69:**2998–3006.

Gai, D., R. Zhao, D. Li, C. V. Finkelstein, and X. S. Chen. 2004. Mechanisms of conformation change for a replicative hexameric helicase of SV40 large tumor antigen. *Cell* **119:**47–60.

Im, D. S., and N. Muzyczka. 1990. The AAV origin-binding protein Rep68 is an ATP-dependent site-specific endonuclease with DNA helicase activity. *Cell* **61:**447–457.

Lechner, R. L., and T. J. Kelly. 1977. The structure of replicating adenovirus 2 DNA molecules. *Cell* **12:**1007–1020.

Li, J. J., and T. J. Kelly. 1984. Simian virus 40 DNA replication in vitro. *Proc. Natl. Acad. Sci. USA* **81:**6973–6977.

Li, J. J., K. W. Peden, R. A. Dixon, and T. J. Kelly. 1986. Functional organization of the simian virus 40 origin of DNA replication. *Mol. Cell. Biol.* **6:**1117–1128.

Link, M. A., L. A. Silver, and P. A. Schaffer. 2007. Cathepsin B mediates cleavage of herpes simplex virus type 1 origin binding protein (OBP) to yield OBPC-1, and cleavage is dependent upon viral DNA replication. *J. Virol.* **81:**9178–9182.

Reese, D. K., K. R. Sreekumar, and P. A. Bullock. 2004. Interaction requiring for binding of simian virus 40 T antigen to the viral origin and molecular modeling of initial assembly events. *J. Virol.* **78:**2921–2934.

Rekosh, D., W. C. Russell, A. J. D. Bellett, and A. J. Robinson. 1977. Identification of a protein linked to the ends of adenovirus. *Cell* **11:**283–295.

Sanders, C. M., and A. Stenlund. 1998. Recruitment and loading of the E1 initiator protein: an ATP-dependent process catalysed by a transcription factor. *EMBO J.* **17:**7044–7055.

Waga, S., and B. Stillman. 1994. Anatomy of a DNA replication fork revealed by reconstitution of SV40 DNA replication in vitro. *Nature* **369:**207–212.

Regulation of Viral DNA Replication

Burkham, J., D. M. Coen, C. B. Hwang, and S. K. Weller. 2001. Interactions of herpes simplex virus type 1 with ND10 and recruitment of PML to replication compartments. *J. Virol.* **75:**2353–2367.

Doncas, V., A. M. Ishor, A. Romo, H. Juguilon, M. D. Weitzman, R. M. Evans, and G. G. Maul. 1996. Adenovirus replication is coupled with the dynamic properties of the PML nuclear structure. *Genes Dev.* **10:**194–207.

Moarefi, L. F., D. Small, I. Gilbert, M. Hopfner, S. K. Randall, T. J. Kelly and E. Farning. 1993. Mutation of the cyclin-dependent kinase phosphorylation site in simian virus 40 (SV40) large T antigen specifically blocks SV40 DNA origin unwinding. *J. Virol.* **67:**4992–5002.

Quinlan, M. P., L. B. Chen, and D. M. Knipe. 1984. The intranuclear location of a herpes simplex virus DNA-binding protein is determined by the status of viral DNA replication. *Cell* **36:**857–868.

Spindler, K. R., C. Y. Eng, and A. J. Berk. 1985. An adenovirus early region 1A protein is required for maximal viral DNA replication in growth-arrested human cells. *J. Virol.* **53:**742–750.

Taylor, T. J., and D. M. Knipe. 2004. Proteomics of herpes simplex virus replication compartments: association of cellular DNA replication, repair, recombination and chromatin remodeling protein with 1CP8. *J. Virol.* **78:**5856–5886.

Whyte, P., J. J. Buchkovich, J. M. Horowitz, S. H. Fried, M. Raybuck, R. A. Weinberg, and E. Harlow. 1988. Association between an oncogene and an anti-oncogene: the adenovirus E1A proteins bind to the retinoblastoma gene product. *Nature* **334:**124–129.

Limited Replication of Viral Genomes

Dhar, S. K., K. Yoshida, Y. Machida, P. Khaira, B. Chaudhuri, J. A. Wohlschlegel, M. Leffak, J. Yates, and A. Dutta. 2001. Replication from oriP of Epstein-Barr virus requires human ORC and is inhibited by geminin. *Cell* **106:**287–296.

Flores, E. R., and P. F. Lambert. 1997. Evidence for a switch in the mode of human papillomavirus type 11 DNA replication during the viral life cycle. *J. Virol.* **71:**7167–7179.

Gilbert, D. M., and S. N. Cohen. 1987. Bovine papilloma virus plasmids replicate randomly in mouse fibroblasts throughout S phase of the cell cycle. *Cell* **50:**59–68.

Kotin, R. M., M. Siniscalco, R. J. Samulski, X. D. Zhu, L. Hunter, C. A. Laughlin, S. McLaughlin, N. Muzyczka, M. Rocchi, and K. I. Berns. 1990. Site-specific integration by adeno-associated virus. *Proc. Natl. Acad. Sci. USA* **87:**2211–2215.

Philpott, N. J., J. Gomos, K. I. Berns, and E. Falck-Pedersen. 2002. A p5 integration efficiency element mediates Rep-dependent integration into AAVS1 at chromosome 19. *Proc. Natl. Acad. Sci. USA* **99:**12381–12385.

Reisman, D., J. L. Yates, and B. Sugden. 1985. A putative origin of replication of plasmids derived from Epstein-Barr virus is composed of two *cis*-acting components. *Mol. Cell. Biol.* **5:**1822–1832.

Schnepp, B. C., R. L. Jensen, C. Chen, P. R. Johnson, and K. R. Clark. 2005. Characterization of adeno-associated virus genomes isolated from human tissues. *J. Virol.* **79:**14793–14803.

Sears, J., M. Ujihara, S. Wong, C. Oh, J. Middeldorp, and A. Aiyar. 2004. The amino terminus of Epstein-Barr virus (EBV) nuclear antigen 1 contains an AT hook that facilitates the replication and partitioning of latent EBV genomes by tethering them to cellular chromosomes. *J. Virol.* **78:**11487–11505.

You, J., J. L. Croyle, A. Nishimura, K. Ozato, and P. M. Howely. 2004. Interaction of the bovine papillomavirus E2 protein with Brd4 tethers the viral DNA to host mitotic chromosomes. *Cell* **117:**349–360.

Repair and Recombination of Viral DNA

Evan, J. D., and P. Hearing. 2005. Relocalization of the Mre11-Rad50-Nbs1 complex by adenovirus E4 Orf3 protein is required for viral replication. *J. Virol.* **79:**6207–6215.

Gaspar, M., and T. Shenk. 2006. Human cytomegalovirus inhibits a DNA damage response by mislocalizing checkpoint proteins. *Proc. Natl. Acad. Sci. USA* **103:**2821–2826.

Hayward, G. S., R. J. Jacob, R. C. Wadsworth, and B. Roizman. 1975. Anatomy of herpes simplex virus DNA: evidence for four populations of molecules that differ in the relative orientations of their long and short segments. *Proc. Natl. Acad. Sci. USA* **72:**4243–4247.

Hwang, Y. T., B.-Y. Liu, D. M. Coen, and C. B. C. Hwang. 1997. Effects of mutations in the ExoIII motif of the herpes simplex virus DNA polymerase gene on enzyme activities, viral replication and replication fidelity. *J. Virol.* **71:**7791–7798.

King, A. J., W. R. Teertstra, L. Blanco, M. Salas, and P. C. van der Vliet. 1997. Processive proofreading by the adenovirus DNA polymerase. Association with the primary protein reduces exonucleolytic degredation. *Nucleic Acids Res.* **9:**1745–1752.

Linden, R. M., E. Winocour, and K. I. Berns. 1996. The recombination signals for adeno-associated virus site-specific integration. *Proc. Natl. Acad. Sci. USA* **93:**7966–7972.

Martin, D. W., and P. C. Weber. 1996. The *a* sequence is dispensable for isomerization of the herpes simplex virus type 1 genome. *J. Virol.* **70:**8801–8812.

Reaven, N. B., A. E. Staire, R. S. Myers, and S. K. Weller. 2003. The herpes simplex virus type 1 alkaline nuclease and single-stranded DNA binding protein mediate strand exchange in vitro. *J. Virol.* **77:**7425–7433.

Stracker, T. H., C. T. Carson, and M. D. Weitzman. 2002. Adenovirus oncoproteins inactivate the Mre11-Rad50-NBS1 DNA repair complex. *Nature* **418:**348–352.

Weitzman, M. D., S. R. M. Kyöstiö, R. M. Kotin, and R. A. Owens. 1994. Adeno-associated virus (AAV) rep proteins mediate complex formation between AAV DNA and its integration site in human DNA. *Proc. Natl. Acad. Sci. USA* **91:**5808–5812.

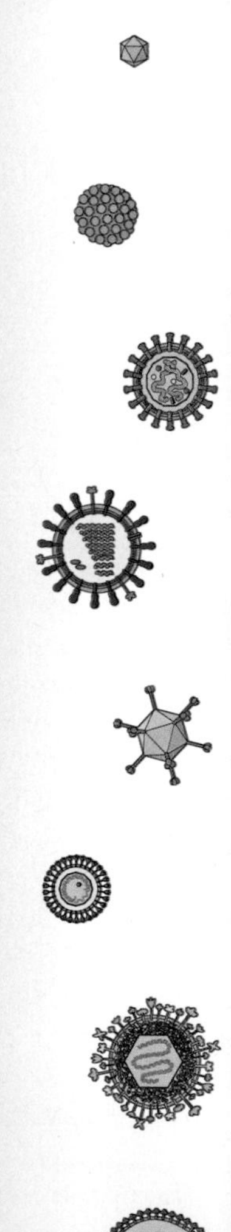

10

Processing of Viral Pre-mRNA

All is flux, nothing stays still.
HERACLITUS

Introduction

Viral messenger RNAs (mRNAs) are synthesized by either viral or cellular enzymes and may be made in the nucleus or the cytoplasm of an infected cell. Regardless of how and where they are made, all must be translated by the protein-synthesizing machinery of the host cell. Consequently, they must conform to the requirements of the cell's translational system. A series of covalent modifications, collectively known as **RNA processing** (Fig. 10.1), endow mRNAs with the molecular features needed for recognition by the protein synthesis machinery, and translation of the protein-coding sequences by cellular ribosomes. Most RNA-processing reactions were discovered in viral systems, primarily because virus-infected cells provide large quantities of specific mRNAs for analysis.

Two modifications important for efficient translation are the addition of m^7GpppN to the 5′ end (**capping**) and the addition of multiple A residues to the 3′ end (**polyadenylation**) (Fig. 10.1). The enzymes that perform these chemical additions (Box 10.1) may be encoded by viral or cellular genes. When an RNA is produced in the nucleus, another chemical rearrangement, called **splicing**, is possible. During splicing, short blocks of noncontiguous coding sequences (**exons**) are joined precisely to create a complete protein-coding sequence for translation, while the intervening sequences (**introns**) are discarded (Fig. 10.1). Splicing therefore dramatically alters the precursor mRNA (pre-mRNA) initially synthesized. As no viral genome encodes even part of the complex machinery needed to catalyze splicing reactions, splicing of viral pre-mRNAs is accomplished entirely by cellular gene products. Some viral pre-mRNAs undergo a different type of internal chemical change, in which a single nucleotide is replaced by another, or one or more nucleotides are inserted at specific positions (Box 10.1). Such **RNA-editing** reactions introduce nucleotides that are not encoded in the genome, and therefore may change the sequence of the encoded protein.

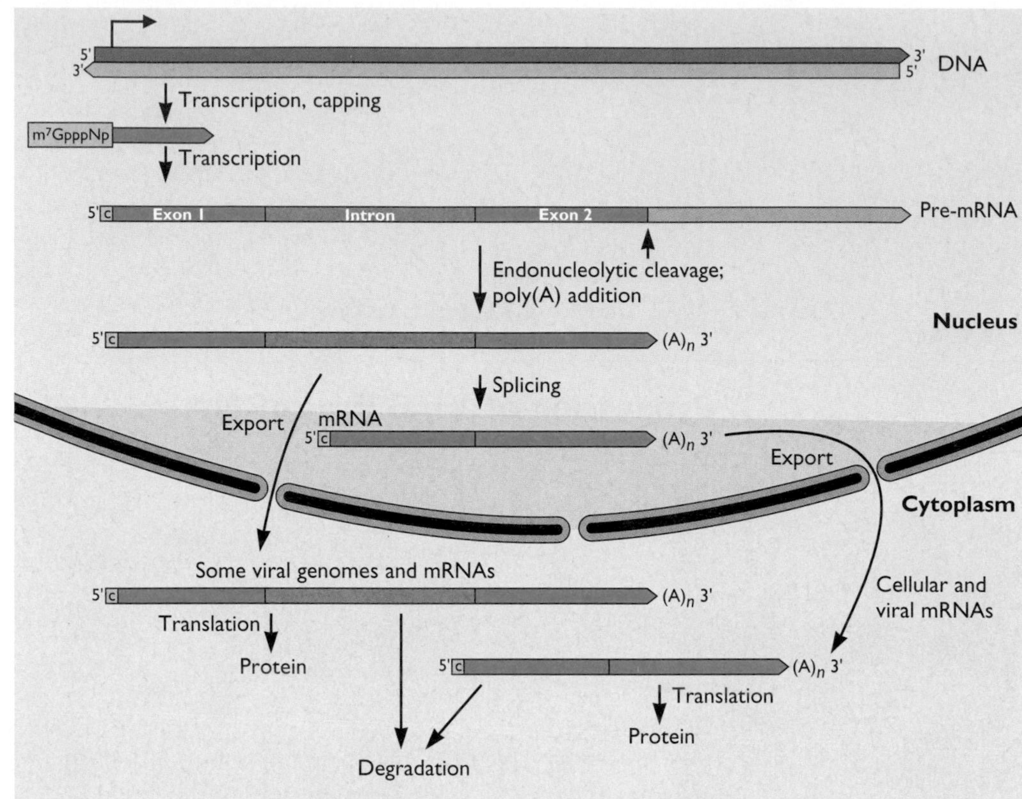

Figure 10.1 Processing of a viral or cellular mRNA synthesized by RNA polymerase II. The reactions by which mature mRNA is made from a typical RNA polymerase II transcript are shown. The first such reaction, capping, takes place cotranscriptionally. For clarity, the exons of a hypothetical, partially processed (i.e., polyadenylated but unspliced) pre-mRNA are depicted, even though polyadenylation and splicing are often coupled, and many splicing reactions are cotranscriptional. Most cellular and viral pre-mRNAs synthesized by RNA polymerase II are processed by this pathway. However, some viral mRNAs that are polyadenylated but not spliced, or are incompletely spliced, are exported to the cytoplasm.

When a viral RNA is produced in the nucleus, it must be exported to the cytoplasm for translation (Fig. 10.1). Such export of mature viral and cellular mRNAs is considered to be part of mRNA processing, even though the RNA is not known to undergo any chemical change during transport. Viral mRNAs invariably leave the nucleus by cellular pathways, but nuclear export mechanisms may be altered in activity or specificity in cells infected by some viruses. Once within the cytoplasm, an mRNA has a finite lifetime before it is recognized and degraded by ribonucleases (RNases). The susceptibilities of individual mRNA species to attack by these destructive enzymes vary greatly, and can be modified in virus-infected cells.

The RNA-processing reactions summarized in Fig. 10.1 not only produce functional mRNAs, but also provide numerous opportunities for posttranscriptional control of gene expression. Regulation of RNA processing can increase the coding capacity of the viral genome, determine when specific viral proteins are made during the infectious cycle, and facilitate selective expression of viral genetic information.

An additional component of the varied repertoire of posttranscriptional mechanisms that regulate viral and cellular gene expression has been recognized much more recently. Cellular and viral genomes encode small RNAs that induce mRNA degradation or inhibition of translation by base pairing to the mRNA. This phenomenon is known as RNA silencing or **RNA interference**. It is mediated by small RNAs 21 to 28 nucleotides in length that are produced by endonucleotytic cleavage of larger precursors in the nucleus and/or the cytoplasm. RNA interference mediated by cellular components is an important component of anti-viral defense. Furthermore, virally encoded small RNAs can alter cellular processes.

In this chapter, we focus on these RNA processing reactions to illustrate both crucial viral regulatory mechanisms, and the seminal contributions of viral systems to the elucidation of essential cellular processes.

BOX 10.1

BACKGROUND

Types of change during pre-mRNA processing

Addition of sequences
- 5′ caps
- 3′ poly(A) tails
- One or a few internal nucleotides by editing

Removal of sequences
- Introns during splicing

Substitution of sequences
- Editing

Relocation
- Export from nucleus to cytoplasm

Covalent Modification during Viral Pre-mRNA Processing

Capping the 5′ Ends of Viral mRNA

The first mRNAs shown to carry the 5′-terminal structure termed the **cap** were those of reovirus and vaccinia virus (Box 10.2). These viral mRNAs are made and processed by virus-encoded enzymes (see "Synthesis of Viral 5′ Cap Structures by Viral Enzymes" below), but subsequent research established that virtually all viral and cellular mRNAs possess the same cap structure, m^7GpppN, where N is any nucleotide (Fig. 10.2A). This structure protects mRNAs from 5′ exonucleolytic attack and allows recognition during splicing of the 5′-terminal exons of pre-mRNAs made by cellular RNA polymerase II. In addition, the cap structure is essential for the efficient translation of most viral and cellular mRNAs, as it is recognized by translation initiation proteins. The principal exceptions are the uncapped mRNAs of certain (+) strand viruses, notably picornaviruses and the flavivirus hepatitis C virus, which are translated by the cap-independent mechanism described in Chapter 11. The cap also blocks recognition of viral RNAs by an antiviral defense mechanism that detects cytoplasmic RNA molecules carrying uncapped 5′ triphosphate groups (see Volume II, Chapter 3).

Although most viral mRNAs carry a 5′-terminal cap structure, there is considerable variation in how this modification is made. Three mechanisms can be distinguished: *de novo* synthesis by cellular enzymes, synthesis by viral enzymes, and acquisition of preformed 5′ cap structures from cellular pre-mRNAs or mRNAs.

Synthesis of Viral 5′ Cap Structures by Cellular Enzymes

Viral pre-mRNA substrates for the cellular capping enzyme are invariably made in the infected cell nucleus by cellular RNA polymerase II (Table 10.1). The formation of cap structures on the 5′ ends of viral and cellular RNA polymerase II transcripts, the first step in their processing, is a cotranscriptional reaction that takes place when the nascent RNA is only 20 to 30 nucleotides in length (Fig. 10.3). Phosphorylation of paired RNA polymerase II at specific serines in the C-terminal domain of the largest subunit is the signal for binding of capping enzyme (see below) and capping of the nascent RNA. The intimate relationship between the cellular capping enzyme and RNA polymerase II ensures that all transcripts made by this enzyme are capped at their 5′ ends. It also explains why RNA species synthesized by other cellular RNA polymerases are not modified in this way: these enzymes do not carry an equivalent of the C-terminal domain of RNA polymerase II.

The 5′ cap structure is assembled by the action of several enzymes as described in Fig. 10.2B. In mammalian cells, a single protein, commonly called **capping enzyme**, contains both the 5′ triphosphatase and the guanylyltransferase required for synthesis of a 5′ cap. Following the action of capping enzyme, the terminal residues are modified by methylation at specific positions (Fig. 10.2B). The cap 1 structure, m^7GpppNm, is common in viral and mammalian mRNAs. However, the sugar of the second nucleotide can also be methylated by a cytoplasmic enzyme, to form the cap 2 structure (Fig. 10.2B).

Synthesis of Viral 5′ Cap Structures by Viral Enzymes

When viral mRNAs are made in the cytoplasm of infected cells, their 5′ cap structures are, of necessity, synthesized by viral enzymes. These enzymes form cap structures typical of those present on cellular mRNA, although with some variations in the mechanism of cap construction. For example, during synthesis of the caps of vesicular stomatitis virus mRNAs, only the α-phosphate group of the initiating nucleotide is retained and guanosine 5′-diphosphate (GDP) is added, and alphaviral mRNAs carry the cap 0 structure (Fig. 10.2B).

Like their cellular counterparts, viral capping enzymes are intimately associated with the viral RNA polymerases responsible for mRNA synthesis. Indeed, in the simplest case, the several enzymatic activities required for synthesis of a 5′ cap structure are supplied by the viral RNA polymerase. This mechanism is exemplified by the vesicular stomatitis virus L protein. This large (>2,000-amino-acid) protein contains discrete domains that catalyze RNA synthesis and cap methylation, and possesses the triphosphatase and guanosine triphosphatase (GTPase) activities necessary for synthesis of the cap. This arrangement presumably facilitates coordination of capping with RNA synthesis. More complex viruses encode dedicated capping enzymes, such as the λ-2 protein of reovirus particles and the VP4 protein of the rotavirus bluetongue

BOX 10.2 EXPERIMENTS
Identification of 5′ cap structures on viral mRNAs

The first clues that the termini of mRNAs made in eukaryotic cells possessed special structures came when viral mRNAs did not behave as predicted from the known structure of bacterial mRNAs. The figure summarizes the identification of 5′ cap structures. The 5′ end of reoviral (or vaccinia virus) mRNA, in contrast to that of a prokaryotic mRNA, could not be labeled by polynucleotide kinase and [γ-^{32}P]ATP (yellow) after alkaline phosphatase treatment. This property established that the 5′ end did not carry a simple phosphate group, but rather was blocked. As indicated below, the structure of the 5′ blocking group (termed the **cap**) was elucidated by differential labeling of specific groups of the viral mRNA, as indicated by colors in the figure, followed by digestion of the mRNA with nucleases with different specificities. In addition to the expected product, 5′pN_{OH}3′, cleavage with P1 nuclease liberated a structure that corresponded to the 5′ terminus of the mRNA (the only position at which β or γ phosphates from nucleoside triphosphates are retained). This structure also contained all the *methyl*-^{3}H label, indicating that it included methyl groups. Its digestion with nucleotide pyrophosphatase produced free phosphate with all the ^{32}P label and two *methyl*-^{3}H-labeled products. Because the terminal structure was known to contain G (from incorporation of ^{32}P from [β,γ-^{32}P]GTP), these nucleotides were identified as m^{7}Gp and 2′-*O*-methylguanosine monophosphate (i.e., methylated at the 2 position of the sugar ring) by comigration with these compounds.

Furuichi, Y., M. Morgan, S. Muthukrishnan, and A. J. Shatkin. 1975. Reovirus messenger RNA contains a methylated, blocked 5′-terminal structure: m^{7}G(5′)ppp(5′)GpCp. *Proc. Natl. Acad. Sci. USA* **72:**362–366.

Wei, C. M., and B. Moss. 1975 Methylated nucleotides block 5′ terminus of vaccinia virus messenger RNA. *Proc. Natl. Acad. Sci. USA* **72:**318–322.

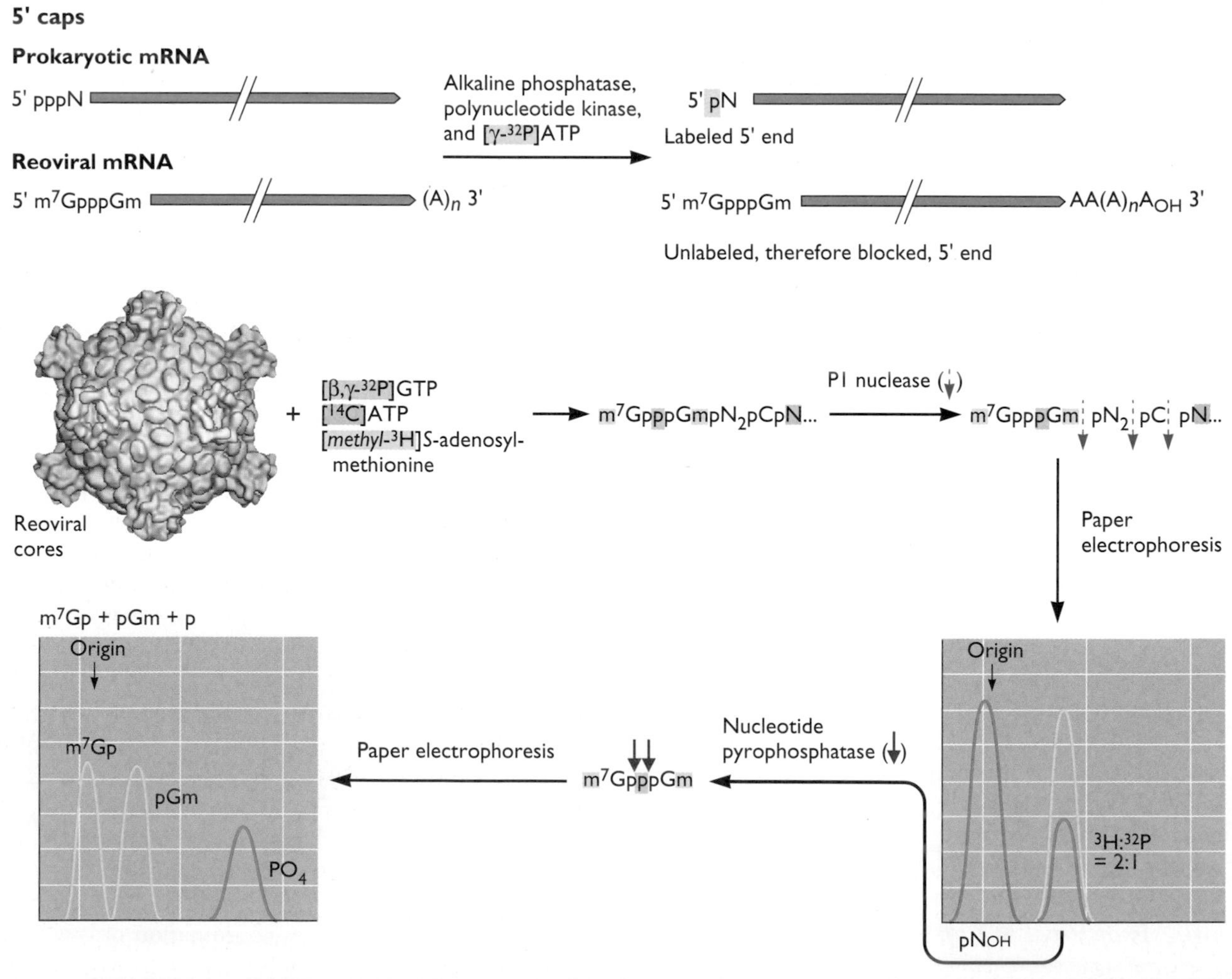

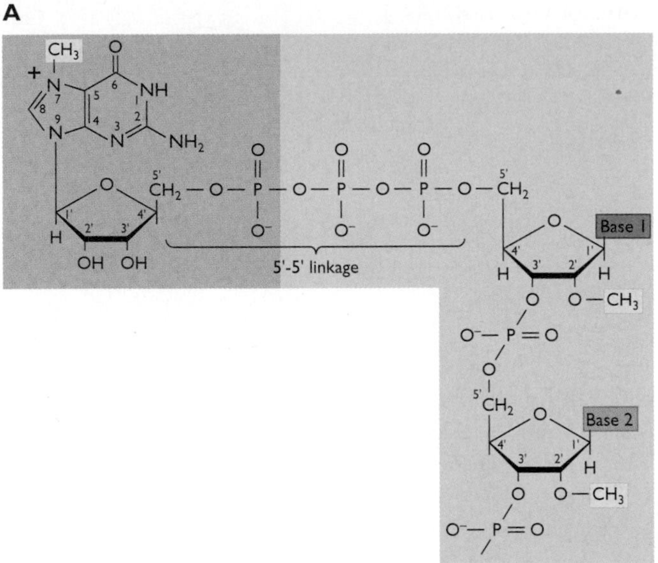

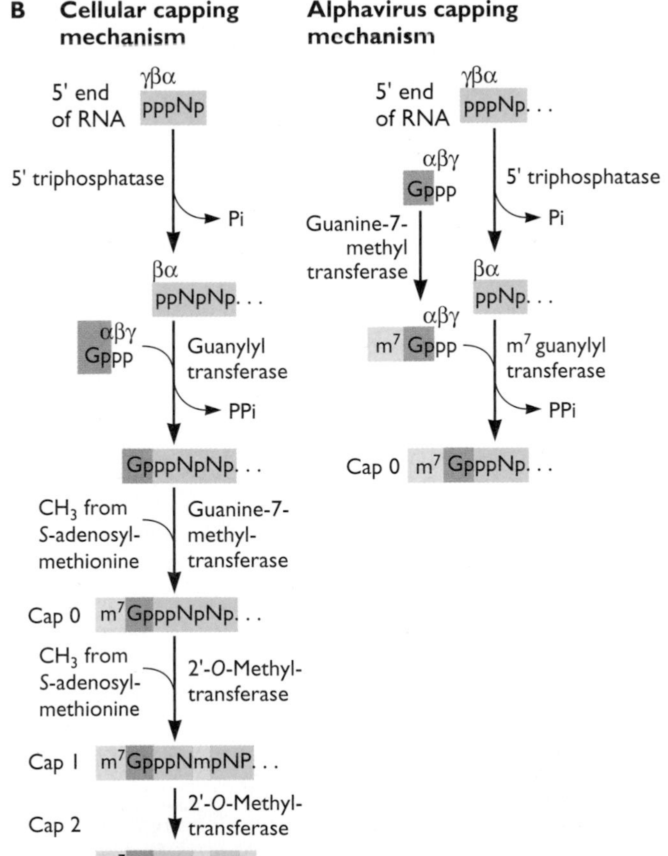

Figure 10.2 The 5′ cap structure and its synthesis by cellular or viral enzymes. (A) In the cap structure shown, Cap 2, the sugars of the two transcribed nucleotides (green) adjacent to the terminal m^{7}G (gray) contain 2′-*O*-methyl groups (yellow). The first and second nucleotides synthesized are methylated in the nucleus and in the cytoplasm, respectively. **(B)** The enzymes and reactions by which this cap is synthesized by cellular enzymes are listed (left) and compared to the synthesis of the caps of Semliki Forest virus (a togavirus) mRNAs by viral enzymes in the cytoplasm of infected cells (right).

virus (both members of the *Reoviridae*). The latter protein catalyses all of the four reactions required for synthesis of the cap 1 structure, and its active sites are organized as a capping "assembly line" (Fig. 10.4). It is closely associated with the viral RNA-dependent RNA polymerase in the core of the virion. One of the first capping enzymes to be analyzed in detail was the vaccinia virus enzyme. This protein binds directly to the viral RNA polymerase and adds 5′ cap structures cotranscriptionally to nascent viral transcripts that are approximately 30 nucleotides in length. The capping enzyme of vaccinia virus therefore displays striking functional similarities to its host cell counterpart.

Most viral capping enzymes cooperate with viral RNA-dependent RNA polymerases that can synthesize both (−) and (+) strand RNAs, but only (+) strand RNAs become capped. The activities of these enzymes must therefore be regulated. The mechanisms that coordinate capping activity with viral mRNA synthesis are not fully understood. In some cases, sequence or structural features of the (+) strand RNA may be recognized by capping enzymes. For example, the methyltransferase of the flavivirus West Nile virus binds specifically to a stem-loop structure at the 5′ end of (+) strand RNA. Substitutions of specific residues within this region inhibit cap methylation and viral replication. In other cases, such as the alphaviruses Sindbis virus and Semliki Forest virus, activation of capping enzymes may be the result of proteolytic processing. As discussed in Chapter 6, the viral P1234 polyprotein is responsible for the initial synthesis of (−) strand RNA from the (+) strand viral genome. This polyprotein includes the sequences of the RNA polymerase and the capping enzyme (Nsp2 and Nsp1, respectively), but the latter is inactive. Cleavage of the polyprotein is necessary for synthesis of viral mRNAs (see Fig. 6.21), and this processing reaction may also activate the capping enzyme.

Acquisition of Viral 5′ Cap Structures from Cellular RNAs

The 5′ cap structures of orthomyxoviral and bunyaviral mRNAs are produced by cellular capping enzymes, but in a unique manner. The 5′ caps of these viral mRNAs are acquired when viral cap-dependent endonucleases cleave cellular transcripts to produce the primers needed for viral mRNA synthesis, a process called **cap snatching** (see Fig. 6.14). The 5′-terminal segments and caps of influenza virus mRNAs are obtained from cellular pre-mRNA in the nucleus. On the other hand, bunyaviral mRNA synthesis

Table 10.1 Mechanisms of synthesis of 5′-terminal cap structures of viral mRNAs

Mechanism	Virus family	Enzyme synthesizing pre-mRNA
Synthesis by host cell enzymes	*Adenoviridae, Hepadnaviridae, Herpesviridae, Papillomaviridae, Parvoviridae, Polyomaviridae, Retroviridae*	Cellular DNA-dependent RNA polymerase II
Synthesis by viral enzymes	*Reoviridae, Rhabdoviridae, Togaviridae*	Viral RNA-dependent RNA polymerase
	Poxviridae	Viral DNA-dependent RNA polymerase
Acquisition from cellular pre-mRNA or mRNA	*Bunyaviridae, Orthomyxoviridae*	Viral RNA-dependent RNA polymerase

is primed with 5′-terminal fragments cleaved from mature cellular mRNAs in the cytoplasm.

Synthesis of 3′ Poly(A) Segments of Viral mRNA

Like the 5′ cap structure, a 3′ poly(A) segment was first identified in a viral mRNA (Box 10.3). This 3′-end modification was soon found to be a common feature of mRNAs made in eukaryotic cells. Like the 5′ cap, the 3′ poly(A) sequence stabilizes mRNA, and it also increases the efficiency of translation. Therefore, it is not surprising that viral mRNAs generally carry a 3′ poly(A) tail. Those that do not, such as reoviral and arenaviral mRNAs, may survive by virtue of a 3′-terminal stem-loop structure that blocks nucleolytic attack. Such structures are also present at the 3′ ends of cellular, poly(A)-lacking mRNAs that encode histones.

The addition of 3′ poly(A) segments to viral pre-mRNAs, like capping of their 5′ ends, can be carried out by either cellular or viral enzymes (Table 10.2). However, cellular and viral polyadenylation mechanisms can differ markedly.

Figure 10.3 Model of cotranscriptional capping of RNA polymerase II transcripts. The elongating RNA polymerase II pauses after 20 to 30 nucleotides have been incorporated into a nascent transcript. The C-terminal domain (CTD) of the largest subunit is then phosphorylated (P) at specific serine residues. Such modification is the signal for binding of the capping enzyme to the CTD and capping (blue) of the nascent RNA.

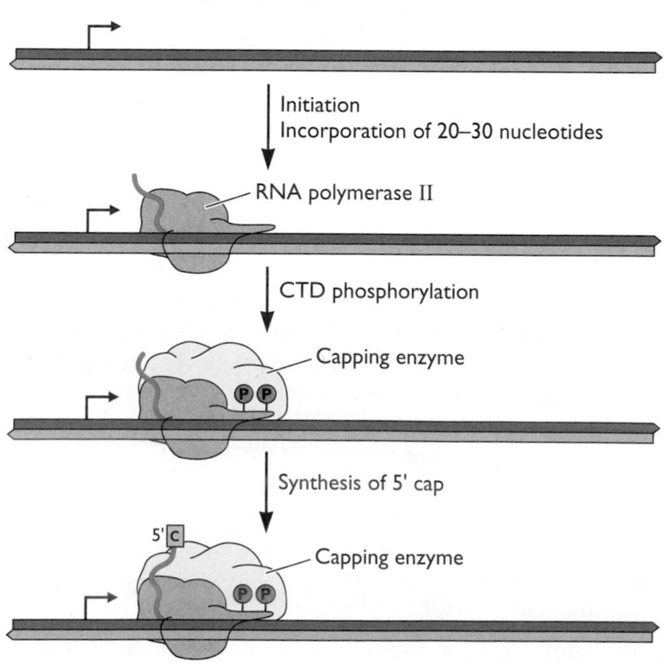

Polyadenylation of Viral Pre-mRNA by Cellular Enzymes

Viral pre-mRNAs synthesized in infected cell nuclei by RNA polymerase II are invariably polyadenylated by cellular enzymes (Table 10.2). Transcription of a viral or cellular gene by RNA polymerase II proceeds beyond the site at which poly(A) will be added. The 3′ end of the mRNA is determined by endonucleolytic cleavage of its pre-mRNA at a specific position. Such cleavage is also required for termination of transcription. Poly(A) is then added to the new 3′ terminus, while the RNA downstream of the cleavage site is degraded (Fig. 10.5). Cleavage and polyadenylation sites are identified by specific sequences, first characterized in simian virus 40 and adenovirus pre-mRNAs. The 3′ end of mature mRNA is formed 10 to 30 nucleotides downstream of a highly conserved and essential polyadenylation signal, the sequence 5′AAUAAA3′. However, this sequence is **not** sufficient to specify poly(A) addition. For example, it is found within mRNAs at internal positions that are never used as polyadenylation sites. Sequences at the 3′ side of the cleavage site, notably a U- or GU-rich sequence located 5 to 20 nucleotides downstream, are also required. In many mRNAs (particularly viral mRNAs), additional sequences 5′ to the cleavage site are also important.

The first reaction in polyadenylation is recognition of the 5′AAUAAA3′ sequence by the protein termed Cpsf (cleavage and polyadenylation specificity protein) (Fig. 10.5). This interaction is stabilized by other proteins (Fig. 10.5). Poly(A) polymerase is then recruited to the complex, prior to cleavage of the pre-mRNA by a subunit of Cpsf. The polymerase then synthesizes a poly(A) segment typically of 200 to 250 nucleotides in a two-stage process (Fig. 10.5).

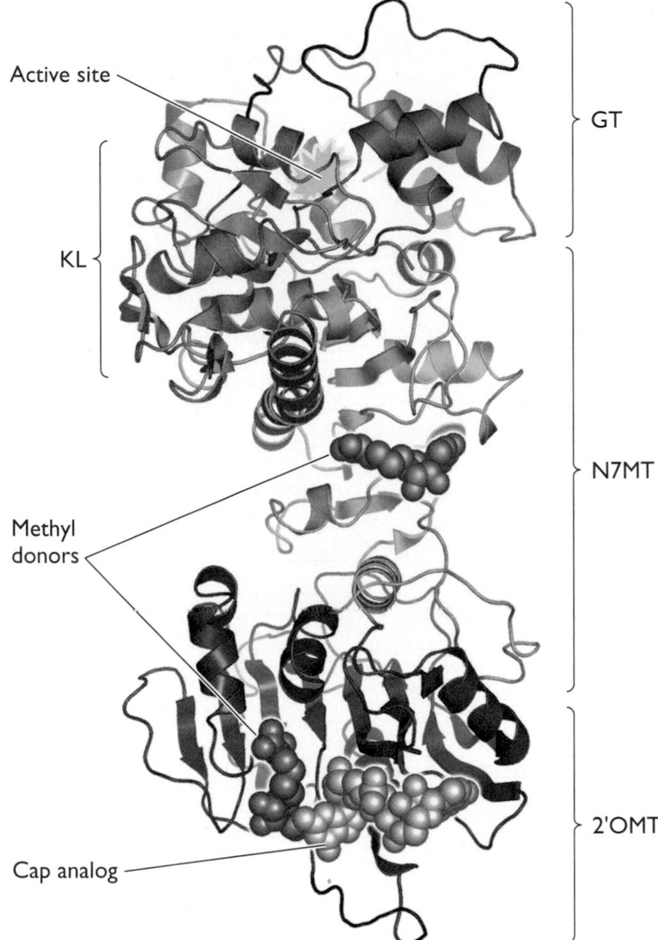

Figure 10.4 A unimolecular assembly line for capping. The structure of the bluetongue virus VP4 protein determined by X-ray crystallography is shown in ribbon form, with each of the four domains in a different color. Localization of the binding sites for substrates and products (e.g., a cap analog) identified the 2′-*O*-methyltransferase (2 O′MT, purple), guanine-7-methyl transferase (N7MT, green), and guanylyltransferase (GT, blue) domains. The latter may also contain the RNA 5′-triphosphatase active site. The linear layout of the active sites in the sequence in which capping reactions take plate (Fig 10.2B) allows efficient coordination of these reactions. The KL domain (orange), which is located on one side of the otherwise linear protein, contains no active sites and is thought to mediate interactions with other proteins, such as the viral RNA-dependent RNA polymerase. Adapted from G. Sutton et al., *Nat. Struct. Mol. Biol.* **14**:449–451, 2007, with permission. Courtesy of Polly Roy, London School of Hygiene and Tropical Medicine.

Components of the polyadenylation machinery, like capping enzymes, are associated with RNA polymerase II, and the polymerase is required for efficient polyadenylation in *in vitro* reactions. Both Cpsf and Csf (Fig. 10.5) bind to the C-terminal domain of the largest subunit of RNA polymerase II. Such interactions are essential for polyadenylation *in vivo,* and are thought to coordinate synthesis of pre-mRNAs with maturation of the 3′ ends.

Polyadenylation of Viral Pre-mRNAs by Viral Enzymes

Formation of 3′ ends by termination of transcription. The 3′ ends of vaccinia virus early mRNAs are formed by termination of transcription by the viral DNA-dependent RNA polymerase at specific sites (Fig. 10.6), a mechanism with no counterpart in cellular or other viral mRNA synthesis systems. Unexpectedly, the vaccinia virus capping enzyme was identified as one protein that is required for termination of transcription. As discussed previously, the capping enzyme binds to the viral RNA polymerase. It is believed to remain with the transcriptional complex until the termination signal 5′UUUUUNU3′ is encountered in the nascent RNA. Termination of transcription, which also requires a specific subunit of the RNA polymerase and a single-stranded DNA-dependent adenosine triphosphatase (ATPase), takes place 30 to 50 nucleotides downstream of this signal. The viral poly(A) polymerase adds poly(A) to the 3′ end thus formed, by a two-step process that is remarkably similar to the mechanism of synthesis of poly(A) by the cellular machinery (compare Fig. 10.6 with Fig. 10.5). The regulatory subunit of the viral poly(A) polymerase that induces processive synthesis of full-length poly(A) after the initial production of short chains is the viral 2′-*O*-methyltransferase. All components of the vaccinia virus capping machinery therefore also participate in formation of the 3′ ends of viral mRNAs. This property seems likely to facilitate coordination of the reactions by which viral mRNAs are synthesized.

Polyadenylation during viral mRNA synthesis. The synthesis of the poly(A) segments of vaccinia virus early mRNAs resembles the cellular mechanism in several respects. In contrast, the poly(A) sequences of mRNAs made by other viral RNA polymerases are produced, during synthesis of the mRNA (Table 10.2). In the simplest case, exemplified by (+) strand picornaviruses, a poly(U) sequence present at the 5′ end of the (–) strand RNA template is copied directly into a poly(A) sequence of equivalent length. The mRNAs of (–) strand RNA viruses like vesicular stomatitis virus and influenza virus are polyadenylated by reiterative copying of short stretches of U residues in the (–) strand RNA template, a mechanism described in Chapter 6.

Splicing of Viral Pre-mRNA

Discovery of Splicing

Between 1960 and the mid-1970s, the study of putative nuclear precursors of mammalian mRNAs established that these RNAs are larger than the mature mRNAs translated in the cytoplasm and are heterogeneous in size.

BOX 10.3

EXPERIMENTS

Identification of poly(A) sequences on viral mRNAs

Polyadenylation of viral mRNAs was first indicated by the observation that an RNA chain resistant to digestion by RNase A, which cleaves after U and C, was produced when vaccinia virus mRNA, but not bacterial mRNA, was treated with this enzyme following labeling with [^{3}H]ATP. The presence of a tract of poly(A) was confirmed by the specific binding of vaccinia virus mRNA, but not of bacterial mRNA, to poly(U)-Sepharose under conditions that allowed annealing of complementary nucleic acids. The position of the poly(A) sequence in viral mRNA was determined by analysis of the products of alkaline hydrolysis, when phosphodiester bonds are broken to produce nucleotides with 5′ hydroxyl and 3′ phosphate (Ap) groups. The liberation of A_{OH3}' by this treatment indicated that the poly(A) was located at the 3′ end of the mRNA.

Kates, J. 1970. Transcription of the vaccinia virus genome and occurrence of polyriboadenylic acid sequences in messenger RNA. *Cold Spring Harbor Symp. Quant. Biol.* **35:**743–752.

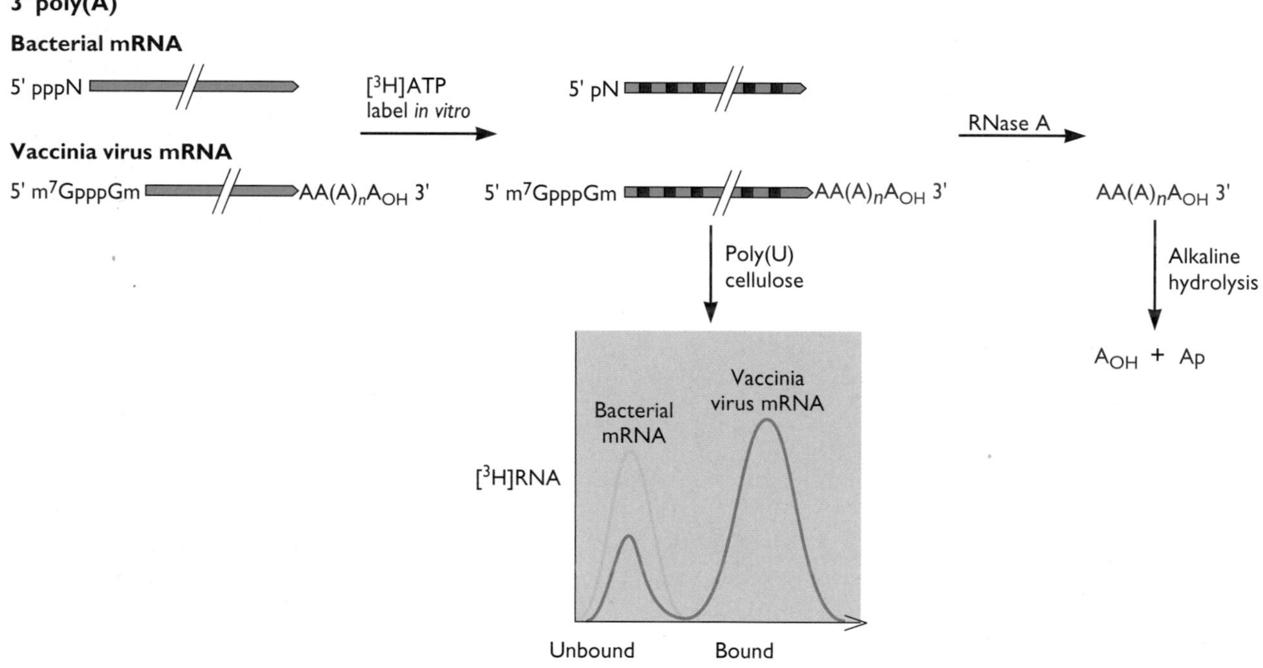

Table 10.2 Mechanisms of addition of poly(A) to viral RNAs

Mechanism	Enzyme	Viruses
During mRNA synthesis		
Reiterative copying of short U stretches in the template RNA	Viral	Orthomyxoviruses, paramyxoviruses, rhabdoviruses
Copying of a long 5′-terminal U stretch in the template RNA	Viral	Picornaviruses, togaviruses
Posttranscriptional		
Cleavage of pre-mRNA followed by polyadenylation	Cellular	Adenovirus, hepadnaviruses, hepatitis delta satellite virus, herpesviruses, papovaviruses, retroviruses
Transcription termination followed by polyadenylation	Viral	Poxviruses

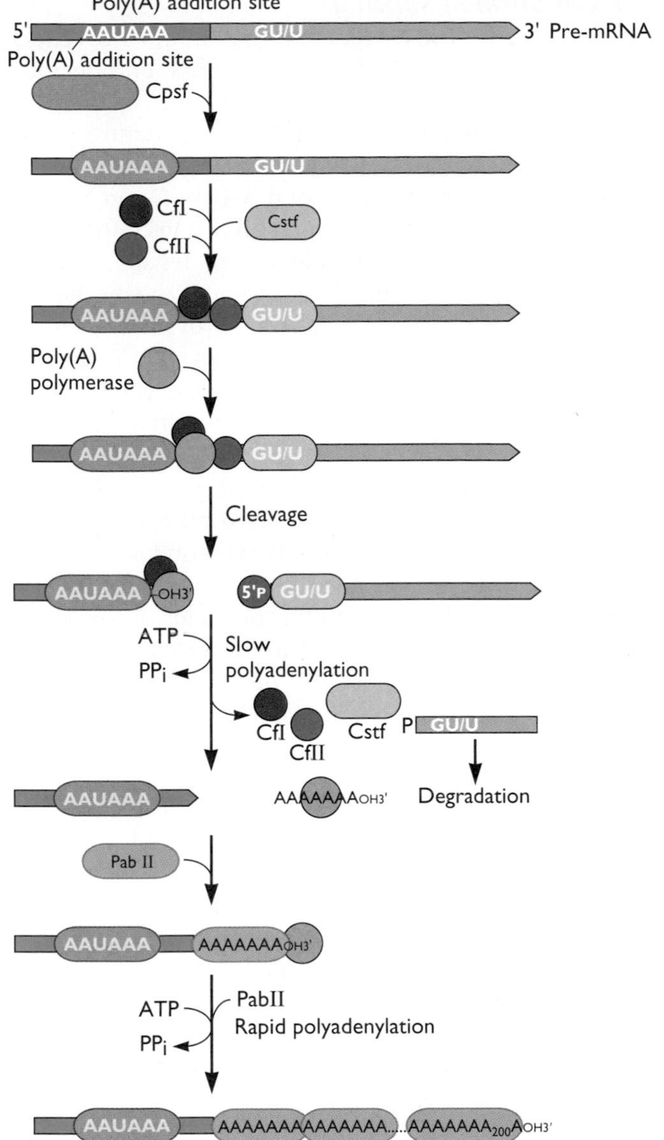

Figure 10.5 Cleavage and polyadenylation of vertebrate pre-mRNAs. The cleavage and polyadenylation specificity protein (Cpsf), which contains four subunits, binds to the 5′AAUAAA3′ poly(A) addition signal that lies upstream of the site at which poly(A) will be added. Cleavage stimulatory protein (Cstf) then interacts with the downstream U/GU-rich sequence to stabilize a complex that also contains the two cleavage proteins, CfI and CfII. Binding of poly(A) polymerase is followed by cleavage at the poly(A) addition site by a subunit of Cpsf, and CfI, CfII, Cstf, and the downstream RNA cleavage product are then released. The polymerase slowly adds 10 to 15 A residues to the 3′-OH terminus produced by the cleavage reaction. Poly(A)-binding protein II (Pab II) then binds to this short poly(A) sequence and, in conjunction with Cpsf, tethers poly(A) polymerase to the poly(A) sequence. This association facilitates rapid and processive addition of A residues until a poly(A) chain of about 200 residues has been synthesized.

They were therefore named **heterogeneous nuclear RNAs** (hnRNAs). Such hnRNAs were shown to carry both 5′-terminal cap structures and 3′ poly(A) sequences, leading to the conclusion that both ends of the hnRNA were preserved in the smaller, mature mRNA. Investigators were faced with the dilemma of deducing how smaller mRNAs could be produced from larger hnRNAs while both ends of the hnRNA were retained.

The puzzle was solved by two groups of investigators, led by Phillip Sharp and Richard Roberts, who shared the 1993 Nobel Prize in physiology or medicine for this work. These investigators showed that adenoviral major late mRNAs are encoded by four **separate** genomic sequences (Box 10.4). The distribution of the mRNA-coding sequences into four separate blocks in the genome, in conjunction with the large size of major late mRNA precursors, implied that these mRNAs were synthesized by excision of noncoding sequences from primary transcripts, with precise joining of coding sequences. The demonstration that nuclear major late transcripts contain the noncoding sequences (introns) confirmed that the mature mRNAs are formed by **splicing** of noncontiguous coding sequences in the pre-mRNA.

This mechanism of mRNA synthesis had great appeal, because it could account for the puzzling properties of hnRNA. Indeed, it was shown within a matter of months that splicing of pre-mRNA is not an obscure, virus-specific device: splicing occurs in all eukaryotic cells. The great majority of mammalian pre-mRNAs, like the adenoviral major late mRNAs, comprise exons interspersed among introns. In all cases, the capped 5′ end and sequences immediately adjacent to the 3′ poly(A) tail are conserved in mature mRNA.

The organization of protein-coding sequences into exons separated by introns has profound implications for the evolution of the genes of eukaryotes and their viruses. Introns are generally much longer than exons, and only short sequences at their ends are necessary for accurate splicing (see "Mechanism of Splicing" below). Consequently, introns provide numerous sites at which DNA sequences can be broken and rejoined without loss of coding information, and greatly increase the frequency with which random recombination reactions can create new functional genes by rearrangement of exons. Evidence of such "exon shuffling" can be seen in the modular organization of many modern proteins. Such proteins comprise combinations of a finite set of structural and functional domains or motifs, or multiple repeats of a single protein domain, each often encoded by a single exon. In similar fashion, the presence of introns is thought to have facilitated the transfer (by recombination) of cellular genetic information into the genomes of DNA viruses and retroviruses.

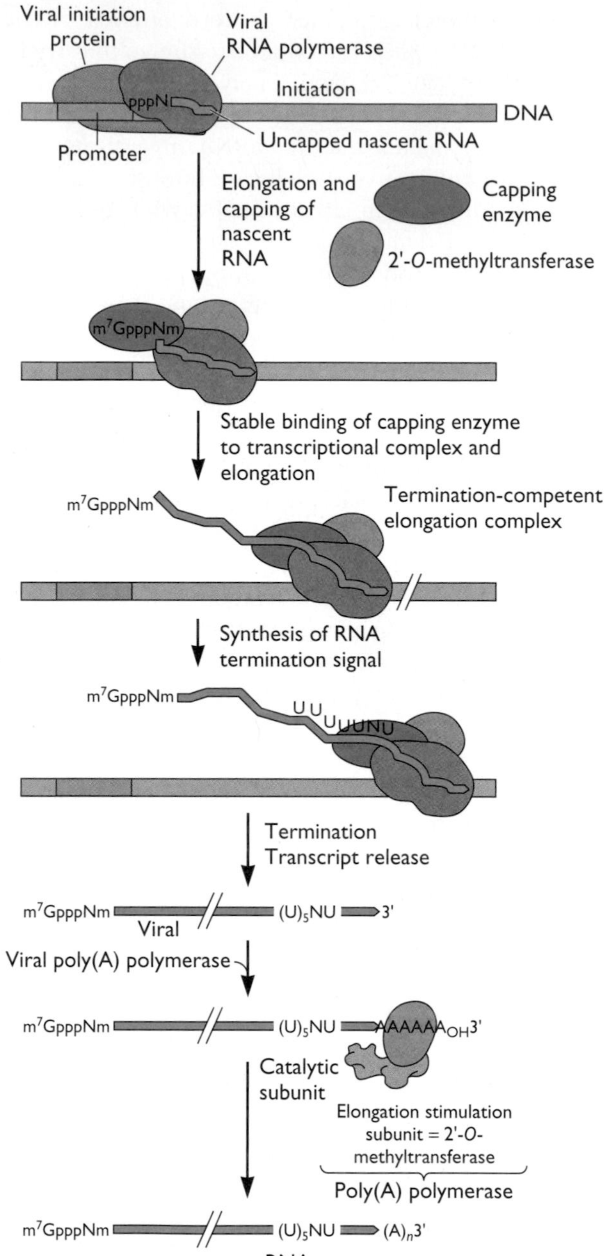

Figure 10.6 The vaccinia virus capping enzyme and 2′-O-methyltransferase process both the 5′ and 3′ ends of vaccinia virus mRNAs. After capping the 5′ ends of nascent viral mRNA chains about 30 nucleotides in length (step 1), the capping enzyme remains bound to the nascent RNA chain and to the RNA polymerase as the latter enzyme transcribes the template DNA. The viral 2′-*O*-methytransferase, which produces a cap 1 structure, also binds to the viral RNA polymerase and stimulates elongation during transcription of viral intermediate and late genes (step 2). This protein is also a subunit of the viral poly(A) polymerase. Termination of transcription (step 3), which takes place 30 to 50 nucleotides downstream of the RNA sequence 5′UUUUUNU3′, is mediated by the termination protein/capping enzyme and the viral nucleoside triphosphate phosphohydrolase I, which is a single-stranded DNA-dependent ATPase. A fraction of the 2′-*O*-methyltransferase molecules act as an elongation stimulation protein for the viral poly(A) polymerase, analogous to cellular poly(A)-binding protein II. This viral enzyme, like its cellular counterpart (Fig. 10.4), adds poly(A) to the 3′ ends of the mRNA in a two-step process (steps 4 and 5).

Mechanism of Splicing

Sequencing of DNA copies of a large number of cellular and viral mRNAs and of the genes that encode them identified short consensus sequences at the 5′ and 3′ **splice sites**, which are joined to each other in mature mRNA (Fig. 10.7A). The conserved sequences lie largely within the introns. The dinucleotides GU and AG are found at the 5′ and 3′ ends, respectively, of almost all introns. Mutation of any one of these four nucleotides eliminates splicing, indicating that all are essential. Elucidation of the mechanism of splicing came with the development of *in vitro* systems in which model pre-mRNAs (initially of viral origin) are accurately spliced. Numerous components needed for splicing have now been identified by both biochemical studies based on such systems and genetic analyses in organisms such as yeast and the fruit fly *Drosophila melanogaster*.

Pre-mRNA splicing occurs by two *trans*-esterification reactions, in which one phosphodiester bond is exchanged for another without the need for an external supply of energy (Fig. 10.7B). The first reaction yields two products, the 5′ exon and the intron-3′ exon **lariat**. In the second *trans*-esterification reaction, the newly formed hydroxyl group at the end of the 5′ exon attacks the phosphodiester bond at the 3′ splice site, yielding the spliced exon product and the intron lariat.

From a chemical point of view, the splicing of pre-mRNA is a simple process. However, each splicing reaction must be completed with a high degree of accuracy to ensure that coding information is not lost or altered. The chemically active hydroxyl groups (Fig. 10.7B) must also be brought into close proximity to the phosphodiester bonds they will attack. Furthermore, some genes contain 50 or more exons separated by **much** larger introns, which must be spliced in the correct order. It is presumably for such reasons that pre-mRNA splicing requires both many proteins and several RNAs, which associate with the pre-mRNA to form the large structure called the **spliceosome**.

Five small nuclear RNAs (snRNAs) participate in splicing: the U1, U2, U4, U5, and U6 snRNAs. In vertebrate cells, these RNAs vary in length from 100 to 200 nucleotides and are associated with proteins to form **small nuclear ribonucleoproteins** (snRNPs). The RNA components of the snRNPs recognize splice sites and other sequences in cellular and viral pre mRNAs. Indeed, they participate in multiple interactions with the pre-mRNA and with each

BOX 10.4

EXPERIMENTS

Discovery of the spliced structure of adenoviral major late mRNAs

(A) Digestion of adenoviral major late mRNAs with RNase T_1, which cleaves after G, and isolation of the capped 5′ oligonucleotides indicated the **same** 11-nucleotide sequence was present at the 5′ ends of several different mRNAs. This observation was surprising, and puzzling. Hybridization studies indicated that these 5′ ends were not encoded adjacent to the main segments of major late mRNAs. Direct visualization of such mRNAs hybridized to viral DNA provided convincing proof that their coding sequences are dispersed in the viral genome. **(B)** Schematic diagram of one major late mRNA (hexon mRNA) hybridized to a complementary adenoviral DNA fragment extending from the left end of the genome to a point within the hexon coding sequence. Three loops of unhybridized DNA (thin lines), designated A, B, and C, bounded or separated by three short segments (1, 2, and 3) and one long segment (hexon mRNA) of DNA-RNA hybrid (thick lines) were observed. Other adenoviral late mRNAs examined yielded the same sets of hybridized and unhybridized viral DNA sequences at their 5′ ends, but differed in the length of loop C, and the length and location of the 3′-terminal RNA-DNA hybrid. It was therefore concluded that the major late mRNAs contain a common 5′-terminal segment (segments 1, 2, and 3) built from sequences encoded at three different sites in the viral genome, and termed the tripartite leader sequence. This sequence is joined to the mRNA body, a long sequence complementary to part of the hexon coding sequence in the example shown. (B) Adapted from S. M. Berget et al., *Proc. Natl. Acad. Sci. USA* **74:**3171–3175, 1977, with permission.

Berget, S. M., C. Moore, and P. A. Sharp. 1977. Spliced segments at the 5′ terminus of adenovirus 2 late mRNA. *Proc. Natl. Acad. Sci. USA* **74:**3171–3175.

Chow, L. T., R. E. Gelinas, T. R. Booker, and R. J. Roberts. 1977. An amazing sequence arrangement at the 5′ ends of adenovirus 2 messenger RNA. *Cell* **12:**1–8.

Gelinas, R. E., and R. J. Roberts. 1977. One predominant undecanucleotide in adenovirus late messenger RNAs. *Cell* **11:**533–544.

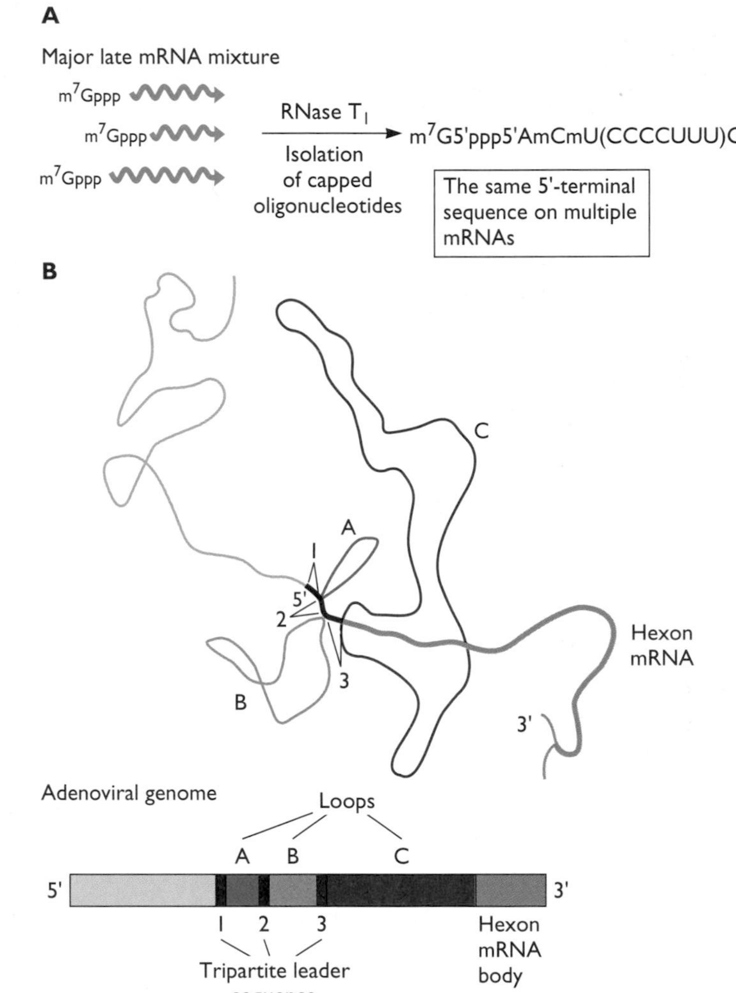

other during splicing. Assembly of the spliceosome results in base pairing of the snRNA components of the U1 and U2 snRNPs with the 5′ splice site and the intronic branch point, respectively (Fig. 10.8). Dramatic rearrangements of the RNA-RNA interactions within the spliceosome then result is juxtaposition of the 5′ splice site and the branch point for the first transesterification reaction (Fig. 10.8). The RNA of the U5 snRNP may juxtapose the 5′ and 3′ exons to facilitate catalysis of their joining in the second transesterification reaction. However, RNAs of the snRNPs do much more than simply organize the pre-mRNA sequences into a geometry suitable for transesterification. It has long been suspected that the spliceosome might be an RNA enzyme (or **ribozyme**), and several observations

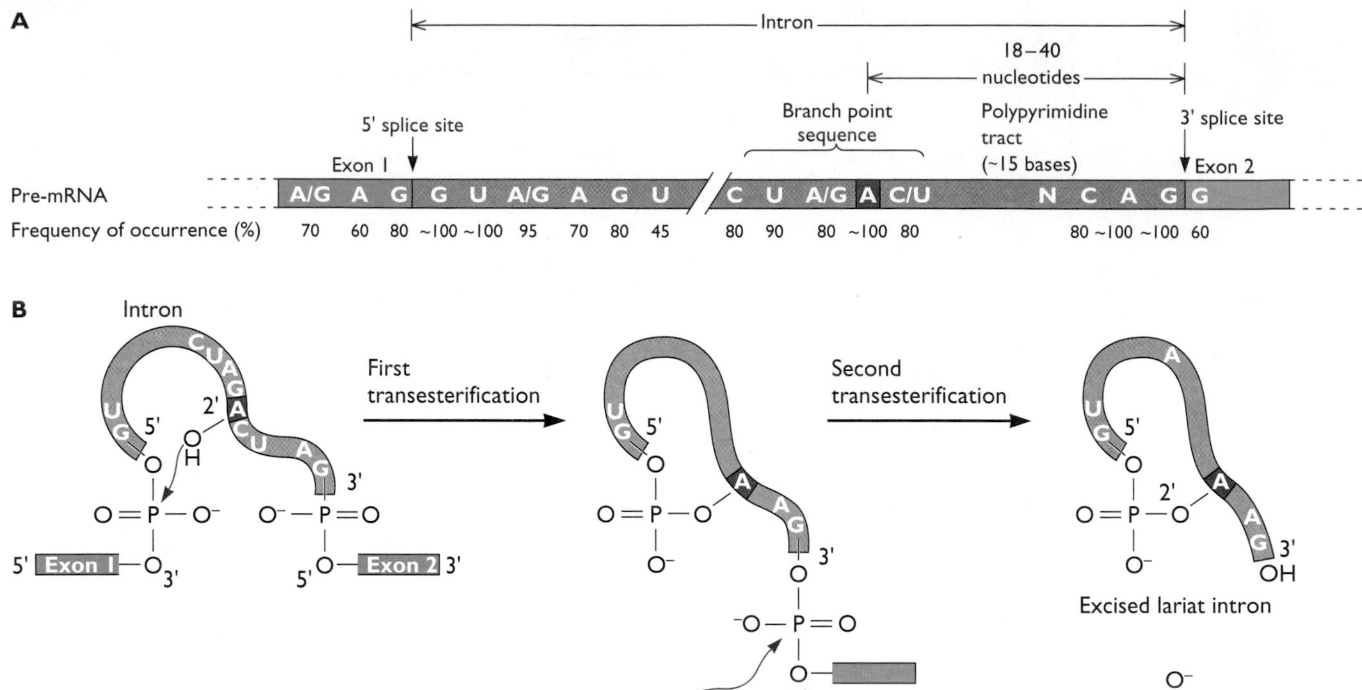

Figure 10.7 Splicing of pre-mRNA. (A) Consensus splicing signals in cellular and viral pre-mRNAs. The most conserved sequences are found at the 5′ and 3′ splice sites at the junctions of exons (green) and introns (pink) and at the 3′ ends of introns. The intronic 5′GU3′ and 5′AG3′ dinucleotides at the 5′ and 3′ ends, respectively, of introns and branch point A (highlighted) are present in all but rare mRNAs made in higher eukaryotes. **(B)** The two transesterification reactions of pre-mRNA splicing. In the first reaction, the 2′ hydroxyl group of the conserved A residue in the intronic branch point sequence makes a nucleophilic attack on the phosphodiester bond at the 5′ side of the GU dinucleotide at the 5′ splice site to produce the intron-3′ exon lariat and the 5′ exon. A second nucleophilic attack by the newly formed 3′ hydroxyl group of the 5′ exon on the phosphodiester bond at the 3′ splice site then yields the spliced exons and the intron lariat.

have provided direct evidence for catalysis by U6 and U2 snRNAs during splicing (Box 10.5).

Although the RNAs of snRNPs play essential roles in splicing as both guides and catalysts, the spliceosome also contains about 150 non-snRNP proteins. One class comprises proteins that package the pre-mRNA substrate, termed heterogeneous nuclear ribonucleoprotein (hnRNP) proteins. Many other splicing proteins contain both RNA-binding and protein-protein interaction domains. Such proteins bind to pre-mRNA sequences within or adjacent to exons, and regulate splicing or exon recognition. Members of one family of such proteins, called the SR proteins because they contain domains rich in serine (S) and arginine (R), act at early stages of splicing to recruit snRNPs to the spliceosome. For example, the SR protein named splicing factor 2 (Sf2, or alternative splicing factor) facilitates binding of U1 snRNP to 5′ splice sites and stabilizes the prespliceosomal complex (Fig. 10.8). Other proteins important for splicing are RNA-dependent helicases, which are thought to catalyze the multiple rearrangements of hydrogen bonding among different snRNAs and the pre-mRNA substrate (Fig. 10.8). Such helicases are generally ATP dependent, and spliceosome assembly and rearrangement depend on energy supplied by ATP hydrolysis.

Splicing of pre-mRNAs is commonly cotranscriptional, and components of the splicing machinery associate with RNA polymerase II. The hyperphosphorylated form of the C-terminal domain of the largest subunit of this enzyme can bind to both SR proteins and spliceosomal snRNPs. Peptide mimics of the C-terminal domain or antibodies raised against it inhibit pre-mRNA splicing *in vivo* or *in vitro*. Furthermore, nontranscribed sequences within RNA polymerase II promoters can dictate whether a particular exon is retained or removed during splicing. As we have

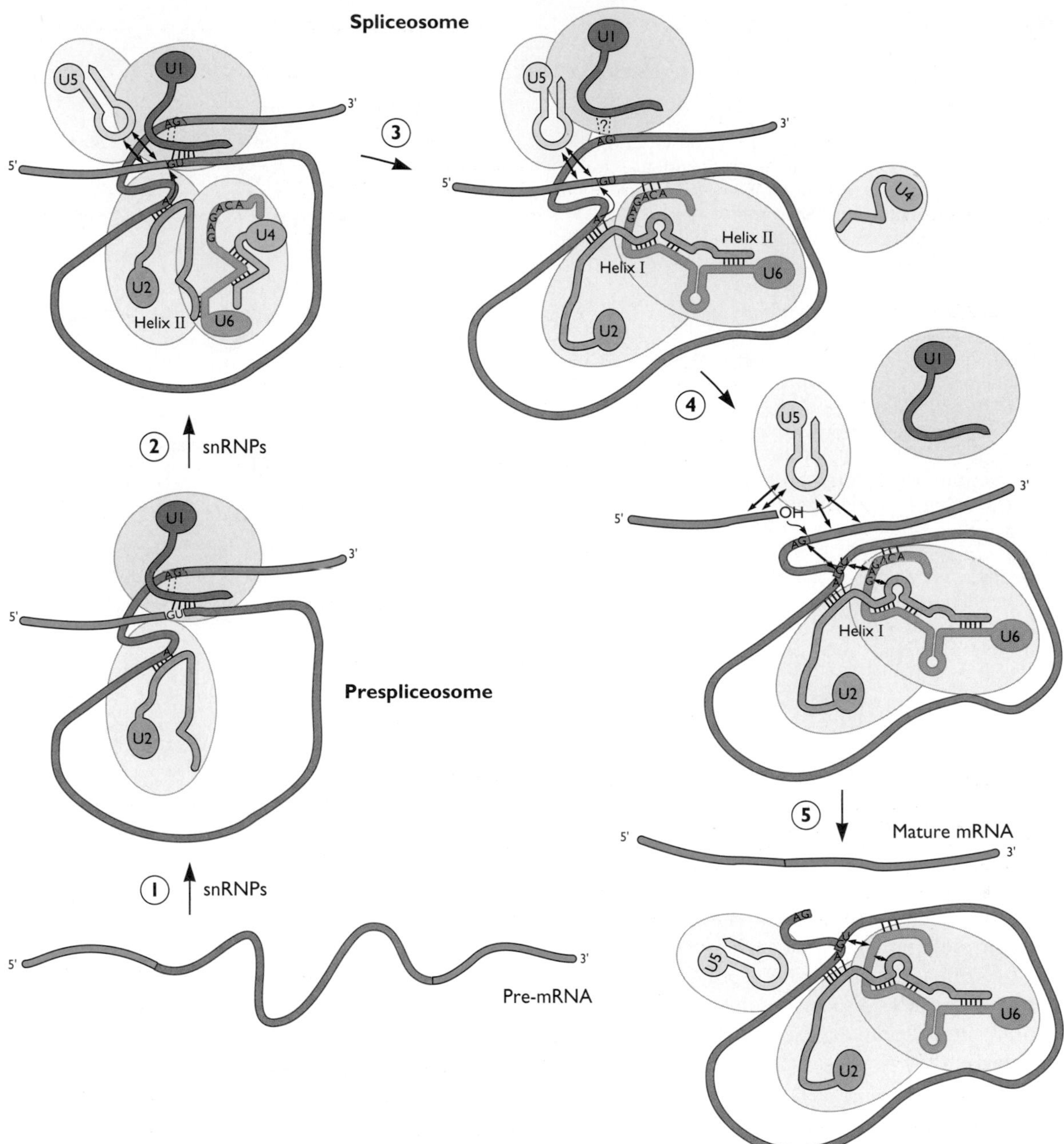

Figure 10.8 RNA-RNA interactions organize substrates and catalysts during splicing. Base pairs are indicated by dashes and experimentally observed or presumed contacts among the RNA molecules by the two-headed arrows. The U1 and U2 snRNAs initially base pair with the 5′ splice site and branch point sequence, respectively, in the pre-mRNA (step 1). The other snRNPs then enter the assembling spliceosome (step 2). The U4 and U6 snRNAs, which are present in a single snRNP, are base paired with one another over an extended complementary region. This snRNP binds to the U5 snRNP, and the snRNP complex associates with that containing the pre-mRNA and U1 and U2 snRNAs to form the spliceosome. RNA rearrangements then activate spliceosomes for catalysis of splicing (step 3). U4 snRNA dissociates from U6 snRNA, which forms hydrogen bonds with both U2 snRNA and the pre-mRNA. The interaction of U6 snRNA with the 5′ splice site displaces U1 snRNA (step 4). One of the U2 sequences hydrogen bonded to U6 snRNA (helix I) is adjacent to the U2 snRNA sequence that is base paired with the pre-mRNA branch point region. The interactions of U2 and U6 snRNAs with each other and with the pre-mRNA therefore juxtapose the branch point and 5′ splice site sequences for the first transesterification reaction (Fig. 10.7B). The U5 snRNA base pairs to sequences in both the 5′ and 3′ exons and may align them for the second transesterification reaction (step 5). The many proteins that participate in spliceosome assembly and activation or that package the pre-mRNA and snRNAs are not shown. Adapted from T. Nilsen, *Cell* **78:**1–4, 1994, with permission.

BOX 10.5

DISCUSSION

RNA catalysis by U2 and U6 snRNAs takes place in the absence of any proteins

In the absence of proteins, U2 and U6 snRNA synthesized *in vitro* base pair to form a stable structure with a three-dimensional conformation similar to that thought to be present at the active site of the spliceosome (shown schematically in the figure). This U2-U6 snRNA complex binds specifically to a short, synthetic RNA molecule containing the sequence of an intron branch point. The branch point adenosine is activated when the substrate RNA is bound to the snRNA complex, and its 2′-OH group attacks a specific phosphodiester bond in U6 RNA to form an unusual phosphotriester bond. This reaction is analogous to the first transesterification during pre-mRNA splicing by the spliceosome. It requires RNA sequences necessary for authentic splicing, including a specific base-pairing interaction between U2 and U6 snRNAs and a particular U6 RNA sequence. These observations therefore provide direct evidence for RNA catalysis of the first splicing reaction.

Valadkhan, S., and J. L. Manley. 2001. Splicing-related catalysis by protein-free snRNAs. *Nature* **413:**701–707.

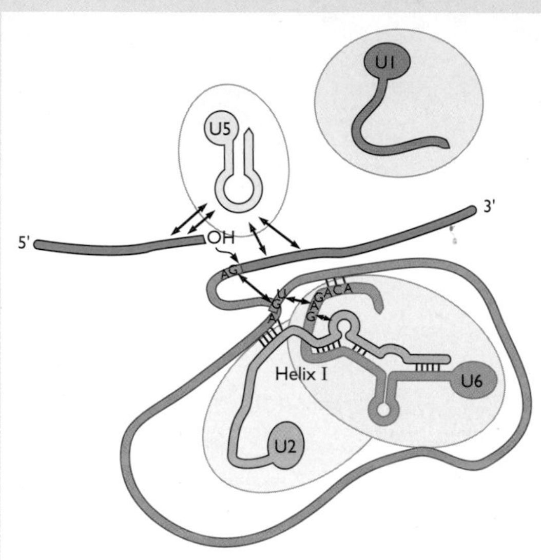

seen, association of components of the 5′ capping and 3′ polyadenylation systems with the C-terminal domain of RNA polymerase is necessary for these processing reactions. The synthesis of pre-mRNA and its complete processing are therefore coordinated as a result of association of specific proteins needed for each processing reaction with the C-terminal domain of RNA polymerase II. Such a transcription and RNA-processing machine is analogous to, but much more complex than, that of vaccinia virus described above.

Production of Stable Viral Introns

Cells infected by certain herpesviruses provide a remarkable exception to the rule that viral intron-containing pre-mRNAs are processed exactly as their cellular counterparts. As discussed in Chapter 8, the most abundant viral RNAs detected in neurons latently infected by herpes simplex viruses are the 2.0-kb major latency-associated transcripts (LATs). These nuclear RNAs not only lack 3′ poly(A) sequences, but also exhibit the properties of excised introns: they are not linear molecules and contain the branch points characteristic of the intron lariats excised from pre-mRNAs during splicing. Although no viral products are necessary for synthesis of the LAT intron, the processing of LAT pre-mRNA is much more efficient in sensory ganglia, in which herpes simplex virus type 1 establishes latent infection (Box 10.6). In contrast to typical introns removed from other viral and cellular pre-mRNAs, these herpesviral RNAs are remarkably stable, with a half-life of 24 h, and accumulate to high concentrations in the nuclei of latently infected cells. The presence of a branch point sequence that is not a close match to the consensus (Fig. 10.7A) is an important determinant of such stability. As discussed in Chapter 8, various functions have been ascribed to this unusual product of viral RNA processing.

Alternative Splicing of Viral Pre-mRNA

Many viral and cellular pre-mRNAs contain multiple exons. Splicing of the majority removes all introns and joins all exons in the order in which they are present in the substrate pre-mRNA. Such **constitutive splicing** (Fig. 10.9A) produces a single mature mRNA. However, numerous cellular and many viral pre-mRNAs yield more than one mRNA as a result of the splicing of different combinations of exons, a process termed **alternative splicing**. Several different types of alternative splicing can be defined (Fig. 10.9B). Alternative splicing generally leads to the synthesis of mRNAs that differ in their protein-coding sequences.

BOX 10.6

EXPERIMENTS

Synthesis of the herpes simplex virus latency-associated transcript intron is tissue specific

The production and processing of the latency-associated transcript (LAT) in different organs and tissues were examined by creation of a line of transgenic mice carrying in their genomes the region of the herpes simplex virus type 1 genome spanning the LAT coding sequence and promoter. This approach allowed LAT RNA synthesis and processing to be examined in any organ or tissues (including those not infected by herpes simplex virus) and in the absence of other viral sequences or promoters. Expression of the LAT gene was detected in several organs and tissues including brain, liver, and kidney. However, as shown in the figure, the processed LAT intron was observed by fluorescent *in situ* hybridization only in neurons of the sensory ganglia, in which the virus establishes latency, such as dorsal root and trigeminal ganglia. Quantitative assays to measure the relative concentrations of unspliced and spliced LAT RNAs confirmed that splicing to produce the LAT intron is much more efficient in sensory ganglia than in other tissues.

Gussow, A. M., N. V. Giorgani, R. K. Tran, Y. Imai, D. L. Kwiatkowski, G. R. Full, T. P. Margolis, and D. C. Bloom. 2006. Tissue-specific splicing of the herpes simplex virus type latency-associated transcript (LAT) intron in LAT transgenic mice. *J. Virol.* **80:**9414–9423.

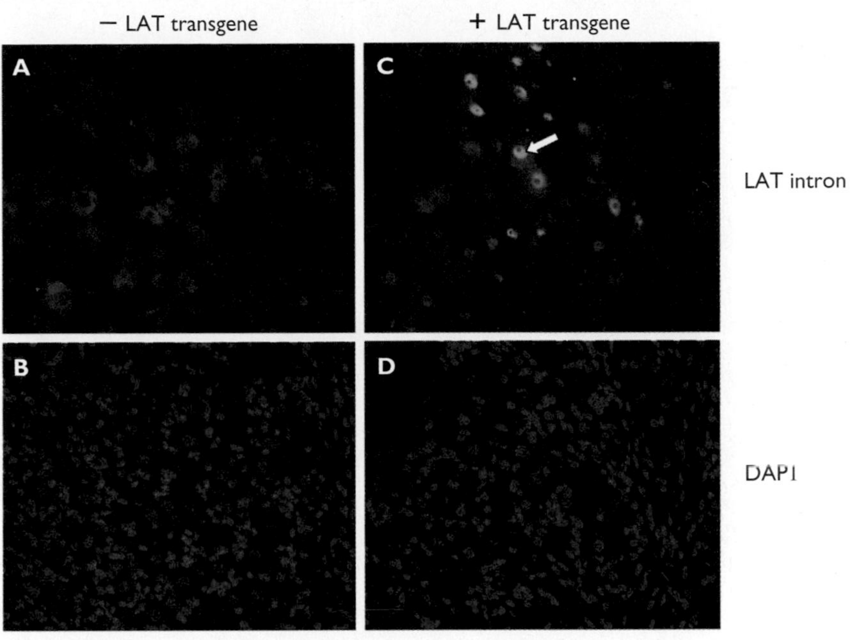

Trigeminal neurons from LAT nontransgenic mice **(A)** or LAT transgene siblings **(C)** were hybridized to a digoxigenin-containing complementary RNA specific for spliced LAT RNA, incubated with fluorescein-labeled anti-digoxigenin antibodies, and examined by fluorescent microscopy. **(B and D)** DNA staining of the fields shown in panels A and C, respectively. The many nuclei that do not contain the LAT intron **(C and D)** are those of glial cells. Reprinted from A. M. Gussow et al., *J. Virol.* **80:**9414–9423, 2006, with permission.

The most obvious advantage of this mechanism is that it can expand the limited coding capacity of viral genomes. For example, the early genes of polyomaviruses, as well as the adenoviral E1A and E4 genes, each specify two or more proteins as a result of splicing of primary transcripts at alternative 5′ or 3′ splice sites. Alternative splicing can also be important for temporal regulation of viral gene expression, or the control of a crucial balance in the production of spliced and unspliced-mRNAs. In many cases, alternative splicing of viral pre mRNAs is coupled with other RNA-processing reactions, or regulated by viral proteins. These more complex phenomena are considered in later sections.

Examples of Alternative Splicing of Viral Pre-mRNAs

Cellular differentiation regulates splicing of papillomaviral late pre-mRNAs. The late proteins of bovine papillomavirus type 1 are synthesized efficiently only in highly differentiated keratinocytes. Productive replication of viral DNA and assembly and release of virions are also restricted to these outer cells in an epithelium. Alterations in splicing are crucial for production of the mRNA encoding the major capsid protein, the L1 mRNA (Fig. 10.10A). *In situ* hybridization studies have shown that this mRNA is made only in fully differentiated cells (Fig. 10.10B). Production of the L1 mRNA requires activation of an alternative 3′ splice site that is not recognized in undifferentiated cells. Several *cis*-acting sequences in the viral late pre-mRNA govern the choice of 3′ splice sites (Fig. 10.10C). These sequences are bound by cellular splicing proteins, including SR proteins. Analogous sequences regulate splicing of human papillomavirus late pre-mRNAs. It has therefore been proposed that terminal differentiation of keratinocytes is accompanied by changes in the activity or abundance of these cellular proteins, such that L1 mRNA can be synthesized.

A Constitutive splicing **B Alternative splicing**

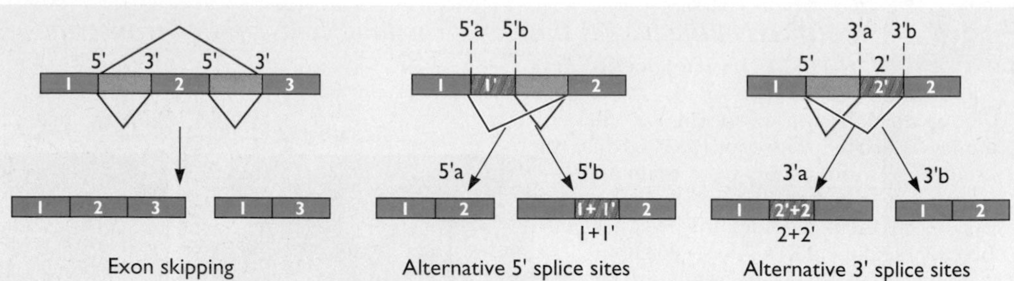

Figure 10.9 Constitutive and alternative splicing. (A) In constitutive splicing all exons (green) are joined sequentially and all introns (pink) are excised. **(B)** Alternative splicing occurs by several mechanisms. In exon skipping, the 3′ splice site of exon 2 is sometimes ignored, so that this exon is not included in some fraction of the spliced mRNA molecules. Alternatively, one of two 5′ splice sites (5′a and 5′b) in exon 1 or one of two 3′ splice sites (3′a and 3′b) in exon 2 are recognized. Recognition of different 5′ and 3′ splice sites produces alternatively spliced simian virus 40 early and adenoviral major late (Fig. 10.11) mRNAs, respectively.

Production of spliced and unspliced RNAs essential for virus replication. The expression of certain coding sequences in retroviral genomes (Gag and Pol) (Appendix, Fig. 19.8) and orthomyxoviruses (M1 and NS1) (Appendix, Fig. 7B) depends on an unusual form of alternative splicing that produces both spliced and unspliced mRNAs: normally all pre-mRNAs that contain introns are fully spliced in mammalian cell nuclei. This phenomenon has been well studied in retrovirus-infected cells, in which the viral genome is produced as unspliced viral transcripts.

In cells infected by retroviruses with simple genomes, such as avian leukosis virus, a full-length, unspliced transcript of proviral DNA serves as both the genome and the mRNA for the capsid proteins and viral enzymes, while a singly spliced mRNA specifies the viral envelope protein (Fig. 10.11A). Retrovirus production depends rather critically on the maintenance of a proper balance between the proportions of unspliced and spliced RNAs: modest changes in splicing efficiencies cause replication defects (Fig. 10.11A). This phenomenon has been used as a genetic tool to select for mutations that affect splicing control. Such mutations arise in different splicing signals at the 3′ splice site, and alter the efficiency of either the first or second step in the splicing reaction. Features that maintain the proper splicing balance include a suboptimal 3′ splice site, and a splicing enhancer in the adjacent exon. A negative regulatory sequence located more than 4000 nucleotides upstream of the 3′ splice site is also important. This sequence, which is bound by both U1 snRNP and specific cellular proteins (Fig. 10.11A), has been proposed to act as a "decoy" 5′ splice site: it forms a complex with the 3′ splice site for production of Env mRNA, but one that does not participate in splicing reactions. This sequence also stimulates polyadenylation.

The splicing of human immunodeficiency virus type 1 pre-mRNA is necessarily much more complex, as more than 40 alternatively spliced mRNAs are made in infected cells. Nevertheless, alternative splicing is also regulated by specific sequences that promote or repress recognition or utilization of splice sites, and by the degree of conformity of 3′ splice sites to the optimal sequence.

Coordination between Polyadenylation and Splicing

Although described separately in previous sections, capping, polyadenylation, and splicing of a pre-mRNA by cellular components are not independent reactions. Rather, one processing reaction governs the efficiency or specificity of another. For example, interaction of the nuclear cap-binding protein with the 5′ end of a pre-mRNA facilitates both removal of the 5′-terminal intron and efficient cleavage at the 3′ poly(A) addition site. Similarly, the presence of a 3′ poly(A) addition signal generally stimulates removal of the intron closest to it. However, splice sites at other positions can suppress recognition of polyadenylation signals. Inhibition of polyadenylation at specific sites is essential for expression of the genes of some retroviruses, and is an important device for the regulation of adenoviral gene expression.

Suppression of recognition of retroviral poly(A) addition sites. The proviral DNAs of retroviruses contain long terminal repeat sequences at either end, both of which contain a poly(A) addition signal (Fig. 10.11B).

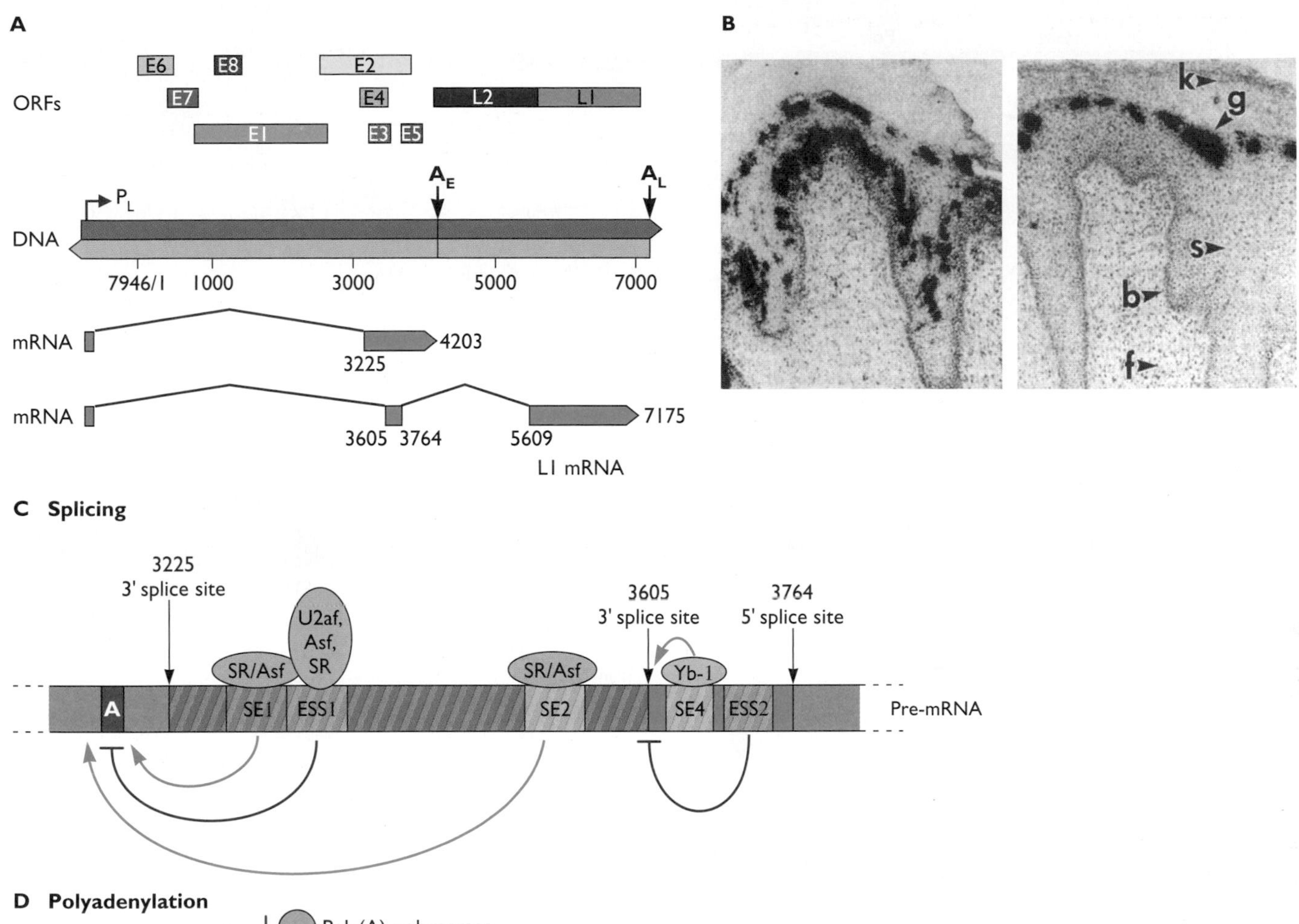

Figure 10.10 Alternative polyadenylation and splicing make active control of the production of bovine papillomavirus type 1 late mRNAs. (A) The circular bovine papillomavirus type 1 genome is represented in linear form, with open reading frames (ORFs) shown above. Two of the many mRNAs made from transcripts from the late promoter (P_L) are shown to illustrate the changes in recognition of splice sites and of poly(A) addition sites necessary to produce the L1 mRNA. Synthesis of this mRNA depends on recognition of a 3′ splice site at position 3605, rather than that at 3225, which is used during the early phase of infection. Polyadenylation of pre-mRNAs must also switch from the early (A_E) to the late (A_L) polyadenylation site. **(B)** *In situ* hybridization of bovine fibropapillomas to probes that specifically detect mRNAs spliced at the 3225 3′ splice site (left) or at the 3605 site (right). The cell layers of the fibropapilloma are indicated in the right panel. Abbreviations: k, keratin horn; g, granular cell layer; s, spinous cell layer; b, basal cell layer; f, fibroma. Note the production of late mRNA spliced at the 3605 3′ splice site only in the outermost layer (g) of fully differentiated cells. From S. K. Barksdale and C. C. Baker, *J. Virol.* **69:**6553–6556, 1995, with permission. Courtesy of C. C. Baker, National Institutes of Health. **(C)** Mechanisms that regulate splicing to produce L1 mRNA, which are specific to highly differentiated keratinocytes of the granular cell layer. The sequences that control alternative splicing at the 3225 and 3605 3′ splice sites are located between these splice sites. The splicing enhancers, SE1, SE2, and SE4, are recognized by cellular SR and other splicing proteins. The SE1 enhancer and the adjacent sequence that inhibits splicing at the 3605 3′ splice site, termed exonic splicing suppressor (ESS1), are thought to facilitate recruitment of U2-associated protein (U2af) and recognition of the branch point sequence upstream of the 3225 3′ splice site. SE2 is located very close to the 3605 3′ splice site and may block access to the branch point for splicing at this site until keratinocytes differentiate. **(D)** Inhibition of polyadenylation at the A_E site by the binding of U1 snRNP to a pseudo-5′ splice site located nearby in the primary transcript (see the text). Such inhibition is the result of binding of the U1 snRNP 70k subunit to poly(A) polymerase.

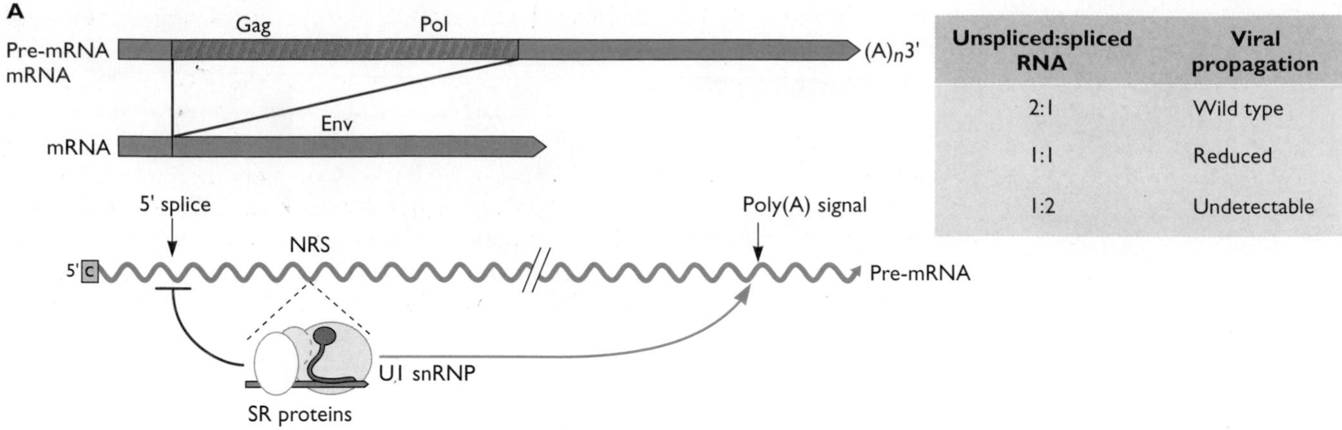

Unspliced:spliced RNA	Viral propagation
2:1	Wild type
1:1	Reduced
1:2	Undetectable

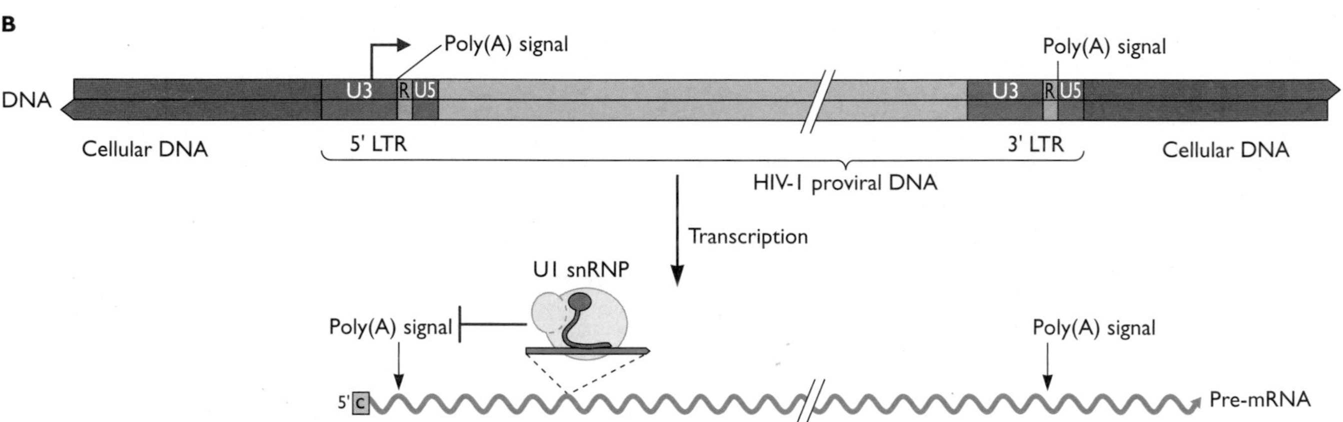

Figure 10.11 Control of RNA-processing reactions during retroviral gene expression. (A) Balanced production of spliced and unspliced mRNAs is illustrated for avian leukosis virus. A single 3′ splice site is recognized in about one-third of the primary transcripts to produce spliced mRNA encoding the Env protein. The Gag and Pol proteins are synthesized from unspliced transcripts. Even a twofold reduction in the ratio of unspliced to spliced mRNA impairs virus reproduction (right). Shown below is the negative regulatory sequence (NRS) located within Gag coding sequences, which is bound by U1 snRNP and SR proteins. This sequence is believed to act as a "decoy" 5′ splice site (to inhibit splicing [red bar]). It also stimulates polyadenylation (green arrow), by an unknown mechanism. **(B)** Suppression of poly(A) site recognition. Utilization of the 5′ polyadenylation site in primary transcripts of human immunodeficiency virus type 1 proviral DNA is inhibited by binding of U1 snRNP to the major 5′ splice site located 195 nucleotides downstream. The ability of the U1 snRNP protein U1a to bind to both poly(A) polymerase and cleavage-polyadenylation specificity protein suggests that the U1 snRNP might inhibit their activity.

Transcription of some proviral DNAs, such as that of Rous sarcoma virus, initiates downstream of the long terminal repeat sequence encoding the essential 5′AAUAAA3′ polyadenylation sequence. Consequently, a functional poly(A) addition site is present only at the 3′ ends. However, many other retroviral transcripts carry complete signals for this modification at both their 5′ and 3′ ends. Nevertheless, poly(A) is added to only the 3′ ends of these pre-mRNAs. At least two mechanisms ensure that the correct poly(A) addition signal of human immunodeficiency virus type 1 pre-mRNA is recognized. Sequences present only at the 3′ end of the pre-mRNA stimulate polyadenylation *in vitro* and *in vivo*, by facilitating binding of Cpsf to the nearby 5′AAUAAA3′ sequence. In addition, recognition of the 5′ poly(A) signal is suppressed by the 5′ splice site lying immediately downstream (Fig. 10.11B). Such inhibition is thought to be mediated by U1 snRNP, which can bind to the 5′ splice site. A related mechanism, binding of U1 snRNP to a pseudo 5′ splice site, appears to suppress recognition of the poly(A) addition site for production of bovine papillomavirus type 1 late mRNAs until the keratinocyte host cell is fully differentiated (Fig. 10.10C).

Alternative polyadenylation and splicing of adenoviral late pre-mRNAs. The production of adenoviral major late mRNAs exemplifies complex alternative splicing and polyadenylation at multiple sites in a pre-mRNA. The major late pre-mRNA contains the sequences of at least 15 different mRNAs. These mRNAs fall into five families (L1 to L5) defined by which of five polyadenylation sites is recognized (Fig. 10.12). The frequency with which each site is used must therefore be regulated to allow production of all major late mRNAs. For example, high efficiency polyadenylation at the L1 site would prevent efficient synthesis of L2 to L5 mRNAs, just as efficient splicing of retroviral pre-mRNAs would preclude production of the essential, unspliced transcripts. During the late phase of adenovirus infection, each of the five polyadenylation sites directs 3′-end formation with approximately the same efficiency. The mechanism responsible for such balanced recognition of multiple poly(A) addition sites is not fully understood. However, the activities of cellular polyadenylation proteins are altered as infection proceeds (see "Posttranscriptional

Figure 10.12 Alternative polyadenylation and splicing of adenoviral major late transcripts. During the late phase of adenoviral infection, major late primary transcripts extend from the major late promoter almost to the right end of the genome. They contain the sequences for at least 15 mRNAs and are polyadenylated at one of five sites, L1 to L5, as a result of decreased activity of cleavage stimulatory protein (see the text). The tripartite leader sequence, present at the 5′ ends of all late mRNAs, is assembled by the splicing of three short exons, l1, l2, and l3. This sequence is then ligated to alternative 3′ splice sites. Such joining of the spliced tripartite leader sequence to an mRNA sequence has been reported to take place after polyadenylation of pre-mRNA. Polyadenylation therefore appears to determine which 3′ splice sites can be utilized during the final splicing reaction.

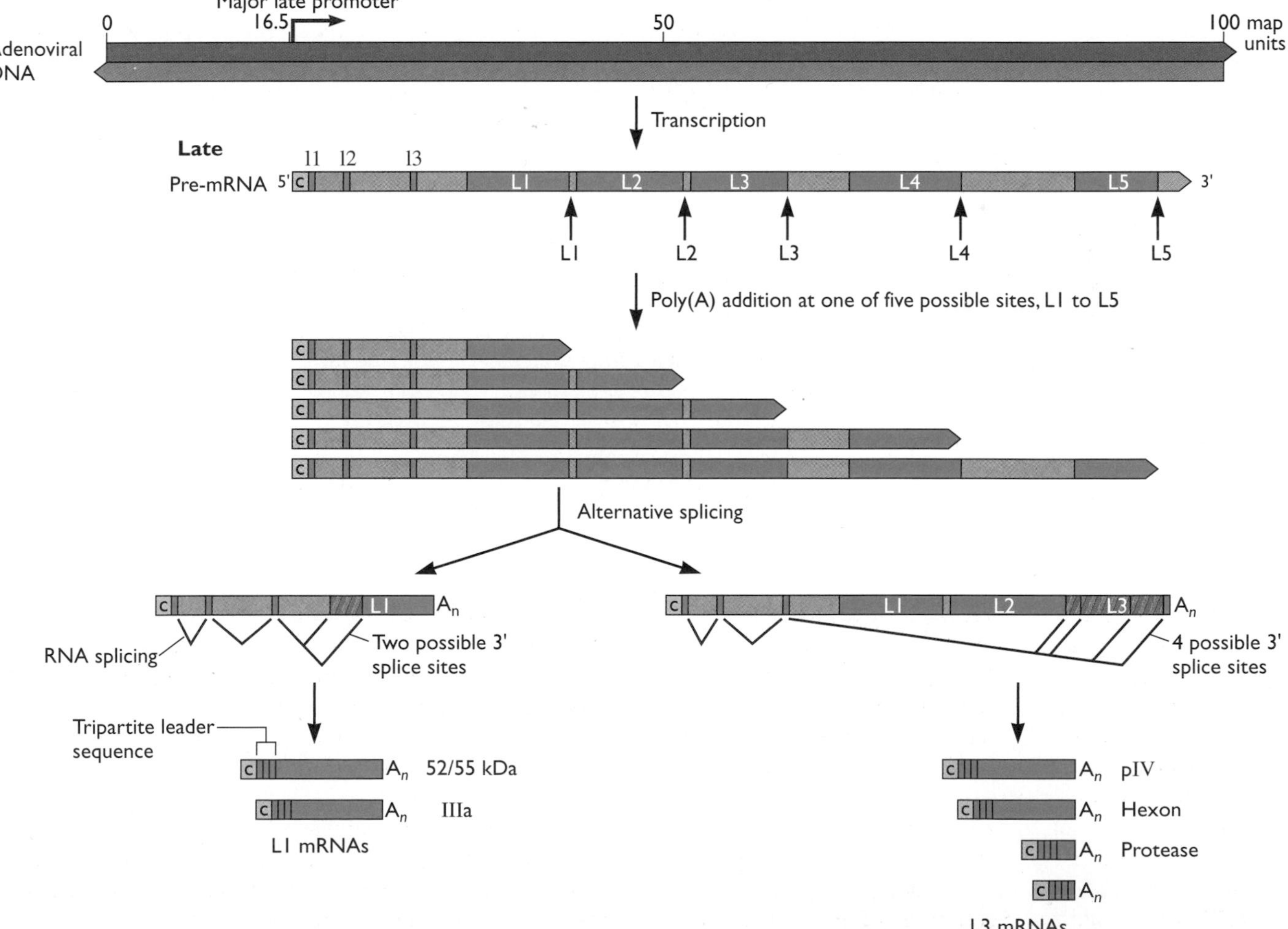

Regulation of Viral or Cellular Gene Expression by Viral Proteins" below).

As discussed previously, all major late mRNAs contain the 5′-terminal tripartite leader sequence. The splicing reactions that produce this sequence from three small exons (Box 10.4) can take place before polyadenylation of the primary transcript. The final splicing reaction joins the tripartite leader sequence to one of many mRNA sequences (Fig. 10.13). Each primary transcript therefore yields only a **single** mRNA, even though it contains the sequences for many, and most of its sequence is discarded. It remains something of a mystery why the majority of adenoviral late mRNAs are made by this bizarre mechanism. However, one contributing factor may be that it ensures that each major late mRNA molecule carries the 5′-terminal tripartite leader sequence, which is important for efficient translation late in the infectious cycle (Chapter 11).

Editing of Viral mRNAs

The term "RNA editing" was first coined in 1980 to describe the origin of uridine residues that are not encoded in the DNA template in a mitochondrial mRNA of trypanosomes. Since this modification was discovered, RNA editing has been identified in many different eukaryotes, as well as in some important viral systems.

Viral mRNAs are edited by either insertion of nucleotides not directly specified in the template during synthesis or alteration of a base *in situ*. Both mechanisms result in an mRNA sequence that differs from that encoded in the viral genome, altering the sequence and function of the protein specified by edited mRNA. Consequently, RNA editing has the potential to make an important contribution to regulation of viral gene expression.

Editing during Pre-mRNA Synthesis

mRNAs of *Paramyxoviridae* (e.g., measles and mumps viruses) and *Filoviridae* (e.g., Ebola virus) are edited during their synthesis. Among paramyxoviruses, this modification occurs only in the mRNAs that encode the RNA-dependent RNA polymerase, the P protein. During mRNA synthesis, the viral RNA polymerase inserts one or two G residues at specific positions in a fraction of the RNA molecules. The genomic RNA template contains a polypyrimidine sequence at the RNA-editing site. Insertion of G residues is therefore thought to occur by a reiterative copying mechanism (Fig. 10.13A), analogous to that by which the viral RNA polymerases synthesize 3′ poly(A) tails. The observation that increased stability of the nascent RNA-template RNA duplex just upstream of the editing site reduces the frequency of editing, and vice versa, is consistent with this mechanism.

Figure 10.13 Cotranscriptional editing of measles virus mRNAs. (A) Proposed mechanism. The viral polymerase pauses near a junction of U_n and C_n sequences in the template virion RNA after two C residues of the template have been copied into G residues in the nascent mRNA. As a result of such pausing, some fraction of viral RNA polymerase molecules and their attached nascent mRNA chains slip backward, such that additional nucleotides are incorporated when RNA synthesis resumes. The most stable structure is formed when the measles virus RNA polymerase slips backward by one (−1) position. Consequently, one G residue is incorporated into the viral mRNA when RNA synthesis resumes. **(B)** The measles virus P gene. The unedited mRNA contains a continuous open reading frame (pink) for the P protein. The addition of one G residue at the editing site changes the translational reading frame to one (yellow) that contains a termination codon. The edited mRNA specifies the V protein, which differs from the P protein in its C-terminal sequence.

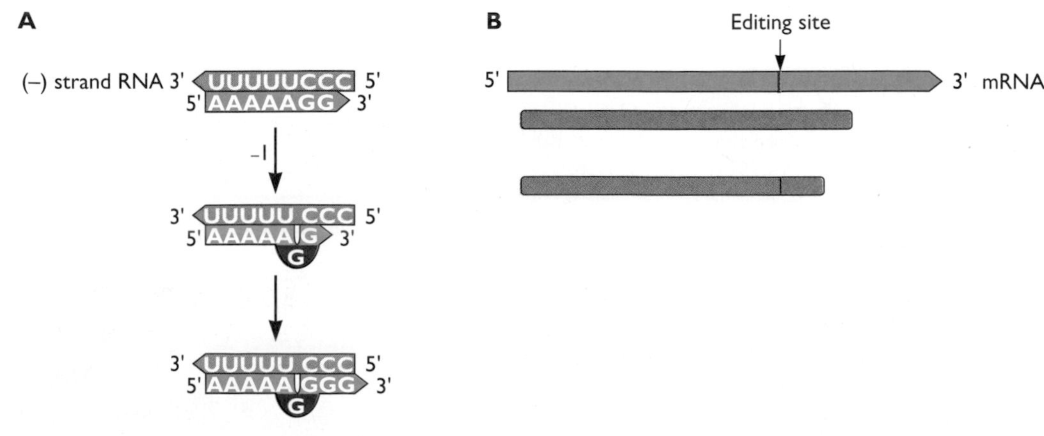

Because insertional editing of both measles and mumps virus mRNAs alters the translational reading frame, paramyxoviral P genes encode two distinct proteins (Fig. 10.13B). Similarly, RNA editing determines whether an Ebola virus gene is expressed as a full-length glycoprotein that is required for virus entry, or as a truncated secreted protein, and modulates the cytotoxicity of the virus (Box 10.7).

Editing following mRNA Synthesis

Production of two proteins from a single coding sequence. One form of posttranscriptional editing of mRNAs is accomplished by cellular enzymes that deaminate adenine bases in double-stranded RNA regions to form inosine (I). One such enzyme has been implicated in editing of the mRNA of hepatitis delta satellite virus. Two forms of the protein encoded in the genome of this agent, called hepatitis delta antigen, are synthesized in infected cells. Both are necessary for satellite virus reproduction. The large delta antigen, which is required for assembly and inhibits RNA replication, is made when editing converts an UAG termination codon to the UGG tryptophan codon (Fig. 10.14). As many as 50% of satellite virus mRNA molecules are modified at this site, but few other sequences in the viral RNA are edited. The mechanisms that restrict the action of the adenosine deaminase, which exhibits little specificity *in vitro*, have not been elucidated. The larger form of hepatitis delta antigen inhibits editing *in vivo*. This property is important for synthesis of the essential, smaller form of the protein, which is translated from unedited mRNA molecules (Fig. 10.14). It also prevents production of too high a concentration of large delta antigen and consequently inhibition of replication of the satellite virus.

Double-stranded RNA adenosine deaminases may play a broader role in the biology of RNA viruses. Before the discovery of these enzymes, it was assumed that nucleotide changes in viral RNA genomes arise solely by the introduction of incorrect nucleotides by RNA polymerases, or by RNA recombination (Chapter 6). However, several changes can now be attributed to editing. For example, many of the genomic RNAs of defective measles viruses

BOX 10.7 DISCUSSION

RNA editing regulates the cytotoxicity of Ebola viruses

The ~19,000-nucleotide (–) strand RNA genome of Ebola virus (a member of the *Filoviridae*) contains an editing site within the coding sequence for glycoprotein (GP) mRNA. This site comprises seven consecutive U residues and resembles the viral polyadenylation signal. Editing is therefore thought to take place cotranscriptionally, by reiterative copying, as during synthesis of paramyxoviral mRNA (Fig. 10.13). As shown in the figure, the products of the edited and unedited GP mRNAs share an N-terminal sequence (blue), but carry different C-terminal sequences, because introduction of an additional A residue by editing changes the reading frame. The protein specified by edited mRNA, GP, is the viral envelope glycoprotein and localizes to the plasma membrane of infected cells. In contrast, the protein synthesized from the unedited mRNA (sGP) is secreted.

When the editing site was eliminated by mutation of cloned DNA from which infectious virus can be recovered, sGP was not made, as expected. The concentration of GP increased, also as anticipated, but most of the protein accumulated as an immature precursor in the endoplasmic reticulum. Furthermore, and unexpectedly, these alterations in production of the GP proteins were accompanied by severe cytotoxicity: the mutant virus formed small plaques because of the earlier than normal death of infected cells.

These observations indicate that maintenance of a proper ratio of the secreted and membrane-associated forms of the viral glycoprotein is necessary to limit the cytotoxicity of Ebola virus. The secreted glycoprotein is therefore likely to make an important contribution to pathogenicity, by indirectly increasing virus reproduction and spread.

Volchkov, V. E., V. A. Volchkova, E. Mühlberger, L. V. Kolesnikova, M. Weik, O. Dolnik and H.-D. Klenk. 2001. Recovery of infectious Ebola virus from complementary DNA: RNA editing of the GP gene and viral cytotoxicity. *Science* **291:**1965–1969.

(Top) The editing site in the GP gene of the viral RNA genome. **(Bottom)** The differences between the sGP and GP proteins specified by unedited and edited mRNAs, respectively.

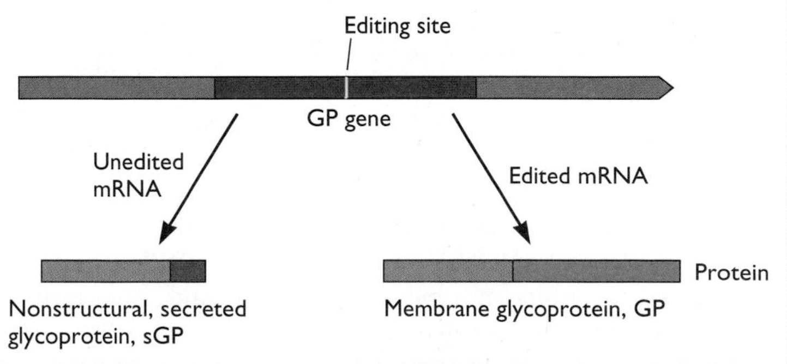

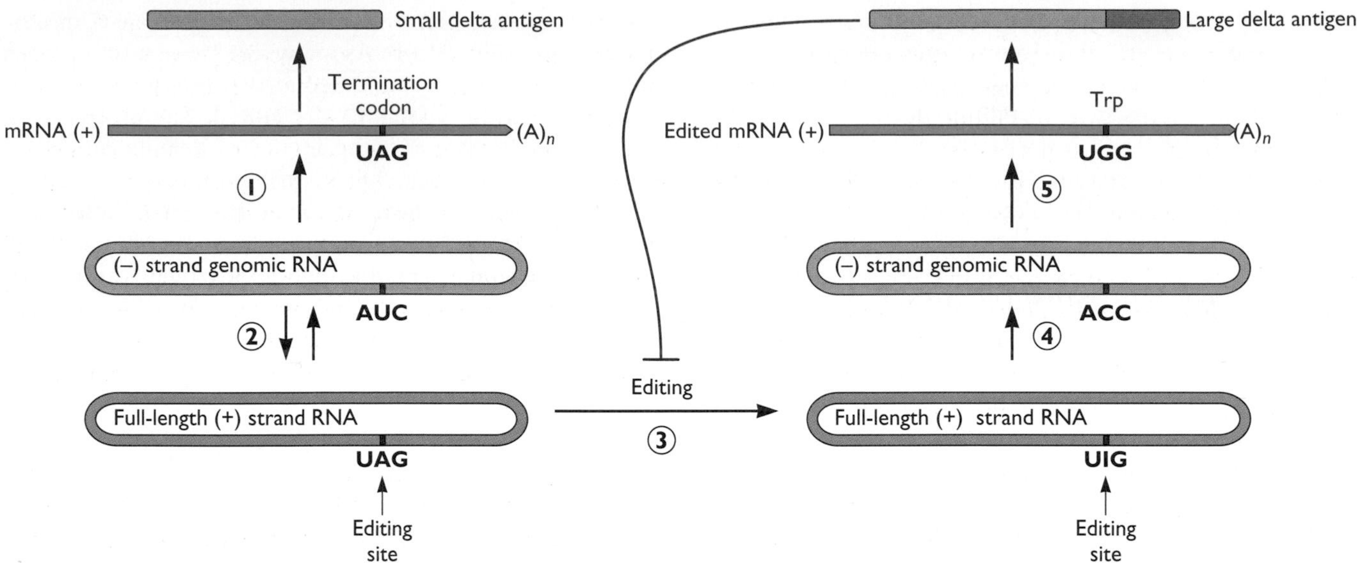

Figure 10.14 Editing of hepatitis delta satellite virus RNA by double-stranded RNA adenosine deaminase. The mRNA synthesized from genomic (–) strand RNA specifies the small delta antigen (step 1), which is required for replication of the genome (2). Double-stranded RNA adenosine deaminase acts on the full-length (+) RNA to convert a specific A residue to I (3). Because I base pairs with C, the genome (–) strands copied from edited full-length (+) strands contain a C residue (4). Such edited (–) RNA is therefore copied into (+) mRNA that contains a UGG codon (tryptophan) at the editing site (5), rather than the UAG stop codon. As a result, the mRNA made from edited RNA specifies the large delta antigen, which contains a 19-amino-acid C-terminal extension (purple). This protein inhibits RNA editing (red bar) and genome replication, and is needed for association of the hepatitis delta satellite viral genome with envelope proteins of its helper virus, hepatitis B virus.

isolated from the brains of patients who died of subacute sclerosing panencephalitis appear to have been edited by cellular adenosine deaminases.

Recent studies suggest that, like their cellular counterparts, transcripts of viral DNA genomes can also be edited by adenine deaminases. Transcripts of the k12 region of the human herpesvirus 8 genome are edited efficiently at just one site, both in infected cells and by adenine deaminase 1 *in vitro*. This modification appears to regulate the function of the kaposin protein that is encoded within the k12 region: this protein exhibits transforming activity only when it is made from the unedited coding sequence.

Editing as a powerful antiviral defense mechanism. Other cellular enzymes, known as Apobec3s, edit RNA by deamination of cytidine to uridine. Inhibition of the activity of such enzymes is important for the successful replication of several viruses. This phenomenon was discovered through efforts to elucidate the function of the human immunodeficiency virus type 1 accessory protein Vif. This protein is required for efficient replication *in vivo*, and in certain types of human cells in culture, which proved to contain Apobec3G. In the absence of Vif, this editing enzyme becomes incorporated into virions. Although it normally edits specific cellular mRNAs, Apobec3G catalyzes cytidine deamination during the first stage of reverse transcription, synthesis of (–) strand complementary DNA (cDNA). Such editing both triggers degradation of the (–) strand, presumably because of the abnormal presence of U residues, and results in hypermutation of proviral coding sequences. It therefore serves as an effective antiviral defense. Vif targets Apobec3G for degradation by the proteasome, and hence prevents its incorporation into assembling virus particles.

Export of RNAs from the Nucleus

Any mRNA made in the nucleus must be transported to the cytoplasm for translation. Other classes of RNA, including small cellular and viral RNAs made by RNA polymerase III, must also enter the cytoplasm permanently (e.g., transfer RNAs [tRNAs]) or transiently (snRNAs). The export of viral mRNAs is mediated by the host cell machinery and, in most cases, is indistinguishable from export of analogous cellular RNAs. However, transport to the cytoplasm of unusual mRNA molecules is a critical step in some viral life cycles. In this section, we describe the cellular export machinery and the mechanisms that ensure export of such viral mRNA substrates.

The Cellular Export Machinery

The substrates for mRNA export are not naked RNA molecules, but rather ribonucleoproteins. Some of the proteins travel to the cytoplasm with the mRNA, but are then displaced and return to the nucleus. Consequently, such hnRNP proteins shuttle between the nucleus and cytoplasm. One such protein, the abundant hnRNP-A1 protein, contains a short sequence that directs both export and import. This sequence was the first nuclear export signal to be identified. A different kind of nuclear export signal is found in proteins that direct export of certain viral mRNAs and small cellular RNAs (see "Export of Viral mRNA" below). In fact, export of RNA molecules (with the exception of tRNAs) is directed by sequences present in the proteins associated with them.

Like proteins entering the nucleus, RNA molecules travel between nuclear and cytoplasmic compartments via the nuclear pore complexes described in Chapter 5. Numerous genetic, biochemical, and immunocytochemical studies have demonstrated that specific nucleoporins (the proteins from which nuclear pore complexes are built) participate in nuclear export. Export of RNA molecules also shares several mechanistic features with import of proteins into the nucleus. Substrates for nuclear export or import are identified by specific protein signals and some soluble proteins, including the small guanosine nucleotide-binding protein Ran, function in both import and export. And RNA export, like protein import, is mediated by receptors that recognize nuclear export signals and direct the proteins, and ribonucleoproteins that contain them, to and through nuclear pore complexes.

Export of Viral mRNA

All viral mRNAs made in infected cell nuclei carry the same 5′- and 3′-terminal modifications as cellular mRNAs that are exported. Furthermore, many viral mRNAs, like their cellular

Figure 10.15 Regulation of export of human immunodeficiency virus type 1 mRNAs by the viral Rev protein. Before the synthesis of Rev protein in the infected cell, only fully spliced (2-kb class) viral mRNAs are exported to the cytoplasm (left). These mRNAs specify viral regulatory proteins, including Rev. The Rev protein enters the nucleus, where it binds to an RNA structure, the Rev-responsive element (RRE) present in unspliced (9-kb class) and singly spliced (4-kb class) viral mRNAs. This interaction induces export to the cytoplasm of the RRE-containing mRNAs, from which virion structural proteins and enzymes are made (right). The Rev protein therefore alters the pattern of viral gene expression as the infectious cycle progresses.

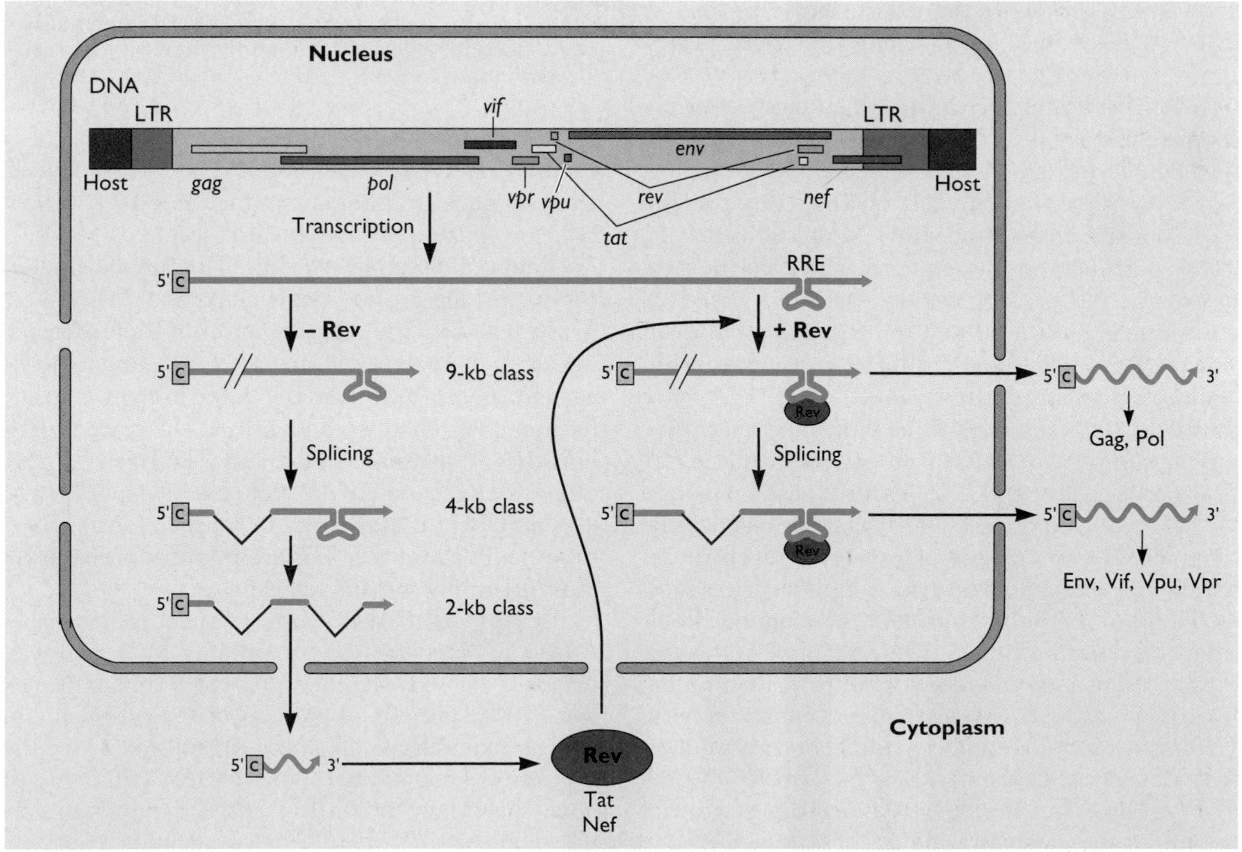

counterparts, are produced by splicing of intron-containing precursors. Cellular pre-mRNAs that contain introns and splice sites ordinarily are retained in the nucleus, at least in part because they remain associated with spliceosomes. Furthermore, a protein complex that marks mature mRNAs for export is assembled on the RNA only during splicing. Therefore the pre-mRNAs are either spliced to completion and exported, or degraded within the nucleus. However, replication of retroviruses, herpesviruses, and orthomyxoviruses requires production of mRNAs that are incompletely spliced, or not spliced at all, and their export from the nucleus for translation and/or virion assembly. Efforts to address the question of how the normal restrictions on cellular mRNA export are circumvented in cells infected by these viruses have provided important insights into the complex process of export of macromolecules from the nucleus.

The Human Immunodeficiency Virus Type 1 Rev Protein Directs Export of Intron-Containing mRNAs

The human immunodeficiency virus type 1 Rev protein is by far the best understood of the viral proteins that modulate mRNA export from the nucleus. This protein and related proteins of other lentiviruses promote export of the unspliced and partially spliced viral mRNAs. Rev binds specifically to an RNA sequence termed the **Rev-responsive element** that lies within an alternatively spliced intron of viral pre-mRNA (Fig. 10.15). The Rev-responsive element is some 250 nucleotides in length and forms several stem-loops (Fig. 10.16A), one of which contains a high-affinity binding site for the arginine-rich RNA-binding domain of Rev (Box 10.8). Export of RNAs that contain the Rev-responsive element depends on the formation of Rev-protein oligomers on this sequence, and a leucine-rich nuclear export signal present in Rev.

The C-terminal domain of Rev contains a short nuclear export signal (Fig. 10.16B) that is sufficient to induce export of heterologous proteins. In the oligomeric Rev-RNA complex, the nuclear export signals of the protein become organized on the surface. One cellular protein that binds to the nuclear export signal of Rev is exportin-1 (also known as Crm-1) (Fig. 10.17). Exportin-1, which binds simultaneously to Rev and the GTP-bound form of Ran, is the **receptor** for Rev-dependent export of the human immunodeficient type 1 RNAs bound to it. The viral protein therefore functions as an **adapter**, directing viral, intron-containing mRNAs to a preexisting cellular export receptor. Translocation of the complex containing viral RNAs, Rev and cellular proteins through the nuclear pore complex to the cytoplasm requires specific nucleoporins. In the cytoplasm, hydrolysis of GTP bound to Ran by a Ran-specific GTPase-activating protein present only in the cytoplasm induces dissociation of the

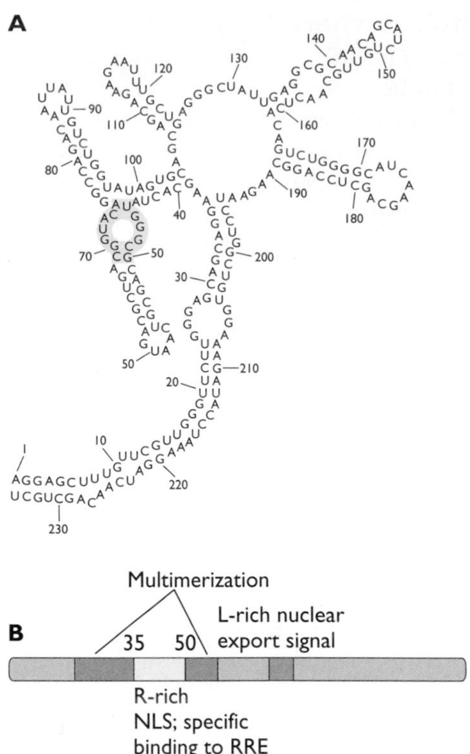

Figure 10.16 Features of the Rev-responsive element and Rev protein. (A) Predicted secondary structure of the 234-nucleotide Rev-responsive element, with the high-affinity binding site for Rev shaded in yellow. **(B)** The functional organization of the Rev protein.

export complex. Rev then shuttles back into the nucleus via a typical nuclear localization signal (Fig. 10.17B), where it can pick up another cargo RNA molecule.

Other cellular proteins required for Rev-dependent RNA export include an ATP-dependent RNA helicase (Ddx3), which may catalyze reorganization of RNA molecules for transit though the nuclear pore, and human Rev-interacting protein (hRip). As the latter protein is needed for release of Rev-associated RNA from the nuclear periphery into the cytoplasm, its discovery identified a previously unknown reaction in Rev-dependent export. This reaction may be a useful target for antiviral drugs, as loss of hRip neither results in mislocalization of cellular mRNAs or proteins, nor impairs cell viability.

Perhaps the most interesting aspect of Rev-dependent RNA export is the exit of mRNAs by a pathway that normally does not handle such cargo, but rather exports small RNA species and proteins of the host cell. The Rev nuclear export signal is similar to, and can be functionally replaced by, that of the cellular protein TfIIIa. This protein binds specifically to 5S rRNA and is required for export of this cellular RNA from the nucleus. Peptides containing the

BOX 10.8

EXPERIMENTS

Structure of a Rev peptide bound to the high-affinity binding site of the Rev-responsive element, determined by nuclear magnetic resonance methods

Comparison of the high-resolution structures of a Rev peptide bound to the high-affinity site and free Rev has provided important mechanistic insights into how specific RNA sequences and structural features are recognized. The Rev peptide corresponding to amino acids 34 to 50 forms an α-helix (red) that binds to the major groove of the RNA (blue), with phosphates contacted by Rev shown as spheres. The bases shown are invariant in RNA molecules selected from a large population for high-affinity binding to Rev *in vitro*. The major groove of an A-form helix is too narrow to accommodate an α-helix. However, in the Rev-RNA complex, the groove is widened by formation of two purine-purine base pairs and distortion of the RNA backbone. The purine-purine pairs are not observed in the solution structure of the RNA alone, indicating that binding of Rev induces a substantial conformational change. The Rev protein penetrates deeply into the major groove, and interacts extensively with the RNA over three to four turns of the α-helix. These interactions include base-specific contacts, for example, via the three arginine and one asparagine residues shown in yellow, and numerous contacts with the phosphate backbone of the RNA, via residues colored orange. From J. L. Battiste et al., *Science* **273:**1547–1551, 1996, with permission. Courtesy of J. R. Williamson, The Scripps Research Institute.

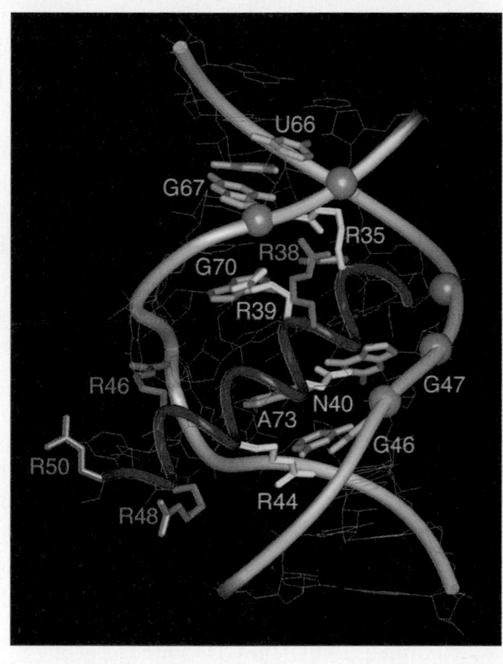

Rev nuclear export signal inhibit export of 5S rRNA, but not of mRNAs. The human immunodeficiency virus type 1 Rev protein therefore circumvents the normal restriction on the export of intron-containing pre-mRNAs from the host cell nucleus by diverting such viral mRNAs to a cellular pathway that handles unspliced RNAs.

RNA Signals Can Mediate Export of Intron-Containing Viral mRNAs by Cellular Proteins

The genomes of many retroviruses do not encode analogous proteins to Rev, even though unspliced viral mRNAs must reach the cytoplasm. These unspliced viral mRNAs contain specific sequences that promote export. Because they must function by means of cellular proteins, such sequences were termed **constitutive transport elements** (CTEs). The first such sequence was found in the 3′ untranslated region of the genome of Mason-Pfizer monkey virus.

Even low concentrations of RNA containing the Mason-Pfizer monkey virus CTE inhibit export of mature mRNAs when microinjected into *Xenopus* oocyte nuclei, but CTE RNA does **not** compete with Rev-dependent export. This observation indicated that this retroviral RNA sequence is recognized by components of a cellular mRNA export pathway. A search for such proteins led to the first identification of a mammalian protein mediating mRNA export: the human Tap protein binds specifically to the CTE and is essential for export from the nucleus of the unspliced viral RNAs and spliced cellular mRNAs. Tap proved to be the human homolog of a yeast protein that is essential for mRNA export.

The pathway of Tap-dependent mRNA export has not yet been fully elucidated, but the Ran-GTP protein does not participate. The direct and specific binding of Tap to the CTE of unspliced retroviral RNAs appears to bypass a cellular process that ensures that export is normally strictly coupled with splicing (Fig. 10.18). The Tap protein can bind only nonspecifically and with low affinity to cellular pre-mRNAs, but it does interact with components of a protein complex that is deposited on cellular mRNAs as they are

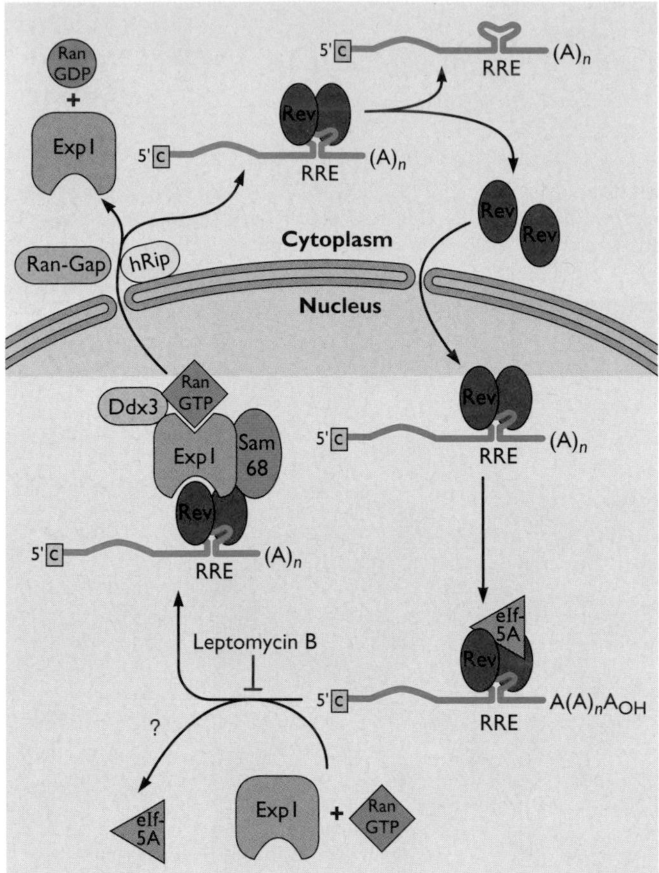

Figure 10.17 Mechanism of Rev protein-dependent export. The cellular nuclear proteins exportin-1 (Exp1), the GTP-bound form of Ran (Ran-GTP), and the 68-kDa Src-associated protein in mitosis (Sam68) have been implicated in Rev-dependent mRNA export, for example by analysis of the effects of dominant negative forms of the proteins. In the presence of Ran-GTP, Rev binds to exportin-1. This protein is related to the import receptors described in Chapter 5, and interacts with nucleoporins. The complex, containing Rev, exportin-1, and Ran-GTP bound to the Rev-responsive element in RNA, is translocated through the nuclear pore complex to the cytoplasm via interactions of exportin-1 with nucleoporins, such as Can/Nup14 and Nup98. Translocation may be facilitated by the action of Ddx3, an ATP-dependent RNA helicase. The Sam68 protein can bind to the Rev nuclear export signal, but does not appear to shuttle between nucleus and cytoplasm. It may therefore act prior to docking of the viral RRE-containing RNA complex at the nuclear pore. The human Rev-interacting protein (hRip) appears to act following translocation, as it is essential for efficient release of Rev-associated RNA into the cytoplasm. Hydrolysis of GTP bound to Ran to GDP induced by the cytoplasmic Ran GTPase-activating protein (Ran-Gap) is presumed to dissociate the export complex, releasing viral RNA for translation or virion assembly, and Ran, exportin-1, and other proteins for reentry into the nucleus.

spliced, such as the Ref/Aly protein. Indirect recruitment of Tap to an mRNA, which couples export to splicing, is circumvented in the case of retroviral pre-mRNAs containing CTEs: such RNAs are recognized directly by Tap, allowing export of unspliced RNAs from the nucleus.

Control of the Balance between Export and Splicing

The successful replication of retroviruses and orthomyxoviruses depends on production of spliced mRNAs in addition to export to the cytoplasm of unspliced RNAs. The relative efficiencies of splicing and export maintain a finely tuned balance in the production of spliced and unspliced viral RNAs. On one hand, splicing of viral pre-mRNA must be inefficient to allow export of the essential, intron-containing mRNAs (Fig. 10.11A). Indeed, increasing the efficiency of splicing of human immunodeficiency virus type 1 pre-mRNA, by replacing the natural, suboptimal splice sites with efficient ones, leads to complete splicing of all pre-mRNA molecules before Rev can recognize and export the unspliced mRNA to the cytoplasm. On the other hand, when unspliced RNAs remain in the nucleus (e.g., before Rev is made in infected cells), they are eventually spliced to completion. Efficient export is therefore required to place unspliced mRNA into the cytoplasm for translation or incorporation into virus particles.

Export of Single-Exon Viral mRNAs

Most of the viral mRNAs made in cells infected by hepadnaviruses, herpesviruses, or orthomyxoviruses are not spliced. In contrast to the retroviral mRNAs described in previous sections, these viral mRNAs do not retain introns. Rather, the viral genes that encode them contain no introns and the RNAs **cannot** be spliced. We therefore designate such mRNAs as **single-exon mRNAs** to distinguish them from those that retain introns.

Because such viral mRNAs do not contain splice sites, they should not become trapped in the nucleus by a spliceosome retention mechanism. On the other hand, they have few counterparts in uninfected mammalian cells growing under normal conditions, a property that raises the question of how viral, single-exon mRNAs are transported efficiently to the cytoplasm. Recent experiments have identified both viral proteins and RNA sequences that promote export of such viral mRNA, analogous to the retroviral Rev protein and constitutive transport elements, respectively.

The herpes simplex virus type 1 early mRNA encoding thymidine kinase contains a sequence that directs export by cellular components. This sequence is sufficient to induce efficient, cytoplasmic accumulation of unspliced β-globin mRNA when fused to it. Furthermore, its recognition by the cellular hnRNP L protein correlates with RNA accumulation in the cytoplasm. It may therefore direct the viral mRNA to a cellular pathway for export of the rare, cellular mRNAs that are not spliced. This same pathway may be responsible for export of hepadnaviral mRNAs. The efficient cytoplasmic accumulation of these viral mRNAs depends on a conserved sequence, termed the **posttranscriptional regulatory element**, which

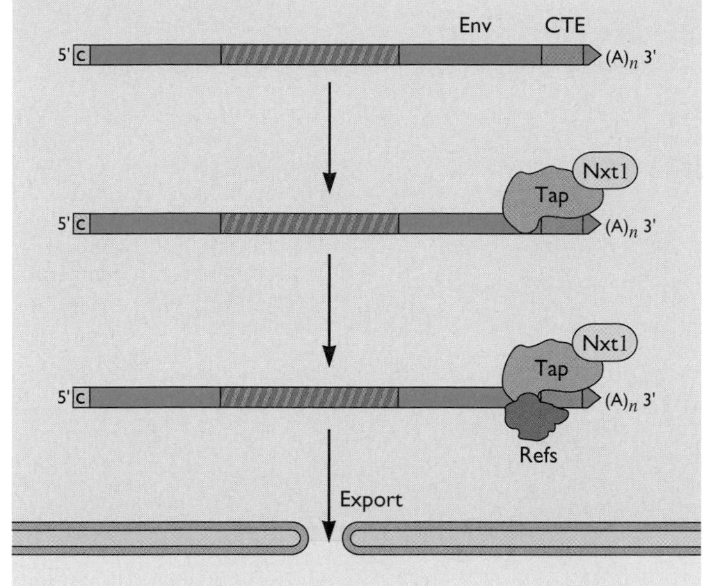

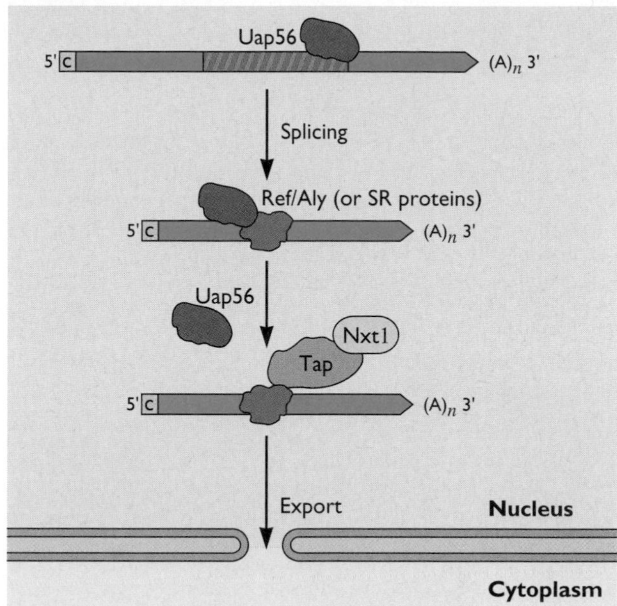

Figure 10.18 Export of unspliced RNA of retroviruses with simple genomes and cellular mRNAs from the nucleus. Export of unspliced, primary transcripts of many retroviruses (left) depends on the constitutive transport element (CTE) in the RNA. This sequence is recognized by the cellular Tap subunit of the export receptor dimer Tap-Nxt1, which is then bound by proteins that mark mRNAs as appropriate substrates for export, such as Ref. A variety of experimental approaches, inducing genetic studies of yeast, have indicated that Tap is an essential component of a cellular mRNA export pathway. However, Tap does not bind to cellular mRNAs (right) with high affinity, but rather becomes associated with them via interactions with specific proteins, such as Ref/Aly and certain SR proteins that are deposited on the cellular mRNA during splicing. In this way, spliced mRNAs are marked for export. The direct interaction of Tap with retroviral CTEs therefore appears to bypass the mechanism(s) that couple splicing of cellular mRNAs with their export.

can function in the absence of viral proteins. Functionally analogous sequences have been identified in one of the uncommon mammalian mRNAs that lacks introns (a histone 2A mRNA), suggesting that hepadnaviral mRNAs leave the nucleus via a cellular pathway that handles a minor fraction of the mRNA traffic in uninfected cells.

The efficient export of the majority of herpesviral single-exon mRNAs requires the viral ICP27 protein. This protein, like the products of all immediate-early genes, is made at the beginning of the infectious cycle. It is a nuclear phosphoprotein that contains a leucine-rich nuclear export signal resembling that of the human immunodeficiency virus Rev protein, and shuttles between the nucleus and the cytoplasm. ICP27 specifically facilitates the export of viral single-exon mRNAs: such mRNAs do not enter the cytoplasm efficiently in cells infected by mutant viruses that do not direct the synthesis of ICP27, but export of viral spliced mRNAs continues normally. This viral protein has been reported to bind to viral RNAs in infected cells. It contains different types of RNA-binding motifs in its N-terminal and C-terminal portions, both of which are necessary for optimal replication of the virus. A considerable body of genetic evidence indicates that the RNA-binding motifs of ICP27 contribute to efficient cytoplasmic accumulation of viral single-exon mRNAs. An unusual property of the ICP27 protein is that it contains both a binding site for Tap and a leucine-rich nuclear export signal that can mediate export by the exportin-1 pathway. The function of the latter is not yet clear, but the results of various experimental approaches indicate that ICP27 functions as an adapter of export of viral single-exon mRNAs via the Tap pathway (Box 10.9). Whether this pathway mediates export of all such viral mRNAs is not clear: it has been reported that export of some mRNAs is inhibited by the exportin-1 inhibitor, leptomycin B.

Posttranscriptional Regulation of Viral or Cellular Gene Expression by Viral Proteins

The genomes of several viruses encode proteins that regulate one or more RNA-processing reactions. These proteins play crucial roles in temporal regulation of viral gene expression, or in inhibition of the production of cellular mRNAs (Table 10.3).

BOX 10.9 DISCUSSION

Multiple lines of evidence indicate that the herpes simplex virus type 1 ICP27 protein engages the cellular Tap-Nxt1 export pathway

The nucleocytoplasmic shuttling of the ICP27 protein under the direction of its leucine-rich nuclear export signal originally suggested that this herpesviral protein, like Rev, engages the exportin-1-mediated export pathway of the host cell. However, shuttling of ICP27 between the nucleus and cytoplasm and export of viral mRNA in microinjected *Xenopus* oocytes are not sensitive to the expotin-1 inhibitor leptomycin B. This drug also had no effect on 1CP27 shuttling in herpes simplex virus type 1-infected cells. Rather, several observations indicate that ICP27 directs viral single exon mRNAs to the Tap-Nxt1 export pathway:

- ICP27 interacts with Tap/Nxt1 in infected cells.
- Overproduction of the Ref/Aly proteins, which are components of the Tap-Nxt1 pathway (Fig. 10.18) in infected cells, increased the efficiency of export of several viral late mRNAs.
- ICP27-dependent export of viral single exon mRNAs from *Xenopus* oocyte nuclei was blocked by a high concentration of a retroviral constitutive transport element, to which Tap binds (Fig. 10.18).
- Overproduction of a dominant negative derivative of Tap, which lacks C-terminal sequences that interact with nucleoporins, inhibited export of viral single exon mRNA in infected cells.

Temporal Control of Viral Gene Expression

Regulation of Alternative Polyadenylation by Viral Proteins

The frequencies with which alternative poly(A) addition sites within specific viral pre-mRNAs are utilized change during infection by adenoviruses, herpesviruses, and papillomaviruses. As discussed previously, the polyadenylation of bovine papillomavirus type 1 late mRNA is activated by a specific complement of cellular proteins found only in fully differentiated cells of the epidermis. In contrast, viral proteins have been implicated in regulation of polyadenylation in cells infected by the larger DNA viruses.

Despite its name, the adenoviral major late promoter is active during the early phase of infection, prior to the onset of viral DNA synthesis. The major late pre-mRNAs made during this period are polyadenylated predominantly at the L1 mRNA site (Fig. 10.12), even though they also contain the L2 and L3 3′ processing sites. Such selective recognition of the L1 site depends on the cellular cleavage stimulatory protein (Cstf), which binds to the U/GU-rich sequence 3′ to the cleavage site (Fig. 10.4). As infection continues, the activity of this cellular protein decreases. The recognition of the other four polyadenylation sites present in major late pre-mRNA synthesized during the late phase (Fig. 10.12) is much less dependent on Cstf. It is therefore likely that these poly(A)-addition signals compete more effectively with the L1 site for components of the polyadenylation machinery later in the infectious cycle. In the case of the L3 polyadenylation site, a downstream sequence is also required.

Experiments in which the synthesis of adenoviral mRNAs from truncated major late transcription units were

Table 10.3 Viral proteins that regulate RNA-processing reactions

Virus	Protein(s)	Functions
Adenovirus		
Human adenovirus type 2	E4 ORF4	Induces dephosphorylation of cellular SR proteins by protein phosphatase 2A; relieves inhibition of L1 pre-mRNA splicing at the IIIa 3′ splice site by phosphorylated SR proteins present early in infection
	E1B 55-kDa–E4 ORF6	The complex inhibits export of fully processed cellular mRNAs and induces selective export of viral late mRNAs
	L4 33 kDa	Promotes alternative splicing to produce L1 IIIa mRNA
Herpesvirus		
Herpes simplex virus type 1	ICP27	Stimulates polyadenylation of viral late mRNAs with suboptimal sequences at their polyadenylation sites; inhibits splicing of intron-containing cellular and probably viral mRNAs; promotes export of viral single-exon mRNAs
Retrovirus		
Human immunodeficiency virus type 1	Rev	Mediates export of unspliced and incompletely spliced viral mRNAs

BOX 10.10

EXPERIMENTS

A single adenoviral protein controls the early-to-late switch in major late RNA processing

In efforts to develop cell lines stably producing adenoviral late proteins, plasmids containing various segments of the major late (ML) transcription unit under the control of an inducible promoter were introduced into human cells. The plasmid ML1-5 supported very efficient expression of all the ML coding sequences and synthesis of the full set of the ML proteins. In contrast, only the L1 52/55-kDa protein was synthesized efficiently in cells containing the ML1-3 plasmid, even through the L2 and L3 coding sequences were present (see figure). Examination of the cytoplasmic concentrations of processed ML mRNAs showed that only the L1 52/55-kDa mRNA was made efficiently in cells containing the ML1-3 plasmid, as is also the case during the early phase of infection. These observations implied that one or more viral proteins encoded in the region of the genome present in the ML1-5 but not in the ML1-3 plasmid induce the early-to-late switch in processing of ML pre-mRNA. In fact, synthesis of the L4 33-kDa protein in cells containing the ML1-3 plasmid allowed production of the L1 IIIa and the L2 and L3 proteins. This viral protein significantly stimulated the synthesis of fully processed hexon mRNA, but did not alter the nuclear concentration of ML pre-mRNA. It was therefore concluded that the L4 33-kDa protein is necessary and sufficient to switch processing of the ML pre-RNA from the early to the late pattern.

Farley, D. C., J. L. Brown, and K. N. Leppard. 2004. Activation of the early-late switch in adenovirus type 5 major late transcription unit expression by L4 gene products. *J. Virol.* **78:**1782–1791.

The major late (ML) coding regions (L1-L5) of the Ad5 genome are shown to scale at the top, with the regions in the ML1-3 and ML1-5 plasmids introduced into human cells shown below. 1, 2, and 3 indicate the positions of the three segments of the tripartite leader sequence. The proteins made in cells containing these plasmids, and the ML1-3 plasmid plus a vector directing synthesis of the viral L4 33-kDa protein, are indicated below.

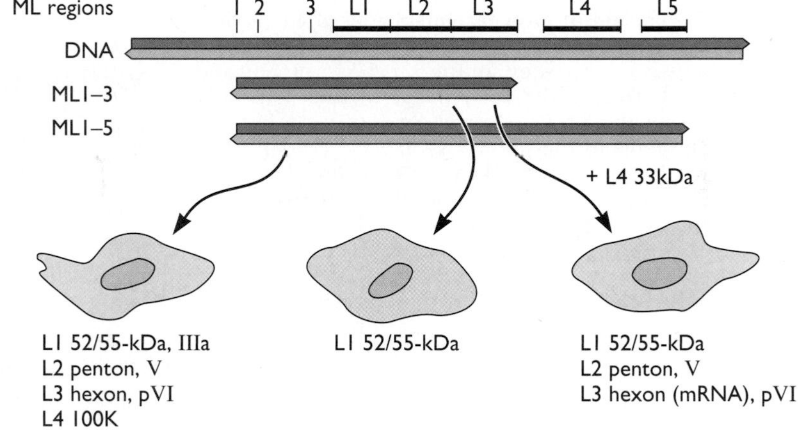

examined (Box 10.10) have established that synthesis of the viral L4 33-kDa protein is essential for the switch to the late pattern of gene expression. It is not yet known whether this protein modulates the activity of Cstf or other components of the polyadenylation machinery. As the L4 33-kDa protein is itself encoded within the major late transcription unit, it is thought that the small quantities produced initially during the late phase activate efficient production of all major late mRNAs.

Some herpesviral late pre-mRNAs contain poly(A) addition signals that function poorly in the absence of the ICP27 protein, at least in part because the downstream U/GU-rich sequences are not optimal. This viral protein appears to stimulate binding of Cstf to such suboptimal sites, and consequently increase the rate of polyadenylation.

Viral Proteins Can Regulate Alternative Splicing

Some viral proteins that regulate pre-mRNA splicing alter the balance among alternative splicing reactions at specific points in the infectious cycle. For example, the ratios of alternatively spliced mRNA products of several adenoviral pre-mRNAs change with the transition into the late phase of infection. This phenomenon has been studied most extensively using the L1 mRNAs. The L1 pre-mRNA, the product of major late transcription and polyadenylation during the early phase of infection, can be spliced at one of two alternative 3′ splice sites (Fig. 10.19). However, only the L1 mRNA that specifies the 52/55-kDa protein is made prior to the onset of viral DNA synthesis. At such early times in the infectious cycle, binding of cellular SR proteins to a negative regulatory sequence located immediately upstream of the branch point for the L1 IIIa mRNA blocks its recognition (Fig. 10.19). A viral early protein encoded within the E4 transcription unit induces dephosphorylation of the SR proteins, which are highly phosphorylated in uninfected cells. Such dephosphorylated SR proteins lose the ability to repress recognition of the 3′ splice site of the L1 IIIa mRNA. Overproduction

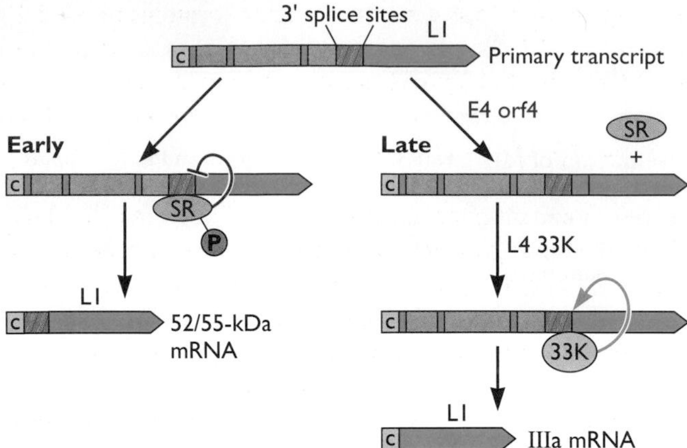

Figure 10.19 Regulation of alternative splicing of adenoviral major late L1 pre-mRNA. The polyadenylated L1 pre-mRNA contains alternative 3′ splice sites, for the 52/55-kDa protein and protein IIIa **(left)**. During the early phase of infection, only the 3′ splice site for the 52/55-kDa protein is utilized, because binding of SR proteins to the pre-mRNA blocks recognition of the 3′ splice site for production of the mRNA for protein IIIa **(right)**. An E4 protein induces dephosphorylation of these cellular proteins by protein phosphatase 2. This modification inhibits binding of the SR proteins to the pre-mRNA. However, efficient utilization of the IIIa mRNA 3′ splice site (during the late phase) requires the viral L4 33-kDa protein, which activates splicing via an infected cell-specific splicing enhancer.

of the SR protein Sf2 in adenovirus-infected cells impairs not only synthesis of the L1 IIIa mRNA, but also the temporal shift in splicing of early E1A and E1B pre-mRNAs, as well as viral replication. This observation indicates that dephosphorylation of cellular SR proteins makes a major contribution to posttranscriptional regulation of adenoviral gene expression. However, other mechanisms are also important. For example, efficient production of the L1 IIIa mRNA depends on a splicing enhancer (Fig. 10.19) that is active only in adenovirus-infected cells (or extracts prepared from them). The viral L4 33-kDa protein stimulates IIIa mRNA production in infected cells, and is sufficient to activate splicing via this enhancer in *in vitro* reactions. This protein also stimulates splicing at other suboptimal 3′ splice sites in major late pre-mRNAs, such as those that produce the L2 mRNAs for proteins V and pre-VII.

Regulation of mRNA Export

Even though all are encoded within a single proviral transcription unit, the regulatory and virion proteins of human immunodeficiency virus type 1 are made sequentially in infected cells, as a result of regulation of mRNA export by the Rev protein. As described previously, this protein regulates a switch in viral gene expression from early production of viral regulatory proteins to a later phase, in which virion components are made (Fig. 10.15).

Viral proteins that modulate mRNA export may play subsidiary, but nonetheless crucial, roles in temporal regulation of viral gene expression. For example, the transcriptional program described in Chapter 8 results in efficient transcription of herpes simplex virus late genes only following initiation of viral DNA synthesis in infected cells. Because all but one of the late mRNAs comprises a single exon, their efficient entry into the cytoplasm and the synthesis of viral late proteins requires ICP27. Consequently, this posttranscriptional regulator is essential for putting the viral transcriptional program into effect. Similarly, as discussed in the previous section, the complete panoply of adenoviral major late gene products can be produced only when the viral L4 33-kDa protein induces the switch to the late pattern of processing of these pre-mRNAs.

Viral Proteins Can Inhibit Cellular mRNA Production

All viral mRNAs are translated by the protein synthesis machinery of the host cell. Inhibition of production of cellular mRNAs can therefore favor this essential step in viral replication. Several mechanisms of selective inhibition of cellular RNA processing operate in virus-infected cells.

Inhibition of Polyadenylation and Splicing

The influenza virus NS1 protein can inhibit both polyadenylation and splicing of cellular pre-mRNAs. This viral protein comprises an N-terminal RNA-binding domain and a C-terminal segment that is required for inhibition of polyadenylation. The C-terminal domain contains binding sites for both Cpsf and PabII (Fig. 10.5). Its interaction with Cpsf inhibits polyadenylation of cellular mRNAs in experimental systems, and NS1 is required for inhibition of cellular mRNA polyadenylation and cytoplasmic accumulation in virus-infected cells. These properties suggest that NS1 could increase the intranuclear concentrations of cellular pre mRNAs, thereby facilitating cap snatching (Chapter 6) and indirectly inhibiting translation of cellular mRNAs. However, cellular protein synthesis is inhibited effectively when NS1 is not made. Under these circumstances, infected cells produce larger quantities of IFN mRNAs and of other cellular mRNAs encoding proteins with antiviral activities. Inhibition of processing of such cellular mRNAs may therefore contribute to the circumvention of host cellular defenses, a critical function of the NS1 protein (see Volume II, Chapter 3).

In addition to its other activities, the herpes simplex virus protein ICP27 inhibits splicing of cellular pre-mRNAs. This protein induces redistribution of spliceosomal snRNPs from a pattern of diffuse speckles in the nucleus to a limited number of sites with a punctate pattern. The viral protein colocalizes with snRNPs at the latter sites, and may bind directly to them. Such abnormal, intranuclear distribution of essential splicing components might contribute to the inhibition of splicing, but ICP27 also inhibits splicing in *in vitro* reactions. Direct interaction of the viral protein with components of the spliceosome inhibits splicing at an early step in assembly. Genetic analyses have shown that disruption of cellular RNA processing by ICP27 leads to inhibition of cellular protein synthesis. Moreover, this function is genetically separable from the role of the protein in efficient production of viral late mRNAs. Because herpesviral genes generally lack introns, inhibition of splicing is an effective strategy for the selective inhibition of cellular gene expression.

Inhibition of Cellular mRNA Export

To facilitate production of viral mRNAs. In contrast to the other viruses considered in this section, adenovirus infection disrupts cellular gene expression by inhibition of export of cellular mRNAs from the nucleus. During the late phase of infection, the great majority of newly synthesized mRNAs entering the cytoplasm are viral in origin. Synthesis and processing of cellular pre-mRNAs are unaffected, but these mRNAs are not exported and are degraded within the nucleus. Viral mRNAs are exported selectively, a phenomenon that is important for efficient synthesis of late proteins. When selective viral mRNA export is prevented by mutations in the viral genome, both the quantities of late proteins made in infected cells and the virus yield are reduced substantially. These same phenotypes are seen in herpes simplex virus-infected cells synthesizing an altered ICP27 protein that is defective only for the inhibition of pre-mRNA splicing. These properties emphasize the importance of posttranscriptional inhibition of cellular mRNA production for efficient virus reproduction.

The preferential export of late mRNAs in adenovirus-infected cells requires two viral early proteins, the E1B 55-kDa and E4 Orf6 proteins. These proteins are found both free and in association with one another in infected cell nuclei. The complex is responsible for the regulation of mRNA export. Both the E1B and the E4 proteins contain leucine-rich nuclear export signals and can shuttle between the nucleus and cytoplasm. Perhaps surprisingly, such shuttling is dispensable for regulation of mRNA export. The E1B 55-kDa and E4 Orf6 proteins associate with several cellular proteins, including cullin 5 and elongins B and C, to form an infected-cell-specific E3 ubiquitin ligase that targets specific proteins for degradation by the proteasome. The results of recent studies indicate that this activity of the viral proteins is required for regulation of mRNA export. However, the cellular export proteins that might be targeted for proteasomal degradation have yet to be identified.

The selectivity of mRNA export in adenovirus-infected cells is especially puzzling, because the viral mRNAs possess all the characteristic features of cellular mRNAs and are made in the same way. One hypothesis is that the viral early-protein complex recruits nuclear proteins needed for export of mRNA to the specialized sites within the nucleus at which the adenoviral genome is replicated and transcribed. As a result of such sequestration, export of cellular mRNAs would be inhibited.

To block cellular antiviral responses. As we have seen, the genomes of many RNA viruses encode all the enzymes necessary for synthesis and processing of viral mRNAs in the cytoplasm. Infection by some of these viruses results in inhibition of nuclear export. One well-characterized example is inhibition of a major mRNA export pathway in cells infected by the rhabdovirus vesicular stomatitis virus. This effect is mediated by binding of the viral matrix (M) protein to the cellular export protein Rae1. This protein normally shuttles between the nucleus and cytoplasm and binds to both Tap/Nxf1 and a cellular nuclear pore protein. As the M protein is made in the cytoplasm, it is likely to sequester Rae1 in that compartment. While the consequent disruption of cellular mRNA export probably facilitates inhibition of host cell protein synthesis, it also appears to block an important anti-viral defense: expression of the cellular Rae1 and Nup98 genes is induced by interferon, a potent antiviral cytokine (see Volume II, Chapter 3).

Picornaviruses also disrupt trafficking from the nucleus to the cytoplasm. The poliovirus 2A protease induces relocation of particular nuclear proteins to the cytoplasm. Such redistribution correlates with loss of structure from the central channel of the nuclear pore, and cleavage of specific nucleoporins (e.g., Nup153). The small leader (L) protein of encephalomyocarditis virus, a member of the cardiovirus group within the *Picornaviridae*, binds to Ran-GTPase, an essential component of pathways of Ran-dependent nuclear export and import. The phenotypes of mutants with deletions in the L gene suggest that inhibition of Ran dependent trafficking by L protein both tempers the interferon antiviral response, and contributes to inhibition of cellular protein synthesis.

Regulation of Turnover of Viral and Cellular mRNAs in the Cytoplasm

Individual mRNAs may differ in the rate at which they are translated, and also in such properties as cytoplasmic location and stability. Indeed, the intrinsic lifetime of an mRNA can be a critical parameter in the regulation of gene expression.

In the cytoplasm of mammalian cells, the lifetimes of specific mRNAs can differ by as much as 100-fold. This property is described in terms of the time required for 50% of the population of the mRNAs to be degraded under conditions in which replenishment of the cytoplasmic pool is blocked, the **half-life** of the mRNA. Many mRNAs are very stable, with half-lives exceeding 12 h. As might be anticipated, such mRNAs encode proteins needed in large quantities throughout the lifetimes of all cells, such as structural proteins (e.g., actin) and ribosomal proteins. At the other extreme are very unstable mRNAs with half-lives of less than 30 min. This class includes mRNAs specifying regulatory proteins that are synthesized in a strictly controlled manner in response to cues from external or internal environments of the cell, such as cytokines, and proteins that regulate cell proliferation. The short lifetimes of these mRNAs ensure that synthesis of their products can be shut down effectively once they are no longer needed. Specific sequences that signal the rapid turnover of the mRNAs in which they reside, such as a 50- to 100-nucleotide AU-rich sequence within the 3′ untranslated region, have been identified, as have several mechanisms of mRNA degradation. The majority of mammalian mRNAs are degraded by the pathway summarized in Figure 10.20, in which deadenylation of the 3′ end of the mRNA triggers decapping of the 5′ end by decapping protein 2. The removal of this protective structure renders the RNA susceptible to rapid 5′→3′ exonucleolytic degradation.

The stabilities of viral mRNAs have not been examined in much detail, in part because many viral infectious cycles are completed within the normal range of mRNA half-lives. Nevertheless, it is clear that viral proteins that induce RNA degradation play an important role in selective expression of viral genes in cells infected by large DNA viruses. Regulation of the stability of specific viral or cellular mRNAs has also been implicated in the permanent changes in cell growth properties (transformation) induced by some viruses. Furthermore, RNA-mediated induction of degradation of specific mRNAs, a widespread phenomenon known as RNA interference, is thought to contribute to host antiviral defense mechanisms (see next section).

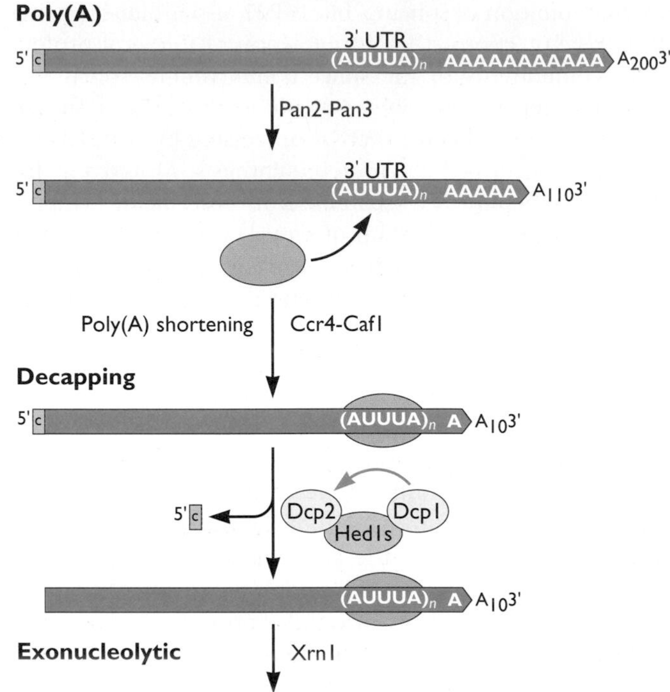

Figure 10.20 Destabilization of cellular mRNAs by 5′AUUUA3′ sequences, repeated in the 3′ untranslated regions (UTR). Such sequences, which are present in short-lived mRNAs encoding cytokines and proteins that regulate cell growth and division, are specifically recognized by one or more proteins. Some proteins stabilize mRNAs upon binding to 5′AUUUA3′ sequences. Binding of others, depicted by the blue oval, appears to induce shortening of the poly(A) tail, in a two-step process catalyzed by different deadenylases, the Pan and Ccr4-Caf1 complexes. Shortening of the poly(A) tail to <110 nucleotides triggers decapping by the enzyme-decapping protein 2 (Dcp2). This reaction is stimulated by Dcp1, which interacts with Dcp2 via the Hedls protein. The exact way in which the decapping enzyme is recruited to target mRNAs, and the contribution of shortening of the poly(A) tail, are not fully understood. The decapped mRNA is then degraded by 5′→3′ exonucleases, such as Xrn1.

Regulation of mRNA Stability by Viral Proteins

The virion host shutoff protein (Vhs) of herpes simplex virus type 1 reduces the stability of mRNAs in infected cells. As its name implies, Vhs is a structural protein of the virion. It is present at low concentrations in the tegument and therefore is delivered to infected cells at the start of the infectious cycle, before viral gene expression begins. It remains in the cytoplasm, where it mediates degradation of some cellular mRNAs. The Vhs protein is an RNase that cleaves mRNA endonucleolytically. The removal of cellular mRNAs at the start of infection facilitates viral gene expression, presumably by reducing or eliminating competition from cellular mRNAs during translation.

The Vhs protein specifically targets mRNAs, probably by virtue of its binding to the translation initiation protein eIF4H, an interaction that stimulates the activity of the viral enzyme. However, Vhs cannot distinguish viral mRNAs from their cellular counterparts. It therefore induces degradation of both viral and cellular mRNAs. Although more Vhs protein is made in infected cells once its coding sequence is expressed during the late phase of infection, the protein is sequestered in the tegument of assembling virus particles by interaction with the viral VP16 protein. As a result, the activity of Vhs decreases as the infection cycle progresses. This mechanism presumably contributes to the efficient synthesis of viral proteins characteristic of the late phase of infection.

The genomes of poxviruses, such as vaccinia virus, also contain the coding sequence for enzymes that induces degradation of viral and cellular mRNAs. The D10 and D9 proteins are not, however, RNases. Rather, they are decapping enzymes that share a motif with the cellular Dcp2 decapping enzyme. Like their cellular counterpart, the viral enzymes hydrolyze the cap to release ^{m7}GDP. It is clear from the results of genetic experiments that the D10 protein induces rapid turnover of viral and cellular mRNAs, and hence facilitates inhibition of cellular protein synthesis in infected cells. It has been suggested that turnover of viral mRNAs may facilitate the production of specific sets of viral proteins during the successive phases of the infectious cycle, with the D9 and D10 enzymes acting early and late in infection.

Regulation of mRNA Stability in Transformation

Stabilization of specific viral mRNAs appears to be important in the development of cervical carcinoma associated with infection by high-risk human papillomaviruses, such as types 16 and 18. As discussed in Volume II, Chapter 7, the E6 and E7 proteins of these viruses induce abnormal proliferation of the cells in which they are made. In benign lesions, the circular human papillomavirus genome is not integrated (Fig. 10.21). The E6 and E7 mRNAs that are synthesized from such templates contain destabilizing, AU-rich sequences in their 3′ untranslated regions and possess short lifetimes. This property, as well as repression of transcription from the viral promoter from which the viral pre-mRNAs are transcribed, results in low, steady-state concentrations of the E6 and E7 mRNAs. In cervical carcinoma cells, the viral DNA is integrated into the cellular genome. Such reorganization of viral DNA frequently disrupts the sequences encoding the E6 and E7 mRNAs, such that their 3′ untranslated regions are copied from cellular DNA sequences (Fig. 10.21). These hybrid mRNAs therefore lack the destabilizing AU-rich sequences and are more stable. The increase in the stability of the viral mRNAs encoding the papillomaviral transforming proteins accounts, at least in part, for their higher concentrations in tumor cells.

Production and Function of Small RNAs That Inhibit Gene Expression

Small Interfering RNAs, Micro-RNAs, and Their Synthesis

In the early 1990s, attempts to produce more vividly purple petunias by creation of transgenic plants carrying an additional copy of the gene for the enzyme that makes

Figure 10.21 Stabilization of human papillomavirus type 16 mRNAs upon integration of the viral genome into cellular DNA. (A) In benign lesions, the viral genome is maintained as an extrachromosomal, circular episome. Transcription of such viral DNA and pre-mRNA processing produce various alternatively spliced mRNAs containing the E6 and/or E7 protein-coding sequences, but, as illustrated, all contain destabilizing 5′AUUUA3′ sequences in their 3′ untranslated regions. **(B)** In cervical carcinoma cells, the viral genome is integrated into cellular DNA (purple) such that the viral genome is disrupted upstream of the E6/E7 mRNA 3′ splice site. The mRNAs encoding these viral proteins are therefore made by using 3′ splice and polyadenylation sites transcribed from adjacent cellular DNA, and they lack the destabilizing sequence.

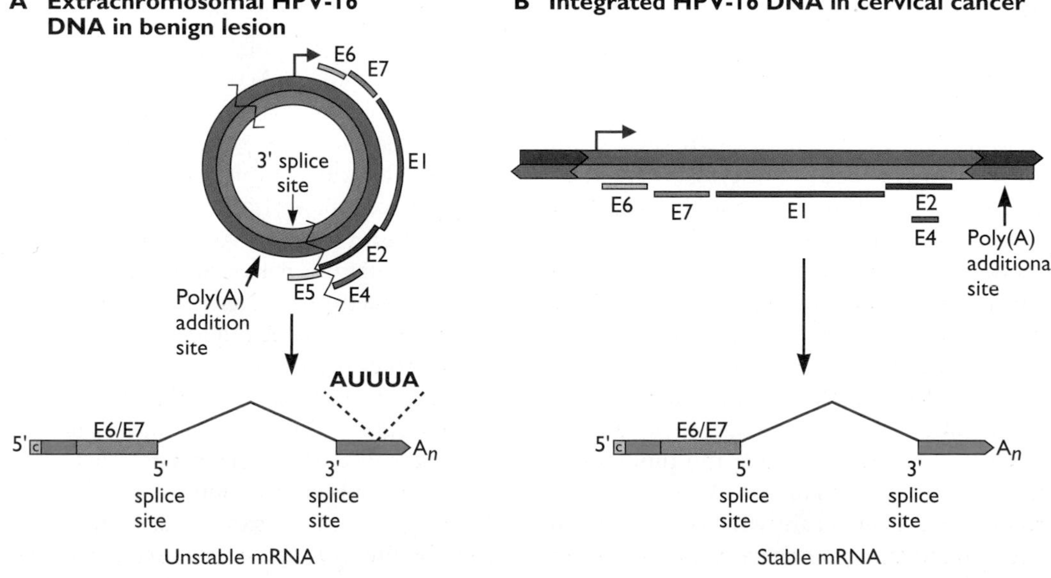

the purple pigment often resulted in white flowers. It is now clear that this seemingly esoteric observation represented the first example of a previously unknown mechanism of posttranscriptional regulation of gene expression, called RNA interference or RNA silencing, that is ancient, and widespread in eukaryotes. Our understanding of the mechanisms and functions of RNA interference, as well as its exploitation as an experimental tool, have advanced at a remarkably rapid pace. Indeed, Andrew Fire and Craig Mello were awarded the Nobel Prize in physiology or medicine in 2006, just 8 years after their groundbreaking study of the mechanism of RNA interference was published.

RNA interference is mediated by small RNA molecules (typically 21 to 28 nucleotides in length) that induce mRNA degradation or inhibit translation. The two main types of these regulatory RNA molecules are distinguished by how they are synthesized. Small interfering RNAs (siRNAs), such as those first discovered in plants, are initially processed by endonucleolytic cleavage of double-stranded RNAs by cytoplasmic dicer enzymes (Fig. 10.22). The double-stranded RNA precursors are formed by base pairing of transcripts that contain complementary sequences, such as the (+) and (−) strand RNAs synthesized in cells infected by many viruses with RNA genomes. Micro-RNAs (miRNAs) are also processed from longer precursors. However, these precursors are capped and polyadenylated transcripts synthesized by RNA polymerase II, in which self-complementary sequences form imperfect hairpin structures (Fig. 10.22). Although the catalog of mammalian miRNAs is far from complete, it is clear that genes encoding them are often clustered. This arrangement allows synthesis of transcripts containing multiple miRNA sequences. Such transcripts are initially processed by endonucleolytic cleavage in the nucleus to liberate pre-miRNAs, imperfect hairpins of 60 to 80 nucleotides. Further processing of pre-miRNAs occurs following export to the cytoplasm, where they are cleaved by dicer enzymes.

In the case of both siRNAs and miRNAs, the products of dicer cleavage are largely double-stranded, with two unpaired bases at the 3′ ends. These RNAs are then unwound from one 5′ end (Box 10.11) and one strand becomes tightly associated with a member of the Argonaute (Ago) family of proteins in the effector ribonucleoprotein, termed the RNA-induced silencing complex, Risc. In these complexes, the small RNA acts as a "guide," identifying the target mRNA by base pairing to specific sequences within it. This complex then induces mRNA cleavage or inhibition of translation, depending in part on the degree of complementarity between the siRNA or miRNA and the mRNA: perfect base pairing usually results in mRNA cleavage. The complement of proteins present in the Risc is also likely to be important. Most organisms that have been examined contain multiple Ago proteins (eight have been identified in humans), and they appear to fulfill different functions. Silencing complexes that contain human Ago2 cleave mRNAs with which the small guide RNA base pairs, but Ago1, Ago3, and Ago4 do not possess such RNase activity.

The introduction of small double-stranded RNAs analogous to the products formed by dicer during production of siRNAs has proved to be a very valuable experimental tool. Such exogenous RNAs are incorporated into Riscs with high efficiency, allowing the experimenter to inhibit expression of particular genes by targeting siRNAs to degrade the corresponding mRNA. However, mammalian cells generally synthesize miRNAs, rather than siRNAs, and the genomes of several DNA viruses contain sequences coding for miRNAs.

Viral Micro-RNAs

The first viral miRNAs were identified in 2004, by cloning and sequencing of small RNAs made in cells latently infected by Epstein-Barr virus. Subsequently, miRNAs of a number of other viruses have been described. Such RNAs are typically identified by combining computational methods that screen viral genomes for sequences with the properties of pre-miRNAs (Fig. 10.22) with assays for detection of viral RNAs, such as hybridization of low-molecular-weight RNAs isolated from infected cells to viral microarrays. Such approaches have identified significant numbers of miRNAs encoded by the genomes of alpha-, beta-, and gammaherpesviruses, as well as by polyomaviral genomes. With the exceptions described below, the functions of these recently recognized viral miRNAs are not known.

Simian Virus 40 miRNAs That Facilitate Immune Modulation

The genome of the polyomavirus simian virus 40 contains the sequence for a single pre-miRNA, which is transcribed as part of the late pre-mRNA (Fig. 10.23). Unusually, both RNA strands produced by processing of this miRNA precursor by the pathway shown in Figure 10.22 are stable, indicating they are both incorporated into effector silencing complexes. The miRNAs are synthesized late in infection, and induce cleavage and degradation of the mRNA for the early-gene product, large T antigen (Fig. 10.23). Mutations designed to disrupt the pre-miRNA secondary structure prevented both viral miRNA synthesis and large-T mRNA degradation, but did not alter the yield of infectious virus. However, the susceptibility of killing of infected cells by cytotoxic T cells specific for large T antigen was reduced significantly. As such cells are important for clearing infected cells from host animals (see Volume II, Chapter 4), this observation suggests that the viral miRNAs facilitate survival of simian virus 40-infected cells in the face of immune surveillance mechanisms.

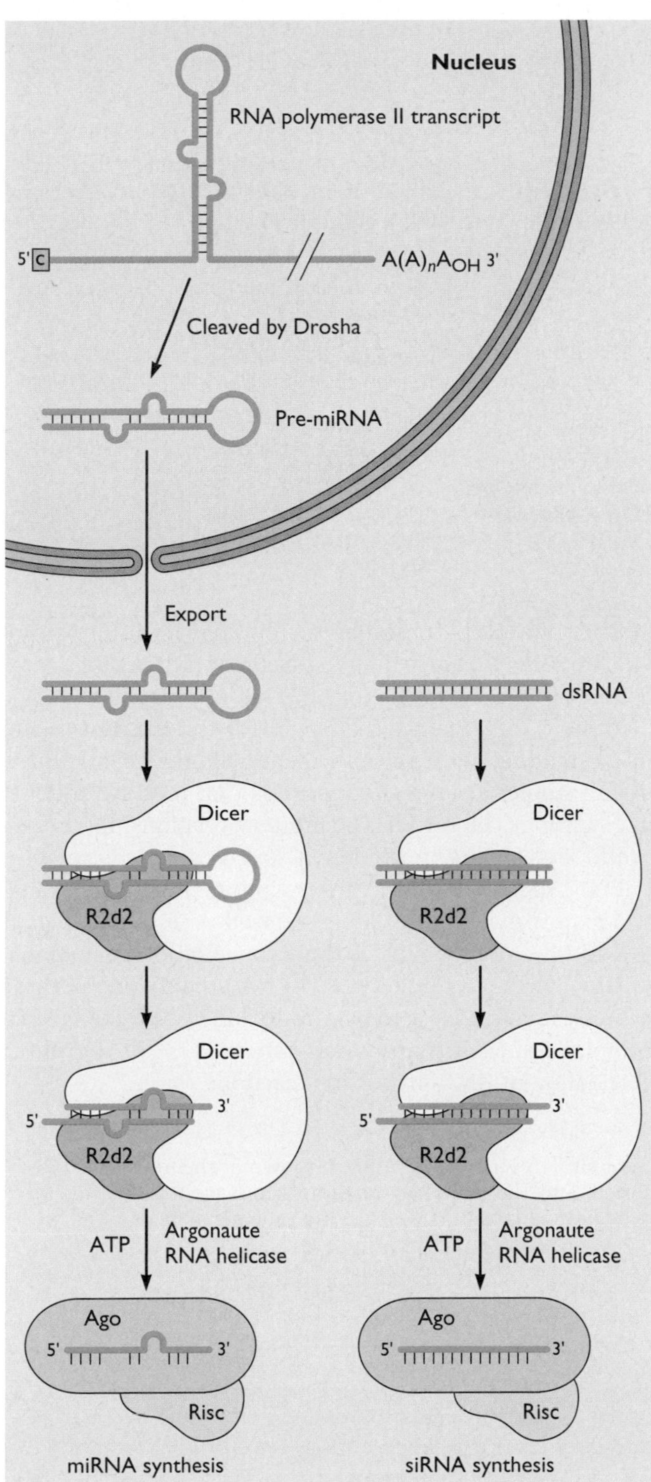

Figure 10.22 Synthesis of siRNAs and miRNAs. As shown, siRNAs are processed from cytoplasmic double-stranded RNAs, whereas the precursors of miRNAs, which are transcripts made by RNA polymerase II, must undergo initial processing in the nucleus. They are then exported to the cytoplasm via the export receptor exportin-5. In both cases, cleavage by Dicer enzymes produces short, largely double-stranded RNAs with unpaired nucleotides at their 3′ ends. Upon unwinding of these duplexes, one RNA strand becomes tightly associated with an Argonaute (Ago) protein (and others) in the RNA-induced silencing complex (Risc). The other strand is degraded.

Latency-Associated miRNAs of Herpesviruses

It is striking that pre-miRNA coding sequences have been identified in regions of several alpha-, beta-, and gammaherpesviral genomes that are expressed in latently infected cells. For example, 12 miRNAs are made in cells latently infected by human herpesvirus type 8. As discussed in Volume II, Chapter 7, this virus is a causative agent of Kaposi's sarcoma and B-cell lymphoma, and the viral gene products made in latently infected transformed cells are of considerable interest. The 12 latency-associated miRNAs are processed from three overlapping transcripts synthesized from viral promoters active in latently infected cells. Indeed, each transcript contains all the pre-miRNA sequences. Ectopic expression of the viral miRNA coding region reduced substantially the concentrations of eight cellular mRNAs for proteins that participate in regulation of proliferation, immune responses and apoptosis. The mRNA encoding thrombospondin 1 is of particular interest, in part because it is a target of several of the latency-associated miRNAs. In addition, an important function of this protein is to block angiogenesis (proliferation of new blood vessels), which is a hallmark of Kaposi's sarcomas.

Viral Gene Products That Block RNA Interference

Recent studies have identified proteins (or RNAs) of a wide range of viruses that counter host cell RNA interference by various mechanisms. Most are RNA-binding proteins that are believed to sequester miRNAs or their precursors. Such viral gene products are discussed in more detail in Volume II, Chapter 3, as they function to subvert an important antiviral defense mechanism.

Perspectives

Many of the molecular processes required for replication of animal viruses, including such virus-specific reactions as synthesis of genomic mRNAs and RNAs from an RNA template, were foretold by the properties of the bacteriophages that parasitize bacterial cells. In contrast, the covalent modifications necessary to produce functional mRNAs in eukaryotic cells were without precedent when discovered in viral systems. Study of the processing of viral RNAs has yielded much fundamental information about the mechanisms of capping, polyadenylation, and splicing. More recently, viral systems have provided equally important insights into export of mRNA from the nucleus to the cytoplasm. Perhaps the most significant lesson learned

BOX 10.11

DISCUSSION

How the guide strand of siRNAs is identified

During formation of RNA-induced silencing complexes, one strand of the double-stranded siRNA, the guide strand, is retained while the second (often called the passenger strand) is destroyed. siRNAs contain many different sequences, raising the question of how guide and passenger strands are distinguished.

The answer came from efforts to identify siRNAs that are most effective in inducing mRNA cleavage when introduced into cells. It was observed that the 5′ end of the guide RNA forms less thermodynamically stable base pairs than the 3′ end. Naturally occurring siRNAs and miRNAs exhibit this same asymmetry. As base pairs at the ends of double-stranded nucleic acids transiently break and re-form (they are said to "breathe"), this sequence might favor recognition of the transiently single-stranded 5′ end of the guide strand. Regardless, the less stable base pairs at the 5′ end of the guide strand favor unwinding from that end, which requires Dicer and the double-stranded RNA-binding protein, R2d2. Subsequent studies demonstrated that R2d2 binds to the siRNA end that contains the most stable base pairs and therefore determines the orientation of the dicer-R2d2 heterodimer on the duplex siRNA.

Khvorava, A., A. Reynolds, and S. D. Juyasena. 2003. Funtional siRNAs and MiRNAs exhibit strand bias. *Cell* **115:**209–216.

Schwarz, D. S., G. Hutvágner, T. Du, Z. Xu, N. Aronin, and P. D. Zamore. 2003. Asymmetry in the assembly of the RNAi enzyme complex. *Cell* **115:**199–208.

Tomari, Y., C. Matranga, B. Haley, N. Martinez, and P. D. Zamore. 2004. A protein sensor for siRNA asymmetry. *Science* **306:**1377–1380.

from the study of viral mRNA processing is the importance of these reactions in the regulation of gene expression.

Regulation of viral RNA processing can be the result of differences in the concentrations or activities of specific cellular components in different cell types, or actively induced by viral gene products. Several mechanisms by which viral proteins or RNAs can regulate or inhibit polyadenylation or splicing reactions, export of mRNA from the nucleus, or mRNA stability have been quite well characterized. However, our understanding of regulation of viral gene expression via RNA-processing reactions is far from complete; the mechanisms of action of several crucial viral regulatory proteins have not been fully elucidated, and many of the specific mechanisms deduced by using experimental systems have yet to be confirmed in virus-infected cells. Furthermore, exploration of the physiological functions of viral miRNAs that target viral or cellular mRNAs for degradation (or inhibit translation) began only very recently.

Among the greatest challenges for the future is a full understanding of the recently identified physical and functional couples among the reactions that produce mRNAs in eukaryotic cell nuclei. The intimate relationships among synthesis, capping, polyadenylation, and splicing of pre-mRNAs suggest that the transcription, RNA processing, and export machineries are organized within the nucleus to optimize all reactions in the production of a functional mRNA. Further study of viral mRNA production via these processes seems likely to be as valuable in addressing such complex issues as many viral systems were in the initial elucidation of RNA-processing reactions.

Figure 10.23 The miRNAs of simian virus 40. The circular simian virus 40 genome is shown at the left, with the positions of the early (P_E) and late (P_L) promoters and the primary transcripts indicated. As shown, the 3′ ends of early and late pre-mRNAs are encoded by opposite strands of the same sequence. Downstream of its polyadenylation site (arrowhead), the late pre-RNA contains an pre-miRNA sequence, that is processed to a 57-nucleotide pre-miRNA and then to two miRNAs, designated 3′ and 5′. Both are perfectly complementary to specific sequences in the early mRNAs which encode T antigen and induce its cleavage.

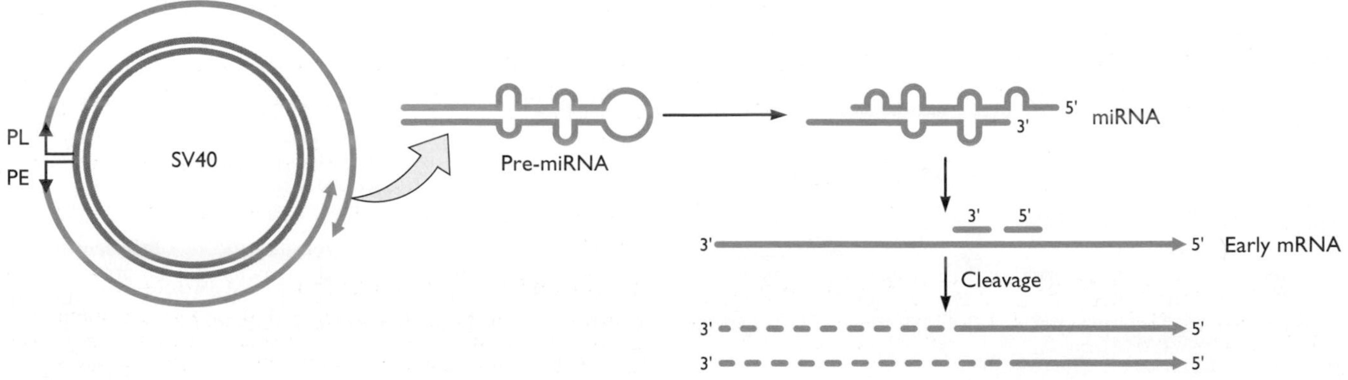

References

Books

Hauber, J., and P. K. Vogt (ed.). 2001. *Nuclear Export of Viral RNAs.* Springer-Verlag, Berlin, Germany.

Reviews

Banks, J. D., K. L. Beemon, and M. F. Linial. 1997. RNA regulatory elements in the genomes of simple retroviruses. *Semin. Virol.* **8:** 194–204.

Bentley, D. L. 2005. Rules of engagement: co-transcriptional recruitment of pre-mRNA processing factors. *Curr. Opin. Cell Biol.* **17:**251–256.

Colgan, D. F., and J. L. Manley. 1997. Mechanism and regulation of mRNA polyadenylation. *Genes Dev.* **11:**2755–2766.

Garneau, N. L., J. Wilusz, and C. J. Wilusz. 2007. The highways and byways of mRNA decay. *Nat. Rev. Mol. Cell Biol.* **8:**113–126.

Maniatis, T., and R. Reed. 2002. An extensive network of coupling among gene expression machines. *Nature* **416:**499–506.

Meister, G., and T. Tuschl. 2004. Mechanisms of gene silencing by double-stranded RNA. *Nature* **431:**343–349.

Mitchell, P., and D. Tollervey. 2000. mRNA stability in eukaryotes. *Curr. Opin. Genet. Dev.* **10:**193–198.

Rodriguez, M. S., C. Dargemont, and F. Stutz. 2004 Nuclear export of RNA. *Biol. Cell* **96:**639–655.

Sandri-Goldin, R. M. 2004. Viral regulation of mRNA export. *J. Virol.* **78:**4389–4396.

Shatkin, A. J., and J. L. Manley. 2000. The ends of the affair: capping and polyadenylation. *Nat. Struct. Biol.* **7:**838–842.

Shuman, S. 2001. Structure, mechanism and evolution of the mRNA capping apparatus. *Prog. Nucleic Acid Res. Mol. Biol.* **66:**1–40.

Smith, H. C., J. M. Gottand, and M. R. Hanson. 1997. A guide to RNA editing. *RNA* **3:**1105–1123.

Sullivan, C. S., and D. Ganem. 2005. MicroRNAs and viral infecton. *Mol. Cell* **20:**3–7.

Staley, J. P., and C. Guthrie. 1998. Mechanical devices of the spliceosome: motors, clocks, springs, and things. *Cell* **92:**315–326.

Zhang, Z. M., and C. C. Baker. 2006. Papillomavirus genome structure, expression and post-transcriptional regulation. *Front. Biosci.* **11:**2286–2302.

Papers of Special Interest

Capping of Viral mRNAs or pre-mRNAs

Dong, H., D. Ray, S. Ren, B. Zhang, F. Paig-Basagoti, Y. Takagi, Z. K. Ho, H. Li, and P.-Y. Shi. 2007. Distinct RNA elements confer specificity to flavivirus RNA cap methylation events. *J. Virol.* **81:**4412–4421.

Lou, Y., X. Mao, L. Deng, P. Cong, and S. Shuman. 1995. The D1 and D12 subunits are both essential for the transcription termination activity of vaccinia virus capping enzyme. *J. Virol.* **69:**3852–3856.

Plotch, S. J., M. Bouloy, I. Ulmanen, and R. M. Krug. 1981. A unique cap (m^7GpppXm)-dependent influenza virion endonuclease cleaves capped RNAs to generate the primers that initiate viral RNA transcription. *Cell* **23:**847–858.

Salditt-Georgieff, M., M. Harpold, S. Chen-Kiang, and J. E. Darnell. 1980. The addition of the 5′ cap structure occurs early in hnRNA synthesis and prematurely terminated molecules are capped. *Cell* 19:69–78.

Polyadenylation of Viral mRNA or Pre-mRNA

Ashe, M. L. P., L. H. Pearson, and N. J. Proudfoot. 1997. The HIV-1 5′ LTR poly(A) site is inactivated by U1 snRNP interaction with the major splice donor site. *EMBO J.* **16:**5752–5763.

Fitzgerald, M., and T. Shenk. 1981. The sequence 5′AAUAAA3′ forms part of the recognition site for polyadenylation of late SV40 mRNAs. *Cell* **24:**251–260.

Gilmartin, G. M., S.-L. Hung, J. D. DeZazzo, E. S. Fleming, and M. J. Imperiale. 1996. Sequences regulating poly(A) site selection within the adenovirus major late transcription unit influence the interaction of constitutive processing factors with the pre-mRNA. *J. Virol.* **70:**1775–1783.

Mohamed, R. M., D. R. Latner, R. C. Condit, and E. G. Niles. 2001. Interaction between the J3R subunit of vaccinia virus poly(A) polymerase and the H4L subunit of the viral RNA polymerase. *Virology* **280:**143–152.

Wilusz, J. E., and K. L. Beemon. 2006. The negative regulator of splicing element of Rous sarcoma virus promotes polyadenylation. *J. Virol.* **80:**9634–9640.

Splicing of Viral Pre-mRNA

Barksdale, S. K., and C. C. Baker. 1995. Differentiation-specific alternative splicing of bovine papillomavirus late mRNAs. *J. Virol.* **69:**6553–6556.

Bouck, J., X.-D. Fu, A. M. Skalka, and R. A. Katz. 1995. Genetic selection for balanced retroviral splicing: novel regulation involving the second step can be mediated by transitions in the polypyrimidine tract. *Mol. Cell. Biol.* **15:**2663–2671.

Kanopka, A., O. Mühlemann, S. Petersen-Mahrt, C. Estmer, C. Öhrmalm, and G. Akusjärvi. 1998. Regulation of adenovirus alternative RNA splicing by dephosphorylation of SR proteins. *Nature* **393:**185–187.

Lamb, R. A., and C. J. Lai. 1980. Sequence of interrupted and uninterrupted mRNAs and cloned DNA coding for the two overlapping non-structural proteins of influenza virus. *Cell* **21:**475–485.

McPhillips, M. G., T. Veerapraditsin, S. A. Cumming, D. Karali, S. G. Milligan, W. Boner, I. M. Morgan, and S. V. Graham. 2004. SF2/ASF binds the human papillomavirus late RNA control element and is regulated during differentiation of virus-infected epithelial cells. *J. Virol.* **78:**10598–10605.

Zabolotny, J. M., C. Krummenacher, and N. W. Fraser. 1997. The herpes simplex virus type 1 2.0-kilobase latency-associated transcript is a stable intron which branches at a guanosine. *J. Virol.* **71:**4199–4208.

Editing of Viral RNA

Gandy, S. Z., S. D. Linnstaedt, S. Muralidhar, K. A. Cashman, L. J. Rosenthal, and J. L. Casey. 2007. RNA editing of the human herpesvirus 8 kaposin transcript eliminates its transforming activity and is induced during lytic replication. *J. Virol.* **81:**13544–13551.

Polson, A. G., B. L. Bass, and J. L. Casey. 1996. RNA editing of hepatitis delta virus antigenome by dsRNA-adenosine deaminase. *Nature* (London) **380:**454–456.

Sanchez, A., S. G. Trappier, B. W. Mahy, C. J. Peters, and S. T. Nichol. 1996. The virion glycoproteins of Ebola viruses are encoded in two reading frames and are expressed through transcriptional editing. *Proc. Natl. Acad. Sci. USA* **93:**3602–3607.

Sato, S., C. Cornillex-Ty, and D. W. Lazinski. 2004. By inhibiting replication, the large hepatitis virus delta antigen can indirectly regulate amber/W editing and its own expression. *J. Virol.* **78:**8120–8134.

Thomas, S. M., R. A. Lamb, and R. G. Paterson. 1988. Two mRNAs that differ by two nontemplated nucleotides encode the amino coterminal proteins P and V of the paramyxovirus SV5. *Cell* **54:**891–902.

Export of Viral mRNAs from the Nucleus

Ernst, R. K., M. Bray, D. Rekosh, and M.-L. Hammarskjöld. 1997. A structured retroviral RNA element that mediates nucleocytoplasmic export of intron-containing RNA. *Mol. Cell. Biol.* **17:**135–144.

Fischer, U., J. Huber, W. C. Boelens, I. W. Mattaj, and R. Lührmann. 1995. The HIV-1 Rev activation domain is a nuclear export signal that accesses an export pathway used by specific cellular RNAs. *Cell* **82:**475–483.

Huang, Z. M., and T. S. Yen. 1995. Role of the hepatitis B virus post-transcriptional regulatory element in export of intronless transcripts. *Mol. Cell. Biol.* **15:**3864–3869.

Pasquinelli, A. E., R. K. Ernst, E. Lund, C. Grimm, M. L. Zapp, D. Rekosh, M.-L. Hammarskjöld, and J. E. Dahlberg. 1997. The constitutive transport element (CTE) of Mason Pfizer monkey virus (MMPV) accesses a cellular mRNA export pathway. *EMBO J.* **16:**7500–7510.

Sanchez-Velar, N., E. B. Udofia, Z. Yu, and M. L. Zapp. 2004. hRIP, a cellular cofactor for Rev function, promotes release of HIV RNAs from the perinuclear region. *Genes Dev.* **18:**23–34.

Zang, W.-Q., B. Li, P.-Y. Huang, M. M. C. Lai, and T. S. B. Yen. 2001. Role of polypyrimidine tract binding protein in the function of the hepatitis B virus posttranscriptional regulatory element. *J. Virol.* **75:**10779–10786.

Regulation of RNA Processing by Viral Proteins

Faria, P. A., P. Chakroborty, A. Levay, G. N. Barber, H. J. Ezelle, J. Enninga, C. Arana, J. van Deursen and B. M. A. Fontoura. 2005. VSV disrupts the Rae1/mrnp14 mRNA export pathway. *Mol. Cell* **17:**93–102.

Gonzalez, R. A., and S. J. Flint. 2002. Effects of mutations in the adenoviral E1B 55-kilodalton protein coding sequence on viral late mRNA metabolism. *J. Virol.* **76:**4507–4519.

McGregor, F., A. Phelan, J. Dunlop, and J. B. Clements. 1996. Regulation of herpes simplex virus poly(A) site usage and the action of immediate-early protein IE63 in the early-late switch. *J. Virol.* **70:**1931–1940.

Molin, M., and G. Akusjärvi. 2000. Overexpression of essential splicing factor ASF/SF2 blocks the temporal shift in adenovirus pre-mRNA splicing and reduces virus progeny formation. *J. Virol.* **74:**9002–9009.

Noah, D. L., K. Y. Twu, and R. M. Krug. Cellular antiviral responses against influenza virus are countered at the post-transcriptional level by the viral NS1A protein via its binding to a cellular protein required for the 3′ end processing of cellular pre-mRNAs.

Pilder, S., M. Moore, J. Logan, and T. Shenk. 1986. The adenovirus E1B 55K transforming polypeptide modulates transport or cytoplasmic stabilization of viral and host cell mRNAs. *Mol. Cell. Biol.* **6:**470–476.

Porter, F. W., Y. A. Bochkov, A. J. Albee, C. Wiese, and A. C. Palmenberg. 2006. A picornavirus protein interacts with Ran-GTPase and disrupts nucleocytoplasmic transport. *Proc. Natl. Acad. Sci. USA* **103:**12417–12422.

Törmänen, H., E. Backström, A. Carlsson, and G. Akusjärvi. 2006. L4-33k, an adenovirus-encoded alternative RNA splicing factor. *J. Biol. Chem.* **281:**36510–36517.

Stability of Viral and Cellular mRNA

Depker, R. C., W.-L. Hsu, H. A. Suffran, and J. R. Smiley. 2004. Herpes simplex virus virion host shutoff protein is stimulated by translation initiation factors eIF4B and eIF4H. *J. Virol.* **78:**4684–4699.

Everly, D. N., Jr., P. Feng, I. S. Mian, and G. S. Read. 2002. mRNA degradation by the virion host shutoff (Vhs) protein of herpes simplex virus: genetic and biochemical evidence that Vhs is a nuclease. *J. Virol.* **76:**8560–8571.

Jeon, S., and P. F. Lambert. 1995. Integration of human papillomavirus type 16 DNA into the human genome leads to increased stability of E6 and E7 mRNAs: implications for cervical carcinogenesis. *Proc. Natl. Acad. Sci. USA* **92:**1654–1658.

Myer, V. E., S. I. Lee, and J. A. Steitz. 1992. Viral small nuclear ribonucleoproteins bind a protein implicated in messenger RNA destabilization. *Proc. Natl. Acad. Sci. USA* **89:**1296–1300.

Parrish, S., W. Resch, and B. Moss. 2007. Vaccinia virus D10 protein has mRNA decapping activity, providing a mechanism for control of host and viral gene expression. *Proc. Natl. Acad. Sci. USA* **104:**2139–2144.

RNA Interference

Cai, X., and B. R. Cullen. 2006. Transcriptional origin of Kaposi's sarcoma-associated herpesvirus microRNAs. *J. Virol.* **80:**2234–2242.

Cui, C., A. Griffiths, G. Li, L. M. Silva, M. F. Kramer, T. Gaasterland, Z.-J. Wang, and D. M. Coen. 2006. Prediction and identification of herpes simplex virus 1-encoded mircoRNAs. *J. Virol.* **80:**5499–5508.

Pfeffer, S., M. Zavolan, F. A. Gässer, M. Chien, J. J. Russo, J. Ju, B. John, A. J. Enright, D. Marks, C. Sander, and T. Tusche. 2004 Identification of virus-encoded microRNAs. *Science* **304:**734–736.

Samols, M. A., R. L. Skalsky, A. M. Maldonado, A. Riva, M. C. Lopez, H. V. Baker, and R. Renne. 2007. Identification of cellular genes targeted by KSHV-encoded microRNAs. *PLOS Pathog.* **3:**611–618.

Sullivan, C. S. A. T. Grandhoff, S. Tevethia, J. M. Pipas, and D. Gahem. 2005. SV40-encoded microRNAs regulate viral gene expression and reduce susceptibility to cytotoxic T cells. *Nature* **435:**682–686.

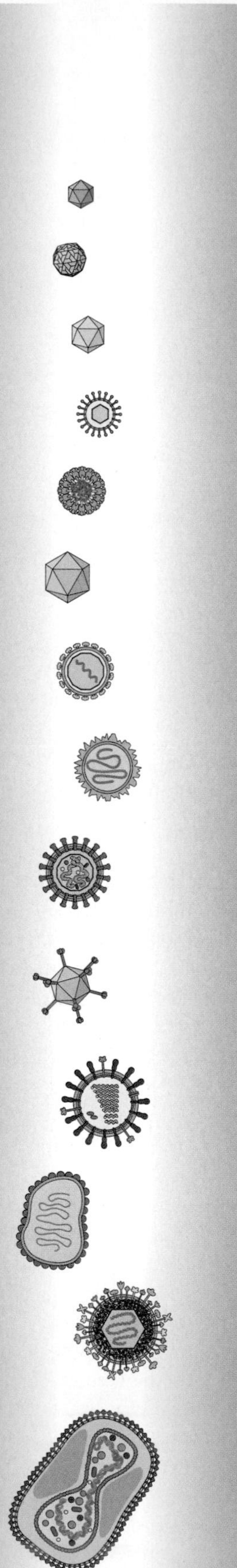

11

Control of Translation

The difficulty lies, not in the new ideas, but in escaping old ones....
JOHN MAYNARD KEYNES

Introduction

Viruses are obligate intracellular parasites; their replication is dependent on components of the host cell. Some aspects of viral multiplication depend more on cellular contributions than do others. For example, all viral genomes encode at least one protein needed for viral nucleic acid synthesis. However, viruses are dependent on the host cell for the translation of their messenger RNAs (mRNAs) because most viral genomes do not encode any part of the translational machinery (Box 11.1). Viral infection often results in modification of the host's translational apparatus so that viral mRNAs can be translated selectively.

Studies of virus-infected cells have contributed much to our understanding of translation and its regulation. Before the advent of recombinant DNA technology, infected cells were a ready source of large quantities of relatively pure mRNA for *in vitro* studies of protein synthesis. The 5′ cap structure was identified on a viral RNA, and new translation initiation mechanisms, such as internal ribosomal entry, were discovered during studies of infected cells. Our understanding of how the activity of the multisubunit cap-binding complex can be regulated originated from the finding that one of its subunits is cleaved in infected cells.

Translation is a universal process in which proteins are produced from mRNA templates read in the 5′ → 3′ direction, and the growing polypeptide chain is synthesized from the amino to the carboxy terminus. Each amino acid is specified by a genetic code consisting of three bases, a **codon**, in the mRNA. Translation takes place on **ribosomes**, and **transfer RNAs (tRNAs)** are the adapter molecules that link specific amino acids with individual codons in the mRNA. This chapter explores the basic mechanisms by which translation occurs in eukaryotic cells, the many ways that viral mRNAs are translated to confer expanded coding capacity in genomes of limited size, and how translation is regulated in infected cells.

BOX 11.1

DISCUSSION

Viral contributions to the translational machinery

Analysis of the nucleic acid of the largest DNA viruses challenges the belief that no viral genomes encode any part of the translational machinery. The 330- to 380-kbp DNA genome of viruses that infect the unicellular green alga *Chlorella* encode 10 to 15 tRNAs. These viral tRNAs are produced in infected cells, and some of them are aminoacylated, suggesting that they function during protein synthesis. These viral genomes also encode a homolog of elongation protein 3 that is produced in infected cells. The 1,180-kbp DNA genome of *Mimivirus*, the largest known virus, encodes six tRNAs, four aminoacyl tRNA synthetases, release protein eRF1, elongation protein eF-Tu, initiation protein 1, and eIF4E.

These remarkable observations show that parts of the cellular translational machinery can be replaced by viral gene products. Why viral genomes encode such gene products is not known. One possibility is that these genes are relics of an ancestral translation machinery that has been lost during viral evolution. An alternative is that these genes were recently acquired and assist in modifying the translation apparatus to favor the production of viral proteins. For example, the use of viral tRNAs may compensate for the low abundance of some tRNAs in host cells, allowing more efficient replication.

Nishida, K., T. Kawasaki, M. Fujie, S. Usami, and T. Yamada. 1999. Aminoacylation of tRNAs encoded by *Chlorella* virus CVK2. *Virology* **263:**220–229.

Raoult, D., S. Audic, C. Robert, C. Abergel, P. Renesto, H. Ogata, B. La Scola, M. Suzan, and J. M. Claverie. 2005. The 1.2 megabase genome sequence of *Mimivirus*. *Science* **308:**1344–1350.

Yamada, T., T. Fukuda, K. Tamura, S. Furukawa, and P. Songsri. 1993. Expression of the gene encoding a translational elongation factor 3 homolog of *Chlorella* virus CVK2. *Virology* **197:**742–750.

Mechanisms of Eukaryotic Protein Synthesis

General Structure of Eukaryotic mRNA

Most eukaryotic mRNAs, with the exception of organelle and certain viral mRNAs, begin with a 5′ 7-methylguanosine (m^7G) **cap structure** (Fig. 11.1; see also Fig. 10.2). It is joined to the second nucleotide by a 5′-5′ phosphodiester linkage, in contrast to the 5′-3′ bonds found in the remainder of the mRNA. The unique cap structure directs pre-mRNAs to processing and transport pathways, regulates mRNA turnover, and is required for efficient translation by the 5′-end-dependent mechanism. Most eukaryotic mRNAs contain **5′ untranslated regions**, which may vary in length from 3 to more than 1,000 nucleotides, although they are typically 50 to 70 nucleotides in length. Such 5′ untranslated regions often contain secondary structures (e.g., hairpin loops [see Fig. 6.2]) formed by base pairing of the RNA. These double-helical regions must be unwound to allow passage of ribosomal 40S subunits during translation.

Figure 11.1 Structure of eukaryotic and bacterial/archaeal mRNAs. UTR, untranslated region; AUG, initiation codon; ORF, open reading frame; Stop, termination codon. Adapted from G. M. Cooper, *The Cell: a Molecular Approach* (ASM Press, Washington, DC, and Sinauer Associates, Sunderland, MA, 1997), with permission.

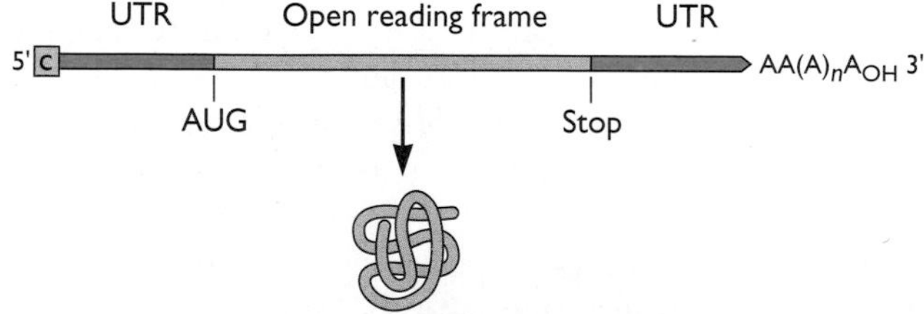

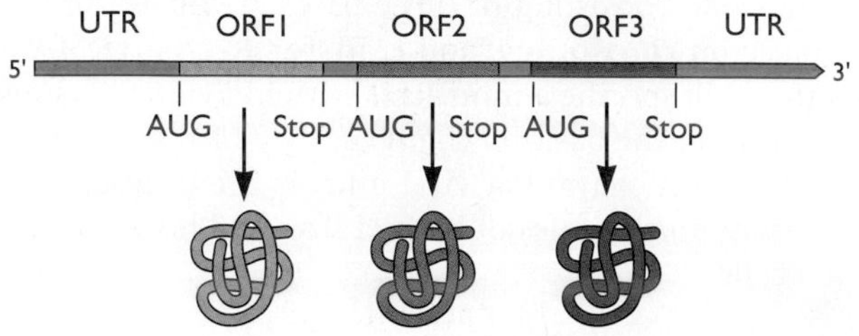

Translation begins at an **initiation codon** and ends at a **termination codon**. The termination codon is followed by a **3′ untranslated region**, which can regulate initiation, translation efficiency, and mRNA stability. At the very 3′ end of the mRNA is a stretch of adenylate residues known as the **poly(A) tail**, which is added to nascent pre-mRNA. The poly(A) tail is necessary for efficient translation, and may promote interactions among proteins that bind both ends of the mRNA.

Most bacterial and archaeal mRNAs are **polycistronic**: they encode several proteins, and each open reading frame is separated from the next by an untranslated spacer region. The vast majority of eukaryotic mRNAs are **monocistronic**; i.e., they encode only a single protein (Fig. 11.1). A small number of eukaryotic mRNAs are functionally polycistronic, and different strategies have evolved for synthesizing multiple proteins from a single mRNA. Members of the virus family *Dicistroviridae* are unique because their virions contain true bicistronic mRNAs.

The Translation Machinery

Ribosomes

Mammalian ribosomes, the sites of translation, are composed of two subunits designated according to their sedimentation coefficients, 40S and 60S (Fig. 11.2). The 40S subunit comprises an 18S rRNA molecule and 30 proteins, while the 60S subunit contains three rRNAs (5S, 5.8S, and 28S rRNAs) and 50 proteins. Actively growing mammalian cells may contain approximately 10 million ribosomes.

Remarkably, the catalytic activity of ribosomes is carried out largely by RNA, not protein. After removal of 95% of the ribosomal proteins, the 60S ribosomal subunit can still catalyze the formation of peptide bonds; the peptidyl transferase center, where peptide bonds are formed, contains only RNA. The ribosome is the largest known RNA catalyst. The protein components of ribosomes are thought to help fold the rRNAs properly, so that they can fulfill their catalytic function, and to position the tRNAs.

tRNAs

tRNAs are adapter molecules that align each amino acid with its corresponding codon on the mRNA. Each tRNA is 70 to 80 nucleotides in length and folds into a highly base-paired L-shaped structure (Fig. 11.2B). This shape is thought to be required for the appropriate interaction between tRNA and the ribosome during translation. The adapter function of tRNAs is carried out by two distinct regions of the molecule. At their 3′ ends, all tRNAs have the sequence 5′-CCA-3′, to which amino acids are covalently linked by **aminoacyl-tRNA synthetase**. Each of these enzymes recognizes a single amino acid and the correct tRNA. At the opposite end of the tRNA is the **anticodon loop**, which base pairs with the mRNA template. The accuracy of protein synthesis is maintained by two different proofreading mechanisms: faithful incorporation of amino acids depends on the specificity of codon-anticodon base pairing, as well as on the correct attachment of amino acids to tRNAs by aminoacyl-tRNA synthetases.

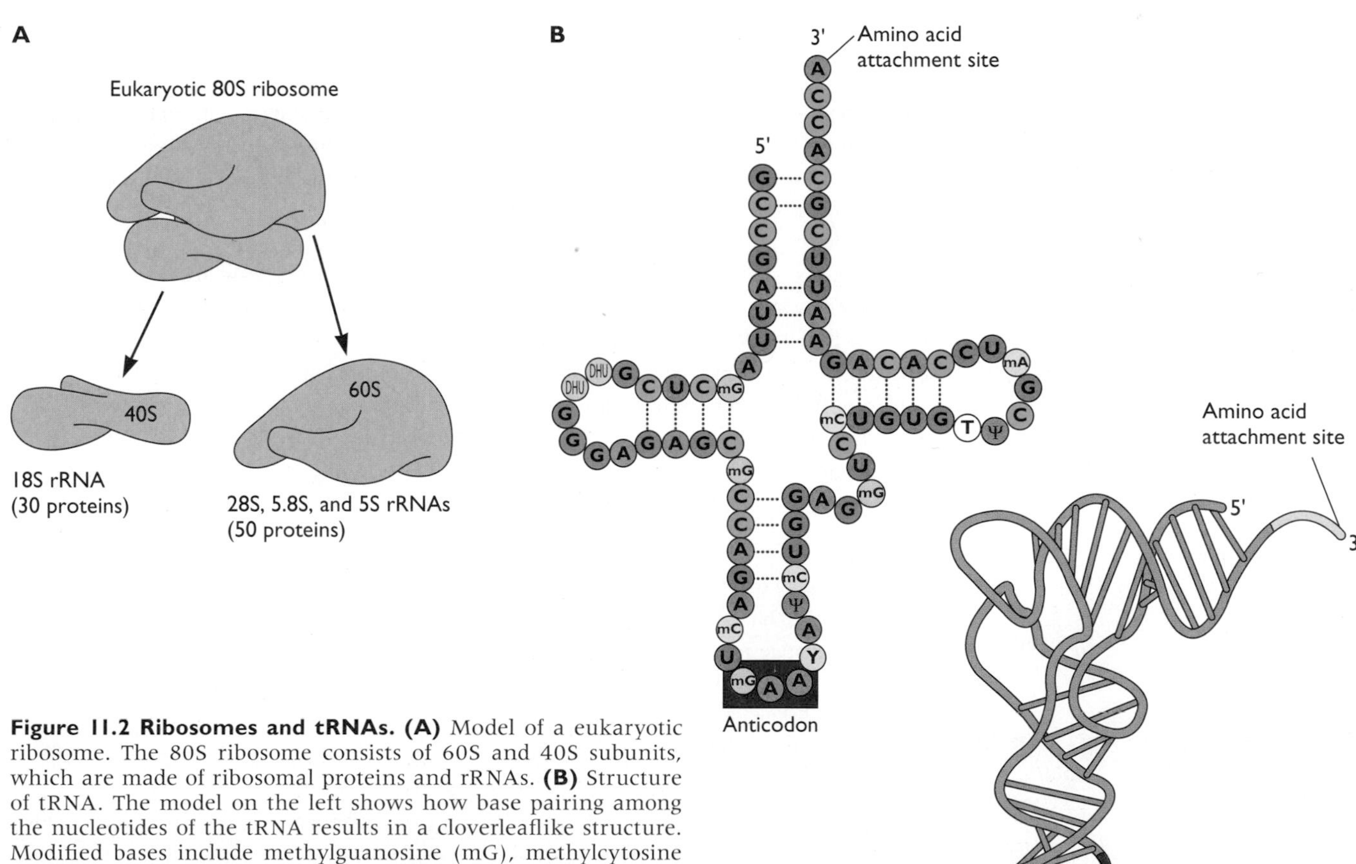

Figure 11.2 Ribosomes and tRNAs. (A) Model of a eukaryotic ribosome. The 80S ribosome consists of 60S and 40S subunits, which are made of ribosomal proteins and rRNAs. **(B)** Structure of tRNA. The model on the left shows how base pairing among the nucleotides of the tRNA results in a cloverleaflike structure. Modified bases include methylguanosine (mG), methylcytosine (mC), dihydrouridine (DHU), ribothymidine (T), a modified purine (Y), and pseudouridine (Ψ). On the right is a folded representation showing the L-shaped structure. Adapted from G. M. Cooper, *The Cell: a Molecular Approach* (ASM Press, Washington, DC, and Sinauer Associates, Sunderland, MA, 1997), with permission.

Table 11.1 Mammalian translation proteins[a]

Name	Subunit	Function
Initiation proteins		
eIF1		Enhances initiation complex formation at the appropriate AUG initiation codon; destabilizes aberrant 48S complexes
eIF1A		Enhances initiation complex formation at the appropriate AUG initiation codon; affects ribosome dissociation and release of 60S subunit; promotes Met-tRNAi binding to 40S, promotes scanning
eIF2		Binds Met-tRNAi, and GTP; associates with eIF3, eIF5
	α	Affects eIF2B binding by phosphorylation
	β	Binds Met-tRNAi; associates with eIF2B, eIF5
	γ	Binds GTP, Met-tRNAi; GTPase
eIF2B		eIF2 recycling
	α	Recognition of P-eIF2
	β	Binds GTP; recognition of P-eIF2
	γ	Binds ATP; recognition of P-eIF2
	δ	Guanine exchange
	ε	Guanine exchange
eIF3		Dissociates 80S ribosomes; promotes Met-tRNAi and mRNA binding to 40S ribosomal subunit
	k	
	j	40S ribosomal subunit, dissociation of mRNA after termination, multifactor complex[b] assembly
	i	
	h	
	g	Binds mRNA, eIF4B
	f	May bind mTor and S6k1[c]
	e	
	d	Binds mRNA
	c	Binds 40S ribosomal subunit; ternary complex; mRNA; AUG recognition; multifactor complex assembly
	b	Binds 40S ribosomal subunit; ternary complex; mRNA; multifactor complex assembly; scanning
	a	Binds 40S ribosomal subunit; eIF4B; ternary complex; mRNA; multifactor complex assembly
eIF4AI		ATPase, RNA helicase
eIF4AII		ATPase, RNA helicase
eIF4B		Binds RNA; promotes mRNA-40S ribosomal subunit interaction; promotes helicase activity of eIF4A
eIF4E		Binds m^7G cap of mRNA
eIF4GI		Binds eIF4E, eIF4A, eIF3, Mnk1, Pab1p, Paip-1, RNA
eIF4GII		Binds eIF4E, eIF4A, eIF3, Mnk1, Pab1p, Paip-1, RNA
eIF4H		Binds RNA; stimulates ATPase and helicase of eIF4A
eIF5		Promotes GTPase activity of eIF2
eIF5B		Ribosome-dependent GTPase; required for 60S ribosomal subunit joining
eIF6		Binds to 60S ribosomal subunit; promotes 80S ribosome dissociation
Elongation proteins		
eEF1A		GTP-dependent binding of aminoacyl-tRNAs; GTPase
eEF1B		Guanine nucleotide exchange on eEF1A
	α	GTP nucleotide exchange activity
	β	GTP nucleotide exchange activity
	γ	Anchors eEF1 to cytoskeleton or membranes
eEF2		GTPase; promotes translocation of peptidyl-tRNA from A site to P site
Termination proteins		
eRF1		Recognizes termination codons; promotes ribosome-catalyzed peptidyl-tRNA hydrolysis; interacts with peptidyl transferase of 60S ribosomal subunit and A site of ribosome
eRF3		GPTase; associates with eRF1 and Pab1p

[a]Modified from N. Sonenberg et al., *Translational Control of Gene Expression* (Cold Spring Harbor Press, Cold Spring Harbor, NY, 2000).
[b]multifactor complex comprises eIF3, eIF1, eIF5, and ternary complex.
[c]S6k1, ribosomal protein S6 kinase.

Translation Proteins

Many nonribosomal proteins are required for eukaryotic translation (Table 11.1). Some form multisubunit complexes containing as many as 11 different proteins, while others function as monomers. Translation can be separated experimentally into three distinct stages: initiation, elongation, and termination. The proteins that participate at each stage are named eukaryotic initiation, elongation, and termination proteins. These proteins are designated in the same way as their bacterial and archaeal counterparts, with the prefix "e" to distinguish them. The amino acid sequences of these proteins are conserved from yeasts to mammals, indicating that the mechanisms of translation are similar throughout eukaryotes.

Initiation

The majority of regulatory mechanisms function during initiation because this is the rate-limiting step in the translation of most mRNAs (see "Regulation of Translation during Viral Infection" below). At least 11 initiation proteins participate in this energy-dependent process. The end result is formation of a complex containing the mRNA, the ribosome, and the initiator Met-tRNA$_i$ in which the reading frame of the mRNA has been set. The 80S ribosome, which is the predominant species in cells, must be dissociated because it is the 40S subunit that participates in initiation. Three initiation proteins, eIF1A, eIF3, and eIF6 (Table 11.1), promote such dissociation.

There are two mechanisms by which ribosomes bind to mRNA in eukaryotes. In 5′-end-dependent initiation, by which the majority of mRNAs are translated, the initiation complex binds to the 5′ cap structure and moves, or scans, in a 3′ direction until the initiating AUG codon is encountered. In 5′-end-dependent initiation, the initiation complex binds at, or just upstream of, the initiation codon. Internal ribosome entry sites were first discovered in picornavirus mRNAs, and are now known to be present in some cellular mRNAs.

5′-End-Dependent Initiation

How ribosomes assemble at the proper end of mRNA. The first step in the 5′-end-dependent initiation pathway is recognition of the m^{7}G cap by the cap-binding protein, eIF4E (Fig. 11.3). eIF4G acts as a scaffold between the cap structure and the 40S subunit, which associates with the mRNA via an interaction of eIF3 with the C-terminal domain of eIF4G. This important adapter molecule was first discovered as the target of proteolytic cleavage in poliovirus-infected cells that results in the abrupt termination of host protein synthesis. The ability of eIF4G to bind RNA is also important for recruitment of the 40S ribosomal subunit. After binding near the cap, the 40S ribosomal subunit moves in a 3′ direction on the mRNA in a process called **scanning**. When the 40S subunit reaches the AUG initiation codon, GTP is hydrolyzed and initiation proteins are released, allowing the 60S ribosomal subunit to associate with the 40S subunit, forming the 80S initiation complex.

Role of the poly(A) tail in initiation. The presence of a poly(A) tail can stimulate mRNA translation. This surprising effect is a consequence of interactions between proteins associated with the 5′ and 3′ ends of the mRNA, promoting 40S subunit recruitment. Such interactions were first demonstrated in the yeast *Saccharomyces cerevisiae*, in which poly(A)-tail-binding protein, Pab1p, is required for efficient mRNA translation. Stimulation of translation by poly(A), which requires Pab1p, occurs by enhancing the binding of 40S ribosomal subunits to mRNA. Pab1p interacts with the N terminus of eIF4G (Fig. 11.4). Alteration of the Pab1p-binding site on eIF4G destroys stimulation of translation by poly(A). These results have led to a model in which Pab1p, bound to the poly(A) tail, interacts with eIF4G bound to the 5′ cap, perhaps stabilizing the interaction and assisting in recruitment of 40S subunits (Fig. 11.4). A consequence of these interactions is that the 5′ and 3′ ends of the mRNA are brought into close proximity.

Many viral mRNAs, such as the mRNA of barley yellow dwarf luteovirus, lack a 5′-terminal cap and 3′ poly(A) sequence. Nevertheless, the ends of these mRNAs are brought together by base pairing between discrete sequences in the 5′ and 3′ untranslated regions. Translation of mRNA of the flavivirus dengue virus, which has a 5′ cap structure but lacks a 3′ poly(A) sequence, may also depend on complementarity between the untranslated regions.

The juxtaposition of mRNA ends might be a mechanism to ensure that only intact mRNAs that contain 5′ cap and poly(A) are translated. Such structures could also stabilize mRNA by preserving the interaction among the translation initiation proteins associated with both ends. Translation reinitiation might also be stimulated by such an arrangement: once the ribosome terminates translation, it might be rapidly repositioned at the AUG initiation codon rather than dissociating into the two subunits.

VPg-dependent ribosomal recruitment. The 40S ribosomal subunit appears to be brought to the mRNAs of members of the *Potyviridae* and the *Caliciviridae* via interactions with VPg, the small protein linked to the first base of the RNA (Fig. 6.11). VPg of the plant virus turnip mosaic virus binds eIF4E, thereby recruiting eIF4G, eIF3, and the 40S ribosomal subunit to the mRNA. In cells infected with members of the *Caliciviridae*, VPg binds both eIF4E and eIF3. Such interactions may also facilitate selective translation of viral mRNAs over cellular mRNAs, although the mechanisms involved have not been elucidated.

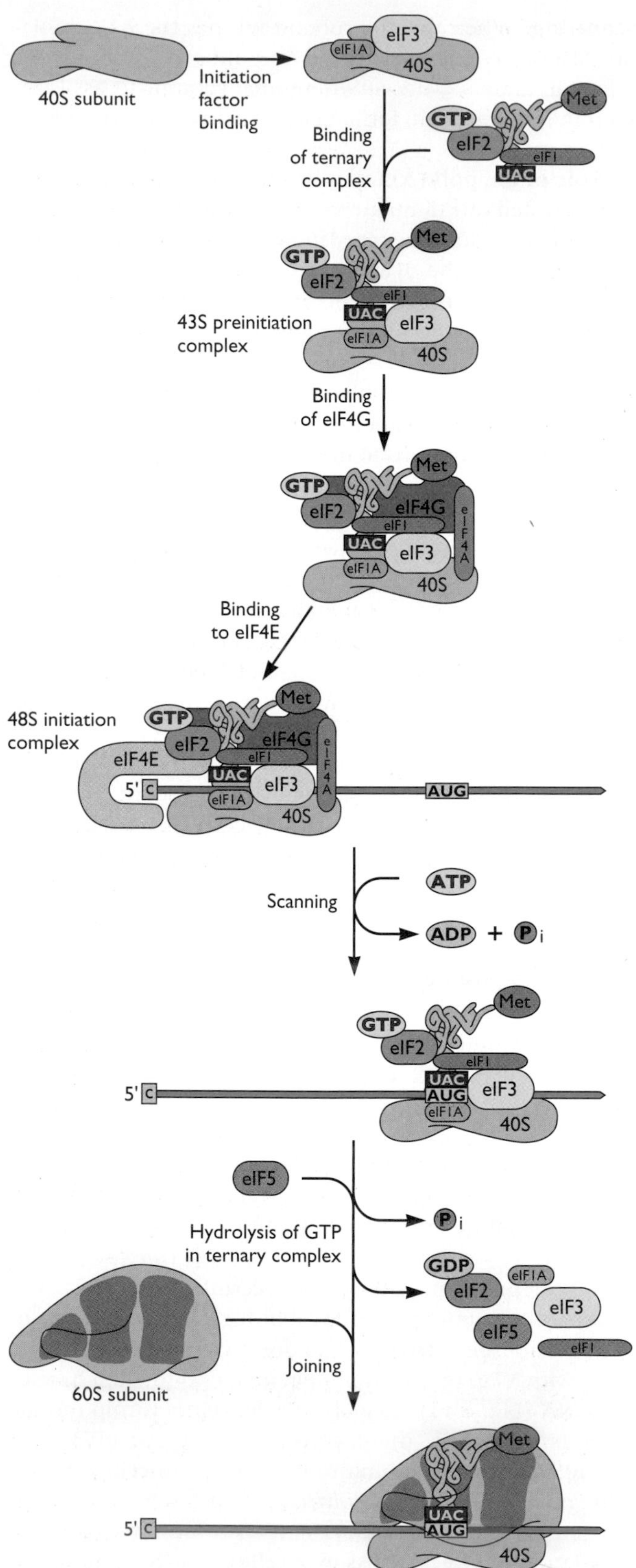

The role of mRNA secondary structure in translation. Translation efficiency is reduced by the presence of a stable secondary structure in the mRNA 5′ untranslated region, particularly when the structure is near the 5′ terminus. There are at least two reasons for this effect. If RNA secondary structure is adjacent to the 5′ cap, it can inhibit binding of the 40S ribosomal subunit. In addition, the presence of secondary structure blocks ribosome movement toward the initiation codon.

The ATP-dependent RNA helicase activity of eIF4A is enhanced by eIF4B and eIF4H (Fig. 11.3). Such activity is thought to unwind regions of double-stranded RNA (dsRNA) near the 5′ end of the mRNA, allowing the 43S preinitiation complex to bind. The helicase may also migrate in a 3′ direction, unwinding dsRNA and enabling ribosomes to scan. mRNAs with less secondary structure in the 5′ untranslated region have a reduced requirement for RNA helicase activity during mRNA translation and hence are less dependent on the cap structure through which the helicase is brought to the mRNA. Dependence of translation on the cap can be measured experimentally by determining the effect on protein synthesis of cap analogs such as m^7GDP and m^7GTP. These compounds competitively inhibit 5′-end-dependent initiation by binding to eIF4E and preventing it from associating with capped mRNAs. For example, the 5′ untranslated region of alfalfa mosaic virus RNA segment 4 is largely free of secondary structure, and translation of this mRNA is quite resistant to inhibition by cap analogs.

Choosing the initiation codon. For over 90% of mRNAs, translation initiation occurs at the 5′-proximal AUG codon. If the 5′-proximal AUG codon is mutated

Figure 11.3 5′-cap-dependent assembly of the initiation complex. Initiation proteins eIF3 and eIF1A bind to free 40S subunits to prevent their association with the 60S subunit, while interaction of eIF6 (not shown) with the larger subunit prevents it from associating with the 40S subunit. eIF4F, which consists of three proteins—eIF4A, eIF4E, and eIF4G—binds the cap via the eIF4E subunit, and the ribosome binds a ternary complex containing eIF2, GTP, and Met-tRNA$_i$, forming a 43S preinitiation complex. The ribosome then binds eIF4G via eIF3. Alternatively, eIF4G may first join the 43S preinitiation complex and then bind the mRNA via eIF4E bound to the cap. The 40S subunit then scans down the mRNA until the AUG initiation codon is reached. eIF1 and eIF1A are required for selection of the correct AUG initiation codon. eIF5 triggers GTP hydrolysis, eIF2 bound to GDP is released along with other initiation proteins, and the 60S ribosomal subunit joins the complex. Adapted from G. M. Cooper, *The Cell: a Molecular Approach* (ASM Press, Washington, DC, and Sinauer Associates, Sunderland, MA, 1997), with permission.

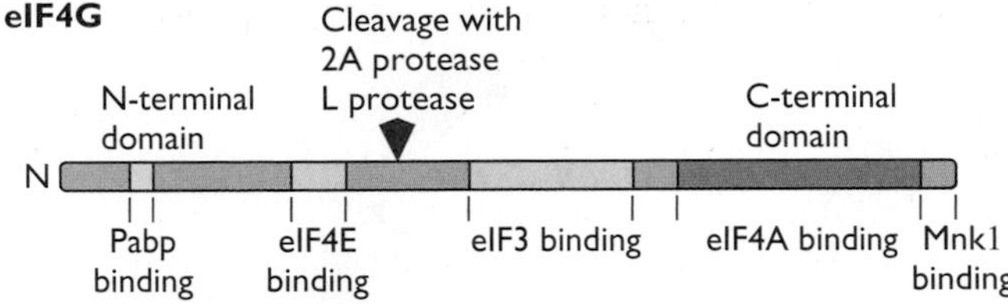

5'-end-dependent initiation

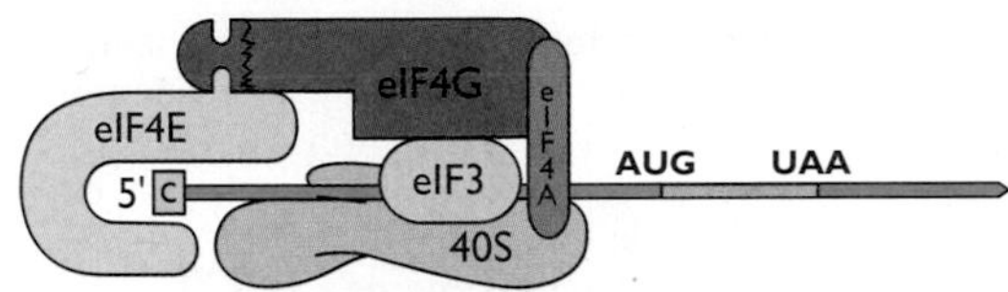

Juxtaposition of mRNA ends

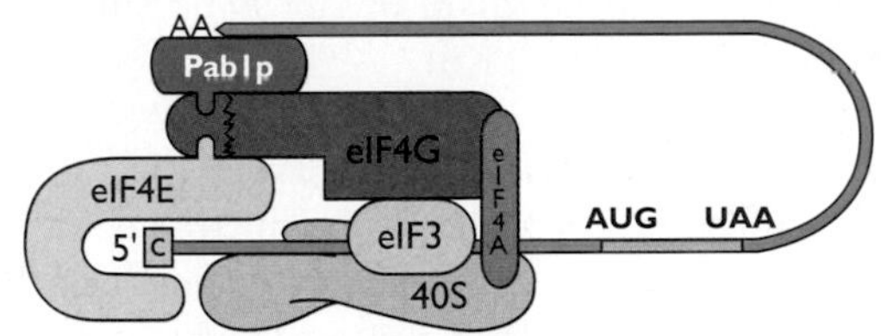

Figure 11.4 5'-end-dependent initiation. (Top) Schematic of eIF4G. Adapted from S. J. Morley et al., *RNA* **3:**1085–1104, 1997, with permission. **(Middle)** Model of initiation complex assembly. eIF4F is brought to the mRNA 5′ end by interaction of eIF4E with the cap structure. The N terminus of eIF4G binds eIF4E, and the C terminus binds eIF4A. The 40S ribosomal subunit binds to eIF4G indirectly via eIF3. **(Bottom)** 5′-end-dependent initiation is stimulated by the poly(A)-binding protein Pab1p, which interacts with eIF4G. This interaction may bring the mRNA ends together and facilitate formation of the initiation complex at the 5′ end. Adapted from M. W. Hentze, *Science* **275:**500–501, 1997, with permission.

so that it cannot serve as an initiation codon, translation starts at the next downstream AUG. Insertion of an AUG codon upstream of the initiating codon causes initiation at the more 5′-proximal site. The efficiency of initiation is influenced by the nucleotide sequence surrounding this codon. Studies of the effects of mutating these sequences have shown that the consensus sequence 5′-GCC**A**CC<u>AUG</u>G-3′ is recognized most efficiently in mammalian cells: the presence of a purine at the −3 position (boldface) is most important for high levels of translation. However, only 5% of eukaryotic mRNAs contain this ideal consensus sequence; most have suboptimal sequences that result in less efficient translation. This finding indicates that not all mRNAs must be translated maximally, but rather only at levels appropriate for the function of the protein product. If a very poor match to this consensus sequence is present, the AUG codon may be passed over by the ribosome and initiation may occur further downstream (see "The Diversity of Viral Translation Strategies" below).

The results of genetic and biochemical experiments demonstrate that eIF1 plays a role in discrimination during selection of the initiation codon. In the absence of eIF1, the 43S preinitiation complex is unable to discriminate between AUG and AUU or GUG codons. Certain alterations of eIF1 of yeasts enhance initiation at a UUG triplet. eIF1 might influence codon-anticodon pairing, either directly or by modulating the conformation of the 43S complex.

Although AUG is the initiation codon for most proteins, synthesis may also begin at ACG, GUG, and CUG codons in viral and cellular mRNAs, although far less frequently. In all cases, the first amino acid of the protein is a methionine. Precisely how ribosomes recognize non-AUG codons so that methionine is inserted is not known. Because the efficiency of initiation at these sites is always low, non-AUG initiation might be another mechanism for regulating translation efficiency. Not all mRNAs support initiation from a non-AUG site, suggesting that there may be structural elements in the mRNA that regulate translation of such a codon.

Methionine-independent initiation. The structural proteins of some viruses begin not with methionine but with glutamine (CAA), proline (CCU), or alanine (GCU or GCA). Initiation of synthesis of these viral proteins does not require initiator tRNA methionine or the ternary complex because the viral mRNA mimics the structure of tRNA (Fig. 11.5A). The tRNA-like structure occupies the P site of the ribosome, allowing initiation to take place within the A site. These mRNAs require no translation initiation proteins, and can bind ribosomes and induce them to enter the elongation phase of translation. Methionine-independent initiation of the mRNA of turnip yellow mosaic virus is accomplished in a similar way, except that the tRNA-like structure is located in the 3′ untranslated region of the viral RNA (Fig. 11.5B). The tRNA-like structure is aminoacylated with valine, which is incorporated as the first amino acid of the viral polyprotein.

Does the 40S subunit move on the mRNA? Translation of most mRNAs begins at the 5′-proximal AUG codon. How does the ribosome reach this location? In one popular model, the 40S ribosomal subunit migrates along the 5′ untranslated region until it reaches the AUG initiation codon. Experimental proof for this long-standing hypothesis is still lacking. Hydrolysis of ATP is necessary for movement of the 43S complex on RNA containing weak

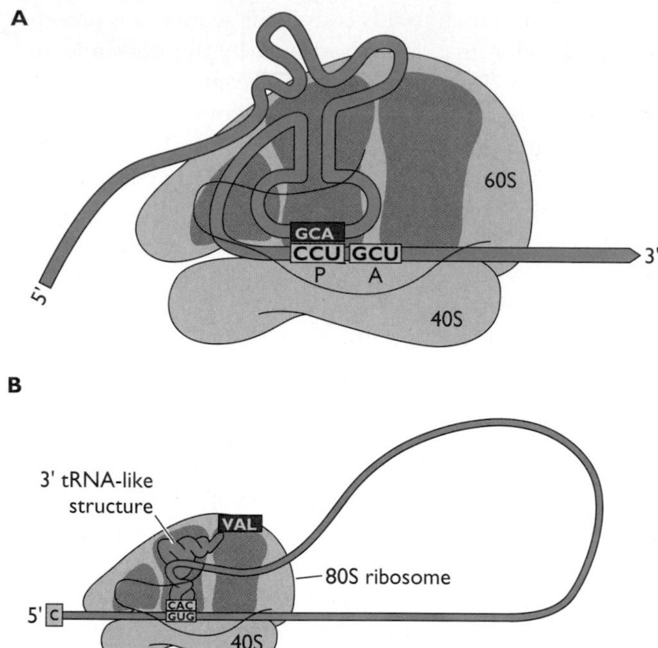

Figure 11.5 Two mechanisms of methionine-independent initiation. (A) The viral mRNA of picornavirus-like viruses of insects mimics the structure of tRNA, which occupies the P site of the ribosome, allowing initiation to take place within the A site. Adapted from M. Bushell and P. Sarnow, *J. Cell Biol.* **158**:395–399, 2002, with permission. **(B)** A tRNA-like structure in the 3′ untranslated region of turnip yellow mosaic virus RNA, aminoacylated with valine, occupies the P site of the ribosome. Adapted from S. Barends et al., *Cell* **112**:123–129, 2003, with permission.

secondary structures. Movement of ribosomes on RNA lacking secondary structure does not require ATP hydrolysis. However, energy is required by the ATP-dependent RNA helicase, not for ribosome movement. In an alternative model, the ribosome does not travel to the initiation codon, but remains at the 5′ cap. The mRNA is threaded through the ribosome, eventually bringing the initiation codon to the P site.

Ribosome shunting. Stable RNA secondary structures in 5′ untranslated regions may inhibit scanning of 40S ribosomes. In some RNAs, such hairpin structures are not inhibitory because ribosomes bypass them. This process, called **ribosome shunting**, may be dependent or independent of viral proteins. Shunting on the 35S cauliflower mosaic virus RNA requires translation of the very short upstream open reading frame on the same viral RNA. In contrast, shunting on adenovirus major late mRNAs occurs in the absence of viral proteins, although its efficiency is increased by the viral L4 100-kDa polypeptide. Translation of cellular inhibitor of apoptosis 2 is also mediated by

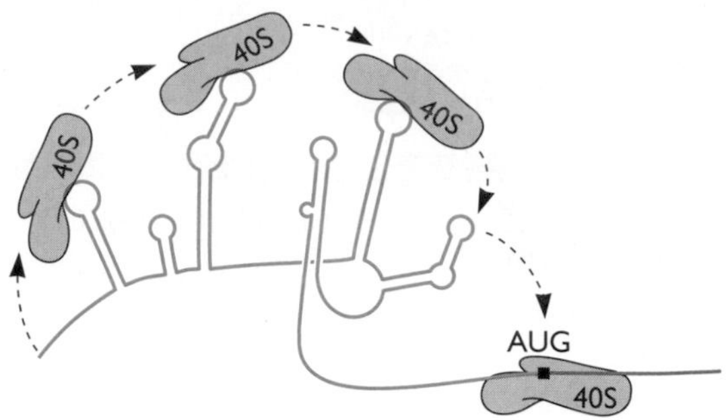

Figure 11.6 Hypothetical model of ribosome shunting. The 40S ribosomal subunit binds to the mRNA by a cap-dependent mechanism and then bypasses large regions of the mRNA with secondary structure to reach the AUG initiation codon. Shunting elements, such as the loops in the figure, and viral or cell proteins may direct ribosome movement.

a ribosome shunt. The mechanism of ribosome shunting remains to be elucidated (Fig. 11.6).

5′-End-Independent Initiation

The internal ribosome entry site. The mRNAs of picornaviruses differ from most host cell mRNAs. They lack the 5′-terminal cap structure, and the 5′ untranslated regions are highly structured and contain multiple AUG codons. Infection of host cells by many picornaviruses results in the inhibition of host cell translation. These observations led to the hypothesis that translation of the mRNA of (+) strand picornaviruses was initiated by an unusual mechanism. It was suggested that the ribosome bound internally, rather than at the mRNA 5′ end. In a key experiment, the 5′ untranslated region of poliovirus mRNA was shown to promote internal binding of the 40S ribosomal subunit, and was termed the **internal ribosome entry site (IRES)** (Box 11.2).

An IRES has been identified in the mRNAs of all picornaviruses, in other viral mRNAs including those of pestiviruses and hepatitis C virus, and in some cellular mRNAs. There is very little nucleotide sequence conservation among these IRESs, with the exception of an oligopyrimidine tract 25 nucleotides upstream of the 3′ end of the IRES. Viral IRESs contain extensive regions of RNA secondary structure (Fig. 11.7). Although such secondary structure is not strictly conserved, it is of extreme importance for ribosome binding. Viral IRESs have been placed in five groups, depending on a variety of criteria, including primary sequence and secondary structure conservation, the location of the initiation codon, and activity in different cell types.

The discovery of the IRES makes even more puzzling the rarity of eukaryotic mRNAs that contain several open reading frames reminiscent of bacterial and archaeal mRNAs (Fig. 11.1). In principle, all the open reading frames of a

BOX 11.2 EXPERIMENTS

Key experiment: discovery of the IRES

The hypothesis that poliovirus mRNA is translated by internal ribosome binding was first tested by examining the translation of mRNAs containing two open reading frames (ORFs) separated by the poliovirus 5′ untranslated region (figure, panel A). These bicistronic mRNAs directed the synthesis of two proteins, but the second ORF was efficiently translated only if it was preceded by the picornavirus 5′ untranslated region. It was concluded that ribosomes bind within the viral 5′ untranslated region, thereby permitting translation of the second ORF. The segment of the 5′ untranslated region that directs internal ribosome entry was called the IRES.

It had long been known that covalently closed circular mRNAs cannot be translated by 5′-end-dependent initiation. Translation by internal ribosome binding, however, should not require a free 5′ end. To test this hypothesis, circular mRNAs with and without an IRES were created. The circular mRNA was translated only if an IRES was present (Figure, panel B). This experiment formally proved that translation initiation directed by an IRES occurs by internal binding of ribosomes and does not require a free 5′ end.

Chen, C. Y., and P. Sarnow. 1995. Initiation of protein synthesis by the eukaryotic translational apparatus on circular RNAs. *Science* **268:**415–417.

Jang, S. K., H. G. Kräusslich, M. J. Nicklin, G. M. Duke, A. C. Palmenberg, and E. Wimmer. 1988. A segment of the 5′ nontranslated region of encephalomyocarditis virus RNA directs internal entry of ribosomes during in vitro translation. *J. Virol.* **62:**2636–2643.

Pelletier, J., and N. Sonenberg. 1988. Internal initiation of translation of eukaryotic mRNA directed by a sequence derived from poliovirus RNA. *Nature* **334:**320–325.

Assays for an IRES. (A) Bicistronic mRNA assay. Plasmids were constructed that encode bicistronic mRNAs encoding the thymidine kinase (tk) and chloramphenicol acetyltransferase (cat) proteins separated by a spacer (light green) or a poliovirus IRES (dark green). Plasmids were introduced into mammalian cells by transformation. In uninfected cells containing either plasmid (top lines), both tk and cat proteins were detected, although without an IRES, cat synthesis was inefficient. Translation of cat from this plasmid probably occurs by reinitiation. In poliovirus-infected cells, 5′-end-dependent initiation is blocked (stop sign), and no proteins are observed without an IRES. cat protein is detected in infected cells when the IRES is present, demonstrating internal ribosome binding. Adapted from J. Pelletier and N. Sonenberg, *Nature* **344:**320–325, 1988, with permission. **(B)** Circular mRNA assay for an IRES. Circular mRNAs containing an ORF (yellow) were produced and translated *in vitro*. No protein product was observed unless an IRES was included in the circular mRNA. Adapted from C. Y. Chen and P. Sarnow, *Science* **268:**415–417, 1995, with permission.

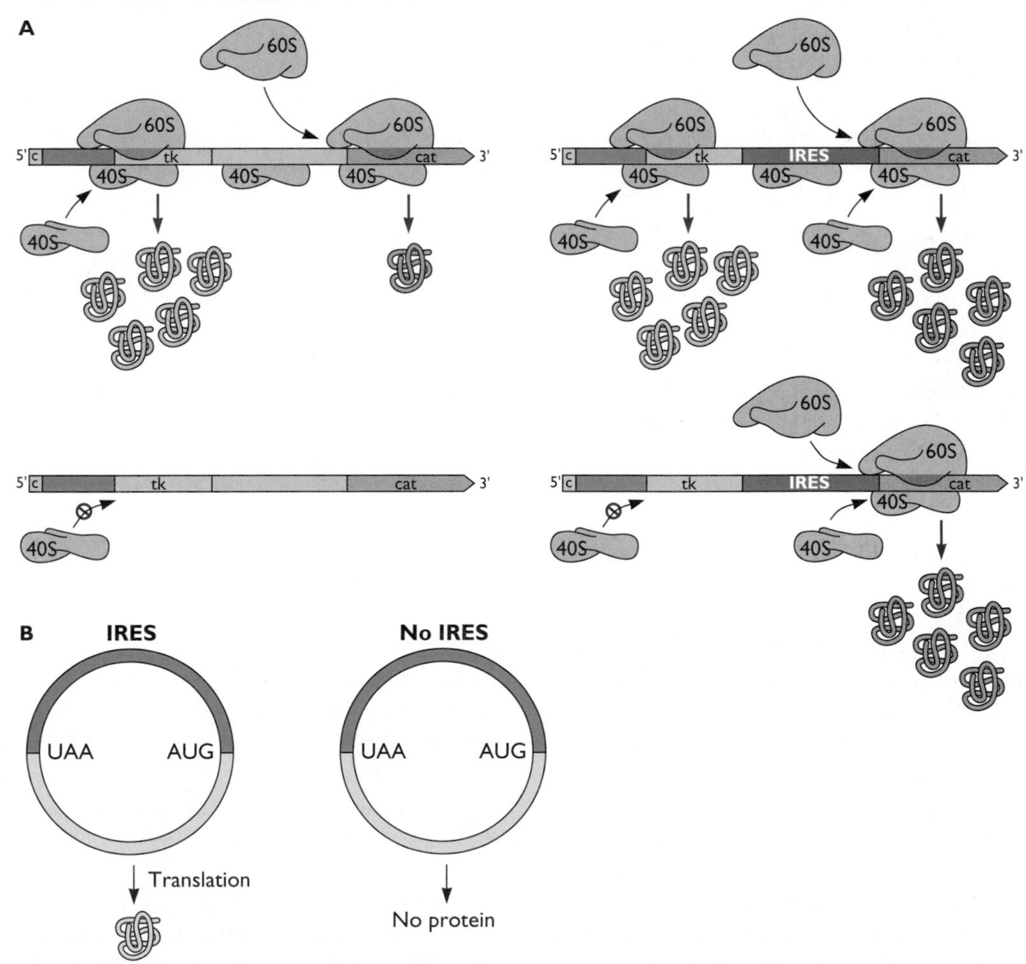

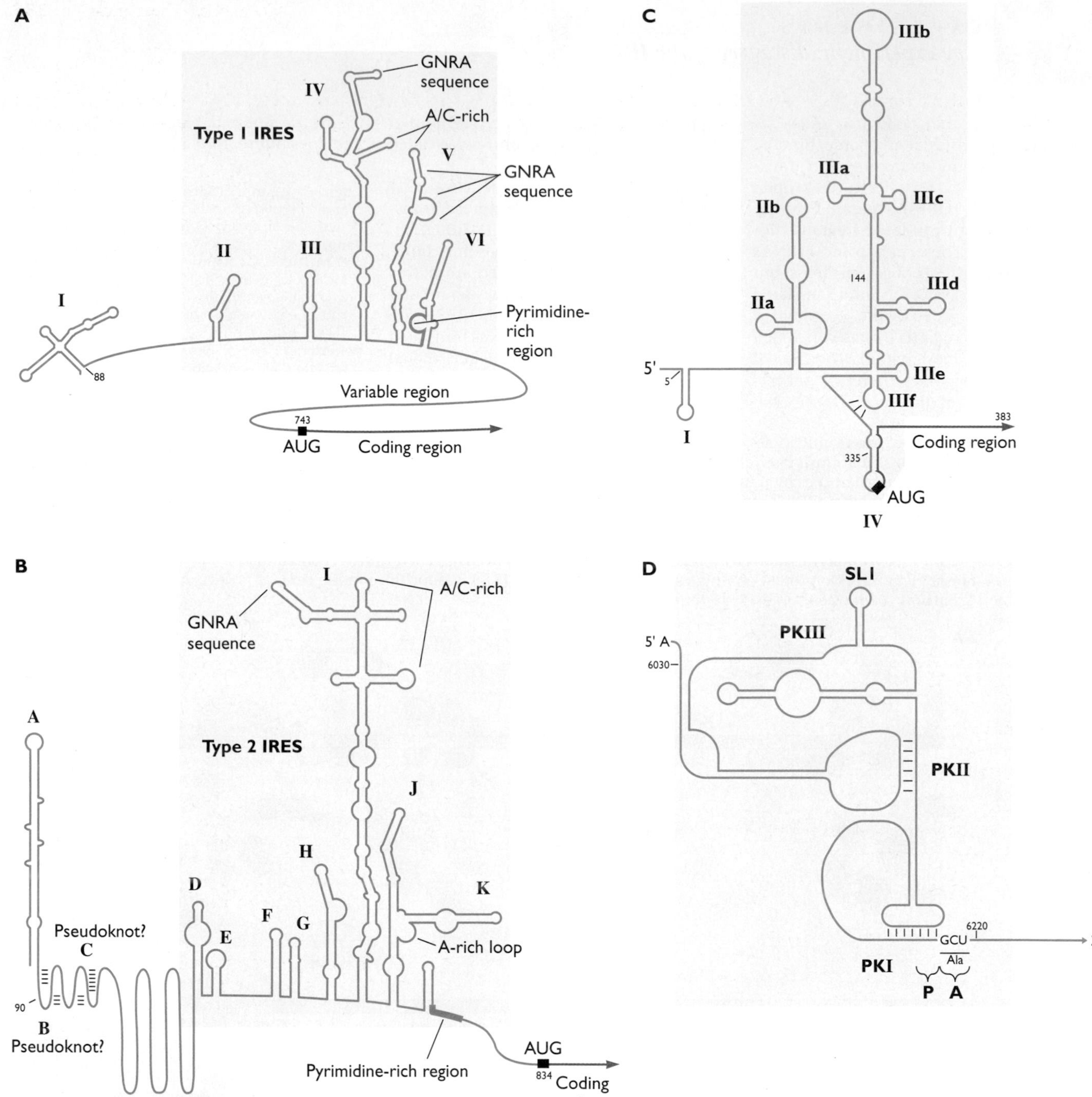

Figure 11.7 Four types of IRES. The 5′ untranslated regions from poliovirus **(A)**, encephalomyocarditis virus **(B)**, hepatitis C virus **(C)**, and cricket paralysis virus **(D)** are shown. The IRES in panels A and B is indicated by yellow shading. Predicted secondary and tertiary RNA structures (RNA pseudoknots) are shown. The poliovirus IRES is a type 1 IRES, which is found in the genomes of enteroviruses and rhinoviruses. The ribosome probably enters the IRES at domains V and VI and scans to the AUG initiation codon, which is located 50 to 100 nucleotides past the 3′ end of the IRES. The type 2 IRES is found in the genomes of aphthoviruses and cardioviruses. The 3′ end of the hepatitis C virus IRES (type 4) extends beyond the AUG initiation codon (black box). The IRES of picornavirus-related viruses of insects (type 5), such as cricket paralysis virus, mimics a tRNA and occupies the P site in the 40S ribosomal subunit. Translation initiates with a non-AUG codon from the A site. THE IRES ends at the initiating codon. A sixth class of IRES element has been identified in the viral RNA of teschoviruses, picornaviruses that infect pigs. These IRESs have similarities to that of hepatitis C virus. The initiation codon of the IRES of hepatitis A virus (type 3, not shown) is located 50 to 100 nucleotides past the 3′ end of the IRES. (A and B) Adapted from S. R. Stewart and B. L. Semler, *Semin. Virol.* **8**:242–255, 1997, with permission. (C) Adapted from S. M. Lemon and M. Honda, *Semin. Virol.* **8**:274–288, 1997, with permission. (D) Adapted from E. Jan and P. Sarnow, *J. Mol. Biol.* **324**:889–902, 2002, with permission.

polycistronic mRNA can be translated in a eukaryotic cell as long as each frame is preceded by an IRES. Nevertheless, only one such naturally occurring polycistronic mRNA has been identified in eukaryotes, and it is not known if an IRES is present. Bicistronic mRNAs produced in the laboratory have been used in the expression and cloning of genes (Box 11.3).

The mechanism of internal initiation. Different sets of translation initiation proteins are required for the function of various IRESs. Internal ribosome binding on the hepatitis A virus IRES requires all of the initiation proteins, including eIF4E. At the other extreme, the intergenic IRES of cricket paralysis virus requires **none** of the translation initiation proteins. However, the

BOX 11.3

BACKGROUND

Use of the IRES in cloning and expression vectors

The IRES has been used widely in the expression and cloning of foreign genes in eukaryotes. One strategy for the expression of genes in mammalian cells is to produce mRNAs in the cytoplasm by using a bacteriophage DNA-dependent RNA polymerase, such as T7 RNA polymerase. Such mRNAs are poorly translated because they are not capped; inclusion of an IRES in the 5′ untranslated region allows them to be translated efficiently.

Another application of the IRES is in the functional cloning of new genes (figure, top panel). A DNA library is made by using a cloning vector that produces a bicistronic mRNA encoding both the desired gene and a selectable marker. The use of a selectable marker on the same mRNA increases the efficiency of screening because most transformants express both genes.

IRESs have also been used in the isolation of mutant mice by homologous recombination in embryonic stem cells. Bicistronic vectors have been designed to produce mRNA encoding the altered protein and β-galactosidase, separated by an IRES (figure, bottom panel). Because β-galactosidase is encoded on the same mRNA as the targeted gene product, it serves as a marker for mRNA expression.

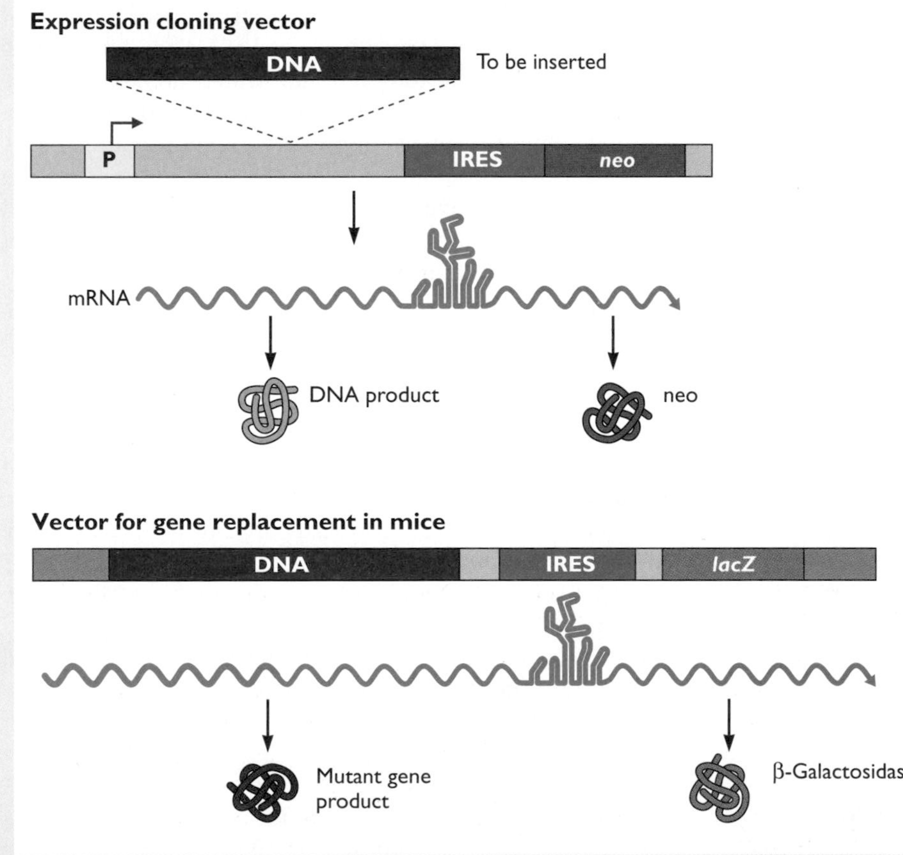

(Top) Design of plasmids for expression cloning. The expression library is created by inserting DNA into a site on the plasmid downstream of a promoter. The foreign DNA is followed by an IRES linked to a selectable marker such as neomycin resistance (neo). The mRNA produced from this plasmid DNA encodes the cloned DNA product and the protein conferring neomycin resistance. The library is introduced into cells that are then screened for expression of the desired gene. **(Bottom)** Vector for gene replacement in mice. In this example, the goal is to replace the gene with a mutant version. The targeting plasmid consists of mutant DNA followed by an IRES and the *lacZ* gene. The flanking light blue bars represent sequences from the mouse gene that mediate homologous recombination. After replacement of the endogenous gene with this synthetic version, mRNA that encodes the mutant gene product as well as the β-galactosidase protein will be produced. The latter can be detected in tissues by staining with the chromogenic substrate X-Gal (5-bromo-4-chloro-3-indolyl-β-D-galactopyranoside).

activity of most IRESs require a subset of translation initiation proteins.

Translation initiation via a type 1 IRES comprises binding of the 40S ribosomal subunit to the IRES, followed by scanning of the subunit to the initiation codon. The 40S subunit may bind directly to the RNA or could be recruited to the IRES by means of interaction with translation initiation proteins (Fig. 11.8). In poliovirus-infected cells, the scaffold eIF4G is cleaved, reducing the translation of most cellular mRNAs. It was therefore assumed that this initiation protein was not required for function of the poliovirus IRES. However, the C-terminal fragment of eIF4G, which contains binding sites for eIF3 and eIF4A (Fig. 11.8), stimulates translation directed by the poliovirus IRES. These results have led to a model in which the 40S ribosomal subunit is recruited to the IRES via interaction with eIF3 bound to the C-terminal domain of eIF4G (Fig. 11.8). It has been reported that IRES function is markedly enhanced in cells in which the poliovirus protease 2A^{pro} is synthesized. Because 2A^{pro} is responsible for cleavage of eIF4G, such stimulation of translation may be due to production of the C-terminal proteolytic fragment of eIF4G.

The IRESs of hepatitis C virus (Fig. 11.7C), pestiviruses such as bovine viral diarrhea virus and classical swine fever virus, and teschoviruses function very differently from those of other picornaviruses. The formation of the 48S initiation complex on the mRNA is independent of eIF4A, eIF4B, and eIF4F and is highly sensitive to secondary RNA structure around and downstream of the AUG initiation codon. Purified 40S ribosomal subunits bind directly to stem-loop IIId of the hepatitis C virus IRES, and single point mutations in this structure abolish the interaction and block internal initiation. Addition of only Met-tRNA$_i$, eIF2, and GTP is required to form 48S complexes. A dramatic conformational change in the 40S ribosomal subunit occurs when it binds the hepatitis C virus IRES (Fig. 11.9), clamping the mRNA in place and setting the AUG initiation codon within the P site of the ribosome. The IRES also contacts the E site of the ribosome, where the deacylated tRNA is harbored after translocation of the 80S ribosome. Initiation of translation from the IRES of hepatitis C virus and related viruses therefore resembles translation initiation of bacterial mRNAs.

The intergenic IRESs of picornavirus-like viruses of insects form complexes with the 40S ribosome independent of initiation proteins, and translation begins at other than an AUG codon. Initiation from these IRESs is uniquely inhibited by ternary complex (Met-tRNA$_i$–eIF2–GTP) and a high concentration of Met-tRNA$_i$. In cells infected with these viruses, recycling of eIF2-GDP is blocked and the concentration of ternary complex is low, inhibiting cellular mRNA translation, but the activity of the intergenic IRES is not reduced. The secondary structure of the IRES of these viruses mimics an uncharged tRNA, and mutations that destabilize the fold abrogate translation. The tRNA-like structure is recognized and bound by the 40S ribosomal subunit, placing the initiation codon within the A site instead of the P site (Fig. 11.5A). Initiation is therefore dependent on elongation proteins eEF1A and eEF2 and the appropriate aminoacylated tRNAs. Like the IRESs of hepatitis C virus and pestiviruses, IRESs of the picornavirus-like viruses of insects also occupy the E site of the 80S ribosome.

As discussed above, translation of cellular mRNAs is enhanced by the juxtaposition of mRNA ends (Fig. 11.4). Translation of viral mRNAs by internal initiation is also stimulated by this arrangement, which may be established in at least two ways. The 5′ and 3′ ends of the RNA genome of foot-and-mouth disease virus are brought together by RNA-RNA interactions. The mRNA of hepatitis C virus is not polyadenylated and therefore cannot bind Pab1p. The interaction between molecules of polypyrimidine-tract-binding protein that bind both the viral 3′ untranslated region and the IRES may bring together the 5′ and 3′ ends of the mRNA.

Figure 11.8 5′-end-independent initiation. (Top) Initiation in the type I or II IRES does not depend on the presence of a cap structure but requires the C-terminal fragment of eIF4G to recruit the 40S ribosomal subunit via its interaction with eIF3. eIF4G probably binds directly to the IRES. **(Bottom)** The ribosomal 40S subunit binds to the hepatitis C virus IRES without the need for translation initiation proteins. eIF3 also binds the IRES and is thought to be necessary for recruitment of the 60S ribosomal subunit.

Type I or II IRES

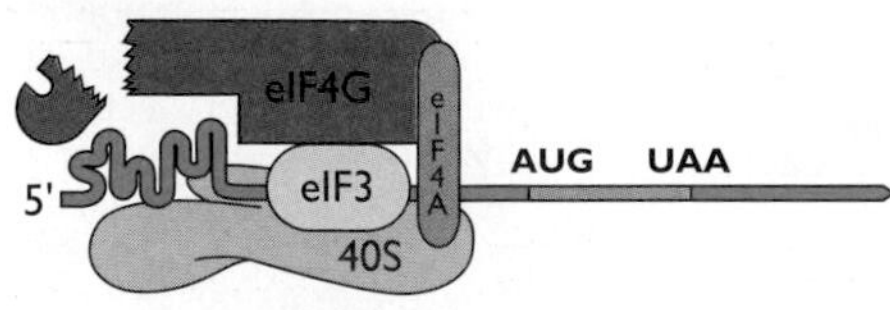

Hepatitis C virus IRES

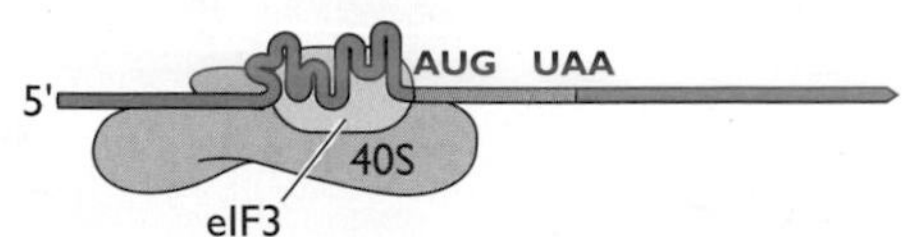

Other host cell proteins that contribute to IRES function. The poliovirus IRES functions poorly in reticulocyte lysates, in which most capped mRNAs are translated efficiently (Box 11.4). Reticulocytes are immature red

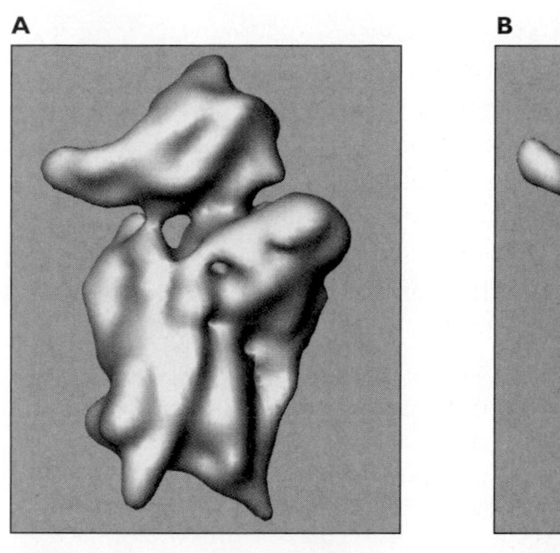

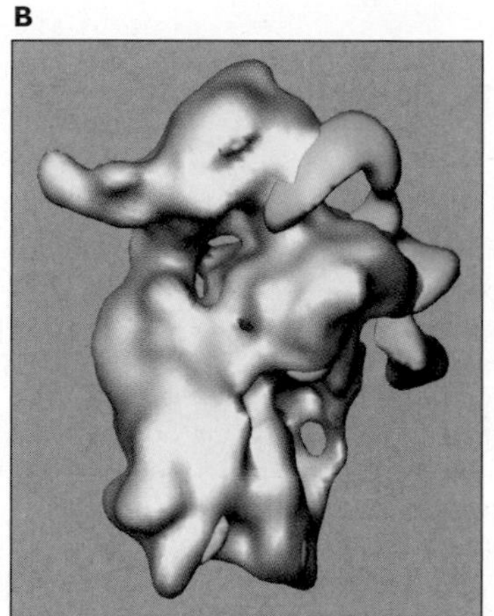

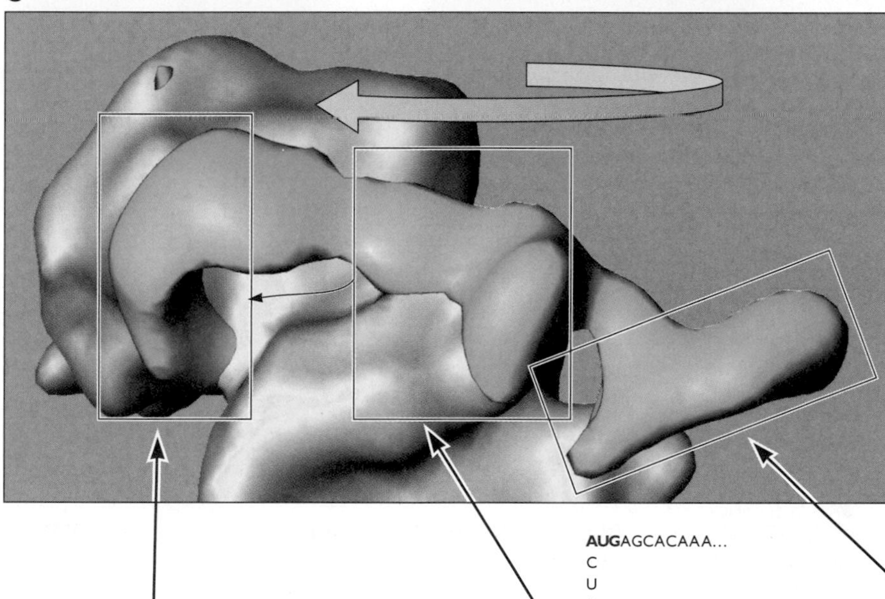

Figure 11.9 Complex of the 40S ribosomal subunit with the hepatitis C virus IRES. **(A)** Structure of 40S ribosomal subunit from reticulocytes. **(B)** Structure of 40S ribosomal subunit bound to 350-nucleotide RNA encompassing the IRES of hepatitis C virus. Binding of the RNA induces major conformational changes in the 40S subunit. The head clamps down on the mRNA and touches the platform. **(C)** Enlarged side view of the hepatitis C virus IRES bound to the 40S ribosomal subunit. The viral RNA lies within the cleft. Domains of the IRES are identified by boxes, and their sequences are shown below. The AUG initiation codon is in red. The arrow indicates the mRNA-binding groove, where the coding region of the viral RNA may be located. Structures were obtained by cryo-electron microscopy and image reconstruction. Adapted from C. M. Spahn, J. S. Kieft, R. A. Grassucci, P. A. Penczek, K. Zhou, J. A. Doudna, and J. Frank, *Science* **291:**1959–1962, 2001, with permission.

BOX 11.4

DISCUSSION

Translation in vitro: *the reticulocyte lysate and wheat germ extract*

Our present understanding of the fundamentals of translation initiation, elongation, and termination, as well as viral translation strategies, would not be possible without the technique of in vitro translation in cell extracts. In this method, cells are lysed and the nuclei are removed by centrifugation. The mRNA is added to the lysate, and the mixture is incubated in the presence of an isotopically labeled amino acid which is incorporated into the translation product.

The ideal extract for *in vitro* translation has two important properties: high translation efficiency and low protein synthesis in the absence of added mRNA. By the early 1970s, cell extracts prepared from Krebs II ascites tumor cells or rabbit reticulocytes were found to translate protein with high efficiency, but the presence of endogenous mRNAs that were also translated complicated the analysis of proteins made from added mRNA. In 1973 a cell extract from commercial wheat germ that had low background levels of protein synthesis, and in which exogenous mRNAs were translated very efficiently, was developed. A few years later, the background in a reticulocyte lysate was eliminated by treatment with micrococcal nuclease, which destroyed the endogenous mRNA. This nuclease requires calcium for its activity, and it was therefore a simple matter of adding a calcium chelator, EGTA [ethylene glycol-bis(β-aminoethylether)-*N,N,N′,N′*-tetraacetic acid], to the reaction to prevent the degradation of exogenously added mRNA.

Wheat germ extract and reticulocyte lysate are still widely used in translation studies, because the cells are abundant and inexpensive and are excellent sources of initiation proteins. Micrococcal nuclease followed by calcium chelation has been successfully used to make mRNA-dependent extracts from many mammalian cell types, although the translation efficiency of such systems does not approach that of wheat germ or reticulocyte lysates. Unfortunately, it has not been possible to prepare consistently translation extracts from normal mammalian tissues, which has hampered the study of tissue-specific translation regulation in virus-infected and uninfected cells.

Pelham, H. R. B., and R. J. Jackson. 1976. An efficient mRNA-dependent translation system from reticulocyte lysates. *Eur. J. Biochem.* **67:**247–256.

Roberts, B. E., and B. M. Patterson. 1973. Efficient translation of tobacco mosaic virus RNA and rabbit globin 9S RNA in a cell-free system from commercial wheat germ. *Proc. Natl. Acad. Sci. USA* **70:**2330–2334.

blood cells that primarily produce hemoglobin; they lack nuclei. Addition of a cytoplasmic extract to reticulocyte lysates restores efficient translation from this IRES. These observations led to the suggestion that ribosome binding to the IRES requires more than translation initiation proteins. Such proteins have been identified by their ability to bind to the IRES and to restore its function in the reticulocyte lysate. One host protein identified by this approach is the La protein, which binds to the 3′ end of the poliovirus IRES. This protein is associated with the 3′ termini of newly synthesized small RNAs, including all nascent transcripts of cellular RNA polymerase III. While predominantly nuclear, La protein is localized to the cytoplasm in poliovirus-infected cells. It is present in low concentrations in reticulocyte lysates, and when added to them it stimulates the function of the poliovirus IRES. Furthermore, depletion of La mRNA by RNA interference reduced poliovirus IRES-dependent translation.

Polypyrimidine-tract-binding protein (Ptb), also called heterogeneous nuclear ribonucleoprotein I (hnRnpI), was also found to bind the poliovirus IRES. This predominantly nuclear protein, a negative regulator of alternative pre-mRNA splicing, is redistributed to the cytoplasm during poliovirus infection. Polypyrimidine-tract-binding protein binds to sequences upstream of the pyrimidine-rich sequence of the poliovirus IRES, and to both the 5′ and 3′ untranslated regions of hepatitis C virus RNA. If this protein is removed from a cell extract, the activity of the poliovirus and hepatitis C virus IRESs is eliminated. However, neither IRES activity is restored when purified Ptb is added to the depleted extracts. These findings suggest that another protein(s) associated with polypyrimidine-tract-binding protein might also be required for IRES function.

Another cellular protein necessary for the activity of the enterovirus and rhinovirus IRES is the cytoplasmic RNA-binding protein, poly(rC)-binding protein 2, originally identified by its ability to bind stem-loop IV of the poliovirus IRES (Fig. 11.7A). Mutations in the poliovirus 5′ untranslated region that abolish binding of poly(rC)-binding protein cause decreased translation *in vitro*. Depletion of poly(rC)-binding proteins from human translation extracts inhibits translation dependent on the IRESs of poliovirus, coxsackievirus B, and rhinovirus, but not on those of encephalomyocarditis virus or foot-and-mouth disease virus. Translation activity of the IRESs was restored by addition of purified poly(rC)-binding protein 2. This protein binds to, and functions cooperatively during internal initiation with, SRp20, a protein that is essential for constituitive splicing and regulation of alternative splice site utilization. Cleavage of poly(rC)-binding protein 2 is thought to enable a switch from translation to replication during poliovirus infection (Chapter 6).

No single host cell protein that is essential for the function of all viral IRESs has been identified. Many of these

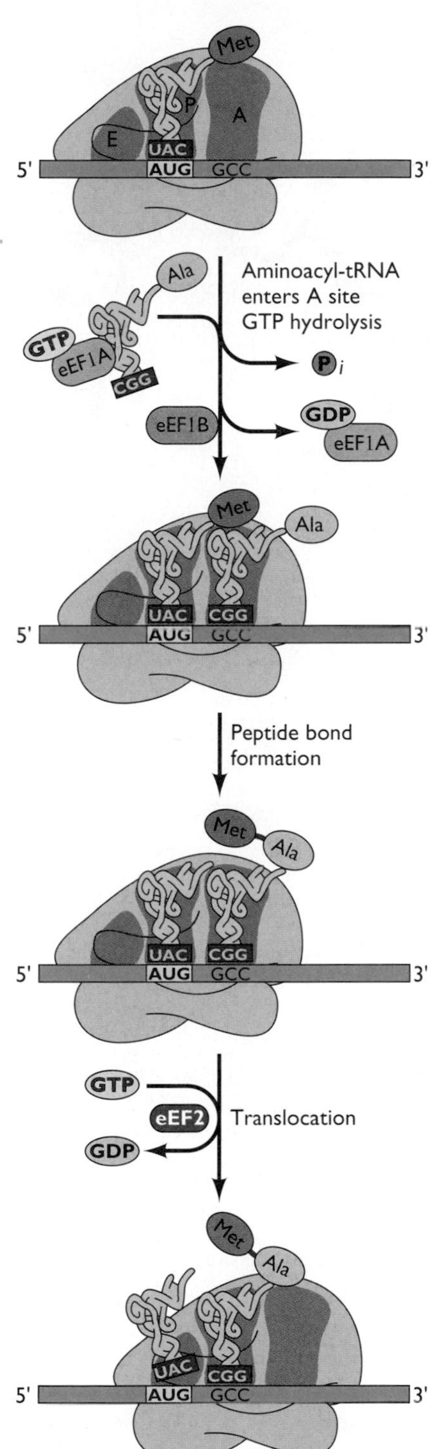

Figure 11.10 Translation elongation. There are three tRNA-binding sites on the ribosome, called peptidyl (P), aminoacyl (A), and exit (E). After the initiating Met-$tRNA_i$ is positioned in the P site, the second aminoacyl-tRNA (alanyl-tRNA is shown) is brought to the A site by eEF1A bound to GTP. After GTP hydrolysis, eEF1A is released. The guanine nucleotide exchange protein eEF1B exchanges GDP of eEF1A-GDP with GTP, allowing eEF1A to interact with a tRNA synthetase and bind a newly aminoacylated tRNA. The peptide bond is then formed; this is followed by movement of the ribosome three nucleotides along the mRNA, a reaction that requires GTP hydrolysis and eEF2. The peptidyl (Met-Ala) tRNA moves to the P site, and the uncharged tRNA moves to the E site. The A site is now empty, ready for another aminoacyl-tRNA. Adapted from G. M. Cooper, *The Cell: a Molecular Approach* (ASM Press, Washington, DC, and Sinauer Associates, Sunderland, MA, 1997), with permission.

proteins normally reside in the nucleus and become relocalized to the cytoplasm during infection. It has been suggested that La, Ptb, poly(rC)-binding proteins, and others act as RNA chaperones, maintaining the IRES in its appropriate three-dimensional structure for binding to ribosomes and translation initiation proteins. In support of this hypothesis is the observation that all are RNA-binding proteins that can form multimers that contact the IRES at multiple points.

Elongation and Termination

During elongation, the ribosome selects aminoacylated tRNA according to the sequence of the mRNA codon, and catalyzes the formation of a peptide bond between the nascent polypeptide and the incoming amino acid. The 40S ribosomal subunit is responsible for both decoding and selection of the cognate tRNA. The RNA of the 60S subunit catalyzes the peptidyl transferase reaction without any soluble nonribosomal proteins or nucleotides. Elongation is assisted by three proteins that maintain the speed and accuracy of translation. In the 80S initiation complex, the Met-$tRNA_i$ is bound to the peptidyl (P) site of the ribosome (Fig. 11.10). Elongation of the peptide chain begins with addition of the next amino acid encoded by the triplet that occupies the acceptor (A) site. An important component of this process is elongation factor eEF1A, which is bound to aminoacylated tRNA, a molecule of GTP, and the nucleotide exchange protein eEF1B. Interaction between the codon and the anticodon leads to a conformational change in the ribosome called **accommodation**, the hydrolysis of GTP and the release of eEF1A-GDP. Accommodation preserves the fidelity of translation, because it can occur only upon proper codon-anticodon base pairing, and is required for GTP hydrolysis. If an incorrect tRNA enters the A site, accommodation does not occur and the aminoacylated tRNA is rejected. The large ribosomal subunit catalyzes the formation of a peptide bond between the amino acids occupying the P and A sites. The 80S ribosome then moves 3 nucleotides along the mRNA. Translocation is dependent upon eEF2 and hydrolysis of GTP. This motion moves the uncharged tRNA to the exit (E) site and the peptidyl-tRNA to the P site, allowing a new aminoacylated tRNA to enter the A site and allowing release of the uncharged

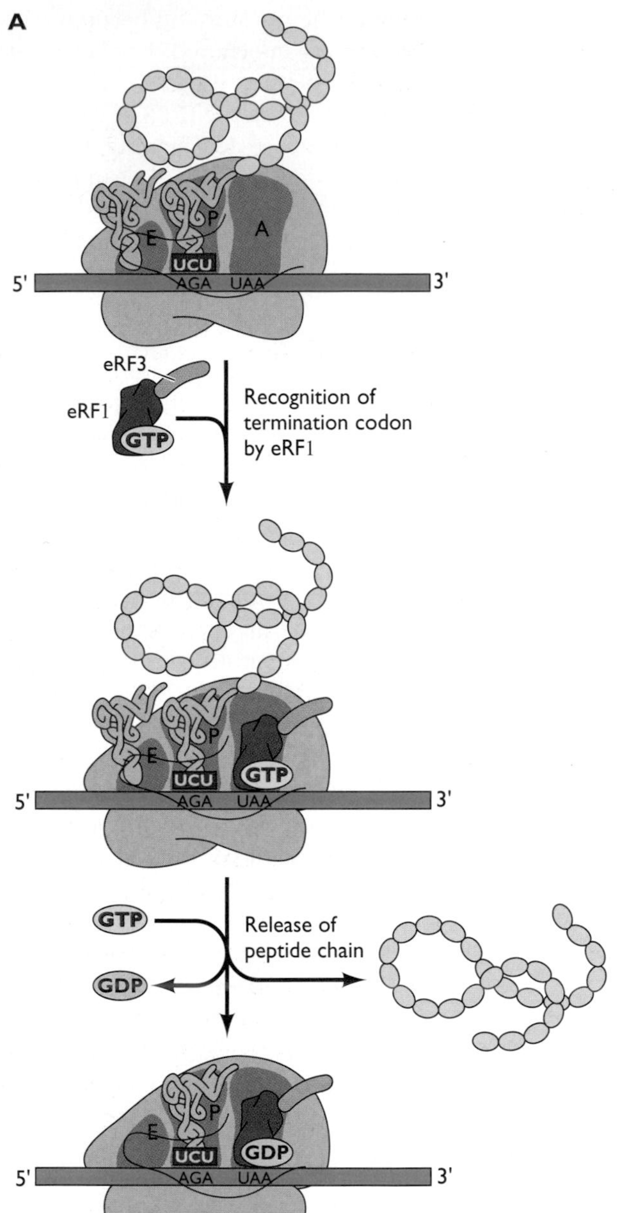

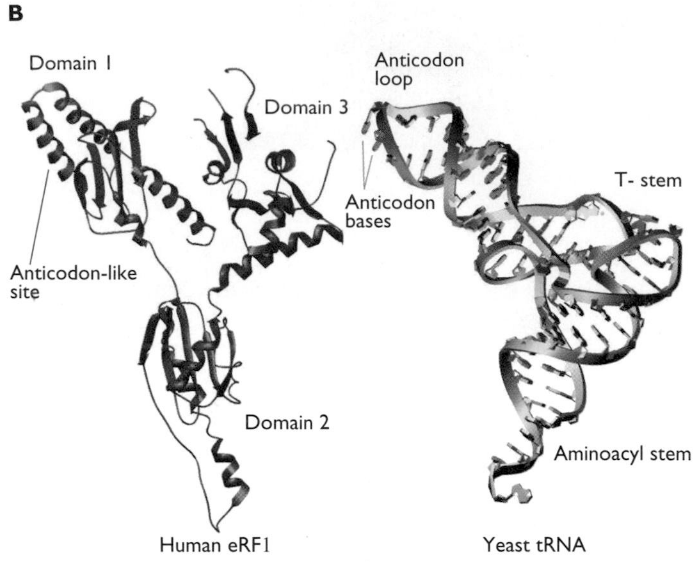

Figure 11.11 Translation termination. (A) Overview of termination. When a termination codon is encountered at the A site, it is usually recognized by a release factor (eRF) instead of a tRNA. The peptide chain is then released, followed by dissociation of tRNA and ribosome from the mRNA. eRF3, which is bound to the C terminus of eRF1, is a GTPase that is required for release of the protein. **(B)** Atomic structure of eRF1 and yeast $tRNA^{Phe}$. The structure of eRF1 mimics that of tRNA, providing a mechanism for recognition of termination codons. (A) Adapted from G. M. Cooper, *The Cell: a Molecular Approach* (ASM Press, Washington, DC, and Sinauer Associates, Sunderland, MA, 1997), with permission. (B) Adapted from H. Song, P. Mugnier, A. K. Das, H. M. Webb, D. R. Evans, M. F. Tuite, B. A. Hemmings, and D. Barford, *Cell* **100**:311–321, 1998, with permission.

tRNA. This cycle is repeated until the ribosome encounters a stop codon. mRNAs are usually bound by many ribosomes **(polysomes)**, with each ribosome separated from its neighbors by approximately 100 to 200 nucleotides, synthesizing a polypeptide chain.

Termination is a modification of the elongation process: once the stop codon enters the A site of the ribosome, it is recognized by the 40S subunit, and the 60S subunit cleaves the ester bond between the protein chain and the last tRNA. Recognition of the three stop codons (UAA, UAG, and UGA) by the 40S ribosomal subunit is facilitated by the **release proteins** eRF1 and eRF3 (Fig. 11.11). The structure of eRF1 mimics that of tRNA, allowing the release protein to occupy the A site of the ribosome. The N terminus of eRF1 recognizes all three stop codons. Once bound in the A site, eRF1 and eRF3 together induce a rearrangement of the 80S ribosome, translocation of the P site codon, and release of the polypeptide. The interaction between eRF1 and the ribosome stimulates the GTPase activity of eRF3, which is bound to the C terminus of eRF1. GTP hydrolysis is required for release of the nascent polypeptide.

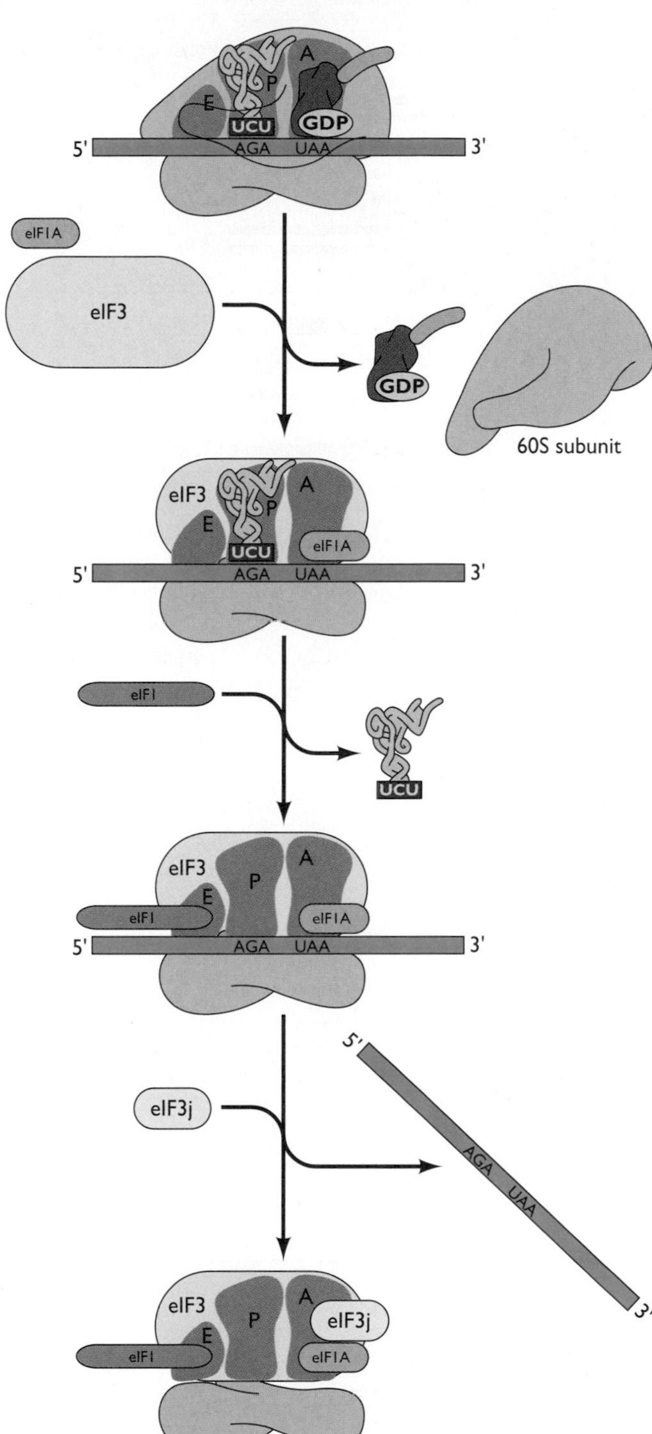

Figure 11.12 Ribosome recycling. After peptide release, eIF1A and eIF3 cause dissociation and release of the 60S ribosomal subunit. Release of the P-site deacylated tRNA is promoted by eIF1 and is followed by dissociation of mRNA mediated by eIF3j binding.

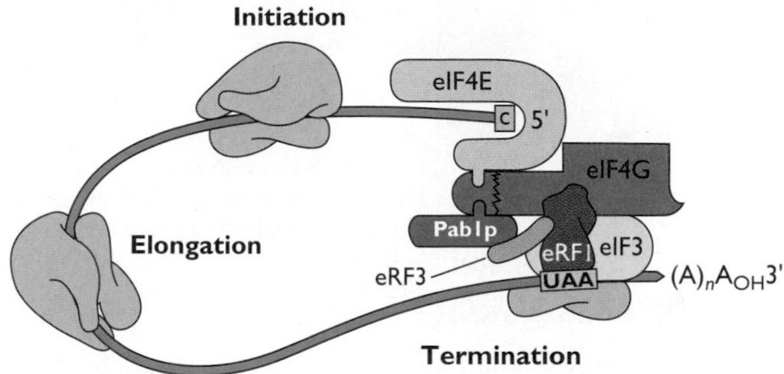

Figure 11.13 Juxtaposition of mRNA ends. Shown is a juxtaposition of mRNA ends by interactions of termination and initiation proteins, Pab1p, and the mRNA 5′ and 3′ ends. eRF3 binds both eRF1 and Pab1p. Adapted from N. Uchida, S. Hoshino, H. Imataka, N. Sonenberg, and T. Katada, *J. Biol. Chem.* **277:**50286–50292, 2002, with permission.

The E site, in addition to accommodation, is another important determinant of the fidelity of protein synthesis. When the E site is occupied by a deacylated tRNA, the affinity of the A site for aminoacyl-tRNA is low. Consequently, incorrect tRNAs are readily rejected. When the E site is empty, the affinity of the A site for aminoacyl-tRNA is significantly higher, making rejection of incorrect tRNAs less likely. An occupied E site also prevents tRNA slippage; when this site is empty, increased ribosomal frameshifting occurs.

Although stop codons are the major determinants of translation termination, other sequences can affect the efficiency of this process. The nucleotide immediately downstream of the stop codon can influence chain termination and ribosome dissociation. In eukaryotes, the preferred termination signals are UAA(A/G) and UGA(A/G).

After release of the polypeptide chain, the 60S ribosomal subunit, tRNA, and mRNA are released by the cooperation of eIF1, eIF1A, and eIF3 (Fig. 11.12). It has been suggested that 40S ribosomal subunits preferentially engage in new rounds of translation initiation on the same mRNA. This hypothesis is supported by the finding that eIF3, which remains bound to the 40S ribosomal subunit after termination, also binds eIF4G (Fig. 11.13). Other observations that are consistent with this model include the ability of eRF3 to bind Pab1p (Fig. 11.13) and the stimulation of 60S ribosomal subunit joining by Pabp1 at initiation. As a result, ribosomes may shuttle from the 3′ end of the mRNA back to the 5′ end, beginning the synthesis of another molecule of the protein.

The Diversity of Viral Translation Strategies

A variety of unusual translation mechanisms have evolved that expand the coding capacity of viral genomes, which are relatively small, and allow the synthesis of multiple polypeptides from a single RNA genome (Fig. 11.14).

Mechanism of translation	Examples	
Polyprotein synthesis	Picornaviruses Flaviviruses Alphaviruses Retroviruses	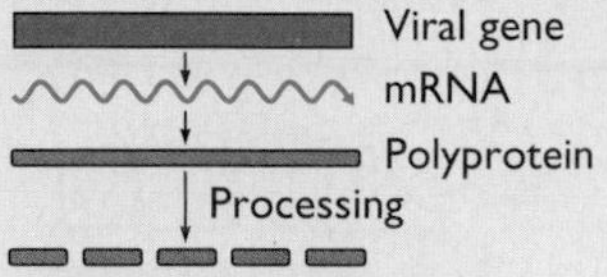 Viral gene mRNA Polyprotein Processing
Leaky scanning	Sendai virus P/C mRNA Influenza B virus RNA 6 Human immunodeficiency virus type 1 Env/Vpu Human T-lymphotropic virus Tax, Rex Simian virus 40 VP2, VP3 Simian virus 40 agnoprotein	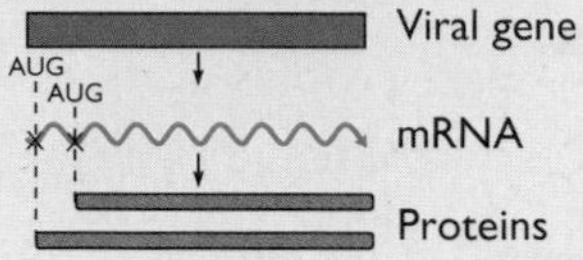 Viral gene AUG AUG mRNA Proteins
Reinitiation	Influenza B virus RNA 7 Cytomegalovirus gp48 mRNA	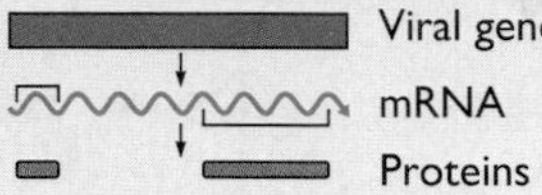 Viral gene mRNA Proteins
Suppression of termination	Alphavirus nsP4 Retrovirus Gag-Pol	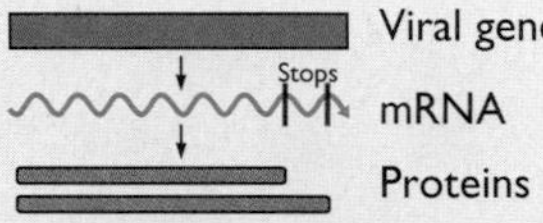 Viral gene Stops mRNA Proteins
Ribosomal frameshifting	Coronavirus ORF1a-ORF1b Human astrovirus type 1 ORF1a-ORF1b Retrovirus Gag-Pol	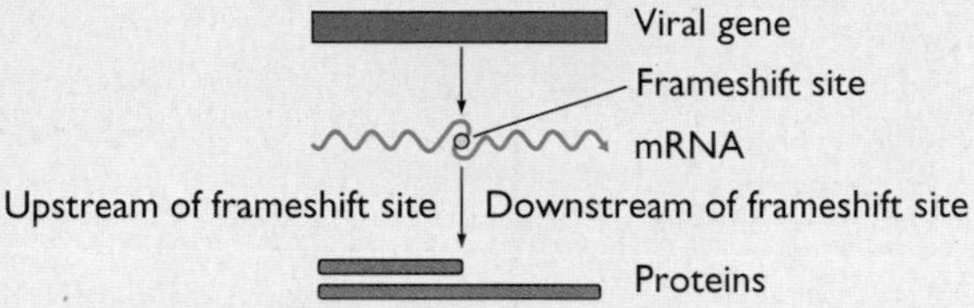 Viral gene Frameshift site mRNA Upstream of frameshift site Downstream of frameshift site Proteins
Internal ribosome entry	Picornaviruses Flaviviruses	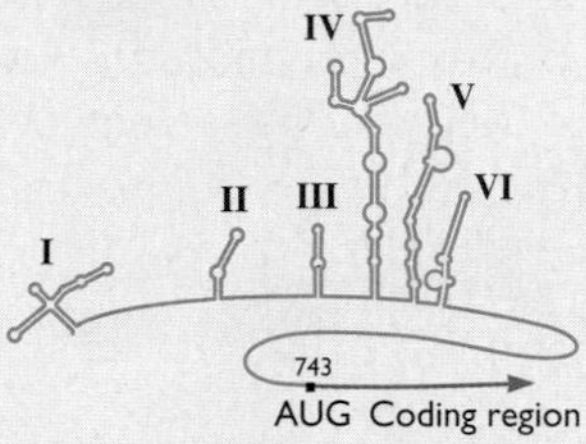 IV V II III VI I 743 AUG Coding region
Ribosome shunting	Adenovirus Cauliflower mosaic virus	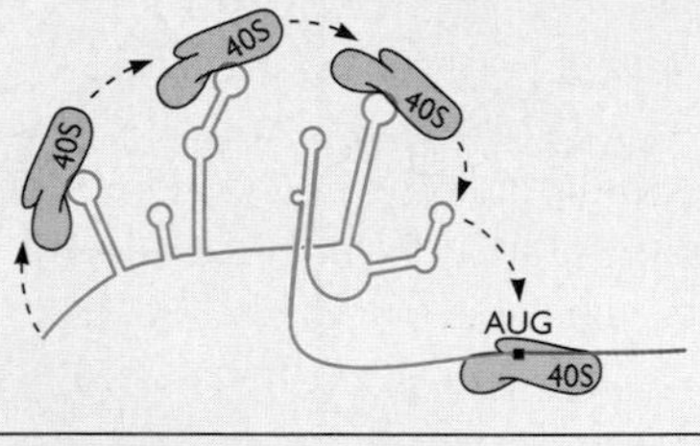 40S 40S 40S AUG 40S
Internal initiation mediated by tRNA-like structure in the 3' untranslated region	Turnip yellow mosaic virus	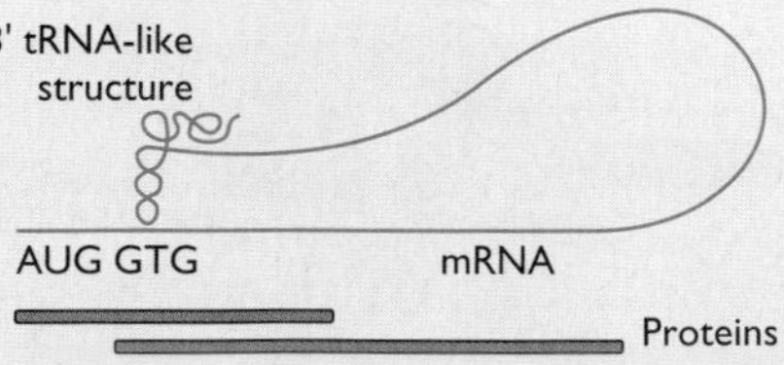 3' tRNA-like structure AUG GTG mRNA Proteins
Bicistronic mRNAs	Cricket paralysis virus *Rhopalosiphum padi* virus	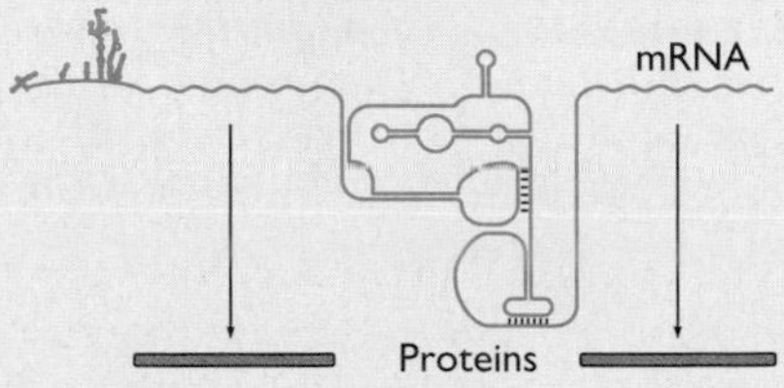 mRNA Proteins

Figure 11.14 The diversity of viral translation strategies.

All were discovered in virus-infected cells and subsequently shown to operate during translation of cellular mRNAs. Nontranslational solutions for maximizing the number of proteins encoded in viral genomes are discussed in other chapters and include the synthesis of multiple subgenomic mRNAs, mRNA splicing, and RNA editing.

Polyprotein Synthesis

One strategy allowing the production of multiple proteins from a single RNA genome is to synthesize from a single mRNA a polyprotein precursor, which is then proteolytically processed to form functional viral proteins.

A dramatic example of protein processing occurs in picornavirus-infected cells: nearly the entire (+) strand RNA is translated into a single large polyprotein (Fig. 11.15A). Processing of this precursor is carried out by two virus-encoded proteases, $2A^{pro}$ and $3C^{pro}$, which cleave between Tyr and Gly and between Gln and Gly, respectively. In both cases, flanking amino acid residues control the efficiency of cleavage so that not all Tyr-Gly and Gln-Gly pairs in the polyprotein are processed. These two proteases are active in the nascent polypeptide and release themselves from the polyprotein by self-cleavage. Consequently, the polyprotein is not observed in infected cells because it is processed as soon as the protease coding sequences have been translated. After the proteases have been released, they cleave other polyprotein molecules.

Protein production can be controlled by the rate and extent of polyprotein processing. In addition, alternative utilization of cleavage sites can produce proteins with different activities. For example, the poliovirus protease $3C^{pro}$ does not process the capsid protein precursor P1 efficiently. Rather, the $3C^{pro}$ precursor, 3CD, is required for processing of P1. By regulating the amount of 3CD produced, the extent of capsid protein processing can be controlled. Because 3CD and $3C^{pro}$ process Gln-Gly pairs in the remainder of the polyprotein with the same efficiency, an interesting question is why 3CD, which also contains $3D^{pol}$ protein, is further processed to produce $3C^{pro}$ (Fig. 11.15A). The answer is that 3CD protein, while active as a protease, does not possess RNA polymerase activity and therefore some molecules must be cleaved to allow RNA replication.

Some viral precursor proteins are processed by cellular proteases. The genome of flaviviruses contains an open reading frame of more than 10,000 bases (Fig. 11.15B). This mRNA is translated into a polyprotein precursor that is processed by a viral serine protease and by host signal peptidase. The latter enzyme is located in the endoplasmic reticulum, where it removes the signal sequence from proteins translocated into the lumen (Chapter 12). The viral proteins processed by the cellular signal peptidase must therefore be inserted into the endoplasmic reticulum.

Leaky Scanning

Although the vast majority of eukaryotic mRNAs are monocistronic (Fig. 11.1), some viral mRNAs that encode two proteins in overlapping reading frames have been identified. The P/C gene of Sendai virus is the model for genes that encode mRNAs with such translational flexibility (Fig. 11.16). P protein is translated from an open reading frame beginning with an AUG codon at nucleotide 104. C proteins are produced from a different reading frame, which begins at nucleotide 81, and are completely different from P proteins. No less than four C proteins (called C′, C, Y1, and Y2) are produced by translation beginning at four in-frame initiation codons. The first start site is an unusual ACG codon, and the third, fourth, and fifth are at AUG codons; the result is a nested set of proteins with a common C terminus.

The first three initiation sites on P/C mRNA are likely to be arranged to permit translation by **leaky scanning**, when ribosomes skip the first AUG codon and initiate at subsequent AUG triplets. The first start site, $ACG^{81/C'}$, is surrounded by a good initiation context but is inefficient because of the unusual start codon. Some ribosomes bypass this initiator codon and initiate at the second, (CGC<u>AUG</u>G). Although the second is an AUG codon, the context is poor, and some ribosomes bypass it and find their way to the third initiation codon, which has a better context (AAG<u>AUG</u>C). Consistent with this hypothesis, mutagenesis of $ACG^{81/C'}$ to AUG abolishes initiation at $AUG^{104/P}$ and $AUG^{114/C}$. When successive initiation codons are used in leaky scanning, they are increasingly efficient as start sites.

The last two C-protein initiation codons, $AUG^{183/Y1}$ and $AUG^{201/Y2}$, are not likely to be translated by leaky scanning because they are in the poorest contexts of the five. Furthermore, mutagenesis of $ACG^{81/C'}$ to AUG has no effect on the synthesis of Y1 and Y2 proteins. Rather, translation of Y1 and Y2 proteins is believed to be initiated by ribosome shunting. An interesting question is how the different translation strategies of P/C mRNA are coordinated such that, for example, shunting does not dominate at the expense of translation of upstream AUG codons. The answer to this question is not known, but Y protein synthesis relative to that of the other C proteins varies in different cell lines. This result suggests that cellular proteins might regulate ribosome shunting on P/C mRNA, although no such protein has been identified.

Translation of overlapping reading frames also occurs in other viral mRNAs (Fig. 11.17). Influenza viruses are classified into three types, A, B, and C; most of the previous discussion in this textbook has concerned work on

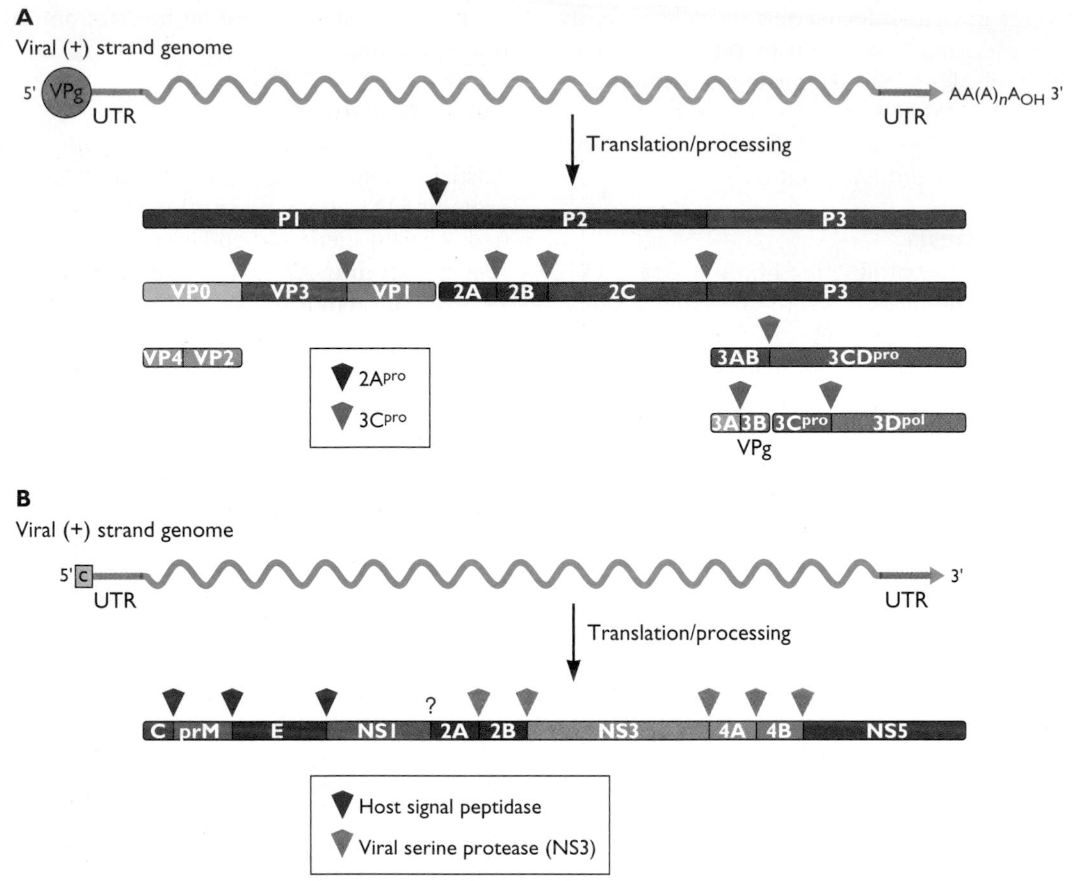

Figure 11.15 Polyprotein processing of picornaviruses and flaviviruses. (A) Processing map of protein encoded by the poliovirus genome. The viral RNA is translated into a long precursor polyprotein that is processed by two viral proteases, $2A^{pro}$ and $3C^{pro}$, to form viral proteins. Cleavage sites for each protease are shown. **(B)** Cleavage map of protein encoded in the flavivirus genome. Processing of the flavivirus precursor polyprotein is carried out either by the host signal peptidase or by the viral protease NS3.

Figure 11.16 Leaky scanning and mRNA editing in the Sendai virus P/C gene. P and C protein open reading frames are shown as brown and blue boxes, respectively. An enlargement of the 5′ end of the mRNA is shown below, indicating the different start sites for four of the C proteins. aa, amino acids. Adapted from J. Curran et al., *Semin. Virol.* **8**:351–357, 1997, with permission.

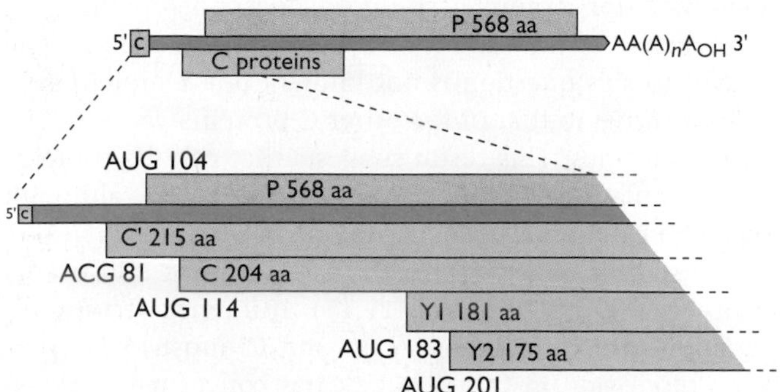

influenza A virus. The mRNA synthesized from influenza B virus RNA segment 6 encodes two different proteins, NB and NA, in overlapping reading frames. NB protein synthesis is initiated at the 5′-proximal AUG codon, while initiation at an AUG codon 4 nucleotides downstream produces the NA protein.

Reinitiation

Reinitiation is another strategy for producing two proteins from a single mRNA. About 10% of mRNAs contain an additional short open reading frame upstream of the main open reading frame (Fig. 11.17). These open reading frames may be translated, with reinitiation occurring at the downstream AUG. For example, a 22-amino-acid peptide is synthesized from an open reading frame in the 5′ region of human cytomegalovirus gp48 mRNA. Other examples

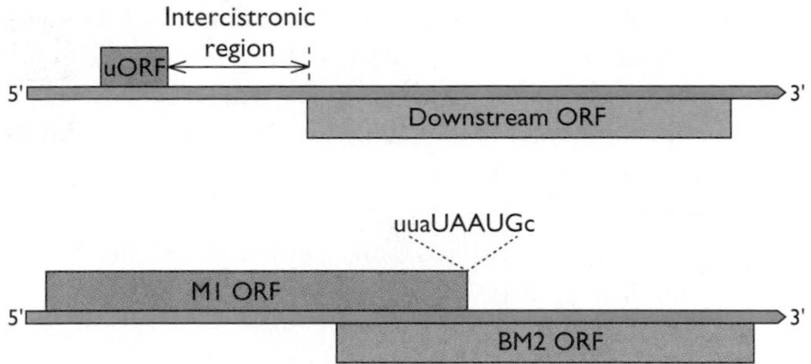

Figure 11.17 Reinitiation of translation. (Top) Some mRNAs contain one or more short upstream open reading frames (uORFs) that may be translated. Expression of the longer, downstream ORF depends on reinitiation. **(Bottom)** mRNA produced from influenza B virus RNA segment 7 encodes two proteins, M1 and BM2. The initiation AUG codon for BM2 overlaps the termination codon of M1. Synthesis of BM2 occurs by reinitiation.

are found in mRNAs of retroviruses and in cellular mRNAs, such as those encoding *S*-adenosylmethionine decarboxylase and fibroblast growth factor 5. These short open reading frames can affect the translation of downstream open reading frames. The extent of regulation depends on many factors, such the as sequence context of the upstream open reading frame AUG initiation codon, the presence of RNA secondary structure, and the distance between the upstream and downstream open reading frames.

In most cases, translation reinitiation involves short upstream open reading frames that precede the main open reading frame. In contrast, reinitiation of translation of longer, overlapping reading frames occurs on mRNA of influenza B virus RNA 7, which encodes two proteins, M1 protein and BM2 protein (Fig. 11.17). M1 protein is translated from the 5′-proximal AUG codon, while the BM2 protein AUG initiation codon is part of the termination codon for M1 protein (UAAUG). Translation of the BM2 open reading frame is dependent on the synthesis of M1 protein, as deletion of the M1 AUG codon abrogates BM2 synthesis.

Suppression of Termination

Suppression of termination occurs during translation of many viral mRNAs as a means of producing a second protein with an extended C terminus. The Gag and Pol genes of Moloney murine leukemia virus are encoded in a single mRNA and separated by an amber termination codon, UAG (Fig. 11.18). Infected cells contain a polyprotein precursor called Gag-Pol. This precursor is synthesized by translational suppression of the amber termination codon. The efficiency of suppression is about 4 to 10%. The Gag-Pol precursor is subsequently processed proteolytically to liberate the Gag and Pol proteins. Without this suppression mechanism, the viral enzymes reverse transcriptase (RT) and integrase (IN) could not be produced. In a similar way, translational suppression of a different termination codon, UGA, is required for the synthesis of nsP4 of alphaviruses (Fig. 11.18). In this example, the efficiency of synthesis is about 10% of that of the normally terminated nsP3 protein. Because nsP4 encodes the RNA-dependent RNA

Figure 11.18 Suppression of termination codons of alphaviruses and retroviruses. (A) Structure of the termination site between Gag and Pol of Moloney murine leukemia virus. The stop codon that terminates synthesis of Gag is underlined; it is followed by a pseudoknot that is important for suppression of termination. Adapted from J. H. Strauss and E. G. Strauss, *Microbiol. Rev.* **58:**491–562, 1994, with permission. **(B)** Suppression of termination during the synthesis of alphavirus P123 to produce nsP4, the RNA-dependent RNA polymerase. The termination codon is shown on the RNA as a box.

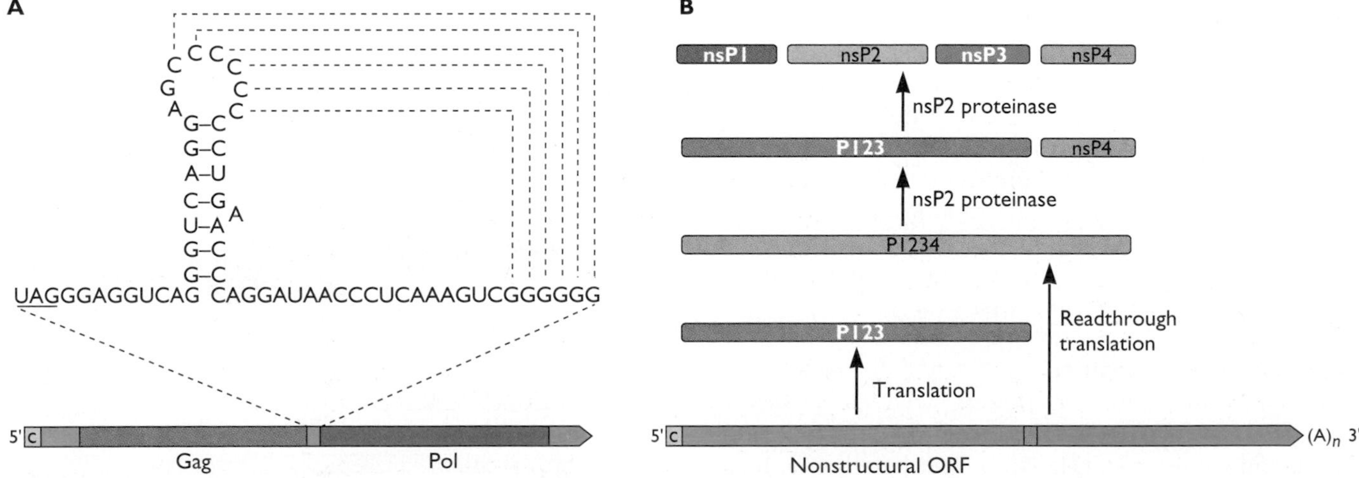

polymerase, suppression is essential for viral RNA replication. Translational suppression in eukaryotic mRNAs is extremely rare.

Most translational suppression occurs when normal tRNAs misread termination codons. The misreading of the amber codon in Moloney murine leukemia virus Gag protein for a Gln codon is an example. More rare are suppressor tRNAs that can recognize termination codons and insert a specific amino acid. One example is a suppressor tRNA that inserts selenocysteine, the 21st amino acid, in place of a UGA codon.

The nucleotide sequence 3′ of the termination codon plays an important role in the efficiency of translational suppression. In bacteria, the nucleotide next to this codon is highly influential. In eukaryotes, the signals range from very simple to complex. In Sindbis virus, efficient suppression of the UGA codon requires only a single C residue 3′ of the termination codon. The effect of this nucleotide on Sindbis virus suppression may be explained in three ways. The 3′ nucleotide might influence the recognition of termination codons by release proteins; it might affect the interaction of misreading tRNAs with the codon by increasing the energy of base pairing; or the misreading tRNA and a tRNA recognizing the next codon might interact, affecting the efficiency of elongation and suppression.

In contrast, readthrough of the UAG codon in Moloney murine leukemia virus mRNA requires a purine-rich sequence 3′ of the termination codon, as well as a pseudoknot structure further downstream (see Chapter 6 for a description of pseudoknots). Because of these differences, it is likely that readthrough can be mediated by different mechanisms. It has been suggested that the pseudoknot of Moloney murine leukemia virus RNA causes the ribosome to pause and allow the suppressor tRNA to compete with eRF1 at the suppression site. The mechanism involved, and the role of the eight-nucleotide, purine-rich segment of the suppression signal, is not known. Maximal readthrough efficiency also requires the interaction of viral reverse transcriptase with eRF1.

Suppression of termination is far more prevalent during translation of viral mRNAs than of cellular mRNAs. The RNA sequences and structures required for suppression are not found in most cellular mRNAs. For example, there is a strong bias against cytidine residues at the 3′ end of UGA termination codons in cellular mRNAs. Suppression by tRNAs charged with selenocysteine has been found in fewer than 50 eukaryotic mRNAs.

Ribosomal Frameshifting

Ribosomal frameshifting is a process by which, in response to signals in mRNA, ribosomes move into a different reading frame and continue translation in that new frame. It was discovered in cells infected with Rous sarcoma virus and has since been described for many other viruses, including additional retroviruses, eukaryotic (+) strand RNA viruses, and herpes simplex virus. There are several examples of frameshifting during translation of mammalian mRNAs. Frameshifting may occur by shifting the reading frame 1 base toward the 5′ end (–1 frameshifting) or the 3′ end (+1 frameshifting) of the mRNA.

In the genome of retroviruses, the *gag* and *pol* genes may be separated by a stop codon (Fig. 11.18), or they may be in different reading frames, with *pol* overlapping *gag* in the –1 direction (Fig. 11.19). During synthesis of Rous sarcoma virus Gag, ribosomes frameshift before reaching the Gag stop codon and continue translating Pol, such that a Gag-Pol fusion is produced at about 10% of the frequency of Gag. Studies on the requirements for frameshifting in retroviruses and coronaviruses have identified two essential components: a "slippery" homopolymeric sequence, which is a heptanucleotide stretch with two homopolymeric triplets of the form X-XXY-YYZ (e.g., in Rous sarcoma virus A-AAU-UUA), and an RNA secondary structure, usually a pseudoknot, five to eight nucleotides downstream. These observations led to the proposal of the tandem shift model for frameshifting, in which two tRNAs in the zero reading frame (X-XXY-YYZ) slip back by one nucleotide during the frameshift to the –1 phase (XXX-YYY). Each tRNA base pairs with the mRNA in the first two nucleotides of each codon (Fig. 11.20). The peptidyl-tRNA is transferred to the P site, the –1 frame *pol* codon is decoded, and translation continues to produce the fusion protein. In this model, slippage occurs before peptide transfer, with the peptidyl- and aminoacyl-tRNAs bound to the P and A sites. However, it is possible that the shift occurs after peptide transfer but before translocation of the tRNAs, or when the aminoacyl-tRNA occupies the A site. These models cannot

Figure 11.19 Frameshifting on a retroviral mRNA. The structure of open reading frames is illustrated. Rous sarcoma virus mRNA encodes Gag and Pol proteins in reading frames that overlap by –1. Normal translation and termination produce the Gag protein; ribosomal frameshifting to the –1 frame results in the synthesis of a Gag-Pol fusion protein.

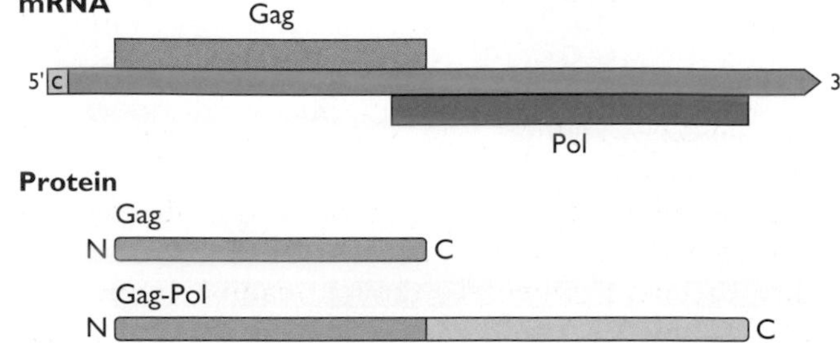

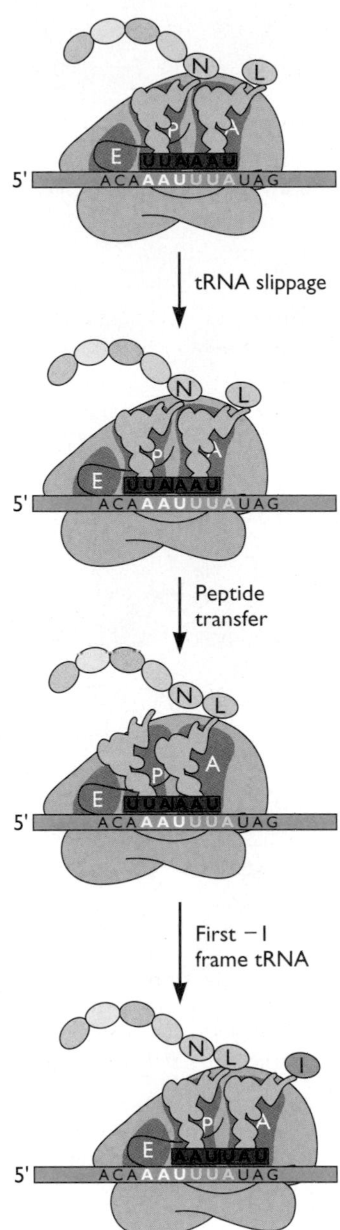

Figure 11.20 A model for −1 frameshifting. Slippage of the two tRNAs occurs after aminoacyl-tRNA enters the A site but before peptide transfer. Slippage allows the tRNAs to form only two base pairs with the mRNA. The site shown is that of Rous sarcoma virus. One-letter amino acid codes are used. Adapted from P. J. Farabaugh, *Microbiol. Rev.* **60:**103–134, 1996, with permission.

be distinguished by mutagenesis or by the sequence of the protein products.

The pseudoknot is thought to cause the ribosome to pause over the slippery sequence, increasing the probability that realignment to the −1 reading frame will occur. Some viral mRNAs contain simple stem-loop structures rather than pseudoknots at this position, but the highest frequencies of frameshifting are associated with the presence of the latter structure. However, when the pseudoknot of a coronavirus is replaced with a stem-loop structure, frameshifting is abolished but ribosomal pausing still occurs. Therefore, pausing is not sufficient for frameshifting; the pseudoknot might also interact with the ribosome to promote frameshifting. It has been suggested that cellular proteins might interact with the pseudoknot to influence frameshifting. For example, eEF1A, eEF2, and a ribosome-associated chaperone complex all have been shown to regulate frameshifting.

Bicistronic mRNAs

The mRNAs of members of the *Dicistroviridae*, including cricket paralysis virus and *Rhopalosiphum padi* virus, are bicistronic (Fig. 11.14). The upstream open reading frame begins with an AUG codon and is preceded by an IRES similar to those of picornaviruses. The downstream open reading frame, which encodes the viral capsid proteins, is translated independently from a completely different IRES. The 40S ribosomal subunit binds directly to the intergenic region that is partially folded to mimic a tRNA. The tRNA-like structure occupies the P site of the ribosome, and initiation occurs from the A site at a nonmethionine codon. Translation of this cistron is therefore dependent on ribosomes and elongation and termination proteins. Because initiation proteins are not required, translation regulation may occur at the stages of elongation and termination.

Regulation of Translation during Viral Infection

Alterations in the cellular translation apparatus are commonplace in virus-infected cells. As part of the antiviral defense, or in response to stress caused by virus infection, the cell initiates measures designed to inhibit protein synthesis and limit virus production. Many viral genomes encode proteins or nucleic acids that neutralize this response, restore translation, and maximize virus replication. In addition, many viral gene products modify the host translation apparatus to favor synthesis of viral proteins over those of the cell. As a result, the entire synthetic capability of the cell can be turned to the production of new virus particles, which should enhance virus yield and perhaps accelerate replication. This supposition is supported by the growth defects of different viral mutants that cannot inhibit cellular translation. These cellular and viral modifications of the translation apparatus usually affect the initiation stages, which are rate-limiting.

Inhibition of Translation Initiation after Viral Infection

Phosphorylation of eIF2α

Members of a large family of secreted proteins, including interferons, are produced as part of the rapid innate immune response of vertebrates in response to viral infection (discussed in Volume II, Chapter 3). Interferons diffuse to neighboring cells, bind to cell surface receptors, and activate signal transduction pathways that result in transcription of hundreds of cellular genes and the establishment of an **antiviral state**. Several interferon-induced genes encode enzymes that effectively prevent association of mRNA with polysomes and hence translation, including RNA-activated protein kinase (Pkr) and RNase L. RNase L degrades RNA and is not considered further here, while Pkr phosphorylates eIF2α, thereby inhibiting translation initiation. Because the block to translation is global, the infected cell may die, but by slowing down viral replication the organism may be spared.

Pkr is a serine-threonine protein kinase composed of an N-terminal regulatory domain and a C-terminal catalytic domain (Fig. 11.21; see also Volume II, Chapter 3). Small quantities of an inactive form of Pkr are present in most uninfected mammalian tissues. Transcription of its gene is induced 5- to 10-fold by interferon. Pkr is activated by the binding of dsRNA to two dsRNA-binding motifs at the N terminus of the protein (Fig. 11.22). Such dsRNA is produced in cells infected by either DNA or RNA viruses. Activation is accompanied by autophosphorylation of Pkr. Low concentrations of dsRNA activate Pkr, but high concentrations are inhibitory, leading to the suggestion that one molecule of Pkr phosphorylates another while both are bound to the same molecule of dsRNA (Fig. 11.22). Phosphorylation is thought to induce conformational changes in the enzyme, thereby rendering the kinase active without further need for dsRNA. Pkr molecules may also be activated by a cell protein, Pact, independently of dsRNA (Fig. 11.22).

Two other eIF2α protein kinases regulate translation during virus infection. In mammalian cells, Gcn2p is activated during amino acid starvation when uncharged tRNA binds a histidyl-tRNA synthetase-like domain in the protein (Fig. 11.21). During infection with Sindbis virus, Gcn2p is activated upon binding of viral RNA, leading to phosphorylation of eIF2α and restriction of virus replication.

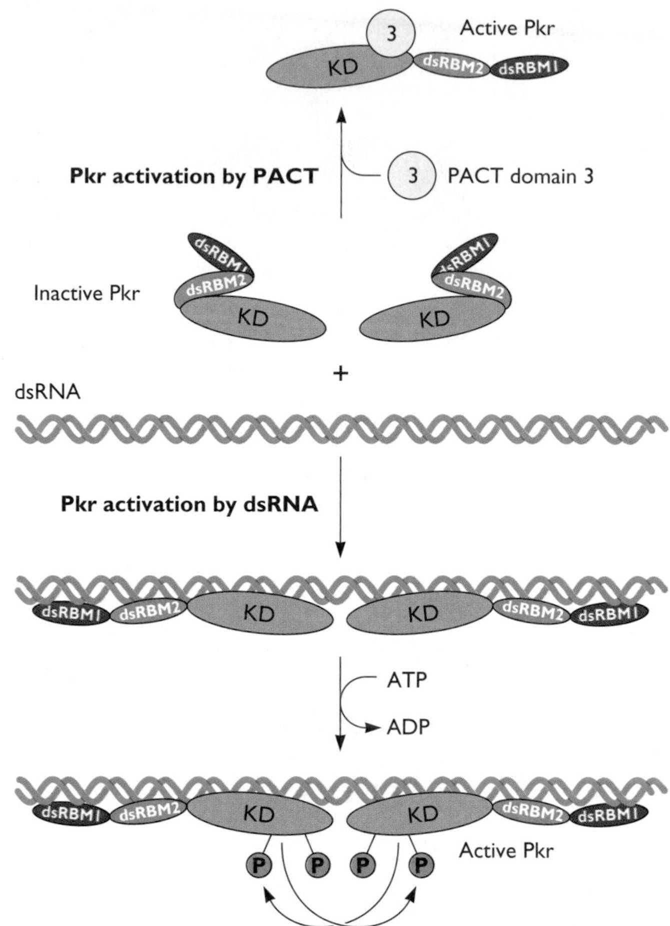

Figure 11.22 Model of activation of Pkr. Pkr is maintained in an inactive monomer by the interaction between a Pact domain 3-binding sequence in Pkr and dsRBM2. Pkr is activated when it binds Pact or dsRNA. When two or more molecules of inactive Pkr bind to one dsRNA molecule, cross-phosphorylation occurs because of the physical proximity of the molecules. Phosphorylation is thought to cause a conformational change in the kinase domain (KD) to allow phosphorylation of other substrates, including eIF2α. dsRBM, double-stranded RNA-binding motif. Adapted from J. W. B. Hershey et al. (ed.), *Translational Control* (Cold Spring Harbor Laboratory Press, Cold Spring Harbor, NY, 1996), with permission.

Figure 11.21 Schematic structures of three eIF2α kinases. IRE1, inositol-requiring enzyme 1; ψ-kinase, pseudokinase domain; HisRS domain, histidyl-tRNA synthetase-like domain. Adapted from C. G. Proud, *Semin. Cell Dev. Biol.* **16:**3–12, 2005, with permission.

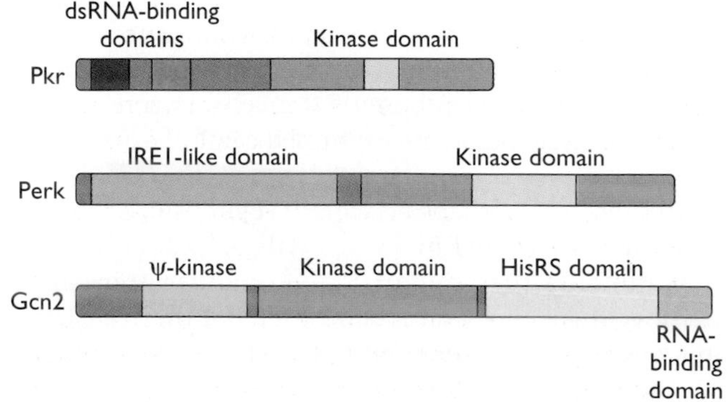

Consistent with a role in mediating antiviral responses, Sindbis virus replication is more efficient in cells lacking Gcn2p. Pkr-like ER kinase (Perk), a transmembrane protein of the endoplasmic reticulum, is a component of the unfolded protein response. Its lumenal domain senses the equilibrium between unfolded and misfolded proteins and chaperone proteins. Under conditions of intracellular stress (Box 11.5), such as occurs during virus infection, Perk oligomerizes within the membrane, is activated, and phosphorylates eIF2α in the cytoplasm.

Activated Pkr, Gcn2p, and Perk phosphorylate eIF2α on serine 51. This initiation protein is part of the complex that also contains GTP and Met-tRNA$_i$ (Fig. 11.3). After GTP hydrolysis, the bound GDP must be exchanged for GTP to permit the binding of another Met-tRNA$_i$. This exchange is carried out by eIF2B (Fig. 11.23). When the alpha subunit of eIF2 is phosphorylated, eIF2-GDP binds eIF2B with such high affinity that it is effectively trapped; recycling of eIF2 stops, and ternary complexes are depleted. eIF2B is less abundant than eIF2, and phosphorylation of about 10 to 40% of eIF2 (depending on the cell type and the relative concentrations of eIF2 and eIF2B) results in the complete sequestration of eIF2B, leading to a block in protein synthesis. As viral translation is also impaired, the production of new virus particles is diminished.

Viral Regulation of Pkr

Most viral infections induce activation of eIF2α kinases and consequent phosphorylation of eIF2α. Global inhibition of translation is clearly a threat to successful viral replication. At least five different viral mechanisms to block Pkr activation or to prevent activated Pkr from inhibiting translation can be distinguished (Table 11.2).

Inhibition of dsRNA binding. The 166-nucleotide adenovirus VA-RNA I, which accumulates to massive concentrations (up to 10^9 copies per cell) late in infection following transcription of the viral gene by RNA polymerase III, is a potent inhibitor of Pkr. An adenovirus mutant that cannot express the VA-RNA I gene grows poorly. In cells infected with this mutant virus, eIF2α becomes extensively phosphorylated, causing global translational inhibition. VA-RNA I binds the dsRNA-binding region of Pkr and blocks activation. It has been suggested that binding of VA-RNA I to Pkr blocks interaction with authentic dsRNA and hence prevents activation of the kinase. Epstein-Barr virus and human immunodeficiency virus type 1 genomes encode small RNAs that inhibit Pkr activation *in vitro*, but whether they function in a similar manner in infected cells is not clear.

While adenovirus VA-RNA I binds Pkr and blocks activation by dsRNA, the vaccinia virus genome encodes a protein that sequesters dsRNA. The viral E3L protein contains the same dsRNA-binding motif as Pkr; it binds dsRNA and prevents it from activating the kinase. Deletion of the gene encoding the E3L protein renders the virus more sensitive to interferon and results in larger quantities of active Pkr in infected cells. The influenza virus NS1 protein and the reovirus σ3 protein also sequester dsRNA.

Inhibition of kinase function. The genomes of several viruses encode proteins that directly inhibit the kinase activity of Pkr, and some do so by acting as pseudosubstrates. For example, vaccinia virus K3L protein has amino acid homology to the N terminus of eIF2α. The protein binds tightly to Pkr within the catalytic cleft and blocks autophosphorylation. The growth of vaccinia virus mutants lacking the K3L gene is severely impaired by interferon. Human immunodeficiency virus type 1 Tat protein may inhibit Pkr by a similar mechanism.

In uninfected cells, Pkr is associated with the chaperone proteins Hsp40 and Hsp70. Influenza virus infection induces the release of Hsp40. Such dissociation may allow aberrant refolding of Pkr by Hsp70, leading to inactivation of the enzyme. Virus infection may also activate a cellular protein, p58^{IPK}, that binds Pkr and prevents autophosphorylation. Activated p58^{IPK} can also block phosphorylation of eIF2α by Perk. In cells infected with herpes simplex virus type 1, acute endoplasmic reticulum stress occurs, but Perk is not activated. The viral glycoprotein gB associates with the lumenal domain of Perk and prevents its activation and subsequent phosphorylation of eIF2α. Another herpes simplex virus protein, Us11, binds to both Pkr and Pact and blocks Pkr activation.

Dephosphorylation of eIF2α. Another mechanism for reversing the consequences of Pkr activation is dephosphorylation of its target. In herpes simplex virus-infected cells, Pkr is activated but eIF2α is not phosphorylated. During infection with viruses lacking the viral ICP34.5 gene, Pkr is activated and eIF2α becomes phosphorylated, causing global inhibition of protein synthesis. This viral protein product associates with a type 1a protein phosphatase and acts as a regulatory subunit, redirecting the enzyme to dephosphorylate eIF2α. The effects of activated Pkr are reversed, ensuring continued protein synthesis. In a similar fashion, the E6 protein of human papillomavirus activates a phosphatase, leading to dephosphorylation of eIF2α.

Beneficial Effects of eIF2α Phosphorylation on Viral Replication

Phosphorylation of eIF2α is not always detrimental to virus replication. Reovirus replication is more efficient in the presence of phosphorylated eIF2α. Such conditions promote the synthesis of a transcriptional regulator (Atf4)

BOX 11.5

BACKGROUND

Viruses and cellular stress

In response to environmental stresses such as heat, starvation, or oxidation, mammalian cells produce cytoplasmic granular RNA structures called stress granules. These structures are aggregates of stalled translational complexes containing intact mRNAs, 40S ribosomal subunits, and a variety of translation initiation proteins such as eIF3 and eIF4G. They are thought to form when protein synthesis is interrupted, specifically by phosphorylation of eIF2α, or inhibition of eIF4A helicase activity. During conditions of stress, mRNAs are protected from degradation within the stress granule. Depending on the condition of the cell, the mRNAs may be subsequently routed to RNA processing bodies (P bodies) for degradation, or to the polysome pool for translation.

Viral infections also induce the formation of stress granules, but different mechanisms have evolved to block or reverse their formation and allow the production of viral proteins. For example, during rotavirus infection, eIF2α is phosphorylated, but stress granule components are relocated and these structures are not formed. Early in cells infected with poliovirus, stress granules form, but they are subsequently dispersed. One of the key proteins for nucleation of stress granule formation, Ras-Gap SH3 domain-binding protein, is cleaved by the poliovirus protease 3C^pro^. Viral interference with stress granule formation demonstrates a critical role for this cellular response in limiting virus infection.

Schütz, S., and P. Sarnow. 2007. How viruses avoid stress. *Cell Host Microbe* **2:**284–285.

White, J. P., A. M. Cardenas, W. E. Marissen, and R. E. Lloyd. 2007. Inhibition of cytoplasmic mRNA stress granule formation by a viral proteinase. *Cell Host Microbe* **2:**295–305.

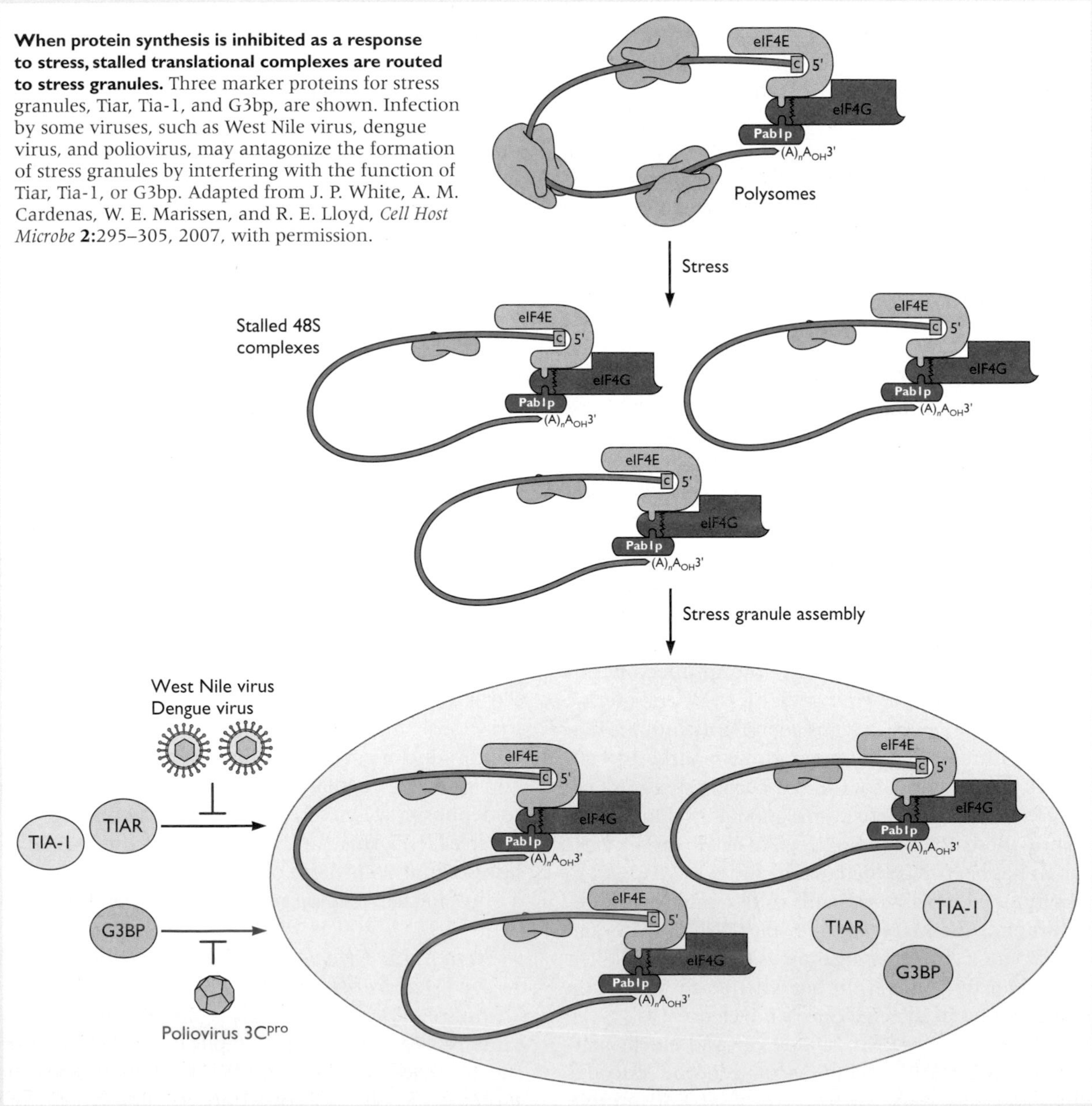

When protein synthesis is inhibited as a response to stress, stalled translational complexes are routed to stress granules. Three marker proteins for stress granules, Tiar, Tia-1, and G3bp, are shown. Infection by some viruses, such as West Nile virus, dengue virus, and poliovirus, may antagonize the formation of stress granules by interfering with the function of Tiar, Tia-1, or G3bp. Adapted from J. P. White, A. M. Cardenas, W. E. Marissen, and R. E. Lloyd, *Cell Host Microbe* **2:**295–305, 2007, with permission.

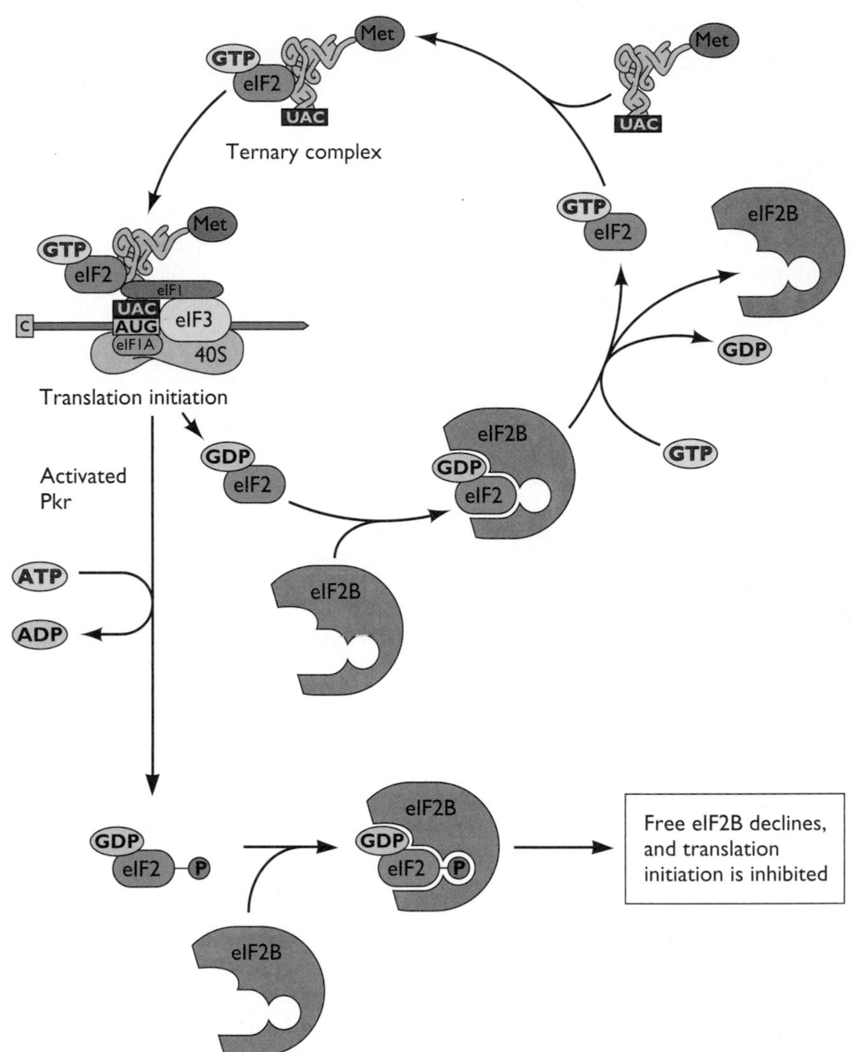

Figure 11.23 Effect of eIF2α phosphorylation on catalytic recycling. eIF2-GTP and tRNA-Met_i form the ternary complex required for translation initiation. During initiation, GTP is hydrolyzed to GDP, and in order for initiation to continue, eIF2 must be recharged with GTP. This recycling is accomplished by eIF2B, which exchanges GTP for GDP on eIF2. When eIF2 is phosphorylated on the alpha subunit, it binds irreversibly to eIF2B, preventing the latter from carrying out its role in recycling active eIF2. As a result, the concentration of eIF2-GTP declines and translation initiation is inhibited.

that controls the synthesis of a variety of proteins important for cellular recovery from stress. Furthermore, inhibition of cellular translation favors translation of the more abundant viral mRNAs. Atf4 is thought to induce the synthesis of mRNAs whose protein products benefit viral replication in cells infected with human cytomegalovirus. The latent membrane protein 1 oncogene of Epstein-Barr virus also induces synthesis of Atf4 by activating Perk. This transcriptional activator in turn increases the synthesis of latent membrane protein 1 mRNA. The increased levels of the protein lead to signaling required for B-cell proliferation.

Regulation of eIF4F

The eIF4F protein plays several important roles during 5′-end-dependent initiation, including recognition of the cap, recruitment of the 40S ribosomal subunit, and unwinding of RNA secondary structure. It is not surprising, therefore, that several viral proteins modify the activity of this protein. The cap-binding subunit eIF4E is frequently a target, probably because its activity can be modulated in at least two ways and because it is present in limiting quantities in cells. The cap-binding complex can also be inactivated by cleavage of eIF4G.

Table 11.2 Targets of viral inhibitors of eIF2α phosphorylation[a]

Target	Virus	Inhibitor	Mechanism
dsRNA	Herpes simplex virus	US11	Binds and sequesters dsRNA
	Influenza virus	NS1	Binds and sequesters dsRNA
	Reovirus	σ3	Binds and sequesters dsRNA
	Vaccinia virus	E3L	Binds and sequesters dsRNA
Pkr	Adenovirus	VA-RNA I	Blocks activation by dsRNA
	Epstein-Barr virus	EBER	Blocks activation by dsRNA
	Human immunodeficiency virus type 1	TAR RNA	Blocks activation by dsRNA
	Herpes simplex virus	US11	Binds Pkr
	Kaposi sarcoma herpesvirus	vIRF-2	Binds Pkr
	Baculovirus	PK2	Inhibits dimerization
	Hepatitis C virus	NS5A	Inhibits dimerization
	Human immunodeficiency virus type 1	Tat	Reduces Pkr expression
eIF2α	Hepatitis C virus	E2	Pseudosubstrate, blocks Pkr-eIF2α interaction
	Human immunodeficiency virus type 1	Tat	Pseudosubstrate, blocks Pkr-eIF2α interaction
	Vaccinia virus	K3L	Pseudosubstrate, blocks Pkr-eIF2α interaction
Phosphatase	Herpes simplex virus	γ34.5	Binds phosphatase, directs to eIF2α
	Simian virus 40	T antigen	Downstream of eIF2α?
Pact	Herpes simplex virus	US11	Binds Pact

[a]RBM, RNA-binding motif.

Cleavage of eIF4G

Poliovirus infection of mammalian cells in culture results in a dramatic inhibition of cellular protein synthesis. By 2 h after infection, polyribosomes are disrupted and translation of nearly all cellular mRNAs dramatically declines (Fig. 11.24). Translationally competent extracts from infected cells can readily translate poliovirus mRNA but not capped mRNAs. Studies of these extracts have demonstrated that they lack functional eIF4F. In poliovirus-infected cells, both isoforms of eIF4G are proteolytically cleaved. As the N-terminal domain of eIF4G binds eIF4E, which in turn binds the 5′ cap of cellular mRNAs, such cleavage prevents eIF4F from recruiting 40S ribosomal subunits to capped mRNAs (Fig. 11.25). Poliovirus mRNA is uncapped and is translated by internal ribosome binding, a process that does not require intact eIF4G. In fact, IRES-mediated initiation function appears to require the C-terminal fragment of eIF4G, which, as discussed above, is necessary to recruit 40S ribosomal subunits to the IRES. Consequently, cleavage of eIF4G not only inhibits translation of cellular mRNAs but also is a strategy for stimulating IRES-dependent translation. Although both eIF4GI and eIF4GII are cleaved in poliovirus- and rhinovirus-infected cells, the kinetics of shutoff of host translation correlates only with cleavage of eIF4GII.

Cleavage of eIF4G is carried out by viral proteases such as $2A^{pro}$ of poliovirus, rhinovirus, and coxsackievirus and the L protease of foot-and-mouth disease virus. Purified $2A^{pro}$ of rhinovirus cleaves eIF4G directly *in vitro*, although very inefficiently unless this protein is bound to eIF4E; that is, eIF4F appears to be the target of $2A^{pro}$ cleavage. The binding of eIF4E to eIF4G might induce conformational changes in the latter protein that make it a more efficient substrate for the protease. Poliovirus $2A^{pro}$ efficiently cleaves eIF4GI but not eIF4GII *in vitro*, consistent with the differential processing of these proteins during viral infection.

Modulation of eIF4E Activity by Phosphorylation

The regulated phosphorylation of eIF4E at Ser209 has been recognized for many years. Inhibition of cellular translation during mitosis and heat shock correlates with reduced phosphorylation of eIF4E. Two protein kinases, Mnk1 and Mnk2, that are associated with eIF4G phosphorylate Ser209 of eIF4E. However, the effect of phosphorylation on the function of eIF4E is unclear. It has been suggested that phosphorylation of eIF4E allows tighter binding to the 5′-terminal cap.

A decrease in eIF4E phosphorylation may be responsible for the inhibition of mRNA translation in cells infected with some viruses. For example, cellular protein synthesis is inhibited at late times in cells infected with adenovirus, a result of virus-induced underphosphorylation of eIF4E. The viral L4 100-kDa protein binds to the C terminus of eIF4G, preventing binding of Mnk1, and hence presumably

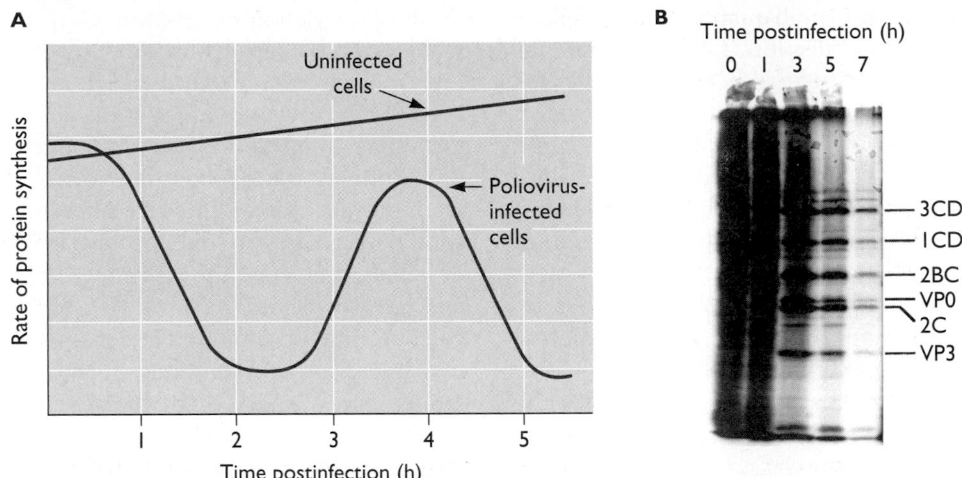

Figure 11.24 Inhibition of cellular translation in poliovirus-infected cells. (A) Rate of protein synthesis in poliovirus-infected and uninfected cells. During poliovirus infection, host cell translation is inhibited by 2 h after infection and is replaced by translation of viral proteins. Adapted from H. Fraenkel-Conrat and R. R. Wagner (ed.), *Comprehensive Virology* (Plenum Press, New York, NY, 1984), with permission. **(B)** Sodium dodecyl sulfate-polyacrylamide gel electrophoresis of [^{35}S]methionine-labeled proteins at different times after poliovirus infection. In this experiment, host translation was shut off by 5 h postinfection and was replaced by the synthesis of viral proteins, some of which are labeled at the right.

blocks phosphorylation of eIF4E. Adenoviral late mRNAs continue to be translated because they possess a reduced requirement for eIF4E. The majority of these viral mRNAs contain the tripartite leader (Fig. 10.12), a common 5′ noncoding region that mediates translation by ribosome shunting. Initiation by this mechanism is less dependent on eIF4F, presumably because the shunting of part of the 5′ untranslated region reduces the requirement for RNA-unwinding (helicase) activity associated with initiation by cap binding and scanning. Furthermore, adenovirus late

Figure 11.25 Regulation of eIF4F activity. The illustration shows regulation of eIF4F activity, and inhibition of translation, by dephosphorylation of eIF4E, interaction with two eIF4E-binding proteins, and proteolytic cleavage of eIF4G.

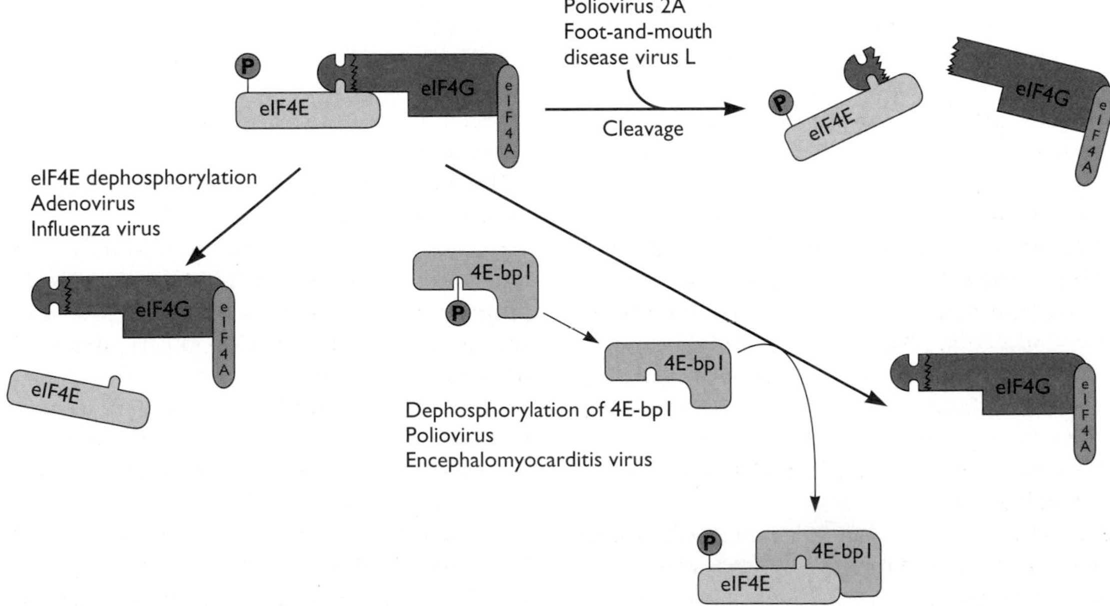

mRNAs efficiently recruit the small quantities of phosphorylated eIF4E present late in infection, a feature of mRNAs with little RNA secondary structure near the 5′ cap. The tripartite leader therefore confers selective translation of viral over cellular mRNAs under conditions in which eIF4E is underphosphorylated. Adenovirus-induced translation inhibition not only boosts viral late mRNA translation but also enhances cytopathic effects and consequently release of virus from cells.

Despite the correlation between reduced phosphorylation of eIF4E and translation inhibition in virus-infected cells, it has also been reported that phosphorylation of this protein may have little or no effect on its affinity for the 5′-terminal cap structure and the rate of protein synthesis. The role of eIF4E phosphorylation during translation remains controversial.

Modulation of eIF4E Activity by Specific Binding Proteins

Three related low-molecular-weight cellular proteins, 4E-bp1, 4E-bp2, and 4E-bp3, can bind to eIF4E and inhibit translation following 5′-end-dependent scanning, but not by internal ribosome entry (Fig. 11.25). The first was found to be identical to a previously described protein, called phosphorylated heat- and acid-stable protein regulated by insulin (Phas-I). This protein was known to be an important substrate for phosphorylation in cells treated with insulin and growth factors. Phosphorylation of 4E-bp *in vitro* prevents it from associating with eIF4E. It was subsequently shown that when bound to 4E-bp, eIF4E cannot bind to eIF4G. As a result, active eIF4F is not formed. eIF4G and 4E-bp proteins carry a common sequence motif that binds eIF4E. Treatment of cells with hormones and growth factors leads, through signal transduction pathways, to the phosphorylation of 4E-bp and its release from eIF4E. Translation of mRNAs with extensive secondary structure in the 5′ untranslated region is preferentially sensitive to the phosphorylation state of 4E-bp.

Infection with several viruses results in alteration of the phosphorylation state of 4E-bp (Fig. 11.25). In contrast to the shutoff that occurs in poliovirus-infected cells, inhibition of cellular protein synthesis in encephalomyocarditis virus-infected cells occurs late in infection and is not mediated by cleavage of eIF4G. Rather, encephalomyocarditis virus infection induces dephosphorylation of 4E-bp1. As a result, translation of cellular mRNAs is inhibited, but, because the viral mRNA contains an IRES, its translation is unaffected. Phosphorylation of 4E-bp1 is also observed in cells infected with vesicular stomatitis virus, poxviruses, and herpes simplex virus type 1, but how this modification favors translation of viral mRNAs is not known.

Phosphorylation of 4E-bp is carried out by the mammalian target of rapamycin kinase (mTor). Rapamycin is an immunosuppressant that binds to an immunophilin protein (Fkbp). The latter binds to mTor and blocks phosphorylation of 4E-bp. As expected, treatment of cells with this drug inhibits translation initiation by 5′-end-dependent scanning but not by internal ribosome entry. The mTor complex regulates protein synthesis in response to a variety of signals, such as amino acid concentrations, energy state, and growth factors. As would be expected, mTor is activated during infection by a wide variety of viruses, leading to increased protein synthesis under conditions (e.g., virus-induced stress) that would otherwise limit translation.

Regulation of Poly(A)-Binding Protein Activity

The poly(A)-binding protein Pab1p plays a crucial role in mRNA translation, bringing together the ends of the mRNA (Fig. 11.4). In cells infected with rotaviruses, inhibition of host translation is a consequence of blocking the function of Pab1p (Fig. 11.26). The 3′ ends of rotaviral mRNAs are not polyadenylated and therefore cannot interact with this protein. Instead, these 3′ untranslated regions contain a conserved sequence that binds the viral protein nsP3. This viral protein associates with eIF4G, bringing together the viral mRNA ends. Therefore, nsP3 assumes the function of Pab1p in translation of rotavirus mRNAs. The nsP3 protein occupies the Pab1p-binding site of eIF4G, thereby evicting Pab1p and preventing juxtaposition of the mRNA ends. The binding of nsP3 to eIF4G is the molecular basis for rotavirus inhibition of host cell translation.

Regulation of eIF3

Three interferon-induced human genes, *ISG54*, *ISG56*, and *ISG60*, encode proteins (P54, P56, and P60) that bind subunits of eIF3 and block translation. The P56 protein binds the e subunit of eIF3, while P54 binds to the c and e subunits. Both P54 and P56 interfere with stabilization of the ternary complex (eIF2-GTP-tRNAi-Met), and P54 also blocks formation of the 48S initiation complex (Fig. 11.3). Both 5′-end-dependent and internal initiation are inhibited by P56. The inhibition of internal initiation by the hepatitis C virus IRES by P56 is probably one of the main reasons for the antiviral effect of interferon on replication of this virus.

Another mechanism for regulation of the activity of eIF3 is binding of one of the subunits, eIF3f, by the spike glycoprotein of severe acute respiratory syndrome (SARS) coronavirus. This interaction leads to reduced translation of cellular genes. It is not known how viral mRNAs are translated. The eIF3a and eIF3b subunits are cleaved by the

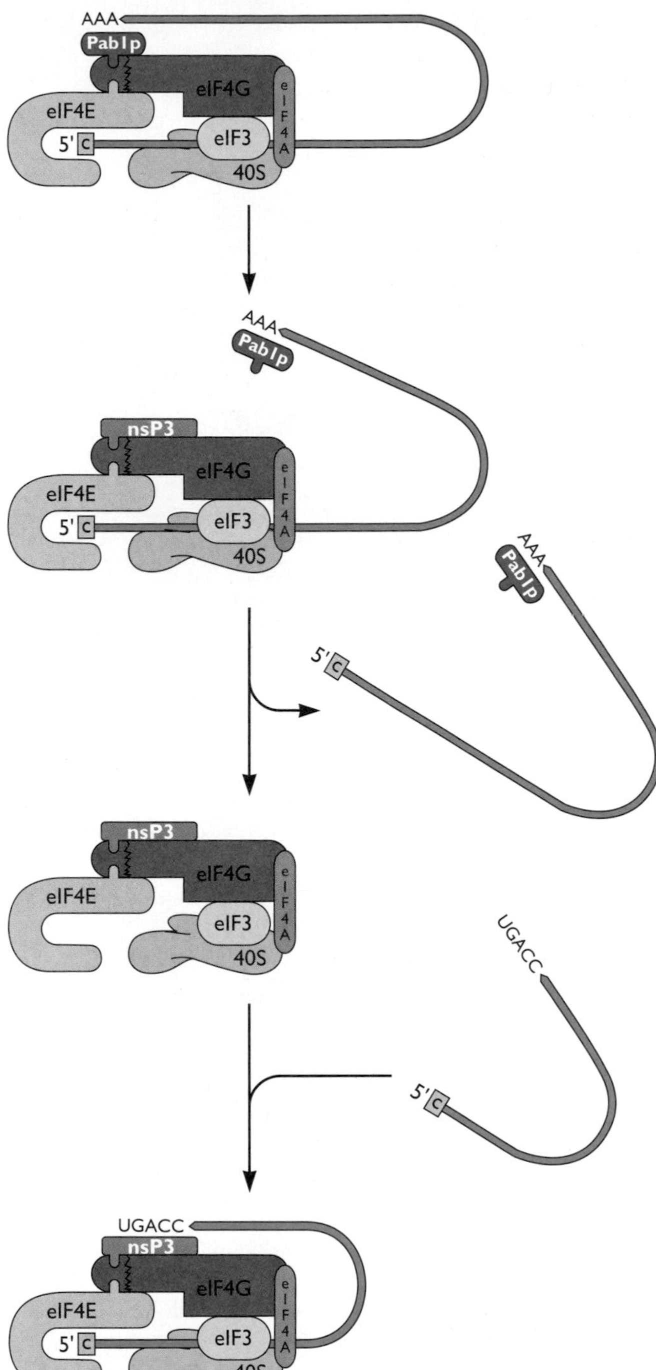

Figure 11.26 Eviction of Pab1p from eIF4G by rotavirus nsP3. The ends of host cell mRNA brought together by the interaction of Pab1p with poly(A) and eIF4G are shown at the top. Rotavirus nsP3 associates with eIF4G at the Pab1p-binding site, and also binds the 3′ untranslated region of the viral mRNA. As a result, host mRNAs are replaced by viral mRNAs in initiation complexes, and translation occurs in a 5′-end-dependent manner. Adapted from M. Piron, P. Vende, J. Cohen, and D. Poncet, *EMBO J.* **17**:5811–5821, 1998.

viral protease in cells infected with the picornavirus foot-and-mouth disease virus, further contributing to inhibition of host protein synthesis caused by cleavage of eIF4G.

Regulation by miRNA

Micro-RNAs (miRNAs) (Chapter 10) regulate the expression of ~30% of all mammalian protein-coding genes by influencing either mRNA stability or translation. They function as part of a ribonucleoprotein (miRNP or RNA-induced silencing complex [Risc]) which includes members of the Argonaute family of proteins. The target site on an mRNA is most often within the 3′ noncoding region and is present in multiple copies (Fig. 11.27). There is evidence that binding of miRNPs to the 3′ noncoding region inhibits translation at the initiation step. In one study, miRNA blocked cap-dependent but not IRES-dependent initiation. It has been suggested that the Argonaute 2 protein competes with eIF4E for binding to the 5′ cap. Translational repression depends not only on the presence of a 5′ cap but also on a poly(A) tail. Alternative mechanisms of action of miRNAs include inhibition of joining of the 60S ribosomal subunit, and interference with elongation and termination. The mechanisms have not yet been established.

miRNAs clearly play important roles in virus infections. They may modulate inflammatory responses to virus infections. As might be expected, viral gene products, such as adenovirus VA-RNA I, that block production of miRNAs have been identified. miRNAs may also influence cell susceptibility: a liver-specific miRNA has been identified which markedly enhances hepatitis C virus replication. The genomes of some mammalian viruses have been shown to encode miRNAs. One miRNA encoded in the genome of simian virus 40 reduces viral gene expression, thus reducing elimination of infected cells by cytotoxic T lymphocytes. miRNAs will probably be found to have enormous influence on viral pathogenesis, and may also present new opportunities for interfering with viral growth (Volume II, Chapter 9).

Perspectives

The study of protein synthesis in virus-infected cells has revealed a great deal about how proteins are made and how the process is regulated. Very early in infection, intrinsic defense responses are mounted, and protein synthesis is inhibited in an attempt to limit viral replication. Should infection proceed, cellular stress responses, which cause further reduction in translation, are activated. As viral proteins and RNAs are produced, modifications to the cellular translation apparatus take place to favor the production of viral proteins. The interplay of cellular and viral modifications is an important determinant of the outcome of infection.

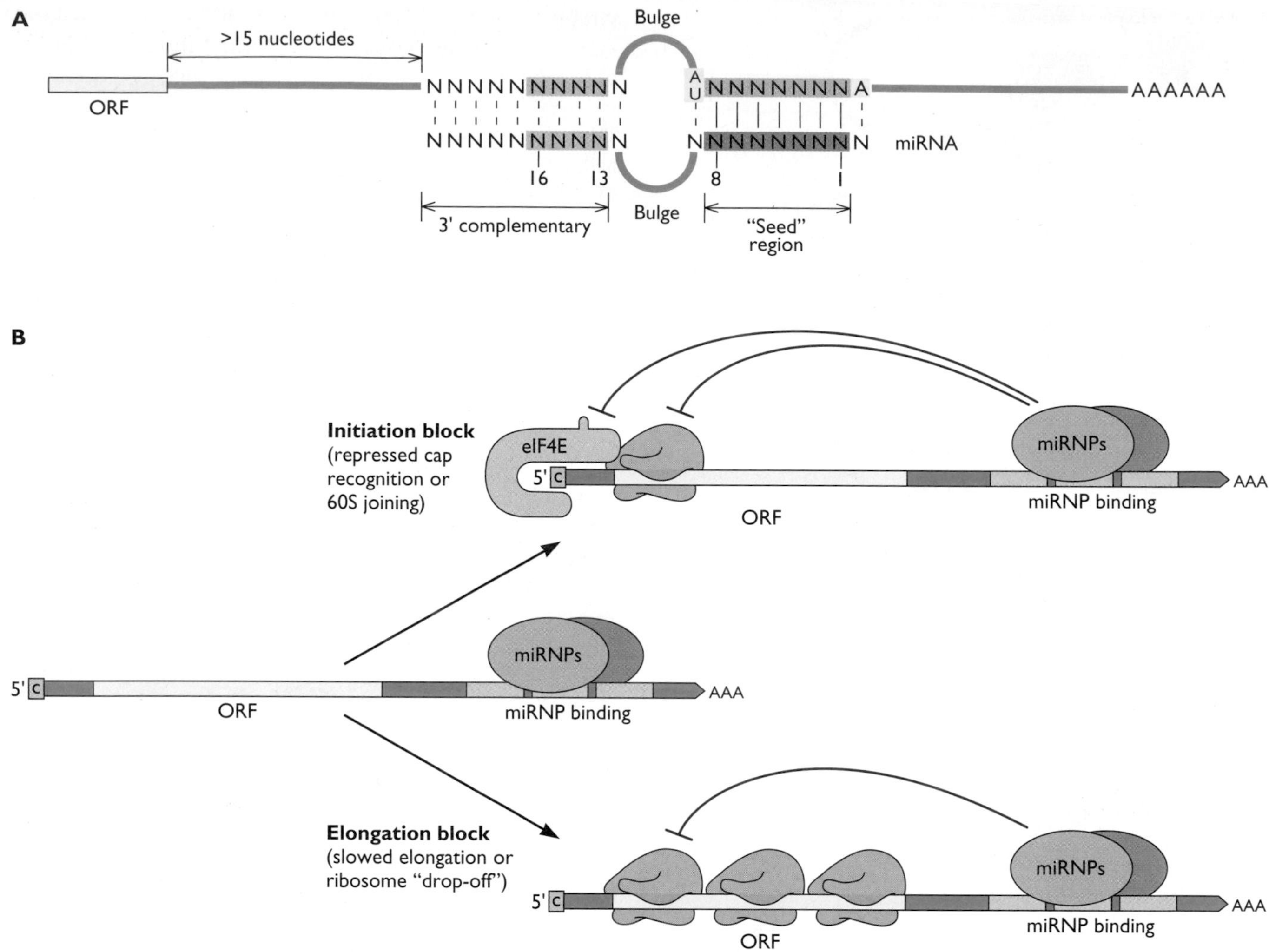

Figure 11.27 Models for miRNA inhibition of translation. (A) Base pairing of miRNA in the 3′ untranslated region of mRNA. Perfect base pairing of nucleotides 2 to 8 (purple/red) in the seed region nucleates the association of the two RNAs. Mismatches in this region decrease repression, with the exception of the two bases shown in yellow. If the central bulge is not present, endonucleolytic cleavage by Argonaute occurs. Good base pairing is required in the 3′ complementarity region, especially in the bases shown in pink. **(B)** Two possible mechanisms of translational repression mediated by miRNA. Binding of miRNPs can impair initiation (top) or elongation (bottom).

Despite years of work in many laboratories, many aspects of translational regulation during virus infection remain unclear. Unanswered questions include the role of phosphorylation in activity of eIF4E, why translation is made more efficient when mRNA ends are juxtaposed, and how viral mRNAs are preferentially translated. What are the signals that lead to phosphorylation of eIF4E-binding proteins through mTor? How are stress responses counteracted so that virus infection can proceed? How do eIF2α kinases benefit viral replication?

The recent discovery of miRNA in mammalian cells has led to the realization that these small RNAs may have significant impact on the outcome of viral infection. An important question is whether miRNAs block protein synthesis at one or more steps. Although results from many laboratories suggest that miRNAs interfere at many steps in translation, it is too early to make general conclusions, because so many different experimental systems have been examined. It is possible that translation initiation is the main target for inhibition, which then leads to alterations in elongation, termination, mRNA, and protein stability. Understanding how miRNAs inhibit translation is likely to improve our understanding of this essential process, and reveal new ways to limit viral infections.

References

Monograph

Sonenberg, N., J. W. B. Hershey, and M. B. Mathews (ed.). 2000. *Translational Control of Gene Expression*. Cold Spring Harbor Laboratory Press, Cold Spring Harbor, NY.

Review Articles

Curran, J., P. Latorre, and D. Kolakofsky. 1997. Translational gymnastics on the Sendai virus P/C mRNA. *Semin. Virol.* **8:**351–357.

Filipowicz, W., S. N. Bhattacharyya, and N. Sonenberg. 2008. Mechanisms of post-transcriptional regulation by microRNAs: are the answers in sight? *Nat. Rev. Genet.* **9:**102–104.

Fraser, C. S. and J. A. Doudna. 2007. Structural and mechanistic insights into hepatitis C viral translation initiation. *Nat. Rev. Microbiol.* **5:**29–38.

Garcia, M. A., E. F. Meurs, and M. Esteban. 2007. The dsRNA protein kinase Pkr: virus and cell control. *Biochimie* **89:**799–811.

Kahvejian, A., G. Roy, and N. Sonenberg. 2001. The mRNA closed-loop model: the function of PABP and PABP-interacting proteins in mRNA translation. *Cold Spring Harb. Symp. Quant. Biol.* **66:**293–300.

Kedersha, N., and P. Anderson. 2007. Mammalian stress granules and processing bodies. *Methods Enzymol.* **431:**61–81.

Lewis, S. M., and M. Holcik. 2008. For IRES trans-acting factors, it is all about location. *Oncogene* **27:**1033–1035.

Martinez-Salas, E., A. Pacheco, P. Serrano, and N. Fernandez. 2008. New insights into internal ribosome entry site elements relevant for viral gene expression. *J. Gen. Virol.* **89:**611–626.

Namy, O., J.-P. Rousset, S. Napthine, and I. Brierley. 2004. Reprogrammed genetic decoding in cellular gene expression. *Mol. Cell* **13:**157–168.

Nierhaus, K. H. 2006. Decoding errors and involvement of the E-site. *Biochimie* **88:**1013–1019.

Pisarev, A. V., N. E. Shirokikh, and C. U. Hellen. 2005. Translation initiation by factor-independent binding of eukaryotic ribosomes to internal ribosome entry sites. *Crit. Rev. Biol.* **328:**589–605.

Prévot, D., J. L. Darlix, and T. Ohlmann. 2003. Conducting the initiation of protein synthesis: the role of eIF4G. *Biol. Cell* **95:**141–156.

Scheper, G. C., and C. G. Proud. 2002. Does phosphorylation of the cap-binding protein eIF4E play a role in translation initiation? *Eur. J. Biochem.* **269:**5350–5359.

Sen, G. C., and G. A. Peters. 2007. Viral stress-inducible genes. *Adv. Virus Res.* **70:**233–263.

Spriggs, K. A., M. Stoneley, M. Bushell, and A. E. Willis. 2008. Re-programming following cell stress allows IRES-mediated translation to predominate. *Biol. Cell* **100:**27–38.

Papers of Special Interest

Translation Termination

Alkalaeva, E. Z., A. V. Pisarev, L. Y. Frolova, L. L. Kisselev, and T. V. Pestova. 2006. In vitro reconstitution of eukaryotic translation reveals cooperativity between release factors eRF1 and eRF3. *Cell* **125:**1125–1136.

Pisarev, A. V., C. U. T. Hellen, and T. V. Pestova. 2007. Recycling of eukaryotic posttermination ribosomal complexes. *Cell* **131:**286–299.

Juxtaposition of mRNA Ends

Guo, L., E. M. Allen, and W. A. Miller. 2001. Base-pairing between untranslated regio-ns facilitates translation of uncapped, nonpolyadenylated viral RNA. *Mol. Cell* **7:**1103–1109.

Hinton, T. M., M. J. Coldwell, G. A. Carpenter, S. J. Morley, and V. M. Pain. 2007. Functional analysis of individual binding activities of the scaffold protein eIF4G. *J. Biol. Chem.* **282:**1695–1708.

Serrano, P., M. R. Pulido, M. Sáiz, and E. Martínez-Salas. 2006. The 3′ end of the foot-and-mouth disease virus genome establishes two distinct long-range RNA-RNA interactions with the 5′ end region. *J. Gen. Virol.* **87:**3013–3022.

Svitkin, Y. V., M. Costa-Mattioli, B. Herdy, S. Perreault, and N. Sonenberg. 2007. Stimulation of picornavirus replication by the poly(A) tail in a cell-free extract is largely independent of the poly(A) binding protein (PABP). *RNA* **13:**2330–2340.

Tarun, S. Z., S. E. Wells, J. A. Deardorff, and A. B. Sachs. 1997. Translation initiation factor eIF4G mediates in vitro poly(A) tail-dependent translation. *Proc. Natl. Acad. Sci. USA* **94:**9046–9051.

Uchida, N., S. Hoshino, H. Imataka, N. Sonenberg, and T. Katada. 2002. A novel role of the mammalian GSPT/eRF3 associating with poly(A)-binding protein in cap/poly(A)-dependent translation. *J. Biol. Chem.* **277:**50286–50292.

Ribosome Shunting

Pooggin, M. M., J. Futterer, K. G. Skryabin, and T. Hohn. 2001. Ribosome shunt is essential for infectivity of cauliflower mosaic virus. *Proc. Natl. Acad. Sci. USA* **98:**886–891.

Sherrill, K. W., and R. E. Lloyd. 2008. Translation of cIAP2 mRNA is mediated exclusively by a stress-modulated ribosome shunt. *Mol. Cell. Biol.* **28:**2011–2022.

Xi, Q., R. Cuesta, and R. J. Schneider. 2005. Regulation of translation by ribosome shunting through phosphotyrosine-dependent coupling of adenovirus protein 100k to viral mRNAs. *J. Virol.* **79:**5676–5683.

Internal Ribosome Entry

Bedard, K. M., S. Daijogo, and B. L. Semler. 2007. A nucleo-cytoplasmic SR protein functions in viral IRES-mediated translation initiation. *EMBO J.* **26:**459–467.

Blyn, L. B., J. S. Towner, B. L. Semler, and E. Ehrenfeld. 1997. Requirement of poly(C) binding protein for translation of poliovirus RNA. *J. Virol.* **71:**6243–6246.

Costantino, D. A., J. S. Pfingsten, R. P. Rambo, and J. S. Kieft. 2008. tRNA-mRNA mimicry drives translation initiation from a viral IRES. *Nat. Struct. Mol. Biol.* **15:**57–64.

Elroy-Stein, O., T. R. Fuerst, and B. Moss. 1989. Cap-independent translation of mRNA conferred by encephalomyocarditis virus 5′ sequence improves the performance of the vaccinia virus/bacteriophage T7 hybrid expression system. *Proc. Natl. Acad. Sci. USA* **86:**6126–6130.

Fujimura, K., F. Kano, and M. Murata. 2008. Identification of PCBP2, a facilitator of IRES-mediated translation, as a novel constituent of stress granules and processing bodies. *RNA* **14:**425–431.

Jang, S. K., H.-G. Kräusslich, M. J. H. Nicklin, G. M. Duke, A. C. Palmenberg, and E. Wimmer. 1988. A segment of the 5′ nontranslated region of encephalomyocarditis virus RNA directs internal entry of ribosomes during in vitro translation. *J. Virol.* **62:**2636–2643.

Kaminski, A., S. L. Hunt, J. G. Patton, and R. J. Jackson. 1995. Direct evidence that polypyrimidine tract binding protein (PTB) is essential for internal initiation of translation of encephalomyocarditis virus RNA. *RNA* **1:**924–938.

Kolupaeva, V. G., T. V. Pestova, C. U. Hellen, and I. N. Shatsky. 1998. Translation eukaryotic initiation factor 4G recognizes a specific structural element within the internal ribosome entry site of encephalomyocarditis virus RNA. *J. Biol. Chem.* **273:**18599–18604.

Macejak, D. G., and P. Sarnow. 1991. Internal initiation of translation mediated by the 5′ leader of a cellular mRNA. *Nature* **353:**90–94.

Meerovitch, K., Y. V. Svitkin, H. S. Lee, F. Lejbkowicz, D. J. Kenan, E. K. L. Chan, V. I. Agol, J. D. Keene, and N. Sonenberg. 1993. La autoantigen enhances and corrects aberrant translation of poliovirus RNA in reticulocyte lysate. *J. Virol.* **67:**3798–3807.

Mokrejs, M., V. Vopálensky, O. Kolenaty, T. Masek, Z. Feketová, P. Sekyrová, B. Skaloudová, V. Kríz, and M. Pospísek. 2006. IRESite: the database of experimentally verified IRES structures (www.iresite.org). *Nucleic Acids Res.* **34:**D125–D130.

Pestova, T. V., I. N. Shatsky, S. P. Fletcher, R. J. Jackson, and C. U. Hellen. 1998. A prokaryotic-like mode of cytoplasmic eukaryotic ribosome binding to the initiation codon during internal translation initiation of hepatitis C and classical swine fever virus RNAs. *Genes Dev.* **12:**67–83.

Siridechadilok, B., C. S. Fraser, R. J. Hall, J. A. Doudna, and E. Nogales. 2005. Structural roles for human translation factor eIF3 in initiation of protein synthesis. *Science* **310:**1513–1515.

Svitkin, Y. V., B. Herdy, M. Costa-Mattioli, A. C. Gingras, B. Raught, and N. Sonenberg. 2005. Eukaryotic translation initiation factor 4E availability controls the switch between cap-dependent and internal ribosome entry site-mediated translation. *Mol. Cell. Biol.* **25:**10556–10565.

Non-AUG Initiation

Barends, S., H. H. Bink, S. H. van den Worm, C. W. Pleij, and B. Kraal. 2003. Entrapping ribosomes for viral translation: tRNA mimicry as a molecular Trojan horse. *Cell* **112:**123–129.

Costantino, D., and J. S. Kieft. 2005. A preformed compact ribosome-binding domain in the cricket paralysis-like virus IRES RNAs. *RNA* **11:**332–343.

Sasaki, J., and N. Nakashima. 2000. Methionine-independent initiation of translation in the capsid protein of an insect RNA virus. *Proc. Natl. Acad. Sci. USA* **97:**1512–1525.

Wilson, J. E., T. V. Pestova, C. U. Hellen, and P. Sarnow. 2000. Initiation of protein synthesis from the A site of the ribosome. *Cell* **102:**511–520.

Leaky Scanning

Williams, M. A., and R. A. Lamb. 1989. Effect of mutations and deletions in a bicistronic mRNA on the synthesis of influenza B virus NB and NA glycoproteins. *J. Virol.* **63:**28–35.

Reinitiation

Horvath, K. M., M. A. Williams, and R. A. Lamb. 1990. Eukaryotic coupled translation of tandem cistrons: identification of the influenza B virus BM2 polypeptide. *EMBO J.* **9:**2639–2647.

Racine, T., C. Barry, K. Roy, S. J. Dawe, M. Shmulevitz, and R. Duncan. 2007. Leaky scanning and scanning-independent ribosome migration on the tricistronic S1 mRNA of avian reovirus. *J. Biol. Chem.* **282:**25613–25622.

Translational Suppression

Feng, Y.-X., H. Yuan, A. Rein, and J. G. Levin. 1992. Bipartite signal for read-through suppression in murine leukemia virus mRNA: an eight-nucleotide purine-rich sequence immediately downstream of the gag termination codon followed by an RNA pseudoknot. *J. Virol.* **66:**5127–5132.

Li, G., and C. M. Rice. 1993. The signal for translational readthrough of a UGA codon in Sindbis virus RNA involves a single cytidine residue immediately downstream of the termination codon. *J. Virol.* **67:**5062–5067.

Yoshinaka, Y., I. Katoh, T. D. Copeland, and S. Oroszlan. 1985. Murine leukemia virus protease is encoded by the gag-pol gene and is synthesized through suppression of an amber termination codon. *Proc. Natl. Acad. Sci. USA* **82:**1618–1622.

Frameshifting

Hwang, C. B. C., B. Horsburgh, E. Pelosi, S. Roberts, P. Digard, and D. M. Coen. 1994. A net −1 frameshift permits synthesis of thymidine kinase from a drug-resistant herpes simplex virus mutant. *Proc. Natl. Acad. Sci. USA* **91:**5461–5465.

Jacks, T., H. D. Madhani, F. R. Masiarz, and H. E. Varmus. 1988. Signals for ribosomal frameshifting in the Rous sarcoma virus gag-pol region. *Cell* **55:**447–458.

Kim, Y. G., L. Su, S. Maas, A. O'Neill, and A. Rich. 1999. Specific mutations in a viral RNA pseudoknot drastically change ribosomal frameshifting efficiency. *Proc. Natl. Acad. Sci. USA* **96:**14234–14239.

Léger, M., D. Dulude, S. V. Steinberg, and L. Brakier-Gingras. 2007. The three transfer RNAs occupying the A, P and E sites on the ribosome are involved in viral programmed −1 ribosomal frameshift. *Nucleic Acids Res.* **35:**5581–5592.

Muldoon-Jacobs, K. L., and J. D. Dinman. 2006. Specific effects of ribosome-tethered molecular chaperones on programmed −1 ribosomal frameshifting. *Eukaryot. Cell* **5:**762–770.

Namy, O., S. J. Moran, D. I. Stuart, R. J. Gilbert, and I. Brierley. 2006. A mechanical explanation of RNA pseudoknot function in programmed ribosomal frameshifting. *Nature* **441:**244–247.

eIF2α kinases

Berlanga, J. J., I. Ventoso, H. P. Harding, J. Deng, D. Ron, N. Sonenberg, L. Carrasco, and C. de Haro. 2006. Antiviral effect of the mammalian translation initiation factor 2α kinase GCN2 against RNA viruses. *EMBO J.* **25:**1730–1740.

Goodman, A. G., J. A. Smith, S. Balachandran, O. Perwitasari, S. C. Proll, M. J. Thomas, M. J. Korth, G. N. Barber, L. A. Schiff, and M. G. Katze. 2007. The cellular protein P58IPK regulates influenza virus mRNA translation and replication through a Pkr-mediated mechanism. *J. Virol.* **81:**2221–2230.

He, B., M. Gross, and B. Roizman. 1997. The γ(1)34.5 protein of herpes simplex virus 1 complexes with protein phosphatase 1α to dephosphorylate the alpha subunit of the eukaryotic translation initiation factor 2 and preclude the shutoff of protein synthesis by double-stranded RNA-activated protein kinase. *Proc. Natl. Acad. Sci. USA* **94:**843–848.

Kitajewski, J., R. J. Schneider, B. Safer, S. M. Munemitsu, C. E. Samuel, B. Thimmappaya, and T. Shenk. 1986. Adenovirus VAI RNA antagonizes the antiviral action of the interferon-induced eIF2 alpha kinase. *Cell* **45:**195–200.

Leib, D. A., M. A. Machalek, B. R. Williams, R. H. Silverman, and H. W. Virgin. 2000. Specific phenotypic restoration of an attenuated virus by knockout of a host resistance gene. *Proc. Natl. Acad. Sci. USA* **97:**6097–6101.

Li, S., G. A. Peters, K. Ding, X. Zhang, J. Qin, and G. C. Sen. 2006. Molecular basis for Pkr activation by PACT or dsRNA. *Proc. Natl. Acad. Sci. USA* **103:**10005–10010.

McKenna, S. A., D. A. Lindhout, T. Shimoike, C. E. Aitken, and J. D. Puglisi. 2007. Viral dsRNA inhibitors prevent self-association and autophosphorylation of Pkr. *J. Mol. Biol.* **372:**103–113.

Control of eIF4F Activity

Cuesta, R., Q. Xi, and R. J. Schneider. 2000. Adenovirus-specific translation by displacement of kinase Mnk1 from cap-initiation complex eIF4F. *EMBO J.* **19:**3465–3474.

Feigenblum, D., and R. J. Schneider. 1996. Cap-binding protein (eukaryotic initiation factor 4E) and 4E-inactivating protein BP-1 independently regulate cap-dependent translation. *Mol. Cell. Biol.* **16:**5450–5457.

Gingras, A.-C., and N. Sonenberg. 1997. Adenovirus infection inactivates the translational inhibitors 4E-bp1 and 4E-bp2. *Virology* **237:**182–186.

Gingras, A.-C., Y. Svitkin, G. J. Belsham, A. Pause, and N. Sonenberg. 1996. Activation of the translational suppressor 4E-bp1 following infection with encephalomyocarditis virus and poliovirus. *Proc. Natl. Acad. Sci. USA* **93:**5578–5583.

Gradi, A., Y. V. Svitkin, H. Imataka, and N. Sonenberg. 1998. Proteolysis of human eukaryotic translation initiation factor eIF4GII, but not eIF4GI, coincides with the shutoff of host protein synthesis after poliovirus infection. *Proc. Natl. Acad. Sci. USA* **95:**11089–11094.

Haghighat, A., Y. Svitkin, I. Novoa, E. Kuechler, T. Skern, and N. Sonenberg. 1996. The eIF4G-eIF4E complex is the target for direct cleavage by the rhinovirus 2A protease. *J. Virol.* **70:**8444–8450.

Mathonnet, G., M. R. Fabian, Y. V. Svitkin, A. Parsyan, L. Huck, T. Murata, S. Biffo, W. C. Merrick, E. Darzynkiewicz, R. S. Pillai, W. Filipowicz, T. F. Duchaine, and N. Sonenberg. 2007. MicroRNA inhibition of translation initiation in vitro by targeting the cap-binding complex eIF4F. *Science* **317:**1764–1767.

Ohlmann, T., M. Rau, V. M. Pain, and S. J. Morley. 1996. The C-terminal domain of eukaryotic protein synthesis initiation factor eIF4G is sufficient to support cap-independent translation in the absence of eIF4E. *EMBO J.* **15:**1371–1382.

Walsh, D., C. Perez, J. Notary, and I. Mohr. 2005. Regulation of the translation initiation factor eIF4F by multiple mechanisms in human cytomegalovirus-infected cells. *J. Virol.* **79:**8057–8064.

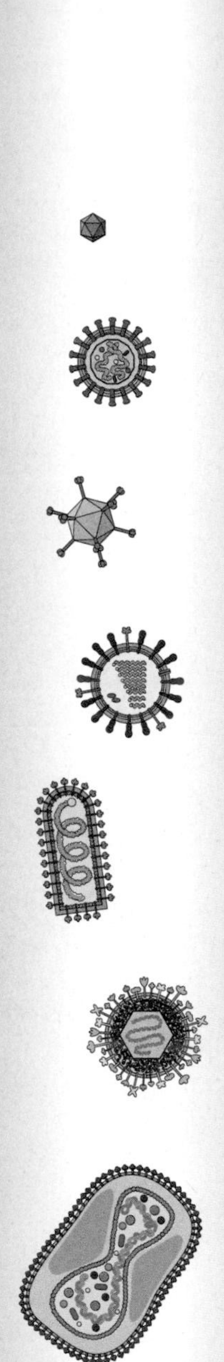

12

Intracellular Trafficking

Anatomy is destiny.
SIGMUND FREUD

Introduction

Successful viral reproduction requires the intracellular assembly of progeny virions from their protein, nucleic acid, and, in many cases, membrane components. In preceding chapters, we have considered molecular mechanisms that ensure the synthesis of the macromolecules from which virions are constructed in the host cell. Because of the structural and functional compartmentalization of eukaryotic cells, virion components are generally produced at multiple intracellular locations, and must be brought together for assembly. Intracellular trafficking, or sorting, of viral nucleic acids, proteins, and glycoproteins to the appropriate sites is therefore an essential prelude to the assembly of all animal viruses.

From our point of view, animal cells are very small, with typical diameters of 10 to 30 μm. However, in the microscopic world inhabited by viruses, an animal cell is large: the distances over which virion components must be transported within a cell are roughly equivalent to up to a mile on the macroscopic, human scale. The properties of the intracellular milieu indicate that viral particles, genomes, or subassemblies could not reach the appropriate intracellular destinations during entry or egress within reasonable periods simply by diffusion (Boxes 5.2 and 12.1). Their movement therefore requires transport systems and a considerable expenditure of energy, supplied by the host cell. The cellular highways most commonly used for movement of viral components for assembly are those formed by microtubules (as is also true during entry). These filaments are polarized and highly organized within the cell, with (−) ends at the centrosome (near the nucleus) and (+) ends at the cell periphery. They are traveled by cellular (−) end- and (+) end-directed motor proteins that convert the chemical energy of ATP into kinetic energy, and carry cargo.

The intracellular trafficking of viral macromolecules must be appropriately directed, so that individual virion components are delivered to the correct assembly site. Virion assembly can occur at any one of several intracellular addresses, depending on whether the particles are enveloped or naked,

BOX 12.1

DISCUSSION

Getting from point A to point B in heavy traffic

Major problems in cell biology are directional movement and coordination of such movements in space and time. Concentrations of high-molecular-weight reactants and products are rarely controlled by diffusion, as they are *in vitro*. Indeed, the inside of a cell is so tightly packed with organelles and cytoskeletal structures (panel A in the figure) that it is simply inappropriate to think of the contents of the cytoplasm, the nucleus, or organelle lumens as a "gel" or a "suspension."

Directional movement in cells is achieved by two general processes (panel B). Short-distance movement across membranes, or in and out of capsids, is measured in angstroms to nanometers and is accomplished primarily via protein channels.

- Common channels are transporters, translocons, pores, and portals.
- Movement generally requires energy.
- Diffusion in the classical sense contributes little to the process.

Long-distance movement of proteins, viral particles or their components, and organelles inside cells is measured in micrometers to meters. It

- invariably requires energy
- is mediated by molecular motors moving on cytoskeletal tracks; myosin motors move cargo on actin fibers, while dynein and kinesin motors move cargo on microtubules.

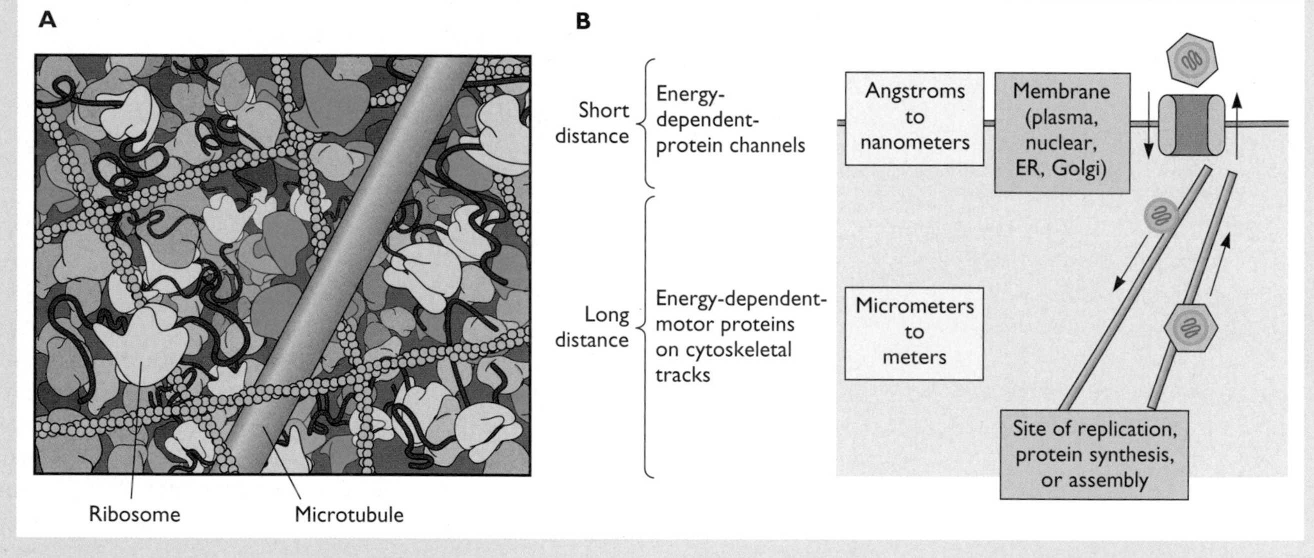

and on the site of and mechanism of genome replication (Table 12.1). All viral envelopes are derived from one of the host cell's membranes, which are modified by insertion of viral proteins. Many virus particles assemble at the plasma membrane, but some envelopes are derived from membranes of internal compartments. Assembly of enveloped viruses therefore requires delivery of some viral proteins to the appropriate membrane, as well as transport of other virion proteins and the nucleic acid genome to the modified membrane. Other common assembly sites are the cell nucleus and within the cytoplasm, where all virion components are also made. These strategies impose less complex trafficking problems than does assembly of enveloped viruses at membrane sites, but additional mechanisms may be required for egress of progeny virions from the cell. In some cases, genome-containing nucleocapsids are formed in infected cell nuclei, but virion assembly is completed at a cellular membrane. Such spatial and temporal separation of assembly reactions depends on appropriate coordination among multiple transport processes.

The need for movement of proteins and nucleic acids from one cellular compartment to another, or for insertion of proteins into specific membranes, is not unique to viruses. The majority of cellular proteins are made by translation of messenger RNAs (mRNAs) by cytoplasmic polyribosomes and must then be transported to their sites of operation. Similarly, most cellular RNA species are exported from the nucleus, in which transcription takes place. Eukaryotic cells are therefore constantly engaged in transport of macromolecules among their compartments via intracellular trafficking systems. The cellular systems that sort macromolecules to each of many possible intracellular sites are just as indispensable for viral replication

Table 12.1 Intracellular trafficking requirements for virus assembly

Assembly site(s)	Virus(es)	Trafficking requirements
Within the nucleus	Adenovirus, papovavirus	Transport of structural proteins from cytoplasm to nucleus
Within the cytoplasm	Picornavirus	Transport of structural proteins to specialized vesicles in which genome replication and assembly take place
At the plasma membrane	Alphavirus, retrovirus, rhabdovirus	Transport of viral glycoproteins to the plasma membrane; transport of other virion proteins made in the cytoplasm to the plasma membrane; transport of viral RNA genomes from nuclear or cytoplasmic sites of synthesis to the plasma membrane
At an internal cellular membrane	Bunyavirus, coronavirus, poxvirus	Transport and sorting of viral glycoproteins to the appropriate internal membrane; transport of other virion proteins and genomes to the internal membrane
Within the nucleus and at a cellular membrane	Herpesvirus, orthomyxovirus	Transport of structural proteins to the nucleus for assembly of the nucleocapsid; transport of the nucleocapsid from the nucleus to the membrane site of assembly; transport of internal virion proteins from cytoplasmic sites of synthesis to the membrane assembly site; transport and sorting of viral glycoproteins to the membrane assembly site

as the cellular biosynthetic machineries responsible for transcription, DNA synthesis, or translation. Indeed, the advances in our understanding of cellular trafficking mechanisms can be traced to initial studies with viral membrane or nuclear proteins. In the following sections, the cellular transport pathways required during viral reproduction are described in the context of the site at which virion assembly takes place.

Assembly within the Nucleus

Assembly of the majority of viruses with DNA genomes, including adenoviruses, papillomaviruses and polyomaviruses, takes place within infected cell nuclei, the site of viral DNA synthesis. All structural proteins of these nonenveloped viruses are imported into the nucleus following synthesis in the infected cell cytoplasm (Fig. 12.1), allowing complete assembly within this organelle (Table 12.1). In contrast, assembly of the more complex herpesviruses, which contain a DNA-containing nucleocapsid assembled within the nucleus, is completed at extranuclear sites (Table 12.1), as is that of some enveloped RNA viruses, such as orthomyxoviruses. In these cases, a subset of virion proteins must be imported into the nucleus.

As far as we know, all viral structural proteins that enter the nucleus do so via the normal cellular pathway of nuclear protein import. As discussed in Chapter 5, these same pathways are responsible for import of both viral genomes (or nucleoproteins) and viral nonstructural proteins that function in the nucleus early in the infectious cycle. Proteins destined for the nucleus are so labeled by the presence of nuclear localization signals (see Fig. 5.21). These signals are recognized by components of the cellular nuclear import machinery for subsequent transport into the nucleus.

Import of Viral Proteins for Assembly

Viral proteins that are known to be localized to the nucleus generally contain specific sequences that target them for import (see Fig. 5.21). The nuclear localization signals of viral proteins cannot be distinguished from those of cellular proteins. Indeed, many of the efforts to identify the cellular receptors that initiate nuclear import employed the well-characterized nuclear localization signal of simian virus 40 large T antigen. It is therefore thought that viral proteins that enter the nucleus contain typical nuclear localization signals. Nevertheless, such signals have been verified for only a small fraction of the many viral proteins that are imported into the nucleus. Furthermore, there are a large number of cellular, nuclear localization signal receptors, some of which are expressed in tissue-specific fashion. These receptors may recognize different types of viral (and cellular) nuclear localization signals.

The cellular components that mediate import of viral proteins into the nucleus are present at finite concentrations. A typical mammalian cell contains on the order of 3,000 to 4,000 nuclear pore complexes, each with a very high translocation capacity (a mass flow of up to 80 MDa/s with 10^3 translocation events/s). However, nuclear import also depends on the limited supply of soluble transport proteins. As large quantities of viral structural proteins must enter the nucleus prior to assembly, there is potential for competition among viral and cellular proteins for access to receptors, the nuclear pore complex, or transport proteins. Such competition is minimized in cells infected by the larger DNA viruses, such as adenoviruses and herpesviruses: by the time structural proteins are made during the late phase of infection, cellular protein synthesis is inhibited severely. The proteins of viruses that do not induce inhibition of cellular protein synthesis, such as those of the

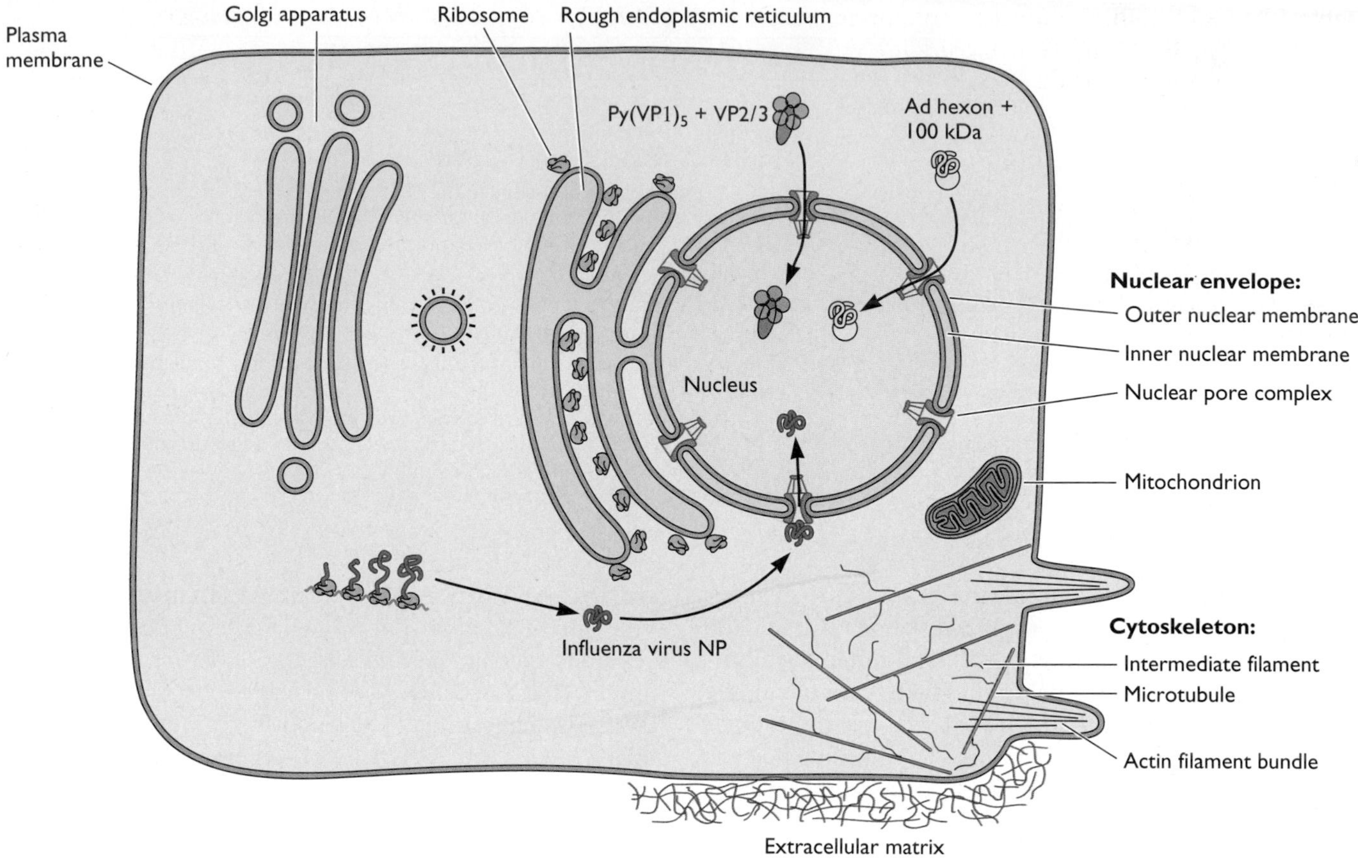

Figure 12.1 Localization of viral proteins to the nucleus. The nucleus and major membrane-bound compartments of the cytoplasm, as well as components of the cytoskeleton, are illustrated schematically and not to scale. Viral proteins destined for the nucleus are synthesized by cytoplasmic polyribosomes, as illustrated for the influenza virus NP protein. They engage with the cytoplasmic face of the nuclear pore complex and are translocated into the nucleus by the protein import machinery of the host cell. Some viral structural proteins enter the nucleus as preassembled structural units (polyomaviral [Py] VP1 pentamers associated with one molecule of either VP2 or VP3) or in association with a viral chaperone (adenoviral [Ad] hexon monomers bound to the L4 100-kDa protein).

polyomaviruses, must enter the nucleus despite continual transport of cellular proteins. Whether import of viral proteins is favored in such circumstances, for example, by the presence of particularly effective nuclear localization signals, is not known.

Many viral structural proteins that enter infected cell nuclei form multimeric capsid components. In some cases, structural units of the virion are assembled in the cytoplasm prior to import into the nucleus. Pentamers of the major capsid protein (VP1) of simian virus 40 and polyomavirus specifically bind a monomer of either VP2 or VP3, the minor virion proteins, which share C-terminal sequences (Appendix, Fig. 15B). Such heteromeric assemblies are the substrates for import into the nucleus. Indeed, efficient nuclear localization of polyomavirus VP2 and VP3 proteins occurs only in cells in which VP1 is also made. Assembly of the heteromeric complex facilitates import of the minor structural proteins, even though each contains a nuclear localization signal. The increased density of these signals may allow more effective competition for essential components of the import pathway, or the nuclear localization signals may be more accessible in the complex.

Despite such potential advantages as increased efficiency of import of viral proteins and transport of the structural proteins in the appropriate stoichiometry, import of preassembled capsid components is not universal. For example, adenoviral hexons, which are trimers of viral protein II, are found only in the nucleus of the infected cell. Association of newly synthesized hexon monomers with a late, nonstructural protein, the L4 100-kDa protein, is essential for both the entry of hexon subunits into the nucleus and their assembly into trimers. The monomeric hexon-L4 protein complex must be the import substrate. The viral L4 100-kDa protein might supply the nuclear localization signal for hexon import, or promote nuclear retention, by facilitating assembly of hexon trimers within the nucleus.

Assembly at the Plasma Membrane

Assembly of enveloped viruses frequently takes place at the plasma membrane of infected cells (Table 12.1). Before such virions can form, viral integral membrane proteins must be transported to this cellular membrane (Fig. 12.2). The first stages of the pathway by which viral and cellular proteins are delivered to the plasma membrane were identified more than 30 years ago, and the process is now quite well understood. Viruses with envelopes derived from the plasma membrane also contain internal proteins, which may be membrane associated, and, of course, nucleic acid genomes. These internal components must also be sorted to appropriate plasma membrane sites for assembly (Fig. 12.2).

Transport of Viral Membrane Proteins to the Plasma Membrane

Viral membrane proteins reach their destinations by the highly conserved, cellular **secretory pathway**. Many of the steps in the pathway have been studied by using viral membrane glycoproteins, such as the vesicular stomatitis virus G and influenza virus hemagglutinin (HA) proteins. These viral proteins offer several experimental advantages: they are frequently synthesized in large quantities, their synthesis is initiated in a controlled fashion following infection, and their transport can be studied readily by genetic, biochemical, and imaging methods.

Entry into the first staging post of the secretory pathway, the endoplasmic reticulum (ER), is accompanied by membrane insertion of integral membrane proteins. Viral envelope proteins generally span the cellular membrane into which they are inserted only once, and therefore contain a single transmembrane domain. In viral proteins, transmembrane segments (described in Chapter 5) usually

Figure 12.2 Localization of viral proteins to the plasma membrane. Viral envelope glycoproteins (red) are cotranslationally translocated into the ER lumen and folded and assembled within that compartment. They travel via transport vesicles to and through the Golgi apparatus and from the Golgi apparatus to the plasma membrane. The internal proteins of the particle (purple) and the genome (green) are also directed to plasma membrane sites of assembly.

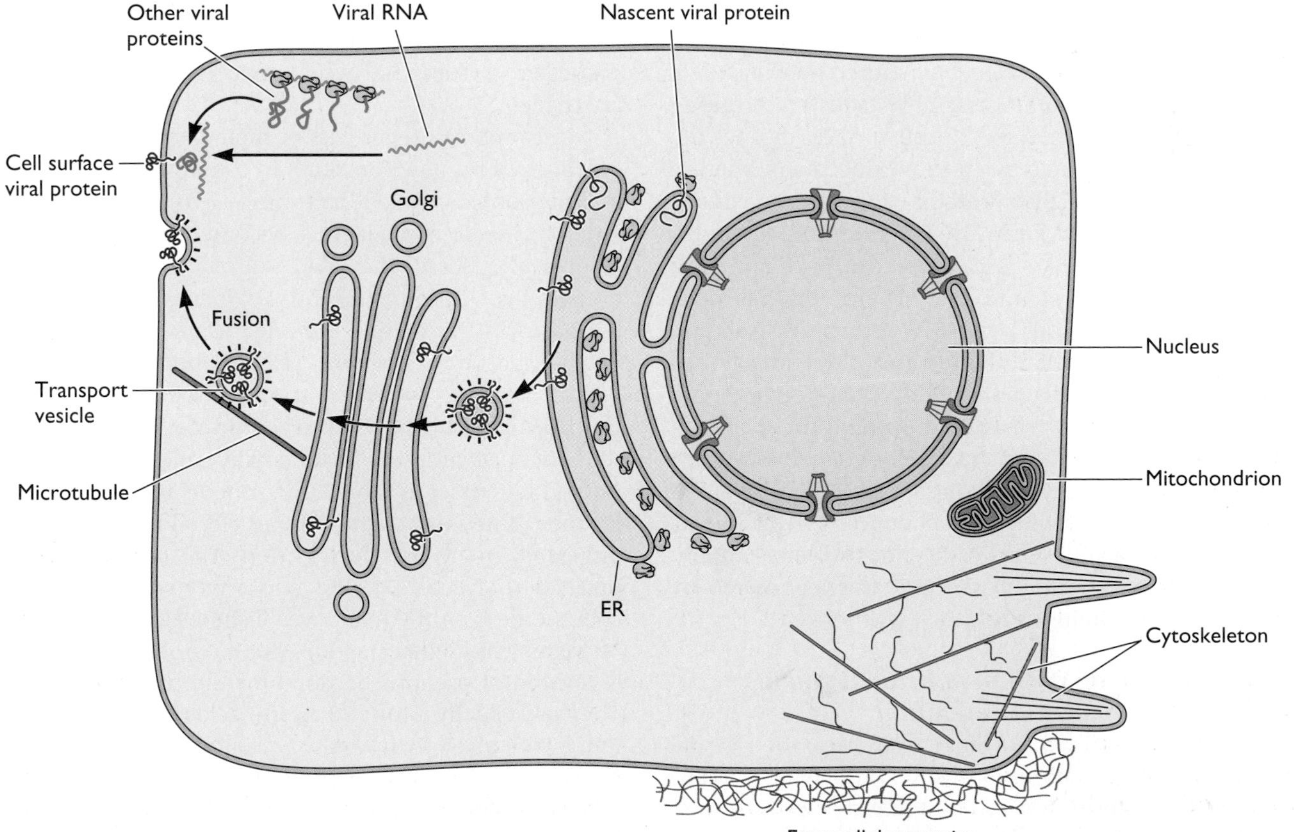

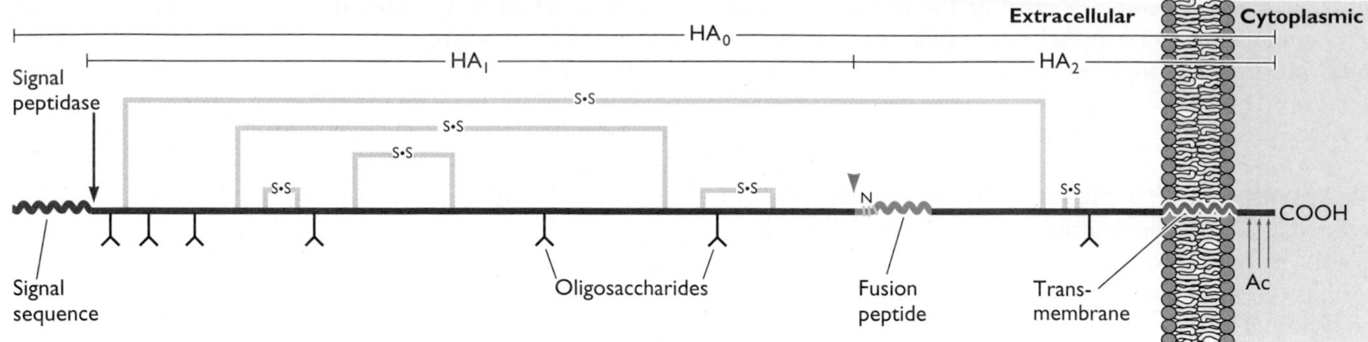

Figure 12.3 Primary sequence features and covalent modifications of the influenza virus HA protein. The primary sequence of the HA0 protein is depicted by the red line in the center, with the orange arrowhead indicating the site of the proteolytic cleavages that produce the HA1 and HA2 subunits. The fusion peptide, the N-terminal signal sequence which is removed by signal peptidase in the ER, and the C-terminal transmembrane domain are hydrophobic. Disulfide bonds, one of which maintains covalent linkage between the HA1 and HA2 proteins following HA0 cleavage, are indicated, as are sites of N-linked glycosylation (oligosaccharides) and palmitoylation (Ac).

separate large extracellular from smaller cytoplasmic domains (Fig. 12.3). The former include the binding sites for cellular receptors, crucial for initiation of the infectious cycle, whereas the latter are important in virus assembly. As discussed in Chapter 4, viral membrane proteins are usually oligomers. Most interactions among the subunits of viral membrane proteins are noncovalent, but some examples of association via covalent interchain disulfide bonds are known. Oligomer assembly takes place during transit from the cytoplasm to the cell surface, as does the proteolytic processing necessary to produce some mature (functional) envelope glycoproteins from the precursors that enter the secretory pathway. For example, the influenza virus HA0 precursor is cleaved into HA1 and HA2 subunits (Fig. 12.3). Viral (and cellular) proteins that travel the secretory pathway also possess distinctive structural features, including disulfide bonds and covalently linked oligosaccharide chains (Fig. 12.3). As illustrated in Figure 12.4 for the influenza A virus HA0 protein, these characteristic covalent modifications (as well as oligomerization) take place as proteins travel through a series of specialized compartments that provide the chemical environments and enzymatic machinery necessary for their maturation. The first such compartment, the ER, is encountered by viral membrane proteins as they are synthesized.

Translocation of Viral Membrane Proteins into the Endoplasmic Reticulum

All proteins destined for insertion into the plasma membrane, or the membranes of such intracellular organelles as the Golgi apparatus, enter the ER as they are translated (Fig. 12.2). This membranous structure appears as a basketwork of tubules and sacs extending throughout the cytoplasm (Fig. 12.5A). The ER membrane demarcates a geometrically convoluted but continuous internal space, the **ER lumen**, from the remainder of the cytoplasm. The ER lumen has a chemically distinctive environment, topologically equivalent to the outside of the cell. Proteins that enter the ER during their synthesis are therefore sequestered from the cytoplasmic environment as they are made.

Most proteins, regardless of their final ports of call, are synthesized on polyribosomes in the cytoplasm. However, polyribosomes engaged in synthesis of proteins that will enter the secretory pathway become associated with the cytoplasmic face of the ER membrane soon after translation begins. Areas of the ER to which polyribosomes are bound form the **rough ER** (Fig. 12.5B). The association of polyribosomes with the ER membrane is directed by a short sequence in the nascent protein termed the **signal peptide**. It is now taken for granted that the primary sequences of proteins include "zip codes" specifying the cellular addresses at which the proteins must reside to fulfill their functions, such as the nuclear localization signals discussed in the previous section. The signal peptides of proteins that enter the ER lumen were the first such zip codes to be identified, and established this paradigm some 30 years ago. Signal peptides are commonly found at the N termini of proteins destined for the secretory pathway. They are usually about 20 amino acids in length and contain a core of 15 hydrophobic residues. Signal peptides are often transient structures that are removed enzymatically during protein translocation into the ER by a protease located in the lumen, signal peptidase.

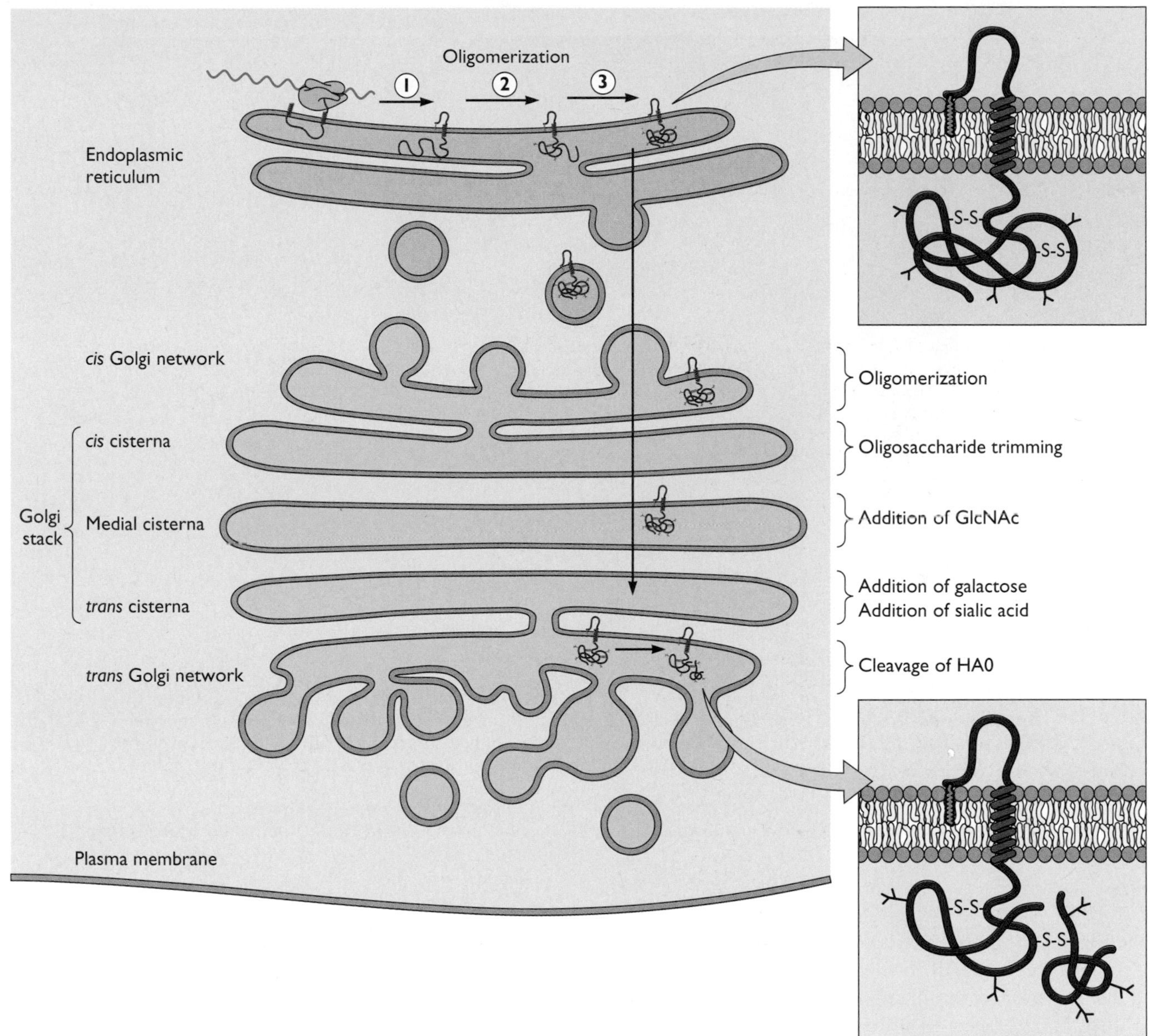

Figure 12.4 Maturation of influenza virus HA0 protein during transit along the secretory pathway. The modifications that occur during transit of the influenza virus HA0 protein through the various compartments of the secretory pathway are illustrated. In the ER, these are translocation and signal peptide cleavage (1), disulfide bond formation, and addition of N-linked core oligosaccharides (2), as the protein folds (3). The cytoplasmic domain acquires palmitate (orange) (see insert, top) while the protein travels to the plasma membrane, but it has not been established when this modification takes place. For simplicity, the protein is depicted as a monomer, although oligomerization also takes place in the ER lumen. Note that the protein domain initially introduced into the ER lumen, in this case the N-terminal portion of the protein (type I orientation), corresponds to the extracellular domain of the cell surface protein.

Translation of a protein that will enter the ER begins in the normal fashion and continues until the signal peptide emerges from the ribosome (Fig. 12.6). This signal then directs binding of the translation machinery to the ER membrane by means of two components: the signal peptide is recognized by the **signal recognition particle (SRP)**, which in turn binds to the cytoplasmic domain of an integral ER membrane protein termed the **SRP receptor**. Binding of the signal recognition particle to the ribosome temporarily halts translation, to allow the

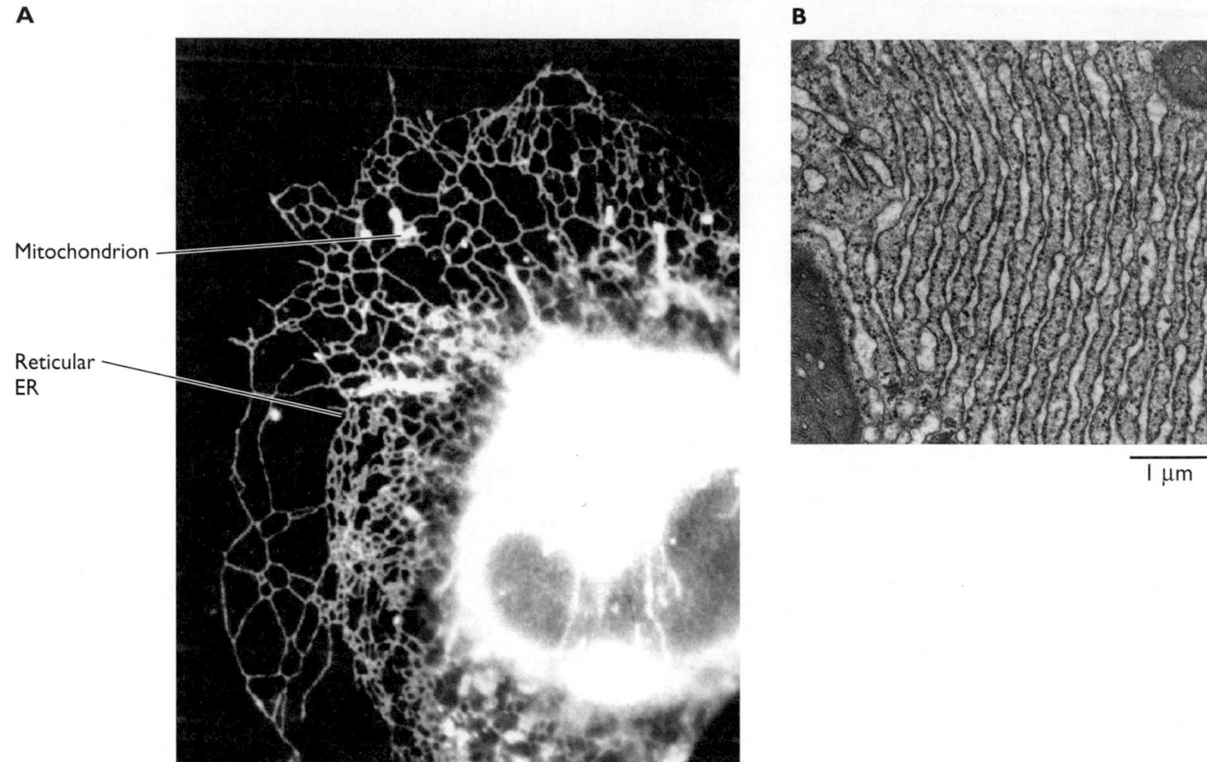

Figure 12.5 The endoplasmic reticulum. (A) The ER of a mammalian cell in culture. The reticular ER, which extends throughout the cytoplasm, was visualized by fluorescence microscopy of fixed African green monkey kidney epithelial cells stained with the lipophilic fluorescent dye 3,3′-dihexyloxacarbocyanine iodide. This dye also stains mitochondria. The ER membrane accounts for over half of the total membrane of a typical animal cell and possesses a characteristic lipid composition. Courtesy of M. Terasaki, University of Connecticut Health Center. **(B)** Electron micrograph of the rough ER in rat hepatocytes. Note the many ribosomes associated with the cytoplasmic surface of the membrane. From R. A. Rodewald, Biological Photo Service.

stalled translation complex to bind to the ER membrane via the SRP receptor. Both the signal recognition particle and its receptor contain subunits that bind guanosine triphosphate (GTP), and efficient targeting requires the presence of this nucleotide. Following the initial docking of the complex at the membrane, the ribosome becomes tightly bound to the membrane, and engaged with a protein translocation channel, which forms a gated, aqueous pore through the ER membrane. This interaction is coordinated with release of the signal recognition particle, association of the signal peptide with the translocation channel, and resumption of translation. Because the ribosome remains bound to the membrane upon release of signal recognition particle, continued translation facilitates movement of the growing polypeptide chain through the membrane. Such coupling of translation and translocation ensures that the protein crosses the membrane as an unfolded chain that can be accommodated within the translocation channel. Movement of the growing polypeptide through the membrane channel is facilitated by binding of the lumenal chaperone Grp78 (Bip) to the nascent protein.

When the protein entering the ER is destined for secretion from the cell, translocation continues until the entire polypeptide chain enters the lumen. During translocation, the signal peptide is proteolytically removed by signal peptidase, releasing the soluble protein into the ER. In contrast, translocation of integral membrane proteins, such as viral envelope proteins, halts when a hydrophobic **stop transfer signal** is encountered in the nascent protein. This sequence may be the signal peptide itself or a second, internal hydrophobic sequence. The number, location, and orientation of such sequences within a protein determine the topology with which it is organized in the ER membrane (Box 12.2). The programming of insertion of proteins into the ER membrane by signals built into their primary sequences ensures that every molecule of a particular protein adopts the identical topology in the

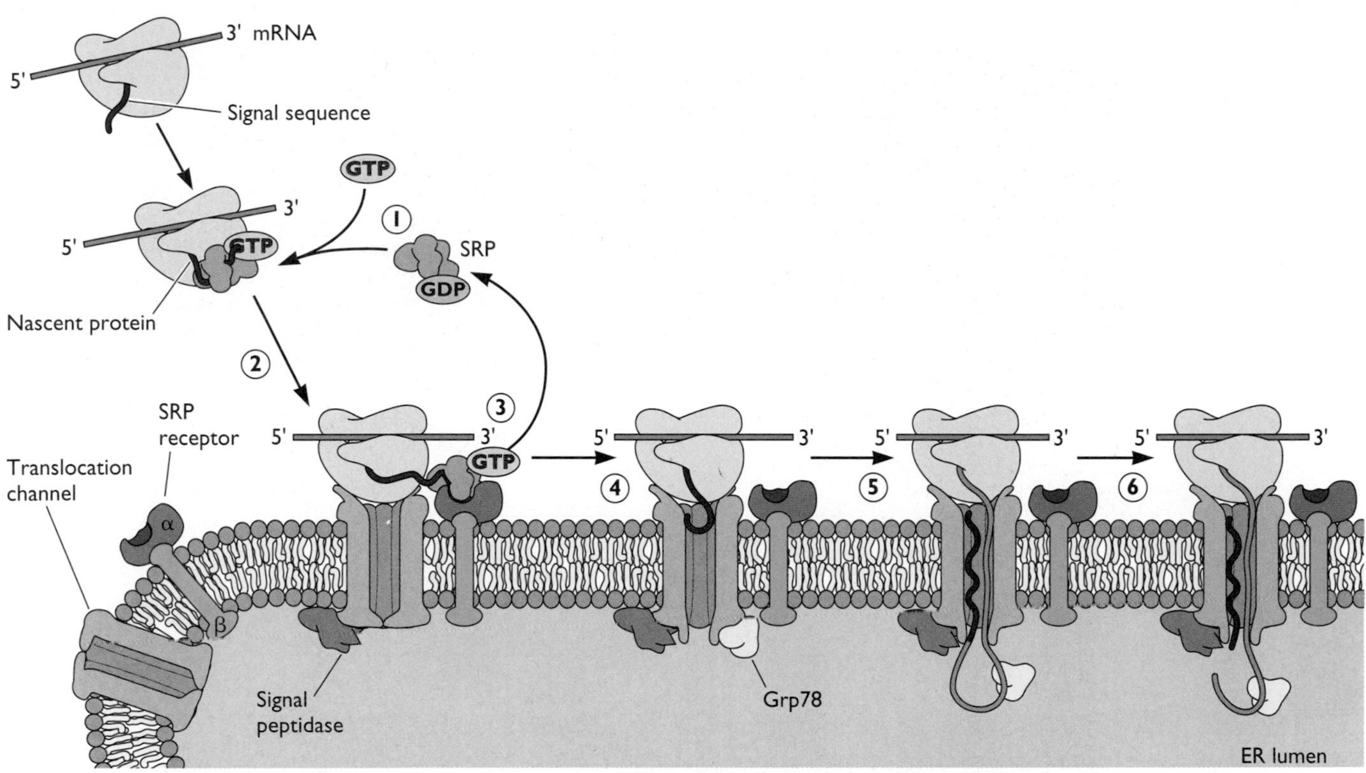

Figure 12.6 Targeting of a nascent protein to the ER membrane. Translation of an mRNA encoding a protein that will enter the ER lumen proceeds until the signal peptide (purple) emerges from the ribosome. The signal recognition particle (SRP), which contains a small RNA molecule and several proteins, binds to both the signal peptide and the ribosome to halt or pause translation, upon binding of GTP to one of the protein subunits (step 1). The nascent polypeptide-SRP-ribosome complex then binds to the SRP receptor in the ER membrane (step 2). This interaction triggers hydrolysis of GTP bound to both the SRP and its receptor, release of SRP (step 3), and close association of the ribosome with, and binding of the hydrophobic signal peptide to, the heterotrimeric protein translocation channel (step 4). These interactions trigger opening of the cytosolic end of the channel. The lumenal end of the translocation channel is also initially closed. Translation is then resumed, and the seal maintained at the lumenal end of the channel early in translocation is reversed by binding of the chaperone Grp78. The growing polypeptide chain is transferred through the membrane as its translation continues (step 5). In some cases, signal peptidase removes the signal peptide cotranslationally (step 6). A lateral gate in the channel opens within the membrane for transfer of the transmembrane domain(s) of translocated proteins into the ER membrane.

membrane. As this topology is maintained during the several membrane budding and fusion reactions by which proteins reach the cell surface, the way in which a protein is inserted into the ER membrane determines its orientation in the plasma membrane.

Reactions within the ER

The folding and initial posttranslational modification of proteins that enter the secretory pathway take place within the ER. The lumen contains many enzymes that catalyze chemical modifications, such as disulfide bond formation and glycosylation, or that promote folding and oligomerization (Table 12.2).

Glycosylation. Viral envelope proteins, like cellular proteins that travel the secretory pathway, are generally modified by the addition of oligosaccharides to either asparagine (N-linked glycosylation) or serine or threonine (O-linked glycosylation). The presence of oligosaccharides on a protein can be detected, as changes in the protein's electrophoretic mobility, following exposure of cells to inhibitors of glycosylation, or of the protein to enzymes that cleave the oligosaccharide (Fig. 12.7A). The first steps in N-linked glycosylation take place as a polypeptide chain emerges into the ER lumen. Oligosaccharides rich in mannose preassembled on a lipid carrier are added to asparagine residues by an oligosaccharyltransferase (Fig. 12.7B).

BOX 12.2

BACKGROUND

Establishing the topology of integral membrane proteins that contain a single transmembrane domain

Many proteins with a single transmembrane domain, including the influenza virus HA0 protein, contain not only an N-terminal signal peptide, but also a second hydrophobic sequence. The signal peptide initiates cotranslational translocation of the nascent polypeptide through the protein translocation channel, as shown in more detail in Fig. 12.6. As shown in the figure, the resident ER enzyme signal peptidase (tan) cleaves off the signal peptide, forming the N terminus of the mature protein. Translocation continues until the second hydrophobic sequence is encountered. This sequence acts as a stop transfer signal and halts translocation, with lateral discharge of the protein into the ER membrane. This mechanism results in type I orientation (i.e., N terminus within the lumen), as in the influenza virus HA0 protein (Fig. 12.4). The position of the stop transfer sequence within the polypeptide determines the sizes of its lumenal (extracellular) and cytoplasmic domains. In proteins that contain multiple hydrophobic sequences of this kind, the first one encountered by the signal recognition particle starts the transfer, the second is a stop-transfer sequence, and so on.

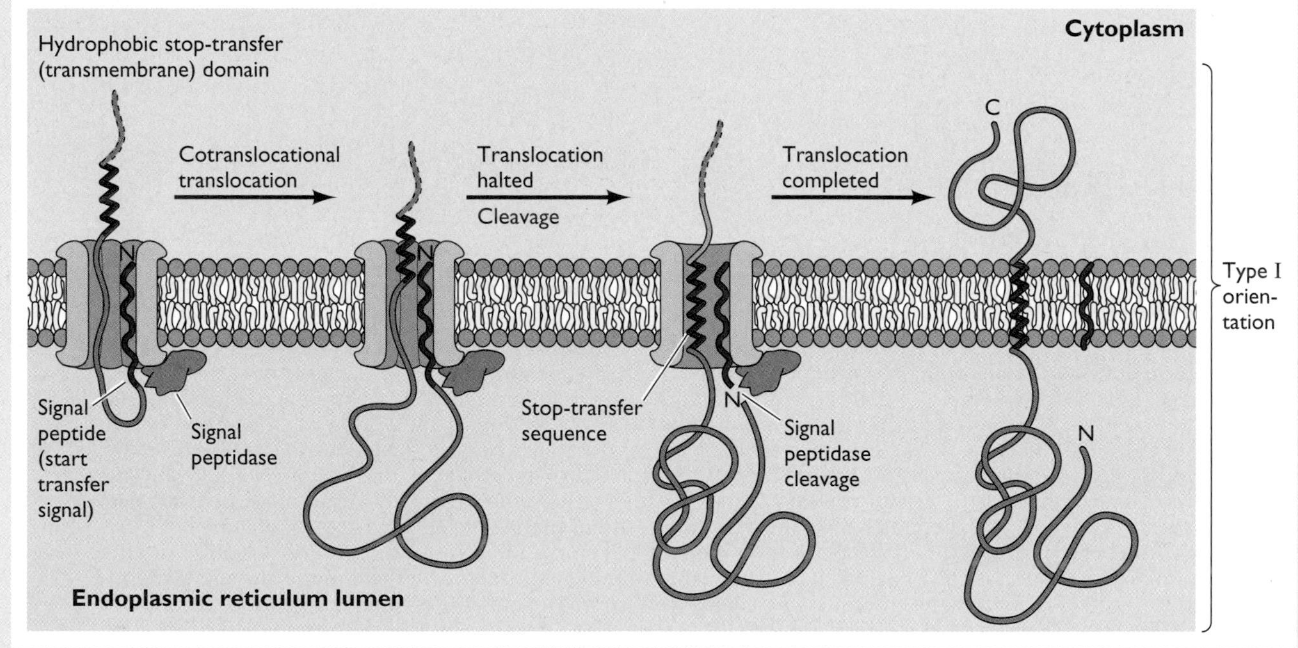

Subsequently, several sugar residues are trimmed from N-linked core oligosaccharides in preparation for additional modifications that take place as the protein travels from the ER to the plasma membrane.

Sites of N-linked glycosylation are characterized by the sequence NXS/T (where X is any amino acid except proline), but not every potential glycosylation site is modified. Even a single specific site within a protein is not necessarily modified with 100% efficiency. Each glycoprotein population therefore comprises a heterogeneous mixture of **glycoforms**, varying in whether a particular site is glycosylated, as well as in the composition and structure of the oligosaccharide present at each site. As many viral and cellular proteins contain a large number of potential N-linked glycosylation sites, particular proteins can exist in an extremely large number of glycoforms. This property complicates investigation of the physiological functions of oligosaccharide chains present on glycoproteins. Nevertheless, this modification has been assigned a wide variety of roles.

As essential components of receptors and ligands, oligosaccharides participate in many molecular recognition reactions. These processes include binding of certain hormones to their cell surface receptors, interactions of cells with one another, binding of virions, such as influenza A virus and herpesviruses, to their host cells, and later steps in virus entry. Some sugar units serve as signals, targeting proteins to specific locations, in particular to lysosomes.

Table 12.2 Some ER enzymes and chaperones that participate in quality control

Protein	Properties and functions
Calnexin (Cnx)	Integral membrane lectin that binds to almost all glycoproteins in the ER; promotes proper folding; retains incompletely folded glycoproteins in the ER
Erp57	Soluble thioloxidoreductase; as cochaperone, binds to and functions with calnexin/calreticulin
Glucosidases I and II	Catalyze removal of terminal glucose residues from high-mannose, N-linked oligosaccharides; glucosidase II promotes release of substrates from Cnx
Grp78 (Bip)	Lumenal protein; binds to exposed hydrophobic surfaces on incompletely folded proteins; retains such proteins in the ER
Protein disulfide isomerase	Catalyzes formation and isomerization of disulfide bonds; retains incompletely folded or assembled proteins in the ER; appears to operate with Grp78/Bip
UDP-glucose-glycoprotein transferase	Lumenal ER enzyme that specifically recognizes unfolded proteins; adds glucose residues to their N-linked oligosaccharides; promotes recognition by Cnx and thus ER retention

Glycosylation has also been suggested to fulfill more general functions, such as protecting proteins (and virus particles) that circulate in body fluids from degradation and host immune defenses. Many proteins contain such a large number of glycosylation sites that carbohydrate can contribute more than 50% of the mass of the mature protein, for example, the poliovirus receptor and the respiratory syncytial virus G protein. The hydrophilic oligosaccharides are present on the surface of such proteins, where they can form a sugar "shell," masking much of the proteins' surfaces, and epitopes recognized by antiviral antibodies (Box 12.3).

Studies of viral glycoproteins have established that glycosylation is absolutely required for the proper folding of some of these proteins. For example, elimination (by mutagenesis) of all sites at which the vesicular stomatitis virus G or influenza virus HA proteins are glycosylated blocks the folding of these proteins and their exit from the ER (see "Protein folding and quality control" below). Before a protein folds, its hydrophobic amino acids, which are ultimately buried in the interior, are exposed. Such exposed hydrophobic patches on individual unfolded polypeptide chains tend to interact with one another nonspecifically, leading to aggregation. The hydrophilic oligosaccharide chains are thought to counter this tendency.

Figure 12.7 Detection and structure of N-linked oligosaccharides. (A) Detection of N-linked oligosaccharides using inhibitors or specific enzymes. Addition to cells of tunicamycin, an inhibitor of the first step in synthesis of the oligosaccharide precursor, prevents N-linked glycosylation, so that the mobility of glycoproteins is altered (left). *In vitro* treatment of glycoproteins with enzymes that cleave within the oligosaccharide, such as endoglycosidase H (Endo H) or *N*-glycanase, can also alter glycoprotein mobility (right). Glycosylation of a protein can also be assayed by incorporation of radioactively labeled monosaccharides. **(B)** The branched, mannose-rich oligosaccharide added via an N-glycosidic bond to asparagine residues of proteins is shown at the left. This common precursor is transferred to N-linked glycosylation sites as proteins are translocated into the ER, from the lipid carrier dolichol phosphate upon which the oligosaccharide is assembled. While within the ER, three glucose residues and one mannose residue are trimmed from the core oligosaccharide.

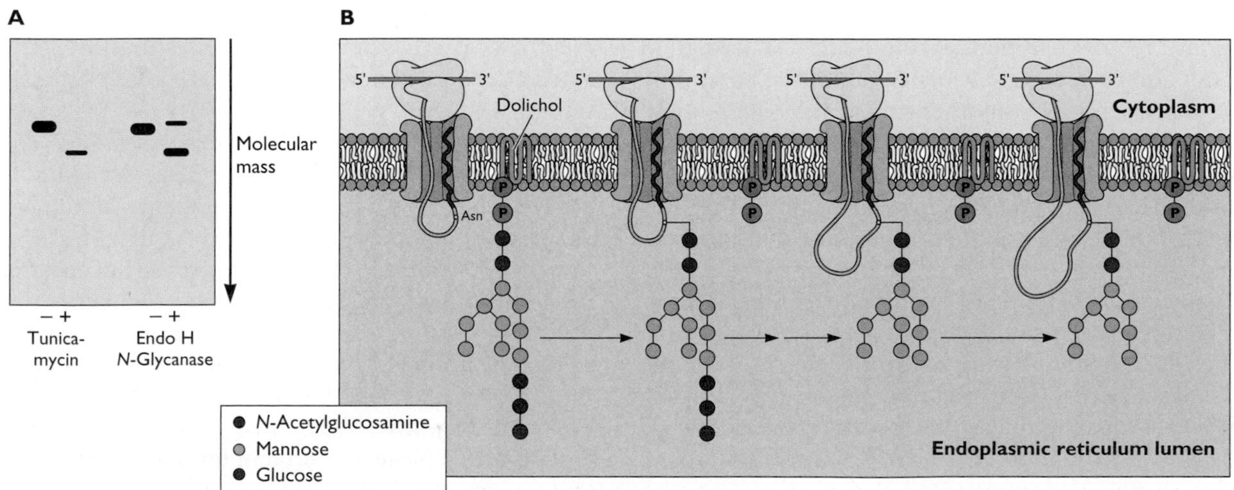

BOX 12.3

DISCUSSION

The evolving sugar "shield" of human immunodeficiency virus type 1

Mutational studies have implicated N-linked glycosylation at specific sites in the envelope proteins of several viruses in protection against host neutralizing antibodies. The Env protein of human immunodeficiency virus type 1 (HIV-1) provides a dramatic example of this phenomenon.

The SU (gp120) subunit of the HIV-1 Env protein carries a large number of oligosaccharide chains, which form a dense shell that masks much of the protein's surface (see the figure). These oligosaccharides govern several properties of HIV-1. For example, the tropism of the virus for CCr5 or CXCr4 coreceptors correlates with specific patterns of glycosylation in the variable loops of the SU subunit. However, a major function of such modification is to block access of host anti-HIV-1 antibodies to SU protein epitopes: high-resolution structural studies of the SU protein core have confirmed that N-linked oligosaccharides cover much of the protein's surface. Furthermore, the sugar chains are highly ordered, forming the outer surface of the Env spike. As predicted from this arrangement, N-linked glycosylation at specific sites blocks binding of monoclonal antibodies that recognize nearby sequences in the protein.

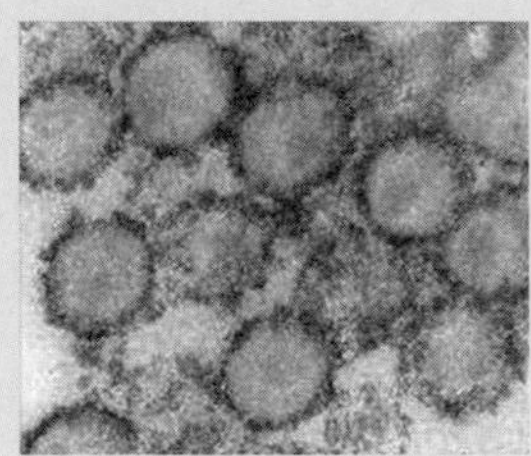

Electron micrograph of HIV-1 particles, showing carbohydrates stained with ruthenium red (dark). Courtesy of Edwin P. Ewing, Jr., Centers for Disease Control and Prevention (CDC), Atlanta, GA (CDC Public Health Image Library).

Several observations have led to the hypothesis that HIV-1 carries an evolving carbohydrate "shield" that enhances immune evasion. For example, the number of N-linked oligosaccharides added to SU tends to increase during the course of an HIV-1 infection, and the sites of N-linked glycosylation also change.

Chen, B., E. M. Vogan, H. Gong, J. J. Skehel, D. C. Wiley, and S. C. Harrison. 2005. Structure of an unliganded simian immunodeficiency virus gp120 core. *Nature* **433:**834–841.

Scanlan, C. N., J. Offer, N. Zitman, and R. A. Derek. 2007. Exploiting the defensive sugars of HIV-1 for drug and vaccine design. *Nature* **446:**1038–1045.

Disulfide bond formation. A second chemical modification that generally is restricted to proteins entering the secretory pathway, and essential for the correct folding of many, is the formation of intramolecular disulfide bonds between pairs of cysteine residues (Fig. 12.3). These bonds can make important contributions to the stability of a folded protein. However, they rarely form in the reducing environment of the cytoplasm. The more oxidizing ER lumen provides an appropriate chemical environment for disulfide bond formation. This compartment contains high concentrations of protein disulfide isomerase and other thioloxidoreductases, enzymes that catalyze the formation, reshuffling, or even breakage of disulfide bonds under appropriate redox conditions (Table 12.2). As formation of the full and correct complement of disulfide bonds in a protein is often the rate-limiting step in its folding, these enzymes are important catalysts of this process.

The cellular enzymes that promote formation of disulfide bonds are present in the ER lumen. Consequently, this modification typically is limited to proteins that enter, or protein domains exposed to, this compartment. Remarkably, however, several viral membrane proteins present in mature virions of the poxvirus vaccinia virus have stable disulfide bonds in their **cytoplasmic** domains: the genome of this virus encodes all the enzymes necessary to catalyze the formation of disulfide bonds in the cytoplasm (Box 12.4).

Protein folding and quality control. A number of other cellular proteins assist the folding of the extracellular domains of viral membrane glycoproteins as they enter the lumen of the ER (Table 12.2). In contrast to the enzymes described above, these proteins do not alter the covalent structures of proteins. Rather, their primary function is to facilitate folding, largely by preventing improper associations among unfolded, or incompletely folded, polypeptide chains, such as the nonspecific, hydrophobic interactions described above. Such **molecular chaperones** play essential roles in the folding of individual polypeptides and in the oligomerization of proteins. The ER chaperones, which include Grp78 and calnexin, are also crucial for **quality control** processes that determine the sorting and fate of newly synthesized proteins translocated into the ER.

Grp78 is a member of the Hsp70 family of stress response proteins. It associates transiently with incompletely folded viral and cellular proteins. Binding of this chaperone, generally at multiple sites in a single nascent protein molecule, is thought to protect against misfolding and aggregation by sequestering sequences prone to nonspecific interaction, such as hydrophobic patches. The release of unfolded proteins from Grp78 is controlled by the hydrolysis of ATP bound to the chaperone. Multiple cycles of association with, and dissociation from, Grp78 probably take place as a protein folds. Once the sequences

BOX 12.4

EXPERIMENTS

A viral thiol oxidoreductase system that operates in the cytoplasm

The intracellular mature virion of the poxvirus vaccinia virus is the first of two infectious particles assembled in infected cells. This particle carries an envelope containing viral membrane proteins surrounding an internal core in which the DNA genome is packaged. In 1999, it was reported that some viral core proteins synthesized in the cytosol, as well as the cytoplasmic domains of some membrane proteins, contain stable disulfide bonds. This property explained the previously reported sensitivity of vaccinia virions to disruption by reducing agents. In addition, it raised the intriguing question of how disulfide bonds could be introduced into viral proteins or domains that are never exposed to the cellular site of thiol oxidation, the ER lumen. Within a few years, viral genes were shown to encode all the components necessary to catalyze formation of disulfide bonds.

This viral thioloxidoreductase system comprise three components, and the final substrates, which include the mature virions proteins L1R and F9L. The order in which the three viral enzymes act, summarized in the figure, was deduced from a variety of experimental observations. For example,

- when expression of the E10R gene was repressed in infected cells, only reduced A2.5L was detected
- conversely, inhibition of synthesis of G4L did not prevent oxidation of A2.5L
- E10R and A2.5 were shown form stable, disulfide-linked heterodimers when synthesized in the absence of other viral proteins
- covalent interactions between A2.5L and G4L or between G4L and substrates, were also identified, but only when thiol-disulfide exchange was prevented
- formation of the A2.5 and G4L heterodimer required synthesis of the E10R protein, as predicted by the pathway shown in the figure

The proteins that comprise the viral thiol oxidoreductase pathway are conserved among all poxvirus, and all three vaccinia virus proteins are essential for assembly of intracellular mature virions.

Lockner, J. K., and G. Griffiths. 1999. An unconventional role for cytoplasmic disulfide bonds in vaccinia virus proteins. *J. Cell Biol.* **144:**267–279.

Senkevich, T. G., C. L. White, E. V. Koonin, and B. Moss. 2002. Complete pathway for disulfide bond formation encoded by poxviruses. *Proc. Natl. Acad. Sci. USA* **99:**6667–6672.

The coupled oxidation-reduction (thiol-exchange) reactions among the proteins of the vaccinia virus disulfide bond formation are depicted in order (left to right). The transfer of electrons to oxygen via flavin adenine dinucleotide FAD (left) is based on homology of E10R with members of a family of flavin adenine dinucleotide (FAD)-containing sulfhydryl oxidases, and has not been demonstrated experimentally. Adapted from T. G. Senkevich et al., *Proc. Natl. Acad. Sci. USA* **99:**6667–6672, 2002, with permission.

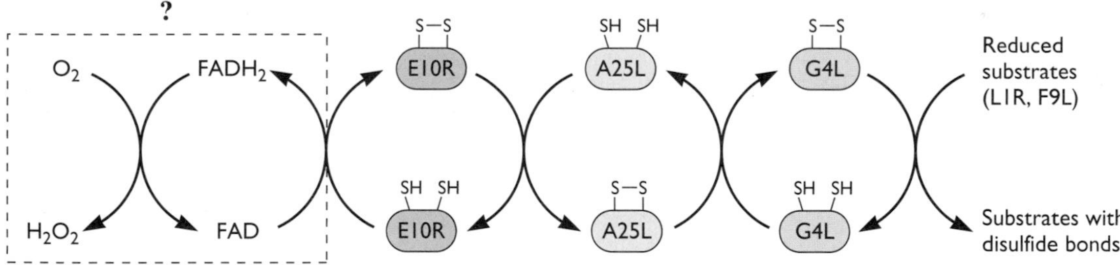

to which the chaperone binds are buried in the interior of the protein, such interactions cease. For example, molecules of vesicular stomatitis virus G, Semliki Forest virus E1, or influenza virus HA0 proteins that have acquired the full complement of correct disulfide bonds can no longer associate with Grp78.

Calnexin is an integral membrane protein of the ER that also binds transiently to immature proteins, as does its soluble relative, calreticulin, present in the ER lumen. In contrast to Grp78, which recognizes protein sequences directly, calnexin and calrecticulin distinguish newly synthesized glycoproteins by binding to immature oligosaccharide chains. For example, the vesicular stomatitis virus G and influenza virus HA0 proteins bind to calnexin only when their oligosaccharide chains retain terminal glucose residues (Fig. 12.7B). In fact, formation of the mature oligosaccharide is intimately coupled with folding of glycoproteins and their retention within the ER (Fig. 12.8). Proteins with monoglucosylated sugars are recognized by calnexin (or calreticulin), which forms a complex with an ER thiol-oxidoreductase, but they are released upon removal of the glucose by the enzyme glucosidase II (Table 12.2). An enzyme that re-adds terminal glucose (uridine diphosphate [UDP]-glucose-glycoprotein transferase [Table 12.3]) appears to be the "sensor" of the folded state of the glycoprotein: it recognizes incompletely folded

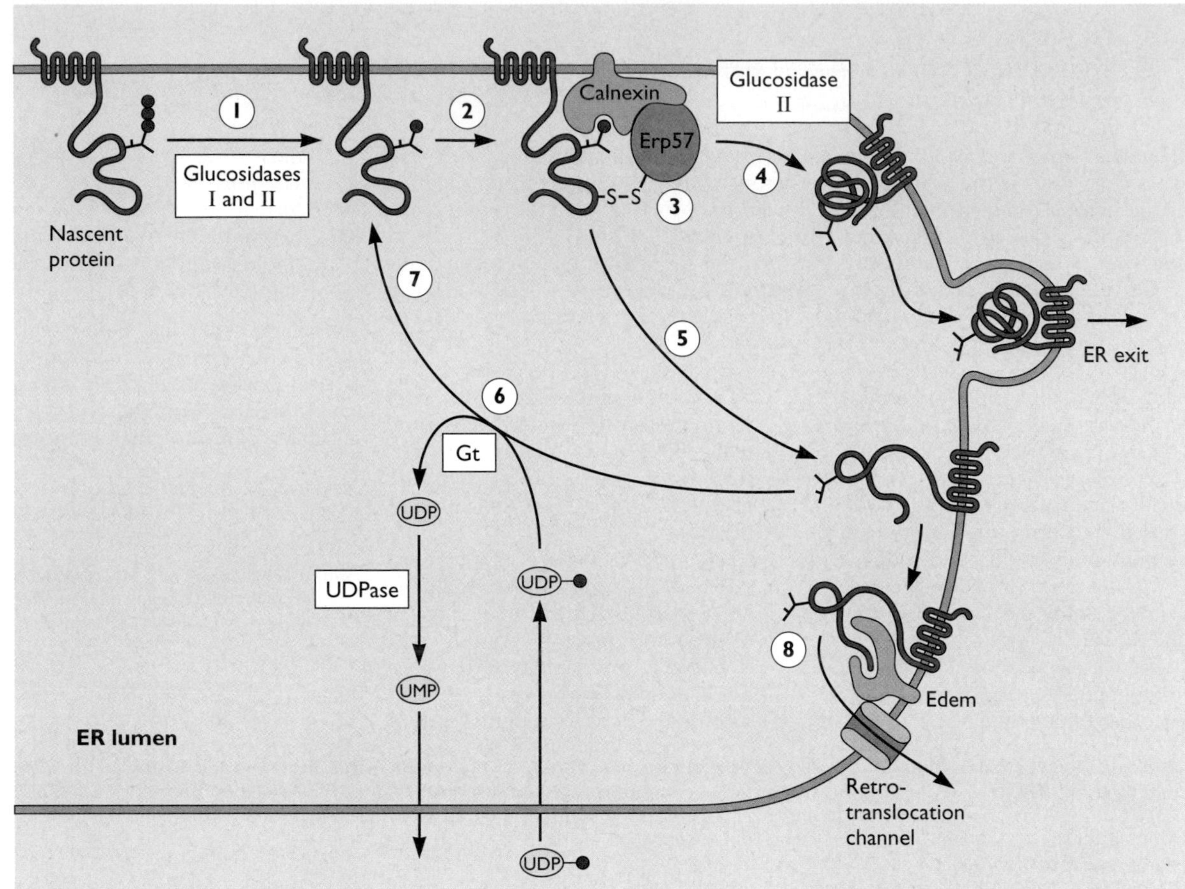

Figure 12.8 Integration of folding and glycosylation in the ER. The model illustrates the coordination of ER retention by calnexin with glycosylation and folding of a newly synthesized glycoprotein (red) containing an N-linked oligosaccharide, depicted as in Fig. 12.3. Trimming of terminal glucose residues by glucosidases I and II (1) yields a monoglucosylated chain, to which calnexin (or calreticulin) binds (2). Because the thioloxidoreductase Erp57 forms a complex with calnexin, the newly synthesized protein is brought into contact with Erp57, with which transient intermolecular disulfide bonds (-S-S-) can form. When the remaining glucose is removed by glucosidase II (3), the protein dissociates from the calnexin-Erp57 complex. If it has attained its native structure, the protein can leave the ER (4). However, if it is incompletely (or incorrectly) folded (5), the protein is specifically recognized by UDP-glucose glycoprotein transferase (Gt) (6), which re-adds terminal glucose residues to the oligosaccharide (7) and therefore allows rebinding to calnexin. Cycles of binding and modification are repeated until the protein is either folded properly or targeted for degradation (8), via binding to Edem and retrotranslocation to the cytosol. Adapted from L. Ellgaard et al., *Science* **286**:1882, 1999, with permission.

proteins by virtue of exposed hydrophobic amino acids and specifically reglucosylates the proteins, controlling cycles of substrate binding and release from calnexin or calreticulin (Fig. 12.8). This specificity ensures that only fully folded proteins can escape these chaperones and travel along the secretory pathway.

The ER contains numerous other folding catalysts and chaperones, many specific for particular proteins. Relatively little is known about the parameters that determine the chaperone(s) to which a newly synthesized protein binds, and the order in which chaperones operate. However, studies of specific viral glycoproteins in living cells indicate that the positions of oligosaccharides within the protein chain are one important determinant of chaperone selection (Box 12.5).

Proteins that are misfolded or not modified correctly cannot escape covalent or noncovalent associations with ER enzymes or molecular chaperones. For example, a temperature-sensitive vesicular stomatitis virus G protein remains bound to calnexin, and hence to the ER membrane, at a restrictive temperature. Consequently, egress of nonfunctional proteins from the ER to subsequent compartments

Table 12.3 Viral envelope glycoprotein precursors processed by signal peptidase or furin-family proteases

Virus family	Precursor glycoprotein	Membrane-associated cleavage products
Signal peptidase		
Alphavirus	Envelope polyprotein precursor	E1, pE2
Bunyavirus	Translation product of M mRNA	G1, G2
Furin family proteases		
Alphavirus	pE2	E2
Flavivirus	PrM	M
Hepadnavirus	preC	C antigen[a,b]
Herpesvirus[b,c]	pre-gB	gB
Orthomyxovirus[d]	HA0	HA_1, HA_2
Paramyxovirus	F0	F1, F2
Retrovirus	Env	TM, SU

[a]This cleavage product is largely secreted into the extracellular medium, but is also associated with plasma membrane of infected cells.

[b]Cleavage is not necessary for production of infectious virus particles in cells in tissue culture.

[c]Some alphaherpesviruses (e.g., varicella-zoster virus), and all known betaherperviruses.

[d]Virulent strains of avian influenza A virus.

in the secretory pathway is prevented. These interactions also target misfolded (and misassembled) proteins for degradation. The first step in this process is translocation of such proteins from the ER into the cytoplasm **(retrotranslocation)**. The mechanisms responsible for specific recognition of misfolded or incompletely folded proteins, and induction of transport from ER to cytosol, are not yet fully understood. However, the stress-inducible ER membrane protein Edem (ER degradation-enhancing 1,2-mannosidase-like protein) has been implicated in diversion of misfolded glycoproteins for retrotranslocation. Furthermore, studies of herpesviral proteins that induce translocation of major histocompatibility complex (MHC) class I molecules from the ER to the cytosol have identified components of a membrane channel that appear to be dedicated to removal of misfolded proteins from the ER (Box 12.6). Once the proteins reenter the cytoplasm, oligosaccharides are removed and multiple copies of the protein ubiquitin (polyubiquitin) are added. Such modification targets the proteins for degradation by the proteasome. The quality control functions of resident ER chaperones therefore ensure that nonfunctional proteins are cleared from the secretory pathway at an early step.

Oligomerization. Most viral membrane proteins are oligomers that must assemble as their constituent protein chains are folded and covalently modified. Such assembly generally begins in the ER, as the surfaces that mediate interactions among protein subunits adopt the correct conformation. For many proteins, these reactions are completed within the ER. For example, influenza virus HA0 protein monomers are restricted to the ER lumen, whereas trimers are found in this and all subsequent compartments of the secretory pathway. Indeed, several viral and some cellular heteromeric membrane proteins must oligomerize to exit the ER, for folding of one subunit depends on association with the other(s). This requirement has been characterized in some detail for the glycoproteins of alphaviruses, such as Sindbis virus. These proteins are translated from the 26S subgenomic mRNA (Appendix, Fig. 25B) as a precursor that contains the E1 and E2 protein sequences, but are liberated from the precursor by signal peptidase cleavage as they enter the ER. The E1 protein associates with Grp78 and folds via three disulfide-bonded intermediates. Folding beyond the second intermediate, and release from chaperone association, requires dimerization of the E1 with the E2 protein. Furthermore, the E2 protein misfolds when synthesized in the absence of E1 protein. The association of these two envelope proteins within the ER therefore is essential for the productive folding, and exit, of both. Similarly, the herpes simplex virus type 1 envelope glycoproteins gH and gL must interact with one another for the transport of either from the ER, and in the absence of gL, gH cannot fold correctly.

Assembly of other viral membrane proteins is completed following exit from the ER: disulfide-linked dimers of the hepatitis B virus surface antigen form higher-order complexes in the next compartment in the pathway, and oligomers of the human immunodeficiency virus type 1 Env protein can be detected only in the Golgi apparatus.

BOX 12.5 EXPERIMENTS
Selectivity of chaperones for viral glycoproteins entering the ER

The parameters that determine which of the many ER chaperones operate on individual proteins are not fully understood. However, analysis of the folding and chaperone association of viral glycoproteins suggests that the position of glycosylation sites can determine chaperone selection. This conclusion is based on the following observations.

The Semliki Forest virus E1 and p62 (pre-E2) glycoproteins enter the ER cotranslationally, and are cleaved from a precursor by signal peptidase. However, E1, which folds via three intermediates differing in their disulfide bonding, initially associates with Grp78, whereas nascent p62 molecules bind to calnexin.

This difference is not a trivial result of a lack of Grp78-binding sites in p62: when access to calnexin was blocked, this protein did bind to Grp78.

One major difference between E1 and p62 is that the latter contains glycosylation sites close to the N terminus. Addition of oligosaccharides at such sites could allow recognition of nascent p62 by calnexin and preclude association with Grp78.

This hypothesis was tested by elimination of N-linked glycosylation sites of the influenza virus HA0 protein, either close to the N terminus or in more C-terminal positions. The wild-type protein does not bind to Grp78, but inhibition of glycosylation at N-terminal sites (positions 8, 22, and 38) led to association with this chaperone.

As summarized in the figure, these observations suggest that nascent proteins that carry N-linked glycosylation sites close to the N terminus (p62, HA0) enter the calnexin folding pathway directly **(A).** Others, such as the Semliki Forest virus E1, and the vesicular stomatitis virus G, proteins associate initially with Grp78 and protein disulfide isomerase and are transferred to the calnexin pathway as they mature **(B).**

Molinari, M., and A. Helenius. 2000. Chaperone selection during glycoprotein translocation into the endoplasmic reticulum. *Science* **288:**331–333.

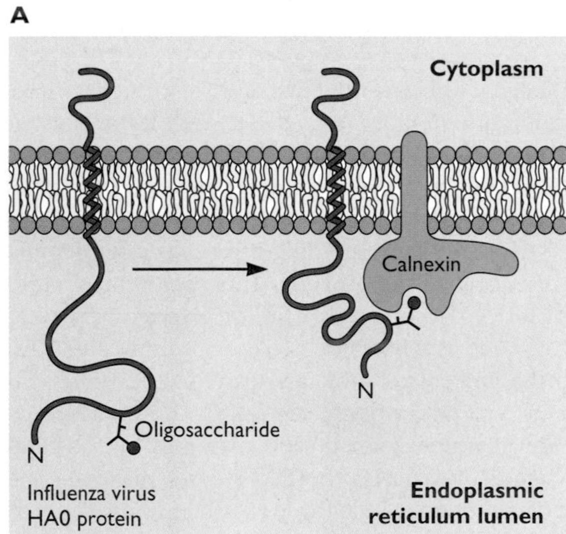

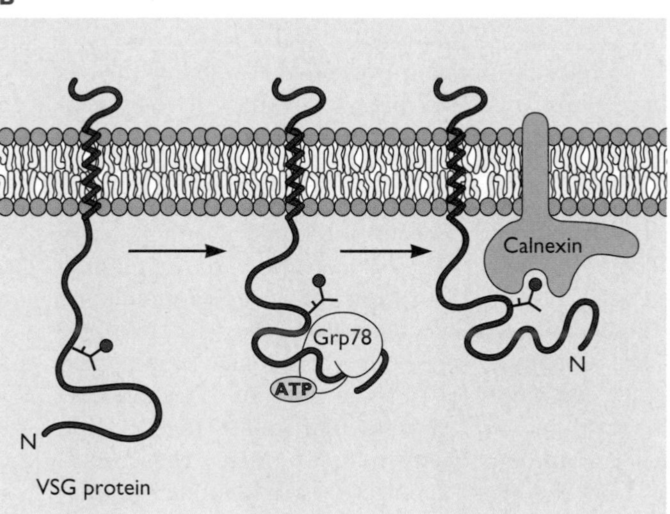

At present, we can discern no simple rules describing the relationship of oligomer assembly and transport of membrane proteins from the ER. Nevertheless, oligomerization begins, and in some cases must be completed, within the ER, where it can be facilitated by the folding catalysts and chaperones characteristic of this compartment.

Vesicular Transport to the Cell Surface

Mechanism of vesicular transport. Viral membrane proteins, like their cellular counterparts, travel to the cell surface through a series of membrane-bound compartments and vesicles. The first step in this pathway, illustrated schematically in Fig. 12.9, is transport of the folded protein from the ER to the Golgi apparatus. Within the Golgi apparatus (Fig. 12.10), proteins are sorted according to the addresses specified in their primary sequences or covalent modifications. **Transport vesicles**, and larger vesicular structures (Box 12.7), which bud from one compartment and move to the next, carry cargo proteins between compartments of the secretory pathway. Fusion of the vesicle membrane with that of the target compartment releases the cargo into the lumen of that compartment. Consequently, proteins that enter the secretory pathway upon translocation into the ER (and are correctly folded) are never again

BOX 12.6

EXPERIMENTS

How a herpesviral glycoprotein was exploited in indentification of an ER retrotranslocation channel

Two human cytomegalovirus (a betaherpesvirus) membrane glycoproteins, US2 and US11, were known to insert into the ER membrane and induce rapid transfer of MHC class I heavy chains from the ER to the cytosol (retro-translocation or dislocation). These cellular proteins then become polyubiquitmated and degraded by the cytosolic proteasome. In the case of the viral US11 protein, a glutamine residue (Glu192) in the transmembrane domain is essential for retrotranslocation of MHC class I proteins.

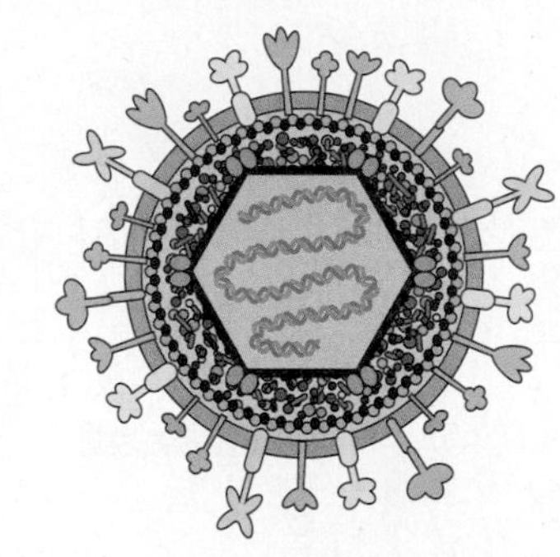

In one set of experiments, this property was exploited to purify human proteins that bound specifically to wild-type US11, but not to the viral protein carrying a Glu192 → Leu substitution. Several ER proteins bound to both US11 proteins, but one associated only with wild-type US11. This protein, identified by mass spectrometry, showed some similarity to the yeast Der1p protein that is known to participate in degradation of misfolded ER proteins, and was named derlin-1. Overproduction of a dominant-negative derivative of derlin-1 inhibited US11-mediated retrotranslocation of MHC class I proteins.

In an alternative approach, components of a canine ER retrotranslocation channel were identified by virtue of their interaction with a cytoplasmic ATPase (ATPase p97) known to be essential for degradation of both misfolded ER proteins in yeast and MHC class I molecules in US11-producing human cells. The protein complex identified in this way contained derlin-1 and a second ER membrane protein. These ER proteins were shown by immunoprecipitation to interact with both US11 and MHC class I proteins.

The US2-mediated degradation of MHC class I proteins was not inhibited by the dominant negative form of derlin that blocked degradation induced by the US11 protein. This observation indicates that US2 must function with a distinct retrotranslocation channel, perhaps one of the other two human derlin proteins or the channel through which nascent polypeptides enter the ER. The results of genetic experiments in yeast support the view that this channel is required for degradation of several ER proteins.

Lilley, B. W., and H. Ploegh. 2004. A membrane protein required for dislocation of misfolded membrane proteins from the ER. *Nature* **429:** 834–840.

Ye, Y., Y. Shibata, C. Yun, D. Ron, and T. A. Rapoport. 2004. A membrane protein complex mediates retro-translocation from the ER lumen to the cytosol. *Nature* **429:**841–847.

exposed to the cytosol of the cell. This strategy effectively sequesters proteins that might be detrimental, such as secreted or lysosomal proteases, and avoids exposure of disulfide-bonded proteins to a reducing environment.

Many soluble and membrane proteins that participate in vesicular transport have been identified and characterized by biochemical, molecular, and genetic methods. The properties of these proteins suggest that similar mechanisms control the budding and fusion of different types of transport vesicle, despite the stringent specificity requirements of vesicular transport. The general mechanism of vesicular transport is quite well understood (Fig. 12.11). Budding of transport vesicles from the membranes of compartments of the secretory pathway requires proteins that form external coats of the vesicles, such as the protein complex called CopII, which mediates ER-to-Golgi transport, and small GTPases (Fig. 12.11A). The coat proteins induce membrane curvature and vesicle budding, and are subsequently removed by various mechanisms. The vesicle then moves to the next compartment, by either passive diffusion or active transport via microtubule-associated motor proteins.

When a transport vesicle encounters its target membrane, it docks as a result of specific interactions among Snare proteins present in the vesicle and target membranes. Docking requires a member of each of the very large Rab and Arf families of small GTPases, which are important determinants of fusion specificity. A complex containing proteins necessary to prepare for membrane fusion then assembles in a regulated manner and juxtaposes the membranes that will fuse (Fig. 12.11B). One component of this complex binds to the **Snares** present in the membranes of the transport vesicle (v-Snare) and of the compartment with which it will fuse (t-Snare). The final reaction of vesicular transport, membrane fusion, requires the Snare proteins, which promote membrane fusion. A v-Snare and an appropriate t-Snare are sufficient to induce membrane fusion, but there is evidence that additional proteins may facilitate or accelerate this reaction.

The high density of intracellular protein traffic requires considerable specificity during vesicle formation and fusion. For example, vesicles that transport proteins from the ER to the first compartment of the Golgi apparatus, the *cis*-Golgi network (Fig. 12.9), must take up only the appropriate proteins when budding from the ER, and must fuse only with the membrane of the *cis*-Golgi network. Specificity of cargo loading during formation of these (and other) transport vesicles is achieved

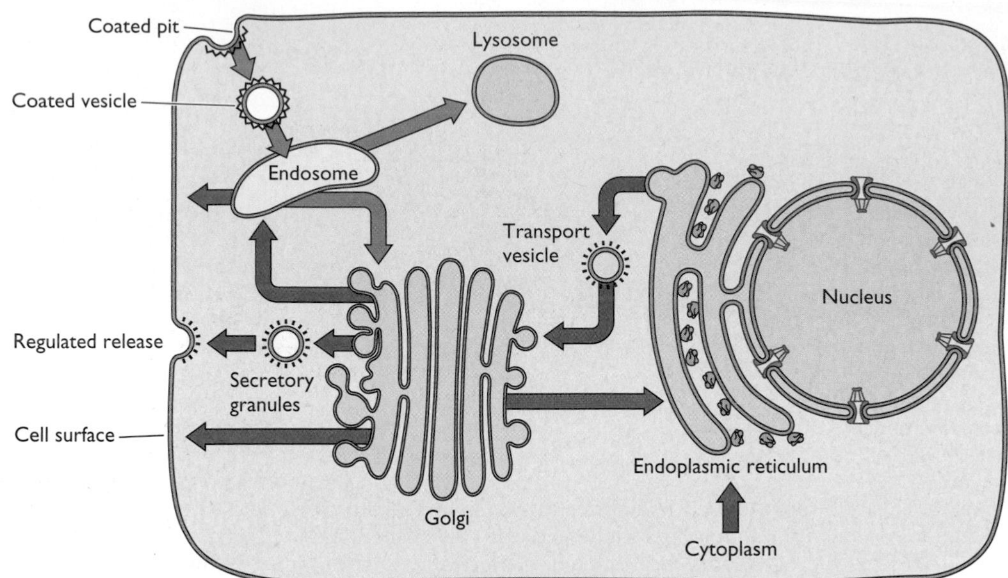

Figure 12.9 Compartments in the secretory pathway. The lumen of each membrane-bound compartment shown is topologically equivalent to the exterior of the cell. Proteins destined for secretion or for the plasma membrane travel from the ER to the cell surface via the Golgi apparatus, as indicated by the red arrows. However, proteins can be diverted from this pathway to lysosomes, or to secretory granules that carry proteins to the cell surface for regulated release. The return of proteins from the Golgi apparatus to the ER is indicated (purple arrow). The endocytic pathway (blue arrows), discussed in Chapter 5, and the secretory pathway intersect in endosomes and the Golgi apparatus.

by both direct association of cargo proteins with proteins of the Cop II vesicular coat and indirect interactions via transmembrane export receptors. Several types of protein establish the specificity of vesicular transport, that is, fusion of particular transport vesicles with the appropriate compartment, including the Snare proteins resident in the vesicle and target compartment membranes. Small GTP-binding proteins of the Rab and Arf families, each of which is associated with a specific organelle, provide, in the GTP-bound form, an "identity signal" recognized by proteins that participate in vesicle budding or fusion. Specific phosphoinositides (lipids) are also important determinants of the identity of some organelles, for example of Golgi compartments.

Reactions within the Golgi apparatus. One of the most important staging posts in the secretory pathway is the Golgi apparatus, which is composed of a series of membrane-bound compartments located near the cell nucleus. Proteins enter the Golgi apparatus from the ER via the *cis*-Golgi network, which is composed of connected tubules and sacs (Fig. 12.9 and 12.10). A similar structure, the *trans*-Golgi network, forms the exit face of this organelle. The *cis*- and *trans*-Golgi networks are separated by a variable number of cisternae termed the *cis*, medial, and *trans* compartments. Each of the compartments of the Golgi apparatus, which can comprise multiple cisternae, is the site of specific processing reactions (Fig. 12.12).

A number of viral envelope glycoproteins are also processed proteolytically by cellular enzymes resident in late Golgi compartments. Retroviral Env glycoproteins are cleaved in the *trans*-Golgi network to produce the TM (transmembrane) and SU (surface unit) subunits from the Env polyprotein precursor (Fig. 12.13A). The larger SU protein associates with TM noncovalently and in many simple retroviruses also via a disulfide bond (Fig. 12.13B). Similarly, the HA0 protein of certain avian influenza A viruses is cleaved into the HA1 and HA2 chains in the *trans*-Golgi network (Fig. 12.4). These and other viral membrane proteins (Table 12.3) are processed by members of a family of resident Golgi proteases that cleave after pairs of basic amino acids. The members of this family, which in mammalian cells includes furins found in the *trans*-Golgi network, are serine proteases related to the bacterial enzyme subtilisin. Various furin family members have been shown by genetic and molecular methods to process viral glycoproteins; their normal function is to process cellular polyproteins, such as certain hormone precursors.

These proteolytic cleavages typically are essential for formation of infectious particles, although they are not

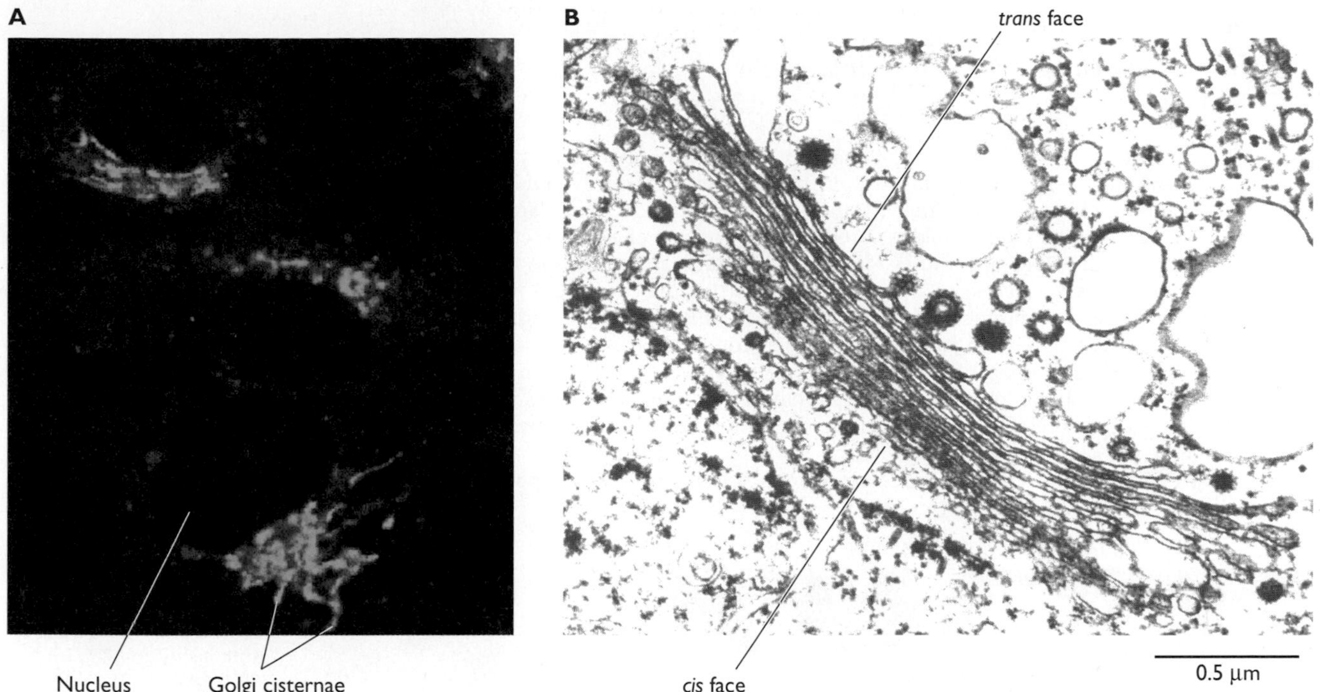

Figure 12.10 The Golgi apparatus. (A) Immunofluorescence of pig kidney epithelial cells following staining with antibody against a Golgi protein (p115). Courtesy of A. Brideau, Princeton University. **(B)** Electron micrograph of negatively stained sections of the alga *Ochromonas danica*, showing the flattened membranous sacs, termed cisternae. These are bordered on either side by more convoluted structures of connected tubules and sacs, the *cis*- and *trans*-Golgi networks. Because the cisternae lie on top of one another, they are often called Golgi stacks. The number of cisternae varies considerably from one cell type to another and under different physiological conditions. From G. T. Cole, Biological Photo Service.

necessary for assembly. For example, proteolytic processing of envelope proteins of retroviruses and alphaviruses is necessary for infectivity, probably because sequences important for fusion and entry become accessible. As discussed in more detail in Volume II, Chapter 1, virulent strains of avian influenza A virus encode HA0 proteins that can be processed by the ubiquitous furin family proteases, such that virus particles carrying fusion-active HA protein are released. It seems likely that the common dependence on furin family proteases (Table 12.3), which act on proteins relatively late in the secretory pathway, helps minimize complications that would arise if viral glycoproteins were initially synthesized with their fusion peptides in an active conformation. Furthermore, exposure to the low pH environment of *trans*-Golgi network compartments (pH ~6.0) can be a prerequisite for processing of viral envelope proteins. This requirement is exemplified by the envelope proteins of flaviviruses, such as dengue virus. The auxillary membrane protein (M) of this virus is synthesized as a precursor (prM) that forms heterodimers with the envelope (E) protein in the ER membrane. Dengue virus particles bud into the ER and travel the secretory pathway. When they reach the *trans*-Golgi network, the prM-E heterodimers undergo extensive conformational rearrangement that renders prM accessible to proteolysis by furin.

For at least one influenza A virus, the fowl plague virus, the ion channel activity of the viral M2 protein helps to maintain HA in a fusion-incompetent conformation following cleavage. This HA protein switches to the fusion-competent conformation at a pH higher than that required by HA proteins of human influenza A viruses. The M2 protein, which forms a proton channel, is found in infected cells at quite high concentrations in the membranes of secretory pathway compartments. By increasing the pH of normally acidic compartments, such as those of the *trans*-Golgi network, this protein prevents premature switching of proteolytically processed HA to the fusion-active conformation described in Chapter 5.

Although all the envelope proteins of viruses that assemble at the plasma membrane travel via the cellular secretory pathway, there is considerable variation in the rate and efficiency of their transport. A champion is the influenza virus HA protein, which folds and assembles

BOX 12.7 EXPERIMENTS
ER-to-Golgi transport in living cells

The vesicular stomatitis virus G protein made in cells infected with the mutant virus *ts*045 misfolds and is retained in the ER at high temperature (40°C). It refolds and is transported to the Golgi apparatus when the temperature is reduced to 32°C. This temperature-sensitive protein has therefore been used extensively to study transport through the secretory pathway. To allow examination of this process in living cells, the green fluorescent protein was attached to the cytoplasmic tail of the viral G protein. Control experiments established that this modification did not alter the temperature-sensitive folding or transport of the G protein. Time-lapse fluorescence microscopy of cells shifted from high to low temperature (see the figure) demonstrated that the chimeric G protein rapidly left the ER at multiple peripheral sites. The protein appeared in membranous structures, which were often larger than typical transport vesicles. These structures moved rapidly toward the Golgi in a stop-start manner, with maximal velocities of 1.4 μm/s. Such transport, but not formation of post-ER structures, was blocked when microtubules were depolymerized by treatment with nocodazole, or when the (–) end-directed microtubule motor dynein was inhibited. It was therefore concluded that vesicles and other membrane-bound structures that emerge from the ER at peripheral sites are actively transported to the Golgi complex.

Presley, J. F., N. B. Cole, T. A. Schroer, K. Hirschberg, K. J. M. Zaal, and J. Lippincott-Schwartz. 1997. ER-to-Golgi transport visualized in living cells. *Nature* **389:**81–85.

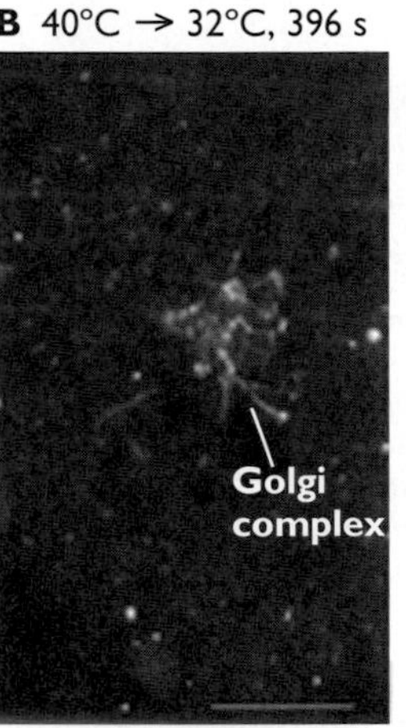

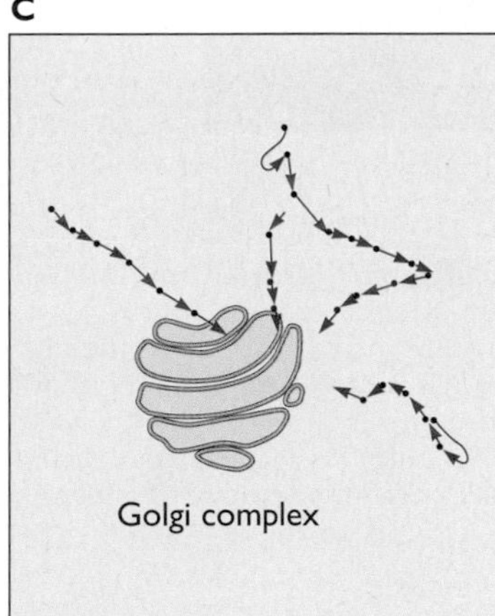

Shown are images of the vesicular stomatitis virus ts045 G protein-green fluorescent protein observed in cells 5 min after a shift to 32°C from 40°C **(A)** and 396 s after the transfer to 32°C **(B)**. Fluorescence associated with the Golgi complex was photobleached upon the shift to 32°C to allow visualization of delivery of the fluorescent protein from intermediates to the Golgi. **(C)** The path of such movement is shown schematically.

with a half time of only 7 min, with more than 90% of the newly synthesized molecules reaching the cell surface. Many other viral proteins do considerably less well. Parameters determining the rate and efficiency of transport may include the complexity of the protein and the inherent asynchrony of protein folding. With some exceptions (see "Inhibition of Transport of Particular Cellular Proteins" below), cellular proteins continue to enter and traverse the secretory pathway as enveloped viruses assemble at the plasma membrane. Competition among viral and cellular proteins, which may vary with the nature and physiological state of the host cell, is also likely to affect the transport of viral proteins to the cell surface.

We have focused our discussion of viral envelope proteins that travel the cellular secretory pathway on the well-understood maturation of their extracellular domains. However, the cytoplasmic portions of these proteins are also frequently modified. Many are **acylated** by the covalent linkage of the fatty acid palmitate to cysteine residues in their cytoplasmic domains (Table 12.4). This modification can be necessary for optimal production of progeny virions. For example, inhibition of palmitoylation of the Sindbis virus E2 glycoprotein or of the human immunodeficiency virus type 1 Env protein impairs virus assembly and budding. The bulky fatty acid chains attached to the short cytoplasmic tails may regulate envelope protein conformation or association with specific membrane domains (see "Signal Sequence-Independent Transport of Viral Proteins to the Plasma Membrane" below).

Sorting of Viral Proteins in Polarized Cells

Proteins that are not specifically targeted to an intracellular address travel from the Golgi apparatus to the plasma membrane. However, the plasma membrane is not uniform in all animal cells: differentiated cells often devote different parts of their surfaces to specialized functions, and the plasma membranes of such **polarized cells** are divided into correspondingly distinct regions. During infection by many enveloped viruses, the asymmetric surfaces of such cells are distinguished during entry, and by the sorting of virion components to a specific plasma membrane region. In this section, we describe the final steps in the transport of proteins to specialized plasma membrane regions in two types of polarized cell in which animal viruses often replicate, epithelial cells and neurons (Fig. 12.14).

Epithelial Cells

As discussed in Chapter 2, epithelial cells, which cover the external surfaces of vertebrates and line all their internal cavities (such as the respiratory and gastrointestinal tracts), are primary targets of virus infection. The cells of an epithelium are organized into close-knit sheets, by both the tight contacts they make with one another and their interactions with the underlying basal lamina, a thin layer of extracellular matrix (Fig. 2.3). Within the best-characterized epithelia, such as those that line the intestine, each cell is divided into a highly folded **apical domain** exposed to the outside world and a **basolateral domain** (Fig. 12.14). The former performs more specialized functions, whereas the latter is associated with cellular housekeeping. These two domains differ in their protein and lipid content, in part because they are separated by specialized cell-cell junctions (tight junctions) (Fig. 2.3), which prevent free diffusion and mixing of components in the outer leaflet of the lipid bilayer. However, such physical separation does not explain how the polarized distribution of plasma membrane proteins is established and maintained.

Viruses have been important tools in efforts to elucidate the molecular mechanisms responsible for the polarity of typical epithelial cells, because certain enveloped viruses bud asymmetrically. For example, in the epithelial cells of all organs studied, influenza A virus buds apically and vesicular stomatitis virus buds basolaterally. As discussed in Volume II, Chapter 1, polarized assembly and release of virus particles can facilitate virus spread within or among host organisms. The polarity of virus budding is generally the result of accumulation of envelope proteins, such as HA and G, in the apical and basolateral domains, respectively. The most common mechanism for selective localization appears to be signal-dependent sorting of proteins in the *trans*-Golgi network, for packaging into appropriately targeted transport vesicles. Signals necessary for basolateral targeting comprise short amino acid sequences located in the cytoplasmic domains of membrane proteins. Many basolateral targeting signals overlap with those that direct proteins to clathrin-coated pits, the first staging post of the endocytic pathway. Indeed, in polarized epithelial cells, certain proteins are transferred through endosomes from one membrane domain to the other, a process termed transcytosis (see Volume II, Fig. 4.18). The sorting of viral glycoproteins to basolateral membrane domains can also be governed by additional viral proteins. The two envelope proteins of measles virus (F and H) possess signals that direct them to the basolateral membrane. However, when the viral matrix protein binds to the cytoplasmic tails of F and H, these proteins are redirected from the default basolateral sorting pathway, and accumulate at the apical surface of epithelial cells.

The signals that direct proteins to the apical membrane are currently a matter of debate, but both N-glycosylation and association of proteins with specific lipids may be important. The apical membranes of epithelial cells are enriched in sphingolipids and cholesterol, which form microdomains called **lipid rafts**. Such rafts are dynamic assemblies that can incorporate specific proteins selectively, and were initially shown to mediate apical transport of glycosylphosphatidylinositol-anchored proteins. The influenza virus HA and NA proteins associate specifically with lipid rafts via their transmembrane domains, which determine apical sorting. Cellular proteins known to participate in apical trafficking of viral glycoproteins, such as caveolin-1, annexin XIIIb, myelin, and lymphocyte-binding protein, are also associated with these rafts: inhibition of the activity or synthesis of these proteins disrupts transport of the influenza virus HA protein (and other proteins) from the Golgi complex to the apical membrane. Lipid rafts seem likely to be more generally important in targeting of viral membrane proteins and assembly: measles virus glycoproteins are selectively enriched in lipid rafts in nonpolarized cells, and association of the human immunodeficiency virus type 1 Gag polyprotein with these membrane domains promotes production of virus particles.

Neurons

Neurons are probably the most dramatically specialized of the many polarized cells of vertebrates. The axon is typically long and unbranched, whereas the dendrites form an extensive branched network of projections (Fig. 12.14). Axons are specialized for the transmission of nerve impulses, ultimately via the formation of synaptic vesicles and release of their contents. In contrast, dendrites provide a large surface area for the receipt of signals from other neurons. The nucleus, the rough ER, and the Golgi apparatus are also located in the dendritic region and the cell body of a neuron. Although axonal and dendritic surfaces are not separated by tight junctions, proteins must be distributed asymmetrically in neurons. Several mechanisms contribute to the establishment and maintenance of neuronal polarity, including transport of vesicles in specific directions along the highly organized

A

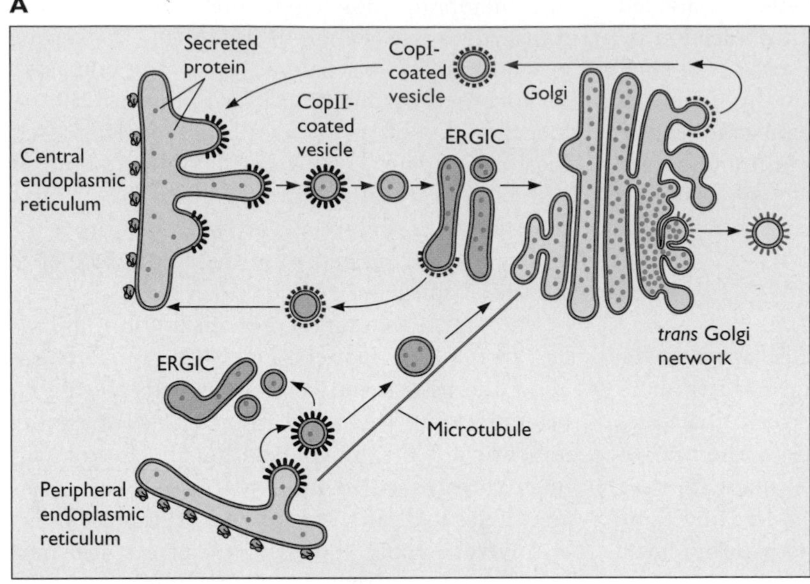

B

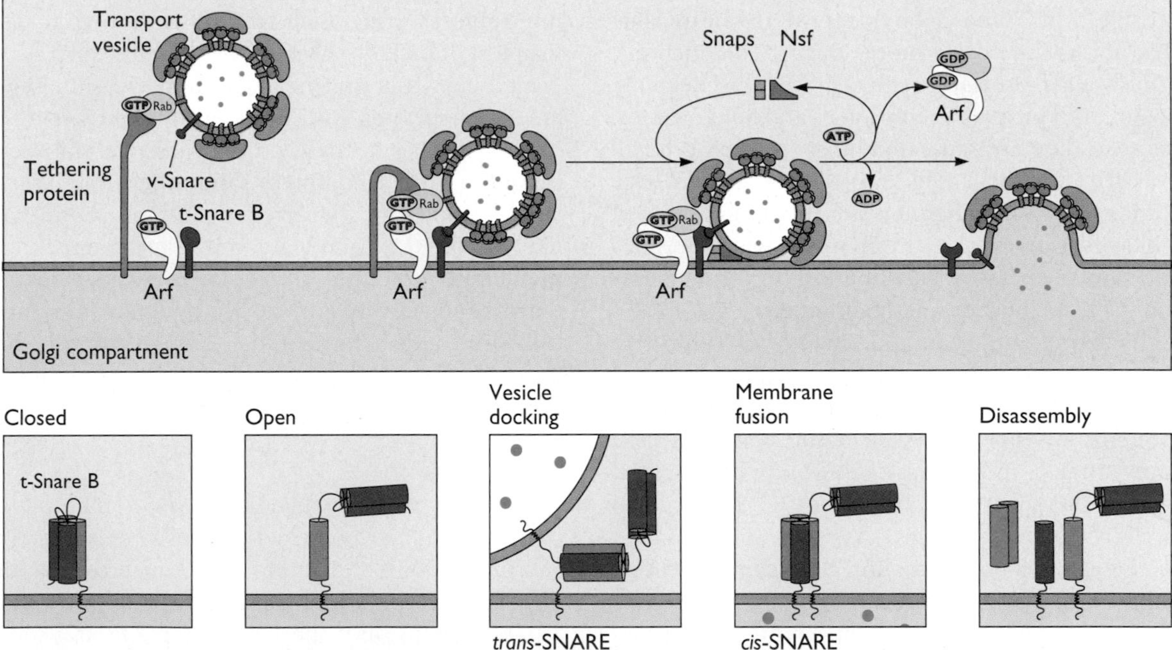

Figure 12.11 Protein transport from the ER to the Golgi. (A) Proteins leave the ER in transport vesicles at regions free of ribosomes in both the center of the cell (that is, near the Golgi) and peripheral regions. Vesicle formation is initiated by binding of cytoplasmic coat protein complex II (CopII), which contains a small GTPase (Sarl) and several other proteins, to the membrane. The vesicle membranes also carry proteins that direct them to appropriate destinations, such as the v-Snares syntaxin 5 and membrin. Cargo is loaded by interactions with proteins of the CopII coat and export receptors. The CopII coat induces budding and pinching off of vesicles, which then lose their coats and fuse to form vesicular tubular clusters. These clusters are also known as the ER-Golgi intermediate compartment (ERGIC). When formed at sites far from the Golgi, vesicles and vesicular-tubular compartments move to this organelle along microtubules (Box 12.7). Vesicles coated by the COP I coat bud off from these compartments and mediate transport of proteins back to the ER. This process is thought to make an important contribution to sorting proteins for forward transport. Whether vesicular-tubular compartments fuse with a fixed *cis*-Golgi network (stable

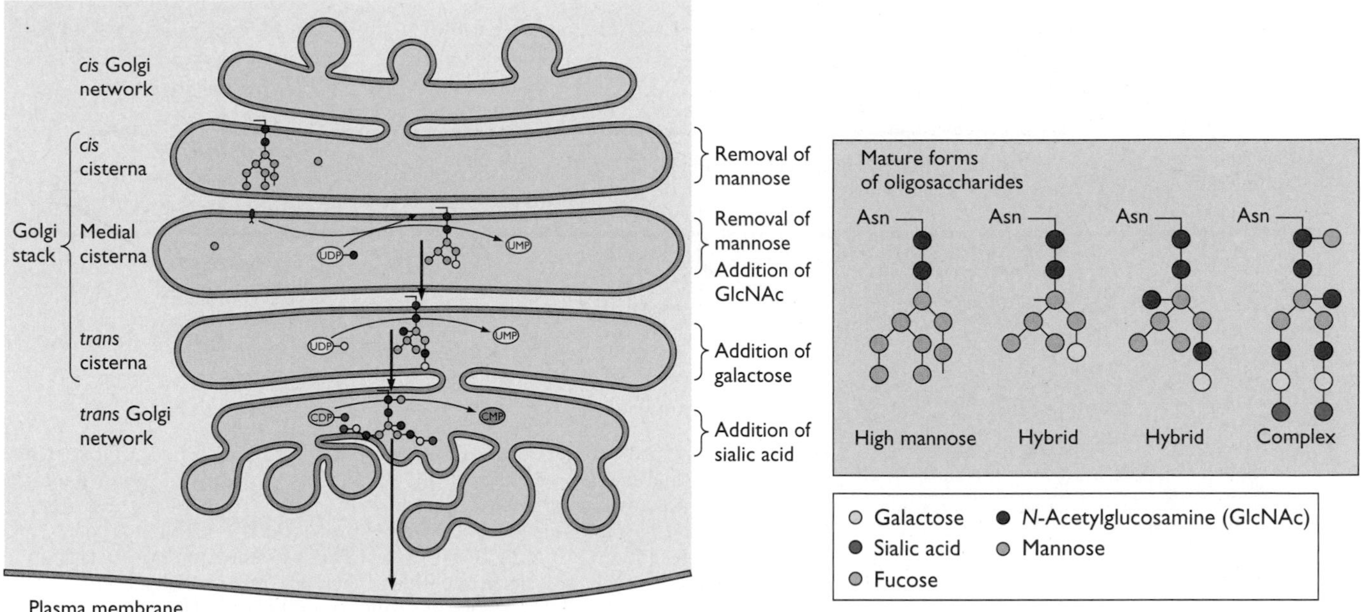

Figure 12.12 Compartmentalization of processing of N-linked oligosaccharides. The reactions by which mature N-linked oligosaccharide chains are produced from the high-mannose core precursor added in the ER (Fig. 12.7B) are shown in the Golgi compartment in which they take place. Trimming of terminal glucose and mannose residues of the common core precedes stepwise addition of the sugars found in the mature chains. The enzymes responsible for early reactions in maturation of oligosaccharides are located in *cis* cisternae, whereas those that carry out later reactions are present in the medial and *trans* compartments. Such spatial separation ensures that oligosaccharide processing follows a strict sequence as proteins pass through the compartments of the Golgi apparatus. Synthesis of O-linked oligosaccharides by glycosyltransferases, which add one sugar unit at a time to certain serine or threonine residues, also takes place in the Golgi apparatus.

microtubules of the axon **(axonal transport)** and transport of mRNA to dendrites. Nevertheless, it is thought that sorting and targeting of membrane proteins for delivery to axonal or dendritic surfaces is also essential for polarization.

The directional movement of vesicles and many cellular organelles in neurons is dependent on polarized microtubules and motor molecules that travel toward either their (−) or (+) ends. Such motors therefore mediate transport both toward the cell body from axons and dendrites and away from the cell body (Fig. 12.15). Infection, assembly, and egress of particles of viruses that infect neurons depend on these mechanisms. An important example is provided by the neurotropic alphaherpesviruses, a group

---

Golgi compartment model) or form *cis* cisternae upon fusion (cisternal maturation model) and whether forward transport through the Golgi cisternae is mediated by transport vescles or maturation of Golgi cisternae have been questions of intense debate. Clear evidence for both vesicle transport and cisternal maturation has been obtained. It is generally accepted that both mechanisms operate, but that one or the other may predominate in specific cell types, or under some conditions. **(B)** Fusion of transport vesicle and target compartment membranes. Both vesicle (v-Snare and specific Rabs bound to GTP) and target compartment (t-Snare and specific Arfs and Rabs bound to GTP) proteins govern the specificity of membrane fusion. The first step is tethering of an uncoated vesicle by interaction of a tethering protein with the GTP-bound Rab. Tethering proteins (e.g., p119 and giantin) generally are extended α-helical proteins that form coiled coils. Docking then takes place by formation of specific v-Snare–t-Snare complexes. As shown below, this process is initiated by a t-Snare adopting an open conformation, which allows formation of a trans-Snare α-helical bundle with helices also contributed by the v-Snare and a second t-Snare. Docking is followed by recruitment of the soluble proteins *N*-ethylmaleimide-sensitive factor (Nsf) (an ATPase) and soluble Nsf-attachment proteins (Snaps) to form a membrane fusion complex. Fusion is accompanied by ATP hydrolysis and disassembly of the fusion complex.

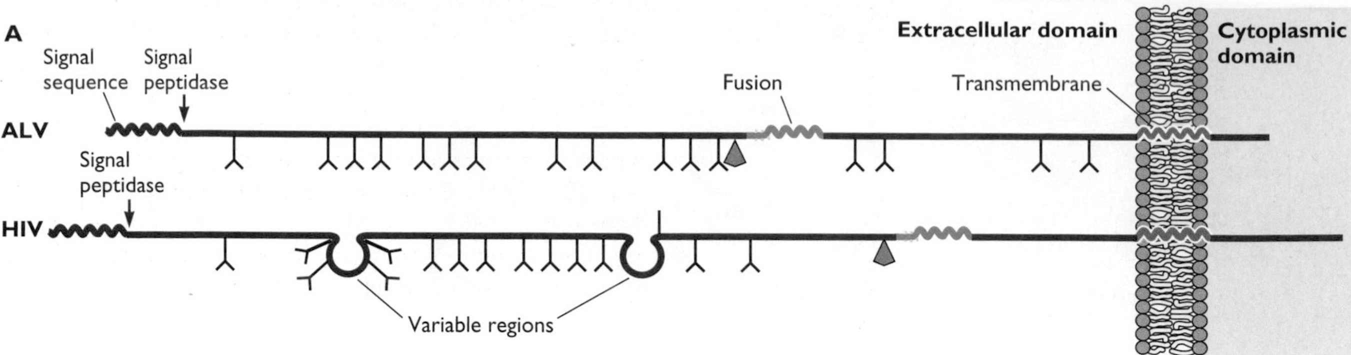

B

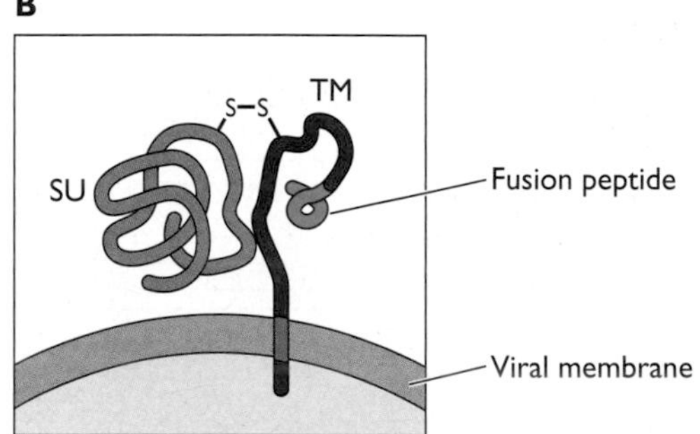

Figure 12.13 Modification and processing of retroviral Env polyproteins. **(A)** Sequence features and modifications of the Env proteins of avian leukosis virus (ALV) and human immunodeficiency virus type 1 (HIV) are depicted as in Fig. 12.3. The variable regions of human immunodeficiency virus type 1 Env differ greatly in sequence among viral isolates. The translocation products shown here are cleaved by the ER signal peptidase (red arrows) and by furin family proteases in the *trans*-Golgi network (orange arrowheads). The latter liberates the transmembrane (TM) and surface unit (SU) subunits from the Env precursor. **(B)** Both noncovalent interactions and a disulfide bond (-S-S-) can maintain the association of the SU and TM proteins.

Table 12.4 Examples of acylated or isoprenylated viral proteins

Virus	Protein	Lipid	Probable function
Envelope proteins			
Alphavirus			
Sindbis virus	E2	Palmitate	Required for efficient budding of virus particles
Coronavirus			
Severe acute respiratory syndrome virus	S	Palmitate	Fusion
Hepadnavirus			
Hepatitis B virus	L (pre-S1)	Myristate	Initiation of infection
Orthomyxovirus			
Influenza A virus	HA	Palmitate	Essential for fusion and infectivity
Retrovirus			
Human immunodeficiency virus type 1, Moloney murine leukemia virus	Env (TM)	Palmitate	Virion budding
Other viral proteins			
Hepatitis delta satellite virus	Large delta antigen	Geranylgeranol	Interaction with HBV L protein; assembly; inhibition of HDV RNA replication
Papovavirus			
Simian virus 40	VP2	Myristate	Virion assembly
Picornavirus			
Poliovirus	VP_0, VP_4	Myristate	Virion assembly; uncoating
Retrovirus			
Human immunodeficiency virus type 1, murine leukemia virus	Gag, MA	Myristate	Membrane association, assembly and budding
Rous sarcoma virus	pp60*src*	Myristate	Membrane association, transformation

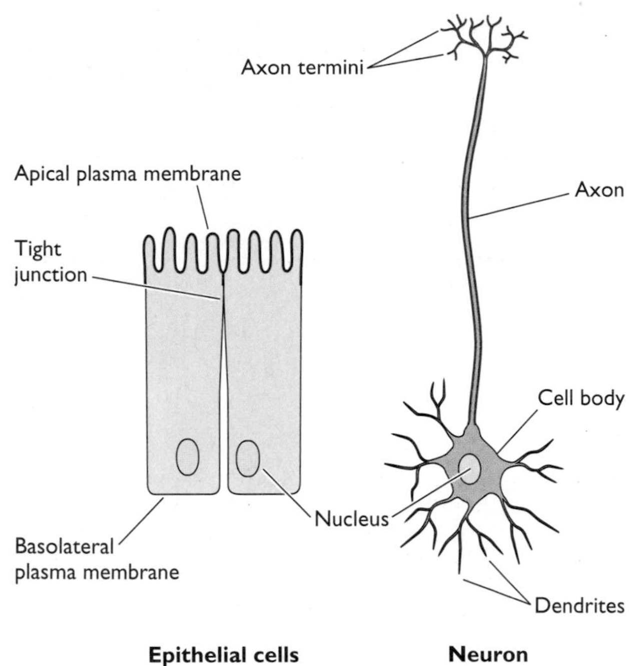

Figure 12.14 Polarized epithelial cells and neurons.

that includes herpes simplex virus type 1 and varicella-zoster virus. Upon infection of sensory neurons, herpesvirus capsids are transported to the nucleus by microtubule-based transport, probably mediated by (–) end-directed motors such as dynein (Fig. 12.15; Box 12.8). Later in the infectious cycle, virion components must be moved in the opposite direction upon association with protons of the kinesin family. How nucleocapsids and other virion proteins are targeted to exons for anterograde transport remains controversial (Box 12.9). The spread of herpesviruses from neuron to neuron occurs at or near sites of synaptic contact, indicating that virus particles must be targeted to specific areas within neurons for egress. As discussed in Volume II, Chapter 1, this attribute can be exploited to define neuronal connections in a living animal by using the virus as a self-amplifying tracer.

Disruption of the Secretory Pathway in Virus-Infected Cells

Inhibition of Transport of Particular Cellular Proteins

The genomes of several viruses encode proteins that interfere with the transport to the plasma membrane of specific cellular proteins, notably MHC class I molecules (Table 12.5; Volume II, Chapter 4). For example, the adenovirus E3 glycoprotein gp19 (Appendix, Fig. 1B) binds to these important components of immune defense in the ER and prevents their exit from this compartment, probably by inhibiting folding and assembly. Several herpesviral proteins also block transport of MHC class I molecules to the cell surface. The human cytomegalovirus US11 and US2 gene products induce transport of the cellular proteins from the ER to the cytosol for rapid degradation by the cellular proteasome. When this activity was first discovered, such retrotranslocation from the ER to the

Figure 12.15 Axonal transport in neurons. Vesicles containing cargo destined for synaptic vesicles (e.g., neurotransmitters) become associated with molecular motors of the kinesin family, and are transported down the axon on the tracks formed by axonal microtubules. The force generated by kinesin family motors allows transport at rates of 1 to 2 μm/s. Transport toward the cell body is thought to be mediated by the cytoplasmic motor dynein. In infected neurons, herpes simplex virus type 1 (HSV-1) components are transported in opposite directions at the beginning and end of the infectious cycle.

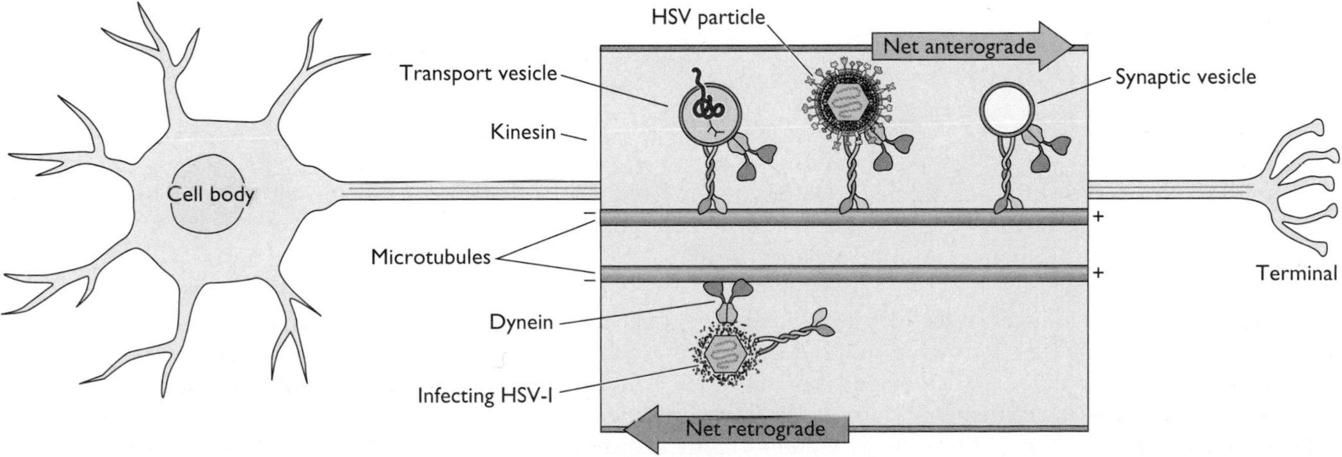

BOX 12.8

DISCUSSION

The dynamics of herpesvirus capsid transport in axons

The core structures of many viruses move within cells by association with host cytoskeletal motor proteins; however, the mechanisms by which intracellular viral particles are transported toward sites of replication or the cell periphery at distinct stages of infections remain to be understood.

In this study, green fluorescence protein-labeled capsids of pseudorabies virus (an alphaherpesvirus) were visualized in living, unfixed axons following infections of chicken sensory dorsal root ganglia neurons. Images of individual green fluorescent protein puncta (capsids) were captured by laser-scanning confocal microscopy at multiple frames per second.

After entry of the capsids into axons, the motion of capsids was bidirectional and saltatory, i.e., proceeding by starts and stops. Nevertheless, the net direction of movement was toward the cell body (retrograde). The dynamics of capsids immediately after entry was unperturbed by the presence of egressing virus particles produced during a prior infection. This observation indicates that transport direction is modulated not by viral gene expression, but rather by a component of the subviral particle. The motion of newly formed capsids later in infection was also bidirectional and saltatory, but the net flow of capsids was reversed, and was predominantly toward the cell body (anterograde). It was suggested that the control of net directional transport of capsids occurs by modulation of plus-end, but not minus-end, motors.

Smith, G. A., L. Pomeranz, S. P. Gross, and L. W. Enquist. 2004. Local modulation of plus-end transport targets herpesvirus entry and egress in sensory axons. *Proc. Natl. Acad. Sci. USA* **101:**16034–16039.

A Nucleocapsid motion: egress versus entry in neurons

Direction	Rate (μm/s)	Run length (μm)
Egress		
Anterograde	1.97	13.1
Retrograde	1.28	6.8
Entry		
Anterograde	0.55	0.5
Retrograde	1.17	8.1

B

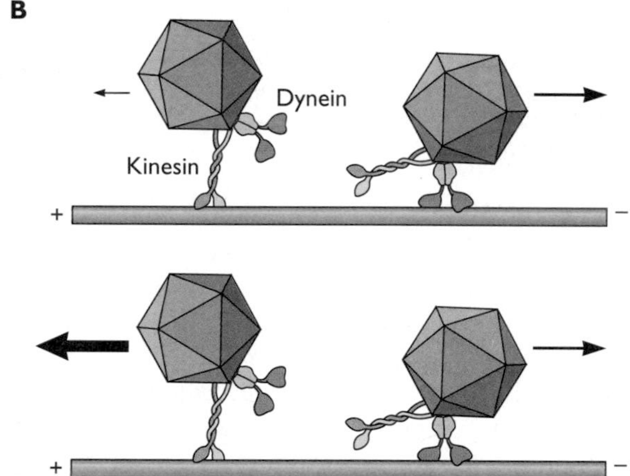

Coordination of a bidirectional motor complex directs capsid transport. The model depicts contrasting transport dynamics of herpesvirus capsids during entry and egress. Dynein and a kinesin family motor are illustrated associated with individual capsids simultaneously as part of a coordinated bidirectional motor complex. A direct interaction of capsids with motor complexes is shown for illustrative purposes only: the interaction of motors with cargo or each other may be direct, or indirect by means of additional proteins and capsids may be inside vesicles during egress. Net retrograde motion required for entry results when kinesin-mediated movement is reduced **(top).** Alternatively, increasing the kinesin contribution results in the net enterograde motion required for egress **(bottom).** In this model, the contribution of the dynein motor does not change during entry and egress.

cytoplasm appeared very unusual. However, as discussed previously, it is now clear that this is a normal activity of uninfected cells. The human immunodeficiency virus type 1 Vpu protein (Appendix, Fig. 19), a transmembrane phosphoprotein, also induces selective degradation of newly synthesized MHC class I proteins, and of CD4, by a similar mechanism (Table 12.5). Such degradation of CD4, the major receptor for this virus, is important for assembly and release: tight binding of this cellular protein to the viral Env glycoprotein in the ER prevents transit of both proteins to the cell surface.

Drastic Effects on Compartments of the Secretory Pathway

Proteins encoded by certain other viruses exert more drastic effects on the cellular secretory pathway. For example, rotaviruses, which lack a permanent envelope but are transiently membrane enclosed during assembly, encode

BOX 12.9

DISCUSSION

How are herpesviral particles targeted to axons for anterograde transport?

Alphaherpesviruses (e.g., herpes simplex virus [HSV] and pseudorabies virus [PRV]) replicate and traffic within polarized neurons, a strategy conducive to their lifestyles in the host peripheral nervous system. Infection begins with virion entry at mucosal surfaces and spread of infection between cells of the mucosal epithelium. The peripheral nervous system is infected via axon termini innervating this region, and subsequent trafficking of nucleocapsids to the cell body. It is here that a reactivatable, latent infection that persists for the life of the host is established. A well-known but poorly understood phenomenon is that, upon reactivation from latent infection, alphaherpesviruses rarely enter the central nervous system, despite having what seems to be two rather similar choices: cross one synapse and infect the central nervous system (rare) or traffic back down the axon and cross to the initial site of (peripheral) infection, mucosal epithelial cells (very common). Inherent in this choice is the fact that viral proteins must be targeted to axons, a highly specialized neuronal compartment restricted to only a subset of neuronal proteins. The primary problem is to identify the mechanisms that gather and sort the many viral structural proteins to this compartment.

The mechanisms by which newly synthesized structural proteins are sorted to axons for anterograde transport have been the subjects of considerable controversy. In fact, different processes have been proposed for herpes simplex virus and pseudorabies virus, separate transport of the nucleocapsid, plus tegument and viral glycoprotein and transport of enveloped virions, respectively. Similar methods have been used to examine anterograde transport of the two viruses in several laboratories. The methods include confocal microscopy, imaging of nucleocapsids and glycoproteins that carry distinguishable fluorescent labels in live cells, and immunoelectron microscopy. Nevertheless, the controversy has not been resolved. Although counterintuitive, it is possible that different processes operate in cells infected by these two herpesviruses, which are the most distantly related among the alphaherpesviruses. This hypothesis implies that envelopment of naked herpes simplex virus nucleocapsids takes place at the membrane of axonal growth cones, whereas unenveloped pseudorabies virus nucleocapsids can travel only in the retrograde direction.

Antinone, S. E., and G. A. Smith. 2006. Two modes of herpesvirus trafficking in neurons: membrane acquisition directs motion. *J. Virol.* **80:**11235–11240.

Diefenback, R. J., M. M. Saksena, M. Douglas, and A. Cunningham. 2007. Transport and egress of herpes simplex virus neurons. *Rev. Med. Virol.* **18:**35–51.

Feierbach, B., M. Bisher, J. Goodhouse, and L. W. Enquist. 2007. In vitro analysis of trans-neuronal spread of an alpha-herpesvirus infection in peripheral nervous system neurons. *J. Virol.*, **81:**6846–6857.

Snyder, A., B. Bruun, H. Browne, and D. Johnson. 2007. A herpes simplex virus gD-YFP fusion glycoprotein is transported seperately from viral capsids in neuronal axons. *J. Virol.* **81:**8337–8340.

Sympathetic neurons from superior cervical ganglia of Sprague-Dawley rats were cultured in a chamber system that separates cell bodies (outside) from axons (in the chamber). Cell bodies outside the chamber were infected with a dually fluorescent derivative of pseudorabies virus: the envelope is labeled green by a gD-GFP fusion protein, and the nucleocapsid is labeled red by fusion of a structural protein with mRFP. Live neurons within the chamber were examined by confocal video microscopy from 15.5 h after infection. Shown are individual images, taken at the intervals indicated. The inset shows an enlargement fo the area outlined in white. The axon runs along the longitudinal axis of the frame. The yellow puncture (yellow arrowhead) represents an enveloped (green) nucleocapsid (red, green plus red = yellow). This particle moved in the opposite direction to a red punctum (orange arrowhead), that is, a naked nucleocapsid. Eventually, the paths of these two particles cross within the same axon. Adapted from B. Feierbach et al., *J. Virol.* **81:**6846–6857, 2007, with permission.

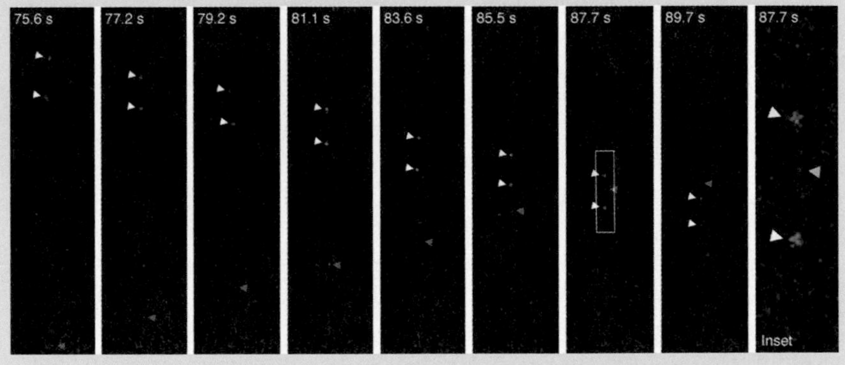

a protein that disrupts the ER membrane. This protein is thought to allow removal of the temporary envelope formed during virion assembly. The replication of most (+) strand RNA viruses takes place in membranous structures derived from various cytoplasmic membranes of the host cell. Such remodeling of cytoplasmic membranes is best understood for poliovirus and other enteroviruses, which induce a dramatic reorganization of compartments of the secretory pathway.

Three functionally distinct alterations of internal cell membranes occur in cells infected with poliovirus (Fig. 12.16). The Golgi complex is disrupted, an effect that is also observed when the viral protein 2B is synthesized in cells. Golgi disassembly is not required for viral replication.

Table 12.5 Viral proteins that inhibit transport of cellular proteins through the secretory pathway

Viral protein	Properties and activities
Adenovirus	
Ad2 E3gp19	ER transmembrane glycoprotein; blocks exit of MHC class I proteins from the ER; probably inhibits folding and assembly
Herpesvirus	
Human cytomegalovirus US2 and US11	ER transmembrane proteins; induce retrotranslocation of MHC class I proteins from ER to cytoplasm for proteasomal degradation
Picornavirus	
Poliovirus 2BC, 3A	Inhibit vesicular transport from ER; induce accumulation of infected-cell-specific vesicles derived from ER and autophagosomes; associated with membranes of the viral replication complex
Retrovirus	
Human immunodeficiency virus type 1	
Vpu	Transmembrane protein with phosphorylated cytoplasmic domain; induces degradation of MHC class I and CD4 proteins
Nef	Induces decreased cell surface accumulation of MHC class I proteins and retention in intracellular compartments; increases rate of endocytosis of CD4

Poliovirus infection also leads to inhibition of protein traffic to the cell surface, a consequence of blocked transport from the ER intermediate compartment of the Golgi complex. This effect may be phenocopied by synthesis of viral protein 3A in cells. Because poliovirus particles lack an envelope, the cellular secretory pathway is not needed for virus reproduction. However, inhibition of protein secretion interferes with some antiviral responses, such as those mediated by MHC class I molecules and cytokines, and may allow evasion of immune surveillance.

Independent of the disruption of the Golgi complex and inhibition of protein secretion is the accumulation in the cytoplasm of double-membrane membranous vesicles 200 to 400 nm in diameter. These vesicles resemble autophagosomes, and serve as the site of viral RNA replication (Chapter 6). They may also play a role in nonlytic release of virus from cells (Chapter 13). The synthesis of double-membrane vesicles can be induced in uninfected cells by the cosynthesis of viral proteins 2BC and 3A. The vesicles that serve as scaffolds for formation of replication complexes in cells infected by coronaviruses also exhibit properties of autophagosomes. The mechanisms by which infection by these viruses override the cellular circuits that normally prevent autophagy are not yet known. Nevertheless, it is clear that formation of autophagosomes facilitates virus replication: virus yield is reduced when synthesis of cellular proteins required for autophagy is prevented.

Signal Sequence-Independent Transport of Viral Proteins to the Plasma Membrane

As discussed in Chapter 4, many enveloped viruses contain matrix or tegument proteins lying between, and making contact with, the inner surface of the membrane of the particle and the capsid or nucleocapsid. In contrast to the integral membrane proteins of enveloped viruses, such internal virion proteins do not enter the secretory pathway, but are synthesized in the cytoplasm of an infected cell and directed to membrane assembly sites by specific signals.

Lipid-plus-Protein Signals

It has been known for many years that cytoplasmic proteins can be modified by the covalent addition of lipid chains (Table 12.4). Best characterized are the addition of the 14-carbon saturated fatty acid myristate to N-terminal glycine residues, or of unsaturated polyisoprenes, such as farnesol (C_{15}) or geranylgeranol (C_{20}), to a specific C-terminal sequence (Fig. 12.17). Palmitate is also added to some viral proteins that do not enter the secretory pathway. The discovery that transforming proteins of oncogenic retroviruses, the Src and Ras proteins, are myristoylated and isoprenylated, respectively, led to a resurgence of interest in these modifications. In this section, we focus on myristoylation and isoprenylation of viral structural proteins.

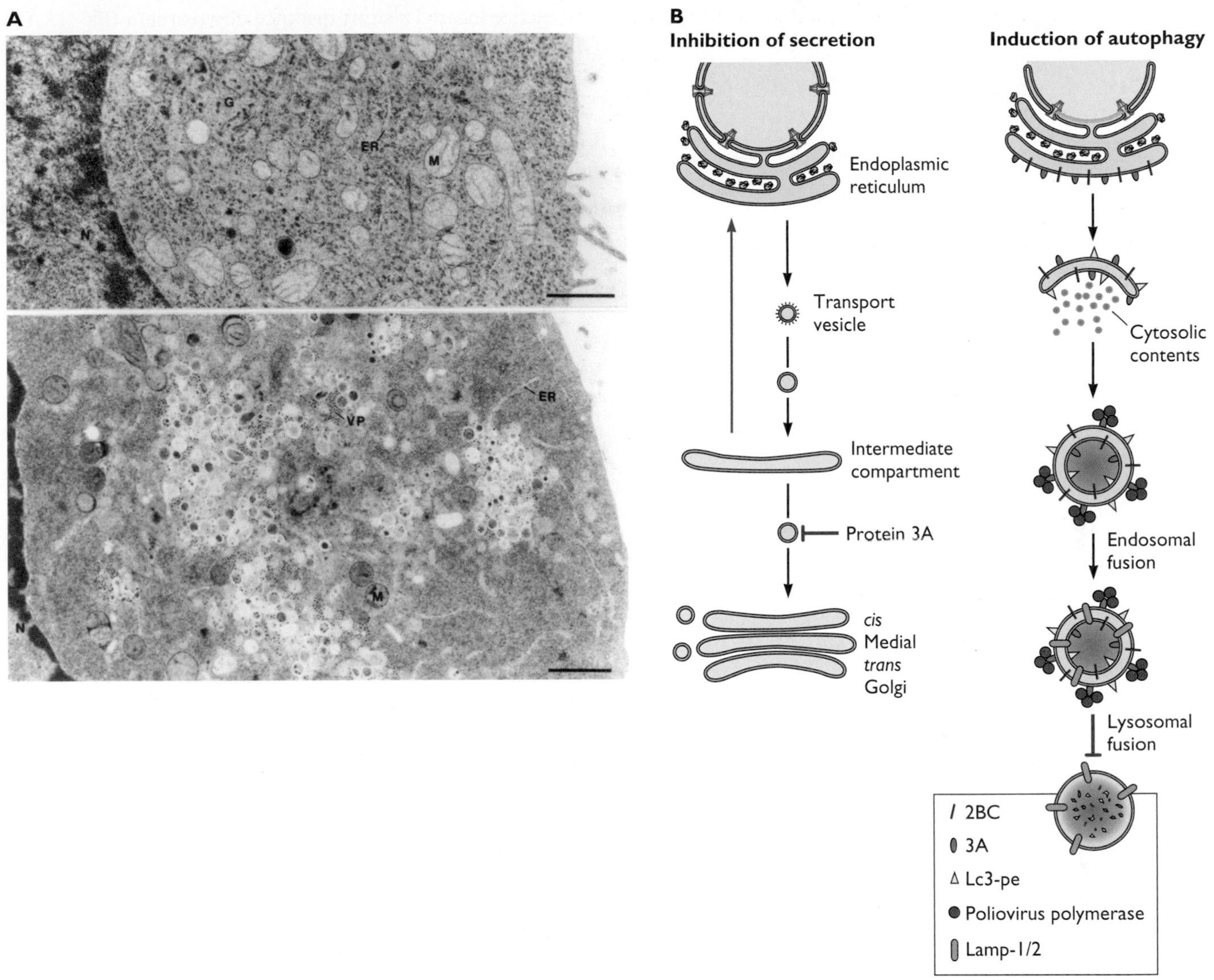

Figure 12.16 Inhibition of the cellular secretory pathway in poliovirus-infected cells. (A) Electron micrographs of uninfected HeLa cells (left) and HeLa cells 5 h after poliovirus infection (right) preserved by high-pressure freezing. Many infected cell-specific vesicles can been seen in the infected cells. G, Golgi apparatus; M, mitochondrion; VP, virus particles. The bars indicate 1 µm. Adapted from A. Schlegel et al., *J. Virol.* **70:**6576–6588, 1996, with permission. Courtesy of Karla Kirkegaard, University of Colorado, Boulder. **(B)** Models for inhibition of vesicular transport from the ER to the cell surface (left) and formation of poliovirus-induced vesicles (right). The viral 3A protein blocks transport between the ER-Golgi intermediate compartment and *cis*-Golgi compartments, and trafficking of proteins to the cell surface via the secretory pathway. The Golgi is also fragmented during polio infection, but this is not needed for viral replication or inhibition of secretion. When Golgi dispersion is prevented by nocodazole treatment in poliovirus-infected cells, the secretory pathway is still blocked. As shown at the right, the viral 3A and 2BC proteins induce the formation of double-membrane vesicles that resemble autophagosomes. These infected cell specific vesicles carry cellular proteins that are required for autophagy, such as Lc3-pe and Lamp1/2.

Myristoylation of the cytoplasmic Gag proteins of retroviruses and its consequences have been examined in detail. The internal structural proteins of these viruses, MA (matrix), CA (capsid), and NC (nucleocapsid), are produced by proteolytic cleavage of the Gag polyprotein following virus assembly. The Gag proteins of the majority of retroviruses are myristoylated at their N-terminal glycines. Mutations that alter the sequence at which murine leukemia virus or human immunodeficiency virus type 1 Gag proteins are myristoylated prevent interaction

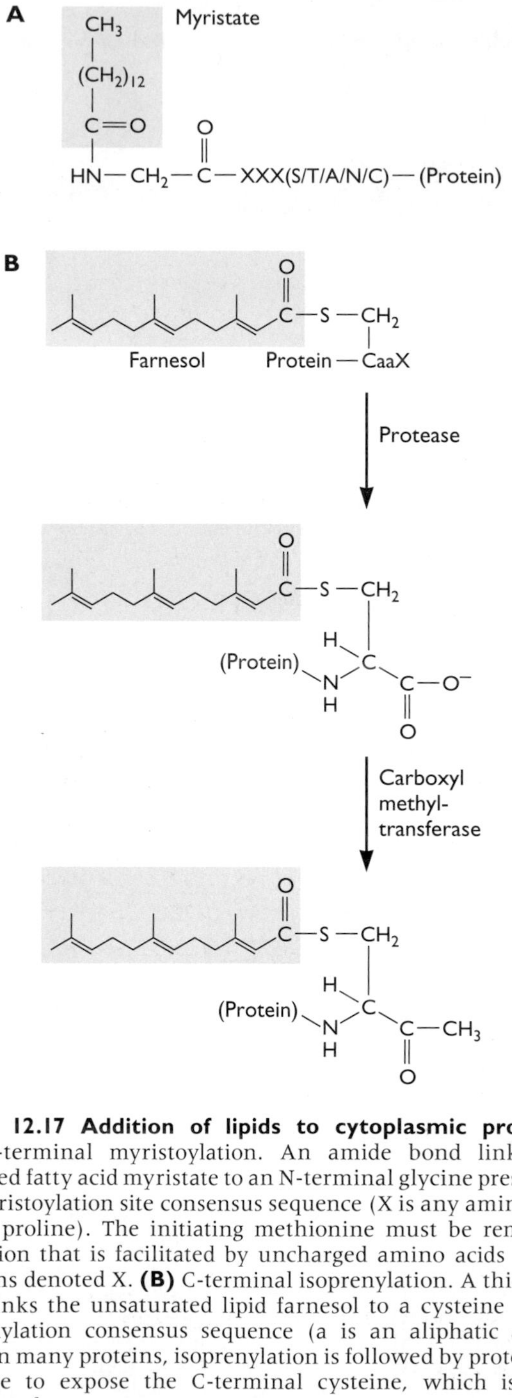

Figure 12.17 Addition of lipids to cytoplasmic proteins. **(A)** N-terminal myristoylation. An amide bond links the saturated fatty acid myristate to an N-terminal glycine present in the myristoylation site consensus sequence (X is any amino acid except proline). The initiating methionine must be removed, a reaction that is facilitated by uncharged amino acids in the positions denoted X. **(B)** C-terminal isoprenylation. A thioether bond links the unsaturated lipid farnesol to a cysteine in the isoprenylation consensus sequence (a is an aliphatic amino acid). In many proteins, isoprenylation is followed by proteolytic cleavage to expose the C-terminal cysteine, which is then methylated.

of the protein with the cytoplasmic face of the plasma membrane, induce cytoplasmic accumulation of Gag, and inhibit virus assembly and budding. In the case of the human immunodeficiency virus type 1 Gag protein, the myristoylated N-terminal segment and a highly basic sequence located a short distance downstream (Fig. 12.18) form a bipartite signal, which allows membrane binding *in vitro* and virus assembly and budding *in vivo*. The MA domain of the Gag protein of this protein binds to phosphatidylinositol-(4,5)-bisphophate, a lipid found only in the plasma membrane. This interaction accounts for the preferential association of Gag with the plasma membrane. It also induces a conformational change that leads to exposure of the N-terminal myristate, and presumably tighter association of Gag with the membrane.

The hepatitis B virus large surface (L) protein is also myristoylated at its N terminus. However, in contrast to retroviral Gag, the L protein is present in the envelope of the virion. Modification of its N terminus must therefore occur while it traverses the secretory pathway. In this case, myristoylation is not necessary for assembly or release of virions, but is required for infection of primary hepatocytes, presumably because it contributes to the initial interaction of the virus with, or its entry into, the host cell. More surprising is the myristoylation of structural proteins of poliovirus (VP4) and polyomavirus (VP2): although these virions do not contain an envelope, such modification is necessary for efficient assembly of both viruses. In mature poliovirus particles, the myristate chain at the N terminus of VP4, which is processed from VP0, interacts with the VP3 protein on the inside of the capsid (Fig. 4.13B). The hydrophobic lipid chain must therefore facilitate protein-protein interactions necessary for the assembly of virions. The fatty acid is also important during entry into cells of poliovirus particles and their uncoating at the beginning of an infectious cycle (see Chapter 5).

Among viral structural proteins, only the large delta protein of the hepatitis delta satellite virus has been found to be isoprenylated. Formation of the particles of this satellite virus depends on structural proteins provided by the helper virus, hepatitis B virus. The isoprenylation of large delta protein is necessary, but not sufficient, for its binding of the hepatitis B virus S protein during assembly of the satellite. This hydrophobic tail of large delta protein seems likely to facilitate interaction with the plasma membrane adjacent to the helper envelope of S protein in cells infected by the two viruses.

The addition of lipid chains allows the association of virion proteins with the cytoplasmic face of the plasma membrane, providing opportunities for interactions with viral membrane glycoproteins. Therefore, we can now explain (at least in principle) how cytoplasmic structural proteins, such as the Gag and Gag-Pol polyproteins of most retroviruses, associate with the plasma membrane.

Protein Sequence Signals

The matrix proteins of members of several families of (−) strand RNA viruses lie between the nucleocapsid and

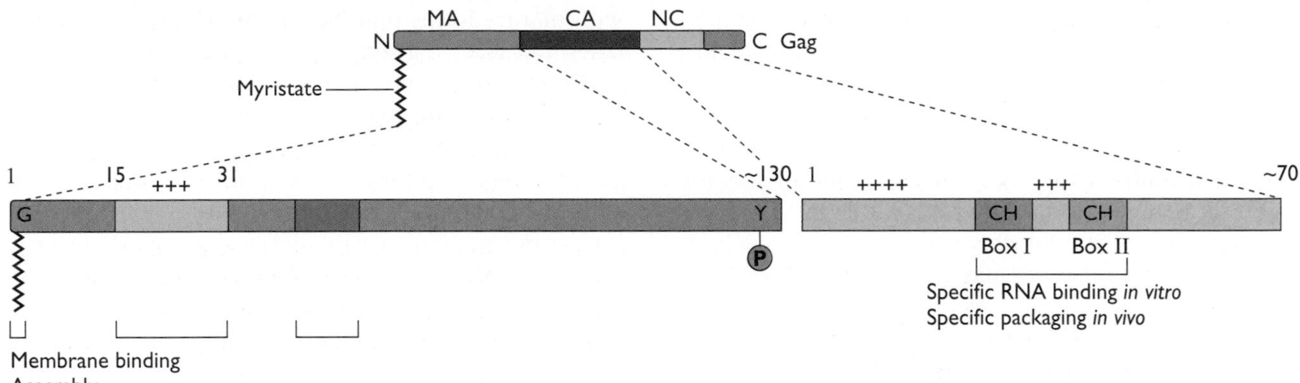

Figure 12.18 Human immunodeficiency virus type 1 Gag proteins and their targeting signals. The locations of the internal structural proteins MA (matrix), CA (capsid), and NC (nucleocapsid) in the Gag polyprotein are shown at the top. Sequence features, localization signals (MA), and the RNA-binding domain (NC) are shown below. The lengths of the MA and NC proteins are listed as approximate because of the variation among virus isolates. Specific amino acids are given in the single-letter code in the boxes, and a plus sign indicates a basic amino acid. The basic region of MA of simple retroviruses, such as avian sarcoma virus, is not required for membrane binding. The CH boxes of NC contain three cysteines and one histidine, and each coordinates one Zn^{2+} ion. CH box I is conserved among retroviruses, but CH box II is not.

the envelope, and are essential for correct localization and packaging of RNA genomes. During assembly, matrix proteins, such as M of vesicular stomatitis virus and M1 of influenza A virus, must bind to the plasma membrane of infected cells. These proteins are produced in the cytoplasm but, as far as is known, receive no lipid after translation. Direct membrane binding *in vivo* appears to be an intrinsic property of matrix proteins. For example, when the influenza virus M1 protein is synthesized in host cells in the absence of any other viral proteins, a significant fraction is tightly associated with cellular membranes. Both this protein and the vesicular stomatitis virus M protein contain specific sequences that are necessary for their interaction with the plasma membrane *in vivo* or with lipid vesicles *in vitro*. This region of the influenza A virus M1 protein contains two hydrophobic sequences (Fig. 12.19A), which might form a hydrophobic surface structure in the folded protein. In addition to hydrophobic segments, membrane association of the vesicular stomatitis virus M protein requires a basic N-terminal sequence (Fig. 12.19B). This latter segment might participate directly in membrane binding, like the basic

Figure 12.19 Targeting signals of matrix proteins of influenza virus (A) and vesicular stomatitis virus (B). Sequence features of specific segments of the proteins and the boundaries of targeting and RNP-binding domains are shown. Amino acids are written in the one letter code, and a plus sign indicates a basic amino acid. VSV, vesicular stomatitis virus; NLS, nuclear localization signal.

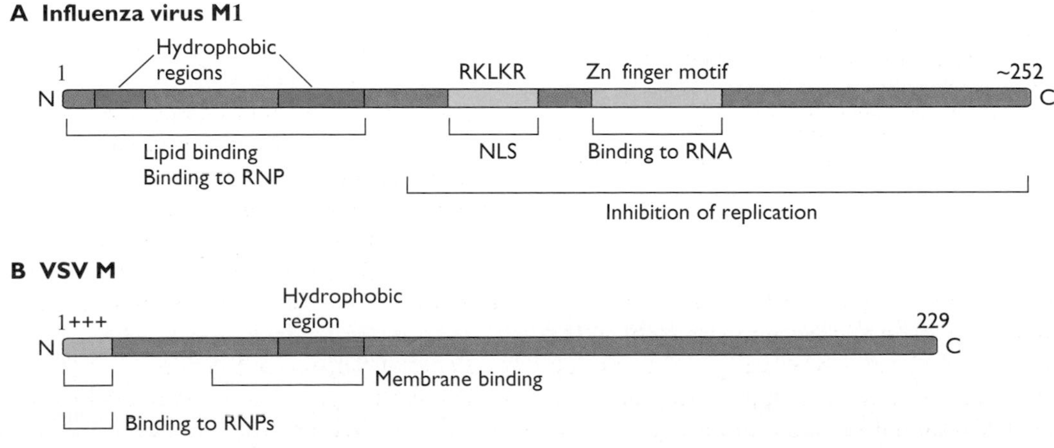

sequence of the human immunodeficiency virus type 1 Gag membrane-targeting signal, or it might stabilize a conformation of the internal sequence favorable for interaction with the membrane.

Assembly of (–) strand RNA viruses requires association of their matrix proteins with the plasma membrane. In some cases, such as the vesicular stomatitis virus M protein, specificity for the plasma membrane is an intrinsic property. However, binding of matrix proteins to the cytoplasmic tails of viral envelope glycoproteins can also be an important determinant of membrane association. The cytoplasmic domains of both the NA and HA proteins of influenza A virus stimulate membrane binding by M1 protein. Similarly, membrane binding by the matrix protein of Sendai virus (a paramyxovirus) is independently stimulated by the presence of either of the two viral glycoproteins (F or HN) in the membrane.

Interactions with Internal Cellular Membranes

The envelopes of a variety of viruses are acquired from internal membranes of the infected cell, rather than from the plasma membrane. The majority assemble at the cytoplasmic faces of compartments of the secretory pathway (Table 12.6). Although a single budding reaction is typical, the more complex herpesviruses and poxviruses interact with multiple internal membranes of the cell during assembly and exocytosis. And components of the enveloped hepatitis C virus associate with both the ER and the membrane-bound lipid droplets, in which cells store triacylglycerols and cholesterol (Box 12.10). Even some viruses with mature forms that lack an envelope, such as rotaviruses, can enter the secretory pathway transiently.

The diversity of the internal membranes with which these viruses associate during envelope acquisition and exocytosis is the result of variations on a single mechanistic theme: the site of assembly is determined by the intracellular location of viral envelope proteins (Fig. 12.20), just as assembly at the plasma membrane is the result of transport of such proteins to that site. Assembly of viruses at internal membranes therefore requires transport of envelope proteins to, and their retention within, appropriate intracellular compartments.

Localization of Viral Proteins to Compartments of the Secretory Pathway

The bunyaviruses, a family that includes Uukuniemi and Hantaan viruses, are among the best-studied viruses that assemble by budding into compartments of the secretory pathway. Bunyavirus particles contain two integral membrane glycoproteins, called G1 and G2, which are encoded within a single open reading frame of the M genomic RNA segment. Like alphaviral envelope proteins, the bunyaviral polyprotein containing G1 and G2 is processed cotranslationally by signal peptidase as the precursor enters the lumen of the ER. However, association of G1 with G2 is required for transport to, and retention in, Golgi compartments. For example, when synthesized alone, G1 accumulates in the Golgi complex in normal fashion, but G2 fails to leave the ER. When both glycoproteins are made, the G2 protein is now transported to the Golgi complex. The G1 protein may therefore contain signals necessary for transport along the secretory pathway, or it may be required for correct

Table 12.6 Interactions of viruses with internal cellular membranes

Virus family	Example	Integral membrane protein(s)	Intracellular membrane(s)	Mechanism of envelopment
Bunyaviruses	Uukuniemi virus Hantaan virus	G1, G2	*cis*-medial Golgi cisternae	Budding into Golgi cisternae
Coronavirus	Mouse hepatitis virus	M, S	ER, *cis*-Golgi network	Budding into ER and *cis*-Golgi network
Flavivirus	Dengue virus West Nile virus	E, prM	ER	Budding into ER
Hepadnavirus	Hepatitis B virus	L, M, S	ER and other compartments	Budding into ER?
Herpesvirus	Herpes simplex virus type 1	gB, gH, UL34	Nuclear membrane	Primary envelopment; budding of capsids from inner nuclear membrane
		gE-gI	*trans*-Golgi cisternae	Budding at *trans*-Golgi membrane
Poxvirus	Vaccinia virus, immature	A14L, A13L, A17L	ER	Formation of mature virion
	Vaccinia virus, intracellular mature virus	A56R (HA), F13L, B5R	Late *trans*-Golgi cisternae and post-Golgi vesicles	Wrapping of mature virion
Rotavirus	Simian rotavirus	VP7, NS28	ER	Budding into ER

BOX 12.10

EXPERIMENTS

An unexpected site of viral genome replication and assembly: lipid droplets

Mammalian cells store uncharged lipids in lipid droplets. These structures contain a core of cholesterol esterified to triacylglycerols, bounded by a single phospholipid leaflet and an external protein layer. Recent studies have highlighted a crucial role for lipid droplets in replication of hepatitis C virus, a member of the family *Flaviviridae*.

The hepatitis C virus (+) strand RNA genome contains a single open reading frame that encodes a large polyprotein. N-terminal sequences of the polyprotein become inserted into the ER membrane during translation, when cleavage by signal peptidase liberates the envelope proteins E1 and E2. The major structural protein, the core protein, is released on the cytoplasmic side of the ER membrane by cleavage by signal peptide peptidase. The several nonstructural proteins also remain in the cytoplasm, and interact directly or indirectly with the ER membrane. In contrast, the core protein becomes associated with lipid droplets.

Remarkably, lipid droplets appear to be the sites of viral genome replication (see, for example, the figure). Furthermore, substitutions in the core protein that prevent its association with lipid droplets block production of progeny virus particles. It is therefore thought that the core protein nucleates assembly of viral replication complexes on lipid droplets. The localization of this structural protein and newly synthesized RNA genomes to the same structure seems likely to facilitate the assembly of capsids. This arrangement would also be expected to promote acquisition of the envelope by budding into the ER: lipid droplets are attached to the ER membrane.

Boulant, S., P. Targett-Adams, and J. McLauchlan. 2007. Disrupting the association of hepatic C virus core protein with lipid droplets correlates with a loss of production of infectious virus. *J. Gen. Virol.* **88:**2204–2213.

Targett-Adams, P., S. Boulant, and J. McLauchlan. 2008. Visualization of double-stranded RNA in cells supporting hepatitis C virus replication. *J. Virol.* **82:**2182–2195.

Colocalization of double-stranded RNA, viral core protein and lipid droplets in hepatitis C virus-infected cells. Human cells infected with hepatitic C virus were examined by immunofluoresence with antibodies specific for double-stranded RNA (dsRNA), the lipid droplet restricted protein adipocyte differentiation-related protein (Adrp), or the viral core protein (core). The enlargements show the area boxed in white in panel iv. Scale bar, 10 µm. A variety of control experiments indicated that the anti-double-stranded RNA antibody detected sites of viral genome replication. All three molecules can be seen to be colocalized (merge), indicating that both viral RNA and the core protein are associated with lipid droplets. Adapted from P. Targett-Adams et al., *J. Virol.* **82:**2182–2195, 2008, with permission. Courtesy of P. Targett-Adams, Pfizer PGRD, Sandwich, Kent, United Kingdom.

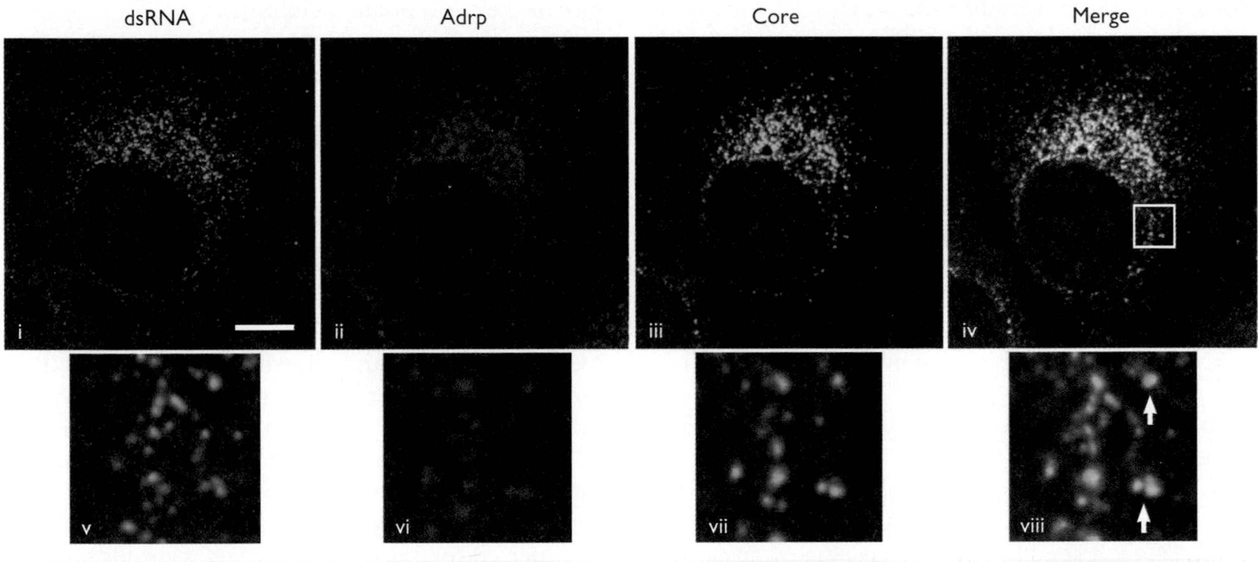

folding of G2 and its exit from the ER. Both visualization of the intracellular distribution of the G1 and G2, and the structures of their N-linked oligosaccharides, indicate that they do not reach the *trans*-Golgi network, but accumulate in *cis* or medial compartments. The signals that specify such locations have not been identified precisely, but are included within the transmembrane domain and the adjacent segment of the cytoplasmic domain of the G1 protein.

Golgi cisternae are by no means the only compartments of the secretory pathway at which virus budding can occur. For example, rotaviruses transiently aquire an evelope by budding into and out of the ER, whereas coronaviruses

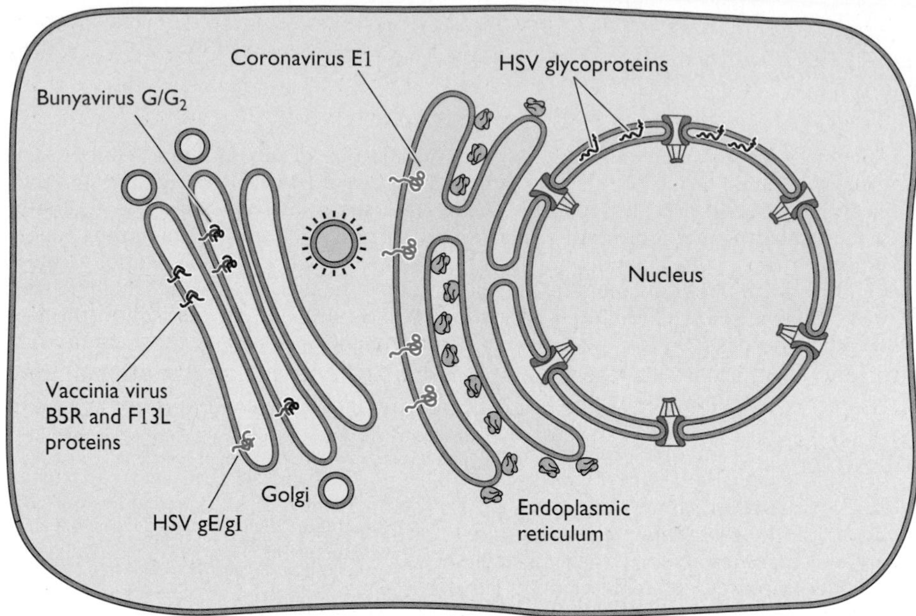

Figure 12.20 Sorting of viral glycoproteins to internal cell membranes. The destinations of membrane glycoproteins of viruses that bud into compartments of the secretory pathway (bunyaviruses and coronaviruses) or from the inner nuclear membrane and compartments of the *trans*-Golgi network (herpesviruses [HSV]) or are wrapped by cellular membranes during assembly (poxviruses) are indicated.

bud into both the ER and the *cis*-Golgi network. In these and other cases, it is the presence of viral membrane glycoproteins in specific cellular membranes (Table 12.6) that determines the site of assembly and budding.

Localization of Viral Proteins to the Nuclear Membrane

Herpesviruses such as herpes simplex virus type 1 are the only enveloped viruses that are known to assemble initially within, and bud from, the nucleus. The first association of an assembling herpesvirus with a cellular membrane is therefore budding of the nucleocapsid through the inner membrane of the nuclear envelope. This process, which is described in Chapter 13, depends on association of particular viral proteins with the inner nuclear membrane (Table 12.6).

Transport of Viral Genomes to Assembly Sites

Like the structural proteins and enzymes of the virion, progeny genomes must be transported to the intracellular site at which assembly takes place. Such transport is frequently coupled to mechanisms that prevent participation of the genome in biosynthetic reactions, i.e., replication, transcription, mRNA synthesis, or translation. The synthesis and packaging of most DNA genomes takes place in the infected cell nucleus, so no transport of progeny DNA molecules from one cellular compartment to another is required. For other DNA viruses, as well as for many with RNA genomes, both synthesis of the nucleic acid genome and assembly of virus particles take place in the cytoplasm of the host cell. However, many of these viruses assemble at membrane sites, necessitating transport of genomic nucleic acid from its site of synthesis to the cytoplasmic face of the appropriate membrane.

The membranes of both poxviruses and hepadnaviruses originate from internal compartments of the host cell (Table 12.6), and genome-bound proteins appear to mediate association of the genome with the appropriate membrane. For example, the hepadnaviral DNA genome is synthesized from the pregenomic RNA in the cytoplasm within cores containing the viral C protein. When infected cells contain a sufficient concentration of the large (L) viral glycoprotein in the ER membrane, core and L proteins can bind to one another, bringing the genome to the appropriate cell membrane and allowing budding of particles into the appropriate intracellular compartment. Similarly, many RNA genomes made in the cytoplasm must be transported to the plasma membrane for assembly of virions. Yet other RNA genomes must travel even further: both influenza virus and retroviral genomic RNAs are synthesized within the infected cell nucleus, but progeny virions bud from the plasma membrane.

Transport of Genomic and Pregenomic RNA from the Nucleus to the Cytoplasm

Retroviral RNA genomes are unspliced RNA transcripts synthesized in infected cell nuclei by host cell RNA polymerase II, as are hepadnaviral pregenomic RNAs. These RNAs must be exported to the cytoplasm for assembly, a process which requires that the inefficient export of unspliced mRNAs characteristic of host cells be circumvented. As discussed in Chapter 10, the genomes of complex retroviruses encode sequence-specific viral RNA-binding proteins that promote export of unspliced RNA, such as human immunodeficiency virus type 1 Rev protein. Specific sequences in the genomes of simple retroviruses and hepadnaviruses also direct export of genomic RNA, but these are recognized by cellular proteins.

Perhaps the most elaborate requirements for transport of viral RNA species between nucleus and cytoplasm are found in influenza virus-infected cells: both the direction of transport of genomic RNA and the nature of the viral RNA exported from the nucleus change as the infectious cycle progresses. When the cycle is initiated, viral genomic ribonucleoproteins (RNPs) released into the cytoplasm enter the nucleus under the direction of the nuclear localization signal of the NP protein. The mechanisms that ensure export of viral (+) strand mRNAs for translation are not fully understood. With the switch to replication, genomic (−) strand RNA segments are synthesized in infected cell nuclei, where they accumulate as viral RNPs containing the NP protein and the three P proteins. These RNPs must be exported to allow virus assembly and completion of the infectious cycle, a reaction that requires the viral M1 and NEP proteins. NEP provides the nuclear export signal, but its interaction with viral RNPs depends on the prior binding of M1 (Fig. 12.21). Because the M1 protein is not available in large quantities until the later stages of infection, such

Figure 12.21 Transport of influenza A virus genomic RNA segments from the nucleus to the plasma membrane. Genomic RNA segments are bound by the NP protein as they are synthesized (see Chapter 6) and subsequently by the M1 protein. M1 is the most abundant protein of the virus particle and enters the nucleus by means of a typical nuclear localization signal (Fig. 5.21). Binding of M1 to genomic RNPs (Fig. 12.19A) both inhibits RNA synthesis and promotes genomic RNP export. M1-continuing RNPs are directed to the cellular exportin-1 export pathway upon binding of NEP, which contains an N-terminal nuclear export signal. The NEP protein possesses no intrinsic RNA-binding activity, but includes a C-terminal M1-binding domain. This domain is thought to allow recognition of RNPs to which the M1 protein is bound. Once within the cytoplasm, viral RNPs are transported to the plasma membrane by a mechanism that is not well understood. The M1 protein interacts with both the membrane itself and the cytoplasmic domains of the HA and NA glycoproteins to initiate assembly and release of enveloped particles.

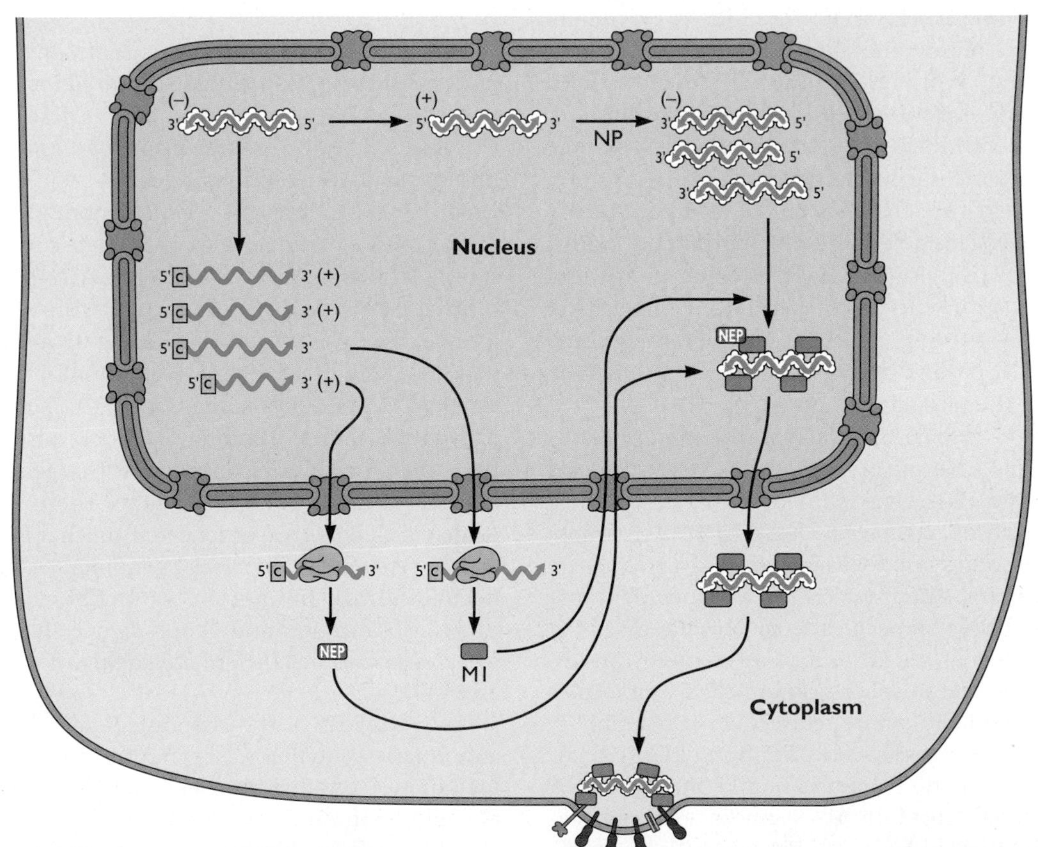

an indirect recognition mechanism probably ensures that genomic RNPs are not exported from the nucleus before sufficient quantities of viral mRNA and progeny genomic RNA have been synthesized. Furthermore, the RNPs enter the cytoplasm only when they have associated with the protein (M1) that is necessary for guiding them to the plasma membrane.

Transport of Genomes from the Cytoplasm to the Plasma Membrane

Like nuclear export, transport of the RNA genomes of enveloped viruses to the appropriate cellular membrane depends on signals present in viral proteins bound to the RNA. As we have seen, the influenza virus M1 protein contains an N-terminal sequence that allows it to bind to lipids (Fig. 12.19A). This membrane-binding domain of M1 allows association of the genomic RNPs with the plasma membrane. Although more limited in its RNA transport functions, the M protein of vesicular stomatitis virus shares several properties with the influenza virus M1 protein.

Replication of vesicular stomatitis virus RNA takes place in the cytoplasm, where newly synthesized genomic RNA molecules assemble with the N, L, and NS proteins to form helical RNPs. Genomic RNA molecules within RNPs can serve as templates for additional cycles of replication or for mRNA synthesis. However, these RNPs eventually must travel to the plasma membrane for association with the G protein and incorporation into virions. Entry into the latter pathway is determined by the viral M protein, which associates with RNPs containing genomic RNA (Fig. 12.19B). This interaction induces formation of a tightly coiled RNA-protein "skeleton," the final structure of the rhabdoviral nucleocapsid (Fig. 12.22). Formation of this structure precludes replication and mRNA synthesis and allows transport of nucleocapsids to the plasma membrane via microtubules (Box 12.11). The membrane-binding domains of the M protein described above, then mediate association with the plasma membrane.

The retroviral proteins that mediate membrane association of genomic RNA are similar to the matrix proteins of these (–) strand RNA viruses in several respects. Once within the cytoplasm, unspliced retroviral RNA is translated on polyribosomes into the Gag and, at low frequency, Gag-Pol polyproteins. The functions of Gag include transport of unspliced RNA molecules to membrane assembly sites and packaging of the RNA into assembling capsids. Consequently, whether unspliced genomic RNA molecules continue to be translated, or are redirected for assembly, is controlled by the cytoplasmic concentration of Gag. Accumulation of Gag may inhibit translation of the viral RNA genome directly: binding of the MA segment to the translational elongation protein EF1α inhibits translation *in vitro*.

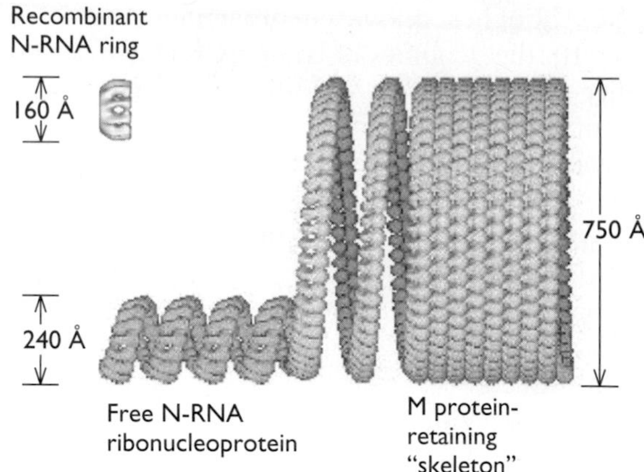

Figure 12.22 Models of the rabies virus nucleocapsid, showing the free nucleocapsid and the nucleocapsid present in virions. The models are based on cryo-electron microscopy and image reconstruction of the two forms of the nucleocapsid, as well as of rings of 9 or 10 molecules of the viral N protein and RNA assembled when the protein is produced in insect cells. The free nucleocapsid, which is the template for viral RNA synthesis, is a loosely coiled helix with a variable pitch and diameter of 240 Å. In contrast, the nucleocapsid helix incorporated into virus particles is tightly wound, with a small pitch and a much larger diameter. These structural transitions are induced by binding of the M protein to the free nucleocapsid. Adapted from G. Schoehn et al., *J. Virol.* **75**:490–498, 2001, with permission.

The NC (nucleocapsid) segment of Gag (Fig. 12.18) contains an RNA-binding domain required for specific recognition of the RNA-packaging sequences described in Chapter 13. The NC region, which functions as an independent protein in mature virions, is very basic and contains at least one copy of a zinc-binding motif (Fig. 12.18). This "zinc knuckle" domain makes a major contribution to the specificity with which Gag or NC proteins bind to unspliced retroviral RNA and, in conjunction with basic amino acids located nearby, is responsible for the RNA-packaging activity of Gag. The N-terminal MA portion of Gag contains the signals described above that target the polyprotein to the plasma membrane. Binding of Gag to unspliced retroviral RNA therefore allows delivery of the genome to assembly sites at the plasma membrane. There is accumulating evidence that more complicated mechanisms govern the fate of cytoplasmic retroviral genomes. For example, trafficking through the nucleus of Gag, which contains both nuclear localization and export signals, has been reported to be essential for efficient packaging of viral RNA during assembly.

Retroviral RNA and associated proteins travel to the appropriate cellular membrane (in most cases the plasma membrane) on recycling endosomal vesicles that move along microtubules. Cellular motor proteins that bind to Gag, such as kinesin, are thought to mediate such

BOX 12.11 EXPERIMENTS
Movement of vesicular stomatitis virus nucleocapsids within the cytoplasm requires microtubules

Vesicular stomatitis virus nucleocapsids that must be transported to the plasma membrane for assembly and budding of virus particles contain the (–) strand RNA genome and several viral proteins, including the P protein (see the text). To examine intracellular trafficking of nucleocapsids, a sequence coding for a green fluorescent protein (eGFP) was inserted into that for the hinge region of the P protein. Control experiments established that the P-eGFP fusion protein catalyzed both viral mRNA synthesis and genome replication, although it exhibited somewhat reduced activity. Furthermore, mutant particles containing P-eGFP in place of the P protein were infectious.

In infected cells, P-eGFP colocalized with newly synthesized viral RNA, as well as with the N and L proteins, in cytoplasmic structures of the size predicted for nucleocapsids. Time-lapse imaging of these P-eGFP-containing structures indicated that nucleocapsids move toward the cell periphery and, unexpectedly, do so in close association with mitochondria. The significance of this association is not yet clear. However, as mitochondria are known to move on microtubules, it seemed likely that viral nucleocapsids also do so. Indeed, nucleocapsids were observed to be distributed throughout the cytoplasm in close association with microtubules (panel A below). Treatment of infected cells with drugs that disrupt microtubules, such as nocodazole, dramatically altered this pattern: nucleocapsids became clustered in the absence of microtubules in large aggregates and did not reach the plasma membrane (panel B). Such drugs also reduced virus yield significantly, confirming the importance of microtubules in the transport of vesicular somatitis virus nucleocapsids to sites of assembly.

Das, S. C., D. Nayak, Y. Zhou, and A. K. Pattnaik. 2006. Visualization of intracellular transport of vesicular stomatitis virus nucleocapsids in living cells. *J. Virol.* **80:**6368–6377.

Localization of P-eGFP-containing nucleocapsids (green) and microtubules (red) in cells infected by the mutant virus VSV-PeGFP and untreated (A) or treated with nocodazole prior to infection (B). Nuclei are in blue. Courtesy of Asit Pattnaik, University of Nebraska—Lincoln.

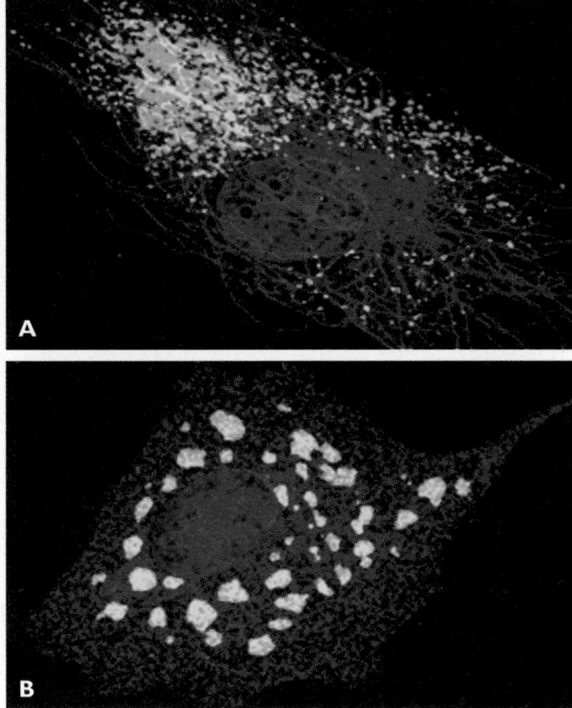

trafficking. Vesicle-associated transport of the viral RNA requires the packaging signal and the NC domain of Gag, which contains the RNA-binding motif described previously. Recruitment to vesicles appears to be via association of Gag with the viral Env protein, which is present in endosomal membranes. These observations suggest that the majority of the interactions among the retroviral components required for assembly are established during transport from the various sites of synthesis to the plasma membrane.

The influenza virus M1, vesicular stomatitis virus M, and retroviral Gag proteins each possess the ability to bind both to RNPs containing genomic RNA and to membranes. Such interactions commit genomic RNA to the assembly pathway, direct genomic RNA to the plasma membrane, and promote interactions among internal and envelope components of virions. These properties are essential at the end of an infectious cycle, when the primary task is assembly of progeny virions. On the other hand, they would be disastrous if the interactions could not be reversed before or at the beginning of a new cycle, when the infecting genome must reach nuclear (influenza virus) or cytoplasmic (vesicular stomatitis virus and retroviruses) sites distant from the plasma membrane. In the case of (–) strand viruses, matrix proteins are removed during virus entry. Matrix-free RNPs can then enter the nucleus for mRNA synthesis (influenza virus) or begin this process in the cytoplasm (vesicular stomatitis virus). Retroviruses exhibit a more elegant mechanism: following virus assembly and budding, Gag (and Gag-Pol) polyproteins are processed by the viral protease to the individual structural proteins shown in Fig. 12.18. Such cleavages place the RNA-binding domain of NC in protein molecules separate from membrane-binding signals of MA, so that matrix-free core RNPs can be released into the cell to initiate a new infectious cycle.

Perspectives

The cellular trafficking systems described in this chapter are just as crucial for virus reproduction as are the host cell's biosynthetic capabilities. The trafficking requirements during the infectious cycle can be quite complex, with transport of viral macromolecules (or structures built from them) over large distances, or in opposite directions during different periods of the infectious cycle. Assembly of progeny particles of all viruses depends on the prior sorting of virion components by at least one cellular trafficking system.

This property has provided important tools (viral proteins or nucleic acids synthesized in large quantities in infected cells) with which to study these essential cellular processes. Indeed, the fundamental principle of protein sorting, that a protein's final destination is dictated by specific signals within its amino acid sequence and/or covalently attached sugars or lipids, was established by analyses of viral proteins. Furthermore, the study of viral proteins that enter the secretory pathway has provided much of what we know of the reactions by which proteins are folded and processed within the ER, as well as those that clear misfolded proteins from the pathway. It therefore seems certain that viral systems will provide equally important insights into the signals and sorting mechanisms that are presently less well characterized, such as those responsible for the direction of proteins to specialized membrane regions of polarized cells.

One of the greatest current challenges in this field remains the elucidation of the mechanics of the movement of proteins, nucleic acids, nucleoproteins, or transport vesicles from one cellular compartment or site to another. The development and application of techniques that exploit fluorescent proteins to visualize transport in living cells is providing important new insights into these processes. Motor protein-mediated movement of various viral proteins, nucleic acids, and assemblies has been observed, and seems likely to emerge as a major theme in the trafficking of viral components for assembly.

Infection by some viruses interferes with the normal transport of specific cellular proteins to the cell surface. Such inhibition can facilitate modulation of the host's immune defense systems, for example, because the display of viral antigens bound to MHC class I proteins on the surface of the infected cell is impaired or prevented. In the decade since the first edition of this book was published, we have come to appreciate that unanticipated and radical alterations in the structure of cellular organelles and transport processes also take place in cells infected by some viruses, to facilitate assembly or egress. Regardless of the degree to which replication and assembly of different viruses impinges on normal cellular processes, transport of components of virus particles to sites of assembly results in formation of microenvironments containing high concentrations of viral structural proteins and the nucleic acid genome. Such microenvironments are ideal niches for the assembly of progeny virions from their multiple parts.

References

Reviews

Behnia, R., and S. Munno. 2005. Organelle identity and the signposts for membrane traffic. *Nature* **438:**597–604.

Braakman, I., and E. van Anken. 2000. Folding of viral envelope viral glycoproteins in the endoplasmic reticulum. *Traffic* **1:**533–539.

Drubin, D. G., and W. J. Nelson. 1996. Origins of cell polarity. *Cell* **84:**335–344.

Ellgard, L., and A. Helenius. 2003. Quality control in the endoplasmic reticulum. *Nat. Rev. Mol. Cell. Biol.* **4:**181–191.

Gahmberg, C. G., and M. Tolvanen. 1996. Why mammalian cell surface proteins are glycoproteins. *Trends Biochem. Sci.* **21:**308–311.

Ikonen, E. 2001. Roles of lipid rafts in membrane transport. *Curr. Opin. Cell Biol.* **13:**470–477.

Jackson, W. T., T. H. Giddings, M. P. Taylor, S. Mulinyawe, M. Rabinovitch, R. R. Kopito, and K. Kirkegaard. 2005. Subversion of cellular autophagosomal machinery by RNA viruses. *PLoS Biol.* **3:**861–871.

Jahn, R., and R. H. Scheller. 2006. SNARES—engines for membrane fusion. *Nat. Rev. Mol. Cell Biol.* **7:**631–643.

Marsh, B. J., and K. E. Howell. 2002. The mammalian Golgi-complex debates. *Nat. Rev. Mol. Cell Biol.* **3:**789–795.

Mostov, K. E., M. Verges, and Y. Altschuler. 2000. Membrane traffic in polarized epithelial cells. *Curr. Opin. Cell Biol.* **12:**483–490.

Pelham, H. R. B. 2001. Traffic through the Golgi apparatus. *J. Cell Biol.* **155:**1099–1101.

Ploubidou, A., and M. Way. 2001. Viral transport and the cytoskeleton. *Curr. Opin. Cell Biol.* **13:**97–105.

Rapoport, T. A. 2007. Protein translocation across the eukaryotic endoplasmic reticulum and bacterial plasma membranes. *Nature* **450:**663–669.

Rothman, J. E., and T. H. Sollner. 1997. Throttles and dampers: controlling the engine of membrane fusion. *Science* **276:**1212–1213.

Suomalainen, M. 2002. Lipid rafts and the assembly of enveloped viruses. *Traffic* **3:**705–709.

Swanson, C. M., and M. H. Malim. 2006. Retrovirus RNA trafficking: from chromatin to invasive genomes. *Traffic* **7:**1440–1450.

Tsai, B., Y. Ye, and T. A. Rapoport. 2002. Retro-translocation of proteins from the endoplasmic reticulum into the cytosol. *Nat. Rev. Mol. Cell Biol.* **3:**246–255.

Wileman, T. 2006. Autophagosomes and autophagy generate sites for virus replication. *Science* **312:**875–878.

Papers of Special Interest

Import of Viral Proteins into the Nucleus

Forstova, J., N. Krauzewicz, S. Wallace, A. J. Street, S. M. Dilworth, S. Beard, and B. E. Griffin. 1993. Cooperation of structural proteins during late events in the life cycle of polyomavirus. *J. Virol.* **67:** 1405–1413.

Lombardo, E., J. C. Ramirez, M. Agbandje-McKenna, and J. M. Almendral. 2000. A beta-stranded motif drives capsid protein oligomers of the parvovirus minute virus of mice into the nucleus for viral assembly. *J. Virol.* **74:**3804–3814.

Zhao, L. J., and R. Padmanabhan. 1988. Nuclear transport of adenovirus DNA polymerase is facilitated by interaction with preterminal protein. *Cell* **55:**1005–1015.

Transport of Viral and Cellular Proteins via the Secretory Pathway

Balch, W. E., W. G. Dunphy, W. A. Braell, and J. E. Rothman. 1984. Reconstitution of the transport of proteins between successive compartments of the Golgi measured by the coupled incorporation of N-acetyl glucosamine. *Cell* **39:**405–416.

Chackerjan, B., L. M. Rudensey, and J. Overbaugh. 1997. Specific N-linked and O-linked glycosylation modifications in the envelope V1 domain of simian immunodeficiency virus variants that evolve in the host alter recognition by neutralizing antibodies. *J. Virol.* **71:** 7719–7727.

Hammond, C., and A. Helenius. 1994. Folding of VSV G protein: sequential interaction with BiP and calnexin. *Science* **266:**456–458.

Horimoto, T., K. Nakayama, S. P. Smeekens, and Y. Kawaoka. 1994. Preprotein-processing endoproteases PC6 and furin both activate hemagglutinin of virulent avian influenza viruses. *J. Virol.* **68:** 6074–6078.

Li, Z., A. Pinter, and S. C. Kayman. 1997. The critical N-linked glycan of murine leukemia virus envelope protein promotes both folding of the C-terminal domains of the precursor polypeptide and stability of the postcleavage envelope complex. *J. Virol.* **71:**7012–7019.

Ono, A., A. A. Waheed, and E. O. Freed. 2007. Depletion of cholesterol inhibits membrane binding and higher-order multimerization of human immunodeficiency virus type 1 Gag. *Virology* **360:**27–35.

Rothman, J. E., and H. F. Lodish. 1977. Synchronized transmembrane insertion and glycosylation of a nascent membrane protein. *Nature* **269:**775–780.

Rousso, I., M. B. Mixon, B. K. Chen, and P. S. Kim. 2000. Palmitoylation of the HIV-1 envelope glycoprotein is critical for viral infectivity. *Proc. Natl. Acad. Sci. USA* **97:**13523–13525.

Takeda, M., G. P. Leser, C. J. Russell, and R. A. Lamb. 2003 Influenza virus hemagglutin concentrates in lipid raft microdomains for efficient viral fusion. *Proc. Nat. Acad. Sci. USA* **100:**14610–14617.

Takeuchi, K., and R. A. Lamb. 1994. Influenza virus M2 protein ion channel activity stabilizes the native form of fowl plague virus hemagglutinin during intracellular transport. *J. Virol.* **65:**911–919.

Yu, I.-M., W. Zhang, H. A. Holdaway, L. Li, V. A. Kostyuchenko, P. R. Chipman, R. Z. Khun, M. G. Rossmann, and J. Chen. 2008. Structure of the immature dengue virus at low pH primes proteolytic maturation. *Science* **319:**1834–1837.

Transport of Viral Proteins to Intracellular Membranes

Alconada, A., U. B. B. Sodeik, and B. Hoflack. 1999. Intracellular traffic of herpes simplex virus glycoprotein gE: characterization of the sorting signals required for its *trans*-Golgi network localization. *J. Virol.* **73:**377–387.

Hobman, T. C., H. F. Lemon, and K. Jewell. 1997. Characterization of an endoplasmic reticulum retention signal in the rubella virus E1 glycoprotein. *J. Virol.* **71:**7670–7680.

Husain, M., and B. Moss. 2001. Vaccinia virus F13L protein with a conserved phospholipase catalytic motif induces colocalization of the B5R envelope glycoprotein in post-Golgi vesicles. *J. Virol.* **75:** 7528–7542.

Law, L. M. J., R. Duncan, A. Esmaili, H. L. Nakhasi, and T. C. Hobman. 2001. Rubella virus E2 signal peptide is required for perinuclear localization of capsid protein and virus assembly. *J. Virol.* **75:**1978–1983.

Melin, L., R. Persson, A. Andersson, A. Bergstrom, R. Ronnholm, and R. F. Pettersson. 1995. The membrane glycoprotein G1 of Uukuniemi virus contains a signal for localization to the Golgi complex. *Virus Res.* **36:**49–66.

Overby, A. K., V. L. Popor, R. F. Peltersson and E. P. A. Neve. 2007. The cytoplasmic tails of Uukuniemi virus (Bungaviridae) GN and Gc glycoproteins are important for intracellular targeting and the budding of virus-like particles. *J. Virol.* **81:**11381–11391.

Saad, J. S., E. Loeliger, P. Luncsford, M. Liriano, J. Tai, A. Kim, J. Miller, A. Joshi, E. O. Freed, and M. F. Summers. 2007. Point mutations in the HIV-1 matrix protein turn off the myristyl switch. *J. Mol. Biol.* **366:**574–585.

Virus-Induced Inhibition of Transport via the Secretory Pathway

Belov, G. A., N. Altan-Bonnet, G. Kovtunovych, C. L. Jackson, J. Lippincott-Schwartz, and E. Ehrenfeld. 2007. Hijacking components of the cellular secretory pathway for replication of poliovirus RNA. *J. Virol.* **81:**558–567.

Beske, O., M. Reichelt, M. P. Taylor, K. Kirkegaard, and R. Andino. 2007. Poliovirus infection blocks EFGIC-to-Golgi trafficking and induces microtubule-dependent disruption of the Golgi complex. *J. Cell Sci.* **120:**3207–3218.

Burgert, H. G., and S. Kvist. 1985. An adenovirus type 2 glycoprotein blocks cell surface expression of human histocompatibility class I antigens. *Cell* **41:**987–997.

Lilley, B. N., and H. L. Ploegh. 2004. A membrane protein required for dissociation of folded proteins from the ER. *Nature* **429:**834–840.

Schubert, U., L. C. Antón, I. Bačík, J. H. Cox, S. Bour, J. R. Bennink, M. Orlowski, K. Strebel, and J. W. Yewdell. 1998. CD4 glycoprotein degradation induced by human immunodeficiency virus type 1 Vpu protein requires the function of proteasomes and the ubiquitin-conjugating pathway. *J. Virol.* **72:**2280–2288.

Suhy, D. A., T. H. Giddings, Jr., and K. Kirkegaard. 2000. Remodeling the endoplasmic reticulum by poliovirus infection and by individual viral proteins: an autophagy-like origin for virus-induced vesicles. *J. Virol.* **74:**8953–8965.

Wiertz, E. J. H., T. R. Jones, L. Sun, M. Bogyo, H. J. Gueze, and H. L. Ploegh. 1996. The human cytomegalovirus US11 gene product dislocates MHC class I heavy chains from the endoplasmic reticulum to the cytosol. *Cell* **84:**709–779.

Transport of Viral RNA Genomes to Sites of Assembly

Chong, L. D., and J. K. Rose. 1994. Interactions of normal and mutant vesicular stomatitis virus matrix proteins with the plasma membrane and nucleocapsids. *J. Virol.* **68:**441–447.

Enami, M., and K. Enami. 1996. Influenza virus hemagglutinin and neuraminidase stimulate the membrane association of the matrix protein. *J. Virol.* **70:**6653–6657.

Martin, K., and A. Helenius. 1991. Nuclear transport of influenza virus ribonucleoproteins: the viral matrix protein (M1) promotes export and inhibits import. *Cell* **67**:117–130.

O'Neill, R. E., J. Tulon, and P. Palese. 1998. The influenza virus NEP (NS2 protein) mediates the nuclear export of viral ribonucleoproteins. *EMBO J.* **17**:288–296.

Tang, Y., U. Winkler, E. O. Freed, T. A. Torrey, W. Kim, H. Li, S. P. Goff, and H. C. Morse III. 1999. Cellular motor protein KIF-4 associates with retroviral Gag. *J. Virol.* **73**:10508–10513.

Ye, Z. P., R. Pal, W. Fox, and R. R. Wagner. 1987. Functional and antigenic domains of the matrix (M1) protein of influenza A virus. *J. Virol.* **61**:239–246.

Zhou, W., L. J. Parent, J. W. Wills, and M. D. Resh. 1994. Identification of a membrane-binding domain within the amino-terminal region of human immunodeficiency virus type 1 Gag protein which interacts with acidic phospholipids. *J. Virol.* **68**:2556–2569.

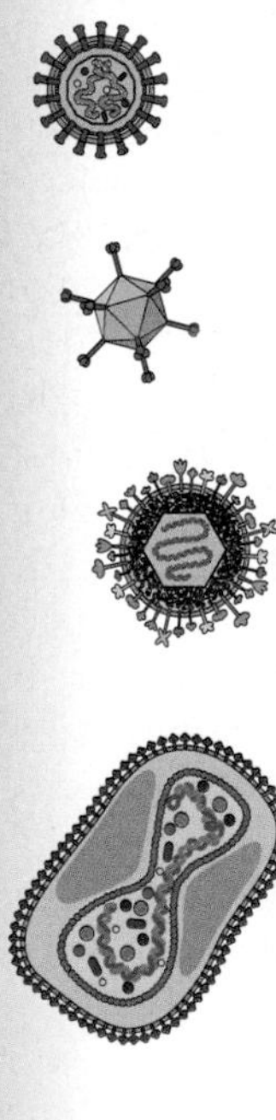

13

Assembly, Exit, and Maturation

The probability of formation of a highly complex structure from its elements is increased, or the number of possible ways of doing it diminished, if the structure in question can be broken down in a finite series of successively smaller substrates!

J. D. Bernal
A. I. Oparin (ed.), *The Origins of Life on Earth* (Pergamon, Oxford, United Kingdom, 1959)

Introduction

Virus particles exhibit considerable diversity in size, composition, and structural sophistication, ranging from those comprising a single nucleic acid molecule and one structural protein to complex structures built from many different proteins and other components. Nevertheless, successful replication of all viruses requires execution of a common set of *de novo* assembly reactions. These processes include formation of the structural units of the protective protein coat from individual protein molecules, assembly of the coat by interaction among the structural units, incorporation of the nucleic acid genome, and release of newly assembled progeny virions (Fig. 13.1). In many cases, formation of internal virion structures must be coordinated with acquisition of a cellular membrane into which viral proteins have been inserted, or additional maturation steps must be completed to produce infectious particles. Assembly of even the simplest viruses is therefore a remarkable process requiring considerable specificity in, and coordination among, each of multiple reactions. Furthermore, virus reproduction is successful only if each of the assembly reactions proceeds with reasonable efficiency, and if the overall pathway is irreversible. The diverse mechanisms by which viruses assemble represent powerful solutions to the problems imposed by *de novo* assembly. Indeed, infectious virus particles are produced in prodigious numbers with great specificity and efficiency.

The structure of a virus particle determines the nature of the reactions by which it is formed (Fig. 13.1), as well as the mechanisms by which it enters a new host cell, reproduces, and promotes pathogenesis within a host animal. For example, the exceptionally stable poliovirus survives passage through the stomach to replicate in the gut. Despite such virus-specific variations in structure and biological properties, all viruses must be metastable structures well suited for protection of the nucleic acid genome in extracellular environments. They must also be built in a way that allows their ready disassembly during entry into a new host cell. A number of elegant mechanisms resolve the apparently paradoxical requirements for very stable associations among virion

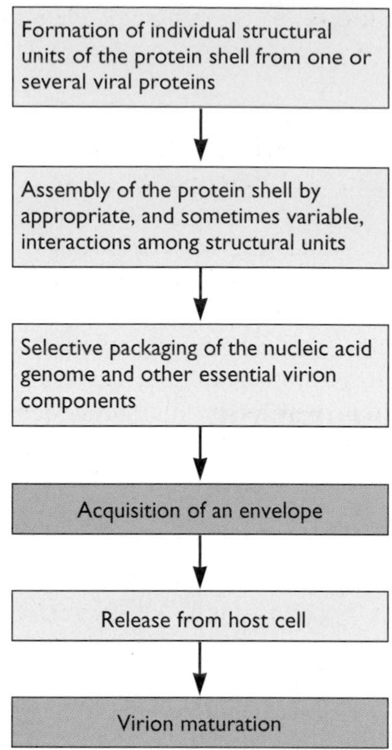

Figure 13.1 Hypothetical pathway of virion assembly and release. Reactions common to all viruses are shown in yellow, and those common to many viruses are shown in blue. The structural units that are often the first assembly intermediates are the homo- or hetero-oligomers of viral structural proteins from which virus particles are built (see Table 4.1). The arrows indicate a general sequence that strictly applies to only some viruses. Packaging of the genome can be coordinated with assembly of the capsid or nucleocapsid, and, for enveloped viruses, assembly of internal components can be coordinated with acquisition of the envelope.

components during assembly and transmission but facile reversal of these interactions when appropriate signals are encountered upon infection of a host cell.

Like synthesis of viral nucleic acids and proteins, production of virus particles depends on host cell components, such as the cellular proteins that catalyze or assist the folding of individual protein molecules. Furthermore, the components from which virions are built are transported to the appropriate assembly site by cellular pathways (see Chapter 12). Such localization of virion components to a specific intracellular compartment or region undoubtedly facilitates virus production: the concentration of virion components in specialized microenvironments must increase the rates of assembly reactions. It is also likely to restrict the number of interactions in which a particular virion component can engage, thereby increasing the specificity of assembly reactions.

A successful virus-host interaction, the survival and propagation of a virus in a host population, generally requires dissemination of the virus beyond the cells infected upon initial contact. Progeny virus particles assembled during the later stages of the infectious cycle must therefore escape from the infected cell for transmission to new host cells within the same animal or to new host animals. The majority of viruses leave an infected cell by one of two general mechanisms: they are released into the external environment (upon budding from, or lysis of, the cell), or they are transferred directly into a new host cell.

Methods of Studying Virus Assembly and Egress

Mechanisms of virus assembly and release can be understood only with the integration of information obtained by structural, biochemical, genetic, and imaging approaches. These methods are introduced briefly in this section.

Structural Studies of Virus Particles

The mechanisms by which viruses form within, and leave, their host cells are intimately related to their structural properties. Our understanding of these processes therefore improves dramatically whenever the structure of a virus particle is determined. An atomic-level description of the contacts among the structural units that maintain the integrity of the virus particle identifies the interactions that mediate assembly, and the ways in which these interactions must be regulated. For example, the X-ray crystal structure of the polyomavirus simian virus 40 particle described in Chapter 4 solved the enigma of how VP1 pentamers could be packed in hexameric arrays. Three distinct modes of interpentamer contact, mediated by conformationally flexible N- and C-terminal arms of VP1 subunits, were also identified. Assembly of the simian virus 40 capsid therefore must require specific variations in the way in which pentamers associate, depending on their position in the capsid shell. Such subtle, yet sophisticated, regulation of association of structural units during assembly was certainly not anticipated and could be revealed only by high-resolution structural information. Such information can also provide important insights into specific features of individual assembly pathways or the mechanisms which ensure that assembly is irreversible, as discussed for picornaviruses in subsequent sections.

The atomic structures of larger naked particles and enveloped viruses cannot yet be determined by X-ray crystallography. However, the newer methods for structural analysis, notably, cryo-electron microscopy and difference imaging, are rapidly improving our understanding of the mechanisms by which these more complex virions assemble.

Visualization of Assembly and Exit by Microscopy

While high-resolution structural studies of purified virions or virion proteins provide a molecular foundation for describing virus assembly, they offer no clues about how assembly (or exit) actually proceeds in an infected cell. Electron microscopy can be applied to investigation of these processes. Examination of thin sections of cells infected by a wide variety of viruses has provided important information about intracellular sites of assembly, the nature of assembly intermediates, and mechanisms of envelope acquisition and release of particles. This approach can be particularly useful when combined with immunocytochemical methods for identification of individual viral proteins, or the structures they form, via binding of specific antibodies attached to electron-dense particles of gold.

The labeling of viral proteins by fusion with the green fluorescent protein (Chapter 2) (or of membranes with fluorescent lipophilic dyes) allows direct visualization of assembly and egress, an approach inconceivable even a few years ago. Such chimeric proteins and virus particles containing them can be observed in living cells, and their associations and movements can be recorded by video microscopy. Consequently, these techniques overcome the limitations associated with traditional microscopic methods, which provide only static views of populations of proteins or virus particles.

Biochemical and Genetic Analysis of Assembly Intermediates

Although of great value, the information provided by X-ray crystallography or microscopy is not sufficient to describe the dynamic processes of virus assembly and release. An understanding of virus assembly requires identification of the intermediates in the pathway by which individual viral proteins and other virion components are converted to mature infectious virions.

When extracts are prepared from the appropriate compartment of infected cells under conditions that preserve protein-protein interactions, a variety of viral assemblies often can be detected by techniques that separate on the basis of mass and conformation (velocity sedimentation in sucrose gradients or gel filtration) or of density (equilibrium centrifugation). These assemblies range from structural units of the capsid or nucleocapsid (see Table 4.1 for the definition of structural units) to empty capsids and mature virions. Similar methods have identified various subcomplexes formed by viral structural proteins in *in vitro* reactions. Furthermore, such structures can be organized into a sequence logical for assembly, from the least to the most complex. On the other hand, it is often quite difficult to **prove** that structures identified by these approaches, such as empty capsids, are true intermediates in the pathway.

By definition, the intermediates in any pathway do not accumulate, unless the next reaction is rate limiting. For this reason, assembly intermediates are generally present within infected cells at low concentrations against a high background of the starting material (mono- or oligomeric structural proteins) and the final product (virus particles). This property makes it difficult to establish precursor-product relationships by pulse-chase experiments: the large pools of structural proteins initially labeled are converted only slowly and inefficiently into subsequent intermediates in the pathway. Genetic methods of analysis provide one powerful solution to this problem. Mutations that confer temperature-sensitive or other phenotypes and block a specific reaction have been invaluable in the elucidation of assembly pathways. A specific intermediate may accumulate in mutant virus-infected cells, and can often be purified and characterized more readily. The judicious use of temperature-sensitive mutants can allow the reactions in a pathway to be ordered, and second-site suppressors of such mutations can identify proteins that interact with one another. Of even greater value is the combination of genetics with biochemistry, an elegant approach pioneered more than 35 years ago during studies of assembly of the complex bacteriophage T4 with the development of *in vitro* complementation (Box 13.1).

The difficulties inherent in kinetic analyses are compounded by the potential for formation of dead-end products, and the unstable nature of some assembly intermediates. Dead-end assembly products are those that form by off-pathway (side) reactions. Because they are not true intermediates, they may accumulate in infected cells and be identified incorrectly as components in the pathway. By definition, such dead-end products differ from true intermediates in some structural property that prevents them from completing assembly. Furthermore, some authentic intermediates may be fragile structures, because they lack the complete set of intermolecular interactions that stabilize the virus particle. Less obvious is the conformational instability of some intermediates: such structures do not fall apart during isolation and purification but, rather, undergo irreversible conformational changes so that the structures studied experimentally do not correspond to **any** present in the infected cell. This kind of instability is not easy to detect, because monoclonal antibodies that distinguish specific structural features, rather than a simple linear sequence of amino acids, are needed. Consequently, such conformational change may well escape notice, as was initially the case for poliovirus empty capsids.

BOX 13.1

BACKGROUND

Late steps in T4 assembly

As illustrated, the head, tail, and tail fibers of this morphologically complex bacteriophage first form separately and then assemble with one another. The many genes encoding products that participate in building the T4 particle are listed by the reaction for which they are required. These gene products and the order in which they act were identified by genetic methods, including identification of second-site suppressors of specific mutations (see Chapter 3). The development of *in vitro* systems in which specific reactions were reconstituted was also of the greatest importance, allowing the development of biochemical complementation. For example, noninfectious T4 particles lacking tail fibers accumulate in infected cells when the tail fiber pathway (right part of figure) is blocked by mutation. These incomplete particles can be converted to infectious phage when mixed *in vitro* with extracts prepared from cells infected with T4 mutated in the gene encoding the major "head" protein. The fact that the phages formed in this way were infectious established that assembly was accurate. This type of system was used to identify the genes encoding proteins required for assembly of heads or tails, as well as scaffolding proteins essential for assembly of the head, but not present in the virus particle. Adapted from W. B. Wood, *Harvey Lect.* **73**:203–223, 1978, and W. B. Wood et al., *Fed. Proc.* **27**:1160–1166, 1968, with permission.

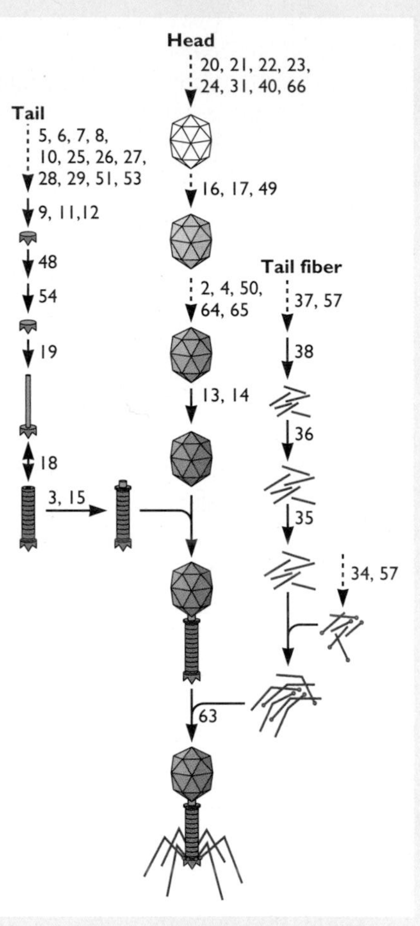

Methods Based on Recombinant DNA Technology

Modern methods of molecular biology and the application of recombinant DNA technology have greatly facilitated the study of virus assembly. Especially valuable is the simplification of this complex process that can be achieved by the synthesis of an individual viral protein or small sets of proteins in the absence of other viral components (Box 13.2).

Assembly of Protein Shells

Although virus particles are far simpler in structure than any cell, they are built from multiple components, such as a capsid (or nucleocapsid), a nucleoprotein core containing the genome, and a lipid envelope carrying viral glycoproteins. The first steps in assembly are therefore the formation of the various components of the virion from their parts. Such intermediates must then associate in ordered fashion, in some cases following transport to the appropriate intracellular site, to complete construction of the virus particle. Application of the techniques described in the previous section has allowed us to sketch the pathways by which many viruses are assembled, and describe some specific reactions in exquisite detail. In this section, we draw on this large body of information to illustrate mechanisms for the efficient assembly of protective protein coats for genomes, the first reaction listed in Fig. 13.1.

Formation of Structural Units

For some viruses, fabrication of a protein shell is coordinated with binding of structural proteins to the viral genome, as

BOX 13.2

METHODS

Assembly of herpes simplex virus type 1 nucleocapsids in a simplified system

The assembly and egress of herpesviruses from infected cells are complex processes that comprise multiple steps (Fig. 13.5 and 13.21). To facilitate analysis of the initial reactions that lead to assembly of the protein shell, the constituent proteins were produced by using baculovirus vectors. Formation of the nucleocapsid was examined by electron microscopy of insect cells infected with various combinations of the recombinant baculoviruses. Empty capsids indistinguishable from those formed in herpes simplex virus type 1-infected cells were observed when six viral genes were expressed together. Four of these encode the structural proteins VP5 (hexons and pentons), VP19C and VP23 (triplexes that link VP5 structural units), and VP26. By

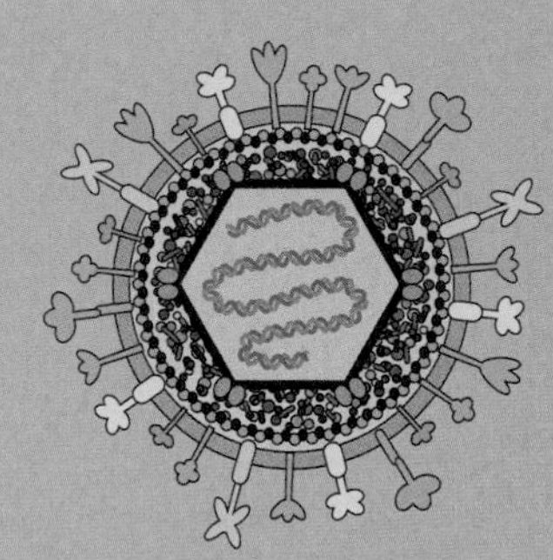

omission of individual recombinant baculoviruses, it was shown that VP26 is not necessary for capsid assembly. Furthermore, only partial or deformed structures assemble in the absence of VP24, VP21, and VP22a, the protease and scaffolding proteins (see "Viral Scaffolding Proteins: Chaperones for Assembly").

Synthesis of subsets of viral proteins is also used widely to identify and characterize protein-protein and protein-membrane interactions critical for assembly.

Tatman, J. D., V. G. Preston, P. Nicholson, R. M. Elliot, and F. J. Rixon. 1994. Assembly of herpes simplex virus type 1 capsids using a panel of recombinant baculoviruses. *J. Gen. Virol.* **75:**1101–1113.

during assembly of the ribonucleoproteins of (−) strand RNA viruses. Consequently, structures built entirely from proteins do not accumulate. In other cases, the first assembly reaction is formation of the structural units from which the protein shell is constructed (Fig. 13.1). Compared to many other steps, this process is relatively simple: individual structural units contain a small number of protein molecules, typically two to six, that must associate appropriately following (or during) their synthesis. Nevertheless, several mechanisms have evolved for formation of structural units, and in some cases additional proteins are required to assist the reactions (Fig. 13.2).

Assembly from Individual Proteins

The structural units of some protein shells, including the VP1 pentamers of simian virus 40, assemble from their individual protein components (Table 13.1; Fig. 13.2A). This straightforward mechanism is analogous to formation of cellular structures containing multiple proteins, such as nucleosomes. In some cases, exemplified by adenoviral pentons, assembly is a two-step process (Fig. 13.2A). In this kind of reaction, the surfaces of individual protein molecules that contact either other molecules of the same protein or a different protein are formed prior to assembly of the structural unit. This arrangement facilitates specific binding when appropriate protein molecules encounter one another: no energetically costly conformational change is required, and subunits that come into contact can simply interlock. Production of these structural units generally can be reconstituted *in vitro*, or in cells that synthesize the component proteins. Such experiments confirm that all information necessary for accurate assembly is contained within the primary sequence, and hence the folded structure, of the protein subunits. On the other hand, the individual protein subunits must find one another in an intracellular environment, in which the concentration of irrelevant (cellular) proteins is as high as that attained in protein crystals (20 to 40 mg/ml). Such a milieu offers uncountable opportunities for nonspecific binding of viral proteins to unrelated cellular proteins. This problem may account for the synthesis of viral structural proteins in quantities far in excess of those incorporated in virus particles, a common feature of virus-infected cells. Such high concentrations must facilitate the encounter of viral proteins with one another by random diffusion, and would provide a sufficient reservoir to compensate for any loss by nonspecific binding to cellular components. Another benefit of high protein concentration is that formation of structural units proceeds efficiently (Fig. 13.2A), driving the assembly pathway in the productive direction.

Assembly from Polyproteins

An alternative mechanism for production of structural units is assembly while covalently linked in a polyprotein precursor (Fig. 13.2B). This mechanism circumvents the need for protein subunits to meet by random diffusion, and avoids competition from nonspecific binding reactions.

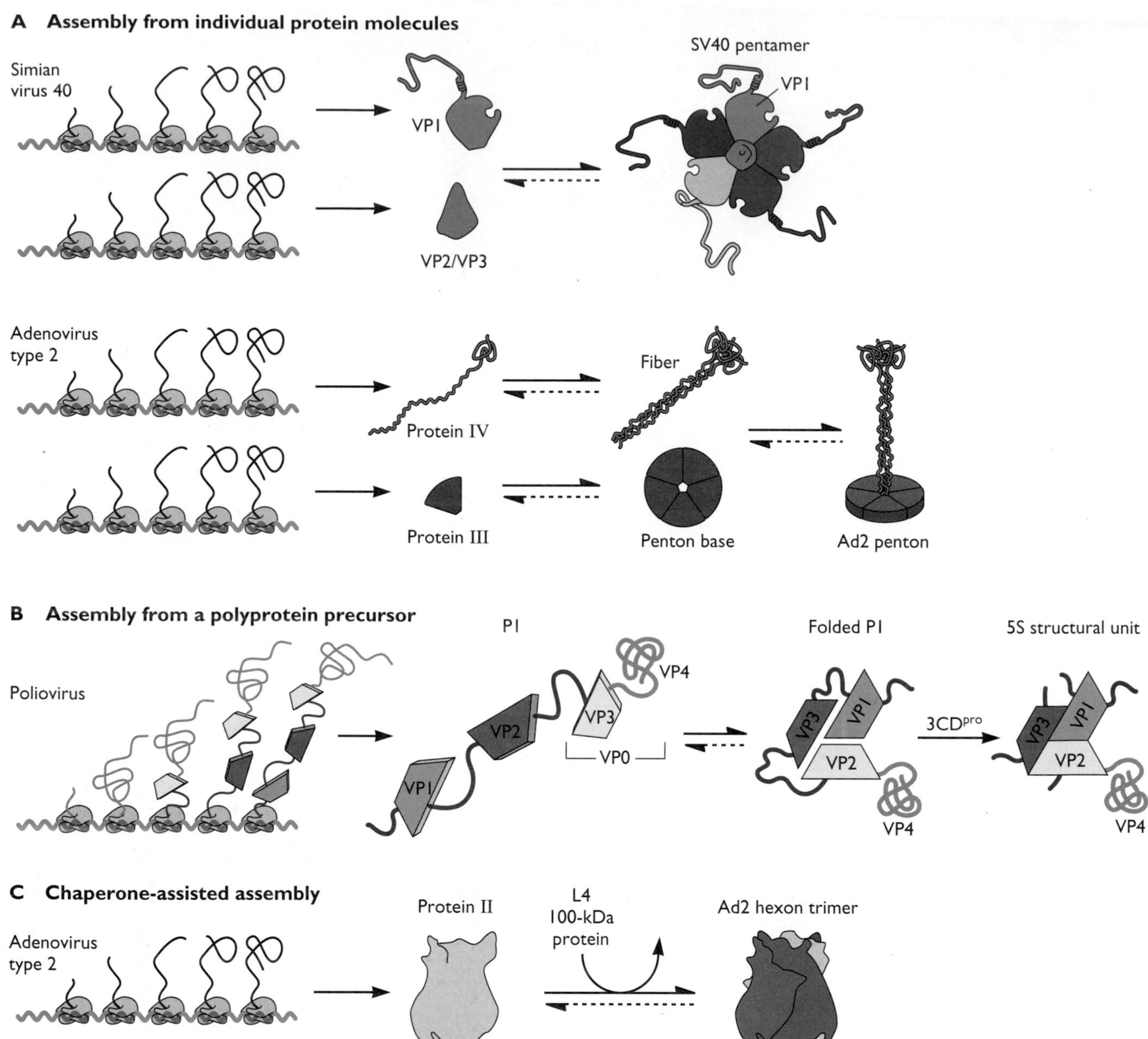

Figure 13.2 Mechanisms of assembly of the structural units of virion protein shells. (A) Assembly from folded protein monomers, illustrated with simian virus 40 (SV40) VP1 pentamers and adenovirus pentons. The assembly reactions shown are the result of specific interactions among the proteins that form structural units. In many cases, for example those among VP1 molecules in the simian virus 40 pentamer, the interactions have been described at atomic resolution (Chapter 4). These assembly reactions are driven in a forward direction by the high concentrations of protein subunits synthesized in infected cells, as indicated by the solid arrows. **(B)** Assembly from a polyprotein precursor, illustrated with the poliovirus polyprotein that contains the four proteins that form the heteromeric structural unit. The latter proteins are synthesized as part of the single polyprotein precursor from which all viral proteins are produced by proteolytic processing. For simplicity, only the P1 capsid protein precursor and its cleavage by the viral 3CD protease following folding and assembly of the proteins of the immature structural unit (VP0, VP3, and VP1) are shown. The flexible covalent connections between VP1, VP3, and VP0 in the P1 precursor, which are exaggerated for clarity, are severed by the protease to form the 5S structural unit. However, VP4 remains covalently linked to VP2 in VP0 until assembly is completed (see the text). **(C)** Assisted assembly. Some structural units are assembled only with the assistance of viral chaperones, such as the adenoviral L4 100-kDa protein, which is required for formation of the hexon trimer from the protein II monomer.

Table 13.1 Mechanisms of assembly of structural units

Mechanism	Virus	Structural unit
Association of individual protein molecules	Adenovirus (adenovirus type 2)	Protein IV trimer (fiber) and protein III pentamer (penton base) that forms pentons
	Hepadnavirus (hepatitis B virus)	C (capsid) protein dimers
	Papovavirus (simian virus 40)	VP1 pentamer, with one molecule of VP2 or VP1 in its central cavity
	Reovirus (reovirus type 1)	λ, σ2 (inner capsid protein) homo-oligomers; σ3-μ, (outer capsid protein) hetero-oligomers
Assisted assembly of protein subunits	Adenovirus (adenovirus type 2)	Hexon trimers of protein II, formed with assistance of the L4 100-kDa protein
	Herpesviruses (herpes simplex virus type 1)	VP5 pentamers and hexamers, formed with assistance of VP22a
Assembly from polyprotein precursors	Alphavirus (Sindbis virus)	Capsid (C) protein folds in, and cleaves itself from, a nascent polyprotein also containing glycoprotein sequences
	Picornavirus (poliovirus)	Immature 5S structural units, VP0-VP3-VP1
	Retrovirus (avian sarcoma virus)	NC, CA, and MA protein shells assembled via Gag polyprotein

The structural units of several viruses are assembled by this mechanism (Table 13.1), which is exemplified by formation of picornaviral capsids. The first poliovirus intermediate, which sediments as a 5S particle, is the immature structural unit that contains one copy each of VP0, VP3, and VP1 (Fig. 13.2B). These three proteins are liberated from the capsid protein precursor P1 upon cleavage by the viral 3CD^{pro} protease. However, it is thought that folding of their central β-barrel domains (Fig. 4.12) takes place during synthesis of P1. The poliovirus structural unit can then form by intramolecular interactions among the surfaces of these β-barrel domains, before the covalent connections that link the proteins are severed.

Retrovirus assembly illustrates an elegant and effective variation on the polyprotein theme. Mature retrovirus particles contain three protein layers. An inner coat of NC protein, which packages the dimeric RNA genome, is enclosed within the capsid built from the CA protein. The capsid is in turn surrounded by the MA protein, which lies beneath the inner surface of the viral envelope (Appendix, Fig. 19). These three structural proteins are synthesized as the Gag polyprotein precursor, which contains their sequences in order of the protein layers they form in virus particles, with MA at the N terminus (Fig. 13.3). Retrovirus particles assemble from such Gag polyprotein molecules by a unique mechanism that allows orderly construction of the three protein layers and, as we shall see, coordination of this reaction with encapsidation of the genome and acquisition of the envelope.

Participation of Cellular and Viral Chaperones

The assembly of viral proteins into structural units is often assisted by cellular **chaperones**. These specialized proteins facilitate protein folding by preventing nonspecific, improper associations among exposed, sticky patches on nascent and newly synthesized proteins. The first chaperone to be identified, the product of the *Escherichia coli*

Figure 13.3 Radial organization of the Gag polyprotein in immature human immunodeficiency virus type I particles. The model for the arrangement of the Gag polyprotein shown below a cryo-electron micrograph of a virus-like particle assembled from Gag was deduced from radial density measurements of digitized images of the particles. The plot shows density as a function of distance from the particle center, in angstroms. Courtesy of T. Wilk, European Molecular Biology Laboratory.

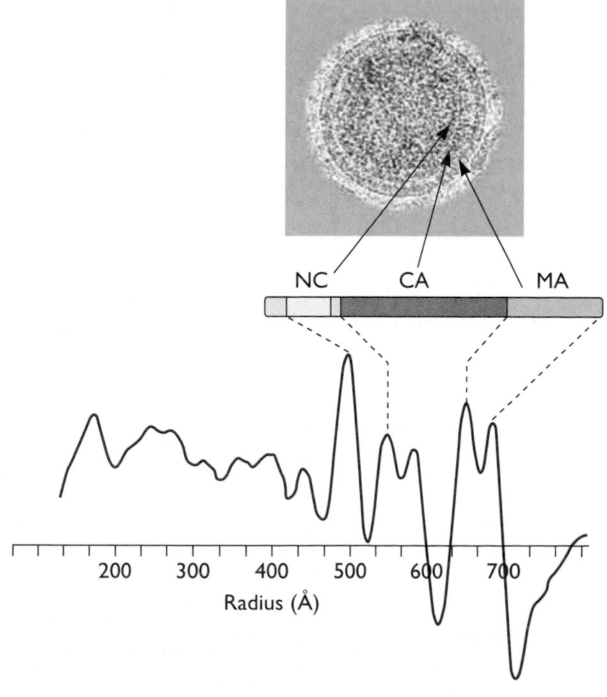

groEL gene, was discovered because it is essential for reproduction of bacteriophages T4 and lambda. As discussed in Chapter 12, the participation of chaperones resident in the lumen of the endoplasmic reticulum (ER) in folding and assembly of oligomeric viral glycoproteins is well established. Cytoplasmic and nuclear chaperones seem likely to play equally important roles in the formation of structural units, or later reactions in assembly. A number of viral structural proteins have been shown to interact with one or more cellular chaperones (Table 13.2). In most cases, a role for these proteins in viral assembly is based on "guilt by association." However, some cellular chaperones have been directly implicated in assembly reactions (Table 13.2). For example, alterations in the Gag protein of the betaretrovirus Mason-Pfizer monkey virus that prevent binding to a cytoplasmic chaperone (Table 13.2) reduce the accumulation of stable Gag and capsids in cells transiently synthesizing the viral protein. These observations, and the dependence of association of Gag molecules with one another on the chaperone, indicate that the latter protein facilitates proper folding of the Gag polyprotein.

Chaperones are abundant in all cells, and some accumulate to concentrations even greater than that of ribosomes. Nevertheless, the genomes of several viruses encode proteins with chaperone activity, some with sequences and functions homologous to those of cellular proteins (Table 13.2). Some viral chaperones are essential participants in the reactions by which structural units are formed. For

Table 13.2 Cellular and viral proteins implicated in viral assembly reactions

Chaperone	Properties and function(s)	Viral protein(s) bound
Cellular chaperones		
Bacterial		
Chaperonins GroEL and GroES	GroEL comprises two rings of 8 identical subunits to which the single heptameric GroES ring binds and dissociates, regulated by the ATPase of GroEL; nonnative proteins enter the GroEL cavity, where they fold; required for assembly of phage/T4 and λ heads	Phage B protein, phage T4 gene 31 protein
Mammalian		
Chaperonin Tri C	Large, double-ring complexes of ~800 kDa surrounding a central cavity; encapsulates substrates upon ATP binding	Mason-Pfizer monkey virus Gag, implicated in productive folding of Gag
Hsp70 proteins	Cytoplasmic proteins synthesized constitutively and in response to stress; in conjunction with Hsp40 cochaperones, participate in ATP-dependent cycles of binding to, and release from, nascent proteins to prevent misfolding or nonspecific aggregation	Adenovirus protein IV Hepatitis B virus L protein Human immunodeficiency virus type 1 Gag Poliovirus P1 capsid protein precursor Simian virus 40 VP1
Hsp68	Contains ATP-binding motifs and an epitope present in a subunit of TriC; RNase L inhibitor	Human immunodeficiency virus type 1 Gag; facilitates a late reaction in assembly of Gag capsids *in vitro*
Viral chaperones		
Adenovirus type 2 L4 100-kDa protein	Formation of hexon trimers	Hexon monomer (protein II)
African swine fever virus CAP80	Productive folding of the major capsid protein, p73	Capsid protein p73
Herpes simplex virus type 1 VP22a	Formation of VP5 pentamers	VP5
Simian virus 40 LT antigen	N-terminal J-domain necessary for assembly of virions; binds to and stimulates activity of cellular Hsc70 proteins	None known
Viral scaffolding proteins		
Adenovirus type 2 L1 52/55-kDa proteins	Necessary for formation of capsids; present in immature particles, but not virions; may be required for encapsidation	IVa_2
Herpes simplex virus type 1 VP22a	Self-associates to form scaffold-like structure that organize assembly of the empty nucleocapsid	VP5 VP19-VP23 triplexes

example, production of adenoviral hexon trimers, which form the faces of the icosahedral capsid, depends on such an accessory protein, the viral L4 100-kDa protein (Fig. 13.2C). The latter protein facilitates folding of monomeric hexon subunits or their assembly into trimers, although its biochemical activity has not been identified.

Capsid and Nucleocapsid Assembly

The accumulation of virion structural units within the appropriate compartment of an infected cell sets the stage for the assembly of more complex capsids or nucleocapsids (see Table 4.1 for nomenclature). For reasons discussed previously, the reactions by which these structures are formed are often not understood in detail. Nevertheless, several different mechanisms for their assembly can be distinguished.

Intermediates in Assembly

A striking feature of well-characterized pathways of bacteriophage assembly (Box 13.1) is the sequential formation of progressively more complex structures: heads, tails, and tail fibers are each assembled in stepwise fashion via defined intermediates. Such an assembly line mechanism appears ideally suited for orderly formation of virus particles, which can be large and morphologically complex. Discrete intermediates also form during assembly of some icosahedral animal viruses. A stepwise assembly mechanism has been well characterized for poliovirus: the 5S structural unit described in the previous section is the immediate precursor of a 14S pentamer, which in turn is efficiently incorporated into virus particles (Fig. 13.4). The pentamer is stabilized by extensive protein-protein contacts and by interactions mediated by the myristate chains present on the five VP0 N termini (see Fig. 4.13C). The contribution of the lipids to pentamer stability is so great that this structure does not form at all when myristoylation of VP0 is prevented. The extensive interactions among its components result in molecular interlocking of the five structural units of the pentamer and impart great stability. Consequently, formation of the 14S assembly intermediate is irreversible under normal conditions in an infected cell, a property that imposes the appropriate directionality on the entire assembly pathway (Fig. 13.4).

For many viruses, discrete assembly intermediates like the poliovirus pentamer have been difficult to identify. In some cases, the absence of discrete intermediates can be attributed to coordination of assembly of protein shells with binding of the structural proteins to the nucleic acid genome. This mode of assembly is exemplified by the ribonucleoproteins of (−) strand RNA viruses, which assemble as genomic RNA is synthesized. Nucleocapsid formation depends on interactions of the protein components with both the nascent RNA and other protein molecules

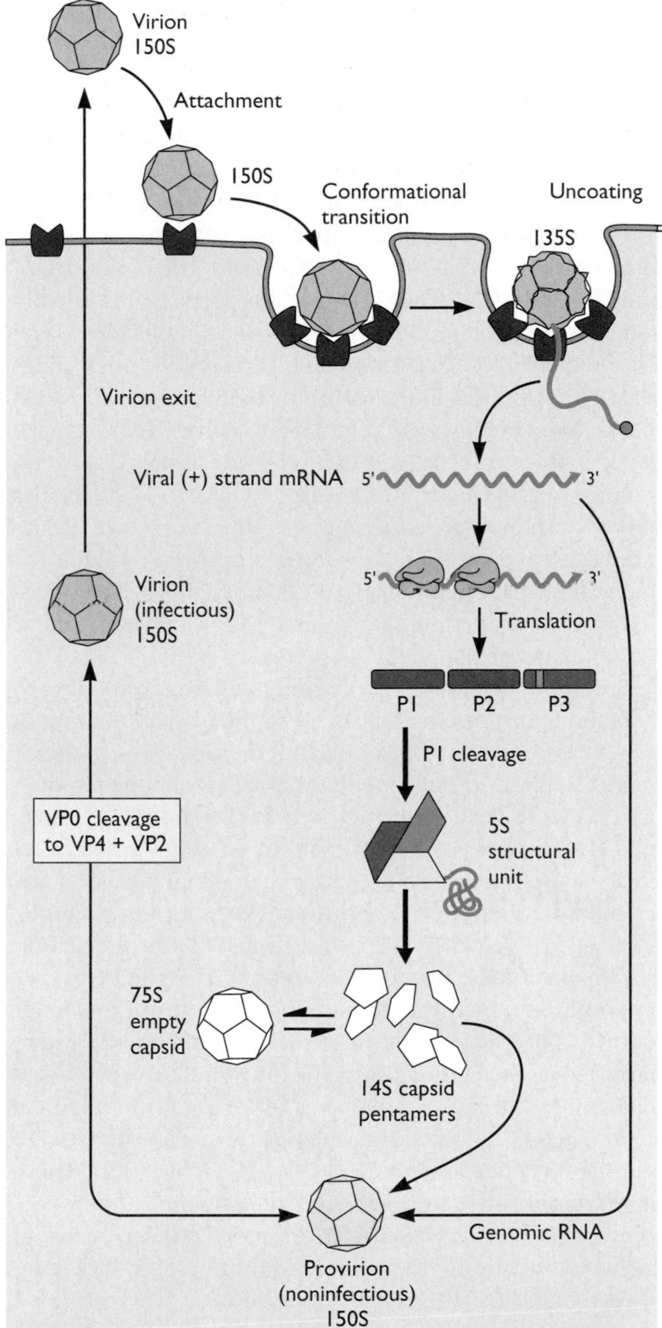

Figure 13.4 Assembly of poliovirus in the cytoplasm of an infected cell. Most of the assembly reactions are essentially irreversible, because of proteolytic cleavage (formation of 5S structural units and mature virions) or extensive stabilizing interactions in the assembled structure (formation of 14S pentamers and of provirions). Stable, empty capsids, originally considered the precursors of provirions, do not possess the same conformation as the mature virion, as symbolized by the white color, and are dead-end products. Formation of the capsid shell from 14S pentamers is coordinated with genome encapsidation, and requires replication of genomic RNA. The conformational transition upon attachment to the poliovirus receptor, for which the virion is primed by cleavage of VP0 to VP2 and VP4, is also illustrated.

previously bound to the RNA. Because the synthesis of genomic RNA molecules is an all-or-none process, so too is the assembly of ribonucleoproteins in infected cells.

Methods that permit synthesis of subsets of virion proteins have begun to provide insights into how such ribonucleoproteins assemble. The vesicular stomatitis virus N protein, which is a dimer in the helical nucleocapsid, aggregates when synthesized alone in *E. coli*. However, when the viral P protein is also made, aggregation does not occur, and discrete, disk-like oligomers assemble. The assembly contains 10 molecules of the N protein, 5 molecules of the P protein, and an RNA molecule of some 90 nucleotides (Fig. 4.7). The disk-like oligomer is equivalent to one turn of the ribonucleoprotein helix formed in vesicular stomatitis virus-infected cells, and the RNA (presumably of bacterial origin) is of the length required to bind to 10 molecules of the N protein. As no further assembly takes place in bacterial cells, even though they contain numerous, long RNA molecules, it has been proposed that assembly of the ribonucleoprotein containing genomic RNA requires a viral or mammalian cell assembly chaperone.

For many enveloped viruses, including retroviruses, assembly of a protein shell is coordinated with binding of structural proteins to a cellular membrane. This property makes isolation of intermediates a technically demanding task. Nevertheless, new methods for separation of intermediates make it possible to examine assembly reactions of these viruses. Specific assembly reactions can also be studied by using simplified experimental systems. For example, when synthesized in a cell-free transcription-translation system, the human immunodeficiency virus type 1 Gag protein multimerizes through a series of discrete intermediates to form 750S particles. These structures resemble virus-like particles released when Gag is the only viral gene expressed in mammalian cells. Conversion of an early intermediate to the 750S particle requires adenosine triphosphate (ATP) and the ATPase cellular Abcel (Hsp68) (Table 13.2). These observations illustrate the power of simplified approaches to the study of virus assembly. An important caveat is that such experimental systems must faithfully reproduce reactions that take place within infected cells. There is good reason to conclude that the *in vitro* assembly of Gag particles meets this crucial criterion, because assembly phenotypes exhibited by altered Gag proteins *in vitro* correspond closely to those observed *in vivo*. Furthermore, binding of Gag to Abcel is required for assembly of later intermediates in cells in culture.

The general dearth of structures simpler than empty capsids in cells infected by nonenveloped viruses might simply be a consequence of the properties that make such intermediates difficult to detect, notably low concentration and instability. In addition, the formation of assembly intermediates may be rate limiting, allowing stable structural units to be stockpiled before the final assembly reactions begin. While these possibilities cannot be excluded, it is more likely that assembly of many capsids is a highly cooperative, all-or-none process. Both simple and more complex icosahedral capsids are built by the repetition of interactions among multiple copies of one or a small number of structural units. Consequently, once the first few structural units were associated in the correct manner, assembly of the capsid would proceed rapidly to completion.

Self-Assembly and Assisted Assembly Reactions

The primary sequences of virion proteins contain sufficient information to specify assembly, including complex reactions like the alternative five- and sixfold packing of VP1 pentamers in the simian virus 40 capsid: when synthesized in *E. coli*, VP1 is isolated as pentamers that assemble into capsid-like structures *in vitro*. Such self-assembly is mediated by interactions between complementary surfaces in individual structural units, or in intermediates assembled from them. Self-assembly of structural proteins is the primary mechanism for formation of protein shells, but other viral components or cellular proteins can assist assembly.

Viral and Cellular Components That Regulate Self-Assembly

Self-assembly of viral structural proteins may be the mortar for construction of virus particles, but other components of the virion often provide an essential foundation, or the blueprint for correct assembly. As we have seen, assembly of the nucleocapsids of (−) strand RNA viruses is both coordinated with, and dependent on, synthesis of genomic RNA. The RNA must serve as a template for productive and repetitive binding of nucleocapsid proteins to one another. Interactions of retroviral Gag proteins with RNA mediated by the NC RNA-binding domain also appear to be essential to initiate assembly of the Gag protein shell (Box 13.3). In other cases, the viral genome plays a more subtle yet equally important role, ensuring that the interactions among structural units are those necessary for infectivity. For example, poliovirus empty capsids lack internal structural features characteristic of the mature virion, because VP0 is not cleaved to VP4 and VP2. The RNA genome is thought to participate in the autocatalytic cleavage of this precursor, which is essential for the production of infectious particles.

Association of structural proteins with a cellular membrane is essential for the assembly of some protein shells, a situation exemplified by many retroviruses: the sequences of MA that specify Gag myristoylation and

BOX 13.3

DISCUSSION

A scaffolding function for RNA

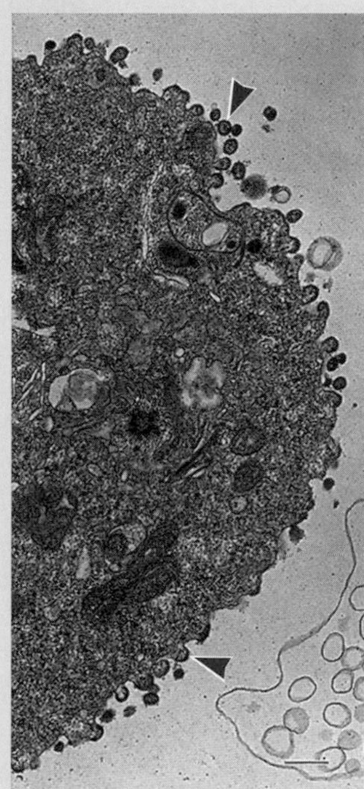

Electron micrograph showing a thin section (fixed and stained) of a human T cell synthesizing the viral Gag-Pol protein. Prior to electron microscopy, viral particles (red arrowheads) were labeled with polyclonal antibodies (attached to gold beads) recognizing the CA protein. Bar, 1.0 μm. N, nucleus; M, mitochondrion. Courtesy of J. J. Wang, Institute of Biomedical Sciences, Academica Sinica, Taipei, Taiwan, and B. Horton and L. Ratner, Washington University School of Medicine, St. Louis, MO.

When synthesized in the absence of any other viral component, retroviral Gag polyproteins direct assembly and release of the virus-like particles shown in the figure. It was therefore assumed for many years that this protein contains all information necessary and sufficient for assembly of particles. However, the results of more recent experiments indicate that RNA acts as a scaffold during Gag assembly.

In vitro studies of the ability of altered Gag proteins to multimerize with the full-length protein initially underscored the importance of the nucleocapsid (NC) RNA-binding domain for efficient assembly. Association of Gag with RNA is also required for multimerization in this system. The apparent contradiction between these findings and efficient assembly of Gag in mammalian cells in the absence of genomic RNA was subsequently resolved: virus-like particles contain cellular RNAs when they form in cells infected by a Moloney murine leukemia virus mutant with a deletion in the signal that directs packaging of the RNA genome. Furthermore, when the Gag coding sequence was expressed from an alphavirus vector, which directs very efficient synthesis of genomic and subgenomic viral RNAs, these alphaviral mRNAs were readily detected in the Gag virus-like particles. Finally, RNase digestion of cores assembled from Gag in wild-type Moloney murine leukemia virus-infected cells was shown to dissociate these structures. This observation indicates that interactions of Gag molecules with RNA, as well as with one another, are required for assembly, and maintain particle stability.

In infected cells, interaction of the MA domain of Gag with the plasma membrane is necessary for efficient assembly (see the text). The high concentrations of Gag typically used in *in vitro* assembly reactions appears to overcome this requirement.

Campbell, S., and V. M. Vogt. 1995. Self assembly in vitro of purified CA-NC proteins from Rous sarcoma virus and human immunodeficiency virus type 1. *J. Virol.* **69:**6487–6497.

Muriaux, D., J. Mirro, D. Harvin, and A. Rein. 2001. RNA is a structural element in retrovirus particles. *Proc. Natl. Acad. Sci. USA* **98:**5246–5251.

binding to the cytoplasmic surface of the plasma membrane (described in Chapter 12) are also required for assembly of the core.

Binding of structural proteins to the genome or to a cellular membrane might simply raise their local concentration sufficiently to drive self-assembly, might organize the proteins in such a way that their interactions become cooperative, or might induce conformational changes necessary for productive association of structural units. These mechanisms, which are not mutually exclusive, have not been distinguished experimentally, but there is evidence for induction of conformational transitions in specific cases. We do not understand adequately the molecular mechanisms by which binding of structural proteins to other virion components directs or regulates particle assembly. However, such a requirement offers the important advantage of integration of assembly of protein shells with acquisition of other essential parts of the virion.

Cellular components can also modulate the fidelity with which viral structural proteins bind to one another. The capsid-like structures assembled when simian virus 40 VP1 is made in insect or mammalian cells are much more regular in appearance than those formed *in vitro* by bacterially synthesized VP1. Modification of VP1 (by acetylation and phosphorylation) or the participation of chaperones, such as Hsp70, must therefore improve the accuracy with which VP1 pentamers associate to form capsids. Similarly, *in vitro* self-assembly of poliovirus structural proteins is very slow, proceeding at a rate at least 2 orders of magnitude lower than that observed in infected cells. Furthermore, the empty capsids that form have the altered conformation described previously, unless the reaction is seeded by 14S pentamers isolated from infected cells. This property indicates that the appropriate folding, modification, and/or interaction of the viral proteins that form the pentamer are critical for subsequent assembly reactions to proceed productively. Within infected cells, these crucial reactions are likely to be modulated by cellular chaperones, such as Hsp70, which is associated with the polyprotein during its folding to form 5S structural units. It is clear from such examples that host cells provide a hospitable environment for productive virus assembly, one that is not necessarily reproduced when viral structural proteins are made and assemble *in vitro*.

Viral Scaffolding Proteins: Chaperones for Assembly

Accurate assembly of some large icosahedral protein shells, such as those of adenoviruses and herpesviruses, is mediated by proteins that are not components of mature virions. Because these proteins participate in reactions by which the capsid or nucleocapsid is constructed, but are then removed, they are termed **scaffolding proteins**. Among the best characterized of such proteins is the precursor of the herpes simplex virus type 1 VP22a protein.

This protein is the major component of an interior core present in assembling nucleocapsids (Fig. 13.5A). In the absence of other viral proteins, it forms specific scaffold-like structures. In immature nucleocapsids isolated from infected cells, it appears as an ordered sphere that lacks the icosahedral symmetry of the mature nucleocapsid. Self-association of pre-VP22a stimulates binding of the scaffolding protein to VP5, the protein that forms the hexameric and pentameric structural units of the nucleocapsid. The VP5 and pre-VP22a proteins form a core via hydrophobic interactions, to which are added additional VP5 hexamers and the triplex structures formed by VP19 and VP23 (Table 4.7). The latter are also required for capsid assembly, which occurs by sequential formation of partial dome-like structures and the spherical immature nucleocapsid. The interactions of VP5 with the scaffolding protein guide and regulate the intrinsic capacity of VP5 hexamers (and other nucleocapsid proteins) for self-assembly: omission of the scaffolding protein from a simplified assembly system (Box 13.2) leads to the production of unclosed and deformed nucleocapsid shells.

One of the 12 vertices of the herpesviral nucleocapsid is formed by the portal through which the DNA enters, rather than a VP5 pentamer (Fig. 13.5A, see also Fig. 4.26). Consequently, this unique structural unit, a dodecamer of the UL6 protein, must be incorporated at just one vertex during assembly. This reaction requires interaction of the portal with the scaffolding protein: a small molecule that blocks this interaction prevents assembly of portal-containing nucleocapsids in infected cells. Although the portal is dispensable for formation of procapsids or nucleocapsids, the results of *in vitro* studies indicate that it can be incorporated only during the initial stages of assembly. The mechanism which ensures that each nucleocapsid contains only one portal remains an enigma.

Once nucleocapsids have assembled, scaffolding proteins must be disposed of, so that viral genomes can be accommodated (Fig. 13.5A). The virion protease (VP24), which is also present in the core of assembling nucleocapsids, is essential for such DNA encapsidation. This protein is incorporated into the assembling nucleocapsid as a precursor (Fig. 13.5B). The protease precursor possesses some activity and initiates cleavage to produce VP24. The protease cleaves a short C-terminal sequence from the scaffolding protein that is required for binding of the scaffolding protein to VP5. Such processing presumably disengages scaffolding from structural proteins, once assembly of the nucleocapsid is complete. The protease also degrades the scaffolding protein so that encapsidation of the genome can begin.

The proteolytic cleavages that liberate the VP5 structural units from their association with the scaffold also induce major changes in the organization of the nucleocapsid shell. Assembly of the nucleocapsid from its constituent proteins *in vitro* proceeds via a short-lived, spherical precursor to the mature structure. This intermediate possesses the $T = 16$ icosahedral symmetry characteristic of the mature nucleocapsid. However, it is unstable and dissociates at low temperature, because the strong interactions among VP5 structural units that form the floor of the nucleocapsid shell are absent. Similarly cold-sensitive particles accumulate at nonpermissive temperatures in cells infected by a mutant virus encoding a temperature-dependent viral protease, and can form infectious virions following shift to a permissive temperature. These properties suggest that the open structures initially assembled *in vitro* may correspond to *bona fide*, but short-lived, intermediates in herpesvirus assembly *in vivo*, analogous to the well-characterized **procapsids** formed during assembly of certain DNA-containing bacteriophages (Box 13.4).

Assembly of simpler protein shells can also depend critically on a viral protein. In addition to its many other functions, the simian virus 40 large T antigen (LT) participates in virion assembly. This protein does not form a scaffold, but an N-terminal domain of LT appears to be essential to

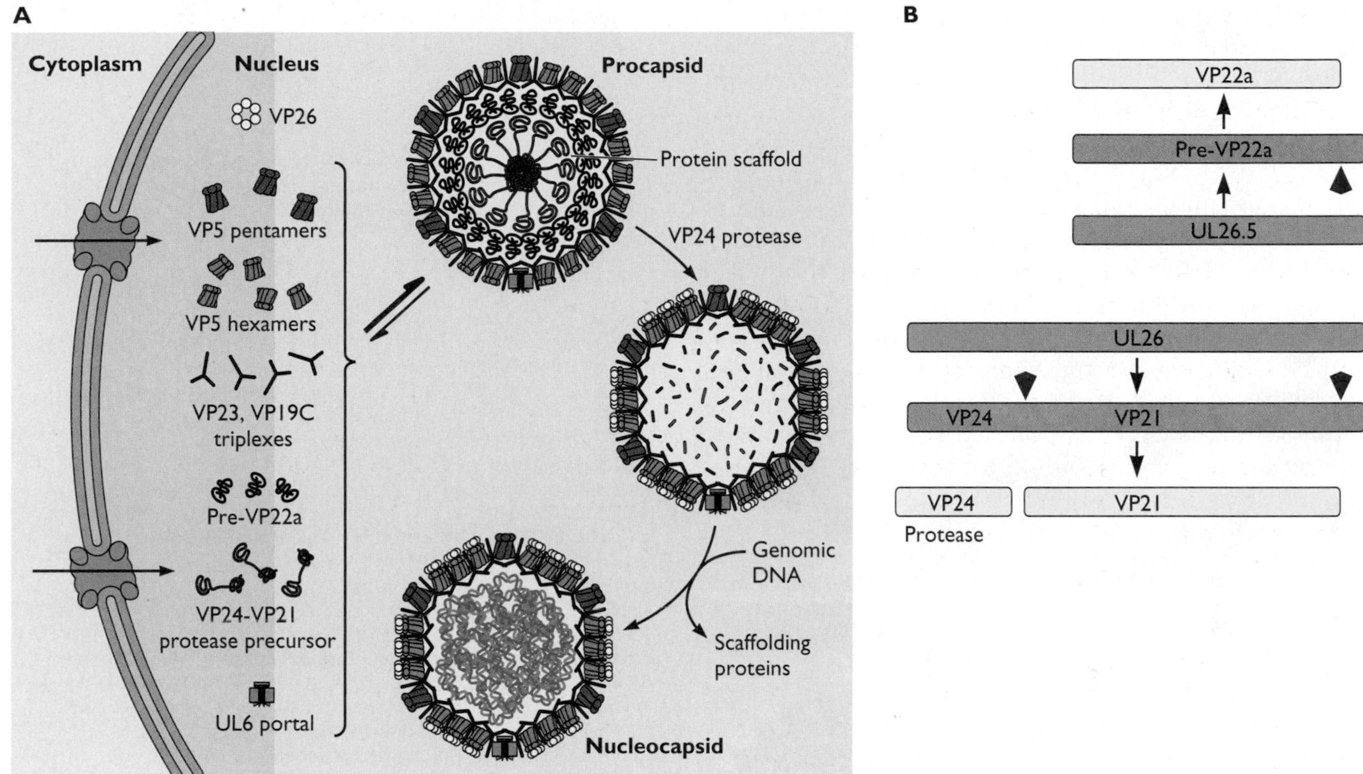

Figure 13.5 Assembly of herpes simplex virus type 1 nucleocapsids. (A) Assembly begins as soon as nucleocapsid proteins accumulate to sufficient concentrations in the infected cell nucleus. Intermediates include pentamers and hexamers of the major capsid protein VP5, which form pentons and hexons in the capsid, and triplexes of the minor proteins VP23 and VP19C. Whether structural units assemble prior to transport into the nucleus is not clear. Viral proteins essential for assembly of the nucleocapsid but not present in mature virions, the scaffolding protein (pre-VP22a) and the viral protease precursor (VP24-VP21), must also enter the nucleus. Assembly of nucleocapsids depends on the formation of an internal scaffold around which the protein shell assembles. Subsequent reactions require the viral protease to remove the scaffolding protein, allowing entry of the DNA genome and morphological transitions. As discussed in the text, encapsidation is concurrent with cleavage of the concatemeric products of herpesviral DNA replication. **(B)** Overlapping sequences of scaffolding proteins. The UL26 and UL26.5 reading frames are shown in purple, and their primary translation products are shown in light brown. The initiating methionine of VP22a protein is within the larger reading frame that encodes the VP24-VP21 polyprotein. Consequently, VP21 and VP22a are identical in sequence, except that the former contains a unique N-terminal segment. All proteolytic cleavages, at the sites indicated by the red arrowheads, including those that liberate the protease itself from the VP24-VP21 precursor, are carried out by the VP24 protease. The cleavage at the C-terminal site in VP22a disengages the scaffolding from the capsid proteins.

organize the capsid shell: alterations within this domain block production of virions and induce accumulation of an incomplete structure, not normally observed during virus assembly, that contains the viral chromatin and VP1. The N-terminal segment of LT possesses chaperone activity (Table 13.2). It is similar in sequence to a specific domain of cellular chaperones of the DnaJ family and, like these cellular proteins, stimulates the activity of Hsp70 chaperones. The chaperone activity of LT may ensure the productive binding of VP1 pentamers to one another, and to other components of the virion during assembly.

Selective Packaging of the Viral Genome and Other Virion Components

Concerted or Sequential Assembly

Incorporation of the viral genome into assembling particles is often called **packaging**. This process requires specific recognition of genomic RNA or DNA molecules (see below). It is clear that all viral genomes are packaged by one of two general mechanisms, concerted or sequential assembly.

In concerted assembly, the structural units of the protective protein shell assemble productively only in association

BOX 13.4

EXPERIMENTS

Visualization of structural transitions during assembly of DNA viruses

The assembly of viruses that package double-stranded DNA genomes into a preformed protein shell exhibits several common features, regardless of the host organism. These include the presence of a portal for DNA entry in the capsid or nucleocapsid precursor and probably the mechanism of DNA packaging (see the text). In addition, as illustrated for bacteriophage λ and herpes simplex virus type 1, assembly of DNA-containing structures is accompanied by major reorganizations of the protein shell.

A Phage λ

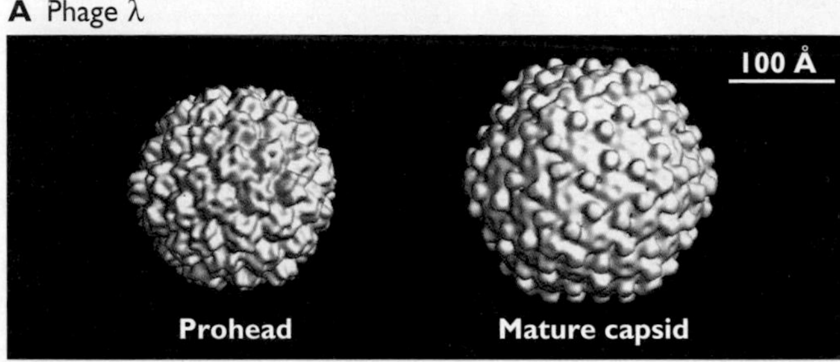

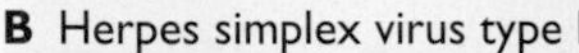
B Herpes simplex virus type I

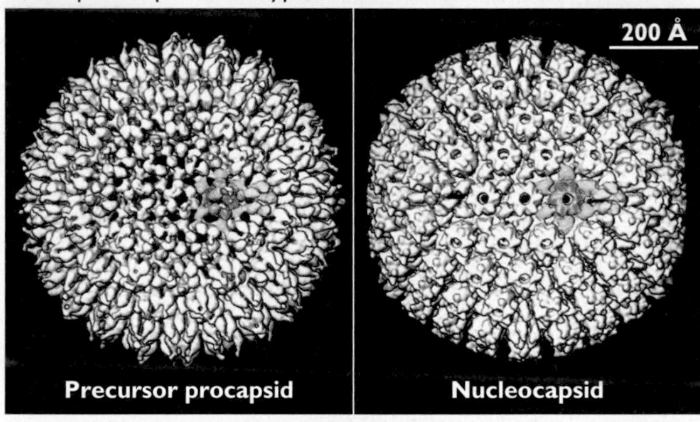

(A) Cryo-electron micrographs of the phage lambda prohead and the DNA-containing mature capsid. The former comprises hexamers and pentamers of the capsid protein gpE organized with $T = 7$ icosahedral symmetry, and is assembled prior to encapsidation of the DNA genome. It is smaller than the mature capsid (270 and 315 Å in diameter, respectively), but its protein shell is considerably thicker. Packaging of the DNA genome leads to an expansion of the capsid shell, as a result of reorganization of gpE hexamers. This change is accompanied by binding of the gpD protein, which contributes to capsid stabilization. Adapted from T. Dokland and H. Murialdo, *J. Mol. Biol.* **233:**682–694, 1993, with permission. **(B)** Cryo-electron micrographs of herpes simplex virus type 1 nucleocapsid precursor and mature nucleocapsid, viewed along a twofold axis of icosahedral symmetry. Some copies of the proteins that form the particles' surfaces are colored as follows: VP5 hexons, red; VP5 pentons, yellow; triplexes containing one molecule of VP19C and two of VP23, green. The precursor nucleocapsid is spherical and less angular than the mature, DNA-containing structure, and its protein shell is thicker. Furthermore, the VP5 hexamers are not organized in a highly regular, symmetric manner in the precursor, resulting in a more open protein shell. The precursor nucleocapsid also lacks the VP26 protein, which binds to the external surface of VP5 hexamers, but not pentamers, in the mature nucleocapsid. Adapted from A. C. Steven et al., *FASEB J.* **10:**733–742, 1997, with permission.

with the genomic nucleic acid. The nucleocapsids of (−) strand RNA viruses form by a concerted mechanism (Fig. 13.6), as do retrovirus particles (Fig. 13.7). In the alternative mechanism, sequential assembly, the genome is inserted into a preformed protein shell. The formation of herpesviral nucleocapsids provides a clear example of this type of assembly (Fig. 13.5). Neither mutations that inhibit viral DNA synthesis nor those that prevent DNA packaging block assembly of capsid-like structures that lack DNA. These phenotypes establish unequivocally that the DNA genome must enter preformed protein shells. In contrast to concerted assembly, encapsidation of the genome in a preformed structure requires specialized mechanisms to pull or push the genome into the capsid (see the next section), as well as to maintain or open a portal for entry of the nucleic acid. The herpesviral portal UL6, which is present at only 1 of the 12 vertices of the nucleocapsid (Fig. 13.5; see also Fig. 4.26), fulfills the latter function.

Despite the clear differences between concerted and sequential assembly pathways, it can be quite difficult to decide which mechanism applies to some viruses with icosahedral symmetry. A classic case in point is poliovirus assembly, which has been studied for more than 30 years. In the first scheme to be proposed, an empty capsid containing 60 copies of the VP0-VP3-VP1 structural unit but lacking the RNA genome was viewed as the precursor of the **provirion** (Fig. 13.4), the RNA-containing but immature virion. Such 75S empty capsids, initially called procapsids, are detected readily in infected cell extracts and also form by self-assembly of 14S pentamers *in vitro*. In the alternative pathway, pentamers are proposed to condense around the RNA genome (Fig. 13.4). In this

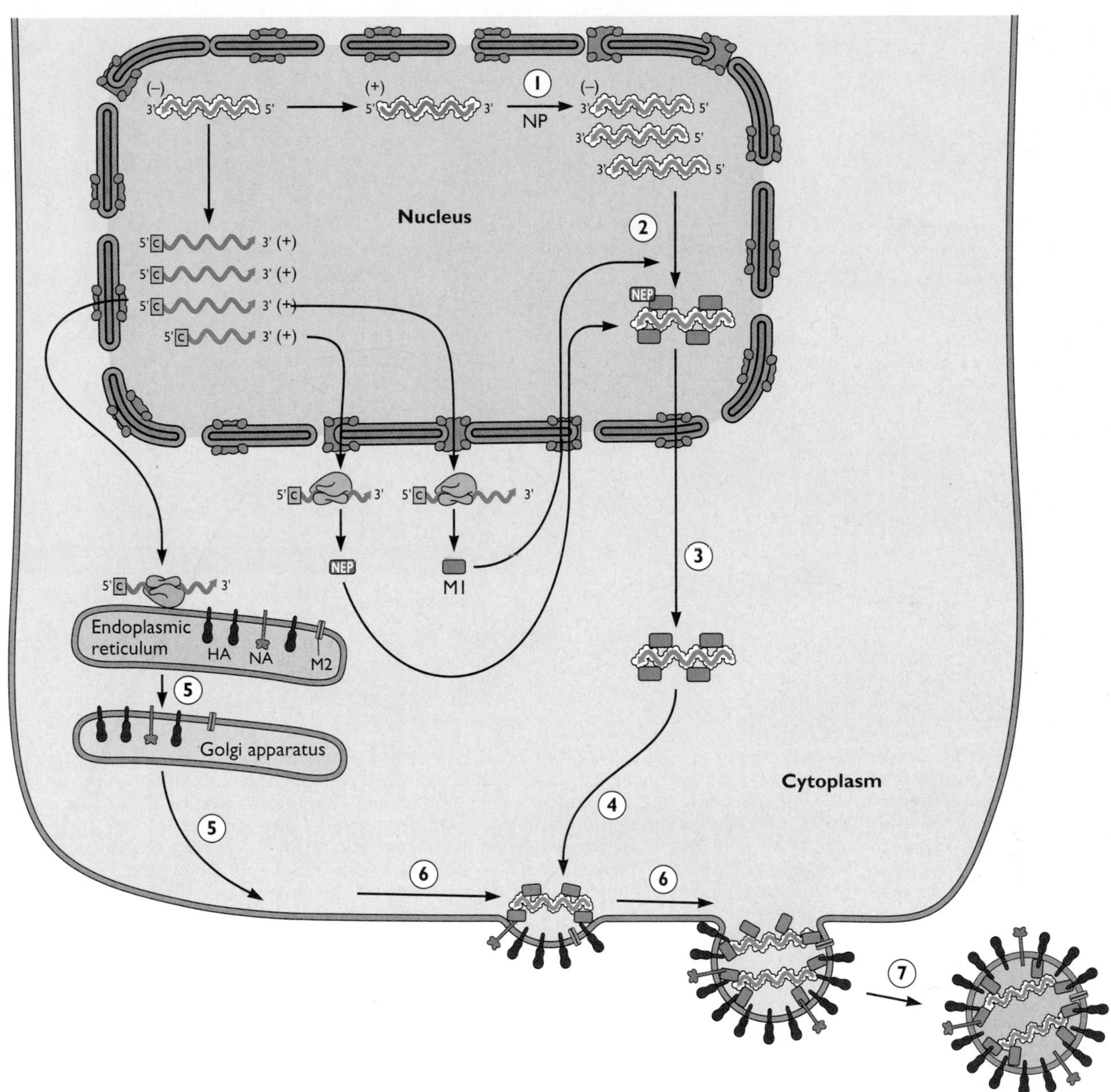

Figure 13.6 Assembly of influenza A virus. Assembly proceeds in stepwise fashion within different compartments of an infected cell. As (–) strand genomic RNA is synthesized in the nucleus, it is packaged by the NP RNA-binding protein (step 1). These ribonucleoproteins may serve as templates for mRNA synthesis, participate in further cycles of replication, or bind the M1 protein (step 2). The latter interaction prevents further RNA synthesis, and allows binding of NEP and export of the nucleocapsid to the cytoplasm (step 3). The M1 protein also binds to the cytoplasmic face of the plasma membrane via specific sequences and directs the nucleocapsid to the plasma membrane (step 4). The plasma membrane carries the viral HA, NA, and M2 proteins, which reach this site via the cellular secretory pathway (step 5). The M1 protein probably controls budding (step 6) via recruitment of cellular components (see the text). Fusion of the membrane bud releases the enveloped particle (step 7). Only two of the eight genome segments are illustrated for clarity.

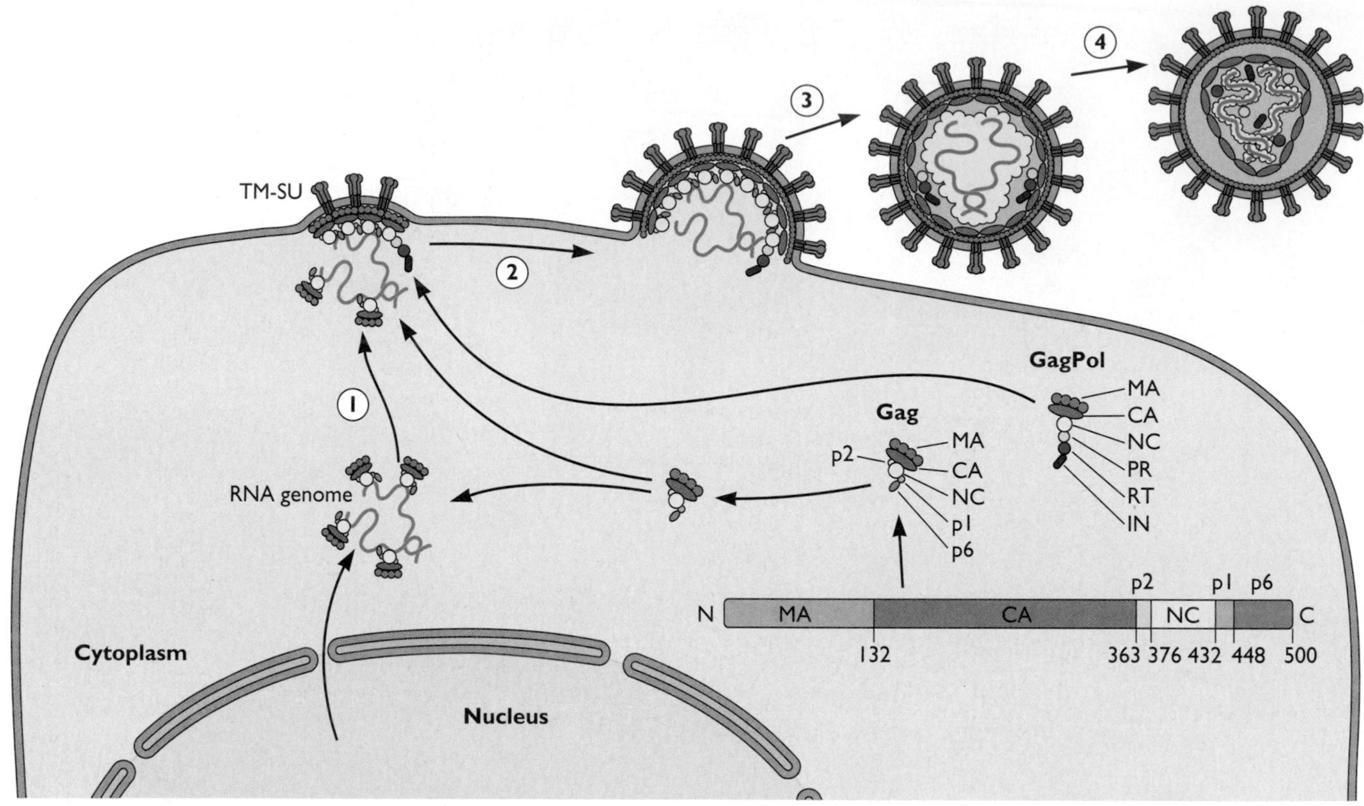

Figure 13.7 Assembly of a retrovirus from polyprotein precursors. The Gag polyprotein of all retroviruses contains the MA, CA, and NC proteins linked by spacer peptides that are variable in length and position. The proteins are in the order (from N to C terminus) of the protein shells of the virus particle, from the outer to the inner. The organization of human immunodeficiency virus type 1 Gag is summarized on the right. A minor fraction, about 1 in 10, of Gag translation products carry the retroviral enzymes, denoted by PR, RT, and IN, at their C termini. The association of Gag molecules with the plasma membrane, with one another, and with the RNA genome via binding of NC segments initiates assembly at the inner surface of the plasma membrane (step 1). In some cases, such as human immunodeficiency virus type 1, the MA segment also binds specifically to the internal cytoplasmic domain of the TM-SU glycoprotein. Assembly of the particle continues by incorporation of additional molecules of Gag (step 2). This pathway is typical of many retroviruses, but some (e.g., betaretroviruses) complete assembly of the core in the interior of the cell prior to its association with the plasma membrane. The dimensions of the assembling particle are determined by interactions among Gag polyproteins. Eventually, fusion of the membrane around the budding particle (step 3) releases the immature noninfectious particle. Cleavage of Gag and Gag-Pol polyproteins by the viral protease (PR) produces infectious particles (step 4) with a morphologically distinct core (see Fig. 13.23).

mechanism, capsid assembly and packaging of the RNA genome are coupled, and empty capsids are considered dead-end. The concerted pathway is supported by several experimental observations. These include the ability of radioactively labeled pentamers to form virions without the appearance of a procapsid intermediate, and the demonstration that stable empty capsids are produced by irreversible, conformational changes that take place during their extraction from infected cells. Furthermore, in a cell-free system for synthesis of infectious poliovirus particles, exogenously added 14S pentamers assemble with newly synthesized viral (+) strand RNA to form virions with antigenic sites characteristic of virions produced in infected cells. In contrast, exogenously added empty capsids undergo no further assembly, even when genomic RNA is synthesized, confirming that they are dead-end products. In this case, there is now strong evidence for the concerted pathway, but in others, such as adenovirus assembly (Fig. 13.8), there is still debate about which mechanism is used.

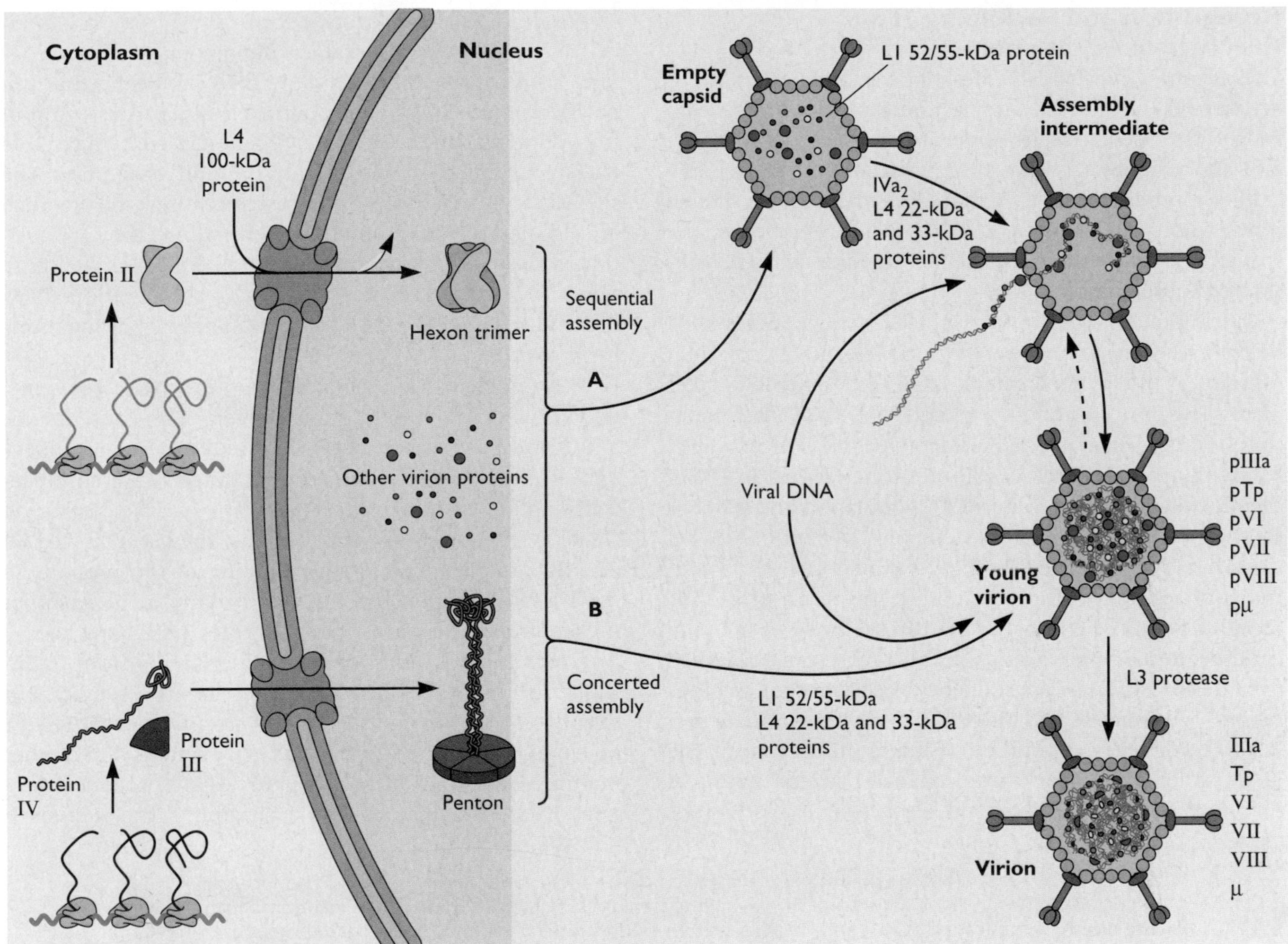

Figure 13.8 Adenovirus assembly pathways. Synthesis and assembly of hexons and pentons and their transport into the nucleus set the stage for assembly. The L4 100-kDa protein is required for formation of hexons, but its molecular function is not known. **(A)** In the pathway originally proposed, these structural units and the proteins that stabilize the capsid assemble into empty capsids. The L1 52/55-kDa proteins are necessary for the formation of structures that can complete assembly, and decrease in concentration as assembly proceeds. The DNA is then inserted into this structure via the packaging signal located near the left end of the genome. The viral IVa_2 and L4 22-kDa proteins bind specifically to this sequence *in vitro* and are required for assembly in infected cells. Premature breakage of DNA in the process of insertion would yield the structure designated "Assembly intermediate," in which an immature capsid is associated with a DNA fragment derived from the left end of the viral genome. Core proteins are encapsidated with the viral genome to yield noninfectious young virions. Mature virions are produced upon cleavage of the precursor proteins listed to the right of the young virion. **(B)** The alternative pathway is based on the failure of any capsid-like structures to assemble in cells infected by viruses with mutations that eliminate the packaging signal, prevent synthesis of the IVa_2 protein, or remove the C-terminal segment of the L4 33-kDa protein. In this model, capsid assembly and encapsidation of the genome are concerted reactions. Empty capsids and the assembly intermediate would then be viewed as dead-end products, or artifacts of the methods by which particles are extracted from infected cells. For example, the structure designated "Assembly intermediate" in pathway A could represent unstable structures with DNA genomes that were partially extruded and broken during extraction of intermediates (dashed arrow).

Recognition and Packaging of the Nucleic Acid Genome

The assembly mechanisms described in previous sections are fruitless unless the viral genome is packaged within progeny virions. Despite diversity in size, composition, and morphology, specific incorporation of the genome during assembly of virus particles is achieved by a limited repertoire of mechanisms, which are discussed below. The special problems imposed by segmented genomes are considered subsequently.

During encapsidation, viral nucleic acid genomes must be distinguished from the cellular DNA or RNA molecules present in the compartment in which assembly takes place. This process requires a high degree of discrimination among similar nucleic acid molecules. For example, retroviral genomic RNA constitutes much less than 1% of an infected cell's cytoplasmic RNA population and bears all the hallmarks of cellular messenger RNAs (mRNAs). Yet it is **the** RNA packaged in the great majority of retrovirus particles. Such discrimination is the result of specific recognition of sequences or structures unique to the viral genome, termed **packaging signals**. These can be defined by genetic analysis as the sequences necessary for incorporation of the nucleic acid into the assembling virion, or sufficient to direct incorporation of foreign nucleic acid. The organization of the packaging signals of several viruses is therefore quite well understood.

Nucleic Acid Packaging Signals

DNA signals. The products of polyomaviral or adenoviral DNA synthesis are genomic DNA molecules that can be incorporated into assembling virus particles without further modification. These DNA genomes contain discrete packaging signals with several common properties. The signals comprise repeats of short sequences, some of which are also part of viral promoters or enhancers; they are positioned close to an origin of replication, and their ability to direct DNA encapsidation depends on this location. They differ in whether they are recognized directly or indirectly by viral proteins.

The encapsidation signal of the adenoviral genome, which is located close to the left inverted repeat sequence and origin, comprises a set of repeated sequences. Several of these sequences overlap enhancers that stimulate transcription of viral genes (Fig. 13.9A). The packaging signal is recognized by viral proteins, the late IVa_2 and L4 22-kDa proteins. Cooperative binding of the proteins to the repeated sequence is thought to form a higher-order nucleoprotein structure that promotes packaging of the genome. The results of genetic experiments have established the importance of these proteins in assembly. For example, genomes and structural proteins are synthesized efficiently in cells infected by IVa_2 null mutants, but neither mature virions nor any other particles are produced. This phenotype also suggests that packaging of the genome is

Figure 13.9 Viral DNA-packaging signals. (A) Human adenovirus type 5 (AD5). The location of the repeated sequences (blue arrows) of the packaging signal relative to the left inverted terminal repeat (ITR), the origin of replication (Ori), and the E1A transcription unit is indicated. The repeated sequences are AT rich and functionally redundant. The viral IVa_2 protein binds directly to the 3′ portion of the sequence that is conserved to each of the repeats. Once the IVa_2 protein is associated, the L22-kDa protein interacts with the 5′ segment of the conserved sequence. The positions of transcriptional enhancers within this region are also shown. Enhancer 1 stimulates transcription of the immediate-early E1A gene, whereas enhancer II increases the efficiency of transcription of all viral genes. **(B)** Simian virus 40 (SV40). The region of the genome containing the enhancer, origin of replication (Ori) and packaging signal is shown, with positions (bp) in the circular genome indicated below. The Sp1-binding sites within the packaging sequence are required for genome packaging.

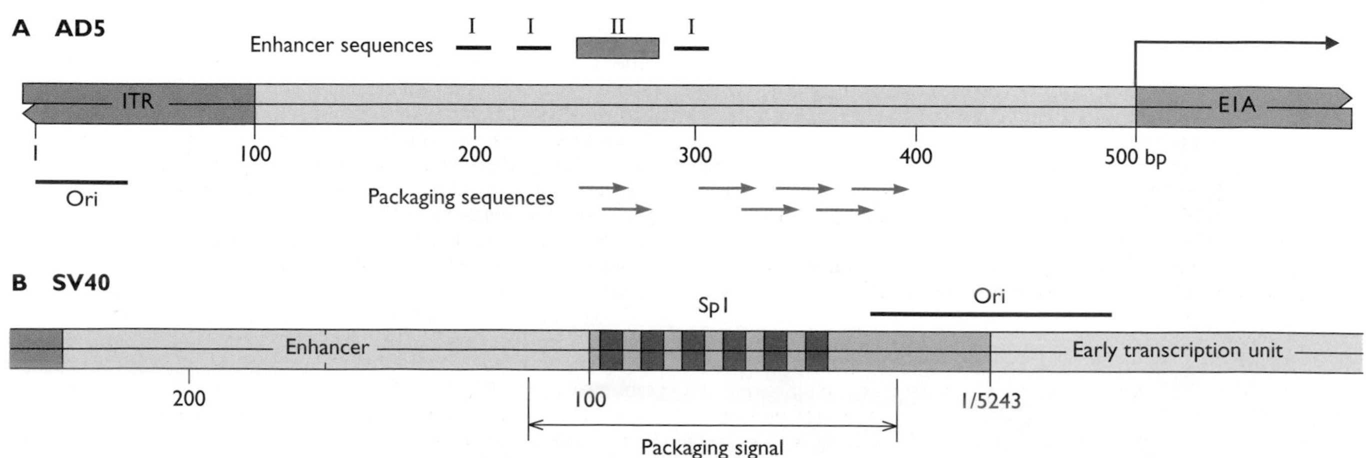

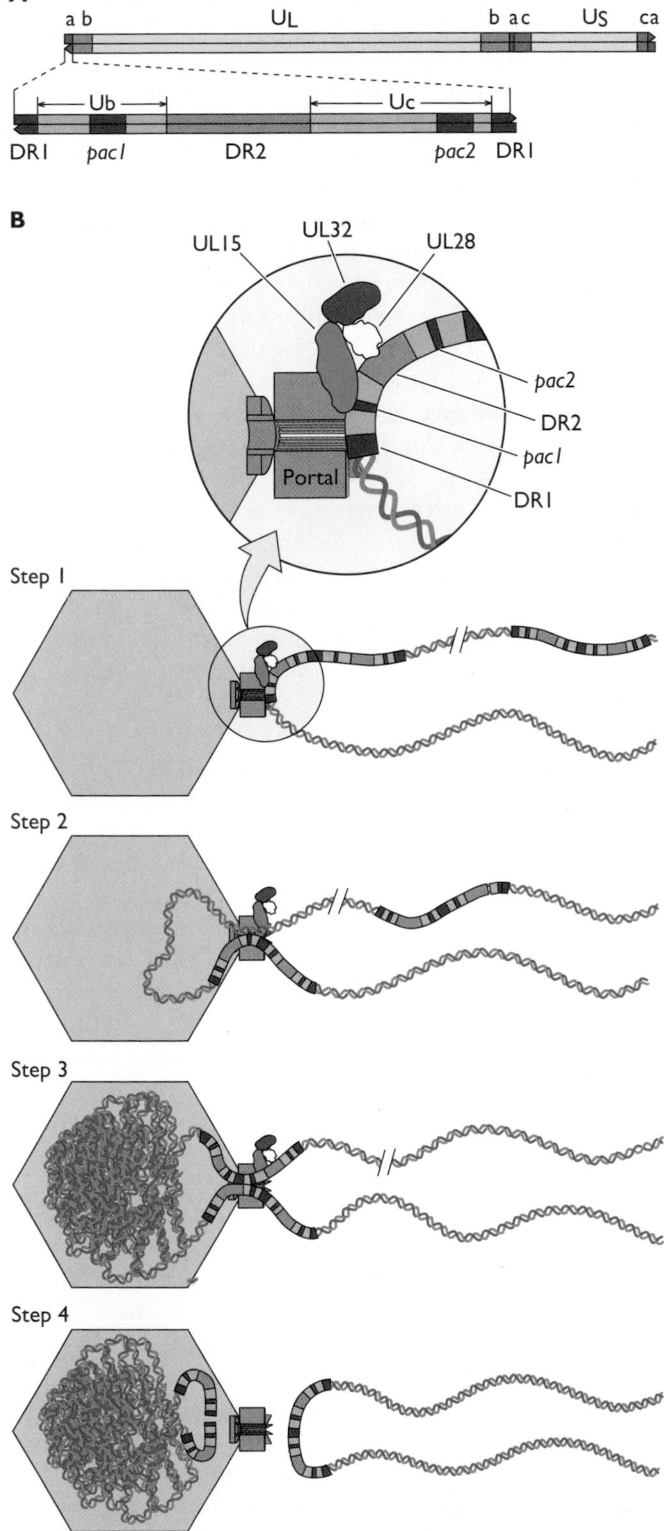

Figure 13.10 Packaging of herpes simplex virus type 1 DNA. (A) Organization of the *a* repeats of the viral genome, showing the location of the *pac1* and *pac2* sequences within the nonrepeated sequences Ub and Uc, and relative to the flanking direct repeats DR1 and DR2. One to several copies of the *a* sequence are present at the end of the unique long (U_L) segment and at the internal L-S junction, but only one copy lies at the end of the U_S region. **(B)** Model of herpes simplex virus type 1 DNA packaging, in which encapsidation is initiated by formation of a terminase complex, which includes the proteins indicated, on the packaging sequence. This protein-DNA complex is oriented to interact with the portal of the nucleocapsid (step 1). The DNA is then reeled into the capsid (steps 2 and 3) until a headful threshold is reached and an *a* sequence in the same orientation (i.e., one genome equivalent) is encountered (step 3), when cleavage in DR1 sequences takes place (step 4). When *a* sequences are tandemly repeated, adjacent copies share a single intervening DR1 sequence. The unit-length DNA genomes packaged in virions are therefore assumed to be released from the concatemeric products of viral DNA synthesis by cleavage at a specific site within shared DR1 sequences.

the result of a concerted assembly mechanism, but exactly how such assembly proceeds is not known. At least one additional viral protein, the L1 52/55-kDa protein mentioned previously, is associated with the packaging signal in infected cells, although it does not directly to the repeat sequence *in vitro*.

The simian virus 40 DNA-packaging signal is located in the regulatory region of the genome containing the origin of replication, the enhancer, and early and late promoters. Multiple sequences within this region contribute to the encapsidation signal. It includes multiple binding sites for the cellular transcriptional regulator Sp1. Although the cellular genome contains numerous binding sites for Sp1, the particular arrangement of sequences recognized by Sp1 in the packaging signal is unique to the viral genome. Neither the internal proteins of the simian virus 40 capsid (VP2/3) nor the major capsid protein (VP1) binds specifically to DNA. However, VP2/3 and Sp1 bind cooperatively to the packaging signal with high affinity and specificity. This cellular protein stimulates the *in vitro* assembly of infectious virus particles by an order of magnitude, consistent with a role in mediating indirect recognition of the packaging signal by capsid proteins. Subsequently, highly cooperative interactions among the structural units appear to drive converted assembly of the capsid, concomitant with displacement of Sp1 and nonspecific binding of capsid proteins to viral minichromosomes.

The replication of herpesviruses produces not genomic DNA molecules but, rather, concatemers containing many head-to-tail copies of the viral genome. Individual genomes must therefore be liberated from such concatemers. The herpes simplex virus type 1 packaging signals *pac1* and *pac2*, which lie within the terminal *a* repeats of the genome, are necessary for both recognition of the viral DNA and its cleavage within the adjacent DR repeats (Fig. 13.10A). It is generally thought that cleavage is concomitant with genome encapsidation. In one model (Fig. 13.10B), it is proposed that a protein complex formed on the unique short *pac* sequence interacts with the portal in the nucleocapsid. Following the first DNA cleavage, a unit-length genome is reeled into the nucleocapsid prior to the second DNA cleavage. This mechanism is analogous to that by which concatemeric DNA products of bacteriophage T4 replication are cleaved and packaged by a terminase complex, which hydrolyzes ATP and associates transiently with the portal protein of the preformed capsid. The products of at least seven herpes simplex virus type 1 genes are dedicated to stable encapsidation of the viral genome.

The UL15, UL28, and UL33 proteins, which interact with one another and with the portal protein, exhibit the properties predicted for the terminase. For example, the UL15 protein contains a sequence motif characteristic of ATPases that is essential for encapsidation of viral genomes, while the UL28 gene product binds to *pac* sequences required for DNA cleavage. The UL17 protein is also essential for DNA cleavage, but its function in this process is not known. In addition, it is necessary for recruitment of the UL15 protein to the nucleocapsid, resulting in formation of a complex present only on mature DNA-containing nucleocapsids (Box 13.5). In the absence of the UL25 protein, DNA cleavage does occur, but fewer nucleocapsids are formed. It has therefore been suggested that one function of this protein complex is to stabilize the protein shell so that it can withstand the pressure exerted by the encapsidated genome.

RNA signals. Because it is also an mRNA, the retroviral genome must be distinguished during encapsidation from both cellular mRNA and subgenomic viral mRNA. In addition, two genomic RNA molecules must interact with one another, for the retroviral genome is packaged as a dimer. This unusual property is thought to help retroviruses survive extensive damage to their genomes (see Chapter 7). In virions, the dimeric genome is in the form of a 70S complex held together by many noncovalent interactions between the RNA molecules. However, most attention has focused on sequences that allow formation of stable dimers, termed the **dimer linkage sequence**. *In vitro* experiments with human immunodeficiency virus type 1 RNA have provided evidence for base pairing between loop sequences of a specific hairpin (SL1) within the dimer linkage sequence (Fig. 13.11A) and the formation of an intermolecular four-stranded helical structure (known as a G tetrad or G quartet). The effects of mutations in, or duplication of, this sequence indicate that it nucleates formation of genome RNA dimers *in vivo*, and that dimerization is required for efficient genome packaging. Indeed, the dimer linkage sequence lies with the RNA-packaging signal (Fig. 13.11A), and it has been proposed that dimerization results in conformational change to expose sequences recognized by the RNA-binding portion of Gag (see below).

Sequences necessary for packaging of retroviral genomes, termed Psi (ψ), vary considerably in their complexity and location. In some cases, exemplified by Moloney murine leukemia virus, a contiguous ψ sequence of about 350 nucleotides (Fig. 13.11B) is both necessary and sufficient for RNA encapsidation. As this sequence lies downstream of the 5′ splice site, only unspliced genomic RNA molecules are recognized for packaging. The human immunodeficiency virus type 1 genome also contains a primary RNA-packaging sequence (Fig. 13.11A) that distinguishes the full-length genome from spliced viral RNA molecules. However, this sequence fails to direct packaging of heterologous RNA species into retrovirus particles, indicating that it is not sufficient. Additional sequences required for genomic RNA encapsidation lie within the TAR sequence and adjacent sequences (Fig. 13.11A), and at more distant locations. Presumably, such dispersed packaging sequences form a distinctive structural feature in folded RNA molecules.

The NC domain of Gag, a basic region that contains at least one copy of a zinc-binding motif (Fig. 13.12A), mediates selective and efficient encapsidation of genomic RNA during retroviral assembly. The central region of NC containing the zinc-binding motif(s) and adjacent basic sequences binds specifically to RNAs that contain ψ sequences *in vitro* and is necessary for selective packaging of the genome in infected cells. It is clear from structural studies of NC proteins bound to RNA-packaging signals that NC binds specifically to short RNA sequences, UCUG in the case of Moloney murine leukemia virus (Fig. 13.12B). The zinc-binding motifs form a compact structure, termed a zinc knuckle, which makes specific contacts with bases and is complementary in charge and shape to the bound RNA. Two features of the Moloney murine leukemia virus ψ signal appear to promote selection of dimeric RNA genomes. First, this region contains 13 copies of the UCUG sequence recognized specifically by the NC protein, a higher frequency than elsewhere in the genome. Second, the UCUG sequences within the regions most important for efficient packaging are not bound by NC when the RNA is monomeric: they are sequestered within base-paired regions of stem-loops, which have been shown to undergo substantial changes in base pairing upon dimerization of the genome. It is therefore thought that dimerization induces conformational transitions that expose high-affinity NC-binding sites (Fig. 13.12C). Whether analogous mechanisms promote selective packaging of dimeric RNAs of retroviruses with more complex

BOX 13.5 EXPERIMENTS
An allosteric transition that may stabilize the herpesviral nucleocapsid

The herpes simplex virus type 1 UL17 and UL25 proteins are not required for assembly of nucleocapsid protein shells, but are necessary for stable encapsidation of the DNA genome. Biochemical studies indicated that approximately equal quantities of these two proteins are exposed on the outer surface of mature, DNA-containing nucleocapsids. Cryo-electron microscopy of such nucleocapsids and shells that contain no DNA and retain or lack the assembly scaffold (so-called B and A capsids, respectively) revealed a structural component present only on nucleocapsids (see the figure). As mature nucleocapsids are also called C capsids, this structure was called the C capsid-specific component (CCSC). Formation of the CCSC correlated with the presence of the UL17 and UL25 proteins. For example, treatment of nucleocapsids with 0.5 M guanidinium hydrochloride for 30 min induced both dissociation of the UL17 and UL25 proteins, and disappearance of CCSCs. The CCSC is most likely to be a heterodimer of these two proteins.

The absence of the CCSC from A and B capsids suggested that its formation depends on encapsidation of the DNA genome. Consistent with this view, the genome is also released by the treatment described above that dissociates the UL17 and UL25 proteins. It has therefore been proposed that pressure exerted on the inner surface of the nucleocapsid as DNA enters induces a structural transition on the outer surface to create a site for stable binding of the CCSC.

Genetic analyses of the functions of the UL25 protein indicate that this protein has no role in cleavage of viral DNA concatemers, but is necessary for stable genome encapsidation, presumably because it stabilizes the protein shell. It may also permit binding of specific tegument proteins.

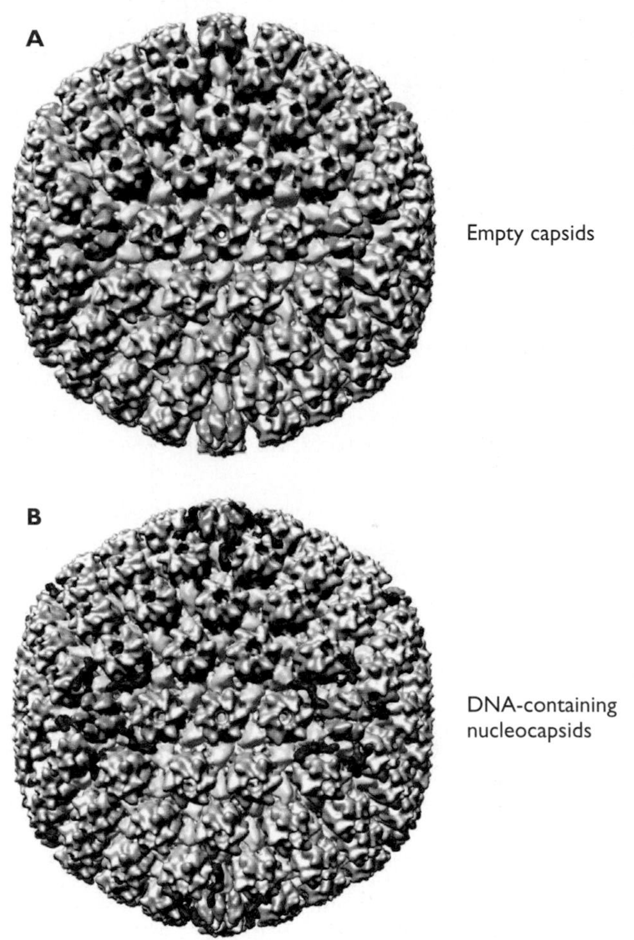

Cryo-electron microscopy reconstructions of the surfaces of mature nucleocapsids (A) and UL25-lacking A capsids (B) and viewed along a twofold axis of icosahedral symmetry. Hexons plus associated VP26 molecules, pentons, and triplexes are in light blue, blue, and green, respectively. The CCSCs present only on nucleocapsids, which are organized around the pentamers, are in purple. Bar, 20 nm. Adapted from B. L. Trus et al., *Mol. Cell* **26:**479–489, 2006, with permission. Courtesy of A. Steven, National Institutes of Health.

Stow, N. D. 2001 Packaging of genomic and amplicon DNA by the herpes simplex virus type 1 UL25-null mutant KUL25NS. *J. Virol.* **75:**10755–10765.

Trus, B. L., W. W. Newcomb, N. Cheng, G. Gardone, L. Marekov, F. L. Huma, J. C. Brown, and A. C. Steven. 2006. Allosteric signaling and nuclear exit strategy: binding of UL25/U17 heterodimers to DNA-filled HSV-1 capsids. *Mol. Cell* **25:**479–489.

packaging signals, such as human immunodeficiency virus type 1, is not yet clear.

As noted previously, binding of Gag to retroviral RNA is coordinated with assembly of the protein shell. One domain of the polyprotein of several retroviruses that is required for polyprotein assembly is located within the NC portion, and RNA binding by NC is necessary for efficient assembly of virus particles. Furthermore, particles that do form when RNA binding is prevented by alterations in NC are of low density, indicating that they contain fewer molecules

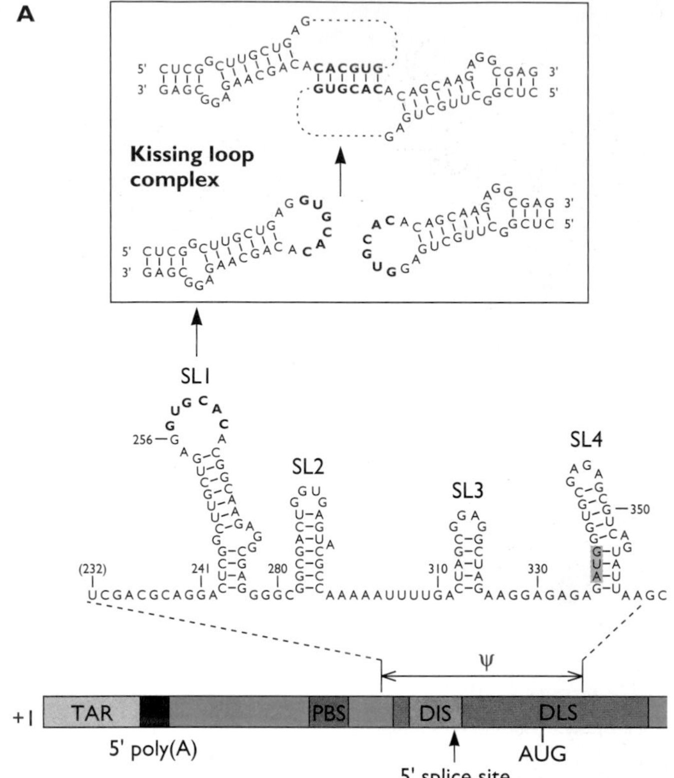

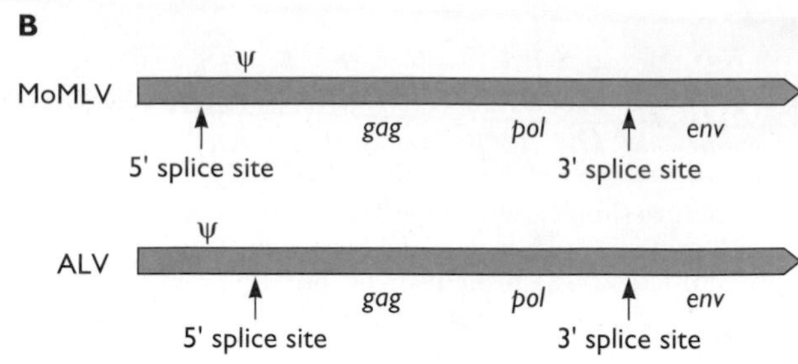

Figure 13.11 Sequences important in packaging of retroviral genomes. (A) The 5′ end of the human immunodeficiency virus type 1 genome is shown to scale at the bottom, indicating the positions of TAR, the 5′ polyadenylation signal [5′ poly (A)], the tRNA primer-binding site (PBS), the 5′ splice site, a packaging signal designated ψ, the sequence that forms the dimer linkage structure (DLS), and the dimerization initiation site (DIS), which can initiate dimerization *in vitro*. The four hairpins (SL1 to SL4) formed by the ψ sequence are shown above. The SL1 hairpin is the dimer initiation sequence. The loop-loop "kissing" complex proposed to form when two genomic RNA molecules dimerize via the self-complementary sequence shown in red is depicted at the top. The ψ sequence, which includes intronic sequences and therefore is present only in unspliced RNA, appears to be necessary but not sufficient for encapsidation of genomic RNA. A hairpin containing the 5′ poly(A)-addition signal and the bottom of the TAR hairpin, as well as more distantly located sequences, are also required. **(B)** Locations within the RNA genomes of sequences necessary for encapsidation of Moloney murine leukemia virus (MoMLV) and avian leukosis virus (ALV) RNAs, designated ψ. The latter ψ signal resides only upstream of the 5′ splice site. Even though both genomic and subgenomic RNAs contain this sequence, spliced mRNA molecules are not encapsidated efficiently.

of Gag. It has been demonstrated that nonspecific binding of Gag to nucleic acids induces Gag dimerization. The current model proposes that this reaction is followed by conformational change in RNA-bound Gag dimers to a form competent for multimerization and hence assembly (Box 13.6).

Other parameters that govern genome encapsidation. Specific signals may be required to mark a viral genome for encapsidation, but their presence does not guarantee packaging. The fixed dimensions of the closed icosahedral capsids or nucleocapsids of many viruses impose an upper limit on the size of viral nucleic acid that can be accommodated. Consequently, nucleic acids that are more than 5 to 10% larger than the wild-type genome cannot be encapsidated, even when they contain appropriate packaging signals. This property has important implications for the development of viral vectors. In some cases, the length of the DNA that can be accommodated in the particle (a "headful") is a critical parameter. This mechanism is exemplified by the coupled cleavage and encapsidation of genomic herpes simplex virus type 1 DNA molecules from the concatemeric products of replication, when both specific sequences and a headful of DNA are recognized. Indeed, the packaging of some viral DNA genomes, such as T4 DNA, depends solely on the latter parameter (Box 13.7).

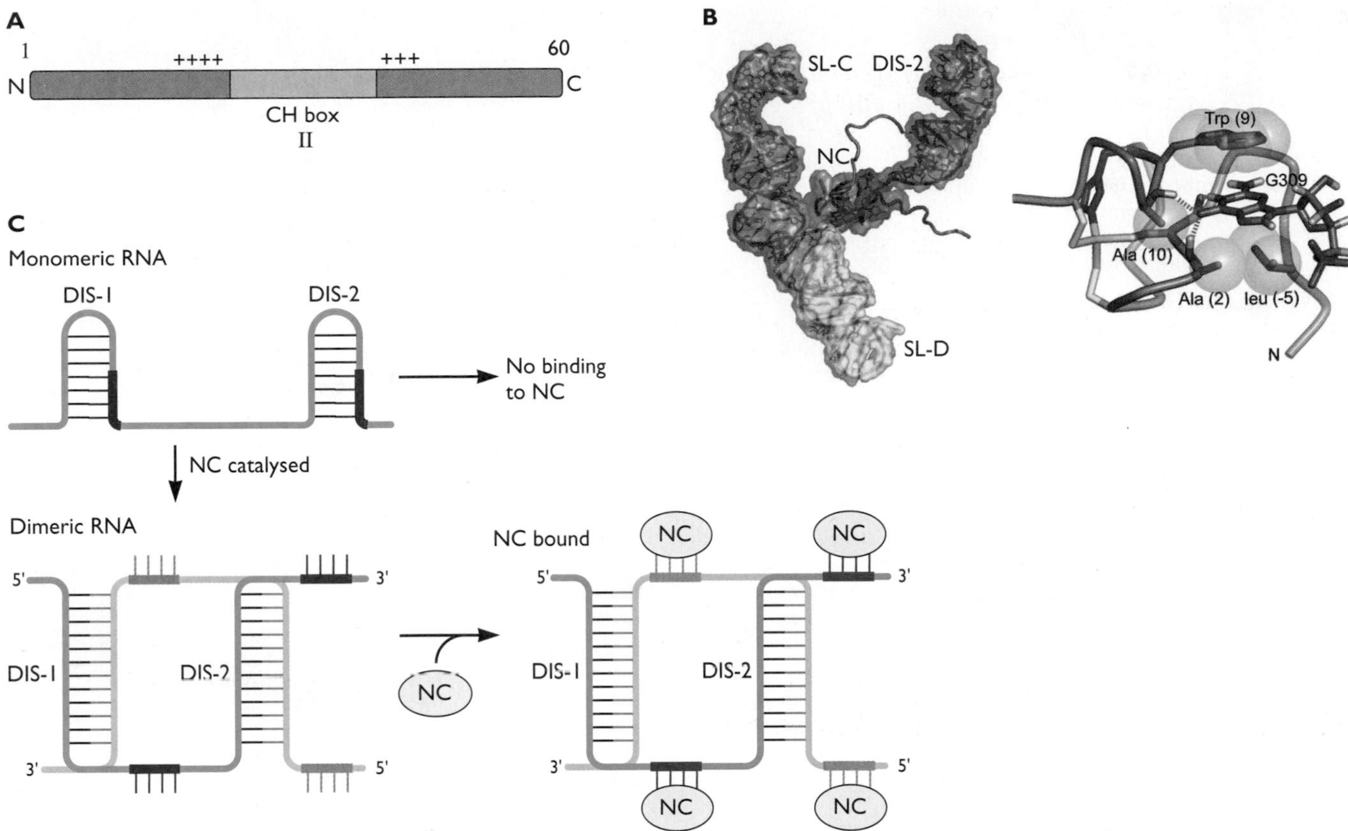

Figure 13.12 The retroviral NC RNA-packaging protein. (A) Schematic illustration of the Moloney murine leukemia virus NC protein, showing the locations of the Cys-His (CH) box and basic regions necessary for both nonspecific binding to RNA and specific recognition of the packaging signal. **(B)** Structural model of the Moloney murine leukemia virus NC protein bound to the 101-nucleotide core element of the packaging signal determined by nuclear magnetic resonance methods (left). Stem-loops are shown in purple, orange, and yellow, and the UCUG with which the NC zinc knuckle interacts is shown in red. The close-up view of the UCUG sequence bound to NC (right) shows hydrogen bonding of the exposed G residue with backbone atoms of the proteins. Adapted from V. D'Souza and M. F. Summers, *Nature* **431**:586–590, 2004, with permission. Courtesy of M. F. Summers, University of Maryland, Baltimore County. **(C)** Model for selective recognition of the dimeric RNA genome. The secondary structures of the monomeric and dimeric Moloney murine leukemia virus packaging signal are shown schematically. These structures are based on mapping the accessibility of RNA sequences to various chemicals, phylogenetic comparisons and free energy calculations, as well as structural studies. Adapted from V. D'Souza and M. F. Summers, *Nat. Rev. Microbiol.* **3**:643–655, 2005, with permission.

The coupling of encapsidation of a viral nucleic acid with its synthesis may also contribute to the specificity with which the viral genome is incorporated into assembling structures. As mentioned previously, such coordination is typical of the assembly of (−) strand RNA viruses (see, e.g., Fig. 13.6). However, the mechanisms by which nascent genomic RNA molecules are recognized by RNA-binding proteins are not well understood. Coordination of replication and encapsidation may also contribute to the great specificity with which picornaviral genomes are packaged: not only abundant cytoplasmic cellular RNA species (transfer RNAs [tRNAs], ribosomal RNAs [rRNAs], and mRNAs), but also (−) strand poliovirus RNA and poliovirus mRNA lacking VPg are excluded from virions. No packaging signal has yet been identified in the poliovirus genome. As discussed in Chapter 12, poliovirus infection induces the formation of cytoplasmic vesicles, with which both viral replication and virion assembly are associated. Such sequestration of genomic RNA molecules with the viral proteins that must bind to them clearly could make a major contribution to packaging specificity, by reducing competition from cytoplasmic cellular RNAs. There is accumulating evidence that encapsidation of the genome is coupled with RNA

BOX 13.6

DISCUSSION

Dimerization-induced conformational change forms assembly-competent Gag polyproteins

The model for higher-order assembly of the Gag polyprotein shown in the figure is based largely on *in vitro* studies of assembly of Rous sarcoma virus and human immunodeficiency virus type 1 Gag.

- Assembly of Gag synthesized in *E. coli* or made by *in vitro* translation into virus-like particles requires the NC domain and nucleic acid
- Heterologous RNA is present in virus-like particles released from Gag-producing cells (see the text), and DNA oligonucleotides can support Gag assembly in *in vitro* reactions
- Oligonucleotides that support such Gag assembly **must** contain at least two binding sites for the NC domain of Gag

These properties suggested that nonspecific binding of the NC domain of Gag to nucleic acid (the viral RNA genome during assembly in infected cells) promotes Gag dimerization. The hypothesis is consistent with the fact that basic sequences of NC, but not the zinc finger domain(s) that binds specifically to packaging sequences, are necessary for assembly and release of virus-like particles. The demonstrations that replacement of NC by either a heterologous, leucine zipper dimerization domain or cysteine residues that can be covalently linked (by chemical crosslinkers or oxidation) allows efficient assembly *in vitro* provides strong evidence for the crucial role of dimerization.

It has been proposed that efficient assembly also depends on conformational change following dimerization. Such assembly

- requires not only NC and the CA dimerization interface (see the text), but also more N-terminal sequences, notably the N-terminal domain of CA
- has been reported to be temperature dependent: Gag assembly *in vitro* take place at 37°C, but not at 23°C. As protein rearrangements occur more readily at the higher temperature, this property implicated conformational change as a prerequisite for Gag multimerization. The temperature dependence can be eliminated by artificial tethering of N-terminal segments of CA to one another.

These observations argue that dimerization of RNA-bound Gag induces conformational change to form dimers competent for assembly into higher-order multimers, via alignment of N-terminal domains of CA.

Alfadhli, A., T. C. Dhenub, A. Still, and E. Barklis. 2005. Analysis of human immunodeficiency virus type 1 Gag dimerization-induced assembly. *J. Virol.* **79:**14498–14506.

Ma, Y. M., and V. M. Vogt. 2004. Nucleic acid binding-induced Gag dimerization in the assembly of Rous sarcoma virus particles *in vitro*. *J. Virol.* **78:**52–60.

The sequential formation of Gag molecules competent for assembly into virus-like particles (VLPs) is shown schematically.

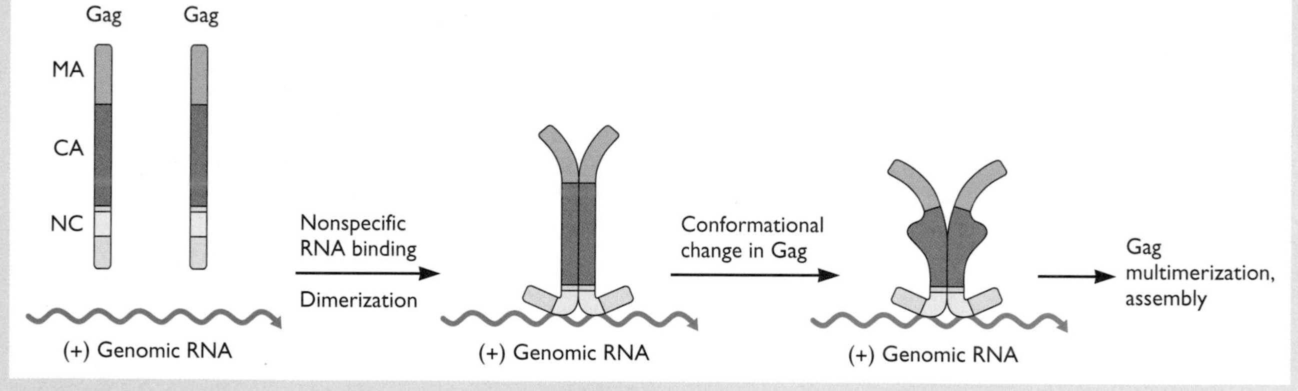

replication and that this step in assembly is assisted by the poliovirus 2C protein. This protein is a nucleoside triphosphatase that might facilitate release of RNA from membrane-bound replication complexes. As packaging of flavivirus RNA also depends on replication of genomic RNA, coincident genome synthesis and assembly may be a general feature of (+) strand RNA viruses.

Packaging of Segmented Genomes

Segmented genomes pose an intriguing packaging problem. The best-studied example among animal viruses is the influenza A virus genome, which comprises eight molecules of RNA. It has been appreciated for many years that production of an infectious virus particle requires incorporation of at least one copy of each of the eight genomic segments. However, it has proved difficult to distinguish random packaging from a selective mechanism for inclusion of a full complement of genomic RNAs.

Packaging of the bacteriophage $\phi 6$ genome provides clear precedent for a selective mechanism. The genome of this bacteriophage comprises one copy of each of three double-stranded RNA segments designated S, M, and L.

BOX 13.7

BACKGROUND

Packaging a headful of viral DNA

During assembly of herpesvirus and several bacteriophages with large, double-stranded DNA genomes, including T4, the linear genome is cleaved from concatemeric products of viral genome replication during insertion into a preformed protein shell, the head in the case of T4. Furthermore, encapsidation of T4 DNA is coordinated with cleavage of concatemers. However, the T4 genome exhibits several unusual features.

- The linear T4 genomes do not have unique terminal sequences
- The genetic map is circular, even though the genome is linear
- The terminal sequences, which are different in each DNA molecule, are repeated at each end of DNA

It was deduced from these properties that the T4 genome is circularly permuted and terminally redundant. These properties can be accounted for by essentially random cleavage of head-to-tail concatemers (the preferred substrate for DNA packaging) that results in encapsidation of DNA molecules that are **longer** than the unique sequence in the genome (see the figure). No specific DNA sequence dictates the cleavages that liberate linear DNA during encapsidation. Rather, the first cleavage occurs randomly, and the second takes place once the phage T4 head has been filled with DNA. As predicted by the "headful" packaging mechanism, when head size is increased or decreased by mutation in specific genes (or other manipulations), longer and shorter DNA molecules, respectively, are encapsidated. Furthermore, when sequences are deleted from, or inserted into, the genome, the length of the terminal repeats increases or decreases to the corresponding degree: when *x* bp are removed, the total length of the terminal repeats increases by *x* bp. These properties demonstrate directly that a fixed length of DNA, a headful, is incorporated during assembly.

A headful of DNA is packaged during assembly of other bacteriophage and animal viruses with double-stranded DNA genomes, including herpesviruses. Structural studies of bacteriophage P22 virions revealed that tight spooling of DNA in the nucleocapsid induces major conformational change in the portal, through which DNA enters. It has therefore been proposed that the change in portal structure provides the signal that the nucleocapsid is full, to activate termination of DNA encapsidation.

Lander, G. C., L. Tang, S. R. Casjens, E. B. Gilcrease, P. Prevelige, A. Poliakov, C. S. Potter, B. Carragher, and J. E. Johnson. 2006. The structure of infectious P22 virion shows the signal for headful DNA packaging. *Science* **312:**1791–1795.

A head-to-tail concatemer, in which the unique genome sequence is represented by ABCDEFGH. Initial cleavage between H and A is followed by packaging of a headful length that is longer than the length of the unique genome sequence, and the second cleavage. Repetition of this process yields a population of particles with encapsidated DNA molecules of the same length, but that are circularly permuted and terminally redundant.

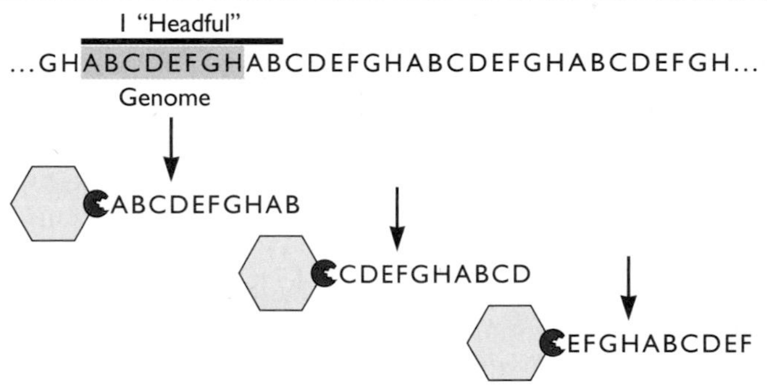

The (+) strand of each segment is packaged prior to synthesis of complementary (−) strands within particles, a mechanism analogous to synthesis of the double-stranded RNA segments of the reovirus genome (Chapter 6). The particle-to-plaque-forming-unit ratio of ϕ6 is close to 1, indicating that essentially all particles contain a complete complement of genome segments. Such precise packaging appears to be the result of the serial dependence of packaging of the (+) strand RNA segments. In *in vitro* reactions, the S segment packages alone, but entry of M RNA requires the presence of S RNA within particles, and packaging of the L segment is dependent on prior entry of both S and M RNAs.

A random packaging mechanism, in which any eight RNA segments of the influenza virus genome were incorporated into virions, would yield a maximum of 1 infectious particle for every 400 or so assembled ($8!/8^8$). This ratio might seem impossibly low, but is within the range of ratios of noninfectious to infectious particles found in virus preparations. Furthermore, if packaging of more than eight RNA segments were possible, the proportion of infectious particles would increase significantly. For example, with 12 RNA molecules per virion, 10% of the particles would contain the complete viral genome. Particles containing more than eight RNA segments have been isolated, consistent with random packaging. In a plasmid-based system for production of infectious influenza A virus particles that included differentially marked versions of three of the genomic RNA segments, some of the virions produced contained two copies of the **same** RNA segment.

This result can be best explained by a random packaging mechanism. On the other hand, the results of more recent studies favor selective packaging.

This mechanism implies that each of the eight (−) strand genome RNAs (vRNAs) carries a unique signal that ensures its packaging. In the past few years, RNA sequences required for incorporation of the individual genome segments have been identified. Such packaging signals comprise the untranslated regions and adjacent coding sequences at both the 5′ and 3′ ends of the vRNAs. Furthermore, these sequences appear to interact to form structures unique to each segment: artificial vRNAs carrying a 5′ end from one genome segment and a 3′ end from a second are packaged with only low efficiency. The observation that defective derivatives of a particular vRNA compete for packaging of only the corresponding normal RNA segment is also consistent with selective packaging of genome RNA segments. The mechanisms by which vRNA packaging signals are recognized and that ensure incorporation of one copy of each segment are not clear. However, there are hints that a multisegment vRNA complex (analogous to the retroviral RNA genome dimer) may form prior to assembly.

Incorporation of Virion Enzymes and Other Nonstructural Proteins

In many cases, the production of infectious particles requires incorporation into the assembling virion of essential viral enzymes, or other proteins that are important in establishing an efficient infectious cycle. Some of these proteins are also structural proteins of the virion. For example, the herpes simplex virus type 1 VP16 protein is both a major component of the virion tegument and the activator of transcription of viral immediate-early genes.

A simple, yet elegant, mechanism ensures entry of retroviral enzymes (protease [PR], reverse transcriptase [RT], and integrase [IN]) into the assembling core. In most cases, these enzymes are synthesized as C-terminal extensions of the Gag polyprotein. The organization and complement of these translation products, here designated Gag-Pol, varies among retroviruses, but the important point is that they contain not only Pol but also the sequences specifying Gag-Gag interactions. These are presumed to direct incorporation of Gag-Pol molecules into assembling particles (Fig. 13.8). The low efficiency with which Gag-Pol polyproteins are translated determines their concentrations relative to Gag in the cell and in virions (1:9).

The enzymes present in other virus particles, such as the RNA-dependent RNA polymerases of (−) strand RNA viruses (see Table 4.8), are synthesized as individual molecules and therefore must enter assembling particles by noncovalent binding to the genome or to structural proteins.

All retroviral cores also contain the cellular tRNA primer for reverse transcription, brought into particles by its base pairing with a specific sequence in the RNA genome and by specific binding to RT. In some cases, including human immunodeficiency virus type 1, the host amino acyl tRNA synthetase that aminoacylates the particular tRNA used as primer is also encapsidated. The absence from virions of other amino acid tRNA synthetases, and the similar concentrations of the enzyme and its tRNA substrate in human immunodeficiency virus type 1 particles, suggest that the synthetase may be recognized by viral proteins during packaging.

Acquisition of an Envelope

Formation of many types of virus particle requires envelopment of capsids or nucleocapsids by a lipid membrane carrying viral proteins. Most such enveloped viruses assemble by virtue of specific interactions among virion components at a cellular membrane before budding and pinching off of a new virus particle. However, they vary in the intracellular site at which the envelope is acquired and therefore in the relationship of envelopment to release of the virus particle. As discussed in Chapter 12, whether particles assemble at the plasma membrane or internal membranes is determined by the destination of viral proteins that enter the cellular secretory pathway. Enveloped viruses assemble by one of two mechanisms, distinguished by whether acquisition of the envelope follows assembly of internal structures, or whether these processes take place simultaneously.

Sequential Assembly of Internal Components and Budding from a Cellular Membrane

For most enveloped viruses, the assembly of internal structures of the virion and their interaction with a cellular membrane modified by insertion of viral proteins are spatially and temporally separated. This class of assembly pathways is exemplified by (−) strand RNA viruses, such as influenza A virus (Fig. 13.6) and vesicular stomatitis virus. Influenza A virus ribonucleoproteins containing individual genomic RNA segments, NP protein, and the polymerase proteins are assembled in the infected cell nucleus as genomic RNA segments are synthesized. They are then transported to the cytoplasm in the M1- and NEP-dependent reactions described in Chapter 12. The viral glycoproteins HA and NA and the M2 membrane protein travel separately to specialized regions in the plasma membrane (lipid rafts) via the cellular secretory pathway (Fig. 13.6; see Chapter 12). The M1 protein interacts with both viral nucleocapsids and the inner surface of the plasma membrane to direct assembly of progeny particles at that membrane. Vesicular stomatitis virus assembles in a similar fashion, although no transport of the ribonucleoprotein from nucleus to cytoplasm is required. The matrix proteins of these (–) strand RNA viruses therefore provide the links among ribonucleoproteins and the modified cellular membrane necessary for assembly and budding.

The cellular membranes destined to form the envelopes of virus particles contain viral integral membrane proteins that play essential roles in the attachment of virus particles to, and their entry into, host cells. In simple enveloped alphaviruses, direct binding of the cytoplasmic portions of the viral glycoproteins to the single nucleocapsid protein (see Fig. 4.23) is necessary for acquisition of the envelope during budding from the plasma membrane. The crucial role and specificity of these interactions in the final steps in assembly are illustrated by the failure of a chimeric Sindbis virus containing the coding sequence for the E1 glycoprotein of Ross River virus, a second togavirus, to bud efficiently. The pE2 and E1 glycoproteins of the chimeric virus form heterodimers that are correctly processed (by cleavage of pE2 to E2) and transported to the plasma membrane. However, such chimeric glycoproteins exhibit an altered conformation and fail to bind to nucleocapsids at the plasma membrane. Binding of viral glycoproteins to internal components also appears to be important for production of more complex enveloped viruses. Interactions between the influenza virus M1 protein and the cytoplasmic tails of the HA and NA glycoproteins are necessary for formation of virus particles with normal size and morphology (Fig. 13.13), and with the appropriate concentration of genomic RNA. The cytoplasmic domain of the M2 ion channel protein is also necessary for efficient incorporation of viral ribonucleoproteins and production of infectious particles.

Coordination of the Assembly of Internal Structures with the Acquisition of the Envelope

The alternative pathway of acquiring an envelope, in which assembly of internal structures and budding from a cellular membrane are largely coincident in space and time, is exemplified by many retroviruses. Assembling cores of the majority first appear as crescent-shaped patches at the inner surface of the plasma membrane. These structures extend to form a closed sphere as the plasma membrane wraps around and eventually pinches off the assembling particle (Fig. 13.7). Formation of the assembling particles depends on the interaction of Gag polyprotein molecules with one another to form the protein core, with the RNA genome via the NC portion, and with the plasma membrane via the MA segment.

Figure 13.13 Deformed influenza A virus particles. These particles assemble when the cytoplasmic (internal) domains of the NA glycoprotein **(B)** or both NA and HA glycoproteins **(C)** are truncated. Purified wild-type **(A)** and mutant virions were negatively stained with phosphotungstic acid and examined by electron microscopy. Bars, 100 μm. The mutant particles are also considerably larger than wild-type virus particles. It is therefore assumed that binding of the internal portions of HA, and particularly of NA, to internal virion components (M1 protein and/or the ribonucleoproteins) determines the spherical shape and characteristic size of these virus particles, as well as the efficiency of budding. From H. Jin et al., *EMBO J.* **16:**1236–1247, 1997, with permission. Courtesy of R. A. Lamb, Northwestern University.

A

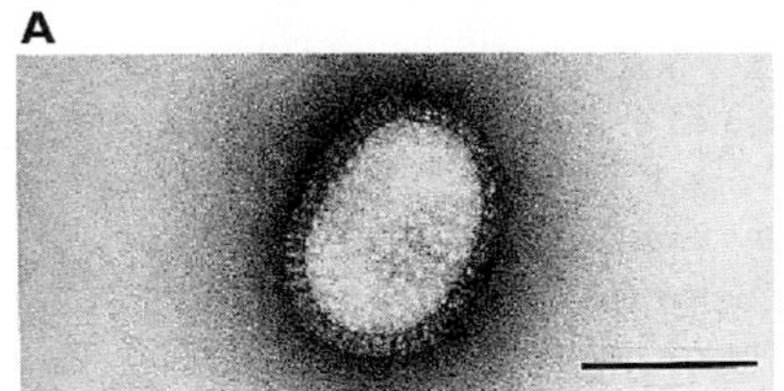

B

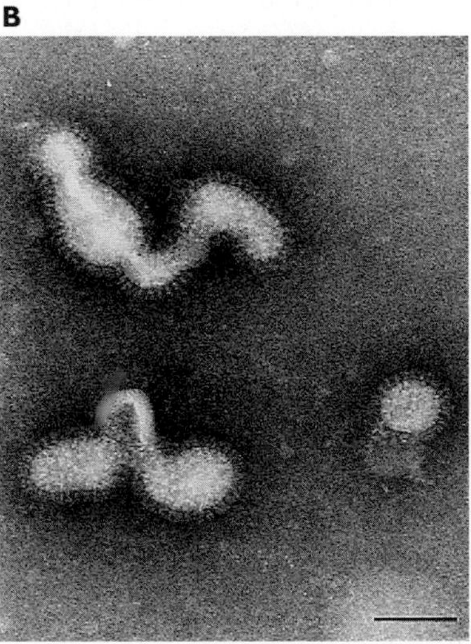

C

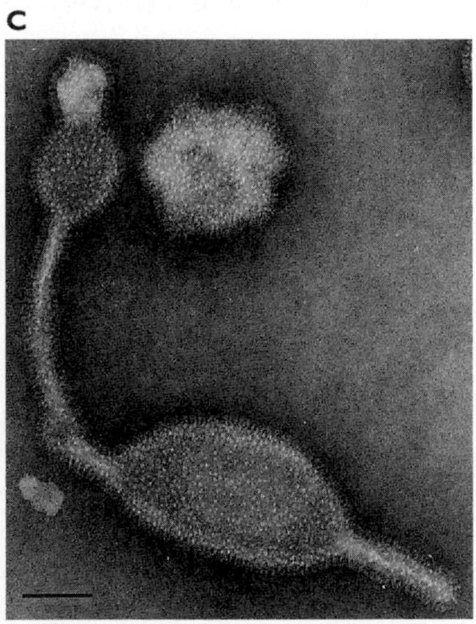

Specific segments of Gag mediate the orderly association of polyprotein molecules with one another, and are required for proper assembly. These sequences include an essential C-terminal multimerization domain of the CA segment: substitutions that disrupt the CA dimer interface block assembly of the CA protein *in vitro* and severely inhibit Gag assembly and formation of virus particles *in vivo*. The capsids of retroviruses can be spherical, conical or cylindrical (Fig. 4.19). Specific CA sequences that determine the morphology of mature particles or of structures formed by CA *in vitro* have also been identified. For example, insertions within an N-terminal CA sequence that forms a loop at the outer surface of CA hexamers in crystals block the formation of spherical Moloney murine leukemia virus cores. Certain sequences present only in the Gag polyprotein also govern morphology, for their removal results in assembly of misshapen particles.

As discussed previously, Gag multimerization during particle assembly is regulated by binding of the NC domain to the RNA genome. This process is also promoted by interaction of Gag with the plasma membrane via the MA membrane-binding signals (Fig. 12.18). Elimination of the signal for myristoylation prevents assembly, as does alteration of the sequence predicted to lie at the interfaces of the MA trimers formed in crystals (Fig. 4.24). It has been suggested that MA trimerization increases the accessibility of the myristate chain. Conversely, efficient membrane binding of Gag depends on sequences other than the membrane-binding region of MA, such as a sequence in the N-terminal portion of NC. Because this sequence is not required for production of stable Gag or its transport to the plasma membrane, it may promote Gag-Gag or Gag-RNA interactions that lead to cooperative and stable binding of Gag molecules to the membrane. In this context, it is noteworthy that in recent studies of assembly in live cells, Gag-Gag interactions were observed only at the plasma membrane.

In some cases, the MA segment of Gag also binds to the cytoplasmic tail of the viral envelope glycoprotein. For example, association of the assembling human immunodeficiency virus type 1 core with the TM-SU glycoprotein requires the N-terminal 100 amino acids of MA. Such Gag-Env interactions ensure specific incorporation of viral glycoproteins into virions. Nonetheless, they do not appear to be universal: glycoprotein-containing virions are produced even when the C-terminal tails of TM of other retroviruses (e.g., avian sarcoma virus) are deleted. Nor can a model based solely on Gag-Env interactions account for the ease with which "foreign" viral and cellular glycoproteins are included in the envelopes of all retroviruses. It is therefore thought that as a particle assembles at the plasma membrane, Gag-membrane interactions displace cellular membrane proteins connected to internal components of the cell. Such displacement would allow lateral diffusion of viral (and cellular) glycoproteins that are not connected in this way into these regions of the cellular membrane, and hence their passive incorporation into assembling virus particles. The final reaction, fusion of membrane regions juxtaposed as the particle assembles (Fig. 13.7), is shared with other viruses that assemble at the plasma membrane. This process is considered in the next section.

Release of Virus Particles

Many enveloped viruses assemble at, and bud from, the plasma membrane. Consequently, the final assembly reaction, fusion of the bud membrane, releases the newly formed virus particle into the extracellular milieu. When the envelope is derived from an intracellular membrane, the final step in assembly, budding, is also the first step in egress, which must be followed by transport of the particles to the cell surface. The assembly of enveloped viruses is therefore both mechanistically coupled and coincident with (or at least shortly followed by) their exit from the host cell. The egress of some viruses without envelopes from certain types of host cell also occurs by specific mechanisms. However, replication of such viruses more commonly results in destruction (lysis) of the host cell. Large quantities of assembled virions may accumulate within infected cells for hours, or even days, prior to release of progeny on cell lysis. The release of many enveloped viruses also destroys the cells that support their replication. However, in some cases nondestructive budding permits a long-lasting relationship with the host cell. The progeny of many simple retroviruses are released throughout the lifetime of an infected cell, which is not harmed (but may be permanently altered; see Volume II, Chapter 7).

Release of Nonenveloped Viruses

The most usual fate of host cells permissive for reproduction of nonenveloped viruses is death and destruction (but see Volume II, Chapter 5). In natural infections, the host defenses are an important cause of infected-cell destruction. However, infection by these viruses destroys host cells more directly: they are cytopathic to cells in culture. In general, we are remarkably ignorant about the mechanisms by which replication of nonenveloped viruses induces death and lysis of host cells.

Infection by many viruses, including poliovirus and adenovirus, leads to inhibition of expression of cellular genetic information by specific effects on cellular transcription, RNA export from the nucleus, or translation. In the case of adenovirus, with a one-step infectious cycle that is a relatively long 1 to 2 days, the shutdown of production of cellular proteins during the late phase of infection makes an important contribution to the eventual destruction of the infected cell. However, more specific mechanisms have also been implicated, as several viral proteins induce degradation of structural components of the host cell. For example, late in the infectious cycle, cytoplasmic intermediate filament components (specific cytokeratins) are cleaved by the L3 protease into polypeptides that can no longer polymerize. Such disruption of intermediate filament networks, which requires inhibition of cellular protein synthesis so that the network cannot be rebuilt, seems likely to damage the structural integrity of the cell and facilitate virus release. In addition, a small E3-encoded glycoprotein is necessary for efficient nuclear disruption and lysis of cells in culture. This viral glycoprotein is made in large quantities late in infection and accumulates in the nuclear envelope, but its mechanism of action is not yet known. One or more poliovirus nonstructural proteins have been implicated in host cell lysis, but this process occurs more rapidly (within approximately 8 h) than can be explained by inhibition of synthesis of cellular macromolecules. One idea currently under investigation is that cells infected by poliovirus, and other viruses with short infectious cycles, succumb to apoptotic lysis.

While cell lysis is the most common means of escape of naked viruses, there is evidence that some are released in the absence of any cytopathic effect. Under certain conditions, polioviruses and other picornaviruses are released without lysis of the infected cell. When poliovirus replicates in polarized epithelial cells resembling those lining the gastrointestinal tract (a natural site of infection), progeny virions are released exclusively from the apical surface by a nondestructive mechanism. The viral 2BC and 3A proteins induce the formation of infected cell-specific vesicles that closely resemble autophagosomes (see Chapter 12). It has been proposed that these vesicles, which contain two membranes and virions (late in infection), provide a route for nonlytic release of particles assembled to the cytoplasm (Fig. 13.14). Simian virus 40 is also released from permissive cells before induction of cytopathic effects and leaves polarized epithelial cells at their apical surfaces, via the secretory transport pathway of a cell. It will be of considerable interest to learn how such nonenveloped virions enter the membrane-bound compartments of this pathway.

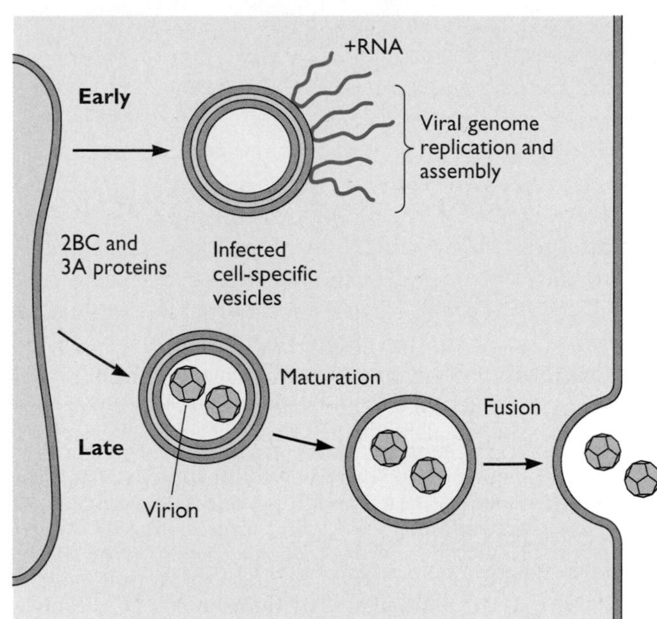

Figure 13.14 Model for nonlytic release of poliovirus particles. As discussed in Chapter 12, synthesis of the viral 2BC and 3A proteins leads to formation of infected cell-specific vesicles that resemble autophagosomes. The surfaces of these vesicles, which are bound by two lipid bilayers, are the sites of genome replication and assembly (top). It has been proposed that as autophagosome-like vesicles form later in infection, they enclose virions present in the cytoplasm. Maturation of such virion-containing vesicles analogous to maturation of autophagosomes would result in complete or partial degradation of the inner membrane. Subsequent fusion of the mature vesicle with the plasma membrane would release virions. This model is based largely on the observation that RNA interference-mediated knockdown of proteins required for formation of autophagosome-like vesicles reduced the yield of extracellular virus particles to a greater degree than the yield of intracellular particles.

Assembly at the Plasma Membrane: Budding of Virus Particles

The release of enveloped virus particles from the plasma membrane is a complex process that comprises induction of membrane curvature by viral components (bud formation), bud growth, and fusion of the bud membrane. Although budding has been visualized repeatedly, often in striking images, the mechanisms that underlie this crucial process are not fully understood.

For many years, it was generally accepted that bud formation was driven by interactions among viral envelope glycoproteins and internal components of viral particles, a mechanism exemplified by alphaviruses such as Sindbis virus. Such a mechanism (Fig. 13.15) can account nicely for how a bud site is determined, by initial binding of the internal nucleocapsid to a cluster of the heterodimeric E1-E2 glycoproteins within the membrane. The bud then expands by cooperative, lateral interactions among glycoprotein spikes and between their internal segments and the capsid protein. Nevertheless, as described in the previous section and summarized in Fig. 13.15, it is clear that there is greater variety in the viral proteins required

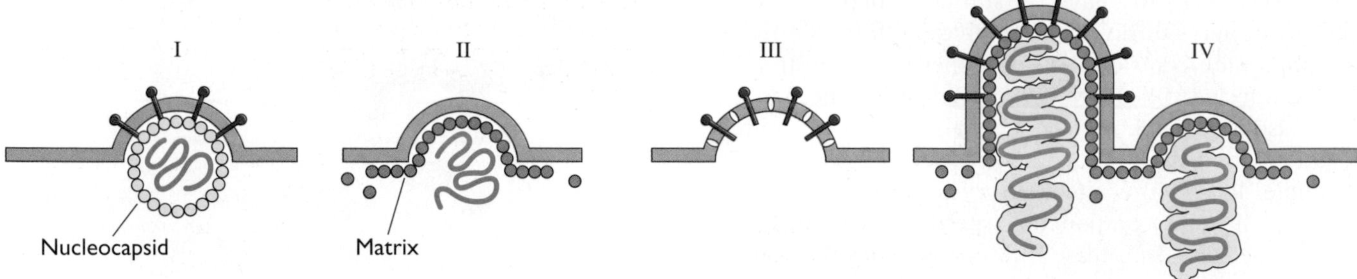

Figure 13.15 Interaction of viral proteins responsible for budding at the plasma membrane. Four distinct budding strategies have been identified. In type I budding, exemplified by alphaviruses such as Sindbis virus, both the envelope glycoproteins and the internal capsid are essential. Quite detailed structural pictures of alphaviruses are now available (Chapter 4). Certain altered or chimeric envelope proteins that reach the membrane normally do not support budding. These properties indicate that lateral interactions among the envelope heterodimers, as well as those of the heterodimers with the capsid, cooperate to drive budding. Type II budding, such as Gag-dependent budding of many retroviruses, requires only the internal matrix protein. For other viruses, type II budding requires only capsid proteins. Conversely, budding can be driven solely by envelope proteins (type III), a mechanism exemplified by the envelope proteins of the coronavirus mouse hepatitis virus. Type IV budding is driven by matrix proteins, but its proper functioning depends on additional components. For example, in the case of rhabdoviruses and orthomyxoviruses, internal matrix proteins alone can drive budding. However, this process is inefficient, or results in deformed or incomplete particles in the absence of envelope glycoproteins or the internal ribonucleoprotein. Adapted from H. Garoff et al., *Mol. Microbiol. Rev.* **62:**1171–1190, 1998, with permission.

for budding than was originally appreciated. Despite such diversity, recent studies of viral protein sequences required for budding have identified mechanistic themes common to different virus families.

Common Sequence Motifs Are Required for Budding at the Plasma Membrane

A major breakthrough in our understanding of how virus particles bud from the plasma membrane came with the identification of mutants of human immunodeficiency virus type 1 with an unusual assembly phenotype: mutations in the coding sequence for the p6 region unique to the Gag polyprotein did not impair assembly of immature particles, but the particles remained attached to the host cell by a thin membrane stalk (Fig. 13.16A). It was therefore concluded that these Gag sequences are required for the fusion reaction that separates the viral envelope from the plasma membrane. Subsequently, functionally analogous sequences, termed late-assembly (L) domains, were identified in Gag proteins of several other retroviruses. These L domains are not conserved in their location within Gag or in amino acid sequence, but nevertheless can substitute for one another to promote budding.

Retroviral L domains contain a small number of short, core sequence motifs, such as PTAP and PPXY (Table 13.3). The recognition of such motifs, and their ability to function independently of position or sequence context, led to definition of L domains containing such motifs in the proteins required for budding of viruses of several different families (Table 13.3). Furthermore, it is now clear that **L domain sequences** promote budding by recruitment of cellular proteins that participate in specific steps in vesicular trafficking.

The Activity of Viral L Domains Depends on Vascular Sorting Proteins

The autonomous activity (and in some cases the sequence) of L domains suggested that these sequences mediate protein-protein interactions. Cellular proteins that bind to each of the prototype sequences have now been identified. The PTAP motif was first shown to recruit the product of tumor susceptibility gene 101, Tsg101, an interaction that is essential for budding of human immunodeficiency virus type 1. This discovery provided the key to identification of the cellular machinery that promotes budding of this and several other enveloped viruses.

Mammalian Tsg101 participates in sorting and trafficking of cellular proteins from late endosomes to structures called multivesicular bodies, which fuse with lysosomes. As their name implies, multivesicular bodies contain vesicles within vesicles. Formation of these structures and budding of virus particles are topologically equivalent processes: in both cases, membranes invaginate away from the cytoplasm and fusion releases vesicles with cytoplasmic contents into a lumen or the extracellular space. Recruitment of Tsg101

A

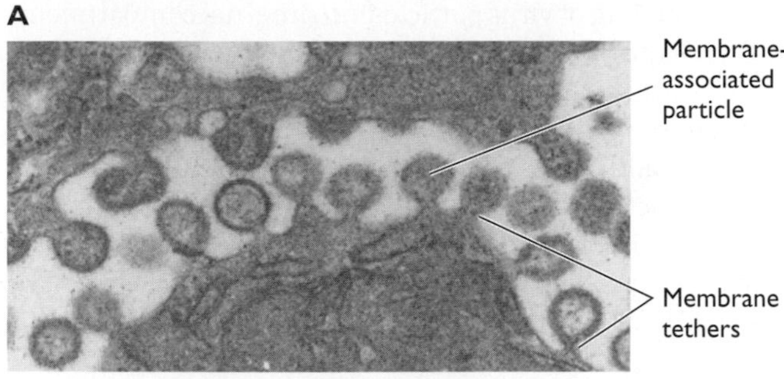

B

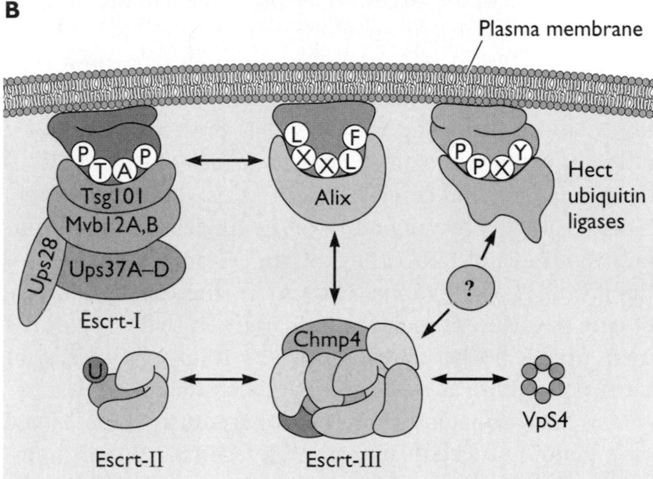

Figure 13.16 Release of retroviral particles. (A) Electron micrograph of monkey Cos-7 cells containing a human immunodeficiency virus type 1 mutant provirus from which the Gag p6 cannot be expressed. The plasma membrane-associated particles exhibit normal morphology, but remain tethered to the membrane. Adapted from H. G. Göttlinger et al., *Proc. Natl. Acad. Sci. USA*, **88**:3195–3199, 1991, with permission. Courtesy of H. Göttlinger, University of Massachusetts Medical Center. **(B)** Summary of association of cellular trafficking proteins with core sequence motifs of L domains in viral proteins required for release of viral particles. Interactions are shown by direct contact between motifs and proteins, and by double-headed arrows. It is thought that an unidentified adapter protein links Hect ubiquitin ligases with EscrtIII. This complex and Vps4 (vacuolar protein sorting 4) are thought to mediate membrane reorganization and fusion during release of enveloped virus particles. Alix binds to endophilins, proteins that induce membrane curvature. Its interaction with Chmp4 (charged multivesicular body protein 4) is required for budding of human immunodeficiency virus type 1.

by the PTAP L domain therefore suggested that the cellular machinery that mediates sorting and trafficking of endocytic vesicles is diverted to promote budding and release of virus particles. In fact, Tsg101 proved to be the human homolog of one subunit of a complex required for sorting of yeast proteins to the vacuole/lysosome, termed endosomal sorting complex required for transport-I (Escrt-I). The other subunits of human Escrt-I are also required

Table 13.3 Common sequence motifs required for budding of enveloped virus particles

Core motif	L domain sequence[a]	Viral protein
PT/SAP	...PEPTAPPEE...	Human immunodeficiency virus type 1 Gag
	...VEPTAPQV...	
	...ILPTAPEY...	Human immunodeficiency virus type 1 Gag
	...YAPSAP...	Ebola virus M
		Vesicular stomatitis virus M
PPXY	...APPPPYVG...	Rous sarcoma virus Gag
	...IAPPPYEE...	Vesicular stomatitis virus M
LXXLF	...ASLRSLFG...	Human immunodeficiency virus type 1 Gag
	...VRLDLLLL...	Sendai virus M

[a]The amino acids of the L domain motifs are underlined.

for release of human immunodeficiency virus type 1. As summarized in Fig. 13.16B, Escrt-I is associated with two other Escrt assemblies, via adapter proteins such as Alix. Importantly, the other protypical L domain sequences (Table 13.3) recruit this **same** cellular machine, although they bind directly to different components (Fig. 13.16B). It therefore appears that formation and release of virus particles with very different structures, genomes, and composition are driven by the same cellular components and mechanism.

Many mechanistic questions about the precise molecular functions of the many proteins shown in Figure 13.16B remain to be addressed. However, Escrt-III is thought to provide the machinery for membrane budding and fusion. Consistent with this view, dominant negative derivatives of Escrt-III subunits inhibit budding of all retroviruses examined, irrespective of which L domain sequence is present in Gag, and of the rhabdovirus vesicular stomatitis virus.

A small fraction of retroviral Gag is ubiquitinylated, and the PPXY L domain sequence recruits specific ubiquitin ligases (Fig. 13.16B). A catalytically active ubiquitin ligase is necessary for release of retrovirus with this Gag L domain sequence. Furthermore, ubiquitinylation of human immunodeficiency virus type 1 Gag at sites C-terminal to the CA domain is necessary for efficient release: substitutions that prevent modification at these sites lower the rate of release, and induce the accumulation of virus particles tethered to the plasma membrane. As ubiquitin is recognized by several of the endocytic trafficking proteins, this modification might promote assembly of the machine that mediates budding and release of retroviruses. A significant fraction of human immunodeficiency virus type 1 Gag localizes to late endosomes and multivesicular bodies, particularly in macrophages. It has therefore been suggested that Gag is sorted to such intracellular compartments prior to transport to the plasma membrane. However, analysis of Gag trafficking by pulse-chase labeling has shown recently that the polyprotein is sorted primarily to the plasma membrane. Much Gag is then released in assembled particles, and the remainder is subsequently internalized, accumulating in late endosomes and multivesicular bodies.

Release from the plasma membrane can also depend on viral proteins other than major structural proteins. For example, in some cell types, efficient release of human immunodeficiency virus type 1 requires the viral Vpu protein. This function of Vpu is distinct from the induction of proteolysis of CD4 in the ER described in Chapter 12. In the absence of Vpu, particles accumulate in intracellular vacuoles, or we tethered to the infected cell surface. The Vpu protein was shown to counteract the action of an antiviral protein that tethers virus particles to the cell surface, and is produced when cells are exposed to alpha interferon. This protein was subsequently identified, based on the pattern of expression of its gene, and termed tetherin. Its organization suggests that interactions between tetherin molecules inserted in the plasma membrane and the viral envelope are responsible for retaining virus particles at the cell surface. How this interaction is disrupted by Vpu is not yet known.

Assembly at Internal Membranes: the Problem of Exocytosis

Cytoplasmic Compartments of the Secretory Pathway

Several enveloped viruses are assembled at the cytoplasmic surfaces of compartments of the secretory pathway under the direction of specifically located viral glycoproteins (Table 12.4). The final step in virion production is budding of the particle into the lumen of one of these compartments, for example, of bunyaviruses into Golgi cisternae. These particles therefore lie within membrane-bound organelles. It is generally assumed that such virus particles must be packaged within cellular transport vesicles for travel along the secretory pathway to the cell surface. Despite almost universal acceptance of this mechanism, we have little direct evidence for this mode of transport. Indeed, in some cases there is evidence indicating that the secretory pathway is not required. Immature capsids of the betaretrovirus Mason-Pfizer monkey virus assemble at internal cytoplasmic sites, near the centrioles. Their transport to the plasma membrane depends on the viral Env protein, implicating some type of vesicular transport. However, this process is refractory to inhibition of ER-to-Golgi transport. Rather, it is blocked by disruption of endocytic trafficking.

The budding of virus particles into internal compartments of the secretory pathway is initiated by interactions among the cytoplasmic domains of viral membrane proteins and internal components of the particle. Consequently, this process generally begins as soon as the integral membrane and cytoplasmic viral proteins attain sufficient concentrations in the infected cell. For example, the concentration of viral membrane proteins (surface proteins) determines the fate of hepadnaviral cores, which contain the capsid (C) protein (Table 13.1), a DNA copy of the pregenomic RNA, and the viral polymerase (Appendix, Fig. 3). Early in infection, the concentration of the large surface protein (L) in membranes is too low for efficient envelopment of cores (Fig. 13.17); these structures therefore enter the nucleus, where they contribute to the pool of viral DNA templates for transcription. As the concentration of the L protein increases, it interacts with cores, and enveloped particles form. The ability of hepadnaviral cores to bind to this viral glycoprotein is also regulated by the nature of the nucleic acid they contain: synthesis of DNA from the pregenomic RNA induces significant conformational changes in the exterior surface of core particles. These induce changes in the shape of the hydrophobic pocket that is formed by residues required for normal assembly and release of virus particles.

An intriguing question is why some viruses formed by a budding mechanism assemble at intracellular membranes rather than at the plasma membrane, where budding ensures release of the virion from the host cell. One possible advantage of intracellular budding is that the concentration of viral glycoproteins exposed on the surface of the infected cell is reduced. This property would decrease the likelihood that the infected cell would be recognized by components of the immune system before the maximal number of progeny virions were assembled and released. Alternatively, the simpler cytoplasmic surfaces of internal membranes, which are not burdened with cytoskeletal structures and the proteins that attach them to the extracellular matrix, may make for more facile assembly or budding reactions. Or the distinctive lipid composition of internal membranes may confer some (as yet unknown) special property that is advantageous to these viruses.

Envelopment by a Virus-Specific Mechanism

The interaction of components of the poxvirus vaccinia virus with internal cellular membranes during assembly is most unusual. One remarkable feature is the assembly of two **different** infectious particles, the intracellular mature and the extracellular enveloped virions, that differ in the number and origin of lipid membranes. Furthermore, the initial acquisition of a membrane early in assembly occurs by an apparently unique, virus-specific mechanism.

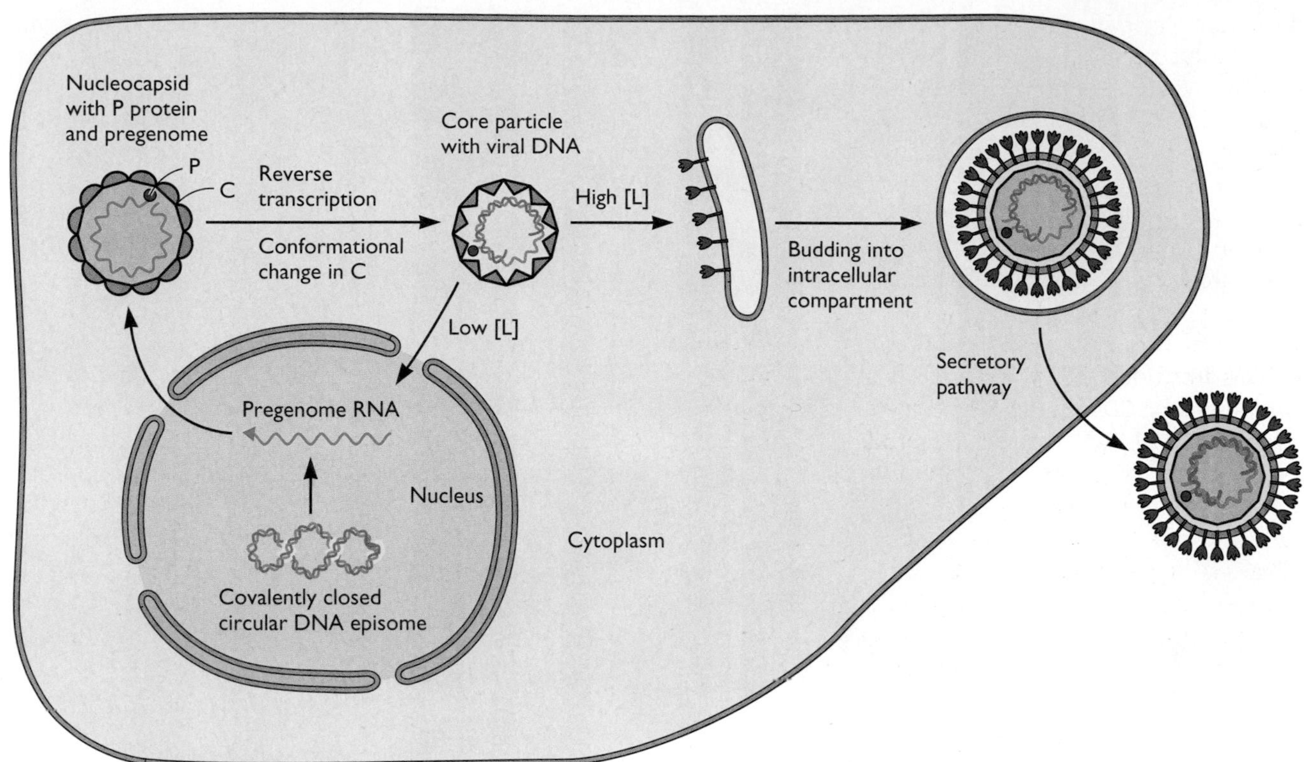

Figure 13.17 Model of hepatitis B virus envelopment. The pregenome RNA synthesized in infected cell nuclei (see Chapter 7) is exported to the cytoplasm, where it is incorporated into particles built from the capsid (C) protein. Reverse transcription to produce the DNA genome induces a conformational change in C protein that allows interaction of C with the large surface protein (L) inserted into internal membranes. Whether core particles containing DNA enter the nucleus or become enveloped by budding into compartments of the secretory pathway is determined by the concentration of the L protein. The L and middle (M) and small (S) envelope glycoproteins accumulate in membranes of the ER-Golgi intermediate compartment, into which subviral particles that contain only lipid and envelope proteins (primarily S) appear to bud. As the S, as well as the L, protein is required for envelopment, it is generally accepted that virions are also formed by budding into this same compartment of the secretory pathway. However, the results of recent experiments implicate the cellular proteins that participate in endocytic trafficking described previously in hepatitis B virus budding and release.

Finally, infectious particles leave the host cell by at least three distinct routes.

Vaccinia virus assembly is a complex process that includes the formation of several intermediates, such as crescents (see below) and immature virions, and major morphological rearrangements as infectious particles are formed (Fig. 13.18A). The assembly pathway was elucidated initially by electron microscopy in some of the earliest studies of vaccinia virus. Numerous viral proteins that participate in the various assembly reactions have been identified by genetic experiments (Table 13.4). Synthesis of viral DNA genomes and structural proteins takes place in discrete cytoplasmic domains termed viral factories. The first morphological sign of assembly is the appearance within viral factories of rigid, curved structures 10 to 15 nm thick (Fig. 13.18B). There has never been any doubt that these structures, termed crescents, contain at least one lipid bilayer. In contrast, the origin of this membrane, and whether a second bilayer is also present, have been subjects of much debate. The current consensus is that crescents contain a single lipid membrane that is derived from the ER membrane, but by a unique, perhaps virus-specific mechanism (Box 13.8). As the crescents enlarge, they retain their original curvature and therefore eventually form spheres surrounding viral macromolecules present in viral factories, including the DNA genome. Such immature virions then undergo major morphological transitions to form brick-shaped mature virions (Fig. 13.18B). This maturation process requires several distinct reactions. These include proteolytic cleavage of several structural proteins by viral protease(s) (Table 13.3), the action of the viral redox system (Chapter 12), and removal of at least one crescent-associated protein. These changes

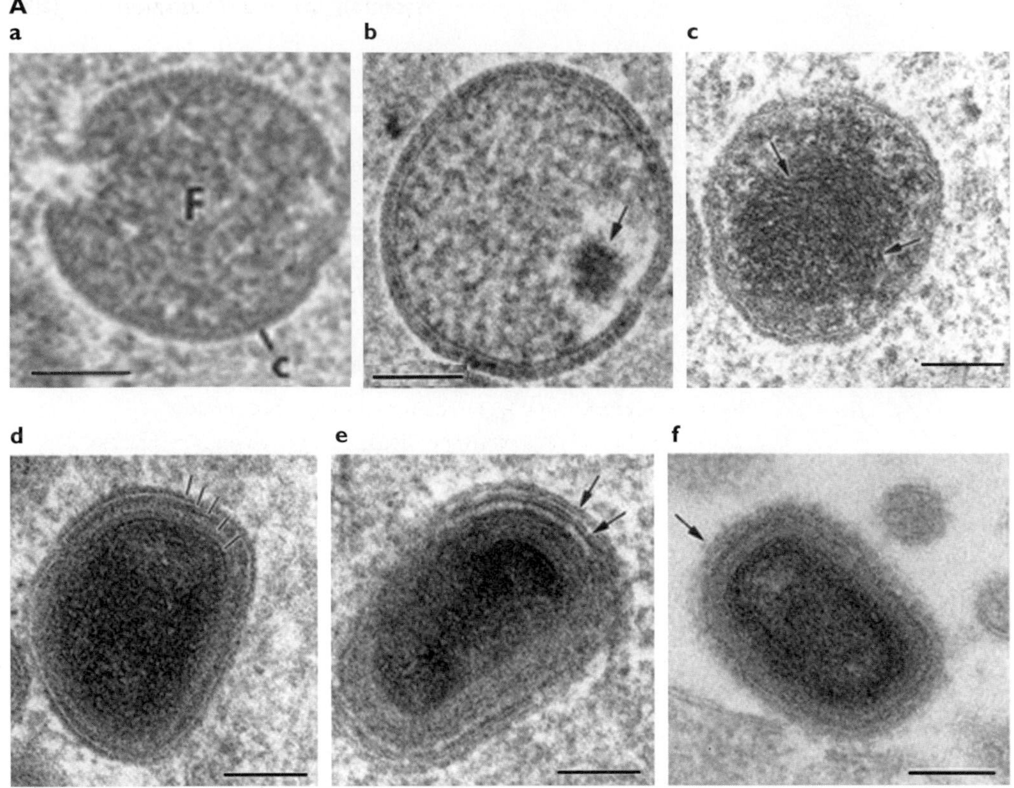
A
a
b
c
F
c
d
e
f
B
Crescent
DNA
Immature virion
Mature virion
ER
Proteins
Post-Golgi
or early
endosome
Cytoplasm
Wrapped
virion
Actin tail
Plasma
membrane
Extracellular
enveloped
virion
Cell-associated
enveloped virion

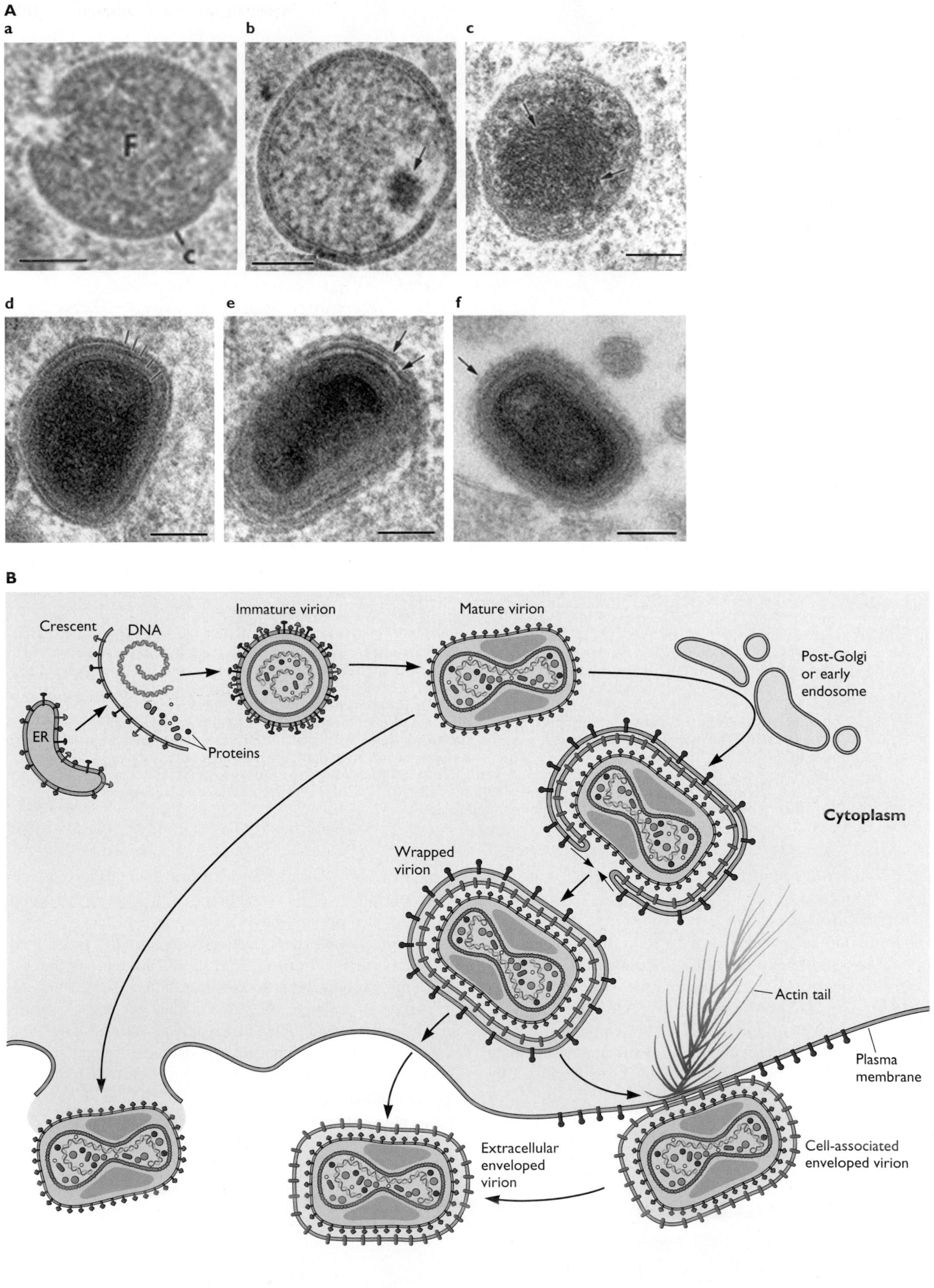

Table 13.4 Some proteins implicated in vaccinia virus assembly

Assembly reaction	Protein	Function/properties
Crescent formation	A14, A17	Essential for this step; integral membrane proteins; phosphorylated by F10; form a disulfide-bond-stabilized lattice
	D13	Imparts crescent curvature and rigidity; interacts with A17
	F10	Essential for ER membrane remodeling and appearance of crescents; dual-specificity protein kinase
Assembly of immature virion (IV)	Complex of 7 proteins, including F10	Association of viroplasm with crescents
	A32	Genome encapsidation; required for packaging DNA and I6
	I6	Genome encapsidation; binds specifically to terminal hairpins in DNA
Formation of mature virion (MV)	A4	Core assembly during morphogenesis; present in outer palisade layer of core wall
	A3	Formation of morphologically normal and transcriptionally active cores; proteolytically processed during morphogenesis
	G1	IV-to-MV transition; metalloprotease
	I7	IV-to-MV transition; cysteine protease, required for processing A3 and other proteins
	E10, A2.5, G4	IV-to-MV transition; thiol redox proteins (see Chapter 12)
Formation of wrapped virion	A27	Essential for this step: disulfide-bonded trimer bound to MV membrane
	B5, F13	Required for efficient wrapping; transmembrane proteins sorted to intracellular wrapping site(s)

resemble those that occur during assembly of herpesviruses and adenoviruses but are undoubtedly more complex.

The mature virus is released only upon lysis of the infected cell. However, a proportion of this population (that varies with the time of infection) becomes engulfed by the membranes of a second intracellular compartment, probably a *trans*-Golgi or early endocytic compartment, to form the wrapped virion (Fig. 13.18B). The mature virion is transported to the site(s) of wrapping via microtubules. The remodeling of organelle membranes to form the wrapped virion depends on a number of viral proteins that are present only in this type of particle (Table 13.4), and that appear to be sorted to wrapping sites via the secretory pathway. The mechanism(s) by which such proteins interact with mature virions to induce formation of the three membrane-containing wrapped virion is not known. This particle can be released from the cell as the two-membrane-containing extracellular enveloped virion following transport to the cell surface and fusion of its outer membrane with the plasma membrane (Fig. 13.18B). As the mature virion and the extracellular enveloped virion bind to different cell surface receptors,

Figure 13.18 Vaccinia virus assembly and exocytosis. (A) Viral structures observed in infected cells. HeLa cells infected with vaccinia virus for 10 or 24 h were prepared for electron microscopy by quick freezing and negative staining while frozen. These procedures preserve fine structural detail. The following structures are shown. (a) Viral factory comprising a viral crescent (C) around a viroplasm focus (F). (b) Immature virions, in this example associated with DNA (arrows) entering via a pore in the particle. (c) Spherical, dense particles containing DNA-like material (arrows). (d) Intracellular mature virion. (e) Intracellular enveloped virion with the additional double membrane (arrows). (f) Extracellular enveloped virion, which carries an external fuzzy layer (arrow). Bars, 100 nm. Adapted from C. Risco et al., *J. Virol.* **76:**1839–1855, 2000, with permission. **(B)** Schematic model of assembly and exocytosis. Assembly begins with the formation of crescents by diversion of membrane from the ER. The viral D13 protein, which is associated with the outer leaflets of crescents, maintains the curvature and rigidity of the crescent membrane as it enlarges and eventually closes with the incorporation of viral DNA and proteins from viral factories. As the D13 protein is lost during the morphological transitions that form the brick-shaped mature virion, it is considered a scaffolding protein. The mature virion is released from infected cells only upon lysis. However, a significant proportion of these structures acquire additional membranes by wrapping, in membranes derived from a late or post-Golgi compartment to form the wrapped virion. This particle is transported to the plasma membrane, where fusion with this cellular membrane forms the cell-associated enveloped virus, which induces formation of actin tails. Adapted from B. Sodeik and J. Krijnse-Locker, *Trends Microbiol.* **10:**15–24, 2002, with permission.

BOX 13.8 DISCUSSION
How is the vaccinia virus crescent membrane formed? An unsolved mystery

It is simple to visualize how reorganization and fusion of internal cellular membranes can "wrap" structures in a double membrane, as during formation of wrapped virions of vaccinia virus (Fig. 13.18B). In contrast, it is not at all obvious how viral structures containing a single lipid bilayer, the crescent (and immature and mature virions), form by a nonbudding mechanism. This conundrum led to the early proposal that the crescent membrane is synthesized *de novo* from cellular lipids. No mechanism for such *de novo* assembly has been identified, and it is generally agreed that crescents are derived from preexisting cellular membranes.

There is accumulating evidence that the ER is the source of the crescent membrane. For example, several of the major viral membrane proteins are inserted into the ER membrane *in vitro* or *in vivo*, and one (A9) is present near sites of assembly in tubular structures that contain the ER lumenal enzyme, protein disulfide isomerase. Furthermore, when a heterologous signal sequence was added to the N terminus of A9, the signal sequence was cleaved off and only the truncated protein was detected in immature and mature virions. As signal peptidase, which removes signal sequences, resides in the ER (see Chapter 12), this observation provides compelling support for the view that the ER membrane containing viral membrane proteins is the origin of the crescent membrane.

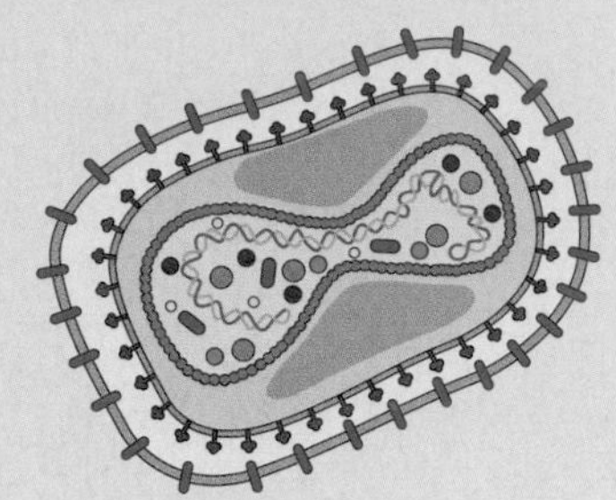

How a single lipid bilayer structure is produced from the ER remains an enigma.

Hussain, M., A. S. Weisberg, and B. Moss. 2006. Existence of an operative pathway from the endoplasmic reticulum to the immature poxvirus membrane. *Proc. Natl. Acad. Sci. USA* **103:**19506–19511.

the release of two types of infectious particle may increase the range of cell types that can be infected. A significant proportion of enveloped virions are not released following membrane fusion but, rather, remain attached to the host cell surface as cell-associated enveloped virions (Fig. 13.18B). The mechanisms of transport and egress that produce cell-associated virions are amazing processes that depend on major reorganization of components of the host cell cytoskeleton.

Wrapped virions initially travel from sites of assembly to the plasma membrane on microtubules, carried by the cellular motor protein kinesin. The interaction of these

Figure 13.19 Movement of vaccinia virus on actin tails. (A) Immunofluorescence micrograph of virus particles (red) at the ends of the cell surface projections containing actin tails (green). The coincidence of the tips of the projecting actin and viral particles gives yellow-orange signals (marked by white arrowheads), indicating that the particles are projected from the cell surfaces on the tips of actin tails. When infected cells are plated with uninfected cells, such actin-containing structures to which virus particles are attached can be seen extending from the former into the latter. **(B)** An electron micrograph of a virus particle attached (arrowhead) to an actin tail. From S. Cudmore et al., *Nature* **378:**636–638, 1995, with permission. Courtesy of S. Cudmore and M. Way, European Molecular Biology Laboratory.

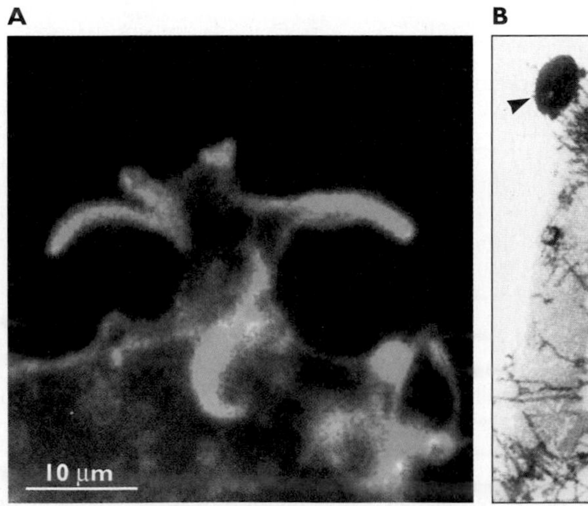

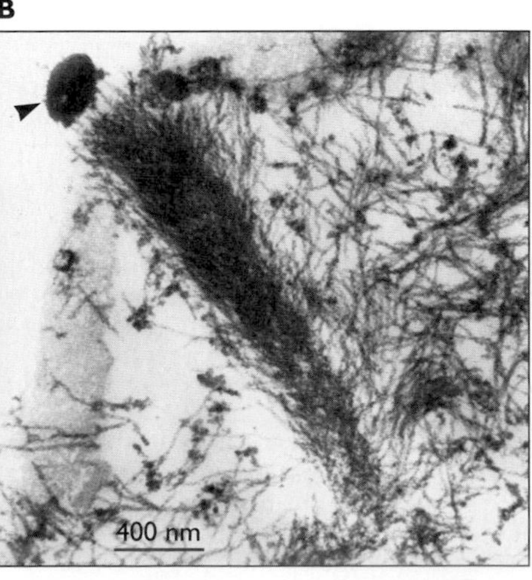

particles with the motor depends on the viral A36 protein present in their outer membrane, which interacts with the light chain of the kinesin motor. Such active transport allows movement of the large wrapped virion to the cell periphery in less than 1 min (compared to an estimated 10 h that would be required by passive diffusion!). Remodeling of the dense layer of cortical actin that lies beneath the plasma membrane (Fig. 2.4) is also required to deliver these particles to the plasma membrane. This phenomenon is induced by a viral protein that modulates the cellular signaling pathway that regulates the dynamics of cortical actin. Actin remodeling is required for efficient release of wrapped virions, presumably in part because the movement of virus particles near the cell surface depends on the actin cytoskeleton.

The particles formed by fusion of wrapped virions with the plasma membrane remain cell associated because of a remarkable activity: they induce a further, and dramatic reorganization of the actin cytoskeleton just below the site of fusion. The number of typical actin stress fibers is significantly decreased, because the virus induces the formation of new, filamentous actin-containing structures. Each of these, which are termed actin tails, is in contact with a single virus particle (Fig. 13.19A and B). Viral particles attached to the tips of actin tails are propelled at an average speed of 0.18 nm/s, by polymerization of actin at the front end of the tail and its depolymerization at the back end. As the infection progresses, they can be seen on large microvilli induced by the actin tails (Fig. 13.19B). Remarkably, formation of actin tails in vaccinia virus-infected cells requires the same viral protein that allows transport of wrapped virions along microtubules, the A36 protein. This protein is phosphorylated at specific positions by the cellular tyrosine kinase Src, which plays an important role in regulation of actin dynamics in uninfected cells. Phosphorylation of A36 triggers its dissociation from kinesin and allows binding of cellular proteins that promote actin polymerization (Fig. 13.20).

The formation of vaccinia-actin tails is necessary for efficient spread of the virus, for mutants that cannot induce these structures form only small plaques on cultured cells. The viral particles attached to the outer surface of cellular projections containing actin tails can dissociate to become extracellular enveloped virions. However, such projections can extend from infected cells into neighboring uninfected cells, suggesting that they may also facilitate direct cell-to-cell spread of infectious particles.

Intranuclear Assembly

The problem of egress is especially acute for the enveloped herpesviruses, because the nucleocapsids assemble in the nucleus. The pathway by which the virus leaves the cell has been a topic of fierce controversy, centered on where and when the viral envelope is acquired. A large body of evidence now favors the less intuitive (and to some, inefficient) double-envelopment model summarized in Figure 13.21.

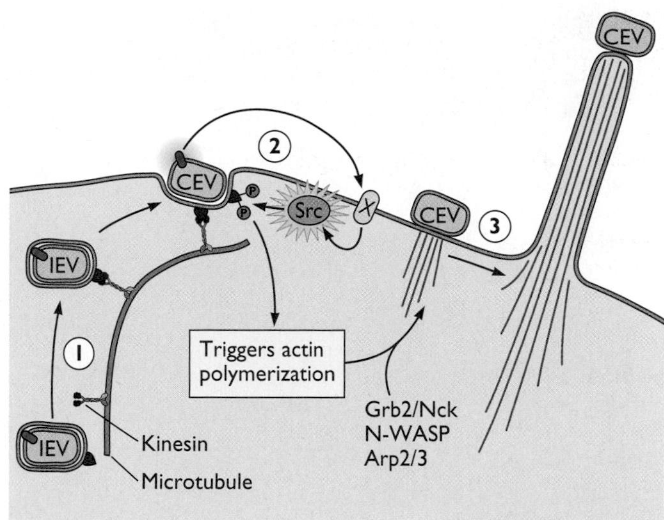

Figure 13.20 Model for the switch from microtubule- to actin-dependent transport of vaccinia virus particles. The A36R protein present in the outer membrane of wrapped virions binds to the light chain of the kinesin motor, which then transports the particles to the cell periphery. Remodeling of cortical actin by viral proteins allows close approach of the virions to the plasma membrane. Fusion of the outer membrane of wrapped virions with the plasma membrane releases cell-associated virions, which carry the B5R protein in their new outer membrane. This viral protein activates the cellular Src tyrosine kinase, presumably via interaction with one or more cellular membrane proteins (X). Src then phosphorylates the membrane-associated A36R protein, a modification shown by genetic experiments to be essential for formation of actin tails. Furthermore, A36 remains bound to kinesin in vaccinia virus-infected cells that lack Src or that are treated with inhibitors of this kinase. Phosphorylated A36R binds via adapter (Grb and Nck) and scaffolding (N-Wasp) to cellular proteins that induce actin polymerization. Such polymerization drives the formation of actin-tail containing protrusions that project cell-associated virions away from the host cell. Adapted from A. Hall, *Science* **306:**65–67, 2006, with permission.

The first step in egress is exit of nucleocapsids from the nucleus, which is achieved not by transport through nuclear pore complexes but, rather, by a unique budding mechanism (Fig. 13.21). In the case of herpes simplex virus type 1, a subset of the tegument proteins, including VP16, associate with the nucleocapsid prior to budding. Late in infection, the dense meshwork of protein filaments that abuts the inner nuclear membrane (the nuclear lamina) is dramatically reorganized and perforated (Fig. 13.22), presumably to allow juxtaposition of the nucleocapsid and membrane. Such disruption of the nuclear

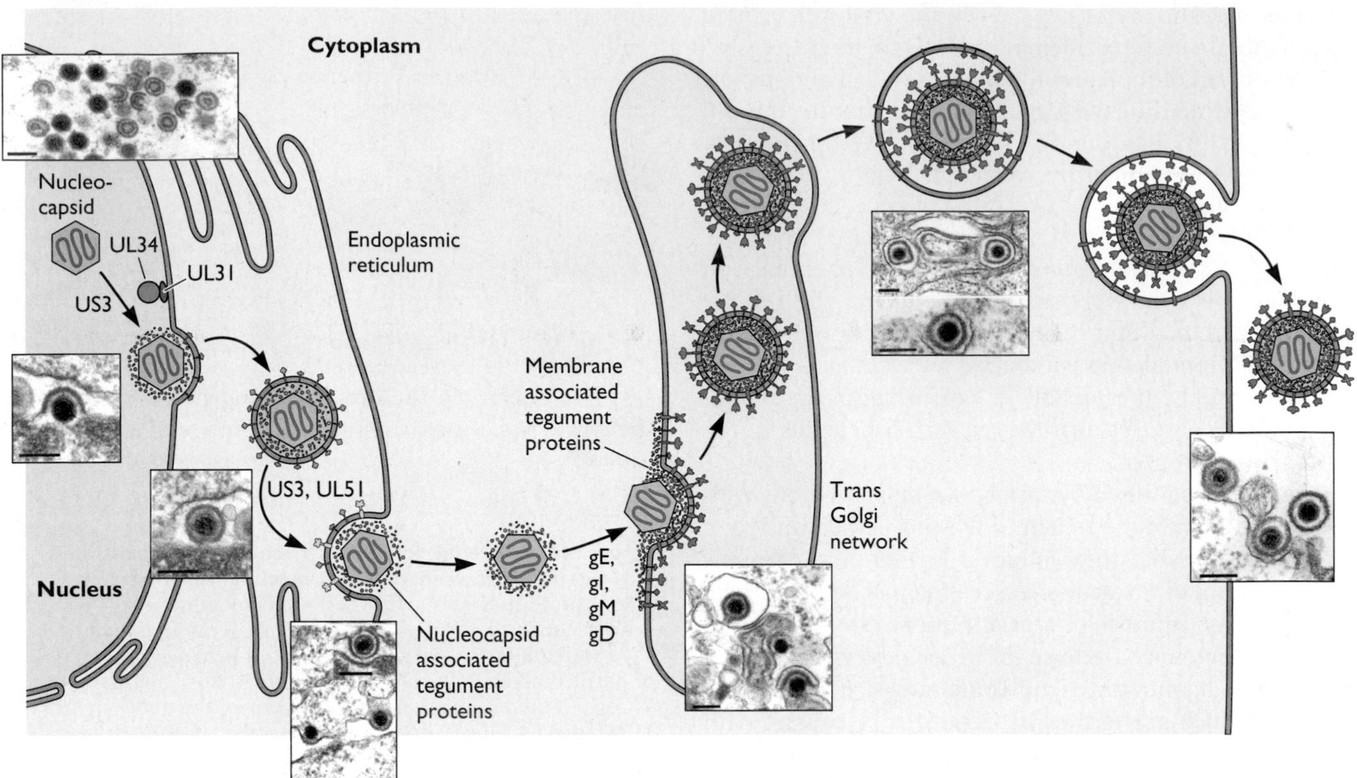

Figure 13.21 Pathway of herpesvirus egress. The mature nucleocapsid assembled within the nucleus (Fig. 13.5 and 13.10) initially acquires an envelope by budding through the inner nuclear membrane. Upon fusion with the outer nuclear membrane, this membrane is lost as unenveloped nucleocapsids are released into the cytoplasm. Some tegument proteins interact with the nucleocapsid in the cytoplasm, whereas others, including the UL11, UL46, and UL49 proteins, concentrate at sites of secondary envelopment. The latter are presumably localized at membranes of *trans*-Golgi compartments, by interactions with the cytoplasmic domains of viral glycoproteins, such as the binding of the UL11 and UL49 proteins to the cytoplasmic domains of gE and gD. The myristyolated UL11 protein accumulates at the membranes of *trans*-Golgi compartments and directs other tegument proteins to sites of secondary envelopment. The viral envelope is acquired upon budding of tegument-containing structures into compartments of the *trans*-Golgi network. Virions formed in this way are thought to be transported to the plasma membrane in secretory transport vesicles and released upon membrane fusion, as illustrated. Viral gene products implicated in specific reactions are indicated. The reactions are illustrated in the corresponding electron micrographs of cells infected by the alphaherpesvirus pseudorabies virus. Bar, 150 nm. Adapted from T. C. Mettenleiter, *J. Virol.* **76:**1537–1547, 2002, with permission. Courtesy of T. C. Mettenleiter, Federal Research Center for Virus Diseases of Animals, Insel Riems, Germany.

lamina requires the viral UL31 phosphoprotein and the UL34 transmembrane protein. When phosphorylated by the viral US3 protein kinase, UL31 and UL34 associate with one another at the inner surface of the inner nuclear membrane, and bind the proteins that form the lamina (lamins A/C and B) and cellular protein kinase C. This enzyme phosphorylates the lamins, while the US3 kinase phosphorylates the nuclear membrane protein emerin, which binds to the lamins and has been implicated in maintenance of nuclear integrity. These modifications are thought to disrupt the interactions that form the nuclear lamina. The mechanisms that then drive budding and fusion of the inner nuclear membrane are not understood, not least because the nucleocapsid component(s) that interacts with the membrane has not been identified. The deenvelopement reaction that subsequently releases nucleocapsids into the cytoplasm (Fig. 13.21) requires either the gB or the gH glycoproteins.

The second envelopment, in which particles acquire the virion envelope, takes place at the cytoplasmic surface of compartments of the *trans*-Golgi network. Viral membrane proteins, including those necessary for secondary envelopment (e.g., gD, gE/gI, gM, and the UL20 protein), are sorted to these cellular compartments via the

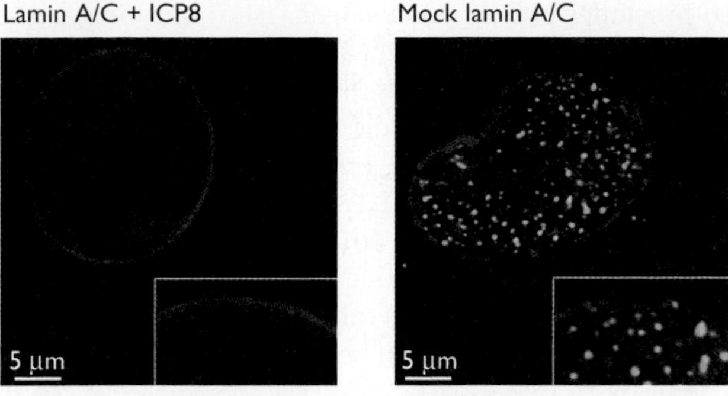

Figure 13.22 Disruption of the nuclear lamina in herpes simplex virus type 1- infected cells. Human cells mock-infected or infected with herpes simplex virus type 1 for 16 h were examined by indirect immunofluoresence. The cellular lamin A/C and viral ICP8 proteins are in red and green, respectively. The insets show magnified regions of equal size. Adapted from M. Simpson-Holley et al., *J. Virol.* **79:**12840–12851, with permission. Courtesy of D. Knipe, Harvard University Medical School.

secretory pathway (see Chapter 12). Some tegument proteins accumulate at the sites of secondary envelopment, and are required for this step. Others associate with the nucleocapsid in the cytoplasm. These proteins include the US3 kinase described previously and the UL36 and UL37 proteins, which are required for transport of nucleocapsids through the cytoplasm. Once the nucleocapsid reaches the *trans*-Golgi network, interactions between these two classes of tegument protein must take place prior to envelopment. Nevertheless, the proteins that mediate such final assembly of the tegument have not yet been identified, nor have those that induce membrane budding and fusion. Some recent observations hint that such viral proteins may function via the cellular budding machinery that mediates the release of simpler enveloped viruses.

It is thought that enveloped virions released into the lumen of a *trans*-Golgi compartment travel to the plasma membrane in secretory vesicles, and are released upon fusion of such vesicles with that membrane (Fig. 13.21).

Maturation of Progeny Virions

Proteolytic Processing of Virion Proteins

The products of assembly of several viruses are noninfectious particles, often called **immature virions.** In all cases, proteolytic processing of specific proteins with which the particles are initially built converts them to infectious virions. The maturation reactions are carried out by virus-encoded enzymes, and take place late in assembly of particles, or following release of immature virions from the host cell. Proteolytic cleavage of virion proteins introduces an irreversible reaction into the assembly pathway, driving it in a forward direction. This modification can also make an important contribution to resolving the contradictory requirements of assembly and virus entry. One consequence of proteolytic processing is the exchange of covalent linkages between specific protein sequences for much weaker noncovalent interactions, which can be disrupted in a subsequent infection. A second is the liberation of a new N terminus and a new C terminus at each cleavage site, and hence opportunities for additional protein-protein contacts. Such changes in chemical bonding among virion proteins clearly facilitate virus entry, for the proteolytic cleavages that introduce them are necessary for infectivity. Accordingly, viral proteases and the structural consequences of their action are of considerable interest. Moreover, these enzymes are excellent targets for antiviral drugs, a property exemplified by the success of therapeutic agents that inhibit the human immunodeficiency virus type 1 protease.

Cleavage of Polyproteins

As we have seen, the protein shells of several viruses, including picornaviruses and many retroviruses, are assembled from polyproteins. The liberation of individual structural proteins is essential for formation of mature particles.

The alterations in the structure of the virus particle and their functional correlates are best understood for small RNA viruses, such as the picornavirus poliovirus. A single cleavage to liberate VP4 and VP2 from VP0 converts noninfectious provirions to mature infectious virions (Fig. 13.4). As the viral proteases are not incorporated into particles, VP0 cleavage may be catalyzed by a specific feature of the virion itself, with internal genomic RNA participating in the reaction. The structural changes induced by such maturation cleavage can be described in great detail, for the structures of mature virions and empty particles in which VP0 have been determined at high resolution. Cleavage of VP0 allows the extensive internal structures of the particle (Fig. 4.13C), to be established in its final form. Cleavage of VP0 therefore allows additional protein-protein interactions that make important contributions to the stability of the virion.

Cleavage of VP0 to VP4 and VP2 is also necessary for release of the RNA genome into a new host cell. The conformational transitions that mediate entry of the genome following attachment of the virus to its receptor are not fully understood. However, many alterations that impair receptor binding and entry map to just those regions of the capsid proteins that participate in the structures that adopt their final organization only upon VP0 cleavage. Cleavage of VP0 therefore not only further stabilizes the virion, but also "spring-loads" it for the conformational transitions that take place during entry and release of the genome.

Following release of most retrovirus particles, the Gag polyprotein is processed by the viral protease, concomitant with substantial morphological and conformational rearrangements (Fig 13.23; Box 13.9). However, the mature, infectious particles of members of the *Spumaretrovirinae*, which resemble hepadnaviruses in some aspects of their infectious cycles, contain uncleaved Gag.

Processing of Gag plays an essential part in the mechanisms by which most infectious retroviruses are assembled and released. As we have seen, interactions among Gag polyproteins, and of their NC and MA domains with the viral RNA and the plasma membrane, respectively, build and organize assembling retrovirus particle. Efficient and orderly assembly also depends on "spacer" peptides that are removed during proteolysis. Furthermore, the membrane-binding signal of MA is exposed when MA is part of Gag, but is blocked by a C-terminal α-helix of MA in the mature protein. It is therefore very unlikely that retrovirus particles could be constructed correctly from mature Gag proteins. Indeed, alterations that increase the catalytic activity of the viral protease inhibit budding and production of infectious particles, indicating that premature processing of the polyproteins is detrimental to assembly. On the other hand, the covalent connection of the virion proteins that is so necessary during assembly is incompatible with release of the virion core following fusion of the viral envelope with the membrane of a new host cell. Such covalent linkage also precludes efficient activity of virion enzymes, which are incorporated as Gag-Pol proteins. In some virions, including those of Moloney murine leukemia virus, the protease also removes a short C-terminal segment of the cytoplasmic tail of the TM envelope protein to activate the fusionogenic activity of TM. The retroviral proteases that sever such connections therefore are absolutely necessary for production of infectious virions, even though they are dispensible for assembly.

Figure 13.23 Morphological rearrangement of retrovirus particles upon proteolytic processing of the Gag polyprotein. These two cryo-electron micrographs show the maturation of human immunodeficiency virus type 1 virions. **(Left)** The immature particles contain a Gag polyprotein layer below the viral membrane and its external spikes. **(Right)** Processing of Gag converts such particles to mature virions with elongated internal capsids. Courtesy of T. Wilk, European Molecular Biology Laboratory.

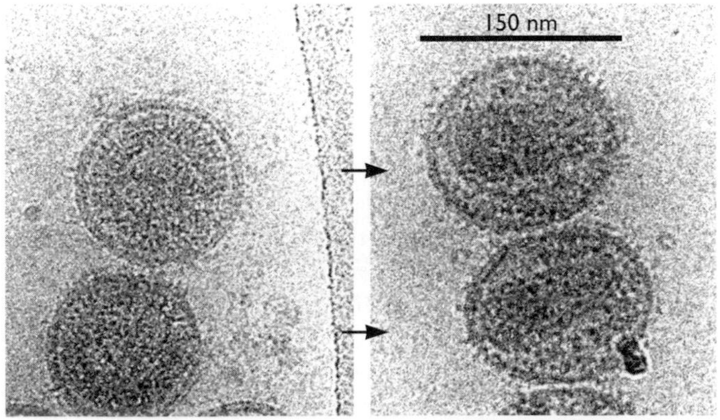

The retroviral proteases belong to a large family of enzymes with two aspartic acid residues at the active site (aspartic proteases). The viral and cellular members of this family are similar in sequence, particularly around the active site, and are also similar in three-dimensional structure. All aspartic proteases contain an active site formed between two lobes of the protein, each of which contributes a catalytic aspartic acid. The retroviral proteases are homodimers, in which each monomer corresponds to a single lobe of their cellular cousins. Consequently, the active site is formed only upon dimerization of two identical subunits. This property undoubtedly helps avoid premature activity of the protease within infected cells, in which the low concentration of their polyprotein precursors mitigates against dimerization. Indeed, dimerization of the protease appears to be rate limiting for maturation of virions. Fusion of the protease to the NC domain of Gag also inhibits dimerization. Consequently, synthesis of the protease as part of a polyprotein precursor not only allows incorporation of the enzyme into assembling virions, but also contributes to regulation of its activity. These properties raise the question of how the protease is activated, a step that requires its cleavage from the polyprotein. Polyproteins containing the protease (e.g., made in bacteria) possess some activity, sufficient to liberate fully active enzyme at a very low rate *in vitro*. It is therefore thought that such activity of the polyproteins initially releases protease molecules within the particle. Furthermore, it has been shown, using Gag-Pol proteins yielding distinguishable cleavage products, that the initial proteolytic cleavages are intramolecular. The high local concentrations of protease molecules within the assembling particle would facilitate subsequent dimerization of protease molecules to form the fully active enzyme.

Cleavage of Precursor Proteins

Like its retroviral counterpart, the adenoviral protease converts noninfectious particles to infectious virions, in this case by cleavage at multiple sites within six virion proteins (Fig. 13.8). Although the adenoviral enzyme does not process polyprotein precursors, the cleavage of so many proteins also appears to alter protein-protein interactions necessary for virion assembly, in preparation for early steps in the next

BOX 13.9

DISCUSSION

Model for refolding of the human immunodeficiency virus type 1 CA protein on proteolytic processing of Gag

The model for the radial organization of the human immunodeficiency virus type 1 Gag polyprotein (left) is based on cryo-electron micrographs like that in Fig. 13.3. The three-dimensional structures of the processed proteins, the MA trimer (red), the CA dimer (blue), and monomeric NC (violet) bound to the SL3 packaging signal (green), shown on the right, are derived from high-resolution structures discussed in this and preceding chapters.

In the X-ray crystal structure of the N-terminal portion of mature CA (right), the charged N terminus is folded back into the protein by a β hairpin formed by amino acids 1 to 13 and forms a buried salt bridge with the carboxylate of Asp51. The lack of a charged N terminus prior to cleavage of CA from MA, and the steric difficulties of burying the N terminus of CA attached to an MA extension (left), indicate that the β hairpin and buried salt bridge can form only after proteolytic cleavage. Furthermore, the viral protease recognizes the cleavage site between MA and CA in an extended conformation. As the N-terminal β hairpin of mature CA forms a CA-CA interface, it has been proposed that proteolytic cleavage and the consequent refolding of the N terminus of CA facilitate the rearrangements to form the conical core during maturation of virus particles. The inhibition of core assembly and formation of infectious viral particles *in vivo* by alteration of amino acids in this interface support this model.

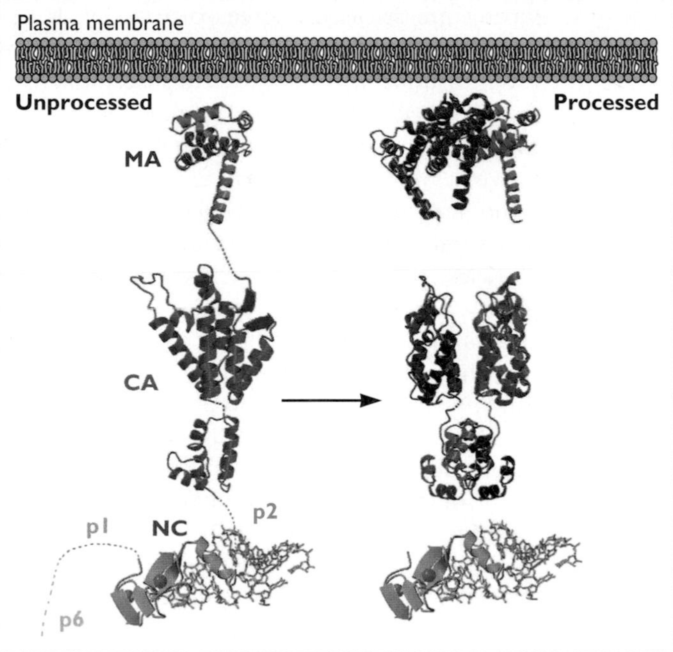

Conformational changes in segments of Gag during assembly may regulate proteolytic processing. For example, efficient cleavage to liberate human immunodeficiency virus type 1 NC depends on the binding of NC to RNA, at least *in vitro*. Figure courtesy of T. L. Stemmler and W. Sundquist, University of Utah.

von Schwedler, U. K., T. L. Stemmler, V. Y. Klishko, S. Li, K. H. Albertine, D. R. Davis, and W. I. Sundquist. 1998. Proteolytic refolding of the HIV-1 capsid protein amino-terminus to facilitate viral core assembly. *EMBO J.* **17:**1555–1568.

infectious cycle. The enzyme is a cysteine protease containing an active-site cysteine and two additional cysteines, all highly conserved. One mechanism by which its activity is regulated is by interaction with a small peptide, a product of cleavage of the virion protein pVI, or with pVI itself (Fig. 13.24). The pVI peptide binds covalently via a disulfide bond to the proteases both *in vitro* and in virions to increase the catalytic efficiency of the enzyme over 1,000-fold. Its binding site is not located close to the active site of the protease, suggesting that pVI peptide induces or stabilizes an active-site conformation that is optimal for catalysis.

Other Maturation Reactions

Newly assembled virus particles appear to undergo few maturation reactions other than proteolytic processing. However, the trimming of certain oligosaccharides, or formation of disulfide bonds, is required in some cases. Moreover, a surprising extracellular assembly process has been identified recently (Box 13.10).

Terminal sialic acid residues are removed from the complex oligosaccharides added to the envelope HA and NA glycoproteins of influenza A virus during their transit to the plasma membrane. The influenza A virus receptor is sialic acid, which is specifically recognized by the HA protein. Consequently, newly synthesized virus particles have the potential to aggregate with one another, and with the surface of the host cell, by binding of an HA molecule on one particle to a sialic acid present in an envelope protein of another particle, or in cell surface proteins. Such aggregation is observed when the viral neuraminidase is inactivated. The neuraminidase eliminates such binding of newly synthesized virions to one

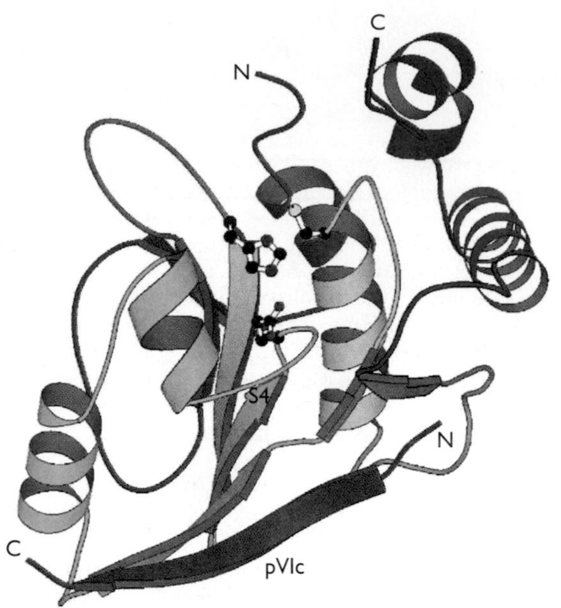

Figure 13.24 Three-dimensional structure of the human adenovirus type 2 protease bound to the pVI peptide. This structure, determined by X-ray crystallography of the complex and shown as a ribbon diagram, allowed identification of the active site of the enzyme (at which side chains are shown) by comparison with the secondary and tertiary structures of cellular cysteine proteases. This structure closely resembles that of papain, suggesting that the adenoviral protease employs the same catalytic mechanism. The β-strand formed by the activating pVIc peptide (deep red) is covalently linked to the enzyme by a disulfide bond, but also makes many noncovalent interactions. Adapted from J. Ding et al., *EMBO J.* **15:**1778–1783, 1995, with permission. Courtesy of W. Mangel, Brookhaven National Laboratory.

another and to cell surface proteins. The activity of this enzyme, which removes terminal sialic acid residues from oligosaccharide chains, is essential for effective release of progeny virus particles from the surface of a host cell. This property has been exploited to develop new drugs (e.g., Tamiflu) designed specifically to inhibit the viral neuraminidase.

The capsids of nonenveloped papillomaviruses, which are built from 72 pentamers of the major structural protein L1, are stabilized by intermolecular disulfide bonds between specific L1 cysteine residues. This protein does not travel the secretory pathway, raising the question of how such cysteines become oxidized. When the human papillomavirus type 16 (or 18) L1 and the minor capsid (L2) proteins are made in mammalian cells, they assemble to form particles that are less stable and less infectious than mature capsids, and that lack disulfide bonds. Such bonds form spontaneously at a low rate, when immature particles are incubated 37°C, and more quickly in the presence of oxidizing agents. Disulfide bond formation is accompanied by increased stability and infectivity, and the appearance of more regularly structured particles. Papillomaviruses are thought to be released slowly during natural infections, as the outer layers of the epithelia in which they replicate are shed. It is therefore likely that newly assembled capsids are exposed to an oxidizing environment for a considerable period (several days) prior to release.

Cell-to-Cell Spread

The raison d'être of all progeny virions is to infect a new host cell in which the infectious cycle can be repeated. Many viruses are released as free particles by the mechanisms described in preceding sections, and must travel within the host until they encounter a susceptible cell. The new host cell may be an immediate neighbor of that originally infected, or a distant cell reached via the circulatory or nervous systems of the host. Virions are designed to withstand such intercellular passage, but they are susceptible to several host defense mechanisms that can destroy virus particles (Volume II, Chapter 4). Localized release of virus particles only at points of contact between an infected cell and its uninfected neighbor(s) can minimize exposure to these host defense mechanisms. Furthermore, some viruses can spread from one cell to another by mechanisms that circumvent the need for release of progeny virions into the extracellular environment.

In some cases, viruses can be transferred directly from an infected cell to its neighbors (Box 13.11), a strategy that avoids exposure to host defense mechanisms targeted against extracellular virus. Such cell-to-cell spread, which is defined operationally as infection that still occurs when released virus particles are neutralized by addition of antibodies, depends on the viral fusion machinery. In the case of herpes simplex virus type 1, the glycoproteins gB, gH, and gL, which promote fusion during entry, are required, as is gD. The latter protein binds to the cell surface protein nectin-1, which is localized to cell-cell junctions. Two additional proteins are also essential for efficient cell-to-cell spread, but have no known role in entry of extracellular particles: mutant viruses that lack the gE or gI genes form only small plaques when transfer of free virus particles from one cell to another is prevented. They are also defective for both lateral spread of infection in polarized epithelial cells and spread of infection from an axon terminal to an uninfected neuron in animals. Such cell-to-cell spread of herpesviruses is thought to occur at specialized junctions, such as tight junctions of epithelial cells, and sites of synaptic contact between individual neurons. Experiments with herpes simplex virus gE mutants in polarized cells demonstrate that in the absence of gE, virions do not accumulate at lateral surfaces with tight junctions, and mutations that

BOX 13.10 EXPERIMENTS

A notable example of virus maturation: extracellular assembly of specific structures

Acidianus two-tailed virus was discovered in an acidic hot spring (pH 1.5, 85 to 93°C) at Pozzuoli, Italy, where it replicates in the thermophilic archaeon *Acidianus convivator*. The virus particles isolated from this source have a lemon-shaped body with filamentous tails of different length protruding from each end **(A)**. However, when the virus replicated in host cells grown in culture at 75°C, the released particles lacked such tails **(B)**. Remarkably, tails formed over 1 week when such particles were incubated at 75°C in the **absence** of host cells **(B, right to left)**. Moreover, this extracellular assembly reaction was complete in less than 1 h when particles were incubated at the temperatures optimal for host cell growth, 85 to 90°C.

Although the morphological changes that accompany maturation of virus particles are well documented (see the text), *Acidianus* two-tailed virus represents the first example of extracellular assembly. This property implies that the tailless particles released from host cells contain all the components and information necessary for tail assembly.

Häring, M., G. Vestergaard, R. Rachel, L. Chen, R. A. Garrett, and D. Prangishvili. 2005. Independent virus development outside a host. *Nature* **436:**1101–1102.

Electron micrographs of *Acidianus* two-tailed virus particles isolated from a hot spring (A) or released from host cells infected in culture at 75°C and maintained in cell-free medium at 75°C for 0, 2, 5, 6, and 7 days (B, right to left). Scale bars, 0.5 µm (A) and 0.1 µm (B). From M. Häring et al., *Nature* **436:**1101–1102, 2005, with permission. Courtesy of David Prangishvili, Institut Pasteur, Paris, France.

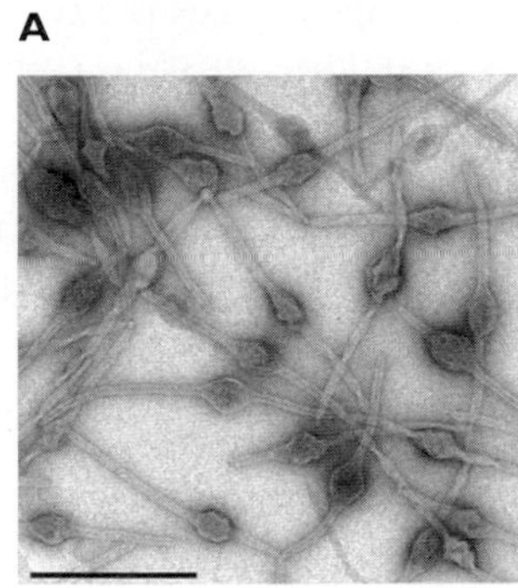

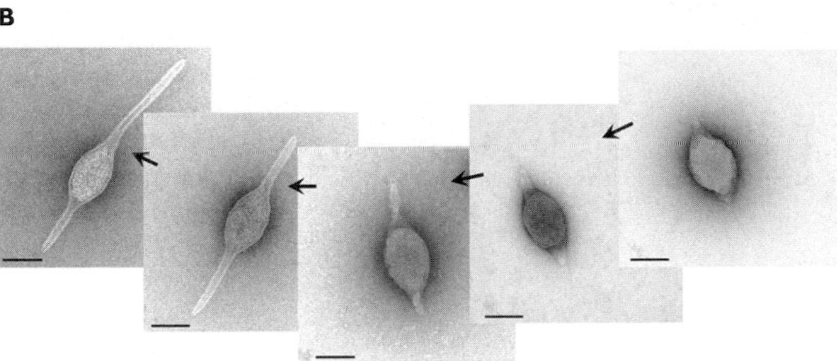

block trafficking of gE from the *trans*-Golgi network to cell junctions impair cell-to cell spread. Exactly how gE-gI glycoprotein oligomers promote cell-to-cell spread via these specialized junctions is not yet clear.

Recent studies indicate that direct cell-to-cell spread is the predominant mechanism for transmission of human immunodeficiency virus type 1 and other retroviruses. Specialized intercellular structures called **virological synapses** assemble when an infected cell contacts an uninfected neighbor. Virological synapses form at lipid raft regions of the plasma membrane that are enriched in cholesterol and sphingomyelin (see Chapter 12), and also the sites of release of viral particles by budding into the extracellular milieu. The viral Env protein, as well as the CD4 and CxCr4 coreceptors, accumulate in virological synapses, and Env-CD4 interactions are required for intercellular transfer of human immunodeficiency virus type 1. This mode of transmission is some 2 to 3 orders of magnitude more efficient than infection via entry of extracellular virions. It depends on formation of stable filopodia between uninfected and infected cells as a result of strong association of the viral Env protein of released virus particles with its receptor on the uninfected cells. Virus particles then travel toward the uninfected cell along the outer surface of the filopodial bridge.

Persistent measles virus (a paramyxovirus) infection of the brain is associated with subacute sclerosing panencephalitis (Volume II, Appendix A, Fig. 14). Spread of this virus between neurons occurs by a mechanism different from that in nonneuronal tissue: little infectious virus can be recovered from brain tissue of patients with this disease, although the genomic RNA and viral proteins are present. Indeed, budding of virus particles does not take place from the surfaces of infected mouse or human neurons in culture, which contain nucleocapsids accumulating at presynaptic membranes. Nor does spread of measles virus between cultured neurons require the viral receptors. Rather, cell-cell contact and the fusion protein are necessary. In neurons, measles virus therefore spreads without release and attachment to infected cells of free

BOX 13.11

BACKGROUND

Extracellular and cell-to-cell spread

Many viruses spread from one host cell to another as extracellular virions released from an infected cell **(A)**. Such extracellular dissemination is necessary to infect another naive host. Some viruses, notably alphaherpesviruses and some retroviruses, can also spread from cell to cell without passage through the extracellular environment **(B)** and can therefore spread by both mechanisms **(C)**.

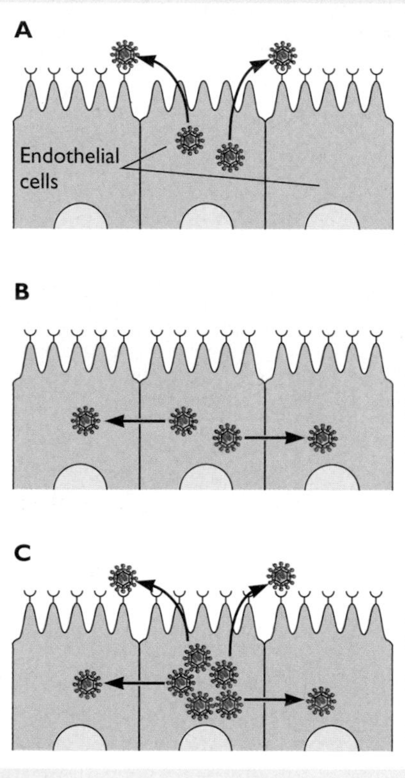

virus particles: rather, it spreads from cell to cell, perhaps through synapses.

There are other examples of more radical mechanisms of transfer. In astrocyctes (supporting cells of the central nervous system), measles virus spreads by inducing the formation of syncytia, sheets of neighboring cells fused to one another (Fig. 13.25A). Certain cell types infected by human immunodeficiency virus type 1 also form syncytia when they would not normally do so (Fig. 13.25B). Perhaps most unusual is the direct transfer of vaccinia virus from an infected cell to its neighbors, by the novel actin-containing structures induced by the virus, as described above.

The production of "decoys," noninfectious particles released in large quantities, is one alternative strategy to avoid host defense mechanisms during transmission. The vast majority of particles detected in hepatitis B virus-infected humans are empty. Another strategy would be to disguise virus particles with normal products of a host cell. Some viral envelopes retain cellular proteins, such as major histocompatibility complex class II proteins and the adhesion receptor Icam-1, in their membranes. The latter protein substantially increases the infectivity of human immunodeficiency virus type 1 particles. However, the importance of such a "wolf in sheep's clothing" strategy for the spread of a virus from one cell to another in the host has yet to be documented.

Perspectives

The assembly of even the simplest virus is a complex process, in which multiple reactions must be completed in the correct sequence and coordinated in such a way that the overall pathway is irreversible. These requirements for efficient production and release of stable structures must be balanced with the fabrication of virus particles primed for ready disassembly at the start of a new infectious cycle. The integration of information collected by the application of structural, imaging, biochemical, and genetic methods of analysis has allowed an outline of the dynamic processes of assembly, release, and maturation for many viruses. Despite the considerable structural diversity of virions, the repertoire of mechanisms for successful completion of the individual reactions is limited. Furthermore, we can identify common mechanisms that ensure that assembly proceeds efficiently and irreversibly, or that resolve the apparent paradox of great particle stability during assembly and release but facile disassembly at the start of the next infectious cycle. These mechanisms include high concentrations of virion components at specific sites within the infected cell, and proteolytic cleavage of virion proteins at one or more steps in the production of infectious particles. Indeed, for some smaller viruses, the structural changes that accompany the production of infectious virions from noninfectious precursor particles can be described in atomic detail. Such information has revealed unanticipated relationships between structures that stabilize virions, and interactions that prime them for conformational rearrangements during entry.

On the other hand, the pathways for assembly, production, and release of even the simplest virions cannot be fully described. These reactions are difficult to study in infected cells, and even the simplest proved more difficult to reconstitute *in vitro* than originally anticipated. The latter property emphasizes the crucial contributions to virus assembly that can be made by cellular proteins that assist

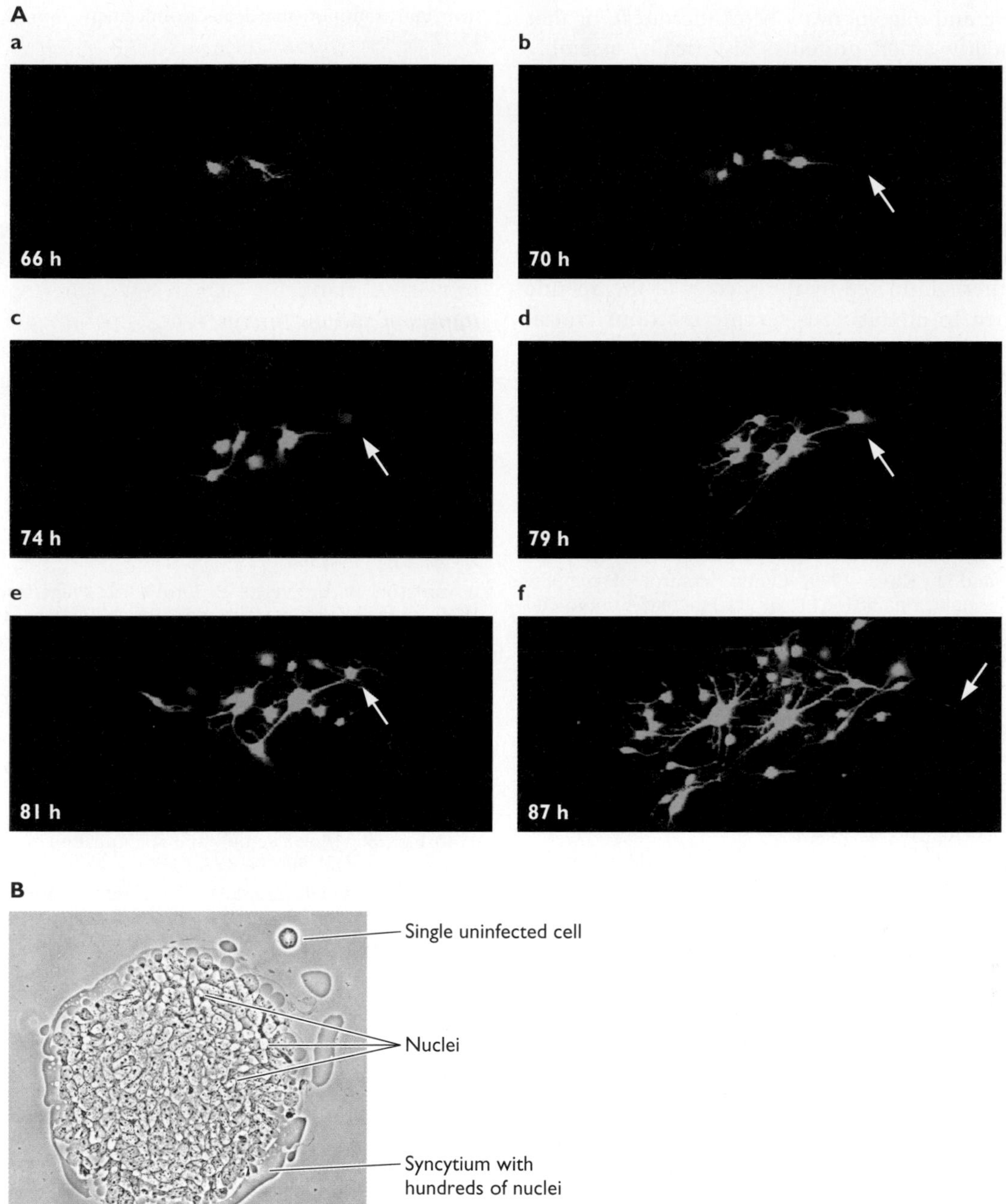

Figure 13.25 Formation of syncytia. (A) Cell-to-cell spread by measles virus. Human astrocytoma cells were infected at low multiplicity with a recombinant measles virus carrying a green fluorescent protein-coding sequence in its genome. The autofluorescence of this protein identifies infected cells (a) and allows spread of the virus to be monitored in living cells. With increasing time, the virus spreads to cells neighboring those initially infected and can be clearly seen in the processes connecting the cells that become infected (b to f). The arrows point to an extended astrocyte process of a newly infected cell (b), the weak autofluorescence of a nucleus of a cell in a very early phase of infection (c), the nucleus from the same cell 5 and 7 h later (d and e, respectively), and an extended astrocytic process issuing from the cell shown in panels d and e (f). From W. P. Duprex et al., *J. Virol.* **73**:9568–9575, 1999. Courtesy of W. D. Duprex, Queen's University, Belfast, United Kingdom. **(B)** Syncytia formed by human immunodeficiency virus type 1 in a T-cell line. The photograph shows a large syncytium of SupT1 cells (a $CD4^+$ T-lymphotropic cell line) that had been infected with a viral vector that expresses the *env* gene. Large quantities of this Env protein accumulate at the cell surfaces mediating fusion. A single cell is indicated for comparison. Courtesy of Matthias Schnell, Philip McKenna, and Joseph Kulkosky, Thomas Jefferson University School of Medicine, Philadelphia, PA.

protein folding and oligomerization (chaperones), or that covalently modify virion proteins. Historically, assembly reactions have received less attention than mechanisms of viral gene expression or replication of viral genomes. However, the development of new structural and imaging methods, coupled with the experimental power and flexibility provided by modern molecular biology, has revitalized investigation of the crucial processes of assembly, release, and maturation of virus particles. This renaissance has been further stimulated by the success of therapeutic agents designed to inhibit virus-specific reactions crucial for the production of infectious particles. Consequently, the advances in our understanding of assembly and egress illustrated in this chapter seem certain to continue at an accelerated pace.

References

Chapters in Books

Wood, W. B., and J. King. 1979. Genetic control of complex bacteriophage assembly, p. 581–633. *In* H. Fraenkel-Conrat and R. R. Wagner (ed.), *Comprehensive Virology*, vol. 13. Plenum Press, New York, NY.

Reviews

Ansardi, D. C., D. C. Porter, M. J. Anderson, and C. D. Morrow. 1996. Poliovirus assembly and encapsidation of genomic RNA. *Adv. Virus Res.* **46:**1–68.

Bieniasz, P. D. 2005. Late budding domains and host proteins in enveloped virus release. *Virology* **344:**55–63.

Brandenburg, B., and X. Zhuang. 2007. Virus trafficking-learning from single-virus tracking. *Nat. Rev. Microbiol.* **5:**197–208.

Bruss, V. 2006. Envelopment of the hepatitis B virus nucleocapsid. *Virus Res.* **106:**199–209.

Condit, R., N. Moussatche, and P. Traktman. 2006. In a nutshell: strucutre and assembly of the vaccinia viron. *Adv. Virus Res.* **66:**31–124.

D'Souza, D., and M. F. Summers. 2005. How retroviruses seclect their genomes. *Nat. Rev. Microbiol.* **3:**643–655.

Enquist, L. W., P. J. Husak, B. W. Banfield, and G. A. Smith. 1999. Infection and spread of alphaherpesviruses in the nervous system. *Adv. Virus Res.* **51:**237–347.

Freed, E. O. 2002. Viral late domains. *J. Virol.* **76:**4679–4687.

Harrison, S. C. 1995. Virus structures and conformational rearrangements. *Curr. Opin. Struct. Biol.* **15:**157–164.

Johnson, D. C., and M. T. Huber. 2002. Directed egress of animal viruses promotes cell-to-cell spread. *J. Virol.* **76:**1–8.

Kirkegaard, K., and W. T. Jackson. 2005. Topology of double-membraned vesicles and the opportunity for non-lytic release of cytoplasm. *Autophagy* **1:**182–184.

Martin-Serrano, J. 2007. The role of ubiquitin in retroviral egress. *Traffic* **8:**1297–1303.

Mettenleiter, T. C. 2002. Herpesvirus assembly and egress. *J. Virol.* **76:**1537–1547.

Moss, B., and B. M. Ward. 2001. High-speed mass transit for poxviruses on microtubules. *Nat. Cell. Biol.* **3:**E245–E246.

Steven, A. C., J. B. Heymann, N. Cheng, B. L. Trus, and J. F. Conway. 2005. Virus maturation: dynamics and mechanism of a stabilizing structural transition that leads to infectivity. *Curr. Opin. Struct. Biol.* **15:**227–236.

Strauss, J. H., E. G. Strauss, and R. J. Kuhn. 1995. Budding of alphaviruses. *Trends Microbiol.* **3:**346–350.

Sullivan, C. S., and J. M. Pipas. 2001. The virus-chaperone connection. *Virology* **287:**1–8.

Weber, J. 1995. The adenovirus endopeptidase and its role in virus infection. *Curr. Top. Microbiol. Immunol.* **199:**227–235.

Wileman, T. 2007. Aggresomes and pericentriolar sites of virus assembly: cellular defense or viral design? *Annu. Rev. Microbiol.* **61:**149–167.

Papers of Special Interest

Assembly of Protein Shells

Basavappa, R., R. Syed, O. Flore, J. P. Icenogle, D. J. Filman, and J. M. Hogle. 1994. Role and mechanism of the maturation cleavage of VP0 in poliovirus assembly: structure of the empty capsid assembly intermediates at 2.9Å resolution. *Protein Sci.* **3:**1651–1669.

Desai, P., N. A. DeLuca, J. C. Glorioso, and S. Person. 1993. Mutations in herpes simplex virus type 1 genes encoding VP5 and VP23 abrogate capsid formation and cleavage of replicated DNA. *J. Virol.* **67:**1357–1364.

Edvardsson, B., E. Everitt, E. Jörnvall, L. Prage, and L. Philipson. 1976. Intermediates in adenovirus assembly. *J. Virol.* **19:**533–547.

Khromykh, A. A., A. N. Varnavski, P. L. Sedlak, and E. G. Westaway. 2001. Coupling between replication and packaging of flavivirus RNA: evidence derived from the use of DNA-based full-length cDNA clones of Kunjin virus. *J. Virol.* **75:**4633–4640.

Li, H., J. Dou, L. Ding, and P. Spearman. 2007. Myristoylation is required for human immunodeficiency virus type 1 Gag-Gag multimerization in mammalian cells. *J. Virol.* **81:**12899–12910.

Ng, S. C., and M. Bina. 1984. Temperature-sensitive BC mutants of SV**40:** block in virion assembly and accumulation of capsid-chromatin complexes. *J. Virol.* **50:**471–477.

Reicin, E. S., A. Ohagen, L. Yin, S. Hoglund, and S. P. Goff. 1996. The role of Gag in human immunodeficiency virus type 1 virion morphogenesis and early steps of the viral life cycle. *J. Virol.* **70:**8645–8652.

Verlinden, Y., A. Cuconati, E. Wimmer, and B. Rombaut. 2000. Cell-free synthesis of poliovirus: 14S subunits are the key intermediates in the encapsidation of poliovirus RNA. *J. Gen. Virol.* **81:**2751–2754.

Yuen, L. K. C., and R. A. Consigli. 1985. Identification and protein analysis of polyomavirus assembly intermediates from infected primary mouse embryo cells. *Virology* **144:**127–136.

Assembly Chaperones and Scaffolds

Cepko, C. L., and P. A. Sharp. 1982. Assembly of adenovirus major capsid protein is mediated by a non-virion protein. *Cell* **31:**407–415.

Chromy, L. R., J. M. Pipas, and R. L. Garcia. 2003. Chaperone-mediated *in vitro* assembly of polyomavirus capsids. *Proc. Natl. Acad. Sci. USA* **100:**10477–10482.

Desai, P., S. C. Watkins, and S. Person. 1994. The size and symmetry of B capsids of herpes simplex virus type 1 are determined by the gene products of the UL26 open reading frame. *J. Virol.* **68:**5365–5374.

Dokland, T., R. McKenna, L. L. Hag, B. R. Bowman, N. L. Incardona, B. A. Fane, and M. G. Rossmann. 1997. Structure of a viral procapsid with molecular scaffolding. *Nature* **389:**308–313.

Dooher, J. E., and J. R. Lingappa. 2004. Conservation of a stepwise, energy-sensitive pathway involving HP68 for assembly of primate lentivirus capsids in cells. *J. Virol.* **78:**1645–1656.

**Gao, M., L. Matusick-Kumar, W. Hurlburt, S. F. DiTusa, W. W. Newcomb, J. C. Brown, P. C. McCann III, I. Deckma, and

R. J. Colonno. 1994. The protease of herpes simplex virus type 1 is essential for functional capsid formation and viral growth. *J. Virol.* **68:**3702–3712.

Hasson, T. B., P. D. Soloway, D. A. Ornelles, W. Doerfler, and T. Shenk. 1989. Adenovirus L1 52- and 55-kilodalton proteins are required for assembly of virions. *J. Virol.* **63:**3612–3621.

Zhou, Z. H., S. J. Macnab, J. Jakana, R. Scott, W. Chiu, and F. J. Rixon. 1998. Identification of the sites of interaction between the scaffold and outer shell in herpes simplex virus type 1 capsids by difference imaging. *Proc. Natl. Acad. Sci. USA* **95:**2778–2783.

Packaging the Viral Genome

Adelman, K., B. Salmon, and J. D. Baines. 2001. Herpes simplex virus DNA packaging sequences adopt novel structures that are specifically recognized by a component of the cleavage and packaging machinery. *Proc. Natl. Acad. Sci. USA* **98:**3086–3091.

Ansardi, D. C., and C. D. Marrow. 1993. Poliovirus capsid proteins derived from P1 precursors with glutamine-valine sites have defects in assembly and RNA encapsidation. *J. Virol.* **67:**7284–7297.

Beard, P. M., N. S. Taus, and J. D. Baines. 2002. DNA cleavage and packaging proteins encoded by genes U_L28, U_L15, and U_L33 of herpes simplex virus type 1 form a complex in infected cells. *J. Virol.* **76:**4785–4791.

Frilander, M., and D. H. Bamford. 1995. In vitro packaging of the single stranded RNA genomic precursors of the segmented double-stranded RNA bacteriophage **S6:** the three segments modulate each other's packaging efficiency. *J. Mol. Biol.* **246:**418–428.

Gordon-Shaag, A., O. Ben-Nun-Shaul, V. Roitman, Y. Yosef, and A. Oppenheim. 2002. Cellular transcription factor Sp1 recruits simian virus 40 capsid proteins to the viral packaging signal, *ses. J. Virol.* **76:**5915–5924.

Liang, Y., T. Huang, H. Ly, T. G. Parslow, and Y. Liang. 2008. Mutational analysis of packaging signals of influenza virus PA, PB1, and PB2 genome RNA segments. *J. Virol.* **82:**229–236.

Muramoto, Y., A. Takada, T. Noda, K. Iwatsuki-Horimoto, S. Watanabe, T. Horimoto, H. Kida and Y. Kawaoka. 2006. Hierachy among viral RNA (vRNA) segments in their role in vRNA incorporation into influenza A virions. *J. Virol.* **80:**2318–2325.

Nugent, C. I., K. L. Johnson, P. Sarnow, and K. Kirkegaard. 1999. Functional coupling between replication and packaging of poliovirus replicon RNA. *J. Virol.* **73:**427–435.

Sakuragi, J.-I., S. Sakuragi, and T. Shioda. 2007. Minimal region sufficient for genome dimerization in the human immunodeficiency virus type 1 virion and its potential roles in the early stage of viral replication. *J. Virol.* **81:**7985–7992.

Schmid, S. I., and P. Hearing. 1997. Bipartite structure and functional independence of adenovirus type 5 packaging elements. *J. Virol.* **71:**3375–3384.

Zhang, J., G. P. Leser, A. Pekosz, and R. A. Lamb. 2000. The cytoplasmic tails of the influenza virus spike glycoproteins are required for normal genome packaging. *Virology* **269:**325–334.

Zhang, W., and M. J. Imperiale. 2000. Interaction of the adenovirus IVa2 protein with viral packaging sequences. *J. Virol.* **74:**2687–2693.

Acquisition of an Envelope

Finzi, A., A. Orthwein, J. Mercier, and E. A. Cohen. 2007. Productive human immunodeficiency virus type 1 assembly takes place at the plasma membrane. *J. Virol.* **81:**7476–7490.

Freed, E. O., and M. A. Martin. 1996. Domains of the human immunodeficiency virus type 1 matrix and gp41 cytoplasmic tail required for envelope incorporation into virions. *J. Virol.* **70:**341–351.

Justice, P. A., W. Sun, Y. Li, Z. Ye, P. R. Grigera, and R. R. Wagner. 1995. Membrane vesiculation function and exocytosis of wild-type and mutant matrix proteins of vesicular stomatitis virus. *J. Virol.* **69:**3156–3160.

Lenthoff, R., and J. Summers. 1994. Coordinate regulation of replication and virus assembly by the large envelope protein of an avian hepadnavirus. *J. Virol.* **68:**4565–4571.

Roseman, A. M., J. A. Beriman, S. A. Wyane, J. G. Bulter, and R. A. Crowther. 2005. A structural model for maturation of the hepatitis B virus core. *Proc. Natl. Acad. Sci. USA* **102:**15821–15826.

Vangenderen, I. L., R. Brandimarti, M. R. Torrisi, G. Campadelli, and G. van Meer. 1994. The phospholipid-composition of extracellular herpes-simplex virions differs from that of host-cell nuclei. *Virology* **200:**831–836.

Watanabe, T., E. M. Sorensen, A. Naito, M. Scholt, S. Kim, and P. Ahlquist. 2007. Involvement of host cellular multivesicular body functions in hepatitis B virus budding. *Proc. Natl. Acad. Sci. USA* **104:**10205–10210.

Yao, E., E. G. Strauss, and J. H. Strauss. 1998. Molecular genetic study of the interaction of Sindbis virus E2 with Ross River virus E1 for virus budding. *J. Virol.* **72:**1418–1423.

Virion Maturation

Bernstein, H., D. Bizub, and A. M. Skalka. 1991. Assembly and processing of avian retroviral Gag polyproteins containing linked protease dimers. *J. Virol.* **65:**6165–6172.

Buck, C. B., C. D. Thompson, Y.-Y. S. Pang, D. R. Lowry, and J. T. Schiller. 2005. Maturation of papillomavirus capsids. *J. Virol.* **79:**2839–2846.

Lee, W.-M., S. S. Monroe, and R. R. Rueckert. 1993. Role of maturation cleavage in infectivity of picornaviruses: activation of an infectosome. *J. Virol.* **67:**2110–2122.

Webster, A., R. T. Hay, and G. Kemp. 1993. The adenovirus protease is activated by a virus-coded disulfide-linked peptide. *Cell* **72:**97–104.

Transport, Release and Spread of Virions

Chen, P., W. Hübner, M. A. Spinelli, and B. K. Chen. 2007. Predominant mode of human immuno deficiency virus transfer between T cells is mediated by sustained Env-dependent neutralization-resistant virological synapses. *J. Virol.* **81:**12582–12595.

Clayson, E. T., L. V. Brando, and R. W. Compans. 1989. Release of simian virus 40 virions from epithelial cells is polarized and occurs without cell lysis. *J. Virol.* **63:**2278–2288.

Farnsworth, A., T. W. Wisner, M. Webb, R. Roller, G. Cohen, R. Eisenberg and D. C. Johnson. 2007. Herpes simplex glycoproteins gB and gH function in fusion between the virion envelope and the outer nuclear membrane. *Proc. Natl. Acad. Sci. USA* **104:**10187–10192.

Johnson, D. C., M. Webb, T. W. Wisner, and C. Brunetti. 2001. Herpes simplex virus gE/gI sorts nascent virions to epithelial cell junctions, promoting virus spread. *J. Virol.* **75:**821–833.

Neil, S. J. D., T. Zang, and P. D. Beiniasz. 2008. Tetherin inhibits retrovirus release and is antagonized by HIV-1 Vpu. *Nature* **451:**425–431.

Roper, R. L., E. J. Wolfe, A. Weisberg, and B. Moss. 1998. The envelope protein encoded by the A33R gene is required for formation of actin-containing microvilli and efficient cell-to-cell spread of vaccinia virus. *J. Virol.* **72:**4192–4204.

Sfakianos, J. N., and E. Hunter. 2003. M-PMV capsid transport is medicated by Env/Gag interactions at the preicentriolar recycling endosome. *Traffic* **4:**671–680.

Sherer, N. M., M. J. Lehmann, L. F. Siminez-Soto, C. Horensavitz, M. Pypaert, and W. Mothes. 2007. Retroviruses can establish filopodial bridges for efficient cell-to-cell transmission. *Nat. Cell Biol.* **9:**310–316.

Strack, B., A. Calistri, M. A. Accola, G. Palu, and H. G. Gottlinger. 2000. A role for ubiquitin ligase recruitment in retrovirus release. *Proc. Natl. Acad. Sci. USA* **97:**13063–13068.

Tollefson, A. E., A. Scaria, T. W. Hermiston, J. S. Ryerse, L. H. Wold, and W. S. Wold. 1996. The adenovirus death protein (E3-11.6K) is required at very late stages of infection for efficient cell lysis and release of adenovirus from infected cells. *J. Virol.* **70:**2296–2306.

Zhang, Y., and R. J. Schneider. 1995. Adenovirus inhibition of cell translation facilitates release of virus particles and enhances degradation of the cytokeratin network. *J. Virol.* **68:**2544–2555.

APPENDIX
Structure, Genome Organization, and Infectious Cycles

Adenoviruses

Family *Adenoviridae*

Genus	Type species
Mastadenovirus	Human adenovirus C
Aviadenovirus	Fowl adenovirus A

Human serotypes are very widespread in the population. Infection by these viruses is often asymptomatic but can result in respiratory disease in children (members of subgroup B and C), conjunctivitis (members of subgroup B and D), and gastroenteritis (subgroup F serotypes 40 and 41). Human adenoviruses 40 and 41 are the second leading cause (after rotaviruses) of infantile viral diarrhea. These viruses share capsid morphology and linear double-stranded DNA genomes, but the members of the two genera differ in size, organization, and coding sequences. The *Mastadenovirinae* comprise 51 human adenoviruses and adenoviruses of other mammals, including mice, sheep, and dogs, and some are oncogenic in rodents. Study of human adenovirus transformation of cultured cells has provided fundamental information about mechanisms that control progression through the cell cycle and oncogenesis. Characteristic features of replication of these viruses include precise temporal control of viral gene expression and an unusual mechanism of initiation of viral DNA synthesis (protein priming). Mastadenoviral genomes also include genes transcribed by cellular RNA polymerase III.

Figure 1 Structure and genome organization of human adenovirus type 5. (A) Virion structure. The electron micrograph shows a negatively stained human adenovirus type 5 particle (courtesy of M. Bisher, Princeton University, Princeton, NJ). Bar = 50 nm. **(B) Genome organization.** The DNA genome length is 36 to 38 kbp. Green and tan arrows represent RNA polymerase II and III transcription products, respectively. Hatched lines show splicing of the major late (ML) tripartite leader. Ori, origin of replication.

Figure 2 Single-cell reproductive cycle of human adenovirus type 2. The virus attaches to a permissive human cell via interaction between the fiber and (with most serotypes) the coxsackie-adenovirus receptor on the cell surface. The virus enters the cell via endocytosis **(1** and **2)**, a step that depends on the interaction of a second virion protein, penton base, with a cellular integrin protein (red cylinder). Partial disassembly takes place prior to entry of particles into the cytoplasm **(3)**. Following further uncoating, the viral genome associated with core protein VII is imported into the nucleus **(4)**. The host cell RNA polymerase II system transcribes the immediate-early E1A gene **(5)**. Following alternative splicing and export of E1A mRNAs to the cytoplasm **(6)**, E1A proteins are synthesized by the cellular translation machinery **(7)**. These proteins are imported into the nucleus **(8)**, where they regulate transcription of both cellular and viral genes. The larger E1A protein stimulates transcription of the viral early genes by cellular RNA polymerase II **(9a)**. Transcription of the VA genes by host cell RNA polymerase III also begins during the early phase of infection **(9b)**. The early pre-mRNA species are processed, exported to the cytoplasm **(10)**, and translated **(11)**. These early proteins include the viral replication proteins, which are imported into the nucleus **(12)** and cooperate with a limited number of cellular proteins in viral DNA synthesis **(13)**. Replicated viral DNA molecules can serve as templates for further rounds of replication **(14)** or for transcription of late genes **(15)**. Some late promoters are activated simply by viral DNA replication, but maximally efficient transcription of the major late transcription unit (Fig. 1, ML) requires the late IVa2 and L4 proteins. Processed late mRNA species are selectively exported from the nucleus as a result of the action of the E1B 55-kDa and E4 Orf6 proteins **(16)**. Their efficient translation in the cytoplasm **(17)** requires the major VA RNA, VA RNA-I, which counteracts a cellular defense mechanism, and the late L4 100-kDa protein. The latter protein also serves as a chaperone for assembly of trimeric hexons as they and the other structural proteins are imported into the nucleus **(18)**. Within the nucleus, capsids are assembled from these proteins and the progeny viral genomes to form noninfectious immature virions **(19)**. Assembly requires a packaging signal located near the left end of the genome, as well as the IVa2 and L4 22/33-kDa proteins. Immature virions contain the precursors of the mature forms of several proteins. Mature infectious virions are formed **(20)** when these precursor proteins are cleaved by the viral L3 protease, which enters the virion core. Progeny virions are released **(21)**, usually upon destruction of the host cell via mechanisms that are not well understood.

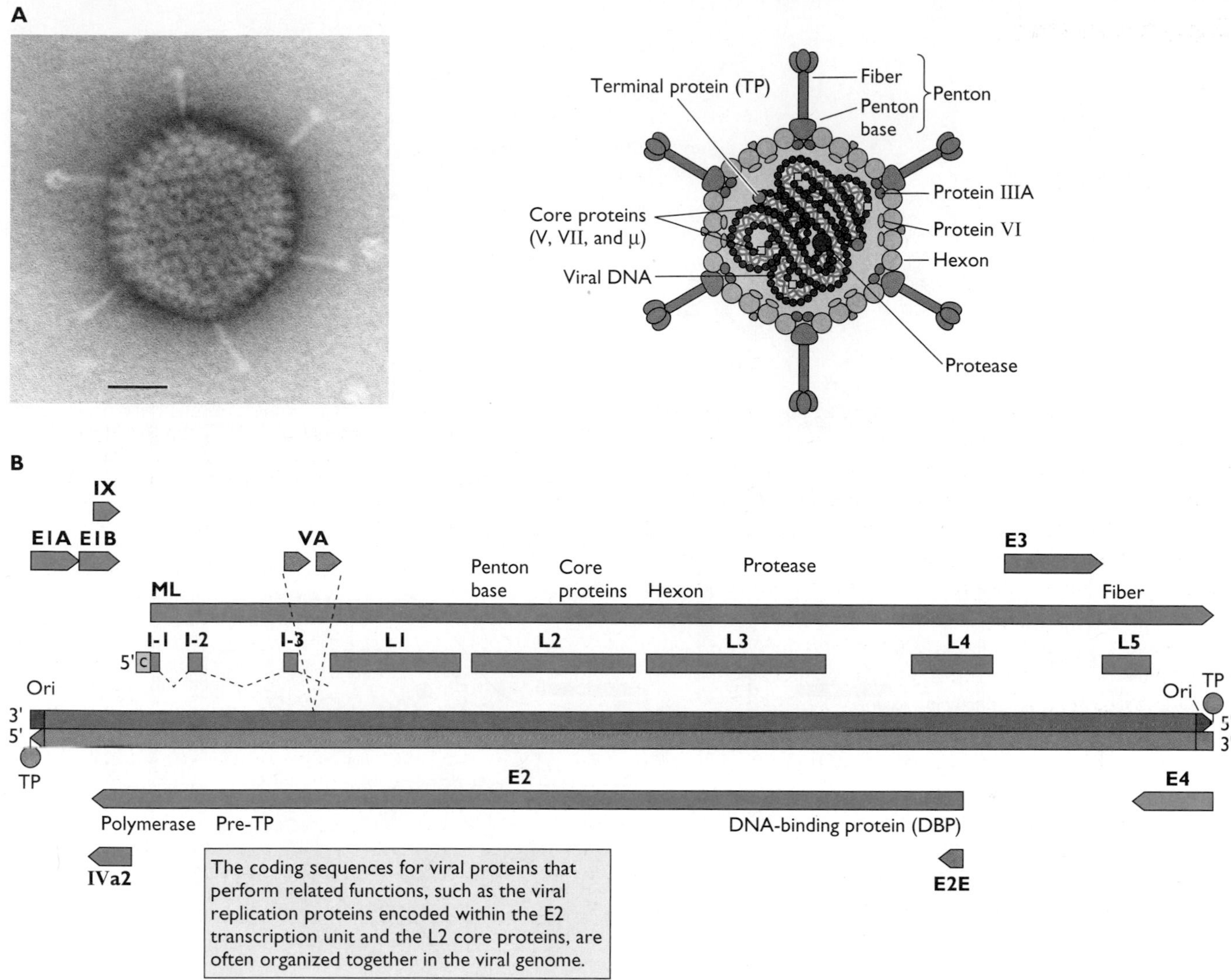

Figure 1 Structure and genome organization of human adenovirus type 5.

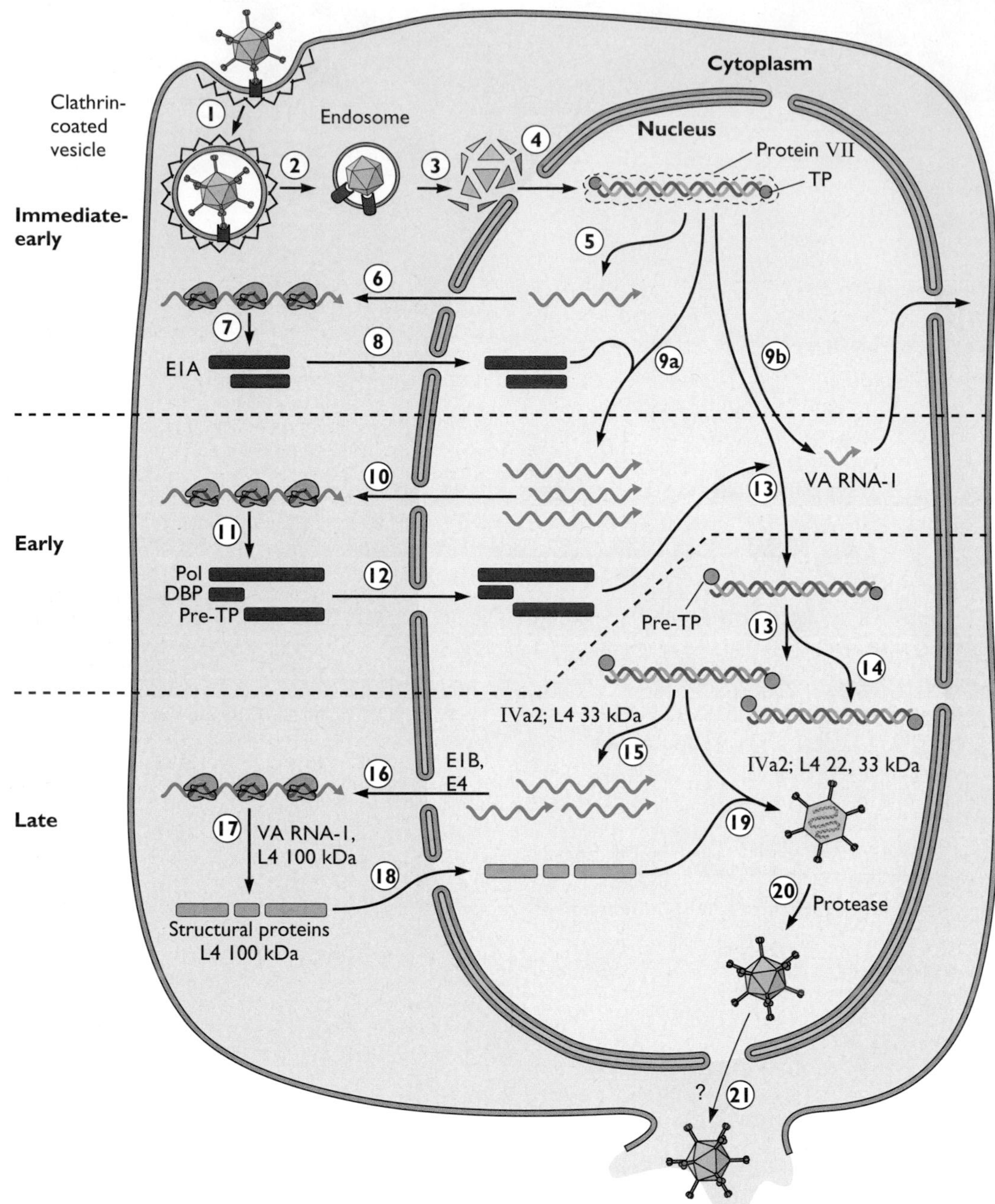

Figure 2 Single-cell reproductive cycle of human adenovirus type 2.

Hepadnaviruses

Family *Hepadnaviridae*

Genus	Type species
Orthohepadnavirus	Human hepatitis B virus
Avihepadnavirus	Duck hepatitis B virus

The hepadnaviruses all show very narrow host specificity and marked tropism for liver tissue. Hepadnaviruses can replicate following inoculation of primary hepatocytes with virus-containing serum, but most hepadnaviruses cannot be propagated in established cell lines. However, replication can be initiated by transfection of liver cell lines with cloned viral DNA. Hepadnaviruses replicate via an RNA intermediate and, like the retroviruses, encode a reverse transcriptase. Both families are included in the group called **retroid viruses.** Natural infections may be acute or persistent, depending on host age, inoculum dose, and other (undefined) factors that influence the host immune response. Approximately 5% of the world's population has been infected with human hepatitis B virus; 250 million to 300 million are persistently infected. Persistent infection with the orthohepadnaviruses but not the avihepadnaviruses confers an increased risk for hepatocellular carcinoma.

Figure 3 Structure and genome organization of orthohepadnaviruses. (A) Virion structure. The electron micrograph shows negatively stained woodchuck hepatitis virus, a mammalian hepadnavirus related to human hepatitis B virus (courtesy of W. Mason and T. Gales, Fox Chase Cancer Center, Philadelphia, PA). **(B) Genome organization**. The (−) strand of the human hepatitis virus DNA genome is 3,227 nucleotides long.

Figure 4 Single-cell reproductive cycle of hepatitis B virus. The virion attaches to a susceptible hepatocyte **(1)** through recognition of a cell surface receptor(s) that has yet to be identified. The mechanism of virus uptake **(2)** is also unknown, and repair of the gapped (+) DNA strand is accomplished **(3)** by as yet unidentified enzymes. The DNA is translocated to the nucleus **(4)**, where it is found in a covalently closed circular form called CCC DNA. The (−) strand of such CCC DNA is the template for transcription by cellular RNA polymerase II **(5)** of a longer-than-genome-length RNA called the pregenome and shorter, subgenomic transcripts (Fig. 3B), all of which serve as mRNAs. Viral mRNAs are transported from the nucleus **(6)**. Subgenomic viral mRNAs encoding the viral envelope protein are translated by ribosomes bound to the endoplasmic reticulum (ER) **(7)**, and the proteins destined to become surface antigens (HBsAg) in the viral envelope enter the secretory pathway. The pregenome RNA is translated to produce capsid protein **(8)** and, at low efficiency, the 90-kDa P protein **(9)**, which possess reverse transcriptase activity. P then binds to a specific site, the packaging signal, at the 5′ end of its own transcript, where viral DNA synthesis is eventually initiated. Concurrently with capsid formation, the RNA-P protein complex is packaged **(10)** and priming of DNA replication occurs from a tyrosine residue on the polymerase **(11)**. Following synthesis of a few nucleotides, there is a template exchange in which the 3′ end of the pregenome is engaged by the polymerase and DNA synthesis continues to the 5′ end of this RNA template **(12)**. At early times after infection, the DNA is transported to the nucleus **(13)**, where the process is repeated, resulting in the eventual accumulation of 10 to 30 molecules of CCC DNA and a concomitant increase in viral mRNA concentrations. At later times, and possibly as a conseqence of the accumulation of sufficient envelope proteins, mature DNA-containing nucleocapsids acquire envelopes as they bud into the ER **(14)**, where viral maturation is completed. Progeny enveloped virions are released from the cell by exocytosis **(15)**.

A

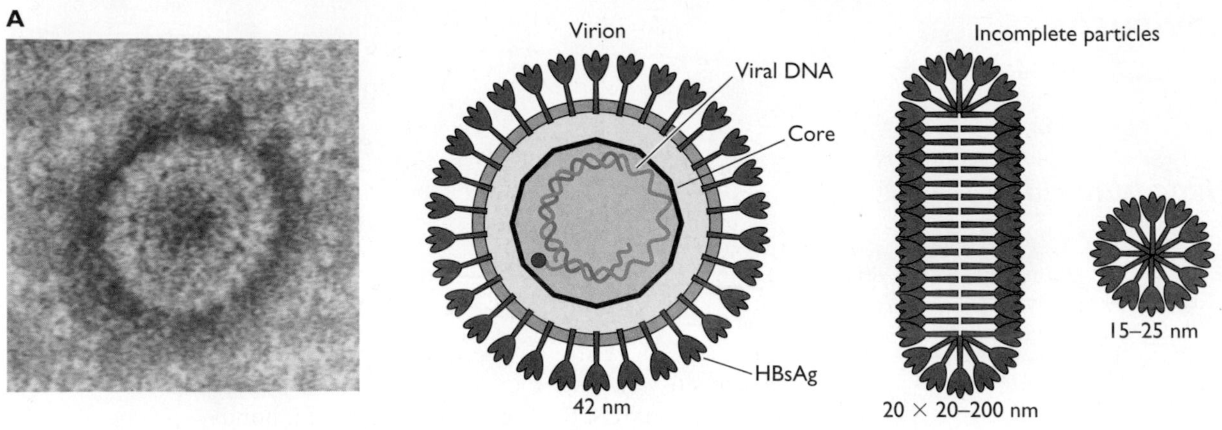

B

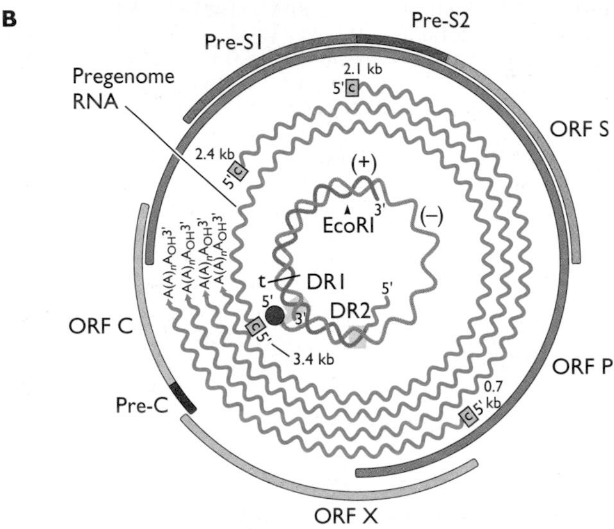

Figure 3 Structure and genome organization of orthohepadnaviruses.

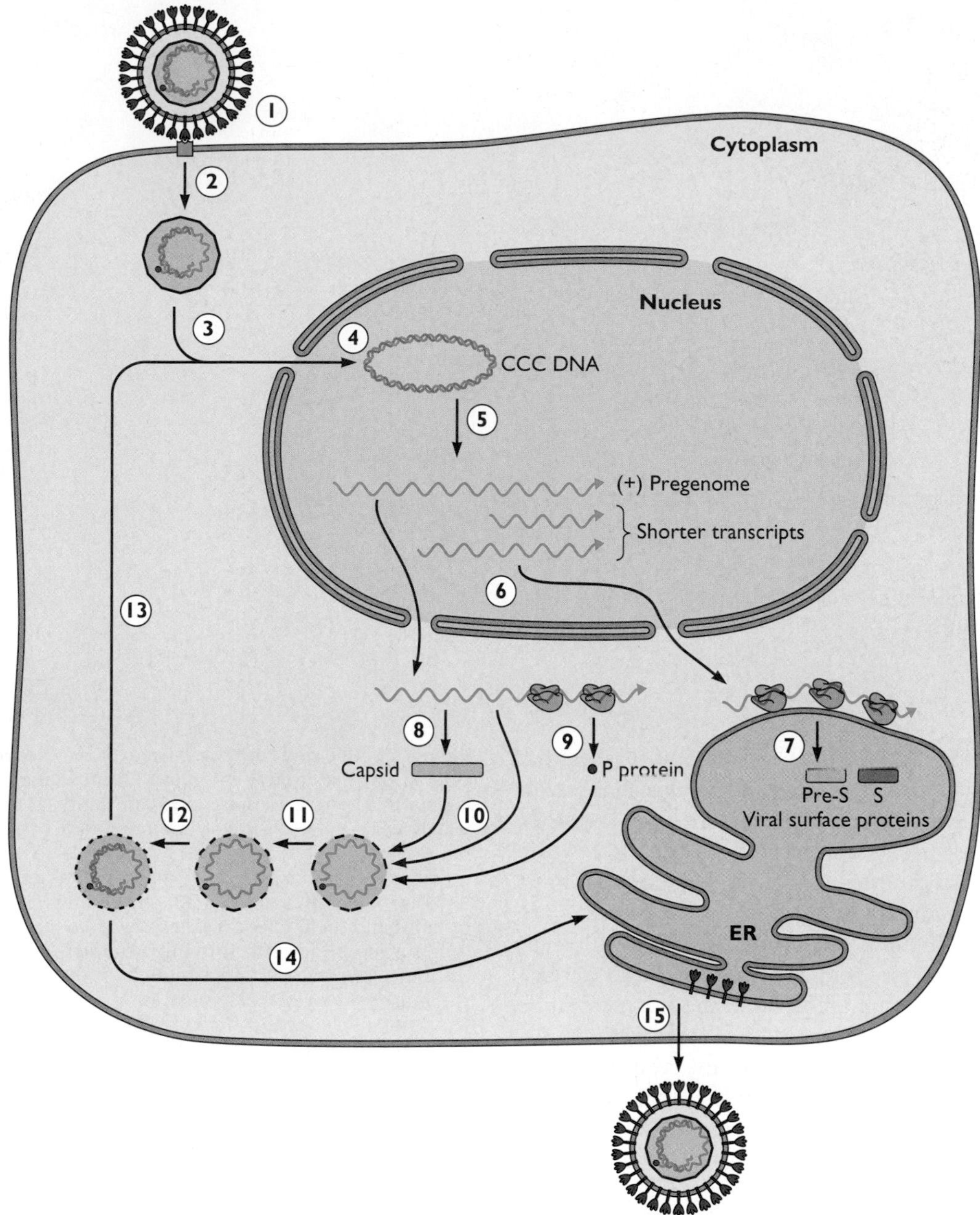

Figure 4 Single-cell reproductive cycle of hepatitis B virus.

Herpesviruses

Family *Herpesviridae*

Genus	Type species
Subfamily *Alphaherpesvirinae*	
Simplexvirus	Human herpes simplex virus type 1
Varicellovirus	Varicella-zoster virus
Subfamily *Betaherpesvirinae*	
Cytomegalovirus	Human cytomegalovirus
Muramegalovirus	Murine cytomegalovirus 1
Roseolovirus	Human herpesvirus 6
Subfamily *Gammaherpesvirinae*	
Lymphocryptovirus	Epstein-Barr virus
Rhadinovirus	Herpesvirus saimiri
Undefined subfamily: channel catfish virus, oyster herpesvirus	

The family *Herpesviridae* comprises over 120 viruses that infect a wide range of vertebrates and at least one invertebrate (the oyster). While some herpesviruses have broad host ranges, most are restricted to infection of a single species and spread in the population by direct contact or aerosols. The hallmark of herpesvirus infections is the establishment of a lifelong, latent infection that can reactivate to cause one or more rounds of disease. Many herpesvirus infections are not apparent, but if the host's immune defenses are compromised, infections can be devastating. Humans are the natural hosts of eight different herpesviruses: herpes simplex virus types 1 and 2, varicella-zoster virus, Epstein-Barr virus, human cytomegalovirus, and human herpesviruses 6, 7, and 8 (the last of these is also known as Kaposi's sarcoma-associated herpesvirus). Agricultural pathogens include the alphaherpesviruses (pseudorabies virus, bovine herpes virus type 1, equine herpesvirus type 1, and avian Marek's disease virus types 1 and 2) and the gamma herpesviruses (bovine herpesvirus 4 and equine herpesvirus 2).

Figure 5 Structure and genome organization of alphaherpesviruses. (A) Virion structure. The electron micrograph shows a negatively stained pseudorabies viron (courtesy of T. Mettenleiter, Federal Research Center for Virus Diseases of Animals, Insel Riems, Germany). **(B) Genome organization.** The herpes simplex virus type 1 genome can "isomerize" or recombine via the inverted repeat sequences (TRL and IRL, or IRS and TRS) such that all populations consist of four equimolar isomers in which the UL and US sequences are inverted with respect to each other. There are at least 84 open reading frames in this genome, as well as three replication origins (Ori). The approximate locations of some genes are noted. Genes encoding related functions have the same color shading to illustrate their dispersed distribution in herpesviral genomes.

Figure 6 Single-cell reproductive cycles of herpes simplex virus type 1. (A) Productive infection. Virions bind to the extracellular matrix (heparan sulfate or chondroitin sulfate proteoglycans) via gB and gC **(1)**. Another viral membrane protein (gD) interacts with a second cellular receptor (nectin-1) **(2)**. Viral and plasma membrane fusion is then mediated by viral membrane glycoproteins (gD, gB, gH, and gL) **(3)**. On membrane fusion, tegument proteins and the nucleocapsid are released into the cytoplasm **(4)**. Viral nucleocapsids attach to microtubules and are transported to the nucleus **(5a)**. Certain tegument proteins are transported to the nucleus **(5b)**. Others, such as Vhs, remain in the cytoplasm **(6)**. Viral nucleocapsid docks at the nuclear pore **(7)**, releasing DNA into the nucleus. VP16 **(8)** interacts with host transcription proteins to stimulate transcription of immediate-early genes by host cell RNA polymerase II. Some immediate-early mRNAs are spliced and transported to the cytoplasm **(9)** where they are translated. The immediate-early proteins (α proteins) are transported to the nucleus, where they activate transcription of early genes and regulate transcription of immediate-early genes **(10)**. Early-gene transcripts, which are rarely spliced, are transported to the cytoplasm **(11)** where they are translated. The early proteins (β proteins) function primarily in DNA replication and production of substrates for DNA synthesis. Some β proteins are transported to the nucleus **(12)**, and some function in the cytoplasm. Viral DNA synthesis is initiated from viral origins of replication **(13)**. DNA replication and recombination produces long, concatemeric DNA, the template for late-gene expression **(14)**. Most late mRNAs are not spliced, but nevertheless are transported to the cytoplasm **(15)**, where they are translated. Late proteins (γ proteins) are primarily virion structural proteins and additional proteins needed for virus assembly and particle egress. Some γ proteins are made on, and inserted into, membranes of the rough endoplasmic reticulum **(16a)**. Many of these membrane proteins are modified by glycosylation. Some precursor viral membrane proteins are thought to be localized both to the outer and inner nuclear membranes, as well as membranes of the endoplasmic reticulum **(16b)**. The precursor glycoproteins are also transported to the Golgi apparatus for further modification and processing **(16c)**. Mature glycoproteins are transported to the plasma membrane of the infected cell **(16d)**. Some γ proteins are transported

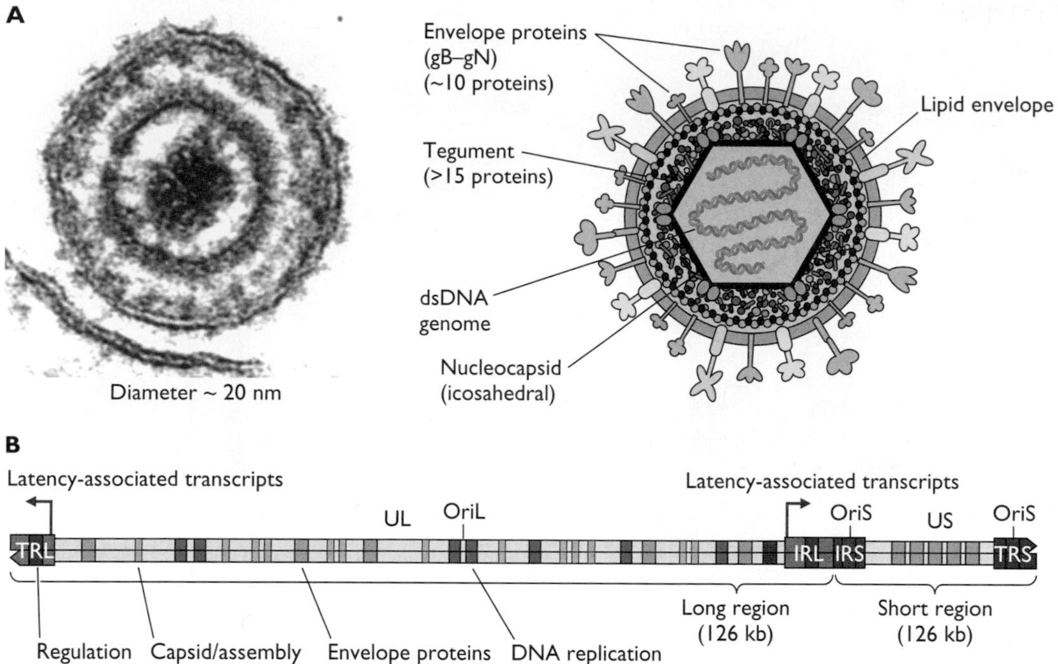

Figure 5 Structure and genome organization of alphaherpesviruses.

to the nucleus for assembly of the nucleocapsid and DNA packaging, while some remain in the cytoplasm **(17)**. Newly replicated viral DNA is packaged into nucleocapsids **(18)**. DNA-containing nucleocapsids, together with some tegument proteins, bud from the inner nuclear membrane into the perinuclear lumen, acquiring an envelope thought to contain precursors to viral membrane proteins **(19)**. Immature enveloped virions fuse with the outer nuclear membrane from within **(20)**, leaving behind viral membrane proteins while releasing the nucleocapsid into the cytoplasm. This structure is transported to a late *trans*-Golgi compartment or an endosome that contains mature viral membrane proteins **(21)**. Tegument proteins added in the nucleus remain with the nucleocapsid, and others are added in the cytoplasm. As nucleocapsids bud into the late Golgi-endosome compartment, they acquire an envelope containing mature viral envelope proteins and the complete tegument layer (secondary envelopment **[22]**). The enveloped virus particle then buds into a vesicle **(23)** that is transported to the plasma membrane for release by exocytosis **(24)**. **(B) Latent infection.** Latent infection occurs primarily in neurons found in ganglia of the peripheral nervous system. In the simplest model, the initiation of latent infection occurs as in **steps 1 to 7** of the productive infection (A), except that the viral DNA circularizes in the nucleus and is wrapped around nucleosomes. It is unclear if tegument proteins are transported into the nucleus during the establishment of the latent infection **(8)**. The latent genome is transcriptionally silent, and only a single pre-mRNA is produced from the latency-associated transcript (LAT) promoter **(9)**. Low-level or sporadic transcription of immediate-early and early genes can occur, but is not sufficient to initiate a productive infection. The LAT RNA is spliced, and a stable intron in the form of a lariat, called the 2-kb LAT, is maintained in the nucleus **(10)**. The spliced LAT mRNA is transported to the cytoplasm, where several small open reading frames (e.g., OrfO and OrfP) may be translated into proteins **(11)**. The function of LAT RNAs and the production of LAT proteins are controversial. After months to years, and in the absence of viral tegument and IE proteins, cellular proteins produced in response to changes in neuronal physiology induced by trauma, hormonal changes, and other stressful conditions render some latently infected neurons permissive for viral replicaton. The entire genome is then transcribed and replicated as during productive infection, and progeny virions are produced as shown in panel A.

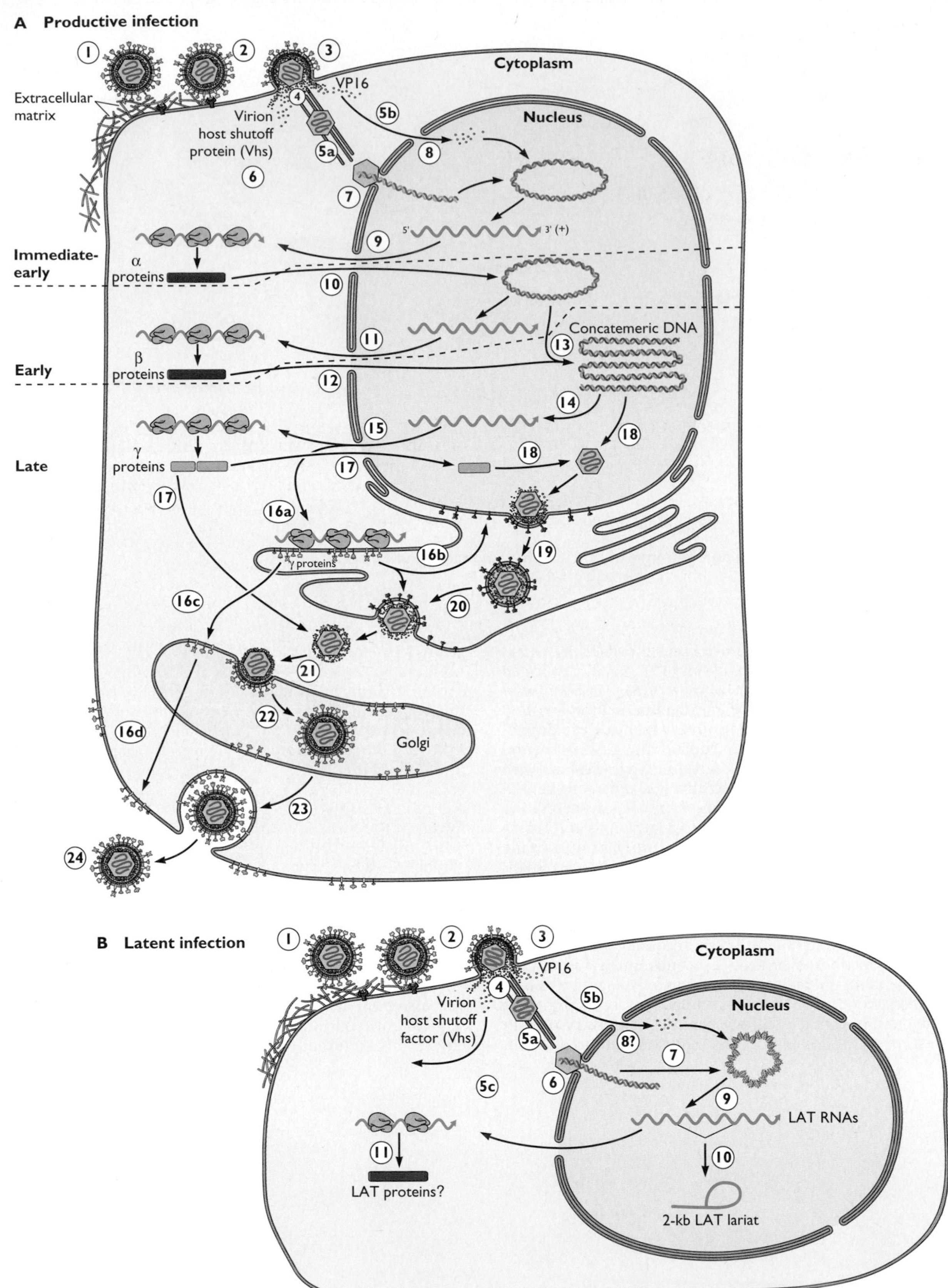

Figure 6 Single-cell reproductive cycles of herpes simplex virus type 1.

Orthomyxoviruses

Family *Orthomyxoviridae*

Genus	Type species
Influenza A virus	A/PR/8/34(H1N1)
Influenza B virus	B/Lee/40
Influenza C virus	C/California/78
Thogotovirus	Thogoto virus
Isavirus	Infectious salmon anemia virus

Influenza viruses are the causative agents of a highly contagious and often serious acute respiratory illness. Influenza viruses are unusual among RNA viruses in that all viral RNA synthesis occurs in the cell nucleus. Initiation of viral mRNA synthesis with a capped primer derived from host cell mRNA was first observed in cells infected with influenza viruses. The viral genomes undergo extensive reassortment and variation, and are expressed via a remarkable panoply of unusual strategies, including RNA splicing, overlapping reading frames, and leaky scanning.

Figure 7 Structure and genomic organization of the orthomyxovirus influenza A virus. (A) Virion structure. The electron micrograph shows negatively stained influenza A virus particles (courtesy of P. Palese, Mount Sinai School of Medicine, New York, NY). **(B) Genome organization.** The (–) strand RNA genome comprises eight segments, each of which encodes at least one viral protein as shown. Some fraction of the (+) strand mRNA of the smallest genomic RNA segments, 7 and 8, is spliced by host cell enzymes, allowing the production of two proteins from each. The NS1 (nonstructural) protein is so named because it is not incorporated into virus particles. An accessory protein with proapoptotic activity, PB1-F2, is produced from the PB1 RNA, by translation of an overlapping open reading frame.

Figure 8 Single-cell reproductive cycle of influenza A virus. The virion binds to a sialic acid-containing cellular surface protein or lipid and enters the cell via receptor-mediated endocytosis **(1)**. Upon acidification of the vesicle, the viral membrane fuses with the membrane of the vesicle, releasing the eight viral nucleocapsids into the cytoplasm (for simplicity, only one is shown) **(2)**. The viral nucleocapsids containing (–) strand genomic RNA, multiple copies of the NP protein, and the P proteins are transported into the nucleus **(3)**. The (–) strand RNAs are copied by virion RNA polymerase into viral mRNA, using the capped 5′ ends of host pre-mRNAs (or mRNAs) as primers to initiate synthesis **(4)**. The mRNAs are transported to the cytoplasm **(5)**, following splicing in the case of the mRNAs encoding NEP and M2 **(6)**. The mRNAs specifying the viral membrane proteins (HA, NA, and M2) are translated by ribosomes bound to the endoplasmic reticulum (ER) **(7)**. These proteins enter the host cell's secretory pathway, where HA and NA are glycosylated. All other mRNAs are translated by ribosomes in the cytoplasm **(8** and **9)**. The PA, PB1, PB2, and NP proteins are imported into the nucleus **(10a)**, where they participate in the synthesis of full-length (+) strand RNAs **(11)** and then of (–) strand genomic RNAs **(12)**, both of which are synthesized in the form of nucleocapsids. Some of the newly synthesized (–) strand RNAs enter the pathway for mRNA synthesis **(13)**. The M1 protein and the NS1 protein are transported into the nucleus **(10b)**. Binding of the M1 protein to newly synthesized (–) strand RNAs shuts down viral mRNA synthesis and, in conjunction with the NEP protein, induces export of progeny nucleocapsids to the cytoplasm **(14)**. The HA, NA, and M2 proteins are transported to the cell surface **(15)** and become incorporated into the plasma membrane **(16)**. The virion nucleocapsids associated with the M1 protein **(17)** and the NEP protein **(18)** are transported to the cell surface and attach to regions of the plasma membrane that contain the HA, NA, M1, and M2 proteins. Assembly of virions is completed at this location by budding from the plasma membrane **(19)**.

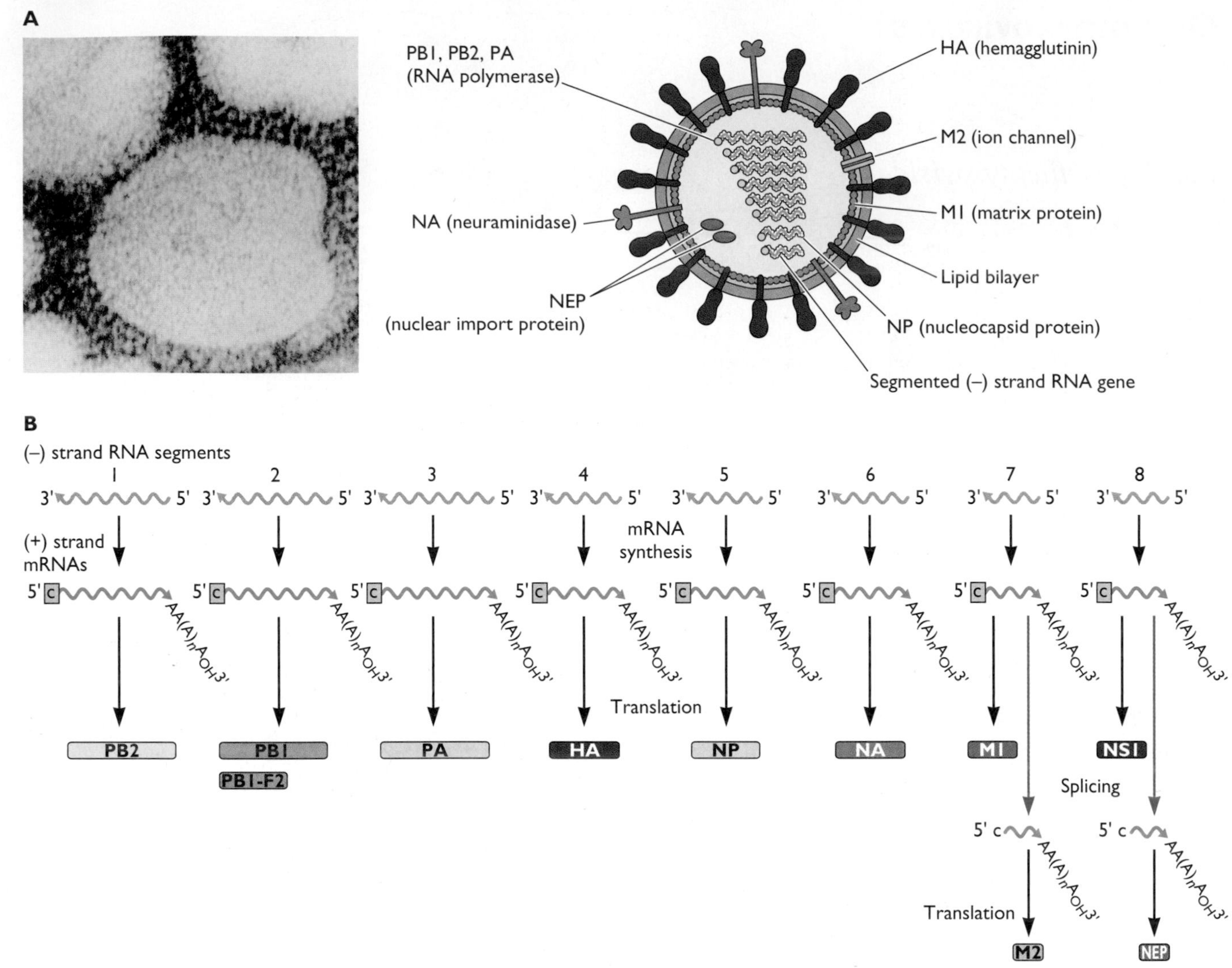

Figure 7 Structure and genomic organization of the orthomyxovirus influenza A virus.

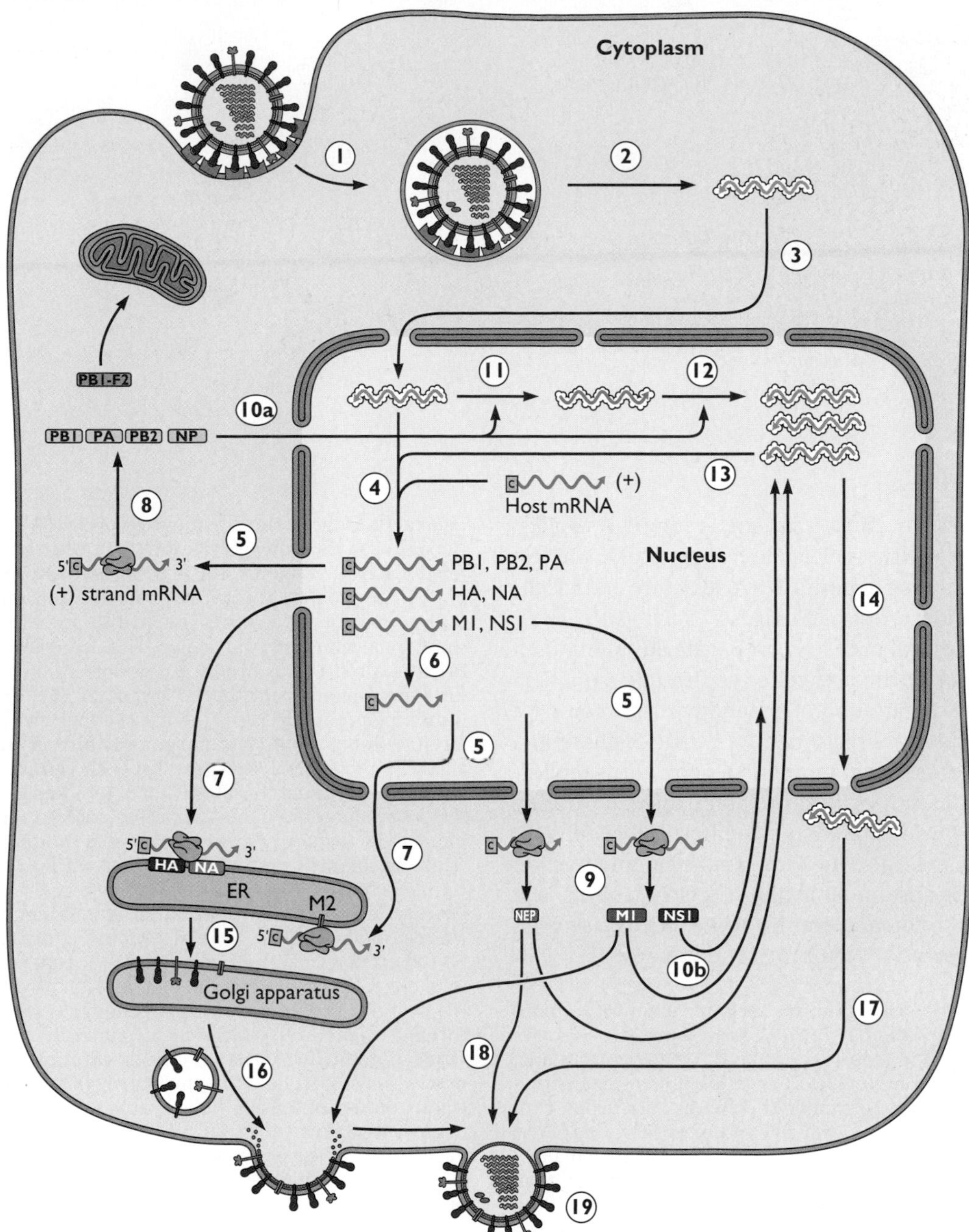

Figure 8 Single-cell reproductive cycle of influenza A virus.

Parvoviruses

Family *Parvoviridae*

Genus	Type species
Subfamily *Densovirinae* (3 genera; insect viruses)	
Subfamily *Parvovirinae*	
Parvovirus	Minute virus of mice
Erythrovirus	Human B19 virus
Dependovirus	Human adeno-associated viruses

Members of the family *Parvoviridae* are among the smallest of the DNA animal viruses. They are of particular interest because of their unique genomic DNA structure and mechanism of replication. Most parvoviruses, such as the well-studied minute virus of mice, can propagate autonomously, although they require the host cell to go through S phase in order to replicate. Replication of dependoviruses requires a helper, adenovirus or herpesvirus, to induce S phase and to provide components that promote dependovirus replication. These viruses can establish a latent infection during which their DNA is integrated into the host cell genome in an inactive state, to be activated upon subsequent infection with a helper. Because of its ability to persist and its lack of pathogenicity, human adeno-associated virus has been developed as a potential vector for gene therapy.

Figure 9 Structure and genome organization of adeno-associated virus (AAV). (A) Virion structure. The electron micrograph shows AAV4 (courtesy of Mavis Agbandje-McKenna, University of Florida, Gainesville). The three-dimensional structure of canine parvovirus is from cryo-electron microscopy with the icosahedral two-, three-, and fivefold axes indicated. The illustration was prepared by Agbandje-McKenna and obtained from N. Muzyczka and K. I. Berns, *in* D. M. Knipe et al. (ed.), *Fields Virology*, 4th ed. (Lippincott Williams & Wilkins, Philadelphia, PA, 2001), with permission. The capsid comprises 60 protein subunits, primarily (~90%) VP3. Virions contain (+) or (–) single-stranded DNA in separate particles. **(B) Genome organization.** The best-characterized DNA genome, that of AAV 2, comprises ca. 4,600 nucleotides and includes terminal repeats (TR) of 145 nucleotides, the first 125 of which form a palindromic sequence. The TR is required in *cis* for genome replication, transcription, and encapsidation, and plays a role in integration into the host DNA during establishment of a latent infection. Use of multiple initiation codons and alternative splicing results in synthesis of multiple Rep (tan bars) and structural proteins (purple bars), respectively. ORF, open reading frame. Adapted from R. M. Linden and K. Berns, p. 68–84, *in* S. Faisst and J. Rommelaere (ed.), *Contributions to Microbiology*, vol. 4, *Parvoviruses: from Molecular Biology to Pathology and Therapeutic Uses* (S. Karger, Basel, Switzerland, 2000), with permission.

Figure 10 Single-cell reproductive cycle. (A) Latent infection. Heparan sulfate proteoglycans are the primary cell surface receptors for AAV2. However, the processes of adsorption **(1)**, uncoating **(2)**, and entry of the DNA into the nucleus **(3)** are poorly understood for all *Parvoviridae*. In the absence of helper virus, some replication of the single-stranded genome occurs prior to integration **(4)** to produce a double-stranded DNA template for transcription from the p5 promoter **(5)**. Translation of this transcript produces Rep proteins **(6)** that are required to both facilitate integration **(7)** and suppress further transcription from all three promoters. The junction with cellular DNA is usually within the terminal repeat (TR) or near a Rep-binding site within the p5 promoter. Neither the actual site in the viral DNA nor the cellular sequence at the junction is unique. However, 70 to 100% of the integration events occur within a region of several hundred nucleotides in chromosome 19q13.3-qter. **(8)**. In most latently infected cells the viral DNA is integrated as a tandem (head-to-tail) array of several genome equivalents. This unexpected arrangement indicates that the mechanism of integration is likely to be distinct from that of viral DNA replication [see (B), step 7]. The integrated viral genome(s) can remain dormant through many cell cycles. Upon superinfection with a helper virus, a productive infection is initiated; molecular details of this process have not yet been elucidated. **(B) Productive infection.** Upon coinfection with a helper virus **(1)**, AAV undergoes a productive infection. With an adenovirus helper, this response is dependent on the expression of early genes E1A, E1B, E4, and E2A **(2)**, which induce S phase and the concomitant production of cellular DNA replication proteins needed for viral DNA synthesis **(3)**. The adenovirus EIA transcriptional activator also induces transcription from the p5 promoter **(4)**, leading to the production of Rep78/68 mRNA and proteins **(5)**. These proteins then function as powerful transcriptional activators (rather than repressors as in latency), and induce transcription from both the p5 and p19 promoters **(6a)**. Viral DNA is replicated by a single-strand displacement mechanism that is initiated by recognition of the terminal resolution site (*trs*) by the Rep78/68 proteins, which remain linked covalently to the DNA through subsequent steps of DNA synthesis **(6b)**. A very large number of replicating forms (ca. 10^6 double-stranded genomes/cell) can be produced within a short time **(7)**. The capsid proteins produced in the cytoplasm **(8)** self-associate in the nucleus during virion assembly **(9)**. As with the adenovirus helper, progeny virions are released **(10)**, usually upon destruction of the cell. The (+) or (–) strand genomes are encapsidated in equal numbers in progeny virions.

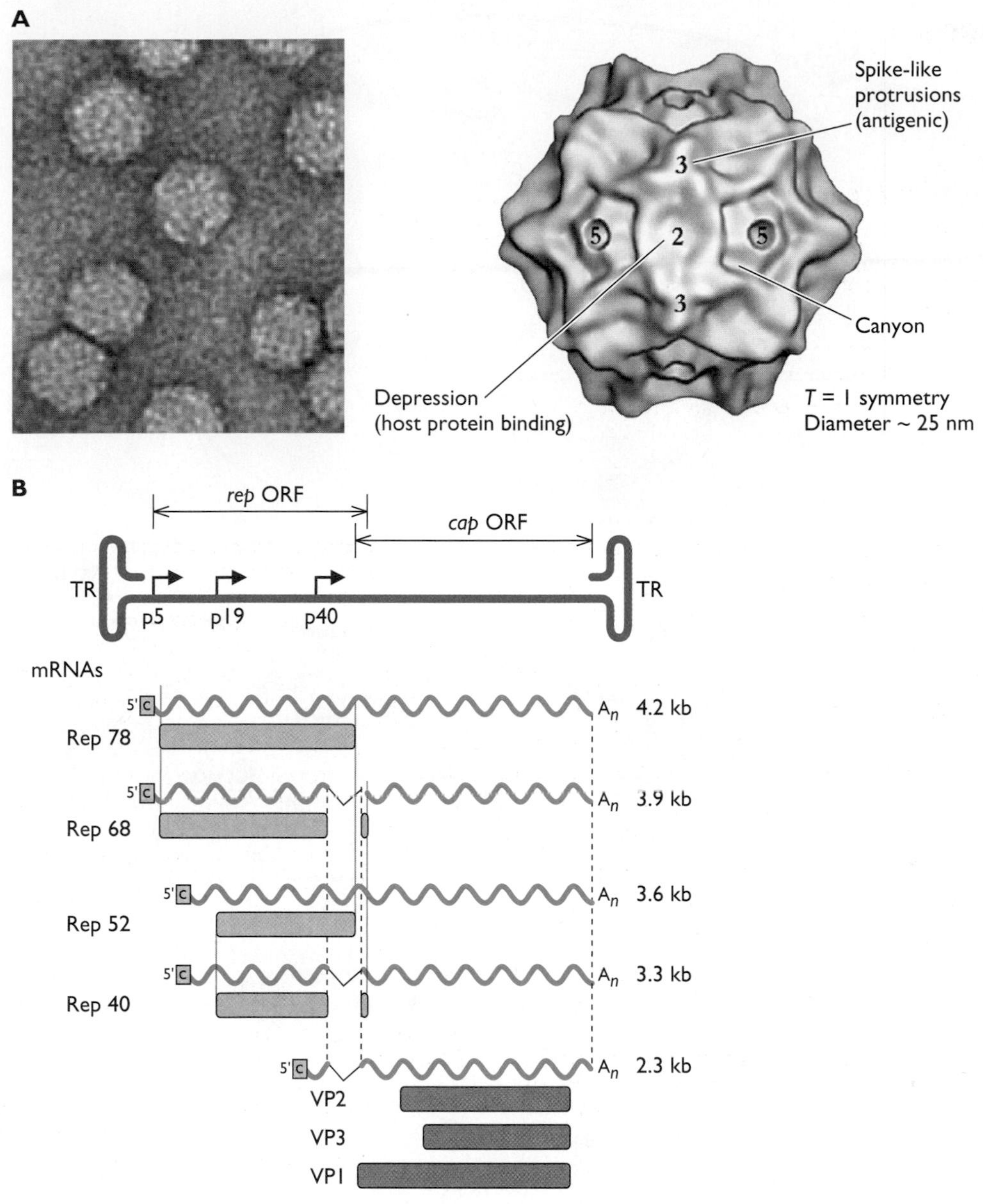

Figure 9 Structure and genome organization of adeno-associated virus (AAV).

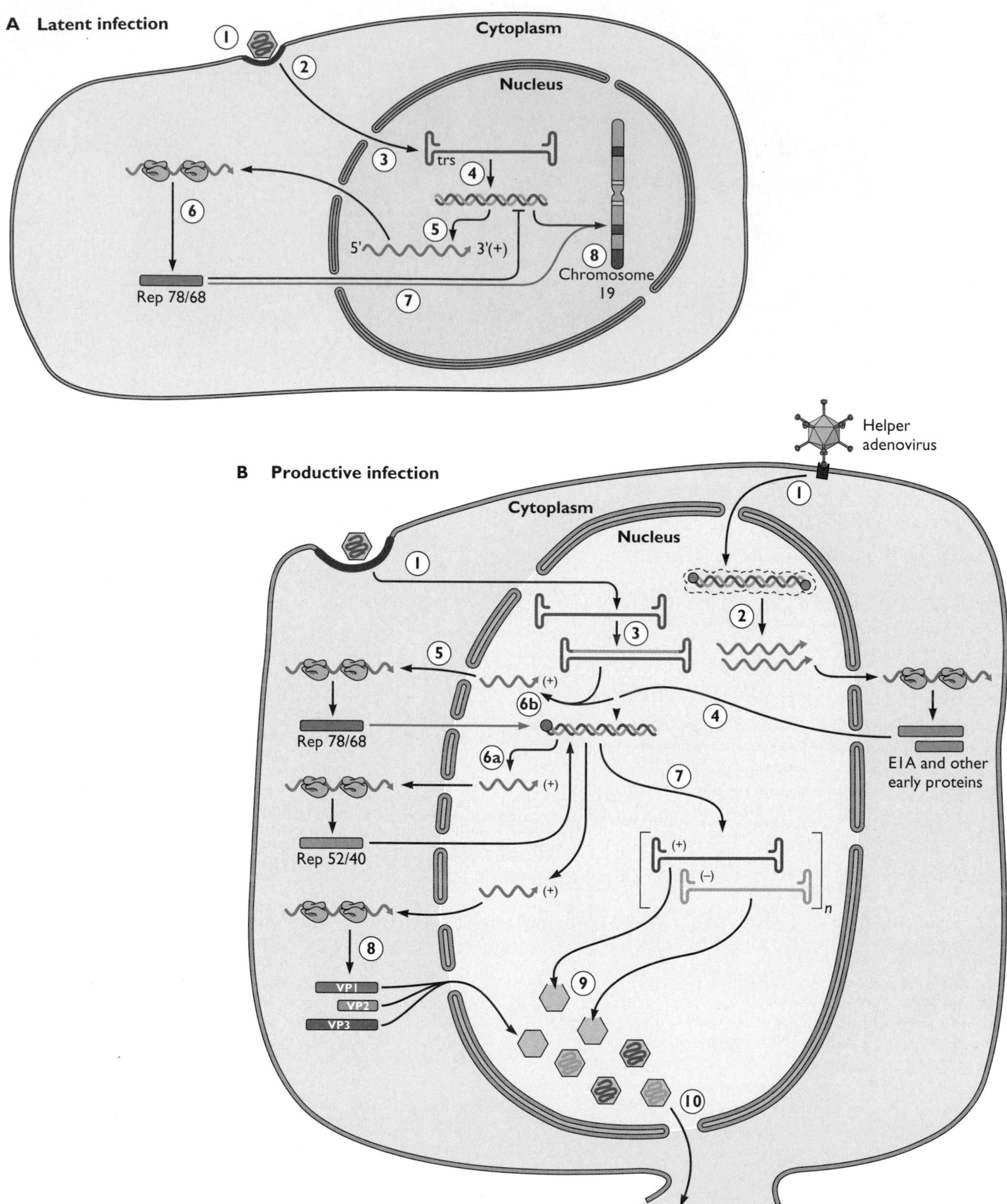

Figure 10 Single-cell reproductive cycle.

Picornaviruses

Family *Picornaviridae*

Genus	Type species
Enterovirus	Poliovirus
Rhinovirus	Human rhinovirus A
Cardiovirus	Encephalomyocarditis virus
Aphthovirus	Foot-and-mouth disease virus
Hepatovirus	Hepatitis A virus
Parechovirus	Human parechovirus
Erbovirus	Equine rhinitis B virus
Kobuvirus	Aichi virus
Teschovirus	Porcine teschovirus 1

The family *Picornaviridae* includes many important human and animal pathogens, such as poliovirus, hepatitis A virus, foot-and-mouth disease virus, and rhinovirus. Because they cause serious disease, poliovirus and foot-and-mouth disease viruses are the best-studied picornaviruses. These two viruses have had important roles in the development of virology. The first animal virus discovered, in 1898, was foot-and-mouth disease virus. The plaque assay was developed using poliovirus, and the first RNA-dependent RNA polymerase discovered was poliovirus $3D^{pol}$. Polyprotein synthesis was discovered in experiments with poliovirus-infected cells, as was translation by internal ribosome entry. The first infectious DNA clone of an animal RNA virus was that of the poliovirus genome, and the first three-dimensional structures of animal viruses determined by X-ray crystallography were those of poliovirus and rhinovirus.

Figure 11 Structure and genomic organization. (A) Virion structure. The electron micrograph shows negatively stained poliovirus (courtesy of N. Cheng and D. M. Belnap, National Institutes of Health, Bethesda, MD). The capsid consists of 60 structural units (each made up of a single copy of VP1, VP2, VP3, and VP4, colored blue, yellow, red, and green, respectively) arranged in 12 pentamers. One of the icosahedral faces has been removed in the diagram to illustrate the locations of VP4 and the viral RNA. **(B) Genome organization.** Polioviral RNA is shown with the VPg protein covalently attached to the 5′ end. The genome is of (+) polarity and encodes a polyprotein precursor. The polyprotein is cleaved during translation by two virus-encoded proteases, $2A^{pro}$ and $3C^{pro}$, to produce structural and nonstructural proteins, as indicated.

Figure 12 Single-cell reproductive cycle. The virion binds to a cellular receptor **(1)**; release of the poliovirus genome occurs from within early endosomes located close (within 100 to 200 nm) to the plasma membrane **(2)**. The VPg protein, depicted as a small orange circle at the 5′ end of the virion RNA, is removed, and the RNA associates with ribosomes **(3)**. Translation is initiated at an internal site 741 nucleotides from the 5′ end of the viral mRNA, and a polyprotein precursor is synthesized **(4)**. The polyprotein is cleaved during and after its synthesis to yield the individual viral proteins **(5)**. Only the initial cleavages are shown here. The proteins that participate in viral RNA synthesis are transported to membrane vesicles **(6)**. RNA synthesis occurs on the surfaces of these infected-cell-specific membrane vesicles. The (+) strand RNA is transported to these membrane vesicles **(7)**, where it is copied into double-stranded RNAs **(8)**. Newly synthesized (–) strands serve as templates for the synthesis of (+) strand genomic RNAs **(9)**. Some of the newly synthesized (+) strand RNA molecules are translated after the removal of VPg **(10)**. Structural proteins formed by partial cleavage of the P1 precursor **(11)** associate with (+) strand RNA molecules that retain VPg to form progeny virions **(12)**, which are released from the cell upon lysis **(13)**.

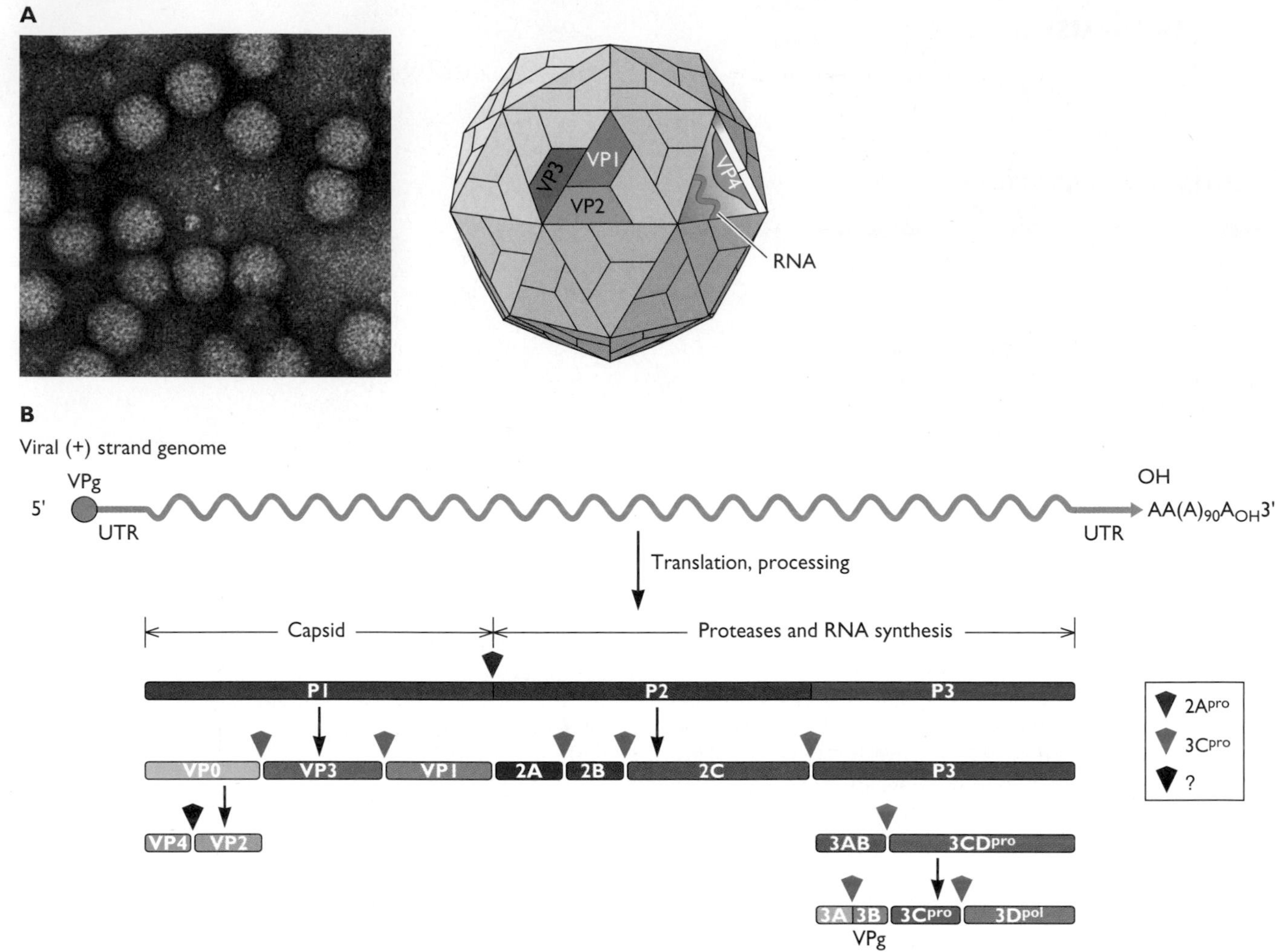

Figure 11 Structure and genomic organization.

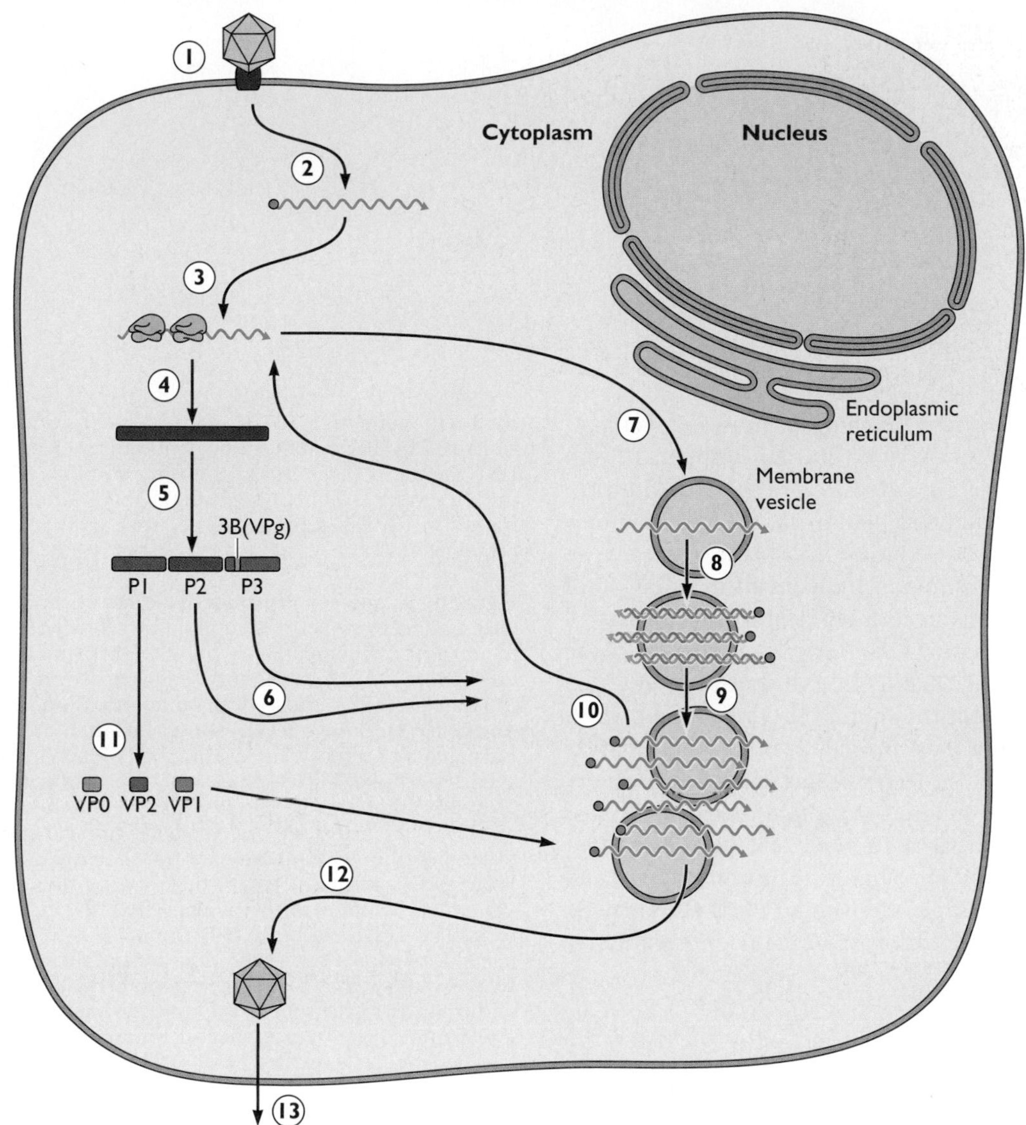

Figure 12 Single-cell reproductive cycle.

Polyomaviruses

Family *Polyomaviridae*

Genus	Type species
Polyomavirus	Simian virus 40

The family *Polyomaviridae* includes mouse polyomaviruses, simian virus 40, and two human viruses, JC and BK viruses, which were isolated from a patient with progressive multifocal leukoencephalopathy and an immunosuppressed recipient of a kidney transplant, respectively. Under some conditions, mouse polyomavirus infection of the natural host results in formation of a wide variety of tumors (hence the name). A characteristic property of the members of this family is an ability to transform cultured cells or to induce tumors in animals. Investigation of such transforming activity has provided much information about mechanisms of oncogenesis, including the discovery of the cellular tumor suppressor protein p53. These viruses, particularly simian virus 40, have also been important in elucidation of cellular mechanisms of transcription and its regulation. For example, the simian virus 40 enhancer was the first member of this class of regulatory sequences to be identified.

Figure 13 Structure and genome organization. (A) Virion structure. The electron micrograph shows negatively stained simian virus 40 virions (from F. A. Andered et al., *Virology* **32**:511–523, 1967, with permission). The double-stranded DNA genome is organized into approximately 25 nucleosomes by the cellular core histones. One molecule of either VP2 or VP3, which possess a common C-terminal sequence, is associated with each VP1 pentamer. **(B) Genome organization.** The 5,243-bp simian virus 40 genome showing locations of the origin of viral DNA synthesis (Ori) and of the early and late mRNAs. The late mRNA species generally contain additional open reading frames in their 5′-terminal exons, such as that encoding leader protein 1 (LP1).

Figure 14 Single-cell reproductive cycle of simian virus 40. The virion attaches to permissive monkey cells upon binding of VP1 to a major histocompatibility complex (MHC) class I molecule on the surface. The virion is then endocytosed in caveolae **(1** and **2)**, transported to the endoplasmic reticulum, and enters that organelle **(3)**. Subsequently, it is transported to the nucleus and uncoated by unknown mechanisms **(4)**. The viral genome packaged by cellular nucleosomes is found within the nucleus **(5)**. The early transcription unit is transcribed by host cell RNA polymerase II **(6)**. After alternative splicing and export to the cytoplasm **(7)**, the early mRNAs are translated to produce the early proteins LT and sT **(8)**. The former is imported into the nucleus **(9)**, where it binds to the simian virus 40 origin of replication to initiate DNA synthesis **(10)**. Apart from LT, all components needed for viral DNA replication are provided by the host cell. As they are synthesized, daughter viral DNA molecules associate with cellular nucleosomes to form the viral nucleoproteins often called minichromosomes. LT also stimulates transcription of the late gene from replicated viral DNA templates **(11)**. Processed late mRNAs are exported to the cytoplasm **(12)**, and translated to produce the virion structural proteins VP1, VP2, and VP3 **(13)**. These structural proteins are imported into the nucleus **(14)** and assemble around viral minichromosomes to form progeny virions **(15)**. Virions are released by an unknown mechanism **(16)**.

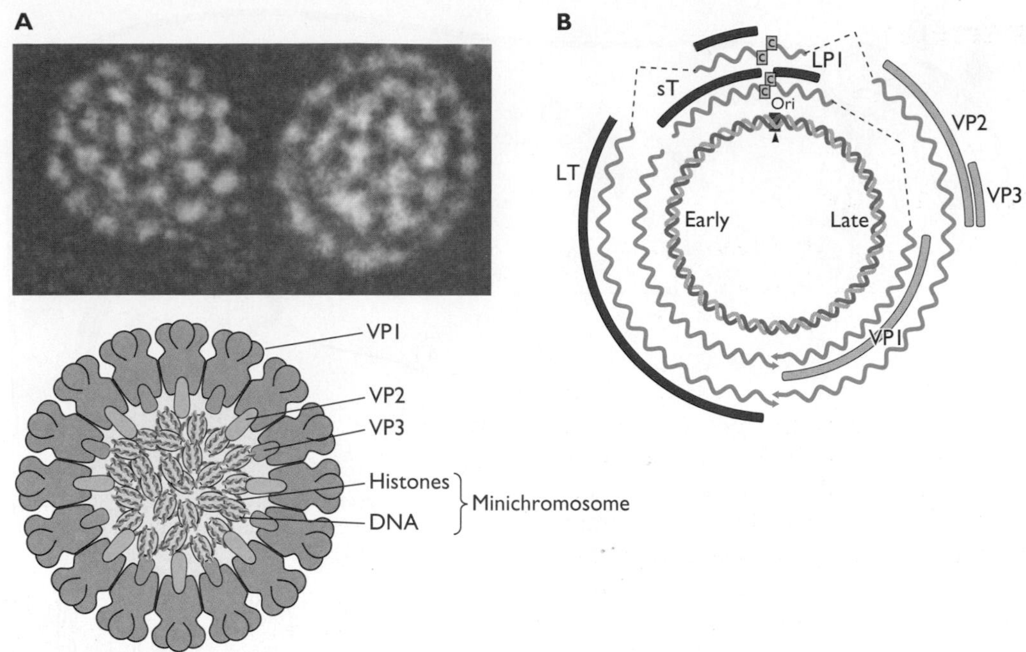

Figure 13 Structure and genome organization.

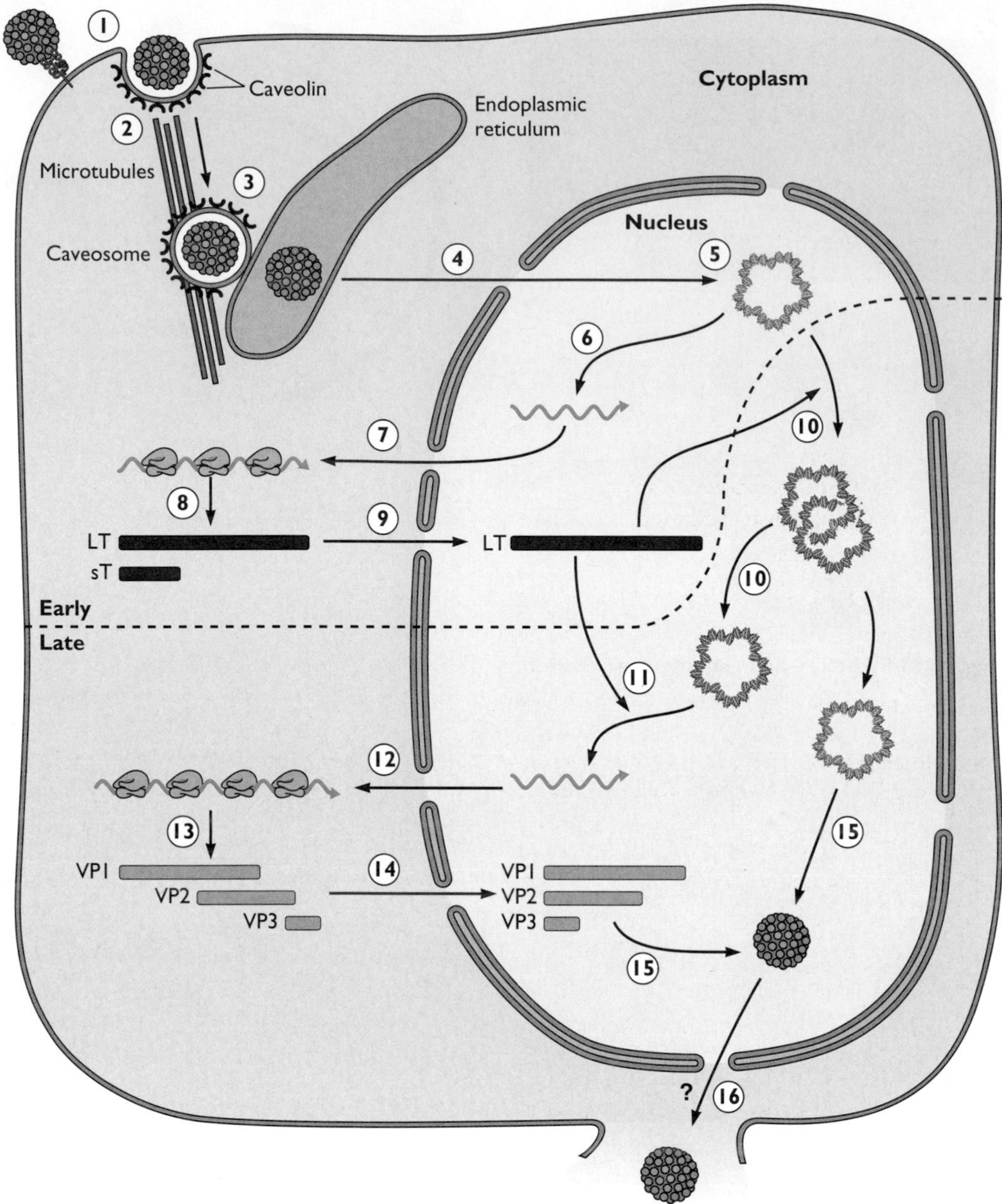

Figure 14 Single-cell reproductive cycle of simian virus 40.

Poxviruses

Family *Poxviridae*

Genus	Type species
Subfamily *Chordopoxvirinae*	
Orthopoxvirus	Vaccinia virus
Parapoxvirus	Orf virus
Avipoxvirus	Fowlpox virus
Capripoxvirus	Sheeppox virus
Leporipoxvirus	Myxoma virus
Suipoxvirus	Swinepox virus
Molluscipoxvirus	Molluscum contagiosum virus
Yabapoxvirus	Yaba monkey tumor virus
Subfamily *Parvovirinae*	
Three genera of viruses that infect insects	

Poxviruses infect most vertebrates and invertebrates, causing a variety of diseases of veterinary and medical importance. The best-known poxviral disease is smallpox, a devastating human disease that has been eradicated by vaccination. The origins of modern vaccinia virus, the virus used in smallpox virus vaccine, are obscure, but this virus is widely studied as a model poxvirus in the laboratory. Myxoma virus, which causes an important disease of domestic rabbits, was described in 1896. Rabbit fibroma virus, which was first described by Shope in 1932, was the first virus proven to cause tissue hyperplasia. The genomes of poxviruses are large DNA molecules that include genes for all proteins needed for DNA synthesis and production of viral mRNAs. These viruses replicate in the cytoplasm and are minimally dependent on the host cell.

Figure 15 Structure and genome organization of the poxvirus vaccinia virus. (A) Virion structure. The electron micrograph shows the mature virion in cross section (courtesy of David J. Vaux, Sir William Dunn School of Pathology, Oxford University, Oxford, United Kingdom). **(B) Genome organization.** Shown are details for the 191-kb genome of the Copenhagen strain of vaccinia virus, with open reading frames in a small section of the genome. This genome includes ~185 unique protein-coding sequences. Those that encode structural proteins and essential enzymes are clustered in the center; those that affect virulence, host range, or immunomodulation are predominantly near the ends.

Figure 16 Single-cell reproductive cycle of vaccinia virus. After fusion of viral and plasma membranes, or fusion following endocytosis, the viral core is released into the cytoplasm **(1)**. Early mRNAs are synthesized **(2)** and are translated by the cellular protein-synthesizing machinery **(3)**. Some early proteins have sequence similarity to cellular growth factors, and can induce proliferation of neighboring host cells following their secretion **(4)**. Other early proteins counteract host immune defense mechanisms. Some early viral proteins induce a second uncoating reaction in which the viral genome is released from the core in a nucleoprotein complex **(5)**, and others mediate replication of the genome **(6)**. Newly synthesized viral DNA molecules can serve as templates for additional cycles of genome replication **(7)** and for transcription of viral intermediate genes **(8)**. Transcription of intermediate genes requires viral initiation proteins, which are products of early genes, and a cellular protein (Vitf2), which relocates from the infected cell nucleus to the cytoplasm. Proteins made upon translation of intermediate mRNAs **(9)** include those necessary for transcription of late genes **(10)**. Late mRNAs are translated to produce virion structural proteins, virion enzymes, and other essential proteins that are needed early in subsequent infections and must be incorporated into virus particles during assembly. Assembly of progeny particles begins in specialized perinuclear sites, termed viral factories, that form upon viral DNA synthesis. These sites contain cellular membranes, probably derived from the endoplasmic reticulum, which are initially reorganized by specific viral proteins to form crescents **(12)**, the precursor to spherical DNA-containing particles, called **immature virions (13)**. Upon proteolysis and release from viral factories, these particles then mature into brick-shaped intracellular mature virions **(14)**, which are released only upon cell lysis **(15)**. However, these particles can acquire a second, double membrane from a *trans*-Golgi or early endosomal compartment to form intracellular **wrapped virions (16)**. The latter particles move to the cell surface on microtubules where fusion with the plasma membrane forms **cell-associated virions (17)** that induce actin polymerization for direct transfer to surrounding cells **(18)** or dissociate from the membrane as the **extracellular virion**. Association of the extracellular virion with a host cell in the next cycle of infection is thought to result in rupture of the outer membrane, giving rise to the **mature virion (MV)**.

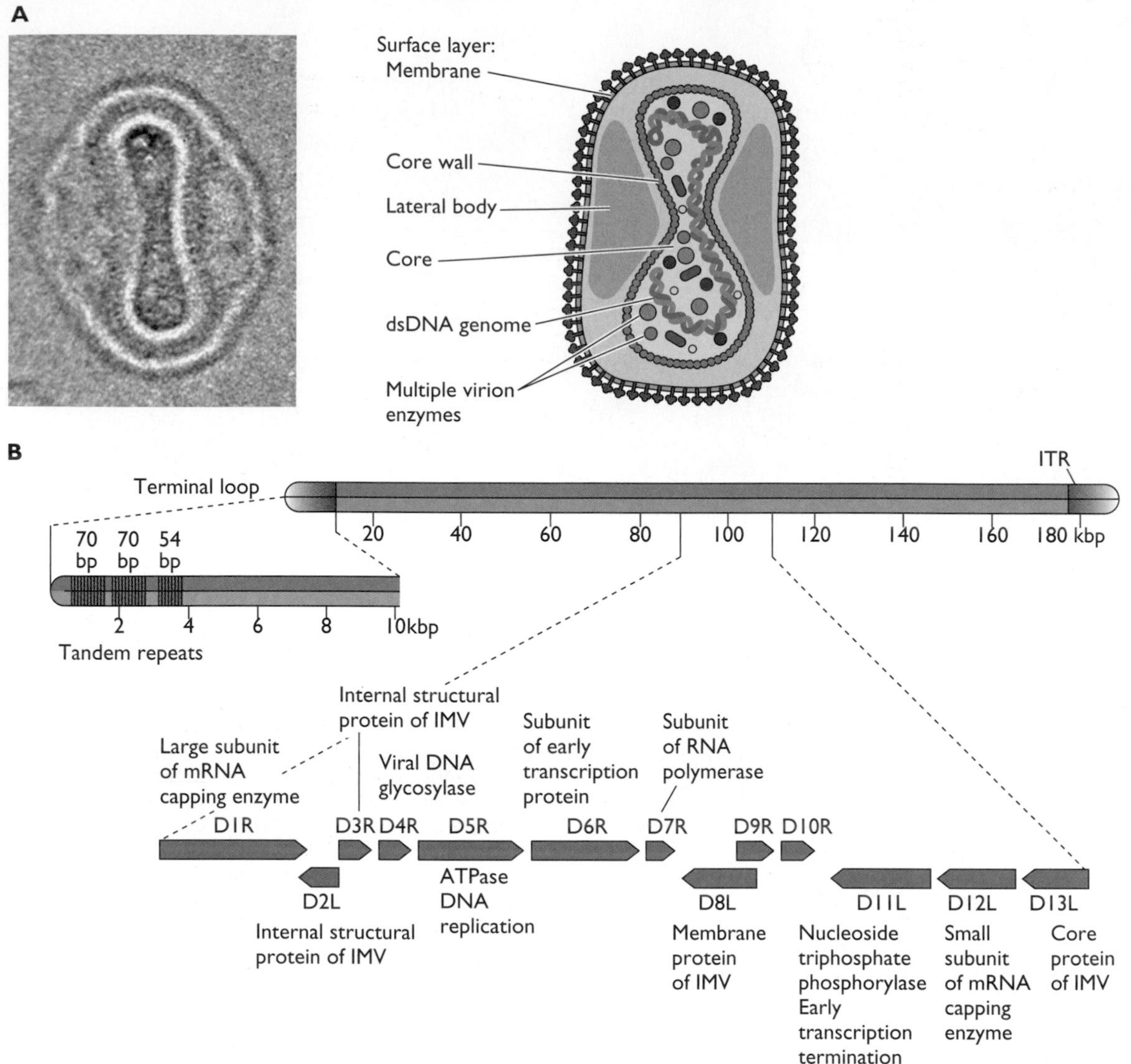

Figure 15 Structure and genome organization of the poxvirus vaccinia virus.

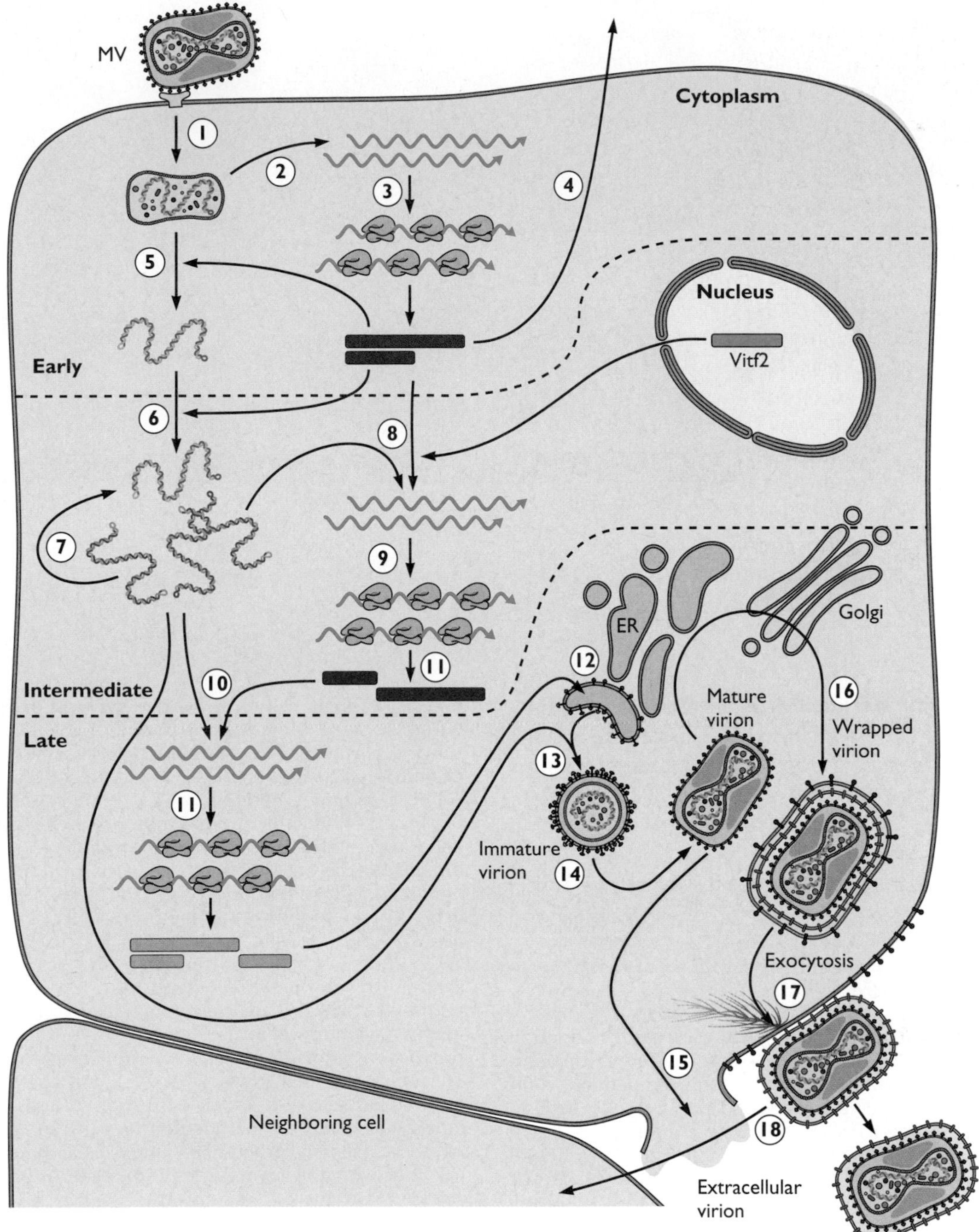

Figure 16 Single-cell reproductive cycle of vaccinia virus.

Reoviruses

Family *Reoviridae*

Genus	Type species
Orthoreovirus	Mammalian orthoreovirus
Orbivirus	Bluetongue virus
Rotavirus	Rotavirus A
Coltivirus	Colorado tick fever virus
Aquareovirus	Aquareovirus A
Cypovirus	Cypovirus 1
Fijivirus	Fijivirus 1
Phytoreovirus	Rice dwarf virus
Oryzavirus	Rice ragged stunt virus
Seadornavirus	Banna virus
Idnoreovirus	Idnoreovirus 1
Mycoreovirus	Mycoreovirus 1

Reoviridae is one of six families of viruses with double-stranded RNA genomes. Included in this family are the human pathogens rotaviruses and Colorado tick fever virus. Reoviruses are the best studied of all the double-stranded RNA viruses. Some of the first *in vitro* research on RNA synthesis was done using reoviruses, and the 5′-terminal cap structure of mRNA was discovered in studies of reovirus mRNAs.

Figure 17 Structure and genomic organization of an orthoreovirus. (A) Virion structure. Electron micrograph of negatively stained reovirus particles, courtesy of S. McNulty, Queen's University, Belfast, United Kingdom. The locations of six virion proteins are indicated on the three-dimensional reconstructions. **(B) Genome organization.** The double-stranded genome comprises 10 segments, named according to size: large (L), medium (M), and small (S). The S1 RNA encodes two proteins: σ1s protein is translated from a second initiation codon in a different reading frame from σ1. Two proteins are also produced from the M3 RNA: protein μNSC is produced by translation at a second initiation codon in the same reading frame as μNS.

Figure 18 Single-cell reproductive cycle of orthoreovirus. The virion **(1)** or a proteolytic derivative **(3)** called the infectious subviral particle (ISVP) **(4)** binds to the same cellular receptor **(2** and **5)** and enters the cell via receptor-mediated endocytosis **(6)**. Under some conditions (e.g., in the lumen of the intestine) the ISVP may be produced by extracellular proteolysis. In endosomes and lysosomes, the virion undergoes acid-dependent proteolytic cleavage **(7)** and can then penetrate the vacuolar membrane **(8** and **9)**. However, the entry of ISVPs may not depend on endocytosis **(10)**. Synthesis of 10 capped viral mRNAs begins within the core particle **(11)**, which is derived from the ISVP. These mRNAs are translated and associate with newly synthesized viral proteins **(12)** to form RNase-sensitive subviral particles in which reassortment may occur **(13)**. Each of the 10 mRNAs is a template for (−) strand RNA synthesis, leading to the production of an RNase-resistant subviral particle that contains 10 double-stranded RNAs **(14)**. Viral mRNAs produced within subviral particles **(15)** are used for the synthesis of viral proteins and the assembly of additional virus particles. In the final steps of capsid assembly, preformed complexes of outer capsid proteins are added to subviral particles **(16)**. Mature virus particles are released from the cell by lysis **(17)**.

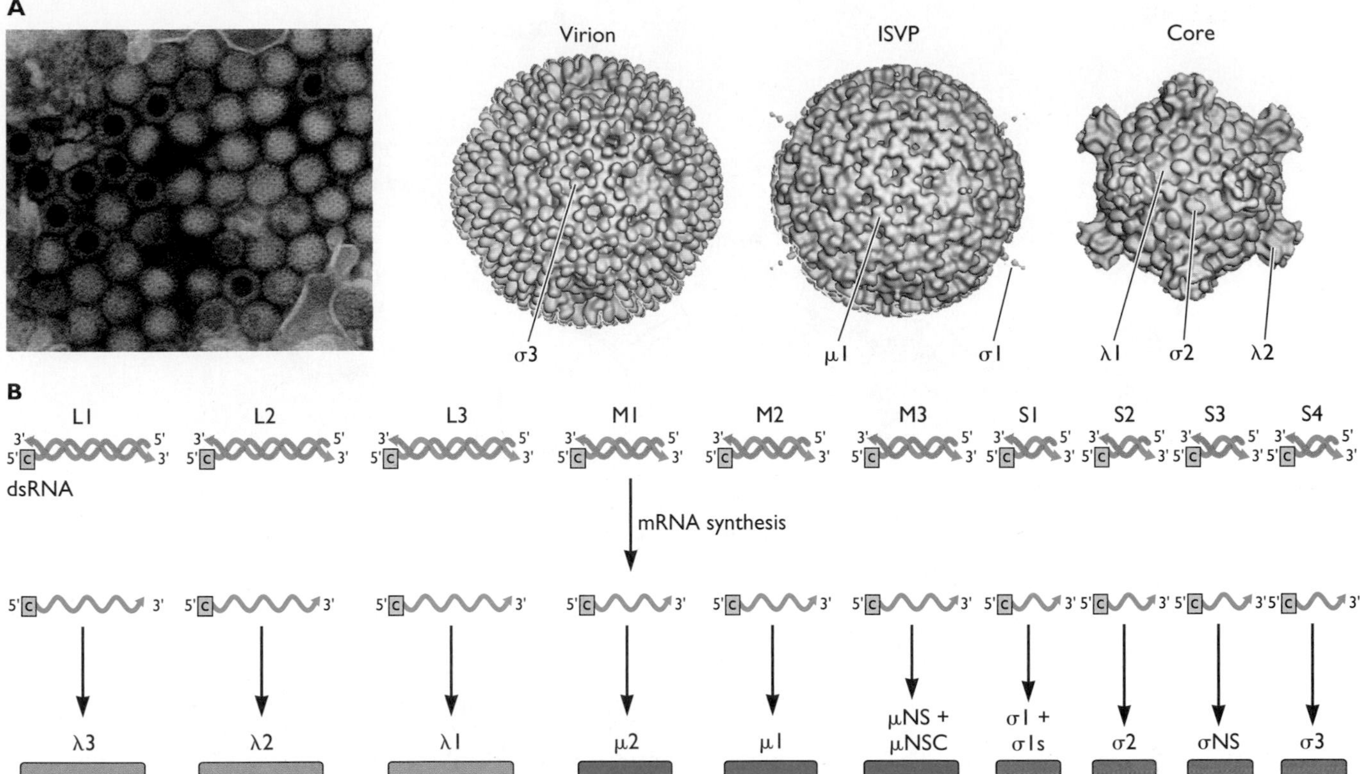

Figure 17 Structure and genomic organization of an orthoreovirus.

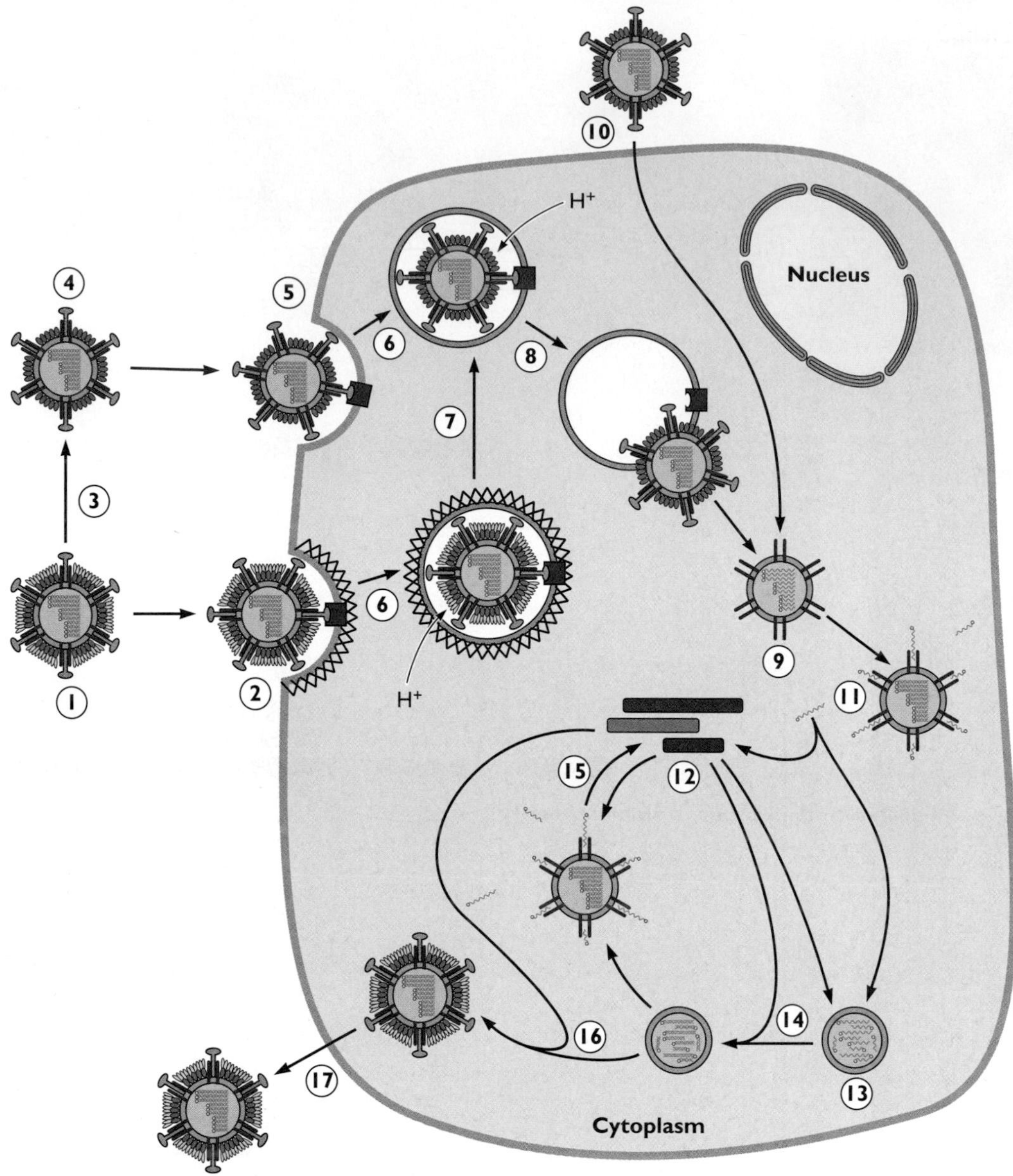

Figure 18 Single-cell reproductive cycle of orthoreovirus.

Retroviruses

Family *Retroviridae*

Genus	Type species
Subfamily *Alphaherpesviridae*	
Alpharetrovirus	Avian leukosis virus
Betaretrovirus	Mouse mammary tumor virus
Gammaretrovirus	Murine leukemia virus
Deltaretrovirus	Bovine leukemia virus
Epsilonretrovirus	Walleye dermal sarcoma virus
Subfamily *Orthoretrovirinae*	
Lentivirus	Human immunodeficiency virus type 1
Subfamily *Spumavirinae*	
Spumavirus	Chimpanzee foamy virus

Retrovirus particles contain the enzyme reverse transcriptase, which mediates synthesis of a double-stranded DNA copy of the viral RNA genome. Although once thought to be unique to this family, similar enzymes are now known to be encoded in other viral genomes (i.e., hepadnaviruses and caulimoviruses), and the term **retroid viruses** has been coined to include these families. Retrovirus particles contain a second enzyme, integrase, that catalyzes the insertion of the viral DNA into essentially random sites in host DNA. The retroviruses can be propagated as integrated elements (called proviruses) that are transmitted in the germ line or as exogenous infectious agents. They infect a wide range of animal hosts and can cause cancer by multiple mechanisms. Some retroviruses, i.e., alpha-, beta-, and gammaretroviruses, have **simple** genomes that encode only the three genes common to all retroviruses—*gag*, *pol*, and *env*. All of the others have more **complex** genomes, which include auxiliary or accessory genes that encode nonstructural proteins that affect viral gene expression and/or pathogenesis.

Figure 19 Structure and genomic organization. (A) Virion structure. The electron micrograph shows a negatively stained alpharetrovirus, Rous sarcoma virus (courtesy of R. Katz and T. Gales, Fox Chase Cancer Center, Philadelphia, PA). Envelope protein projections are not visible in this image. **(B) Genome organization.** (Left) A retrovirus with a simple genome (avian leukosis virus [ALV]), here denoted "Simple retrovirus." Proviral genes are encoded in different reading frames (indicated by different horizontal positions) and are also overlapping. Colored boxes delineate open reading frames; LTR, long terminal repeats that include transcription signals. Origins of RNA and protein products are shown below. U, sequences unique to the 5′ (U5) or 3′ (U3) RNA ends. (Right) A retrovirus with a more complex genome, here denoted "Complex retrovirus" and illustrated with the lentivirus human immunodeficiency virus type 1 (HIV-1). Proviral genes are located in all three reading frames, as indicated by the overlaps. Human immunodeficiency virus type 1 mRNAs fall into one of three classes. The first type is an unspliced transcript of 9.1 kb, identical in function to that of the simple retrovirus shown at the left. The second type comprises singly spliced mRNAs (average length, 4.3 kb) that result from splicing from a 5′ splice site upstream of *gag* to any one of a number of 3′ splice sites near the center of the genome. One of these mRNAs specifies the Env polyprotein precursor, as illustrated for the simple retrovirus. The others specify the human immunodeficiency virus type 1 accessory proteins. The third type comprises a complex class of mRNAs (average length, 1.8 kb) derived by multiple splicing from 5′ and 3′ splice sites throughout the genome. They include mRNAs that specify the regulatory proteins Tat and Rev and are the first to accumulate after infection.

Figure 20 Single-cell reproductive cycle of a simple retrovirus. The virus attaches by binding of the viral envelope protein to specific receptors on the surface of the cell **(1)**. The identities of receptors are known for several retroviruses. The viral core is deposited into the cytoplasm **(2)** following fusion of the virion and cell envelopes. Entry of some beta- and gammaretroviruses may involve endocytic pathways. The viral RNA genome is reverse transcribed by the virion reverse transcriptase (RT) **(3)** within a subviral particle. The product is a linear double-stranded viral DNA with ends that are shown juxtaposed in preparation for integration. Viral DNA and integrase (IN) protein gain access to the nucleus with the help of intracellular trafficking machinery or, in some cases, by exploiting nuclear disassembly during mitosis **(4)**. Integrative recombination, catalyzed by IN, results in site-specific insertion of the viral DNA ends, which can take place at virtually any location in the host genome **(5)**. Transcription of integrated viral DNA (the **provirus**) by host cell RNA polymerase II **(6)** produces full-length RNA transcripts, which are used for multiple purposes. Some full-length RNA molecules are exported from the nucleus and serve as mRNAs **(7)**, which are translated by cytoplasmic ribosomes to form the viral Gag and Gag-Pol polyprotein precursors **(8)**. Some full-length RNA molecules are destined to become encapsidated as progeny viral genomes **(9)**. Other full-length RNA molecules are spliced within the nucleus **(10)** to form mRNA for the Env polyprotein precursor proteins. Env mRNA is translated by ribosomes bound to the endoplasmic reticulum (ER) **(11)**. The Env proteins are transported through

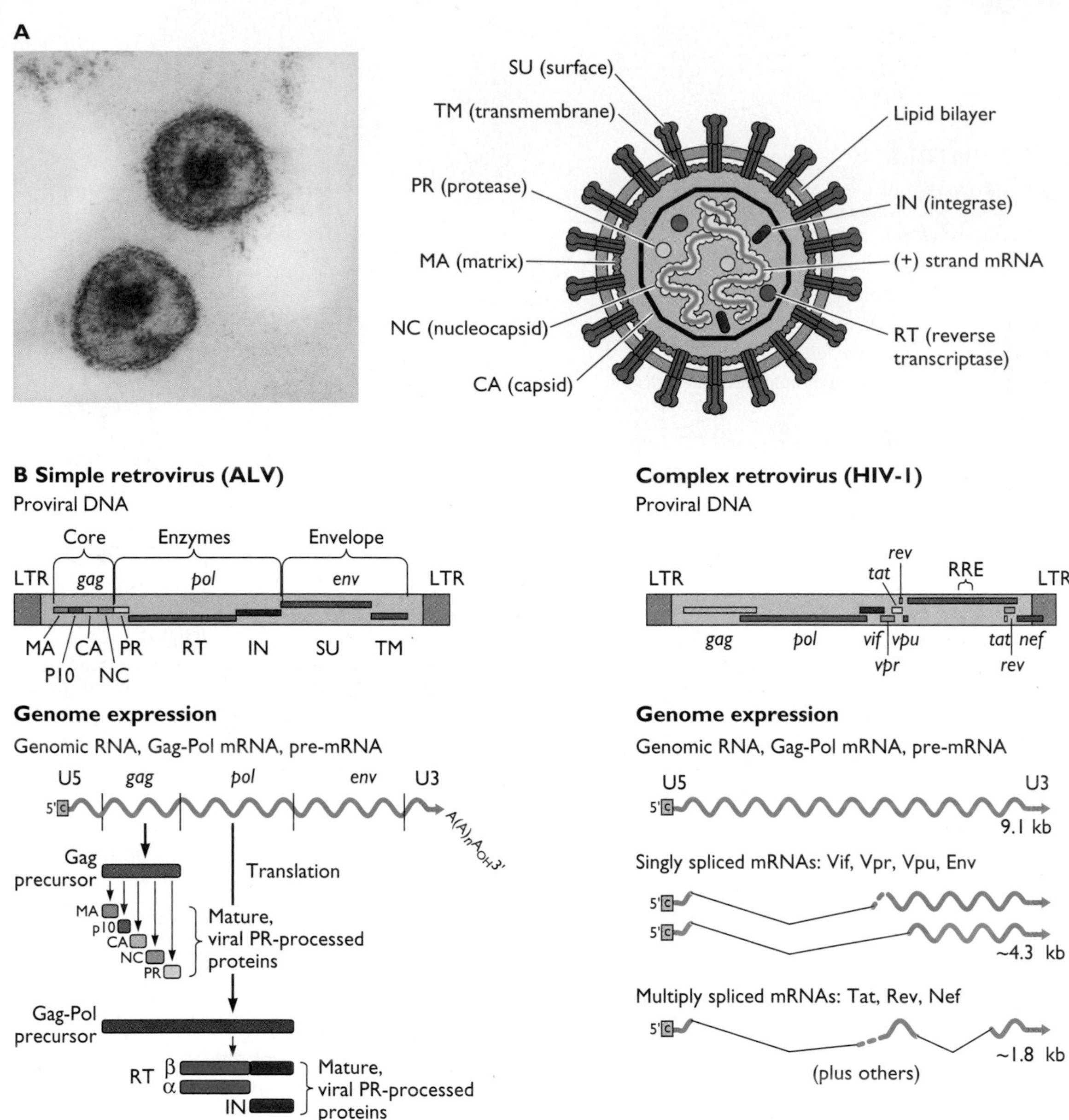

Figure 19 Structure and genomic organization.

the Golgi apparatus **(12)**, where they are glycosylated and cleaved by cellular enzymes to form the mature SU-TM complex. Mature envelope proteins are delivered to the surface of the infected cell **(13)**. Virion components (viral RNA, Gag and Gag-Pol precursors, and SU-TM) assemble at budding sites **(14)** with the help of *cis*-acting signals encoded in each that exploit intracellular vesicular trafficking machinery. Type C retroviruses (e.g., alpharetroviruses and lentiviruses) assemble at the inner face of the plasma membrane, as illustrated. Other types (A, B, and D) assemble on internal cellular membranes. The nascent virions bud from the surface of the cell **(15)**. Maturation (and infectivity) requires the action of the virus-encoded protease (PR), which is itself a component of the core precursor polyprotein. During or shortly after budding, PR cleaves at specific sites within the Gag and Gag-Pol precursors **(16)** to produce the mature virion proteins. This process causes a characteristic condensation of the virion cores.

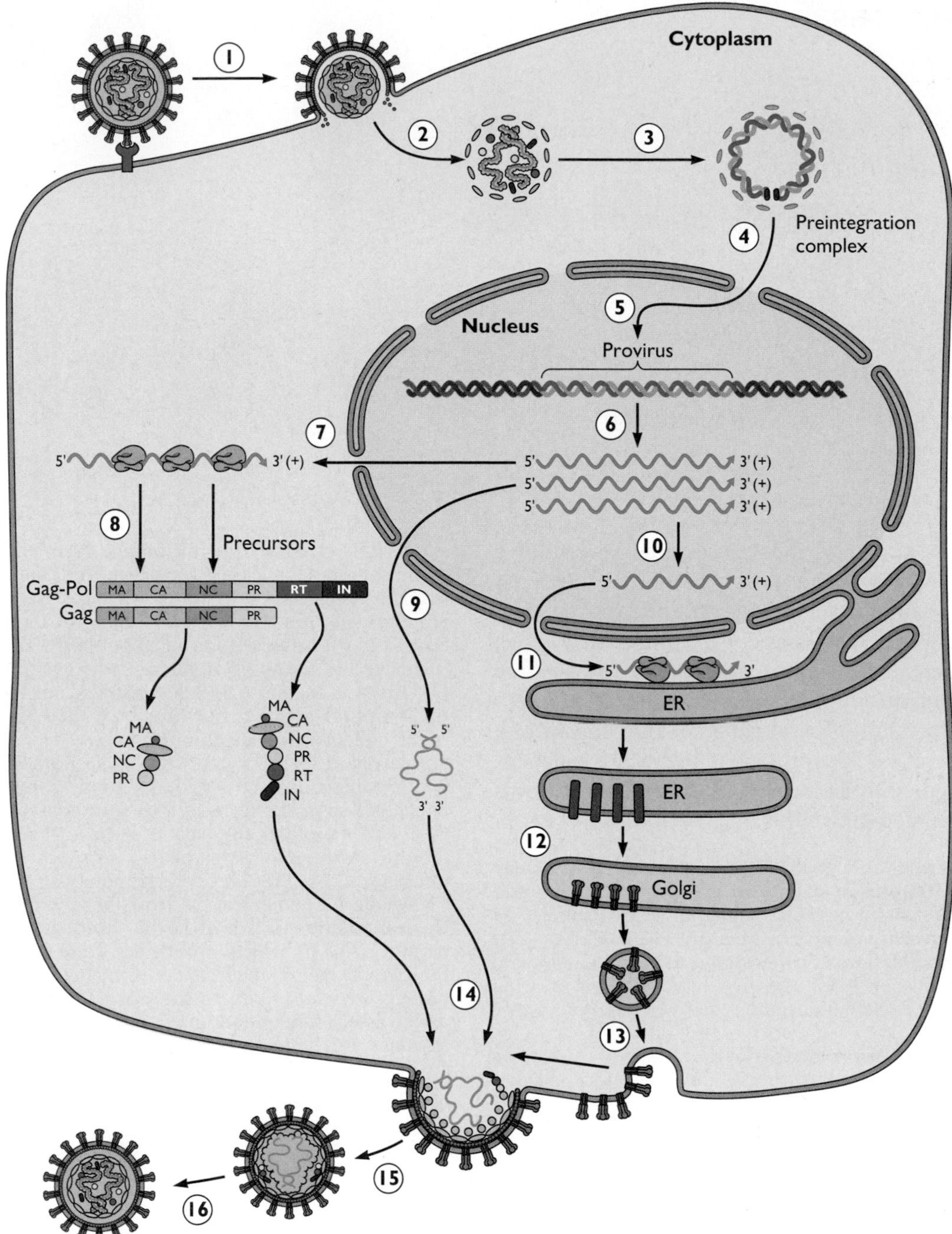

Figure 20 Single-cell reproductive cycle of a simple retrovirus.

Rhabdoviruses

Family *Rhabdoviridae*

Genus	Type species
Vesiculovirus	Vesicular stomatitis virus Indiana
Lyssavirus	Rabies virus
Ephemerovirus	Bovine ephemeral fever virus
Novirhabdovirus	Infectious hematopoietic necrosis virus
Cytorhabdovirus	Lettuce necrotic yellows virus
Nucleorhabdovirus	Potato yellow dwarf virus

Among the 175 known rhabdoviruses are the causative agents of rabies, one of the oldest recognized infectious diseases, and economically important diseases of fish. The host range of these viruses is very broad: they infect many vertebrates, invertebrates, and plants. The genome of vesicular stomatitis virus has been a model for the replication and expression of viral genomes that consist of a single molecule of (–) strand RNA. The first RNA-dependent RNA polymerase discovered in a virus particle was that of vesicular stomatitis virus.

Figure 21 Structure and genomic organization of vesicular stomatitis virus. (A) Virion structure. The electron micrograph shows negatively stained vesicular stomatitis virus (courtesy of J. Rose, Yale University School of Medicine, New Haven, CT). **(B) Genome organization.** The (–) strand RNA is the template for synthesis of leader RNA and five monocistronic mRNAs (capped and polyadenylated) encoding the five viral proteins.

Figure 22 Single-cell reproductive cycle. The virion binds to a cellular receptor and enters the cell via receptor-mediated endocytosis **(1)**. The viral membrane fuses with the membrane of the endosome, releasing the helical viral nucleocapsid **(2)**. This structure comprises (–) strand RNA coated with nucleocapsid protein molecules and a small number of L and P protein molecules, which catalyze viral RNA synthesis. The (–) strand RNA is copied into five subgenomic mRNAs by the L and P proteins **(3)**. The N, P, M, and L mRNAs are translated by free cytoplasmic ribosomes **(4)**, while G mRNA is translated by ribosomes bound to the endoplasmic reticulum **(5)**. Newly synthesized N, P, and L proteins participate in viral RNA replication. This process begins with synthesis of a full-length (+) strand copy of genomic RNA, which is also in the form of a ribonucleoprotein containing the N, L, and P proteins **(6)**. This RNA in turn serves as a template for the synthesis of progeny (–) strand RNA in the form of nucleocapsids **(7)**. Some of these newly synthesized (–) strand RNA molecules enter the pathway for viral mRNA synthesis **(8)**. Upon translation of G mRNA, the G protein enters the secretory pathway **(9)**, in which it becomes glycosylated and travels to the plasma membrane **(10)**. Progeny nucleocapsids and the M protein are transported to the plasma membrane **(11** and **12)**, where association with regions containing the G protein initiates assembly and budding of progeny virions **(13)**.

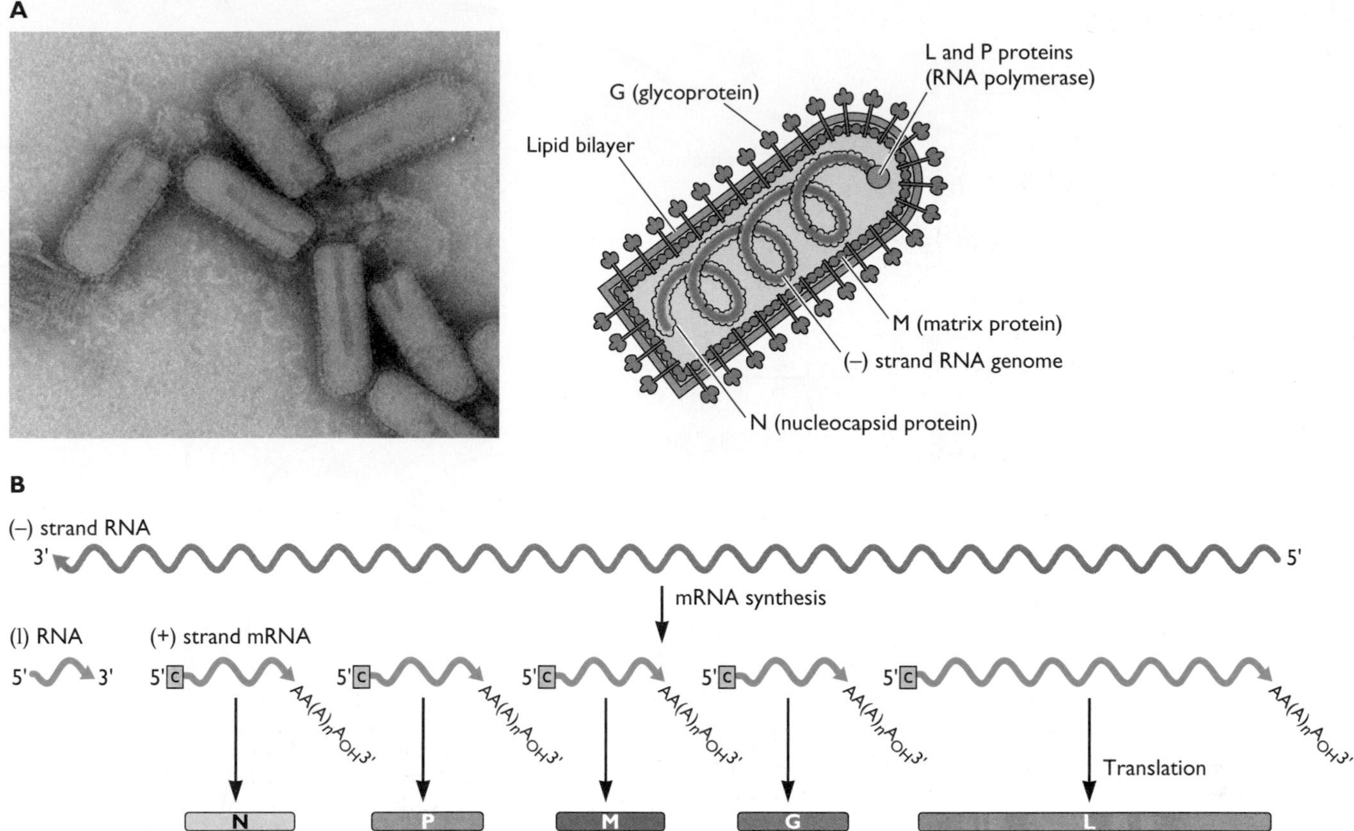

Figure 21 Structure and genomic organization of vesicular stomatitis virus.

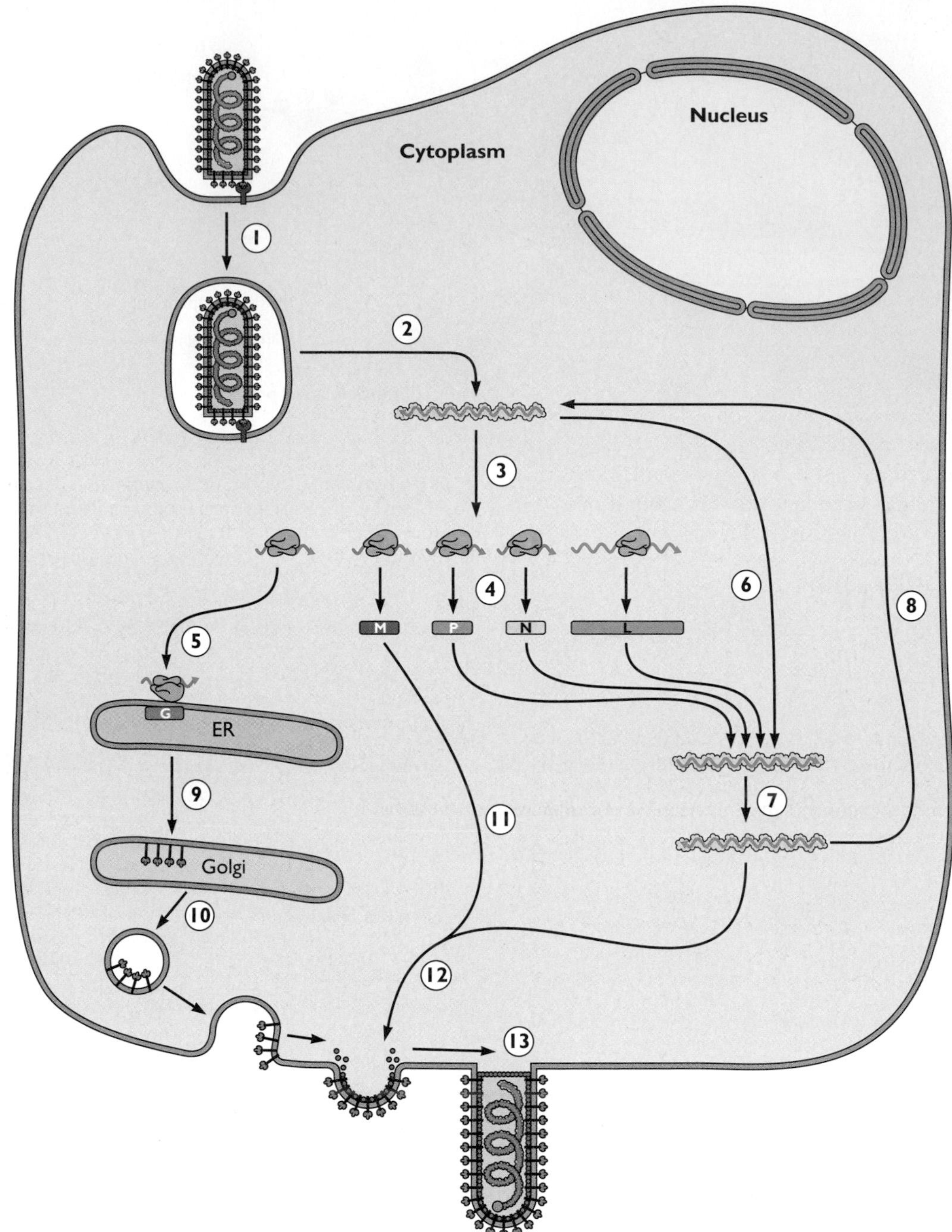

Figure 22 Single-cell reproductive cycle.

Togaviruses

Family *Togaviridae*

Genus	Type species
Alphavirus	Sindbis virus
Rubivirus	Rubella virus

Members of the *Togaviridae* are responsible for two very different kinds of human disease. The alphaviruses are all transmitted by arthropods and cause encephalitis, arthritis, and rashes in humans. Rubella virus is the agent of a mild rash disease that can also cause congenital abnormalities in the fetus when acquired by the mother early in pregnancy. Because these virions have a lipid envelope, they have been important models for studying the synthesis, posttranslational modification, and localization of membrane glycoproteins.

Figure 23 Structure and genomic organization. (A) Virion structure. The cryo-electron micrograph shows the alphavirus Ross River virus (courtesy of N. Olson, Purdue University, West Lafayette, IN). The three-dimensional image reconstructions (courtesy of B. V. V. Prasad, Baylor College of Medicine, Houston, TX) show the intact Sindbis virus (yellow, with glycoprotein spikes visible), its nucleocapsid (blue), and a cross section of the particle (green), and illustrate the relationship among the spike glycoproteins (S), the lipid membrane (M), the capsid (C), and the viral RNA genome (RNA). **(B) Genome organization.** The first two-thirds of togavirus genomic RNA, which is of (+) polarity and carries a 5′ cap, is translated to produce the polyproteins P123 and P1234. The latter is the precursor of the RNA polymerase. For some alphaviruses, the P1234 polyprotein is produced by translational suppression of a stop codon located at the end of the nsP3 coding region. The proteins encoded in the 3′-terminal one-third of the genome are produced from a subgenomic mRNA that is copied from a full-length (−) strand RNA intermediate. The subgenomic mRNA encodes the structural proteins.

Figure 24 Single-cell reproductive cycle. The virion binds to a cellular receptor and enters the cell via receptor-mediated endocytosis **(1)**. Upon acidification of the vesicle, viral RNA is uncoated **(2)** and translated to form the polyprotein P1234. Sequential cleavage of this polyprotein at different sites produces RNA polymerases with different specificities. **(3)** These viral enzymes then copy (+) strands into full-length (−) and (+) strands **(4)** and catalyze synthesis of the subgenomic mRNA **(5)**. This mRNA is translated by free cytoplasmic ribosomes to produce the capsid protein **(6)**; proteolytic cleavage to liberate the capsid protein exposes a hydrophobic sequence of PE2 that induces the ribosomes to associate with the endoplasmic reticulum (ER) **(7)**. As a result, the PE2-6K-E1 polyprotein enters the secretory pathway. The glycoproteins are transported to the cell surface (**8** and **9**). The capsid protein and (+) strand genomic RNA assemble to form capsids **(10)** that migrate to the plasma membrane and associate with viral glycoproteins **(11)**. The capsid acquires an envelope by budding at this site **(12)**, and mature virions are released **(13)**.

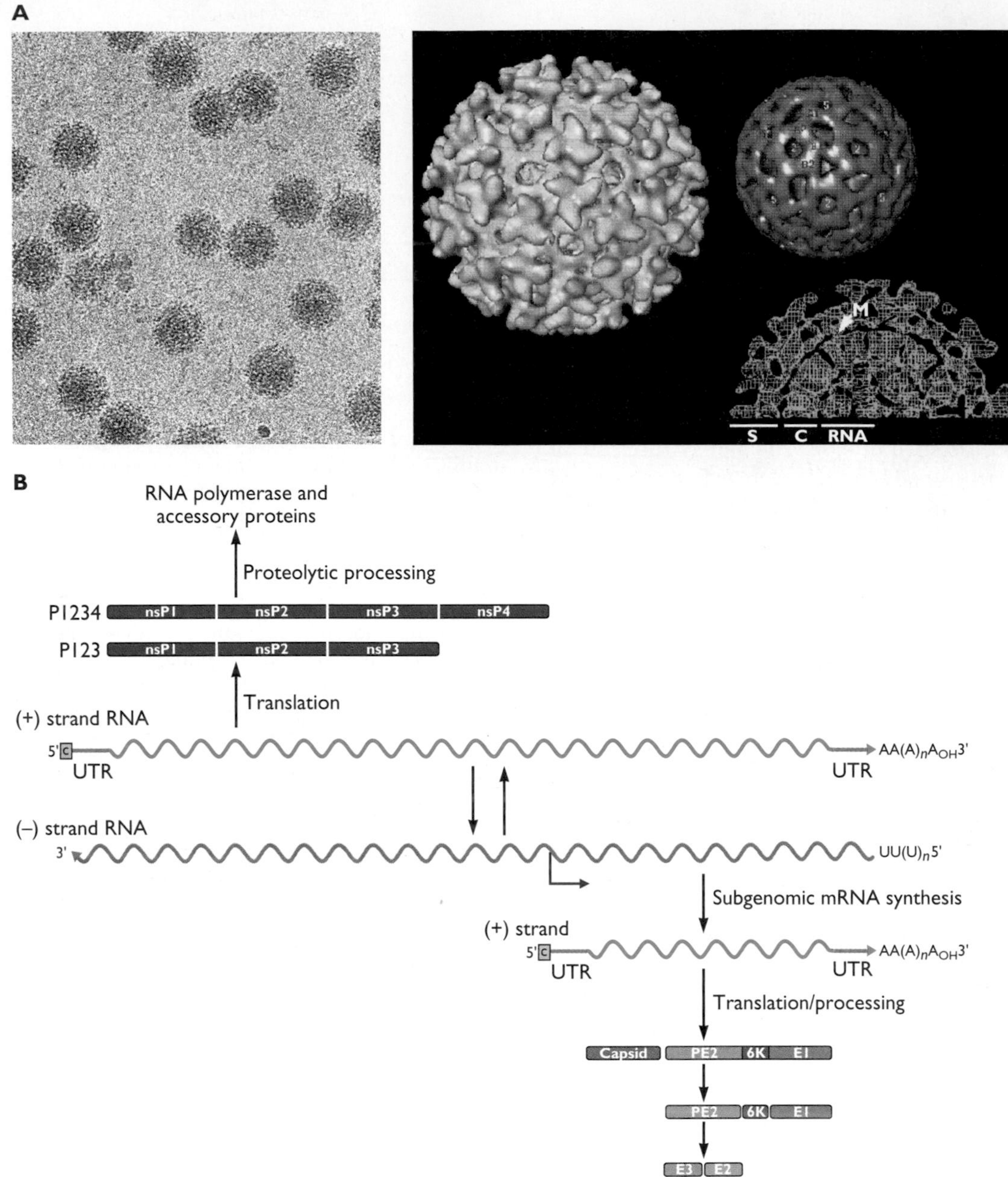

Figure 23 Structure and genomic organization.

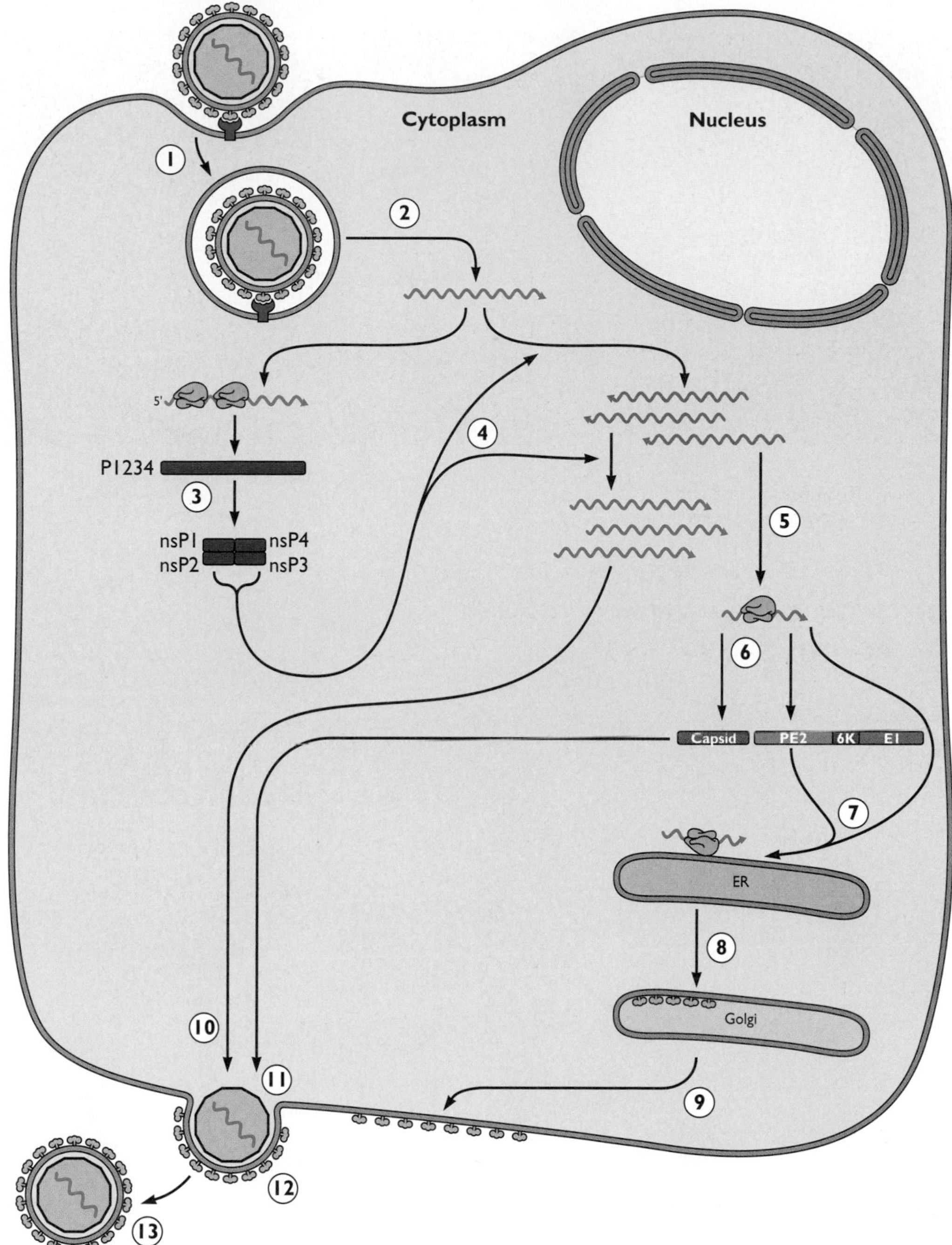

Figure 24 Single-cell reproductive cycle.

Glossary

Absolute efficiency of plating The plaque titer divided by the number of virus particles in the sample. *(Chapter 2)*

Acylation Posttranslational addition of saturated or unsaturated fatty acids to a protein. *(Chapters 9 and 12)*

Allele specific Complementing only a specific change; refers to suppressor mutations. *(Chapter 3)*

Alternative splicing Splicing of different combinations of exons in a pre-mRNA, generally leading to synthesis of mRNAs with different protein-coding sequences. *(Chapter 10)*

Ambisense Producing mRNAs from both (−) strand genomic RNA and the complementary (+) strand; refers to viral genomes. *(Chapter 6)*

Amphipathic Having both hydrophilic and hydrophobic portions. *(Chapter 2)*

Aneuploid Abnormal in chromosome morphology and number. *(Chapter 2)*

Apical surface The specialized surface of a epithelial cell exposed to the environment. Also called apical domain. *(Chapters 2, 9, and 12)*

Asymmetric unit The unit from which capsids or nucleocapsids of a virus particle are built. Also called protomer or structural unit. *(Chapter 4)*

Attenuated An infection in which normally severe symptoms or pathology are mild or inconsequential; a state of reduced virulence. *(Chapter 1)*

Bacteriophages Viruses that infect bacteria; derived from the Greek word *phagein,* meaning "to eat." *(Chapter 1)*

Basal lamina A thin layer of extracellular matrix bound tightly to the basolateral surface of cells; the basal lamina is linked to the basolateral membrane by integrins. *(Chapter 2)*

Basolateral surface The nonspecialized surface of an epithelial cell that contacts an internal basal lamina or adjacent or underlying cells in the tissue. Also called basolateral domain. *(Chapters 2, 9, and 12)*

Burst The yield of viruses from one cell. *(Chapter 2)*

Capping The addition of m^7G via a 5′-5′ phosphodiester bond to the 5′ ends of cellular and viral transcripts made in eukaryotic cells. *(Chapter 10)*

Capsid The outer shell of viral proteins that surrounds the genome in a virus particle. *(Chapters 1, 3, and 4)*

Cap snatching Cleavage of cellular RNA polymerase II transcripts by a viral endonuclease to produce capped primers for viral mRNA synthesis. *(Chapter 10)*

Caveolae Flask-shaped invaginations of the plasma membrane of many types of cells that contain the protein caveolin and are rich in lipid rafts; caveolae internalize membrane components, extracellular ligands, bacterial toxins, and some animal viruses. *(Chapter 5)*

Centrosome An organelle that is the main microtubule-organizing center. *(Chapter 5)*

Chaperone A protein that facilitates the folding of other polypeptide chains, the assembly of multimeric proteins, or the formation of macromolecular assemblies (e.g., chromatin). Also called molecular chaperone. *(Chapters 4, 12, and 13)*

Chemokines Small proteins that attract and stimulate cells of the immune defense; produced by many cells in response to infection. Also called chemotactic cytokines. *(Chapter 5)*

Coactivator A protein that stimulates transcription by RNA polymerase II without binding to a specific DNA sequence; generally interacts with sequence-specific transcriptional activators. *(Chapter 8)*

Codon Three contiguous bases in an mRNA template that specify the amino acids incorporated into protein. *(Chapter 11)*

Complementation The ability of gene products of two different, nonreplicating mutants to interact functionally in the same cell to permit viral replication. *(Chapter 3)*

Constitutive splicing Splicing of a pre-mRNA that removes all introns and joins all exons. *(Chapter 10)*

Constitutive transport elements Sequences in certain unspliced viral mRNAs that direct export from the nucleus by host cell proteins. *(Chapter 10)*

Continuous cell lines Cultures of a single cell type that can be propagated indefinitely in culture. *(Chapter 2)*

Copy choice A mechanism of recombination in which an RNA polymerase first copies the 3′ end of one parental strand and then exchanges one template for another at the corresponding position on a second parental strand. *(Chapters 6 and 7)*

Coreceptor A cell surface molecule that is required, in addition to the receptor, for entry of virus particles into cells. *(Chapter 5)*

Core promoter The minimal set of DNA sequences required for accurate initiation of transcription by RNA polymerase II. *(Chapter 8)*

Culling Removing and destroying diseased or potentially exposed animals to prevent further spread of infection. *(Chapter 1)*

Cytopathic effects The morphological changes induced in cells by viral infection. *(Chapter 2)*

Cytoskeleton The intracellular structural network composed of actin filaments, microtubules, and intermediate filaments. *(Chapter 2)*

Defective interfering RNAs Subgenomic RNAs that replicate more rapidly than full-length RNA and therefore compete for the components of the RNA synthesis machinery and interfere with the replication of full-length RNAs. *(Chapter 6)*

Deletion mutation Loss of one or more bases in a nucleic acid. *(Chapter 3)*

Diploid cell strains Cell cultures that consist of a homogeneous population of a single type and that can divide up to 100 times before dying. *(Chapter 2)*

Eclipse period The phase of viral infection during which the viral nucleic acid is uncoated from its protective shell and no infectious virus can be detected inside cells. *(Chapter 2)*

Elongation Stepwise incorporation of ribodeoxynucleoside monophosphates or deoxynucleoside monophosphates into the 3′-OH end of the growing RNA or DNA chain in the 5′ → 3′ direction. *(Chapter 6)*

Endemic Having a disease pattern typical of a particular geographic area; persisting in a population for a long period without reintroduction of the causative virus from outside sources. *(Chapter 1)*

Endogenous proviruses Proviruses that enter the germ line at some point in the history of an organism and are thereafter inherited in normal Mendelian fashion by every cell in that organism and by its progeny. *(Chapter 7)*

Endosome A vesicle that transports molecules from the plasma membrane to the cell interior. *(Chapter 5)*

Enhancer A DNA sequence containing multiple elements that can stimulate RNA polymerase II transcription over long distances, independently of orientation or location relative to the site of transcriptional initiation. *(Chapter 8)*

Envelope The host cell-derived lipid bilayer carrying viral glycoproteins that forms the outer layer of many virus particles. *(Chapter 4)*

Epidemic A pattern of disease characterized by rapid and sudden appearance of cases spreading over a wide area. *(Chapter 1)*

Epitope A short contiguous sequence or unique conformation of a macromolecule that can be recognized by the immune system; also called an antigenic determinant; a T-cell epitope is a short peptide recognized by a particular T-cell receptor, while a B-cell epitope is recognized by the antigen-binding domain of antibody and is part of an intact protein. *(Chapter 2)*

Exons Blocks of noncontiguous coding sequences (generally short) present in many cellular and viral pre-mRNAs. *(Chapter 10)*

Foci Clusters of cells that are derived from a single progenitor and share properties, such as unregulated growth, that cause them to pile up on one another. *(Chapter 2)*

Fusion peptide A short hydrophobic amino acid sequence (20 to 30 amino acids) that is thought to insert into target membranes to initiate fusion. *(Chapter 5)*

Fusion pore An opening between two lipid bilayers formed by the action of fusion proteins; it allows exchange of material across membranes. *(Chapter 5)*

Glycoforms The total set of forms of a protein that differ in the number, location, and nature of oligosaccharide chains. *(Chapter 12)*

Glycoprotein A protein carrying covalently linked sugar chains (oligosaccharides). *(Chapter 4)*

G_0 state A state in which the cell has ceased to grow and divide and has withdrawn from the cell cycle. Also called resting state. *(Chapter 9)*

Half-life The time required for decay to half of the original value. *(Chapter 10)*

Helical symmetry The symmetry of regularly wound structures defined by the relationship $P = \mu \times \rho$, where P = pitch of the helix, μ = the number of structural units per turn, and ρ = the axial rise per unit. *(Chapter 4)*

Helper virus A virus that provides viral proteins needed for the replication of a coinfecting defective virus. *(Chapter 6)*

Hemagglutination A method for measuring virus concentrations via the linking of multiple red blood cells by virus particles, resulting in a lattice. *(Chapter 2)*

Heterogeneous nuclear RNAs Nuclear precursors to mRNAs that are larger than mRNAs and heterogeneous in size. *(Chapter 10)*

Homologous recombination The exchange of genetic information between any pair of related DNA sequences. *(Chapter 9)*

Host range A listing of species and cells (hosts) that are susceptible to and permissive for infection. *(Chapter 5)*

Icosahedral symmetry The symmetry of the icosahedron, the solid with 20 faces and 12 vertices related by axes of two-, three-, and fivefold rotational symmetry. *(Chapter 4)*

Indirectly anchored proteins Proteins that are indirectly bound to the plasma membrane by interacting with either integral membrane proteins or the charged sugars of membrane glycolipids. *(Chapter 2)*

Infectious DNA clone A double-stranded DNA copy of the viral genome carried on a bacterial plasmid or other vector. *(Chapter 3)*

Initiation site The site at which transcription of a gene begins. *(Chapter 8)*

Initiator A short DNA sequence that is sufficient to specify the site at which RNA polymerase II initiates transcription. *(Chapter 8)*

Insertion mutation Addition of a nucleic acid sequence. *(Chapter 3)*

Integral membrane proteins Proteins that are embedded in a lipid bilayer, with external and internal domains connected by one or more membrane-spanning domains. *(Chapters 2 and 4)*

Introns Noncoding sequences that separate coding sequences (exons) in many cellular and viral pre-mRNAs. *(Chapter 10)*

Inverted terminal repetitions Sequences that are present in the opposite orientation at the ends of certain linear viral DNA genomes. *(Chapter 9)*

Koch's postulates Criteria developed by the German physician Robert Koch in the late 1800s to determine if a given agent is the cause of a specific disease. *(Chapter 1)*

Lagging strand The daughter DNA strand made by discontinuous synthesis during DNA replication. *(Chapter 9)*

Lariat An intermediate in pre-mRNA splicing containing the intron and 3′ exon, with the branch point A residue of the intron linked via a 2′-5′ phosphodiester bond to the nucleotide at the 5′ end of the intron. *(Chapter 10)*

Latency-associated transcript RNA produced specifically during a latent infection by herpes simplex virus. *(Chapter 8)*

Latent period The phase of viral infection during which no extracellular virus can be detected. *(Chapter 2)*

L domain sequences Short amino acid sequences required for membrane fusion during budding of enveloped viruses. *(Chapter 13)*

Leading strand The daughter DNA strand made by continuous synthesis during DNA replication; its synthesis begins before that of the lagging strand. *(Chapter 9)*

Lipid raft A microdomain of the plasma membrane that is enriched in cholesterol and saturated fatty acids and is more densely packed and less fluid than other regions of the membrane. *(Chapters 2 and 12)*

Long terminal repeat A direct repeat of genetic information that is present in the proviral DNA of retroviruses; it is formed by reverse transcription of the RNA template and includes *cis*-acting elements required for viral DNA integration and its subsequent transcription. *(Chapter 7)*

Lysogenic Pertaining to a bacterium that carries the genetic information of a quiescent bacteriophage, which can be induced to reproduce, and subsequently lyse, the bacterium. *(Chapter 1)*

Lysogeny The phenomenon by which the lysogenic state is established and maintained in bacteria. *(Chapter 1)*

Lysosome A vesicle in the cell that contains enzymes that degrade sugars, proteins, nucleic acids, and lipids. *(Chapter 5)*

Marker rescue Replacement of all local nucleic acids that include a mutation with wild-type nucleic acid. *(Chapter 3)*

Marker transfer Introduction of a mutation by replacement of a segment of viral nucleic acid with one containing the mutation. *(Chapter 3)*

Membrane-spanning domain A segment of an integral membrane protein that spans the lipid bilayer; often α-helical. *(Chapters 2 and 4)*

Metastable structure A structure that has not attained the lowest free energy state. *(Chapter 4)*

Microdomains Regions of the plasma membrane with distinct lipid and protein composition. *(Chapter 2)*

Minichromosome maintenance element A sequence of papillomavirus genomes required for maintenance of episomal viral genomes in a host cell population. *(Chapter 9)*

Missense mutation A change in a single nucleotide or codon that results in the production of a protein with a single amino acid substitution. *(Chapter 3)*

Molecular chaperone *See* Chaperone.

Monocistronic Encoding one polypeptide; refers to mRNA. *(Chapter 11)*

Monoclonal antibody An antibody of a single specificity made by a clone of antibody-producing cells. *(Chapter 2)*

Monolayer A layer of cultured cells growing in a cell culture dish. *(Chapter 2)*

Multiplicity of infection The number of infectious viruses added per cell. *(Chapter 2)*

Mutagen An agent that causes base changes in nucleic acids. *(Chapter 3)*

Negative [(−)] strand The strand of DNA or RNA that is complementary in sequence to the (+) strand. *(Chapter 1)*

Neutralize To block (by antibodies) the infectivity of viruses. *(Chapter 2)*

Nonsense mutation A substitution mutation that produces a translation termination codon. *(Chapter 3)*

Nuclear localization signal Amino acid sequence that is necessary and sufficient for import of a protein into the nucleus. *(Chapter 5)*

Nucleocapsid A nucleic acid-protein assembly packaged within the virion; the term is used when this complex is a discrete substructure of a complex particle. *(Chapter 4)*

Obligate parasites Organisms that are dependent on another living organism for reproduction. *(Chapter 1)*

Oligomerization Association of protein subunits, which may be the same or different, to form a protein with multiple subunits. *(Chapter 8)*

Oligosaccharide A short linear or branched chain of sugar residues (monosaccharides). *(Chapter 4)*

One-hit kinetics A linear relationship between plaque count and virus concentration that indicates that one infectious particle is sufficient to initiate infection. *(Chapter 2)*

One-step growth curve A single replication cycle that occurs synchronously in every infected cell. *(Chapter 2)*

Origin for plasmid maintenance The origin of Epstein-Barr virus DNA active in latently infected cells but not in productively infected cells. *(Chapter 9)*

Origins (of replication) Specific sites at which replication of DNA begins. *(Chapter 9)*

Packaging Incorporation of the viral genome during assembly of virus particles. *(Chapter 13)*

Packaging signal Nucleic acid sequence or structural feature directing incorporation of a viral genome into a virus particle. *(Chapter 13)*

Particle–to–plaque-forming-unit (PFU) ratio The inverse value of the absolute efficiency of plating. *(Chapter 2)*

Pathogen Disease-causing virus or other microorganism. *(Chapter 1)*

Permissive Able to support virus replication when viral nucleic acid is introduced; refers to cells. *(Chapter 2)*

Plaque A circular zone of infected cells that can be distinguished from the surrounding monolayer. *(Chapter 2)*

Plaque-forming units per milliliter A measure of virus infectivity. *(Chapter 2)*

Plaque purified Prepared from a single plaque (refers to virus stock); when one infectious virus particle initiates a plaque, the viral progeny within the plaque are clones. *(Chapter 2)*

Polarized cells Differentiated cells with surfaces divided into functionally specialized regions. *(Chapter 12)*

Polyadenylation The addition of ~200 A residues to the 3′ ends of cellular and viral transcripts made in eukaryotic cells. *(Chapter 10)*

Polycistronic Encoding several polypeptides; refers to mRNA. *(Chapter 11)*

Polyclonal antibodies The antibody repertoire against the many epitopes of an antigen produced in an animal. *(Chapter 2)*

Portal A specialized structure for entry of a viral genome into a preassembled protein shell. *(Chapter 4)*

Positive [(+)] strand The strand of DNA or RNA that corresponds in sequence to that of the messenger RNA. Also known as the sense strand. *(Chapter 1)*

Pregenomic mRNA The hepadnaviral mRNA that is reverse transcribed to produce the DNA genome. *(Chapter 7)*

Preinitiation complex A promoter-bound complex of an RNA polymerase and initiator proteins competent to initiate transcription. *(Chapter 8)*

Primary cell cultures Cell cultures prepared from animal tissues; these cultures include several cell types and have a limited life span, usually no more than 5 to 20 cell divisions. *(Chapter 2)*

Primary cells Cells that have been freshly derived from an organ or tissue. *(Chapter 1)*

Primase An enzyme that synthesizes RNA primers for DNA synthesis. *(Chapter 9)*

Primer A free 3′-OH group required for initiation of synthesis of DNA from DNA or RNA templates and initiation of synthesis of some viral RNA genomes. *(Chapters 6 and 9)*

Prions Infectious agents comprising an abnormal isoform of a normal cellular protein but no nucleic acid; implicated as the causative agents of transmissible spongiform encephalopathies. *(Chapter 1)*

Procapsid A closed, protein-only structure into which viral genomes are inserted; precursor to a capsid or nucleocapsid. *(Chapter 13)*

Processivity The ability of an enzyme to copy a nucleic acid template over long distances from a single site of initiation. *(Chapters 7, 8, and 9)*

Promoter A set of DNA sequences necessary for initiation of transcription by a DNA-dependent RNA polymerase. *(Chapter 8)*

Promoter occlusion The mechanism by which access to a promoter is blocked by passage of a transcribing RNA polymerase. *(Chapter 8)*

Proofreading Correction of mistakes made during chain elongation by exonuclease activities of DNA-dependent DNA polymerases. *(Chapter 6)*

Prophage The genome of the quiescent bacteriophage in a lysogenic bacterium. *(Chapter 1)*

Proteasome A complex containing multiple proteases that is responsible for degradation of proteins tagged with polyubiquitin. *(Chapter 8)*

Proteoglycans Proteins linked to glycosaminoglycans, which are unbranched polysaccharides made of repeating disaccharides. *(Chapter 2)*

Protomer *See* Asymmetric unit.

Proviral DNA *See* Provirus.

Provirion A noninfectious precursor of a mature virion. *(Chapter 13)*

Provirus Retroviral DNA that is integrated into its host cell genome and is the template for formation of retroviral mRNAs and genomic RNA. Also called proviral DNA. *(Chapter 7)*

Pseudodiploid Having two RNA genomes per virion that give rise to only one DNA copy, as is the case for retroviruses. *(Chapter 7)*

Pseudoreversion Phenotypic reversion caused by second-site mutation; also known as suppression. *(Chapter 3)*

Quasiequivalence The arrangement of structural units in a virus particle such that similar interactions among them are allowed. *(Chapter 4)*

Quasispecies Virus populations that exist as dynamic distributions of nonidentical but related replicons. *(Chapter 6)*

Reactivation A switch from a latent to a productive infection; usually applied to herpesviruses. *(Chapter 8)*

Reassortants Viral genomes that have exchanged segments after coinfection of cells with viruses with segmented genomes. *(Chapter 3)*

Reassortment The exchange of entire RNA molecules between genetically related viruses with segmented genomes. *(Chapters 3 and 6)*

Receptor The cellular molecule to which a virus attaches to initiate replication. *(Chapter 5)*

Receptor-mediated endocytosis The uptake of molecules into the cell from the extracellular fluid; in this process, the molecule binds a cell surface receptor, and the complex is taken into the cell by invagination of the membrane and formation of a vesicle. *(Chapter 5)*

Relative efficiency of plating A ratio of viral titers obtained on two different cell types: this number may be more or less than 1, depending on how well the virus replicates in the different host cells. *(Chapter 2)*

Replication centers Specialized nuclear structures in which viral DNA genomes are replicated. Also called replication compartments. *(Chapter 9)*

Replication forks The sites of synthesis of nascent DNA chains that move away from origin as replication proceeds. *(Chapter 9)*

Replication intermediate An incompletely replicated DNA molecule containing newly synthesized DNA. *(Chapter 9)*

Replication licensing Mechanisms which ensure that replication of cellular DNA is initiated at each origin once, and only once, per cell cycle. *(Chapter 9)*

Replicon A unit of replication in large genomes, defined by discrete origin and termini. *(Chapter 9)*

Resolution The minimal size of an object that can be distinguished by microscopy or other methods of structural analysis. *(Chapter 4)*

Resting state *See* G_0 state.

Retroid viruses Viruses that replicate their genomes via reverse transcription. *(Chapter 7)*

Revert To change to the parental, or wild-type, genotype or phenotype. *(Chapter 3)*

Ribosome A complex of RNAs and proteins that is the site of translation. *(Chapter 11)*

Ribozyme An RNA molecule with catalytic activity. *(Chapter 10)*

RNA-dependent RNA polymerase The protein assembly required to carry out RNA synthesis. *(Chapter 6)*

RNA editing The introduction into an RNA molecule of nucleotides that are not specified by a cellular or viral gene. *(Chapter 10)*

RNA interference A mechanism of posttranscriptional regulation of gene expression by small RNA molecules that induce mRNA degradation or inhibition of translation. *(Chapters 3 and 10)*

RNA processing The series of co- or posttranscriptional covalent modifications that produce mature mRNAs from primary transcripts. *(Chapter 10)*

RNA pseudoknot An RNA secondary structure formed when a single-stranded loop region base pairs with a complementary sequence outside the loop. *(Chapter 6)*

Rule of six The requirement that the (–) strand RNA genome of a paramyxovirus be copied efficiently only when its length in nucleotides is a multiple of 6. *(Chapter 6)*

Satellites Small, single-stranded RNA molecules that lack genes required for their replication but do replicate in the presence of another virus (the **helper virus**). *(Chapter 1)*

Satellite virus A satellite with a genome that encodes one or two proteins. *(Chapter 1)*

Scaffolding protein A viral protein that is required for assembly of an icosahedral protein shell but is absent from mature virions. *(Chapter 13)*

Secretory pathway The series of membrane-demarcated compartments (e.g., the endoplasmic reticulum and Golgi apparatus), tubules, and vesicles through which secreted and membrane proteins travel to the cell surface. *(Chapter 12)*

Self-priming A mechanism by which some viral DNA genomes serves as primers, as well as templates, for DNA synthesis. *(Chapter 9)*

Semiconservative replication Production of two daughter DNA molecules, each containing one strand of the parental template and a newly synthesized complementary strand. *(Chapter 9)*

Serotype A virus type as defined on the basis of neutralizing antibodies. *(Chapter 2)*

Signal peptide A short sequence (generally hydrophobic) that directs nascent proteins to the endoplasmic reticulum. The signal may be removed, or retained as a transmembrane domain. *(Chapter 12)*

Single-exon mRNAs mRNAs produced without splicing, because their precursors lack introns and splice sites. *(Chapter 10)*

siRNAs *See* Short interfering RNAs.

Site-specific recombination Exchange of DNA sequences at short DNA sequences that are specifically recognized by proteins that catalyze recombination. *(Chapter 9)*

Small interfering RNAs Small RNA molecules that induce mRNA cleavage or inhibition of translation. Abbreviated siRNAs. *(Chapter 3)*

Small nuclear ribonucleoproteins Structures that contain small nuclear RNAs and several proteins; several participate in pre-mRNA splicing. *(Chapter 10)*

S phase The phase of the cell cycle in which the DNA genome is replicated. *(Chapter 9)*

Spliceosome The large complex that assembles on an intron-containing pre-mRNA before splicing; in mammalian cells, it comprises the small nuclear ribonucleoproteins containing U1, U2, U4, U5, and U6 small nuclear RNAs and ~150 proteins. *(Chapter 10)*

Splice sites Sites at which pre-mRNA sequences are cleaved and ligated during splicing; defined by short consensus sequences. *(Chapter 10)*

Splicing The precise ligation of blocks of noncontiguous coding sequences (exons) in cellular or viral pre-mRNAs with excision of the intervening noncoding sequences (introns). *(Chapter 10)*

Stop transfer signal A hydrophobic sequence that halts translocation of a nascent protein across the endoplasmic reticulum membrane; serves as a transmembrane domain. *(Chapter 12)*

Structural unit *See* Asymmetric unit.

Substitution mutation Replacement of one or more nucleotides in a nucleic acid. *(Chapter 3)*

Subunit A single folded protein of a multimeric protein. *(Chapter 4)*

Supercoiling The winding of one duplex DNA strand around another. *(Chapter 9)*

Suppression *See* Pseudoreversion.

Susceptible Producing the receptor(s) required for virus entry; refers to cells. *(Chapter 2)*

Suspension cultures Cells propagated in suspension, in which a spinning magnet continuously stirs the cells. *(Chapter 2)*

Tegument The layer interposed between the nucleocapsid and the envelope of herpesvirus particles. *(Chapter 4)*

Termini Sites at which DNA replication stops. *(Chapter 9)*

Tight junctions The areas of contact between adjacent epithelial cells, circumscribing the cells at the apical edges of their lateral membranes. *(Chapter 2)*

Topology The geometric arrangement of, and connections among, secondary-structure units in a protein. *(Chapter 4)*

Transcriptional control region Local and distant DNA sequences necessary for initiation and regulation of transcription. *(Chapter 8)*

Transcytosis A mechanism of transport in which material in the intestinal lumen is endocytosed by M cells, transported to the basolateral surface, and released to the underlying tissues. *(Chapter 2)*

Transfection Introduction of viral nucleic acid into cells by transformation, resulting in the infection of cells. *(Chapter 3)*

Transfer RNAs Adapter molecules that align each amino acid with its corresponding codon on the mRNA. Abbreviated tRNAs. *(Chapter 11)*

Transport vesicles Membrane-bound structures with external protein coats that bud from compartments of the secretory pathway and carry cargo in anterograde or retrograde directions. *(Chapter 12)*

Triangulation Division of the triangular face of a large icosahedral structure into smaller triangles. *(Chapter 4)*

tRNAs *See* Transfer RNAs.

Tropism The predilection of a virus to invade, and replicate, in a particular cell type. *(Chapter 5)*

Tumor suppressor gene A cellular gene encoding a protein that negatively regulates cell proliferation; mutational inactivation of both copies of the genes is associated with tumor development. *(Chapter 9)*

Two-hit kinetics A parabolic relationship between plaque count and virus concentration which indicates that two different types of virus particle must infect a cell to ensure replication. *(Chapter 2)*

Type-specific antigens Epitopes, defined by neutralizing antibodies, which define viral serotypes (e.g., poliovirus types 1, 2, and 3). *(Chapter 2)*

Uncoating The release of viral nucleic acid from its protective protein coat or lipid envelope; in some cases, the liberated nucleic acid is still associated with viral proteins. *(Chapter 5)*

Vaccination Inoculation of healthy individuals with attenuated or related microorganisms, or their antigenic products, to elicit an immune response that will protect against later infection by the corresponding pathogen. *(Chapter 1)*

Variolation Inoculation of healthy individuals with material from a smallpox pustule, or in modern times from a related or attenuated cowpox (vaccinia) virus preparation, through a scratch on the skin (called scarification). *(Chapter 1)*

Viral pathogenesis The processes by which viral infections cause disease. *(Chapter 2)*

Virion An infectious virus particle. *(Chapters 1 and 4)*

Viroids Unencapsidated, small, circular, single-stranded RNAs that replicate autonomously when inoculated into plant cells. *(Chapter 1)*

Viruses Submicroscopic, obligate parasitic pathogens comprising genetic material (DNA or RNA) surrounded by a protective protein coat. *(Chapter 1)*

Virus replication The sum total of all events that occur during the infectious cycle. *(Chapter 2)*

Virus titer The concentration of a virus in a sample. *(Chapter 2)*

Wild type The original (often laboratory-adapted) virus from which mutants are selected and which is used as the basis for comparison. *(Chapter 3)*

Zoonotic Transmitted among humans and other vertebrates; refers to infections and diseases. *(Chapter 1)*

Index

B

C

D

H

I

Q

R

S

T

U

W

X

Y

Z